# Advanced Structured Materials

Volume 140

**Series Editors**

Common engineering materials reach in many applications their limits and new developments are required to fulfil increasing demands on engineering materials. The performance of materials can be increased by combining different materials to achieve better properties than a single constituent or by shaping the material or constituents in a specific structure. The interaction between material and structure may arise on different length scales, such as micro-, meso- or macroscale, and offers possible applications in quite diverse fields.

This book series addresses the fundamental relationship between materials and their structure on the overall properties (e.g. mechanical, thermal, chemical or magnetic etc.) and applications.

The topics of *Advanced Structured Materials* include but are not limited to

- classical fibre-reinforced composites (e.g. glass, carbon or Aramid reinforced plastics)
- metal matrix composites (MMCs)
- micro porous composites
- micro channel materials
- multilayered materials
- cellular materials (e.g., metallic or polymer foams, sponges, hollow sphere structures)
- porous materials
- truss structures
- nanocomposite materials
- biomaterials
- nanoporous metals
- concrete
- coated materials
- smart materials

Advanced Structured Materials is indexed in Google Scholar and Scopus.

More information about this series at http://www.springer.com/series/8611

Dilipkumar Pal · Amit Kumar Nayak
Editors

# Bioactive Natural Products for Pharmaceutical Applications

Springer

*Editors*
Dilipkumar Pal
Institute of Pharmaceutical Sciences
Guru Ghasidas Vishwavidyalaya
(A Central University)
Bilaspur, Chhattisgarh, India

Amit Kumar Nayak
Department of Pharmaceutics
Seemanta Institute of Pharmaceutical
Science
Mayurbhanj, Odisha, India

ISSN 1869-8433 ISSN 1869-8441 (electronic)
Advanced Structured Materials
ISBN 978-3-030-54026-5 ISBN 978-3-030-54027-2 (eBook)
https://doi.org/10.1007/978-3-030-54027-2



This Springer imprint is published by the registered company Springer Nature Switzerland AG
The registered company address is: Gewerbestrasse 11, 6330 Cham, Switzerland

# Foreword

Bioactive products derived from natural resources possess several biological and pharmacological significances. Due to numerous health-promoting potential, these bioactive natural products are widely used by human as a source of medication since ancient times. For example, many secondary metabolites have already found huge significance in nutrition and therapeutics. In this perspective, the exploitation of natural materials possessing different pharmacological actions (such as antioxidant, antitumor, immunomodulatory, anti-inflammatory, antidiabetic, cardioactivity, diuretic, osteogenic, wound healing, etc.) presents a wide range of therapeutic activities and hence they are being used in many pharmaceutical applications. In addition, the assessment of naturally derived bioactive molecules for their inclusive therapeutic potential has led to the discovery of numerous drug leads in recent times. Even, these products are also being used as functional additives in many pharmaceutical preparations. Therefore, the discovery of such bioactive natural products has inspired many scientists and researchers to explore their potential pharmaceutical applications. In this regard, many natural resources are being explored to produce and accomplish the demand for bioactive natural products for various ranges of pharmaceutical applications.

A thorough understanding of different bioactive natural products is very much required in order to promote the drug discovery research and their utility in pharmaceutical fields. I believe this book *Bioactive Natural Products for Pharmaceutical Applications* edited by **Dr. Dilipkumar Pal** and **Dr. Amit Kumar Nayak** surely provides updated information on the various aspects of bioactive natural products for pharmaceutical and pharmacological uses to graduate and undergraduate students, academicians, industry persons, researchers, pharmaceutical formulators, and healthcare professionals involved in natural product and pharmaceutical fields. This new book is a reflection of the rapid development in this area, contains **25** important chapters presenting the latest research updates on the recent innovations relating to various bioactive natural products and their uses in various pharmaceutical fields. I congratulate the editors: Dr. Dilipkumar Pal and Dr. Amit Kumar Nayak and all contributing authors for bringing the collection of their noble piece of works.

I shall be happy if this book receives wide attention.

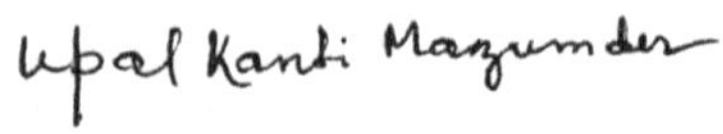

Prof. (Dr.) Upal Kanti Mazumder
Ex-Professor
Department of Pharmaceutical Technology
Jadavpur University
Kolkata, India

# Preface

In the present day, there are tough arguments and huge significance in terms of the safety concerns of naturally derived materials for their potential uses in various pharmaceutical applications. In this perspective, the exploitation of various bioactive natural products possessing a variety of biological and pharmacological activities has been discussed and these bioactive natural products are also considered to have beneficial effects in nutrition and health. Thus, bioactive natural products are a rich source of novel therapeutics. Natural materials are also currently utilized in different pharmaceutical preparations mainly as functional additives. Therefore, the exploration of bioactive molecules from the natural resources continues to play a significant role in fashioning new pharmaceutical uses. In this context, bioactive natural products and their utility in pharmaceutical fields need to be thoroughly understood.

The book entitled *Bioactive Natural Products for Pharmaceutical Applications* contains **25** important chapters, which present the latest research updates on the recent innovations relating to various bioactive natural products (such as alkaloids, glycosides, flavonoids, anthraquinones, steroids, polysaccharides, tannins and polyphenolic compounds, volatile oils, fixed oils, fats and waxes, proteins and peptides, vitamins, marine products, camptothecin, piperines, carvacrol, gedunin, GABA, ginsenosides, etc.) and their applications in various pharmaceutical fields. Chapter 1 gives a brief account of secondary metabolites isolated from plant sources. This is followed by a discussion of bioactive natural products and their general applications in Chap. 2. Chapter 3 presents various pharmaceutical applications of polysaccharides derived from plant resources. The role of phytochemicals in cancer prevention and cure has been discussed in Chap. 4. Further, the role of stress and defense in plant secondary metabolites production is also important, which has been discussed in Chap. 5. Chapter 6 presents a brief discussion of various natural compounds extracted from medicinal plants and their immunomodulatory activities. Biological activities of marine products and nutritional importance are accounted in Chap. 7. Chapter 8 presents a comprehensive review on the capillary electrophoresis as a new evolutionary platform of plant secondary metabolites. The occurrence, chemistry, and mode of action of camptothecin as anticancer drug have been presented in Chap. 9. Chapter 10 accounts a brief discussion on elicitor signal transduction leading to the production of plant secondary metabolites. Sources, properties, applications, and biotechnological production of piperine as an important bioactive molecule have been discussed in Chap. 11. Chapter 12 presents the discussion of the antimicrobial applications of phytoconstituents from turmeric and garlic. In Chap. 13, therapeutic properties and molecular mechanisms of carvacrol (*Origanum vulgare*) have been discussed. Current findings and future directions of pharmaceutical application of bioactives from *Alstonia* genus have been presented in Chap. 14. The role of natural bioactive compounds as antidiabetic agents has been discussed in Chap. 15. In Chap. 16, bioactivities of gedunin have been comprehensively reviewed. Antibacterial and antifungal plant metabolites isolated from the tropical medicinal plants have been accounted in Chap. 17. The role of natural biomaterial in cardiac tissue engineering has been discussed in Chap. 18. Chapter 19 presents a discussion on the importance of natural products in cosmetics. In Chap. 20, the encapsulation of bioactive compounds and their therapeutic potential have been reviewed. Advances and perspectives of gamma-aminobutyric acid as a bioactive compound in food have been accounted in Chap. 21. Chapter 22 gives a brief account of pharmaceutical and therapeutic applications of fenugreek gum. Protein and enzymes isolated from plant sources and their utilization in pharmaceutical field have been reviewed in Chap. 23. Chapter 24 presents a comprehensive discussion on tannins and polyphenols extracted from natural plants and their versatile application. In the last chapter, the medicinal attribution of ginsenoside as a huge source of plant bioactive compound has been discussed.

We would like to convey our sincere thanks to all the authors of the chapters for providing timely and valuable contributions. We specially thank the publisher **Springer International Publishing** and **Dr. Mayra Castro** for their invaluable support in organization of the editing process. We specially thank **Mr. Ashok Arumairaj**, Project Coordinator, Books Production, **Springer Nature** for his priceless support right through the beginning to finishing point of this book. We gratefully acknowledge the permissions to reproduce copyright materials from various sources. Finally, we would like to thank our family members, respected teachers, friends, colleagues, and dear students for their continuous encouragements, inspirations, and moral supports during the preparation of the current book. Together with our contributing authors and the publishers, we will be extremely pleased if our efforts fulfill the needs of graduate and postgraduate students, academicians, industry persons, researchers, pharmaceutical formulators, and healthcare professionals involved in natural products and pharmaceutical fields.

Bilaspur, India Dr. Dilipkumar Pal
Mayurbhanj, India Dr. Amit Kumar Nayak

# About This Book

A thorough review of the natural bioactive compounds collected from plants, microbes, and algae has been presented in this book. It contributes detailed and specialized information on different types of phytoconstituents involving alkaloids, glycosides, tannins, phenolic compounds, polysaccharides, gum, proteins, enzymes, etc. and their versatile applications in the field of pharmaceutical and medical sciences, cosmetics, nutrition, tissue engineering, food, immunology, and biomedicine. The role of different phytocompounds in cancer preventions and cure; diabetic control; and antimicrobial applications including antibacterial and antifungal properties, cardiac tissue engineering, and cosmetics has been discussed in this book. Some special topics related to secondary metabolites production, stress and defense mechanism, and elicitor signal transduction in plant secondary metabolites, camptothecin, carvacrol, GABA, ginsenoside, etc. are presented in this book in detail. In a summary, the present book possesses a valuable resource with respect to its exclusivity on various bioactive natural products and their applications in pharmaceutical fields. This book is a valuable resource for research scholars, academics, students, industrialists, and subject experts working in the multidisciplinary fields like medicinal chemistry, biochemistry, pharmacology, natural product chemistry, and other areas related to drug discovery and research.

# Contents

# About the Editors

**Dr. Dilipkumar Pal**, (born in West Bengal, India) Ph.D., M.Pharm., Chartered Chemist, PostDoc (Australia), is an Associate Professor in the Department of Pharmaceutical Sciences, Guru Ghasidash Vishwavidyalaya (A Central University), Bilaspur, Chhattisgarh, India. He received his Master's and Ph.D. degrees from Jadavpur University, Kolkata and performed postdoctoral research as "Endeavor Post-Doctoral Research Fellow" in University of Sydney, Australia. His areas of research interest include "Isolation, structure Elucidation and biological evaluation of indigenous plants" and "natural biopolymers." He has published 172 full research papers in peer-reviewed reputed national and international scientific journals, having good impact factor and contributed 114 abstracts in different national and international conferences. He has written 1 book, 52 book chapters, and edited 6 books published by reputed international publishers. His research publications have acquired a highly remarkable cited record in Scopus and Google Scholar (H-Index: 40; i–10–index: 97; total citations 5286 till date). He, in his 21 years of research-oriented teaching profession, received 13 prestigious national and international professionals awards also. He has received Endeavour Post-Doctoral Research Fellowship Award (Australia) 2010; ISTE Rajalaxmi Award, 2009; APTI young Pharmacy Teacher Award, 2010; SPER Innovative Researcher Award-2016. He has guided 8 Ph.D. and 41 Master's students for their dissertation thesis. He is the reviewer and editorial board member of 30 and 29 scientific journals, respectively. He is the member and life member of 15 professional organizations.

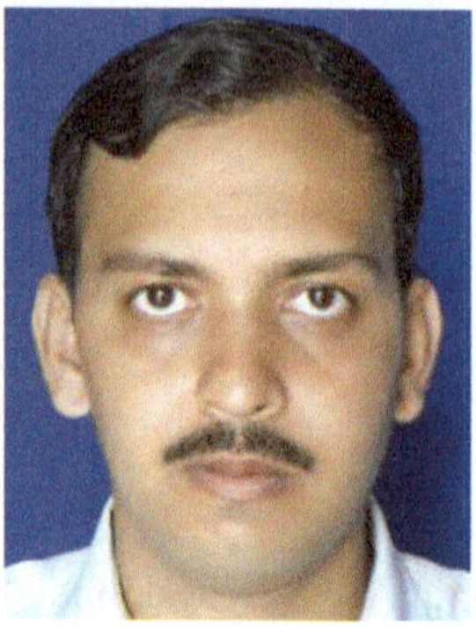

**Dr. Amit Kumar Nayak** (M.Pharm., Ph.D.) is currently working as Associate Professor at Seemanta Institute of Pharmaceutical Sciences, Odisha, India. He has earned his Ph.D. from IFTM University, Moradabad, Uttar Pradesh, India. He worked as Senior Research Associate at IIT, Kanpur in a CSIR sponsored project. He has over 12 years of research experiences in the field of pharmaceutics, especially in the development and characterization of novel biopolymeric and nanostructured drug delivery systems made of natural biopolymers. Till date, he has authored over 120 research and review publications in various high-impact peer-reviewed journals and 76 book chapters and 7 books to his credit. Overall, he has earned highly impressive publishing and cited record in Scopus (*h*-index: 35) and in Google Scholar (*h*-index: 39, i10-Index: 104). He has presented his research work at several conferences. He has received University Foundation Day Research Award-2019 by Biju Patnaik University of Technology, Odisha. He is life member of Association of Pharmaceutical Teachers of India (APTI) and a Registered Pharmacist.

# Chapter 1
# Elicitor Signal Transduction Leading to the Production of Plant Secondary Metabolites

**Supriyo Saha and Dilipkumar Pal**

**Abstract** Plant metabolites are highly effective as medicine with a higher efficacy and lower adverse effect. Two basic metabolites are obtained from nature, namely primary metabolites and secondary metabolites. Alkaloids, glycosides, terpenoids, flavonoids are the principal secondary metabolites, and also the primary source for the drug discovery and development. Elicitors are the substances which under stress conditions induce the biosynthesis of secondary metabolites of plants. Both biotic and abiotic elicitors are used in the process. Most common secondary metabolites Ferulic acid, cinnamic acid, vanillin, coumaric acid, silymarin, affinin, hypocrellin A, steroiside, menthone, piperitone, glycyrrhizic acid, colchicine, thiocolchicoside, phenolic acid, gymnemic acid, flavonoids are utilized the elicitation technique. Elicitors are two types such as: abiotic and biotic. Abiotic elicitors such as salicylic acid, methyl jasmonate, hydrogen peroxide, lanthanum, different hormones, light, gamma rays and controlled temperature are used to generate secondary metabolites of wheat grass, *Thymus vulgaris*, *Silybum marianum*, *Shiraia bambusicola*, *Ajuga bracteosa*, broccoli plant, etc. Biotic elicitors like chitosan, rhizobacteria, *Rhizobium leguminosum*, *Aspergillus tenius*, *Agrobacterium tumefacians*, carrageenan, Streptomyces, *Rhizopus*, dextran, yeast are used to develop or improvise secondary metabolites of Khus, *Mentha pulegium*, *Tavernia cuneifolia*, chickpea, -*Vitis vinifera*, *Rumex gmelini* Turcz, *Cupressus lusitanica*, etc. Some secondary metabolites of *Coleus aromaticus* Benth, *Rhododendron tomentosum*, *Fagonia indica*, *Rauwolfia serpentine*, *Solanum khasianum*, *Ocimum tenuiflorum*, *Stevia rebaudiana* etc. are used both abiotic and biotic elicitors.

**Keywords** Secondary metabolites · Abiotic elicitor · Biotic elicitor · Phenolic compounds · Silymarin · Affinin · Hypocrellin A

---

S. Saha (✉)
School of Pharmaceutical Sciences & Technology, Sardar Bhagwan Singh University, Dehradun, Uttarakhand 248161, India
e-mail: supriyo9@gmail.com

D. Pal
Department of Pharmaceutical Sciences, Guru Ghasidas Vishwavidyalaya (A Central University), Bilaspur, Chhattisgarh 495009, India



D. Pal and A. K. Nayak (eds.), *Bioactive Natural Products for Pharmaceutical Applications*, Advanced Structured Materials 140,
https://doi.org/10.1007/978-3-030-54027-2_1

## 1.1 Introduction

Nowadays, plants are the primary sources of food and medicine for mankind. High efficacy and lesser side effects are the main advantage of plant metabolites (Sato and Matsui 2012). Not only that, synthetic molecules are carry various side and adverse effects to our body system (Hesketh et al. 2002). Natural sources contain diversified medicinal component such as alkaloids, glycosides, terpenoids, flavonoids; those are essential for maintaining body immune system. Plants are capable to cultivate a diversified source of natural metabolites; those are important to link up the behavior of other organism. Also, in present time exposure to air pollution, water pollution, ultraviolet ray exposure and deforestation are the most common fact to our nature and our life style. Natural sources are also helps to protect us from these abiotic changes (Hiroaki et al. 2012). In this situation ancient folkloric knowledge of plants and their metabolites is the primary weapon to conquer against various viruses, pathogens, bacteria and their mutant strains. Also in this era different complex structure of secondary metabolites is developed using recombinant DNA technology. As per the definition, metabolites are compounds produced by plants for essential functions (Buchanan et al. 2015); as growth and development (primary metabolites), as well as specific functions as pollinator attraction or defense mechanism. Also metabolites are organic compounds procured from enzyme-dependent chemical reactions of organism known as metabolic pathways (Lena 2012). Nowadays there is a tremendous demand of various plant derived product as health supplement, natural flavors, natural colors, but due to natural calamity and lesser growth of the secondary metabolites the producer does not able to fulfil the demand of the market. Plant tissue culture is one of the way to cope up with the market demand of secondary metabolites (such as alkaloids, glycosides, terpenoids, flavonoids etc.) and helps to over produce the secondary metabolites (Pichersky and Gang 2000). Actually plant tissue culture comprised of clonal propagation, callus formation and formation of germplasm using a perfect combination of plant hormones, temperature, humidity and elicitors. Tissue culture is not only helps to generate secondary metabolites but also participate in seed germination, improvement of crop quality, immobilization, biomass accumulation, micropropagation and existence of some rare plants (those are lost during evolution) (Jensen et al. 2014). Tissue culture mainly focused on shortening of biosynthetic process of plant cell, higher cell division and metabolism; these all factors are cumulatively increase the growth of secondary metabolites. This chapter mainly focus on the role of elicitors on the production of secondary metabolites.

## 1.2 Types of Plant Metabolites

Plant metabolites were two types such as primary metabolite and secondary metabolite.

1. Primary metabolites were the basic requirement for growth and development and it was present in all plants, organisms or cells. Low molecular weight primary metabolites were important components of crop plants for both consumers and producers. Primary metabolites such as ethanol, lactic acid, and certain amino acids were the most primary one. Primary metabolism was referred as trophophase, which was characterized by balanced growth of microorganisms. It occurred when all the nutrients needed by the organisms were provided in the medium. Primary metabolism was essential for plant existence and reproduction of cells. In the trophophase, the cells possessed with optimal concentrations of all the macromolecules such as proteins, DNA, RNA etc. In during trophophase, growth of microorganisms was followed the exponential nature. During trophophase, the collective metabolic products were known as primary metabolites (Croteau et al. 2000).
   Basically primary metabolites were subdivided into two groups, such as:
   a. Primary essential metabolites: These were the compounds produced in adequate quantizes to sustain cell growth e.g. vitamins, amino acids, nucleosides. Primary microorganisms usually do not overproduce essential primary metabolites because it was wasted. In case of industrial overproduction, the regulatory mechanisms were taken.
   b. Primary metabolic end products: These were the normal and traditional end products of fermentation process of primary metabolism. The end products had many industrial applications such as in the form of ethanol, acetone and lactic acid. Carbon dioxide was a metabolic end product of *Saccharomyces cerevisiae*, which was essential for leavening of dough in baking industry.

   Growth limitation: Due to insufficient supply of any nutrient (substrate or even Oxygen), the growth of microorganisms was slows down. However, the metabolism does not stop, continues until cell lives with the differentiate product formation (Pal et al. 2019a).
2. Secondary Metabolites: These are sometimes colored, fragrant and flavorful compounds, which typically conjugate with plant metabolite and organisms. Interactions were included pollination through animal such as butterfly, bees, moths, flies as well as by fungi, bacteria, nematodes, and viruses and by foliage, gastropods and caterpillar. After the ending of exponential growth of microorganisms, idiophase was initiated. Idiophase was characterized by secondary metabolism. Secondary metabolites (produced in abundance) were not required by the microorganisms, but it had some industrially very important mainly in the biotechnology to produce antibiotics, steroids, alkaloids, glycosides, plant hormones and toxins (Pal et al. 2017).

Characteristics of secondary metabolites:

a. Secondary metabolites were specifically produced by selected microorganisms.
b. Secondary metabolites were less essential for development and reproduction of organisms.
c. Environmental factors were influenced the production of secondary metabolites (Pal et al. 2019b).
d. Certain microorganisms can produce structurally related secondary metabolites as a group of compounds not a single one such as anthracyclines, produced by *Streptomyces*.
e. The biosynthetic pathways of were not clearly established for most secondary metabolites.
f. The regulation of the formation of secondary metabolites was more complex than primary metabolites (Pal et al. 2018).

Functions of secondary metabolites: There were two possible hypotheses for the justification of secondary metabolite function as:

a. Secondary metabolites were beneficial for the cells to survive.
b. The secondary metabolites were important for the cell development.

The metabolic regulation was equally complex to achieve overproduction of secondary metabolites. Some regulatory mechanisms were as follows:

i. Induction: Methionine addition induced certain enzymes, which enhanced the production of cephalosporin. Also tryptophan regulated ergot alkaloid biosynthesis.
ii. ii. End product regulation: Secondary metabolites inhibited their own biosynthesis by negative feed regulation such as penicillin, streptomycin, puromycin and chloramphenicol (Pal et al. 2019c).
iii. Catabolite regulation: In this process, a key enzyme was participated in the catabolic pathway was inactivated, inhibited or repressed the process. Catabolic repression was obtained by carbon or nitrogen sources. Where the most common source of carbon was glucose, which was participated in the inhibition process of several antibiotics, such as: penicillin, streptomycin, bacitracin, chloramphenicol, puromycin; whereas ammonia was the primary source of nitrogen which was catabolite regulators for the overproduction of certain antibiotics (Pal et al. 2019d).
iv. Phosphate regulation: Inorganic phosphate was essential for the growth and multiplication of prokaryotes and eukaryotes. Concentration of inorganic phosphate (up to 1 mM) was directly correlated with the concentration of secondary metabolites e.g. streptomycin, tetracycline, alkaloids, gibberellins.
v. Auto regulation: Certain microorganisms such as: actinomycetes were associated with the self-regulation for the production of secondary metabolites. A compound (factor A) was a derivative of a hormone was suggested to be closely involved in auto regulation for the production streptomycin by *Streptomyces griseus*.

vi. Bioconversions: Biotransformation process using microorganisms was very important for the production of several compounds e.g. vinegar, sorbose, steroid hormones and certain amino acids. In this process, microorganisms were responsible for the conversion of a compound to another structurally related product in one or a few enzymatic reactions. The bioconversions can be performed with resting cells, spores or even killed cells. Non-growing cells were preferred for bioconversions due to its high substrate concentration (Pal et al. 2019e).

## 1.3 Elicitor and Its Type

Elicitors are the substances which under stress conditions induce the biosynthesis of secondary metabolites of plants. So elicitation is the process to create stress on the plant, which relates with the chemical composition and growth of the plant. The process of elicitation mainly increase the growth of the plant and its metabolites.

Elicitors are two types such as abiotic and biotic elicitor. Abiotic elicitors are obtained from non-living source. Abiotic elicitors are mainly three types such as: physical, chemical and hormonal abiotic elicitors. Biotic elicitors are obtained from biological source and it has four types such as: saccharides, yeast, fungal and bacterial biotic elicitors (Fig. 1.1).

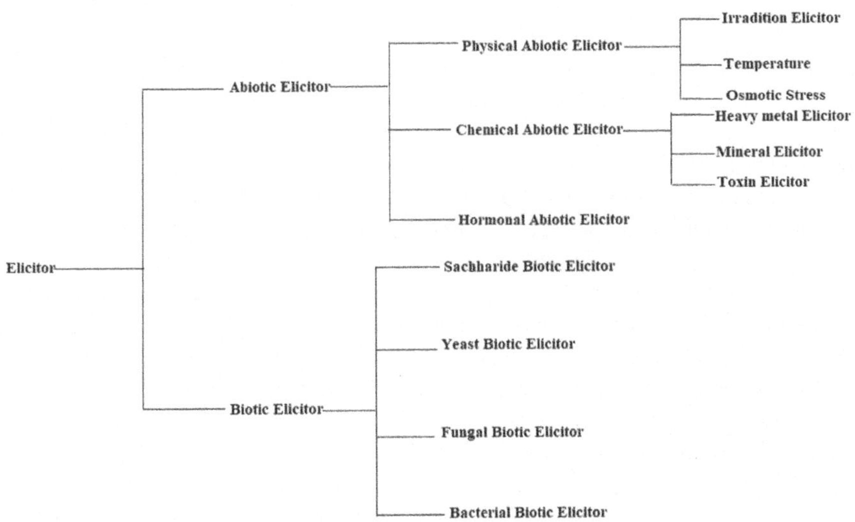


**Fig. 1.1** Elicitor and its types

## 1.4 Application of Elicitors on the Production of Secondary Metabolites

### *1.4.1 Production of Secondary Metabolites Using Abiotic Elicitors*

#### 1.4.1.1 Effect of Arachidonic and Jasmonic Acid Elicitors on Secondary Metabolites of Wheatgrass

Zlotek et al. was scientifically tested the effect of abiotic elicitors as arachidonic acid and jasmonic acid on the phenolic and flavonoid content of wheatgrass (*Triticum aestivum* L.). The experiment started with slightly grown wheatgrass plant treated with different concentration of arachidonic acid and jasmonic acid (arachidonic acid: 0.01, 1.0 and 100 μM; jasmonic acid: 0.01, 1 and 100 μM) at a relative humidity of 70%, 18 °C of temperature and 200 μmol $m^{-2}$ $s^{-1}$ of photon flux density. Thereafter the growth of the plants were prominent, then make it dry followed by centrifuged at 9000 g force for 30 min. The total phenolic content, flavonoid content, Antioxidant and Anti-inflammatory activities were evaluated. The outcomes revealed that amount of principle flavonoids (luteolin and apigenin) and phenolic compounds (ferulic acid and syringic acid) were remain unchanged after elicitation, some polyphenolic compounds observed with increased amount; but the anti-inflammatory activity was markedly improved. The outcomes concluded that 0.01 μM of arachidonic acid showed better elicitation behavior than jasmonic acid (Złotek et al. 2019).

#### 1.4.1.2 Effect Abiotic Elicitors on Secondary Metabolites of Thyme, Greater Celandine and Parsley

Kleinwachter et al. was experimentally proved the effects of slight drought and abiotic elicitors on the production of secondary metabolites of thyme (Thymus vulgaris), greater celandine (*Chelidonium majus*) and parsley (*Petroselinum crispum*). The experimental procedure was started with the minimization of water supply to the plant by 50% and continue until the evaporation done by plants were 80% reduced than usual. These conditions leads to the abrupt ratio of soil and water by (6–10)% as compare to the control plant (17–20)%. Then the samplings of thyme and greater celandine were treated with methyl jasmonate (MJ) (0.2 mM) and parsley sampling was treated with (2 mM) concentration, followed by addition of salicylic acid solution. Then the samplings were quantified the production of monoterpenes in thyme, benzylisoquinoline alkaloids in greater celandine, flavones and essential oil in in parsley. The outcomes revealed that the quantity of benzylisoquinoline in greater celandine was increased by 46% and flavones were increased by 70% in parsley; but the minimized water supply reduced the overall content of the plants due to its lesser growth. The presence of salicylic acid does not create any remarkable effect and also the different ratios of MJ regulated the plant secondary metabolites growth depends upon their species characteristics (Kleinwachter et al. 2015).

#### 1.4.1.3 Effect of Abiotic Elicitor on Triterpenoid Accumulation in *Centella asiatica*

Buraphaka et al. was scientifically tested the effects of abiotic elicitor (MJ and salicylic acid) on the accumulation of triterpenoid in *Centella asiatica* plant. The experiment was started with the reaction between different concentrations of MJ and salicylic acid (1, 2 and 4 mM) with the leaves of the plant under a shaker for 20 min, 40, 60 and 120 min. Then the treated leaves were divided into two groups: first group for anti-inflammatory was dried at 50 °C whereas another group was stored at (−) 80 °C for messenger RNA level expression analysis. The outcomes revealed that after the treatment with salicylic acid (2 mM), the amount of triterpenoid was doubled; whereas treatment with MJ the amount of increased content was 1.4 times than normal. Also the elicited leaves were remarkably inhibited the nitric oxide production in lipopolysaccharide induced RAW 264.7 macrophage cells with increased activity of phenylalanine ammonia lyase, peroxidase and catalase enzymes. So these data correlated with the positive effect of abiotic elicitor on the aforementioned plant (Buraphaka and Putalun 2020).

#### 1.4.1.4 Effect of Abiotic Elicitor on Affinin Contents in *Heliopsis longipes* (Chilcuague)

Parola-Contreras et al. was experimentally proved the effects of salicylic acid and hydrogen peroxide as aboiotic elicitor on the amount of affinin content in *Heliopsis longipes* (chilcuague). The experiment was started with the treatment of the samplings of chilcuague and salicylic acid (5 and 10 mM) at 150 days after transplanting as well as hydrogen peroxide (200 and 400 mM) at 157 days after transplanting. Then the elicited plants were evaluated with the enzymatic activity for superoxide dismutase, catalase, phenylalanine ammonia lyase, valine decarboxylase and also quantified the amount of affinin present in the plant. The outcomes revealed that after 150 days maximum superoxide dismutase, catalase, phenylalanine ammonia lyase and valine decarboxylase activities were observed with 10 mM of salicylic acid, 200 mM of hydrogen peroxide, both salicylic acid and hydrogen peroxide, 200 mM of hydrogen peroxide as abiotic elicitor. After 157 days, maximum superoxide dismutase, catalase, phenylalanine ammonia lyase were observed with 10 mM of salicylic acid whereas maximum valine decarboxylase activity was observed with 200 mM of hydrogen peroxide elicitor

**Fig. 1.2** Typical morphological aspect at day 164 post-transplanting of *H. longipes* treated with hydrogen peroxide and salicylic acid. Copyright permission obtained from Parola-Contreras et al. @ 2020 Elsevier B.V

(Fig. 1.2). After 150 and 157 days, maximum affinin content was observed with 10 mM of salicylic acid and 200 mM of hydrogen peroxide. These data confirmed the positive effects of abiotic elicitors on chilcaugaue (Parola-Contreras et al. 2020).

#### 1.4.1.5 Effects of Abiotic Elicitor on Hypocrellin a Content in *Shiraia bambusicola*

Lu et al. was scientifically proved the effect of abiotic elicitor (lanthum $La^{3+}$) on the content of hypocrellin A (anticancer agent mainly against lung adenocarcinoma) present in *Shiraia bambusicola* (parasitic fungi present in bamboo). The experiment was started with mycelium of *S. bambusicola* and lanthum ($La^{3+}$) with a concentration range from (0.0–1.4 g/L). The elicited fungus was evaluated by membrane permeabilization assay with fluorescent dye SYTOX green, generation of reactive oxygen species using 7-dichloro dihydro fluoresceindiacetate dye followed by antioxidative activity using NADPH oxidase produce superoxide, superoxide dismutase and catalase enzymes. The outcomes showed that after lanthanum treatment for 4 h, the hyphae became more strong and intense. The bioactivities showed that after lanthum treatment the amount of hydrogen peroxide was higher with many fold increased in antioxidative enzymatic activity. Also vitamin C suppressed the up regulation of major facilitator superfamily transporter and O-methyltransferase, polyketide synthase, ATP-binding cassette transporter, O-methyltransferase/FAD-dependent monooxygenase and FAD/FMN-dependent oxidoreductase genes. These data correlated with the greater production of hypocrellin A in lanthanum elicitor induced fungi (Lu et al. 2019a).

#### 1.4.1.6 Effects of Abiotic Elicitor on Steviosides Production in *Stevia rebaudiana* Bertoni Calli

Mejia-Espejel et al. was experimentally proved the effects of abiotic elicitor (salicylic acid, MJ, citric acid, ascorbic acid) with different light (red, blue and white) and temperature environment on the production of steviosides (diterpene glycoside) present in *Stevia rebaudiana*. The experiment was started with the reaction between 3 gm of calii (grown in Murashige and Skoog (MS) medium containing 2,4-dichlorophenoxyacetic acid, BA, citric acid and ascorbic acid) and different elicitors such as salicylic acid (10 and 100 mM), MJ (10 and 100 mM), antioxidants, growth regulator, different light (white light, red light, blue light and a combination of red light and blue light) and temperature (25 °C and 28 °C) as physical abiotic elicitor. The outcomes revealed that maximum stevioside was obtained at a combination of white light, 28 °C, 100 mM of salicylic acid and 10 mM of MJ. These data strictly adhere with the importance of abiotic elicitor on the production of steviosides (Mejia-Espejel et al. 2018).

#### 1.4.1.7 Effect of Light as Abiotic Elicitor on the Secondary Metabolite Growth in *Stevia rebaudiana*

Ahmad et al. was experimentally proved the effect of light of the production of secondary metabolites (total phenolic and flavonoid content) and antioxidative properties of the *Stevia rebaudiana* plant. The experiment was started with the development of callus from the leaves of the plant in MS medium contained with 2,4-dichlorophenoxyacetic acid, BA; then the callus was exposed to different light medium such as green light, yellow light, blue light, red light and white fluorescent light environments at a 25 °C temperature flowed by antioxidative property assessment using DPPH antioxidant method. The outcomes revealed that maximum fresh callus weight was observed in white light with greater growth kinetic. Total phenolic and flavonoid contents were observed in blue light (Fig. 1.3). Also blue light elicitor increased the antioxidative property of the callus. These data confirmed the importance of blue light on the growth of secondary metabolites of *Stevia rebaudiana* (Ahmad et al. 2016).

**Fig. 1.3** Effect of different spectral lights on callus morphological features in *S. rebaudiana*. **a** red light induced callus **b** blue light **c** yellow light **d** green light and **e** control white light. Copyright permission obtained from Ahmad et al. @ 2016 Elsevier B.V

#### 1.4.1.8 Effect of Gamma Radiation on Secondary Metabolites of *Hypericum triquetrifolium* Turra

Azeez et al. was scientifically proved the effect of abiotic elicitor (gamma radiation) on the production of biomass and secondary metabolites (phenolic compounds and naphtodiantrones) in *Hypericum triquetrifolium* Turra plant. The experiment was started with induction of explants (obtained from leaf, stem and root) with indole acetic acid (IAA) and thidiazuron (synthetic cytokinin) in MS medium followed by irradiated with gamma rays of 10, 20, 30 and 40 Gy unit; then finally quantify its average growth index based on the values of callus biomass at initial and final point. The outcomes revealed that at 10 Gy scale maximum growth index was observed as well as maximum accumulation of 4-hydroxybenzoic acid, chlorogenic acid and epicatechin. At 10 Gy of gamma irradiation the accumulation of naphtodiantrones (hypericina and pseudohypericin) were observed (Fig. 1.4). So it was quite resemble that 10 Gy scale of gamma rays directly helps to accumulate the principle secondary metabolites present in the plant (Azeez et al 2017).

#### 1.4.1.9 Effects of Abiotic Elicitors on Secondary Metabolites of *Vitis vinifera* Suspension Culture

Cai et al. was experimentally proved the importance of abiotic elicitor (streptomycin, activated charcoal, etephon) and pressure on the growth of secondary metabolites of *Vitis vinifera* plant. The experiment was started with the reaction between plant cell cultures with the above mentioned abiotic elicitors along with a diverse pressure treatment from (40–50) Mega Pascal unit for a seven days regime. Here streptomycin was used to reduce the load of contamination, activated charcoal

**Fig. 1.4** The morphological characteristics of *Hypericum triquetrifolium* Turra callus mass accumulated from leaf explant and irradiated with 10 Gy dose after the third successive subculture (**a**). *Hypericum triquetrifolium* T. callus mass from leaf irradiated with 30 Gy dose and harvested after the third regular subculture (**b**). Copyright permission obtained from Azeez et al. @ 2017 Elsevier B.V

was utilized as source of carbon and etephon was a plant growth regulator. The outcomes revealed that after 5th day of treatment maximum fresh and dry weight of the callus was observed with activated charcoal and etephon treatments. Also the amount of anthocyanins were increased with etephon and activated charcoal treatments. But the amount of extracellular 3-*O*-glucosyl resveratrol and phenolic compound were increased with the combination treatment of etephon and high pressure. These data confirmed the importance of abiotic elicitors on the growth and accumulation of secondary metabolites present in *Vitis vinifera* plant (Cai et al. 2011).

#### 1.4.1.10 Effect of Salicylic Acid on Alkaloid Biosynthesis in Marine Microalgae *Arthrospira platensis*

Hadizadeh et al. experimentally proved the importance of salicylic acid as abiotic elicitor on the biosynthesis of pharmaceutical alkaloid present in marine microalgae *Arthrospira platensis*. The experiment as started with addition of salicylic acid (0, 5, 20 and 100) μM concentration to *A. platensis* suspension culture after 3, 7, 10 and 14 days of culture; followed by estimation of dry weight of biomass and total alkaloid content. The outcomes revealed that maximum growth was observed between 8 to 10 days of culture and at 15 days maximum dry weight of biomass was observed with salicylic acid (5 and 20) μM as well as at 15th day of treatment maximum content of alkaloid was obtained with 5 μM of salicylic acid. These data confirmed the importance of salicylic acid on the production of alkaloid in the marine microalgae (Hadizadeh et al. 2019).

#### 1.4.1.11 Effect of Abiotic Elicitor on Growth of Secondary Metabolites in Broccoli Plant

Hassini et al. scientifically proved the importance of abiotic elicitors (salicylic acid, MJ and methionine) on the growth of secondary metabolites present in broccoli plant. The experiment was started with treating the seeds with potassium chloride, potassium sulfate and sodium chloride followed by elicited the roots and shoots with methionine (10 mM), salicylic acid (200 μM) and MJ (100 μM) concentrations. The capacity of roots to conduct water between roots to xylem was measured by root hydraulic conductivity using root fresh weight and pressure, followed by assessment of plant defense mechanism using myrosinase activity and total phenolic content present in the plant was also evaluated. The outcomes showed that maximum dry weight of plant was obtained from roots using potassium sulfate and from shoots using methionine as elicitor. The root hydraulic conductivity showed that methionine positively culminated the plant. The myrosinase activity showed that methionine increase the defense mechanism while sodium chloride, salicylic acid were suppressed the defense of the plant against insect and herbivores. Two phenolic acid component as chlorogenic acid and sinapic acid were increased in presence of potassium chloride and potassium sulfate whereas flavonol was maximum obtained through methionine elicitation. These data confirmed the importance of abiotic elicitors on the growth of secondary metabolites present in broccoli plant (Hassini et al. 2019).

#### 1.4.1.12 Effect of Abiotic Elicitors on Steviol and Adventitious Root Growth in *Stevia rebaudiana* Plant

Kazmi et al. scientifically proved the importance of abiotic elicitors (MJ, phenyl acetic acid and melatonin) on the growth of steviol glycoside and adventitious roots in *Stevia rebaudiana* plant. The experiment was started with culturing the small pieces of root, stem and leaf portions of the plant in MS medium with BA, IAA and naphthalene acetic acid (NAA) as growth regulators, then the explants were elicited using different concentrations of melatonin, phenyl acetic acid and MJ for a period of 15, 30 and 45 min. The elicited dry mass of the explants were estimated with total phenolic and flavonoid content as well as evaluated by antioxidative assay methods. The outcomes revealed that in case of root maximum morphogenic response was given by combination of 2.0 mg/L of benzylaminopurine (BA) and 1.0 mg/L of NAA; in case of leaf maximum morphogenic response was obtained from 0.5 mg/L of NAA and from stem maximum morphogenic response was obtained using 2.0 mg/L of BA plant regulators. Maximum adventitious roots were obtained from 0.5 mg/L of MJ after 30 min dipping time. Total phenolic and flavonoid contents were highly obtained from in vitro plants but among the elicitors MJ-adventitious root combination wins the race and MJ-adventitous root combination was also observed with higher antioxidative effect (Fig. 1.5). So these data confirmed the importance of the abiotic elicitors on *Stevia rebaudiana* plant (Kazmi et al. 2019).

#### 1.4.1.13 Effect of Ultrasound on Secondary Metabolite Accumulation in Tomato Plant

Lu et al. experimentally proved the importance of high intensity ultrasound on secondary metabolite accumulation and antioxidative efficiency of *Solanum lycopersicum* (tomato) plant. The experiment was started with ultrasound treatment of tomatoes with ultrasound (25 kHz) for (1–4) min at room temperature with 26 W/L of acoustic power density. After 0 h, 24 h and 48 h of treatment, the elicited tomatoes were checked for firmness, total phenolic content, lycopene content, total carotenoid content and ascorbic acid followed by antioxidative efficiency evaluation. The outcomes showed that firmness of tomatoes were not remarkably changed so it can stored for longer time; total phenolic content was maximum after 48 h of storage with 2 min of ultrasound treatment; maximum lycopene and carotenoid contents were obtained after 48 h of storage with (1–4) min of treatment as well as ascorbic acid content was higher with 3 min of

**Fig. 1.5** In vitro morphogenesis and adventitious root (AR) formation **a** In vitro germinated plants, **b** and **c** response of root and stem explants to 6-benzyladenine (BA), **d–f** different morphological responses by leaf explants, **d** callus formation, **e** AR induction from callus, **f** direct AR formation, **g–j** AR formation in leaf explants pretreated with elicitors for defined time periods, **g** Melatonin (Mel) induced AR, **h**: Phenyl acetic acid (PAA) induced AR, **i** and **j** Methyl jasmonate (Me-J) induced AR. Copyright permission obtained from Kazmi et al. @ 2019 Elsevier B.V

ultrasound treatment after 48 h. Maximum phenylalanine ammonia-lyase activity was also observed after 48 h of storage with (1–4) min of ultrasound treatment. These data confirmed the importance of abiotic elicitors on secondary metabolite accumulation in tomato plant (Lu et al. 2020).

#### 1.4.1.14 Effect of Abiotic Elicitors on Metabolite Accumulation of *Trifolium resupinatum*

Twaij et al. experimentally proved the importance of abiotic elicitors and precursors on the accumulation of secondary metabolites in *Trifolium resupinatum*. The experiment was started with culture propagation of *T. resupinatum* shoot in MS medium containing BA (2.0 mg/L), IAA (0.5 mg/L), sucrose and phytage at pH 7.0 until calli was full grown. Then the calli further transferred into the MS medium with sucrose, NAA and kinetin as growth regulators. Then the grown explants were divided into four categories such as shade grown plant, somatic embryos, light induced and dark induced followed by treated with MJ, salicylic acid and glutathione with concentration range (0, 10, 20, 40, 80 and 160) μM for a period of (0–4) weeks. Then total phenolic content, total flavonoid and antioxidative efficiency using DPPH radical scavenging and ferric reducing antioxidant potential assay activity were checked and evaluated for/by elicited explant. The outcomes revealed that for shade grown plant, somatic embryos and light induced and dark induced conditions antioxidative efficiencies against DPPH and ferric reducing antioxidative processes were observed with greater efficiency in MJ, salicylic acid and glutathione elicitors. Maximum effects as antioxidative and enzymatic activities against glutathione peroxidase, glutathione reductase and glutathione-S-transferase were observed with salicylic acid as elicitor. These data correlated abiotic elicitor and secondary metabolites accumulation in *Trifolium resupinatum* plant (Twaij et al. 2019).

#### 1.4.1.15 Effect of MJ Abiotic Elicitor on Anthraquinone Production in *Rubia tinctorum*

Perassolo et al. experimentally proved the effects of combined culture medium and MJ as abiotic elicitor on the production of anthraquinone in *Rubia tinctorum* plant. The process was started with culturing of young small clones of leaf margin or aerial stem of the plant with *Agrobaterium rhizogenes* in Gamborg B5 medium with sucrose and ampicillin as source of carbon and strengthening of natural defense system. Then after two to four weeks roots were developed for cloning in Gamborg B5 and Llyod-McCown woody plant medium with thiamine, pyridoxine, nicotinic acid and myoinositol as growth regulators. After fourteen days, inoculum cultures were elicited in presence of MJ (100 μM) concentration. After a time interval of 0, 2, 4 and 7 days, the amount of anthraquinone and growth of biomass were evaluated. The outcomes revealed that after six weeks of treatment biomass in woody plant medium was higher than Gamborg medium and for anthraquinone production Gamborg medium was perfectly suited. But after seven days treatment with MJ, anthraquinone content was remarkably good as compare to dimethylsulfoxide. These information stated the importance of MJ and growth regulators for the production of anthraquinone (Perassolo et al. 2017).

#### 1.4.1.16 Effect of Polyunsaturated Fatty Acids on Gymnemic Acid Production in *Gymnema Sylvestre*

Praveen et al. scientifically proved the importance of polyunsaturated fatty acids (oleic acid and linoleic acid) on the production of gymnemic acid present in *Gymnema sylvestre* plant. The experiment was started with formation of hairy root culture of the plant with the plantation in MS medium with sucrose as carbon source and these were sub-cultured for fortnight period. After the incubation period, cultures were elicited with oleic acid and linoleic acid with concentration range (0, 1, 5, 10 and 50) μM. The dry biomass, amount of gymnemic acid, total phenolic content, total flavonoid content were checked as well as antioxidative property was also evaluated. The outcomes revealed that the growth ratio between fresh and dry biomass was higher with linoleic acid (1.0 μM) elicitation. Highest amount of gymnemic acid and greater antioxidative property were observed with linoleic acid (5.0 μM) elicitation. These data confirmed the importance of linoleic acid for the production of gymnemic acid (Praveen et al. 2014).

#### 1.4.1.17 Effect of Magnesium Oxide Nanoparticle on Secondary Metabolite in *Atropa belladonna*

Tian et al. experimentally proved the importance of magnesium oxide nanoparticle on the growth and accumulation of antioxidative metabolites present in *Atropa belladonna* plant. The experiment was started with germination of the plant in MS medium containing sodium hypochlorite, then the clones of root and shoot were elicited with magnesium oxide nanoparticle of (25, 50, 100 and 200) mg/L concentration without any presence of cytokinin and auxin. Then the shoot/root number, length and fresh weights were evaluated; evaluation of relative water content, chlorophyll content, malondialdehyde and membrane stability index, enzymatic activities against superoxide dismutase, ascorbate peroxidase were also evaluated. Also the total phenolic flavonoid content, total alkaloid and antioxidative efficiencies were measured. The outcomes revealed that shoot/root number, length and fresh weight, relative water, chlorophyll and membrane stability index were high with 25 mg/L of magnesium oxide nanoparticle presence. Total phenolic and flavonoid contents were observed with (100 mg/L) magnesium oxide nanoparticle; maximum alkaloid content was observed with (25 mg/L) of nanoparticle was well as maximum antioxidative effect of the elicited plant was observed with (200 mg/L) magnesium oxide nanoparticle (Tian et al. 2018).

#### 1.4.1.18 Effect of Temperature on Secondary Metabolite Accumulation in Cold Environment Soil Fungi

Ulaganathan et al. experimentally suggested the effect of temperature on secondary metabolite accumulation in forty soil isolated fungal strains. The experiment was started with cultivation of isolated fungi into potato dextrose agar plate for the mycelia formation followed by antimicrobial assessment against gram positive (*Bacillus subtilis*, *Enterococcus facaellis* and *Bacillus cereus*) and gram negative (*Pseudomonas aeruginosa* and *Escherichia coli*) strains. As per the first screening test, the pass over fungal strains [HND 10 (*Atradidymella sp*), AK 102 (*Pseudogymnoascus* sp.) and HND 11 (*Penicillium flavigenum*)] were evaluated against bacterial strains (*E. coli*, *B. subtilis*, *S. aureus*, *P. aeruginosa* and *Candida albicans*) at different temperature modules as 4, 10, 15 and 28 °C. The outcomes showed that the growth of *E. coli*, *B. subtilis*, *S. aureus* and *C. albicans* were highly inhibited by AK 102 at 4 and 15°C temperature; *P. aeruginosa* was not observed with any susceptibility against strains. So it was quite justified the importance of temperature on microbial growth inhibition (Ulaganathan et al. 2017).

#### 1.4.1.19 Effect of Polyunsaturated Fatty Acids on Secondary Metabolite Production in *Panax ginseng*

Wu et al. experimentally proved the effect of polyunsaturated fatty acids (linoleic and alpha linolenic acid) on secondary metabolite accumulation and biomass production of *Panax ginseng* plant. The experiment was started with culturing of adventitious root of ginseng into MS medium containing indole butyric acid and sucrose. When the roots were 5 fresh weight per litre, these were inoculated into the same medium with addition of linoleic acid and alpha-linolenic acid with concentrations of (1.0, 2.5, 5.0, 10.0 and 20.0) μmol per litre. Total ginsenosides, diol ginsenoside, triol ginsenoside, total phenolic and total flavonoid contents were checked for polyunsaturated acid elicited ginseng plant adventitious roots. The outcomes showed that maximum amount of total ginsenosides, diol ginsenoside, triol ginsenoside and total flavonoid contents were observed with 5.0 μmol/l of fatty acid elicitation, whereas total phenolic content was maximum with 5.0 μmol/l of linoleic acid and 10 μmol/l of alpha linolenic acid. The enzymatic activity data expressed that maximum superoxide dismutase activity was observed with 20 μmol/l of linoleic acid and 5.0 μmol/l of alpha linolenic acid, maximum catalase activity was obtained with 2.5 μmol/l of linoleic acid and 5.0 μmol/l of alpha linolenic acid, maximum ascorbate peroxidase activity was obtained with 20 μmol/l of linoleic acid and 5.0 μmol/l of alpha linolenic acid as well as glutathione peroxidase was higher in case of 2.5 μmol/l of linoleic acid and alpha linolenic acid. These data confirmed the importance of fatty acids on the secondary metabolite production and antioxidative properties of ginseng plant (Wu et al. 2009).

### *1.4.2 Production of Secondary Metabolites Using Biotic Elicitors*

#### 1.4.2.1 Plant Growth Regulating Rhizobacteria Stimulated Secondary Metabolite Growth in Pennyroyal

Asghari et al. experimentally proved the effect of plant growth regulating rhizobacteria on the biosynthesis of secondary metabolites in pennyroyal (*Mentha pulegium*) plant under droght situation. The experiment was started with culturing the seeds of pennyroyal plant with rhizobacteria in four different categories such as control group without rhizobacteria, treated with *Azotobacter chroococcum*, treated with *Azospirillum brasilense* and final group treated with a combination of *A. chroococcum* and *A. brasilense* followed by treat in the environment with enough water, moderate water and less water condition. Relative water content, chlorophyll fluorescence, total phenolic content, total flavonoid content, antioxidative activities using glutathione peroxidase, catalase and superoxide dismutase and radical scavenging activity against DPPH were checked and evaluated by the elicited plant. The outcomes showed that relative water and chlorophyll contents were highest in enough water condition and lowest with less water condition; the glutathione peroxidase, catalase and superoxide dismutase activities were maximum less water condition with *A. chroococcum* elicitated condition followed by *A. brasilense* and its combination. Abscisic acid content, total phenolic content, total flavonoid content, radical scavenging activities as well as presence of essential oils such as 1,8-cineloe, menthone and pulegone were maximum in case of less water condition with *A. chroococcum* and *A. brasilense* combination; whereas piperitone was maximum obtained with moderate water condition with *A. chroococcum* and *A. brasilense* combination. So these data directly stated the importance of rhizobacteria on pennyroyal plant (Asghari et al. 2020).

#### 1.4.2.2 Effect of Carrageenan on Secondary Metabolite Growth in Chickpea and Maize Plant

Bi et al. scientifically proved the importance of carrageenan (polysaccharide) on the growth of secondary metabolite present in chick pea and maize plants. The process was started with formation of two types of elicitors as liquid carrageenan elicitor of 100 μg glucose concentration and solid carrageenan elicitor (mixer of hot aqueous extract of carrageenan with soil). The treatment module had three types: (i) 50 ml of liquid elicitor (ii) 5 g of solid elicitor (iii) 5 ml of liquid elicitor. These modules were applied to both chickpea and maize plants. Plant height, number of pods/branches/leaves per plant, and plant height, number of cobs per plant, stem diameter, number of leaves per plant, days to flowering, secondary metabolite production were evaluated for maize plant. Maximum results were obtained from treatment with 50 ml of liquid elicitor. So these data confirmed the importance of carrageenan in liquid form for the growth and production of secondary metabolite present in chickpea and maize plants (Bi et al. 2011).

#### 1.4.2.3 Effect of Fungal Elicitor on Phenylalanine Ammonia Lyase Activity in French Bean Cells

Bolwell et al. scientifically proved the importance of fungal pathogen obtained from *Colletotrichum lindemuthianum* fungal pathogen on phenylalanine ammonia lyase activity of cultured french bean cells. The experiment was started with treatment of cultured french bean cells with calcium ionophore, calmodulin inhibitor (trifluroperazine), calcium channel blocker (verapamil), pertussis toxin, cholera toxin, polyether antibiotic (monensin), labdane diterpene (forskolin), local anesthetic (procaine), caffeine and fungal elicitor. The outcomes revealed that on the activity of elicitor, caffeine and procaine had no direct effect whereas most effect on elicitation was observed with ionophore and most negative effect on elicitation was observed with monensin (Bolwell et al. 1991).

#### 1.4.2.4 Effect of Endophytic Fungi on Secondary Metabolite Accumulation in *Rumex gmelini* Turcz

Ding et al. scientifically proved the importance endophytic fungi (*Aspergillus sp.*, *Fusarium* sp., and *Ramularia* sp.) on the secondary metabolite accumulation of *R. gmelini* Turcz plant. The experiment was started with the development of explant using rhizomes of the plant in MS medium. After fifteen days, roots were co-cultured with endophytic fungi with three concentration range (1000, 10,000 and 100,000 $mL^{-1}$) followed by measured the amount of secondary metabolites (polydatin, resveratrol, chrysophaein, musizin, emodin, chrysophanol and physcion) accumulation. The outcomes revealed that *Aspergillus* sp. was used to increase the production of resveratrol, chrysophaein, musizin, emodin, chrysophanol and physcion whereas *Fusarium sp.* was used for polydatin. In case of combination treatment, *R. gmelini* seedling cultured with three endophytic fungi (*Aspergillus* sp. = 1000 $mL^{-1}$; *Fusarium* sp. = 10,000 $mL^{-1}$ and *Ramularia* sp. = 100,000 $mL^{-1}$) for a period of 20 days created positive effects on polydatin, resveratrol, chrysophaein, musizin, chrysophanol and physcion whereas productivity of emodin was increased many folds with combination of *R. gmelini* seedling cultured with two endophytic fungi *Aspergillus* sp. = 1000 $mL^{-1}$; *Fusarium* sp. = 10,000 $mL^{-1}$ after 15 days of treatment. These data cumulatively stated the importance of endophytic fungi on the production of secondary metabolites present in *R. gmelini* (Ding et al. 2018).

#### 1.4.2.5 Effect of Chitosan on Flavonoid Productivity in *Isatis tinctoria* L. Hairy Root Cultures

Jiao et al. experimentally proved the importance of chitosan on the increased production of flavonoid and antioxidative efficiency of *Isatis tinctoria* L. The experiment was started with formation of hairy root of *I. tinctoria* in MS medium with sucrose as principle constituent. After 24 days, chitosan in acetic acid (concentration = 50, 100, 150, 200 and 400 mg/l) were added to hairy roots of the plant followed by incubated for a period of 0, 6, 12, 18, 24, 30, 36, 48, 60, 72, and 96 h. Total flavonoid content especially (rutin, neohesperidin, buddleoside, liquiritigenin, quercetin, isorhamnetin, kaempferol, and isoliquiritigenin) were checked after the chitosan treatment. The outcomes revealed that total flavonoid content was increased with increasing concentration of chitosan (upto 200 mg/l); incase of 400 mg/l chitosan, the productivity of total flavonoid was decreased. Among other flavonoids, rutin, followed quercetin and isorhamnetin were highly expressed (Fig. 1.6). Antioxidative efficiencies of elicited hairy root cultures of *I. tinctoria* (150 mg/l) were good as compare to ascorbic acid (as per percent radical scavenging activity) and butylated hydroxytoluene (as per percent bleaching inhibition activity). These data confirmed the effects of chitosan on *I. tinctoria* secondary metabolite production (Jiao et al. 2018).

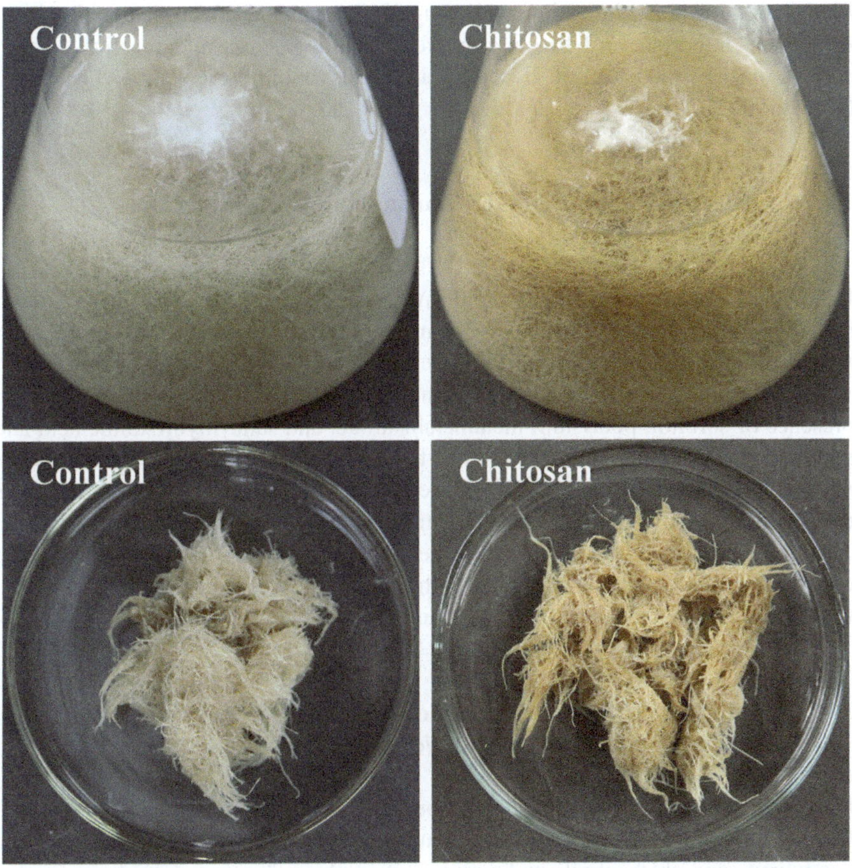


**Fig. 1.6** Phenotype comparison of hairy root tissues from control and ITHRCs elicited by 150 mg/L chitosan for 36 h. Copyright permission obtained from Jiao et al. @ 2018 Elsevier B.V

#### 1.4.2.6 Effect of Biotic Elicitor on Glyceollin in Soybean (*Glycine max*)

Kalli et al. scientifically proved the cumulative effects of reactive oxygen species and a mixture of fungal strains (*Rhizopus oligosporus* and *Rhizopus oryzae*) on the production of natural antimicrobial prenylated pterocarpan glyceollin and isoflavonoid present in soybean plant. The experiment was started with the formation of seedling using two different cultivar (I and II) in three phase (soaking, germination and priming). Soaking was done for one day followed by germination for two days. Priming was done by two phase such as early priming with reactive oxygen species and stress priming with slicing and sonication of seed for two days as well as late priming by mixture of fungal strains for five days. The outcomes revealed that the amount of glyceollin, glycinol (source of glyceollin) and isoflavonoid were maximum with primed and elicited (reactive oxygen species with fungal mixture) and sonication with fungal mixture dual effect. These data confirmed the importance of fungal mixture, wounding stress and reactive oxygen species on the production of glyceollin and isoflavonoid present in soybean plant (Kalli et al. 2020).

#### 1.4.2.7 Effect of Fungal Biotic Elicitor in Sign-Al Transduction in Potato Tubers

Kawakita et al. experimentally confirmed the positive role of fungal elicitor (*Phytophthora infestans*) in the involvement of guanosine triphosphate and guanosine triphosphatase activity in the tubers of potato. The experiment was started with reaction between potato tuber and hyphae wall component of *P. infestans* in presence of sorbitol, potassium metametabisulfate, and salicylhydroxamic acid and phenylmethylsulfonyl fluoride followed by centrifugation at 14,000 rotations per minute for fifteen minutes and again recentrifuged at one hour. The outcomes revealed that relative gamma guanosine triphosphatase activity was increased with 1 micromolar concentration of adenosine triphosphate upto 5 h with *P. infestans* elicitor. So these outcomes confirmed the importance of fungal strains in signal transduction in potato tuber (Kawakita and Doke 1994).

#### 1.4.2.8 Effect of Carbohydrate on Secondary Metabolite Accumulation in *Fagonia indica*

Khan et al. scientifically proved the importance of carbohydrates (sucrose, glucose, fructose and maltose) on the accumulation of secondary metabolites (phenolic compounds and chlorophyll) in *Fagonia indica* plant. The experiment was started with the formation of aforementioned plant callus culture by treating the explant of plant stem portion with different concentrations (1, 3 and 5%) of sucrose, glucose, fructose and maltose in MS medium followed by estimation of callus dimension, maximum production of biomass, total phenolic compound production, chlorophyll and estimation of free radical scavenging activity. The outcomes revealed the maximum width and height of callus were obtained with 5% and 3% sucrose concentration, respectively; maximum biomass production from fresh weight and dry weight were obtained from 3 and 5% sucrose concentration, respectively; maximum phenolic compounds were gained from 5% of maltose; phloroglucinol was the most accumulated phenolic compound obtained from all the carbohydrate callus as well as maximum radical scavenging activity was observed with 3% glucose elicitation (Fig. 1.7). So these data confirmed the importance of different carbohydrates on secondary metabolite accumulation in *Fagonia indica* (Khan et al. 2018).

**Fig. 1.7** Callus cultures as affected by various concentrations of Sucrose (S), Glucose (G), Fructose (F) and Maltose (M) 42 days after culture initiation. Copyright permission obtained from Khan et al. @ 2018 Elsevier B.V

#### 1.4.2.9 Effect of Dextran on Secondary Metabolite Accumulation and Improve Defense in Tomato Fruit

Lu et al. experimentally proved the importance of dextran on the accumulation of phenylpropanoid and flavonoids along with improved defense mechanism against grey mold infection in tomato (*Solanum lycopersicum*) fruit. The experiment was started with treatment between sterilized wounded tomato fruit (sodium hypochlorite treated) and dextran (0.1%, 0.5% and 1.0%) followed by ruthenium red treatment on the site of wound to view the effect of calcium channel blockers on site of wound as well as activity against grey mold infection *Botrytis cinerea* inoculation with dextran elicitor. The checked activities enlisted with percent disease incidence and development of lesion with dextran elicitor alone or along with ruthenium red after one and three days of inoculation on original and synthetic wounds after two days; total phenolic and flavonoid accumulation; phenylalanine lyase activity along with percent grey mold infection germination rate. The outcomes revealed that development of disease incidence and induction lesion were decreased with gradual increase in dextran concentration but the incidence was slightly higher with dextran and ruthenium red combination; accumulation of total phenolic and flavonoid compounds were higher with dextran elicitation after one day and two days of inoculation, respectively; the phenylalanine lyase activity was higher with dextran treatment after twelve hour of inoculation but it hits the lowest point after two of inoculation also the rate of spore germination in presence of *B. cinerea* was gradually decreased with dextran treatment. So these data confirm the importance of dextran as biotic elicitor to strengthening the plant against grey mold and greater accumulation of phenolic compound as well as flavonoid content in tomato fruit (Lu et al. 2019b).

#### 1.4.2.10 Effect of Yeast on Vincristine and Vinblastine Production in *Catharanthus roseus* Plant

Maqsood et al. scientifically proved the importance yeast as biotic elicitor on the production of vincristine and vinblastine obtained from protoplast sourced tissues and plant of *Catharanthus roseus* (commonly known as periwinkle). The experiment was commenced from protoplast culture involving incubation between suspended cells and enzymes treatment (cellulose, pectinase, macroenzyme and driselase) in MS medium followed by treatments with NAA, dichlorophenoxyacetic acid, BA and gibberellin within the same medium to culture the callus and embryo formation followed by elicitation using different concentrations of yeast (0.5, 1.0, 1.5 and 2.0 g/L). The proliferation, maturation and germination of vincristine and vinblastine after different yeast elicitation. The outcomes showed that maximum observations were observed with 1.5 and 2.0 g/L yeast concentrations, so these elicitations were selected for enzymatic activities (catalase, superoxide dismutase, ascorbate peroxidase and glutathione reductase) in the medium with different plant regulators. The results showed that maximum catalase and ascorbate peroxidase activities were observed with leaf harvested tissues with 2.0 g/L yeast in the medium contained BA and NAA; the superoxide dismutase and glutathione peroxidase activities were observed with 2.0 g/L yeast in the medium contained BA and NAA in embryo germination stage (Fig. 1.8). So these outcomes confirmed the importance of yeast on production of vincristine and vinblastine in *Catharanthus roseus* plant (Maqsood and Mujib 2017).

**Fig. 1.8** **a** Isolated protoplasts; **b** development of embryogenic callus from protoplasts; **c** individual embryo developed from PDEC treated with T3 treatment of YE and **d** regenerated plantlet from protoplast derived embryo, grown in MS medium containing yeast. Copyright permission obtained from Maqsood et al. @ 2017 Sociedade Brasileira de Farmacognosia

#### 1.4.2.11 Effect of Chitosan on Curcumin Production and Improved Defense in *Curcuma longa* Plant

Sathiyabama et al. experimentally proved the importance of chitosan polysaccharide on the curcumin production, increased biomass production from leaf and rhizome followed by estimation of protein and enzymatic activities against beta glucanase, peroxidase and polyphenol oxidase enzymes. The experiment was started with the reaction between new root/leaf of turmeric obtained from sodium hypochlorite treated turmeric rhizomes and chitosan (0.1% weight/volume) followed by elicited for a time period of six month. During the period, proper dosing of chitosan elicitation was provided after seven month leaves and roots were evaluated for the activities. The outcomes revealed that rhizome biomass was higher with fresh weight culture; chitosan elicitation was increased the curcumin production followed by elicitated leaves produced higher amount of protein as well as greater enzymatic activities against beta glucanase, peroxidase and polyphenol oxidase enzymes than elicitated rhizome. So these data confirmed the importance of chitosan on the accumulation of curcumin and enzymatic activity of turmeric plant (Sathiyabamaa et al. 2016).

### *1.4.3 Production of Secondary Metabolites Using Abiotic-Biotic Dual Elicitors*

#### 1.4.3.1 Effects of Microbial Elicitors on Glycyrrhizic Acid Production in *Taverniera cuneifolia* Culture

Awad et al. experimentally proved the importance of microbial elicitors on the production of glycyrrhizic acid from root cultures of *Taverniera cuneifolia* (Indian liquorice) plant. The experiment was started with the development of root from synthetic roots of the plant followed by treatment with fungal (*Aspergillus niger, Aspergillus tenius, Penicillium fellutanum, Fusarium moniliforme, Mucor hiemalis*) and bacterial (*Bacillus aminovorans, A. rhizogenes, A. tumefacians, B. cereus, Rhizobium leguminosarum*) cultures. Another elicitation process include treatment of root cultures (after 6th week) with MJ (1, 2.5, 5, 10, 100 and 1000) μM concentration. After three days treatment both elicited root cultures were stored and measured for glycyrrhizic acid production. The outcomes revealed that maximum glycyrrhizic acid was produced from *Fusarium moniliforme* and *R. leguminosarum* microbial elicitors. Another outcomes showed that glycyrrhizic acid was maximum obtained with MJ from plant biomass with 100 μM concentration (Fig. 1.9) These data confirmed the importance of microbial elicitors and MJ on the production of glycyrrhizic acid present in *T. cuneifolia* plant (Awad et al. 2014).

#### 1.4.3.2 Effects of Yeast and MJ on β-thujaplicin in *Cupressus lusitanica* Culture

Zhao et al. scientifically proved the importance of yeast as biotic elicitor on the production of natural antimicrobial agent generated from geranyl pyrophosphate

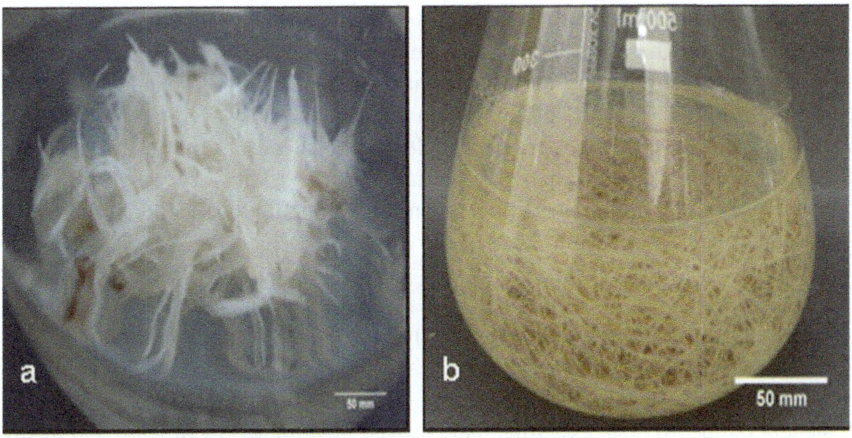


**Fig. 1.9** Root cultures of *T. cuneifolia*. **a** Mother culture used for root culture initiation and **b** six weeks old root culture used for elicitation. Copyright permission obtained from Awad et al. @ 2014 Elsevier B.V

present in *Cupressus lusitanica* cell culture. The process was stared with the formation of fresh suspension culture of the plant from fresh cells (4 g) followed by addition yeast (1 mg/ml) on to the five days old culture. Then in a single treatment all plant growth regulators such as calcium ionophore, cholera toxin, mastopaoran and actinomycin were added alone or along with twenty minutes prior to elicitation. The outcomes revealed that β-thujaplicin was greater in production with yeast elicitor within 2–4 days of incubation. The enzymatic activities against isopentenyl pyrophosphate isomerase, geranyl pyrophosphate synthase and monoterpene synthase showed that maximum activity within 1–2 days incubation with yeast elicitor. In all the cases MJ was just one step down than yeast elicitation but greater than dimethylsulfoxide control. These data confirmed the importance of yeast on the natural antimicrobial β-thujaplicin production from *C. lusitanica* callus culture (Zhao et al. 2006).

#### 1.4.3.3 Effects of Yeast, Chitosan, MJ and Heat on Secondary Metabolite Accumulation in Khus Root Extracts

Moon et al. scientifically proved the importance of yeast, chitosan, MJ and heat as dual abiotic-biotic elicitor on the production of para hydroxyl benzoic acid, vanillin, para coumaric acid, ferulic acid, total phenolic content followed by antioxidative effects using DPPH (Diphenyl picrylhydrazyl), ABTS (Azino bisethyl benzothiazoline sulfonic acid) and FRAP (Ferric reducing ability of plasma) methods as well as acetycholinesterase activity. The experiment was started with reaction between (5–6) long root of khus plant and chitosan biotic elicitor with 100, 200 and 300 mg/l or MJ with 25, 50 and 75 micromolar concentration or yeast elicitor with 5, 10 and

15 mg/ml concentration with or without treatment with 100 °C for twenty minute period. The outcomes revealed that maximum para hydroxyl benzoic acid, vanillin, para coumaric acid, ferulic acid, total phenolic content were obtained from heat treated chitosan (200 mg/l) concentration elicited khus dry weight. Antioxidative efficiency and acetycholinesterase data depicted that best activity observed with heat treated chitosan (200 mg/l) concentration elicited khus dry weight. So these data confirmed the importance chitosan on the production of secondary metabolite from khus plant (Moon et al. 2020).

#### 1.4.3.4 Effects of Yeast and MJ on Silymarin Production in *Silybum marianum* Culture

Sanchez-Sampedro et al. experimentally proved the importance of yeast and methyl as dual abiotic-biotic elicitor on the production and accumulation of silymarin obtained from *Silybum marianum* cell culture. The process was started with development of cell suspension culture from hypocotyl callus of the plant cultured in MS medium containing sucrose, dichlorophenoxyacetic acid, benzyladenine, then the suspension culture was elicited with 1 mg/ml concentration of yeast and 100 micromolar concentration of MJ for a period of 3 days. Also the silymarin production was monitored in presence of calcium antagonist, calcium effectors and inhibitors of protein kinase and protein phosphatase enzymes. The outcomes showed that MJ increased silymarin production many folds than yeast. It was also observed that silymarin production was exponentially increased with ruthenium red, neomycin and diphenylene iodonium treated both MJ and yeast elicitors. These data confirmed the importance of MJ and yeast elicitors on silymarin production present in *S. marianum* suspension culture (Sanchez-Sampedro et al. 2008).

#### 1.4.3.5 Effects of MJ, Phenylacetic Acid and Light Elicitors on *Ajuga bracteosa*

Ali et al. scientifically proved the importance of MJ, phenylacetic acid and light as dual abiotic and biotic elicitors on the production of volatile oils (α-phelendrene, Sabinene, α-terpinene, limonene, 1,8-cineole, γ-terpinene, β-pinene, β-myrcene, D-limonene, β-phelendrene, β-ocimene, Thujyl alcohol, cis-sabinol, β-linalool, 1-terpinene-4-ol, Cis-geraniol, α-terpineol, myrtenol, myrtenal, nerol, caryophyllene, β-farnesene, carvone, citronellyl acetate, p-cymen-7-ol, bornyl acetate) from cell cultures of *Ajuga bracteosa*. The process was started with the reaction between callus culture and plant growth regulators (kinetin, BA, dichlorophenoxyacetic acid, indole butyric acid) and elicitors (MJ and phenyl acetic acid) with 0.5, 1.0, 1.5 g/l of concentrations in three different luminous condition such as dark for one day, 16 h light with 8 h dark and complete light for one day. The amount of dry biomass, total phenolic content, total flavonoid content, free radical scavenging activity as well as volatile oils productions were evaluated. The outcomes showed that amount of dry biomass was maximum (11.3 g/l) with BA (1.0 g/l) in complete dark condition; total phenolic content was maximum (7.0 mg of gallic acid equivalent/g) with MJ (0.5 g/l) in complete light condition; total flavonoid content was maximum (3.8 mg quercetin equivalent/g) with MJ (0.5 g/l) in complete light condition; percent free radical scavenging activity was greater (86%) with MJ (0.5 g/l) in complete light condition; higher amount of volatile oils was observed with MJ (0.5 g/l) and BA (1.0 g/l) in dark period. So these data confirmed the importance of MJ, phenylacetic acid, other plant regulators and light treatment on the production and accumulation of secondary metabolites obtained from *A. bracteosa* cell culture (Ali et al. 2018).

#### 1.4.3.6 Effects of Salicylic Acid, Yeast and Casein Hydrolysate on Colchicine and Thiocolchicoside Production from *Gloriosa superba* Plant

Mahendran et al. scientifically proved the importance of salicylic acid, yeast, casein hydrolysate and silver nitrate on the accumulation of colchicine and thiocolchicoside in *Gloriosa superba* (Calihari) plant. The process was started with reaction between cell suspension culture (generated from plant rhizomes cultured in MS medium contained sucrose, dichlorophenoxyacetic acid, NAA) and salicylic acid (13.812, 27.624, 41.436, 55.248 and 69.060 mg/l), yeast, casein hydrolysate and silver nitrate with (100, 200, 300, 400 and 500 mg/l) followed by estimation of colchicine and thiocolchicoside after fifteen and thirty days of elicitation. The outcomes showed that a proper callus induction was happen with a combination of 2.0 mg/l of dichlorophenoxyacetic acid and 0.5 mg/l of NAA as well as maximum colchicine was obtained from casein hydrolysate (300 mg) and (27.624 mg) of salicylic acid after fifteen and thirty days of elicitation respectively whereas maximum thiocolchicoside was obtained from silver nitrate (200 mg) and (300 mg) of silver nitrate after fifteen and thirty days of elicitation respectively. So these data confirmed the importance of mixed abiotic and biotic elicitors on the accumulation of colchicine and thiocolchicoside in *G. superba* plant (Mahendran et al. 2018).

#### 1.4.3.7 Effects of Yeast and MJ on Polyphenolic Compound in *Aster scaber* Plant

Ghimire et al. scientifically proved the importance of biotic elicitor yeast and abiotic elicitor MJ on the production of polyphenolic compounds (flavonols, hydroxy cinnamic acid derivative, hydroxybenzoic acid derivative, vanillin, resveratrol and homogentisic), biomass production, total phenolic and flavonoid content as well as effects on antioxidative properties of the hairy root culture of the plant using diphenylpicraylhydrazyl and ferric reducing potential processes. The experiment was started with the formation of hairy root culture of *A. scaber* plant (Chwinamul) by the germination of seeds in MS medium contained with sucrose, cefotaxime antibiotic and excised with *A. rhizogenes* followed by elicited with yeast with 50 mg/l, 100 mg/l, 200 mg/l concentration and MJ with 50, 100, 200 micromolar concentration. The outcomes revealed that maximum biomass was obtained with MS medium, Sucrose (3%) with twenty seven days of interval. Maximum flavonol, hydroxy cinnamic acid derivative, hydroxybenzoic acid derivative, vanillin, resveratrol and homogentisic were obtained with MJ elicitation (Fig. 1.10). Total phenolic and flavonoid contents were higher with increased dose of both elicitors as well as MJ (100 micromolar concentration) was showed maximum antioxidative property. So these data confirmed the importance of MJ and yeast as elicitors on the accumulation of polyphenols in *A. scaber* plant (Ghimire et al. 2019).

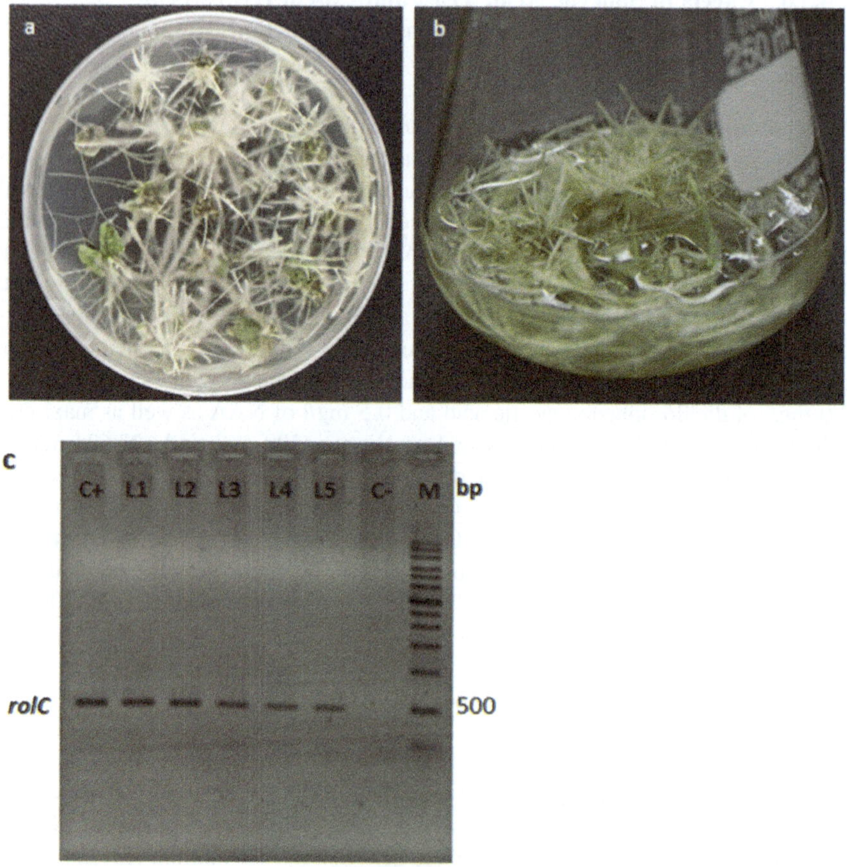


**Fig. 1.10** Agrobacterium rhizogenes-mediated hairy root cultures in *Aster scaber*. **a** Hairy roots induction, **b** Hairy root cultures in hormone-freeMS liquid medium, **c** PCR analysis of the rolC gene in the transgenic root lines. DNA ladder marker Lane M, pRiKCTC2703 DNA C(+), transgenic root lines induced by A. rhizogenes L1–L5, roots from a non-transgenic plant C(−). Copyright permission obtained from Ghimire et al. @ 2019 Elsevier B.V

#### 1.4.3.8 Effects of Phenylalanine, Salicylic Acid and Chitosan on Secondary Metabolites of *Coleus aromaticus* Benth

Govindaraju et al. experimentally confirmed the importance of phenylalanine, salicylic acid and chitosan as dual abiotic-biotic elicitor on the accumulation of alkaloid, flavonoid, saponin, terpenoids, total phenolic content followed by expression of phenylalanine messenger ribonucleic acid in the root/shoot culture and regenerated *Coleus aromaticus* Benth plant. The experiment was started from the induction of shoot followed by development of explants in the MS medium with activated charcoal and ascorbic acid as carbon and antioxidant source. The effect of BA and kinetin were also tested. The outcomes revealed that healthy explants were maximized with (7.5 g/l) activated charcoal and (1.0 mg/l) ascorbic acid. The percent induction of explants and number of shoots were maximum with a combination of (1 mg/l) of BA and kinetin. Different concentration of phenylalanine (0.5, 1.0, 1.5, 2.0, and 2.5) mg/l, salicylic acid (0.2, 0.4, 0.6, 0.8, and 1.0) mg/l and chitosan (20, 40, 60, and 80) mg/l were used as elicitors with BA (1 mg/l) concentration. The outcomes showed that maximum number of shoots were observed with 40 mg/l of chitosan cultured with BA as well as maximum development of roots was observed with (0.5 mg/l) of a combination NAA and IAA (Fig. 1.11). The outcomes also showed that regenerated explants observed with greater amounts of alkaloid, flavonoid, saponin, terpenoids, total phenolic content as well as higher

**Fig. 1.11** In vitro propagation of *C. aromaticus*, **a** shoot bud induction, **b** multiple shoot bud development from in vitro derived nodal explants, **c** multiple shoots (1.0 mg BAP + 40 mg/l Ch), **d** root initiation and **e**, **f** rooted plant growing in the plastic cup with soil and sand in the ratio of 1:1. Copyright permission obtained from Govindaraju et al. @ 2016 Elsevier B.V

expression of phenylalanine messenger ribonucleic acid. These data confirmed the effects of phenylalanine, salicylic acid and chitosan on secondary metabolites of *C. aromaticus* Benth (Govindaraju and Arulselvi 2018).

#### 1.4.3.9 Effects of MJ, Chitosan and Microbial Lysates on Essential Oil Accumulation in *Rhododendron tomentosum*

Jesionek et al. experimentally proved the importance of abiotic elicitors (copper, nickel, methyl jasmoante in dimethyl sulfoxide and ethanol, dimethylsulfoxide, ethanol) and biotic elicitors (chitosan hydrochloride, ergosterol and aphid extract in ethanol, lysates of *C. albicans*, *E. coli*, *Enterobacter sakazaki*, *Pectobacterium Carotovorum*, *Dickeya dadantii*) on the quantity of essential oils (monoterpene hydrocarbons, oxygenated monoterpenes, sesquiterpene hydrocarbons, oxygenated sesquiterpenes, aromadendrane-type sesquiterpenes) in *R. tomentosum* plant. The experiment was started with formation of in vitro culture of the plant on Schenk-Hildebrandt medium contained ammonium nitrate and isopentenyladenine within a period of one month. The formation of biomass was done using micro-shoots of the plant within sterilized bioreactor with proper illumination followed by elicitation using the aforementioned abiotic and biotic elicitors. The fresh weight and dry weight of biomass with growth index reflected that nickel, MJ in ethanol, ergosterol in ethanol and lysate of *Enterobacter sakazaki* showed greater production. The productions of monoterpene hydrocarbons, oxygenated monoterpenes, sesquiterpene hydrocarbons, oxygenated sesquiterpene and aromadendrane-type sesquiterpenes were maximized with copper and moderate with *Escherichia coli* lysates, nickel and ethanolic ergosterol and ethanolic MJ, respectively. The outcomes stated the importance of both abiotic and biotic elicitors on the production of volatile oils in *R. tomentosum* plant (Jesionek et al. 2018).

#### 1.4.3.10 Effects of Yeast, MJ and Salicylic Acid on Ursolic Acid and Eugenol Production in *Ocimum tenuiflorum* L.

Sharan et al. experimentally proved the importance of abiotic elicitors [(MJ: 0, 3, 60 and 120 mg/l) and (salicylic acid: 0, 30, 60 and 120 mg/l)] and biotic elicitor (yeast: 0, 25, 50 and 100 mg/l) on the production of ursolic acid and eugenol in *Ocimum tenuiflorum* L (holy basil) plant. The experiment was started with formation of hairy root culture from explants cultured in MS medium with cefotaxime and sucrose as antibiotic and sugar source, respectively followed by elicitation using aforementioned elicitors and evaluated the effects of *A. rhizogenes* bacterial strains (LBA 9402 and A4) on the transformation into leaf and stems as well as evaluated the production of biomass, ursolic acid and eugenol contents for a time period of thirty five days and finally quantified the effects of elicitors on seventeen and twenty one old hairy roots tendency to accumulate ursolic acid and eugenol within four to twelve days period. The outcomes revealed that LBA 9402 strain of *A. rhizogenes* facilitated the transformation of explants into leaf and stem; highest amounts of biomass, ursolic acid and eugenol accumulations were observed within twenty to twenty five days of incubation as well as maximum ursolic acid was accumulated with (50 mg/l) yeast, (60 mg/l) of methyl jasmoante and (60 mg/l) of salicylic acid from seventeen day old hairy roots and (50 mg/l) yeast, (0 mg/l) of methyl jasmoante and (0 mg/l) of salicylic acid from twenty one day old hairy roots; whereas maximum eugenol was accumulated with (50 mg/l) yeast, (60 mg/l) of methyl jasmoante and (0 mg/l) of salicylic acid from seventeen day old hairy roots and (0 mg/l) yeast, (0 mg/l) of methyl jasmoante and (0 mg/l) of salicylic acid from twenty one day old hairy roots. These data confirmed the importance of yeast, MJ and salicylic acid contents present in *O. tenuiflorum* L plant (Sharan et al. 2019).

## 1.5 Conclusion

This chapter provides a detailed information about What is Elicitation? Types of Elicitors along with applications of elicitors to increase productivity of secondary metabolites. This chapter also includes the information about secondary metabolites and its types such as alkaloid, glycoside, flavonoid and terpenoids etc. Elicitors are the physical/chemical/biological/microbial substances which under stress conditions induce the biosynthesis of secondary metabolites of plants. Both biotic and abiotic elicitors are used in the process. Most common secondary metabolites Ferulic acid, cinnamic acid, vanillin, coumaric acid, silymarin, affinin, hypocrellin A, steroiside, menthone, piperitone, glycyrrhizic acid, colchicine, thiocolchicoside, phenolic acid, gymnemic acid, flavonoids are utilized the elicitation technique. Elicitors are two types such as: abiotic and biotic. Abiotic elicitors such as salicylic acid, methyl jasmonate, hydrogen peroxide, lanthanum, different hormones, light, gamma rays and controlled temperature are used to generate secondary metabolites of wheat grass, *Thymus vulgaris*, *Silybum marianum*, *Shiraia bambusicola*, *Ajuga bracteosa*, broccoli plant, etc. Biotic elicitors like chitosan, rhizobacteria, *Rhizobium leguminosum*, *A. tenius*, *A. tumefacians*, carrageenan, Streptomyces, *Rhizopus*, dextran, yeast are used to develop or improvise secondary metabolites of Khus, *M. pulegium*, *T. cuneifolia*, chickpea, *V. vinifera*, *R. gmelini* Turcz, *C. lusitanica*, etc. Some secondary metabolites of *C. aromaticus* Benth, *R. tomentosum*, *F. indica*, *R. serpentine*, *S. khasianum*, *O. tenuiflorum*, *S. rebaudiana* etc. are used both abiotic and biotic elicitors. In the Table 1.1 detailed tabulated information of elicitors used to accumulate and increased productivity of secondary metabolites. Elicitors are not only increase secondary metabolite production as well as increase the defense mechanism, antioxidative, antimicrobial and enzymatic activities. This data collectively express the detailed knowledge about the elicitors and its signal transduction behavior. So if scientific minds hunts for secondary metabolites with greater activity but with very lesser abundance, then elicitation is the future.

**Table 1.1** Detailed effects of elicitors on different plants and secondary metabolites

| SN | Types of elicitation | Description of elicitors | Accumulated secondary metabolites | Reference |
|---|---|---|---|---|
| 1 | Abiotic | Arachidonic acid, Jasmonic acid | Flavonoid and polyphenolic compounds in *Triticum aestivum* L.) | Hassini et al. (2019) |
| 2 | | Methyl jasmonate, Salicylic acid | Benzylisoquinoline flavonoids in *Thymus vulgaris*, *Chelidonium majus* and *Petroselinum crispum* | Hesketh et al. (2002) |
| 3 | | Methyl jasmonate, Salicylic acid | Triterpenoid in *Centella asiatica* | Hiroaki et al. (2012) |
| 4 | | Salicylic acid, Hydrogen peroxide | Affinin contents in *Heliopsis longipes* | Jensen et al. (2014) |
| 5 | | Lanthum ($La^{3+}$) | Hypocrellin A content in *Shiraia bambusicola* | Jesionek et al. (2018) |
| 6 | | Salicylic acid, Methyl jasmonate, Citric acid, Ascorbic acid | Steviosides in *Stevia rebaudiana* Bertoni calli | Jiao et al. (2018) |
| 7 | | Light | Total phenolic and flavonoid contents in *Stevia rebaudiana* | Kalli et al. (2020) |
| 8 | | Gamma rays | Phenolic compounds and naphtodiantrones in *Hypericum triquetrifolium* Turra | Kawakita and Doke (1994) |
| 9 | | Streptomycin, Activated charcoal and Etephon | 3-*O*-glucosyl resveratrol and phenolic compound in *Vitis vinifera* | Kazmi et al .(2019) |
| 10 | | Salicylic acid | Alkaloid in marine microalgae *Arthrospira platensis* | Khan et al. (2018) |
| 11 | | Salicylic acid, Methyl jasmonate and Methionine | Chlorogenic acid and Sinapic acid in Broccoli plant | Kleinwachter et al. (2015) |
| 12 | | Methyl jasmonate, Phenyl acetic acid and Melatonin | Steviol glycoside in *Stevia rebaudiana* | Lena (2012) |
| 13 | | High intensity ultrasound | Total phenolic content, lycopene and carotenoid contents in *Solanum lycopersicum* | Lu et al. (2020) |

(continued)

**Table 1.1** (continued)

| SN | Types of elicitation | Description of elicitors | Accumulated secondary metabolites | Reference |
|---|---|---|---|---|
| 14 | | Methyl jasmonate, Salicylic acid and Glutathione | Total phenolic content and total flavonoid in *Trifolium resupinatum* plant | Lu et al. 2019.a |
| 15 | | Methyl jasmonate | Anthraquinone in *Rubia tinctorum* plant | Lu et al. (2019b) |
| 16 | Abiotic | Oleic acid and Linoleic acid | Gymnemic acid in *Gymnema sylvestre* | Mahendran et al. (2018) |
| 17 | | Magnesium oxide nanoparticle | Total phenolic and flavonoid contents in *Atropa belladonna* | Maqsood and Mujib 2017 |
| 18 | | Temperature | *E. coli*, *B. subtilis*, *S. aureus* *C. albicans* and *P. aeruginosa* in soil | Mejia-Espejel et al. (2018) |
| 19 | | Linoleic and Alpha linolenic acid | Total ginsenosides, diol ginsenoside, triol ginsenoside and total flavonoid contents in *Panax ginseng* | Moon et al. (2020) |
| 20 | Biotic | *Azotobacter chroococcum*, *Azospirillum brasilense*, *Azotobacter chroococcum* and *Azospirillum brasilense* | Relative water content, Chlorophyll fluorescence, Total phenolic content, Total flavonoid contents in *Mentha pulegium* | Pal et al. (2017) |
| 21 | | Carrageenan | Plant height, number of pods/branches/leaves per plant, and plant height, number of cobs per plant, stem diameter, number of leaves per plant, days to flowering, secondary metabolite productions in Chickpea and maize plant | Pal et al. (2019a) |
| 22 | | *Colletotrichum lindemuthianum* fungal | Phenylalanine ammonia lyase activity | Pal et al. (2018) |
| 23 | | *Aspergillus sp.*, *Fusarium sp.*, and *Ramularia sp.* | Resveratrol, Chrysophaein, Musizin, Emodin, Chrysophanol and Physcion in *Rumex gmelini* | Pal et al. (2019b) |

(continued)

**Table 1.1** (continued)

| SN | Types of elicitation | Description of elicitors | Accumulated secondary metabolites | Reference |
|---|---|---|---|---|
| 24 | | Chitosan | Rutin, Neohesperidin, Buddleoside, Liquiritigenin, Quercetin, Isorhamnetin, Kaempferol, and Isoliquiritigenin in *Isatis tinctoria* L | Pa and Saha (2019a) |
| 25 | | *Rhizopus oligosporus* and *Rhizopus oryzae* fungal strains | Prenylated pterocarpan glyceollin and Isoflavonoid present in soybean plant | Pa and Saha (2019b) |
| 26 | Biotic | Fungal elicitor (*Phytophthora infestans*) | Guanosine triphosphatase activity | Pal et al. (2019c) |
| 27 | | Sucrose, Glucose, Fructose and Maltose | Phenolic compounds and chlorophyll in *Fagonia indica* plant | Parola-Contreras et al. (2020) |
| 28 | | Dextran | Phenylpropanoid and flavonoids with improved defense mechanism against grey mold infection in tomato (*Solanum lycopersicum*) fruit | Perassolo et al. (2017) |
| 29 | | Yeast | Vincristine and Vinblastine in *Catharanthus roseus* | Pichersky and Gang (2000) |
| 30 | | Chitosan | Curcumin in *Curcuma longa* plant | Praveen et al. (2014) |
| 31 | Abiotic-Biotic | Fungal (*Aspergillus niger, Aspergillus tenius, Penicillium fellutanum, Fusarium moniliforme, Mucor hiemalis*) and bacterial (*Bacillus aminovorans, Agrobacterium rhizogenes, Agrobacterium tumefacians, Bacillus cereus, Rhizobium leguminosarum*) and Methyl jasmonate | Glycyrrhizic acid from root cultures of *Taverniera cuneifolia* | Sanchez-Sampedro et al. (2008) |
| 32 | | Yeast and Methyl jasmonate | β-thujaplicin in *Cupressus lusitanica* | Sathiyabamaa et al. (2016) |

(continued)

**Table 1.1** (continued)

| SN | Types of elicitation | Description of elicitors | Accumulated secondary metabolites | Reference |
|---|---|---|---|---|
| 33 | | Yeast, Chitosan, Methyl jasmonate and Heat | Para hydroxyl benzoic acid, Vanillin, Para coumaric acid, Ferulic acid, Total phenolic contents in Khus plant | Sato and Matsui (2012) |
| 34 | | Yeast and Methyl jasmonate | Silymarin production in *Silybum marianum* culture | Sharan et al. (2019) |
| 35 | | Methyl jasmonate, Phenylacetic acid and Light | Total phenolic and total flavonoid contents in *Ajuga bracteosa* | Tian et al. (2018) |
| 36 | | Salicylic acid, Yeast and Casein hydrolysate | Colchicine and Thiocolchicoside in *Gloriosa superba* plant | Twaij et al. (2019) |
| 37 | | Yeast and Methyl jasmonate | Polyphenolic compounds in *Aster scaber* plant | Ulaganathan et al. (2017) |
| 38 | Abiotic-Biotic | *Phenylalanine, Salicylic acid and Chitosan* | *Alkaloid, Flavonoid, Saponin, Terpenoids, Total phenolic contents in Coleus aromaticus Benth* | Wu et al. (2009) |
| 39 | | Abiotic elicitors (copper, nickel, methyl jasmoante in dimethyl sulfoxide and ethanol, dimethylsulfoxide, ethanol) and biotic elicitors (chitosan hydrochloride, ergosterol and aphid extract in ethanol, lysates of *Candida albicans, Escherichia coli, Enterobacter sakazaki, Pectobacterium Carotovorum, Dickeya dadantii* | Essential oils (monoterpene hydrocarbons, oxygenated monoterpenes, sesquiterpene hydrocarbons, oxygenated sesquiterpenes, aromadendrane-type sesquiterpenes) in *Rhododendron tomentosum* plant | Zhao et al. (2006) |
| 40 | | Yeast, Methyl jasmonate and Salicylic acid | Ursolic acid and Eugenol production in *Ocimum tenuiflorum* L | Złotek et al. (2019) |

## References


Ahmad N, Rab A, Ahmad N (2016) Light-induced biochemical variations in secondary metabolites production and antioxidant activity in callus cultures of *Stevia rebaudiana* (Bert.). J Photochem Photobiol B Biol 154:51–56

Ali H, Khan MA, Ullah N, Khan RS (2018) Impacts of hormonal elicitors and photoperiod regimes on elicitation of bioactive secondary volatiles in cell cultures of *Ajuga bracteosa*. J Photochem Photobiol B Biol 183:242–250

Asghari B, Khademian R, Sedaghati B (2020) Plant growth promoting rhizobacteria (PGPR) confer drought resistance and stimulate biosynthesis of secondary metabolites in pennyroyal (*Mentha pulegium* L.) under water shortage condition. Scientia Horticulture 263:109132

Awad V, Kuvalekar A, Harsulkar A (2014) Microbial elicitation in root cultures of *Taverniera cuneifolia* (Roth) Arn. for elevated glycyrrhizic acid production. Ind Crops Prod 54:13–16

Azeez H, Ibrahim K, Pop R, Pamfil D, Harta M, Bobis O (2017) Changes induced by gamma ray irradiation on biomass production and secondary metabolites accumulation in *Hypericum triquetrifolium* Turra callus cultures. Ind Crops Prod 108:183–189

Bi F, Iqbal S, Arman M, Ali A, Hassan M (2011) Carrageenan as an elicitor of induced secondary metabolites and its effects on various growth characters of chickpea and maize plants. J Saudi Chem Soc 15:269–273

Bolwell GP, Coulson V, Rodgers MW, Murphy DL, Jones D (1991) Modulation of the elicitation response in cultured french bean cells and its implication for the mechanism of signal transduction. Phytochemistry 30(2):397–405

Buchanan BB, Gruissem W, Jones RL (2015) Signal Transduction, In: Buchanan (ed). Biochemistry and molecular biology of plants. Wiley, London, pp 834–871

Buraphaka H, Putalun W (2020) Stimulation of health-promoting triterpenoids accumulation in *Centella asiatica* (L.) Urban leaves triggered by postharvest application of methyl jasmonate and salicylic acid elicitor. Ind Crops Prod 146: 112171

Cai Z, Riedel H, Saw NMMT, Mewis I, Reineke K, Knorr D, Smetanska I (2011) Effects of elicitors and high hydrostatic pressure on secondary metabolism of *Vitis vinifera* suspension culture. Process Biochem 46:1411–1416

Croteau R, Kutchan TM, Lewis NG (2000) Natural products (secondary metabolites). Biochem Mol Biol Plants 24:1250–1319

Ding CH, Wang QB, Guo S, Wang ZY (2018) The improvement of bioactive secondary metabolites accumulation in *Rumex gmelini* Turcz through co-culture with endophytic fungi. Braz J Microbiol 49(2):362–369

Ghimire BK, Thiruvengadam M, Chung IM (2019) Identification of elicitors enhances the polyphenolic compounds and pharmacological potential in hairy root cultures of *Aster scaber*. South African J Botany 125:92–101

Govindaraju S, Arulselvi IP (2018) Effect of cytokinin combined elicitors (L-phenylalanine, salicylic acid and chitosan) on in vitro propagation, secondary metabolites and molecular characterization of medicinal herb *Coleus aromaticus* Benth (L). J Saudi Soci Agric Sci 17 (4):435–444

Hadizadeh M, Ofoghi H, Kianirad M, Amidi Z (2019) Elicitation of pharmaceutical alkaloids biosynthesis by salicylic acid in marine microalgae *Arthrospira platensis*. Algal Res 42:101597

Hassini I, Rios JJ, Garcia-Ibanez P, Baenas N, Carvajal M, Moreno DA (2019) Comparative effect of elicitors on the physiology and secondary metabolites in broccoli plants. J Plant Physiol 239:1–9

Hesketh AR, Chandra G, Shaw AD, Rowland JJ, Kell DB, Bibb MJ, Chater KF (2002) Primary and secondary metabolism, and post-translational protein modifications, as portrayed by proteomic analysis of *Streptomyces coelicolor*. Mol Microbiol 46(4):917–932

Hiroaki S, Akira O, Ryo N, Fumio M, Kazuki S, Seiichi T (2012) Changes in primary and secondary metabolite levels in response to gene targeting-mediated site-directed mutagenesis of the anthranilate synthase gene in rice metabolites. 2(4):1123–1138

Jensen LM, Wallis IR, Marsh KJ, Moore BD, Wiggins NL, Foley WJ (2014) Four species of arboreal folivore show differential tolerance to a secondary metabolite. Oecologia 176:251–258

Jesionek A, Kokotkiewicz A, Krolicka A, Zabiegala B, Luczkiewicz M (2018) Elicitation strategies for the improvement of essential oil content in *Rhododendron tomentosum* (Ledum palustre) bioreactor-grown microshoots. Ind Crops Prod 123:461–469

Jiao J, Gaia QY, Wang X, Qin QP, Wang ZY, Liu J, Fu YJ (2018) Chitosan elicitation of *Isatis tinctoria* L. hairy root cultures for enhancing flavonoid productivity and gene expression and related antioxidant activity. Ind Crops Prod 124:28–35

Kalli S, Cloutier CA, Lin Y, Bruijn WJC, Chapman J, Vincken JP (2020) Enhanced biosynthesis of the natural antimicrobial glyceollins in soybean seedlings by priming and elicitation. Food Chem 317:126389

Kawakita K, Doke N (1994) Involvement of a GTP-binding protein in signal transduction in potato tubers treated with the fungal elicitor from *Phytophthora infestans*. Plant Sci 96:81–86

Kazmi A, Khan MA, Mohammad S, Ali A, Kamil A, Arif M, Ali H (2019) Elicitation directed growth and production of steviol glycosides in the adventitious roots of *Stevia rebaudiana* Bertoni. Ind Crops Prod 139:111530

Khan T, Abbasi BH, Zeb A, Ali GS (2018) Carbohydrate-induced biomass accumulation and elicitation of secondary metabolites in callus cultures of *Fagonia indica*. Ind Crops Prod 126:168–176

Kleinwachter M, Paulsen J, Bloem E, Schnug E, Selmar D (2015) Moderate drought and signal transducer induced biosynthesis of relevant secondary metabolites in thyme (*Thymus vulgaris*), greater celandine (*Chelidonium majus*) and parsley (*Petroselinum crispum*). Ind Crops Prod 64:158–166

Lena BB (2012) Metabolic defenses against plant allelochemicals. In: Gerald AR, May RB (eds) Herbivores: their interactions with secondary plant metabolites: ecological and evolutionary processes. Academic Press, London, pp 175–242

Lu C, Ding J, Park HK, Feng H (2020) High intensity ultrasound as a physical elicitor affects secondary metabolites and antioxidant capacity of tomato fruits. Food Control 113:107176

Lu C, Ma Y, Wang J (2019a) Lanthanum elicitation on hypocrellin A production in mycelium cultures of *Shiraia bambusicola* is mediated by ROS generation. J Rare Earths 37(8):895–902

Lu L, Ji L, Shi R, Li S, Zhang X, Guo Q, Wang C, Qiao L (2019b) Dextran as an elicitor of phenylpropanoid and flavonoid biosynthesis in tomato fruit against gray mold infection. Carbohydrate Polymer 225:115236

Mahendran D, Kishor PBK, Sreeramanan S, Venkatachalam P (2018) Enhanced biosynthesis of colchicine and thiocolchicoside contents in cell suspension cultures of *Gloriosa superba* L. exposed to ethylene inhibitor and elicitors. Ind Crops Prod 120:123–130

Maqsood M, Mujib A (2017) Yeast extract elicitation increases vinblastine and vincristine yield in protoplast derived tissues and plantlets in *Catharanthus roseus*. Revista Brasileira de Farmacognosia 27(5):549–556

Mejia-Espejel L, Robledo-Paz A, Lozoya-Gloria E, Pena-Valdiviac CB, Carrillo-Salazar JA (2018) Elicitors on steviosides production in Stevia rebaudiana *Bertoni calli*. Scientia Horticultura 242:95–102

Moon UR, Durge AA, Kumar M (2020) Value addition of khus (*Vetiveria zizanioides* L. Nash) root extracts by elicitor and heat treatment. Ind Crops Prod 144:112037

Pal D, Saha S (2019a) Chondroitin: a natural biomarker with immense biomedical applications. RSC Adv 9(48):28061–28077

Pal D, Saha S (2019b) Current status and prospects of chitosan-metal nanoparticles and their applications as nanotheranostic agents. In: M Rai, B Jamil (eds.) Nanotheranostics. Springer Nature, Switzerland, pp. 79–114

Pal D, Kumar S, Saha S (2017) Antihyperglycemic activity of phenyl and ortho-hydroxy phenyl linked imidazolyl triazolo hydroxamic acid derivatives. Int J Pharm Pharmaceut Sci 9(12):247–251

Pal D, Nayak AK, Saha S (2018) Interpenetrating polymer network hydrogels of chitosan: applications in controlling drug release. In: Mondal MdIH (ed) Cellulose-Based Superabsorbent Hydrogels. Polymers and Polymeric Composites, A Reference Series. Springer Cham, pp 1–41

Pal D, Nayak AK, Hasnain MS, Saha S (2019a) Pharmaceutical applications of chondroitin. In: Nayak AK, Hasnain MS, Pal D (eds) Natural polymers for pharmaceutical applications: volume 3: animal-derived polymers. Apple Academic Press, New York, pp 1–31

Pal D, Nayak AK, Saha S (2019b) Cellulose based hydrogel. In: Akhtar MS et al (eds) Natural bio-active compounds, vol 1: production and applications. Springer Nature, Singapore, pp 285–332

Pal D, Saha S, Nayak AK, Hasnain MS (2019c) Marine-derived polysaccharides: pharmaceutical applications. In: AK Nayak, Md Saquib Hasnain, D Pal (eds) Natural polymers for pharmaceutical applications: volume 2: marine and microbiologically derived polymers. Apple Academic Press, New York, pp 1–42

Parola-Contreras I, Tovar-Perez EG, Rojas-Molina A, Luna-Vazquez FJ, Torres-Pacheco I, Ocampo-Velazquez RV, Guevara-Gonzalez RG (2020) Changes in affinin contents in *Heliopsis longipes* (chilcuague) after a controlled elicitation strategy under greenhouse conditions. Ind Crops Prod 148:112314

Perassolo M, Cardillo AB, Mugas ML, Montoya SCN, Giulietti AM, Talou JR (2017) Enhancement of anthraquinone production and release by combination of culture medium selection and methyl jasmonate elicitation in hairy root cultures of *Rubia tinctorum*. Ind Crops Prod 105:124–132

Pichersky E, Gang DR (2000) Genetics and biochemistry of secondary metabolites in plants: an evolutionary perspective. Trends Plant Sci 5:439–445

Praveen N, Thiruvengadam M, Yang YS, Kim SH, Murthy HN, Chung IM (2014) Production of gymnemic acid from hairy root cultures of *Gymnema sylvestre* R. Br. as influenced by polyunsaturated fatty acids (PUFAs) and their antioxidant activity. Ind Crops Prod 54:54–61

Sanchez-Sampedro MA, Tarrago JF, Corchete P (2008) Some common signal transduction events are not necessary for the elicitor-induced accumulation of silymarin in cell cultures of *Silybum marianum*. J Plant Physiology 165:1466–1473

Sathiyabamaa M, Bernstein N, Anusuya S (2016) Chitosan elicitation for increased curcumin production and stimulation of defence response in turmeric (*Curcuma longa* L.). Ind Crops Prod 89:87–94

Sato F, Matsui K (2012): Engineering the biosynthesis of low molecular weight metabolites for quality traits (essential nutrients, health-promoting phytochemicals, volatiles, and aroma compounds). Plant Biotechnol Agric Prospects 21st Century 2012:443–461

Sharan S, Sarin NB, Mukhopadhyay K (2019) Elicitor-mediated enhanced accumulation of ursolic acid and eugenol in hairy root cultures of *Ocimum tenuiflorum* L. is age, dose, and duration dependent. South African J Botany 124:199–210

Tian H, Ghorbanpour M, Kariman K (2018) Manganese oxide nanoparticle-induced changes in growth, redox reactions and elicitation of antioxidant metabolites in deadly nightshade (*Atropa belladonna* L.). Ind Crops Prod 126:403–414

Twaij BM, Jazar ZH, Hasan MN (2019) The effects of elicitors and precursor on in-vitro cultures of *Trifolium resupinatum* for sustainable metabolite accumulation and antioxidant activity. Biocatal Agric Biotechnol 22:101337

Ulaganathan Y, Weber JFF, Convey P, Rizman-Idid M, Siti Aisyah Alias SA (2017) Antimicrobial properties and the influence of temperature on secondary metabolite production in cold environment soil fungi. Polar Sci 14:60–67

Wu CH, Popova EV, Hahn EJ, Paek KY (2009) Linoleic and α-linolenic fatty acids affect biomass and secondary metabolite production and nutritive properties of *Panax ginseng* adventitious roots cultured in bioreactors. Biochem Eng J 47:109–115

Zhao J, Matsunaga Y, Fujita K, Sakai K (2006) Signal transduction and metabolic flux of b-thujaplicin and monoterpene biosynthesis in elicited *Cupressus lusitanica* cell cultures. Metab Eng 8:14–29

Złotek U, Szymanowska U, Jakubczyk A, Sikora M, Swieca M (2019) Effect of arachidonic and jasmonic acid elicitation on the content of phenolic compounds and antioxidant and anti-inflammatory properties of wheatgrass (*Triticum aestivum* L.). Food Chem 288:256–261

# Chapter 2
# An Introduction to Bioactive Natural Products and General Applications

**Tijjani Ahmadu and Khairulmazmi Ahmad**

**Abstract** Nature has been a powerful source of potential medicinal plants with natural bioactive products for a long time period. The plants are known to contain quantum of various active principles of therapeutic value and possess biological activity against a number of diseases. These plants synthesized phytochemicals, serving as their natural defence system and also used in medicine, dye colours, fragrant, pharmaceutical, agrochemicals and flavouring. They also possess antimicrobial properties that are correlated with their ability to manufacture several secondary metabolites with antimicrobial properties like phenols, phenolic acids, flavonoids, alkanoids, tannins, quinones, coumarins, saponins, terpenoids, triterpenoids, glycosides and organic acids. Today, an increasing research interest in the herbal medicine field has been gaining popularity in both developed and developing countries. The bioactive compounds are provided by chemical diversity and natural plant products as purified compounds or as plant crude extracts. The extracts (crude) existed as a combination of different bioactive phytocompounds combined with various polarities and their partition still remains a challenge in the process of identification and characterization. Since bioactive natural products have diverse range of uses, the present book chapter constitutes a review on their distribution and geographical sources, phytochemistry, delivery technology, medicinal properties and their general potential applications in agricultural plant protection and pharmaceutical field.

T. Ahmadu (✉) · K. Ahmad
Department of Plant Protection, Faculty of Agriculture, University Putra Malaysia, 43400 UPM Serdang, Selangor Darul Ehsan, Malaysia
e-mail: tijjaniahmadu72@gmail.com

K. Ahmad
e-mail: tijjaniahmadu72@gmail.com

T. Ahmadu
Department of Crop Production, Faculty of Agriculture and Agricultural Technology, Abubakar Tafawa Balewa University, Bauchi, Bauchi State P. M. B 0248, Nigeria

K. Ahmad
Institute of Plantation Studies, University Putra Malaysia, 43400 UPM Serdang, Selangor Darul Ehsan, Malaysia



D. Pal and A. K. Nayak (eds.), *Bioactive Natural Products for Pharmaceutical Applications*, Advanced Structured Materials 140,
https://doi.org/10.1007/978-3-030-54027-2_2

**Keywords** Antimicrobial properties · Bioactive compounds · Biological activity · Medicinal plants · Phenolic acids

## 2.1 Introduction

Since times immemorial many countries across the planet have accumulated a rich body of empirical knowledge tracing the occurrence of bioactive natural compounds with medicinal properties from medicinal plants for the treatment of different illnesses throughout their long history. Natural bioactive compounds are metabolites manufactured by plants and their uses in different fields like medicine in form of pharmaceutical drugs and plant based bio-pesticides in agricultural plant protection are well known and documented. Approximately, 85% of preparation of traditional medicine involves the use of bioactive natural compounds extracted from plant (Meenupriya et al. 2014). Medicinal plants play a key role in world health and global economy (Meenupriya et al. 2014). Plants generate numerous categories of bioactive natural compounds, thus, making them rich source of various types of drugs and biopesticides (Meenupriya et al. 2014). All parts of these plants (roots, flowers, stem, bark, leaves, essential oils and seeds have microbial qualities and are therefore used for many purposes incuding traditional medicine (Dwivedi and Enespa 2012; Anwar et al. 2007). The choice and use of these potential plants and their general applications in pharmaceuticals and agriculture is due to: (1) their safety for human consumption, (2) non environmental pollution. (3) more acceptable to the pharmaceuticals and local farmers because they are indigenous (4) their biological activity and dispersion in harvested tissue (5) our ability to develop formulations that allow the delivery of non-toxic concentrations but at the same time interfere with antimicrobial development (Sogvar et al. 2016; Bhattacharjee and Dey 2014).

Today, research interest on medicinal plants is focus in recent time in such areas such as pharmacognosy, phytochemistry and horticulture. Bioactive compounds have been validated and detailed of their structural analysis have been present in the field or area phytochemistry. Ríos et al. (2015) and Jacob and Narendhirakannan (2019) have reported the use of such plants as fenugreek, caper, aloe, soybean, banaba, green tea, bitter melon, turmeric, cinnamon, walnut, coffee, guava, cocoa, garlic, sage, gymnema, nettle and yerba mate for treating illness like diabetes and its alike and the mechanisms of natural bioactive products as antidiabetic agents, with attention to compounds of high interest. The compounds are gurmarin, phlorizin, berberine, gymnemic acids, palmatine, honokiol, amorfrutins, trigonelline and fukugetin. The review described by (Jacob and Narendhirakannan 2019) has categorized 81 plants from literature that are native to Asian countries with anti-lipidemic, antidiabetic, insulin-mimetic, hypoglycemic, and antihyperglycemic properties. The presence of some bioactive compounds in the leaves, stem bark and flowers of Eugenia jambolana Lam. such as *n*-dotricontanol, anthocyanins, acylated flavonol glycosides delphinidin, friedelan-3-α-ol, friedelin, corilagin, petunidin, ellagic acid, garlic acid, jambosine, β-sitoterol, *n*-hentriacontane, gallotannin, myricitrin, quercetin, myricetin, *n*-nonacosane, ellagicacids, *n*-hepatcosane, *n*-triacontanol, isoquercetin, malvidin-diglucosides, myricetin, betulinic acid, pinocarveol, cineole, oleanolic acid, β-sitosterol-D-glucoside, muurolol kaempferol, myrtenol, eucarvone, noctacosanol, α-myrtenal, geranyl acetone, α-cadinol, 3-galloylglucose, pinocarvone, 3,6-hexahydroxy diphenoylglucose, 1-galloylglucose, 4,6-hexahydroxydiphenoylglucose, crategolic acid and flavonol glycosides, myricetin 3-O-(4″-acetyl)-α-L-rhamnopyranosides, α-terpeneol, were also reported (Baliga et al. 2011). Bio-assays and mechanisms of action of medicinal plants has been the interest in the area of pharmacognosy. The interest in horticultural research on medicinal plants has been on optimal growth and yield in cultivation. In this regards, bioactive compounds have been found to be good in improving postharvest quality, crop immunity and growth parameters (Freire et al. 2015; Mustafa et al. 2014). Advancement in search for alternatives to synthetic drugs and pesticides led to the wild harvesting of these plants yet the growth conditions have not been optimized. The consequences could be loss of biodiversity, improper identification of plant and potential variation in medicinal plant quality. Additionally, the feature of medicinal plants is seemingly being endangered and by complacency regarding their preservation. Stocks and herbs of these plants are in menace of extinction and diminishing in developing countries due to the demands for the growing market for cheaper alternative healthcare products and new plant-based therapeutic markets in preference to more expensive target-specific drugs and biopharmaceuticals.

Extracts of plant origin are found to be the most pressing sources of natural bioactive compounds and can be screened from local traditional plants (Mari et al. 2016; Freire et al. 2015). These plant bioactive compounds are extracted using solvents of different polarities and are shown to acquire antidiabetic, antifungal, antimicrobial, antioxidant and antibacterial properties etc. (Bhattacharjee and Dey 2014). Bioactive compounds obtainable from these plants have been shown to be the best alternative for pharmaceutical drugs and chemical fungicides (Bahare et al. 2019; Mari et al. 2016; Sogvar et al. 2016). Extraction of these bioactive natural compounds from species of medicinal plants is done by using various solvents and methods of extraction. Bioactive natural compounds are mostly aromatic or organic compounds that are saturated in nature. As a result, they are collected frequently through initial extraction with ethanol or methanol. Ethyl aceted, hexane, dicholro-methane, chloroform, acetone, butanol among others, and/or their appropriate ratio combination have been suggested by researchers (Gurjar et al. 2012) to get the best solvent systems for their extraction. Filtration is a common method for the extraction of bioactive natural compounds from plants which involves dissolving fresh ground sample (wet or dried) into certain volume of solvent, followed by vigorous shaking and filtration after sometimes usually after 24 h. Serial exhaustive is another extraction method for the bioactive compounds suggested by Green and Beestman (2007) which includes successive extraction with solvents of different polarity beginning from non-polar to higher polar solvents in order to make necessary extraction of crucial quantum compounds with enough and excellent polarity range. Gurjar et al. (2012) reported the extraction of phytochemical compounds from dried plant material with soxhlet method using organic solvent. High Pressure Liquid Chromatography is an analytical method that have been commonly used with other chromatography methods such as gas chromatography mass spectrometry (GC–MS) and liquid chromatography mass spectrometry/mass spectrometry (LC–MS/MS) for the separation and identification of bioactive natural compounds (Snyder et al. 2012; Villas-Boas et al. 2005). GC–MS had high ability in the partitioning and identification of compounds from multiple biological mixtures (Villas-Boas et al. 2005). More so, GC–MS technique can identify various bioactive natural compounds with lower molecular weight, volatile in nature and have different functions (Huang et al. 2012; Wang et al. 2012a, b). LC–MS is another chromatography technique apart from GC–MS where liquid chromatography is attached to spectrometry together with electrospray ionization and atmospheric pressure chemical ionization (Villas-Boas et al. 2005). LC–MS can analyse and identify huge number of non-volatile compounds even in smaller quantities. Nuclear magnetic resonance ($^{1}H$ NMR) and thin layer chromatography (TLC) are other methods utilized in the validation, characterization and separation of active compounds (Sharma and Paliwal 2013).

## 2.2 Bioactive Natural Compounds: Distribution and Geographical Sources

Table 2.1 has 251 plants species belonging to 98 genera and 56 families distributed in virtually 96 different continents and countries across the planet which includes Africa, Algeria, America, Asia, Bangladesh, Brazil, Cameroon, central America, China, Chinese medicines, Cuba, east Asia, Egypt, Ethiopia, Ghana, India, India (Ayurveda), India (Ayurveda, Unani), India (Ayurveda, Unani, Siddha and homeopathy), Japan, Ivory Coast, Iraq, Iran, Indonesia, Jordan, Kenya, Korea, Laos, Latin America, Marshall Islands, Mali, Mauritius, Mexico, Malaysia, New Guinea, Morocco, Nigeria, Pakistan, Papua, Paraguay, Philippines, Puerto Rico, Sierra Leone, Saudi Arabia, South Africa, Senegal, South East Asia, Nepal, Sri lanka, Sudan, South East Asian Countries,Taiwan, Tanzania, Thailand, Tobago, Togo, Trinidad and Tobago, Trinidad, Turkey, Uganda, Vietnam, West Indies. Of the 98 genera given, some have more than 8 species giving different biological activity. Eighteen (18) species were observed in *Ficus*, 14 in *Artemisia*, 11 in *Terminalia*, while *Cinnamomum, Phyllanthus and Ziziphus* with 9 each and *Acacia* and *Zizigium* with 8 each in that order. Of the 18 species in the *Ficus* genus, the most prominent and important species are, *Ficus elastic*, *Ficus benghalensis and F. hispida*. Indian Banyan tree also known as *F. benghalensis* is a frequent used plant to avert most illness diabetes inclusive (Jaya Kumari et al. 2016) and is used in folk medicines, Ayurveda, Unani, Siddha (Gopukumar and Praseetha 2015), and homeopathy (Deepa et al. 2018). *F. benghalensis*, *F. glomerata*, *F. carica*, *F. glumosa*, *F. religiosa* and

**Table 2.1** Distribution and Geographical Sources of Bioactive Natural Compounds

| Species | Bioactive compounds | Geographic zone | References |
|---|---|---|---|
| *Allium cepa* | Allyl propyl disulphide, S-methyl cysteine sulphoxide | Mauritius, Algeria | Mootoosamy and Fawzi Mahomoodally Ý(2014) |
| *Allium porrum* | S-methyl cysteine sulphoxide, Allyl propyl disulphide | Turkey | Bahare et al. (2019) |
| *Allium sativum* | Allyl propyl disulphide | China, India (Ayurveda), Iran, Indonesia, Cuba, Mauritius, Togo | García Mesa (2014), Sukandar et al. (2015) |
| *Allium stipitatum* | Allyl propyl disulphide | Iran | Bahare et al. (2019) |
| *Amaranthus hybridus* | Allyl propyl disulphide | Mauritius | Mondal et al. (2015) |
| *Amaranthus spinosus* | Allyl propyl disulphide | Taiwan | Bahare et al. (2019) |
| *Mangifera mekongensis* | Mangiferin, Flavonoid, Phenolics, | Vietnam | Nguyen et al. (2016) |
| *Panax notoginseng* | Saponin | China | Bahare et al. (2019) |
| *Panax quinquefolius* | Saponin | China, Korea | Bahare et al. (2019) |
| *Aloe marlothii* | Lophenol, Luteolin | South Africa | Sudha et al. (2011) |
| *Aloe ferox* | 24-methyl-lophenol, Luteolin, 24ethyllophenol | India (Ayurveda) | Kamel et al. (2017) |
| *Aloe vera* | 24-methyl-lophenol, Luteolin, 24ethyllophenol, Lophenol, cycloartanol, 24-methylene-cycloartanol | Saudi Arabia, Ghana, Philippines, Uganda Tanzania India, Pakistan, Iran, Ghana, Mauritius, Chinese medicines | Mina and Mina (2017), Asase and Yohonu (2016), Ssenyange et al. (2015) |
| *Artemisia absinthium* | Polysaccharide | Africa | Islam et al. (2014) |
| *Artemisia capillaris* | Polysaccharide | Africa | Islam et al. (2014) |

(continued)

**Table 2.1** (continued)

| Species | Bioactive compounds | Geographic zone | References |
|---|---|---|---|
| *Artemisia campestris* | Polysaccharide | Morocco | Dib et al. (2017) |
| *Artemisia dracunculus* | Polysaccharide | Morocco | Ota and Ulrih (2017) |
| *Artemisia judaica* | Polysaccharide | Jordan | Abu-Darwish et al. (2016) |
| *Artemisia ludoviciana* | Polysaccharide | Mexico | Anaya-Eugenio et al. (2014) |
| *Artemisia herba-alba* | Polysaccharide | Algeria, Iraq, Jordan | Nedjimi and Beladel (2015) Abu-Darwish et al. (2016) |
| *Artemisia parviflora* | Polysaccharide | India | Stanifer et al. (2015) |
| *Artemisia pallens* | Polysaccharide | Asia | Bahare et al. (2019) |
| *Artemisia princeps* | Polysaccharide | Asia | Bahare et al. (2019) |
| *Artemisia roxburghiana* | Polysaccharide | China | Shah et al. (2016a; b) |
| *Artemisia sacrorum* | Polysaccharide | China | Yuan et al. (2010) |
| *Artemisia sphaerocephala* | Polysaccharide | Jordan | Shah et al. (2016a, b) |
| *Eugenia jambolana* | Pandanus odorus | India (Ayurveda) | Kumar et al. (2013) |
| *Eugenia uniflora* | Tarpen | Paraguay | Bahare et al. (2019) |

(continued)

Table 2.1 (continued)

| Species | Bioactive compounds | Geographic zone | References |
|---|---|---|---|
| *Eugenia polyantha* | Tarpen | Indonesia, India | Jerang et al. (2015) |
| *Brassica rapa* | Isorhamnetin diglucoside | India | Jokar et al. (2016) |
| *Brassica oleracea* | Isorhamnetin diglucoside | India | Jokar et al. (2016) |
| *Terminalia alata* | Phenolics | Vietnam | Nguyen et al. (2016) |
| *Terminalia bellirica* | Phenolics | Bangladesh, Sri lanka, South East Asia, Vietnam, India | Nguyen et al. (2016), Tanaka et al. (2016) |
| *Terminalia arjuna* | Phenolics | Bangladesh, India (Ayurveda), | Nguyen et al. (2016) |
| *Terminalia chebula* | Phenolics | Bangladesh, Iran, India (Ayurveda), Thailand | Sudha et al. (2011) Kadir et al. (2012) |
| *Terminalia citrina* | Phenolics | Bangladesh | Biswas et al. (2014) |
| *Terminalia corticosa* | Phenolics | Vietnam | Nguyen et al. (2016) |
| *Terminalia macroptera* | Phenolics | Africa | Pham et al. (2014) |
| *Terminalia glaucescens* | Phenolics | Cameroon | Biswas et al. (2014) |
| *Terminalia superba* | Phenolics | India | Padmashree and Pandey (2010) |
| *Terminalia sericea* | Phenolics | Thailand | Nkobole et al. (2011) |

(continued)

**Table 2.1** (continued)

| Species | Bioactive compounds | Geographic zone | References |
|---|---|---|---|
| *Tremella mesenterica* | Phenolics | India (Ayurveda) | Singh et al. (2014), Mamun-or-Rashid et al. (2014) |
| *Calendula officinalis* | isorhamnetin | India | Sudha et al. (2011) |
| *Coccinia indica* | B-amyrin, Cucurbitacin B, Lupeol | India (Ayurveda) | Singh et al. (2014), Mamun-or-Rashid et al. (2014) |
| *Coccinia grandis* | B-amyrin, Cucurbitacin B, Lupeol | India(Ayurveda), Sri Lank | Pulbutr et al. (2017) |
| *Cucumis sativus* | B-carotne, Fatty acid | Malaysia | Panwar et al. (2014) |
| *Cucumis metuliferus* | Fatty acid, B-carotne | India | Jamal et al. (2011) |
| *Cucumis callosus* | B-carotne | India | Jamal et al. (2011) |
| *Momordica balsamina* | Luteolin, Saponin, Cucurbitacin, Alkaloid, Carotene, Fixed oil, Vitamin C, Resin acid, Momordicin | South Africa | Attanayake et al. (2015) |
| *Momordica cymbalaria* | Cucurbitacin, Glycoside, Steroidal glycoside or phenolics, Luteolin | Nigeria, South Africa | Balkhande and Surwase (2013) |
| *Momordica charantia* | Charantin, Cucurbitacin glycoside lanosterol, Luteolin, Momordicin, Galactosebinding lectin Non-bitter, Cholesterol, Diosgenin, β-sitosterol, | Bangledash, Nigeria, India (Ayurveda), Taiwan, Vietnam, Philippines, Mauritius, central America, Trinidad and Tobago | Akharaiyi et al. (2017), Giovannini et al. (2016) |
| *Momordica grosvenori* | Phenolics or Steroidal glycoside, Luteolin | China | Di et al. (2011) |
| *Momordica foetida* | Cucurbitacin, Saponin, Alkaloid Luteolin | South Africa | Mahomoodally and Ramalingum (2015) |

(continued)

**Table 2.1** (continued)

| Species | Bioactive compounds | Geographic zone | References |
|---|---|---|---|
| *Ibervillea sonorae* | Fatty acid, Monoglyceride (MG) | India | Mamun-or-Rashid et al. (2014) |
| *Diospyros peregrina* | Lupeol, Betulin, Betulinic acid, Gallic acid, Hexacosane, Hexacosanol | India | Kiran Kumar et al. (2012) |
| *Diospyros melanoxylon* | Phenolics | Sri Lanka, India | Dewanjee et al. (2011) |
| *Diospyros canaliculata* | Betulinic acid | Cameroon | Kuete and Efferth (2011) |
| *Diospyros crassiflora* | Phenolics, Hexacosane, Hexacosanol | Cameroon | Kuete and Efferth (2011) |
| *Diospyros lotus* | Phenolics, Sitosterol | India | Singh et al. (2014), Mamun-or-Rashid et al. (2014) |
| *Vaccinium angustifolium* | Anthocyanoside, Phenolic | India | Sánchez-Villavicencio et al. (2017) |
| *Vaccinium bracteatum* | Anthocyanoside, Phenolic | China | Qian et al. (2017) |
| *Vaccinium arctostaphylos* | Anthocyanoside, Phenolic | Iran | Nickavar and Amin (2010 |
| *Vaccinium vitis* | Anthocyanoside, Phenolic | | Beaulieu et al. (2010) |
| *Vaccinium myrtillus* | Anthocyanoside, Phenolic | China | Kellogg et al. (2010) |
| *Vaccinium ovalifolium* | Anthocyanoside, Phenolic | China, India | Mishra et al. (2013) |
| *Emblica officinalis* | Tannoid | India, Vietnam | Mamun-or-Rashid et al. (2014), Singh et al. (2014) |

(continued)

**Table 2.1** (continued)

| Species | Bioactive compounds | Geographic zone | References |
|---|---|---|---|
| *Jatropha curcas* | Diterpene | Nigeria, India, China | Mamun-or-Rashid et al. (2014), Singh et al. (2014) |
| *Phyllanthus acidus* | Tannin | India | Mamun-or-Rashid et al. (2014) |
| *Phyllanthus emblica* | Tannin | Thailand, Southeast Asia, India (Ayurveda) | Mamun-or-Rashid et al. (2014), Singh et al. (2014) |
| *Phyllanthus amarus* | Tannin | Malaysia, Nigeria, Vietnam, India(Ayurveda, Unani, Siddha and homeopathy) | Sarin et al. (2014), Adedapo and Ofuegbe (2014) |
| *Phyllanthus engleri* | Tannin | Tanzania | Bahare et al. (2019) |
| *Phyllanthus gardnerianus* | Tannin | India | Muthulakshmi et al. (2014) |
| *Phyllanthus urinaria* | Tannin | Vietnam | Bharati et al. (2016) |
| *Phyllanthus watsonii* | Tannin | Asia, Vietnam | Ramasamy et al. (2013) |
| *Phyllanthus niruri* | Tannin | Asia, Vietnam | Bharati et al. (2016) |
| *phyllanthus virgatus* | Tannin | Asia, Vietnam | Hashim et al. (2013) |
| *Acacia modesta* | Polyphenol, Tannin | Pakistan, India | Rao et al. (2015) |
| *Acacia senegal* | Polyphenol, Tannin | Sudan | Hilmi et al. (2014) |
| *Acacia ferruginea* | Polyphenol, Tannin | Sudan | Deb and Dash (2013) |
| *Acacia tortilis* | Polyphenol, Tannin | Sudan | Vadivel and Biesalski (2012) |

(continued)

**Table 2.1** (continued)

| Species | Bioactive compounds | Geographic zone | References |
|---|---|---|---|
| *Acacia farnesiana* | Polyphenol, Tannin | Bangladesh | Kingsley et al. (2013) |
| *Acacia catechu* | Polyphenol, Tannin | India, Nepal | Rao et al. (2015) |
| *Acacia nilotica* | Polyphenol, Tannin | India | Rao et al. (2015) |
| *Butea frondosa* | Palasonin, Butein, Genistein | India | Rao et al. (2015) |
| *Butea monosperma* | Stigmasterol-3, Genistein, Butein, β-D-glucopyranoside | India | Kumar et al. (2012) |
| *Bauhinia monandra* | Quercetin-3-O-rutinoside | India | Bahare et al. (2019) |
| *Cassia auriculata* | Triterpenoid, Sterol, Tannin, Flavonoid | Tanzania, India | Mamun-or-Rashid et al. (2014) |
| *Cassia sieberiana* | Triterpenoid, Sterol, Tannin, Flavonoid | Nigeria | Salihu Shinkafi et al. (2015) |
| *Cassia obtusifolia* | Triterpenoid, Sterol, Tannin, Flavonoid | China | He et al. (2014) |
| *Cassia fistula* | Triterpenoid, Sterol, Tannin, Flavonoid | India | Thakur et al. (2016) |
| *Cassia spectabilis* | Triterpenoid, Sterol, Tannin, Flavonoid | India | Mamun-or-Rashid et al. (2014) |
| *Clitoria ternatea* | kaempferol-3-neohesperidoside | Asia | Singh et al. (2014) |
| *Glycine max* | 3-*O*-methyl-D-chiro-inositol | Africa, Asia, Latin America | Singh et al. (2014) |
| *Tamarindus indica* | Polysaccharide, Flavonoid | Africa, Asia | Mamun-or-Rashid et al. (2014) |
| *Xanthocercis zambesiaca* | Castanospermine, Fagomine, 4-*O*-beta-D-glucopyranosylfagomine | Africa, Asia | Mamun-or-Rashid et al. (2014), Singh et al. (2014) |
| *Pongamia pinnata* | Pongaflavonol | India | Bahare et al. (2019) |

(continued)

**Table 2.1** (continued)

| Species | Bioactive compounds | Geographic zone | References |
|---|---|---|---|
| *Tephrosia purpurea* | Purpurin | India | Bahare et al. (2019) |
| *Cinnamomum burmannii* | Cinnamaldehyde | | Orhan et al. (2013) |
| *Cinnamomum iners* | Cinnamaldehyde | Malaysia | Mustaffa et al. (2014) |
| *Cinnamomum japonicum* | Cinnamaldehyde | Korea | Seo et al. (2013) |
| *Cinnamomum tamala* | Cinnamaldehyde | India (Ayurveda) | Bahare et al. (2019) |
| *Cinnamomum verum* | Cinnamaldehyde | India (Ayurveda) | Sudha et al. (2011) |
| *Cinnamomum obtusifolium* | Cinnamaldehyde | Bangladesh | Grover et al. (2012) |
| *Cinnamomum impressinervium* | Cinnamaldehyde | India | Zaidi et al. (2015) |
| *Cinnamomum cassia* | Cinnamaldehyde | India (Ayurveda, Unani), China, Japan, South Africa | Zaidi et al. (2015) |
| *Persea americana* | Mineral, fat, Protein, Vitamin | America | Mamun-or-Rashid et al. (2014) |
| *Mentha arvensis* | Flavonoid, Terpen, Essential oil, Zinc, Vanadium, Chromium, Iron, Copper, Sodium, Potassium, Nickel, Luteolin-7-*O*-glycoside | India | Rao et al. (2015) |
| *Mentha longifolia* | Flavonoid, Terpen, Essential oil, Zinc, Vanadium, Chromium, Iron, Copper, Sodium, Potassium, Nickel, Luteolin-7-*O*-glycoside | India | Rao et al. (2015) |

(continued)

**Table 2.1** (continued)

| Species | Bioactive compounds | Geographic zone | References |
|---|---|---|---|
| *Mentha piperita* | Flavonoid, Terpen, Essential oil, Zinc, Vanadium, Chromium, Iron, Copper, Sodium, Potassium, Nickel, Luteolin-7-O-glycoside | India | Bahare et al. (2019) |
| *Ocimum canum* | Saponin, Phenolics, Tannin, Eugenol (1-hydroxy-2-methoxy-4-allylbenzene), | Ghana | Berhow and Affum (2012) |
| *Ocimum campechianum* | Saponin, Tannin, Eugenol (1-hydroxy-2-methoxy-4-allylbenzene), Phenolics | Trinidad and Tobago | Bahare et al. (2019) |
| *Ocimum sanctum* | Saponin, Phenolics, Tannin, Eugenol (1-hydroxy-2-methoxy-4- allylbenzene), | Bangladesh, China, India (Ayurveda) | Upadhyay (2017) |
| *Ocimum gratissimum* | Saponin, Tannin, Eugenol (1-hydroxy-2-methoxy-4-allylbenzene), Phenolics | Nigeria, Bangladesh | Akharaiyi et al. (2017) |
| *Ocimum tenuiflorum* | Saponin, Tannin, Phenolics | India (Ayurveda) | Mousavi et al. (2016) |
| *Salvia officinalis* | Flavonoid | India | Mamun-or-Rashid et al. (2014) |
| *Glyccheriza glabra* | Mimosoidea | India | Sudha et al. (2011) |
| *Thespesia populnea* | Populnetin, Quercetin, Populneol, Herbacetin | India | Singh et al. (2014), Mamun-or-Rashid et al. (2014) |
| *Melia azadirachta* | Liminoid | India | Ranganathan et al. (2012) |
| *Melia orientalis* | Liminoid | India (Ayurveda) | Marimuthu et al. (2013) |
| *Melia dubia* | Liminoid | Mexico | Manosroi et al. (2011) |
| *Moringa peregrina* | Phenolics, Tannin, Saponin, Flavonoid, Steroid, Glycoside, Quercetin, Stearic acid, Malic acid | West Indies | Ullah et al. (2015) |

(continued)

**Table 2.1** (continued)

| Species | Bioactive compounds | Geographic zone | References |
|---|---|---|---|
| *Moringa stenopetala* | Phenolics, Tannin, Saponin, Flavonoid, Steroid, Glycoside, Quercetin, Stearic acid, Malic acid | Ethiopia | Geleta et al. (2016) |
| *Artocarpus communis* | Sapogenin | Nigeria | Bahare et al. (2019) |
| *Artocarpus altilis* | Sapogenin | Indonesia, Mauritius Trinidad and Tobago | Mahomoodally and Ramalingum (2015), Wahyudin et al. (2017) |
| *Artocarpus mariannensis* | Sapogenin | Marshall Islands | Englberger et al. (2014) |
| *Ficus bengalensis* | Leucopelargonidin | Southeast Asia, India (Ayurveda, Siddha, Unani, homoeopathy) | Gopukumar and Praseetha (2015) |
| *Ficus religiosa* | Leucopelargonidin | India (Ayurveda) | Singh et al. (2011) |
| *Ficus virens* | Leucopelargonidin | India (Ayurveda) | Babu et al. (2010) |
| *Ficus racemosa* | Leucopelargonidin | Bangladesh, Southeast Asia, India | Shah et al. (2016a, b) |
| *Ficus cunia* | Leucopelargonidin | India | Sheikh et al. (2015) |
| *Ficus carica* | Leucopelargonidin | India (Ayurveda, Siddha, Unani, homoeopathy) | Badgujar et al. (2014) |
| *Ficus thonningii* | Leucopelargonidin | Africa | Ahmed et al. (2012) |
| *Ficus sansibarica* | Leucopelargonidin | Africa | Awolola et al. (2014) |
| *Ficus lutea* | Leucopelargonidin | Africa | Olaokun et al. (2014) |
| *Ficus palmata* | Leucopelargonidin | Africa | Shahreen et al. (2012) |
| *Ficus microcarpa* | Leucopelargonidin | south Asia | Akhtar et al. (2016), Chan et al. (2017) |
| *Ficus hispida* | Leucopelargonidin | Bangladesh | Singh et al. (2014) |

(continued)

**Table 2.1** (continued)

| Species | Bioactive compounds | Geographic zone | References |
|---|---|---|---|
| *Ficus glumosa* | Leucopelargonidin | Cameroon, Nigeria | Zayyanu et al. (2015) |
| *Ficus elastica* | Leucopelargonidin | Philippines | Kamel et al. (2017) |
| *Ficus deltoidea* | Leucopelargonidin | Southeast Asia, Malaysia | Fidele et al. (2015) |
| *Ficus exasperata* | Leucopelargonidin | Nigeria, Ivory Coast, Cameroon, Sierra Leone | Salihu Shinkafi et al. (2015) |
| *Ficus glomerata* | Leucopelargonidin | India (Ayurveda, Siddha, Unani, homoeopathy) | Vaishnav et al. (2015) |
| *Ficus amplissima* | Leucopelargonidin | India (Ayurveda, Siddha, Unani) | Arunachalam and Parimelazhagan (2013) |
| *Musa sapientum* | Flavonoid, Glycoside, Steroid, Pectin | Africa, India | Jayamurthy et al. (2013) |
| *Musa acuminata* | Flavonoid, Glycoside, Steroid, Pectin | Africa, India | Venkatesh et al. (2013) |
| *Eucalyptus globules* | Calytoside | Iran | Asgharpour et al. (2012) |
| *Psidium guajava* | Polysaccharide, Flavonoid, Terpen, Pedunculagin, Isostrictinin, Strictinin | China, Togo, New Guinea, Sri Lanka, Japan, Mauritius, Central America, Papua, | Deguchi and Miyazaki (2010) |
| *Psidium cattleianum* | Polysaccharide, Flavonoid, Terpen, Pedunculagin, Isostrictinin, Strictinin | East Asia | Im et al. (2012) |
| *Syzygium aromaticum* | Gallic acid, Malic acid, Citric acid, Anthocyanin | India | Adefegha et al. (2014) |
| *Syzygium densiflorum* | Gallic acid, Malic acid, Citric acid, Anthocyanin | India | Muthusamy and Krishnasamy (2016) |
| *Syzygium jambosa* | Gallic acid, Malic acid, Citric acid, Anthocyanin | Puerto Rico | Gavillán-Suárez and Aguilar-Perez (2012) |
| *Syzygium alternifolium* | Gallic acid, Malic acid, Citric acid, Anthocyanin | Puerto Rico | Kasetti et al. (2012) |

(continued)

**Table 2.1** (continued)

| Species | Bioactive compounds | Geographic zone | References |
|---|---|---|---|
| *Syzygium samarangense* | Gallic acid, Malic acid, Citric acid, Anthocyanin | Bangladesh | Shahreen et al. (2012) |
| *Syzygium cumini* | Gallic acid, Malic acid, Citric acid, Anthocyanin | Bangladesh, Brazil, India (Ayurveda), | Bansode and Salalkar (2015), Sharma et al. (2017) |
| *Syzygium jambolanum* | Gallic acid, Malic acid, Citric acid, Anthocyanin | India (Ayurveda) | Baliga et al. (2013) |
| *Syzygium cordatum* | Gallic acid, Malic acid, Citric acid, Anthocyanin | India | Mamun-or-Rashid et al. (2014) |
| *Biophytum sensitivum* | Phenol | India | Mamun-or-Rashid et al. (2014), Singh et al. (2014) |
| *Butea monosperma* | Flavonoid | India | Rao et al. (2015) |
| *Piper guineense* | Saponin, Tannin, Terpen, Terpenoid, Flavonoid | Nigeria | van de Venter et al. (2008) |
| *Piper betle* | Saponin, Tannin, Terpen, Terpenoid, Flavonoid | Asia | Srividya et al. (2015) |
| *Piper sarmentosum* | Saponin, Tannin, Terpen, Terpenoid, Flavonoid | South East Asia | Sh Ahmed et al. (2017) |
| *Piper cubeba* | Saponin, Tannin, Terpen, Terpenoid, Flavonoid | India | Sh Ahmed et al. (2017) |
| *Piper crocatum* | Saponin, Tannin, Terpen, Terpenoid, Flavonoid | India | Sh Ahmed et al. (2017) |
| *Piper nigrum* | Saponin, Tannin, Terpen, Terpenoid, Flavonoid | India | Mishra et al. (2011) |
| *Triticum vulgare* | Albumin | India, Nigeria | Singh et al. (2014) Mamun-or-Rashid et al. (2014) |
| *Panax quinquefolius* | Tannin | India | Bahare et al. (2019) |
| *Panax ginseng* | Tannin | Korea | Lee et al. (2016) |

(continued)

**Table 2.1** (continued)

| Species | Bioactive compounds | Geographic zone | References |
|---|---|---|---|
| *Ziziphus xylopyrus* | Fatty acid, Christinin-A, Alkaloid | China, India (Ayurveda), Pakistan | Modi et al. (2014) |
| *Ziziphus oxyphylla* | Fatty acid, Christinin-A, Flavonoid, Alkaloid, Tannin, Saponin | Pakistan | Ahmad et al. (2017) |
| *Ziziphus mucronata* | Fatty acid, Christinin-A, Flavonoid, Alkaloid, Tannin, Saponin | Nigeria | Ibrahim and Islam (2017) |
| *Ziziphus nummularia* | Fatty acid, Christinin-A, Flavonoid, Alkaloid, Tannin, Saponin | India | Goyal et al. (2011) |
| *Ziziphus lotus* | Fatty acid, Flavonoid, Alkaloid, Tannin, Saponin | Algeria | Benammar et al. (2010) |
| *Ziziphus mauritiana* | Christinin-A, Tannin, Flavonoid, Alkaloid, Saponin | Mali, Southeast Asia | Marwat et al. (2014) |
| *Ziziphus jujuba* | Flavonoid, Alkaloid, Tannin, Saponin | Turkey | Sadegh-Nejadi et al. (2018) |
| *Ziziphus amole* | Fatty acid, Christinin-A, Flavonoid, Alkaloid, Tannin, Saponin | Mali | Romero-Castillo et al. (2013) |
| *Mucuna pruriens* | Alkaloid, Carbazole | India (Ayurveda) | Bahare et al. (2019) |
| *Limonia acidissima* | Polysaccharide | India | Singh et al. (2014), Mamun-or-Rashid et al. (2014) |
| *Aegle marmelos* | Alkaloid, Coumarin, Aegeline 2, Flavonoid | India | Singh et al. (2014), Mamun-or-Rashid et al. (2014) |
| *Citrus aurantium* | Sterol, Essential oil | China | Choi et al. (2012) |
| *Citrus sinensis* | Sterol, Essential oil | India | Shakthi Deve et al. (2014) |

(continued)

**Table 2.1** (continued)

| Species | Bioactive compounds | Geographic zone | References |
|---|---|---|---|
| *Citrus reticulata* | Sterol, Essential oil | China | Choi et al. (2012) |
| *Citrus grandis* | Sterol, Essential oil | China | Tzeng et al. (2011) |
| *Citrus paradisi* | Sterol, Essential oil | Nigeria, Trinidad and Tobago, Cuba | García Mesa (2014), Adeneye (2008) |
| *Citrus medica* | Hesperidin | China | Sudha et al. (2011) |
| *Feronia elephantum* | Bergapten,Triterpenoid, Bioflavonoid, Stigma sterol | India | Singh et al. (2014), Mamun-or-Rashid et al. (2014) |
| *Bacopa moneirra* | Tannin, Luteolin | India | Sudha et al. (2011) |
| *Physalis alkekengi* | Polysaccharide | India | Ramasamy et al. (2013) |
| *Physalis peruviana* | Polysaccharide | India | Maobe et al. (2013) |
| *Physalis mnima* | Polysaccharide | India | Ayyanar and Ignacimuthu (2011) |
| *Physalis angulata* | Polysaccharide | India | Ramasamy et al. (2013) |
| *Withania somnifera* | Alkaloid, Withanolide | India, Ayurveda | Rehman et al. (2015), Mishra et al. (2013), Maurya et al. (2008) |
| *Withania coagulans* | Alkaloid, Fatty oil, Milk-coagulating enzyme, Esterase, Essential oil, | Pakistan, India, Ayurveda | Jonathan et al. (2015) |
| *Lycium barbarum* | Polysaccharide | China | Potterat (2010) |

(continued)

**Table 2.1** (continued)

| Species | Bioactive compounds | Geographic zone | References |
|---|---|---|---|
| *Lycium ruthenicum* | Polysaccharide | China | Dhar et al. (2011) |
| *Lycium chinense* | Polysaccharide | China | Potterat (2010), Wang and Ye (2016), Chan et al. (2009) |
| *Helicteres hirsuta* | Terpenoid, Steroid, Alkaloid, Phenolics, Carbohydrate, | Southeast Asia | Varghese et al. (2012) |
| *Urtica dioica* | Lectin, Flavonoid, Coumarin | Turkey, Iran, Kenya | Rezaei Aref et al. (2012), Nickavar and Yousefian (2011) |
| *Urtica angustifolia* | Coumarin | Africa | Zhang et al. (2012) |
| *Curcuma xanthorrhiza* | Curcuminoid | Bangladesh, Indonesia, Laos | Peltzer et al. (2016) |
| *Curcuma longa* | Curcuminoid | China, Laos Indonesia, Bangladesh, India (Ayurveda) | Yadav and Chaudhury (2016), Peltzer et al. (2016), Mishra et al. (2011) |
| *Curcuma domestica* | Curcuminoid | India | Sushma et al. (2015) |
| *Zingiber striolatum* | Ethanol, Gingerol | China | Chen et al. (2016) |
| *Zingiber officinale* | Ethanol, Gingerol | Africa, Latin America India (Ayurveda) | Sudha et al. (2011), Ranilla et al. (2010), Morakinyo et al. (2011) |
| *Andrographis paniculata* | 5-hydroxy-7,8-dimethoxyflavone | India, Nigeria | Sudha et al. (2011) |

(continued)

**Table 2.1** (continued)

| Species | Bioactive compounds | Geographic zone | References |
|---|---|---|---|
| *Allium ampeloprasum* | S-allyl cysteine, Ajoene, Allyl propyl disulfide,, S-allyl mercaptocysteine, Diallyl disulphide oxide | Iran | Rahimi-Madiseh et al. (2017) |
| *Amaranthus cruentus* | Allyl propyl disulphide | Kenya | Mahomoodally and Ramalingum (2015) |
| *Mangifera indica* | Mangiferin, Flavonoid, Phenolics, | Nigeria, India | Sudha et al. (2011) |
| *Cuminum cyminum* | Aldehyde | India | Mnif and Aifa (2015) |
| *Catharanthus roseus* | Alkaloid, Vinculin | South Africa, China, Malaysia, Trinidad, South East Asian countries, India | Marwat et al. (2014), Rasineni et al. (2010) |
| *Panax ginseng* | Saponin | Korea | Xia et al. (2014) |
| *Aloe barbadensis* | Lophenol, 24-methyl-lophenol, 24ethyllophenol, Luteolin | India (Ayurveda) | Bhaludra et al. (2013) |
| *Artemisia afra* | Polysaccharide | Africa | Bahare et al. (2019) |
| *Betula pendula* | Quercetrin | Indonesia, India | Sudha et al. (2011) |
| *Oroxylum indicum* | Chrysin | India | Bahare et al. (2019) |
| *Brassica juncea* | Isorhamnetin diglucoside | India (Ayurveda) | Bahare et al. (2019) |
| *Opuntia dillenii* | Polysaccharide | India | Mamun-or-Rashid et al. (2014), Singh et al. (2014) |
| *Viburnum opulus* | Tanin | Africa, India | Singh et al. (2014) |
| *Carica papaya* | Alkaloid, Flavonod, Saponin, Tannin, Glycoside, Reducing sugar, Anthraquinone | Africa, Asia | Mamun-or-Rashid et al. (2014) |
| *Beta vulgaris* | Polydextrose, Sugar beet pectin | India | Mamun-or-Rashid et al. (2014) |

(continued)

**Table 2.1** (continued)

| Species | Bioactive compounds | Geographic zone | References |
|---|---|---|---|
| *Terminalia catappa* | Phenolics | India | Venkatalakshmi et al. (2014) |
| *Cannabis sativa* | Quercetin | Latin America, India | Sudha et al. (2011) |
| *Coccinia cordifolia* | B-amyrin, Cucurbitacin B, Lupeol | India | Singh et al. (2014) |
| *Vaccinium uliginosum* | Anthocyanoside, Phenolic | Iran | Singh et al. (2014) |
| *Acalypha indica* | Kaempferol glycoside | India | Sudha et al. (2011) |
| *Acacia arabica* | Polyphenol, Tannin | India | Yasir et al. (2010) |
| *Ganoderma lucidum* | Polysaccharide Fraction | India | Singh et al. (2014) |
| *Leonotis leonurus* | Terpen | India | Mamun-or-Rashid et al. (2014) |
| *Cinnamomum zeylanicum* | Cinnamaldehyde | | Orhan et al. (2013) |
| *Cajanus cajan* | 7R*,9as*)-7-phenyloctahydroquinolizin-2-one | Africa, Asia | Mamun-or-Rashid et al. (2014) |
| *Lyophyllum decastes* | Polysaccharide | India | Mamun-or-Rashid et al. (2014) |
| *Brysoni macrassa* | (+)-catechin | India | Geleta et al. (2016) |
| *Abelmoschus esculentus* | Carbohydrate, Flavonoid, Gum, Phenolics, Mucilage, Tannin Protein, Phytosterol, Volatile oil | India | Mamun-or-Rashid et al. (2014) |
| *Azadirachta indica* | Nimbidin, Quercetin | India, Africa | Mamun-or-Rashid et al. (2014) |

(continued)

**Table 2.1** (continued)

| Species | Bioactive compounds | Geographic zone | References |
|---|---|---|---|
| *Grifola frondosa* | Disaccharide | India | Mamun-or-Rashid et al. (2014) |
| *Artocarpus heterophyllus* | Sapogenin | Mauritius, India (Ayurveda) | Bahare et al. (2019) |
| *Moringa oleifera* | Phenolics, Tannin, Saponin, Flavonoid, Steroid, Glycoside, Quercetin, Stearic acid, Malic acid | India, Nigeria, Kenya, South Africa, Senegal, Mexico, Mauritius | Geleta et al. (2016) |
| *Musa paradisiaca* | Pectin, Dietary fibre | Africa, India | Venkatesh et al. (2013) |
| *Eucalyptus torreliana* | Calytoside | Nigeria | Guillen et al. (2015) |
| *Nelumbo nucifera* | Tolbutamide | India | Mamun-or-Rashid et al. (2014) |
| *Averrhoa bilimbi* | Tannin, Terpen | China | Singh et al. (2014), Mamun-or-Rashid et al. (2014) |
| *Lodoicea sechellarum* | Flavonoid, Carbohydrate | India | Mamun-or-Rashid et al. (2014), Singh et al. (2014) |
| *Butea frondosa* | Flavonoid | India | Kumar and Malik (2012) |
| *Passiflora incarnate* | Vitexin | India | Sankaranarayanan et al. (2010) |
| *Piper angustifolium* | Saponin, Tannin, Terpen, Terpenoid, Flavonoid | Latin America | Ranilla et al. (2010) |
| *Hordeum vulgare* | Beta-glucan, Carbohydrate, Flavonoids | Asia, Africa | Mamun-or-Rashid et al. (2014) |
| *Panax notoginseng* | Tannin | China | Xia et al. (2014), Yang et al. (2010) |

(continued)

**Table 2.1** (continued)

| Species | Bioactive compounds | Geographic zone | References |
|---|---|---|---|
| *Nigella sativa* | Thymoquinone | India | Mamun-or-Rashid et al. (2014) |
| *Ziziphus spina-christi* | Fatty acid, Christinin-A, | Egypt | Singh et al. (2014) |
| *Morinda citrifolia* | Flavonoid, Saponin, steroid, Triterpene | India | Singh et al. (2014), Mamun-or-Rashid et al. (2014) |
| *Murraya panicutata* | Alkaloid, Carbazole | Nigeria | Mamun-or-Rashid et al. (2014) |
| *Scoparia dulcis* | Tannin | India | Singh et al. (2014) |
| *Capsicum frutescens* | Capsaicin | Africa, Asia | Singh et al. (2014) |
| *Helicteres isora* | Terpenoid, Steroid, Alkaloid, Phenolics, Carbohydrate, | India (Ayurveda) | Varghese et al. (2012) |
| *Tilia cordata* | Hyperoside | India | Bahare et al. (2019) |
| *Turnera diffusa* | Terpen, Flavonoid | Africa | Mamun-or-Rashid et al. (2014) |
| *Urtica urens* | Lectin, Flavonoid, Coumarin | China, India | Nencu et al. (2015) |
| *Clerodendrum phlomidis* | Pectolinarigenin | | Sushma et al. (2015) |
| *Curcuma angustifolia* | Curcuminoid | India | Sushma et al. (2015) |

*F. racemosa* have shown excellent biological activity with different modes of action and are reported as candidates for controlling diabetes disease (Deepa et al. 2018).

## 2.3 Phytochemistry of Bioactive Natural Compounds

The leaves, roots, stem bark, flowers, seeds, pulp, rhizomes and the essential oils of medicinal plants are known to possess various bioactive natural compounds (Bahare et al. 2019; Baliga et al. 2011). Essentially, these phytochemicals could be produced by these plants as flavonoids, saponins, phenolic acids, alkaloids, tannin, glycosides, terpenoids, polysaccharides and stilbenes which are investigated intensively for their chemical constituents. Flavonoids are a group of natural bioactive compounds with more than 4000 varieties that have been identified (Shashank and Abhay 2013). Flavonoids are chemically a structure with a skeleton of fifteen-carbon that exist as 2 rings of benzene (Fig. 2.1a, b) connected through a pyrane heterocyclic ring (Fig. 2.1c). Therefore on the basis of chemical structure, flavonoids can be categorized into different number of classes that includes flavonols (i.e., quercetin, myricetin, kaempferol and fisetin), flavones (i.e., apigenin, Chrysin, flavone and luteolin), flavanones (i.e., hesperetin, flavanone and naringenin), isoflavone (i.e., daidzin and Genistin), Anthocyanidin (i.e., Apigenidin and cyanidin), flavanonol (i.e., taxifolin) and others. The different classes of flavonoids differ from one another in the level of oxidation and pattern of substitution of the C ring, while individual compounds within a class differ in the pattern of substitution of the A and B rings.

Normally aglycones, glycosides, and methylated derivatives are the most common forms that flavonoids exist. Aglycone is the structural base for flavonoid (Fig. 2.1). A six-member ring usually condensed or compressed with the benzene ring which is either a α-pyrone or its dihydroderivative. The site of the benzenoid substitution categorizes flavonoid into flavonoids and isoflavonoids (i.e., 2-position and 3-position). At the 3 position and a C2–C3 double bond there is a hydroxyl group that differentiates flavonols from flavanones structurally (Narayana et al. 2001). Naturally, acetylesters and methylethers of the alcohol group are known to

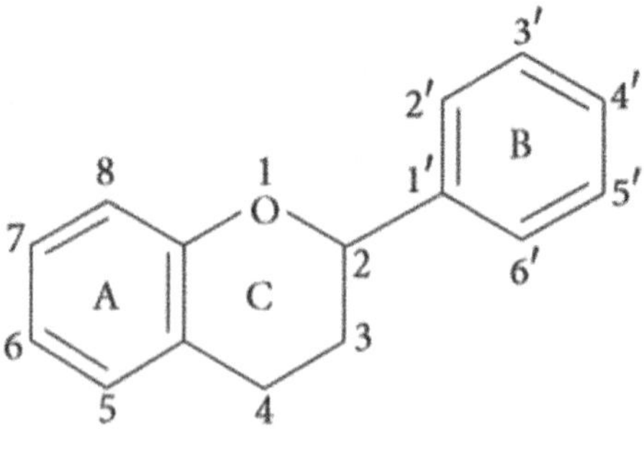


**Fig. 2.1** Basic chemical structure of flavonoids

occur. Flavonoids are often hydroxylated in positions 3, 5, 7, 2, 3′, 4′, and 5′. When glycosides are formed, the glycosidic linkage is normally placed in positions 3 or 7 and the carbohydrate can be glucorhamnose, galactose, L-rhamnose, D-glucose, or arabinose (Narayana et al. 2001).

The presence of bioactive natural compounds (phenolic compounds) in plant products like vegetables, fruits and legumes offer protection afforded by their consumption. Phenolic compounds are well-known phytochemicals found in all plants. Phenolic compounds are manufactured by plants as a result of reaction to physiological or ecological pressures such as pests and diseases attack, wounding and UV radiation (Ali et al. 2013; Kennedy and Wightman 2011) The aromatic ring having one or more hydroxyl groups forms the basic in the structures of phenolic compounds (Fig. 2.2) (Ali et al. 2013). The location and number of hydroxyl groups on the aromatic ring gave rise to the variation in phenolic acids (Ali et al. 2013). Phenolic compounds in plants are divided into polyphenols or simple phenols depending phenol units present in the molecule. This therefore shows that phenolics in plant exist as benzoic and cinnamic acid, simple phenols, coumarins, lignans, lignins, condensed and hydrolysable tannins, phenolic acids and flavonoids (Ali et al. 2013; Soto-Vaca et al. 2012) Phenolic acids as one class of phenols have hydroxycinnamic and hydroxybenzoic acid as their parent structures. The derivatives of hydroxybenzoic acid exit as protocatechuic acids, syringic, gallic and vanillic. The derivatives of hydroxycinnamic acid derivatives include p-coumaric, sinapic acids, ferulic and caffeic. Cell wall phenolics is a phenolic compound group that are found to be insoluble and although capable of making complexes with other components of the cell. Vanholme et al. (2010) indicated that lignins and hydroxycinnamic acids are the main cell wall phenolics. At the time plant growth, the compounds play a key role in the cell wall against such stresses like infection and wounding (Ali et al. 2013).

The structure of saponins consists of water soluble glucidic chain and liposoluble structure in that they are glycosylated compounds (Chaieb 2010; Eskandar and Somayeh 2015). The structure of saponin is given in Fig. 2.3 (Chaieb 2010). The aglycone and glycone are the non-sugar and sugar components respectively. The steroid backbone or triterpenoid are the main component of aglycone portion (Abid Ali Khan et al. 2012). Some of the main components of saponin include among others L-arabinose, D-galactose, D-glucose, D-xylose, D-glucuronic acid, D-fructose, L-rhamnose (Chaieb 2010). The aglycone has glycosylation sites where the sugar moiety is linked to it via one or two ether glycosidic linkage. It is important to note

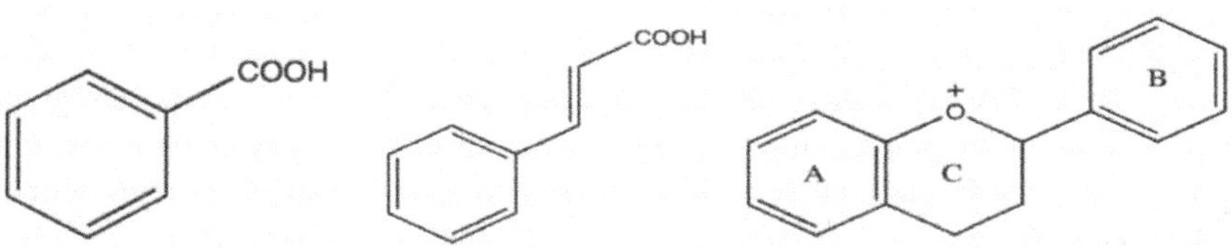


**Fig. 2.2** **a** hydroxybenzoic acid, **b** hydroxycinnamic acid, **c** flavonoids

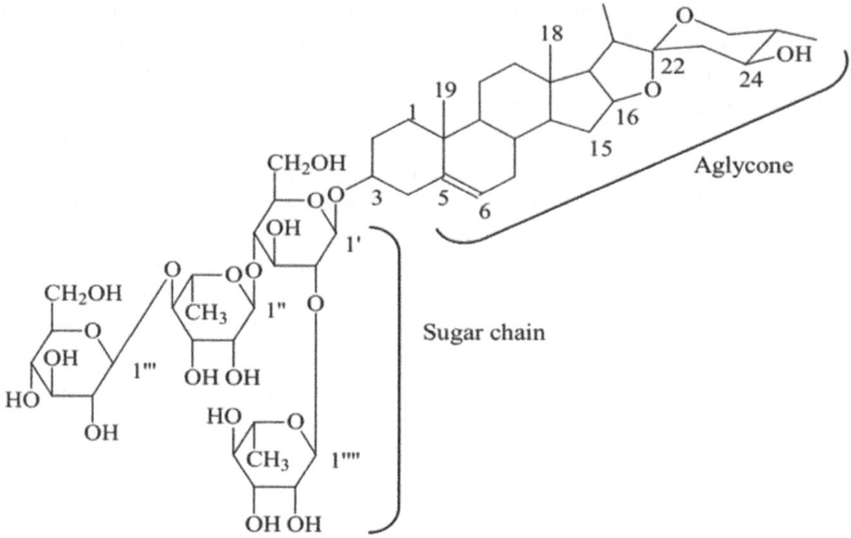


**Fig. 2.3** Chemical structure of saponin

that the aglycone may consist in it one or more unsaturated carbon to carbon (C–C) bonds. If the chain (oligosaccharide) is connected to the site of $C_3$, then the molecule is known as monodesmosidic, whereas if additional sugar moiety were present at the site of $C_{26}$ or $C_{28}$ of the saponin, such saponins are called bidesmosidic. The saponins structure is affected by the amount and types of sugars present, as well as the composition of steroid ring. Abid Ali Khan et al. (2012) observed high saponin composition in plants at tender age than plants at their peak age, even though many factors like environmental factors and physiological state affect the saponin contents. Classification of saponins depends on the nature of their aglycone and as result two classes were identified as steroidic aglycone saponosides and triterpenic aglycone saponosides. The triterpenic aglycones come from the cyclization of the (3S)-2,3-epoxy-2,3 dihydrosqualene. The skeleton of steroidic aglycones was observed to have 27 carbon atoms. The molecules were from cetalisation (intramolecular) that intervenes after oxidation in $C_{16}$, $C_{22}$ and $C_{26}$ of a cholestanic precursor taking into consideration spironature of $C_{22}$; which is usually shown by the spirostane term. It is known as furostane, if the structure is pentacyclic. This process of cyclization gives rise to pentacyclic compounds such as ursanes, dammaranes, hopanes and oleananes. The majority of triterpenic sapogenins belong to these four basic skeletons.

It is well known that bioactive natural compounds are many and we are giving an introduction, therefore, in this section, some bioactive natural compounds, their chemical structures and sources will be presented as are available in the literature (Table 2.2).

**Table 2.2** Some bioactive natural compounds, their chemical structures and sources

| Compound | Structure | Source | Reference |
|---|---|---|---|
| Fisetin | 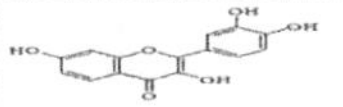 | *Acacia berlandieri*<br>*Acacia greggii*<br>*Butea fronds*<br>*Gleditschia triacanthow*<br>*Quebracho colorado*<br>*Gleditsia triacanthos*<br>*Rhus cotinus*<br>*Cotinus coggygria*<br>*Rhus vemiciflua* | Prasath et al. (2014) |
| Gambogic acid | 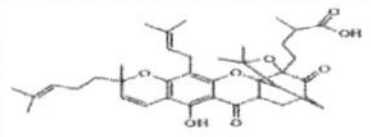 | *Garcinia hanburyi*<br>*Garcinia cambogia*<br>*Garcinia indica* | Cui et al. (2018) |
| Chrysin | 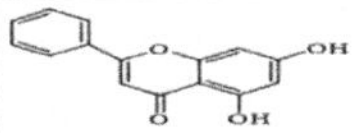 | *Passiflora caerulea*<br>*Oroxylum indicum*<br>*Passiflora incarnata* | Amjid et al. (2014), Oza and Kulkarni (2016) |
| Berberine | 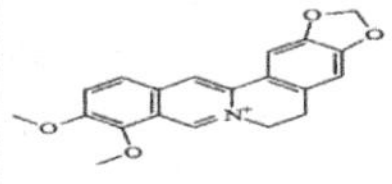 | *Argemone mexicana*<br>*Berberis vulgaris*<br>*Berberis aristata*<br>*Berberis aquifolium*<br>*Eschscholzia californica*<br>*Coptis chinensi*,<br>*Hydrastis canadensis*<br>*Xanthorhiza simplicissima*<br>*Tinospora cordifolia*<br>*Phellodendron amurense* | Cui et al. (2018) |
| Butein | 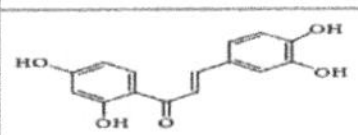 | *Toxicodendron vernicifluum*<br>*Dalbergia odorifera*,<br>*Cyclopia subternata*,<br>*Semecarpus anacardium Creopsis tungtoria* | Benzler et al. (2015) |
| Boldine | 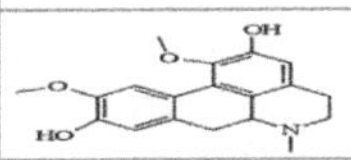 | *Peumus boldus* | Oza and Kulkarni (2016) |
| Baicalein | 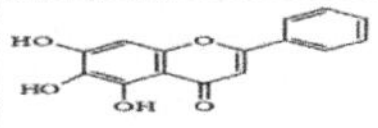 | *Scutellaria baicalensis*<br>*Oroxylum indicum*, | Oza and Kulkarni (2016), Ahad et al (2014) |
| Embelin | 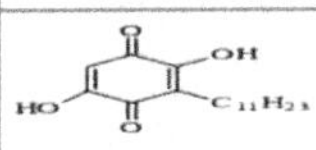 | *Embelia ribes*,<br>*Lysimachia punctata*,<br>*Lysimachia erythrorhiza* | Ahad et al. et al. (2014) |

(continued)

**Table 2.2** (continued)

| Compound | Structure | Source | Reference |
|---|---|---|---|
| Curcumin | 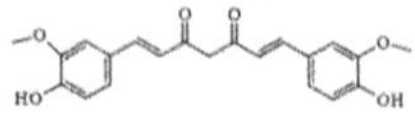 | *Curcuma longa*<br>*Zingiberaceae plants* | Kunwar and Priyadarsini (2016) |
| Withanolides | 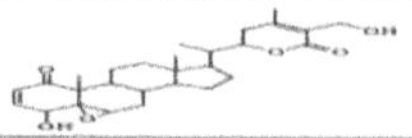 | *Withania somnifera* | Ahad et al. (2014) |
| Piperine | 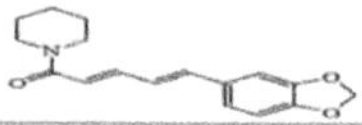 | *Piper nigrum*<br>*Piper longum* | Atal et al. (2016) |
| Embelin | 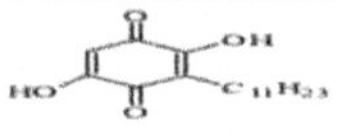 | *Embelia ribes*<br>*Lysimachia erythrorhiza*<br>*Lysimachia punctata* | Durg et al. (2017)<br>Cui et al. (2018) |
| Erianin | 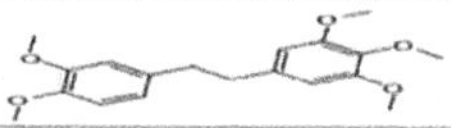 | *Dendrobium chrysotoxum* | Yu et al. (2016) |
| Luteolin | 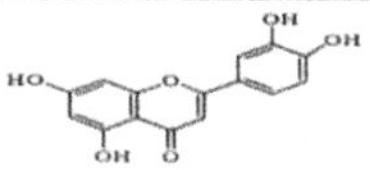 | *Leonotis leonurus*<br>*Mentha arvensis*<br>*Mentha longifolia*<br>*Mentha piperita*<br>*Ocimum canum*<br>*Ocimum campechianum*<br>*Ocimum sanctum*<br>*Ocimum gratissimum*<br>*Ocimum tenuiflorum*<br>*Salvia officinalis* | Wang et al. (2012a, b) |
| Silymarin | 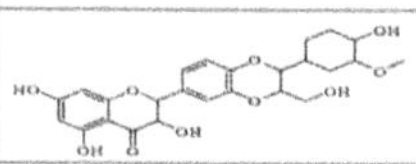 | *Silybum marianum* | Oza and Kulkarni (2016) |
| Ursolic acid | 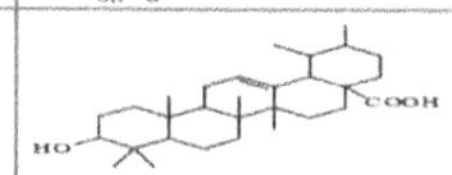 | *Calluna vulgaris*<br>*Crataegus laevigata*<br>*Eriobotrya japonica*<br>*Eugenia jambolana*<br>*Melissa officinalis*<br>*Mentha piperita*<br>*Ocimum sanctum*<br>*Rosmarinus officinalis Thymus vulgaris*<br>*Dracocephalum heterrophyllum*<br>*Hyssopus seravshanicus* | Lee et al. (2014), Castro et al. (2015) |
| Triptolide | 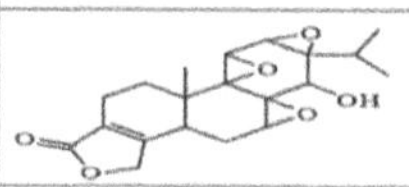 | *Tripterygium wilfordii* | Lee et al. (2014) |
| Sanguinarine | 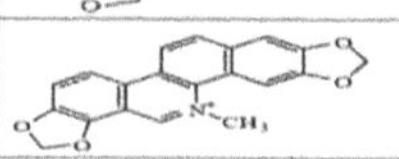 | *Sanguinaria canadensis* | Lee et al. (2014) |

(continued)

**Table 2.2** (continued)

| Compound | Structure | Source | Reference |
|---|---|---|---|
| Resveratrol | 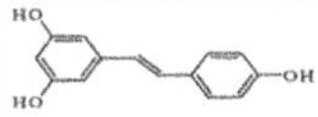 | *Vitis vinifera*<br>*Vaccinium corymbosum*<br>*Pistacia vera* | Do et al. (2012) |
| Quercetin | 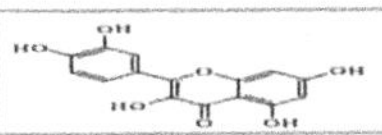 | in many grains, fruits, leaves, vegetables | Li et al. (2013) |
| Piceatannol | 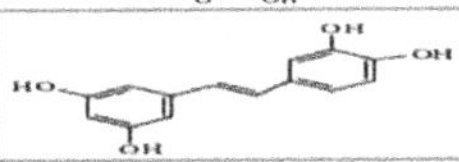 | *Passiflora edulis* | Uchida-Maruki et al. (2015) |
| Neferine | 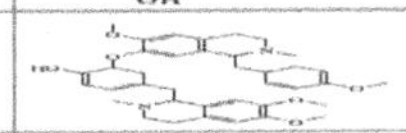 | *Nelumbo nucifera* | Oza and Kulkarni (2016) |
| Oxymatrine | 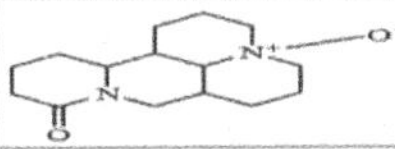 | *Sophora flavescens* | Oza and Kulkarni (2016) |
| Naringenin | 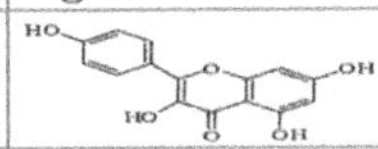 | *Citrus paradisi* | Tsai et al. (2012) |
| Catechins | 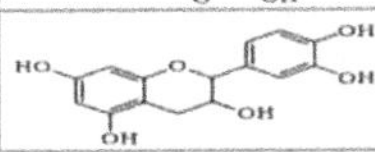 | *Theobroma cacao*<br>*Camellia sinensis* | Oza and Kulkarni (2016) |
| Garcinol | 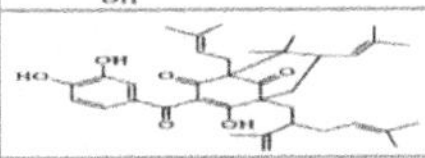 | *Garcinia indica* | Mali et al. (2017) |
| Honokiol | 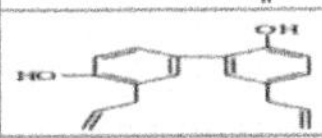 | *Magnolia officinalis* | Sun et al. (2015) |
| Lupanine | 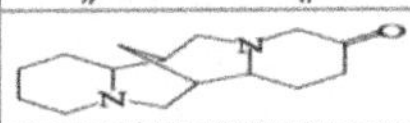 | *Lupinus perennis* | Madhuri and Naik (2017) |
| Morin | 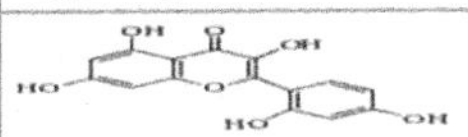 | *Morus alba*<br>*Psidium guajava*<br>*Maclura pomifera*<br>*Prunus dulcis*<br>*Maclura*<br>*Chlorophora tinctoria tinctoria*<br>*Castanea sativ* | Pandey et al. (2019) |
| Inulin | 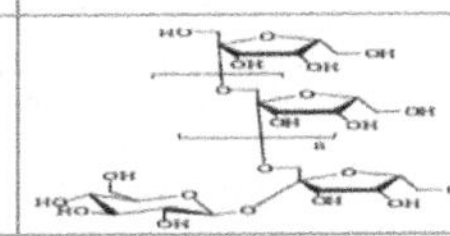 | *Helianthus tuberosus* | Ma et al. (2011) |
| Boswellic acids | 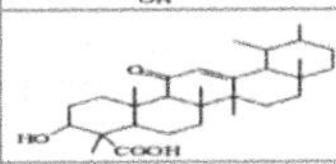 | *Boswellia carteri*<br>*Boswellia serrata* | Razavi et al. (2018) |
| Kaempferol | 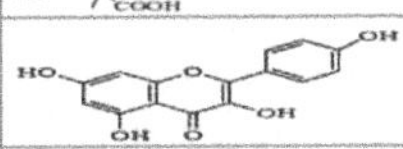 | *Acalypha indica* | Alkhalidy et al. (2015) |
| Genistein | 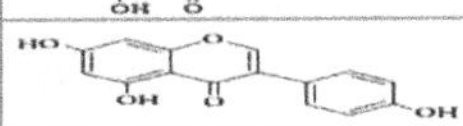 | *Butea monospermea*<br>*Butea frondosa* | Shashank and Abhay (2013) |
| Mangiferin | 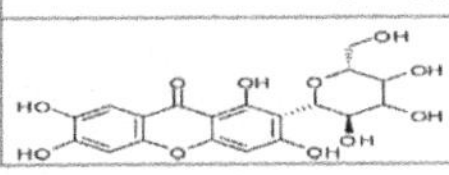 | *Anemarrhena asphodeloides* | Razavi et al. (2018) |

(continued)

**Table 2.2** (continued)

| Compound | Structure | Source | Reference |
|---|---|---|---|
| Oleanoic acid | 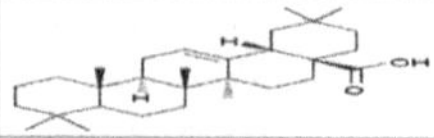 | *Aralia cortex*<br>*Phellodendron cortex* | Taguchi et al. (2016) |
| Oleanoic acid-28-*O*-beta-D-glucopyranoside | 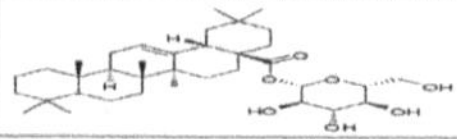 | *Aralia cortex*<br>*Phellodendron cortex* | Taguchi et al. (2016) |
| Kaempferitrin | 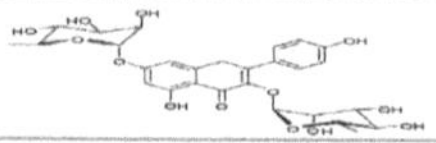 | *Bauhinia forficata* | Taguchi et al. (2016) |
| Apigenin | 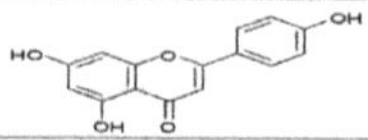 | *Buddleja officinalis* | Alkhalidy et al. (2015) |
| Bassic acid | 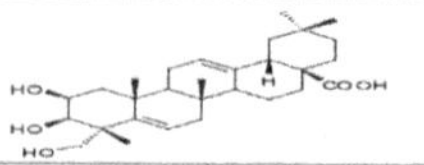 | *Bumelia sartorum* | Alkhalidy et al. (2015) |
| Ellagic acid | 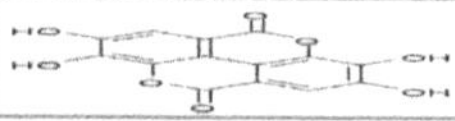 | *Caesalpinia ferrea* | Uchida-Maruki et al. (2015) |
| Alpha-homonojirimycin | 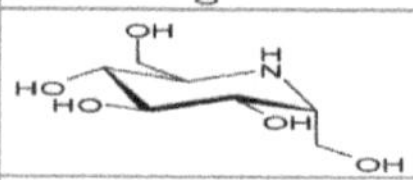 | *Commelina communis* | Pandey et al. (2019) |
| Ursolicacid | 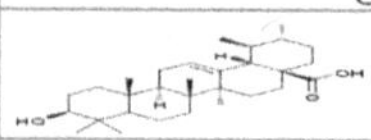 | *Cornus officinalis* | Uchida-Maruki et al. (2015) |
| Cryptolepine | 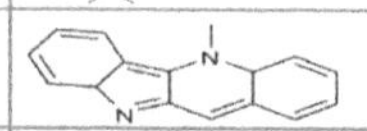 | *Cryptolepis sanguinolenta* | Pandey et al. (2019) |
| Dioscoretine | 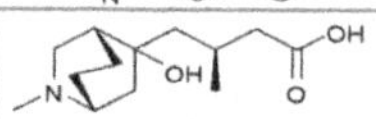 | *Dioscorea dumetorum* | Alkhalidy et al. (2015) |
| Leucopelargonidin | 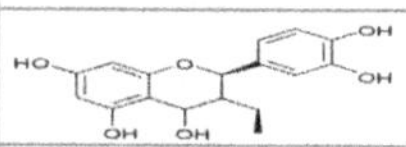 | *Ficus bengalensis* | Alkhalidy et al. (2015) |
| Achyrophuran | 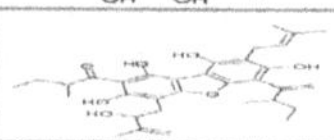 | *Achyrocline satureioides* | Taguchi et al. (2016) |
| Maturine | 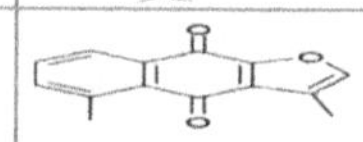 | *Psacalium decompositum*<br>*Psacalium peltatum*<br>*Acourtia thurberi* | Baileyand Day (2004) |
| Metformin | 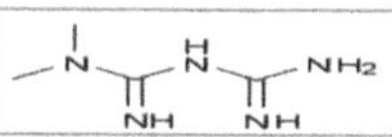 | *Galega officinalis* | Bailey and Day (2004) |
| Isoliquiritigenin | 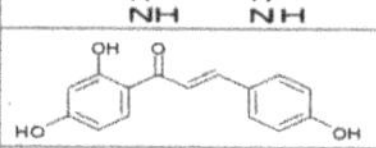 | *Glycyrrhizae radix* | Taguchi et al. (2016) |
| Coutaraegenin | 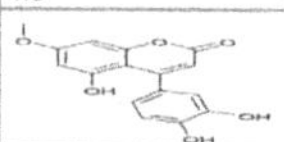 | *Hintonia latiflora* | Mali et al. (2017) |
| Hydnocarpin | 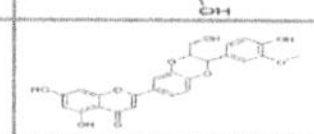 | *Hydnocarpus wightiana* | Reddy et al. (2005) |
| Hedearagenin | 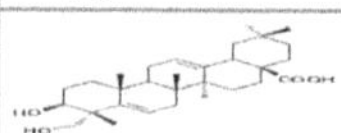 | *Kalopanax pictus* | Taguchi et al. (2016) |
| Masopropol | 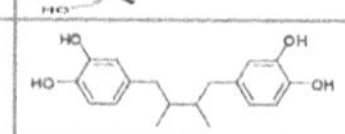 | *Larrea tridentata* | Taguchi et al. (2016) |

(continued)

**Table 2.2** (continued)

| Compound | Structure | Source | Reference |
|---|---|---|---|
| Maesanin |  | *Maesa lancelata* | Taguchi et al. (2016) |
| Flaviolin | 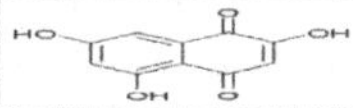 | *Maesa lancelata* | Taguchi et al. (2016) |
| Caffeic acid | 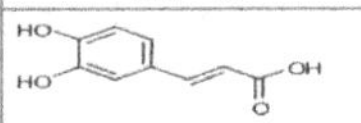 | *Origanum vulgare* | Kim et al. (2005) |
| Rosmarinicacid | 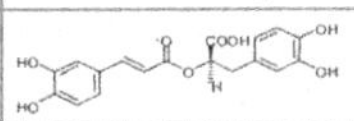 | *Origanum vulgare* | Mali et al. (2017) |
| Bakuchiol | 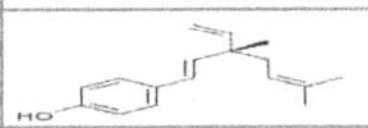 | *Otholobium pubescens* | Mali et al. (2017) |
| Maurine | 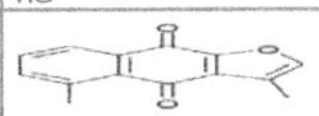 | *Psacalium decompositum* | Mali et al. (2017) |
| Psoralidin | 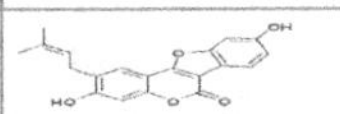 | *Psoralea corylifolia* | Kim et al. (2005) |
| Marsupin | 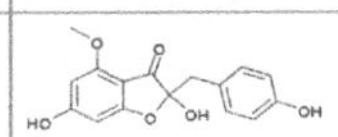 | *Pterocarpus marsupium Trigonella foenumgraecum* | Pandey et al. (2019) |
| Swerchirin | 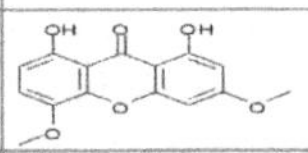 | *Swertia chirayita* | Pandey et al. (2019) |
| Fagomine | 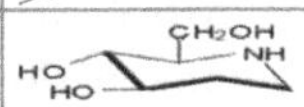 | *Xanthocercis zambesiaca* | Pandey et al. (2019) |

## 2.4 General Applications of Bioactive Natural Products

### *2.4.1 Traditional Medicine*

The practise of traditional medicine is worldwide in spread, China, Japan, India, Pakistan, Thailand and Sri Lanka inclusive. Folklore and traditional systems of medicine bequeathed from generation to generation is rich in domestic recipes and communal practice. Medicinal plants have the great potential used in various traditional systems of medicine like in the Ayurveda, Siddha, Unani, in the Tibetan, in the Srilankan, and in the Homeopathy systems of alternative and complementary medicine (Meenupriya et al. 2014). The best known examples of traditional medicine, differing in concept and protocol, are well-developed systems such as ayurvedic and acupuncture medicine that have been widely used to conserve human health in India, China and native Indians from Brazil in the Amazonian region (Meenupriya et al. 2014). In Chinese medicinal folk practice, the search for new medicines to treat emergent and old diseases such as AIDS and malaria, attention is now being geared towards discovering the active ingredients encountered in the treasury of over 5,000 Chinese plants, herbs and roots that have been used routinely and traditionally. *Chaihu* and *Quinghaosu* are two such examples. Whereas the former, called *artemisinin* and obtained from *Artemisia annua* is expected to yield, in the coming millennium, a potent new class of antimalarials, the latter, obtained from *Bupleurum chinense* and used as a popular remedy for hepatitis is the focus of intense research by the Japanese pharmaceutical industry (Baliga et al. 2011).

The practise of traditional medicine using medicinal plants such as *Panax quinquefolium*, *Podophyllum peltatum* and *Eupatorium perfoliatum* has long been associated with the American Indians in the United States of America. These medicinal plants have also been recognised and appreciated for their aesthetic and ornamental value. In Central America medicinal plants have been widely used by the *Maya* Indians in Mexico, the *Pipiles* in El Salvador, the *Miskitos* and *Sumus* in Nicaragua and Honduras, the *Pech*, *Xicaques* and *Lencas* in Honduras, the *Guaymis* and *Kunas* in Panama and the *Talamancas* in Costa Rica. In Europe, more than 1,500 species of medicinal plants are widely used in Albania, Croatia, Bulgaria, France, Germany, Poland, Hungary, Turkey, Spain, and the United Kingdom. The islands of Maltese constitute a good example where medicinal plants are widely used in everyday life as part of folk medicinal remedies.

Africa is another rich source of medicinal plants rich in bioactive natural compounds. Perhaps, the best known species include that of *Moringa oleifera* reported by different researchers in the literatures to have important biological potentials include antimicrobial activities (Arora and Onsare 2014; Arora et al. 2013), antifungal (Kadhim and AL-Shammaa 2014), antioxidant (Satish et al. 2014), antibacterial and antiulcer (Belay and Sisay 2014), anti-inflamatory, diuretic, antispasmodic (Krishnamurthy et al. 2015; Araujo et al. 2013), antimutagenic and antioxidant (Satish et al. 2014), antistress (Luqman et al. 2012), anticancer (Krishnamurthy et al. 2015; Pinto et al. 2015), cytotoxic activities (Araujo et al. 2013; Asare et al. 2012), antitumour, antipyretic, antiepileptic, antinociceptive and antidiabetic (Vinoth et al. 2012). Other notable examples are *Catharanthus roseus*, which yields anti-tumour agents such as vinblastine and vincristine; and *Ricinus communis*, which yields the laxative castor oil. In Botswana, Namibia, Lesotho and South Africa, *Harpagophytum procumbens* is produced as a crude drug for export. Similarly, *Hibiscus sabdariffa* is exported from Egypt and Sudan. Other exports are *Pausinystalia yohimbe* from Cameroon, Nigeria and Rwanda, which yields *yohimbine;* and *Rauwolfia vomitoria*, from Madagascar, Mozambique and Zaire, which is exploited to yield ajmaline and reserpine.

### 2.4.2 Plant Based Pesticides and Agrochemical Industry

The plant world is rich store house of natural chemicals that could be exploited for use as biopesticides (Dwivedi and Enespa 2012). Medicinal plants are now emerging as safer and more compatible approach to chemical control of pests and diseases. All parts of these plants including roots, flowers, bark, stem, leaves, seeds and essential oils possess antimicrobial properties and are therefore used for medicinal and other purposes (Dwivedi and Enespa 2012). The use of these potential plants as biopesticides has emerged as a result of the drawbacks associated with the use of synthetic chemical methods notably among them are their high cost, their carcinogenicity, teratogenicity, high and acute residual toxicity, long degradation period, environmental pollution and possible side-effects on human health through the food chain (Ai-ying et al. 2011). These drawbacks coupled with public concern have increased interest in developing further alternative control methods, particularly those that are eco-friendly, biodegradable, feasible to the farmers, non-toxic to human and animals, specific in their action and have a broad spectrum of antimicrobial activity (Marino and Bersani 2001; Abhishek et al. 2013).

Scientific investigations on different morphological parts of medicinal plants indicated the presence of biologically active compounds which possesses an array of antimicrobial properties acting on different types of microorganisms. These bioactive compounds were evaluated in vitro and further subjected to in vivo testing to validate their efficacy in controlling the incidence and severity of diseases in crops and insect infestations. Recently, Batista et al. (2014) reported the antifungal inhibitory activity of Moringa-Chitin Binding Protein (*Mo*-$CBP_3$) purified from the seeds of *M. oleifera* Lam. against mycelial growth and spore germination of *Fusarium solani* at 0.05 mg $mL^{-1}$. Similarly, in another study, Gifoni et al. (2012) also showed the antifungal efficacy of *Mo*-$CBP_3$ chitin-binding protein purified also from *M. oleifera* seed against phyopathogenic fungi *Fusarium oxysporium*, *F. solani*, *Colletotrichum gloesporiodes* and *Pythium oligandrum* at 0.05 mg $mL^{-1}$ and 0.1 mg $mL^{-1}$ respectively. Dwivedi and Enespa (2012) reported that *M. oleifera* extracts (leaves, bark and seeds) at 75% (v/v) give significant inhibitory effect on the mycelial growth of *F. solani* and *Fusarium oxysporum* f. sp. *Lycopersici*. The fungicidal effect of *Moringa* extracts on some soil-borne fungi such as *Rhizoctonia, Pythium and Fusarium* causing tomato rots was also reported (Moyo et al. 2012). Again, El-Mohamedy and Abdalla (2014) reported the fungicidal effect of *M. oleifera* against *F. oxysporm*, *F. solani*, *Alternaria alternate*, *A. solani*. *Rhizoctonia solani*, *Sclerotium rolfsii* and *Macrophomina phaseolina* causing rots on fruit and other perishables. Ukeh and Chiejina (2012) reported that phytopesticides from two plants including *Afromomium meleguata* and *Zingiber officinale* inhibited *Penicillium digitatum*, *Mucor piriformis*, *Aspergillus niger* and *Heminsthoporium solani* causal agents involved with soft rot of tomato. Ijato et al. (2011) reported the antimicrobial activity of *Vernonia amygdalina* and *Tridax procumbens* each with two varying formulations and concentrations (aqueous extracts: 80 and 60% and ethanol extracts: 30 and 20%) against *A. niger*, *F. oxysporum*, *G. candidum* and *Rhizopus stolonifer*. Again, the aqueous and organic solvents (water and ethanol) extracts from leaves of *Chromolaena odorantum* and *Azachirachta indica* were reported to have antifungal activity against fungal pathogens that cause tomato rots (*A. niger*, *F. oxysporum*, *R. stolonifer* and *G. candidium*) by poisoned food method (Ijato et al. 2011).They showed that among the various extracts with varying concentrations, ethanol extracts of 30% *A. indica* had the best inhibitory effect (83.30%) against *A. niger* followed by 30% ethanol extract of *C. odorata* (80.00%) against *G. candidium* which proved the potentiality of the plant extracts for the control of postharvest and transit fungal rot of tomato fruit. Ijato et al. (2011) reported that these plants extracts in addition to their ability to retard mycelial growth of the fungi, they also inhibit their spore germination. Other reports on the antifungal activity of plant-based pesticides include that of Prapassom et al. (2012) who reported the efficacy of 14 crude leaf extracts including *Piper sarmentosum*, *Cymbopogon citratus*, *Citrus hystrix*, *Murraya paniculata*, *Ocimum basilicum*, *Ocimum canum*, *Annona squamosal*, *M. oleifera*, *Psidium guajava*, *Ocimum sanctum*, *Eucalyptus camaldulensis*, *Artocarpus heterophyllus*, *Cassia siamea*, *Mentha cordifolia*, using ethanol, methanol, and chloroform (80%) as solvents against *C. gloesporioides* (Penz.) and found that crude methanol extract of *P. sarmentosum* leaves effectively inhibited the growth of fungal mycelium (100%), followed by crude chloroform extract(81.85%). Similarly, extracts from the bark, roots, and leaves of *A. indica* at 400 and 500 mg/ml concentrations tested against *C. gloesporioides*, a causal agent of field soft rot of fruit completely inhibit the growth of the fungus (Ijato et al. 2011).

The challenge in agrochemical industry today is not only to formulate pesticides with high efficiency, but also need to developed better brands with improved safety to the users and have less impact to the environment (Polychniatou and Tzia 2014). Not only that, such formulations should be an adequate delivery system that will resolve the problem of inconsistency associated with bipesticides that reduce their rate of competition with long standing synthetic pesticides in the market (Su et al. 2014). Since bipesticides are formulated with active ingredients and inert ingredients (Jiang et al. 2011), the primary aim of formulation process is to make the active ingredient easy to handle, use, ensure that that it is stable during storage and transport, and achieved an adequate shelf life in the formulation. The active ingredient should possess antimicrobial or chemical properties that can control the target pest. Specifically, the active ingredient should function as antifungal, antibacterial, antiviral, antioxidant, cytotoxic, repellent, destroyer, killer, or mitigate pest, or as plant regulator, desiccant or defoliant. Bioactive compounds (active ingredients) of biopesticides come from various sources that include those extracted from botanicals (plants) such as rotenone, *Mo*$CBP_3$, nicotine, pyrethrum, saponins for the preparation of plant-based biopesticides. Typical example of plant-based biopesticide is that of MILSANA and REGALIA formulated from anthraquinone containing extracts of giant knotweed (*Reynoutria sachalinensis*). Both biopesticides are commercially developed and marketed by Marrone Bio Innovatives Inc sold as MILSANA® and REGALIA®. Both have antifungal and antibacterial properties against various fungal and bacterial pathogens. In addition, they also serve as plant defence inducers and act in the accumulation of fungistatic phenolic compounds in the plants (Huang and Campbell 2016; Su et al. 2014).

### *2.4.3 Pharmacological Applications in Drugs Development*

Another application of bioactive natural compounds is in the field of drugs development as a result of their efficacy and safety concerns of the hundreds of millions of individuals which are currently seeking better management of diseases (Mancha-Ramirez and Slaga 2016). Medicinal plants produce a wide array of phytoconstituents which include alkaloids, flavonoids, phenolic acids, saponins, glycosides, polysaccharides and tannin, which are intensively investigated for their bioactive effects. They are today, the main quarry for discovering promising lead candidates and play an imperative role in the upcoming drug development programs (Sharifi-Rad et al. 2018a, b, c; Salehi et al. 2018). They have low cost, ease of availability and least side effects which make plant-based preparations the main key player of all available therapies, especially in rural areas (Arya et al. 2011). Moreover, many plants provide a rich source of bioactive phytochemicals, which are free from undesirable side effects and possess powerful pharmacological actions (Abdolshahi et al 2018; Mishra et al. 2018a, b; Sharifi-Rad et al. 2018a, b, c; Singab et al. 2014;). These plants also have been an exemplary source of drugs with many of the currently available drugs being obtained directly or indirectly from them (Abdolshahi et al. 2018; Mishra et al. 2018a, b; Arumugam et al. 2013). Despite the presence of drugs in the pharmaceutical market, the treatment of most diseases with medicinal plants is often successful. Herbal medicines and plant components with insignificant toxicity and no side effects are notable therapeutic options for the treatment of diseases around the world (Bahare et al. 2019). Furthermore, reliance on the use of medicinal plants that is increasing in the industrialised societies has been traced to the extraction of bioactive natural compound from medicinal plants and development of several drugs and chemotherapeutics from these plants as well as from traditionally used rural herbal remedies (Baliga et al. 2011). Validated pharmacological properties of the bioactive natural compounds are antiviral, antifungal, antibacterial, free radical scavenging and antitoxic effect, Gastroprotective effects, anti-inflamatory, Hepatoprotective effects, antidiabetic activities, Cardioprotective effects, Radioprotective effects, Chemopreventive effects, Anticlastogenic effects, Antineoplastic effects, Neuropsychopharmacological effects, Antipyretic effects, Anti-diarrheal effects, Antifertility activity, Anti-allergic effects, Hypolipidemic effects (Baliga et al. 2011; Jabeen and Javaid 2010).

### *2.4.4 Cosmetics*

Bioactive natural compounds used in the production of herbal products have today gained increase acceptance and popularity, expanding which are sold as herbal cosmetics (Ribeiro et al. 2015; Lee et al. 2013). Phenolic components such as quercetin, kaempherol and carbohydrates such as glucose, galacturonic acid, arabinose and rhamnose are some of the substances present in the chemical composition of many medicinal plants that the cosmetic industry has interest in them as anti-aging and moisturizers products (Zhang et al. 2011; Ribeiro et al. 2015). Dry skin is a common problem among the human populace. Its main characteristic is a phenomenon called xerosis, which is a rough peeling and appearance. To accomplish the main function of the human skin as a barrier provider between the internal contents external environment, moisturizers made from bioactive natural products are needed. Moisturizers are cosmetic products that prevent the skin against xerosis and delaying of premature ageing and for their use to help dermatological therapies in a wide variety of skin disorders (Capitani et al. 2012). The mode of action of moisturizers should not be overemphasized as 1. They may act by an occlusive mode thereby impairing the evaporation of skin moisture by forming an epicutaneous lipidic film that prevents water loss, when oils and lipids are used as moisturizers; or 2. As humectants that act by attracting water from the other layers of the epidermis to the stratum corneum i.e., glycerine and sodium pyrrolidone carboxylic acid; are used as moisturizers. Other moisturizing modes of action include active hydration by rearranging the stratum corneum and aquaporin formation. The aquaporins are transmembrane proteins which form water channels and facilitate water flux through the cell plasma membrane thus being important to maintain aconstant water content in viable epidermis (Capitani et al. 2012).

## 2.5 Delivery Technology for Bioactive Natural Products

Nanotechnology has been considered as a promising delivery system with respect to transdermal drugs delivery and eco-friendly pesticide formulation, including those with natural plant products incorporated as active ingridients (Angajala et al. 2014; Da Costa et al. 2014; Rodríguez-Rojo et al. 2012). Nanotechnology is a useful system for delivering chemical compounds across the cuticle and it has the potential of application in the food industry, disease treatment delivery system, food packaging and bioactive compound delivery to target sites. The widespread commercial application of nanoemulsion technology especially in pharmaceutical and agrochemical industries is due to its small droplet size (10–100 nm) as suggested by some researchers (Rai et al. 2015; Polychniatou and Tzia 2014), that causes a reduction in the Brownian motion and gravitational strength (Mishra et al. 2014; Tadros et al. 2004). Similarly, the small size of droplets spread uniformly on the surface and helps to enhanced wetting, spreading and penetration as a result of the low surface tension. Additionally, it can prevent flocculation and coalescence of the droplets and enables the system to remain dispersed without separation. Furthermore, the surfactant film thickness prevents disruption and thinning of liquid film between the droplets. Nanoemulsion can encapsulate active ingredients within their droplets and this help in reducing chemical degradation (McClements and Decker 2000) and hence expand cell wall penetration of the fungus, due to their smaller size (Zahid et al. 2012)). Other benefits include, less amount of energy requirement, increase rate of absorption and eliminates variation on absorption, and increases bioavailability (Mishra et al. 2014).

## 2.5.1 Components of Nanoemulsion

Basically, a nanoemulsion consists of an active ingredient, an oil phase, surfactant and aqueous phase. The oil phase is usually the carrier(s) which are mineral oils and/or vegetable oils (Xu et al. 2011). Today, vegetable oils and their esterified derivatives are gaining interest compared to the mineral oils because of their biodegradability and renewability (Xu et al. 2011). Castor oil, copaiba oil, olive oil, soy oil, Agnique® AMD 10, Edenol SP 100 is few examples of vegetable oils used in nanoemulsion formulations. The oil phase in nanoemulsion formulations enhance the spread and penetration of droplets on the surface of the skin as well as plants; split open the cuticle to increase both the fluidity of cuticular components and pesticide diffusion rates (Xu et al. 2011).

### 2.5.1.1 Surfactants

Surfactants, also called surface-active agents are organic compounds that are amphiphilic in nature (having both hydrophobic groups and hydrophilic groups) that reduce surface and interfacial tensions by accumulating at the interface of immiscible fluids and increase solubility, bioavailability, mobility and subsequent biodegradation of hydrophobic or insoluble organic compounds (Singh et al. 2011). They can be found and used as emulsifiers, de-emulsifiers, wetting agents, forming agents, detergents in petroleum, petrochemicals, functional food ingredients, environmental management, foods and beverages, agrochemicals, cosmetic and pharmaceuticals, and in the mining and metallurgical industries (Singh et al. 2011). Surfactant emulsifiers are added to nanoemusion formulations to ensure spontaneous emulsification with good stability qualities in the spray tank. Surfactants can be classified as ionic and non-ionic. Non-ionic surfactants are preferred in pesticide formulation system (Mehmood 2015) than the ionic ones due to their less toxicity, enhanced solubility, spreading, adsorption, translocation and penetration of active ingredients into the target. Myers (2005) define non-ionic surfactant as a surfactant that carries no electric charge, as its water solubility is derived from the occurrence of polar functionalities capable of significant hydrogen bonding relations with water (e.g. polyglycidols and polyoxyethylenes) (Shafiq et al. 2007). Although stable nanoemulsions are best formulated with surfactants or a combination of surfactants having hydrophile-lipophile balance (HLB) values close to that required for the oil phase, it is important to know that there are no specified rules to resolve the ratio of surfactants in the blend surfactants. However, guidance can be obtained from the (HLB) system.

## 2.6 Conclusions

The current book chapter tries to review the global efforts by scientists regarding their investigation on natural bioactive compounds from medicinal plants that will serve as a novel chemotherapeutants and their general applications in various fields. Alongside, this provides a prospect for further research in order to find novel antimicrbial agents, their potentials, modes of action and synergistic effects for the eventual formulation of herbal mixtures and their combination with synthetic medicines. Such efforts will be more lucrative if their toxicological effects are confirmed by necessary and careful studies. It will be more scientific to standardize techniques of extraction, in vitro and in vivo antimicrobial efficacy testing so that the search for new biologically active compounds could be more systematic and results interpretation would be facilitated. The phytochemicals responsible for the potential properties of the bioactive natural compounds are mainly phenolic acids, alkaloids, flavonoids, saponins, glycosides, polysaccharides, tannins and stilbenes. Notwithstaanding, the phytochemical composition in plant depend highly on several factors, including plant specie, genetic traits, plant organs used, and the growing, drying, and storing condition. Moreover, new techniques must be employed to get excellent plant products in high amounts and to determine their potential applications more accurately in food industry, traditional medicine, cosmetics, drugs development and plant based pesticides and agrochemical industry.

## References

Abdolshahi A, Naybandi-Atashi S, Heydari-Majd M, Salehi B, Kobarfard F, Ayatollahi SA, Ata A, Tabanelli G, Sharifi-Rad M, Montanari C (2018) Antibacterial activity of some lamiaceae species against *Staphylococcus aureus* in yoghurt-based drink (doogh). Cell Mol Biol (Noisy Le Grand) 64

Abhishek T, Sharma N, Sharma V, Afroz AA (2013) review on conventional and non-conventional methods to manage postharvest diseases of perishables. Researcher 5:6–19

Abid Ali Khan MM, Naqvi TS, Naqvi MS (2012) Identification of phytosaponins as novel biodynamic agents: an updated overview. Asian J Exp Biol Sci 3:459–467

Abu-Darwish MS, Cabral C, Gonçalves MJ, Cavaleiro C, Cruz MT, Zulfiqar A, Khan IA, Efferth T, Salgueiro L (2016) Chemical composition and biological activities of *Artemisia judaica* essential oil from southern desert of jordan. J. Ethnopharmacol 191:161–168

Adedapo AA, Ofuegbe SO (2014) The evaluation of the hypoglycemic effect of soft drink leaf extract of *Phyllanthus amarus* (euphorbiaceae) in rats. J Basic Clin Physiol Pharmacol 25:47–57

Adefegha SA, Oboh G, Adefegha OM, Boligon AA, Athayde ML (2014) Antihyperglycemic, hypolipidemic, hepatoprotective and antioxidative effects of dietary clove (*Szyzgium aromaticum*) bud powder in a high-fat diet/streptozotocin-induced diabetes rat model. J Sci Food Agric 94:2726–2737

Adeneye AA (2008) Methanol seed extract of *Citrus paradisi* macfad lowers blood glucose, lipids and cardiovascular disease risk indices in normal wistar rats. Nig Q J Hosp Med 18:16–20

Ahad A, Mujeeb M, Ahsan H, Siddiqui WA (2014) Prophylactic effect of baicalein against renal dysfunction in type 2 diabetic rats. Biochimie 106:101–110

Ahmad R, Ahmad N Naqvi AA (2017) *Ziziphus oxyphylla*: Ethnobotanical, ethnopharmacological and phytochemical review. Biomed Pharmacother 91:970–998

Aiying WY, Lou BG, Xu T, Lin C (2011) Defense responses in tomato fruit induced by oligandrin against *Botrytis cinerea*. Afr J Biotechnol 10:4596–4601

Akharaiyi FC, Akinyemi AJ, Isitua CC, Ogunmefun OT, Opakunle SO, Fasae JK (2017) Some antidiabetic medicinal plants used by traditional healers in ado ekiti, Nigeria. Bratisl Med 118:504–505

Akhtar N, Syed DN, Khan MI, Adhami VM, Mirza B, Mukhtar H (2016) The pentacyclic triterpenoid, plectranthoic acid, a novel activator of ampk induces apoptotic death in prostate cancer cells. Oncotarget 7:3819–3831

Ali K, Meredith AW, Thomas HR (2013) Techniques for analysis of plant phenolic compounds. Molecules 18:2328–2375

Alkhalidy H, Moore W, Zhang Y, McMillan R, Wang A, Ali M, Suh KS, Zhen W, Cheng Z, Jia Z (2015) Small molecule kaempferol promotes insulin sensitivity and preserved pancreatic b-cell mass in middle-aged obese diabetic mice. J Dia Res 532984

Amjid A, Ajaz AG, Mohd M (2014) Chrysin, an antiinflammatory molecule, abrogates renal dysfunction in type 2 diabetic rats. Toxicol Appl Pharm 279:1–7

Anaya-Eugenio GD, Rivero-Cruz I, Rivera-Chávez J, Mata R (2014) Hypoglycemic properties of some preparations and compounds from *Artemisia ludoviciana* nutt. J Ethnopharmacol 155:416–425

Angajala G, Ramya R, Subashini R (2014) In-vitro antiinflammatory and mosquito larvicidal efficacy of nickel nanoparticles phytofabricated from aqueous leaf extracts of *Aegle marmelos* Correa. Acta Trop 135:19–26

Anwar P, Latif S, Ashraf M, Gyan A (2007) Moringa oleifera A food plant with multiple medicinal uses. Phytother Res 21:17–25

Araujo LCC, Aguiar JS, Napoleao TH, Mota FVB, Barros ALS, Moura MC (2013) Evaluation of cytotoxic and anti-inflammatory activities of extracts and lectins from *Moringa oleifera* seeds. PLoS One 8:e0081973

Arora DS, Onsare JG (2014) In vitro antimicrobial evaluation and phytoconstituents of *Moringa oleifera* pod husks. Ind Crops Prod 52:125–135

Arora DS, Onsare JM, Kuar H (2013) Bioprospecting of *Moringa* (Moringaceae): microbiological perspective. J Pharmacog Phytochem 1:193–215

Arumugam G, Manjula P, Paari N (2013) A review: anti diabetic medicinal plants used for diabetes mellitus. J Acute Dis 5:196–200

Arunachalam K, Parimelazhagan T (2013) Antidiabetic activity of *Ficus amplissima* smith. Bark extract in streptozotocin induced diabetic rats. J Ethnopharmacol 147:302–310

Arya V, Gupta V, Ranjeet K (2011) A review on fruits having anti-diabetic potential. J Chem Pharmace Res 3:204–212

Asare GA, Gyan B, Bugyei K, Adjei S, Mahama R, Addo P, Nyarko A (2012) Toxicity potentials of the nutraceutical *Moringa oleifera* at supra-supplementation levels. J Ethnopharmacol 139:265–272

Asase A, Yohonu DT (2016) Ethnobotanical study of herbal medicines for management of diabetes mellitus in dangme west district of southern Ghana. J Herb Med 6:204–209

Asgharpour F, Pouramir M, Moghadamnia AA (2012) Evaluation of viscosity of traditional medicinal antihyperglycemic plant extracts and relationship with glucose diffusion in vitro. J Med Plants 11:166–176

Atal S, Atal S, Vyas S, Phadnis P (2016) Bio-enhancing effect of piperine with metformin on lowering blood glucose level in alloxan induced diabetic mice. Pharmacogn Res 8:56–60

Attanayake AP, Jayatilaka KAPW, Pathirana C, Mudduwa LKB (2015) Antihyperglycemic activity of *Coccinia grandis* (L.) voigt in streptozotocin induced diabetic rats. Indian J Trad Knowl 14:376–381

Awolola GV, Koorbanally NA, Chenia H, Shode FO, Baijnath H (2014) Antibacterial and anti-biofilm activity of flavonoids and triterpenes isolated from the extracts of *Ficus sansibarica* warb. Subsp. Sansibarica (moraceae) extracts. Afr J Tradit Complement Altern Med 11:124–131

Ayyanar M, Ignacimuthu S (2011) Ethnobotanical survey of medicinal plants commonly used by kani tribals in tirunelveli hills of westernghats. India J Ethnopharmacol 134:851–864

Babu K, Gokul SS, Rai S (2010) Comparative pharmacognostic studies on the barks of four *Ficus* species. Turk J Bot 34:215–224

Badgujar SB, Patel VV, Bandivdekar AH, Mahajan RT (2014) Traditional uses, phytochemistry and pharmacology of *Ficus carica*: a review. Pharm Biol 52:1487–1503

Bahare S, Athar A, Nanjangud VA, Farukh S, Karina R, Ana R, Seyed AA, Patrick VT, Farzad K, Zainul Amiruddin Z, Marcello I, Yasaman T, Miquel M, Antoni S, William NS, Alessandra D, Massimo L, Antonello S, Raffaele C, Elise AO, Atta-ur-Rahman O, Muhammad IC, William CC, Javad S (2019) Antidiabetic potential of medicinal plants and their active components. Biomolecules 9:551

Bailey CJ, Day C (2004) Metformin: its botanical background. Pract Diab Int 21:115–117

Baliga MS, Bhat HP, Baliga BRV, Wilson R, Palatty PL (2011) Phytochemistry, traditional uses and pharmacology of *Eugenia jambolana* Lam. (black plum): a review. Food Res Int 44:1776–1789

Baliga MS, Fernandes S, Thilakchand KR, D'Souza P, Rao S (2013) Scientific validation of the antidiabetic effects of *Syzygium jambolanum* dc (black plum), a traditional medicinal plant of india. J Altern Complement Med 19:191–197

Balkhande SV, Surwase BS (2013) Antimicrobial activity of tuberous root extracts of *Momordica cymbalaria* hook. Asian J Pharm Clin Res 6:201–203

Bansode TS, Salalkar BK (2015) Phytochemical analysis of some selected Indian medicinal plants. Intl J Pharma Bio Sci 6:550-P556

Batista AB, Oliveira JT, Gifoni JM, Pereira ML, Almeida MG, Gomes VM, Lopes JLS (2014) New insights into the structure and mode of action of Mo-CBP3, an antifungal chitin-binding protein of *Moringa oleifera* seeds. PloSone 9:0111427

Beaulieu LP, Harris CS, Saleem A, Cuerrier A, Haddad PS, Martineau LC, Bennett SAL, Arnason JT (2010) Inhibitory effect of the cree traditional medicine wiishichimanaanh (*Vaccinium vitis-idaea*) on advanced glycation endproduct formation: Identification of active principles. Phytother Res 24:741–747

Belay K, Sisay M (2014) Phytochemical constituents and physicochemical properties of medicinal plant (*Moringa Oleifera*) Around Bule Hora. Int J Innov Sci Res 8:88–98

Benammar C, Hichami A, Yessoufou A, Simonin A, Belarbi M, Allali H, Khan NA (2010) *Zizyphus lotus* (desf.) modulates antioxidant activity and human t-cell proliferation. BMC Complement Altern Med 10

Benzler J, Ganjam GK, Pretz D, Oelkrug R, Koch CE, Legler K, Stöhr S, Culmsee C, Williams LM, Tups A (2015) Central inhibition of ikkβ/nf-κb signaling attenuates high-fat diet–induced obesity and glucose intolerance. Diabetes 64:2015–2027

Berhow MA, Affum AO (2012) Gyan BA (2012) Rosmarinic acid content in antidiabetic aqueous extract of *Ocimum canum* sims grown in Ghana. J Med Food 15:611–620

Bhaludra CSS, Bethapudi RR, Murugulla AC, Pullagummi C, Latha T, Venkatesh K, Bheemagani AJ, Pudutha A, Rani AR (2013) Cultivation, phytochemical studies, biological activities and medicinal uses of *Aloe ferox*, grandfather of aloes an important amazing medicinal plant. Int J Pharmacol 9:405–415

Bharati D, Rawat S, Sharma P, Shrivastava B (2016) Comparative evaluation of antidiabetic antihypertensive activity of cynodon dactylon l. and *Phyllanthus niruri* l in ratswith simultaneous type 2 diabetic and hypertension. Der Pharm Lett 8:255–263

Bhattacharjee R, Dey U (2014) An overview of fungal and bacterial biopesticides to control plant pathogens/diseases. Afr J Microbiol Res 8:1749–1762

Biswas M, Kar B, Bhattacharya S, Kumar RBS, Gosh AK, Haldar PK (2014) Antihyperglycemic activity and antioxidant role of *Terminalia arjuna* leaf in streptozotocin-induced diabetic rats. Pharm Biol 49:335–340

Capitani RD, Mercúrio DG, Camargo Júnior FB, Maia Campos PMBG (2012) Stability and clinical efficacy of moisturizing cosmetics formulations containing vitamins C and E. Biomed Biopharm Res 9:215–224

Castro AJ, Frederico MJ Cazarolli LH Mendes CP, Bretanha LC, Schmidt EC, Bouzon ZL, de Medeiros Pinto VA, da Fonte Ramos C, Pizzolatti MG (2015) The mechanism of action of ursolic acid as insulin secretagogue and insulinomimetic is mediated by cross-talk between calcium and kinases to regulate glucose balance. Biochim Biophys Acta 1850:51–61

Chaieb I (2010) Saponins as insecticides: A review. Tunis J Plant Prot 5:39–45

Chan EWC, Tangah J, Inoue T, Kainuma M, Baba K, Oshiro N, Kezuka M, Kimura N (2017) Botany, uses, chemistry and pharmacology of *Ficus microcarpa*: a short review. Syst Rev Pharm 8:103–111

Chan JYW, Lam FC, Leung PC, Che CT, Fung KP (2009) Antihyperglycemic and antioxidative effects of a herbal formulation of *Radix astragali*, *Radix codonopsis and Cortex lycii* in a mouse model of type 2 diabetes mellitus. Phytother Res 23:658–665

Chen T, Cai J, Ni J, Yang F (2016) An uplc-ms/ms application to investigate chemical compositions in the ethanol extract with hypoglycemic activity from *Zingiber striolatum* diels. J Chin Pharm Sci 25:116–121

Choi EK, Kim KS, Yang HJ, Shin MH, Suh HW, Lee KB, Ahn KS, Um JY Lee SG, Lee BC (2012) Hexane fraction of citrus aurantium l. Stimulates glucagon-like peptide-1 (glp-1) secretion via membrane depolarization in nci-h716 cells. Bioch J 6:41–47

Cui J, Gong R, Hu S, Cai L, Chen L (2018) Gambogic acid ameliorates diabetes-induced proliferative retinopathy through inhibition of the hif-1α/vegf expression via targeting pi3k/akt pathway. Life Sci 192:293–303

Da Costa S, Basri M, Shamsudin N, Basri H (2014) Stability of positively charged nanoemulsion formulation containing steroidal drug for effective transdermal application. J Chem 1:1–8

Deb J, Dash GK (2013) Review on *Acacia ferruginea* dc. (Mimosaceae): an endangered medicinal plant. Intl J Pharm Res 5:1–3

Deepa P, Sowndhararajan K, Kim S, Park SJ (2018) A role of ficus species in the management of diabetes mellitus: a review. J Ethnopharmacol 215:210–232

Deguchi Y, Miyazaki K (2010) Anti-hyperglycemic and anti-hyperlipidemic effects of guava leaf extract. Nutr Metab 7:12–28

Dewanjee S, Maiti A, Sahu R, Dua TK, Mandal V (2011) Effective control of type 2 diabetes through antioxidant defense by edible fruits of *Diospyros peregrina*. Evid Based Complement Altern Med 2011

Dhar P, Tayade A, Ballabh B, Chaurasia OP, Bhatt RP, Srivastava RB (2011) *Lycium ruthenicum* murray: a less-explored but high-value medicinal plant from trans-himalayan cold deserts of ladakh, india. Plant Arch 11:583–586

Di R, Huang MT, Ho CT (2011) Anti-inflammatory activities of mogrosides from *Momordica grosvenori* in murine macrophages and a murine ear edema model. J Agric Food Chem 59:7474–7481

Dib I, Tits M, Angenot L, Wauters JN, Assaidi A, Mekhfi H, Aziz M, Bnouham M, Legssyer A, Frederich M (2017) Antihypertensive and vasorelaxant effects of aqueous extract of *Artemisia campestris* (L.) from eastern Morocco. J Ethnopharmacol 206:224–235

Do GM, Jung UJ, Park HJ, Kwon EY, Jeon SM, McGregor RA, Choi MS (2012) Resveratrol ameliorates diabetes-related metabolic changes via activation of amp-activated protein kinase and its downstream targets in db/db mice. Mol Nutr Food Res 56:1282–1291

Durg S, Veerapur VP, Neelima S, Dhadde SB (2017) Antidiabetic activity of *Embelia ribes*, embelin and its derivatives: a systematic review and meta-analysis. Biomed Pharmacother 86:195–204

Dwivedi SK, Enespa A (2012) Effectiveness of extract of some medical plants against soil borne fusaria causing diseases on *Lycopersicon esculantum* and *Solanum melongena*. Int J Pharm Bio Sci 3:1171–1180

EL-Mohamedy RSR, Abdalla AM (2014) Evaluation of antifungal activity of *Moringa oleifera* extracts as natural fungicide against some plant pathogenic fungi in-vitro. J Agr Technol 10:963–982

Englberger L, Lorennij R, Taylor M, Tuia VS, Aalbersberg W, Dolodolotawake U, Tibon L, Tibon J, Alfred J (2014) Carotenoid content and traditional knowledge of breadfruit cultivars of the Republic of the Marshall Islands. J Food Compos Anal 34:192–199

Eskandar M, Somayeh H (2015) Saponin: properties, methods of evaluation and applications. Ann Res Rev Biol 5:207–220

Fidele N, Abakar D, Emmanuel T, Sélestin SD, Paulin N, Hamadjida A, Marcel NR, Christian B, Samuel G, Nicolas NY (2015) Hypolipidemic and anti-atherogenic effect of aqueous extract leaves of *Ficus glumosa* (Moraceae) in rats. Exp Gerontol 62:53–62

Freire JEC, Vasconcelos IM, Moreno FBMB, Batista AB, Lobo MDP, Pereira ML (2015) *Mo*-$CBP_3$, an antifungal chitin-binding protein from *Moringa oleifera* seeds, is a member of the 2S albumin family. PLoS ONE 10:e0119871

García Mesa M (2014) Hypolipidemic potential of plants used in Cuba. Pharmacologyonline 1:73–80

Gavillan-Suarez J, Aguilar-Perez A, Rivera-Ortiz N, Rodríguez-Tirado K, Figueroa-Cuilan W, Morales-Santiago L, Maldonado-Martínez G, Cubano LA, Martínez-Montemayor MM (2015) Chemical profile and in vivo hypoglycemic effects of *Syzygium jambos*, *Costus speciosus* and *Tapeinochilos ananassae* plant extracts used as diabetes adjuvants in puerto rico. BMC Complement Altern Med 2015:15

Geleta B, Makonnen E, Debella A, Abebe A, Fekadu N (2016) In vitro vasodilatory activity and possible mechanisms of the crude extracts and fractions of *Moringa stenopetala* (baker f.) cufod. Leaves in isolated thoracic aorta of guinea pigs. J Exp Pharm 8:35–42

Gifoni JM, Oliveira JT, Oliveira HD, Batista AB, Pereira ML, Gomes AS, Vasconcelos IM (2012) A novel chitin-binding protein from *Moringa oleifera* seed with potential for plant disease control. Peptide Sci 98:406–415

Giovannini P, Howes MJR, Edwards SE (2016) Medicinal plants used in the traditional management of diabetes and its sequelae in Central America: a review. J Ethnopharmacol 184:58–71

Gopukumar ST, Praseetha PK (2015) *Ficus benghalensis* Linn.—the sacred Indian medicinal tree with potent pharmacological remedies. Int J Pharm Sci Rev Res 32:223–227

Goyal M, Sasmal D, Nagori BP (2011) Review on medicinal plants used by local community of jodhpur district of thar desert. Int J Pharmacol 7:333–339

Green JM, Beestman GB (2007) Recently patented and commercialized formulation and adjuvant technology. Crop Prot 26:320–327

Grover JK, Vats V, Yadav SS (2012) *Pterocarpus marsupium* extract (Vijayasar) prevented the alteration in metabolic patterns induced in the normal rat by feeding an adequate diet containing fructose as sole carbohydrate. Diabetes Obes Metab 7:414–417

Guillen A, Granados S, Rivas KE, Estrada O, Echeverri LF Balcázar N (2015) Antihyperglycemic activity of *Eucalyptus tereticornis* in insulin-resistant cells and a nutritional model of diabetic mice. Adv Pharmacol Sci

Gurjar MS, Ali S, Akhtar M, Singh KS (2012) Efficacy of plant extracts in plant disease management. Agr Sci 3:425–433

Hashim A, Khan MS, Khan MS, Baig MH, Ahmad S (2013) Antioxidant and α-amylase inhibitory property of *Phyllanthus virgatus* L.: an in vitro and molecular interaction study. BioMed Res Int 2013

He ZW, Wei W, Li SP, Ling Q, Liao KJ, Wang X (2014) Anti-allodynic effects of obtusifolin and gluco-obtusifolin against inflammatory and neuropathic pain: possible mechanism for neuroinflammation. Biol Pharm Bull 37:1606–1616

Hilmi Y, Abushama MF, Abdalgadir H, Khalid A, Khalid H (2014) A study of antioxidant activity, enzymatic inhibition and in vitro toxicity of selected traditional sudanese plants with anti-diabetic potential. BMC Complement Altern Med 2014:14

Huang H, Campbell B (2016) U.S. Patent No. 9,380,778. U.S. Patent and Trademark Office, Washington, DC

Huang HC, Wang HF, Yih KH, Chang LZ, Chang TM (2012) Dual bioactivities of essential oil extracted from the leaves of *Artemisia argyi* as an antimelanogenic versus antioxidant agent and chemical composition analysis by GC/MS. Int J Mol Sc 13:14679–14697

Ibrahim MA, Islam MS (2017) Effects of butanol fraction of *Ziziphus mucronata* root ethanol extract on glucose homeostasis, serum insulin and other diabetes-related parameters in a murine model for type 2 diabetes. Pharm Biol 55:416–422

Ijato JY, Adebiyi AO, Ijadunola JA (2011) Antifungal effects of four tropical plant aqueous and thanol extracts on post harvest rot of tomato in Ado-Ekiti, Nigeria. New York Sc J 4:64–68

Im I, Park KR, Kim SM, Kim C, Park JH, Nam D, Jang HJ, Shim BS, Ahn KS, Mosaddik A (2012) The butanol fraction of guava (psidium cattleianum sabine) leaf extract suppresses mmp-2 and mmp-9 expression and activity through the suppression of the ERK1/2 MAPK signaling pathway. Nutr Cancer 64:255–266

Islam MN, Ishita IJ, Jung HA, Choi JS (2014) Vicenin 2 isolated from artemisia capillaris exhibited potent anti-glycation properties. Food Chem Toxicol 69:55–62

Jabeen K, Javaid A (2010) Antifungal activity of *Syzygium cumini* against *Ascochyta rabiei*—the cause of chickpea blight. Nat Prod Res 24:1158–1167

Jacob B, Narendhirakannan R (2019) Role of medicinal plants in the management of diabetes mellitus: A review. 3 Biotech 9:4.

Jamal P, Barkat AA, Amid A (2011) Response surface optimization of the process conditions for anti-diabetic compounds from *Cucumis sativus*. Afr J Biotechnol 10:18788–18794

Jaya Kumari S, Sangeetha M, Pavithra RA (2016) retrospective review on Indian traditional herbs and its biocompounds in diabetes. Int J Pharm Res 9:444–460

Jayamurthy P, Aparna B, Gayathri G, Nisha P (2013) Evaluation of antioxidant potential of inflorescence and stalk of plantain (*Musa sapientum*). J Food Biochem 37:2–7

Jerang G, Vrushabendra Swamy BM, Kotagiri S, Dey T, Fariyaz SM (2015) Indian medicinal plants with antidiabetic and related beneficial effects: a review. Res J Pharm Biol Chem Sci 6:31–38

Jiang LC, Basri M, Omar D, Rahman MBA, Salleh AB, Rahman RNZRA (2011) Self-assembly behaviour of alkylpolyglucosides (APG) in mixed surfactant-stabilized emulsions system. J Mol Liquids 158:175–181

Jokar A, Masoomi F, Sadeghpour O, Nassiri-Toosi M, Hamedi S (2016) Potential therapeutic applications for terminalia chebula in Iranian traditional medicine. J Tradit Chin Med 36:250–254

Jonathan G, Rivka R, Avinoam S, Lumír H, Nirit B (2015) Hypoglycemic activity of withanolides and elicitated *Withania somnifera*. Phytochemistry 116:283–289

Kadhim EJ, AL-Shammaa DA (2014) Phytochemical characterisation using GC-MS analysis of methanolic extract of *Moringa oleifera* (family Moringaceae) plant cultivated in Iraq. Chem Mater Res 6:9-26

Kadir, M.F.; Bin Sayeed, M.S.; Mia, M.M.K. Ethnopharmacological survey of medicinal plants used by indigenous and tribal people in rangamati, Bangladesh. J Ethnopharmacol 144:627–637

Kamel FO, Magadmi RM, Hagras MM, Magadmi B, AlAhmad RA (2017) Knowledge, attitude, and beliefs toward traditional herbal medicine use among diabetics in Jeddah Saudi Arabia. Complement Ther Clin Pract 29:207–212

Kasetti RB, Nabi SA, Swapna S, Apparao C (2012) Cinnamic acid as one of the antidiabetic active principle(s) from the seeds of *Syzygium alternifolium*. Food Chem Toxicol 50:1425–1431

Kellogg J, Wang J, Flint C, Ribnicky D, Kuhn P, De Mejia EG, Raskin I, Lila MA (2010) Alaskan wild berry resources and human health under the cloud of climate change. J Agric Food Chem 58:3884–3900

Kennedy DO, Wightman EL (2011) Herbal extracts and phytochemicals: plant secondary metabolites and the enhancement of human brain function. Adv Nutr 2:32–50

Kim YC, Oh H, Kim BS, Kang TH, Ko EK, Han YM, Kim BY, Ahn JS (2005) In vitro protein tyrosine phosphatase 1B inhibitory phenols from the seeds of *Psoralea corylifolia*. Planta Med 71:87–89

Kingsley B, Aravinth Vijay Jesuraj S, Brindha P, Subramoniam A, Atif M (2013) Anti-diabetes activity of *Acacia farnesiana* (L.) willd in alloxan diabetic rats. Int J Pharm Res 5:112–118

Kiran Kumar ANDE, Gowrishankar NL, Manju Bhargavi V, Nagarjuna M, Rajani G, Swetha Y, Vinay Reddy P (2012) Evaluation of anti ulcer activity of ethanol extract of *Diospyros melanoxylon* (roxb). Bark. Int J Pharmcy Pharm Sci 4:537–539

Krishnamurthy PT, Vardarajalu A, Wadhwani A, Patel V (2015) Identification and characterization of a potent anticancer fraction from the leaf extracts of *Moringa oleifera* Lam. Indian J Exp Biol 53:98–103

Kuete V, Efferth T (2011) Pharmacogenomics of Cameroonian traditional herbal medicine for cancer therapy. J Ethnopharmacol 137:752–766

Kumar M, Malik J (2012) Pharmacognostical studies and evaluation of quality parameters of *Butea frondosa* leaves. Int J Pharmcy Pharm Sci 4:610–614

Kumar D, Gaonkar RH, Ghosh R, Pal BC (2012) Bio-assay guided isolation of α-glucosidase inhibitory constituents from Eclipta alba. Nat Pro 7:989–990

Kumar P, Mehta M, Satija S (2013) Garg M (2013) Enzymatic in vitro anti-diabetic activity of few traditional Indian medicinal plants. J Biol Sci 13:540–544

Kunwar A, Priyadarsini KI (2016) Curcumin and its role in chronic diseases. In: Gupta SC, Prasad S, Aggarwal BB (eds) Advances in experimental medicine and biology. Anti-inflammatory nutraceuticals and chronic diseases. Springer International Publishing, Switzerland, pp 1–26

Lee DH, Seo ES, Hong JT, Lee GT, You YK, Lee KK, Jo GW, Kim NK (2013) The efficacy and safety of proposed herbal moisturizing cream for dry skin and itch relief: a randomized, doubled-blind, placebo-controlled treial-study protocol. BMC Complement Altern Med 13:330

Lee H, Choi J, Shik Shin S, Yoon M (2016) Effects of korean red ginseng (*Panax ginseng*) on obesity and adipose inflammation in ovariectomized mice. J Ethnopharmacol 178:229–237

Lee J, Lee HI, Seo KI, Cho HW, Kim MJ, Park EM, Lee MK (2014) Effects of ursolic acid on glucose metabolism, the polyol pathway and dyslipidemia in non-obese type 2 diabetic mice. Indian J Exp Biol 52:683–691

Li XH, Xin X, Wang Y (2013) Pentamethylquercetin protects against diabetes-related cognitive deficits in diabetic goto-kakizaki rats. J Alzheimers Dis 34:755–767

Luqman S, Srivastava S, Kumar R, Maurya AK, Chanda D (2012) Experimental assessment of *Moringa oleifera* Leaf and fruit for its antistress, antioxidant, and scavenging potential using in vitro and in vivo assays. Evid Based Complement Alternat Med. https://doi.org/10.1155/2012/519084

Ma XY, Zhang LH, Shao HB, Xu G, Zhang F, Ni FT, Brestic M (2011) Jerusalem artichoke (*Helianthus tuberosus*), a medicinal salt-resistant plant has high adaptability and multiple-use values. J Med Plant Res 5:1272–1279

Madhuri K, Naik PR (2017) Modulatory effect of garcinol in streptozotocin-induced diabetic wistar rats. Arch Physiol Biochem 123:322–329

Mahomoodally MF, Ramalingum N (2015) An investigation into the consumption patterns, attitude, and perception of mauritians towards common medicinal food plants. J Herb Med 5:99–112

Mali KK, Dias RJ, Havaldar VD, Yadav SJ (2017) Antidiabetic effect of garcinol on streptozotocin-induced diabetic rats. Indian J Pharm Sci 79:463–468

Mamun-or-Rashid ANM, Hossain MS, Hassan N, Biplab KD, Ashrafuzzaman MS, Kumer MS (2014) A review on medicinal plants with antidiabetic activity. J Pharmacog Phytochem 3:149–159

Mancha-Ramirez AM, Slaga TJ (2016) Ursolic acid and chronic disease: an overview of ua's effects on prevention and treatment of obesity and cancer. In: Gupta SC, Prasad S, Aggarwal BB (eds) Advances in experimental medicine and biology. anti-inflammatory nutraceuticals and chronic diseases. Springer International Publishing: Switzerland, p 928

Manosroi J, Zaruwa MZ, Manosroi A (2011) Potent hypoglycemic effect of nigerian anti-diabetic medicinal plants. J Complement Integr Med 2011:8

Maobe MAG, Gitu L, Gatebe E, Rotich H, Karanja PN, Votha DM, Nderitu IW, Kungu W (2013) Antifungal activity of eight selected medicinal herbs used for the treatment of diabetes, malaria and pneumonia in Kisii region, southwest Kenya. World J Med Sci 8:74–78

Mari M, Bautista-Baños S, Sivakumar D (2016) Decay control in the postharvest system: role of microbial and plant volatile organic compounds. Postharv Biol Technol 122:70–81

Marimuthu S, Padmaja B, Nair S (2013) Phytochemical screening studies on melia orientalis by gc-ms analysis. Pharmacogn Res 5:216–218

Marino M, Bersani CCG (2001) Impedancemeasurement to study antimicrobial activity of essential oils from Lamiaceae and Compositae. Int J Food Microbiol 67:187–195

Marwat SK, Fazal UR, Khan EA, Khakwani AA, Ullah I, Khan KU, Khan IU (2014) Useful ethnophytomedicinal recipes of angiosperms used against diabetes in south east asian countries (india, Pakistan & Sri lanka). Pak J Pharma Sci 27:1333–1358

Maurya R, Akanksha J, Singh AB, Srivastava AK (2008) Coagulanolide, a withanolide from *Withania coagulans* fruits and antihyperglycemic activity. Bioorg Med Chem Lett 18:6534–6537

Meenupriya J, Sahaya VA, Priya P (2014) *Cassia alata* and *Cassia auriculata* – Review of their bioactive potential. World J Pharm Sci 2:1760–1769

Mehmood T (2015) Optimization of the canola oil based vitamin E nanoemulsions stabilized by food grade mixed surfactants using response surface methodology. Food Chem 183:1–7

Mina EC, Mina JF (2017) Ethnobotanical survey of plants commonly used for diabetes in tarlac of central Luzon Philippines. Int Med J Malaysia 16:21–28

Mishra AP, Saklani S, Salehi B, Parcha V, Sharifi-Rad M, Milella L, Iriti M, Sharifi-Rad J, Srivastava M (2018a) *Satyrium nepalense*, a high altitude medicinal orchid of Indian himalayan region: chemical profile and biological activities of tuber extracts. Cell Mol Biol (Noisy Le Grand) 2018:64

Mishra AP, Sharifi-Rad M, Shariati MA, Mabkhot YN, Al-Showiman SS, Rauf A, Salehi B, Župunski M, Sharifi-Rad M, Gusain P (2018b) Bioactive compounds and health benefits of edible Rumex species—a review. Cell Mol Biol (Noisy Le Grand) 2018:64

Mishra J, Dash AK, Mishra SN, Gupta AK (2013) Withania coagulans in treatmen of diabetics and some other diseases: a review. Res J Pharm Biol Chem Sci 4:1251–1258

Mishra R, Shuaib M, Mishra PS (2011) A review on herbal antidiabetic drugs. J Appl Pharm Sci 1:235–237

Mishra RK, Soni GC, Mishra RP (2014) A review article: on nanoemulsion. World J Pharm Pharm Sci 3:258–274

Mnif S, Aifa S (2015) Cumin (*Cuminum cyminum* l.) from traditional uses to potential biomedical applications. Chem Biodivers 12:733–742

Modi A, Jain S, Kumar V (2014) *Zizyphus xylopyrus* (retz.) willd: a review of its folkloric, phytochemical and pharmacological perspectives. Asian Pac J Trop Dis 4:S1–S6

Mondal A, Guria T, Maity TK (2015) A new ester of fatty acid from a methanol extract of the whole plant of *Amaranthus spinosus* and its α-glucosidase inhibitory activity. Pharm Biol 53:600–604

Mootoosamy A, Fawzi Mahomoodally M (2014) Ethnomedicinal application of native remedies used against diabetes and related complications in Mauritius. J Ethnopharmacol 151:413–444

Morakinyo AO, Akindele AJ, Ahmed Z (2011) Modulation of antioxidant enzymes and inflammatory cytokines: possible mechanism of anti-diabetic effect of ginger extracts. Afr J Biomed Res 14:195–202

Mousavi L, Salleh RM, Murugaiyah V, Asmawi MZ (2016) Hypoglycemic and anti-hyperglycemic study of *Ocimum tenuiflorum* (L.) Leaves extract in normal and streptozotocin-induced diabetic rats. Asian Pac J Trop Biomed 6:1029–1036

Moyo B, Masika PJ, Muchenge V (2012) Antimicrbial activities of *Moringa oleifera* Lam leaf extracts. Afr J Biotechnol 11:2797–2802

Mustafa MA, Ali A, Manickam S, Siddiqui Y (2014) Ultrasound-assisted chitosan-surfactant nanostructure assemblies: towards maintaining postharvest quality of tomatoes. Food Biopr Technol 7:2102–2111

Mustaffa F, Hassan Z, Yusof NA, Razak KNA, Asmawi MZ (2014) Antidiabetic and antihyperlipidemic potential of standardized extract, fraction and subfraction of *Cinnamomum iners* leaves. Int J Pharmcy Pharm Sci 6:220–225

Muthulakshmi S, Bhavani K, Manju R, Mohamed Shahila NA (2014) Hepatoprotective activity of *Phyllanthus gardnerianus* (wight) baill. Against d-galactosamine Induced Hepatotoxicity. Biomedicine 34:36–44

Muthusamy K, Krishnasamy G (2016) A computational study on role of 6-(hydroxymethyl)-3-[3,4,5-trihydroxy-6-[(3,4,5-trihydroxyoxan-2-yl)oxymethyl]oxan-2-yl]oxyoxane-2,4,5 -triol in the regulation of blood glucose level. J Biomol Struct Dyn 34:2599–2618

Myers D (2005) Surfactant science and technology, 3rd, ed. Wiley, New Jersey, 400pp. ISBN: 978-0-471-68024

Narayana KR, Reddy S, Chaluvadi MR, Krishna DR (2001) Bioflavonoids classification, pharmacological, biochemical effects and therapeutic potential. Indian J Pharmacol 33:2–16

Nedjimi B, Beladel B (2015) Assessment of some chemical elements in wild shih (*Artemisia herba-alba*) using inaa technique. J Appl Res Med Aromat Plants 2:203–205

Nencu I, Vlase L, Istudor V, Mircea T (2015) Preliminary research regarding *Urtica urens* and *Urtica dioica*. Farmacia 63:710–715

Nguyen HX, Le TC, Do TNV, Le TH, Nguyen NT, Nguyen MTT (2016) A-glucosidase inhibitors from the bark of *Mangifera mekongensis*. Chem Cent J 2016:10

Nickavar B, Amin G (2010) Bioassay-guided separation of an α-amylase inhibitor anthocyanin from *Vaccinium arctostaphylos* berries. Z Naturforsch Sect C J Biosci 65:567–570

Nickavar B, Yousefian N (2011) Evaluation of α-amylase inhibitory activities of selected antidiabetic medicinal plants. J Verbrauch Lebensmittelsicherh 6:191–195

Nkobole N, Houghton PJ, Hussein A, Lall N (2011) Antidiabetic activity of *Terminalia sericea* constituents. Nat Pro Comm 6:1585–1588

Olaokun OO, McGaw LJ, Awouafack MD, Eloff JN, Naidoo V (2014) The potential role of glut4 transporters and insulin receptors in the hypoglycaemic activity of *Ficus lutea* acetone leaf extract. BMC Complement Altern Med 2014:14

Orhan N, Aslan M, Şüküroğlu M, Deliorman Orhan D (2013) In vivo and *in vitro* antidiabetic effect of *Cistus laurifolius* and detection of major phenolic compounds by uplc-tof-ms analysis. J Ethnopharmacol 146:859–865

Ota A, Ulrih NP (2017) An overview of herbal products and secondary metabolites used for management of type two diabetes. Front Pharmacol 2017:8

Oza MJ, Kulkarni YA (2016) Phytochemical and complication in type 2 diabetes—an update. Int J Pharm Sci Res 7:14–24

Padmashree PPP, Pandey S (2010) Anti diabetic activity of methanol/methylene chloride extract of terminalia superba leaves on streptozotocin induced diabetes in rats. Int J Pharm Res 2:2415–2419

Pandey VK, Mathur A, Khan MF, Kakkar P (2019) Activation of perk-EIF2α-atf4 pathway contributes to diabetic hepatotoxicity: attenuation of er stress by morin. Cell Signal 2019

Panwar NS, Pradheep K, Bhatt KC, Deswal RPS (2014) Ethnobotany of a threatened medicinal plant "indravan" (*Cucumis callosus*) from central India. Med Plants 6:307–309

Peltzer K, Sydara K, Pengpid S (2016) Traditional, complementary and alternative medicine use in a community population in Lao PDR. Afr J Trad Complement Altern Med 13:95–100

Pham AT, Malterud KE, Paulsen BS, Diallo D, Wangensteen H (2014) A-glucosidase inhibition, 15-lipoxygenase inhibition, and brine shrimp toxicity of extracts and isolated compounds from *Terminalia macroptera* leaves. Pharm Biol 52:1166–1169

Pinto CE, Farias DF, Carvalho AF, Oliveira JT, Pereira ML, Grangeiro TB, Vasconcelos IM (2015) Food safety assessment of an antifungal protein from *Moringa oleifera* seeds in an agricultural biotechnology perspective. Food Chem Toxicol 83:1–9

Polychniatou V, Tzia C (2014) Study of formulation and stability of co-surfactant free water-in-olive oil nano-and submicron emulsions with food grade non-ionic surfactants. J Amer Oil Chemists' Soc 91:79–88

Potterat O (2010) Goji (*Lycium barbarum* and *L. Chinense*): Phytochemistry, pharmacology and safety in the perspective of traditional uses and recent popularity. Planta Med 76:7–19

Prapassom P, Piyarat N, Paweena R, Chandrapatya A (2012) Effect of crude leaf extracts on *Colletotrichum gloeosporioides* (Penz.) Sacc. Pysche, ID 309046. 1–6. https://doi.org/10.1155/2012/309046

Prasath GS, Pillai SI, Subramanian SP (2014) Fisetin improves glucose homeostasis through the inhibition of gluconeogenic enzymes in hepatic tissues of streptozotocin induced diabetic rats. Eur J Pharmacol 740:248–254

Pulbutr P, Saweeram N, Ittisan T, Intrama H, Jaruchotikamol A, Cushnie B (2017) In vitro α-amylase and α-glucosidase inhibitory activities of *Coccinia grandis* aqueous leaf and stem extracts. J Biol Sci 17:61–68

Qian HF, Li Y, Wang L (2017) *Vaccinium bracteatum* Leaves' polysaccharide alleviates hepatic gluconeogenesis via the downregulation of mir-137. Biomed Pharmacother 95:1397–1403

Rahimi-Madiseh M, Heidarian E, Kheiri S, Rafieian-Kopaei M (2017) Effect of hydroalcoholic *Allium ampeloprasum* extract on oxidative stress, diabetes mellitus and dyslipidemia in alloxan-induced diabetic rats. Biomed Pharmacother 86:363–367

Rai M, Jogee PS, Ingle AP (2015) Emerging nanotechnology for detection of mycotoxins in food and feed. Int J Food Sci Nutr 66:363–370

Ramasamy S, Abdul Wahab N, Zainal Abidin N, Manickam S (2013) Effect of extracts from *Phyllanthus watsonii* airy shaw on cell apoptosis in cultured human breast cancer MCF-7 cells. Exp Toxicol Pathol 65:341–349

Ranganathan R, Vijayalakshmi R, Parameswari P (2012) Ethnomedicinal survey of Jawadhu Hills in Tamil Nadu. Asian J Pharm Clin Res 5:45–49

Ranilla LG, Kwon YI, Apostolidis E, Shetty K (2010) Phenolic compounds, antioxidant activity and in vitro inhibitory potential against key enzymes relevant for hyperglycemia and hypertension of commonly used medicinal plants, herbs and spices in Latin America. Bioresour Technol 101:4676–4689

Rao PK, Hasan SS, Bhellum BL, Manhas RK (2015) Ethnomedicinal plants of kathua district, j&k, India. J Ethnopharmacol 171:12–27

Razavi T, Kouhsari SM, Abnous K (2018) Morin exerts anti-diabetic effects in human hepg2 cells via down-regulation of mir-29a. Exp Clin Endocrinol Diabetes 2018.

Reddy SV, Tiwari AK, Kumar US, Rao RJ, Rao JM (2005) Free radical scavenging, enzyme inhibitory constituents from antidiabetic Ayurvedic medicinal plant *Hydnocarpus wightiana* Blume. Phytother Res 19:277–281

Rehman K, Mashwani ZUR, Khan MA, Ullah Z (2015) Chaudhary HJ (2015) An ethno botanical perspective of traditional medicinal plants from the Khattak tribe of Chonthra Karak, Pakistan. J Ethnopharmacol. 165:251–259

Rezaei Aref T, Minaii Zangii B, Latifpour M (2012) Protective effects of urtica dioica extract on the damage of rat small intestinal mucosa caused by diabetes. J Babol Univ Med Sci 14:31–37

Ribeiro RCD, Barreto SMA, Ostrosky EA, Rocha-Filho PAD, Veríssimo LM, Ferrari M (2015) Production and characterization of cosmetic nanoemulsions containing Opuntia ficus-indica (L.) mill extract as moisturizing agent. Molecules 20:2492–2509

Ríos JL, Francini F, Schinella GR (2015) Natural products for the treatment of type 2 diabetes mellitus. Planta Med 81:975–994

Rodríguez-Rojo S, Varona S, Nunez M, Cocero MJ (2012) Characterization of rosemary essential oil for biodegradable emulsions. Ind Crops Prod 37:137–140

Romero-Castillo PA, Pérez Amador Barron MC, Guevara Fefer P, Muñoz Ocotero V, Reyes Dorantes A, Aguirre Garcia F, Amaya Chavez A (2013) Anti-infammatory activity of *Ziziphus amole*. Phyton 82:75–80

Sadegh-Nejadi S, Aberomand M, Ghaffari MA, Mohammadzadeh G, Siahpoosh A, Afrisham R (2018) Inhibitory effect of *Ziziphus jujuba* and *Heracleum persicum* on the activity of partial purified rat intestinal alpha glucosidase enzyme. J Maz Univ Med Sci 25:135–146

Salehi B, Kumar NVA, Şener B, Sharifi-Rad M, Kılıc M, Mahady GB, Vlaisavljevic S, Iriti M, Kobarfard F, Setzer WN (2018) Medicinal plants used in the treatment of human immunodeficiency virus. Int J Mol Sci 19:1459

Salihu Shinkafi T, Bello L, Wara Hassan S, Ali S (2015) An ethnobotanical survey of antidiabetic plants used by hausa-fulani tribes in sokoto, northwest Nigeria. J. Ethnopharmacol 172:91–99

Sánchez-Villavicencio ML, Vinqvist-Tymchuk M, Kalt W, Matar C, Alarcón Aguilar FJ, Escobar Villanueva MC, Haddad PS (2017) Fermented blueberry juice extract and its specific fractions have an anti-adipogenic effect in 3 t3–l1 cells. BMC Complement Altern Med 2017:17

Sankaranarayanan S, Bama P, Ramachandran J (2010) Ethnobotanical study of medicinal plants used by traditional users in Villupuram district of Tamil Nadu, India. J Med Plant Res 4:1089–1101

Sarin B, Verma N, Martín JP, Mohanty A (2014) An overview of important ethnomedicinal herbs of *Phyllanthus* species: Present status and future prospects. Sci World J 2014.

Satish A, Reddy PV, Sairam S, Ahmed F, Urooj A (2014) Antioxidative effect and DNA protecting property of *Moringa oleifera* root extracts. J Herb Spice Med Plants 20:209–220

Seo EJ, Kuete V, Kadioglu O, Krusche B, Schröder S, Greten HJ, Arend J, Lee IS, Efferth T (2013) Antiangiogenic activity and pharmacogenomics of medicinal plants from traditional korean medicine. Evid.-Based Complement Altern Med 2013.

Sh Ahmed A, Ahmed QU, Saxena AK, Jamal P (2017) Evaluation of in vitro antidiabetic and antioxidant characterizations of *Elettaria cardamomum* maton (zingiberaceae), *Piper cubeba* (Piperaceae), and *Plumeria rubra* (Apocynaceae). Pak J Pharma Sci 30:113–126

Shafiq S, Shakeel F, Talegaonkar S, Ahmad FJ, Khar RK, Ali M (2007) Development and bioavailability assessment of ramipril nanoemulsion formulations. Eur J Pharm Biopharm 66:227–243

Shah MR, Hizbullah SM, Habtemariam S, Zarrelli A, Muhammad A, Collina S, Khan I (2016) Protein tyrosine phosphatase 1b inhibitors isolated from *Artemisia roxburghiana*. J Enzym Inhib Med Chem 31:563–567

Shah SK, Garg G, Jhade D, Pandey H (2016) *Ficus racemosa* Linn: Its potentials food security and rural medicinal management. J Pharm Sci Res 8:317–322

Shahreen S, Banik J, Hafiz A, Rahman S, Zaman AT, Shoyeb MA, Chowdhury MH, Rahmatullah M (2012) Antihyperglycemic activities of leaves of three edible fruit plants (*Averrhoa carambola, Ficus hispida* and *Syzygium samarangense*) of bangladesh. Afr J Tradit Complement Altern Med 9:287–291

Shakthi Deve A, Sathish kumar T, Kumaresan K, Rapheal VS (2014) Extraction process optimization of polyphenols from Indian *Citrus sinensis*—as novel antiglycative agents in the management of diabetes mellitus. J Diabetes Metab Disord 13

Sharifi-Rad M, Nazaruk J, Polito L, Morais-Braga MFB, Rocha JE, Coutinho HDM, Salehi B, Tabanelli G, Montanari C, del Mar Contreras M (2018a) Matricaria genus as a source of antimicrobial agents: From farm to pharmacy and food applications. Microbiol Res 2018

Sharifi-Rad M, Roberts TH, Matthews KR, Bezerra CF, Morais-Braga MFB, Coutinho HDM, Sharopov F, Salehi B, Yousaf Z, Sharifi-Rad M (2018b) Ethnobotany of the genus Taraxacum—phytochemicals and antimicrobial activity. Phytother Res 2018

Sharifi-Rad M, Salehi B, Sharifi-Rad J, Setzer WN, Iriti M (2018c) *Pulicaria vulgaris* Gaertn. Essential oil: an alternative or complementary treatment for leishmaniasis. Cell Mol Biol (Noisy-Le-GrandFr.) 64:18–21

Sharma S, Pathak S, Gupta G, Sharma SK, Singh L, Sharma RK, Mishra A, Dua K (2017) Pharmacological evaluation of aqueous extract of *Syzigium cumini* for its antihyperglycemic and antidyslipidemic properties in diabetic rats fed a high cholesterol diet—role of ppar$\gamma$ and ppar$\alpha$. Biomed Pharmacother 89:447–453

Sharma V, Paliwal R (2013) Isolation and characterization of saponins from *Moringa oleifera* (Moringaceae) pods. Int J Pharm Pharm Sci 5:179–183

Shashank K, Abhay KP (2013) Chemistry and biological activities of flavonoids: an overview. Sci World J 16 (Article ID162750)

Sheikh Y, Maibam BC, Biswas D, Laisharm S, Deb L, Talukdar NC, Borah JC (2015) Anti-diabetic potential of selected ethno-medicinal plants of North East India. J Ethnopharmacol 171:37–41

Singab A, Youssef F, Ashour M (2014) Medicinal plants with potential antidiabetic activity and their assessment. Med Aromat Plants 3:1–12

Singh D, Singh B, Goel RK (2011) Traditional uses, phytochemistry and pharmacology of *Ficus religiosa*: a review. J Ethnopharmacol 134:565–583

Singh R, Tasleem A, Imran K, Poonam S (2014) Phytochemicals in antidiabetic drug discovery. J Biomed Ther Sci 1:1–33

Snyder LR, Kirkland JJ, Glajch JL (2012) Practical HPLC method development, 2nd ed. Wiley, New York, 542p

Sogvar OB, Saba MK, Emamifar A (2016) *Aloe vera* and ascorbic acid coatings maintain postharvest quality and reduce microbial load of strawberry fruit. Postharvest Biol Technol 114:29–35

Soto-Vaca A, Losso JN, Xu Z, Finley JW (2012) Review: Evolution of phenolic compounds from color and flavor problems to health benefits. J Agric Food Chem [Epub ahead of print]

Srividya S, Roshana Devi V, Subramanian S (2015) Hypoglycemic and hypolipidemic properties of hydroxychavicol, a major phenolic compound from the leaves of *Piper betle* Linn. Studied in high fat diet fed- low dose stz induced experimental type 2 diabetes in rats. Der Pharm Lett 7:130–140

Ssenyange CW, Namulindwa A, Oyik B, Ssebuliba J (2015) Plants used to manage type II diabetes mellitus in selected districts of Central Uganda. Afr Health Sci 15:496–502

Stanifer JW, Lunyera J, Boyd D, Karia F, Maro V, Omolo J, Patel UD (2015) Traditionalmedicine practices among community members with chronic kidney disease in Northern Tanzania: an ethnomedical survey. BMC Nephrol 2015:16

Su H, Marrone P, Osborne JL (2014) U.S. Patent No. 8,658,567. U.S. Patent and Trademark Office, Washington, DC

Sudha P, Zinjarde SS, Bhargava SY, Kumar AR (2011) Potent $\alpha$-amylase inhibitory activity of Indian ayurvedic medicinal plants. BMC Complement Altern Med 2011:11

Sukandar EY, Adnyana RIK, Nurfitria RS (2015) Antioxidant potential of garlic and turmeric mixture–a traditional Indonesian formulation. Indian J Trad Knowl 14:632–636

Sun J, Fu X, Liu Y, Wang Y, Huo B, Guo Y, Gao X, Li W, Hu X (2015) Hypoglycemic effect and mechanism of honokiol on type 2 diabetic mice. Drug Des Devel Ther 9:6327–6342

Sushma SM, Sharath R, Sujan Ganapathy PS, Sivakamisundari P, Preetham J (2015) Pharmacognostic and phytochemical evaluation of *Curcuma angustifolia* roxb. (Rhizome) indigenous ethno-medicinal plant used by tribal Soliga community of Biligirirangana hills. Int J Pharmacogn Phytochem Res 7:820–824

Tadros TF, Izquierdo P, Esquena J, Solans C (2004) Formation and stability of nano-emulsions. Adv Colloid Interfac 108:303–318

Taguchi K, Hida M, Hasegawa M, Matsumoto T, Kobayashi T (2016) Dietary polyphenol morin rescues endothelial dysfunction in a diabetic mouse model by activating the AKT/ENOS pathway. Mol Nutr Food Res 60:580–588

Tanaka M, Kishimoto Y, Saita E, Suzuki-Sugihara N, Kamiya T, Taguchi C, Iida K, Kondo K (2016) Terminalia bellirica extract inhibits low-density lipoprotein oxidation and macrophage inflammatory response in vitro. Antioxidants 2016:5

Thakur M, Asrani RK, Thakur S, Sharma PK, Patil RD, Lal B, Parkash O (2016) Observations on traditional usage of ethnomedicinal plants in humans and animals of Kangra and Chamba districts of Himachal Pradesh in north-western himalaya, india. J Ethnopharmacol 191:280–300

Tsai SJ, Huang CS, Mong MC (2012) Anti-inflammatory and antifibrotic effects of naringenin in diabetic mice. J Agric Food Chem 60:514–521

Tzeng YM, Rao YK, Lee MJ, Chen K, Lee YC, Wu WS (2011) Insulin-mimetic action of rhoifolin and cosmosiin isolated from *Citrus grandis* osbeck leaves: enhanced adiponectin secretion and insulin receptor phosphorylation in 3T3-L1 cells. Evid.-Based Complement Altern Med11

Uchida-Maruki H, Inagaki H, Ito R, Kurita I, Sai M, Ito T (2015) Piceatannol lowers the blood glucose level in diabetic mice. Biol Pharm Bull 38:629–633

Ukeh N, Chiejina J (2012) Preliminary investigation of the cause of the postharvest fungal rot of tomato. IOSR J Pharmocol Biol Sci 4:36–39

Ullah MF, Bhat SH, Abuduhier FM (2015) Antidiabetic potential of hydro-alcoholic extract of moringa peregrina leaves: Implication as functional food for prophylactic intervention in prediabetic stage. J Food Biochem 39:360–367

Upadhyay RK (2017) Tulsi: A holy plant with high medicinal and therapeutic value. Int J Green Pharm 11:S1–S12

Vadivel V, Biesalski HK (2012) Total phenolic content, in vitro antioxidant activity and type ii diabetes relevant enzyme inhibition properties of methanolic extract of traditionally processed underutilized food legume, *Acacia nilotica* willd ex. Delile. Int Food Res J 19:593–601

Vaishnav R, Agrawal RD, Sandeep S (2015) Medicinal value and future perspective of some therapeutically important plants from indian western region. Int J Pharm Sci Rev Res 34:88–93

van de Venter M, Roux S, Bungu LC, Louw J, Crouch NR, Grace OM, Maharaj V, Pillay P, Sewnarian P, Bhagwandin N (2008) Antidiabetic screening and scoring of 11 plants traditionally used in south africa. J Ethnopharmacol 119:81–86

Vanholme R, Demedts B, Morreel K, Ralph J, Boerjarr W (2010) Lignin biosynthesis and structure. Plant Physiol 153:895–905

Varghese E, Pappachen KL Narayanan SS (2012) Isolation and evaluation of antimicrobial properties of isolated phytoconstituents of fruits of *Helicteres isora* linn. Res J Pharm Biol Chem Sci 3:959–964

Venkatalakshmi P, Brindha P, Saralla RP (2014) Analytical and chemical standardisation studies on *Terminalia catappa* bark. Int J Pharmcy Pharm Sci 6:4–8

Venkatesh KV, Girish Kumar K, Pradeepa K, Santosh Kumar SR (2013) Antibacterial activity of ethanol extract of *Musa paradisiaca* cv. Puttabale and musa acuminate cv. Grand naine. Asian J Pharm Clin Res 6:167–170

Villas-Boas SG, Mas S, Akesson M, Smedsgaard J, Nielsen J (2005) Mass spectrometry in metabolome analysis. Mass Spectrom Rev 24:613–646

Vinoth B, Manivasagaperumal R, Balamurugan S (2012) Phytochemical analysis and antibacterial activity of *Moringa oleifera* Lam. Int J Res Biol Sci 2:98–102

Wahyudin MMN, Natzir R, Alam G, Bukhari AS (2017) Effect of sukun leaf extract [*Artocarpus altilis* (park.) fosberg] on insulin resistance in obese rats (*Rattus norvegicus*): A study of free fatty acid (ffa) levels. Pak J Nutr 16:521–524

Wang D, Ye Z (2016) Cortex *Lycii radicis* extracts protect pancreatic beta cells under high glucose conditions. Curr Mol Med 16:591–595

Wang G, Li W, Lu X, Bao P, Zhao X (2012a) Luteolin ameliorates cardiac failure in type i diabetic cardiomyopathy. J Diabetes Complicat 26:259–265

Wang Y, Chang L, Zhao X, Meng X, Liu Y (2012b) Gas chromatography-mass spectrometry analysis on compounds in volatile oils extracted from Yuan Zhi (*Radix polygalae*) and Shi Chang Pu (*Acorus tatarinowii*) by supercritical $CO_2$. J Trad Chin Med 32:459–464

Xia P, Guo H, Liang Z, Cui X, Liu Y, Liu F (2014) Nutritional composition of sanchi (*Panax notoginseng*) seed and its potential for industrial use. Genet Resour Crop Evol 61:663–667

Xu L, Zhu H, Ozkan HE, Bagley WE, Krause CR (2011) Droplet evaporation and spread on waxy and hairy leaves associated with type and concentration of adjuvants. Pest Manag Sci 67:842–851

Yadav KD, Chaudhury AK (2016) Anti-obesity mechanism of *Curcuma longa* - an over view. Ind J Nat Prod Resour 7:99–106

Yang CY, Wang J, Zhao Y, Shen L, Jiang X, Xie ZG, Liang N, Zhang L, Chen ZH (2010) Anti-diabetic effects of *Panax notoginseng* saponins and its major anti-hyperglycemic components. J Ethnopharmacol 130:231–236

Yasir M, Jain P, Debajyoti Kharya MD (2010) Hypoglycemic and antihyperglycemic effect of different extracts of *Acacia arabica* lamk bark in normal and alloxan induced diabetic rats. Int J Phytomed 2:33–138

Yu Z, Zhang T, Gong C, Sheng Y, Lu B, Zhou L, Ji L, Wang Z (2016) Erianin inhibits high glucose-induced retinal angiogenesis via blocking ERK1/2-regulated HIF-1α-VEGF/VEGFR2 signaling pathway. Sci Rep 6:34306

Yuan HD, Yuan HY, Chung SH, Jin GZ, Piao GC (2010) An active part of artemisia sacrorum ledeb. Attenuates hepatic lipid accumulation through activating amp-activated protein kinase in human hepg2 cells. Biosci Biotechnol Biochem 74:322–328

Zahid N, Ali A, Manickam S, Siddiqui Y, Maqbool M (2012) Potential of chitosan-loaded nanoemulsions to control different *Colletotrichum* spp. and maintain quality of tropical fruits during cold storage. J Appl Microbiol 113:925–939

Zaidi SF, Muhammad JS, Kadowaki M (2015) Diverse pharmacological properties of cinnamomum cassia: A review. Pak J Pharma Sci 28:1433–1438

Zayyanu U, Mohammed UA, Binti Mohamed M (2015) Role of ethanol leaf extracts of *Ficus glumosa* on fasting blood glucose and liver function test results of diabetes treated rats. J Med Biomed Res 14:64–71

Zhang H, Yan X, Jiang Y, Han Y, Zhou Y (2011) The extraction, identification and quantification of hypoglycemic active ingredients from stinging nettle (*Urtica angustifolia*). Afr J Biotechnol 10:9428–9437

Zhang B, Yuan Q, Huang L, Liu X, Li X, Lin S, Chen M, Ge X (2012) Locality identification of Chinese medicinal plant Scutellaria baicalensis (Lamiaceae) based on population-level DNA barcoding. Zhongguo Zhongyao Zazhi 37:1100–1106

# Chapter 3
# Plant Polysaccharides in Pharmaceutical Applications

**Amit Kumar Nayak, Md Saquib Hasnain, Amal Kumar Dhara, and Dilipkumar Pal**

**Abstract** Plant polysaccharides are the by-products of photosynthesis within the plants and are being extracted from different parts of the plants, such as leaves, pods, fruits, seeds, cereals, stems, roots, rhizomes, corms, exudates, etc. The important advantages for the uses of plant polysaccharides include easy availability from the nature as plant resources are abundant, sustainable and low cost production, biodegradability, biocompatibility, water solubility, swelling ability, etc. Since long, numerous plant polysaccharides have already been explored and exploited as excipients in a variety of common pharmaceutical dosage forms, such as suspensions, emulsions, gels, tablets, capsules, beads, microparticles, nanoparticles, liposomes, transdermal formulations, buccal formulations, nasal formulations, ophthalmic formulations, etc. The current chapter presents a brief review on the pharmaceutical applications of various plant polysaccharides.

**Keywords** Plant polysaccharides · Biopolymers · Excipient · Pharmaceutical applications · Drug delivery

---

A. K. Nayak
Department of Pharmaceutics, Seemanta Institute of Pharmaceutical Sciences, Mayurbhanj, Odisha 757086, India

Md. S. Hasnain
Department of Pharmacy, Shri Venkateshwara University, Gajraula, Amroha, UP, India

A. K. Dhara
Department of Pharmacy, Contai Polytechnic, Darua, Contai, Purba Medinipur, West Bengal 721406, India

D. Pal (✉)
Department of Pharmaceutical Sciences, Guru Ghasidas Vishwavidyalaya (A Central University), Koni, Bilaspur, C.G 495009, India
e-mail: drdilip2003@yahoo.co.in

 
D. Pal and A. K. Nayak (eds.), *Bioactive Natural Products for Pharmaceutical Applications*, Advanced Structured Materials 140,
https://doi.org/10.1007/978-3-030-54027-2_3

## 3.1 Introduction

An inspiring array of plant-derived materials and products have already been revealed for the uses in many applications including pharmaceuticals, cosmeceuticals, biomedical, foods, textiles, paints, paper-making, etc. (Ali et al. 2019; George and Suchithra 2019; Hasnain et al. 2019a; Nayak et al. 2020; Pal et al. 2019a; Sinha Mahapatra et al. 2011; Xie et al. 2016). Amongst the different plant-derived materials, plant polysaccharides have already been demonstrated as a useful group of polymeric biomacromolecules possessing some outstanding merits over the synthetic polymers and these merits include easy availability from the nature as plant resources are abundant, sustainable and low cost production, biodegradability, biocompatibility, water solubility, swelling ability, etc. (Nayak and Hasnain 2019a; Nayak et al. 2018a). The most important reason for the rising interest for the uses of various plant polysaccharides is the advantages of easy cultivation and harvesting to offer a constant supply of raw plant materials for the polysaccharide extractions. Plant polysaccharides are the by-products of photosynthesis within the plants and are being extracted from different parts of the plants, such as leaves, pods, fruits, seeds, cereals, stems, roots, rhizomes, corms, exudates, etc. (Prajapati et al. 2013). These are high molecular weight biopolymers possessing many monosaccharidic units as building blocks, which are linked each other by O-glycosidic linkages in different patterns (Nayak and Pal 2016). Similar or different monosaccharidic units are arranged as extremely complex molecular structures with the variations in sequences, linkages, branching patterns and distributions of side chains. In addition, the molecular structural features of plant polysaccharides possess the presence of many functional groups, which can be modified or tailored to produce polysaccharides of desirable quality (Nayak and Pal 2018; Nayak et al. 2018b). During past few decades, an extensive volume of research efforts have been directed to use various plant polysaccharides in many biomedical and medical applications including their pharmaceutical uses as the dosage formulation excipients and dosage performance enhancers in terms of desired pattern of drug releasing, improved drug stability, enhanced bioavailability, desired target specificity, etc. (Nayak and Hasnain 2019a; b; Pal and Nayak 2017).

Various useful plant polysaccharides have already been explored and exploited as biopolymeric agents in many healthcare area including pharmaceutical industry and researches. Some of the widely used plant polysaccharides as efficient excipients in various kinds of pharmaceutical dosage forms are gum Arabic (Nayak and Hasnain 2019c), gum tragacanth (Dhupal et al. 2019), pectin (Nayak and Pal 2016), guar gum (Jana et al. 2019), locust bean gum (Hasnain et al. 2019b; Nayak and Hasnain 2019d), sterculia gum (Bera et al. 2019; Nayak and Hasnain 2019e), tamarind gum (Dey et al. 2019; Nayak, 2016; Nayak and Hasnain 2019f), cashew gum (Nayak et al. 2019), okra gum (Nayak and Hasnain 2019g; Nayak et al. 2018b), gum odina (Samanta et al. 2019), fenugreek seed mucilage (Nayak and Hasnain 2019h; Pal et al. 2019b), linseed polysaccharides (Nayak and Hasnain 2019i), ispaghula mucilages (Guru et al. 2018), plant starches (Nayak and Pal

2017a), etc. Important plant starches used as pharmaceutical excipients are rice starch, potato starch, maize starch, jackfruit seed starch, sago starch, etc. (Nayak and Hasnain 2019i; Nayak and Pal 2017a). The current chapter presents a brief review on the pharmaceutical applications of various plant polysaccharides.

## 3.2 Classifications and Sources of Plant Polysaccharides

Plant polysaccharides possess a wider ranging of structurally diversified biomacromolecules with different useful physicochemical properties (Nayak et al. 2015). These useful biomacromolecules are occurred abundantly in the plant-resources in forms of gums, mucilages and starches (Avachat et al. 2011; Prajapati et al. 2013).

### *3.2.1 Plant Gums*

Plant gums are commonly known as useful class of polysaccharides comprising multiple numbers of sugar units, which are linked together (Nayak and Pal 2016). Upon the hydrolysis process, these yield different sugar units, such as glucose, galactose, arabinose, xylose, mannose or uronic acids, etc., which are occurred in the respective plant gums (Prajapati et al. 2013; Nayak and Pal 2016; Pal and Nayak 2017). This type of plant polysaccharides are commonly occurred in various parts of higher plants, such as leaves, pods, fruits, seeds, cereals, stems, roots, rhizomes, corms, exudates, etc., on account of their own defense mechanisms following natural injury or harvesting via the break-down of cell walls (Prajapati et al. 2013). This gum generation process in plants is commonly known as gummosis (Rana et al. 2011). Plant gums are mainly pathologically generated products containing complex substances, commonly known as polyuronides. These gums are also capable of producing 3-dimensional gel networks like other gums derived from animal as well as microbiological resources (Rana et al. 2011).

The sources of some important plant gums, which have already been studied for pharmaceutical applications, are listed in Table 3.1.

### *3.2.2 Plant Mucilages*

Mucilages originate in the plants either as a part of contents of the cell or as a part of the wall thereof (Malviya et al. 2011). These serve as food reserve and membrane thickener and aid in water storage and seed germination (Pal and Nayak 2017). Chemically, the molecular structural feature of mucilages possess complex structure of polysaccharides containing uronic acid and sugar residues. Principally, plant

**Table 3.1** Sources of some important plant gums for pharmaceutical applications

| Plant gums | Common sources | |
|---|---|---|
| | Scientific name of the plant | Family |
| Gum Arabica/gum acacia | *Acacia Arabica, Acacia Senegal* | Leguminoseae |
| Gum tragacanth | *Astragalus gummifer* | Leguminosae |
| Locust bean gum | *Ceratonia siliqua* | Fabaceae |
| Guar gum | *Cyamompsis tetraganolobus* | Leguminoseae |
| Okra gum | *Hibiscus esculantus* | Malvaceae |
| Tamarind gum | *Tamarindus indica* | Leguminoseae |
| Sterculia (karaya) gum | *Sterculia urens* | Sterculiaceae |
| Gum kondagogu | *Cochlospermum gossypium* | Colchospermaceae |
| Gum odina | *Lannea woodier* | Anacardiaceae |
| Gum cordia | *Cordia obliqua* | Boraginaceae |
| Moringa gum | *Moringa oleifera* | Moringaceae |
| *Albizia* gum | *Albizia procera* | Leguminoseae |
| Khaya gum | *Khaya grandifoliola* | Meliaceae |
| Terminalia gum | *Terminalia randii* | Combretaceae |
| Gum ghatti | *Anogeissus latifolia* | Combretaceae |
| Honey locust gum | *Gleditsia triacanthos* | Leguminosea |
| Abelmoschus gum | *Abelmoschus esculantus* | Malavaceae |
| Gum copal | *Bursera bipinnata* | Burseraceae |
| Gum dammar | *Shorea Wiesneri* | Dipterocarpaceae |
| Tara gum | *Caesalpinia spinosa* | Leguminosae |
| Moi gum | *Lannea coromandelica* | Anacardiaceae |
| Bahera gum | *Terminalia bellerica* roxb | Combretaceae |
| Hakea gum | *Hakea gibbosa* | Proteaceae |
| Leucaena gum | *Leucaena leucocephata* | Fabaceae |
| Grewia gum | *Grewia mollis* | Malvaceae |
| Balangu gum | *Lallemantia royleana* | Labiatae |
| Dillenia fruit gum | *Dillenia indica* L | Dilleniaceae |
| Cashew tree gum | *Anacardium occidentale* | Anacardiaceae |

mucilages are the sulphuric acid esters (Malviya et al. 2011). Due to the high concentration of hydroxyl groups in their structures, these have high water binding capacity and this has led to studies of their role in plant water relations (Nayak and Pal 2016).

The sources of some important plant mucilages, which have already been studied for pharmaceutical applications, are listed in Table 3.2.

**Table 3.2** Sources of some important plant mucilages for pharmaceutical applications

| Plant mucilages | Common sources | |
|---|---|---|
| | Scientific name of the plant | Family |
| Ispaghula mucilage | *Plantago ovata* | Plantaginaceae |
| Fenugreek seed mucilage | *Trigonella foenum-graecum* | Fabaceae |
| Aloe mucilage | *Aloe barbadensis* | Liliaceae |
| Flaxseed mucilage | *Linum usitatissimum* | Linaceae |
| *Mimosa pudica* seed mucilage | *Mimosa pudica* | Mimosaceae |
| *Spinacia oleraceae* leaves mucilage | *Spinacia oleraceae* | Amaranthaceae |
| *Basella alba* leaves and stem mucilages | *Basella alba* | Basellaceae |
| *Prosopis juliflora* seed mucilage | *Prosopis juliflora* | Fabaceae |
| *Hibiscus rosasinensis* L. leaves mucilages | *Hibiscus rosasinensis* | Malvaceae |
| Basil seed mucilage | *Ocimum* basilicum | *Lamiaceae* |
| Clove basil seed mucilage | *Ocimum gratissimum* | *Lamiaceae* |
| *Amaranthus viridis* L. (Leotia) leaves mucilage | *Amaranthus viridis* | Amaranthaceae |
| Yellow mustard seed mucilage | *Sinapis alba* | Brassicaceae, |
| *Lepidium perfoliatum* seed. mucilage | *Lepidium perfoliatum* | Brassicaceae |

### *3.2.3 Plant Starches*

Starches are the widely studied storage carbohydrates. The molecular structure of starches comprises of 2 co-polymers: amylose (20–30%) and amylopectin (70–80%) (Nayak and Pal 2012). Amylose is essentially linear with the glucose units bound together through α-(1, 4)-linkages and very few α-(1, 6)-bonds. Amylose chains have a degree of polymerization up to 6,000 dependent on source and a molecular mass of 105–106 g/mol (Nayak and Pal 2017a). Amylopectin has a molecular mass of 107–109 g/mol. It is highly branched with α-(1, 6)-bonds in the branching points and has an average degree of polymerization of 2 million, making it one of the largest molecules in nature (Malakar et al. 2013a; Nayak and Pal 2017a). Many starches are also being extracted from different plant resources as these are occurred in various parts of these plant resources, such as seeds, cereals, rhizomes, roots, tubers, corms, etc. (Builders and Arhewoh 2016). Starches are exists as microscopic granules with characteristically origin-specific shapes and sizes (Nayak and Pal 2017a).

The sources of some important plant starches, which have already been studied for pharmaceutical applications, are listed in Table 3.3.

**Table 3.3** Sources of some important plant starches for pharmaceutical applications

| Plant starches | Common sources | |
|---|---|---|
| | Scientific name of the plant | Family |
| Rice starch | *Oryza sativa* | Poaceae |
| Potato starch | *Solanum tuberosum* | Solanaceae |
| Sweet potato starch | *Ipomoea batatas* | Convolvulaceae |
| Maize starch | *Zea mays* | Poaceae |
| Tapioca starch | *Manihot esculenta* | Euphorbiaceae |
| Sago starch | *Metroxylon sagu* | Palmae |
| Cocoyam starch or Malanga starch | *Xanthosoma sagittifolium* | Araceae |
| Jackfruit seed starch | *Artocarpus heterophyllus* | Moraceae |
| Sorghum starch | *Sorghum bicolor* | Poaceae |
| Dioscorea starch | *Dioscorea dumetorum, Dioscorea oppositifolia, Dioscorea alata, Dioscorea rotundata*, etc | Dioscoreaceae |
| Arrowroot starch | *Maranta arundinacea* | Marantaceae |
| Ginger starch | *Zingiber officinale* | Zingiberaceae |
| Indian palo rhizome starch | *Curcuma angustifolia* Roxb | Zingiberaceae |
| Tiger nut starch | *Cyperus esculentus* | Cyperaceae |

## 3.3 Applications of Plant Polysaccharides in Pharmaceutical Dosage Forms

Various plant polysaccharides have extensively been utilized as thickeners, suspending agents, emulsifiers, stabilizers, gel forming agents, binders, disintegrating agents, matrix formers, release retardants, film formers, coating materials, mucoadhesive agents, etc., in a variety of common pharmaceutical dosage forms, such as suspensions, emulsions, gels, tablets, capsules, beads, microparticles, nanoparticles, liposomes, transdermal formulations, buccal formulations, nasal formulations, ophthalmic formulations, etc. (Avachat et al. 2011; Nayak and Hasnain 2019a; Nayak and Pal 2017a, b; Nayak et al. 2018c; Prajapati et al. 2013; Rana et al. 2011).

### *3.3.1 Emulsions*

Plant polysaccharides are used as emulsifiers in various emulsions. These materials can efficiently stabilize the emulsion via the interfacial absorption and the minimizations of droplet-coalescence. Porto and Cristianini (2014) investigated the emulsifying property of cashew (*Anacardium occidentale* L.) gum and compared its emulsifying property with that of the gum Arabic. Emulsions prepared using gum

Arabic were found more uniform than the emulsions prepared using cashew gum. The smaller oil droplets were formed in the emulsions prepared using gum Arabic. Figure 3.1a presents the microscopic observations of emulsions prepared using both emulsifiers (gum Arabic and cashew gum). Figure 3.1b presents a confocal microscopy image of emulsions prepared using both emulsifiers (gum Arabic and cashew gum). From the results, a prominent flocculation was noticed in the emulsions prepared using gum Arabic as emulsifier in comparison with emulsions prepared using cashew gum. This flocculation behavior of emulsions was found to be directly influenced the emulsion instability. In a research, gum odina has been studied as an emulsifier in the preparation of primary emulsion, where gum odina at low concentration produced more stable primary emulsion than that of gum acacia (Samanta et al. 2010). In another research, the same research group studied the efficacy of gum odina as a stabilizer for the formulation of W/O/W multiple emulsions of lamivudine (Jena et al. 2018). The in vitro lamivudine releasing from these W/O/W multiple emulsions stabilized by gum odina produced more sustained lamivudine release over 6 h, which was comparable that of the W/O/W multiple emulsions stabilized by Tween 80. Verma and Razdan (2003) studied the use of leucaena gum (extracted from the seeds of *Leucaena leucocephala*) as emulsifier via the preparation of 30% liquid paraffin emulsions (o/w) employing 1–4% w/v leucaena seed gum. The emulsions prepared using leucaena seed gum as emulsifier were compared with the emulsions prepared using another plant-derived gum, gum acacia. The results demonstrated enhanced emulsifying property of leucaena seed gum as emulsifier in comparison with that of gum acacia in liquid paraffin emulsions. Lago et al. (2019) prepared o/w nanoemulsions using mucilage extracted from *Pereskia aculeata* Miller leaves via the ultra-sound assisted technique. These

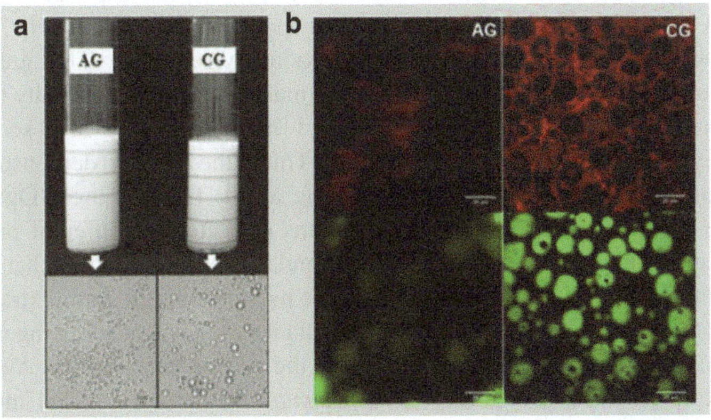


**Fig. 3.1** **a** Optical microscopy images of emulsions prepared with both gum Arabic and cashew gum after 24 h of quiescent storage. Horizontal bar displayed on far right-bottom of figure corresponds to 1 mm extent. **b** Confocal scanning laser microscopy images of o/w emulsions prepared with gum Arabic and cashew gum. Red: emulsifier; green: D-limonene; black: water. Horizontal bar displayed on far right-bottom of figure corresponds to 20 mm extent. (Porto and Cristianini 2014; Copyright @ 2014, with permission from Elsevier Ltd.)

o/w nanoemulsions exhibited good stability when 1–1.5% *Pereskia aculeata* Miller leave mucilage was used as emulsifier. In a research, Gemede et al. (2018) studied the emulsifying property of Ethiopian okra (*Abelmoschus esculentus*) pod mucilage and the prepared emulsions showed good stability. Avlani et al. (2019) explored the emulsifying property of sweet basil (*Ocimum basilicum* L.) seed mucilage to formulate surfactant-free stable sunflower oil emulsions. The sunflower oil emulsions stabilized by 0.3–0.5% w/v sweet basil seed mucilage exhibited good stability. Khunkitti et al. (2006) evaluated the emulsifying property of jackfruit seed starch (1–5% w/v). Although the jackfruit seed starch was found to thicken the external phase of the emulsions prepared, it demonstrated poor emulsifying property. Zhao et al. (2017) formulated and characterized o/w soybean oil emulsions using and gelatinized kudzu starch as emulsifier. 10% (w/w) soybean oil prepared using 3% (w/w) gelatinized kudzu starch showed comparatively good stability.

### *3.3.2 Suspensions*

As suspending agents, numerous plant polysaccharides are being investigated and successfully used in various pharmaceutical suspensions. In suspensions, plant polysaccharides augment the tensile strength of the hydration layer produced around the suspended particles, via hydrogen bonding as well as molecular interactions. These materials are generally hydrophilic colloids and capable of forming aqueous dispersions to enhance the viscosity of continuous phase, in order that the solid particles remain suspended in it over a longer period. In a research, Ogaji and Hoang (2011) prepared ibuprofen pediatric oral suspensions using 0.5% w/v grewia gum as suspending agent. The activity of grewia gum as suspending agent was compared to semi-synthetic conventional suspending agents, namely hydroxylpropyl methylcellulose and sodium carboxymethylcellulose. The grewia gum showed better redispersion property with minimal alterations in viscosity on storage in comparison with that of hydroxylpropyl methylcellulose and sodium carboxymethylcellulose as suspending agent. Thus, the results demonstrated that grewia gum may serve as a good suspending agent. In a research, Ogaji (2011) found the excellent performance of okra gums extracted by different procedures, when used as suspending agents in the formulations of acetaminophen pediatric suspensions. Rao et al. (2005) investigated the use of moringa gum (extracted from the exudate material of *Moringa oleifera* plant) as a suspending agent in sulphamethoxazole suspension and they found the prospective results. Avlani et al. (2019) explored the use of sweet basil (*Ocimum basilicum* L.) seed mucilage to formulate adult and pediatric paracetamol suspensions. The formulated suspensions prepared using 1% w/v sweet basil seed mucilage showed flocculated property with enhanced stability because of the high sedimentation volume as well as good redispersibility. In a research, Pal et al. (2010) evaluated the suitability of *Basella alba* L. leaves mucilage as suspending agent in zinc oxide (20% w/v) suspensions. The *Basella alba* L. leaves mucilage as suspending agent showed better results than that of both gum tragacanth and bentonite. Even, zinc oxide suspensions prepared using *Basella alba* L. leaves mucilage as suspending agent were found easy redispersible. Nayak et al. (2010) assessed the suitability of spinach (*Spinacia oleracea* L.) leaves as suspending agent in zinc oxide (20% w/v) suspensions. The spinach (*Spinacia oleracea* L.) leaves mucilage as suspending agent showed better degree of flocculation and redispersibilty than those of both gum tragacanth and bentonite. The same research group also studied the suitability of fenugreek (*Trigonella foenum-graecum* L.) seed mucilage as suspending agent in zinc oxide (20% w/v) suspensions (Nayak et al. 2012). The fenugreek seed mucilage performed as better suspending agent exhibiting better degree of flocculation and redispersibilty than those of gum tragacanth, gum acacia and bentonite. Piriyaprasarth et al. (2010) studied the uses of arrowroot (*Maranta arundinacea*) starch and yam (Dioscorea sp.) starch as suspending agents in paracetamol suspension. From the results of the study, it was noticed that the optimal concentrations of arrowroot starch was 5–6% and yam starch was 7–8% as suspending agents in paracetamol suspensions. Khunkitti et al. (2006) evaluated the use of jackfruit seed starch as suspending agent and jackfruit seed starch (1–5% w/v) was found able to flocculate titanium dioxide suspensions.

### *3.3.3 Tablets*

Since long different plant polysaccharides are being extensively used in many pharmaceutical tablets as binders, disintegrating agents, matrix formers, and release retardants. The excellent binding property of these materials is due to their adhesive characteristics, which impart cohesiveness to the powdered mass to prepare granules, which are used for further compression in the preparations of tablets. Plant polysaccharides are also employed as disintegrating agents in many tablets (Prajapati et al. 2013). The disintegrating property of plant polysaccharides is attributable to their capability to absorb water and swelling. Plant polysaccharides are also being employed as matrix formers, and release retardants in many sustained drug releasing tablets, especially in matrix tablets. Because of the hydrophilic and high swelling ability, when the matrix tablets containing plant polysaccharides come in contact with water, these get highly hydrated and produce viscous gels onto the surface of the tablets, which produce sustained drug releasing over a longer period.

In a study, Menon et al. (2011) studied the application of orange peel pectin as tablet binder in ibuprofen tablets and they found tablets prepared using 30 mg orange peel pectin exhibited better friability and disintegrating time. Even, these tablets (prepared using 30 mg orange peel pectin as binder) exhibited 82% ibuprofen releasing, in vitro, and were comparable to that of the same amount of starch. In another study, Srivastava et al. (2010) also found the excellent tablet binding property of orange peel pectin in the formulations of paracetamol tablets. Jena et al. (2014) studied the usefulness of gum odina as tablet binder to prepare paracetamol tablets. The in vitro dissolution results showed that 98.55% paracetamol releasing within 30 min from the paracetamol tablets prepared using 0.125% gum odina as binder. However, the in vitro paracetamol releasing was slowed from the paracetamol tablets prepared using 0.25 and 0.375% gum odina. Gum odina was also studied as controlled release matrix former for the preparations of tolterodine tartarate and pioglitazone HCl matrix tablets (Sinha et al. 2011). Both the tolterodine tartarate and pioglitazone HCl matrix tablets exhibited a controlled drug releasing over a prolonged period (Fig. 3.2). Pachuau and Mazumdar (2012) assessed the efficacy of *Albizia procera* gum as release retardant excipients in matrix tablets. They formulated paracetamol tablets using *Albizia procera* gum and observed a controlled sustained releasing of drug from these matrix tablets over12 h. Ofori-Kwakye et al. (2016) prepared matrix tablets of diclofenac sodium and metformin HCl by direct compression using cashew gum, xanthan gum and hydroxypropyl methylcellulose as release retardants. These matrix tablets exhibited extended release of drugs over a longer period. Hasnain et al. (2017a) used cashew

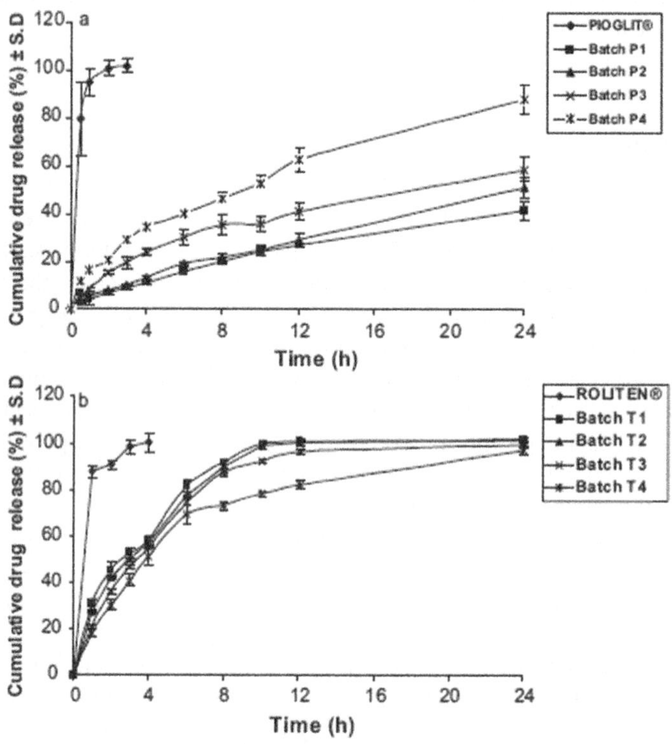


**Fig. 3.2** **a** In vitro mean cumulative percentage release of pioglitazone HCl from matrices containing various proportions of gum odina. **b** In vitro mean cumulative% release of tolterodine tartarate from matrices containing various proportions of gum odina. Each point is the mean value of three samples ($n = 3$) (Sinha et al. 2011; Copyright @ 2010, with permission from Elsevier Ltd.)

gum with hydroxypropylmethylcellulose to prepare matrix tablets of hydralazine HCl. These matrix tablets showed a prolonged sustained drug releasing along with good floating as well as bioadhesive behaviors. Kaleemullah et al. (2017) evaluated the use of *Hibiscus rosa-sinensis* leaves mucilage as matrix former and release retardant excipients in the formulations of sustained release ketoprofen matrix tablets. The in vitro dissolution studies of formulated ketoprofen matrix tablets exhibited sustained ketoprofen releasing up to 24 h. Gaikar and Sandhya (2012) evaluated the tablet binding properties of *Aegle marmelos* fruit mucilage at different concentrations in the formulations of paracetamol tablets. *Aegle marmelos* fruit mucilage of 3% w/w was found to produce comparable results with starch paste of 10% w/w as tablet binder. Ahuja et al. (2013a) studied tablet binding and disintegrant properties of *Mimosa pudica* seed mucilage. The tablet binding property was assessed via the preparation of the paracetamol tablets using *Mimosa pudica* seed mucilage (6, 8 and 10% w/w) as tablet binder and compared with the standard binders (polyvinyl pyrrolidone K25 and gum acacia). *Mimosa pudica* seed mucilage at 10% (w/w) concentration produced tablets possessing adequate hardness as well as friability. The tablet disintegrant property was assessed via the preparation of directly compressed hydrochlorothiazide tablets using *Mimosa pudica* seed mucilage of 1–10% (w/w) as disintegrant and compared with the standard disintegrants (Ac-Di-Sol and starch). The *Mimosa pudica* seed mucilage at 3% (w/w) concentration showed better disintegrant property of the hydrochlorothiazide tablets. In a research, Okunlola and Odeku (2011) evaluated tablet binding properties of starches from four *Dioscorea* species, namely *Dioscorea alata*, *Dioscorea dumetorum*, *Dioscorea rotundata*, *Dioscorea oppositifolia*, etc., in chloroquine phosphate tablet. The results of the research suggested that starches of *Dioscorea alata*, and *Dioscorea rotundata* could be functional for faster disintegration. On the other hand, starches of *Dioscorea oppositifolia* and *Dioscorea dumetorum* could be useful for the minimizations the tablet defects like capping and lamination in tablet formulations. Manek et al. (2012) employed the starch extracted from *Cyperus esculentus* as binder in formulation of metronidazole tablets. Metronidazole tablets prepared using 5, 7.5 and 10% *Cyperus esculentus* starch were compared with tablets prepared using 10% potato starch. Metronidazole tablets prepared using 10% *Cyperus esculentus* starch showed better tablet binding property than others. In a research, Builders et al. (2013) studied the efficacy of tiger nut starch as direct compression excipient in the formulations of acetylsalicylic acid tablets and reported the potential of tiger nut starch as direct compression excipient for the compressible tablets along with enhanced flow properties and improved disintegration. Recently, Peerapattana et al. (2020) evaluated the potential of spray-dried glutinous rice starch as direct compression excipient to prepare hydrophilic matrix tablets of propranolol. The content of spray-dried glutinous rice starch was found to be influenced the propranolol releasing rate from the matrix tablets, significantly.

### 3.3.4 Capsules

Plant polysaccharides are also investigated in the formulations of capsules, where these are used to prepare capsule shells, diluents and capsule matrices. Misale et al. (2008) studied sago starch in the preparation capsule shell as because of its film forming ability. The results of the study indicated that sago starch made capsule shells were moderately comparable to that of gelatin made capsule shells. The in vitro drug releasing rate using sago starch capsule shells exhibited a slower and sustained drug releasing as compared to that of gelatin capsule shells.

### 3.3.5 Beads

Recent years, many plant polysaccharides have been used to formulate polymeric beads loaded with a variety of drugs and the uses of plant polysaccharides to formulate beads is mainly due to the matrix forming and release retarding properties (Nayak and Hasnain 2019k). Most of these polymeric beads made up of plant polysaccharides are intended for oral administrations. Sometimes, mucoadhesive plant polymers are being used to formulate bioadhesive beads, which have been proved useful by facilitating better drug delivery (Nayak and Pal 2016, 2017a, b; Pal and Nayak 2017). In a work, Bera et al. (2015a) developed interpenetrating polymer network beads of ziprasidone HCl employing two plant polysaccharides, namely pectin and sterculia gum via the simultaneous ionotropic crosslinking by zinc acetate and covalent crosslinking by glutaraldehyde. These zinc pectinate-sterculia gum beads were of spherically shaped with characteristic large wrinkles and cracks on the bead surface (Fig. 3.3). These beads showed with controlled drug release over 8 h (Fig. 3.4) with excellent buoyancy and excellent mucoadhesivity onto the goat gastric mucosal membrane. Guru et al. (2018) prepared ispaghula

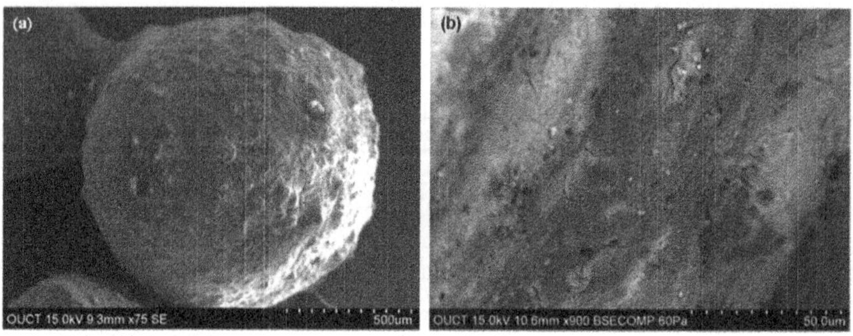


**Fig. 3.3** Scanning electron microphotograph of zinc pectinate-sterculia gum interpenetrating polymer network beads of ziprasidone HCl showing rough surface: **a** 75 × and **b** 900 × (Bera et al. 2015a; Copyright @ 2015, with permission from Elsevier Ltd.)

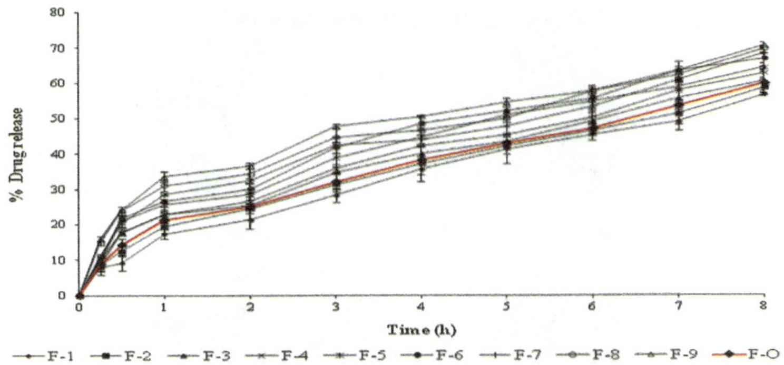


**Fig. 3.4** In vitro drug release from zinc pectinate-sterculia gum interpenetrating polymer network beads of ziprasidone HCl (Bera et al. 2015a; Copyright @ 2015, with permission from Elsevier Ltd.)

husk mucilage-zinc pectinate beads for controlled delivery of aceclofenac. These beads showed a controlled aceclofenac releasing pattern over 10 h with favorable pH dependent swelling. Nayak et al. (2014a; b) developed metformin HCl releasing calcium pectinate-ispagula husk mucilage mucoadhesive beads and calcium pectinate-tamarind seed polysaccharide mucoadhesive beads. Both the mucoadhesive bead formulations showed controlled in vitro drug releasing with good ex vivo mucoadhesion and in vivo antidiabetic activity in diabetic rats. The same group also developed similar kinds of calcium pectinate-fenugreek seed mucilage mucoadhesive beads for oral delivery of metformin HCl (Nayak et al. 2013a). These calcium pectinate-fenugreek (*Trigonella foenum-graecum* L.) seed mucilage mucoadhesive beads of metformin HCl were of spherical in shape with rough surface morphology (Fig. 3.5) and exhibited controlled in vitro drug releasing over 10 h (Fig. 3.6). In addition, these mucoadhesive beads showed significant antidiabetic action in alloxan-induced diabetic rats, in vivo (Fig. 3.7). In another research, an almost similar result of in vivo antidiabetic action in alloxan-induced diabetic rats was noticed by the fenugreek seed mucilage-calcium alginate beads of metformin HCl (Fig. 3.8) (Nayak et al. 2013b). The same research group developed metformin HCl releasing ispaghula husk mucilage-gellan gum mucoadhesive beads for oral administration (Nayak et al. 2014c). These ispaghula husk mucilage-gellan gum mucoadhesive beads exhibited controlled in vitro release of metformin HCl with good ex vivo bioadhesion. They also developed similar kinds of controlled releasing mucoadhesive beads of metformin HCl using polymeric blends of tamarind seed polysaccharide-gellan gum (Nayak et al. 2014b), fenugreek seed mucilage-gellan gum (Nayak and Pal 2014) and jackfruit (*Artocarpus heterophyllus* L.) seed starch-gellan gum (Nayak et al. 2014e). Tamarind seed polysaccharide-calcium alginate beads of metformin HCl were formulated for the use in oral administration (Nayak et al. 2016; Nayak and Pal 2013a). In these biopolymeric beads, tamarind seed polysaccharide was employed as release retardant and bioadhesive polymeric excipients. These beads showed a prolonged

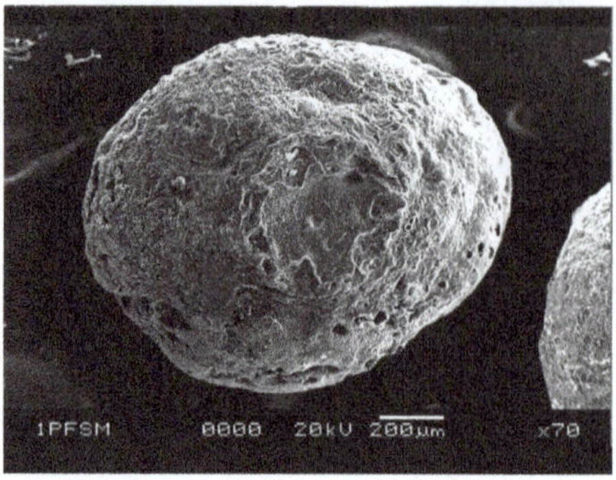


**Fig. 3.5** Scanning electron microphotograph of the optimized calcium pectinate-fenugreek (*Trigonella foenum-graecum* L.) seed mucilage mucoadhesive beads of metformin HCl (Nayak et al. 2013a; Copyright @ 2013, with permission from Elsevier Ltd.)

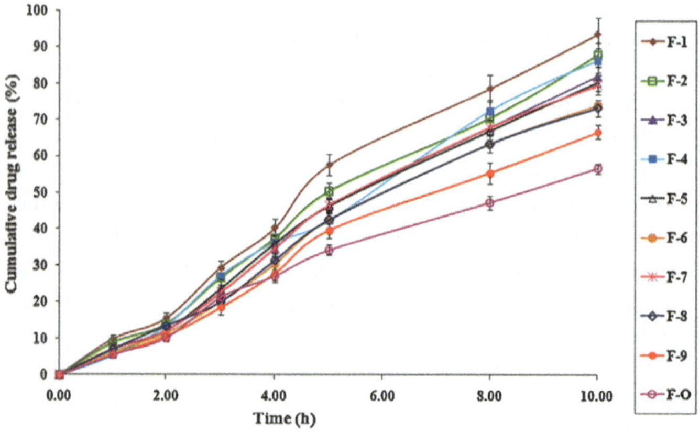


**Fig. 3.6** In vitro drug release from various calcium pectinate-fenugreek (*Trigonella foenum-graecum* L.) seed mucilage mucoadhesive beads of metformin HCl [mean ± S.D., $n = 3$] (Nayak et al. 2013a; Copyright @ 2013, with permission from Elsevier Ltd.)

metformin HCl release over 10 h, in vitro. Even these beads exhibited excellent bioadhesion onto goat intestinal mucosa, ex vivo and significant antidiabetic action in alloxan-induced diabetic rats, in vivo. Sinha et al. (2015a) used okra gum as release retardant polymeric blend with alginate to prepare zinc alginate-okra gum beads for sustained release of diclofenac sodium (over 8 h). The same research group, in another research, formulated okra gum-calcium alginate mucoadhesive beads for controlled releasing glibenclamide (over 8 h) and these beads showed excellent bioadhesion onto goat intestinal mucosa, ex vivo (Sinha et al. 2015b). Hasnain et al. (2018a) prepared mucoadhesive beads of diclofenac sodium using *Linum usitatisimum* mucilage with sodium alginate. In these beads, *Linum*

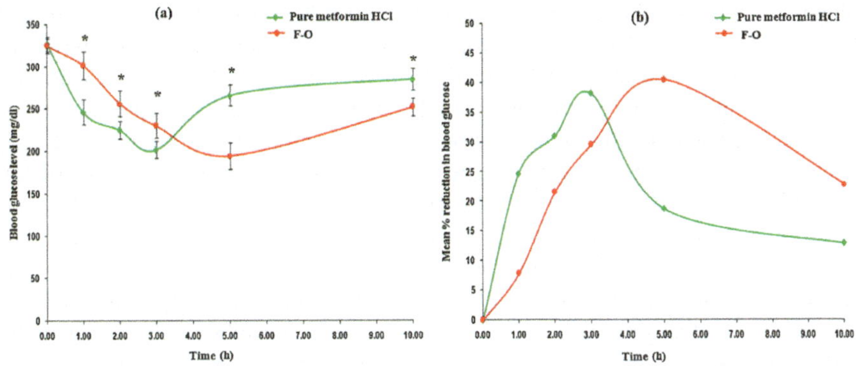


**Fig. 3.7** **a** Comparative in vivo blood glucose level in alloxan-induced diabetic rats after oral administration of pure metformin HCl and optimized calcium pectinate-fenugreek (*Trigonella foenum-graecum* L.) seed mucilage mucoadhesive beads of metformin HCl.The data were analyzed for significant differences (*$p < 0.05$) by paired samples t-test, and **b** comparative in vivo mean percentage reduction in blood glucose level in alloxan-induced diabetic rats after oral administration of pure metformin HCl and optimized calcium pectinate-fenugreek seed mucilage mucoadhesive beads of metformin HCl (Nayak et al. 2013a; Copyright @ 2013, with permission from Elsevier Ltd.)

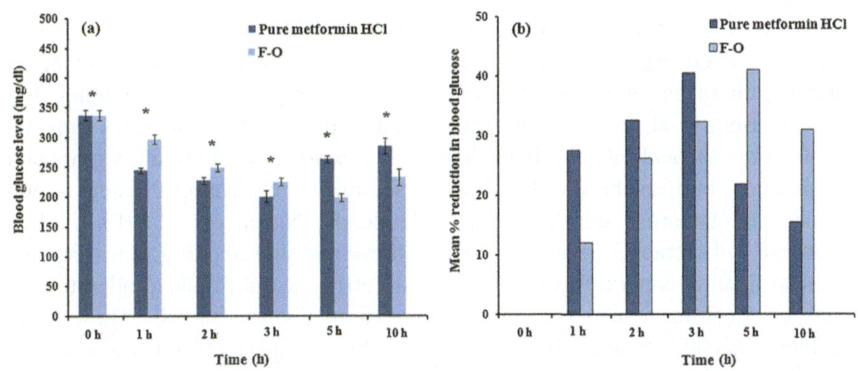


**Fig. 3.8** **a** Comparative in vivo blood glucose level in alloxan-induced diabetic rats after oral administration of pure metformin HCl and fenugreek seed mucilage-calcium alginate beads of metformin HCl. The data were analyzed for significant differences ($p < 0.05$) by paired samples t-test. **b** Comparative in vivo mean percentage reduction in blood glucose level in alloxan-induced diabetic rats after oral administration of pure metformin HCl and fenugreek seed mucilage-calcium alginate beads of metformin HCl (Nayak et al. 2013b; Copyright @ 2012, with permission from Elsevier B.V.)

*usitatisimum* mucilage was used as matrix former, release retardant and mucoadhesive agents. These beads exhibited prolonged in vitro drug releasing with excellent ex vivo biomucoadhesion. The use of jackfruit seed starch-low methoxy pectin blends (Nayak and Pal 2013b) and jackfruit seed starch-sodium alginate blends (Nayak and Pal 2013c) to formulate two different kinds of mucoadhesive beads of metformin HCl by ionotropic gelation was also studied and reported. These beads showed a prolonged metformin HCl release over 10 h, in vitro and significant antidiabetic effects, in vivo. In another research, the applicability of jackfruit seed starch-sodium alginate blends for the preparation of controlled drug releasing was investigated using pioglitazone as a model drug (Nayak et al. 2013c). Malakar et al. (2013b) studied the efficacy of potato starch used as release retardant blends with sodium alginate to formulate potato starch-alginate beads of tolbutamide and these beads showed a controlled tolbutamide releasing pattern, in vitro. Even various plant polysaccharides have been used to formulate buoyant beads for the uses in floating drug delivery. Bera et al. (2015b) developed alginate-sterculia gum gel-coated oil-entrapped calcium alginate beads of resperidone for gastrorentive floating drug delivery. In this work, sterculia gum coat was applied for its bioadhesive nature and these alginate-sterculia gum gel-coated buoyant bioadhesive beads exhibited prolonged release resperidone over 8 h in gastric pH medium. Scanning electron microphotographs exhibited rough surface morphology of the uncoated oil-entrapped calcium alginate beads of resperidone; whereas in case of the alginate-sterculia gum gel-coated oil-entrapped calcium alginate beads of resperidone, comparatively smooth surface morphology was noticed. The cross-sectional view of the alginate-sterculia gum gel-coated oil-entrapped calcium alginate beads of resperidone exhibited a sponge like structural morphology, in which the oil was entrapped with the beads. In another work, the same research group developed risperidone-loaded alginate gel-coated oil-entrapped alginate–tamarind gum–magnesium stearate buoyant beads for gastrorentive floating drug delivery (Bera et al. 2015c). The use of tamarind gum in these beads imparted sustained release and bioadhesive behavior. In a work, emulsion-gelled groundnut oil-entrapped buoyant beads of diclofenac sodium were developed using sodium alginate and tamarind seed polysaccharide-blends (Nayak et al. 2013d). These groundnut oil-entrapped buoyant beads showed sustained drug releasing and excellent floating pattern, in vitro. Guru et al. (2013) also formulated oil-entrapped beads of aceclofenac using sterculia gum-sodium alginate blends. These beads exhibited excellent floating behavior with prolonged sustained release of encapsulated aceclofenac.

### *3.3.6 Microparticles*

Since past few years, many plant polysaccharides have already been exploited for the formulation of microparticles to deliver numerous drugs due to the matrix forming and release retarding properties of plant polysaccharides. Pal and Nayak (2012) formulated gliclazide-loaded tamarind seed polysaccharide-calcium alginate mucoadhesive microspheres for oral adminstration. In these biopolymeric mucoadhesive microspheres, tamarind seed polysaccharide was employed as release retardant and bioadhesive polymeric excipients. These beads showed a prolonged metformin HCl release in vitro and significant antidiabetic activity in alloxan induced diabetic rats, in vivo. Das et al. (2014) developed alginate-based microbeads encapsulated with isoxsuprine HCl using carboxymethyl cashew gum. The scanning electron micrographs of these microbeads demonstrated that these microbeads were of spherically shaped and no agglomeration of particles was noticed. The micrographs also exhibited a rough surface morphology (Fig. 3.9). These microbeads of isoxsuprine HCl showed a prolonged drug releasing pattern (Fig. 3.10). Jana et al. (2013) assessed the use of tamarind seed polysaccharide to prepare chitosan-based interpenetrating polymeric network microparticles of aceclofenac. These microparticles exhibited sustained aceclofenac releasing over 8 h and anti-inflammatory activity was noticed in the carrageenin-induced rats, in vivo, after oral administration (Fig. 3.11) Mohanty et al. (2015) prepared microcapsules of lornoxicam using two plant polysaccharides like *Dillenia indica* pectin and gum dikamali. These formulated microcapsules exhibited sustained release of lornoxicam over a longer period. Nayak et al. (2018d) developed starch-blended $Ca^{2+}$-$Zn^{2+}$-alginate microparticles of aceclofenac. During in vitro release study at pH 1.2 (for initial 2 h), more than 20% aceclofenac was released from these beads and at pH 7.4, it was found to produce sustained release of encapsulated aceclofenac over 7 h Jha and Bhattacharya (2008) investigated the usefulness of sweet potato starch blends with sodium alginate to prepare microbeads for sustained releasing of ibuprofen. These sweet potato starch-alginate microbeads of ibuprofen exhibited sustained drug releasing. Sachan and Bhattyacharya (2009) studied the sustained drug releasing matrix properties of Assam bora rice starch blends with sodium

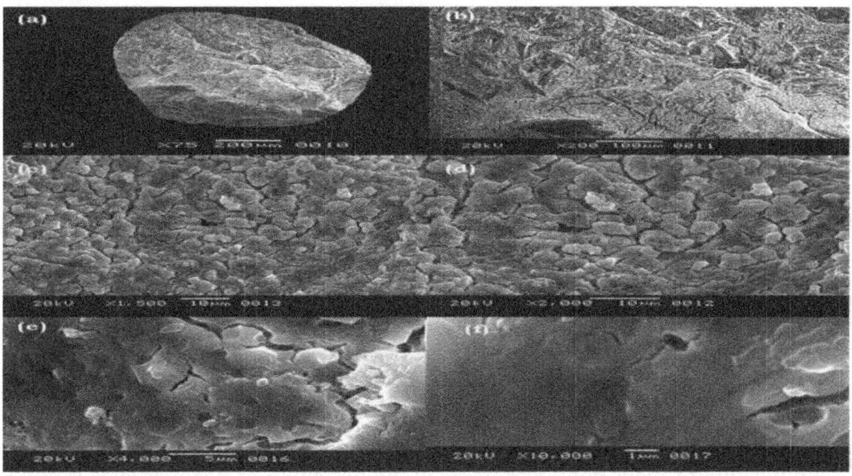


**Fig. 3.9** Scanning electron micrographs of the surface of optimized zinc alginate-carboxymethyl cashew gum microbeads containing isoxsuprine HCl: **a** 75 ×, **b** 200 ×, **c** 1500 ×, **d** 2000 ×, **e** 4000 × and **f** 10,000 × (Das et al. 2014; Copyright @ 2014, with permission from Elsevier B. V.)

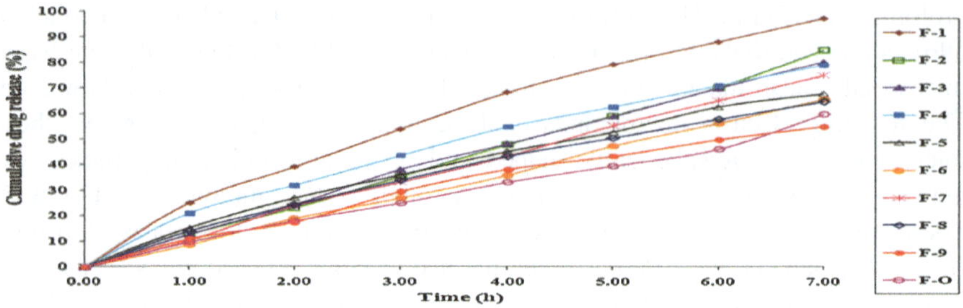


**Fig. 3.10** In vitro drug release from various zinc alginate-carboxymethyl cashew gum microbeads containing isoxsuprine HCl [mean ± S.D., $n$ = 3] (Das et al. 2014; Copyright @ 2014, with permission from Elsevier B.V.)

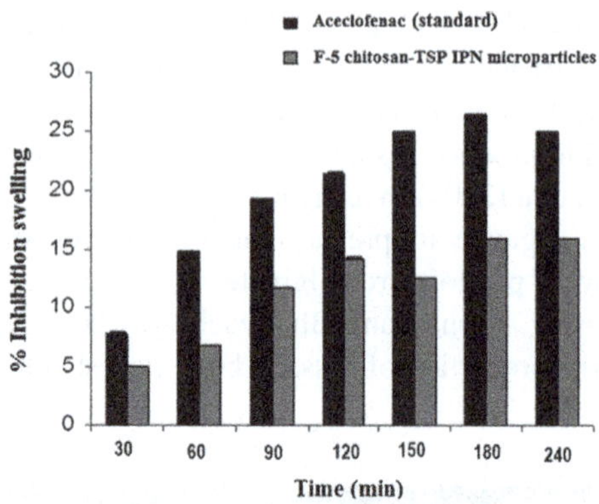


**Fig. 3.11** The percentages inhibition of paw oedema swelling in carageenan induced rat paw oedema model for the standard and chitosan-based interpenetrating polymeric network microparticles of aceclofenac at various time intervals (Jana et al. 2013; Copyright @ 2013, with permission from Elsevier B.V.)

alginate to prepare ionotropically-gelled microbeads of metformin HCl. These Assam bora rice starch-alginate microbeads showed a sustained drug releasing profile over a longer period.

### *3.3.7 Nanoparticles*

The utilization of plant polysaccharides in the designing of nanoparticles is still in its infancy and relatively few studies are reported where plant polysaccharides have been investigated for the preparation of nanoparticles. Pitombeira et al. (2015) prepared nanoparticles using acetylated cashew gum via the self-assembled technique by dialysis. These self-assembled cashew gum-based nanoparticles were loaded with indomethacin. Indomethacin-loaded self-assembled cashew gum-based nanoparticles were of spherically shaped as evidenced in scanning electron microscopy and the particle size characterization demonstrated a unimodal distribution of nanoparticles with an average size of 179 nm (Fig. 3.12). In vitro drug release tests demonstrated a preliminary burst releasing of loaded indomethacin from these cashew gum-based nanoparticles in the initial 2 h followed by a controlled releasing pattern up to 72 h. Burapapadh et al. (2012) prepared pectin-based nanoparticles for delivery of itraconazole via the nanoemulsion templates. The nanoemulsion templates were formed by means of a high-pressure homogenization process employing different types of pectins, such as high methoxyl pectin, low methoxyl pectin, amidated low methoxyl pectin. The results of this investigation indicated that nanoparticles prepared using high methoxyl pectin produced better in vivo absorption than others. Soumya et al. (2010) synthesized the lipase functionalized guar gum-based nanoparticles based via the nanoprecipitation and cross-linking. They tested these formulated nanoparticles as the carrier for anti-hypertensive drug. The drug release evaluation results demonstrated that the rate as well as quantity of encapsulated drug releasing from the lipase functionalized guar gum-based nanoparticles was elevated up to 24 h and subsequently, these were found to be decreased. Sadrjavadi et al. (2018) prepared de-esterified gum tragacanth-chitosan nanoparticles of methotrexate. These gum tragacanth-chitosan nanoparticles of methotrexate were found to be endocytosed via the asialoglycoprotein receptors with sustained release of encapsulated methotrexate for 9 days. Tan et al. (2016) prepared gum Arabic-based nanoparticles via polyelectrolyte complexation with chitosan for the delivery of curcumin. The curcumin loading and curcumin encapsulation efficiency of gum Arabic-chitosan nanoparticles were 3.8 and 90%, respectively. These nanoparticles demonstrated a delayed releasing of curcumin in the simulated gastrointestinal milieu, in vitro.

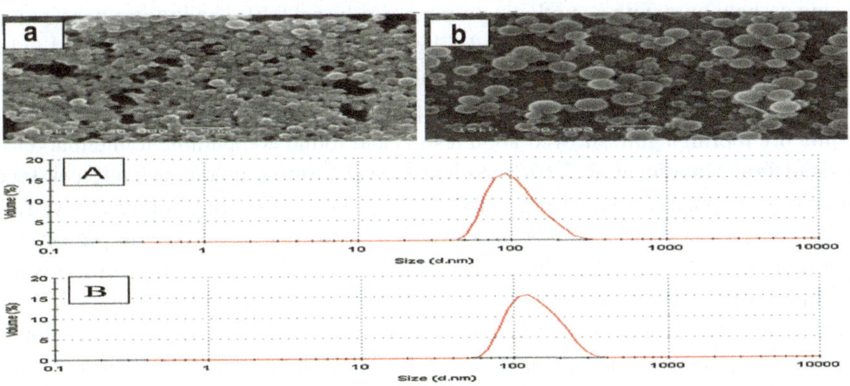


**Fig. 3.12** Scanning electron microscopy and the particle size distribution of self-assembled cashew gum-based nanoparticles without indimethacin (**a**) and with indomethacin (**b**) (Pitombeira et al. 2015; Copyright @ 2014, with permission from Elsevier Ltd.)

### 3.3.8 Liposomes

Plant polysaccharides have also been used to coat liposomes for their better activity. Haghighi et al. (2018) employed coating of pectin onto the nanoliposomes of phloridzin to enhance their targeting property. The pectin-coated nanoliposomes of phloridzin exhibited improved drug entrapment efficiency and storage stability. In another research, Zhou et al. (2014) employed the coating of coated with high methoxyl pectin or low methoxyl pectin onto vitamin C liposomes. These pectin-coated vitamin C liposomes exhibited the enhanced the stability, especially with the coating of high methoxyl pectin.

### 3.3.9 Transdermal Formulations

Plant polysaccharides have already been investigated to formulate different transdermal gels and films due to gel forming and film forming properties, respectively. Hadebe et al. (2014) formulated transdermal patches using amidated pectin for delivery of insulin. The slow releasing insulin from these amidated pectin-based transdermal patches improved various diabetic parameters in streptozotocin-induced diabetic rats. Hasnain et al. ( 2020a) used dillenia (*Dillenia indica* L.) fruit gum as gel-forming agent to prepare lidocaine HCl topical gels. In the ex vivo permeation study, lidocaine HCl was permeated across the porcine ear skin membrane from formulated 4% lidocaine HCl topical gels and the results showed a sustained drug permeation profile over 7 h. The same research group also reported the potential of the application of cashew gum as a potential gel forming material with hydroxypropyl methylcellulose K4M to prepare 4% lidocaine HCl topical gels, which exhibited good skin permeation (Hasnain et al. 2017b). In another research, Das et al. (2013) formulated same type of 4% lidocaine HCl topical gels using cashew gum with a synthetic polymer, Carbopol 940. The in vitro permeation of lidocaine HCl from 4% lidocaine HCl gels through porcine skin was found to be sustained over 7 h (Fig. 3.13). Panda et al. (2006) evaluated gel-forming ability of moringa (*Moringa oleifera*) gum to prepare diclofenac gels. The gels prepared using 8% moringa gum showed good results and found comparable to the marketed formulation Mundhe et al. (2012) studied the gel-forming potential fenugreek seed mucilage to prepare topical gels of diclofenac potassium. In another study, Rao et al. (2010) assessed the *Cocculas hirsutus* leaf powder as gel base to prepare gel of flurbiprofen. From the results of the research, it was found that the in vitro release of flurbiprofen form from the formulated gels and in vivo anti-inflammatory activity were better than that of the marketed gel compared. Nazim et al. (2011) developed hydrotrope potato starch topical gels of rofecoxib. The in vitro rofecoxib releasing results demonstrated that these 1% rofecoxib hydrotrope-gelled potato starch based topical formulation containing 5% w/w potato starch and 15% w/w sodium salicylate exhibited rofecoxib releasing of 16.65% within 6 h; while topical

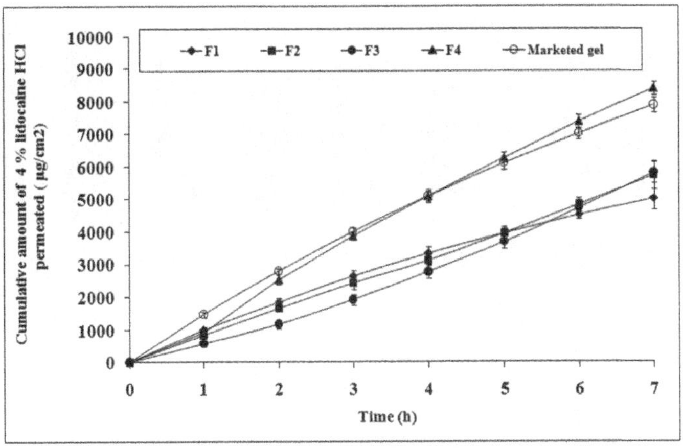


**Fig. 3.13** In vitro lidocaine HCl permeation profile through porcine skin per unit area from 4% lidocaine HCl topical gels containing cashew gum and Carbopol 940 [mean ± S.D., $n = 3$] (Das et al. 2013; Copyright @ 2013, with permission from Elsevier B.V.)

formulation containing 10% w/w potato starch and 15% w/w sodium salicylate exhibited rofecoxib releasing of 16.39% within 6 h.

### *3.3.10 Buccal Formulations*

Due to film forming properties and excellent bioadhesivity, various plant polysaccharides have already been investigated to formulate different buccal drug delivery systems including. Buccal tablets, buccal films and patches (Adhikari and Panda 2017; Hasnain et al. 2020b). Gowthamarajan et al. (2012) prepared buccal tablets of curcumin using cashew gum considering its mucoadhesive potential. The buccal tablets of curcumin formulated using 20% cashew gum, 0.1% methol, 40 mg ethylcellulose (as backing layer forming agent) with a compression force of 2 tons/cm$^2$ for 10 s was identified as an optimized buccoadhesive tablet formulations on the basis of residence time and mucoadhesive strength. The buccal acceptance evaluation of optimized buccoadhesive tablet formulations of curcumin is presented in Fig. 3.14. The in vitro release of curcumin of these buccoadhesive tablets was found to be dependent on the cashew gum concentration used in the formulations. The in vitro release of curcumin from buccoadhesive tablets prepared using 20% cashew gum and 20% hydroxypropyl methylcellulose was compared and the in vitro curcumin release results demonstrated the faster release of curcumin from the buccoadhesive tablets prepared using cashew gum after 8 h (Fig. 3.15). Curcumin release results indicated that cashew gum could be used as a mucoadhesive polymeric excipient to formulate buccoadhesive tablets of curcumin. Avachat et al. (2013) prepared tamarind seed glucan-based buccoadhesive buccal

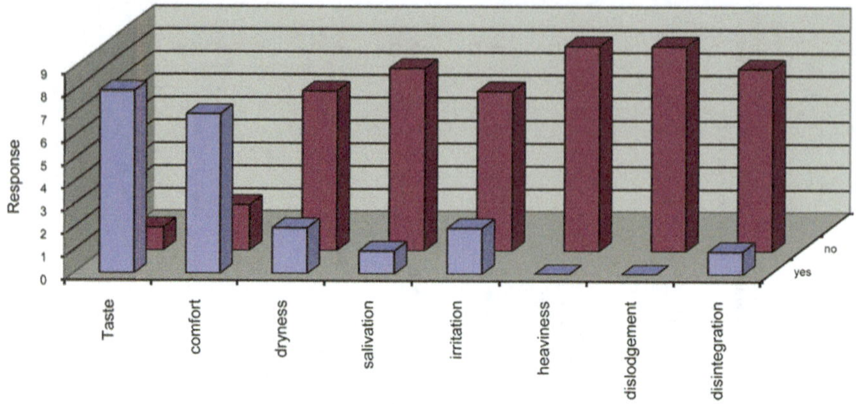


**Fig. 3.14** Comparative results of buccoadhesive tablet acceptance for curcumin buccal tablets (Gowthamarajan et al. 2012; Copyright @ 2012, with permission from Elsevier Ltd.)

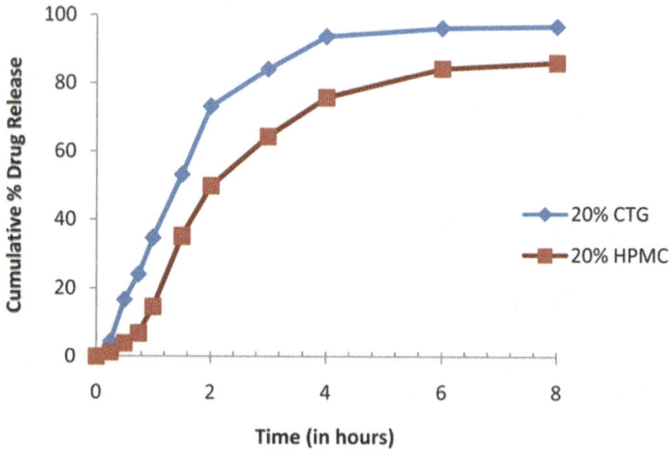


**Fig. 3.15** The comparative in vitro release of curcumin from buccoadhesive tablets prepared using 20% cashew gum and 20% hydroxypropyl methylcellulose (Gowthamarajan et al. 2012; Copyright @ 2012, with permission from Elsevier Ltd.)

films for rizatriptan benzoate delivery. In these buccal films, tamarind seed xyloglucan and Carbopol 934 P were used as mucoadhesive agents. The in vitro rizatriptan benzoate permeation from these buccal films across the porcine buccal mucosal membrane exhibited desirable flux over a prolonged time. Hasnain et al. (2020b) investigated the application of dillenia fruit gum as mucoadhesive film former excipient in the formulation of atenolol buccal patches to treat hypertension. They formulated atenolol buccal patches composed of a mucoadhesive layer of dillenia fruit gum-hydroxypropyl methylcellulose K4M and a backing layer (drug-free) of 1% ethyl cellulose via solvent-casting method. The ex vivo atenolol permeation across the porcine buccal mucosa showed the sustained atenolol permeation over a period of 12 h along with excellent bioadhesion. Ahuja et al. (2013b) evaluated the mucoadhesive property of gum cordia in the preparation of buccal discs containing fluconazole. These gum cordia buccal discs showed good ex vivo buccoadhesion onto the buccal mucosa and it was found to be significantly affected by the compression pressure during preparation.The optimized buccal discs was made by using gum cordia to lactose ratio of 0.66, fluconazole of 20 mg compression pressure of 6600 kg, which showed ex vivo buccoadhesion of 22 h and in vitro fluconazole release of 80% within 24 h. In a research, Mylangam et al. (2016) prepared metoprolol succinate buccoadhesive tablets employing badam gum as mucoadhesive agent by wet granulation technique. These showed good buccoadhesive retention profile. Adhikari and Panda (2017) assessed the mucoadhesivity potential of fenugreek seed mucilage in the formulation of atenolol-releasing buccal patches. They formulated atenolol buccal patches composed of a mucoadhesive layer of fenugreek seed mucilage-hydroxypropyl methylcellulose K4M and a backing layer (drug-free) of 1% ethyl cellulose via solvent-casting method. The sustained atenolol permeation across the porcine buccal mucosa over 12 h was measured along with excellent boaddhesion. Nerkar and Gattani (2011) prepared buccomucoadhesive microspheres of venlafaxine using linseed mucilage as mucoadhesive excipient. They evaluated in vitro as well as in vivo performances of these buccomucoadhesive microspheres. The results exhibited high encapsulation of venlafaxine, higher swelling and good buccoadhesion.

### *3.3.11 Nasal Formulations*

Plant polysaccharides have already been investigated to formulate different nasal drug delivery systems due to gel-forming properties and excellent bioadhesivity. Datta and Bandyopadhyay (2006) assessed the potential of gel-forming property and mucoadhesive property of tamarind seed polysaccharide for the preparation of diazepam nasal gels. The mucoadhesivity and gelling characteristics of tamarind seed polysaccharide was found higher than that of the conventionally used synthetic gel-forming agents, Carbopol 934 and hydroxypropyl methylcellulose. In vitro diazepam releasing from these nasal gels was carried out by Franz-diffusion cell using excised bovine nasal membrane, the results of which was found better than the synthetic gel-forming agents. The same research group evaluated the potential use of *Linum usitatissimum* L. seed mucilage in the formulation of midazolam mucoadhesive nasal gels (Basu et al. 2007). In this work, *Linum usitatissimum* L. seed mucilage was proved as better mucoadhesive agent than Carbopol 934 and hydroxypropyl methylcellulose. The midazolam nasal gels prepared using *Linum usitatissimum* L. seed mucilage demonstrated favorable mucoadhesive characteristics, which facilitate to adhere onto the nasal mucosal surface over a prolonged period. Thus, enhancement of drug absorption can be achieved when administered via the intra-nasal route. Sahu et al. (2011) accessed the use of mucoadhesive agent extracted from *Dillenia* fruits in the formulation of felodipine nasal gels. The nasal gels exhibited controlled releasing of felodipine. In another work, Ketousetuo and Bandyopadhyay (2007) also studied the preparation oxytocin nasal gel using mucoadhesive agent extracted from *Dillenia* fruits.

### 3.3.12 Ophthalmic Formulations

Plant polysaccharides have already been investigated as potential excipient materials in various ocular drug delivery formulations like ophthalmic solutions, gels, and nanoparticles (Dilbaghi et al. 2013; Gheraldi et al. 2004). The high viscosity as well as mucoadhesive characteristics of plant polysaccharides makes these excellent excipients in various ocular formulations for enhancing the ocular residence time. Suzuki and Lim (1994) studied the use of locust bean gum in ocular drug delivery system. They prepared gentamicin-loaded locust bean gum/*i*-carrageenan microparticles by emulsification process, and these microparticles were further incorporated in the poly(vinyl alcohol) gel. The ocular formulations without locust bean gum exhibited an initial burst releasing of gentamicin within the initial 6 h, which was found to decrease by more than 50% by the incorporation of 10% locust bean gum in the microparticle formula. Gheraldi et al. (2000) evaluated the mucoadhesive potential of tamarind gum for the ocular administration of gentamicin and ofloxacin. In this study, the concentrations of gentamicin and ofloxacin were found significantly high in the aqueous humor and cornea in the rabbit eye, when treated with tamarind gum-based ocular formulations containing tamarind gum than that of the free drugs. The same research group also investigated the potential of tamarind gum as mucoadhesive polymer for the improvement of intra-ocular penetration of rufloxacin to treat bacterial keratitis (Gheraldi et al. 2004). Mehra et al. (2010) found the improvement of miotic activity of pilocarpine by the treatment with a tamarind gum based in situ ocular gels, which was found to produce sustained release of pilocarpine up to 12 h Dilbaghi et al. (2013) prepared tamarind seed xyloglucan nanoaggregates loaded with tropicamide for ocular delivery. In this work, they evaluated the ex vivo corneal permeation of tropicamide across the goat cornea and the results demonstrated a significantly increased ex vivo corneal permeation of tropicamide in comparison with that of the marketed formulation.

### 3.3.13 Colon-Targeting Formulations

Some plant polysaccharides have been studied for the development of biodegradable carriers for colon targeting drug releasing. Vivekanandan et al. (2015) prepared guar gum-based matrix tablets of budenoside via the wet granulation. The developed matrix tablets demonstrated 97.12 and 76.86% of budenoside release in rat cecal medium and in the dissolution medium without cecal content, respectively. Dodi et al. (2016) formulated rhodamine-B loaded carboxymethyl guar gum nanoparticles via ionic gelation for colon delivery. These rhodamine-B loaded nanoparticles (208 nm of average diameter) demonstrated a pH-responsive rhodamine-B releasing in simulated gastrointestinal fluids. The MTT assay results demonstrated nontoxicity of the carboxymethyl guar gum nanoparticles prepared by crosslinking using trisodium trimetaphosphate (up to $\sim$0.3 mg/ml). In a study, El-Gibaly, (2002) prepared orally administrable zinc pectinate microparticles for colonic delivery of ketoprofen. These ketoprofen-loaded zinc pectinate microparticles were mixed with mixtures of pectin-dextran to prepare matrix tablets, which were assessed for colonic delivery. These pectinate-tablets exhibited the sigmoidal pattern releasing of ketoprofen with a sustained manner. Odeku and Fell (2005) studied the successful uses of Albizia gum and khaya gum as compression coating material for colon drug targeting. Mishra and Khandare, (2011) prepared tamarind seed polysaccharide-based matrix tablets of ibuprofen via the wet granulation. The in vitro ibuprofen release from these matrix tablets demonstrated that the utmost quantity of ibuprofen was found to be released in simulated colonic fluid pH (6.8) containing rat caecal contents (2 and 4% w/v). On the other hand, less amounts of ibuprofen were found to be released in both simulated gastric fluid (pH 1.2) and simulated intestinal fluid (pH 7.4) before as well as after enzyme induction. Newton et al. (2015) evaluated the chronotherapeutic propranolol HCl delivery matrix tablets for colon targeting therapeutics. These matrix tablets were prepared using tamarind gum and okra gum along with chitosan. The in vitro propranolol HCl release from these matrix tablets demonstrated that the formulations prepared using tamarind gum showed the sustained the propranolol HCl releasing for extended period in comparison with matrix tablets prepared using other polymer based formulations.

### 3.3.14 Dental Formulations

Plant-derived polysaccharides are also used in several dental formulations for loacalized drug releasing. In a research, Hasnain et al. (2018b) used cashew (*Anacardium occidentale*) gum for the preparation of dental pastes containing aceclofenac for the use in periodontitis treatment. These dental pastes of aceclofenac exhibited sustained in vitro aceclofenac releasing over 6 h and excellent bioadhesion onto the oral mucosal membrane. Recently, the same research group investigated the use of dillenia fruit gum to prepare dental pastes containing aceclofenac (Hasnain et al. 2020b). The in vitro aceclofenac releasing from these pastes was found decreased as viscosity increment of these dental pastes due to incorporation more amount of dillenia fruit gum within paste formulations. These dental pastes also exhibited excellent bioadhesion onto the oral mucosal membrane.

## 3.4 Conclusion

Since long, plant polysaccharides have gained much more attention as pharmaceutical as excipients in a variety of pharmaceutical dosage forms due to some important advantages for the uses of plant polysaccharides include easy availability from the nature as plant resources are abundant, sustainable and low cost production, biodegradability, biocompatibility, water solubility, swelling ability, etc. Plant polysaccharides have extensively been utilized as thickeners, suspending agents, emulsifiers, stabilizers, gel forming agents, binders, disintegrating agents, matrix formers, release retardants, film formers, coating materials, mucoadhesive agents, etc., in various common pharmaceutical dosage forms, such as suspensions, emulsions, gels, tablets, capsules, beads, microparticles, nanoparticles, liposomes, transdermal formulations, buccal formulations, nasal formulations, ophthalmic formulations, etc. It is anticipated that a variety of high-quality and multipurpose pharmaceutical dosage forms will possible be formulated in the near future as a result of the continuous research and development in the field of pharmaceutical formulation with the successful uses of new plant polysaccharides as advanced pharmaceutical excipient. Therefore, the future outlook in exploration and exploitation of new plant polysaccharides as useful and multipurpose pharmaceutical excipient is extremely promising.

## References

Adhikari SNR, Panda S (2017) Buccal patches of atenolol formulated using fenugreek (*Trigonella foenum-graecum* L.) seed mucilage. Polim Med 47:5–11

Ahuja M, Kumar A, Yadav P, Singh K (2013) *Mimosa pudica* seed mucilage: Isolation, characterization and evaluation as tablet disintegrant and binder. Int J Biol Macromol 57: 105–110

Ahuja M, Kumar S, Kumar A (2013) Evaluation of mucoadhesive potential of gum cordia, an anionic polysaccharide. Int J Biol Macromol 55:109–112

Ali SA, Nayak AK, Sen KK, Prabhakar T (2019) Preparation and characterization of vetiver oil encapsulated polymeric microcapsules for sedative and hypnotic activity. Int J Res Pharm Sci 10:3616–3625

Avachat A, Gujar KN, Wagh KV (2013) Development and evaluation of tamarind seed glucan-based mucoadhesive buccal films of rizatriptan benzoate. Carbohydr Polym 91: 537–542

Avachat AM, Dash RR, Shrotriya SN (2011) Recent investigations of plant based natural gums, mucilages and resins in novel drug delivery systems. Indian J Pharm Educ Res 45:86–99

Avlani D, Agarwal V, Khattry V, Biswas GR, Majee SB (2019) Exploring properties of sweet basil seed mucilage in development of pharmaceutical suspensions and surfactant-free stable emulsions. Int J Appl Pharm 11:124–129

Basu S, Chakraborty S, Bandyopadhyay AK (2007) Development and evaluation of a mucoadhesive nasal gel of midazolam prepared with *Linum usitatissimum* L. seed mucilage. Sci Pharm 77:899–910

Bera H, Abbasi YF, Hasnain MS, Nayak AK (2019) Sterculia gum in drug delivery applications. In: Hasnain MS, Nayak AK, (eds.,), Natural polysaccharides in drug delivery and biomedical applications. Academic Press, Elsevier Inc., pp 223–247

Bera H, Boddupalli S, Nandikonda S, Kumar S, Nayak AK (2015) Alginate gel-coated oil-entrapped alginate–tamarind gum–magnesium stearate buoyant beads of risperidone. Int J Biol Macromol 78:102–111

Bera H, Boddupalli S, Nayak AK (2015) Mucoadhesive-floating zinc-pectinate-sterculia gum interpenetrating polymer network beads encapsulating ziprasidone HCl. Carbohydr Polym 131:108–118

Bera H, Kandukuri SG, Nayak AK, Boddupalli S (2015) Alginate-sterculia gum gel-coated oil-entrapped alginate beads for gastroretentive risperidone delivery. Carbohydr Polym 120:74–84

Builders PF, Anwunobi AP, Mbah CC, Adikwu MU (2013) New direct compression excipient from tiger nut starch: physicochemical and functional properties. AAPS PharmSciTech 14:818–827

Builders PF, Arhewoh MI (2016) Pharmaceutical applications of native starch in conventional drug delivery. Starch/Stärke 68:1–10

Burapapadh K, Takeuchi H, Sriamornsak P (2012) Novel pectin-based nanoparticles prepared from nanoemulsion templates for improving *in vitro* dissolution and *in vivo* absorption of poorly water-soluble drug. Eur J Pharm Biopharm 82:250–261

Das B, Dutta S, Nayak AK, Nanda U (2014) Zinc alginate-carboxymethyl cashew gum microbeads for prolonged drug release: Development and optimization. Int J Biol Macromol 70:505–515

Das B, Nayak AK, Nanda U (2013) Topical gels of lidocaine HCl using cashew gum and Carbopol 940: Preparation and *in vitro* skin permeation. Int J Biol Macromol 62:514–517

Datta R, Bandyopadhyay AK (2006) A new nasal drug delivery system for diazepam using natural mucoadhesive polysaccharide obtained from tamarind seeds. Saudi Pharm J 14:115–119

Dey S, Nandy BC, De JN, Hasnain MS, Nayak AK (2019) Tamarind gum in drug delivery applications. In: Hasnain MS, Nayak AK, (eds.,), Natural polysaccharides in drug delivery and biomedical applications. Academic Press, Elsevier Inc., pp 285–306

Dhupal M, Gupta MK, Tripathy DR, Kumar M, Yi DK, Nanda SS, Chowdhury D (2019) Recent advances in pharmaceutical applications of natural carbohydrate polymer gum tragacanth, In: Nayak AK, Hasnain MS, Pal D, (eds.,), Natural Polymers for pharmaceutical applications, volume I: plant derived polymers. Apple Academic Press, USA, pp 49–85

Dilbaghi N, Kaur H, Ahuja M, Kumar S (2013) Evaluation of tropicamide-loaded tamarind seed xyloglucan nanoaggregates for ophthalmic delivery. Carbohydr Polym 94:286–291

Dodi G, Pala A, Barbu E, Peptanariu D, Hritcu D, Popa MI, Tamba BI (2016) Carboxymethyl guar gum nanoparticles for drug delivery applications: Preparation and preliminary *in-vitro* investigations. Mater Sci Eng C 63:628–636

El-Gibaly I (2002) Oral delayed-release system based on Zn pectinate gel (ZPG) microparticles as an alternative carrier to calcium pectinate beads for colonic drug delivery. Int J Pharm 232:199–211

Gaikar N, Sandhya P (2012) Evaluation of *Aegle marmelos* fruit mucilage for binding properties. Int J Pharm Sci Res 3:620–622

Gemede HF, Haki GD, Beyene F, Rakshit SK, Woldegiorgis AZ (2018) Indigenous Ethiopian okra (*Abelmoschus esculentus*) mucilage: a novel ingredient with functional and antioxidant properties. Food Sci Nutr 6:563–571

George B, Suchithra TV (2019) Plant-derived bioadhesives for wound dressing and drug delivery system. Fitoterapia 137:104241

Gheraldi E, Tavanti A, Celandroni F, Lupetti A, Blandizzi C, Boldrini E, Campa M, Senesi S (2000) Effect of novel mucoadhesive polysaccharide obtained from tamarind seeds on the intraocular penetration of gentamicin and ofloxacin in rabbits. J Antimicrob Chemother 46:831–834

Gheraldi E, Tavanti A, Davini P, Celandroni F, Salvetti S, Parisio E et al (2004) A mucoadhesive polymer extracted from tamarind seed improves the intraocular penetration and efficacy of rufloxacin in topical treatment of experimental bacterial keratitis. Antimicrob Agents Chemother 48:3396–3401

Gowthamarajan K, Jawahar N, Wake P, Jain K, Sood S (2012) Development of buccal tablets for curcumin using *Anacardium occidentale* gum. Carbohydr Polym 88:1177–1183

Guru PR, Bera H, Das M, Hasnain MS, Nayak AK (2018) Aceclofenac-loaded *Plantago ovata* F. husk mucilage-Zn+2-pectinate controlled-release matrices. Starch/Stärke 70:1700136

Guru PR, Nayak AK, Sahu RK (2013) Oil-entrapped sterculia gum-alginate buoyant systems of aceclofenac: Development and *in vitro* evaluation. Colloids Surf B: Biointerf 104:268–275

Hadebe S, Ngubane P, Serumula M, Musabayane C (2014) Transdermal delivery of insulin by amidated pectin hydrogel matrix patch in streptozotocin-induced diabetic rats: effects on some selected metabolic parameters. *PLoS ONE* 9(7).

Haghighi M, Yarmand MS, Emam-Djomeh Z, McClements DJ, Saboury AA, Rafiee-Tehrani M (2018) Design and fabrication of pectin-coated nanoliposomal delivery systems for a bioactive polyphenolic: Phloridzin. Int J Biol Macromol 112:626–637

Hasnain MS, Guru PR, Rishishwar P, Ali S, Ansari MT, Nayak AK (2020) Atenolol-releasing buccal patches made of *Dillenia indica L.* fruit gum: preparation and *ex vivo* evaluations. SN Appl Sci 2:57

Hasnain MS, Javed MN, Alam MS, Rishishwar P, Rishishwar S, Ali S, Nayak AK, Beg S (2019) Purple heart plant leaves extract-mediated silver nanoparticle synthesis: optimization by Box-Behnken design. Mater Sci Eng C 99:1105–1114

Hasnain MS, Nayak AK, Ansari MT, Pal D (2019b) Pharmaceutical applications of locust bean gum. In: Nayak AK, Hasnain MS, Pal D (eds.,), Natural polymers for pharmaceutical applications, volume I: plant derived polymers. Apple Academic Press, USA, 2019: pp139–162

Hasnain MS, Rishishwar P, Ali S (2017a) Floating-bioadhesive matrix tablets of hydralazine HCl made of cashew gum and HPMC K4M. Int J Phar Pharm Sci 9:124–129

Hasnain MS, Rishishwar P, Ali S (2017b) Use of cashew bark exudate gum in the preparation of 4% lidocaine HCl topical gels. Int J Phar Pharm Sci 9:146–150

Hasnain MS, Rishishwar P, Ali S, Alkahtani S, Tabish M, Milivojevic M, Ansari MT, Nayak AK (2020) Formulation and *ex vivo* skin permeation of lidocaine HCl topical gels using dillenia (*Dillenia indica* L.) fruit gum. Rev Mex Ing Quím 19:1465–1476

Hasnain MS, Rishishwar P, Ali S, Nayak AK (2020) Preparation and evaluation of aceclofenac dental pastes using dillenia fruit gum for periodontitis treatment. SN Appl Sci 2:1–8

Hasnain MS, Rishishwar P, Rishishwar S, Ali S, Nayak AK (2018a) Isolation and characterization of *Linum usitatisimum* polysaccharide to prepare mucoadhesive beads of diclofenac sodium. Int J Biol Macromol 116:162–172

Hasnain MS, Rishishwar P, Rishishwar S, Ali S, Nayak AK (2018b) Extraction and characterization of cashew tree (*Anacardium occidentale*) gum; use in aceclofenac dental pastes. Int J Biol Macromol 116:1074–1081

Jana S, Maiti S, Jana S, Sen KK, Nayak AK (2019) Guar gum in drug delivery applications. In: Hasnain MS, Nayak AK, (eds.,), Natural polysaccharides in drug delivery and biomedical applications. Academic Press, Elsevier Inc., pp 187–201

Jana S, Saha A, Nayak AK, Sen KK, Basu SK (2013) Aceclofenac-loaded chitosan-tamarind seed polysaccharide interpenetrating polymeric network microparticles. Colloids Surf B: Biointerf 105:303–309

Jena AK, Das M, De A, Mitra D, Samanta A (2014) Determination of efficacy of a natural tablet binder: Characterization and *in- vitro* release study. Asian J Pharm Clin Res 7:164–168

Jena AK, Nayak AK, De A, Mitra D, Samanta A (2018) Development of lamivudine containing multiple emulsions stabilized by gum odina. Future J Pharm Sci 4:71–79

Jha AK, Bhattacharya A (2008) Preparation and *in vitro* evaluation of sweet potato starch blended sodium alginate microbeads. Adv Nat Appl Sci 2:122–128

Kaleemullah M, Jiyauddin K, Thiban E, Rasha S, Al-Dhalli S, Budiasih S, Gamal OE, Fadli A, Eddy Y (2017) Development and evaluation of ketoprofen sustained release matrix tablet using *Hibiscus rosa-sinensis* leaves mucilage. Saudi Pharm J 25:770–779

Ketousetuo K, Bandyopadhyay AK (2007) Development of oxytocin nasal gel using natural mucoadhesive agent obtained from the fruits of *Dellinia indica* L. Sci Asia 33:57–60

Khunkitti W, Aromdee C, Vorarat S, Chitropas P (2006) The potential of jackfruit starch for use as suspending agent and emulsifying agent. Songklanakarin J Sci Technol 28:145–155

Lago AMT, Neves ICO, Oliveira NL, Botrel DA, Minim LA, de Resende JV (2019) Ultrasound-assisted oil-in-water nanoemulsion produced from *Pereskia aculeata* Miller mucilage. Ultrason Sonochem 50:339–353

Malakar J, Nayak AK, Das A (2013) Modified starch (cationized)-alginate beads containing aceclofenac: Formulation optimization using central composite design. Starch/Stärke 65:603–612

Malakar J, Nayak AK, Jana P, Pal D (2013) Potato starch-blended alginate beads for prolonged release of tolbutamide: Development by statistical optimization and *in vitro* characterization. Asian J Pharm 7:43–51

Malviya R, Srivastava P, Kulkarni GT (2011) Applications of mucilages in drug delivery-a review. Adv Biol Res 5:1–7

Manek RV, Builders PF, Kolling WM, Emeje M, Kunle OO (2012) Physicochemical and binder properties of starch obtained from *Cyperus esculentus*. AAPS PharmSciTech 13:379–388

Mehra GR, Manish M, Rashi S, Neeraj G, Mishra DN (2010) Enhancement of miotic potential of pilocarpine by tamarind gum based *in situ* gelling ocular dosage form. Acta Pharm Sci 52:145–154

Menon SS, Basavraj BV, Bharath S, Deveswaran R, Madhavan V (2011) Formulation and evaluation of ibuprofen tablets using orange peel pectin as binding agent. Der Pharm Lett 3:241–247

Misale BV, Gavali HBM, Katare SD, Yadav AV (2008) Sago starch capsule shell: A suitable alternative to gelatin capsule shells. Indian J Pharm Educ Res 42:48–52

Mishra MU, Khandare JN (2011) Evaluation of tamarind seed polysaccharide as a biodegradable carrier for colon specific drug delivery. Int J Pharm Pharmaceut Sci 3:139–142

Mohanty S, Mohan GK, Reddy MS (2015) Formulation, evaluation and optimization of sustained release microcapsules of lornoxicam prepared with gum dikamali & pectin extracted from *Dillenia indica*. Int J Chem Sci 13:97–106

Mundhe MR, Pagore RR, Biyani KR (2012) Isolation and evaluation of *Trigonella foenum-gracum* mucilage as gelling agent in diclofenac potassium gel. Int J Ayur Herb Med 2:300–306

Mylangam CK, Beeravelli S, Medikonda J, Pidaparthi JS, Kolapalli VR (2016) Badam gum: a natural polymer in mucoadhesive drug delivery. Design, optimization, and biopharmaceutical evaluation of badam gum-based metoprolol succinate buccoadhesive tablets. Drug Deliv 23:195–206

Narkhede SB, Bendale AR, Jadhav AG, Patel K, Vidyasagar G (2011) Isolation and evaluation of starch of *Artocarpus heterophyllus* as a tablet binder. Int J PharmTech Res 3:836–840

Nayak AK (2016) Tamarind seed polysaccharide-based multiple-unit systems for sustained drug release. In: Kalia S, Averous L (eds) Biodegradable and bio-based polymers: environmental and biomedical applications. WILEY-Scrivener, USA, pp 471–494

Nayak AK, Ansari MT, Sami F, Bera H, Hasnain MS (2019) Cashew gum in drug delivery applications. In: Hasnain MS, Nayak AK, (eds) Natural polysaccharides in drug delivery and biomedical applications. Academic Press, Elsevier Inc., pp 263–283

Nayak AK, Ara TJ, Hasnain MS, Hoda N (2018b) Okra gum-alginate composites for controlled releasing drug delivery, In: Inamuddin, Asiri AM, Mohammad A (eds) Applications of nanocomposite materials in drug delivery. Elsevier Inc., pp 761–785

Nayak AK, Beg S, Hasnain MS, Malakar J, Pal D (2018) Soluble starch-blended $Ca^{2+}$-$Zn^{2+}$-alginate composites-based microparticles of aceclofenac: Formulation development and *in vitro* characterization. Future J Pharm Sci 4:63–70

Nayak AK, Bera H, Hasnain MS, Pal D (2018a) Synthesis and characterization of graft-copolymers of plant polysaccharides, In: Thakur VK (ed) Biopolymer grafting, synthesis and properties. Elsevier Inc., pp 1–62
Nayak AK, Hasnain MS (2019a) Plant polysaccharides in drug delivery applications. In: Nayak AK, Hasnain MS (eds) Plant polysaccharides-based multiple-unit systems for oral drug delivery. Springer, Singapore, pp 19–23
Nayak AK, Hasnain MS (2019b) Some other plant polysaccharide based multiple units for oral drug delivery. In: Nayak AK, Hasnain MS (eds) Plant polysaccharides-based multiple-unit systems for oral drug delivery. Springer, Singapore, pp 123–1128
Nayak AK, Hasnain MS (2019c) Gum Arabic based multiple units for oral drug delivery. In: Nayak AK, Hasnain MS (eds) Plant polysaccharides-based multiple-unit systems for oral drug delivery. Springer, Singapore, pp 25–30
Nayak AK, Hasnain MS (2019d) Locust bean gum based multiple units for oral drug delivery. In: Nayak AK, Hasnain MS (eds) Plant polysaccharides-based multiple-unit systems for oral drug delivery. Springer, Singapore, pp 61–66
Nayak AK, Hasnain MS (2019e) Sterculia gum based multiple units for oral drug delivery. In: Nayak AK, Hasnain MS (eds) Plant polysaccharides-based multiple-unit systems for oral drug delivery. Springer, Singapore, pp 67–82
Nayak AK, Hasnain MS (2019f) Tamarind polysaccharide based multiple units for oral drug delivery. In: Nayak AK, Hasnain MS (eds) Plant Polysaccharides-Based Multiple-Unit Systems for Oral Drug Delivery. Springer, Singapore, pp 31–59
Nayak AK, Hasnain MS (2019g) Okra gum based multiple units for oral drug delivery. In: Nayak AK, Hasnain MS (eds) Plant polysaccharides-based multiple-unit systems for oral drug delivery. Springer, Singapore, pp 83–92
Nayak AK, Hasnain MS (2019h) Fenugreek seed mucilage based multiple units for oral drug delivery. In: Nayak AK, Hasnain MS (eds) Plant polysaccharides-based multiple-unit systems for oral drug delivery. Springer, Singapore, pp 93–112
Nayak AK, Hasnain MS (2019i) Linseed polysaccharide based multiple units for oral drug delivery. In: Nayak AK, Hasnain MS (eds) Plant polysaccharides-based multiple-unit systems for oral drug delivery. Springer, Singapore, pp 117–121
Nayak AK, Hasnain MS (2019j) Potato starch based multiple units for oral drug delivery. In: Nayak AK, Hasnain MS (eds) Plant polysaccharides-based multiple-unit systems for oral drug delivery. Springer, Singapore, pp 113–116
Nayak AK, Hasnain MS (2019k) Background: Multiple units in oral drug delivery. In: Nayak AK, Hasnain MS (eds) Plant polysaccharides-based multiple-unit systems for oral drug delivery. Springer, Singapore, pp 1–17
Nayak AK, Hasnain MS, Pal D (2018c) Gelled microparticles/beads of sterculia gum and tamarind gum for sustained drug release, In: V.K. Thakur, M.K. Thakur (eds) Handbook of Springer on polymeric gel. Springer International Publishing, Switzerland, pp 361–414
Nayak AK, Hasnain MS, Pal K, Banerjee I, Pal D (2020) Gum-based hydrogels in drug delivery. Pal K, Banerjee I, Sarkar P, Kim D, Deng W-P, Dubey NK, Majumder K, (eds) Biopolymer-based formulations. Elsevier Inc., pp 605–645
Nayak AK, Pal D (2012) Natural polysaccharides for drug delivery in tissue engineering. Everyman's Science XLVI: 347–352.
Nayak AK, Pal D (2013a) Ionotropically-gelled mucoadhesive beads for oral metformin HCl delivery: Formulation, optimization and antidiabetic evaluation. J Sci Ind Res 72:15–22
Nayak AK, Pal D (2013b) Blends of jackfruit seed starch-pectin in the development of mucoadhesive beads containing metformin HCl. Int J Biol Macromol 62:137–145
Nayak AK, Pal D (2013c) Formulation optimization of jackfruit seed starch-alginate mucoadhesive beads of metformin HCl. Int J Biol Macromol 59:264–272
Nayak AK, Pal D (2014) *Trigonella foenum-graecum* L. seed mucilage-gellan mucoadhesive beads for controlled release of metformin HCl. Carbohydr Polym 107:31–40

Nayak AK, Pal D (2016) Plant-derived polymers: Ionically gelled sustained drug release systems, In: Mishra M (ed) Encyclopedia of biomedical polymers and polymeric biomaterials, Vol III. Taylor & Francis Group, U.S.A., pp 6002–6017

Nayak AK, Pal D (2017a) Natural starches-blended ionotropically-gelled micrparticles/beads for sustained drug release, In: Thakur VK, Thakur MK, Kessler MR (eds) Handbook of composites from renewable materials, Volume 8, Nanocomposites: advanced applications. WILEY-Scrivener, USA, pp 527–560

Nayak AK, Pal D (2017) Tamarind seed polysaccharide: An emerging excipient for pharmaceutical use. Indian J Pharm Educ Res 51:S136–S146

Nayak AK, Pal D (2018) Functionalization of tamarind gum for drug delivery, In: Thakur VK, Thakur MK (eds) Functional biopolymers. Springer International Publishing, Switzerland, pp 35–56

Nayak AK, Pal D, Das S (2013) Calcium pectinate-fenugreek seed mucilage mucoadhesive beads for controlled delivery of metformin HCl. Carbohydr Polym 96:349–357

Nayak AK, Pal D, Hasnain MS (2013) Development, optimization and *in vitro-in vivo* evaluation of pioglitazone-loaded jackfruit seed starch-alginate beads. Curr Drug Deliv 10:608–619

Nayak AK, Pal D, Malakar J (2013) Development, optimization and evaluation of emulsion-gelled floating beads using natural polysaccharide-blend for controlled drug release. Polym Eng Sci 53:338–350

Nayak AK, Pal D, Pany DR, Mohanty B (2010) Evaluation of *Spinacia oleracea* L. leaves mucilage as innovative suspending agent. J Adv Pharm Technol Res 1:338–341

Nayak AK, Pal D, Pradhan J (2012) Ghorai T (2012) The potential of *Trigonella foenum-graecum* L. seed mucilage as suspending agent. Indian J Pharm Educ Res 46:312–317

Nayak AK, Pal D, Pradhan J, Hasnain MS (2013) Fenugreek seed mucilage-alginate mucoadhesive beads of metformin HCl: design, optimization and evaluation. Int J Biol Macromol 54:144–154

Nayak AK, Pal D, Santra K (2014a) Development of pectinate-ispagula mucilage mucoadhesive beads of metformin HCl by central composite design. Int J Biol Macromol 66:203–221

Nayak AK, Pal D, Santra K (2014b) Development of calcium pectinate-tamarind seed polysaccharide mucoadhesive beads containing metformin HCl. Carbohydr Polym 101:220–230

Nayak AK, Pal D, Santra K (2014c) Ispaghula mucilage-gellan mucoadhesive beads of metformin HCl: Development by response surface methodology. Carbohydr Polym 107:41–50

Nayak AK, Pal D, Santra K (2014d) Tamarind seed polysaccharide-gellan mucoadhesive beads for controlled release of metformin HCl. Carbohydr Polym 103:154–163

Nayak AK, Pal D, Santra K (2014e) *Artocarpus heterophyllus* L. seed starch-blended gellan gum mucoadhesive beads of metformin HCl. Int J Biol Macromol 65:329–339

Nayak AK, Pal D, Santra K (2015) Screening of polysaccharides from tamarind, fenugreek and jackfruit seeds as pharmaceutical excipients. Int J Biol Macromol 79:756–760

Nayak AK, Pal D, Santra K (2016) Swelling and drug release behavior of metformin HCl-loaded tamarind seed polysaccharide-alginate beads. Int J Biol Macromol 82:1023–1027

Nazim S, Dehghan MHG, Shaikh S, Shaikh A (2011) Studies on hydrotrope potato starch gel as topical carrier for rofecoxib. Der Pharm Sin 2:227–235

Nerkar PP, Gattani S (2011) *In vivo, in vitro* evaluation of linseed mucilage based buccal mucoadhesive microspheres of venlafaxine. Drug Deliv 18:111–121

Newton AMJ, Indana VL, Kumar J (2015) Chronotherapeutic drug delivery of tamarind gum, chitosan and okra gum controlled release colon targeted directly compressed propranolol HCl matrix tablets and *in vitro* evaluation. Int J Biol Macromol 79:290–299

Odeku OA, Fell JT (2005) *In vitro* evaluation of khaya and Albizia gums as compression coatings for drug targeting to the colon. J Pharm Pharmacol 57:163–168

Ofori-Kwakye K, Mfoafo KA, Kipo SL, Kuntworbe N, Boakye-Gyasi ME (2016) Development and evaluation of natural gum-based extended release matrix tablets of two model drugs of different water solubilities by direct compression. Saudi Pharm J 24:82–91

Ogaji I (2011) Some physicochemical properties of acetaminophen pediatric suspensions formulated with okra gums obtained from different extraction process as suspending agent. Asian J Pharm 5:15–20

Ogaji IJ, Hoang SW (2011) Effect of grewia gum as a suspending agent on ibuprofen pediatric formulation. AAPS PharmSciTech 12:507–513

Okunlola A, Odeku OA (2011) Evaluation of starches obtained from four *Dioscorea* species as binding agent in chloroquine phosphate tablet formulations. Saudi Pharm J 19:95–105

Pachuau L, Mazumdar B (2012) *Albizia procera* gum as an excipient for oral controlled release matrix tablets. Carbohydr Polym 90:284–295

Pal D, Chandra P, Sachan N, Hasnain MS, Nayak AK (2019b) Pharmaceutical applications of fenugreek seed gum. In: Nayak AK, Hasnain MS, Pal D (eds) *Natural polymers for pharmaceutical applications, Vol. I: plant derived polymers*, Apple Academic Press, USA, pp.203–226.

Pal D, Nayak AK (2012) Novel tamarind seed polysaccharide-alginate mucoadhesive microspheres for oral gliclazide delivery. Drug Deliv 19:123–131

Pal D, Nayak AK (2017) Plant polysaccharides-blended ionotropically-gelled alginate multiple-unit systems for sustained drug release, In: Thakur VK, Thakur MK, Kessler MR (eds) Handbook of composites from renewable materials, volume 6, polymeric composites. WILEY-Scrivener, USA, pp 399–400

Pal D, Nayak AK, Kalia K (2010) Studies on *Basella alba* L. leaves mucilage: Evaluation of suspending properties. Int J Drug Discov Technol 1:15–20

Pal D, Nayak AK, Saha S (2019) Cellulose-based hydrogels: present and future. In: Akhtar MS Swamy MK Sinniah UR (eds) Natural bio-active compounds. Springer, Singapore, pp 285–332

Panda D, Si S, Swain S, Kanungo SK, Gupta R (2006) Preparation and evaluation of gels from gum of *Moringa oleifera*. Indian J Pharm Sci 68:777–780

Peerapattana J, Laovachirasuwan P, Sodata P, Srijesdaruk V, Otsuka M (2020) Evaluation of using spray-dried glutinous rice starch as a direct compression hydrophilic matrix tablet. Biomed Mater Eng. https://doi.org/10.3233/BME-201080

Piriyaprasarth S, Patomchaiviwat V, Sriamornsak P, Seangpongchawal N, Katewongsa P, Akeuru P, Srijarreon P, Suttiphratya P (2010) Evaluation of yam (Dioscorea sp.) starch and arrowroot (*Maranta arundinacea*) starch as suspending agent in suspension. Adv Mater Res 93–94:362–365

Pitombeira NAO, Veras Neto JG, Silva DA, Feitosa JPA, Paula HCB, dePaula RCM (2015) Self-assembled nanoparticles of acetylated cashew gum: characterization and evaluation as potential drug carrier. Carbohydr Polym 117:610–615

Porto BC, Cristianini M (2014) Evaluation of cashew tree gum (*Anacardium occidentale* L.) emulsifying properties. LWT Food Sci Technol 59:1325–1331

Prajapati VD, Jani GK, Moradiya NG, Randeria NP (2013) Pharmaceutical applications of various natural gums, mucilages and their modified forms. Carbohydr Polym 92:1685–1699

Rana V, Rai P, Tiwary AK, Singh RS, Kennedy JF, Knill CJ (2011) Modified gums: Approaches and applications in drug delivery. Carbohydr Polym 83:1031–1047

Rao GC, Prasad PG, Srinivas K, Rao MF (2005) Application of *Moringa oleifera* as a suspending agent in the formulation of sulphamethoxazole. Indian Pharmacist 5:75–77

Rao KM, Gnanaprakash K, Badrinath AV, Chetty CM, Alagusundaram M (2010) Preparation and evaluation of flurbiprofen gel: mucilage of *Cocculas hirsutus* leaf powder as gel base. Int J PharmTech Res 2:1578–1583

Sachan NK, Bhattyacharya A (2009) Feasibility of Assam bora rice based matrix microdevices for controlled release of water insoluble drug. Int J Pharm Pharmaceut Sci 1:96–102

Sadrjavadi K, Shahbazi B, Fattahi A (2018) De-esterified tragacanth-chitosan nano-hydrogel for methotrexate delivery; optimization of the formulation by Taguchi design. Artif Cells Nanomed Biotechnol 46(sup2):883–893

Sahu BP, Sharma HK, Das MK (2011) Development and evaluation of a mucoadhesive nasal gel of felodipine prepared with mucoadhesive substance of *Dillenia indica* L. Asian J Pharm Sci 5:175–187

Samanta A, De A, Hasnain MS, Bera H, Nayak AK (2019) Gum odina as pharmaceutical excipient. In: Hasnain MS, Nayak AK, (eds) Natural polysaccharides in drug delivery and biomedical applications. Academic Press, Elsevier Inc., pp 327–337

Samanta A, Ojha D, Mukherjee B (2010) Stability analysis of primary emulsion using a new emulsifying agent gum odina. Nat Sci 2:494–505

Sinha Mahapatra S, Mohanta S, Nayak AK (2011) Preliminary investigation on angiogenic potential of *Ziziphus oenoplia* M. root ethanolic extract by chorioallantoic membrane model. Sci Asia 37:72–74

Sinha P, Ubaidulla U, Hasnain MS, Nayak AK, Rama B (2015) Alginate-okra gum blend beads of diclofenac sodium from aqueous template using $ZnSO_4$ as a cross-linker. Int J Biol Macromol 79:555–563

Sinha P, Ubaidulla U, Nayak AK (2015) Okra (*Hibiscus esculentus*) gum-alginate blend mucoadhesive beads for controlled glibenclamide release. Int J Biol Macromol 72:1069–1075

Sinha VR, Al-Azaki AA, Kumar RV (2011) Novel *Lannea Woodier* gum matrices for controlled release of drugs. Carbohydr Polym 83:1492–1498

Soumya RS, Ghosh S, Abraham ET (2010) Preparation and characterization of guar gum nanoparticles. Int J Biol Macromol 46:267–269

Srivastava P, Malaviya R, Kulkarni GT (2010) Formulation and evaluation of paracetamol tablets to access binding property of orange peel pectin. Int J Pharm Sci Rev Res 3:982–996

Sukhavasi S, Kishore VS (2012) Formulation and evaluation of fast dissolving tablets of amlodipine besylate by using fenugreek seed mucilage and *Ocimum basilicum* gum. Int Curr Pharm J 1:243–249

Tan C, Xie J, Zhang X, Cai J, Xia S (2016) Polysaccharide-based nanoparticles by chitosan and gum Arabic polyelectrolyte complexation as carriers for curcumin. Food Hydrocolloids 57:236–245

Verma PRP, Razdan B (2003) Studies on *Leucaena leucocephala* seed gum: Emulsifying properties. J Sci Ind Res 62:198–206

Vivekanandan K, Gunasekaran V, Jayabalan G (2015) Design and evaluation of colon specific drug delivery of budenoside. Int J Pharm Pharm Sci 7:261–263

Xie JH, Jin ML, Morris GA, Zha XQ, Chen HQ, Yi Y, Li JE, Wang ZJ, Gao J, Nie SP, Shang P, Xie MY (2016) Advances on bioactive polysaccharides from medicinal plants. Crit Rev Food Sci Nutr 56(Suppl 1):S60-84

Zhao Y, Khalid N, Shu G, Neves MA, Kobayashi I, Nakajima M (2017) Formulation and characterization of oil-in-water emulsions stabilized by gelatinized kudzu starch. Int J Food Prop 20:1329–1341

Zhou W, Liu W, Zou L, Liu W, Liu C, Liang R, Chen J (2014) Storage stability and skin permeation of vitamin C liposomes improved by pectin coating. Colloids Surf B Biointerf 117:330–337

# Chapter 4
# The Role of Phytochemicals in Cancer Prevention and Cure

**Braganza Cilwyn, Soundararajan Vijayarathna, Shanmugapriya, Rameshwar Naidu Jegathambigai, Subramaniam Sreeramanan, Yeng Chen, and Sreenivasan Sasidharan**

**Abstract** The comprehensive search for therapeutic agents, predominantly the green phytochemicals derived from plants, has recently gained high momentum. Phytochemicals are plant's (phyto) chemical referring to various types of compounds that occur naturally in plants. Green phytochemicals from Mother Nature especially from medicinal plants are a rich source of novel therapeutics and prevention agents. Interestingly, their unique mechanism of action as compared with the conventional drug mechanism against cancer cells made them a vital target for novel drug discovery for cancer prevention and cure. Most plants do yield a vast array of phytochemicals that play essential roles in cancer prevention and cure via various biological activities and novel mechanisms of actions. Since cancer has a significant impact on human health and contributes to mortality, it is appropriate to examine the role of phytochemicals in cancer prevention and cure. This chapter provides a comprehensive overview of the role of phytochemicals in cancer prevention and cancer cures via antioxidant activity, pro-oxidant activity, apoptosis induction, necrosis induction, autophagy induction and regulation of miRNA in cancer cells.

**Keywords** Phytochemical · Cancer · Apoptosis · Necrosis · Autophagy · miRNA

---

B. Cilwyn · S. Vijayarathna · Shanmugapriya · S. Sasidharan (✉)
Institute for Research in Molecular Medicine (INFORMM), Universiti Sains Malaysia, 11800 USM, Pulau Pinang, Malaysia
e-mail: srisasidharan@yahoo.com

R. N. Jegathambigai
Department of Biochemistry, Faculty of Medicine, Asia Metropolitan University, No 6 Jalan Lembah, 81750 Bandar Seri Alam, Johor Bahru, Malaysia

S. Sreeramanan
School of Biological Sciences, Universiti Sains Malaysia, 11800 Minden, Penang, Malaysia

Y. Chen
Dental Research & Training Unit, and Oral Cancer Research and Coordinating Centre (OC-RCC), Faculty of Dentistry, University of Malaya, 50603 Kuala Lumpur, Malaysia

 
D. Pal and A. K. Nayak (eds.), *Bioactive Natural Products for Pharmaceutical Applications*, Advanced Structured Materials 140,
https://doi.org/10.1007/978-3-030-54027-2_4

## 4.1 Introduction

Cancer is one of the greatest consequential cause for the death of world human population every year (Zaorsky et al. 2017). The extensive search for therapeutic agents, predominantly the green phytochemicals derived from medicinal plants, has recently gained high momentum especially for cancer prevention and cure (Zhang et al. 2020). Various plants have shown beyond doubt their worth as a source of phytochemicals with therapeutic potential and still represent as an important source for the discovering of new therapeutic agents which leads as the blue print for chemically synthetic compound (Newman and Cragg 2012). Phytochemicals are plant's (phyto) bioactive non-nutrient chemicals referring to various types of compounds that occur naturally in plants (Liu 2004; Mousavi et al. 2018). The utilization of plants has a long-established history in traditional treatment of various diseases. The breakthrough in chemical field with the advances in chemical analysis lead to the isolation and characterization of various purified bioactive compounds of plants, which initiated the exploration of plant sources as chemotherapeutic candidates for cancer (Cragg and Newman 2005). The continuous rise in cancer cases and the failure of synthetic conventional chemotherapies due to drug resistance and excessive toxicity towards healthy normal non-targeted tissues have also pushed to the utilization of naturally occurring phytochemicals of plants, which is evidently shown to improve treatment efficiency with lesser side effects. According to increasing scientific evidences confirming the remarkable anticancer activity induced by phytochemicals derived from plants has prompted to explore the role of phytochemicals in cancer cure. Green phytochemical from Mother Nature especially from medicinal plants are a rich source of novel therapeutics and prevention agents. In reality, some of the commonly used and commercially available anticancer drugs were isolated from medicinal plants, namely vinblastine, and vincristine (Newman et al. 2011). Interestingly, their unique mechanism of action as compared with the conventional drug mechanism against cancer cells made them a vital target for novel drug discovery for cancer prevention and cure. The important characteristic of plant-based phytochemical is that they can kill the cancer cells with least toxicity. Moreover, phytochemical also exhibits a higher selectivity towards cancer cells in comparison to normal non-cancerous cells. Most plants do yield a vast array of phytochemicals that play essential roles in cancer prevention and cure via various biological activities and novel mechanisms of actions. Since cancer has a significant impact on human health and contributes to mortality, it is appropriate to examine the role of phytochemicals in cancer prevention and cure (Vineis and Fecht 2018). The objective of this chapter is to furnish the role of phytochemicals in cancer prevention and cure that addressed various mechanisms of actions of various phytochemicals against the cancer cells as reported in the literature. Hence, this chapter provides a comprehensive overview of the role of phytochemicals in cancer prevention and cancer cures via antioxidant activity, pro-oxidant activity, apoptosis induction, necrosis induction, autophagy induction and regulation of miRNA in cancer cells as shown in Fig. 4.1.

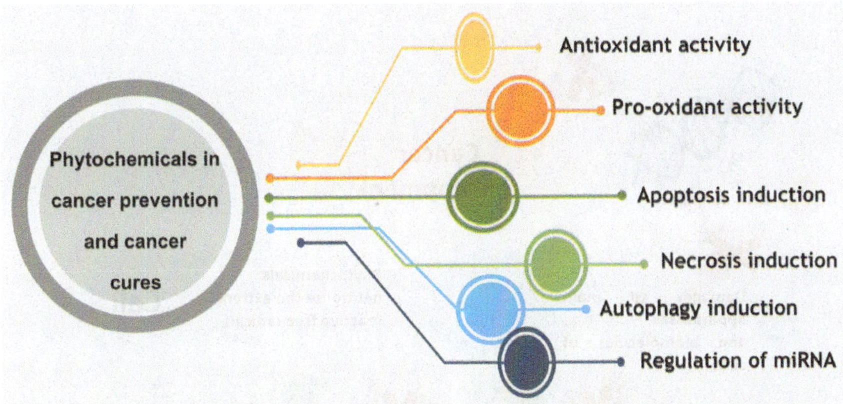


**Fig. 4.1** The role of phytochemicals in cancer prevention and cancer cures via various modes of actions in cancer cells

## 4.2 Role of Phytochemicals in Cancer Prevention via Antioxidant Activity

A free radical is molecules (or atom) comprising one or more unpaired electrons in outer orbit that make them unstable and extremely reactive, which leads them to steals electrons from other molecules to achieve stability. Subsequently, the attacked molecule converts to a free radical itself to start the chain reaction cascade, which eventually injuries the healthy cell (Mukherji and Singh 1984). Free radicals consist of reactive oxygen species, and reactive nitrogen species are made in the human body naturally via numerous endogenous systems according to diverse physiochemical or pathological circumstances (Valko et al. 2007). Free radicals are created in the human body either from usual important metabolic routes or from outside sources such as exposure to irradiation, ozone, pollutants, and various toxic chemicals (Bagchi and Puri 1998). However, the generated free radicals can be dangerous at the excessive level and lead to the damages to the main apparatuses and biomolecules of cells such as DNA, proteins and cell membranes. The various injuries caused by excessive level free radicals to human cells, particularly the damage to DNA, may lead to the development of cancer and other diseases (Dreher and Junod 1996). Antioxidants also identified as "free radical scavengers" are chemicals such as phytochemicals that neutralize the extremely reactive free radicals by donating an electron to the free radical or via quenching chain-initiating catalyst to avoid them from instigating harm to the healthy cell, which can lead to the development of cancer. Hence, the plants' phytochemicals react as a natural antioxidant to block the harmful action of free radicals preventing the development of cancer in humans (Rohman et al. 2006). Plants' phytochemicals are the most commonly known antioxidants, which include ascorbate, tocopherols, polyphenols and terpenoids (Dimitrios 2006).

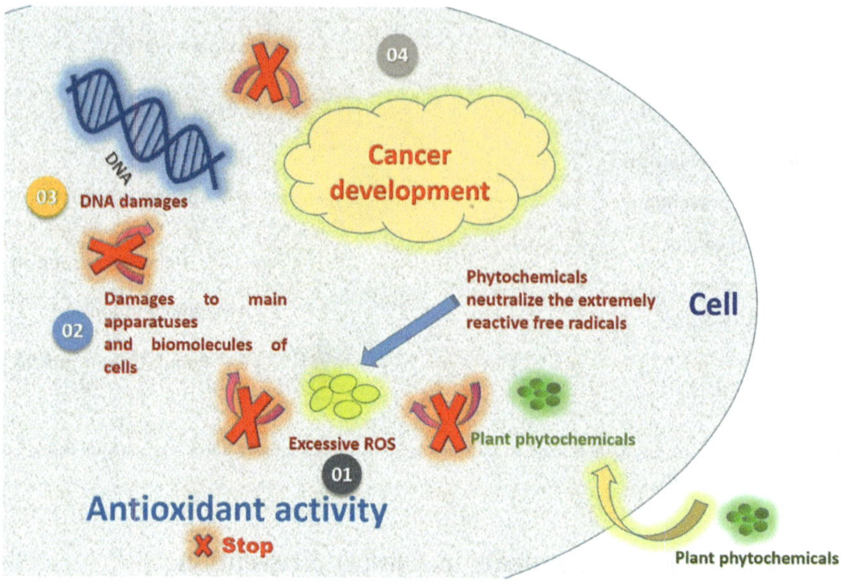


**Fig. 4.2** The antioxidant activity of phytochemicals in cancer prevention

The anticancer activity of phytochemicals via antioxidant activity may depend on the diverse inter-dependent pathways, as shown in Fig. 4.2. Firstly, the phytochemical will be entering the cells to start the pro-oxidant anticancer activity. Firstly, the accumulation of phytochemicals in the cell will reduce the excessive amount of free radicals. Subsequently, the reduced excessive amount of free radicals will not trigger the damages to the main apparatuses and biomolecules of cells, especially to DNA. Eventually, the phytochemicals neutralize the extremely reactive free radicals in the cell which will stop the development of cancer (Fig. 4.2). Conclusively, the various inter-dependent processes exhibit the beneficial effects of the antioxidant activity of phytochemicals that efficiently prevent cancer development in humans.

## 4.3 Role of Phytochemicals in Cancer Prevention Via Pro-Oxidant Activity

Pro-oxidant represents any phytochemicals that persuade oxidative stress either by the generation of free radicals or by hindering the antioxidant systems of a cell (Rahal et al. 2014). The redox level of a cell is determining how well a healthy cell is functioning. Besides, the protection of antioxidant systems in the cells is essential for the survival of the cells. Free radicals play a vital part in cancer. Excessive exposure to free radicals might be linked with a high risk of cancer for a normal cell. Although high levels of free radicals can be a high risk of cancer for a normal cell, a high level of free radicals can also trigger apoptosis and cell death in various types of cancer cells. Decisively, the actions of free radicals in cancer development, inhibition and treatment is tremendously complex and extremely challenging to research. Alteration of redox homeostasis in cancerous and healthy cells recommends that pro-oxidant based upregulation of cellular free radicals would target specifically against cancer cells without damaging the normal healthy cells (Wondrak 2009). Even though the antioxidant activity of various phytochemicals is well researched and commonly applied to stop or cure cancer, multiple phytochemicals also exhibit the pro-oxidant and free radicals generating activities under unique conditions especially in cancer cells. Accordingly, a pro-oxidant activity reached phytochemical might attack the cancerous cells, which is already at an extraordinary level of free radicals with high oxidative stress without affecting the well-tolerated non-cancerous cells (González-Bártulos et al. 2015). Several phytochemicals that target the cellular redox balance produce an exercise amount of reactive oxygen species (ROS) (Kruk et al. 2019) and eventually leads to cell death. Moreover, the transition metal-based phytochemicals could be favourable phytochemicals for pro-oxidant therapies (Rahal et al. 2014). Once, the metal-based phytochemicals accumulate metals, namely iron and copper, they induced the cycling redox reactions in the cancer cells, which will lead to the productions of the excessive amount of free radicals, mainly the extremely damaging hydroxyl radical species via the Fenton reaction. The phytochemicals belong to the flavonoid group, such as quercetin and kaempferol that have been reported to exhibit the pro-oxidant activity when a transition metal is available (Halliwell 2008).

The anticancer activity of phytochemicals via pro-oxidant activity may depend on diverse inter-dependent routes, as shown in Fig. 4.3. Firstly, the phytochemical will be entering the cells to start the pro-oxidant anticancer activity. Secondly, the accumulation of phytochemicals will trigger the cell to produce an excessive amount of free radicals. Subsequently, the excessive amount of free radicals will trigger DNA fragmentation and DNA damages via oxidative mechanisms. An excessive amount of free radicals in the cells will react with the cellular DNA, thus altering its structure, and disturbing the normal function of the DNA is one of the main reasons for DNA damage induced by pro-oxidant activity of the phytochemical (Beckman and Ames 1997). Even though the DNA molecule is an intact molecule, free radicals can act against the DNA and can cause various types of harm, namely alteration of DNA molecule bases, single- and double-strand DNA molecule disruptions, loss of purines in the DNA, destruction to the deoxyribose sugar, cross-link between DNA and protein and destruction of the naturally occurred DNA repair systems (Srinivas et al. 2019). The DNA fragmentation will lead to the induction of cell cycle arrest. Eventually, the cell cycle arrest will lead the cell to apoptotic cell death (Fig. 4.3). Conclusively, the various inter-dependent processes exhibit the beneficial effects of the pro-oxidant activity of phytochemicals that efficiently kills the cancer cells.

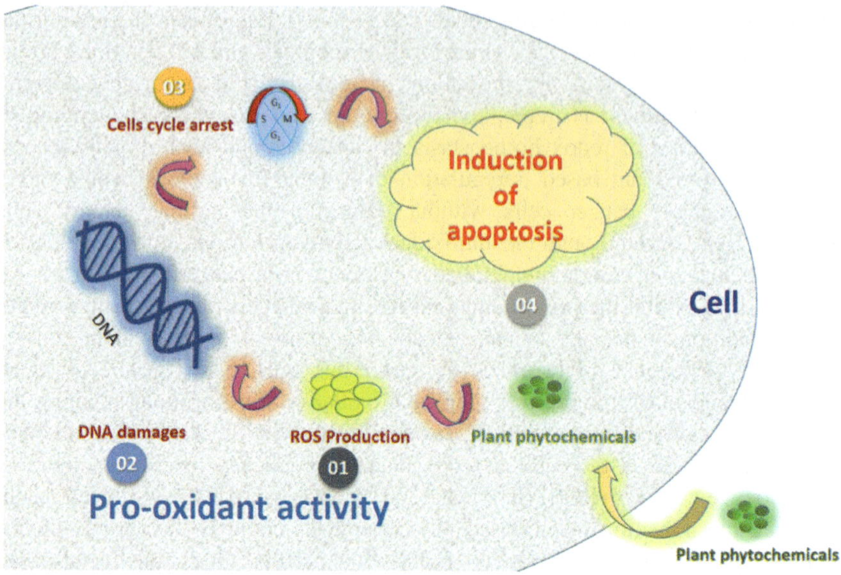


**Fig. 4.3** The pro-oxidant activity of phytochemicals in cancer prevention

## 4.4 Role of Phytochemicals in Cancer Cure Via Apoptosis Induction

Major plant-based phytochemicals such as terpenoids, phenolic acids and alkaloids have shown potentially promising anticancer property by fine-tuning the reactive oxygen species (ROS) signalling pathways (Chirumbolo et al. 2018). In line with the anti-oxidative and ROS-scavenging properties of phytochemicals, numerous phytochemicals have been studied and scientifically demonstrated to induce apoptosis through the ROS generation. An established plant-based anticancer agent can exemplify this, resveratrol, which was shown to induce caspase-8/Caspase-3-dependent apoptosis in human colon cancer cells, HCT29 and COLO201 by significantly increasing the intracellular ROS levels (Miki et al. 2012). Another commonly known phytochemical exhibiting anticancer activity, quercetin, was also inhibiting cancer growth by inducing apoptosis via cyclooxygenase-2 (COX-2)-dependent ROS generation.

Furthermore, numerous phytochemicals showed potentially promising anti-cancer activity by inducing apoptosis via upregulating the expression of caspases. This can be exemplified by a phytochemical called hyperforin which was reported to promote caspase-dependent apoptosis in various leukemia cell lines by upregulating caspase-9, caspase-8 and caspase-3 (Hostanska et al. 2003). One such phytochemical is corosolic acid reported to promoting caspase activation, leading to mitochondria-mediated signalling pathways to induce cell death in HeLa cells (Xu et al. 2009). The purified bioactive compound, Pyranocycloartobiloxanthone A, was also reported to play an imperative role in inducing apoptosis in breast cancer cells by upregulating Bcl-2 expression and downregulating Bax expression, which eventually lead to the release of cytochrome c, initiating the caspase cascade (Mohan et al. 2012).

Interestingly, another phytochemical from the class of phenolics, known as Scutellarin, has been proven to promote apoptosis by activating the p53 pathway (Yang et al. 2017). Scutellarin was found to suppress the anti-apoptotic protein Bcl-2, which eventually activates the pro-apoptotic protein, p53, leading to the upregulation of Bax protein to induce caspase-3 dependent apoptosis in human colon cancer (Yang et al. 2017). Another such phenolic compound, gallic acid, also reported inducing apoptosis in cancer cells via the upregulation of the p53. This, in turn, depolarizes the mitochondrial membrane potential, facilitates the release of caspase-activator, cytochrome c and induces an intrinsic apoptotic pathway (Yang et al. 2018). Role of another important phytochemical, capsaicin to induce p53-mediated apoptosis in various cancer cells have been well elucidated in previous studies (Jin et al. 2014; Clark and Lee 2016; Garufi et al. 2016; Lee and Clark 2016).

Various other phytochemicals have been reported to play an essential role in cancer cure by targeting nuclear factor kappa B (NF-kB), to promote cancer cell death (Kumar et al. 2016). The fact that NF-kB is highly expressed in cancer cells is inevitable due to its function in regulating anti-apoptotic and apoptotic genes (Tse et al. 2007; Manu and Kuttan 2008; Oh et al. 2012; Kumar et al. 2016). Intriguingly, various phytochemicals, including alkaloids and flavonoids, are known to induce apoptosis in cancer cells by specifically targeting NF-kB signalling pathway. For instance, known phytochemicals including xanthohumol (Colgate et al. 2007), Magnolol (Tse et al. 2007), Morusin (Lee et al. 2008), urosolic acid (Manu et al. 2008), Corilagin (Gambari et al. 2012) were ostensibly demonstrated to significantly down-regulate the expression of NF-kB in various cancer cells. The suppression of this apoptosis-inhibitor, NF-kB, eventually leads to tumour necrotic factor-$\alpha$ (TNF-$\alpha$)-induced apoptosis. Recent review collectively elucidated TNF-$\alpha$ induced activation of NF-kB in mitochondria to stimulate programmed cell death by the release of cytochrome c to the cytoplasm, followed by the activation of a caspase cascade (Albensi 2019).

Besides, there are also several plant-based secondary metabolites reported to induce the extrinsic pathway of apoptosis in cancer cells. A form of flavonol can depict this, kaempferol, which has been previously reported to up-regulate the expression of FasL, leading to the activation of caspase-8 in colon cancer cells (Lee et al. 2014). Bid protein, which is cleaved by the activated caspase-8 in the means of extrinsic apoptosis pathway, is then translocated into mitochondria, promoting intrinsic apoptosis (Lim et al. 2014). Another phytochemical known to encourage the extrinsic pathway is a phenolic compound called hispidin. Hispidin was scientifically proven to increase the level of death receptor 3 in colon cancer cells, leading to activation of the caspases involved in the extrinsic apoptosis pathway, namely, caspase-1 and caspase-8, along with the cleavage of PARP to induce cell death (Hengartner 2000).

Cumulatively, it can be suggested that these plant-derived biologically active compounds induce apoptosis mainly through the mitochondria-dependent mechanism. In short, phytochemicals promote ROS generation in cancer cells, causing polarization of mitochondria membrane potential leading to the release of various toxin proteins, including cytochrome c. Eventually, cytochrome c actively binds to apoptotic protease activating factor-1 (Apaf-1), which activates caspase-9 leading to the formation of cytochrome c/Apaf-1/caspase 9 complexes termed as the apoptosome which activates the executioner caspase 3 resulting in apoptosis (Hengartner 2000). On the other hand, phytochemicals were also found to be inducing the extrinsic apoptotic pathway by upregulation of death ligands such as FasL, TNF-α and TRAIL. These ligands are responsible for the activation of caspase-8 by binding to death receptors such as FAS, TNFR and other death receptors. Active caspase-8 cleaves Bid protein into tBid, which is then translocated to mitochondria to promote BAX and BAK proteins, allowing the intrinsic pathway to take place by the activation of the caspase cascade. Figure 4.4 shows the intrinsic and extrinsic apoptotic pathways induced by phytochemicals.

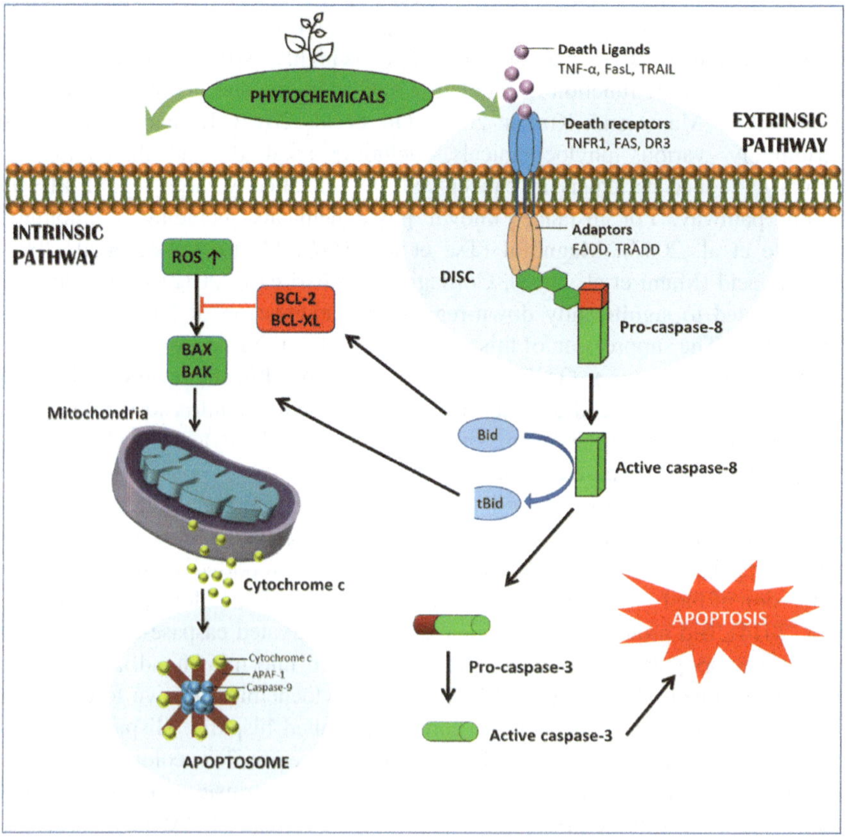


**Fig. 4.4** Intrinsic and extrinsic apoptosis mechanisms induced by phytochemicals

## 4.5 Role of Phytochemicals in Cancer Cure Via Necrosis Induction

Phytochemicals have been shown to protect cells by interfering with their molecular pathways that regulate the cell cycle, survival, angiogenesis and cell death. These properties made phytochemicals an essential source of drug for the prevention and treatment of cancer. Many anticancer drugs such as paclitaxel and vinblastine are derived from phytochemicals, and still, many more are under investigation. Notably, most of these compounds target the apoptotic mechanism mainly by interfering with caspase-dependent pathways (Ashraf 2020). However, other non-apoptotic cell death pathways such as necrosis also play an essential role in checking neoplastic cells and destroying tumour cells. Inducing necrosis is a known mechanism of some anticancer drugs such as DNA-alkylating agents in treating human cancers (Cho and Park 2014). Therefore, understanding the role of phytochemicals in signalling cascades involved in the induction of necrotic cell death will allow us to develop a novel drug to treat cancer.

Depending on the physiological and pathological conditions, a cell would either take the apoptotic or necrotic pathway. Unlike apoptosis, necrosis does not have a dedicated molecular pathway, instead it overlaps with many of those caspase-independent apoptotic pathways, which culminates in disruption of organelle and loss of membrane integrity resulting in the spillover of cellular contents (Lee et al. 2018). In a tumour microenvironment, induction of necrotic pathway would cause more damage as it destroys the cells around, and the contents released from these cells create a pro-inflammatory environment (Lee et al. 2018). Phytochemicals with enhanced necrosis may help to exert more effective tumour suppression property.

Well-controlled, a programmed form of necrosis is known as necroptosis, which is mainly triggered by extracellular stimuli similar to the extrinsic apoptotic pathway. Necroptosis is primarily orchestrated by serine/threonine kinase receptor-interacting protein 1/3 (RIP1 and RIP3) to induce necrotic cell death. RIP1 and RIP3 can be activated by signalling via tumour necrotic factor receptor 1/2 (TNF-R1/2), Toll-like receptor 3/4 (TLR3/4), DNA damage-induced Poly [ADP-ribose] polymerase 1 (PARP1) pathways, especially when caspase-8 is either downregulated, non-responsive or inactivated by other regulators (De Giffoni De Carvalho et al. 2019). Finally, RIP1 and RIP3 form a dimer which is one of the ways of induction of necroptosis by the activation of mixed lineage kinase domain-like protein (MLKL) that destroys the integrity of plasma membrane or via activation of mitochondrial protein phosphatases PGAM5 and Drp1 (mitochondrial fission protein) which lead to mitochondrial dysregulation (Fig. 4.5) (Mishra et al. 2018). However, progression depends on the level of caspase-8 and Fas-associated protein with death domain (FADD) that regulates RIP1/RIP3 levels. Experiments have demonstrated downregulation of Caspase 8 and FADD promotes RIP3-dependent necrosis (Wattanathamsan et al. 2019). Further, reactive oxygen species (ROS), advanced glycation end products (AGE), calcium, cyclophilin D (CypD), NO/NOS, phospholipase A2 (PLA2), calpains, cathepsin B, ceramide, methylglyoxal and high mobility group box 1 (HMGB1), act as important mediators in the necrotic pathway. In addition, the necrotic pathway can be triggered by oncogenic metabolic stress and hypoxia by inducing transcription factors Snail and Dlx-2 (Lee et al. 2018). Polymorphisms and defects in necroptosis regulators such as RIP3 have been shown to have a positive correlation with tumour progression in non-Hodgkin's lymphoma (Mishra et al. 2018). There is also increasing evidence of impaired necroptosis in cancer cells (Lalaoui and Brumatti, 2017). Hence, targeting necroptosis is very promising to treat various cancers, and trials of repurposing anticancer drugs for inducing necroptosis has been explored (Fulda 2018). Moreover, phytochemicals have been shown to induce necrosis by targeting many of the above mediators. Therefore, it is imperative to evaluate the effects of phytochemicals on the above molecular mediators to understand their role in inducing necrosis/necroptosis.

It is beginning to unravel mounting evidence on the molecular regulation of phytochemicals by RIP1/RIP3 upregulation of mitochondrial disruption by ROS generation, ATP depletion and fragmentation. Several in vitro and in vivo studies have reported the involvement of phytochemicals in necrotic/necroptotic signalling in cancer (Fig. 4.5). For example, Solamargine has shown to induce necrosis in melanoma and non-melanoma skin cell lines by targeting the lysosomal mitochondrial death pathway in lung cancer, breast cancer, squamous cell carcinoma and leukemia cell lines (Al Sinani et al. 2016). Similarly, Phenethyl isothiocyanate and Shikonin induce necroptosis in lung cancer cells via ROS, and β-Lapachone induces necroptosis in human hepatocellular carcinoma SK-Hep1 cells through the RIP1-PARP-AIF-dependent pathway (Diederich and Cerella 2016).

Green tea polyphenol is shown to induce necroptosis in p53-deficient Hep3B cells through mitochondrial-associated signalling via activation of Bax/Bak translocation (Lin and Tongyi 2014). Polyphenols resveratrol and analogs from roots of Fallopia japonica has shown to induce necrosis in MCF-7 breast cancer and C6 glioma cell lines. Curcumin and analogs have shown to induce necrosis in prostate (DU-145) cancer cells by generating ROS, and in bladder cancer xenografts by downregulating NF-kB, cyclin D1 with increased p21 expression. Genistein from Genista tinctorial has shown to cause necrosis in cervical cancer (HeLa) cell lines (Gali-Muhtasib et al. 2015). Alkaloids such as berberine and analogs showed necrosis in melanoma (B16) Prostate (RM-1) cell lines. Colchicin and analogs inhibit cell division and cause necrosis in Lung, colorectal, ovarian, prostate and breast mouse model. Terpenoids parthenolide and analogs were shown to cause necrosis in leukemia (HL60, Jurkat), breast (MDA-MB-231) cancer cell lines by ROS generation, induce dissolution of mitochondria membrane potential and RIP1 activation. Organosulfur sulforaphane and analogs induce necrosis in the breast (MCF-7), Colorectal (Caco-2) cancer cells by regulating CDK1 (Gali-Muhtasib H et al. 2015). Flavonoids such as quercetin shown to promote necroptosis in MCF-7 cells. Artemisinin (ARS) derivatives such as artemether (ARM) has been shown to exert necrosis in PG 100 gastric cancer cell line and artesunate (ART) induce necroptosis in RT4 schwannoma cell line (Efferth 2017).

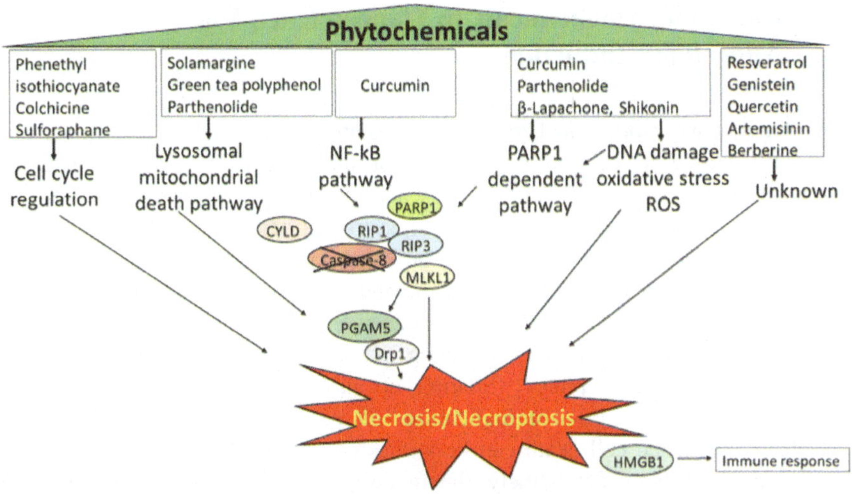


**Fig. 4.5** Mechanisms of phytochemical induced necrosis and necroptosis. Various phytochemicals follow different molecular pathways to induce necrosis and necroptosis. Molecular mechanism of NF-kB induced necrosis is well understood with RIP1, RIP3, PARP1 which are key players while CYLD is the regulator of RIP1. Phosphorylation of MLKL1 by RIP3 will induce necrosis either by direct disintegration of plasma membrane or via PGAM5/Drp1 causing mitochondrial damage leading to necrosis/necroptosis. HMGB1 proteins are released by necrotic cells that stimulate immune response

Even the extracts of plants have been demonstrated to induce necrosis in liver cancer cell line Huh7it (Ficus carica leaves and fruits) (Purnamasari et al. 2019). Further, Hymenocallis speciosa extract in acute myeloid leukaemia cell line showed necroptosis and Jacaranda decurrens extract induced necrosis in K562 erythroleukaemia cells (De Giffoni De Carvalho et al. 2019).

Overall, the role of phytochemicals in necrosis induction and molecular mechanisms are less explored and has the potential to contribute to cancer cure. Drugs targeting necrotic pathways are mainly useful to treat tumour cells that are resistant to apoptotic drugs either by upregulation of anti-apoptotic proteins (Bcl-2) or impaired pro-apoptotic proteins such as p53. This resistance, either intrinsic such as mutations in p53 or acquired to anticancer drugs, is a significant reason for poor treatment outcomes. Besides, necrosis-inducing phytochemicals could be used in combination therapy to simultaneously hit both apoptosis and necrosis pathways and thus provide an effective cure for cancer.

## 4.6 Role of Phytochemicals in Cancer Cure Via Autophagy Induction

Autophagy, a word originated from Greek meaning self-eating, is a self-digestion approach leading to cellular degradation of the subcellular materials such as organelles and proteins to generate energy and metabolic precursors for prolonging cell survival (Glick et al. 2010). The cytoplasmic proteins and cellular organelles will be enveloped in autophagosomes and degraded by fusion with lysosomes during this process. The cellular stress response that usually serves as a quality control mechanism leads to apoptosis-independent cell death. Various forms of autophagy have been described, including macroautophagy, microautophagy and chaperone-mediated autophagy (Sharma et al. 2014). It is a morphological description, and no decisive indication of precise mechanisms underlying autophagy-persuaded cell death can be witnessed (Tsujimoto and Shimizu 2005). Numerous phytochemicals, namely phenols, alkaloids, flavones and organic acids are stated to be autophagy regulators, and it was revealed that autophagy might exhibit either cytoprotective or cytotoxic role in natural products phytochemicals treated cancer cells (Wang and Feng 2014).

Characteristics of cancer cells, including enhanced cell proliferation, altered apoptotic pathways, and reprogrammed cellular metabolism, all together have shown to influence the autophagic route (Linder and Kögel 2019). Most chemotherapeutic drugs promote autophagy, which is generally considered a cytoprotective response in that its inhibition frequently promotes apoptotic cell death (Sharma et al. 2014). In some circumstances, both autophagy and apoptosis are required in parallel pathways to contribute to cell death (Mettlin 1997). Under stress conditions, it can induce programmed cell death, called "autophagy-dependent cell death" (ADCD). Several reports have indicated that a variety of naturally occurring compounds play roles in the prevention or therapy of cancer, and their bioactive compounds lead to autophagy. Recent studies have suggested the role of phytochemicals in modulating the autophagy pathway. The Resveratrol, found abundantly in grape skins and red wine, induced cell death and growth inhibition in ovarian cancer cell lines, through autophagocytosis (Kim et al. 2011). Resveratrol induced molecular features of apoptosis, including the mitochondrial release of cytochrome c and caspase activation, resveratrol-treated cells exhibited the morphologic and ultrastructural changes indicative of autophagocytic death (Opipari et al. 2004). Soy-derived isoflavone, genistein, was also shown to induce both apoptosis and autophagy (Kueck et al. 2007). Genistein induced autophagy in ovarian cancer cells, indicating recruitment and localization of LC3-II to autophagosomes (Gossner et al. 2007).

## 4.7 Role of Phytochemicals in Cancer Cure Via Regulation of miRNA

Cancer has become an extensive malicious disease that relies greatly on the treatment of chemotherapy. The chemotherapy drugs, however, not only contrives the intensity to cause deleterious side effects, but also triggers tumour regression after a temporary period of improvement. Strange enough, these relapsed cancer cells enhance their survival propensity by being chemoresistance. Hence, critical measurements are indispensable to pursue an agent without or with minimal side effects. Nature endows vegetables, fruits and herbs with chemical properties called phytochemicals possessing the aptitude in treating cancer, infections and healing proneness. These phytochemicals are found copious from a wide range of plant products and of these, 10,000 compounds were documented and described (Russo et al. 2010). Some of these chemicals displayed anticancer potential with almost zero or minimal cytotoxicity to normal cell physiology (Rao et al. 2007). Enigmatically, 47% of drugs ratified by FDA are of plant origin that can be treated as single chemotherapeutic agent usage or by merging with other standard anticancer drugs (Newman and Cragg 2007).

When phytochemicals are exerting anticancer properties, various questions emerged in connection to microRNA's role and regulation. The miRNAs some time back had been linked with oncogenic and tumour suppressor action, serving a perfect goal in cancer deterrence and therapy. The explicate mechanism that involves anomalous expression of miRNAs discovered in many cancer types has yet to be determined. Mounting data implicates anomalous transcription machinery, epigenetic and miRNA biosynthesis alteration, mutations or the presence of different DNA number to be prompting miRNA dysregulation in human cancer (Deng et al. 2008). Nonetheless, researchers exposed the potentiality of phytochemicals in the modulation miRNA expression and their effect in cancer pathobiology.

There were several studies revealing the modus operandi of phytochemicals in regulating miRNA expression with an exception to transcriptional regulation. Epigallocatechin-3-gallate (EGCG) is a rich constituent and the most efficient catechin in green tea where its role has been contributed to cancer treatment (Mukherjee et al. 2015). This compound was reported to incite the binding of hypoxia inducible factor-1$\alpha$ (HIF-1$\alpha$) to the promoter region of miR-210, eventually deceiving the cells to stop cell proliferation and anchorage-independent growth as noted in human non-small cell lung cancer cell lines, H1299, H460 and A549 (Wang et al. 2011). Coincidingly, Yamada et al. (2016) had remarkable discovery on EGCG upregulating miRNA-let-7b which then activated 67 kDa laminin receptor to inhibit cancer growth in B16 melanoma cells. Of late, ECGC was also demonstrated to down-regulate the expression of miR-25 and escalate PARP, pro-caspase-3 and pro-caspase-9 inducing cell apoptosis in MCF- 7 cells (Zan et al. 2019).

The resveratrol has long been insinuated in the treatment of cancer and other diseases. In human colon cancer study, resveratrol significantly reduced the level of miR-17, miR-21, miR-25, miR-92a-2, miR-103–1 and miR-103–2, where these miRNAs were deliberately known to behave as oncomiRs (Tili E, J-J Michaille 2011). The biological effect of resveratrol was studied in prostate cancer (Kumar et al. 2017) where it was found to down-regulate the expression of miR-221 while in lung cancer, a high surge was reported in the expression of miR-200c (Bai et al. 2014). Notably, in estrogen-responsive breast cancer cells, resveratrol was reported to dysregulated two miRNAs mainly miR-542-3p which observes a reduction while miR-122-5p that records an increment all in together to promote apoptosis cell death. Comparatively, the triple-negative breast cancer cells with the treatment of resveratrol manage to only display an increase in miR-122-5p (Venkatadri et al. 2016). Osteosarcoma (OS) is considered as an aggressive cancer and when the cells U20S and MG63 were introduced to resveratrol, impressive effects in response to apoptosis were perceived. Expression of miR-139-5p was reduced where antecedently it has the aptitude in binding to 3′UTR region of NOTCH1 and mediate the progression of osteosarcoma (Xiao et al. 2020).

Curcumin is an all-natural occurring phytochemical dominating the root and rhizome of Curcuma longa which has extensive established records on antioxidant, anti-inflammatory, and anticancer properties (Gupta et al. 2013; Park et al. 2013). Coker-Gurkan et al. demonstrated the anti-proliferation activities of curcumin in T47D cells through the downregulation of miR-183, miR-96 and miR-182 along with NF-κB signalling inhibition (Coker-Gurkan et al. 2019). Competently, treatment with curcumin has demonstrated suppressed proliferation and increased apoptosis of NSCLS cells by mediating the downregulating of miR-21 via PTEN axis (Bai et al. 2014; Zhang and Bai 2014). PTEN gene implicated as the target of miR-21 is a tumour suppressor and the suppression of the oncomiR has resulted in upregulation of PTEN, hence resuming back to mediating the anticancer properties as exerted by curcumin (Liu et al. 2013). In another case, an expression of elevated miR-130a directed to target Wnt/β-catenin pathway was halted by curcumin which may lead to an inhibition of cell proliferation in colon cancer SW480 cells (Dou et al. 2017). Curcumin also attested to induce apoptosis in bladder cancer cell lines (T24 and SV-HUC-1) via upregulation of miR-7641 levels, which attenuated cell proliferation and invasion and, thereafter, inducing apoptosis by repressing p16 (Wang et al. 2018). The activity of curcumin was further investigated utilizing SCID mice xenograft tumour model of glioblastoma multiforme. Intriguing results exhibited a surge in the level of miR-378 through target mediator p38. This, however, resulted in the inhibition of cellular growth thus substantiating potential effects of curcumin (Li et al. 2017).

Similar to other polyphenols, quercetin can be detected present in apples, onions and broccoli (Nam et al. 2016) where it is prominently studied for its anticancer mechanism. Notably, when quercetin is applied to pancreatic cancer cells, it disclosed a regulated plethora of 105 miRNAs at which 25 miRNAs were down-regulated (chiefly; miR-103a-3p, miR-125b, and miR-1202) whereas 80 miRNAs were up-regulated (namely; let-7c, miR-200a-3p, and miR-200b-3p). Additional studies were performed on one of the most expressed let-7c with connection to pancreatic cancer anti-proliferation mechanism (Nwaeburu et al. 2016).

Outstandingly, let-7c behaves in an unusual manner than the common miRNAs by binding to 3′UTR of Numbl and evidently increasing the expression of Numbl, an inhibitor of Notch signalling (Miele et al. 2006). As a fact, tumour progression and apoptosis were observed in these pancreatic ductal adenocarcinoma cells. The influence of quercetin was also studied using in vitro oral cancer cells (HSC-6 and SCC-9). Cells treated with quercetin expressed higher miR-16 expression which was then shown to mediate the inhibition of HOXA10 level (Zhao et al. 2019). An earlier in vitro study revealed HOXA10 to be positively related to the outcome of cancer cell proliferation on head and neck squamous cell carcinoma (Guo et al. 2018). Collectively, the activated expression of miR-16 acted as a tumour suppressor in oral cancer by interrupting HOXA10 (Zhao et al. 2019). Likewise, quercetin mediating expression of distinct microRNA: miR-16, miR-217 and miR-145 were discovered through many studies conducted in lung adenocarcinoma, osteosarcoma, and ovarian cancer cells, respectively (Sonoki et al. 2015; Zhang et al. 2015; Zhou et al. 2015).

Sulforaphane (SFN) originates as an isothiocyanate derivative of cruciferous plants such as cauliflower and broccoli known to be effective as anticancer agent based on reputable in vivo studies (Zhang and Tang 2007). Importantly, SFN treated non-small cell lung (H1299, 95C and 95D) cancer cells which disclosed a suppression of miR-616, an oncomiR modulating GSK3β/β-catenin signalling pathway commonly implicated in cancer invasiveness (Wang et al. 2017). In particular, the SFN targets the EGFR signalling which is associated with EMT (epithelial mesenchymal transition) of lung cancer. Studies, however, indicate EMT largely influencing the progression of metastasis, chemoresistance and other types of tumours (Xiao and He 2010). Another study uniformly related to EMT was recorded in SFN-treated bladder cancer cells, (T24 and UMUC-3). Consequently, expression of (miR-200c) was greatly enhanced where it inhibits mesenchymal phenotype (vimentin) and upregulated epithelial phenotype (E-cadherin), thus bringing the mechanism of EMT down to a halt (Huang et al. 2018). Combination treatment of SFN and iberin in colorectal cancer cell lines (NCM460 and NCM356) exhibited better potential via upregulation of miR-23b and downregulation of oncomiR, miR-27b. The performance of these miRNAs is contributed to the anti-invasive, anti-angiogenesis and anti-proliferation activities in cancer cells (Slaby et al. 2013). Of the reasons why pancreatic cancer cells are resistant to treatment is due to the loss of gap junction intercellular communication and connexin 43 expression. However, PANC-1 cells treated with SFN seem to restore these features by making cell sensitive again to anticancer drugs. This is achieved by reducing the expression of miR-30a-3p thus enhancing the expression of Cx43, a protein associated with gap junction (Georgikou et al. 2020). In another study of breast cancer (MCF-7, MDA-MB-231 and SK-BR-3) cell lines, the introduction of SFN dramatically reduced the expression of miR-23, miR-92b, miR-381 which eventually promoted cell cycle arrest and senescence in these cells (Lewinska et al. 2017).

Genistein is a natural soy isoflavone that has featured exceeding intensity as an anticancer agent (Spagnuolo et al. 2015; Varinska et al. 2015). In MCF-7 and

MDA-MB-435 breast cancer cells, downregulation of oncogenic miR-155 was observed after treatment with genistein which led to the expression FOXO3, PTEN, casein kinase, and p27. Interestingly, the inhibition of invasiveness and metastasis were achieved as miR-155 has been demonstrated to escalate in various tumours (De La Parra et al. 2016). When oncogenic miR-23b-3p was reduced in renal cancer cells treated with genistein, expression of target PTEN begin to enhance followed by reduced expression of cancer-related genes PI3K (phosphatidylinositol-3-kinase), Akt and IL-32 (interleukin-32) (Zaman et al. 2012). Genistein increased the levels of miR-145 in retinoblastoma cell (Y79) cell, thereby inhibiting ABCE1 target gene to demonstrate anti-proliferation and growth impediment. The gene ABCE1 is known as a member of ATP-binding cassette (ABC) transporters which carries the responsibility for actively transferring molecules across phospholipid bilayers (Wei et al. 2017). Genistein exhibited apoptotic in laryngeal cancer when the expression of miR-1469 was greatly up-regulated followed by a suppression in myeloid cell leukemia 1, an anti-apoptotic member of Bcl-2 family (Ma et al. 2018). In the case of pancreatic cancer cells, it was reported that following treatment with genistein, the expression of miR-223 was significantly decreased. Further investigation also indicated the up-surge of F-box/WD repeat-containing protein 7 gene, a tumour suppressor gene which was one of the targets of miR-223 (Ma et al. 2013). The expression of miR-29b in U266 multiple myeloma cells were revealed with the administration of genistein. The miR-29b observes a significant twofold increase that targets the NF-κB expression, hence promoting and restoring apoptosis in U266 multiple myeloma cells (Xie et al. 2016).

Indole-3-carbinol (I3C) is a constituent of cruciferous vegetables that has the prospecting value as a chemo-preventive agent (Beier 1990) according to many studies (Chinni and Sarkar 2002; Rahman et al. 2004). It mainly works by modifying estrogen metabolism upon consumption to aid its cancer-preventive measurements. Concurrent researches have provided substantiate evidence indicating I3C in cell growth arrest, cell proliferation prevention and apoptosis (Takada et al. 2005; Weng et al. 2007; Omar et al. 2009). In human breast cancer cells (MCF-7), I3C has demonstrated an upregulation of miR-34a (Hargraves et al. 2016). The higher expression of miR-34a targets Bcl-2 and other anti-apoptotic proteins hinders chromatin silencers of p53 genes and assist the repression of CDK4, CDK6, and cyclin D1 (He et al. 2007; Hermeking, 2010). Taken together, the expression of miR-34a modulates p53 tumour suppressor escalation and its downstream gene 21 to achieve apoptosis (He et al. 2007). In an in vivo lung cancer study, female mice were observed after exposing to vinyl carbamate (VC). A plethora of miRNAs were down-regulated which includes miR-21, miR-31, miR-130a, miR-146b, and miR-377 where reversed changes were observed in mice treated with carcinogen only. Further analysis of miR-21 indicated PTEN, PDCD4, and reversion-inducing-cystein-rich protein with Kazal motifs as potential targets for the oncogenic effect of miR-21 and the chemopreventive activity of I3C (Melkamu et al. 2010). Identically, in hepatocellular carcinoma (HCC) cell, I3C was also revealed to reduce and prevent the progression of tumorigenicity by arresting miR-21 (Wang et al. 2015).

Ellagic acid (EA) is a phenol with anticancer property found rich in countless fruits and vegetables (Zhang et al. 2014; González-Sarrías et al. 2016). Human colon cancer cells (Caco-2, HT-29, and SW480) that were subjected to EA displayed downregulated oncogenic miR-224 while tumour suppressor miR-215 was up-regulated. These miRNAs induced facilitated modulation of p53 via p21 accumulation (Munagala et al. 2013). In a different case, EA treatment was conducted on female ACI rats which had been previously exposed to estrogen to develop mammary tumorigenesis. The observation of this treatment includes upregulation of miR-182, miR-375, miR-183, miR-34c, miR-196c and miR-429 and downregulation of miR122, miR-127, miR-335, miR-205, and miR-206 expression in tumour cells. The overall inhibition of the tumour was reached with the modulation of key proteins; ERa, cyclin D1, RASD1, FoxO3a, FoxO1, cyclin G1, Bcl-w and Bcl-2 (Ravindranath and Chandrasekhara, 1980).

Though there are many phytochemicals which have entered clinical trials for the treatment of cancer, lower bioavailability and poor potency of dietary plant-compounds still pose as a great challenge to scientists (Ravindranath et al. 1980; GonzáLez-Barrio et al. 2010). Based on evidence here, phytochemicals are shown to induce miRNAs in various cancers both in vivo and in vitro (Fig. 4.6). Hence, the limitation of the bioavailability can be resolved by imitating a synthetic analog of the miRNA of interest and by encapsulating their formulation into nanoparticle component. Convincingly, phytochemicals deliver as promising agents in regulating cancers and with the discovery of corresponding miRNA activities together with their target gene mechanism that proposes a magnificent future approach to combat this disease.

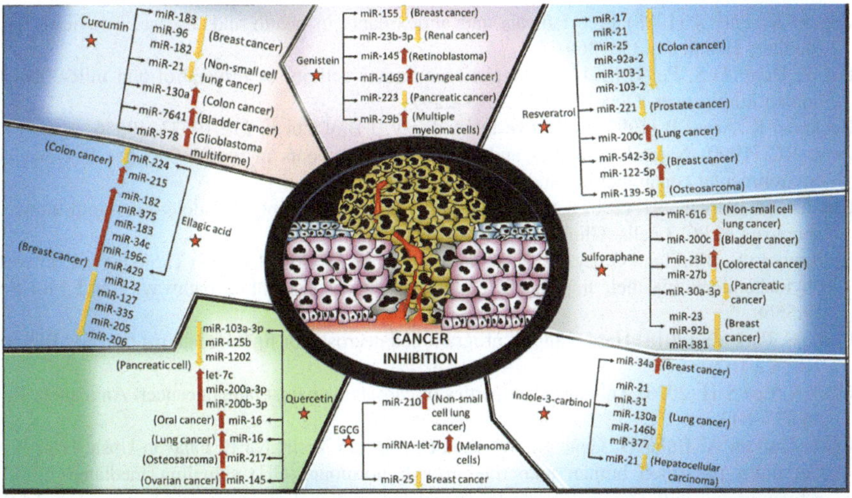


**Fig. 4.6** Molecular targets of different phytochemicals in curbing various tumour growths/ malignancies through microRNA regulation. The red arrow (↑) indicates upregulation, yellow arrow (↓) represents downregulation and (*) marks the phytochemical in subject

## 4.8 Conclusions

Phytochemical from plants is an essential source of novel therapeutics and prevention agents against cancer. This chapter discusses the various mechanism of actions of different types of phytochemicals against the cancer cells for understanding the critical role played by phytochemicals in cancer prevention and treatment. In conclusion, this chapter discussed in detail the role of phytochemicals in cancer prevention and cures via antioxidant activity, pro-oxidant activity, apoptosis induction, necrosis induction, autophagy induction and regulation of miRNA in cancer cells, as reported by the worldwide researcher.

**Acknowledgements** This project was funded by the Fundamental Research Grant Scheme (FRGS; Grant No.: 203/CIPPM/6711723) from the Ministry of Education Malaysia, Government of Malaysia, Malaysia. Braganza Cilwyn is supported by Fundamental Research Grant Scheme (FRGS; Grant No.: 203/CIPPM/6711723) from the Ministry of Education Malaysia, Government of Malaysia, Malaysia.

## References

Al Sinani SS, Eltayeb EA, Coomber BL, Adham SA (2016) Solamargine triggers cellular necrosis selectively in different types of human melanoma cancer cells through extrinsic lysosomal mitochondrial death pathway. Cancer Cell Int 16:11

Albensi BC (2019) What is nuclear factor kappa B (NF-kB) doing in and to the Mitochondrion? Front Cell Dev Biol 7:154

Ashraf MA (2020) Phytochemicals as potential anticancer drugs: time to ponder nature's bounty. Biomed Res Int 2020:1–7

Bagchi K, Puri S (1998) Free radicals and antioxidants in health and disease: a review. East Mediterr Health J 4:350–360

Bai T, Dong D-S, Pei L (2014) Synergistic antitumor activity of resveratrol and miR-200c in human lung cancer. Oncol Rep 31:2293–2297

Beckman K, Ames B (1997) Oxidative DNA decay. J Biol Chem. 272:19633–19636

Beier RC (1990) Natural pesticides and bioactive components in foods. Springer Reviews of environmental contamination and toxicology, 47–137

Chinni SR, Sarkar FH (2002) Akt inactivation is a key event in indole-3-carbinol-induced apoptosis in PC-3 cells. Clin Cancer Res 8:1228–1236

Chirumbolo S, Bjørklund G, Lysiuk R, Vella A, Lenchyk L, Upyr T (2018) Targeting cancer with phytochemicals via their fine tuning of the cell survival signaling pathways. Int J Mol Sci 19:3568

Cho YS, Park SY (2014) Harnessing of programmed necrosis for fighting against cancers. Biomol Ther 22:167

Clark R, Lee S-H (2016) Anticancer properties of capsaicin against human cancer. Anticancer Res 36:837–843

Coker-Gurkan A, Bulut D, Genc R, Arisan E-D, Obakan-Yerlikaya P, Palavan-Unsal N (2019) Curcumin prevented human autocrine growth hormone (GH) signaling mediated NF-kB activation and miR-183-96-182 cluster stimulated epithelial mesenchymal transition in T47D breast cancer cells. Mol Biol Rep 46:355–369

Colgate EC, Miranda CL, Stevens JF, Bray TM, Ho E (2007) Xanthohumol, a prenylflavonoid derived from hops induces apoptosis and inhibits NF-kappaB activation in prostate epithelial cells. Cancer Lett 246:201–209

Cragg GM, Newman DJ (2005) Plants as a source of anti-cancer agents. J Ethnopharmacol 100:72–79

de Giffoni de Carvalho JT, da Silva Baldivia D, Leite DF, de Araújo LCA, de Toledo Espindola PP, Antunes KA, Rocha PS, de Picoli Souza K, dos Santos EL (2019) Medicinal plants from Brazilian cerrado: antioxidant and anticancer potential and protection against chemotherapy oxicity. Oxid Med Cell Longev 1–16

De La Parra C, Castillo-Pichardo L, Cruz-Collazo A, Cubano L, Redis R, Calin GA, Dharmawardhane S (2016) Soy isoflavone genistein-mediated downregulation of miR-155 contributes to the anticancer effects of genistein. Nutr Cancer 68:154–164

Deng S, Calin GA, Croce CM, Coukos G, Zhang L (2008) Mechanisms of microRNA deregulation in human cancer. Cell Cycle 7:2643–2646

Diederich M, Cerella C (2016) Non-canonical programmed cell death mechanisms triggered by natural compounds. Semin Cancer Biol 40:4–34

Dimitrios B (2006) Sources of natural phenolic antioxidants. Trends Food Sci Technol 17:505–512

Dou H, Shen TJ, Huang L, Shi H, Chen H, Wang Y, Wang T (2017) Curcumin suppresses the colon cancer proliferation by inhibiting Wnt/β-catenin pathways via miR-130a. Front Pharmacol 8:877

Dreher D, Junod AF (1996) Role of oxygen free radicals in cancer development. Eur J Cancer 32:30–38

Efferth T (2017) From ancient herb to modern drug: artemisia annua and artemisinin for cancer therapy. Semin Cancer Biol 46:65–83

Fulda S (2018) Repurposing anticancer drugs for targeting necroptosis. Cell Cycle 17:829–832

Gali-Muhtasib H, Hmadi R, Kareh M, Tohme R, Darwiche N (2015) Cell death mechanisms of plant-derived anticancer drugs: beyond apoptosis. Apoptosis 20:1531–1562

Gambari R, Borgatti M, Lampronti I, Fabbri BE, Bianchi N, Piccagli L, Yuen MC-W, Kan C-W, Hau DK-P (2012) Corilagin is a potent inhibitor of NF-kappaB activity and downregulates TNF-alpha induced expression of IL-8 gene in cystic fibrosis IB3-1 cells. Int Immunopharmacol 13:308–315

Garufi A, Pistritto G, Cirone M, D'Orazi G (2016) Reactivation of mutant p53 by capsaicin, the major constituent of peppers. J Exp Clin Cancer Res 35:136

Georgikou C, Yin L, Gladkich J, Xiao X, Sticht C, de la Torre C, Gretz C, Gross W, Schäfer M, Karakhanova S (2020) Inhibition of miR30a-3p by sulforaphane enhances gap junction intercellular communication in pancreatic cancer. Cancer Lett 469:238–245

Glick D, Barth S, Macleod KF (2010) Autophagy: cellular and molecular mechanisms. J Pathol 221:3–12

González-Barrio R, Borges G, Mullen W, Crozier A (2010) Bioavailability of anthocyanins and ellagitannins following consumption of raspberries by healthy humans and subjects with an ileostomy. J Agric Food Chem 58:3933–3939

González-Bártulos M, Aceves-Luquero C, Qualai J, Cussó O, Martínez MA, de Mattos SF, Menéndez JA, Villalonga P, Costas M, Ribas X (2015) Pro-oxidant activity of amine-pyridine-based iron complexes efficiently kills cancer and cancer stem-like cells. PLoS ONE 10:e0137800

González-Sarrías A, Núñez-Sánchez MÁ, Tomé-Carneiro J, Tomás-Barberán FA, García-Conesa MT, Espín JC (2016) Comprehensive characterization of the effects of ellagic acid and urolithins on colorectal cancer and key-associated molecular hallmarks: MicroRNA cell specific induction of CDKN1A (p21) as a common mechanism involved. Mol Nutr Food Res 60:701–716

Gossner G, Choi M, Tan L, Fogoros S, Griffith KA, Kuenker M, Liu JR (2007) Genistein-induced apoptosis and autophagocytosis in ovarian cancer cells. Gynecol Oncol 105:23–30

Guo L-m, Ding G-f, Xu W, Ge H, Jiang Y, Chen X-j, Lu Y (2018) MiR-135a-5p represses proliferation of HNSCC by targeting HOXA10. Cancer Biol Ther 19:973–983

Gupta SC, Patchva S, Aggarwal BB (2013) Therapeutic roles of curcumin: lessons learned from clinical trials. AAPS J 15:195–218

Halliwell B (2008) Are polyphenols antioxidants or pro-oxidants? What do we learn from cell culture and in vivo studies? Arch Biochem Biophys 476:107–112

Hargraves KG, He L, Firestone GL (2016) Phytochemical regulation of the tumor suppressive microRNA, miR-34a, by p53-dependent and independent responses in human breast cancer cells. Mol Carcinog 55:486–498

He L, He X, Lim LP, De Stanchina E, Xuan Z, Liang Y, Xue W, Zender L, Magnus J, Ridzon D (2007) A microRNA component of the p53 tumour suppressor network. Nature 447:1130–1134

Hengartner MO (2000) The biochemistry of apoptosis. Nature 407:770–776

Hermeking H (2010) The miR-34 family in cancer and apoptosis. Cell Death Differ 17:193–199

Hostanska K, Reichling J, Bommer S, Weber M, Saller R (2003) Hyperforin a constituent of St John's wort (*Hypericum perforatum* L.) extract induces apoptosis by triggering activation of caspases and with hypericin synergistically exerts cytotoxicity towards human malignant cell lines. Eur J Pharm Biopharm 56:121–132

Huang L, Li B-L, He C-X, Zhao Y-J, Yang X-L, Pang B, Zhang X-H, Shan Y-J (2018) Sulforaphane inhibits human bladder cancer cell invasion by reversing epithelial-to-mesenchymal transition via directly targeting microRNA-200c/ZEB1 axis. J Funct Foods 41:118–126

Jin J, Lin G, Huang H, Xu D, Yu H, Ma X, Zhu L, Ma D, Jiang H (2014) Capsaicin mediates cell cycle arrest and apoptosis in human colon cancer cells via stabilizing and activating p53. Int J Biol Sci 10:285

Kim M-K, Kim K, Han JY, Lim JM, Song YS (2011) Modulation of inflammatory signaling pathways by phytochemicals in ovarian cancer. Genes Nutr 6:109–115

Kruk J, Aboul-Enein HY, Kładna A, Bowser JE (2019) Oxidative stress in biological systems and its relation with pathophysiological functions: the effect of physical activity on cellular redox homeostasis. Free Radic Res 53:497–521

Kueck A, Opipari AW Jr, Griffith KA, Tan L, Choi M, Huang J, Wahl H, Liu JR (2007) Resveratrol inhibits glucose metabolism in human ovarian cancer cells. Gynecol Oncol 107:450–457

Kumar A, Rimando AM, Levenson AS (2017) Resveratrol and pterostilbene as a microRNA-mediated chemopreventive and therapeutic strategy in prostate cancer. Ann N Y Acad Sci 1403:15–26

Kumar M, Kaur V, Kumar S, Kaur S (2016) Phytoconstituents as apoptosis inducing agents: strategy to combat cancer. Cytotechnology 68:531–563

Lalaoui N, Brumatti G (2017) Relevance of necroptosis in cancer. Immunol Cell Biol 95:137–145

Lee HS, Cho HJ, Yu R, Lee KW, Chun HS, Park JHY (2014) Mechanisms underlying apoptosis-inducing effects of Kaempferol in HT-29 human colon cancer cells. Int J Mol Sci 15:2722–2737

Lee J-C, Won S-J, Chao C-L, Wu F-L, Liu H-S, Ling P, Lin C-N, Su C-L (2008) Morusin induces apoptosis and suppresses NF-κB activity in human colorectal cancer HT-29 cells. Biochem Biophys Res Commun 372:236–242

Lee S, Clark R (2016) Anti-tumorigenic effects of capsaicin in colon cancer. J Food Chem Nanotechnol 2:162–167

Lee SY, Ju MK, Jeon HM, Jeong EK, Lee YJ, Kim CH, Park HG, Han SI, Kang HS (2018) Regulation of tumor progression by programmed necrosis. Oxidative Med Cell Longev 2018:3537471

Lewinska A, Adamczyk-Grochala J, Deregowska A, Wnuk M (2017) Sulforaphane-induced cell cycle arrest and senescence are accompanied by DNA hypomethylation and changes in microRNA profile in breast cancer cells. Theranostics 7:3461

Li W, Yang W, Liu Y, Chen S, Chin S, Qi X, Zhao Y, Liu H, Wang J, Mei X (2017) MicroRNA-378 enhances inhibitory effect of curcumin on glioblastoma. Oncotarget 8:73938
Lim J-H, Lee Y-M, Park SR, Da Hye K, Lim BO (2014) Anticancer activity of hispidin via reactive oxygen species-mediated apoptosis in colon cancer cells. Anticancer Res 34:4087–4093
Lin W, Tongyi S (2014) Role of Bax/Bcl-2 family members in green tea polyphenol induced necroptosis of p53-deficient Hep3B cells. Tumour Biol 35:8065–8075
Linder B, Kögel D (2019) Autophagy in cancer cell death. Biology 8:82
Liu RH (2004) Potential synergy of phytochemicals in cancer prevention: mechanism of action. J Nutr 134:3479S-3485S
Liu Z-L, Wang H, Liu J, Wang Z-X (2013) MicroRNA-21 (miR-21) expression promotes growth, metastasis, and chemo-or radioresistance in non-small cell lung cancer cells by targeting PTEN. Mol Cell Biochem 372:35–45
Ma C-H, Zhang Y-X, Tang L-H, Yang X-J, Cui W-M, Han C-C, Ji W-Y (2018) MicroRNA-1469, a p53-responsive microRNA promotes genistein induced apoptosis by targeting Mcl1 in human laryngeal cancer cells. Biomed Pharmacother 106:665–671
Ma J, Cheng L, Liu H, Zhang J, Shi Y, Zeng F, Miele L, Sarkar FH, Xia J, Wang Z (2013) Genistein down-regulates miR-223 expression in pancreatic cancer cells. Curr Drug Targets 14:1150–1156
Manu K, Kuttan G (2008) Ursolic acid induces apoptosis by activating p53 and caspase-3 gene expressions and suppressing NF-kB mediated activation of bcl-2 in B16F–10 melanoma cells. Int Immunopharmacol 8:974–981
Melkamu T, Zhang X, Tan J, Zeng Y, Kassie F (2010) Alteration of microRNA expression in vinyl carbamate-induced mouse lung tumors and modulation by the chemopreventive agent indole-3-carbinol. Carcinogenesis 31:252–258
Mettlin C (1997) Chemoprevention: will it work? Int J Cancer 71:18–21
Miele L, Miao H, Nickoloff B (2006) NOTCH signaling as a novel cancer therapeutic target. Curr Cancer Drug Targets 6:313–323
Miki H, Uehara N, Kimura A, Sasaki T, Yuri T, Yoshizawa K, Tsubura A (2012) Resveratrol induces apoptosis via ROS-triggered autophagy in human colon cancer cells. Int J Oncol 40:1020–1028
Mishra AP, Salehi B, Sharifi-Rad M, Pezzani R, Kobarfard F, Sharifi-Rad J, Nigam M (2018) Programmed cell death, from a cancer perspective: an overview. Mol Diagn Ther 22:281–295
Mohan S, Abdelwahab SI, Kamalidehghan B, Syam S, May KS, Harmal NSM, Shafifiyaz N, Hadi AHA, Hashim NM, Rahmani M (2012) Involvement of NF-kB and Bcl2/Bax signaling pathways in the apoptosis of MCF7 cells induced by a xanthone compound Pyranocycloartobiloxanthone A. Phytomedicine 19:1007–1015
Mousavi L, Salleh RM, Murugaiyah V (2018) Phytochemical and bioactive compounds identification of *Ocimum tenuiflorum* leaves of methanol extract and its fraction with an anti-diabetic potential. Int J Food Prop 21:2390–2399
Mukherjee S, Ghosh S, Das DK, Chakraborty P, Choudhury S, Gupta P, Adhikary A, Dey S, Chattopadhyay S (2015) Gold-conjugated green tea nanoparticles for enhanced anti-tumor activities and hepatoprotection-Synthesis, characterization and in vitro evaluation. J Nutr Biochem 26:1283–1297
Mukherji SM, Singh SP (1984) Reaction mechanism in organic chemistry. Macmillan Indian Press, Madras
Munagala R, Aqil F, Vadhanam MV, Gupta RC (2013) MicroRNA 'signature'during estrogen-mediated mammary carcinogenesis and its reversal by ellagic acid intervention. Cancer Lett 339:175–184
Nam J-S, Sharma AR, Nguyen LT, Chakraborty C, Sharma G, Lee S-S (2016) Application of bioactive quercetin in oncotherapy: from nutrition to nanomedicine. Molecules 21:E108
Newman DJ, Cragg GM (2007) Natural products as sources of new drugs over the last 25 years. J Nat Prod 70:461–477

Newman DJ, Cragg GM (2012) Natural products as sources of new drugs over the 30 years from 1981 to 2010. J Nat Prod 75:311–335

Newman DJ, Cragg GM, Kingston DG (2011) Anticancer agents from natural products. CRC Press Taylor & Francis, Boca Ratan

Nwaeburu CC, Bauer N, Zhao Z, Abukiwan A, Gladkich J, Benner A, Herr I (2016) Up-regulation of microRNA Let-7c by quercetin inhibits pancreatic cancer progression by activation of Numbl. Oncotarget 7:58367–58380

Oh JH, Ban JO, Cho M-C, Jo M, Jung JK, Ahn B, Yoon D-Y, Han SB, Hong JT (2012) 4-O-methylhonokiol inhibits colon tumor growth via p21-mediated suppression of NF-kB activity. J Nutr Biochem 23:706–715

Omar HA, Sargeant AM, Weng J-R, Wang D, Kulp SK, Patel T, Chen C-S (2009) Targeting of the Akt-nuclear factor-kB signaling network by [1-(4-chloro-3-nitrobenzenesulfonyl)-1H-indol-3-yl]-methanol (OSU-A9), a novel indole-3-carbinol derivative, in a mouse model of hepatocellular carcinoma. Mol Pharmacol 76:957–968

Opipari AW, Tan L, Boitano AE, Sorenson DR, Aurora A, Liu JR (2004) Resveratrol-induced autophagocytosis in ovarian cancer cells. Cancer Res 64:696–703

Park W, Amin AR, Chen ZG, Shin DM (2013) New perspectives of curcumin in cancer prevention. Cancer Prev Res 6:387–400

Purnamasari R, Winarni D, Permanasari AA, Agustina E, Hayaza S, Darmanto W (2019) Anticancer activity of methanol extract of *Ficus carica* leaves and fruits against proliferation, apoptosis, and necrosis in Huh7it cells. Cancer Inform 18:1176935119842576

Rahal A, Kumar A, Singh V, Yadav B, Tiwari R, Chakraborty S, Dhama K (2014) Oxidative stress, prooxidants, and antioxidants: the interplay. BioMed Res Int 2014:761264

Rahman KW, Li Y, Sarkar FH (2004) Inactivation of Akt and NF-kB play important roles during indole-3-carbinol-induced apoptosis in breast cancer cells. Nutr Cancer 48:84–94

Rao YK, Geethangili M, Fang S-H, Tzeng Y-M (2007) Antioxidant and cytotoxic activities of naturally occurring phenolic and related compounds: a comparative study. Food Chem Toxicol 45:1770–1776

Ravindranath V, Chandrasekhara N (1980) Absorption and tissue distribution of curcumin in rats. Toxicology 16:259–265

Rohman A, Riyanto S, Utari D (2006) Antioxidant activities, total phenolic and flavonoid contents of ethyl acetate extract of Mengkudu (*Morinda citrifolia*, L) fruit and its fractions. Indones J Pharm 17:136–142

Russo M, Spagnuolo C, Tedesco I, Russo GL (2010) Phytochemicals in cancer prevention and therapy: truth or dare? Toxins 2:517–551

Sharma K, Le N, Alotaibi M, Gewirtz DA (2014) Cytotoxic autophagy in cancer therapy. Int J Mol Sci 15:10034–10051

Slaby O, Sachlova M, Brezkova V, Hezova R, Kovarikova A, Bischofová S, Sevcikova S, Bienertova-Vasku J, Vasku A, Svoboda M (2013) Identification of microRNAs regulated by isothiocyanates and association of polymorphisms inside their target sites with risk of sporadic colorectal cancer. Nutr Cancer 65:247–254

Sonoki H, Sato T, Endo S, Matsunaga T, Yamaguchi M, Yamazaki Y, Sugatani J, Ikari A (2015) Quercetin decreases claudin-2 expression mediated by up-regulation of microRNA miR-16 in lung adenocarcinoma A549 cells. Nutrients 7:4578–4592

Spagnuolo C, Russo GL, Orhan IE, Habtemariam S, Daglia M, Sureda A, Nabavi SF, Devi KP, Loizzo MR, Tundis R, Nabavi SM (2015) Genistein and cancer: current status, challenges, and future directions. Adv Nutr 6:408–419

Srinivas US, Tan BW, Vellayappan BA, Jeyasekharan AD (2019) ROS and the DNA damage response in cancer. Redox Biol 25:101084

Takada Y, Andreeff M, Aggarwal BB (2005) Indole-3-carbinol suppresses NF-kB and IkBα kinase activation, causing inhibition of expression of NF-kB-regulated antiapoptotic and metastatic gene products and enhancement of apoptosis in myeloid and leukemia cells. Blood 106:641–649

Tili E, Michaille J-J (2011) Resveratrol, microRNAs, inflammation and cancer. J Nucleic Acids 2011:102431

Tse AK-W, Wan C-K, Zhu G-Y, Shen X-L, Cheung H-Y, Yang M, Fong W-F (2007) Magnolol suppresses NF-kB activation and NF-kB regulated gene expression through inhibition of IkappaB kinase activation. Mol Immunol 44:2647–2658

Tsujimoto Y, Shimizu S (2005) Another way to die: autophagic programmed cell death. Cell Death Differ 12:1528–1534

Valko M, Leibfritz D, Moncol J, Cronin MT, Mazur M, Telser J (2007) Free radicals and antioxidants in normal physiological functions and human disease. Int J Biochem Cell B 39:44–84

Varinska L, Gal P, Mojzisova G, Mirossay L, Mojzis J (2015) Soy and breast cancer: focus on angiogenesis. Int J Mol Sci 16:11728–11749

Venkatadri R, Muni T, Iyer A, Yakisich J, Azad N (2016) Role of apoptosis-related miRNAs in resveratrol-induced breast cancer cell death. Cell Death Dis 7:e2104

Vineis P, Fecht D (2018) Environment, cancer and inequalities—the urgent need for prevention. Eur J Cancer 103:317–326

Wang D-X, Zou Y-J, Zhuang X-B, Chen S-X, Lin Y, Li W-L, Lin J-J, Lin Z-Q (2017) Sulforaphane suppresses EMT and metastasis in human lung cancer through miR-616-5p-mediated GSK3β/β-catenin signaling pathways. Acta Pharmacol Sin 38:241–251

Wang H, Bian S, Yang CS (2011) Green tea polyphenol EGCG suppresses lung cancer cell growth through upregulating miR-210 expression caused by stabilizing HIF-1α. Carcinogenesis 32:1881–1889

Wang K, Tan S-L, Lu Q, Xu R, Cao J, Wu S-Q, Wang Y-H, Zhao X-K, Zhong Z-H (2018) Curcumin suppresses microRNA-7641-mediated regulation of p16 expression in bladder cancer. Am J Chinese Med 46:1357–1368

Wang N, Feng Y (2015) Elaborating the role of natural products-induced autophagy in cancer treatment: achievements and artifacts in the state of the art. Biomed Res Int 2015:934207

Wang X, He H, Lu Y, Ren W, Teng K-Y, Chiang C-L, Yang Z, Yu B, Hsu S, Jacob ST (2015) Indole-3-carbinol inhibits tumorigenicity of hepatocellular carcinoma cells via suppression of microRNA-21 and upregulation of phosphatase and tensin homolog. Biochim Biophys 185:244–253

Wei D, Yang L, Lv B, Chen L (2017) Genistein suppresses retinoblastoma cell viability and growth and induces apoptosis by upregulating miR-145 and inhibiting its target ABCE1. Mol Vis 23:385

Weng J-R, Tsai C-H, Kulp SK, Wang D, Lin C-H, Yang H-C, Ma Y, Sargeant A, Chiu C-F, Tsai M-H (2007) A potent indole-3-carbinol–derived antitumor agent with pleiotropic effects on multiple signaling pathways in prostate cancer cells. Cancer Res. 67:7815–7824

Wondrak GT (2009) Redox-directed cancer therapeutics: molecular mechanisms and opportunities. Antioxid Redox Signal 11:3013–3069

Xiao D, He J (2010) Epithelial mesenchymal transition and lung cancer. J Thorac Dis 2:154–159

Xiao X, Zhang Y, Pan W, Chen F (2020) miR-139-mediated NOTCH1 regulation is crucial for the inhibition of osteosarcoma progression caused by resveratrol. Life Sci 242:117215

Xie J, Wang J, Zhu B (2016) Genistein inhibits the proliferation of human multiple myeloma cells through suppression of nuclear factor-kB and upregulation of microRNA-29b. Mol Med Rep 13:1627–1632

Xu Y, Ge R, Du J, Xin H, Yi T, Sheng J, Wang Y, Ling C (2009) Corosolic acid induces apoptosis through mitochondrial pathway and caspases activation in human cervix adenocarcinoma HeLa cells. Cancer Lett 284:229–237

Yamada S, Tsukamoto S, Huang Y, Makio A, Kumazoe M, Yamashita S, Tachibana H (2016) Epigallocatechin-3-O-gallate up-regulates microRNA-let-7b expression by activating 67-kDa laminin receptor signaling in melanoma cells. Sci Rep 6:19225

Yang C, Xie X, Tang H, Dong X, Zhang X, Huang F (2018) Transcriptome analysis reveals GA induced apoptosis in HCT116 human colon cancer cells through calcium and p53 signal pathways. RSC Adv 8:12449–12458

Yang N, Zhao Y, Wang Z, Liu Y, Zhang Y (2017) Scutellarin suppresses growth and causes apoptosis of human colorectal cancer cells by regulating the p53 pathway. Mol Med Rep 15:929–935

Zaman MS, Thamminana S, Shahryari V, Chiyomaru T, Deng G, Saini S, Majid S, Fukuhara S, Chang I, Arora S (2012) Inhibition of PTEN gene expression by oncogenic miR-23b-3p in renal cancer. PLoS ONE 7:e50203

Zan L, Chen Q, Zhang L, Li X (2019) Epigallocatechin gallate (EGCG) suppresses growth and tumorigenicity in breast cancer cells by downregulation of miR-25. Bioengineered 10:374–382

Zaorsky NG, Churilla T, Egleston B, Fisher S, Ridge J, Horwitz E, Meyer J (2017) Causes of death among cancer patients. Ann Oncol 28:400–407

Zhang H-M, Zhao L, Li H, Xu H, Chen W-W, Tao L (2014) Research progress on the anticarcinogenic actions and mechanisms of ellagic acid. Cancer Biol Med 11:92–100

Zhang L, Chinnathambi A, Alharbi SA, Veeraraghavan VP, Mohan SK, Zhang G (2020) Punicalagin promotes the apoptosis in human cervical cancer (ME-180) cells through mitochondrial pathway and by inhibiting the NF-kB signaling pathway. Saudi J Biol Sci 27:1100–1106

Zhang W, Bai W (2014) MiR-21 suppresses the anticancer activities of curcumin by targeting PTEN gene in human non-small cell lung cancer A549 cells. Clin Transl Oncol 16:708–713

Zhang X, Guo Q, Chen J, Chen Z (2015) Quercetin enhances cisplatin sensitivity of human osteosarcoma cells by modulating microRNA-217-KRAS axis. Mol Cells 38:638–642

Zhang Y, Tang L (2007) Discovery and development of sulforaphane as a cancer chemopreventive phytochemical. Acta Pharmacol Sin 28:1343–1354

Zhao J, Fang Z, Wang Zha Z, Sun Q, H, Sun M, Qiao B, (2019) Quercetin inhibits cell viability, migration and invasion by regulating miR-16/HOXA10 axis in oral cancer. Eur J Pharmacol 847:11–18

Zhou J, Gong J, Ding C, Chen G (2015) Quercetin induces the apoptosis of human ovarian carcinoma cells by upregulating the expression of microRNA-145. Mol Med Rep 12:3127–3131

# Chapter 5
# Role of Stress and Defense in Plant Secondary Metabolites Production

**Humberto Aguirre-Becerra, Ma Cristina Vazquez-Hernandez, Diana Saenz de la O, Aurora Alvarado-Mariana, Ramon G. Guevara-Gonzalez, Juan Fernando Garcia-Trejo, and Ana Angélica Feregrino-Perez**

**Abstract** Secondary plant metabolites are natural bioactive compounds which are an important income for the pharmaceutical, food, cosmetic, agriculture, and other sectors due to their health-promoting properties and prevention and treatment of some diseases. The secondary metabolites can be classified into three main groups: phenolic compounds, terpenoids, and nitrogen compounds. The secondary metabolism in plants is a mechanism of adaptation and evolution as a defense to harsh environmental factors that induce stress. According to the hormesis curve of each plant model, the stress can be divided into distress (bad stress that leads to damage and ultimately plant death) or eustress (good stress that leads to activation of secondary metabolism). The environmental factors can be divided into biotic and abiotic which can be artificially induced to activate plant defense responses leading to the production of secondary metabolites. Several approaches to this process called elicitation have been proposed in the last decades with different types of metabolism-inducing factors or elicitors. Novel elicitation using abiotic factors includes electromagnetic waves (including several wavelengths of the light spectra, and electric and magnetic fields), acoustic waves, nanostructures, volatile compounds, nutrient deprivation, and several metals and salt soil pollutants. In the same order, novel elicitation using biotic factors include new bacteria consortium, fungi, phytohormones, and miRNA solutions. In general, the purpose of elicitation is to interact with the biochemical routes in order to produce secondary metabolites in

H. Aguirre-Becerra · D. Saenz de la O · R. G. Guevara-Gonzalez · J. F. Garcia-Trejo · A. A. Feregrino-Perez (✉)
Facultad de Ingenieria, Universidad Autonoma de Queretaro,
C.P. 76010 Santiago de Querétaro, QRO, Mexico
e-mail: feregrino.angge@hotmail.com

M. C. Vazquez-Hernandez
Departamento de Ingeniería, Tecnologico Nacional de Mexico Roque, Carr. Celaya-Juventino Rosas Km. 8, C.P. 38110 Celaya, GTO, Mexico

A. Alvarado-Mariana
Facultad de Ciencias Naturales, Horticultura Ambiental, Av. de la Ciencia s/n, C.P. 76230 Juriquilla, QRO, Mexico



D. Pal and A. K. Nayak (eds.), *Bioactive Natural Products for Pharmaceutical Applications*, Advanced Structured Materials 140,
https://doi.org/10.1007/978-3-030-54027-2_5

high quantities, usually with negative effects in biomass production or morphology, but increasing plant quality in terms of aroma, taste or color. The metabolic profile and general response of elicitation vary greatly depending on the plant model, the elicitor level or concentration, and the stimulation time. Considering these facts, this chapter clearly and concisely discusses the most current strategies of elicitation for the increase of secondary metabolites production in plants.

**Keywords** Secondary metabolites · Eustress · Distress · Defense in plant · Elicitation

## 5.1 Introduction

A variety of substances or compounds called secondary metabolites are mainly responsible for the adaptation of plants to environmental changes. Plants synthesize a wide range of secondary metabolites such as alkaloids, flavonoids, phenolic, steroids, anthocyanins between others, which are used as pharmaceuticals, agro-chemicals, biopesticides, color, additives, etc. Secondary metabolites are not considered to be part of fundamental life processes of plants, however, they play a significant role in protection from insect, pest, herbivores, phytopathogens, and other harsh environmental variables (Thakur et al. 2019). Therefore, the synthesis of secondary metabolites depends on internal and external factors (stressors and stress factors, respectively) that influence plant metabolism, and can affect plant reproduction and productivity (Kranner et al. 2010). The stress factors, that for their nature can be biotic or abiotic, modify positively or negatively the plant metabolism (Cheynier et al. 2013). Biotic stress is the result of the interaction between plant and viral, bacterial, fungi, pheromones, phytohormones, and nucleic acids among others. Meanwhile, abiotic stress can be physical factors (Light spectra, temperature, water stress, acoustic waves, others) and chemical factors (Nanostructures, gasses, nutriment, others) (Vázquez-Hernández et al. 2019).

According to data provided by Agathokleous et al. (2019), exist 109,821 publications that review the topic of "plant stress" only in the period 2000 to 2018. The results indicated that the dose–response is not always linear, observing a biphasic response when the doses allow it. This balance of response between the plant and stress factors provides information on a positive (eustress) or a negative effect (distress), a phenomenon called "Hormesis" (Agathokleous et al. 2019; Vázquez-Hernández et al. 2019). The concept of hormesis is a term used in medicine for the application of toxins in low doses (Calabrese 2004). Paracelsus (1493–1541) defined it as "All things are poison and nothing is without poison, only the dose permits something not to be poisonous" (cited by (Vázquez-Hernández et al. 2019)). Currently, the term is applied in horticultural and agricultural practices as a biphasic response in which doses of a toxic agent could cause inhibition (distress) or can cause stimulation (eustress) (Vargas-Hernandez et al. 2017).

Plants are sessile organisms susceptible to the interaction between various types of stress, which has resulted in an evolved defense system that increases the synthesis of secondary metabolites (Ghorbanpour et al. 2014). The variety of stress factors together can affect the plant physiology, plant-plant interaction, defense type, reproductivity, among others. For example, salinity and low/high-temperature are conditions which restrict plant growth and productivity (Akula and Ravishankar 2011). In the signaling response to pathogens or herbivorous insects, several response pathways are invoked, some of these are induced by infection and some are performed regardless of the antimicrobial nature (Zaynab et al. 2018). Another example is the interaction between plants and herbivory insects that causes the plant to emit volatile organic compounds which influence the plant-to-plant communication, pollinators, and other insects, and increase fluidity of cell membranes for thermo-tolerance and leaf tissue protection from atmospheric oxidants within and around leaves (Faiola and Taipale 2020).

The foregoing indicates that plants can react in various ways in the presence of one or more stress factors and that, in the same way, the response to the stimulus may be the activation of a synthesis pathway of only one metabolite or a series of secondary metabolites. For this reason, this work will focus on stress factors and the production of secondary metabolites in plants.

## 5.2 Abiotic Stress

"*Any unfavorable condition or substance that affects or blocks a plant's metabolism, growth or development*" is the definition of plant stress suggested by Lichtenthaler (1996). Currently, this definition has been modified depending on the stimulus origin, defining as stress factors those stimuli that are external to the plant, biotic (fungi, insects, etc.) or abiotic (temperature, luminosity, nanoparticles, metals and polluting salts, water, etc.) (Kranner et al. 2002; Thakur et al. 2019). Abiotic stress origin is not biological and can be divided into chemical or physical (Vázquez-Hernández et al. 2019). For example, the adaptation to cold environments in some plants is the result of an increase in the synthesis of flavonoids due to acclimatization processes at low temperatures or by the application of UV radiation (Samanta et al. 2011; Nakabayashi et al. 2014). In *Vitis vinifera*, it has been observed that stimulation with heavy metals such as Cadmium ($Cd^{2+}$), Cobalt ($Co^{2+}$) and Silver ($Ag^{+}$), can increase the synthesis of Resveratrol (Cai et al. 2013), while the application of UV-C irradiation induces the synthesis of stilbene (Wang et al. 2010; Liu et al. 2010). This indicates that the synthesis of secondary metabolites will depend on various factors such as the type of stimulus, the concentration, and the form of application. These same observations are appreciated by Feregrino-Perez et al. (2018), where the effect of nanomaterials on germination, development of plants, and synthesis of secondary metabolites is reviewed, concluding that the stress level will depend on the used nanomaterial, the dose and the time of exposition. Low/high-temperature, relative humidity in air, drought, microelements shortage, and $CO_2$ reduction are classical abiotic factors for plants elicitation. The role of novel abiotic stress factors on the production of secondary metabolites is described below.

### 5.2.1 Electromagnetic Sources

Electromagnetic phenomena can be seen as an abiotic stress elicitor to affect plants. In this context, many reports have demonstrated more advantages than disadvantages when strong or weak electric fields, magnetic fields were applied to plants (Dannehl 2018). Electromagnetic sources have been studied as another possibility to increase plant growth and development due to the alteration in the electrostatic balance of the plant system at the cell membrane level (Radhakrishnan 2019). An electromagnetic field is produced by a distribution of electric current and charge. An electric field (EF) can occur under high-voltage lines and the units in the SI are newtons per coulomb or, equivalently volts per meter (V/m); in the same way, a magnetic field (MF) is usually measured in terms of its magnetic flux density whose unit is expressed as Tesla (T) (Dannehl 2018).

Pulsed electric field PEF technology consists of the application of short, high power electrical pulses to products placed in a treatment chamber, confined between electrodes (Soliva-Fortuny et al. 2017). A high electric field can cause cell membrane disruption, whereby inner secondary metabolites are released from intracellular cell compartments, resulting in a high content of bioactive compounds (Odriozola-Serrano et al. 2009; Janositz and Knorr 2010). In the case of MF, several studies have used small boxes with coils, iron bars, a function generator, and a power amplifier. The application of magnetized water to plants is a novel area of indirect application of MF that the scientific community is currently researching (Dannehl 2018).

Light is an electromagnetic wave within the visible spectrum, however, that definition depends upon the sense of sight involving the response of individuals (Koshel 2004), therefore, the UV and infrared parts of the electromagnetic spectrum are roughly included, but will be considered in this chapter due to their importance in the plant production of secondary metabolites. Plants sense light through specific molecules called photoreceptors that trigger specific signals for photomorphogenesis or other defense systems. Depending on the dose rate and exposure time, either insufficient or excess levels, light can become a type of eustress, producing several effects, from damage to cellular components to triggering of defense systems for secondary metabolite production (Alvarado et al. 2019; Muller-Xing et al. 2014; Akula and Ravishankar 2011). The next sections of this chapter will discuss the effect of electromagnetic sources in the production of natural bioactive compounds.

#### 5.2.1.1 Light

Light is one of the most important and obvious requirements for plant growth and development, where the energy of sunlight and artificial light sources is mainly used for photosynthesis. However, light is not only involved in the photosynthesis process but also in the production of natural bioactive compounds, gene expression, and synchronization of the circadian clock in the light/dark cycle (Larner et al. 2018). Changes in the light intensity, quality, direction, and duration are sensed by specialized photoreceptors which are specially designed proteins that sense light, triggering chain reactions that have been studied in terms of photomorphogenesis and primary and secondary metabolites production (Alvarado et al. 2019). Photoreceptors perceive specific light wavelengths of over a continuous spectral range through a small cofactor or chromophore molecule (Burgie et al. 2014). Five photosensory systems have been identified: phytochromes perceiving red (660–700 nm) and far-red (700–750 nm), cryptochromes, phototropins, and members of the Zeitlupe family perceiving blue (495–400 nm) and UV-A (400–315 nm), and UV Resistance Locus 8 (UVR8) perceiving (315–280 nm) (Bantis et al. 2018; Alvarado et al. 2019).

Recent investigation has focused on the effect of light technology in plant growth, developmental traits, and primary and secondary metabolites by using one or more light wavelengths, intensities, and photoperiods. It has been reported that blue light increases phenolic compounds by promoting the production of malonyl CoA and coumaroyl CoA, participating in the synthesis of phenolic compounds (Qian et al. 2016). In addition, red and far-red wavelengths are perceived by the phytochromes photoreceptors, which regulates biosynthetic pathways involved in the synthesis of anthocyanins, molecules that belong to the phenolic compounds known as flavonoids and have many functions in plants including pigmentation (Alokam et al. 2002). In the same way, plants produce secondary metabolites such as flavonoids and anthocyanins to cope with cell damage produced by UV radiation (Jiang et al. 2017b). Serious damage to DNA, membrane, and proteins can be caused by UV-B radiation, whereas UV-A induces DNA damage less efficiently because of the activation of photoreactions forming reactive oxygen species (ROS) (Hideg and Strid 2017; Häder et al. 2015).

Supplemental lighting has been accepted for improving horticultural crops. Light-emitting diode (LED) technology has been linked to controlled environments in horticulture for achieving crop yield, phytochemical content, nutritional value, flowering control, transplant success, pre-harvest and postharvest product quality, and production of regeneration material (Bantis et al. 2018; Alvarado et al. 2019). LEDs have allowed a sustainable and highly efficient use of energy and reproduce true spectral composition of blue, green, red, and far-red wavelengths that matches with plant-specific photoreceptors (Singh et al. 2015). Other light technologies, as high sodium pressure (HSP) and other high-intensity discharge (HID) lamps are still used in greenhouse and plant experimentation, however, LED technology is replacing these devices due to the various advantages LEDs offer. Table 5.1 summarizes some examples of the application of supplemental light on plants or foods with a commercial interest and presents the effect on the production of natural bioactive compounds.

#### 5.2.1.2 Electric and Magnetic Fields

Magnetic fields (MFs) are considered an abiotic factor that can induce eustress with significant effects on the growth and development of plants. The effect of light, gravity, mechanical damage, and electrical signaling on plants has been studied and documented over the past years concluding strong facts relating to phototropism, gravitropism, and thigmotropism (Maffei 2014). The geomagnetic field (GMF) is a natural component of our environment, however, its impact on plant growth and development is not well-understood, moreover, the effects of artificial magnetic fields on plants have been poorly studied (Maffei 2014). Several experiments with lower and higher values than the GMF has been conducted with predominantly positive effects depending on the plant, time of exposure and intensity. For example, an increase in germination or subsequent seedling growth barley, corn, beans, wheat, hornwort, mung bean, pea, chickpea, tomato, and okra, but it was reduced in seeds of rice. In a similar way, the effect on roots, shoots, gravitropism, photosynthesis, and lipid composition present a similar pattern (Maffei 2014).

Several theories and studies about the biological effect on MF have been proposed. A polar structure in various chemical bonds in the organic material may be linked to the polar water molecules and dissociated ions of mineral salts conferring magnetic properties (Chepets et al. 1985). A MF can decrease the disease index of plants due to the modulation of calcium signaling, and proline and polyamines pathways (Radhakrishnan 2019). The plant cells contain about 4500 iron atoms in the ferritin molecules involved in growth and metabolism. The magnetic rotator moment of ultimate iron atoms creates an external MF which collectively generates an atom re-positioning in the direction of MF that leads to an increase of the plant temperature (Vaezzadeh et al. 2006). Photoreceptors have been also proposed to be potential magnetoreceptors since cryptochromes and phytochromes produce radical pairs after the exposure to their corresponding light wavelength triggers (Maffei 2014; Dhiman and Galland 2018). Cryptochrome-dependent responses such as blue-light-dependent anthocyanin accumulation and blue-light-dependent degradation of CRY2 protein were enhanced at higher magnetic intensities in Arabidopsis mutants lacking cryptochromes (Ahmad and Jones 1979). Limited information is available on the molecular basis and the function of the MF receptors and their activation by physiological signals, therefore, their involvement in directing the overall response in different plant organs is yet to be determined (Radhakrishnan 2019).

Static magnetic field (SMF) exposition in plants has been found to be an effective and emerging tool to control diseases and increase tolerance against the adverse environment (Radhakrishnan 2019). However, a small number of studies have been attempted to determine the role of MF on plant tolerance against various

**Table 5.1** Effect on natural bioactive compounds of experiments where supplemental light was the stress factor on plants or food with commercial interest

| Food or cultivar/ Reference | Light condition treatment | Result |
|---|---|---|
| Stored tomato fruit var. *Cappricia*/Panjai et al. (2017) | (T1) Darkness (control), (T2) Darkness + UV, (T3) R, (T4) R + UV. UV of 4.98 kJ $m^{-1}$ per 30 min $day^{-1}$<br>T2/T4: UV tube for 15 min in the morning and at night<br>T3/T4: 60% UV-B, 30% UV-A, 4% UV-C and 6% visible light<br>R (665 nm): applied for the whole storage period, equivalent to PAR of 113 μmol $m^{-2}$ per day | (↑) Lycopene concentration. Sharply increase with T3 and T4<br>(↑) Concentration of β-carotene. The highest at T3 after 10 days of postharvest. T2 had the highest value after 15 days<br>(↑) TFC. The highest with T3 after 10 and 15 days of postharvest<br>(↓) TFC. T2 showed a significant decrease at day 5<br>(↑) TPC. A sharply increase with T3 at day 10 and peaked on day 20<br>(↑) AC – ABTS. T4 showed highest Hydrophilic and Lipophilic AC 20 days after harvesting compared to control |
| Stored habanero pepper (*Capsicum chinense*)/ Pérez-Ambrocio et al. (2018) | Combination of B (0, 1.5, and 3 min) and UV-C (0, 0.5, and 1 min)<br>B: 48 W $m^{-2}$<br>UV-C: 11.3 W $m^{-2}$ | (↑) TPC and TFC. All treatments with B and/or UV-C showed a significant increase compared to control. The highest at 3 min of B + 0.5 min of UV-C<br>(↑) TCC. Increase the first 10 days of storage<br>(↑) Capsaicin. Almost all treatments (1.5 min of B and 1.5 min of B + 1 min UV-C) presented an increase in capsaicin<br>(↑) AC. Statically higher in all treatments with B and UV-C light |
| Green and purple basil (Ocimum basilicum) plants/Dou et al. (2019) | Ten treatments: Combination of two PPFDs and five UV-B radiation doses. PPFDs: 160 and 224 μmol·$m^{-2}$·$s^{-1}$ (high and low) with a 16-h photoperiod provided by cool white fluorescent lamps with UV of 2.2 and 2.5 μmol $m^{-2}$ $s^{-1}$, respectively<br>UV-B (16.0 μmol $m^{-2}$ $s^{-1}$): | (↑) TAC. 9–23% higher after UV-B radiation compared to control for green basil. Greater under high PPFD for purple basil<br>(↑) TPC. 28–126% higher after UV-B compared to control for green basil. Greater in high PPFD for purple and green basil. 29–63% higher under |

(continued)

**Table 5.1** (continued)

| Food or cultivar/ Reference | Light condition treatment | Result |
|---|---|---|
| | (Control) No UV-B<br>(1H2D) 1 h·d$^{-1}$ for 2 days<br>(2H2D) 2 h·d$^{-1}$ for 2 days<br>(1H5D) 1 h·d$^{-1}$ for 5 days<br>(2H5D) 2 h·d$^{-1}$ for 5 days | 2H2D and 2H5D for purple basil<br>(↑) TFC. 80–169% higher after UV-B compared to control for green basil. 37–79% higher under 2H2D and 2H5D for purple basil<br>(↑) AC. Higher in green basil under all supplemental UV-B treatments. Only higher under 2H2D and 2H5D in purple basil plants |
| Coriander (Coriandrum sativum)/Naznin et al. (2016) | R (661 nm) and B (449 nm) ratios (R:B) of LED light<br>Four treatments: 100% R, 5:1, 10:1, and 19:1<br>Photoperiod of 16/8 h (day/ night) and PPFD of 120 μmol m$^{-2}$ s$^{-1}$ in growth chamber | (↑) AC – DPPH. 2.0, 1.6, and 1.5 times higher, under 5:1, 10:1, and 19:1 respectively, compared to the plants under 100% R |
| Soybean (Glycine max L.) sprout/Azad et al. (2018) | Experiment 1: B (450–495 nm), G (510–550 nm) LEDs and florescent lamps (control). PPFD in growth chamber was 150 μmol m$^{-2}$ s$^{-1}$. Samples were harvested at 3rd, 4th, 5th, 6$^{th}$, and 7th days after sowing (DAS)<br>Experiment 2: Far Infrared (FIR) irradiation, photoperiod was same as experiment 1. Exposure time for 30, 60, and 120 min at 90, 110, and 130 °C on harvested sprout | (↑) TPC and isoflavones. Higher under B compared to G and fluorescent light. Isoflavones increased at five and six DAS<br>Significant increase in total isoflavones with FIR of 110 °C/120 min, nearly 2.3 times higher than the control. Further increase in the FIR temperature decreased the isoflavones content<br>(↑) AC – DPPH and FRAP. DPPH 75% higher with B compared to 69% with G and 58% with control. FRAP was more than 2 times higher in B compared to control<br>FIR 110 °C/120 min treatment had the highest FRAP value among the FIR treatments |
| Blueberry Leaves (*Vaccinium corymbosum* L.)/ Routray et al. (2018) | R (661 nm–24 μmol m$^{-2}$ s$^{-1}$) and B (417 nm–6 μmol m$^{-2}$ s$^{-1}$) at 12, 24, and 48 h | (↑) TPC. Higher under 12 h of B light compared to untreated leaves (control)<br>(↓) TPC. 48 h induced deterioration of TPC in all observed cases. R light led to a decrease in TPC compared to control |

(continued)

**Table 5.1** (continued)

| Food or cultivar/ Reference | Light condition treatment | Result |
|---|---|---|
| | | (↑) Monomeric anthocyanins. An increase with time up to 24 h with B light, after which it decreased. Lower than the untreated sample with R light (↑) AC. The extract prepared with leaves treated with R light for 12 h had higher FRAP than control and treated samples with both R and B. For 24 and 48 h, FRAP from treatment with B light was maintained at a higher level as compared to samples treated with R |
| Persimmon (Diospyros kaki L. cv. Vanilla) fruit/Denoya et al. (2020) | Two maturity stages (unripe and ripe) were exposed to a pulsed light treatments of 20 and 60 kJ $m^{-2}$. Time of exposure: 1.2 s, Frequency: 3 Hz, 5.7 kJ $m^{-2}$ per pulse, of polychromatic light in the wavelength range between 200 and 1100 nm | (↑) TPC. Higher than control in unripe fruit exposed to 20 kJ $m^{-2}$<br>In ripe fruit, No significant differences in TPC were found between treatments<br>(↑) AC - ABTS, DPPH, and FRAP. Al treatments presented higher AC than the control treatment in unripe fruit |

Used abbreviations *PAR* Photosynthetic available radiation, *PPFD* Photosynthetic photon flux density, *LED* Light-Emitting diode, *R* Red, *B* Blue, *UV* Ultraviolet, *AC* Antioxidant capacity, *TCC* Total carotenoid content, *TPC* Total phenolic content, *TFC* Total flavonoid content, *TAC* Total anthocyanin content, *DW* dry weight, *FW* fresh weight, (↑) increment of, (↓) decrement of

stress conditions (Radhakrishnan 2019). The effect on secondary metabolites production has been scarcely studied; presenting an opportunity area in this field.

Electrostatic fields (EF) has also been used in the horticultural industry. Most of the applications of this technology are for germination and seedlings improvement due to its importance in the supply chain. For example, drought resistance and removal of free radicals in maize seedlings with electric field intensity 200 kV/m, pulse width 80 ms and frequency 1 Hz were analyzed by He et al. (2017), where the growth of root, the ability of self-organization, and the respiration metabolism of root cells was improved. Another example with potato tubers subjected to pulsed electric field (PEF) through a treatment voltage across the potatoes of 5 kV and a discharge capacitance of 450 pF prior to planting had an increase in the yield of 22–29% (Gachovska et al. 2015). In the same way, winter wheat seeds increased the germination energy up to 32.41% and weight up to 23.8% under PEF (Starodubtseva et al. 2018). In the experiment of Yan et al. (2017), PEF through a high-voltage pulse power supply and arc electrode was studied on cotton seeds vigor with different frequencies of 1, 5, 10, 20, and 50 Hz at the voltage of 16 and 20 kV, and the treatment time was 40 s, finding that when the frequency of the electric field increased, the effects increased and reached the maximum at 10 Hz, and after 10 Hz, as the electric field frequency increased, the effects began to decrease. This last experiment exemplifies the hormetic curve in a vigor treatment where there is a maximal dose at which a maximum response value is reached and then vigor starts decreasing.

In several studies, PEF pretreatments in plants, whole fruits, or other food sources result in an increase in the natural bioactive compounds. When biological cells are exposed to an EF, the charge accumulates along the plasma membrane causing electroporation, a transmembrane potential difference which causes porosity, and thus the diffusion of intracellular components in cellular juice increasing the extractability of natural bioactive compounds by the release of solutes into the solvent (El Kantar et al. 2018; Vicaş et al. 2017; Barba et al. 2015; Hendrawan et al. 2019). The time exposure and intensity of the EF are critical since a lower EF may form smaller pores allowing the ions to pass through, but large molecules may not get out of the cell, however, higher EF are suspected to damage antioxidant compounds due to long exposure to high-voltage electric current (Hendrawan et al. 2019). Table 5.2 summarizes some examples of the application of magnetic or electric fields on plants or foods with a commercial interest and presents the effect on the production of natural bioactive compounds.

### *5.2.2 Acoustic Emissions*

Acoustic emissions (AE) stimulus is one of the recent physical abiotic factors whose beneficial effects on plant growth, development, and health have been discussed. AE from ecological conditions or artificially applied can initiate diverse signals that trigger transduction cascades, similar to other abiotic stress factors (Alvarado et al. 2019). From a bioacoustics perspective, chewing serves as an alarm signal to plants and has been demonstrated that applying recorded insect chewing sounds caused an increase of phytochemical production (Appel and Cocroft 2014). In the same way, Jeong et al. (2014), reported an improvement of natural protection responses in rice plants caused by amplification at 100 decibels of a wide range of frequencies between 0 and 1.5 kHz. Moreover, Hassanien et al. (2014), found a higher disease resistance in pepper, cucumber, and tomato after AE treatments.

The biological mechanism of how sound affects plants is still under discussion. A mechano-stimuli perception of waves has been proposed, but a reliable explanation of sound-specific structure for recognition by plants has not been completely elucidated (Alvarado et al. 2019). This mechanism consists of the second messenger of calcium ion ($Ca^{2+}$) signals. The channels that mediate $Ca^{2+}$ flux are possibly located in the plasmatic membrane where $Ca^{2+}$ is sensed possibly through various $Ca^{2+}$ sensors and/or CDPKs (Calcium-dependent protein kinase), which pass the message through phosphorylation/dephosphorylation to different signaling

**Table 5.2** Effect on natural bioactive compounds of experiments where magnetic or electric fields was the stress factor on plants or food with a commercial interest

| Food or cultivar/Reference | MF or EF treatment | Result |
|---|---|---|
| Apples var. Golden Delicious (Whole fruit)/Ribas-Agustí et al. (2019) | PEF:<br>(a) 0.4 kV $cm^{-1}$, 5 pulses (0.01 kJ $kg^{-1}$, 20 μs total treatment time)<br>(b) 2.0 kV $cm^{-1}$, 35 pulses (1.8 kJ $kg^{-1}$, 140 μs total treatment time)<br>(c) 3.0 kV $cm^{-1}$, 65 pulses (7.3 kJ $kg^{-1}$, 260 μs total treatment time)<br>The system supplied 4 μs monopolar pulses at a fixed frequency of 0.1 Hz | (↑) AC. 24% at (a) just after treatment. (Not significant 24 h after treatment)<br>(↓) AC. 39% at (c) just after treatment. At (b) 46% and (c) 62% after 24 h<br>(↑) TPC and flavan-3-ol. 25–26% in TPC and 43–35% in total flavan-3-ols just after and 24 h after treatment (respectively) with (a)<br>(↓) TPC by 32% at (b) and 43% at (c) just after treatment and 50% at (b) and 66% at (c) after 24 h<br>(↓) Total flavan-3-ols by 19% at (b) and 51% at (c) just after treatment and 52% at (b) and 59% at (c) after 24 h<br>(↓) TFC by 28% at (b) and 50% at (c) just after treatment and 60% at (b) and 68% at (c) 24 h after treatment |
| Orange, pomelo and lemon fruits/El Kantar et al. (2018) | PEF of 3 kV $cm^{-1}$ for whole fruits and 10 kV $cm^{-1}$ for stacks of skins. Time interval between pulses: 2 s. Pulse duration: 70 μs | (↑) by approximately 39% for orange, 66% for pomelo, and 135% for lemon in the release of polyphenols from the inner parts of the cells |
| Apples var. Golden delicious (Whole fruit)/ Soliva-Fortuny et al. (2017) | PEF intensities of 0.4 – 2 kV $cm^{-1}$, using 5–35 monopolar pulses of 4 μs at a frequency of 0.1 Hz, corresponding to an specific energy input of 0.008 – 1.3 kJ $kg^{-1}$ | (↑) TPC (13%) and flavan-3-ol (92%). Best result with 0.008 kJ $kg^{-1}$ in apples stored 24 h at 22 °C<br>(↑) Flavonoids (58%). Stored at 4 °C for<br>(↑) AC enhanced by 43% after 12 h at 4 °C and by 15% after 24 h at 22 °C |
| Tissues of apple var. Ligol and carrot var. Baltimore/ Wiktor et al. (2015) | PEF intensities of 1.85, 3 and 5 kV $cm^{-1}$ with the combination of 10, 50 and 100 exponential shaped pulses of average 10 μs width each. Interval between pulses was set at 2 s | (↑) TCC up to 11.34% in carrots with 1.85 kV $cm^{-1}$ regardless of the applied pulse number<br>(↑) Maximal TPC and antioxidant activity (EC50) in 10 pulses at 1.85 kV $cm^{-1}$ in apple tissue |

(continued)

**Table 5.2** (continued)

| Food or cultivar/Reference | MF or EF treatment | Result |
|---|---|---|
| | | (↓) TCC up to 25.33% with 3 kV $cm^{-1}$<br>(↓) TPC and EC50 up to 35.93 and 32.95% respectively at 5 kV $cm^{-1}$ and 100 pulses |
| Wine of grapes Muscat Ottonel (MO), Pinot Noir (pn), and Merlot (MT)/Vicaş et al. (2017) | Grape mash was treated with PEF and then fermented for wine elaboration<br>Pulses of 150 μs of 7 kV $cm^{-1}$ were applied with a frequency of 178 Hz | (↑) TPC. 1.4 times higher in wine of MO grapes, 2.98 in wine of PN grapes, and 1.72 in wine of MT grapes, than the untreated one<br>(↑) TFC. The highest in PN wine, followed by MT and MO in which the grapes were PEF treated before the fermentation process<br>(↑) TAC. An increase of 11.11% in the case of PN and only 5.22% in the case of MT<br>(↑) AC. The highest in PN wine treated by PEF |
| Blackberries (*Rubus fruticosus*)/Barba et al. (2015) | PEF of 13.3 kV $cm^{-1}$ and High Voltage Electric Discharge (HVED) of 40 kV. Distance between pulses of 2 s. 1 min pause was made after each 100 pulses. Damped oscillations duration of 10 μs. Discharges were applied with a repetition rate of 0.5 Hz, | (↑) TPC yield using HVED (333.8 mg/100 g), followed by PEF (108.0 mg/100 g)<br>Extracts obtained after applying PEF were more clear and stable than HVED assisted extraction<br>TPC relative recovery of PEF after supplementary extraction was also higher (sixfold and fourfold higher after hot water and ethanol extraction, respectively) than for HVED (1.8-fold and 1.5-fold higher for hot water and ethanol extraction, respectively)<br>(↑) Anthocyanin Content, similarly to TPC |
| Basil leaves/(Hendrawan et al. (2019) | PEF. Two factors. Combination of 2, 3, and 4 kV $cm^{-1}$ with an exposure time of 1, 2, and 3 min | (↑) TPC. The highest (115.203 ± 1.115 mg GAE $g^{-1}$) at 3 kV $cm^{-1}$ for 2 min while the lowest (23.507 ± 1.656 mg GAE $g^{-1}$) at 2 kV $cm^{-1}$ for 1 min |

(continued)

**Table 5.2** (continued)

| Food or cultivar/Reference | MF or EF treatment | Result |
|---|---|---|
| Medicinal herb *Dracocephlum polychaetum* Bornm/Taghizadeh et al. (2019) | MF and $Fe_3O_4$ magnetic nanoparticles (MNP). SMF1: (3 days, 3 h $day^{-1}$) and SMF2: (4 days, 5 h $day^{-1}$) both of 30 mT | (↑) TPC. All treatments higher than control. The highest at SMF2 + MNP with 5.9 mg $g^{-1}$ FW (1.73 fold increase compared with control)<br>(↑) TFC. All treatments higher than control. The highest at treatments with MNP with 405.65 µg $g^{-1}$ FW<br>(↑) TAC. Significantly higher by either SMF + MNP or SMF2 as compared to the control samples. The highest at SMF2 + MNP with 37.41 nmol $g^{-1}$ FW corresponding to 1.95 times higher than control |
| Almonds seeds from two plant species (*Amygdalus scoparia* Spach and *A. eburnea* Spach)/ Abdollahi et al. (2019) | MF. 10 mT for 5 h per day, for 4 days | (↑) TPC compared to control group. (About 175 and 240 mg $g^{-1}$ DW for *A. scoparia* and A. *eburnea* respectively)<br>(↑) TAC remarkably compared to control group. (About 0.015 and 0.009 mmol $g^{-1}$ FW for *A. scoparia* and A. *eburnea* respectively)<br>(↑) AC compared to control group. (About 45% and 40% for *A. scoparia* and A. *eburnea* respectively)<br>Note: According to the DPPH test, SMF treated seeds were able to convert the stable radical DPPH into yellow diphenyl picryl hydrazine more readily than those in the control group |
| Grapes. Two cultivars 'Rasha' and 'Sultana'/ (Zareei et al. (2019) | Magnetic Solutions (MS) Two methods for the preparation of MS<br>(a) Prepared Hoagland solution was passed through 0.1 T (T2) and 0.2 T (T3) | (↑) TPC. All higher than control. The highest at T2 and T5, up to 53.25 and 52.1 mg $g^{-1}$ FW, respectively, in Rasha. The highest at T4 with |

(continued)

**Table 5.2** (continued)

| Food or cultivar/Reference | MF or EF treatment | Result |
|---|---|---|
| | (b) Distilled water was first exposed to 0.1 T (T4) and 0.2 T (T5) and then the salts were added to the magnetized water. Magnetized water or nutrient solution was provided by passing them through magnets installed on the pipes at the flow rate of 3 L $min^{-1}$ | 25.57 mg g − 1 FW in Sultana<br>(↑) TAC. The highest at T2 with 0.76 mg $g^{-1}$ FW among all treatments with Rasha. The highest at T2 with 0.58 mg $g^{-1}$ FW among all treatments with Sultana<br>(↑) TFC. All higher than control in Rasha. The highest at T4 and T2, 52.38 and 50.97 mg $g^{-1}$ FW respectively. No statistic difference with Sultana<br>(↓) AC. T4 and T5 on the Sultana grapevines presented 73.63% and 74.83%, respectively, values lower than control<br>(↑) Trans-resveratrol. Based on HPLC–DAD results, the peak area of trans-resveratrol increased in Rasha and Sultana grapevines subjected by T3 and T4, 45.03 and 44.35 ± 0.5 μg $g^{-1}$ FW, respectively, |
| Tea plant (*Camellia sinensis* L.)/Azizi et al. (2019) | MF. Eight different treatments:<br>(a) 1 mT during 30 min<br>(b) 1 mT during 60 min<br>(c) 2 mT during 30 min<br>(d) 2 mT during 60 min<br>(e) 4 mT during 30 min<br>(f) 4 mT during 60 min<br>(g) 6 mT during 30 min<br>(h) 6 mT during 60 min<br>All for 7 continuous days | (↑) TPC. Higher than control with (b), (c), (d), and (e). The highest with (b), about 0.4 gGae $g^{-1}$<br>(↓) TPC. Lower than control with (f), (g), and (h)<br>(↑) TFC. Significantly higher than control with (b) and (d), about 225 and 140 mg $g^{-1}$ of QE respectively<br>(↓) TFC. Lower than control with (h)<br>Note. This study exemplify a hormetic condition, where low doses have a beneficial effect on the production of secondary metabolites and high doses have a negative effect |

(continued)

**Table 5.2** (continued)

| Food or cultivar/Reference | MF or EF treatment | Result |
|---|---|---|
| Seedlings of moringa (Two species *Moringa oleifera* and *Moringa peregrina*)/ Hasan et al. (2018) | MW (Tap water after magnetization through passing in a magnetron of 30 mT, output 4–6 $m^3$ $h^{-1}$) under medium (MS) and severe (SS) drought stress | (↓) TPC lower in both species. *M. olifera* and *M. peregrine* seedlings exposed to MS, SS increased TPC by 20, 30 and 13, 29% under tap water, however, decreased by 11%, 15% and 16%, 21% with MW<br>(↓) TFC and AC. Similar pattern than TPC |
| Lens (*Lens culinaris* L.) plants and seeds/(Azimi et al. (2018) | MW. Tap water was treated by the magnetic field of 110 mT | TPC, TAC, and proline in the leaf of both control and MW-treated plants were identical<br>(↑) AC. The Ferric reducing antioxidant power in plant and seeds leaves was significantly increased by MW<br>(↑) TCC. Noticeably increased in leaf by MW<br>No significant difference was observed between TFC and TAC of seeds of both groups<br>( ↓) TPC and proline. Significantly lower than control in seeds |
| Blackberries. Wild (*Rubus sulcatus* Vest) and cultivated (*Rubus fruticosus* Thornfree)/Răcuciu and Oancea (2018) | MF. Homogenous 50 Hz MF with 3 mT magnetic flux density at 1, 2, 4, 6, and 12 h exposure times | (↑) TAC. 9–33% in wild blackberries and 30–129% in cultivated ones depending of irradiation time. The highest with 1 h, both in wild and cultivated. Long MF exposure (12 h) caused a decrease in wild blackberries by 20% compared to control<br>(↑) TPC. 5–27% in wild blackberries and 6.5–63% in cultivated ones depending of irradiation time. The highest of both was recorded after 1 h<br>Enhanced levels of TAC and TPC of blackberries were obtained for samples exposed to homogeneous magnetic field (50 Hz, 3 mT) for relatively short times (1–6 h) compared to control |

Used abbreviations *MW* Magnetized water, *MF* Magnetic field, *PEF* Pulsed electric field, *AC* Antioxidant capacity, *TCC* Total carotenoid content, *TPC* Total phenolic content, *TFC* Total flavonoid content, *TAC* Total anthocyanin content, *DW* dry weight, *FW* fresh weight, (↑) increment of, (↓) decrement of proteins or to transcription factors (Mishra et al. 2016). In that way, it is strongly suggested that AE can influence the synthesis of secondary metabolites. The most common acoustic emission utilized for the stimulation of bioactive compounds is ultrasound (US). Several mechanisms of how the US interacts within the cell have been proposed. When the US is applied, cavitation bubbles creates a pressure zone change that occurs and increases up to 400 km $h^{-1}$, causing higher porosity, rupture or removal of cell membranes, facilitating the mass transfer from the cells' interior when imploding (Toma et al. 2001; Vinatoru 2001). In that way, the increase of natural bioactive compounds may be caused by better extractability due to rupture of membranes of cell organelles, however, when a decrease is presented, it could be explained by the creation of reactive forms of oxygen (ROS) during cavitation, and that the collapsing bubbles release high doses of energy, raising the temperature (>5000 K) enough to decompose polyphenols (Witrowa-Rajchert et al. 2014; Kentish and Ashokkumar 2011). The second possible explanation is an enhancement of enzymes activity when the US is applied by contact, leading to phenolic compounds' reduction, more significantly after longer treatment time (Wiktor et al. 2016). Ampofo and Ngadi (2020) established that elicitation of common beans with the US, increased the accumulation of stress markers from the onset until the process was arrested, signifying a demand for sprout protection, resulting in an elevated stimulation of defense phenolic triggering enzymes (PAL and TAL), and final biosynthesis of phenolic compounds. In that way, application in food or plants of US treatment should be studied to find the optimal time, frequency, and intensity in order to optimize the production of secondary metabolites. Table 5.3 summarizes some examples of the application of acoustic emissions on plants or foods with a commercial interest and presents the effect on the production of natural bioactive compounds.

### 5.2.3 Nanoparticles

Nanoparticles (NPs) vary in size from 1 to 100 nm and have physicochemical properties, due to their dimensions, which generate a high added value for the nanotechnology industry (Yokel and MacPhail 2011). Nanoscale materials can be found on medical imaging, drug delivery, personal care products, cosmetics, clothing, electronics, agrochemicals, motor vehicles, among other products and applications (Vance et al. 2015; Yokel and MacPhail 2011). The metal-based NPs most commonly studied and found in industrial products are Cd (cadmium) in various complexes, GaAs (gallium arsenide), Au (gold), Ni (nickel), Pt (platinum), Ag (silver), $Al_2O_3$ (aluminum oxide or alumina), $CeO_2$ (cerium dioxide or ceria), $SiO_2$ (Silicon dioxide or silica), $TiO_2$ (titanium dioxide or titania), ZnO (zinc oxide), CuO (copper oxide), and $Fe_3O_4$/$Fe_2O_3$ (iron oxides) (Khot et al. 2012; Yokel and MacPhail 2011). Among the carbon-based nanomaterials often studied are fullerene, single-walled carbon nanotubes (SWCNTs), and multiwalled carbon nanotubes (MWCNTs) (Balbus et al. 2007).

**Table 5.3** Effect on natural bioactive compounds of experiments where the magnetic or electric field was the stress factor on plants or food with a commercial interest

| Food or cultivar/ Reference | AE stimuli | Result |
|---|---|---|
| Romaine lettuce (*Lactuca sativa*, var. *longifolia*)/Yu et al. (2016) | US of 25 kHz and 2 kW nominal power. The acoustic energy delivered 69.4, 138.8, and 208.3 kJ for treatments at 1, 2, and 3 min, respectively | (↑) TPC. After 60 h storage, sample of 1-min had a 22.50% higher TPC than control<br>(↓) TPC. After storage for 30 h, samples treated for 1 min had significantly lower TPC than the control<br>(↑) AC. After 60 h storage, the DPPH inhibition for 1, 2, and 3 min was 97.84, 75.22, and 75.87%, significantly higher than the control, respectively. After 90 h storage, only the AC of samples sonicated for 2 min remained significantly higher than the control<br>Note: The AC with 1 and 2 min US decreased by 50.87 and 64.24% compared with the control respectively during the first 30 h storage, followed by a significant increase 97.07 and 83.67% respectively during the next 30 h |
| Black cumin (*Nigella Sativa*)/Moghimi et al. (2018) | US pretreatment. 30, 60, and 90 W with the constant frequency equal to 25 kHz, and irradiation time of 30, 45, and 60 min | (↑) TPC. With enhancements in US power from 30 to 90 W, TPC increased from 93.21 to 106.6 ppm. 1 h after the extraction process, TPC was 5% higher compared to control<br>(↑) AC. Increment in the AC of the extracted infusions |
| Lavender (*Lavandula Stoechas* L) from Adekar and Keddara regions/ Lilia et al. (2018) | US (pulse system 270, Italy, 26 kHz, 150 W) for different times: 10, 20, 30, 45 and 60 min were applied on the plant materials as a pretreatment before hydrodistillation | (↑) AC. Essential oils obtained by US-HD had a higher AC than those obtained from untreated samples. "Keddara and Adekar of treated and untreated samples revealed a percent of inhibition of DPPH of 20.22 ± 1.72% and 18.88 ± 2.08%, 23.96 ± 3.08% and 20.70 ± 4.41% respectively"<br>Note: Antioxidant activities of the essential oils from aromatic plants are mainly attributed to the active compounds present in them |

(continued)

**Table 5.3** (continued)

| Food or cultivar/ Reference | AE stimuli | Result |
|---|---|---|
| Commercially mature tomato (*Lycopersicon esculentum*) fruit/Lu et al. (2020) | High intensity US (25 kHz frequency with a nominal power of 1 kW) for 1, 2, 3, and 4 min. Storage time of 2, 24, and 48 h | (↑) TPC. Higher than control at all storage times. 2 h storage: maximum at 2 min, and decrease with longer US time. 24 h storage: the highest at 1 min US (17.05% higher than control). 48 h storage: 2 min US presented an increase in TPC<br>(↑) LC. 2 h storage: significantly higher at 3 min US. 24 h storage: the highest at 2 min US (12.21% higher than the control fruit). 48 h storage: the highest at 2 min US<br>(↑) TCC. 24 h storage: an increase in all the US treatments (significantly higher than the control at 2 min). 48 h storage: the highest at 1 min US (17.13% higher compared to the control)<br>(↑) AC. All the US treatments had a higher DPPH than the control in all storage times, except the samples treated with US for 3 min |
| Common bean (*Phaseolus vulgaris*) sprouts/Ampofo and Ngadi (2020) | US at power levels of 360 and 180 W (40 kHz) and time levels of 30, 45 and 60 min | (↑) Total Phenolic Acids. The greatest (216.7 mg 100 $g^{-1}$) with 360 W (60 min) at 96 h of sprouting, 11.65 folds compared to control<br>(↑) TFC. The greatest (203.5 mg 100 $g^{-1}$) with 360 W (60 min) at 96 h of sprouting, 6.6 folds compared to control<br>(↑) TAC. Significantly higher (30.35 mg 100 $g^{-1}$) at 96 h of sprouting with 360 W (60 min), 11.54 folds compared to control<br>(↑) AC. Significantly higher (97.81% with DPPH and 98.34% with ABTS) at 96 h of sprouting with 360 W (60 min), 13.84% and 25.57% folds compared to control, respectively, |

(continued)

**Table 5.3** (continued)

| Food or cultivar/ Reference | AE stimuli | Result |
|---|---|---|
| Apple (Malus domestica var. Ligol) tissue/Wiktor et al. (2016) | US at 21 kHz or 40 kHz and 180 W. Sonication lasted for 0, 5, 10, 20 and 30 min | (↑) TPC. 543.4 mg 100 $g^{-1}$ DW at 21 kHz - 30 min and 1046.5 mg/100 $g^{-1}$ DW at 40 kHz and 5 min, 27.4% and 145.3% higher than control, respectively, (↓) TPC. Only at contact sonication method for 30 min (c_US_30), for which the lowest amount of TPC (298.5 mg/100 g DM) were determined. 30% less than raw apples |
| Black currant fruits (*Ribes nigrum* L.)/ (Oancea et al. (2014) | US of 150 W power and 40 kHz frequency. Three predetermined extraction times (3, 6 and 10 min) and three ultrasonic amplitudes (10, 40 and 70%). Sample preparation: frozen, freeze-dried, and oven air-dried | (↑) TPC. Increased by approximately 4% compared to control at 10 min and 70% amplitude in all three cases of sample preparation (↑) TAC. Frozen: increase by 4% at 3 min at 70% amplitude. Freeze-dried: increase by 20% at 10 min and 70% amplitude. Oven air-dried: increase by 7% at 6 min and 70% amplitude. All compared to control (↑) AC – FRAP method. Frozen: increase by 1% at 6 min and 40% amplitude. Freeze-dried: increase by 28% at 10 min and 70% amplitude. Oven air-dried: increase by 144% at 3 min and 70% amplitude. All compared to control |

Used abbreviations *US* Ultrasound, LC Lycopene content, *AC* Antioxidant capacity, *TCC* Total carotenoid content, *TPC* Total phenolic content, *TFC* Total flavonoid content, *TAC* Total anthocyanin content, *DW* dry weight, *FW* fresh weight, (↑) increment of, (↓) decrement of

In contrast to its benefits, the release of nanomaterial-containing wastes has become a threat, since they cause pollution to air, water, and soil (Oberdörster et al. 2005). Their size, equivalent to that of cellular components, allows them to easily permeate cells, causing adverse biological effects (plants and animals) (Shang et al. 2014; Vecchio et al. 2012). In plant cells, NPs can enter from the apoplast and cross the plasma membrane towards cytosol or other organelles via endocytosis, specific membrane-bound transporter proteins or through induction of new pores by using ion carrier substances; subsequently, they can be transported between cells through symplastic flow (Anjum et al. 2019; Marslin et al. 2017). Despite its potential of toxicity, studies have reported positive effects on plant development and physiology, which are dependent on the nature of the nanomaterial, dose and time of exposure, the plant species, and growth conditions (Cox et al. 2016).

Positive physiological effects using carbon-based NPs include increased water uptake, and enhanced assimilation of $CO_2$ in broccoli (Martínez-Ballesta et al. 2016), promotion of seed germination and root growth in rice (Jiang et al. 2014), enhanced germination and seedling growth in sweet corn, barley, rice soybean, switchgrass, and tomato (Lahiani et al. 2015; Tiwari et al. 2014), increase fruit yield in tomato and bitter melon (Khodakovskaya et al. 2013; Kole et al. 2013), among many others. Studies using metallic nanoparticles have reported similar effects. In wheat, $CeO_2$ particles improved plant growth, shoot biomass, and grain yield (Rico et al. 2014). Au nanoparticles in chinese mustard (Brassica juncea) had positive effects on growth parameters and seed yield (Arora et al. 2012). The response of maize exposed to ZnO nanoparticles showed enhanced germination, seedling vigor, and zinc biofortification of grains (Subbaiah et al. 2016).

The effect of NPs on secondary plant metabolism is still largely unknown compared to physiological and phenotypic responses. However, studies have shown that a constant response between species is the induction of reactive oxygen species (ROS) (Marslin et al. 2017). Studies reporting NPs-elicitation of specialized metabolites often report a reduction in the photosynthetic rate and inhibition of growth. These phytotoxic effects have been linked to the inhibition of Ribulose-1,5-bisphosphate carboxylase/oxygenase (Rubisco) activity and decreased photo-protective capacity of PSII (Jiang et al. 2017a; Wang et al. 2016). When NPs permeate the cells, damage on the photosynthetic apparatus is done because of its accumulation in chloroplasts, at the same time, when they cross the plasma membrane they probably dissociate into ions (rather than stay as intact particles) and bound to NADPH oxidases, causing the production of ROS at the apoplast (Jiang et al. 2017a; Sosan et al. 2016). Besides oxidative burst, it has been reported that NPs also induce reactive nitrogen species (*NO, nitric oxide) (Marslin et al. 2017).

Initial responses also include calcium ion ($Ca^{2+}$) spikes, $Ca^{2+}$ flux movements, and upregulation/phosphorylation of mitogen-activated protein kinase (MAPK) cascades that together with ROS production, ultimately lead to the activation of the pathways of specialized metabolites biosynthesis (Anjum et al. 2019; Marslin et al. 2017). As expected, plants exposed to stressful concentrations of NPs have shown to cope with the oxidative stress and lipid peroxidation through the upregulation of enzymatic antioxidants such as superoxide dismutase (SOD), ascorbate peroxidase (APX), glutathione-S-transferase (GST), and catalase (CAT) (Dimkpa et al. 2012; Fu et al. 2014; Mirzajani et al. 2014; Zhao et al. 2012).

Through these findings, the concept of "nano-elicitors" have recently emerged as a novel alternative to stimulate the production of valuable bioactive compounds that might be used as additives in food, cosmetics, and pharmaceutical products.

The most widely used nano-elicitors with this purpose are carbon nanotubes, silver, gold, copper, zinc oxide, and titanium dioxide (Anjum et al. 2019). The way to supply NPs to plants can be carried out through foliar spray, directly in the soil or through the nutrient solution applied. Table 5.4 summarizes some published studies in recent years reporting the enhanced production of commercially important specialized metabolites using metallic-, metal oxide-, and carbon related-NPs.

**Table 5.4** Summary of the effects of different types of nanoparticles used as elicitors of secondary metabolites in different plant species

| Plant species/reference | Treatment | Result |
|---|---|---|
| Feverfew (*Tanacetum parthenium* L.)/Shahhoseini et al. (2020) | ZnONPs (1000 ppm) sprayed during seedling stage growth in soil | (↑) Essential oil (0.9% V/W) with anti-cancer compounds compared to control (0.56% V/W) |
| Deadly nightshade (*Atropa belladonna* L.)/Tian et al. (2018) | $Mn_2O_3$ (25 mg $L^{-1}$) applied to shoot tip in MS growth media | (↑) Alkaloids (23%)<br>(↑) TPC (12%)<br>(↑) TFC (32%) |
| Aloe vera (*Aloe vera* L.)/Raei et al. (2014) | $TiO_2$ (120 mg $L^{-1}$) in cell suspension using MS media | (↑) Aloin (118%) |
| | Ag (0.625 mg $L^{-1}$) in cell suspension using MS media | (↑) Aloin (127%) |
| Selfheal (*Prunella vulgaris* L.)/Fazal et al. (2019) | $Ag^+$ Au (1:3) supplemented to cell suspension culture in MS media with NAA | (↑) 1.8-Fold in TPC and TFC |
| Cucumber (*Cucumis sativus*)/Zhao et al. (2016) | Cu (20 mg $L^{-1}$) in hydroponic culture at early development stages | (↑) TPC (2.35 mg $g^{-1}$ DW) |
| Salvia (*Salvia verticillata* L.)/Rahmani et al. (2020) | Multi-walled carbon nanotubes (MWCNTs – 50 and 1000 mg $L^{-1}$) foliar sprayed to 2-month-old plants growth in soil and greenhouse conditions | (↑) Rosmarinic acid, nearly four times relative to the control |
| Bitter melon (*Momordica charantia*)/Kole et al. (2013) | Fullerene (10.8 mM) during seed germination using ·B potting mix under greenhouse conditions | (↑) Anticancerous (cucurbitacin-B, 74% and lycopene, 82%) and antidiabetic (charantin, 20% and insulin, 90%) compounds |

Used abbreviations *TPC* Total phenolic content, *TFC* Total flavonoid content, (↑) increment of

### 5.2.4 Metals and Salt Metals

Metals, at high concentrations, act as stress agents to plants, therefore, they can induce changes in the secondary metabolism causing an elicitation effect. Exposure of plants to metals, such as Ni, Ag, Fe, and Co, has shown increased production of secondary metabolites in a variety of plants (Zhao et al. 2001). For instance, cadmium ($Cd^{2+}$) and copper ($Cu^{2+}$) are known for their toxicity and for not having any value for plants (Das et al. 1997). However, Cd and Cu treatments resulted in enhanced phenolic accumulation on the medicinal plant *Gynura procumbens* (Ibrahim et al. 2017). Several factors influence the response of plants to metal exposure, mainly depending on the chemical metal species and concentration, the plant species, climate conditions, growth stage, among others (Lajayer et al. 2017).

The use of nonfood crops with the capacity of absorbing and accumulating heavy metals is an alternative for remediation of contaminated environments. It has been shown in certain medicinal and aromatic plants that this practice can lead to the accumulation of secondary metabolites, which can be phytoextracted to obtain high-value compounds (Lajayer et al. 2017). Metabolic changes by the action of heavy metals can lead to inhibition of enzymes involved in the production of photosynthetic pigments, sugars, proteins, and nonprotein thiols (Naik and Al-Khayri 2016; Nasim and Dhir 2010). To date, many studies have shown increases in medicinal plant performance following exposure to heavy metal stress. For example, in a study where garden mint (*Mentha crispa* L., Lamiaceae) was used for phytoaccumulation of lead (Pb), the chemical composition of the essential oil of the plant was affected by improving the production of carvone, a major component of essential oils (Sá et al. 2015).

Heavy metals have also shown to have a role in stress amelioration through changes in antioxidant balance which often comes hand in hand with increased secondary metabolites. A study subjecting *Camellia sinensis* (L) plants to drought stress was performed to understand the role of Zn in modulating stress conditions. Results showed decreases in hydrogen peroxide ($H_2O_2$) and lipid peroxidation, and at the same time increases in phenolics content and differential expression of antioxidant enzymes, such as superoxide dismutase (SOD), catalase (CAT), peroxidase (POX), polyphenol peroxidase (PPO), glutathione reductase (GR), and ascorbate peroxidase (APX) (Upadhyaya et al. 2013). Similar results were obtained in *Brassica napus* exposed to cadmium (Cd) stress, exogenous application of low concentrations of selenium (Se) increased the tolerance of plants meanwhile concentrations of ascorbic acid and reduced glutathione were increased (Hasanuzzaman et al. 2012).

Metallic salts have also shown enhanced production of secondary metabolites during in vitro root cultures treatments such as two tropane alkaloids, scopolamine and hyoscyamine, by eliciting with silver nitrate ($AgNO_3$) and cadmium chloride ($CdCl_2$) in *Brugmansia candida* (Angelova et al. 2006), increases in tanshinone contents using $AgNO_3$ in *Perovskia abrotanoides* (Zaker et al. 2015) and sesquiterpenoid–defensive compounds using cadmium salts in *Datura stramonium* (Furze et al. 1991).

### 5.2.5 Volatile Organic Compounds

Volatile organic compounds (VOCs) are low-molecular weight compounds that are emitted in the atmosphere in vapors or gaseous form. VOCs are produced as secondary metabolites by micro- (bacteria and fungi) and macro-organisms (animals and plants) and play vital roles, such as regulation of physiological processes and inter-organismal communication (Fincheira and Quiroz 2018; Rakshit et al. 2020). Plants can emit VOCs constitutively to attract pollinators and seed dispersers, or in response to a stimulus as a defense against insects or predators, plant-to-plant communication, thermo-tolerance, and environmental stress adaptation (Vivaldo et al. 2017). Plant VOCs can be classified into terpenoids, fatty acid derivatives, phenylpropanoids/benzenoids, and amino acid derivatives (Dudareva et al. 2013). Under-ground VOC's are emitted by plants through their roots (Rakshit et al. 2020), where they also interact with bacteria and fungi in the rhizosphere zone giving rise to a deep symbiotic plant-microorganisms relation (Dessaux et al. 2016). Microorganisms benefit from root's exudates while they produce nonvolatile metabolites that affect the plant's nutrient assimilation and benefits plant growth (Dotaniya and Meena 2015).

A new plant–microbe interaction involving microbial volatile organic compounds (mVOCs) was discovered by (Ryu et al. 2003). In their study, they identified that volatile compounds from *Bacillus subtilis* act as strong promoters of growth in Arabidopsis thaliana. Since then, studies focusing on mVOCs as potential compounds with practical applications on regulating characteristics of agronomic importance have emerged. Bacterial and fungal volatile compounds may activate defense responses against biotic and abiotic stress, induce systemic resistance, promote growth, and enhance health processes in plants (Kanchiswamy et al. 2015; Piechulla and Degenhardt 2014). There are approximately 1000 mVOCs produced by bacteria and fungi reported in the literature, a few examples include 3-hydroxy-2-butanone (acetoin), 2,3-butanediol, 2-pentylfuran, or dimethylhexadecylmine (Fincheira and Quiroz 2018; Piechulla and Degenhardt 2014).

The idea of using VOCs to elicit secondary metabolites in plants is novel and still very little studied. There are cases of success in the literature to this purpose, which are summarized in Table 5.5, most of them have shown that different volatile compounds from bacterial can increase commercially valued components of essential oils, such as monoterpenes, pulegone, menthone, menthol, limonene, menthyl acetate, terpineol, and eugenol (Banchio et al. 2009; Santoro et al. 2011, 2016; Zhou et al. 2016). These studies are often performed in sterile plastic boxes or petri dishes with divided into two compartments by a physical barrier so that microbial and plant cultures interact without physical contact.

**Table 5.5** Summary of the effects of different sources of VOCs used as elicitors of secondary metabolites in different plant species

| Plant species/ Reference | Treatment | Result |
|---|---|---|
| *Arabidopsis thaliana*/ Sánchez-López et al. (2016) | VCs emitted by the phytopathogen *Alternaria alternata* cultured in petri dishes in MS medium with 14-day-old plants | (↑) Total chlorophyll and TCC |
| Peppermint (*Mentha piperita*)/ Santoro et al. (2011) | VOCs emitted by plant growth-promoting rhizobacteria (*Pseudomonas fluorescens and Azospirillum brasilense*) grown on the same Petri dish with a center partition | (↑) Essential oils: monoterpenes (twofold), pulegone (3.14-fold) and menthone (15.4-fold) in *P. fluorescens*-treated plants (↑) Menthone (13.5-fold) in *A. brasilense*-treated plants |
| Peppermint (*Mentha piperita*)/ Santoro et al. (2016) | VCs from three native rhizospheric bacterial strains (SJ04, SJ25, SJ48) suspended in Hoagland solution positioned on the same partitioned Petri dish with plant's young shoot in MS solid medium | (↑) Total essential oil production (↑) Limonene, menthol and Menthyl acetate |
| *Atractylodes lancea*/Zhou et al. (2016) | Nitrogenous volatiles (formamide and N,N-dimethyl-formamide) and benzaldehyde volatile emitted by *Pseudomonas fluorescens* ALEB7B. Bacterial suspension cultured on MS agar in petri dish was placed beside tissue culture plantlets in MS rooting agar with NAA | (↑) Volatile oils accumulation (1.8-fold) |
| Sweet Basil (*Bacillus subtilis*)/ Banchio et al. (2009) | VCs emitted by soil benefic bacterium *Bacillus subtilis* GB03 by positioning plants and bacteria in separate regions of a partitioned petri dish with MS media in both sides | (↑) Essential oil components content: terpineol (twofold) and eugenol (tenfold) |

Used abbreviations *TCC* Total carotenoid content, (↑) increment of

### 5.2.6 Nutrient Deficiency

The soil provides water and nutrients to plants. Fourteen mineral nutrients are required for plant correct growth and development which are divided into macronutrients (N, P, K, Ca, Mg, and S) and micronutrients (Cl, Fe, B, Mn, Zn Cu, Ni, and Mo). Macronutrients form structural and energy compounds in plants. On the other hand, microelements are related to enzymatic responses. For example, Zn, Fe, Mn, and Cu are components of enzymatic antioxidants, which regulate oxidation processes in the plant (Hajiboland 2012; Nath and Tuteja 2016). Due to the indispensable role of nutrients, plant roots have developed an efficient sensing and signaling system to maintain nutrient homeostasis. Low availability of nutrients in the soil is detected by roots, and in response, chemical signaling and chain reactions are produced. Plants employ signaling players as phytohormones, reactive oxygen species (ROS), sugars, and transcription factors to maintain nutrients homeostasis within the plant (Nath and Tuteja 2016; Isah 2019).

Nutrient deficiency can produce metabolic responses that cause an increased accumulation of secondary metabolites. Natural bioactive compounds are sought in bioproduction processes and improvement of nutritional quality of vegetables and fruits, therefore a nutrient deficiency of specific macro or micronutrients may be an alternative. However, this stress can cause a decrease in crop yields (Hawkesford et al. 2012). To produce secondary metabolites of interest without a considerable loss of growth and biomass, it is necessary to generate eustress in the plant (El-Nakhel et al. 2019). Currently, nutritional eustress is a strategy used in protected production systems, where soilless crops allow greater control of nutrient supply through nutrient solutions.

Nitrogen is a macronutrient constituent of primary metabolites (e.g., protein, peptides, amino acids, and nucleic acids), phytohormones and secondary metabolites. Plants can uptake nitrogen as nitrate and ammonium (mineral form) (Isah 2019). Research has shown an inverse relationship between low nitrogen availability and the synthesis of phenolic compounds. According to this hypothesis, low nitrogen availability increases synthesis of metabolites that contain C, H, and O in their structure. Therefore, terpenes and phenolic compounds synthesis will be favored. On the contrary, metabolites that contain N in its structure such as, alkaloids, nonprotein amino acids, and cyanogenic compounds, will decrease its synthesis (Nath and Tuteja 2016). For example, growing lettuce (Lactuca sativa) increases its content of phenolic compounds and antioxidant capacity in the presence of nitrogen deficiency and drought (Galieni et al. 2015). Table 5.6 shows more examples of the effect on phenolic compounds of experiments where nitrogen deficiency was the stress factor on plants or food with commercial interest.

**Table 5.6** Effect on phenolic compounds of experiments where nitrogen deficiency was the stress factor on plants or food with commercial interest

| Plant/Reference | Treatment or culture conditions | Phenolic compounds |
|---|---|---|
| *Matricaria chamomilla*/Kováčik and Klejdus (2014) | N deficiency and N source ($NH_{4+}$ and $NO_{3-}$) in growth chamber | Phenolic acids |
| *Vitis vinifera* 'Cabernet Suavignon'/Gutiérrez-Gamboa et al. (2017) | Nitrogen application in field (foliar application) | Wine flavonoids |
| *Castilleja tenuiflora*/Medina-Pérez et al. (2015) | N deficiency (1.23 mM $KNO_3$ and 0.09 mM $(NH_2)_2SO4$) in vitro | Phenylethanoid glycosides |
| *Lactuca sativa* cv crispa and cv Satine/Becker et al. (2015) | N deficiency (0.75 and 3 mM under greenhouse | Flavonoids and caffeic acid derivatives |

Phosphorus (P) is a component of molecules such as nucleic acids, lipids and nucleotides with an energy function (ATP and ADP). Plants uptake P in the form of inorganic orthophosphate (Pi, $HPO_4^{2-}$, and $H_2PO_{4-}$) which deficiency promotes anthocyanin synthesis (Jezek et al. 2016). Peng et al. (2019) proposed a model of com-modulation (miR399d) and epigenetic modification as a regulatory mechanism of anthocyanin synthesis that depends on the P availability. Sulfur is a structural component of amino acid precursors of secondary metabolites. Therefore, its deficiency negatively affects the biosynthesis of lycopenes and carotenoids (Mohammed et al. 2015). Micronutrient deficiencies are shown to have a negative impact on the synthesis of phenolic compounds and terpenes. Micronutrients such as Cu, Fe, Mo, and Mn, act as factors for the synthesis of secondary metabolites.

## 5.3 Biotic Stress

Biological stressors are those considered within living organisms (plants or pathogens) including bacteria, insect or herbivores, fungi, phytohormones, and miRNA, among others, that results in biotic stress (Patel and Krishnamurthy 2013). The action mechanism of this factor includes activation or inactivation of enzymes, interaction with receptors, ion channels, stimulation of bioactive compounds, and so forth (Joshi et al. 2019). Some biotic stress factors and their role in the synthesis of secondary metabolites in plants are described below.

### *5.3.1 Bacteria and Viruses*

Plants are exposed to interactions with other living things. The interaction between microorganisms and plants can have positive effects. Microorganism colonization (pathogens and non-pathogens) triggers the resistance mechanism of the plant, conferring resistance against other stressors (Nejat and Mantri 2017; Choudhary et al. 2016). Nonpathogenic microorganisms act as plant biostimulants. A plant biostimulant is defined as any substance or microorganism applied to plants in order to improve nutritional efficiency, tolerance to biotic and abiotic stress, and quality (Van Oosten et al. 2017; Du Jardin 2015). Arbuscular mycorrhizal fungi, Trichoderma, and plant growth-promoting rhizobacteria are biostimulant microorganisms used in crops.

Microorganisms can confer a certain degree of tolerance against abiotic stress conditions. Colonized plants produce a wide range of enzymes and metabolites that allow them to generate tolerance to stress (Miliute et al. 2015). Some genera of bacteria like *Rhizobium*, *Bacillus*, *Pseudomonas*, *Pantoea*, *Paenibacillus*, *Burkholderia*, A*chromobacter*, *Azospirillum*, *Microbacterium*, *Methylobacterium*, *variovorax*, *Enterobacter*, have been shown to induce tolerance to abiotic stress (Choudhary et al. 2016; Naveed et al. 2014; Gururani et al. 2013). Tolerance generated by pathogen attack can induce resistance to abiotic stress factors. The biochemical response generated by the attack of the pathogen is similar to the response generated by abiotic factors. Plants attacked by Verticillium dahliae (pathogenic fungus) develop tolerance to drought due to the formation of xylem but reducing the growth rate (Tani et al. 2018).

Viruses are considered symbiotes. They can behave as pathogens or mutualists depending on the environmental conditions where the host is (Roossinck 2015). Research suggests that the mutualistic behavior of a virus occurs when the titer virus is low and the environmental disturbance is low (Bao and Roossinck 2013). Plant viruses can have a positive effect like other pathogens. The presence of the virus in the plant can increase its ability to cope with biotic and abiotic stress factors because of the activation of the plant defense system. Metabolomic studies in infected plants have shown a significant increase in the quantity and diversity of secondary metabolites. This metabolic effect allows the plant to cope with the stress caused by the infection, as well as other stressors present in the environment. For example, Sade et al. (2015), reported a significant impact on the metabolome in tomato plants infected with Tomato yellow leaf curl virus (TYLCV) where resistant and susceptible cultivars showed a major expression of the phenylpropanoid pathway which is related to the production of antioxidant compounds, among others. In the same research, the expression in resistant cultivars was more significant in terms of the production of flavonoids and other antioxidants. On the other hand, rice plants infected with Brome mosaic virus (BMV) and beet plants (Beta vulgaris) infected with Cucumber mosaic virus (CMV) increased the accumulation of osmoprotectants and antioxidant compounds, conferring drought tolerance to both crops (Xu et al. 2008).

### *5.3.2 Fungi*

Plants have a strong symbiosis relationship with some fungi and bacteria present in the substrate where they are grown. These microorganisms, endophytes or exogenous, can induce eustress to the crop, increasing the production of specialized metabolites, e.g.,*Aspergillus* sp. applied as an elicitor in *Artemisia annua* L. callus culture, enhanced the production of artemisinin, an endoperoxide sesquiterpene lactone and an effective antimalarial agent (Yuliani et al. 2018). Soil-borne beneficial microbes have shown a protecting potential against pathogens and herbivores via the elicitation of plant responses e.g. plant growth-promoting fungi (PGPF) and arbuscular mycorrhizal fungi (AMF) (Pappas et al. 2018). Fungal elicitation (including yeas extract) is one of the most used to enhance the production of secondary metabolites (Singh et al. 2018).

Fungi with a beneficial effect on plant development that associate to plant roots are called PGPF and are considered the first prevention mechanism in the pathogen infection. Plants need to detect PGPFs and take advantage of the presence of microbe-associated molecular patterns (MAMPs) that can be recognized by pattern recognition receptors (PRRs). PGPFs can stimulate the defense system of plants involving the modification of cell walls by the accumulation of lignin, callose, phenols, etc., preventing the growth and proliferation of pathogens. In addition, elicitors such as chitin, chitosan, and β-glucan that are part of the fungal cell wall have been researched (Naziya et al. 2020; Fesel and Zuccaro 2016; Li et al. 2016).

Recent studies have been performed in different plants, e.g., fungal elicitation in Leguminosae increased the accumulation of isoflavonoids and stilbenoids (Araya-Cloutier et al. 2017). Moola and Diana (2019) used *Aspergillus niger*, *Penicillium notatum,* and *Rhizopus oligosporus* as elicitors on hairy root culture of Beta vulgaris to enhance betalain synthesis. *Trichoderma spp.* is one of the most widely used microorganisms as a pathogen biocontrol agent and elicitor. In addition, this fungus has been shown to colonize roots and can also induce systemic resistance (ISR), which favors plant growth, increases nutrient availability and enhances disease resistance. The mechanisms employed by *Trichoderma spp.* are the modulation of plant hormonal mechanisms and the production of secondary metabolites (Nandini et al. 2020; Guzmán-Guzmán et al. 2019; Silva et al. 2019; Martínez-Medina et al. 2017). PGPFs can be used as bio-fertilizers, improving the quality and quantity of products, and reducing the contamination of the agricultural environment by lowering the use of chemical fertilizers (Pereira et al. 2019; Zhou et al. 2018). Table 5.7 shows some of the fungi used to induce the production of secondary metabolites in plants.

**Table 5.7** Plants elicited by fungi application to increase the production of secondary metabolites

| Plant species/Reference | Elicitor | Secondary metabolites |
|---|---|---|
| *Tagetes patula*/Moola and Diana (2019) | *Fusarium conglutinans* | Total thiophenes |
| *Catharanthus roseus*/Moola and Diana (2019) | *Penicillium jasmonate* | Catharanthine (chemical precursor of vinblastine) |
| *Hyoscyamus muticus*/Moola and Diana (2019) | *Rhizoctonia solani* | Phytoalexins, Solavetivone, lubimin |
| *Hyoscyamus muticus*/Moola and Diana (2019) | *Lnonotus obliquus* | Stimulation hyoscyamine |
| *Anoectochilus formosanus*/ Zhou et al. (2018) | *Mycena sp* F-23 | Kinsenoside, flavonoid content |
| *S. miltiorrhiza*/Zhou et al. (2018) | *Alternaria sp.* A-13 | Total phenolic acid, Lithospermic acid A and Lithospermic acid B |
| *Salvia mittiorrhiza*/Halder et al. (2019) | *Trichoderma atroviride* D-16 | Tanshinone |
| *Sinapis alba*/Andini et al. (2019) | *Rhizopus oryzae* | Glucosinolates |

### *5.3.3 Phytohormones*

Phytohormones are molecules synthesized by defined organs that regulate plant growth and have a prominent impact on plant metabolism. Additionally, they play a vital role in the stimulation of response mechanisms of plant defense against stress. Auxins, gibberellins, cytokinins (CK), abscisic acid (ABA), jasmonates (jasmonic acid (JA), methyl jasmonate (MeJA)), salicylic acid (SA), brassinosteroids, strigolactones, cinnamic acid (CA), among others, are examples of phytohormones. Auxins and the auxin índole-3-acetic acid (IAA) act as promoters of growth and developmental events in plants (cell división, elongation, and differentiation). CKs are involved in the maintaining of cellular proliferation, differentiation, and prevention of senescence. ABA has an important role in the plant response to stress and adaptation. Gibberellic acid (GA) is a plant growth regulator and has a vital role in seed dormancy, formation of floral organs, and lateral shoot growth (Egamberdieva et al. 2017). SA has an important role in plant stress tolerance through modulation of antioxidative enzyme activities. JA is a lipid-derived compound synthesized via the octadecanoid pathway and is important in the development, structure, and flowering of plants (Złotek et al. 2020). CA is a bioprecursor of podophyllotoxin and a huge number of plant substances, including tannins, flavonoids, etc. (Kašparová et al. 2018).

Currently, the use of phytohormones in agriculture and research has been implemented as a strategy to induce the production of secondary metabolites (Egamberdieva et al. 2017). The exogenous application of phytohormones by spraying has resulted in increased production of various significant bioactive compounds in plants (Akram et al. 2020, Wang et al. 2018a). For example, Garcia-Ibañez et al. (2019), applied MeJA, SA, and SA + MeJA to Bimi® plants resulting in a differentiated response to each elicitor, all treatments showed an increased content of GLs in leaves and inflorescences. Table 5.8 shows the result of plant elicitation with phytohormones in the production of bioactive compounds of recent studies.

### *5.3.4 miRNA*

MicroRNAs (miRNAs - length of 18 to 28 nucleotides) are noncoding RNAs. The main function of these molecules is to participate in gene expression and regulation at the post-transcriptional level by degrading mRNA or at the translational level by blocking protein biosynthesis at different stages (Fig. 5.1), resulting in regulation of plant development, metabolism, and response to biotic or abiotic stress (Tripathi et al. 2019). In plants miRNAs genes are transcribed by RNA polymerase II producing primary miRNA (Pri-miRNA) which are important for the regulation of genome integrity, primary and secondary metabolism, development, signal transduction, signaling pathways, homeostasis, innate immunity, and environmental stress responses (Vargas-Hernández et al. 2019; Wang et al. 2018b; Samad et al. 2019).

**Table 5.8** Secondary metabolites produced by application of phytohormones to several plant species

| Plant species/Reference | Elicitor | Secondary metabolite |
|---|---|---|
| *Solanum tuberosum*/Egamberdieva et al. (2017) | ABA | Antioxidant enzyme peroxidase |
| *Zea mays*/Egamberdieva et al. (2017) | ABA | Ethylene |
| *Calotropis gigantean*/Halder et al. (2019) | JA | Cardenolide |
| *Stevia rebaudiana*/Vazquez-Hernandez et al. (2019) | SA | Steviol glycosides |
| *Levisticum officinale* Koch cv. Elsbetha/Złotek et al. (2020) | JA | Chlorophylls, vitamin C and phenolic compounds |
| *Artemisia annua*/Qi et al. (2018) | JA | Artemisinin |
| *Mentha canadensis* L./Qi et al. (2018) | JA | Mentol |
| *Ajuga bracteosa* / (Saeed et al. 2017) | MeJA | Phenolic content and flavonoid conten |
| *Bacopa monnieri*/Singh and Dwivedi (2018) | MeJa | Bacoside A |
| *Hyoscyamus niger Vitis vinífera*/Ramirez-Estrada et al. (2016) | MeJa | Scopolamine, hyoscyamine<br>Stilbenes and t-resveratrol |
| *Corylus avellana, T. chinensis, T. bacccata*/ Ramirez-Estrada et al. (2016) | SA | placlitaxel |
| *Brassica oleracea* I. var. Italica/Villarreal-García et al. (2016) | Etilene, MeJA | Glucosinolate<br>Phenolic compounds |
| *Rhazya stricta*/Akhgari et al. (2019) | MeJA | Terpenoid índole alkaloids (vindoline |
| *Juniperus virginiana*/Kašparová et al. (2018) | CA | Podophyllotoxin |

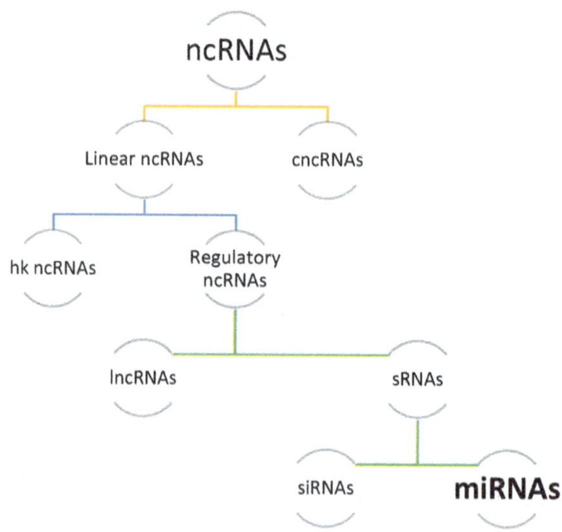


**Fig. 5.1** Clasification of plant noncoding RNAs (ncRNAs). Linear noncoding RNAs (Linear ncRNAs), circular noncoding RNAs (cncRNAs), HouseKeeping noncoding RNAs (hk ncRNAs), Regulatory noncoding RNAs (Regulatory ncRNAs), long noncoding RNAs (lncRNAs), small RNAs (sRNAs), small interfering RNAs (siRNAs), micro RNAs (miRNAs)

The regulation mechanism of gene expression by miRNAs is via RNA interference. These small molecules are transcribed as longer precursors in the nucleus and are then further processed into their mature forms (Wang et al. 2018b; Samad et al. 2017). Most plant miRNA genes are located inside intergenic regions between two adjacent genes and are transcriptionally regulated by their promoters and terminators (Hossain et al. 2019). miRNAs were reported to be involved in plant secondary metabolite regulation such as terpenoid, phenolic, fatty acid, flavonoids, phenolic, and nitrogen-containing compound biosynthesis (Liu et al. 2017; Samad et al. 2019).

There is a very close relationship between miRNAs and the transcription factors (TFs), either to be switched "on" or switched "off". Current research trends have focused on knowing the regulation of miRNA in response to environmental stress and how they interact with transcription factors (Fig. 5.2) (Samad et al. 2019).

The main goal of functional studies on miRNAs has been to understand the biological processes in which the miRNAs are involved. Different technologies

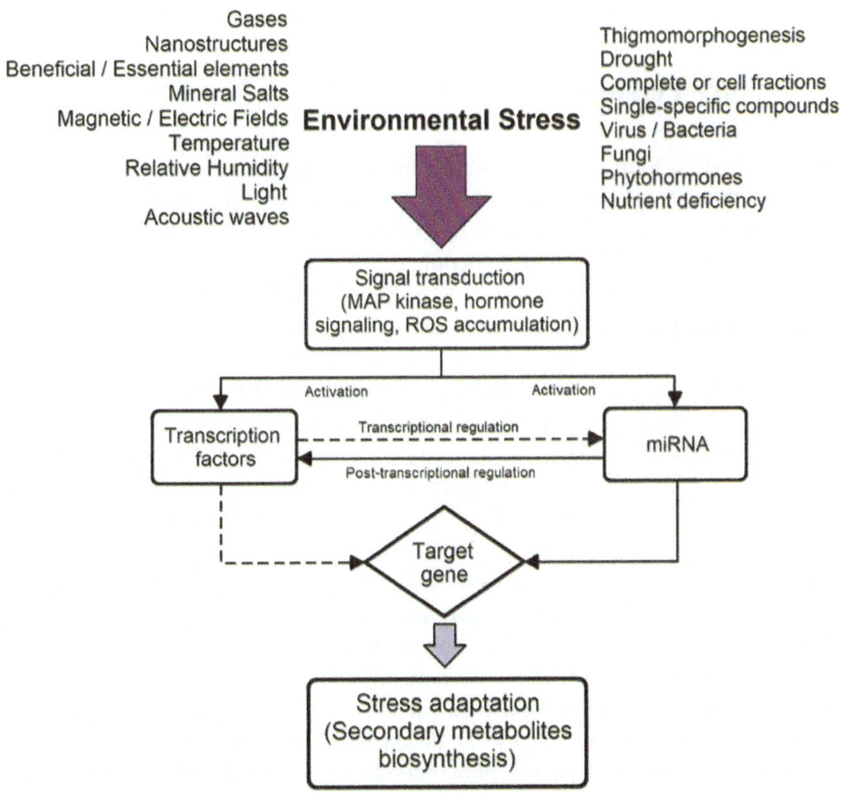


**Fig. 5.2** Stress mechanism and interaction between transcription factors and miRNAs

have been developed to characterize the function and action mechanisms of these small molecules in various plant materials (Liu et al. 2017). Research on the regulation of miRNAs in *Taxus* callus cells has been conducted, concluding that miRNAs are capable of direct regulation of secondary metabolism by modulating transcriptional factors (Chen et al. 2020). The expresión level of miRNAs is mostly governed by temperature and radiation (Tripathi et al. 2019). Studies have demonstrated that miRNAs may act as master regulators of flavonoid biosynthesis, e.g., miR156-SPL9 network directly influences anthocyanin production, miR163 targets S-adenosyl-Met-dependent methyltransferases that methylates secondary metabolites and signaling molecules, and miR397 regulates lignin biosynthesis in Arabidopsis and Populus spp (Sharma et al. 2016).

Studies carried out on different plant species indicate that the cellular levels of miRNA have a high regulation control for optimal spatiotemporal regulation of target genes, adding an additional layer of complexity to the signaling processes (Gupta et al. 2017; Sharma et al. 2016). These results suggest that a strategy to induce the production of secondary metabolites may be the use of miRNAs that promote the expression of genes involved in plant biosynthetic pathways.

## 5.4 Future Perspectives

The use of AE stimuli as a new elicitor in plants has huge sustainable potential. However, this "environmentally friendly" agricultural technology is still under discussion and more studies should be encouraged. Currently, the optimal sound therapy is not known since the effect could differ depending on several factors such as plant model, amplitude, frequency or time and duration of treatment, and application distance among others that are not explained in most studies (Alvarado et al. 2019).

PEF and MF in food, horticulture, and biotechnology asthe postharvest processes have increased substantially during the last few years (Xi-ran and Ting 2017; Gilani et al. 2017; Rusakova et al. 2017). Generally, it can be noted that the application of PEF and MF as pretreatment is a novel method to improve plant development due to a higher production of ROS and an associated activation of antioxidant defense systems, such as POD, SOD, and CAT. However, the effect of electromagnetic sources is plant and treatment specific. It is difficult to have a strong conclusion in the analyzed experiments in this chapter since the experimental details often were incomplete, including an insufficient description of equipment and treatment conditions. It is recommended to consider the description of parameters, such as electric field strength (V/m), frequency (Hz), magnetic flux density (T), electric current (A), PPFD, distance of light source to plant, light decay among the experimental source, photoperiod, light wavelength, the duration of application, and plant or food complete physical description (Dannehl 2018; Alvarado et al. 2019).

Nanoparticles in agriculture have shown benefits related to physiological and growth parameters. Moreover, the new knowledge related to its effect on secondary metabolism has opened a field of possibilities in the production of bioactive compounds with high commercial value using NPs. However, as with many elicitors, there is still a lack of knowledge about the type and size of nanoparticle and appropriate concentrations to use depending on the species of interest. Furthermore, a large part of the existing nanoparticles has not yet been studied and it is also necessary to continue generating knowledge to understand the molecular mechanisms of elicitation with NPs.

Metal ions have been proposed as suitable elicitors of secondary metabolism in cell cultures (Rudrappa et al. 2004), since they can make more efficient the tissue culture techniques during the obtention of valuable secondary metabolites. Many studies have proposed the use of different chemical metal species to enhance bioactive compounds in plants. This tool is very attractive in the sense that, in addition to producing these types of compounds, it can become an environmental remediation technique. Also, the idea of producing specialized metabolites for later extraction eliminates the risks associated with the potential health risks to the consumers.

Among the study of new agricultural tools in the last decade, VOCs stands out for being considered as an eco-friendly, cheap, and effective alternative. Even genetically modified plants with altered VOC emission and synthetic formulations of plant VOCs are been developed as a promising technology for agriculture and horticulture (Rakshit et al. 2020). However, there are still many unknowns, especially about the mode of action of VOCs and about the molecular and biochemical mechanisms related to eliciting responses of interest in plants.

The application of phytohormones in different phenological stages of plants, including in the postharvest stage, induces the production of secondary metabolites and is an effective strategy that uses defense mechanisms to mimic stress caused by various environmental factors. This method can be used at the agronomic level to improve the quality of plant products by increasing the content of bioactive molecules. Defense plant response due to different types of stress depends on the type of crosstalk (positive or negative) between the hormone signaling pathways rather than on the individual contributions of each hormone (Verma et al. 2016). This suggests that for future perspectives it should be taken into account that the results of elicitation will depend on the synergy of the used phytohormones, the concentration, the application conditions, and the type of cultivation. For example, there is a balance between SA and JA to regulate biotic stress in tomato (Verma et al. 2016). In the same way, SA and gibberellins have been used as elicitor and biostimulant to enhance the production of steviol glycosides in stevia, producing tall plants with a greater number of leaves and a larger stem diameter (Vazquez-Hernandez et al. 2019).

Endophytic microbes have been shown to be able to promote plant growth, induce tolerance and production of bioactive compounds (Lata et al. 2018). Endophytic microbes generally reside in tissues and plants without causing symptoms. However, they activate plant defense system and induce secondary metabolites production with potential pharmaceutical use (Jalgaonwala and Mahajan 2014). Therefore, the inoculation of plants with this type of microorganism is a strategy that can be applied to induce and/or increase the production of natural bioactive compounds in plants of commercial interest.

Fungi elicitation is a popular practice among horticultural producers due to the demonstrated increase in crop yield and the production of bioactive compounds. The ability of plant roots to uptake nutrients from the substrate is enhanced by this type of elicitation, with a positive influence on phytohormone production and gene expression reprogramming. Fungi are used as biocontrol treatments, bioremediation agents, and as biostimulants, which can contribute to the development of sustainable agriculture. In the same way, mineral nutrients play an important role in the growth, development, and yield of crops (Nath and Tuteja 2016). Nutrient management is a common agricultural practice, especially in soilless cultivation and controlled systems in order to guarantee high crop yield. On the other hand, consumers demand natural products with high nutritional quality. Both objectives are feasible to be induced by the nutrition of the crops. Therefore, the management of nutrients during the production cycle allows a balance between growth and accumulation of bioactive compounds.

The use of specific miRNAs to enhance the production of metabolites is a novel technique that is still under research and discussion among the scientific community. Research has focused on the identification of the regulatory mechanisms of these molecules that will lead to the design of strategies for direct manipulation, identification, and understanding of the spatial and temporal expression scheme of miRNAs. Innovative tools (e.g., bioinformatics) can be used to predict, by means of algorithms, the modulation of miRNA molecules which are part of a very complex regulatory network of secondary metabolites. Research in this area could help to understand how biosynthetic pathways are modified by these small molecules that have particular target genes and that can influence metabolic plant bioengineering, generating technology to induce the production of secondary metabolites.

## References

Abdollahi F, Amiri H, Niknam V, Ghanati F, Mahdigholi K (2019) Effects of static magnetic fields on the antioxidant system of almond seeds. Russ J Plant Physl 66(2):299–307

Agathokleous E, Kitao M, Calabrese EJ (2019) Hormesis: a compelling platform for sophisticated plant science. Trends Plant Sci 24(4):318–327

Ahmad N, Jones RW (1979) Glycinebetaine, proline and inorganic ion levels in barley seedlings following transient stress. Plant Sci Lett 15(3):231–237

Akhgari A, Laakso I, Maaheimo H, Choi YH, Seppänen-Laakso T, Oksman-Caldentey K-M, Rischer H (2019) Methyljasmonate elicitation increases terpenoid indole alkaloid accumulation in rhazya stricta hairy root cultures. Plants 8(12):534

Akram W, Saeed T, Ahmad A, Yasin NA, Akbar M, Khan WU, Ahmed S, Guo J, Luo W, Wu T (2020) Liquiritin elicitation can increase the content of medicinally important glucosinolates and phenolic compounds in Chinese kale plants. J Sci Food Agric 100(4):1616–1624

Akula R, Ravishankar GA (2011) Influence of abiotic stress signals on secondary metabolites in plants. Plant Signal Behav 6(11):1720–1731

Alokam S, Chinnappa C, Reid DM (2002) Red/far-red light mediated stem elongation and anthocyanin accumulation in *Stellaria longipes*: differential response of alpine and prairie ecotypes. Can J Bot 80(1):72–81

Alvarado AM, Aguirre-Becerra H, Vázquez-Hernández MC, Magaña-Lopez E, Parola-Contreras I, Caicedo-Lopez LH, Contreras-Medina LM, Garcia-Trejo JF, Guevara-Gonzalez RG, Feregrino-Perez AA (2019) Influence of elicitors and eustressors on the production of plant secondary metabolites. In: Natural bio-active compounds. Springer, pp 333–388

Ampofo JO, Ngadi M (2020) Ultrasonic assisted phenolic elicitation and antioxidant potential of common bean (*Phaseolus vulgaris*) sprouts. Ultrason Sonochem 64:104974

Andini S, Dekker P, Gruppen H, Araya-Cloutier C, Vincken J-P (2019) Modulation of glucosinolate composition in brassicaceae seeds by germination and fungal elicitation. J Agric Food Chem 67(46):12770–12779

Angelova Z, Georgiev S, Roos W (2006) Elicitation of plants. Biotechnol Biotechnol Equip 20 (2):72–83

Anjum S, Anjum I, Hano C, Kousar S (2019) Advances in nanomaterials as novel elicitors of pharmacologically active plant specialized metabolites: current status and future outlooks. RSC Adv 9(69):40404–40423

Appel HM, Cocroft R (2014) Plants respond to leaf vibrations caused by insect herbivore chewing. Oecologia 175(4):1257–1266

Araya-Cloutier C, den Besten HM, Aisyah S, Gruppen H, Vincken J-P (2017) The position of prenylation of isoflavonoids and stilbenoids from legumes (Fabaceae) modulates the antimicrobial activity against Gram positive pathogens. Food Chem 226:193–201

Arora S, Sharma P, Kumar S, Nayan R, Khanna P, Zaidi M (2012) Gold-nanoparticle induced enhancement in growth and seed yield of *Brassica juncea*. J Plant Growth Regul 66(3):303–310

Azad MOK, Kim WW, Park CH, Cho DH (2018) Effect of artificial LED light and far infrared irradiation on phenolic compound, isoflavones and antioxidant capacity in soybean (Glycine max L.) sprout. Foods 7(10):174

Azimi N, Majd A, Nejadsattari T, Ghanati F, Arbabian S (2018) Effects of magnetically treated water on physiological characteristics of *Lens culinaris* L. Iran J Sci Technol A 42(2):331–337

Azizi SMY, Sarghein SH, Majd A, Peyvandi M (2019) The effects of the electromagnetic fields on the biochemical components, enzymatic and non-enzymatic antioxidant systems of tea *Camellia sinensis* L. Physiol Mol Biol Plants 25(6):1445–1456

Balbus JM, Maynard AD, Colvin VL, Castranova V, Daston GP, Denison RA, Dreher KL, Goering PL, Goldberg AM, Kulinowski KM (2007) Meeting report: hazard assessment for nanoparticles—report from an interdisciplinary workshop. Environ Health Perspect 115 (11):1654–1659

Banchio E, Xie X, Zhang H, Pare PW (2009) Soil bacteria elevate essential oil accumulation and emissions in sweet basil. J Agric Food Chem 57(2):653–657

Bantis F, Smirnakou S, Ouzounis T, Koukounaras A, Ntagkas N, Radoglou K (2018) Current status and recent achievements in the field of horticulture with the use of light-emitting diodes (LEDs). Sci Hortic-Amsterdam 235:437–451

Bao X, Roossinck MJ (2013) A life history view of mutualistic viral symbioses: quantity or quality for cooperation? Curr Opin Microbiol 16(4):514–518

Barba FJ, Galanakis CM, Esteve MJ, Frigola A, Vorobiev E (2015) Potential use of pulsed electric technologies and ultrasounds to improve the recovery of high-added value compounds from blackberries. J Food Eng 167:38–44

Becker C, Urlić B, Špika MJ, Kläring H-P, Krumbein A, Baldermann S, Ban SG, Perica S, Schwarz D (2015) Nitrogen limited red and green leaf lettuce accumulate flavonoid glycosides, caffeic acid derivatives, and sucrose while losing chlorophylls, β-carotene and xanthophylls. PloS one 10 (11)

Burgie ES, Bussell AN, Walker JM, Dubiel K, Vierstra RD (2014) Crystal structure of the photosensing module from a red/far-red light-absorbing plant phytochrome. Proc Natl Acad Sci U S A 111(28):10179–10184

Cai Z, Kastell A, Speiser C, Smetanska I (2013) Enhanced resveratrol production in Vitis vinifera cell suspension cultures by heavy metals without loss of cell viability. Appl Biochem Biotechnol 171(2):330–340

Calabrese EJ (2004) Hormesis: a revolution in toxicology, risk assessment and medicine. EMBO Rep 5(S1):S37–S40

Chen Y, Zhang M, Jin X, Tao H, Wang Y, Peng B, Fu C, Yu L (2020) Transcriptional reprogramming strategies and miRNA-mediated regulation networks of Taxus media induced into callus cells from tissues. BMC Genomics 21(1):1–15

Chepets A, Morozova Z, Tkacheva N (1985) Wheat and sorghum yield and grain quality as affected by magnetic field.

Cheynier V, Comte G, Davies KM, Lattanzio V, Martens S (2013) Plant phenolics: recent advances on their biosynthesis, genetics, and ecophysiology. Plant Physiol Biochem 72:1–20

Choudhary DK, Kasotia A, Jain S, Vaishnav A, Kumari S, Sharma KP, Varma A (2016) Bacterial-mediated tolerance and resistance to plants under abiotic and biotic stresses. J Plant Growth Regul 35(1):276–300

Cox A, Venkatachalam P, Sahi S, Sharma N (2016) Silver and titanium dioxide nanoparticle toxicity in plants: a review of current research. Plant Physiol Biochem 107:147–163

Dannehl D (2018) Effects of electricity on plant responses. Sci Hortic-Amsterdam 234:382–392

Das P, Samantaray S, Rout G (1997) Studies on cadmium toxicity in plants: a review. Environ Pollut 98(1):29–36

Denoya GI, Pataro G, Ferrari G (2020) Effects of postharvest pulsed light treatments on the quality and antioxidant properties of persimmons during storage. Postharvest Biol Technol 160:111055

Dessaux Y, Grandclément C, Faure D (2016) Engineering the rhizosphere. Trends Plant Sci 21 (3):266–278

Dhiman SK, Galland P (2018) Effects of weak static magnetic fields on the gene expression of seedlings of *Arabidopsis thaliana*. J Plant Physiol 231:9–18

Dimkpa CO, McLean JE, Latta DE, Manangón E, Britt DW, Johnson WP, Boyanov MI, Anderson AJ (2012) CuO and ZnO nanoparticles: phytotoxicity, metal speciation, and induction of oxidative stress in sand-grown wheat. J Nanopart Res 14(9):1125

Dotaniya M, Meena V (2015) Rhizosphere effect on nutrient availability in soil and its uptake by plants: a review. P NATL A SCI INDIA B 85(1):1–12

Dou H, Niu G, Gu M (2019) Pre-harvest UV-B radiation and photosynthetic photon flux density interactively affect plant photosynthesis, growth, and secondary metabolites accumulation in basil (*Ocimum basilicum*) plants. Agronomy 9(8):434

Du Jardin P (2015) Plant biostimulants: definition, concept, main categories and regulation. Sci Hortic-Amsterdam 196:3–14

Dudareva N, Klempien A, Muhlemann JK, Kaplan I (2013) Biosynthesis, function and metabolic engineering of plant volatile organic compounds. New Phytol 198(1):16–32

Egamberdieva D, Wirth SJ, Alqarawi AA, Abd_Allah EF, Hashem A (2017) Phytohormones and beneficial microbes: essential components for plants to balance stress and fitness. Front Microbiol 8:2104

El Kantar S, Boussetta N, Lebovka N, Foucart F, Rajha HN, Maroun RG, Louka N, Vorobiev E (2018) Pulsed electric field treatment of citrus fruits: improvement of juice and polyphenols extraction. Innov Food Sci Emerg 46:153–161

El-Nakhel C, Pannico A, Kyriacou MC, Giordano M, De Pascale S, Rouphael Y (2019) Macronutrient deprivation eustress elicits differential secondary metabolites in red and green-pigmented butterhead lettuce grown in a closed soilless system. J Sci Food Agric 99 (15):6962–6972

Faiola C, Taipale D (2020) Impact of insect herbivory on plant stress volatile emissions from trees: a synthesis of quantitative measurements and recommendations for future research. Atmos Environ X 5:100060

Fazal H, Abbasi BH, Ahmad N, Ali M, Shujait Ali S, Khan A, Wei D-Q (2019) Sustainable production of biomass and industrially important secondary metabolites in cell cultures of selfheal (*Prunella vulgaris* L.) elicited by silver and gold nanoparticles. Artif Cells Nanomed Biotechnol 47(1):2553–2561

Feregrino-Perez AA, Magaña-López E, Guzmán C, Esquivel K (2018) A general overview of the benefits and possible negative effects of the nanotechnology in horticulture. Sci Hortic-Amsterdam 238:126–137

Fesel PH, Zuccaro A (2016) β-glucan: Crucial component of the fungal cell wall and elusive MAMP in plants. Fungal Genet Biol 90:53–60

Fincheira P, Quiroz A (2018) Microbial volatiles as plant growth inducers. Microbiol Res 208:63–75

Fu PP, Xia Q, Hwang H-M, Ray PC, Yu H (2014) Mechanisms of nanotoxicity: generation of reactive oxygen species. J Food Drug Anal 22(1):64–75

Furze JM, Rhodes MJ, Parr AJ, Robins RJ, Withehead IM, Threlfall DR (1991) Abiotic factors elicit sesquiterpenoid phytoalexin production but not alkaloid production in transformed root cultures of Datura stramonium. Plant Cell Rep 10(3):111–114

Gachovska TK, Armyanov N, Palov I (2015) Pre-planting pulsed electric field treatment of potato tubers. In: 2015 IEEE Canada international humanitarian technology conference (IHTC2015). IEEE, pp 1–4

Galieni A, Di Mattia C, De Gregorio M, Speca S, Mastrocola D, Pisante M, Stagnari F (2015) Effects of nutrient deficiency and abiotic environmental stresses on yield, phenolic compounds and antiradical activity in lettuce (*Lactuca sativa* L.). Sci Hortic-Amsterdam 187:93–101

Garcia-Ibañez P, Moreno DA, Nuñez-Gomez V, Agudelo A, Carvajal M (2019) The use of elicitation in the cultivation of Bimi® for food and ingredients. J Sci Food Agric 100(5):2099–2109

Ghorbanpour M, Khavazi K, Hatami M (2014) Chemical compositions and antimicrobial activity of *Salvia officinalis* L. essential oil under rhizobacteria (Pseudomonas fluorescens and Putida) inoculation. Eur J Soil Biol

Gilani A, Kermanshahi H, Gholizadeh M, Golian A (2017) Agricultural water managenemt through magnetizatio of irrigation and drinking water: a review. J Aridland Agric 3:23–27

Gupta OP, Karkute SG, Banerjee S, Meena NL, Dahuja A (2017) Contemporary understanding of miRNA-based regulation of secondary metabolites biosynthesis in plants. Front Plant Sci 8:374

Gururani MA, Upadhyaya CP, Baskar V, Venkatesh J, Nookaraju A, Park SW (2013) Plant growth-promoting rhizobacteria enhance abiotic stress tolerance in Solanum tuberosum through inducing changes in the expression of ROS-scavenging enzymes and improved photosynthetic performance. J Plant Growth Regul 32(2):245–258

Gutiérrez-Gamboa G, Garde-Cerdán T, Gonzalo-Diago A, Moreno-Simunovic Y, Martínez-Gil AM (2017) Effect of different foliar nitrogen applications on the must amino acids and glutathione composition in Cabernet Sauvignon vineyard. LWT 75:147–154

Guzmán-Guzmán P, Porras-Troncoso MD, Olmedo-Monfil V, Herrera-Estrella A (2019) Trichoderma species: versatile plant symbionts. Phytopathology 109(1):6–16

Häder D-P, Williamson CE, Wängberg S-Å, Rautio M, Rose KC, Gao K, Helbling EW, Sinha RP, Worrest R (2015) Effects of UV radiation on aquatic ecosystems and interactions with other environmental factors. Photoch Photobio SCI 14(1):108–126

Hajiboland R (2012) Effect of micronutrient deficiencies on plants stress responses. In: Abiotic stress responses in plants. Springer, pp 283–329

Halder M, Sarkar S, Jha S (2019) Elicitation: A biotechnological tool for enhanced production of secondary metabolites in hairy root cultures. Eng Life Sci 19(12):880–895

Hasan M, Alharby H, Hajar A, Hakeem K, Alzahrani Y, Arabia S (2018) Effects of magnetized water on phenolic compounds, lipid peroxidation and antioxidant activity of moringa species under drought stress. J Anim Plant Sci 28(3):1–6

Hasanuzzaman M, Hossain MA, Fujita M (2012) Exogenous selenium pretreatment protects rapeseed seedlings from cadmium-induced oxidative stress by upregulating antioxidant defense and methylglyoxal detoxification systems. Biol Trace Elem Res 149(2):248–261
Hassanien RH, Hou T, Li Y, Li B (2014) Advances in effects of sound waves on plants. J Integr Agric 13(2):335–348
Hawkesford M, Horst W, Kichey T, Lambers H, Schjoerring J, Møller IS, White P (2012) Functions of macronutrients. In: Marschner's mineral nutrition of higher plants. Elsevier, pp 135–189
He R, Xi G, Liu K (2017) Alleviating effect of extremely low frequency pulsed electric field on drought damage of maize seedling roots. J Lumin 188:441–447
Hendrawan Y, Sabrinauly S, Hawa LC, Argo BD, Rachmawati M (2019) Analysis of the phenol and flavonoid content from basil leaves (*Ocimum Americanum* L) extract using pulsed electric field (PEF) pre-treatment. Agric Eng Int CIGR J 21(2):149–158
Hideg É, Strid Å (2017) The effects of UV-B on the biochemistry and metabolism in plants. In: CABI (ed) UV-B radiation and plant life. Molecular biology to ecology. pp 90–110
Hossain MS, Hoang NTH, Yan Z, Tóth K, Meyers BC, Stacey G (2019) Characterization of the spatial and temporal expression of two soybean miRNAs identifies SCL6 as a novel regulator of soybean nodulation. Front Plant Sci 10:475
Ibrahim MH, Chee Kong Y, Zain M, Amalina N (2017) Effect of cadmium and copper exposure on growth, secondary metabolites and antioxidant activity in the medicinal plant Sambung Nyawa (*Gynura procumbens* (L.) Merr). Molecules 22(10):1623
Isah T (2019) Stress and defense responses in plant secondary metabolites production. Biol Res 52 (1):39
Jalgaonwala R, Mahajan R (2014) A review on microbial endophytes from plants: a treasure search for biologically active metabolites. Glob J Res Med Plants Indig Med 3(6):263
Janositz A, Knorr D (2010) Microscopic visualization of pulsed electric field induced changes on plant cellular level. Innov Food Sci Emerg 11(4):592–597
Jeong M-J, Cho J-I, Park S-H, Kim K-H, Lee SK, Kwon T-R, Park S-C, Siddiqui ZS (2014) Sound frequencies induce drought tolerance in rice plant. Pak J Bot 46:2015–2020
Jezek M, Zörb C, Merkt N, Mühling K-H, Geilfus C-M (2016) The Relationship between Plant nutrition and Anthocyanins and their significance for agriculture. In: Magnesium deficiency in maize and effectiveness of nutrient supply through $MgSO_4$ leaf-application. Kiel, pp 78–118
Jiang HS, Yin LY, Ren NN, Zhao ST, Li Z, Zhi Y, Shao H, Li W, Gontero B (2017) Silver nanoparticles induced reactive oxygen species via photosynthetic energy transport imbalance in an aquatic plant. Nanotoxicology 11(2):157–167
Jiang N, Grundy S, Bian Z, Lu C (2017) Investigation of LED light effects on plant growth in improved protected horticulture system. Greensys 2017—international symposium on new technologies for environment control. Energy-Saving and Crop Production in Greenhouse and Plant Factory, Beijing, China, pp 20–24
Jiang Y, Hua Z, Zhao Y, Liu Q, Wang F, Zhang Q (2014) The effect of carbon nanotubes on rice seed germination and root growth. In: Proceedings of the 2012 international conference on applied biotechnology (ICAB 2012). Springer, pp 1207–1212
Joshi R, Paul M, Kumar A, Pandey D (2019) Role of calreticulin in biotic and abiotic stress signalling and tolerance mechanisms in plants. Gene 714:144004
Kanchiswamy CN, Malnoy M, Maffei ME (2015) Bioprospecting bacterial and fungal volatiles for sustainable agriculture. Trends Plant Sci 20(4):206–211
Kašparová M, Pilařová P, Tůmová L, Siatka T (2018) Effect of precursor and phytohormones on podophyllotoxin production in juniperus virginiana suspension cultures. Nat Prod Commun 13 (11):1934578X1801301129
Kentish S, Ashokkumar M (2011) The physical and chemical effects of ultrasound. In: Ultrasound technologies for food and bioprocessing. Springer, pp 1–12
Khodakovskaya MV, Kim BS, Kim JN, Alimohammadi M, Dervishi E, Mustafa T, Cernigla CE (2013) Carbon nanotubes as plant growth regulators: effects on tomato growth, reproductive system, and soil microbial community. Small 9(1):115–123

Khot LR, Sankaran S, Maja JM, Ehsani R, Schuster EW (2012) Applications of nanomaterials in agricultural production and crop protection: a review. Crop Protect 35:64–70

Kole C, Kole P, Randunu KM, Choudhary P, Podila R, Ke PC, Rao AM, Marcus RK (2013) Nanobiotechnology can boost crop production and quality: first evidence from increased plant biomass, fruit yield and phytomedicine content in bitter melon (*Momordica charantia*). BMC Biotechnol 13(1):37

Koshel RJ (2004) Illumination engineering. Opt Eng 43(7):1478–1480

Kováčik J, Klejdus B (2014) Induction of phenolic metabolites and physiological changes in chamomile plants in relation to nitrogen nutrition. Food Chem 142:334–341

Kranner I, Beckett RP, Wornik S, Zorn M, Pfeifhofer HW (2002) Revival of a resurrection plant correlates with its antioxidant status. Plant J 31(1):13–24

Kranner I, Minibayeva FV, Beckett RP, Seal CE (2010) What is stress? Concepts, definitions and applications in seed science. New Phytol 188(3):655–673

Lahiani MH, Chen J, Irin F, Puretzky AA, Green MJ, Khodakovskaya MV (2015) Interaction of carbon nanohorns with plants: uptake and biological effects. Carbon 81:607–619

Lajayer BA, Ghorbanpour M, Nikabadi S (2017) Heavy metals in contaminated environment: destiny of secondary metabolite biosynthesis, oxidative status and phytoextraction in medicinal plants. Ecotoxicol Environ Saf 145:377–390

Larner VS, Franklin KA, Whitelam GC (2018) Photoreceptors and light signalling pathways in plants. Annual Plant Rev Online 107–131

Lata R, Chowdhury S, Gond SK, White JF Jr (2018) Induction of abiotic stress tolerance in plants by endophytic microbes. Lett Appl Microbiol 66(4):268–276

Li P, Linhardt RJ, Cao Z (2016) Structural characterization of oligochitosan elicitor from *Fusarium sambucinum* and its elicitation of defensive responses in Zanthoxylum bungeanum. Int J Mol Sci 17(12):2076

Lichtenthaler HK (1996) Vegetation stress: an introduction to the stress concept in plants. J Plant Physiol 148(1–2):4–14

Lilia C, Abdelkader A, Karima A-KA, Tarek B (2018) The effect of ultrasound pre-treatment on the yield, chemical composition and antioxidant activity of essential oil from wild Lavandula stoechas L. J Essent Oil-Bear 21(1):253–263

Liu J, Yuan Y, Wang Y, Jiang C, Chen T, Zhu F, Zhao Y, Zhou J, Huang L (2017) Regulation of fatty acid and flavonoid biosynthesis by miRNAs in Lonicera japonica. RSC Adv 7 (56):35426–35437

Liu W, Liu C, Yang C, Wang L, Li S (2010) Effect of grape genotype and tissue type on callus growth and production of resveratrols and their piceids after UV-C irradiation. Food Chem 122 (3):475–481

Lu C, Ding J, Park HK, Feng H (2020) High intensity ultrasound as a physical elicitor affects secondary metabolites and antioxidant capacity of tomato fruits. Food Control 113:107176

Maffei ME (2014) Magnetic field effects on plant growth, development, and evolution. Front Plant Sci 5:445

Marslin G, Sheeba CJ, Franklin G (2017) Nanoparticles alter secondary metabolism in plants via ROS burst. Front Plant Sci 8:832

Martínez-Ballesta MC, Zapata L, Chalbi N, Carvajal M (2016) Multiwalled carbon nanotubes enter broccoli cells enhancing growth and water uptake of plants exposed to salinity. J Nanobiotechnol 14(1):42

Martínez-Medina A, Van Wees SC, Pieterse CM (2017) Airborne signals from Trichoderma fungi stimulate iron uptake responses in roots resulting in priming of jasmonic acid-dependent defences in shoots of *Arabidopsis thaliana* and *Solanum lycopersicum*. Plant Cell Environ 40 (11):2691–2705

Medina-Pérez V, López-Laredo AR, Sepúlveda-Jiménez G, Zamilpa A, Trejo-Tapia G (2015) Nitrogen deficiency stimulates biosynthesis of bioactive phenylethanoid glycosides in the medicinal plant Castilleja tenuiflora Benth. Acta Physiol Plant 37(5):93

Miliute I, Buzaite O, Baniulis D, Stanys V (2015) Bacterial endophytes in agricultural crops and their role in stress tolerance: a review. Zemdirbyste 102(4):465–478

Mirzajani F, Askari H, Hamzelou S, Schober Y, Römpp A, Ghassempour A, Spengler B (2014) Proteomics study of silver nanoparticles toxicity on *Oryza sativa* L. Ecotoxicol Environ Saf 108:335–339

Mishra RC, Ghosh R, Bae H (2016) Plant acoustics: in the search of a sound mechanism for sound signaling in plants. J Exp Bot 67(15):4483–4494

Moghimi M, Farzaneh V, Bakhshabadi H (2018) The effect of ultrasound pretreatment on some selected physicochemical properties of black cumin (*Nigella Sativa*). Nutrire 43(1):18

Mohammed KA, Hellal F, SAA E-S, (2015) Influence of sulfur deprivation on biomass allocation, mineral composition and fruit quality of tomato plants. Middle East J 4(1):42–48

Moola AK, Diana RKB (2019) Elicitation as a means for enhanced plant secondary metabolites through hairy root system. J Adv Sci Res 10(3)

Muller-Xing R, Xing Q, Goodrich J (2014) Footprints of the sun: memory of UV and light stress in plants. Front Plant Sci 5:474. https://doi.org/10.3389/fpls.2014.00474

Naik PM, Al-Khayri JM (2016) Abiotic and biotic elicitors–role in secondary metabolites production through in vitro culture of medicinal plants. Abiotic and biotic stress in plants-recent advances and future perspectives: 247–277

Nakabayashi R, Yonekura-Sakakibara K, Urano K, Suzuki M, Yamada Y, Nishizawa T, Matsuda F, Kojima M, Sakakibara H, Shinozaki K (2014) Enhancement of oxidative and drought tolerance in Arabidopsis by overaccumulation of antioxidant flavonoids. Plant J 77 (3):367–379

Nandini B, Puttaswamy H, Prakash HS, Adhikari S, Jogaiah S, Nagaraja G (2020) Elicitation of novel trichogenic-lipid nanoemulsion signaling resistance against Pearl Millet Downy Mildew disease. Biomolecules 10(1):25

Nasim SA, Dhir B (2010) Heavy metals alter the potency of medicinal plants. In: Reviews of environmental contamination and toxicology. Springer, pp 139–149

Nath M, Tuteja N (2016) NPKS uptake, sensing, and signaling and miRNAs in plant nutrient stress. Protoplasma 253(3):767–786

Naveed M, Hussain MB, Zahir ZA, Mitter B, Sessitsch A (2014) Drought stress amelioration in wheat through inoculation with Burkholderia phytofirmans strain PsJN. J Plant Growth Regul 73(2):121–131

Naziya B, Murali M, Amruthesh KN (2020) Plant growth-promoting fungi (PGPF) instigate plant growth and induce disease resistance in *Capsicum annuum* L. upon infection with Colletotrichum capsici (Syd.) Butler & Bisby. Biomolecules 10(1):41

Naznin M, Lefsrud M, Gravel V, Hao X (2016) Different ratios of red and blue LED light effects on coriander productivity and antioxidant properties. In: VIII international symposium on light in horticulture 1134. pp 223–230

Nejat N, Mantri N (2017) Plant immune system: crosstalk between responses to biotic and abiotic stresses the missing link in understanding plant defence. Curr Issues Mol Biol 23:1–16

Oancea S, Ghincevici D, Ketney O (2014) The effect of ultrasonic pretreatment and sample preparation on the extraction yield of antioxidant compounds and activity of black currant fruits. Acta Chim Slov 62(1):242–248

Oberdörster G, Oberdörster E, Oberdörster J (2005) Nanotoxicology: an emerging discipline evolving from studies of ultrafine particles. Environ Health Perspect 113(7):823–839

Odriozola-Serrano I, Soliva-Fortuny R, Hernández-Jover T, Martín-Belloso O (2009) Carotenoid and phenolic profile of tomato juices processed by high intensity pulsed electric fields compared with conventional thermal treatments. Food Chem 112(1):258–266

Panjai L, Noga G, Fiebig A, Hunsche M (2017) Effects of continuous red light and short daily UV exposure during postharvest on carotenoid concentration and antioxidant capacity in stored tomatoes. Sci Hortic-Amsterdam 226:97–103

Pappas ML, Liapoura M, Papantoniou D, Avramidou M, Kavroulakis N, Weinhold A, Broufas GD, Papadopoulou KK (2018) The beneficial endophytic fungus Fusariumsolani strain K alters tomato responses against spider mites to the benefit of the plant. Front Plant Sci 9:1603

Patel H, Krishnamurthy R (2013) Elicitors in plant tissue culture. Int J Pharmacogn Phytochem 2 (2):60–65

Peng Z, Tian J, Luo R, Kang Y, Lu Y, Hu Y, Liu N, Zhang J, Cheng H, Niu S (2019) MiR399d and epigenetic modification comodulate anthocyanin accumulation in Malus leaves suffering from phosphorus deficiency. Plant Cell Environ 43(5):1148–1159

Pereira M, Morais L, Marques E, Martins A, Cavalcanti V, Rodrigues F, Gonçalves W, Blank A, Pasqual M, Dória J (2019) Humic substances and efficient microorganisms: elicitation of medicinal plants—a review. J Agric Sci 11(7)

Pérez-Ambrocio A, Guerrero-Beltrán J, Aparicio-Fernández X, Ávila-Sosa R, Hernández-Carranza P, Cid-Pérez S, Ochoa-Velasco C (2018) Effect of blue and ultraviolet-C light irradiation on bioactive compounds and antioxidant capacity of habanero pepper (*Capsicum chinense*) during refrigeration storage. Postharvest Biol Technol 135:19–26

Piechulla B, Degenhardt J (2014) The emerging importance of microbial volatile organic compounds. Plant Cell Environ 37(4):811–812

Qi X, Fang H, Yu X, Xu D, Li L, Liang C, Lu H, Li W, Chen Y, Chen Z (2018) Transcriptome analysis of JA signal transduction, transcription factors, and monoterpene biosynthesis pathway in response to methyl jasmonate elicitation in *Mentha canadensis* L. Int J Mol Sci 19 (8):2364

Qian H, Liu T, Deng M, Miao H, Cai C, Shen W, Wang Q (2016) Effects of light quality on main health-promoting compounds and antioxidant capacity of Chinese kale sprouts. Food Chem 196:1232–1238

Răcuciu M, Oancea S (2018) Impact of 50 Hz magnetic field on the content of polyphenolic compounds from blackberries. Bulg Chem Commun 50(3):393–397

Radhakrishnan R (2019) Magnetic field regulates plant functions, growth and enhances tolerance against environmental stresses. Physiol Mol Biol Plants 1–13

Raei M, Angaji SA, Omidi M, Khodayari M (2014) Effect of abiotic elicitors on tissue culture of Aloe vera. Int J Biosci 5(1):74–81

Rahmani N, Radjabian T, Soltani BM (2020) Impacts of foliar exposure to multi-walled carbon nanotubes on physiological and molecular traits of *Salvia verticillata* L. as a medicinal plant. Plant Physiol Biochem 150:27–38

Rakshit A, Singh HB, Singh AK, Singh US, Fraceto L (2020) New frontiers in stress management for durable agriculture. Springer

Ramirez-Estrada K, Vidal-Limon H, Hidalgo D, Moyano E, Golenioswki M, Cusidó RM, Palazon J (2016) Elicitation, an effective strategy for the biotechnological production of bioactive high-added value compounds in plant cell factories. Molecules 21(2):182

Ribas-Agustí A, Martín-Belloso O, Soliva-Fortuny R, Elez-Martínez P (2019) Enhancing hydroxycinnamic acids and flavan-3-ol contents by pulsed electric fields without affecting quality attributes of apple. Food Res Int 121:433–440

Rico CM, Lee SC, Rubenecia R, Mukherjee A, Hong J, Peralta-Videa JR, Gardea-Torresdey JL (2014) Cerium oxide nanoparticles impact yield and modify nutritional parameters in wheat (*Triticum aestivum* L.). J Agric Food Chem 62(40):9669–9675

Roossinck MJ (2015) Plants, viruses and the environment: ecology and mutualism. Virology 479:271–277

Routray W, Orsat V, Lefsrud M (2018) Effect of postharvest LED application on phenolic and antioxidant components of blueberry leaves. Chem Eng 2(4):56

Rudrappa T, Neelwarne B, Aswathanarayana RG (2004) In situ and ex situ adsorption and recovery of betalains from hairy root cultures of Beta vulgaris. Biotechnol Prog 20(3):777–785

Rusakova A, Nosachev I, Lysenko V, Guo Y, Logvinov A, Kirichenko E, Varduny T, Cherednikov S, Chugueva O (2017) Impact of high strength electromagnetic fields generated by Tesla transformer on plant cell ultrastructure. Inf Process Agric 4(3):253–258

Ryu C-M, Farag MA, Hu C-H, Reddy MS, Wei H-X, Paré PW, Kloepper JW (2003) Bacterial volatiles promote growth in Arabidopsis. Proc Natl Acad Sci U S A 100(8):4927–4932

Sá RA, Sá RA, Alberton O, Gazim ZC, Laverde A Jr, Caetano J, Amorin AC, Dragunski DC (2015) Phytoaccumulation and effect of lead on yield and chemical composition of Mentha crispa essential oil. Desalin Water Treat 53(11):3007–3017

Sade D, Shriki O, Cuadros-Inostroza A, Tohge T, Semel Y, Haviv Y, Willmitzer L, Fernie AR, Czosnek H, Brotman Y (2015) Comparative metabolomics and transcriptomics of plant response to Tomato yellow leaf curl virus infection in resistant and susceptible tomato cultivars. Metabolomics 11(1):81–97

Saeed S, Ali H, Khan T, Kayani W, Khan MA (2017) Impacts of methyl jasmonate and phenyl acetic acid on biomass accumulation and antioxidant potential in adventitious roots of Ajuga bracteosa Wall ex Benth., a high valued endangered medicinal plant. Physiol Mol Biol Plants 23(1):229–237

Samad AF, Sajad M, Nazaruddin N, Fauzi IA, Murad A, Zainal Z, Ismail I (2017) MicroRNA and transcription factor: key players in plant regulatory network. Front Plant Sci 8:565

Samad AFA, Rahnamaie-Tajadod R, Sajad M, Jani J, Murad AMA, Noor NM, Ismail I (2019) Regulation of terpenoid biosynthesis by miRNA in Persicaria minor induced by Fusarium oxysporum. BMC Genomics 20(1):586

Samanta A, Das G, Das SK (2011) Roles of flavonoids in plants. Carbon 100(6):12–35

Sánchez-López ÁM, Baslam M, De Diego N, Muñoz FJ, Bahaji A, Almagro G, Ricarte-Bermejo A, García-Gómez P, Li J, Humplík JF (2016) Volatile compounds emitted by diverse phytopathogenic microorganisms promote plant growth and flowering through cytokinin action. Plant Cell Environ 39(12):2592–2608

Santoro MV, Bogino PC, Nocelli N, Cappellari LdR, Giordano WF, Banchio E (2016) Analysis of plant growth-promoting effects of fluorescent Pseudomonas strains isolated from Mentha piperita rhizosphere and effects of their volatile organic compounds on essential oil composition. Front Microbiol 7:1085

Santoro MV, Zygadlo J, Giordano W, Banchio E (2011) Volatile organic compounds from rhizobacteria increase biosynthesis of essential oils and growth parameters in peppermint (Mentha piperita). Plant Physiol Biochem 49(10):1177–1182

Shahhoseini R, Azizi M, Asili J, Moshtaghi N, Samiei L (2020) Effects of zinc oxide nanoelicitors on yield, secondary metabolites, zinc and iron absorption of Feverfew (Tanacetum parthenium (L.) Schultz Bip.). Acta Physiol Plant 42:1–18

Shang L, Nienhaus K, Nienhaus GU (2014) Engineered nanoparticles interacting with cells: size matters. J Nanobiotechnol 12(1):5

Sharma D, Tiwari M, Pandey A, Bhatia C, Sharma A, Trivedi PK (2016) MicroRNA858 is a potential regulator of phenylpropanoid pathway and plant development. Plant Physiol 171 (2):944–959

Silva RN, Monteiro VN, Steindorff AS, Gomes EV, Noronha EF, Ulhoa CJ (2019) Trichoderma/pathogen/plant interaction in pre-harvest food security. Fungal Biol 123(8):565–583

Singh A, Dwivedi P (2018) Methyl-jasmonate and salicylic acid as potent elicitors for secondary metabolite production in medicinal plants: a review. J Pharmacogn Phytochem 7(1):750–757

Singh D, Basu C, Meinhardt-Wollweber M, Roth B (2015) LEDs for energy efficient greenhouse lighting. Renew Sust Energ Rev 49:139–147

Singh NR, Rath SK, Behera S, Naik SK (2018) In vitro secondary metabolite production through fungal elicitation: an approach for sustainability. In: Fungal nanobionics: principles and applications. Springer, pp 215–242

Soliva-Fortuny R, Vendrell-Pacheco M, Martín-Belloso O, Elez-Martínez P (2017) Effect of pulsed electric fields on the antioxidant potential of apples stored at different temperatures. Postharvest Biol Technol 132:195–201

Sosan A, Svistunenko D, Straltsova D, Tsiurkina K, Smolich I, Lawson T, Subramaniam S, Golovko V, Anderson D, Sokolik A (2016) Engineered silver nanoparticles are sensed at the plasma membrane and dramatically modify the physiology of *Arabidopsis thaliana* plants. Plant J 85(2):245–257

Starodubtseva GP, Livinskiy SA, Gabriyelyan SZ, Lubaya SI, Afanacev MA (2018) Process control of Pre-sowing seed treatment by Pulsed electric field. Acta Technol Agric 21(1):28–32

Subbaiah LV, Prasad TNVKV, Krishna TG, Sudhakar P, Reddy BR, Pradeep T (2016) Novel effects of nanoparticulate delivery of zinc on growth, productivity, and zinc biofortification in maize (*Zea mays* L.). J Agric Food Chem 64(19):3778–3788

Taghizadeh M, Nasibi F, Kalantari KM, Ghanati F (2019) Evaluation of secondary metabolites and antioxidant activity in Dracocephalum polychaetum Bornm. cell suspension culture under magnetite nanoparticles and static magnetic field elicitation. Plant Cell Tiss Org 136(3):489–498

Tani E, Kizis D, Markellou E, Papadakis I, Tsamadia D, Leventis G, Makrogianni D, Karapanos I (2018) Cultivar-dependent responses of eggplant (*Solanum melongena* L.) to simultaneous Verticillium dahliae infection and drought. Front Plant Sci 9:1181

Thakur M, Bhattacharya S, Khosla PK, Puri S (2019) Improving production of plant secondary metabolites through biotic and abiotic elicitation. J Appl Res Med Aroma 12:1–12

Tian H, Ghorbanpour M, Kariman K (2018) Manganese oxide nanoparticle-induced changes in growth, redox reactions and elicitation of antioxidant metabolites in deadly nightshade (*Atropa belladonna* L.). Ind Crop Prod 126:403–414

Tiwari D, Dasgupta-Schubert N, Cendejas LV, Villegas J, Montoya LC, García SB (2014) Interfacing carbon nanotubes (CNT) with plants: enhancement of growth, water and ionic nutrient uptake in maize (Zea mays) and implications for nanoagriculture. Appl Nanosci 4 (5):577–591

Toma M, Vinatoru M, Paniwnyk L, Mason TJ (2001) Investigation of the effects of ultrasound on vegetal tissues during solvent extraction. Ultrason Sonochem 8(2):137–142

Tripathi AM, Singh A, Singh R, Verma AK, Roy S (2019) Modulation of miRNA expression in natural populations of A. thaliana along a wide altitudinal gradient of Indian Himalayas. Sci Rep 9(1):1–16

Upadhyaya H, Dutta BK, Panda SK (2013) Zinc modulates drought-induced biochemical damages in tea [*Camellia sinensis* (L) O Kuntze]. J Agric Food Chem 61(27):6660–6670

Vaezzadeh M, Noruzifar E, Faezeh G, Salehkotahi M, Mehdian R (2006) Excitation of plant growth in dormant temperature by steady magnetic field. J Magn Magn Mater 302(1):105–108

Van Oosten MJ, Pepe O, De Pascale S, Silletti S, Maggio A (2017) The role of biostimulants and bioeffectors as alleviators of abiotic stress in crop plants. Chem Biol Technol Agric 4(1):5

Vance ME, Kuiken T, Vejerano EP, McGinnis SP, Hochella MF Jr, Rejeski D, Hull MS (2015) Nanotechnology in the real world: redeveloping the nanomaterial consumer products inventory. Beilstein J Nanotechnol 6(1):1769–1780

Vargas-Hernandez M, Macias-Bobadilla I, Guevara-Gonzalez RG, Romero-Gomez SdJ, Rico-Garcia E, Ocampo-Velazquez RV, Alvarez-Arquieta LdL, Torres-Pacheco I (2017) Plant hormesis management with biostimulants of biotic origin in agriculture. Front Plant Sci 8:1762

Vargas-Hernández M, Vázquez-Marrufo G, Aguilar-Ruiz Carlos A, González-Márquez MA, Rocha O, Cerna-Pantoja D, Cruz-Hernández A (2019) MicroRNAs associated with secondary metabolites production. In: Plant physiological aspects of phenolic compounds. IntechOpen

Vazquez-Hernandez C, Feregrino-Perez AA, Perez-Ramirez I, Ocampo-Velazquez RV, Rico-García E, Torres-Pacheco I, Guevara-Gonzalez RG (2019) Controlled elicitation increases steviol glycosides (SGs) content and gene expression-associated to biosynthesis of SGs in Stevia rebaudiana B. cv. Morita II. Ind Crop Prod 139:111479

Vázquez-Hernández M, Parola-Contreras I, Montoya-Gómez L, Torres-Pacheco I, Schwarz D, Guevara-González R (2019) Eustressors: Chemical and physical stress factors used to enhance vegetables production. Sci Hortic-Amsterdam 250:223–229

Vecchio G, Galeone A, Brunetti V, Maiorano G, Rizzello L, Sabella S, Cingolani R, Pompa PP (2012) Mutagenic effects of gold nanoparticles induce aberrant phenotypes in Drosophila melanogaster. Nanomed Nanotechnol Biol Med 8(1):1–7

Verma V, Ravindran P, Kumar PP (2016) Plant hormone-mediated regulation of stress responses. BMC Plant Biol 16(1):86

Vicaş SI, Bandici L, Teuşdea AC, Turcin V, Popa D, Bandici GE (2017) The bioactive compounds, antioxidant capacity, and color intensity in must and wines derived from grapes processed by pulsed electric field. CYTA-J Food 15(4):553–562

Villarreal-García D, Nair V, Cisneros-Zevallos L, Jacobo-Velázquez DA (2016) Plants as biofactories: Postharvest stress-induced accumulation of phenolic compounds and glucosinolates in broccoli subjected to wounding stress and exogenous phytohormones. Front Plant Sci 7:45

Vinatoru M (2001) An overview of the ultrasonically assisted extraction of bioactive principles from herbs. Ultrason Sonochem 8(3):303–313

Vivaldo G, Masi E, Taiti C, Caldarelli G, Mancuso S (2017) The network of plants volatile organic compounds. Sci Rep 7(1):1–18

Wang W, Li Y, Dang P, Zhao S, Lai D, Zhou L (2018) Rice secondary metabolites: structures, roles, biosynthesis, and metabolic regulation. Molecules 23(12):3098

Wang W, Tang K, Yang H-R, Wen P-F, Zhang P, Wang H-L, Huang W-D (2010) Distribution of resveratrol and stilbene synthase in young grape plants (Vitis vinifera L. cv. Cabernet Sauvignon) and the effect of UV-C on its accumulation. Plant Physiol Biochem 48(2–3):142–152

Wang X, Yang X, Chen S, Li Q, Wang W, Hou C, Gao X, Wang L, Wang S (2016) Zinc oxide nanoparticles affect biomass accumulation and photosynthesis in Arabidopsis. Front Plant Sci 6:1243

Wang X, Zhang D, Cui N, Yu Y, Yu G, Fan H (2018) Transcriptome and miRNA analyses of the response to Corynespora cassiicola in cucumber. Sci Rep 8(1):1–16

Wiktor A, Sledz M, Nowacka M, Rybak K, Chudoba T, Lojkowski W, Witrowa-Rajchert D (2015) The impact of pulsed electric field treatment on selected bioactive compound content and color of plant tissue. Innov Food Sci Emerg 30:69–78

Wiktor A, Sledz M, Nowacka M, Rybak K, Witrowa-Rajchert D (2016) The influence of immersion and contact ultrasound treatment on selected properties of the apple tissue. Appl Acoust 103:136–142

Witrowa-Rajchert D, Wiktor A, Sledz M, Nowacka M (2014) Selected emerging technologies to enhance the drying process: A review. Dry Technol 32(11):1386–1396

Xi-ran L, Ting F (2017) The application of high pulsed electric field technology in extracting natural compounds. Food Ind 1:69

Xu P, Chen F, Mannas JP, Feldman T, Sumner LW, Roossinck MJ (2008) Virus infection improves drought tolerance. New Phytol 180(4):911–921

Yan Y, Zhou S, Song Z, Li F, Zhang C, Zhang X, Song H (2017) Effects of frequency and voltage of high voltage pulsed electric field on improving vigor of aged cotton seed. Trans Chinese Soc Agric Eng 33(13):310–314

Yokel RA, MacPhail RC (2011) Engineered nanomaterials: exposures, hazards, and risk prevention. J Occup Med Toxicol 6(1):7

Yu J, Engeseth NJ, Feng H (2016) High intensity ultrasound as an abiotic elicitor—effects on antioxidant capacity and overall quality of romaine lettuce. Food Bioprocess Tech 9(2):262–273

Yuliani F, Dewi WS, Yunus A, Siswanto U (2018) The study of artemisinin content in callus Artemisia annua L. cultures elicited with endophytic fungi Aspergillus sp. Molekul 13(2):155–161

Zaker A, Sykora C, Gössnitzer F, Abrishamchi P, Asili J, Mousavi SH, Wawrosch C (2015) Effects of some elicitors on tanshinone production in adventitious root cultures of Perovskia abrotanoides Karel. Ind Crop Prod 67:97–102

Zareei E, Zaare-Nahandi F, Oustan S, Hajilou J (2019) Effects of magnetic solutions on some biochemical properties and production of some phenolic compounds in grapevine (Vitis vinifera L.). Sci Hortic-Amsterdam 253:217–226

Zaynab M, Fatima M, Abbas S, Sharif Y, Umair M, Zafar MH, Bahadar K (2018) Role of secondary metabolites in plant defense against pathogens. Microb Pathog 124:198–202

Zhao J, Zhu W-H, Hu Q (2001) Selection of fungal elicitors to increase indole alkaloid accumulation in Catharanthus roseus suspension cell culture. Enzyme Microb Technol 28(7–8):666–672

Zhao L, Huang Y, Hu J, Zhou H, Adeleye AS, Keller AA (2016) 1H NMR and GC-MS based metabolomics reveal defense and detoxification mechanism of cucumber plant under nano-Cu stress. Environ Sci Technol 50(4):2000–2010

Zhao L, Peng B, Hernandez-Viezcas JA, Rico C, Sun Y, Peralta-Videa JR, Tang X, Niu G, Jin L, Varela-Ramirez A (2012) Stress response and tolerance of Zea mays to $CeO_2$ nanoparticles: cross talk among $H_2O_2$, heat shock protein, and lipid peroxidation. ACS Nano 6(11):9615–9622

Zhou J-Y, Li X, Zheng J-Y, Dai C-C (2016) Volatiles released by endophytic Pseudomonas fluorescens promoting the growth and volatile oil accumulation in Atractylodes lancea. Plant Physiol Biochem 101:132–140

Zhou LS, Tang K, Guo SX (2018) The plant growth-promoting fungus (PGPF) Alternaria sp. A13 markedly enhances Salvia miltiorrhiza root growth and active ingredient accumulation under greenhouse and field conditions. Int J Mol Sci 19(1):270

Złotek U, Szymanowska U, Rybczyńska-Tkaczyk K, Jakubczyk A (2020) Effect of jasmonic acid, yeast extract elicitation, and drying methods on the main bioactive compounds and consumer quality of lovage (Levisticum Officinale Koch). Foods 9(3):323

# Chapter 6
# Natural Compounds Extracted from Medicinal Plants and Their Immunomodulatory Activities

**Vinod Kumar Gurjar and Dilipkumar Pal**

**Abstract** The fight against cancer cells in the human body involves a defense system that is comprised of the innate and adaptive immunities which are controlled by a series of immune responses mediated by different immune cells (ICs) and their secretory substances including cytokines and chemokines. Natural substances, synthetic compounds, and antibody elements are used as immunostimulating and immunosuppressive agents. But here are certain restrictions to the overall use of these compounds, such as the increased risk of infection and generalized effect throughout the immune system. The use of plants and plant products as immunomodulators is still in a developing stage. At non-cytotoxic concentrations, the phytoconstituents exhibited three types of immunomodulation including type 1 of PHA, ConA, and quercetin (increased lymphocyte activation and IFN-$\gamma$ secretion); type 2 of isopimpinellin (enhanced lymphocyte activation) and type 3 of rutin, bergapten and xanthotoxin (elevated IFN-$\gamma$ secretion). The augmentation of lymphocyte proliferation was closely correlated to an increase in the number of lymphocyte cells including T-helper lymphocytes ($CD4^+$), $CD8^+$ T cells and activated PBMC, whereas elevation of IFN-$\gamma$ secretion was due to the activated $CD8^+$ T cells. The present chapter revealed the immunomodulating activity, which could be explained the traditional use of medicinal plant extract worldwide.

**Keywords** Immunomodulatory activities · T-helper lymphocytes · IFN-$\gamma$ · $CD8^+$ T cells · $CD4^+$ T cells

V. K. Gurjar
School of Pharmacy, Parul University, P.O. Limda, Ta.Waghodia, 391760 District Vadodara, Gujarat, India

D. Pal (✉)
Institute of Pharmaceutical Sciences, Guru Ghasidas Vishwavidyalaya (A Central University), Bilaspur 495009, Chhattisgarh, India
e-mail: drdilip2003@yahoo.co.in

 
D. Pal and A. K. Nayak (eds.), *Bioactive Natural Products for Pharmaceutical Applications*, Advanced Structured Materials 140,
https://doi.org/10.1007/978-3-030-54027-2_6

**Abbreviations**

| | |
|---|---|
| ABCG2 | ATP-binding cassette super-family G member-2 |
| ADA | Adenosine deaminase |
| ADCC | Antibody-Dependent Cellular Cytotoxicity |
| AINFLCs | Anti-inflammatory Cytokine |
| AL | Alkaline Phosphatase |
| APCs | Antigen-Presenting Cells |
| BAOECs | Bovine Aortic Endothelial Cells |
| BRMs | Biological Response Modifiers |
| CAMs | Cell Adhesion Molecules |
| CMCs | Cortical Microglia Cells |
| COX-2 | Cyclooxygenase |
| Tc | Cytotoxic T Cells |
| DCs | Dendritic cells |
| DP | Dopamine |
| EGCG | Epigallocatechin-3-gallate |
| ELAM-1 | Endothelial Leukocyte Adhesion Molecule |
| elF 2 | Eukaryotic Initiation Factor 2 |
| ERK | Extracellular Signal-Regulated Kinases |
| GM-CSF | Granulocyte-Macrophage Colony-Stimulating Factor |
| HBMECs | Human Brain Microvascular Endothelial Cells |
| HECs | Human Mammary Cells |
| Hsp90 | Heat Shock Protein 90 |
| HMECs | Human Mammary Epithelial Cells |
| HSAECs | Human Primary Small Airway Epithelial Cells |
| HUVECs | Human Umbilical Vein Endothelial Cells |
| ICAM) | Intracellular Adhesion Molecule |
| ICAM-1 | Intercellular Adhesion Molecule 1 |
| ICs | Immune Cells |
| iNOS | Inducible Nitric Oxide Synthase |
| INFLCs | Inflammatory Cytokine |
| LOXs | Lipoxygenases |
| LPS | Lipopolysaccharide |
| MAPKs | Mitogen-Activated Protein Kinase |
| MCs | Mast Cells |
| MC3T3 | E1 Mouse Osteoblastic Cells |
| MCP-1 | Monocyte Chemoattractant Protein-1 |
| MIECs | Murine Intestinal Epithelial Cells |
| MIP-1β | Macrophage Inflammatory Protein |
| MMP-2 | Matrix Metalloproteinase-2 |
| MMPs | Matrix Metallopeptidases |
| MNGCs | Mesencephalic Neuron Glial Cells |
| mPGES-1 | Microsomal Prostaglandin E Synthase-1 |
| MPO | Myeloperoxidase |

| | |
|---|---|
| MSU | Monosodium Urate |
| NFAT | Nuclear Factor Of Activated T-Cells |
| NK | Natural Killer |
| NLGP | Neem Leaf Glycoprotein |
| NO | Nitric Oxide |
| NOX | NADPH oxidase |
| NTP | Nitro Tyrosine-Protein |
| PBMCs | Peripheral Blood Mononuclear Cells |
| PBMCs | Peripheral Blood Mononuclear Cells |
| PCNA | Proliferating Cell Nuclear Antigen |
| PINFLCs | Pro-inflammatory Cytokine |
| PINFLMs | Pro-inflammatory mediators |
| PHA | Pulmonary Arterial Hypertension |
| PhI | Phagocytic Index |
| PMA | Phorbol 12-Myristate 13-Acetate |
| PPARs | Peroxisome proliferator-activated receptors |
| RA | Rheumatoid Arthritis |
| ROS | Reactive Oxygen Species |
| SOCS3 | Suppressor of Cytokine Signaling 3 |
| STAT3 | Signal Transducers And Activators of Transcription 3 |
| TGF-β | Transforming Growth Factor |
| TLR4 | Toll-Like Receptor 4 |
| TREM2 | Triggering Receptor Expressed on Myeloid Cells 2 |
| VCAM-1 | Vascular Cell Adhesion Molecule 1 |
| VEGF-A | Vascular Endothelial Growth Factor A |

## 6.1 Introduction

The immune system is the body's defense mechanism against numerous common pathogens. The elements which activate the immune system include earlier infections, vaccination, and several outside stimuli. Also, immunity is accomplished of discerning between the proteins of the body, cells and foreign invaders. Once the external element is recognized, the cumulative and corresponding counter of precise cells and arbitrator against different elements establishes the immune response (Baxter 2007). Role-based immune mechanisms classified into two main classes, i.e., nonspecific defense mechanisms and the specific immune system, (Vesely et al. 2011). The physicochemical and microbiological hurdles further seldom counted in the nonspecific immune system, but, the key intermediaries of the immune mechanism which transfer immediate shield comprise macrophages, acute phase proteins, monocytes, cytokines, neutrophils and complement. Entire stages of innate immunity comprise APCs and macrophages that have a crucial function in antibody-dependent cellular cytotoxicity (ADCC), the release of cytokines, NO production and expression of antigen (Ag) molecules, dispensation and phagocytosis. The DCs are important for the stimulation of innate immune memory B and T cells. Through several stages of DCs' modification, the modulators of non-specific immunity along with NKCs are controlled, which regulates the adaptive and innate immune system by releasing IFN-γ, GM-CSF and TNF-α (Jantan et al. 2015). The integrated or complement system is the tertiary relevant factor of nonspecific immunity. The key effectors of humoral immunity amongst whole physiological perturbations of host defenses mechanism. The complement system's (CSs) mechanisms/components (C3a and C3b) are triggered by component C9, and increase and intervene immune response (Oh et al. 2012). Adaptive immunity or acquired immune system is developed by producing antigen-specific B and T lymphocytes over a gene reorganization procedure. The introduction of the body to an antigen (pathogen) to produce an acquired immune response that matures in weeks or months or maybe long-lasting or even life-long is termed as active immunity. Active immunity might be natural or artificially developed. The immune response of higher animals is specially equipped with adaptive immunity. The antigen-specific immune reactions are intimately involved in the acquired immune system. The active phagocytic activity of myelogenous cells (MCs) and Tc is increased by Th1 lymphocytes which release IFN-γ, TNF-α, and interleukin-2. The IL-4, IL-10 and IL-5, are composed of Th2 lymphocytes, classified by B lymphocytes-stimulated secretion of ABs. The pathogens, toxins, or bacteria are deactivated when binding with the antibodies (ABs). Besides, ABs can opsonize various impaired microbes, pathogens, and activate bacteria eradication by phagocytes via stimulation of various complement proteins (Puri et al. 1994; Jantan et al. 2015). The type of cells involved in the adaptive and innate immune system are outline in Fig. 6.1.

## 6.2 Immunomodulators

In a healthy body, the immune system maintains equilibrium inside the organism and protect from pathogens. The purpose and efficacy of the immune system are altered by several endogenous and exogenous influences that produce either immunostimulation or immunosuppression. Numerous substances having an activity to regulate or modify pathophysiological progress are termed as immunomodulators (Jantan et al. 2015). In other words, the biomolecules of biological origin or synthetic, capable of modulating, or normalizing, suppressing and stimulating or modify activity of any components of adaptive or innate immunity, decreasing the inflammatory responses are termed as immunoaugmentors, immunorestoratives, immunomodulators or biological response modifiers (BRMs). Immunomodulators are mostly classified into immunosuppressants, immunostimulants and immunoadjuvants in clinical practice. Immunoadjuvants are distinct immune mechanism stimulators that are added to vaccines, to enhance the immune

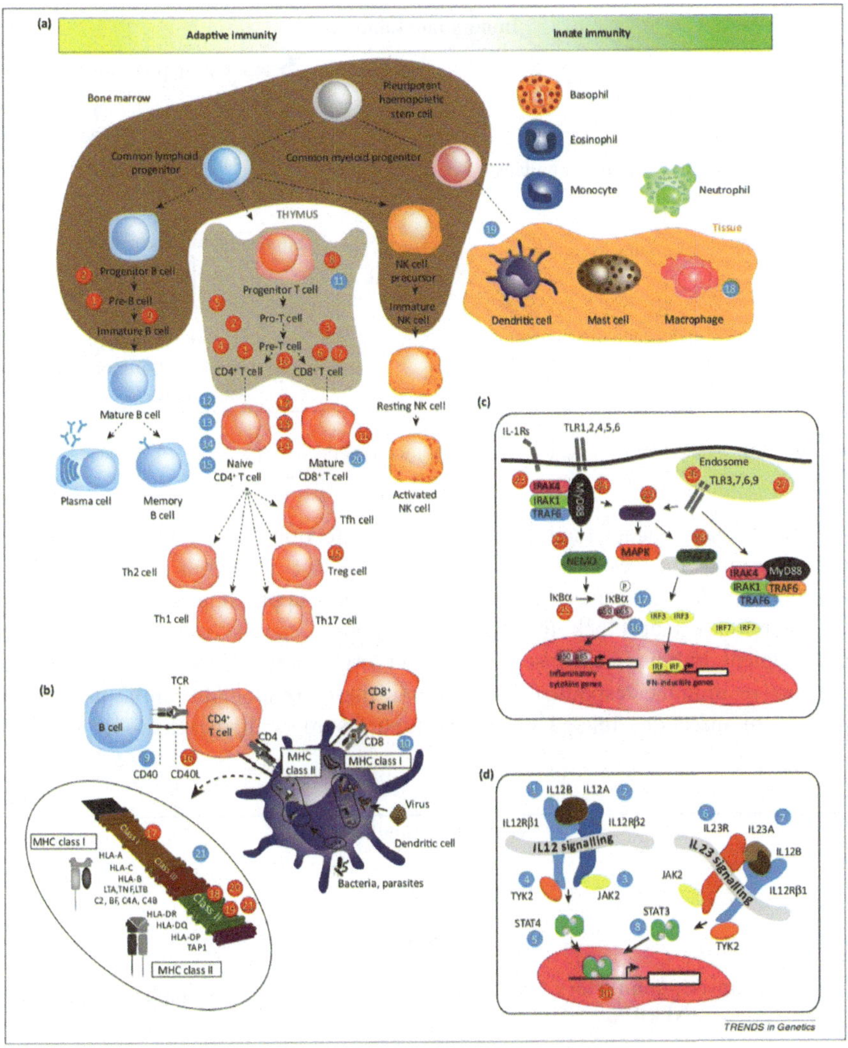


**Fig. 6.1** Overview of key mediators of innate and adaptive immunity, development, and signalling

response. Agents that stimulate or induce the mediators or increase the activity of components of the immune system are called immunostimulants or immunostimulators. Immunostimulators are used to restore the inefficiency in the immune system as observed in the treatment of diseases. They boost the immune system (Fig. 6.2).

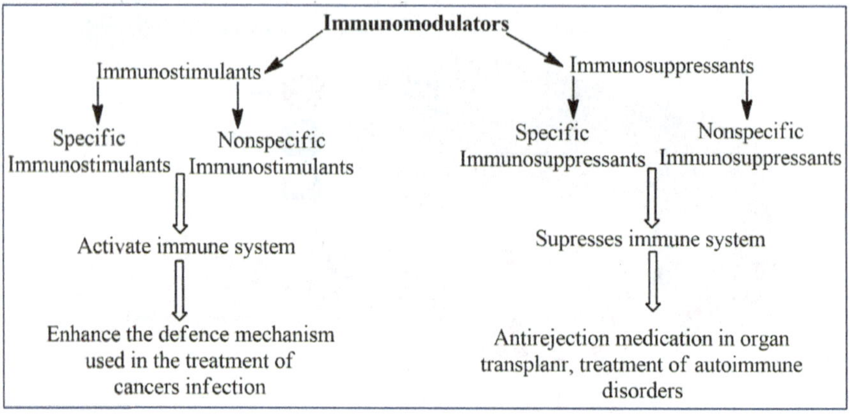


**Fig. 6.2** An overview of the action of immunomodulators

The resistance against autoimmunity, viral infection, tumor, allergy, immunodeficiency and the resistance of the body against infections is increased by immunostimulating agents. Alternatively, immune-suppressing agents are the drug or compounds that block the immune response, can be used to regulate the pathological immune response involved in organ transplants (prevent the graft destructive immune response). Furthermore, these compounds can be used in the therapy of infections related to autoimmune disorders, hypersensitivity reactions

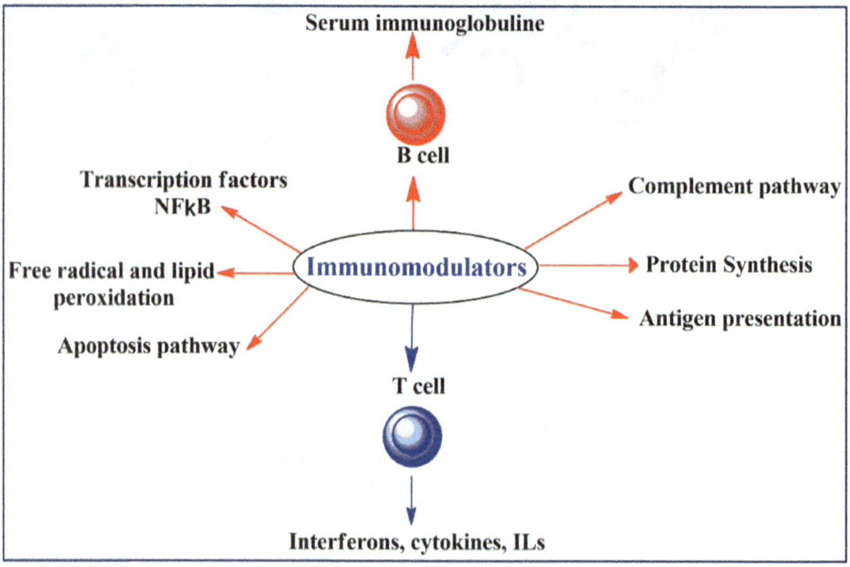


**Fig. 6.3** An overview of various targets and basic principles of functioning of immunomodulators

(HR) and immunopathology (IP) or diseases origin from an autoimmune disorder. Immunoadjuvants, on the other hand, can enhance the efficacy of vaccines, for example, Freud's adjuvant. Immunomodulators can modulate various cellular activities such as protein synthesis, apoptosis, antigen presentation, etc. and target various transcription factors and immune mediators (Fig. 6.3).

Some chemically synthesized compounds and monoclonal antibodies (mAb or moAb) are also being used as immunomodulators. Hence, immunomodulators with further protection and efficacy are still required. Owing to the incidence of chemical drugs-associated adverse reactions and effects, natural immunomodulators are the potential regimens to substitute them in therapeutic immunomodulatory agents (Puri et al. 1994). Presently, the major drug discovery and advancement aiming at biochemicals, biologics, or individual molecules as lead molecules that focus on specific targets thought to be linked to specific diseases. It is difficult to get single lead compounds with strong efficacy, selectivity, low toxicity and favorable drug-like properties for molecular and cellular targets and diseases. Thus, the design and development of a druglike molecule from several traditional or integral and substitute remedies is getting attention. The inhibition and therapy of various diseases by plant-derived drugs reported for most of human history (Puri et al. 1994; Mir et al. 2019; Pal and saha 2019; Pal et al. 2006, 2019; Pal and Mitra 2010; Pal and Dutta 2006; Pal 2015).

The present chapter gives an outline of widely studied plant-origin bioactive compounds, which displayed potent effects on cellular and humoral immune response in the pre-clinical evaluation and highlight the clinical potential. The immunomodulatory properties of plant-derived medicines have seen the interest of investigators. Advanced technologies and the undue study on plant extracts, natural immunomodulators, and their active constituents with immunomodulatory properties, may give us important constituents to get a novel immunomodulators to boost the current chemotherapies. This chapter an overview of the plant-derived bioactive immunomodulators which are under clinical studies. Furthermore, the possible practice as immunomodulators, mechanism of action and plant-derived compounds of various noteworthy plant-origin lead molecules has also been broadly studied (Fig. 6.4). Besides these agents, other phytoconstituents including alkaloids, essential oils, terpenoids, pigments, phenolics, flavonoids and steroids, etc., have shown valuable noted immunomodulatory activity. Plant-based bioactive agents display —Ú*f*d in Table 6.1.

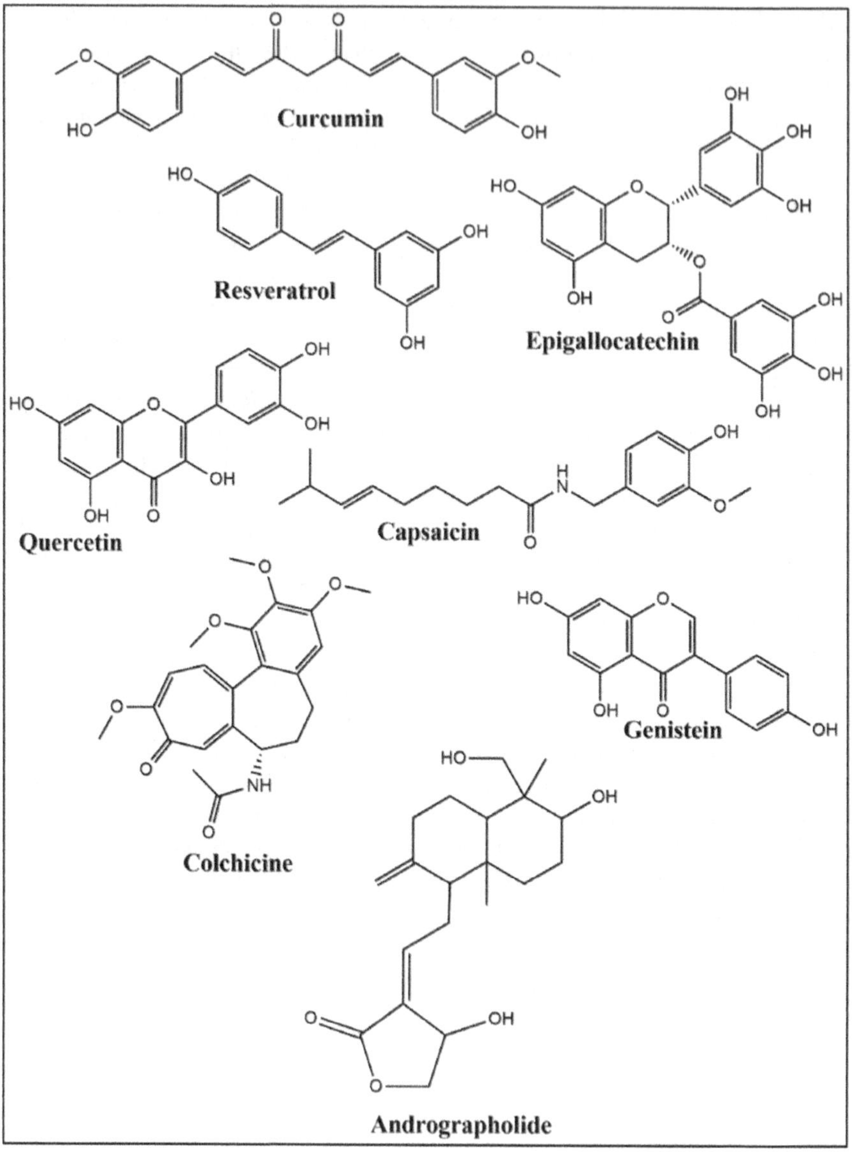


**Fig. 6.4** Chemical structures of selected plant-derived immunomodulators in clinical trials

**Table 6.1** Plant-based immunomodulator (Jantan et al. 2015)

| Chemical category | *Plant source* | Mode of action |
|---|---|---|
| **Alkaloids** | | |
| Berberine | *Coptis chinensis Franch* | Downregulate T-helper cells, Th2 (IL-4)] production, and cytokines TNF-α, Th1, IL-2 |
| Chelerythrine | *Chelidonium majus L* | Block PGE2 secretion by modifying COX-2 activity |
| Gelselegine | *Gelsemium elegans* | Block T lymphocyte propagation |
| Koumine | *Gelsemium elegans* | Block T lymphocyte propagation |
| Leonurine | *Leonurus japonicas Houtt* | Downregulate IL-6, TNF-α, COX-2 and iNOS, and upregulate IL-10 by blocking the appearance of toll like receptors and the stimulation of NF-κB Block the VCAM-1 and ICAM-1 action |
| Lycorine | *Lycoris radiate* | Block COX-2 and iNOS action |
| Piperine | *Piper longum Linn* | Decrease count of PINFCs IL-6, cytokines TNF-α,IL-1β. Down regulate expression of NOS-2, COX-2 and NF-κB. Block eicosanoide production by blocking PLA2 and TXA2 synthase action |
| Pseudocoptisine | *Corydalis turtschaninovii Besser* | Downregulating the phosphorylation of p38 and to ERK block NF-κB stimulation, which leads in lessening of PRINFL mediators counts (TNF-α, COX-2, and iNOS) |
| Rhynchophylline | *Uncaria rhynchophylla(Miq.) Jack* | Block phosphorylation of MAPK. Block release of PINFCs, PGE2, NO, monocyte chemoattractant protein IL-1β, MCP-1, TNF-α |

(continued)

**Table 6.1** (continued)

| Chemical category | *Plant source* | Mode of action |
|---|---|---|
| Sinomenine | *Sinomenium acutum (Thunb.) Rehd.etWils* | Block the release of INFLCs. Block expression of VCAM-1 |
| Sophocarpine | *Sophora alopecuroides L* | Block production of NO and INFLCs. block expression of COX-2 and iNOS |
| Tetrandrine Matrine | *Stephania tetrandra Matrine. Sophora flavescens Ait* | Decreased release of ROS and INFL mediators. Block maleic dialdehyde and myeloperoxidase function |
| *Essential oils* | | |
| Z-ligustilide | *Angelica sinensis (Oliv.) Diels* | Block COX-2 and iNOS initiation by controlling the NF-kB and MAPK signal pathways |
| Tetramethylpyra-zine | *Ligusticum chuanxiong Hort* | Block INFLCs and ROS production. Block macrophages neutrophile infiltration, chemotaxis, and NO synthase action Block the phosphorylation of p38 MAPK |
| **Flavonoids Chalcone** | | |
| Butein | *Semecarpus anacardium, Dalbergia odorifera, Toxicodendron vernicifluum* | Inhibit NO production by reducing iNOS appearance. Block translocation of NF-κB |
| Dihydroxanthohumol | *Humulus lupulus* | Block NO production induced by INF-γ and LPS |
| Licochalcone E | *Glycyrrhiza inflata* | Block release of PINFLCS blocking the activity of NF-κB and activator protein (AP-1) |
| Mallotophilippens C, D, E | *Mallotus, Philippinensis* | Block NO production induced by INF-γ and LPS. mRNA gene expression of IL-1β, COX-2, iNOS and IL-6. Deactivate NF-κB |
| Xanthohumol | *Humulus lupulus* | Block NO production that is induced by INF-γ and LPS |

(continued)

**Table 6.1** (continued)

| Chemical category | *Plant source* | Mode of action |
|---|---|---|
| **Flavones** | | |
| *Luteolin* | *Lonicera Japonica* | Reduce the release of inflammatory mediators, reduce ICAM-1 expression COX-2. Inhibit Hsp90 activity |
| Apigenin | *Cynodon dactylon, Salvia officinalis L., Portulaca oleracea, Mentha longifolia* | Downregulate the expression of cytokines. Decrease response of Th1 and Th17 cells. Downregulate the expression of iNOS and COX-2. Reduced expression of ICAM and VCAM leading to lessened neutrophile Chemotaxis |
| Chrysin | *Picea crassifolia* | Block release of PINFCs by modulating intracellular-calcium decrease histamine release from MCs |
| Nobiletin | *Citrus nobilis Lour, Citrus aurantium L* | Block proinflammatory mediators, iNOS and COX-2 expression by blocking MAPK and NF-κB signaling pathway |
| Baicalein | *Scutellaria altissima L.Georgi* | Block mRNA appearance of TNF-α, COX-2, and iNOS<br>Block release of NO and INFCs by regulating NF-κB and ER-dependent pathway |
| Oroxylin A | *Scutellariae baicalensis* | Block NO production and COX-2 and iNOS proteins expression via inhibiting the nuclear factor-κB pathway |
| Wogonin | *Scutellaria baicalensis Georgi* | Block adhesion and migration of leukocytes by Blocking CAMs expression. Decrease allergic airway inflammation |

(continued)

**Table 6.1** (continued)

| Chemical category | *Plant source* | Mode of action |
|---|---|---|
| **Flavonols** | | |
| Quercetin | *Dysosma veitchii Hemsl. et Wils* | Reduced expression of PINFLCS, iNOS, and NF-κB<br>Decrease expression of E-selectin and VCAM-1 |
| Kaempferol | *Found in various fruits and vegetables. e.g., tea, tomato* | Decrease COX-2 and iNOS activity through suppression of the signaling of AP-1, NF-kappa B, and STAT-1<br>Reduce the gene expression of VCAM-1, ICAM-1, MCP-1 |
| Rutin | *Ruta graveolens* | Block leukocyte relocation. Suppress release of TNF-α and IL-6. Reduce the activation of NF-κB |
| **Flavanols** | | |
| Epigallocatechin-3gallate | *Camellia sinensis L* | Block MAPKs phosphorylation, ROS production, adhesion molecules expression STAT-3 and STAT-2 |
| **Isoflavones** | | |
| Daidzein | *Pueraria mirifica, Pueraria lobata, Glycine max* | Reduce IL-1β, TNF-α, MCP-1, iNOS and NO expression at the mRNA level |
| Genistein | *Glycine max* | Block expression of COX-2 and iNOS<br>Decrease TNF-α and IL-1β production via activation of PPARs |
| Puerarin | *Pueraria lobata (wild)Ohwi* | Reduce cytokines count. Block NF-κB and stimulation of STAT3 |
| ***Phloroglucinols*** | | |
| Myrtucommulone | *Myrtus communis L* | Block the PGE2 release by blocking the mPGES-1 action automatically blocking the COX activity |
| Arzanol | *Helichrysum italicum* | Decrease eicosanoids generation by blocking LOXs and COX action in arachidonic acid metabolism pathways |

(continued)

**Table 6.1** (continued)

| Chemical category | *Plant source* | Mode of action |
|---|---|---|
| Emodin-8-O-β-D glucoside | *Polygonum amplexicaule D. Don var. sinense Forb* | Activate the production and separation of osteoblasts<br>Block PGE2 release by enhancing ALP expression in MC3T3-E1 |
| **Other** | | |
| Apocynin | *Apocynum cannabinum L. (Canadian hemp), Picrorhiza kurroa Royle ex Benth* | Block NOX action. Suppress PINFLCs, $CD4^+$ and $CD8^+$ T cell release |
| **Stilbenes** | | |
| Resveratrol | *Fallopia japonica, grape, nuts* | Reduce MPO action and mPGES-1 to basal levels. Inhibit COX-2 and iNOS appearance. Decrease PINFLCs. Increase the AINFCs IL-10 level |
| Piceatannol | *Fallopia japonica, nuts, grape* | Reduce iNOS expression. Block transcription factors stimulation such as ERK, NF-kB, and STAT3 |
| **Terpenoid** | | |
| 14-deoxy-11,12didehydroandrographolide | *Andrographis paniculata* | Increase the production of lymphocytes. Increase IL-2 initiation in lymphocytes |
| Ginsan | *Panax ginseng* | Increase the release of cytokines and ROS by macrophages. Activate the phagocytic property of macrophages |
| Oleanolic acid | *Luffa cylindrica, Phytolacca americana* | Decrease the count of TNF-α IL-1α and IL-6 along with their effect on the complement pathway by blocking C3 convertase. Blocks ADA action |
| Echinocystic acid | *Luffa cylindrica* | Block PhI of macrophages in antibody and cell-mediated immune responses |

(continued)

**Table 6.1** (continued)

| Chemical category | *Plant source* | Mode of action |
|---|---|---|
| Triptolide | *Tripterygium wilfordii* | Blocks lymphocyte stimulation and PINFLCs gene appearance. Blocks stimulation of STAT3, NFAT and NF-kB |
| Demethylzelasteral | *Tripterygium wilfordii* | Blocks the production of vascular ECs |
| Celastrol | *Tripterygium wilfordii* | Block expression of PINFLCs, topoisomerase II and proteasome action, |
| Asiaticoside | *Centella asiatica* | Reduce NO release |
| Madecassoside | *Centella asiatica* | Decrease spleen cell proliferation Blocking of PINFLMs. Block release of COX-2 and PGE2 |
| 11-keto-β-boswellic acid | *Boswellia carteri* | Reduce PINFLCs through blocking of NF-kB stimulation |
| **Flavonoids Chalcone** | | |
| Butein | *Semecarpus anacardium, Dalbergia odorifera, Toxicodendron vernicifluum* | Inhibit NO production by reducing iNOS appearance. Block translocation of NF-κB |
| Dihydroxanthohumol | *Humulus lupulus* | Block NO production induced by INF-γ and LPS |
| Licochalcone E | *Glycyrrhiza inflata* | Block release of PINFLCS blocking the activity of NF-κB and activator protein (AP-1) |
| Mallotophilippens C, D, E | *Mallotus, Philippinensis* | Block NO production induced by INF-γ and LPS. mRNA gene expression of IL-1β, COX-2, iNOS and IL-6. Deactivate NF-κB |
| Xanthohumol | *Humulus lupulus* | Block NO production that is induced by INF-γ and LPS |
| **Flavones** | | |
| *Luteolin* | *Lonicera Japonica* | Reduce the release of inflammatory mediators, reduce ICAM-1 expression COX-2. Inhibit Hsp90 activity |

(continued)

**Table 6.1** (continued)

| Chemical category | *Plant source* | Mode of action |
|---|---|---|
| Apigenin | *Cynodon dactylon, Salvia officinalis L., Portulaca oleracea, Mentha longifolia* | Downregulate the expression of cytokines. Decrease response of Th1 and Th17 cells. Downregulate the expression of iNOS and COX-2. Reduced expression of ICAM and VCAM leading to lessened neutrophile Chemotaxis |
| Chrysin | *Picea crassifolia* | Block release of PINFCs by modulating intracellular-calcium decrease histamine release from MCs |
| Nobiletin | *Citrus nobilis Lour, Citrus aurantium L* | Block proinflammatory mediators, iNOS and COX-2 expression by blocking MAPK and NF-κB signaling pathway |
| Baicalein | *Scutellaria altissima L.Georgi* | Block mRNA appearance of TNF-α, COX-2, and iNOS<br>Block release of NO and INFCs by regulating NF-κB and ER-dependent pathway |
| Oroxylin A | *Scutellariae baicalensis* | Block NO production and COX-2 and iNOS proteins expression via inhibiting the nuclear factor-κB pathway |
| Wogonin | *Scutellaria baicalensis Georgi* | Block adhesion and migration of leukocytes by Blocking CAMs expression. Decrease allergic airway inflammation |
| **Flavonols** | | |
| Quercetin | *Dysosma veitchii Hemsl. et Wils* | Reduced expression of PINFLCS, iNOS, and NF-κB<br>Decrease expression of E-selectin and VCAM-1 |

(continued)

**Table 6.1** (continued)

| Chemical category | *Plant source* | Mode of action |
|---|---|---|
| Kaempferol | *Found in various fruits and vegetables. e.g., tea, tomato* | Decrease COX-2 and iNOS activity through suppression of the signaling of AP-1, NF-kappa B, and STAT-1<br>Reduce the gene expression of VCAM-1, ICAM-1, MCP-1 |
| Rutin | *Ruta graveolens* | Block leukocyte relocation. Suppress release of TNF-α and IL-6. Reduce the activation of NF-κB |
| **Flavanols** | | |
| Epigallocatechin-3gallate | *Camellia sinensis L* | Block MAPKs phosphorylation, ROS production, adhesion molecules expression STAT-3 and STAT-2 |
| **Isoflavones** | | |
| Daidzein | *Pueraria mirifica, Pueraria lobata, Glycine max* | Reduce IL-1β, TNF-α, MCP-1, iNOS and NO expression at the mRNA level |
| Genistein | *Glycine max* | Block expression of COX-2 and iNOS<br>Decrease TNF-α and IL-1β production via activation of PPARs |
| Puerarin | *Pueraria lobata (wild)Ohwi* | Reduce cytokines count. Block NF-κB and stimulation of STAT3 |
| **Phloroglucinols** | | |
| Myrtucommulone | *Myrtus communis L* | Block the PGE2 release by blocking the mPGES-1 action automatically blocking the COX activity |
| Arzanol | *Helichrysum italicum* | Decrease eicosanoids generation by blocking LOXs and COX action in arachidonic acid metabolism pathways |
| Emodin-8-O-β-D glucoside | *Polygonum amplexicaule D. Don var. sinense Forb* | Activate the production and separation of osteoblasts<br>Block PGE2 release by enhancing ALP expression in MC3T3-E1 |

(continued)

**Table 6.1** (continued)

| Chemical category | *Plant source* | Mode of action |
|---|---|---|
| **Other** | | |
| Apocynin | *Apocynum cannabinum L. (Canadian hemp), Picrorhiza kurroa Royle ex Benth* | Block NOX action. Suppress PINFLCs, $CD4^+$ and $CD8^+$ T cell release |
| **Stilbenes** | | |
| Resveratrol | *Fallopia japonica, grape, nuts* | Reduce MPO action and mPGES-1 to basal levels. Inhibit COX-2 and iNOS appearance. Decrease PINFLCs. Increase the AINFCs IL-10 level |
| Piceatannol | *Fallopia japonica, nuts, grape* | Reduce iNOS expression. Block transcription factors stimulation such as ERK, NF-kB, and STAT3 |
| **Terpenoid** | | |
| 14-deoxy-11,12didehydroandrographolide | *Andrographis paniculata* | Increase the production of lymphocytes. Increase IL-2 initiation in lymphocytes |
| Ginsan | *Panax ginseng* | Increase the release of cytokines and ROS by macrophages. Activate the phagocytic property of macrophages |
| Oleanolic acid | *Luffa cylindrica, Phytolacca americana* | Decrease the count of TNF-α IL-1α and IL-6 along with their effect on the complement pathway by blocking C3 convertase. Blocks ADA action |
| Echinocystic acid | *Luffa cylindrica* | Block PhI of macrophages in antibody and cell-mediated immune responses |
| Triptolide | *Tripterygium wilfordii* | Blocks lymphocyte stimulation and PINFLCs gene appearance. Blocks stimulation of STAT3, NFAT and NF-kB |

(continued)

**Table 6.1** (continued)

| Chemical category | *Plant source* | Mode of action |
|---|---|---|
| Demethylzelasteral | *Tripterygium wilfordii* | Blocks the production of vascular ECs |
| Celastrol | *Tripterygium wilfordii* | Block expression of PINFLCs, topoisomerase II and proteasome action, |
| Asiaticoside | *Centella asiatica* | Reduce NO release |
| Madecassoside | *Centella asiatica* | Decrease spleen cell proliferation Blocking of PINFLMs. Block release of COX-2 and PGE2 |
| 11-keto-β-boswellic acid | *Boswellia carteri* | Reduce PINFLCs through blocking of NF-kB stimulation |

### 6.2.1 Plant-Derived Bioactive as Immunomodulators

Plant extracts and phytocompounds are found to fortify the host's immune system, and numerous plants have been listed in this category. Phytoimmunomodulatory agents can increase the body's immune-responsiveness against pathogens by activating the immune system in a specific or non-specific manner that includes both the innate and adaptive immune systems. Immunomodulatory plants play a pivotal role in the treatment of infection, inflammation, and immunodeficiencies by their effect on various cell types via cytokines and interleukins. The mode of action could be as immunostimulators, immunosuppressors, or immunoadjuvants to boost antigen-specific immune response (Nair et al. 2019). Plant products are widely considered as immunopotentiators and collectively known as biological response modifiers (BRMs). Many dietary phytomolecules, vitamins, and minerals have a protective role in cellular nutrition and the management of diabetic complications. The immunomodulatory function attributed to BRMs is possibly through modulation of the ICs of the body like the macrophages, the lymphocytes (B-cells and T-cells), Dendritic Cells (DCs), etc. For example, Concanavalin A lectin, a carbohydrate-binding protein from *Canavalia ensiformis* can cross-link glycoproteins like TCR/CD3 and thus activate T-lymphocytes. Similarly, *Shorea robusta*, a well-known traditional medicine of India can modulate NO, prostaglandins E2, TNF-α, iNOS expression with anti-inflammatory and wound-healing activity, while one of its major phytochemical Bergenin enhances T helper-1 responses affording antimycobacterial immunity by activating the MAP kinase pathway in macrophages (Nair et al. 2019).

#### 6.2.1.1 Curcumin

Curcumin is one of the natural diarylheptanoid belonging to the group of curcuminoids, which are natural phenols found in the rhizome of turmeric (*Curcuma longa*) and other *Curcuma* species. The medicinal properties of curcumin used for centuries in Ayurvedic medicine. Curcumin has a variety of therapeutic properties like anti-proliferative, anti-cancer, pro-apoptotic, anti-angiogenic, antioxidant, etc. It is one of the most reviewed and studied plant-derived bioactive molecules for its immunomodulatory activity. Curcumin minimizes the inflammatory responses (inflammation) by blocking nitric oxide (NO) production, (Surh et al. 2001) COX-2, iNOS, NK-kB, inducible and LOX in IFN-γ and NK cells or TNF-α-stimulated macrophages. In PMA and hydrogen peroxide activated human acute myeloid leukemia cell lines, curcumin blocked NF-κB stimulation via inhibition of breakdown and phosphorylation of I kappa B alpha (IκB-α). The PKC, which controls the production and survival of the cell, is stimulated by PMA. Furthermore, lipopolysaccharide (LPS) and TNF-α also stimulate PKC, which then stimulates NF-κB (Holden et al. 2008). Hence, curcumin might lessen NF-kB stimulation by the blocking of PKC. The anti-inflammatory property of curcumin moderately facilitated by preventing the activator protein-1 (AP-1) and NF-κB. The AP-1 and NF-κB act together and may increase tumor growth. Curcumin lessened the binding of NF-κB and AP-1 on the treatment of glioma cells (Dhandapani et al. 2007). The AP-1-stimulation also blocked by curcumin in TNF- α-stimulated bovine aortic endothelial cells (BAOECs).

The stimulated ICs to produce PINFLCs that play an important role in various inflammation diseases. The expression of PINFLCs like IL-1, IL-6, IL-12 and TNF-α, blocked by curcumin via LPS or PMA-activated macrophages, monocytes, splenic lymphocytes and DCs (Gao et al. 2004; Kim et al. 2005). The binding of T cells to endothelial and APCs is dependent on CAMs. The binding of monocytes to endothelial cells inhibited by pre-treatment with curcumin. Furthermore, the appearance of ICAM-1, VCAM-1 and ELAM-1 also reduced in TNF-α-activated HUVECs via blocking of NF-κB (Kumar et al. 1998). Curcumin improved the RANKL-mediated differentiation, fusion and development of osteoclasts and has an immunomodulatory result on macrophage polarization. The defensive action of curcumin on osteoclast genesis facilitated by decreasing the up-regulation of Akt and p65 phosphorylation and the stimulation of the downstream transcription factor NFATc1 (Yang et al. 2020). Curcumin treatment reduced activation of the NFkB, MAPK, AKT and pBAD pathways either systemically, or within the inflamed kidneys (Wu et al. 2020). Curcumin is capable of emulating anti-Aβ vaccine in stimulating phagocytic clearance of amyloid by decreasing CD33 and increasing TREM2 and transmembrane immune signaling adaptor (TyroBP) though improving neuroinflammatory systems concerned in neurodegenerative diseases. A low dose of curcumin decreased CD33 (Siglec-3) and increased triggering receptor expressed on myeloid cells 2 (TREM2) expression and also increased TyroBP, which controls a neuroinflammatory gene network implicated in AD, in addition to phagocytosis markers cluster of differentiation 68 (CD68) and Arginase 1. A low dose of curcumin uniformly reestablished closely associated relations among these genes expression levels and decreased expression of genes characteristic of toxic proinflammatory M1 microglia. It stimulated microglial migration to and phagocytosis of amyloid plaques both *in-vivo* and in ex vivo assays of sections of the human AD brain and mouse brain. Curcumin also reduced levels of miR-155, a micro-RNA reported driving a neurodegenerative microglial phenotype. Similarly, it decreased CD33 and increased TREM2. Like curcumin, anti-Aβ antibody increased TREM2 in APPsw mice and decreased amyloid in human AD sections ex vivo (Teter et al. 2019).

#### 6.2.1.2 Resveratrol

Chemically resveratrol is (3,5,4′-trihydroxytrans-stilbene). Resveratrol is a stilbenoid polyphenol present in many nutritional foods and floras including red wine grapevines, and peanuts. Like curcumin, resveratrol shows a broad range of pharmacological activities like anti-inflammatory, anticancer/proapoptotic, chemopreventive, antioxidant and antimicrobial (Malaguarnera 2019). Resveratrol actively blocks inflammatory particles. The immunomodulatory actions of resveratrol comprise the blocking of NF-κB in LPS, PMA or TNF-α-activated epithelial (HeLa), macrophages, Jurkat, DCs and myeloid (U-937). Resveratrol block NF-κB stimulation via blocking of IκB kinase (IKK) activity (Holmes-McNary and Baldwin 2000). Resveratrol also blocks the expression of COX-2 and iNOS in cytokines excited human primary small airway epithelial cells (HSAECs) although it also blocks the transcription of COX-2 in HMECs via PMA stimulation. Resveratrol also significantly blocks the release of TNF-α and NO in LPS-stimulated N9 microglial and CMCs (Bi et al. 2005), the production of PINFCs by splenic macrophages and also block lymphocytes (Kowalski et al. 2005). It also actively inhibits C5 anaphylatoxin (C5a)-activated inflammation *in-vivo*. The secretion of INFCs in C5a-stimulated human and mouse neutrophils block by the pre-incubation with resveratrol. Resveratrol also blocks ERK-phosphorylation, production of glucuronide and C5a mediated oxidative burst. Additionally, resveratrol blocks the production of INFCs and C5a-stimulated neutrophil recruitment in the C5a-activated severe peritonitis mouse model (Issuree et al. 2009). Resveratrol inhibits the expression of CAMs. It is also found that resveratrol decrease IL-6-activated ICAM-1 appearance in ECs (Wung et al. 2005), further to the blocking of *Porphyromonas gingivalis* LPS-activated endothelial dysfunction in HMECs. Resveratrol also, block VCAM-1 and ICAM-1 appearance on HMECs by inhibiting the NF-κB activation (Park et al. 2009). Resveratrol acts by various cellular signaling pathways and shows anti-inflammatory activity. Resveratrol can potentiate its tumor-suppressive effect through modulation of the signaling pathways of cellular components (fibroblasts, macrophages and T cells). Also, studies have shown that resveratrol can suppress malignant phenotypes of cancer cells acquired in response to stresses of the tumor microenvironment, such as hypoxia, oxidative stress and inflammation (Han et al. 2019).

Resveratrol decreases MSU-induced recurrent attacks of gouty arthritis. Despite its demonstrated anti-inflammatory effects, the mechanisms underlying resveratrol-mediated repression of IL-1β production in MSU-activated monocytes remain poorly understood. Resveratrol suppresses the secretion of active IL-1β by human primary monocytes stimulated with MSU crystals through suppression of Syk activation (Chung et al. 2019). Antioxidant or cytotoxic properties, of resveratrol, able to alter the process of differentiation of naive T lymphocytes into Th17 lymphocytes through activation in particular of sirtuin-1, and to reduce the production of pro-inflammatory substances such as interleukins. This action on the cells of the immune system thus highlights a new property of resveratrol as an immunomodulator that could counteract the occurrence or progression of inflammation in various pathological processes (Delmas et al. 2019).

#### 6.2.1.3 Epigallocatechin-3-Gallate

Polyphenols are a big family of phytochemicals that includes a wide range of natural substances with various biological activities. Amongst these activities, the immunomodulatory property has significant importance due to the central and vital roles of the immune system in the human body (Pan et al. 2019; Sobhani et al. 2020). It is abundant in phenolic composition of green tea, *Camellia sinensis* (Fm: Theaceae). Epigallocatechin-3-gallate (EGCG) is the ester of epigallocatechin and gallic acid. It has a broad range of reported in-vivo and in-vitro anti-inflammatory, anti-proliferative, chemopreventive, anti-invasive, and anti-oxidant anti-angiogenic activity (Singh et al. 2011; Yang et al. 2011). It blocks the stimulation of NF-κB by blocking the degeneration of inhibitor of nuclear factor kappa B (Muraoka et al. 2002) and also blocks the MAPK pathways. The downregulation of iNOS transcription and NO release in macrophages depend on blocking of NF-κB. EGCG increases the production of NO to block endothelial exocytosis and to leukocyte adherence ECs (Yamakuchi et al. 2008). Furthermore, that EGCG blocks NF-κB stimulation in HECs and therefore it inhibits the expression of MCP-1 (Hong et al. 2007).

#### 6.2.1.4 Quercetin

The plant flavonol, quercetin, is a polyphenol, describing very broadly spreading secondary plant metabolites. Food like grapevines, apples, broccoli, red onions, berries, and tea, capers are a rich source of quercetin. It shows anti-mutagenic, neuroprotective, anti-oxidative, anti-inflammatory, anticancer/chemopreventive, antihypertensive activities. It also helps to lower blood-glucose-level. It activates several kinase enzymes that phosphorylate elF 2, hence blocking cell translation. The underlying mechanisms behind these activities are broad and extensively categorized. Quercetin scavenges nitrogen and ROS, targets notable proinflammatory signaling pathways including NF-κB, MAPK and STAT1, and block replication of various forms of viruses and infections of target cells (Boots et al. 2008). It inhibits the action of iNOs and COX-2 by suppressing NF-κB, AP-1 and STAT-1 signaling in cytokines or LPS-activated macrophages and HUVECs (Hamalainen et al. 2007). Quercitin decreases the appearance of PINFLCs in calcium ionophore and PMA-activated mast cells. Moreover, quercetin also blocks the TNF-α-activated NF-κB recruitment to proinflammatory gene promoters in MIECs (Ruiz et al. 2007; Park et al. 2008). Quercitin reduces the TNF-α- or PMA-activated expression of ICAM-1 in HECs. Quercetin also inhibits LPS-activated NF-κB and NO production in mice. It modulates Th1/Th2 cytokine dysregulation in asthma and types 1 diabetes (Ravikumar and Kavitha 2020). Quercetin, attenuate the proinflammatory phenotype and function of DCs *in-vitro*. Quercetin is a potent immunomodulatory agent to alter human DC phenotype and function, shifting the immune balance from inflammation to resolution (Jantan et al. 2015; Michalski et al. 2020).

#### 6.2.1.5 Colchicine

Colchicine is a tropolone derivative and it is the main alkaloid of *Colchicum autumnale* (family: Colchicaceae), commonly known as autumn crocus or meadow saffron. The extracts have been used to treat gout for centuries. Colchicine has approved by the US FDA for the prevention and therapy for familial Mediterranean fever and acute gout flares. Colchicine inhibits activation and migration of neutrophils to sites of inflammation it also inhibits microtubule polymerization by binding to its constitutive protein, tubulin (Bhattacharyya et al. 2008; Stanton et al. 2011).

#### 6.2.1.6 Capsaicin

Capsaicin is a hydrophobic alkaloid chemically it is 8-methyl-N-vanillyl-6-nonenamide major active phytoconstituents of chili peppers *Capsicum* sp; belong to family *Solanaceae*, and accountable for the typical spiciness/pungency of chili peppers. It is used in classical and folk medicine as a counter-irritant and topical rubefacient to reduce the pain of joints and muscles (Caterina et al. 1997). Capsaicin suppresses the production of cytokines at low concentrations, and it blocks the release of cytokines at variable concentrations whereas it stimulates IL-6 (Bessler 2016). Furthermore, capsaicin found to inhibit the iNOS countenance and COX-2 expression in the macrophages in a transient receptor potential vanilloid 1 (TRPV1)-independent way. In an inflammation-based experiment, topical capsaicin found to be inactive against osteoarthritis (Cameron and Chrubasik 2013).

#### 6.2.1.7 Andrographolide

It is a labdane diterpenoid obtained from *Andrographis paniculate* (Acanthaceae). Andrographolide exhibit varied biological activities. Several immunomodulatory properties of andrographolide studied *in-vitro* with lessening of cytokines in macrophages and microglia (Maiti et al. 2006). Andrographolide block the LPS-activated COX-2 and iNOS countenance in RAW264.7 cells. Andrographolide also blocks Akt and Erk ½ signaling, subsequently blocking the chemotaxis movement of macrophages on inflammation site (Tsai et al. 2004; Qin et al. 2006). Andrographolide blocks the production of ROS in neutrophils (Shen et al. 2002). It also regulates the secretion of immunoregulatory cytokines and chemokines such as TNF-α, IFN-γ, IL-2 and NK cells. The andrographolide administered at different doses resulting in the enhanced appearance of CD signs and the release of TNF-α, thus raising the cytotoxic efficacy of lymphocytes (Rajagopal et al. 2003). It inhibits IFNγ, IL-2, and IL-6 appearance, thus reducing the cellular and humoral acquired immune response in T cells. It also reduces the antigen-cited action of DCs to T cells. Andrographolide decreases the serum Ig, ILs, and Th2 cytokine, shown in a study on the ovalbumin (OVA)-induced asthma rat model. It decreases movement and ECs proliferation, invasion, and intercellular adhesion molecule 1 (ICAM-1), suggesting its function in angiogenesis. The activity of NF-κB blocked by several immunomodulatory reactions and andrographolide block NF-κB binding to DNA and resulting in decreasing proinflammatory protein expressions (Hidalgo et al. 2005; Iruretagoyena et al. 2005). Andrographolide downregulates iNOS and COX-2 gene countenance by blocking STAT3 and NF-κB countenance via inhibiting the appearance of suppressor of cytokine signaling 1 and 3 (Lee et al. 2011; Zhang et al. 2014) shown in a study to control the role of andrographolide on insulinoma growth. Andrographolide prevents the growth of insulinoma by targeting the TLR4/NF-κB signaling pathway (Zhang et al. 2014). It also inhibits the growth of cancer cells via several mechanisms, such as cytotoxic activity, induction of cell cycle arrest, induction of apoptosis, immunomodulatory effect, anti-inflammatory and anti-angiogenic activities and chemoprotective mechanism. Furthermore, andrographolide also decreases the expression of Bcl-2 studied by immunohistochemistry analysis (Rajaratinam and Nafi 2019).

#### 6.2.1.8 Genistein

It is a naturally occurring phytoestrogen, found in soy and soy-derived products. chemically it is 4,5,7-trihydroxyisoflavone, a prominent tyrosine kinase blocker. It inhibits IL-1b/IFNγ-mediated iNOS and COX-2 expression. Furthermore, producing NO and PGE2 in human islets that can stop the pathogenesis of diabetes and increase insulin resistance. Genistein (Jantan et al. 2015) induces apoptosis similar to other topoisomerase blockers. It actively blocks angiogenesis, exert a blocking effect on proliferating cells. It controls vascular action and protects against atherosclerosis (Si and Liu 2007). Genistein blocked the countenance of

TNF-α-stimulated cell adhesion molecule CD62E and CD106 and monocyte adhesion when administered in HBMECs and HUVECs (Lee and Lee 2008). It also decreases the interaction among the ECs and monocyte via stimulation of PPARs that decrease monocyte adhesion in culture cells and animals. Genistein blocks the LPS-stimulated release of MCP-1 from macrophages that activated in reduced migration of monocyte in-vitro (Nagarajan et al. 2008). It blocks the LPS-mediated countenance of NTP and iNOS in vascular tissue that blocks vascular changes and hypotension. It exerts a potential effect on neurodegenerative diseases, diabetes mellites (DM), metabolic syndromes, rheumatoid arthritis (RA) and chronic colitis by modulating inflammatory response (Jantan et al. 2015). Genistein modulated the Th1-dominant immune reaction by rising IL-4 secretion and inhibit the release of IFN-γ in a collagen-mediated rheumatoid arthritis rat (Wang et al. 2008). Nonalcoholic fatty liver disease (NAFLD) is an obesity-related fatty liver disease initiated by PINFLCs and TNF-α and leads to rising fatty acid uptake and the defunctions of hepatic cells. It decreases the high fat diet-stimulated hepatic steatosis by enhancing liver performance and reducing the level of plasma TNF-α count in rats (Yalniz et al. 2007). Furthermore, genistein decreased LPS-induced DP uptake loss in rat MNGCs by reducing the release of NO, TNF-α and microglia stimulation this might prevent the pathogenesis of Parkinson's disease produced by dopaminergic neuron damage. The growth of astrocytes at Ab accumulation sites is the early neuropathological change that initiates inflammation in Alzheimer's disease (AD). The NOD, LRR- and pyrin domain-containing protein 3 (NLRP3) inflammasome reported being associated with inflammatory bowel disease including colitis due to its potential ability to induce IL-1β secretion. Genistein blocks NLRP3 inflammasome via TGR5-cAMP signaling in macrophages and it can be a possibly effective therapeutic drug for inflammatory bowel diseases (Chen et al. 2019).

### *6.2.2 Classification of Immunomodulators*

Several factors play a deciding role in determining the efficacy of an immunomodulatory compound such as its **mechanism of action** structural conformation, molecular weight, and solubility Table 6.2.

#### 6.2.2.1 Based on the Mechanism of Action

See Table 6.2.

**Table 6.2** Classification of immunomodulators (Konidena et al. 2017)

| | Immunosuppressants | Immunostimulants | |
|---|---|---|---|
| Mechanism of action | Examples | Types | Examples |
| Blocking of lymphocyte gene expression classic (immunosuppressants) | Glucocorticoids | Vaccine | Bacillus Calmette Guerin (BCG) |
| Blocker of lymphocyte signaling (Steroid sparingdrugs) | a. Calcineurin blocker (cyclosporine, tacrolimus)<br>b. mTOR inhibitors (sirolimus, everolimus) | Antihelminthics | Levamisole |
| Cytotoxic agents (Steroid sparing drugs) | Antimetabolites (azathioprine, methotrexate, leflunomid, mycophenolate mofetil)<br>Alkylating agents (cyclophosphamide) | Thalidomide | |
| Cytokine inhibitors (Anticytokine-antibodies) [Biologics] | a. TNF-α blockers (etanercept, infliximab, adalimumab)<br>b. IL-1 blockers (anakinra)<br>c. IL-2 blockers (daclizumab, basiliximab) | Recombinant cytokines | Interferons (α, β, γ)<br>Interleukins (aldesleukin, des-alanyl-1, serine 125 human IL 2) Colony stimulating factors [filgrastim (r metHuG CSF)] |
| Antibodies against specific immune cell molecules [Biologics] | a. Polyclonal antibodies [antithymocyte globulin (ATG)]<br>b. Monoclonal Antibodies [alemutuzmab (antiCD-52 antibodies)<br>c. Muromunab (antiCD-3 antibodies, OKT-3)] | Isoprinosine | |
| Blockers of immune cell adhesion [Biologics] | Efalizumab (LFA-1 inhibitor) | Immunocynin | |
| Tolerogens or blockers of immune cell costimulation Miscellaneous | Rh (D) immune globulin | | |

#### 6.2.2.2 Based on Molecular Weight

Based on the molecular weight, immunomodulators are classified into low molecular weight and high molecular weight compounds, both exhibiting immunostimulatory properties.

### *6.2.3 Low Molecular Weight Immunomodulators*

Several low molecular weight compounds have been reported for immunomodulatory properties. Alkaloids, Aristolochic acid obtained from the plant *Aristolochia clematitis* exhibited immunostimulatory property. Aristolochic acid enhanced the phagocytic activity of peritoneal macrophages and leukocytes, but due to its carcinogenic property, its use as an immunostimulant is limited. Other bioactive immunostimulatory alkaloids obtained from *Actinidia macrosperma*, *Cissampelos pareira*, *Achillea millefolium*, and *Murraya koenigii*. Cepharanthine isolated from *Stephania cepharantha* and Vincristine from *Catharanthus roseus* stimulate the production of antibodies. Cepharanthine also counteracts the effect of cytostatic agents on hematopoiesis. Lower molecular weight compounds like Vincristine and Staurosporine show immunomodulation in a dose-dependent manner, lower doses used as an immunostimulant while higher doses used as an immunosuppressor. The alkaloids isolated from *Uncaria tomentosa* increased the number of granulocytes. A beta-carboline indole alkaloid harmine, isolated from *Ophiorrhiza nicobarica*, inhibits lysine-specific demethylase-1 during immediate-early transcription of herpes simplex virus (HSV) 1 and 2 in-vitro and diseased animal. The interference on early transcription, a decisive factor for HSV lytic cycle or latency, reveals an epigenetic target that may help to develop a nonnucleoside antiherpesvirus drug. Thiosulfinate obtained from *Allium hirtifolium* are potent adaptogens and immunomodulators. Naphthoquinones also showed similar effects on lymphocytes and granulocytes. Plumbagin, a quinoid compound isolated from the roots of *Plumbago zeylanica*, reported for inhibitory action on the growth of hormone-refractory prostate cancers and restrict the growth of *Staphylococcus aureus*. Terpenes and its oxygenated derivatives terpenoids, like Amyrine from *Bauhinia variegata* Eugenol from *Ocimum sanctum*, a diterpene from *Andrographis paniculata*, *Achillea millefolium*, Alternanthera tenella and pentacyclic triterpene from *Cecropia telenitida* showed immunomodulatory activity. The sapogenins triterpenoids and diterpenoids from Gymnema *sylvestre* have diverse immunomodulatory potentials. Immunostimulatory phorbol esters (derivatives of tetracyclic diterpenoids phorbol) shoed anticancer properties at lower doses. Cichoric acid isolated from *Echinacea purpurea* activated phagocytic cells both *in-vitro* and *in-vivo* (Nair et al. 2019).

Plant-derived glycosidase yields sugar on acid or enzymatic hydrolysis. These include iridoid glycosides of *Picrorhiza scrophulariiflora* and anthraquinone glycosides of *Andrographis paniculata*. The Chinese medicinal plant *Dendrobium nobile* yields sesquiterpene glycosides Dendroside A, Dendronobilosides A, and B that exerts proliferative effects on lymphocytes. Flavonoids are another class of phytocompounds that exerts immunomodulatory effects. Lasure et al. (1994) reported the effect of flavonoids on the activation of complements using hemolytic assay and found that quercetin, quercitrin, rutin, myricetin, taxifolin, pelargonidin chloride, and cyanidin chloride inhibited the classical pathway, while hyperoside, myricetin, baicalein, and pelargonidin chloride inhibited the alternative pathway in a dose-dependent manner. Besides other flavonoids viz apigenin, anthocyanidins, flavones, isoflavonoids, and oligomeric proanthocyanidins found in plants like *Terminalia arjuna* exhibit immunomodulatory activity. Furthermore, these phenolic compounds isolated from (Euphorbiaceae) family also stop the classical pathway of complement activation. Coumarin glycosides isolated from *Achillea millefolium*, Citrus *natsudaidai* and *Heracleum persicum* revealed immunomodulatory properties. Hydroxycoumarins like Esculin, Esculetin, and Scopoletin improve complement-mediated hemolysis. Esculentin a class of 6,7-dihydroxycoumarin isolated from *Euphorbia lathyris, Citrus limonia,* and *Artemisia capillaris* attributed to a wide range of immunomodulatory properties like free radical scavenging, protecting DNA against oxidative damage (Leung et al. 2005). It also shows cancer chemopreventive, antitumor, lipoxygenase-inhibitory activity, and tyrosinase-inhibitory activity. Certain plants and plant-derived bioactive compounds like polysaccharides and lipopolysaccharides can activate the alternative pathway of complement activation and hence play a key role in the regulation of inflammation. Moreover, bioactive Rosmarinic acid from *Rosmarinus officinalis* L. inhibits the complement system via blocking both the classical and alternative pathways of complement activation (Fig. 6.5).

### *6.2.4 High Molecular Weight Immunomodulators*

The compound having High molecular weight like polysaccharides show immunomodulatory action primarily on the innate arm of the immune system and specifically by enhancing the phagocytosis of granulocytes and affecting the macrophages functions. Polysaccharide obtained from the plant *Cistanche deserticola* enhanced the proliferation of murine thymus lymphocytes. Compounds isolated from *Salicornia herbacea* show anticancer properties via activating monocytic cells and inducing differentiation of monocytes into macrophages (Rios 2010). Axillary mode of action includes induced interferons, interleukins, and complement activation. The binding sites for several polysaccharides to immune components might vary. Xyloglucans (doRosa'rio et al. 2011), glucuronic acid-containing arabinogalactan and methyl glucuronoxylans, play a critical role in stimulating the phagocytes (Luettig et al. 1989). The complement-activating polysaccharides comprise acidic polygalacturonans isolated from *E. purpurea*. Polygalacturonans are also active for activating macrophages to counteract tumor cells and to destroy intracellular pathogens like *Candida albicans, Leishmania*

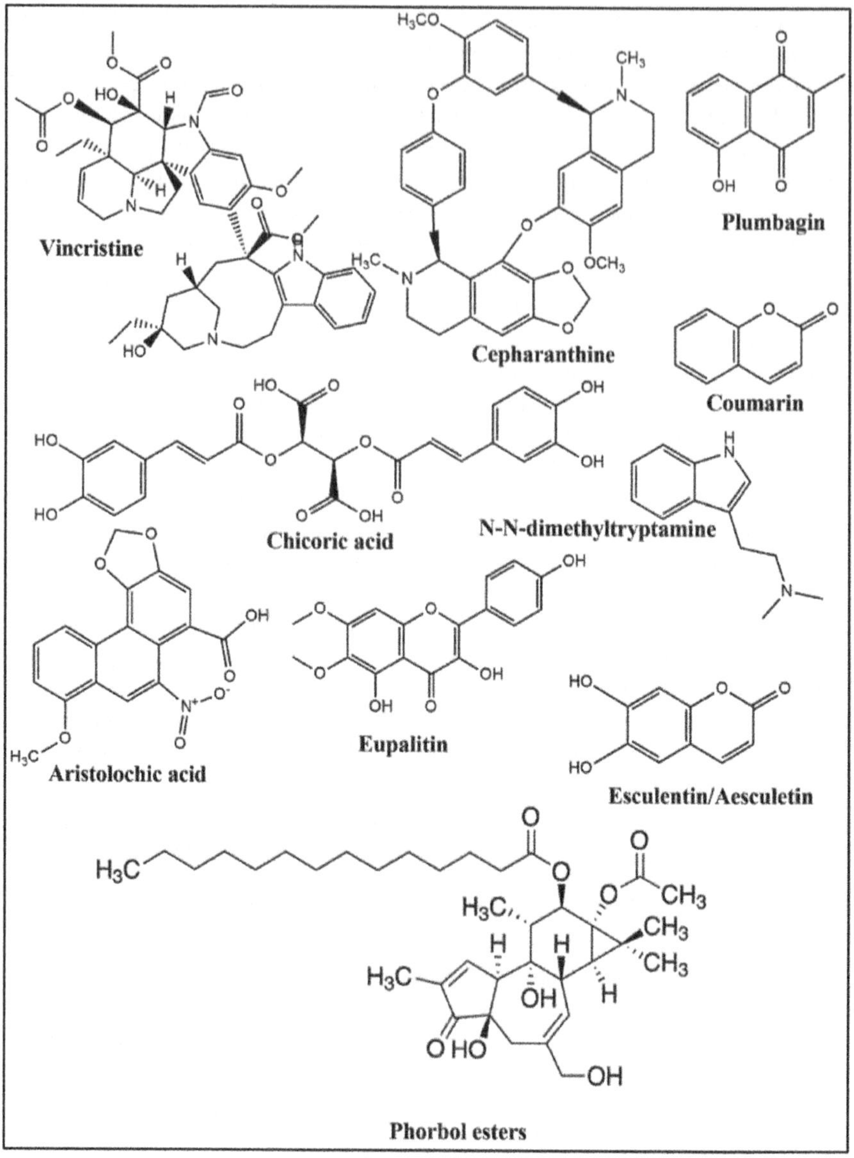


**Fig. 6.5** Structure of low molecular weight immunomodulators

*enrietti*, and *Listeria cytogenes*. Glycanogalacturonans from *Achyrocline satureioides* enhance the phagocytic potential of macrophages and granulocytes and exhibit strong anti-complement and anti-inflammatory activities. Polysaccharides are also known to develop chemotaxis, for instance, the rhizome of *Urtica dioica* produces α-glucan that can stimulate leukocyte migration. Lectins isolated from

*Lens culinaris, Canavalia ensiformis, Ricinus communis, Viscum album, Phaseolus vulgaris*, and *Phytolacca americana* challenged mitosis, inhibited protein synthesis, bind to the lymphocytes, and agglutinate malignant cells. Certain lectins having interferon-inducing properties were isolated from *Viscum album* and *Urtica dioica* (Peumans et al. 1984). Not only lectins but also proteins capable to induce interferons were isolated and analyzed from *Artemisia princeps*. N-Glycosidase that recognizes a conserved stem-loop structure in 23S/25S/28S rRNA and irreversibly block protein translation is known as Ribosome- inactivating proteins (RIPs). RIPs have been reported from over 50 plant species including (Cucurbitaceae) (Zhang and Halaweish 2007), (Euphorbiaceae) (Wu et al. 2015), and (Poaceae) (Loss-Morais et al. 2013). RIPs extracted from *Trichosanthes kirilowii* (Trichosanthin) was found to decrease the retroviral protein and RNA levels in acutely infected T lymphoblast and monocyte/macrophage (Shaw et al. 2005). Other known RIPs are isolated from plants such as *Cucurbita pepo, Ecballium elaterium, Bryonia cretica*, etc. Saponins derived from the plant *Quillaja saponaria* of the family (Rosaceae) are potent immunoadjuvants. Quil A, a purified form of Quillaja extract, is used in many veterinary vaccines (e.g., rabies, and foot and mouth disease), but its adverse reactions makes it unfit for use in humans (Rajput et al. 2007) (Fig. 6.6).

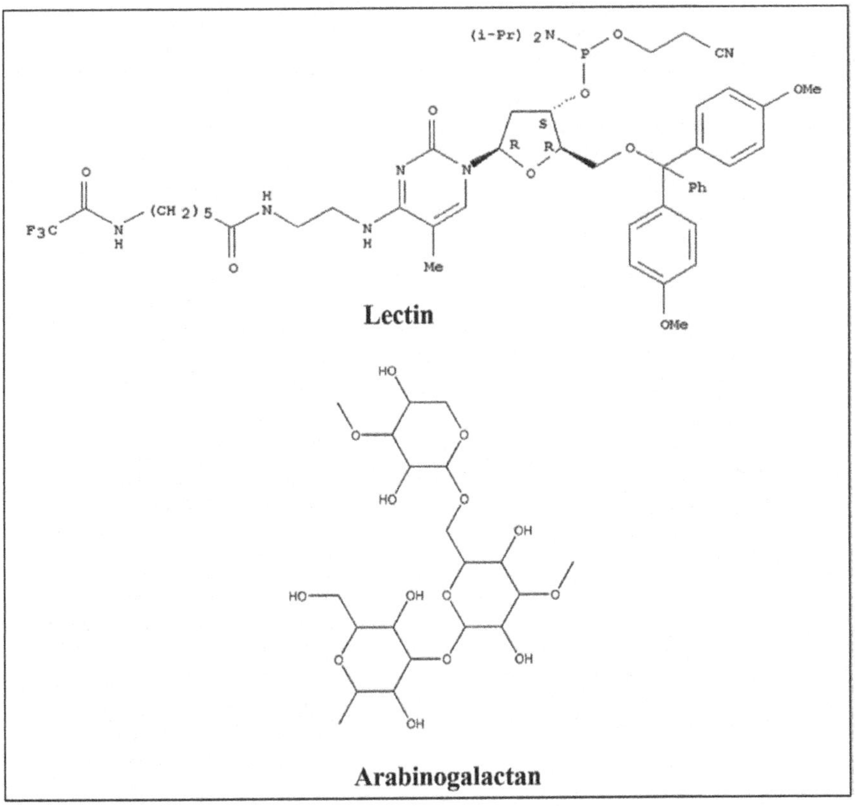


**Fig. 6.6** Structure of high molecular weight immunomodulators

### 6.2.5 *High Throughput Screening (HTS) for Plants and Bioactive Compounds*

Some medicinal plants are currently under high throughput screening for a quick assessment of pharmacologically important hit molecules that could be utilized as a lead molecule in drug development. Some of the medicinally important plants currently under research for their immunomodulatory utility are given in Tables 6.3 and 6.4.

**Table 6.3** List of plants with immunomodulatory potential (Nair et al. 2019; Kumar et al. 2012)

| S. No. | Name of the plant | Parts used | Compounds isolated |
|---|---|---|---|
| 1 | *Acanthopanax sessiliflorus (Rupr. and Maxim.) (Araliaceae)* | Root, berry, leaf | Saponins, coumarins, lignans, syringin, eleutheroside syringaresinol, phenylpropane, flavones, beta-sitosterol |
| 2 | *Achyranthes bidentate (Amaranthaceae)* | Root, leaf, seed | Triterpenoid, alkaloids, ketosteroids, saponins, sterols, flavonoids, anthraquinones, organic acids |
| 3 | *Achyrocline satureioides (Asteraceae)* | Flower | Luteolin, 3-O-methyl quercetin, flavonoids quercetin |
| 4 | *Aconitum carmichaeliim (Ranunculaceae)* | Root | Unesterified-diterpene diester-diterpene alkaloids, alkaloids, monoester-diterpene alkaloids |
| 5 | *Actinidia macrosperma C. F. Liang Actinidiaceae* | Berry | Phenolic compounds, alkaloids |
| 6 | *Aeginetia indica (Orobanchaceae)* | Whole plant | Glycoproteins, terpenoids, cichoric acid, macrocyclic lactones, polysaccharides, alkylamides |
| 7 | *Albizia julibrissin (Fabaceae)* | Bark, flower | Aromatic ethyl esters, dicarboxylic esters |
| 8 | *Aloe vera Tourn. ex Linn. (Liliaceae)* | Leaf | Glucomannans, sterols, anthraquinones, lipids, vitamins, amino acids, glycosides |
| 9 | *Alsophila spinulosa (Cyatheaceae)* | Stem, leaf | Diploptene, glucopyranose, beta-sitosterol, astragalin |
| 10 | *Angelica acutiloba (Apiaceae)* | Root, aerial part | Ferulic acid, Z-ligustilide, tokiaerialide, bergaptol |
| 11 | *Angelica sinensis (Apiaceae)* | Root | Polysaccharides, ferulic acid, Z-ligustilide |
| 12 | *Aralia mandshurica (Araliaceae)* | Root | Aralosides (A, B, C), flavonoids, polysaccharides, saponins, sterols terpenoid acids, terpenoid |
| 13 | *Arnica montana (Asteraceae)* | Flower | Coumarins, lignans, phenolic acids, oligosaccharides, sesquiterpene lactones, flavonoids, alkaloids, carotenoids, pyrrolizidine |

(continued)

**Table 6.3** (continued)

| S. No. | Name of the plant | Parts used | Compounds isolated |
|---|---|---|---|
| 14 | *Artemisia capillaris (Asteraceae)* | Leaf | Capillin, isochlorogenic acid, chlorogenic acid, esculetin, scoparone, scopoletin, artepillin, isorhamnetin-1,6-diglucoside, hyperin |
| 15 | *Artemisia iwayomogi (Asteraceae)* | Aerial parts | Scopoletin, scopolin, β-sitosterol, esculetin 6-methyl ether, quebrachitol, chlorogenic acid, |
| 16 | *Asarum europaeum, (Aristolochiaceae)* | Root, leaf | Trans-isoasarone, methyl ether, trans-isoeugenol |
| 17 | *Astragalus embranaceus (Fabaceae)* | Whole plant | Formononetin, Astragaloside IV |
| 18 | *Atractylodes lancea (Asteraceae)* | Root | Phenolic acids, polyacetylenes, sesquiterpenoids, steroids, atractylodin, β-eudesmol, monoterpenes |
| 19 | *Azadirachta indica (Meliaceae)* | Whole plant | Azadirachtin, quercetin, nimbin, ß-sitosterol, nimbidin |
| 20 | *Benincasa cerifera (Cucurbitaceae)* | Fruit | Isovitexin, cucurbitacin B,β-sitosterol, lupeol |
| 21 | *Bryonia dioica, (Cucurbitaceae)* | Root | Carbohydrates, triterpenes, polyphenols, sterols alkaloids, c-heterosides, saponins |
| 22 | *Bupleurum chinense (Apiaceae)* | Root | Polysaccharides |
| 23 | *Caesalpinia sappan (Fabaceae)* | Heartwood, seed | Brazilin, diterpene caesalsappanin, secang, brazilein |
| 24 | *Calendula Officinalis L. (Asteraceae)* | Flower | Quercetin (1), isorhamnetin (2) |
| 25 | *Camellia sinensis L. (Theaceae)* | Leaf | Theophylline, catechins, flavonoids, saponins |
| 26 | *Carthamus tinctorius (Asteraceae)* | Flower, seed | Safflower polysaccharide, alkaloids, flavonoids, |
| 27 | *Caulophyllumthalictroides (Berberidaceae)* | Root, rhizome | Triterepenes saponins, alkaloids |
| 28 | *Chelidonium majus (Papaveraceae)* | Whole plant | Isoquinoline alkaloids, coptisine, sparteine, chelerythrine,stylopine, chelidonine, bebeerine |
| 29 | *Choerospondias axillaris (Anacardiaceae)* | Folium, fruit peel | Flavones, proanthocyanidins |
| 30 | *Centaurea macrocephala (Asteraceae)* | Flower | Anthocyanins, cyanidin glycoside, cyanidin, patuletin, flavonoids, kaempferol glycoside |
| 31 | *Cimicifuga simplex (Ranunculaceae)* | Root | Triterpenoid glycosides, phenolic compounds |
| 32 | *Cinnamomum cassia (Lauraceae)* | Twig | Coumarin, cinnamate, cinnamyl alcohol, cinnamic acid |

(continued)

**Table 6.3** (continued)

| S. No. | Name of the plant | Parts used | Compounds isolated |
|---|---|---|---|
| 33 | *Cistanche salsa (Orobanchaceae)* | Whole plant | Iridoid glycosides, phenylpropanoid-substituted diglycosides (Cistansalside A) |
| 34 | *Cnidium officinale (Apiaceae)* | Root, rhizome | Falcarindiol |
| 35 | *Coffea arabica (Rubiaceae)* | Berry, leaf | Caffeine, ferulic acid, 5-caffeoylquinic acid, vanillic, protocatechuic acids, p-coumaric |
| 36 | *Combretum micranthum (Combretaceae)* | Leaf | Kinke′loids A, B, C, and D, sorbitol, inositol, vitexin, orientin, myricetin, isovitexin |
| 37 | *Cordycepssinensis (Ophiocordycipitaceae)* | Whole plant | Polysaccharides, sterols, nucleosides, amino acids |
| 38 | *Curcuma longa (Zingiberaceae)* | Root | Curcuminoids |
| 39 | *Daucus carota (Apiaceae)* | Root, seed | Anthocyanin, β-cyclodextrin, Cyaniding 3-xylosyl |
| 40 | *Echinacea purpurea (Asteraceae)* | Whole plant | Alkamides, polysaccharides caffeic acid derivatives |
| 41 | *Echinosophora koreensis (Leguminosae)* | Root | Prenylated flavonoids, alkaloids |
| 42 | *Epimedium alpinum Berberidaceae* | Root, rhizome | Glycosides, lignans, flavonoid, terpenoids, polysaccharides, alkaloids |
| 43 | *Eupatorium cannabinum (Asteraceae)* | Aerial parts | Flavonoids, pyrrolizidine, terpenoids, polysaccharides, sesquiterpene lactones, benzofurans, alkaloids |
| 44 | *Fagopyrum cymosum (Polygonaceae)* | Rhizome | Flavonoids, luteolin, β-sitosterol, phenolic compounds |
| 45 | *Forsythia koreana (Oleaceae)* | Fruit, leaf | Lignans (arctigenin) |
| 46 | *Geranium macrorrhizum (Geraniaceae)* | Leaf | Gallic acid, ellagic acid, quercetin |
| 47 | *Glycyrrhiza glabra (Fabaceae)* | Root | Glycyrrhetinic acid, glabridin, glycyrrhizin, isoflavones, flavonoids, beta-chalcones, triterpenoid |
| 48 | *Gymnema sylvestre (Apocynaceae)* | Leaf | Triterpene, saponins, flavonoids, coumarins, tannins |
| 49 | *Gynostemmapentaphyllum (Cucurbitaceae)* | Whole plant | Saponins, flavones, polysaccharides |
| 50 | *Hedysarum polybotrys (Fabaceae)* | Radix | Heteropolysaccharide, proanthocyanidin |
| 51 | *Herpestis monniera (Plantaginaceae)* | Whole plant | Glycosides (bacosides A and B) |

(continued)

**Table 6.3** (continued)

| S. No. | Name of the plant | Parts used | Compounds isolated |
|---|---|---|---|
| 52 | *Morus alba (Moraceae)* | Berry,Leaf | Flavones, Anthocyanin, oxyresveratrol, flavonoids, pyrrole alkaloids, lignans, fatty acids, polyphenols |
| 53 | *Nyctanthes arbor tristis L. (Oleaceae)* | Flower, leaf | Ursolic acid, flavonoids, phenol |
| 54 | *Ocimum sanctum Linn. (Labiateae)* | Whole plant | Carvacrol, eugenol, ursolic acid, eugenic acid, estragol, anthocyans, sitosterol, linalool |
| 55 | *Ophiopogon japonicus (Asparagaceae)* | Root | Galactan, flavonoids, |
| 56 | *Paeonia albiflora (Paeoniaceae)* | Root | Monoterpenes, glucosides |
| 57 | *Panax ginseng Wall. (Araliaceae)* | Whole plant | Ginsenosides, polyacetylenes, saponins |
| 58 | *Petiveria alliacea (Petiveriaceae)* | Leaf | S-Propyl propanethiosulfinate, S-benzyl phenylmethanethiosulfinate |
| 59 | *Phellodendron (Rutaceae)* | Bark, root | Alkaloids, limonoids, phenolic compounds |
| 60 | *Picrorhiza kurroa (Plantaginaceae)* | Rhizome | Iridoid glycoside (picroside I, picroside II) |
| 61 | *Pinellia ternate (Araceae)* | Tuber | Ferulic acid, lectins, coniferin, p-coumaryl alcohol, lariciresinol, dihydroxy-cinnamyl alcohol |
| 62 | *Pinus strobus(Pinaceae)* | Whole plant | Dihydrobenzofurans, xanthenes |
| 63 | *Polygala tenuifolia (Polygalaceae)* | Roots | Polysaccharides |
| 64 | *Potentilla tormentilla (Rosaceae)* | Rhizome, root | Procyanidin, monogalloylquinic acids, polyphenols |
| 65 | *Pseudostellaria heterophylla (Caryophyllaceae)* | Whole plant | Pectic polysaccharide, peptides, saponins amino acids |
| 66 | *Quillaja Saponaria (Quillajaceae)* | Bark | Triterpenoid saponins |
| 67 | *Rehmannia glutinosa (Orobanchaceae)* | Whole plant | Polysaccharides, acetoside |
| 68 | *Sapium sebiferum (Euphorbiaceae)* | Leaf | Gallic acid, astragalin, kaempferol, quercetin, β-sitosterol glycoside, 6-O-galloyl-d-glucose, methyl gallate, methyl-3,4,5trihydroxybenzoate |
| 69 | *Schisandra chinensis (Schisandraceae)* | Fruit | Schisandrin A, B, C, lignans |
| 70 | *Serenoa repens Arecaceae* | Whole plant | Linoleic acid, myristic acid oleic acid, lauric acid |

(continued)

**Table 6.3** (continued)

| S. No. | Name of the plant | Parts used | Compounds isolated |
|---|---|---|---|
| 71 | *Solenostemma argel (Apocynaceae)* | Leaf | Solenoside A, Stemmosides E K, kaempferol-3-O-glucoside, kaempferol-3-Orutinoside |
| 72 | *Taraxacum platycarpum (Asteraceae)* | Aerial parts | Sesquiterpene glycoside, polysaccharides, desacetylmatricarin, triterpenes |
| 73 | *Tinospora cordifolia (Menispermaceae)* | Stem, Root, Leaf | Polysaccharides, steroids, glycosides, alkaloids, peptides |
| 74 | *Trichosanthes kirilowii (Cucurbitaceae)* | Fruit peel | Polysaccharide, cucurbitacin B |
| 75 | *Tripterygium wilfordii (Celastraceae)* | Root | Glucosides, triptolide, nerolidol-type sesquiterpene triptergosidols A-D, triptonide, celastrol, tripdiolide |
| 76 | *Viscum album (Santalaceae)* | Whole plant | Viscotoxins, lectins, flavonoids, phenylpropanoids, triterpene, phenolic acids, phytosterols |
| 77 | *Uncaria tomentosa (Rubiaceae)* | Bark, leaf | Procyanidin, hydroxybenzoic acid propelargonidin dimers, flavanols, flavalignans, cinchonains, hydroxycinnamic acids, alkaloids |
| 78 | *Zea mays (Poaceae)* | Seed | Diterpenoids, flavonoids, phenolic compounds |
| 79 | *Zingiber officinale (Zingiberaceae)* | Rhizome | 6-Gingerol |
| 80 | *Ziziphus jujube (Rhamnaceae)* | Aerial parts | Phenolics, vitamin C, polysaccharides, triterpenic acids |
| 81 | *Achillea millefolium (Compositae)* | Leaves | Alkaloids, Flavonoids,Coumarins, triterpenes |
| 82 | *Aloe vera Tourn. ex Linn. (Liliaceae* | Gel from leaves | Anthraquinone glycosides |
| 83 | *Andrographis paniculata Nees (Acanthaceae)* | Leaves | Diterpenoids, andrographolide, flavonoids |
| 84 | *Asparagus racemosus Wild. (Liliaceae)* | Roots | Steroidal Saponins, Isoflavones, racemosol, asparagamine, polysaccharides, Vitamins |
| 85 | *Abutilon indicum linn. (Malvaceae)* | Whole plant | Flavonoids triterpenoids |
| 86 | *Alternanthera tenella Colla (Amaranthaceae* | Herb | Flavonoids, triterpenes |
| 87 | *Actinidia macrosperma C. F.Liang (Actinidiaceae* | Fruits | Alkaloids, Flavonoid, saponins |
| 88 | *Acacia catechu Willd. (Leguminosae)* | Leaf | Flavonoids, tannins, quercetin, alkaloids, and |
| 89 | *Allium hirtifolium Boiss. (Alliaceae)* | Herb | Thiosulfinates, flavonoids |

(continued)

**Table 6.3** (continued)

| S. No. | Name of the plant | Parts used | Compounds isolated |
|---|---|---|---|
| 90 | *Acanthopanax sessiliflorus (Araliaceae)* | Shoots and roots | Biopolymers, eleutherosides, flavonoids, phenylpropionic acids, and triterpenic acids |
| 91 | *Agelas mauritianus (Porifera)* | Sponge | Glycolipid |
| 92 | *Aphanothece halophytica (Chroococcales)* | Cyanobacterium | Exopolysaccharide |
| 93 | *Apium graveolens Linn. (Apiaceae)* | Leaves, seeds | Flavonoids, coumarins |
| 94 | *Artemisia annua Linn. (Compositea)* | Herb | Artemisinin |
| 95 | *Bauhinia variegata Linn. (Caesalpiniaceae)* | Roots, bark, buds | Flavonoids, lupeol, betasitosterol |
| 96 | *Botryllus schlosseri (Botryllidae)* | Tunicates | Cytokines |
| 97 | *Bidens pilosa L. (Asteraceae)* | Flowers, leaves | Polyacetylenes |
| 98 | *Boerhaavia diffusa (Nyctaginaceae)* | Herb | Alkaloid |
| 99 | *Bugula neritina L. (Bugulidae)* | Marine invertebrates | Macrocyclic lactones, cholesterol |
| 100 | *Byrsonima crassa Nied. (Malpighiaceae)* | Leaves | Flavonoids, Monoterpenoids tannins, Triterpenoids, Iridoid Glycosides and Phenolic Compounds |
| 101 | *Couroupita guianensis Aubl. (Lecythidaceae)* | Fruits, flowers | Steroids, phenolics, flavonoids |
| 102 | *Cleome gynandra Linn. (Capperdiceae)* | Leaf, seeds, rots | Flavonoids, terpenoids, alkaloids, and steroids Hexacosanol, kaempferol |
| 103 | *Citrus natsudaidai Hayata (Rutaceae)* | Fruits | Auraptene, flavonoids |
| 104 | *Calendula Officinalis L. (Asteraceae)* | Flowers | Polysaccharides, fatty acids, proteins, carotenoids |
| 105 | *Cistanche deserticola (Orobanchaceae)* | Herb | Polysaccharide |
| 106 | *Cliona celata (Clionaidae)* | Sponge | Clionamide, dehydrodopamine |
| 107 | *Cordyceps militaris L. (Clavicipitaceae)* | Fungus | Cordycepin, cordyceps acid, free amino acid |
| 108 | *Crinum latifolium Andr. (Amaryllidaceae)* | Herb | Alkaloids |
| 109 | *Cordia superba Cham. and C. rufescens A. DC. (Boraginaceae)* | Bark, leaf, fruit, | Quercetin, Alpha-amyrin, Linolenic acid |
| 110 | *Cissampelospareira Linn. enispermiaceae)* | Roots | Hayatine alkaloids |

(continued)

**Table 6.3** (continued)

| S. No. | Name of the plant | Parts used | Compounds isolated |
|---|---|---|---|
| 111 | *Chlorophytum borivilianum Sant. F (Liliaceae)* | Roots | Sapogenins |
| 112 | *Camellia sinensis L. (Theaceae)* | Leaves | Alkaloids, glycosides, terpenoids (-) Epigallocatechin gallate, quercetin, gallicacid |
| 113 | *Cannabis sativa (Cannabaceae)* | Leaves | Cannabinoids, volatile terpenes and sesquiterpenes |
| 114 | *Carpobrotus edulis L. (Aizoaceae)* | Flowers, fruit | Alkaloids, flavanoids proanthocyanidins, tannins |
| 115 | *Centella asiatica Linn. (Umbelliferae),* | Herb | Triterpenoid saponins |
| 116 | *Dracocephalum Kotschyi (Lamiaceae)* | Herb | Essential oil, flavonoids, terpinen |
| 117 | *Echinacea angustifolia (Asteraceae)* | Flowers | Polysaccharide, Phenolic compounds, alkamides |
| 118 | *Eclipta alba L. (Compositae)* | Leaves | Triterpenoid glucoside, wedelolactone, eclalbasaponins, ursolic acid, oleanolic acid |
| 119 | *Euphorbia hirta Linn. (Euphorbiaceae)* | Herb | Quercitol, gallic acid, myricitrin, alkanes, triterpenes, tannins, polyphenols phytosterols |
| 120 | *Evolvulus alsinoides Linn. (Convolvulaceae)* | Herb | Alkaloids |
| 121 | *Ganoderma lucidum (Fr.) P. Karst. (Polyporaceae)* | Whole plant | Flavonoids, polysaccharides, triterpenes |
| 122 | *Genus Ardisia (Myrsinaceae)* | Shrub, Leaves | Peptides, Isocoumarins alkyl phenols saponins |
| 123 | *Genus Aristolochia (Aristolochiaceae)* | Leaves | Aristolochic acid |
| 124 | *Genus Aspergillus (Trichocomaceae)* | Fungus | Polyene triazole |
| 125 | *Hibiscus rosa sinensis Linn. (Malvaceae)* | Flowers | Cyclopropanoids |
| 126 | *Hyptis suaveolens (L.) Poit. (Lamaceae)* | Leaf, flowers | Lupeol, beta sitosterol |
| 127 | *Heracleum persicum Desf. (Apiaceae)* | Shurb | Flavonoids furanocoumarins |
| 128 | *Inonotus obliquus Pers. (Hymenochaetaceae)* | Mushroom | Polysaccharide |
| 129 | *Larrea divaricata DC. (Zygophyllaceae)* | Herb | Lignans |
| 130 | *Lycium barbarum Linn. (Solanaceae)* | Fruits | Lycium barbarum polysaccharides, (LBPs), |
| 131 | *Matricaria chamomilla (Rhabdoviridae)* | Flowers | Protein |

(continued)

**Table 6.3** (continued)

| S. No. | Name of the plant | Parts used | Compounds isolated |
|---|---|---|---|
| 132 | *Mollugo verticillata L. (Molluginaceae)* | Herb | Quercetin, triterpenoid glycosides |
| 133 | *Moringa oleifera L. (Moringaceae)* | Leaves | Vitamin A carotenoids saponins |
| 134 | *Nyctanthes arbor tristis L. (Oleaceae)* | Leaf, seeds | Iridoid glucosides |
| 135 | *Ocimum sanctum Linn. (Labiateae)* | Entire plant | Eugenol, cavacrol |
| 136 | *Piper longum L. (Piperaceae)* | Fruits | Alkaloids |
| 137 | *Panax ginseng Wall. (Araliaceae)* | Fruits, root | Saponins (ginsenosides), panaxtriole panaxdiol, |
| 138 | *Picrorhiza scrophulariiflora Benth. (Scrophulariaceae)* | Roots | Iridoid glycosides, amphicoside |
| 139 | *Rhodiola imbricate Gray. (Crassulaceae)* | Rhizomes | Phenolics |
| 140 | *Randia dumetorum Lamk. (Rubiaceae)* | Fruits | Saponins triterpenes Chlorosis |
| 141 | *Silybum marianum L. (Asteraceae)* | Flowers | Flavonoid |
| 142 | *Salicornia herbacea (Chenopodiaceae)* | Herb | Polysaccharides |
| 143 | *Momordica charantia L. (Cucurbitaceae)* | Leaf | Momordicolide,dihydrophaseic, onordicophenoide |
| 144 | *Apios americana (Fabaceae)* | Flowers | Polysaccharide |
| 145 | *Leucas aspera(Willd.) Linn.(Lamiaceae)* | Whole Plant | Triterpenoids, Diterpenes, Ursolic Acid, Nicotine |
| 146 | *Eleutherococcus senticosus (Araliaceae)* | Root, Stem Bark | Syringin, Caffeic Acid, oleanolic acid, Isofraxidin |
| 147 | *Ziziphora tenuior (Lamiaceae)* | Aerial parts | Essential oil |
| 148 | *Litchi chinensis Sapindaceae* | Fruit | Flavonoids |
| 149 | *Terminalia chebula (Combretaceae)* | Fruit | Alkaloids |
| 150 | *Trichilia glabra (Meliaceae)* | Leaf | Trichin, Monadelphin |

**Table 6.4** Indian medicinal plants used for immunomodulatory activity (Kumar et al. 2012)

| S. No. | Plant name | Parts used | Chemical constituents |
|---|---|---|---|
| 1 | *Glycyrrhiza uralensis Fisch (Leguminosae)* | Dried roots | Polysaccharides |
| 2 | *Aesculus indica (Sapindaceae)* | Leaf | Alkaloids, tannins saponins |
| 3 | *Argyreia speciosa (Convolvulaceae)* | Roots | Glycosides |
| 4 | *Abrus precatorius (Fabaceae)* | Seeds | Alkaloids, saponins phenolics, tannins |
| 5 | *Adhatoda vasica Linn. (Acanthaceae)* | Leaf | Quinazoline vasicinone and essential oils |
| 6 | *Balanites roxburghii Zygophyllaceae* | Leaf | Alkaloids, saponins, tannins, and flavonoids |
| 7 | *Clitoria ternatea (Linn.) (Fabaceae)* | Aerial Part | β-sitosterol and kaempferol |
| 8 | *Citrus aurantifolia (Rutaceae)* | Fruits | Volatile oils |
| 9 | *Capparis zeylanica Capp (Araceae)* | Leaf | Flavonoids |
| 10 | *Caesalpinia bonducella (Fabaceae)* | Seeds | Flavonoids, amino acids alkaloids, and tannins |
| 11 | *Habenaria intermedia (Orchidaceae)* | Tubers | Alkaloids and phenolic compounds |
| 12 | *Murraya koenigii (Rutaceae)* | Leaf | Coumarins, glucoside, carbazole alkaloids, |
| 13 | *Mangifera indica (Anacardiaceae)* | Stem bark | Alkaloids, flavonoids, tannins |
| 14 | *Nyctanthes arbortristis (Oleaceae)* | Leaf | Iridoid glucosides |
| 15 | *Salacia chinensis (Celastraceae)* | Root | Flavonoids, alkaloids, carbohydrates, tannins |
| 16 | *Syzygium cumini (Myrtaceae)* | Seeds | Alkaloids, phytosterols, glycosides, flavonoids |
| *17* | *Terminalia arjuna (Combretaceae)* | Leaves and bark | Flavonoids, tannins, and oligomeric proanthocyanidins |
| 18 | *Trapa bispinosa (Lythraceae)* | Fruits | Flavonoids, carbohydrates, proteins, |
| 19 | *Tridax procumbens (Asteraceae)* | Aerial parts | Tannins, steroids, flavonoids and alkaloids |
| 20 | *Urena lobata (Malvaceae)* | Fruits | Flavonoids and glycosides |
| 21 | *Withania somnifera (Solanaceae)* | Roots | Withanolides |

## 6.3 Immunomodulatory Plants

### 6.3.1 *Acacia Catechu/Senegalia Catechu* (Family: (Fabaceae))

This plant commonly known as Cutch Tree, Black Catechu, Cachou, found in Asia. The heartwood and bark are used in traditional medicine to treat sore throat and diarrhea. Several bioactive compounds are isolated from *A. catechu* like free radical scavenging catechin (polyhydroxylated benzoic acid), rutin, isorhamnetin (Li et al. 2011), and other compounds like epicatechin, epicatechin-3-O-gallate, epigallocatechin-3-O-gallate (Stohs and Bagchi 2015), 4-hydroxybenzoic acid, ophioglonin, quercetin, afzelechin, kaempferol, 3,4,7-trihydroxyl-3,5-dimethoxy-flavone, epiafzelechin, mesquitol, aromadendrin, and phenol (Li et al. 2010). Naik et al. (2003) demonstrated that the aqueous extracts of *A. catechu*, in rat liver microsomal preparation, could inhibit radiation-induced lipid peroxidation. Flavonoids isolated from *A. catechu* reduce the production of pro-inflammatory eicosanoids (Burnett et al. 2007). Of note, 70% methanolic extracts of *A. catechu* is found to have DNA protective properties. The acetone, ethyl acetate, and methanolic extracts of the heartwood, leaves, and bark of *A. catechu* not only scavenges free radicals but also protects DNA against strand breaks (Guleria et al. 2011) and has antimicrobial properties (Negi and Dave 2010). *A. catechu* shows its immunomodulatory effect on both cell-mediated and humoral immunity. The aqueous extract of *A. catechu* exhibited an increase in the neutrophil adhesion to the nylon fibers, produced a significant increase in the phagocytic index, and significant protection against cyclophosphamide-induced neutropenia indicating its effect on cell-mediated immunity. *A. catechu* extract produced a significant increase in the serum immunoglobulin levels, an increase in the hemagglutination titer values, and a decrease in the mortality ratio in mice, suggesting enhanced humoral immunity (Ismail and Asad 2009).

Methanolic and hexane extracts of *A. catechu* bark have reported for antiproliferative, cytotoxic, and anticancer properties against various cancer cell lines but do not show any effect on human peripheral lymphocytes (Nadumane and Nair 2011). Methanolic extract of *A. catechu* heartwood exhibited cytotoxic activity in breast cancer cell line MCF-7, which is due to the enhancement in Bax/Bcl2 ratio leading to the activation of caspases and subsequent cleavage of polyadeno ribose polymerase (Ghate et al. 2014). The aqueous extracts of heartwood also show antidiabetic and antinociceptive action in a dose-dependent manner (Rahmatullah et al. 2013). Various extract of *A. catechu* has a chemo-protective role in chemical (carbon tetrachloride, t-butyl hydrogen peroxide, 7,12-dimethylbenz[a]anthracene, DMBA) induced hepatocytic damage, breast and squamous cells cancers (Monga et al. 2011).

The bark extracts of *A. catechu* showed potent anti-HIV effects, owing to its effect on viral protease and via hampering the interaction of Viral Tat protein to its HIV-1 promoter sequence of LTR (Modi et al. 2013). The antiviral compounds isolated from *A. catechu* can overcome the conventional problem of generation of a drug-resistant HIV-1 strain (Ma'rquez et al. 2005) has hence is a promising candidate in drug discovery.

### 6.3.2 *Acorus Calamus* (Family: Acoraceae)

This plant is commonly known as Sweet Flag, Calamus, found in Central Asia, Southern Russia, Siberia and Eastern Europe. *Acorus calamus* shows varied pharmacological properties including antibacterial, insecticidal, anti-ulcerative, etc. (Pandit et al. 2011). It is a potent adaptogenic drug. The key bioactive compounds present in *A. calamus* are flavonoid, monoterpene, quinone, sesquiterpene, and phenylpropanoid. The ethanolic extract has antiproliferative and immunosuppressive properties and is found to inhibit the growth of murine and human cell lines, inhibit mitogen-induced proliferation of peripheral blood mononuclear cells (PBMCs), and the generation of IL-12 and TNF-α (Mehrotra et al. 2003). The volatile oil, petroleum ether, and alcoholic extracts of the leaves of *A. calamus* enhance the phagocytic activity of neutrophils (Ravichandiran and Patil Vishal 2015). A D-galacturonic acid-containing pectic polysaccharide isolated from the rhizomes of *A. calamus* at low concentrations can stimulate murine macrophages to produce NO and IL-12 similar to those induced by LPS, thus promoting a Th1 and suppressing the Th2 response. It also lowers serum levels of IgG1 and IgE and induces the secretion of TNF-α secretion by human PBMCs. Thus, the polysaccharide activates the macrophages into M1 type (classically activated macrophages) (Belska et al. 2010). The polysaccharide possibly acts via binding to certain receptors on APCs, releasing immunoregulatory cytokines, and adhesion molecules (Retini et al. 2001). Its properties can be used for treating oncological and allergic diseases. *A. calamus* exhibited hepatoprotective activity, it restores the hepatic enzymes in acetaminophen-induced liver damage and lowers free radical-induced oxidative stress (Palani et al. 2011). Chronic stress can be detrimental to the immune system. Noise can activate the pituitary-adrenal-cortical axis and the sympathetic adrenal medullary axis thereby increasing the secretion from adrenal glands that directly correlate to stress (Babisch 2003) and hence can have detrimental effects on the immune status of the body. Noise-induced stress diminishes the number of $CD4^+$ and $CD8^+$ T-cells, which is reversed by *A. calamus* and its active compound α-asarone. The free radical-induced oxidation of lipids is also prevented by the extract of *A. calamus* and α-asarone (Dharini et al. 2012).

That the leaf extracts inhibit inflammatory reactions in HaCaT cells via various mechanisms (Kim et al. 2009). Also, β-asarone showed a neuroprotective role. It suppresses neuronal apoptosis by downregulating Bcl2, Bcl-w, and caspase-3 and preventing JNK phosphorylation. It is currently under investigation as a drug in rat models of Alzheimer (Geng et al. 2010). Lectins isolated from the roots of *A. calamus* are potent mitogenic agents for human lymphocytes and murine splenocytes. These lectins significantly inhibited the growth of a J774, a murine macrophage cancer cell line, and B-cell lymphoma (Bains et al. 2005).

### *6.3.3 Allium Sativum* (Family: Amaryllidaceae)

This plant is commonly known as garlic found in worldwide. Garlic is an essential medicinal spice and dietary element attributed to immunomodulatory properties having certain proteins mainly agglutinins, Alliinase, antifungal Allivin and the antimicrobial protein Alliumin. The aged garlic constituents show antiallergic and antitumor properties (Kyo et al. 2001). Alliin isolated from *A. sativum* enhances the expression of PINFCs genes like IL-6, MCP-1, and EGR-1 in LPS-stimulated 3T3-L1 adipocytes. It also modulates the cytokine generation. Low doses of garlic extract increase IL-10 while decreasing the IL-12. IL-1α, IFN-γ, TNF-α, IL-6, and IL-8 on treatment with the extract (Quintero-Fabia'n et al. 2013). Other *A. sativum* derived bioactive compounds like allitridin, S-allyl-L-cysteine, Caffeic acid, uracil, and diallyl sulfide inhibit transcription of several PINFLCs genes like IL-6, MCP-1, TNF-α, IL-1β, and IL-12 by inhibiting the transcription factor NF-κB (Fu et al. 2015). The lectin isolated from garlic non-specifically activates mast cells and basophils. Thus, lectins and agglutinins are potent mitogens and have potential utility in immunomodulation (Clement et al. 2010). Allicin reduced parasitemia when administered in *Plasmodium yoelii*-infected mice due to the generation of PINFLCs like IFN-γ. Allicin treatment also activated the macrophages and stimulates the expansion of $CD4^+$ T-cells (Feng et al. 2012). Not only the bioactive compounds of *A. sativum* affect the T-cells but also oil macerated extracts containing active ingredient Z-ajoene affects the B-cells and increased the levels of IgA and interleukins (Washiya et al. 2013). Allicin promotes the maturation of DCs by increasing the expression of CD40: a costimulatory molecule, thus inducing a proinflammatory response in rodent malaria models. Aged garlic soaked, sliced, and extracted (AGE) in ethanol inhibit the antigen-specific generation of histamines in rat basophil cell line RBL-2H3. Orally administered AGE significantly decreased the IgE-mediated skin reactions (Kyo et al. 2001) and induced the PINFLCs IL-12, INF-γ, and iNOS in *Leishmania*-infected murine macrophages (Gharavi et al. 2011). Differing to this, AGE upregulated IL-10 and decreased IL-12 production in PBMCs, which in effect reduced IFN-γ, IL-2, TNF-α, and IL-6 (Oft 2014; Gazzinelli et al. 1992). Garlic extracts exert an anti-inflammatory effect on monocytes and lymphocyte proliferation by upregulating IL-10 and downregulating the production TNF-α on LPS stimulation. Diallyl disulfide from *A. sativum* decreased PINFLCS, NO production in murine macrophages, and leukemic monocyte cell lines (Shin et al. 2013). AGE also affects NK cells and increases its activity against various cancer cell lines. Fructooligosaccharides present in AGE show mitogenic potential, activated macrophages and increased phagocytic activity, comparable to effects shown by mitogens like zymosan and mannan (Chandrashekar et al. 2011). Immunoproteins isolated from garlic-like lectins and agglutinins are also known for their mitogenic properties similar to those of ConA and PHA (Clement et al. 2010). AGE also affects the unique subset of the T-cell population by increasing the proliferation of $\gamma\delta$-T-cells that plays a crucial role in the recognition of pathogen-associated molecular patterns (Nantz et al. 2012). Bioactive compounds isolated from garlic also exhibit antiviral activity. Allitridin or diallyl tri-sulfide inhibits T-reg cells in-vivo (Li et al. 2013) and thus mounts an antiviral immune response against murine cytomegalovirus. Fresh garlic extracts stimulate the proliferation and activation of $CD8^+$ T-cells and cause a delayed-type hypersensitivity (DTH) response (Ebrahimi et al. 2013).

### *6.3.4 Andrographis Paniculate* (Family: Acanthaceae)

This plant is commonly known as King of Bitters, Kalmegh and found in Southeast Asia, China, America, West Indies. *Andrographis paniculata*, a well-known medicinal plant in various parts of the world, contains polyphenols, terpenoids (diterpene lactones, entalabdane) xanthones, nocardioides, and flavonoids (flavones) as key bioactive compounds. Ethanolic (Pongiuluran and Rofaani 2015) and dichloromethane extracts (Chao and Lin 2010) of A. paniculata augment the proliferation of lymphocytes at low concentrations. Andrographolide has diverse immunoregulatory effects. On administration of Andrographolide in animals bearing metastatic tumors, antibody-dependent cytotoxicity (ADCC) enhanced when compared to untreated controls. Serum from andrographolide treated mice in the presence of complement showed higher cytotoxicity suggesting that the extract can activate the humoral immune system and generate tumor-specific antibodies that can mediate ADCC (Sheeja and Kuttan 2010). Andrographolide significantly increased the mean $CD4^+$ T-cells and inhibited HIV-induced dysregulation of cell cycle tested against HIV-1 infection. Andrographolide and its derivatives also inhibited the fusion of HL2/3 cells with TZM-bl cells, which occurs via the interaction of gp120 with CD4 and CCR5, CXCR4, thus inhibiting the HIV (Uttekar et al. 2012). Andrographolide also inhibits the expression of Epstein-Barr virus (EBV) (Lin et al. 2008) and HSV-1. It increases secretion of IL-2, IFN-$\gamma$ by T-cells and the activity of NK cells, thus inhibiting the growth of the tumor. It also plays a role in autoimmune disorders such as encephalomyelitis, wherein it interferes with the maturation of DCs. Andrographolide can modulate the innate immune response, regulate the production of antibodies and the generation of antigen-specific splenocytes, and can activate macrophages both via classical and alternative pathways (Wang et al. 2010). The immunomodulating property of the purified andrographolide and neoandrographolide is low than that of ethanolic extracts this finding reveals that multiple components may bring about such combinatorial immunomodulatory effects (Puri et al. 1993). Andrographolide, 7-O-methylwogonin is andrographolide, and skullcapflavone-1 significantly inhibited INFLCs IL-6, NO, IL-1β in LPS-stimulated macrophages, and inflammatory mediators like PGE2 and TXB2 in activated promyelocytic leukemia cells. Andrographolide, dehydroandrographolide, and neoandrographolide exhibit its anti-inflammatory effect by inhibiting the cyclooxygenase (COX) enzyme (Parichatikanond et al. 2010). One more bioactive compound Andrograpanin extracted from the leaves of A. paniculate also inhibits PINFLCs in LPS-stimulated macrophages (Liu et al. 2008). *A. paniculata* is also known for its antioxidant properties. Andrographolide and 14-deoxy-11,12-didehydroandrographolide exhibited free radical scavenging effect and inhibited lipid peroxidation under DPPH-induced oxidative stress (Akowuah et al. 2009).

### *6.3.5 Azadirachta Indica* (Family: Meliaceae)

This tree is commonly known as the Neem tree found in Asia. *Azadirachta indica* extract found to increases the number and the activity of peripheral blood lymphocytes in various studies. It also increases the $CD4^+$, $CD8^+$ T-cells, and the markers for T-cell and macrophage activation, namely, CD25 and MAC-3, respectively. *A. indica* extracts also resulted in lesser lung and liver metastases in the sarcoma model of Balb/c mice (Beuth et al. 2006). The nimbolide a triterpenoid isolated from the leaves exhibits antiapoptotic and antiproliferative properties via downregulation of bcl2/bax and upregulation of Apaf-1 and caspase-3 (Kumar et al. 2006). Nimbolide inhibits tumor cell proliferation and exerts its growth inhibitory effects through alterations of cyclins, cdks, PCNA and p53 levels. Nimbolide also retards tumor cell migration, invasion and angiogenesis by downregulating MMP-2/9, VEGF-A expression via inhibiting ERK1/2, reducing the nuclear translocation and DNA-binding activity of NF-κB in cancer cells (Bodduluru et al. 2014). *A. Indica* shows anticancer properties, and it fortifies the body's immune system. Neem leaf preparation (NLP) also acts as an adjuvant to generate antigen (B16MelSAg) specific antiserum in C57BL/6 mice (Baral and Chattopadhyay 2004). NLP also stimulates the CD40-CD40L interaction that leads to the activation of p38MAPK and generation of PINFLCs IL-12, thus the activation of NK cells (Bose and Baral 2007). Glycoprotein isolated from *A. Indica* leaves enhances the expression of IFN-γ that downregulates CXCR3B (responsible for apoptosis in lymphocytes) and thus restored the impaired chemotactic movement of PBMCs toward tumor (Chakraborty et al. 2008). NLGP leads to the activation of T-cells and generates a Th1 type of cytokine IFN-γ and can effectively activate erythroleukemia and oral cancer cells (Bose et al. 2009). Administration of aqueous extracts of *A. Indica* significantly enhanced the activity of macrophages and facilitates tumor antigen presentation by macrophages and DCs to T and B-lymphocytes and generation of an effector and memory response (Tsang et al. 2011).

### 6.3.6 *Boerhavia Diffusa* (Family: Nyctaginaceae)

This plant is commonly known as Tarvine, Punarnava, Red Spiderling found in Asia, Africa, North America, Caribbean, South America, South Pacific region. Major compounds isolated from *Boerhavia diffusa* are Boerhavia acid, isoflavonoids (rotenoids), Punarnavine, sitosterol, Boeravinone, palmitic acid, steroids (ecdysteroid), lignan glycosides, and esters of sitosterol. *Boerhavia diffusa* showed anti-inflammatory and antioxidant properties and scavenges free radicals during oxidative stress (Gacche and Dhole 2006). Two well-characterized immunostimulants isolated from the roots of *B. diffusa* are Punarnavine and syringaresinol. Aqueous extracts exhibited significant leukocytosis and lowered the mortality in *E. coli* induced abdominal sepsis in mice and reduced stress-induced increase in the level of glucose and cholesterol. Alkaloid fraction normalizes the plasma cortisol levels and reduced DTH reactions in animals (Mungantiwar et al. 1999). *B. diffusa* extracts enhanced the phagocytic activity of macrophages (Sumanth and Mustafa 2007). *B. diffusa* is also shown certain adaptogenic effects. The immunomodulatory activity is attributed to compounds like quercetin, Punarnavine, syringaresinol mono-β-D-glucoside, etc. Ethanolic extracts of *B. diffusa* exhibit immunosuppressive properties and reduces the production of TNF-α and IL-2 in human PBMCs, suppresses human NK cells and the generation of NO in murine macrophages (Mehrotra et al. 2003). Immunosuppressive action is possibly due to the alkaloid/lignin compounds. The chloroform and ethanolic extract treatment in RAW 264.7 cell line under LPS stimulation induced NO production, inhibited PHA-induced proliferation, and production of TNF-α and IL-2 from PBMCs. Eupalitin-3-O-β-D-galactopyranoside the bioactive compound isolated from ethanolic extracts found to be most effective (Pandey et al. 2005).

Antiosteoporotic properties are also attributed to Eupalitin (O-methylated flavone) in addition to its anti-inflammatory and immunosuppressive effects. Therefore, its anti-inflammatory effects are extrapolated for the treatment of rheumatic disorders. Various studies have been performed that show the anticancer properties of *B. diffusa* extract. Punarnavine downregulates the expression ERK-1/2, MMP-2, MMP-9, and VEGF (Manu and Kuttan 2009) and prophylactic as well as simultaneous administration of punarnavine-reduced lung melanoma metastasis. It also enhances the production of IL-2 and IFN-γ; the activity of NK cells and showed enhanced ADCC. The level of PINFLCs like IL-1β, IL-6, and TNF-α was reduced on Punarnavine administration. Extracts of *B. diffusa* also showed Cytotoxic effects in tumor cells line Hela (Srivastava et al. 2011). Methanolic extracts of the whole plant reduced cell viability in MCF-7 cell line, and the cells were arrested in G0-G1 phase (Sreeja and Sreeja 2009). The methanolic extracts also inhibit metastasis in B16F10 melanoma in C57BL/6 mice and reduced the serum parameters of metastasis. Retinoids isolated from the root; Boeravinones G and H are efflux inhibitors of cancer-resistance protein (ABCG2) (Ahmed-Belkacem et al. 2007).

### 6.3.7 *Clerodendrum Splendens* (Family: Lamiaceae)

This plant is commonly known as Flaming Glorybower, Pagoda Flower, Bleeding Heart Vine and found in tropical Western Africa. The type II arabinogalactan isolated from *Clerodendrum splendens* exhibited potent immunomodulatory activity. Specifically, the high molecular weight monosaccharides induced NO and cytokine [interleukin (IL)-1α, -1β, -6, -10], TNF-α, and GM-CSF production by human peripheral blood mononuclear cells (PBMCs) and monocyte/macrophages. CSP-AU1-induced secretion of TNF prevented by Toll-like receptor 4 (TLR4) antagonist LPS-RS, indicating a role for TLR4 signaling. It also induced phosphorylation of many MAPKs in human PBMC and activated AP-1/NF-κB. In-vivo administration in mice resulted in increased serum IL-6, IL-10, TNF, monocyte chemoattractant protein-1 (MCP-1), macrophage inflammatory protein (MIP)-1α/CCL3, and MIP-1β/CCL4. It also significantly reduced disease severity in this experimental model of multiple sclerosis. Levels of IL-13, TNF, interferon (IFN)-γ, IL-17, and GM-CSF also significantly decreased, whereas transforming growth factor (TGF)-β increased in LN cells in mice (Kouakou et al. 2013).

### 6.3.8 *Curcuma Longa* (Family: Zingiberaceae)

This plant is commonly known as Turmeric mainly found in Southeast Asia. The main bioactive compounds isolated from *C. longa* are Curcuminoids, sesquiterpenoids, and turmerones. The active component curcuminoid, a mixture of demethoxycurcumin, curcumin, and bisdemethoxycurcumin, having anti-inflammatory activity. It has various activity and block COX-2 (Plummer et al. 1999) and stimulating factor NF-κB (Singh and Aggarwal 1995). Extracts of *C. longa* increases the action of NK cells and increases the amount of TNF-α and IL-6 (Xia et al. 2006). It also regulates the propagation and sensitivity of NK cells macrophages, lymphocytes, and DCs (Jagetia and Aggarwal 2007). Curcumin exhibits immunosuppressive activity and blocks PMA propagation of human spleen-derived lymphocytes and IL-2, accountable for leading this circulation (Ranjan et al. 2004). Immunosuppressive activity of curcumin also examines in human PBMCs, where it blocks expression of NF-κB, IL-2 and PHA-induced propagation (Yadav et al. 2005). Curcumin also influences the phosphantigen-mediated human Vγ9Vδ2 T cell propagation (Cipriani et al. 2001). Polar extracts of *C. longa* primarily contain polysaccharides (lack curcuminoids) that have mitogenic activity and increases the splenocytes count equivalent to LPS and ConA. The NR-INF-02 increases the IFN-γ, TNF-α, NO, IL-2, IL-6 and MCP-1 in deactivated splenocytes and murine macrophages (Chandrasekaran et al. 2013).

### 6.3.9 *Cynodon Dactylon* (Family: Poaceae)

This plant is commonly known as Bermuda Grass Mainly found in Tropical and Subtropical Regions. The active compound isolated in ethanolic extracts of *C. dactylon*, are, Cynodin, beta-carotene, hydrocyanic acid, and triticin show powerful antioxidants activity (Santhi and Annapoorani 2010; Pal et al. 2008; Pal 2008, 2009; Pal and Pandav 2010). Intraperitoneal administration of *C. dactylon* extract significantly increased the adhesion of neutrophils to nylon fibers. Furthermore, the antibody responsiveness in mice to sheep red blood cells (SRBC) significantly increased owing to heightened responsiveness of macrophages and antibody-synthesizing B-cells (Mungantiwar et al. 1999). Not only humoral but also cellular immunity affected the administration of protein fraction of *C. dactylon*. DTH that correlates to cell-mediated immunity increased due to increased sensitivity of T-cells (Santhi and Annapoorani 2010). Oral administration of *C. dactylon* juice increased humoral antibody response against an antigen. *C. dactylon* inhibits the release of TNF-α, IL-6, and IL-1 and promoted anti-inflammatory IL-10 that in turn inhibits PINFLCs (Moore et al. 2001). Studies in *Catla catla* showed that ethanolic extracts of *C. dactylon* incorporated into the diet of the fish activated nonspecific immune mechanism and afforded resistance against *A. hydrophila* infection (Kaleeswaran et al. 2011).

### 6.3.10 *Ficus Benghalensis* (Family: Moraceae)

This tree is commonly known as Banyan Tree, Banyan Fig mainly found in Indian Subcontinent. The root extract used in medicine since ages to boost the immune system. The active phytoconstituents isolated from *F. benghalensis* contain glucosides (Bhattacharjee 2008), flavonoids, etc. The aqueous extract (methanolic and water) shows immunostimulatory activity and increases the phagocytic action of peripheral blood mononuclear cells (PBMC). It also activates the proliferation of lymphocytes and thus the production of cytokines that activate further ICs (Gabhe et al. 2006). The hydroalcoholic leaf extracts significantly increase the phagocytic action of neutrophils and thus engulfment and clearance of microbes by leukocytes, together with free radical scavenging activity and decrease of oxidative stress, exhibited immunomodulatory and antioxidant action (Bhanwase and Alagawadi 2016).

### 6.3.11 *Glycyrrhiza Uralensis Fisch* (Family: Fabaceae)

This plant is commonly known as Chinese liquorice and found in Europe, Asia, Russia and Turkey. Polysaccharides isolated from Licorice exhibited immunomodulatory activities in CT 26 tumor-bearing BALB/c mice. The polysaccharides significantly suppressed tumor growth and increased immune organ index. Furthermore, the immunomodulatory effect was evident with the activation of $CD4^+$ and $CD8^+$ ICs population. The polysaccharides also affected the production of various cytokines, by increasing IL 2, IL 6, IL 7 levels and decreasing TNFα levels. Low molecular weight exhibit polysaccharides anticancer and immunomodulatory activities by suppressing tumor growth and improving the general health of mice. They also augment the thymus/spleen index and population of T lymphocytes. Furthermore, the polysaccharides enhance the levels of serum antitumor cytokines, IL 2, IL 6 and IL 7 while decreasing pro-tumor cytokine TNFα (Ayeka et al. 2017).

### *6.3.12 Murraya Koenigii* (Family: Rutaceae)

This plant is commonly known as Curry Leaves found in Asia and South Africa, *Murraya koenigii* has anti-inflammatory, antioxidant, antitumor properties due to the presence of bioactive ingredients like carbazole alkaloid. Many bioactive compounds are present in *M. koenigii* like koenigin, koenine, koenimbine, girinimbin, iso-mahanimbin, koenidine, cyclomahanimbine, tetrahydromahanimbine, carbazole alkaloids, murrayazoline, mahanimbine, murrayastine, etc. Aqueous extracts of *M. koenigii* exhibit antioxidant activity and protects cardiac tissue against cadmium-induced oxidative stress (Mitra et al. 2012). Leaf extracts also reduced lipid peroxidation in ethanol-induced toxicity in hepatocytes (Sathaye et al. 2011). Methanolic extracts reduce carrageenan-induced paw edema in albino rats, (Bhandari 2012). Methanolic extracts of *M. koenigii* also significantly increased the NO production from macrophages, hence an indicator of its enhanced cytotoxic activity. It also increased the phagocytic index in the carbon-clearance test. Moreover, the humoral antibody response to ovalbumin also increased on treatment with the methanolic extract, thus indicating an overall elevation in humoral response and immunostimulatory effect of the extract on B-cells (Shah et al. 2008).

### *6.3.13 Ocimum Sanctum* (Family: Lamiaceae)

This plant is commonly known as Holy Basil mainly found in Asia, Europe and the USA. *O. sanctum* has anti-inflammatory, analgesic, immunostimulatory activity and has multidirectional therapeutic uses. Key active compounds are isolated are eugenol-2 and its methylated derivatives, sitosterol ursolic acid, limatrol, methyl carvicol, etc. Aqueous extract of *O. sanctum* mainly contains tannins, flavonoids, and alkaloids (Gupta et al. 2002). Leaf extract *O. sanctum* shows immunostimulatory activity. It induces an innate or nonspecific immunity, increases the counts of WBCs, RBCs, phagocytes and lymphocytes and therefore can be used to manage the immunity against several diseases (Nahak and Sahu 2014). A study demonstrates that *O. sanctum* rise a innate immune system against *A. hydrophilia* contamination and increases the total Ig count and lysosomal action, and having a specific activity on biochemical and hematological parameters (Das et al. 2015). The immunostimulatory property of the leaf extract is likely owing to the presence of bioactive constituents like methyleugenol, eugenol, ursolic acid, caryophyllene, salrigenin and oleanolic acid (Mukherjee et al. 2005). The aqueous leaf extracts exhibit immunomodulatory property in subclinical trails in bovines and rise the count and action of lymphocytes and neutrophils and decrease the count of bacteria. *O. sanctum* regulates several cytokines such as TNF-$\alpha$, IL-2 and IFN-$\gamma$, at the time of *S. typhimurium* contamination which blocks bacterial population and is required for stimulation of macrophages and therefore supporting unproductive clearance of the organism (Goel et al. 2010). Oil of *O. sanctum* seed modulates both humoral and cell-mediated immune reactions induced by the GABAergic pathway (Vaghasiya et al. 2010). It increases the count of antibody, number of RBCs, WBCs, and hemoglobin (Jeba et al. 2011). It exhibits free radical scavenging activity and restricts certain categories of cancer cell growth. The active component of ursolic acid show anti-inflammatory and anticancer activity and block COX enzyme action. Alcoholic and aqueous extracts *O. sanctum* shows antitumor activity and decreases the size of Sarcoma-180 solid tumors (Prashar et al. 1998).

### 6.3.14 *Panax Ginseng* (Family: Araliaceae)

This plant is commonly known as Ginseng found in Asia. Panax ginseng is a well-known traditional medicine and immunomodulator. The main isolated components are saponins, polyphenolic compounds, polyacetylenes, triterpenoids, etc. Root, stem, and leaves are commonly made use of to boost the immune system. Ginseng extracts containing polysaccharides increased the phagocytic activity of macrophages. On treatment with red ginseng acidic polysaccharides (RGAPs), peritoneal macrophages showed increased phagocytic activity (Shin et al. 2002). Ginseng enhances the generation of NO from activated peritoneal macrophages and RAW 264.7 cells that aids in microbe clearance. Co-treatment of macrophages with RGAPs and IFN-$\gamma$ further leads to an increase in the production of IL-1$\beta$, TNF-$\alpha$, and NO (Choi et al. 2008). Polysaccharides obtained from *P. ginseng* enhanced the excretion of factors like TNF-$\alpha$ and IL-1$\beta$ from macrophages *in-vitro* (Lim et al. 2002). Treatment of murine macrophage cells J774A.1 with ginseng extract increased the production of IL-12 (Wang et al. 2003). *P. ginseng* also has immunostimulatory effects on DCs (Kim et al. 2009). Administration of ginseng aqueous extracts has immunostimulatory effects on NK cells in both immunocompetent and immunosuppressed mice (Jie et al. 1984; Kim et al 1990) A study using blood samples of immunodeficient patients (AIDS and chronic fatigue syndrome) showed that *P. ginseng* enhanced the NK cell activity in PBMCs (See et al. 1997). The metabolic end products (M1 and M4) of steroidal ginseng saponin, can drive DCs into maturation and increased the expression of cell-surface markers like MHC class II, CD80, CD83, and CD86. M4 primed mature DCs boosted antigen-presentation ability as evident from the production of IFN-γ and 51Cr by the DCs. A reciprocal effect is seen on the treatment of dendritic cells with whole saponins of *Pinex ginseng* followed by oxidized LDLp. It inhibits the maturation of DCs and downregulates maturation markers like HLA-DR, CD1a, CD40, CD86, etc. It also inhibits TNF-α and IL-12. These reciprocal effects of P. ginseng might be due to the activation of different signaling pathways (Su et al. 2010). Ginsan also shows immunostimulatory effects on the maturation of DCs (Kim et al. 2009). Ginsan not only modulates innate immune response but also affects adaptive immune response. Orally administered with ginseng before infection with Salmonella in mice showed higher serum IgG1 and IgG2 and IgA against Salmonella (Na et al. 2010). Ginsenoside injected subcutaneously increases serum antibodies specific against *Toxoplasma gondii* and ginsenoside likewise accordingly increase serum special immunoglobulins like IgG1, IgG, IgG2a and IgG2b mechanism against H3N2 influenza virus (Yoo et al. 2012). Both ginseng and ginsenoside stimulated T-cell proliferation and affect cell-mediated immune response. It was also found that splenocytes cultured with ginseng produce pro-inflammatory IL-2, IFN-γ, IL-1α, and GMSF (Kim et al. 1998). This effect was reversed in mice infected with *S. aureus*. Macrophages treated with ginseng radix extract (GRE) produce PINFLCs such as IFN-γ, IL-1β, TNF-α, and IL-6 *in-vitro* (Liou et al. 2006) possibly through activation of TLR4 signaling (Nakaya et al. 2004) through a non-LPS agent present in GRE. The active compound of ginseng can also downregulate TLR2 and its downstream Myd88 and inhibit PINFLCs production (Ahn et al. 2005).

### 6.3.15 *Picrorhiza Scrophulariiflora* (Family: Scrophulariaceae)

This plant is commonly named as Kutki found in Alpine Himalayas, Tibet. The rhizomes are rich in iridoid glycosides phenolic glycosides, phenylethanoid glycosides and terpenoids. Extracts of *P. scrophulariiflora* inhibits lipid peroxidation and possess free radical scavenging activity. Scrocaffeside A isolated from *P. scrophulariiflora* is an immunostimulant. In response to Con A and LPS, it significantly increased the number of splenocytes. It also enhanced the activity of macrophages and NK cells even in the absence of stimulation (Coico et al. 2015). The number of mature $CD4^+$ and $CD8^+$ cells and both Th1 and Th2 cytokines also significantly increased. Activated $CD4^+$ T-cells differentiate either into Th1 type cells and secrete TNF-α, IL-12, IL-2, IFN-γ, or into Th2 cells and secrete IL-5, IL-10, IL-4, IL-13, and this balance is critical in deciding the immune response and ultimately the disease outcome. Picrogentiosides isolated from *P. scrophulariiflora* enhanced the proliferation of splenocytes comparable to Con A and LPS and enhanced both humoral and cell-mediated immunity (Zou et al. 2010). Picracin and deacetylbaccatin isolated from *P. scrophulariiflora* has antiproliferative effects and inhibited proliferation of T-cells that inhibits the release of IL-2 and subsequent IL-2-IL-2 receptor interaction. Picracin inhibited the release of PINFLC such as IL-1β and TNFα in human monocytes, whereas deacetylbaccatin failed to do so. Moreover, systemic administration of both picracin and deacetylbaccatin decreased paw edema, suggesting its immunosuppressive activity. These compounds have dualistic effects and can induce an inflammatory response on local administration whereas an immunosuppressive effect on systemic administration.

### *6.3.16 Syzygium Aromaticum* (Family: Myrtaceae)

This plant is commonly known as clove and used as a spice and commercially found in Indonesia, as well as in India, Pakistan, Sri Lanka, Comoro Islands, Madagascar, Seychelles, and Tanzania. Eugenol (4-allyl-2-methoxyphenol) is a phenolic compound representing the most important component of the clove. Eugenol inhibits IL-1b, IL-6 and IL-10 production and exerts an efficient action with lipopolysaccharide (LPS) challenge for cytokines. Eugenol affects IL-1b production but inhibits IL-6 and IL-10 production. The action of eugenol on IL-6 production prevented efficient effects of LPS either before or after its addition, whereas on IL-10 production it counteracted significantly LPS action when added after LPS incubation (Bachiega et al. 2012).

### *6.3.17 Terminalia Arjuna* (Family: Combretaceae)

This plant is commonly Arjuna found in the Pantropical region. *T. arjuna* is commonly used in the traditional medicinal system to treat cardiac ailments. Treatment of peritoneal macrophages with various doses of *T. arjuna* methanolic bark extracts and gemmo-modified (embryonic tissue) extracts increased its phagocytic potential as evaluated by increased sheep red blood cells (sRBC) engulfed by macrophages. Both the extracts also enhanced the antibody-mediated humoral immune response and cell-mediated phagocytosis. This immunomodulatory property of the *T. arjuna* extract is attributed to the presence of polyphenols and flavonoids (Chiang et al. 2003). Polyphenols attenuate PINFLCs, MMPs, SOCS3 and downregulate TLR2 and NLRP1-an inflammasome component-and upregulate AINFLCs, thus inhibiting the proliferation of lymphocytes and reducing the activity of NK cells (Jung et al. 2012). In a study, the *T. arjuna* extracts significantly decreased formalin-induced paw edema due to its anti-inflammatory properties (Halder et al. 2009). Arjunolic acid a triterpene is a strong cardioprotective and antioxidant in rat models. Ethanolic extracts of *T. arjuna* decreases the level of endothelin-1 and INFLCs like IL-6 and TNF-α in diabetic rats (Khaliq et al. 2013).

### *6.3.18 Tinospora Cordifolia* (Family: Menispermaceae)

This plant is commonly known as Guduchi, Moonseed, Giloy found in Indian Subcontinent, China. *Tinospora cordifolia* has a broad spectrum of immunotherapeutic properties ranging from antipyretic anti-inflammatory, antiallergic antidiabetic, antihepatotoxic, antibacterial properties and has relatively low toxicity. It also helps the cell repair and rejuvenation process (Sharma et al. 2020). The key active components isolated from the plant include phenyl propanoid glycosides such as Cordifolioside A, Cordifolioside B and syringin (Maurya et al. 1996), and immunostimulatory compound-D-glucan (Nair et al. 2004, 2006). The immunomodulatory property of the Guduchi extract was validated by assessing its effect on activating resting macrophages and by estimating the generation of secretory factors like NO and lysosomes (More and Pai 2011). Water and ethyl acetate extracts of *T. cordifolia* stem increased the phagocytic activity of human neutrophils. The immunostimulatory action may be attributed to the synergistic action of two or more than two active components like Cordifolioside A and syringin. *T. cordifolia* extract is also shown to stimulate the proliferation of stem cells. It increases the total number of WBCs and alpha-esterase-positive cells an indicator of an increase in bone marrow cells. The extract also increases the number of antibody-producing cells, hence implying its role in fortifying the humoral immune system. The extracts of *T. cordifolia* are also found to be effective against tumors and reduced tumor growth comparable to that of treatment with cyclophosphamide (Mathew and Kuttan 1999).

## 6.4 Conclusion and Future Perspective

Phytochemicals and phytomedicine have been widely used in medicine for centuries. Various traditional and folk medicines were based on the unexplored character of these phytochemicals. Scientific investigations in the last several decades have uncovered the molecular roles of these phytochemicals. Recently *in-vitro* and *in-vivo* studies have highlighted their role in immune system modulation. The role of phytochemicals in treating infectious, acute as well as chronic diseases was studied extensively. Though, their efficacy, biosafety and bioavailability have not been fully studied. Application of a systems biology approach to model the multidimensional targets of these chemicals would be a much effective approach. This would further pave the path for the application of phytochemicals in personalized medicine to minimize the adverse/side effects and maximize effectiveness.

Immunomodulatory therapeutics are the agents that could modulate the immune system of an organism if it increases or stimulate the immune response are called as immunostimulants or if it decreases or suppresses immune response are called immunosuppressants. These drugs are most commonly used in autoimmune diseases, inflammatory conditions, cancer, allergic reactions, AIDS, and some viral infections. Only a few plants have been screened for immunomodulatory activities. From the above chapter, it is evident that there are several medicinal plants and marine products which have immunomodulatory activity but insufficient evidence does not allow their use in clinical practice. Therefore immunomodulatory agents will gain more importance in the future research of herbal medicine.

## References

Ahmed-Belkacem A, Macalou S, Borrelli F, Capasso R, Fattorusso E, Taglialatela-Scafati O, Di Pietro A (2007) Nonprenylated rotenoids, a new class of potent breast cancer resistance protein inhibitors. J Med Chem 50:1933–1938

Ahn JY, Song JY, Yun YS, Jeong G, Choi IS (2005) Protection of Staphylococcus aureus-infected septic mice by suppression of early acute inflammation and enhanced antimicrobial activity by ginsan. FEMS Immunol Med Microbiol 46:187–197

Akowuah GA, Zhari I, Mariam A, Yam MF (2009) Absorption of andrographolides from Andrographis paniculata and its effect on CCl4-induced oxidative stress in rats. Food Chem Toxicol 47:2321–2326

Amirghofran Z, Azadbakht M, Karimi MH (2000) Evaluation of the immunomodulatory effects of five herbal plants. J Ethnopharmacol 72:167–172

Ayeka PA, Bian Y, Githaiga PM, Zhao Y (2017) The immunomodulatory activities of licorice polysaccharides (*Glycyrrhiza uralensis* Fisch.) in CT 26 tumor-bearing mice. BMC Complement Altern Med 15:536

Babisch W (2003) Stress hormones in the research on cardiovascular effects of noise. Noise Health 5:1

Baral R, Chattopadhyay U (2004) Neem (Azadirachta indica) leaf mediated immune activation causes prophylactic growth inhibition of murine Ehrlich carcinoma and B16 melanoma. Int Immunopharmacol 4:355–366

Bachiega TF, de Sousa JP, Bastos JK, Sforcin JM (2012) Clove and eugenol in noncytotoxic concentrations exert immunomodulatory/anti-inflammatory action on cytokine production by murine macrophages. J Pharm Pharmacol 64:610–616

Bafna A, Mishra S (2009) Antioxidant and immunomodulatory activity of the alkaloidal fraction of *Cissampelos pareira* Linn. Sci Pharm 78:21–31

Bains JS, Dhuna V, Singh J, Kamboj SS, Nijjar KK, Agrewala JN (2005) Novel lectins from rhizomes of two Acorus species with mitogenic activity and inhibitory potential towards murine cancer cell lines. Int Immunopharmacol 5:1470–1478

Ballarin L (2008) Immunobiology of compound ascidians, with particular reference to *Botryllus schlosseri*: state of art. Inv Surv J 5:54–74

Baxter D (2007) Active and passive immunity, vaccine types, excipients and licensing. Occup Med 57:552–556

Belska NV, Guriev AM, Danilets MG, Trophimova ES, Uchasova EG, Ligatcheva AA, Belousov MV, Agaphonov VI, Golovchenko VG, Yusubov MS, Belsky YP (2010) Water-soluble polysaccharide obtained from *Acoruscalamus* L. classically activates macrophages and stimulates Th1 response. Int Immunopharmacol 10:933–942

Benencia F, Courrèges MC, Coulombié FC (2000) In-vivo and in-vitro immunomodulatory activities of *Trichilia glabra* aqueous leaf extracts. J Ethnopharmacol 69:199–205

Bessler H, Djaldetti M (2016) Capsaicin modulates the immune cross talk between human mononuclears and cells from two colon carcinoma lines. Nutr Cancer 69:14–20

Beuth J, Schneider H, Ko HL (2006) Enhancement of immune responses to neem leaf extract (Azadirachta indica) correlates with antineoplastic activity in BALB/c-mice. Vivo 20:247–251

Bhandari PR (2012) Curry leaf (*Murraya koenigii*) or cure leaf: review of its curative properties. J Med Nutr Nutraceuticals 1:92

Bhanwase AS, Alagawadi KR (2016) Antioxidant and immunomodulatory activity of hydroalcoholic extract and its fractions of leave of *Ficus benghalensis* Linn. Pharmacognosy Res 8:50–55

Bhatt RP, Pandya BK, Sheth RN (2010) *Camellia sinensis* L: the medicinal beverage: a review. IJPSRR 3:6–9

Bhattacharjee S (2008) Handbook of medicinal plants. Pointer publishers, Jaipur, Raj, India, p 118

Bhattacharyya B, Panda D, Gupta S, Banerjee M (2008) Anti-mitotic activity of colchicine and the structural basis for its interaction with tubulin. Med Res Rev 28:155–183

Bhaumik S, Jyothi MD, Khar A (2000) Differential modulation of NO production by curcumin in host macrophages and NK cells. FEBS Lett 483:78–82

Bi XL, Yang JY, Dong YX, Wang JM, Cui YH, Ikeshima T, Zhao YQ, Wu CF (2005) Resveratrol inhibits NO and TNF-α production by lipopolysaccharide-activated microlia. Int Immunopharmacol 5:185–193

Bodduluru LN, Kasala ER, Thota N, Barua CC, Sistla R (2014) Chemopreventive and therapeutic effects of nimbolide in cancer: the underlying mechanisms. Toxicol In-vitro 28:1026–1035

Bonacorsi C, Raddi MSG, Carlos IZ, Sannomiya M, Vilegas W (2009) Anti-Helicobacter pylori activity and immunostimulatory effect of extracts from *Byrsonima crassa* Nied. (Malpighiaceae). BMC Complement Altern Med 9:1–7

Boots AW, Haenen GR, Bast A (2008) Health effects of quercetin: from antioxidant to nutraceutical. Eur J Pharmacol 585:325–337

Bopana N, Saxena S (2007) *Asparagus racemosus* ethnopharmacological evaluation and conservation needs. J Ethnopharmacol 110:1–15

Bose A, Baral R (2007) Natural killer cell mediated cytotoxicity of tumor cells initiated by neem leaf preparation is associated with CD40-CD40L-mediated endogenous production of interleukin-12. Hum Immunol 68:823–831

Bose A, Chakraborty K, Sarkar K, Goswami S, Chakraborty T, Pal S, Baral R (2009) Neem leaf glycoprotein induces perforin-mediated tumor cell killing by T and NK cells through differential regulation of IFNγ signaling. J Immunother 32:42–53

Burnett BP, Jia Q, Zhao Y, Levy RM (2007) A medicinal extract of *Scutellaria baicalensis* and *Acacia catechu* acts as a dual inhibitor of cyclooxygenase and 5-lipoxygenase to reduce inflammation. J Med Food 10:442–451

Cameron M, Chrubasik S (2013) Topical herbal therapies for treating osteoarthritis. Cochrane Database Syst Rev 5:CD010538

Caterina MJ, Schumacher MA, Tominaga M, Rosen TA, Levine JD, Julius D (1997) The capsaicin receptor: a heat-activated ion channel in the pain pathway. Nature 389:816–824

Chakraborty K, Bose A, Pal S, Sarkar K, Goswami S, Ghosh D, Laskar S, Chattopadhyay U, Baral R (2008) Neem leaf glycoprotein restores the impaired chemotactic activity of peripheral blood mononuclear cells from head and neck squamous cell carcinoma patients by maintaining CXCR3/CXCL10 balance. Int Immunopharmacol 8:330–340

Chandrasekaran CV, Kannan Sundarajan JRE, Gururaja GM, Mundkinajeddu D, Agarwal A (2013) Immune stimulatory and anti-inflammatory activities of *Curcuma longa* extract and its polysaccharide fraction. Pharmacognosy Res 5:71–79

Chandrashekar PM, Prashanth KV, Venkatesh YP (2011) Isolation, structural elucidation and immunomodulatory activity of fructans from aged garlic extract. Phytochemistry 72:255–264

Chang SL, Chiang YM, Chang CLT, Yeh HH, Shyur LF, Kuo YH, Wu TK, Yang WC (2007) Flavonoids, centaurein and centaureidin, from *Bidens pilosa*, stimulate IFN expression. J Ethnopharmacol 112:232–236

Chao WW, Lin BF (2010) Isolation and identification of bioactive compounds in *Andrographis paniculata* (Chuanxinlian). Chin Med 5:17

Chen C, Zheng W, Gao X, Xiang X, Sun D, Wei J, Chu C (2007) Aqueous extract of *Inonotus obliquus* (Fr.) Pilat (Hymenochaetaceae) significantly inhibits the growth of sarcoma 180 by inducing apoptosis. Am J Pharmacol Toxicol 2:10–17

Chen Y, Le TH, Du Q, Zhao Z, Liu Y, Zou J, Hua W, Liu C, Zhu Y (2019) Genistein protects against DSS-induced colitis by inhibiting NLRP3 inflammasome via TGR5-cAMP signaling. Int Immunopharmacol 71:144–154

Cherng JM, Chiang W, Chiang LC (2007) Immunomodulatory activities of common vegetables and spices of Umbelliferae and its related coumarins and flavonoids. Food Chem 106:944–950

Chiang LC, Ng LT, Chiang W, Chang MY, Lin CC (2003) Immunomodulatory activities of flavonoids, monoterpenoids, triterpenoids, iridoid glycosides and phenolic compounds of Plantago species. Planta Med 69:600–604

Cho Q, Zhang Y, Chen W, Jia R, Yu X, Wang Y, Li Y, Liu Y, Ye X, Yu L, Zheng X (2020) *Apios americana* Medik flowers polysaccharide (AFP) alleviate Cyclophosphamide-induced immunosuppression in ICR mic. Int J Biol Macromol 144:829–836

Choi HS, Kim KH, Sohn E, Park JD, Kim BO, Moon EY, Rhee DK, Pyo S (2008) Red ginseng acidic polysaccharide (RGAP) in combination with IFN-γ results in enhanced macrophage function through activation of the NF-κB pathway. Biosci Biotechnol Biochem 72:1817–1825

Chung YH, Kim HY, Yoon BR, Kang YJ, Lee WW (2019) Suppression of Syk activation by resveratrol inhibits MSU crystal-induced inflammation in human monocytes. J Mol Med (Berl) 97:369–383

Cipriani B, Borsellino G, Knowles H, Tramonti D, Cavaliere F, Bernardi G, Battistini L, Brosnan CF (2001) Curcumin inhibits activation of Vγ9Vδ2 T cells by phosphoantigens and induces apoptosis involving apoptosis-inducing factor and large scale DNA fragmentation. J Immunol 167:3454–3462

Clement F, Pramod SN, Venkatesh YP (2010) Identity of the immunomodulatory proteins from garlic (*Allium sativum*) with the major garlic lectins or agglutinins. Int Immunopharmacol 10:316–324

Coico R, Sunshine G, Benjamini E (2015) Immunology. Wiley-Liss, New Jersey, p 201

Cooper JC, Turcasso N (1999) Immunostimulatory effects of b-1,3 glucan and acemannan. JANA 2:5–11

Costa JFO, David JPL, David JM, Giulietti AM, Queiroz LP, Santos RR, Soares MBP (2008) Immunomodulatory activity of extracts from *Cordia superba Cham.* and *Cordia rufescens A.* DC. (Boraginaceae), plant species native from Brazilian semiarid. Rev Bras Farmaacogn 18:11–15

Das R, Raman RP, Saha H, Singh R (2015) Effect of *Ocimum sanctum* Linn (Tulsi) extract on the immunity and survival of *Labeo rohita* (Hamilton) infected with Aeromonas hydrophila. Aquacult Res 46:1111–1121

Dashputre NL, Naikwade NS (2010) Immunomodulatory activity of *Abutilon indicum* Linn. On Albino Mice IJPSR 1:178–184

Davicino R, Mattar A, Casali Y, Porporatto C, Correa SG, Micalizzi B (2007) In-vivo immunomodulatory effects of aqueous extracts of *Larrea divaricata* Cav. Immunopharmacol Immunotoxicol 29:351–366

Delmas D, Limagne E, Ghiringhelli F, Aires V (2019) Immune Th17 lymphocytes play a critical role in the multiple beneficial properties of resveratrol. Food Chem Toxicol 137:111091

Dhandapani KM, Mahesh VB, Brann DW (2007) Curcumin suppresses growth and chemoresistance of human glioblastoma cells via AP-1 and NFκB transcription factors. J Neurochem 102:522–538

Dharini S, Gajalakshmi G, Devi RS (2012) Effect of *Acorus calamus* Linn and its active principle on noise stress-induced changes on immune system. Int J Biomed Adv Res 3:494–501

do Rosa´rio MMT, Kangussu-Marcolino MM, do Amaral AE, Noleto GR, de Oliveira Petkowicz CL (2011) Storage xyloglucans: potent macrophages activators. Chem Biol Interact 189:127–133

Ebrahimi M, Hassan ZM, Mostafaie A, Mehrjardi NZ, Ghazanfari T (2013) Purified protein fraction of garlic extract modulates cellular immune response against breast transplanted tumors in BALB/c mice model. Cell J (Yakhteh) 15:65–74

Ebringerova A, Hromadkova Z, Hribalova V, Hirsch J (2002) An immunomodulating pectic arabinogalactan from roots *Cistanche deserticola*. Chem Pap 56:320–325

Feng Y, Zhu X, Wang Q, Jiang Y, Shang H, Cui L, Cao Y (2012) Allicin enhances host proinflammatory immune responses and protects against acute murine malaria infection. Malar J 11:268

Ferreira AP, Soares GLG, Salgado CA, Goncalves LS, Teixeira FM, Teixeira HC, Kaplan MAC (2003) Immunomodulatory activity of *Mollugo verticillata* L. Phytomedicine 10:154–158

Fu E, Tsai MC, Chin YT, Tu HP, Fu MM, Chiang CY, Chiu HC (2015) The effects of diallyl sulfide upon Porphyromonas gingivalis lipopolysaccharide stimulated proinflammatory cytokine expressions and NKk B activation in human gingival fibroblasts. J Periodontal Res 50:380–388

Gabhe SY, Tatke PA, Khan TA (2006) Evaluation of the immunomodulatory activity of the methanol extracts of *Ficus benghalensis* roots in rats. Indian J Pharmacol 38:271–275

Gacche RN, Dhole NA (2006) Antioxidant and possible anti-inflammatory potential of selected medicinal plants prescribed in the Indian traditional system of medicine. Pharm Biol 44:389–395

Gan L, Zhang SH, Liu Q, Xu HB (2003) A polysaccharide-protein complex from *Lycium barbarum* upregulates cytokine expression in human peripheral blood mononuclear cells. Eur J Pharmacol 471:217–222

Ganju L, Karan D, Chanda S, Srivastava KK, Sawhney RC, Selvamurthy W (2003) Immunomodulatory effects of agents of plant origin. Biomed Pharmacother 57:296–300

Gao X, Kuo J, Jiang H, Deeb D, Liu Y, Divine G, Chapman RA, Dulchavsky SA, Gautam SC (2004) Immunomodulatory activity of curcumin: suppression of lymphocyte proliferation, development of cell-mediated cytotoxicity, and cytokine production in-vitro. Biochem. Pharmacol 68:51–61

Gaur K, Kori ML, Nema RK (2009) Comparative screening of immunomodulatory activity of hydro- alcoholic extract of *Hibiscus rosa sinensis* Linn. and ethanolic extract of *Cleome gynandra* Linn. Global J Pharmacol 3:85–89

Gazzinelli RT, Oswald IP, James SL, Sher A (1992) IL-10 inhibits parasite killing and nitrogen oxide production by IFN-gamma-activated macrophages. J Immunol 148:1792–1796

Geng Y, Li C, Liu J, Xing G, Zhou L, Dong M, Li X, Niu Y (2010) Beta-asarone improves cognitive function by suppressing neuronal apoptosis in the beta-amyloid hippocampus injection rats. Biol Pharm Bull 33:836–843

Ghaisas MM, Saikh SA, Deshpande AD (2009) Evaluation of immunomodulatory activity of ethanolic extract of stem bark of *Bauhinia variegata* Linn. Int J Green Pharm 3:70–74

Gharavi MJ, Nobakht M, Khademvatan SH, Bandani E, Bakhshayesh M, Roozbehani M (2011) The effect of garlic extract on expression of INF$\gamma$ and iNOS genes in macrophages infected with Leishmania major. Iranian J Parasitol 6:74–81

Ghate NB, Hazra B, Sarkar R, Mandal N (2014) Heartwood extract of *Acacia catechu* induces apoptosis in human breast carcinoma by altering bax/bcl-2 ratio. Pharmacognosy Mag 10:27–33

Goel A, Kumar S, Bhatia AK (2010) Effect of *Ocimum sanctum* on the development of protective immunity against *Salmonella typhimurium* infection through cytokines. Asia Pac J Trop Med 3:682–686

Guerra RNM, Pereira HAW, Silveria LMS, Olea RSG (2003) Immunomodulatory properties of Alternanthera tenella Colla aqueous extracts in mice. Braz J Med Biol Res 36:1215–1219

Guleria S, Tiku AK, Singh G, Vyas D, Bhardwaj A (2011) Antioxidant activity and protective effect against plasmid DNA strand scission of leaf, bark, and heartwood extracts from *Acacia catechu*. J Food Sci 76:C959–C964

Gupta A, Gautam MK, Singh RK, Kumar MV, Rao CV, Goel RK, Anupurba S (2010) Immunomodulatory effect of *Moringa oleifera* extract on cyclophosphamide induced toxicity in mice. Ind J Exp Biol 48:1150–1167

Gupta SK, Prakash J, Srivastava S (2002) Validation of claim of Tulsi, *Ocimum sanctum* Linn as a medicinal plant. Indian J Exp Biol 40:765–773

Habijanic J, Berovic M, Wraber B, Hodzar D, Boh B (2001) Immunostimulatory effects of fungal polysaccharides from *Ganoderma lucidum* submerged biomass cultivation. Food Technol Biotechnol 39:327–331

Hamalainen M, Nieminen R, Vuorela P, Heinonen M, and Moilanen E (2007) Anti-inflammatory effects of flavonoids: genistein, kaempferol, quercetin, and daidzein inhibit STAT-1 and NF-κB activations, whereas flavone, isorhamnetin, naringenin, and pelargonidin inhibit only NF-κB activation along with their inhibitory effect on iNOS expression and NO production in activated macrophages. Mediators Inflamm 45673

Halder S, Bharal N, Mediratta PK, Kaur I, Sharma KK (2009) Anti-inflammatory, immunomodulatory and antinociceptive activity of Terminalia arjuna Roxb bark powder in mice and rats. Indian J Exp Biol 47:577–583

Han Y, Jo H, Cho JH, Dhanasekaran DN, Song YS (2019) Resveratrol as a tumor-suppressive nutraceutical modulating tumor microenvironment and malignant behaviors of cancer. Int J Mol Sci 20:E925

Hidalgo MA, Romero A, Figueroa J, Cortes P, Concha II, Hancke JL, Burgos RA (2005) Andrographolide interferes with binding of NKk B to DNA in HL-60-derived neutrophilic cells. Br J Pharmacol 144:680–686

Holden NS, Squires PE, Kaur M, Bland R, Jones CE, Newton R (2008) Phorbol ester-stimulated NF-κB-dependent transcription: roles for isoforms of novel PKC. Cell Signal 20:1338–1348

Holmes-McNary M, Baldwin AS Jr (2000) Chemopreventive properties of trans-resveratrol are associated with inhibition of activation of the IκB kinase. Cancer Res 60:3477–3483

Hong MH, Kim MH, Chang HJ, Kim NH, Shin BA, Ahn BW, Jung YD (2007) (-)-Epigallocatechin-3-gallate inhibits monocyte chemotactic protein-1 expression in endothelial cells via blocking NF-κB signaling. Life Sci 80:1957–1965

Hsu CH, Sun HL, Sheu JN, Ku MS, Hu CM, Chan Y, Lue KH (2008) Effects of the immunomodulatory agent *Cordyceps militaris* on airway inflammation in a mouse asthma model. Pediatr Neonatol 495:171–178

Hung LC, Lin CC, Hung SK, Wu BC, Jan MD, Liou SH (2007) A synthetic analog of a galactosylceramide induces macrophage activation via the TLR4-signaling pathways. Biochem Pharmacol 73:1957–1970

Im SA, Kim K, Lee CK (2006) Immunomodulatory activity of polysaccharides isolated from *Salicornia herbacea*. Int Immunopharmacol 6:1451–1458

Iruretagoyena MI, Tobar JA, Gonzalez PA, Sepulveda SE, Figueroa CA, Burgos RA, Hancke JL, Kalergis AM (2005) Andrographolide interferes with T cell activation and reduces experimental autoimmune encephalomyelitis in the mouse. J Pharmacol Exp Ther 312:366–372

Ismail S, Asad M (2009) Immunomodulatory activity of *Acacia catechu*. Indian J Physiol Pharmacol 53:25–33

Issuree PD, Pushparaj PN, Pervaiz S, Melendez AJ (2009) Resveratrol attenuates C5a-induced inflammatory responses in-vitro and in-vivo by inhibiting phospholipase D and sphingosine kinase activities. FASEB J 23:2412–2424

Jafarian A, Ghannadi A, Elyasi A (2003) The effects of *Allium hirtifolium* Boiss. on cell-mediated immune response in mice. IJPR 2:51–55

Jagetia GC, Aggarwal BB (2007) "Spicing up" of the immune system by curcumin. J Clin Immunol 27:19–35

Jain V, Bhagwat D, Jat RC, Bhardwaj S (2005) The immunomodulation potential of *Hyptis suaveolens*. IJPRD 1:1–6

Jantan I, Ahmad W, Bukhari SN (2015) Plant-derived immunomodulators: an insight on their preclinical evaluation and clinical trials. Front Plant Sci 25:655

Jayathirtha MG, Mishra SH (2004) Preliminary immunomodulatory activities of methanol extracts of *Eclipta alba* and *Centella asiatica*. Phytomedicine 11:361–365

Jeba RC, Vaidyanathan RAMA, Rameshkumar G (2011) Efficacy of *Ocimum basilicum* for immunomodulatory activity in Wistar albino rat. Int J Pharm Pharm Sci 3:199–203

Jeong SC, Jeong YT, Yang BK, Song CH (2005) Chemical characteristics and immune stimulating properties of biopolymers extracted from *Acanthopanax sessiliflorus*. J Biochem Mol Biol 39:84–90

Jie YH, Cammisuli S, Baggiolini M (1984) Immunomodulatory effects of *Panax ginseng* CA Meyer in the mouse. Inflamm Res 15:386–391

Jin L, Schmiech M, El Gaafary M, Zhang X, Syrovets T, Simmet T (2020) A comparative study on root and bark extracts of *Eleutherococcus senticosus* and their effects on human macrophages. Phytomedicine 68:153181

Jung MK, Ha S, Son JA, Song JH, Houh Y, Cho E, Chun JH, Yoon SR, Yang Y, Bang SI, Kim M, Park HJ, Cho D (2012) Polyphenon-60 displays a therapeutic effect on acne by suppression of TLR2 and IL-8 expression via down-regulating the ERK1/2 pathway. Arch Dermatol Res 304:655–663

Kaleeswaran B, Ilavenil S, Ravikumar S (2011) Dietary supplementation with *Cynodon dactylon* (L.) enhances innate immunity and disease resistance of Indian major carp, *Catla catla* (Ham.). Fish Shellfish Immunol 31:953–962

Kannan M, Ranjit Singh AJA, Ajith Kumar TT, Jegatheswari P, Subburayalu S (2007) Studies on immuno-bioactivities of *Nyctanthes arbortristis* (Oleaceae). Afr J Microbiol Res 1:88–91

Khaliq F, Parveen A, Singh S, Gondal R, Hussain ME, Fahim M (2013) Improvement in myocardial function by *Terminalia arjuna* in streptozotocin-induced diabetic rats: possible mechanisms. J Cardiovasc Pharm Ther 18:481–489

Khare CP (2007) Indian medicinal plants. An illustrated dictionary. New York: Springer Publications. Panax Ginseng Monograph Altern Med Rev 14:172–176

Killestein J, Hoogervorst ELJ, Reif M, Blauw B, Smits M, Uitdehaag BMJ, Nagelkerken L, Polman CH (2003) Immunomodulatory effects of orally administered cannabinoids in multiple sclerosis. J Neuroimmunol 137:140–143

Kim GY, Kim KH, Lee SH, Yoon MS, Lee HJ, Moon DO, Lee CM, Ahn SC, Park YC, Park YM (2005) Curcumin inhibits immunostimulatory function of DCs: MAPKs and translocation of NF-κB as potential targets. J Immunol 174:8116–8124

Kim JY, Germolec DR, Luster MI (1990) *Panax ginseng* as a potential immunomodulator: studies in mice. Immunopharmacol Immunotoxicol 12:257–276

Kim KH, Lee YS, Jung IS, Park SY, Chung HY, Lee IR, Yun YS (1998) Acidic polysaccharide from Panax ginseng, ginsan, induces Th1 cell and macrophage cytokines and generates LAK cells in synergy with rIL-2. Planta Med 64:110–115

Kim MH, Byon YY, Ko EJ, Song JY, Yun YS, Shin T, Joo HG (2009) Immunomodulatory activity of ginsan, a polysaccharide of *Panax ginseng*, on DCs. Korean J Physiol Pharmacol 13:169–173

Kobayashi H, Mejia ED (2005) The genus Ardisia: a novel source of health-promoting compounds and phytopharmaceuticals. J Ethnopharmacol 96:347–354

Konidena A, Sharma S, Patil DJ, Dixit A, Gupta R, Kaur M (2017) Immunosuppressants in Oral Medicine: A Review. J Indian Acad Oral Med Radiol 29:306–313

Kouakou K, Schepetkin IA, Jun S, Kirpotina LN, Yapi A, Khramova DS, Pascual DW, Ovodov YS, Jutila MA, Quinn MT (2013) Immunomodulatory activity of polysaccharides isolated from *Clerodendrum splendens*: beneficial effects in experimental autoimmune encephalomyelitis. BMC Complement Altern Med 28:149

Kowalski J, Samojedny A, Paul M, Pietsz G, Wilczok T (2005) Effect of apigenin, kaempferol and resveratrol on the expression of interleukin-1beta and TNF-alpha genes in J774.2 macrophages. Pharmacol Rep 57:390–394

Kumar D, Arya V, Kaur R, Bhat ZA, Gupta VK, Kumar V (2012) A review of immunomodulators in the Indian traditional health care system. J Microbiol Immunol Infect 45:165–184

Kumar A, Dhawan S, Hardegen NJ, Aggarwal BB (1998) Curcumin (Diferuloylmethane) inhibition of TNF (TNF)-mediated adhesion of monocytes to endothelial cells by suppression of cell surface expression of adhesion molecules and of NKk B activation. Biochem. Pharmacol 55:775–783

Kumar S, Suresh PK, Vijayababu MR, Arunkumar A, Arunakaran J (2006) Anticancer effects of ethanolic neem leaf extract on prostate cancer cell line (PC-3). J Ethnopharmacol 105:246–250
Kurian A, Van Doan H, Tapingkae W, Elumalai P (2020) Modulation of mucosal parameters, innate immunity, growth and resistance against Streptococcus agalactiae by enrichment of Nile tilapia (*Oreochromis niloticus*) diet with *Leucas aspera*. Fish Shellfish Immunol 97:165–172
Kyo E, Uda N, Kasuga S, Itakura Y (2001) Immunomodulatory effects of aged garlic extract. J Nutr 131:1075S-1079S
Lasure A, Van Poel B, Chimanga K, Pieters L, Berge DV, Vlientnik AJ (1994) Modulation of complement system by flavonoids. Pharm Pharm Lett 4:32–35
Lee KC, Chang HH, Chung YH, Lee TY (2011) Andrographolide acts as an anti-inflammatory agent in LPS-stimulated RAW264.7 macrophages by inhibiting STAT3-mediated suppression of the NF-κB pathway. J Ethnopharmacol 135:678–684
Lee YW, Lee WH (2008) Protective effects of genistein on proinflammatory pathways in human brain microvascular endothelial cells. J Nutr Biochem 19:819–825
Leung KN, Leung PY, Kong LP, Leung PK (2005) Immunomodulatory effects of esculetin (6,7-dihydroxycoumarin) on murine lymphocytes and peritoneal macrophages. Cell Mol Immunol 2:181–188
Li X, Wang H, Liu C, Chen R (2010) Chemical constituents of Acacia catechu. Zhongguo Zhong yao za zhi Zhongguo zhongyao zazhi. China J Chinese Mater Med 35:1425–1427
Li XC, Yang LX, Wang HQ, Chen RY (2011) Phenolic compounds from the aqueous extract of *Acacia catechu*. Chin Chem Lett 22:1331–1334
Li YN, Huang F, Liu XL, Shu SN, Huang YJ, Cheng HJ, Fang F (2013) *Allium sativum* derived allitridin inhibits treg amplification in cytomegalovirus infection. J Med Virol 85:493–500
Lim DS, Bae KG, Jung IS, Kim CH, Yun YS, Song JY (2002) Anti-septicaemic effect of polysaccharide from *Panax ginseng* by macrophage activation. J Infect 45:32–38
Lin TP, Chen SY, Duh PD, Chang LK, Liu YN (2008) Inhibition of the Epstein-Barr virus lytic cycle by andrographolide. Biol Pharm Bull 31:2018–2023
Liou CJ, Huang WC, Tseng J (2006) Short-term oral administration of ginseng extract induces type-1 cytokine production. Immunopharmacol Immunotoxicol 28:227–240
Liu J, Wang ZT, Ge BX (2008) Andrograpanin isolated from *Andrographis paniculata*, exhibits anti-inflammatory property in lipopolysaccharide-induced macrophage cells through down-regulating the p38 MAPKs signaling pathways. Int Immunopharmacol 8:951–958
Loss-Morais G, Turchetto-Zolet AC, Etges M, Cagliari A, Korbes AP, Maraschin FDS, Margis-Pinheiro M, Margis R (2013) Analysis of castor bean ribosome-inactivating proteins and their gene expression during seed development. Genet Mol Biol 36:74–86
Lu Y, Fan J, Zhao Y, Chen S, Zheng X, Yin Y, Fu C (2007) Immunomodulatory activity of aqueous extract of Actinidia macrosperma. Asia Pac J Clin Nutr 16:261–265
Luettig B, Steinmuller C, Gifford GE, Wagner H, Lohmann-Matthes ML (1989) Macrophage activation by the polysaccharide arabinogalactan isolated from plant cell cultures of Echinacea purpurea. J Natl Cancer Inst 81:669–675
Ma'rquez N, Sancho R, Bedoya LM, Alcamı' J, Lo'pez-Pe'rez JL, San Feliciano A, Fiebich BL, Muñoz E (2005) Mesuol, a natural occurring 4-phenylcoumarin, inhibits HIV-1 replication by targeting the NF-κB pathway. Antivir Res 66:137–145
Mahamat O, Flora H, Tume C, Kamanyi A (2020) Immunomodulatory activity of *Momordica Charantia* l. (cucurbitaceae) leaf diethyl ether and methanol extracts on *salmonella typhi*-infected mice and lps-induced phagocytic activities of macrophages and neutrophils. Evid Complement Alternat Med 5248346
Maiti K, Gantait A, Mukherjee K, Saha B, Mukherjee PK (2006) Therapeutic potentials of andrographolide from *Andrographis paniculata*: a review. J Nat Remed 6:1–13
Malaguarnera L (2019) Influence of resveratrol on the immune response. Nutrients 2019(11):946
Mali RJ, Hatapakki BC (2008) An in-vitro study of effect of *Centella asiatica* on phagocytosis by human neutrophils. IJPSN 1:297–302
Mohan MR, Rao GBS, Narender B, Kumar CA, Rao PV, Bakshi V (2019) Indian medicinal plants used as immunomodulatory agents: a review. Int J Green Pharma 13:312

Manu KA, Kuttan G (2009) Anti-metastatic potential of Punarnavine, an alkaloid from *Boerhaavia diffusa* Linn. Immunobiology 214:245–255

Mathew S, Kuttan G (1999) Immunomodulatory and antitumour activities of *Tinospora cordifolia*. Fitoterapia 70:35–43

Maurya R, Wazir V, Kapil A, Kapil RS (1996) Cordifoliosides A and B, two new disaccharides from *Tinospora cordifolia* possessing immunostimulant activity. Nat Prod Lett 8:7–10

Medina EJ, Lora AG, Paco L, Algarra I, Collado A, Garrido F (2006) A new extract of the plant *Calendula officinalis* produces a dual in-vitro effect: cytotoxic anti-tumor activity and lymphocyte activation. BMC Complement Altern Med 6:1–14

Meeran SM, Katiyar S, Elmets CA, Katiyar SK (2006) Silymarin inhibits UV radiation-induced immunosuppression through augmentation of interleukin-12 in mice. Mol Cancer Ther 5:1161–1168

Mehrotra S, Mishra KP, Maurya R, Srimal RC, Yadav VS, Pandey R, Singh VK (2003) Anticellular and immunosuppressive properties of ethanolic extract of *Acorus calamus* rhizome. Int Immunopharmacol 3:53–61

Michalski J, Deinzer A, Stich L, Zinser E, Steinkasserer A, Knippertz I (2020) Quercetin induces an immunoregulatory phenotype in maturing human DCs. Immunobiology 24:151929

Mir MA, Hamdani SS, Sheikh BA, Mehraj U (2019) Recent advances in metabolites from medicinal plants in cancer prevention and treatment. Curr Immunol Rev 15:185–201

Mishra KP, Padwad YS, Dutta A, Ganju L, Sairam M, Banerjee PK, Sawhney RC (2008) Aqueous extract of *Rhodiola imbricata* rhizome inhibits proliferation of an erythroleukemic cell line K-562 by inducing apoptosis and cell cycle arrest at G2/M phase. Immunobiology 213:125–131

Mitra E, Ghosh AK, Ghosh D, Mukherjee D, Chattopadhyay A, Dutta S, Pattari SK, Bandyopadhyay D (2012) Protective effect of aqueous Curry leaf (*Murraya koenigii*) extract against cadmium-induced oxidative stress in rat heart. Food Chem Toxicol 50:1340–1353

Modi M, Dezzutti CS, Kulshreshtha S, Rawat AKS, Srivastava SK, Malhotra S, Verma A, Ranga U, Gupta SK (2013) Extracts from *Acacia catechu* suppress HIV-1 replication by inhibiting the activities of the viral protease and Tat. Virol J 10:309

Monga J, Chauhan CS, Sharma M (2011) Human epithelial carcinoma cytotoxicity and inhibition of DMBA/TPA induced squamous cell carcinoma in Balb/c mice by *Acacia catechu* heartwood. J Pharm Pharm 63:1470–1482

More P, Pai K (2011) Immunomodulatory effects of *Tinospora cordifolia* (Guduchi) on macrophage activation. Biol Med 3:134–140

Moore KW, de Waal Malefyt R, Coffman RL, O'Garra A (2001) Interleukin-10 and the interleukin-10 receptor. Annu Rev Immunol 19:683–765

Mukherjee R, Dash PK, Ram GC (2005) Immunotherapeutic potential of *Ocimum sanctum* (L) in bovine subclinical mastitis. Res Vet Sci 79:37–43

Mungantiwar AA, Nair AM, Shinde UA, Dikshit VJ, Saraf MN, Thakur VS, Sainis KB (1999) Studies on the immunomodulatory effects of *Boerhaavia diffusa* alkaloidal fraction. J Ethnopharmacol 65:125–131

Muraoka K, Shimizu K, Sun X, Tani T, Izumi R, Miwa K, Yamamoto K (2002) Flavonoids exert diverse inhibitory effects on the activation of NF-κB. Transplant Proc 34:1335–1340

Na HS, Lim YJ, Yun YS, Kweon MN, Lee HC (2010) Ginsan enhances humoral antibody response to orally delivered antigen. Immune Netw 10:5–14

Nadkarni KM, Nadkarni AK (2005) Indian Materia medica, 3rd edn. Popular Prakashan, Mumbai, p 623

Nadumane V, Nair S (2011) Evaluation of the anticancer and cytotoxic potentials of *Acacia catechu* extracts in-vitro. J Nat Pharm 2:190

Naeini A, Khosravi A, Tadjbakhsh HG, T, Yaraee R, Shokri H, (2010) Evaluation of the immunostimulatory activity of *Ziziphora tenuior* extracts. Comp Clin Pathol 19:459–463

Nagarajan S, Burris RL, Stewart BW, Wilkerson JE, Badger TM (2008) Dietary soy protein isolate ameliorates atherosclerotic lesions in apolipoprotein E-deficient mice potentially by inhibiting monocyte chemoattractant protein-1 expression. J Nutr 138:332–337

Nahak G, Sahu R (2014) Immunostimulatory effects of *Ocimum sanctum* Linn. leaf extracts in *Clarias batrachus* Linn. Asian J Pharm Clin Res 7:15

Naik GH, Priyadarsini KI, Satav JG, Banavalikar MM, Sohoni DP, Biyani MK, Mohan H (2003) Comparative antioxidant activity of individual herbal components used in ayurvedic medicine. Phytochemistry 63:97–104

Nakaya TA, Kita M, Kuriyama H, Iwakura Y, Imanishi J (2004) Panax ginseng induces production of proinflammatory cytokines via toll-like receptor. J Interferon Cytokine Res 24:93–100

Nair A, Chattopadhyay D, Saha B (2019) Plant-derived immunomodulators. In: Ahmad Khan MS, Ahmad I, Chattopadhyay D (eds) New look to phytomedicine, 1st edn. Academic Press, Cambridge, pp 435–499

Nair PK, Melnick SJ, Ramachandran R, Escalon E, Ramachandran C (2006) Mechanism of macrophage activation by (1, 4)-α-D-glucan isolated from Tinospora cordifolia. Int Immunopharmacol 6:1815–1824

Nair PR, Rodriguez S, Ramachandran R, Alamo A, Melnick SJ, Escalon E, Garcia PI, Wnuk SF Jr, Ramachandran C (2004) Immune stimulating properties of a novel polysaccharide from the medicinal plant *Tinospora cordifolia*. Int Immunopharmacol 4:1645–1659

Nantz MP, Rowe CA, Mulle CE, Creasy RA, Stanilka JM, Percival SS (2012) Supplementation with aged garlic extract improves both NK and γδ-T cell function and reduces the severity of cold and flu symptoms: a randomized, double-blind, placebo-controlled nutrition intervention. Clin Nutr 31:337–344

Negi BS, Dave BP (2010) In-vitro antimicrobial activity of *Acacia catechu* and its phytochemical analysis. Indian J Microbiol 50:369–374

Noori S, Naderi GA, Hassan ZM, Habibi Z, Bathaie SZ, Mahmoud S (2004) Immunosuppressive activity of a molecule isolated from *Artemisia annua* on DTH responses compared with cyclosporin A. Int Immunopharmacol 4:1301–1306

Oh YC, Kang OH, Kim SB, Mun SH, Park CB, Kim YG, Kim YI, Lee YS, Han SH, Keum JH, Shin DW, Ma JY, Kwon DY (2012) Anti-inflammatory effect of sinomenine by inhibition of proinflammatory mediators in PMA plus A23187-stimulated HMC-1 Cells. Eur Rev Med Pharmacol Sci 16:1184–1191

Ordway D, Hohmann J, Viveiros M, Viveiros A, Molnar J, Leandro C, Arroz MJ, Gracio MA, Amaral L (2003) Carpobrotus edulis methanol extract inhibits the MDR efflux pumps, enhance killing of phagocytosed *S aureus* and promotes immune modulation. Phytother Res 17:512–519

Pal D, Saha S (2019) Chondroitin: a natural biomarker with immense biomedical applications. RSC Adv 9(48):28061–28077

Pal D, Ahuja C, Mukherjee S (2019) Celsia coromandeliane Vahl-A New biomarker with immense pharmaceutical applications. J Drug Deliv Ther 9(3):1109–1115

Pal D, Mitra S (2010) A preliminary study on the in vitro antioxidant activity of the stems of Opuntia vulgaris. JAPTR 2:268–270

Pal DK, Dutta S (2006) Evaluation of the antioxidant activity of the roots and rhizomes of cyperus rotundus L. Indian J Pharm Sci 68(2):212–220

Pal DK, Mandal M, Senthilkumar GP, Padhiari A (2006) Antibacterial activity of Cuscuta reflexa stem and Corchorus olitorius seed. Fitoterapia 77(7):589–591

Pal D (2015) A review on Cyperus rotundus as a tremendous source of pharmacologically active herbal medicine. IJGP 9(4):13–18

Pal D (2009) Determination of brain biogenic amines in ethanolic extract of *Cynodon dactylon* Pers and *Cyperus rotundus* L.treated mice. Int J Pharm Pharm Sci 1(1):190–197

Pal D, Pandav K (2010) Evaluation of anthelmintic activity of *Cynodon dactylon*. Ancient Sci Life 30(1):12–13

Pal D, Kumar M, Chakrabarty P, Kumar S (2008) Evaluation of the antioxidant activity of aerial parts of *Cynodon dacvtylon*. Asian J Chem 20(3):2479–2481

Pal D (2008) Evaluation of CNS Activities of *C. dactylon* Pers in mice. Acta Pol Pharm Drug-Res 65(1):37–43

Palani S, Raja S, Kumar SN, Kumar BS (2011) Nephroprotective and antioxidant activities of *Salacia oblonga* on acetaminophen-induced toxicity in rats. Nat Prod Res 25:1876–1880

Pan P, Huang YW, Oshima K, Yearsley M, Zhang J, Arnold M6, Yu J, Wang LS (2019) The immunomodulatory potential of natural compounds in tumor-bearing mice and humans. Crit Rev Food Sci Nutr 59:992–1007

Pandey R, Maurya R, Singh G, Sathiamoorthy B, Naik S (2005) Immunosuppressive properties of flavonoids isolated from *Boerhaavia diffusa* Linn. Int Immunopharmacol 5:541–553

Pandit S, Mukherjee PK, Ponnusankar S, Venkatesh M, Srikanth N (2011) Metabolism mediated interaction of α-asarone and Acorus calamus with CYP3A4 and CYP2D6. Fitoterapia 82:369–374

Parichatikanond W, Suthisisang C, Dhepakson P, Herunsalee A (2010) Study of anti-inflammatory activities of the pure compounds from *Andrographis paniculata* (burm. f.) Nees and their effects on gene expression. Int Immunopharmacol 10:1361–1373

Park HH, Lee S, Son HY, Park SB, Kim MS, Choi EJ, Singh TS, Ha JH, Lee MG, Kim JE, Hyun MC, Kwon TK, Kim YH, Kim SH (2008) Flavonoids inhibit histamine release and expression of PINFLCS pro-inflammatory cytokines in mast cells. Arch Pharm Res 31:1303–1311

Park HJ, Jeong SK, Kim SR, Bae SK, Kim WS, Jin SD, Koo TH, Jang HO, Yun I, Kim KW, Bae MK (2009) Resveratrol inhibits *Porphyromonas gingivalis* lipopolysaccharide induced endothelial adhesion molecule expression by suppressing NF-κB activation. Arch Pharm Res 32:583–591

Patil SB, Naikwade NS, Magdum CS (2009) Review on phytochemistry and pharmacological aspects of *Euphorbia hirta* Linn. JPRHC1:113–133

Peumans WJ, De Ley M, Broekaert WF (1984) An unusual lectin from stinging nettle (*Urtica dioica*) rhizomes. FEBS Lett 177:99–103

Plummer SM, Holloway KA, Manson MM, Munks RJ, Kaptein A, Farrow S, Howells L (1999) Inhibition of cyclo-oxygenase 2 expression in colon cells by the chemopreventive agent curcumin involves inhibition of NF-κB activation via the NIK/IKK signalling complex. Oncogene 18:6013–6020

Pongituluran OB, Rofaani E (2015) Antiviral and immunostimulant activities of *Andrographis paniculate*. HAYATI J Biosci 22:67–72

Pradhan D, Panda PK, Tripathy G (2009) Evaluation of immunomodulatory activity of methanolic extract of *Couroupita guianensis Aubl* flowers in rat. IJNPR 8:37–42

Prashar R, Kumar A, Hewer A, Cole KJ, Davis W, Phillips DH (1998) Inhibition by an extract of *Ocimum sanctum* of DNA-binding activity of 7,12-dimethylbenz [a] anthracene in rat hepatocytes in-vitro. Cancer Lett 128:155–160

Puri A, Saxena R, Saxena RP, Saxena KC, Srivastava V, Tandon JS (1994) Immunostimulant activity of *Nyctanthes arbor-tristis* L. J Ethnopharmacol 42:31–37

Puri A, Saxena R, Saxena RP, Saxena KC, Srivastava V, Tandon JS (1993) Immunostimulant agents from *Andrographis paniculate*. J Nat Prod 56:995–999

Qin LH, Kong L, Shi GJ, Wang ZT, Ge BX (2006) Andrographolide inhibits the production of TNF-alpha and interleukin-12 in lipopolysaccharide stimulated macrophages: role of mitogen-activated protein kinases. Biol Pharm Bull 29:220–224

Quintero-Fabian S, Ortuno-Sahagun D, Vazquez-Carrera M, Lopez-Roa RI (2013) Alliin, a garlic (*Allium sativum*) compound, prevents LPS-induced inflammation in 3T3-L1 adipocytes. Mediators Inflamm 2013:381815

Rahmatullah M, Hossain M, Mahmud A, Sultana N, Rahman SM, Islam MR, Khatoon MS, Jahan S, Islam F (2013) Antihyperglycemic and antinociceptive activity evaluation of 'khoyer' prepared from boiling the wood of *Acacia catechu* in water. Afr J Trad Complement Altern Med 10:1–5

Rajagopal S, Kumar RA Deevi DS, Satyanarayana C, Rajagopalan R (2003) Andrographolide, a potential cancer therapeutic agent isolated from *Andrographis paniculate*. J Ex. Ther Oncol 3:147–158

Rajaratinam H, Nafi SNM (2019) Andrographolide is an alternative treatment to overcome resistance in ER-positive breast cancer via cholesterol biosynthesis pathway *Malays*. J Med Sci 26:6–20

Rajput ZI, Hu SH, Xiao CW, Arijo AG (2007) Adjuvant effects of saponins on animal immune responses. J Zhejiang Univ Sci B 8:153–161

Ranjan D, Chen C, Johnston TD, Jeon H, Nagabhushan M (2004) Curcumin inhibits mitogen stimulated lymphocyte proliferation, NFκB activation, and IL-2 signaling. J Surg Res 121:171–177

Ravichandiran V, Patil Vishal S (2015) In-vitro evaluation for immunomodulatory activity of *Acorus calamus* on human neutrophils. Int Res J Pharm 6(7):450–452

Ravikumar N, Kavitha CN (2020) Immunomodulatory effect of Quercetin on dysregulated Th1/ Th2 cytokine balance in mice with both type 1 diabetes and allergic asthma. J Appl Pharm Sci 10:80–87

Rios JL (2010) Effects of triterpenes on the immune system. J Ethnopharmacol 128:1–14

Reis LSLS, Frazatti-Gallina NM, Paoli RL, Giuffrida AA, Oba E, Pardo PE (2008) Efficiency of Matricaria chamomilla CH12 and number of doses of rabies vaccine on the humoral immune response in cattle. J Vet Sci 9:433–435

Retini C, Kozel TR, Pietrella D, Monari C, Bistoni F, Vecchiarelli A (2001) Interdependency of interleukin-10 and interleukin-12 in regulation of T-cell differentiation and effector function of monocytes in response to stimulation with Cryptococcus neoformans. Infect Immun 69:6064–6073

Ruiz PA, Braune A, Holzlwimmer G, Quintanilla-Fend L, Haller D (2007) Quercetin inhibits TNF-induced NF-κB transcription factor recruitment to proinflammatory gene promoters in murine intestinal epithelial cells. J Nutr 137:1208–1215

Santhi R, Annapoorani S (2010) Efficacy of *Cynodon dactylon* for immunomodulatory activity. Drug Invention Today 2:112–114

Sathaye S, Bagul Y, Gupta S, Kaur H, Redkar R (2011) Hepatoprotective effects of aqueous leaf extract and crude isolates of *Murraya koenigii* against in-vitro ethanolinduced hepatotoxicity model. Exp Toxicol Pathol 63:587–591

Satpute KL, Jadhav MM, Karodi RS, Katare YS, Patil MJ, Rub R, Bafna AR (2009) Immunomodulatory activity of fruits of *Randia dumetorum* Lamk. JPP 1:36–40

Senchina DS, McCann DA, Asp JM, Johnson JA, Cunnick JE, Kaiser MS, Kohut ML (2005) Changes in immunomodulatory properties of *Echinacea* spp root infusions and tinctures stored at 4 degrees C for four days. Clin Chim Acta 355:67–82

See DM, Broumand N, Sahl L, Tilles JG (1997) In vitro effects of Echinacea and ginseng on natural killer and antibody-dependent cell cytotoxicity in healthy subjects and chronic fatigue syndrome or acquired immunodeficiency syndrome patients. Immunopharmacology 35:229–235

Shah AS, Wakade AS, Juvekar AR (2008) Immunomodulatory activity of methanolic extract of *Murraya koenigii* (L) Spreng Leaves. Indian J Exp Biol 46:505–509

Shahnazi M, Azadmehr A, Andalibian A, Hajiaghaee R, Saraei M, Alipour M Protoscolicidal and immunomodulatory activity of *Ziziphora tenuior* extract and its fractions. Asian Pac J Trop Med 9:1062–1068

Sharififar F, Pournournohammadi S, Arabnejad M (2009) Immunomodulatory activity of aqueous extract of *Achiella wilhelmsii* C. Koch in mice. Indian J Exp Biol 47:668–671

Sharma A, Kalotra S, Bajaj P, Singh H, Kaur G (2020) Butanol extract of *Tinospora cordifolia* ameliorates cognitive deficits associated with glutamate-induced excitotoxicity: a mechanistic study using hippocampal neurons. Neuromolecular Med 22:81–99

Shaw PC, Lee KM, Wong KB (2005) Recent advances in trichosanthin, a ribosome inactivating protein with multiple pharmacological properties. Toxicon 45:683–689

Sheeja K, Kuttan G (2010) *Andrographis paniculata* downregulates proinflammatory cytokine production and augments cell mediated immune response in metastatic tumorbearing mice. Asian Pac J Cancer Prev 11:723–729

Shen YC, Chen CF, Chiou WF (2002) Andrographolide prevents oxygen radical production by human neutrophils: possible mechanism(s) involved in its anti-inflammatory effect. Br J Pharmacol 135:399–406

Shin IS, Hong J, Jeon CM, Shin NR, Kwon OK, Kim HS, Kim JC, Oh SR, Ahn KS (2013) Diallyl-disulfide, an organosulfur compound of garlic, attenuates airway inflammation via activation of the Nrf-2/HO-1 pathway and NF-kappaB suppression. Food Chem Toxicol 62:506–513

Shin JY, Song JY, Yun YS, Yang HO, Rhee DK, Pyo S (2002) Immunostimulating effects of acidic polysaccharides extract of *Panax ginseng* on macrophage function. Immunopharmacol Immunotoxicol 24:469–482

Shivaprasad HN, Kharya MD, Rana AC, Mohan S (2006) Preliminary immunomodulatory activities of the aqueous extract of *Terminalia-chebula*. Pharmaceutical Biology 44:32–34

Si H, Liu D (2007) Phytochemical genistein in the regulation of vascular function: new insights. Curr Med Chem 14:2581–2589

Sikarwar MS, Patil MB, Sharma S, Bhat V (2010) Aloe vera: plant of immortality. IJPSR 1:7–10

Singh BN, Shankar S, Srivastava RK (2011) Green tea catechin, epigallocatechin-3-gallate (EGCG): mechanisms, perspectives and clinical applications. Biochem Pharmacol 82:1807–1821

Singh S, Aggarwal BB (1995) Activation of transcription factor NF-κB is suppressed by curcumin (diferuloylmethane). J Biol Chem 270:24995–25000

Sobhani M, Farzaei MH, Kiani S Khodarahmi R (2020) Immunomodulatory; anti-inflammatory/antioxidant effects of polyphenols: a comparative review on the parental compounds and their metabolites. Food Rev Int 1–53

Sredni B, Kalechman Y, Albeck M, Gross O, Aurbach D, Sharon P, Sehgal SN (1990) Cytokine secretion effected by synergism of the immunomodulator AS101 and the PKC inducer bryostatin. Immunology 70:473–477

Sreeja S, Sreeja S (2009) An in-vitro study on antiproliferative and antiestrogenic effects of *Boerhaavia diffusa* L. extracts. J Ethnopharmacol 126:221–225

Srivastava R, Saluja D, Dwarakanath BS, Chopra M (2011) Inhibition of human cervical cancer cell growth by ethanolic extract of Boerhaavia diffusa Linn. (punarnava) root. Evid Based Complement Alternat Med 427031

Stanton RA, Gernert KM, Nettles JH, Aneja R (2011) Drugs that target dynamic microtubules: a new molecular perspective. Med Res Rev 31:443–481

Steinbach WJ, Stevens DA (2003) Review of newer antifungal and immunomodulatory strategies for invasive Aspergillosis. ClinInfect Dis 37:157–187

Stohs SJ, Bagchi D (2015) Antioxidant, anti-inflammatory, and chemoprotective properties of Acacia catechu heartwood extracts. Phytother Res 29:818–824

Su W, Sun AJ, Xu DL, Zhang HQ, Yang L, Yuan LY, Jia JG, Zou YZ, Wu YL, Wang KQ, Ge JB (2010) Inhibiting effects of total saponins of *panax ginseng* on immune maturation of DCs induced by oxidized-low density lipoprotein. Cell Immunol 263:99–104

Sugumaran M, Robinson WE (2007) Bioactive dehydrotyrosyl and dehydrodopyl compounds of marine origin. Mar Drugs 8:2906–2935

Sumanth M, Mustafa SS (2007) Antistress, adoptogenic and immunopotentiating activity roots of *Boerhaavia diffusa* in mice. Int J Pharmacol 3:416–420

Sunila ES, Kuttan G (2004) Immunomodulatory and antitumor activity of *Piper longum* Linn and piperine. J Ethnopharmacol 90:339–346

Surh YJ, Chun KS, Cha HH, Han SS, Keum YS, Park KK, Lee SS (2001) Molecular mechanisms underlying chemopreventive activities of anti-inflammatory phytochemicals: down-regulation of COX-2 and iNOS through suppression of NF-κB activation. Mutat Res 481:243–268

Syed I, Mohammed A (2009) Immunomodulatory activity of *Acacia catechu*. Indian J Physicol Pharmacol 53:25–33

Tanaka T, Sugiura H, Inaba R, Nishikawa A, Murakami A, Koshimizu K, Ohigashi H (1999) Immunomodulatory action of Citrus auraptene on macrophage functions and cytokine production of lumphocytes in female BALB/c mice. Carcinogenesis 20(8):1471–1476

Teter B, Morihara T, Lim GP, Chu T, Joness MR, Zuo X, Paul RM, Frautschy SA, Cole GM (2019) Curcumin restores innate immune Alzheimer's disease risk gene expression to ameliorate Alzheimer pathogenesis. Neurobiol Dis 127:432–448

Thakur M, Bhargava S, Dixit VK (2006) Immunomodulatory activity of *Chlorophytum borivilianum* Sant. F, Evid Based Complement Alternat Med 4:419–423

Tsai HR, Yang LM, Tsai WJ, Chiou WF (2004) Andrographolide acts through inhibition of ERK1/2 and Akt phosphorylation to suppress chemotactic migration. Eur J Pharmacol 498: 45–52

Tsang JYS, Li D, Ho D, Peng J, Xu A, Lamb J, Chen Y, Tam PK (2011) Novel immunomodulatory effects of adiponectin on dendritic cell functions. Int Immunopharmacol 11:604–609

Uttekar MM, Das T, Pawar RS, Bhandari B, Menon V, Nutan Gupta SK, Bhat SV (2012) Anti-HIV activity of semisynthetic derivatives of andrographolide and computational study of HIV-1 gp120 protein binding. Eur J Med Chem 56:368–374

Vaghasiya J, Datani M, Nandkumar K, Malaviya S, Jivani N (2010) Comparative evaluation of alcoholic and aqueous extracts of *Ocimum sanctum* for immunomodulatory activity. Int J Pharm Biol Res 1:25–29

Varma A, Padh H, Shrivastava N (2011) Andrographolide: a new plant-derived antineoplastic entity on horizon. Evid Complement Alternat Med 2011:815390

Vesely MD, Kershaw MH, Schreiber RD, Smyth MJ (2011) Natural innate and adaptive immunity to cancer. Annu Rev Immunol 29:235–271

Wang H, Actor JK, Indrigo J, Olsen M, Dasgupta A (2003) Asian and Siberian ginseng as a potential modulator of immune function: an in vitro cytokine study using mouse macrophages. Clin Chim Acta 327:123–128

Wang C, Tao W, Wang Y, Bikow J, Lu B, Keating A, Verma S, Parker TG, Han R, Wen XY (2010) Rosuvastatin, identified from a zebrafish chemical genetic screen for antiangiogenic compounds, suppresses the growth of prostate cancer. Eur Urol 58:418–426

Wang J, Zhang Q, Jin S, He D, Zhao S, Liu S (2008) Genistein modulate immune responses in collagen-induced RA model. Maturitas 59:405–412

Washiya Y, Nishikawa T, Fujino T (2013) Enhancement of intestinal IgA production by ajoene in mice. Biosci Biotechnol Biochem 77:2298–2301

Wu T, Marakkath B, Ye Y, Khobahy E, Yan M, Hutcheson J, Zhu J, Zhou X, Mohan C (2020) Curcumin attenuates both acute and chronic immune nephritis. Int J Mol Sci 4:E1745

Wu Y, Mao Y, Jin S, Hou J, Du H, Yang M, Wu L (2015) Identification, characterization and structure analysis of a type I ribosome-inactivating protein from *Sapium sebiferum* (Euphorbiaceae). Biochem Biophys Res Commun 463:557–562

Wung BS, Hsu MC, Wu CC, Hsieh CW (2005) Resveratrol suppresses IL-6-induced ICAM-1 gene expression in endothelial cells: effects on the inhibition of STAT3 phosphorylation. Life Sci 78:389–397

Xia X, Pan Y, Zhang WY, Cheng G, Kong LD (2006) Ethanolic extracts from *Curcuma longa* attenuates behavioral, immune, and neuroendocrine alterations in a rat chronic mild stress model. Biol Pharm Bull 29:938–944

Xia YF, Ye BQ, Li YD, Wang JG, He XJ, Lin X, Yao X, Ma D, Slungaard A, Hebbel RP, Key NS, Geng JG (2004) Andrographolide attenuates inflammation by inhibition of NF-κB activation through covalent modification of reduced cysteine 62 of p50. J Immunol 173:4207–4217

Yadav VS, Mishra KP, Singh DP, Mehrotra S, Singh VK (2005) Immunomodulatory effects of curcumin. Immunopharmacol Immunotoxicol 27:485–497

Yalniz M, Bahcecioglu IH, Kuzu N, Poyrazoglu OK, Bulmus O, Celebi S, Ustundag B, Ozercan IH, Sahin K (2007) Preventive role of genistein in an experimental nonalcoholic steatohepatitis model. J Gastroenterol Hepatol 22:2009–2014

Yamakuchi M, Bao C, Ferlito M, Lowenstein CJ (2008) Epigallocatechin gallate inhibits endothelial exocytosis. Biol Chem 389:935–941

Yang C, Zhu K, Yuan X, Zhang X, Qian Y, Cheng T (2020) Curcumin has immunomodulatory effects on RANKL-stimulated osteoclastogenesis in-vitro and titanium nanoparticle-induced bone loss in-vivo. J Cell Mol Med 24:1553–1567

Yang CS, Wang H, Li GX, Yang Z, Guan F, Jin H (2011) Cancer prevention by tea: evidence from laboratory studies. Pharmacol Res 64:113–122

Yoo DG, Kim MC, Park MK, Park KM, Quan FS, Song JM, Wee JJ, Wang BZ, Cho YK, Compans RW (2012) Protective effect of ginseng polysaccharides on influenza viral infection. PLoS ONE 7:e33678

Zhang D, Halaweish FT (2007) Isolation and characterization of ribosome inactivating proteins from Cucurbitaceae. Chem Biodivers 4:431–442

Zhang QQ, Ding Y, Lei Y, Qi CL, He XD, Lan T, Li JC, Gong P, Yang X, Geng JG, Wang LJ (2014) Andrographolide suppress tumor growth by inhibiting TLR4/NF-κB signaling activation in insulinoma. Int J Biol Sci 10:404–414

Zhao M, Yang B, Wang J, Liu Y, Yu L, Jiang Y (2007) Immunomodulatory and anticancer activities of flavonoids extracted from litchi (*Litchi chinensis Sonn*) pericarp. Int Immunopharmacol 7:162–166

Zheng W, Chen C, Cheng Q, Wang Y, Chu C (2006) Oral administration of exopolysaccharide from *Aphanothece halophytica Chroococcales* significantly inhibits influenza virus H1N1-induced pneumonia in mice. Int Immunopharmacol 6:1093–1099

Zou LC, Yan ZH, Zhu TF, Xiang H, Deng XM (2010) Two new phenylethanoid glycosides and a new secoiridoid glycoside from the roots of *Picrorhiza scrophulariiflora*. Chin Chem Lett 21:1103–1106

Zvetkova E, Wirleitner B, Tram NT, Schennach H, Fuchs D (2001) Aqueous extracts of *Crinum latifolium* L and *Camellia sinensis* show immunomodulatory properties in human peripheral blood mononuclear cells. Int Immunopharmacol 1:2143–2150

# Chapter 7
# Antibacterial and Antifungal Plant Metabolites from the Tropical Medicinal Plants

**Luiz Everson da Silva, Camila Confortin, and Mallappa Kumara Swamy**

**Abstract** A wide-ranging organism, such as plants, fungi, insects, marine organisms, and bacteria are the main sources of bioactive substances. Among them, medicinal plants offer harmless elusive means to improve our health conditions. Moreover, plant-derived bioactive is an inspiration for developing several therapeutic agents having extensively diverse chemo-structures and exhibit superior biological properties like antimicrobial, anticancer, and antioxidant, anti-inflammatory properties. Hence, many of them are used in therapeutic applications to treat various human ailments. Markedly, tropical countries harbor the maximum global biodiversity and constitute several vital plant species with pharmacological potential. Hence, they are being explored for bioactive compounds, and other raw materials to manufacture herbal medicines or chemo-drugs. In the present time, the emergence of microbial resistance to conventional antibiotics has become a major threat in treating microbial infections. Further, side effects caused by classical antibiotics have forced scientists to work toward exploring novel antimicrobial agents to overcome these hitches. Phytoconstituents have been shown to have prospective antimicrobial properties against both sensitive and resistant pathogenic microbes by exerting diverse mode of action. In this chapter, the potential of tropical plant species with antibacterial and antifungal activities are reviewed in detail.

**Keywords** Tropical medicinal flora · Natural products · Antifungal and antibacterial activities · Plant metabolites · Phytotherapies

L. E. da Silva (✉) · C. Confortin
Post-Graduate Program in Sustainable Territorial Development, Universidade Federal do Paraná, Matinhos 83260-000, Brazil
e-mail: luiever@gmail.com

M. K. Swamy
Department of Biotechnology, East West First College, Bengaluru, Karnataka 560091, India

 
D. Pal and A. K. Nayak (eds.), *Bioactive Natural Products for Pharmaceutical Applications*, Advanced Structured Materials 140,
https://doi.org/10.1007/978-3-030-54027-2_7

## 7.1 Introduction

Many medicinal plants contain large amounts of bioactive compounds, such as phenolic compounds, terpenoids, nitrogen compounds, vitamins, and various other secondary metabolites that are being used for various therapeutic purposes, since the dawn of mankind (Arumugam et al. 2016; Swamy et al. 2017; Kirubakari et al. 2019). The nature is a treasure of several medications, and provides assistance in the treatment of different health issues, affecting humans. Indian, Egyptian, Roman, Chinese, and Greek traditional medicines have documented the use of herbs against numerous health problems. Currently, despite the development of medicine, technological progress, and the creation of new synthetic substances with medicinal properties, the use of natural products is gradually growing, mainly in developed countries. Thus, medicinal plants are present as an alternative in health care (Pandey and Kumar 2013; Swamy et al. 2016; Mohanty et al. 2017; Anand et al. 2019). Natural products contribute immensely to the development of numerous drug molecules having several therapeutic applications (Beutler 2009). A wide-ranging organism, such as plants, fungi, insects, marine organisms, and bacteria are the main sources of natural products. Among them, medicinal plants offer harmless elusive means to improve our health conditions. Moreover, plant-derived bioactives are an inspiration for developing several therapeutic agents having extensively diverse chemo-structures and exhibit superior biological properties like antimicrobial, anticancer, antioxidant, and anti-inflammatory properties. Hence, many of them are used in therapeutic applications to treat various human ailments (Swamy et al. 2016; Kirubakari et al. 2019).

In this context, the tropical region is home to almost a third of the world's flora represented in many different biomes with exuberant biodiversity (Brooks et al. 2002; Sobrinho et al. 2014). However, even in the face of the vast biodiversity and even having great potential for study and knowledge of new perspectives on the issue of genetic resources, the environmental degradation and the intrusion of new cultural elements accompanied by the breakdown of traditional life systems threaten, in addition to a collection of empirical knowledge, the heritage invaluable genetics for future generations (Chazdon 2019). Therefore, it is essential that scientific and technological innovation provides advances in order to add value to the products of the tropical biodiversity, and we can make use of these assets ensuring the sovereignty and vanguard of our nations.

At present, microbial infections due to pathogenic microbes, in addition to their resistance to conventional antibiotics are posing a major challenge and threatening the health of humans and other animals. Globally, microbial infections cause millions of human demises, and are increasing every year (Rudramurthy et al. 2016; Khameneh et al. 2019). Bacterial and fungal infections, in general, have increased significantly at an accelerated rate. They are being considered as a wide public health problem that affects the whole world, affecting countless people. Due to this advancement, it is necessary to put aside many conventional methods of combating such diseases, which are often not having the expected effect, and to seek new sources or methods that can significantly affect these infections (Swamy and Rudramurthy 2016; Anand et al. 2019; Kirubakari et al. 2019). Plant-derived bioactives are known to be relatively safer compared to the presently available antibiotic agents. Further, phytocompounds exert numerous tonic advantages linked with their extraordinary effectiveness. Various plant metabolites have demonstrated synergistic bioactivity when combined with antibiotic agents against several multi-drug resistant pathogenic microbes. These plant metabolites comprise phenolics, flavonoids, alkaloids, tannins, terpenoids, and many more (Arumugam et al. 2016; Mohanty et al. 2017; Shin et al. 2018; Kirubakari et al. 2019).

In view of these aspects, the discovery of new active substances from natural sources is the most priority to researchers, considering several advantages. The pharmacological potential of medicinal plants occurring in the Tropical region has attracted to develop novel antimicrobials (Calixto 2019). Further, side effects caused by classical antibiotics have forced scientists to work toward exploring novel antimicrobial agents to overcome these hitches. Phytoconstituents have shown to have prospective antimicrobial properties against both sensitive and resistant pathogenic microbes by exerting diverse mode of action. In this chapter, the potential of tropical plant species with antibacterial and antifungal activities is reviewed in detail.

## 7.2 Target for Antimicrobial Agents and Resistance Mechanisms

Microbial infections are affecting the lives of several thousands of populaces around the globe, and are accounted for a major cause of human deaths. Nearly 17% of total deaths in the year 2013 were because of microbial infections. Further, microbes are evolving toward resistance to existing antimicrobials, making the present way of treating pathogens less effective. In recent times, clinically pathogenic bacteria and fungi are being classified on the basis of microbial resistance to either a single or multi-drug. Majorly, microbial resistance is due to many reasons, including overdose, misuse, and continuous use of antibiotics (Swamy and Rudramurthy 2016; Rudramurthy et al. 2016; Khameneh et al. 2019). On the other hand, bacterial resistance has become an increasing health problem with great impact on the pharmaceutical industry, as many antibiotics have become resistant. More recently, different approaches are being recommended to overcome the multi-drug-resistance mechanisms by pathogenic microbes. One such endorsed approach is the combination therapy, involving two or more molecules with the weakened antibiotics (Inui et al. 2007; Kirubakari et al. 2019). These molecules can be other than antibiotic drugs having antibacterial potentials, for example, phytochemicals. Particularly, these molecules function either individually or together with antibiotic drugs to improve the antimicrobial property, and thus can be very effective against wide-ranging bacterial pathogens (Khameneh et al. 2019).

This approach effectively restores the desired antimicrobial property (Brown 2015; Fazly Bazzaz et al. 2018; Rana et al. 2018).

Antimicrobial agents exhibit their actions through various ways, including the interfering with the biosynthesis or inhibiting bioactivities of bacterial components. The established targets for antibiotics may include (a) biosynthesis of bacterial proteins; (b) biosynthesis of bacterial cell wall; (c) damage of bacterial cell membrane; (d) interfering with bacterial DNA replication and/or repair mechanisms, and (e) suppressing metabolic pathways. Some classes of antibiotic drugs like tetracyclines, macrolides, and aminoglycosides exhibit their antibacterial properties, which particularly inhibit protein biosynthesis via targeting the ribosomal subunits (Swamy et al. 2016; Khameneh et al. 2019). A protein is generally synthesized inside the cells by several means of the molecular processes, including initiation, elongation, termination, and assembly of protein-mediated by ribosomes. Hence, bacterial pathogens can be targeted by inhibiting the actions of ribosomes and affecting protein synthesis (Walsh 2003). Further, some antibiotics can modify the permeability of the bacterial exterior cell membrane, and later disrupt the structural alterations of cell membrane, leading to a rapid bacterial death by creating osmotic imbalances. The polymyxin class of antibiotics attach to the lipid A constituent of lipopolysaccharide, and results to cause cell membrane structural modifications (McBain et al. 2003; Tenover 2006). Several classes of antibiotics inhibit cell wall synthesis. Bacterial cell wall has covalently cross-linked strands of peptide and glycan, and can be responsible for increased mechanical strength, and prevents cell lysis due to osmotic pressure. The enzymes transglycosylases and transpeptidases are responsible for forming this layer. Antibiotics such as penicillins and cephalosporins are shown to target the cell wall assembly, and exhibit bactericidal properties. Antibiotics such as vancomycin is proved to, particularly disrupting the peptidoglycan layer, and weaken the cell wall structure to result in bacterial cell death (Schneider and Sahl 2010). The DNA gyrase enzyme is accounted to perform the uncoiling and supercoiling of DNA strands, and controls DNA replication process. Thus, targeting this enzyme can be another target for antibacterial drugs/ antibiotics. The antibiotics, ciprofloxacin (a fluoroquinolone), and nalidixic acid suppress the replication of DNA by attaching to DNA Gyrase enzyme bound DNA complex (Maxwell 1997). Bacteria might exhibit drug-resistance property to one or more antibiotics via different types of mechanisms. The resistance property may vary from one bacterial species to another and to different classes of antimicrobial agents. So, knowing about the resistance mechanisms exhibited by bacterial strains can be very useful in designing novel antibacterial drugs (Walsh 2003; Tenover 2006; Khameneh et al. 2016).

Noteworthy to mention here that the drug-resistance could be linked to a single or more types of mode of actions together (Fig. 7.1). Some of the major proved mechanisms of antibacterial drug-resistance by bacteria includes the destruction of the antibiotic drugs by producing damaging enzymes, modifying antibiotic drugs by secreting modifying enzymes, stimulation of efflux pumps, and changing the structure of target in the bacterial cells, so that it will have less affinity for recognizing antibacterial agents (Khameneh et al. 2016; Munita and Arias 2016; Peterson

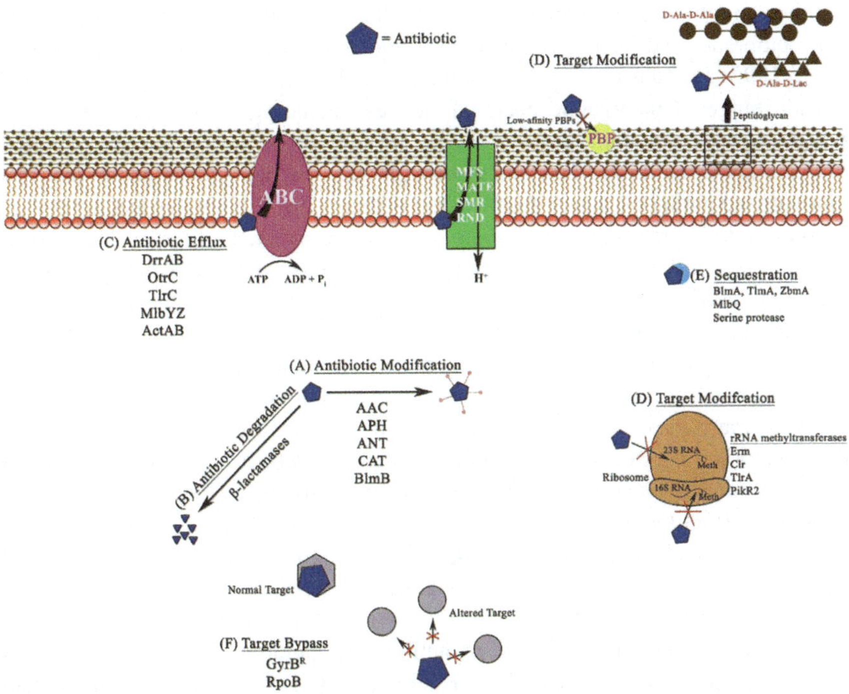


**Fig. 7.1** Schematic representation of different antibiotic resistance mechanisms in bacteria, shown with examples. **A** Antibiotic modification involves the addition of acetyl, phosphate, or adenyl groups to aminoglycosides by *N*-acetyl transferases (AAC), *O*-phosphotransferases (APH), and *O*-adenyltransferases (ANT). Other examples include chloramphenicol acetyl transferases (CAT) and bleomycin *N*-acetyltransferases (BlmB). **B** Antibiotic degradation is observed with β-lactamases, which hydrolyze the antibiotic. **C** Antibiotic efflux pumps remove the antibiotic from the cell using energy from ATP hydrolysis in ABC pumps like DrrAB, OtrC, TlrC, and MlbYZ, or proton gradients in MFS, MATE, SMR, and RND family pumps. **D** Target modification includes various target alterations, such as 23S rRNA or 16S rRNA methylation, alterations in the peptidoglycan precursors (for example, in the case of glycopeptides), or synthesis of alternate low-affinity targets (PBPs) that reduce or completely block antibiotic (penicillins) from associating with the target. **E** Antibiotic sequestration involves proteins that can associate with the antibiotic and block them from reaching their targets. **F** Target bypass involves generation of additional antibiotic targets or subunits that are not susceptible to binding of the antibiotic. Meth, methylation ( Adapted from Peterson and Kaur 2018)

and Kaur 2018; Khameneh et al. 2019). In general, antibacterial drugs will be more effective at definite concentrations and after it reaches to the precise site of action. Efflux pumps may function by acting as an efflux or transfer system, where antibacterial drugs are quickly pushed out of cells than the time required for drug molecule to get diffused into the bacterial cell. This condition will subsequently lead to a condition, where the intra-bacterial concentration will be much lesser as compared to the concentration required for effective actions. For instance, the cytoplasm consists of ribosomes required for proteins. When protein synthesis inhibitors, including antibiotics are treated, they are enforced to pass through the cell membrane, and later accumulate at a higher level to inhibit protein synthesis (Levy 1992; Paulsen et al. 1996; Munita and Arias 2016; Shriram et al. 2018). A wide range of pathogenic bacterial species, including *Staphylococcus aureus*, *Acinetobacter baumannii*, *P. aeruginosa*, in addition to fungal species, *Candida albicans* show antibiotic resistance through this mechanism. Efflux pumps can eliminate the antibiotic agents from bacterial strains, for example, trimethoprim and fluoroquinolones resistance in *Pseudomonas aeruginosa*. Further, making use of efflux pump inhibitors along with antibacterial drugs could be a better option of treatment in overcoming the threatening microbial infections due to multi-drug resistant microbes (Blair et al. 2015; Munita and Arias 2016; Khameneh et al. 2019). The structural modifications of porins (protein channels) can control the diffusion of intracellular molecules, including antibiotics. Through this way, few bacterial strains prevent the antibiotics influx by altering the structure of porins, and also membrane permeability to exhibit resistance mechanism. This mechanism of antimicrobial resistance is mediated by several microbes, including gram-negative pathogenic *Pseudomonas* spp. and *Acinetobacter* spp. (De et al. 2001; Vila et al. 2007; Pages et al. 2008). Bacterial species belonging to Enterobacteriaceae can degrade the β-lactam antibiotics (cephalosporins, penicillins, and carbapenems) (Olsen et al. 2015; Blair et al. 2015). *P. aeruginosa* produces modifying enzymes that destroy or alter the structure of antibiotics, including chloramphenicol and fosfomycin (Munita and Arias 2016; Khameneh et al. 2019). In addition, a study carried out with α-Bisabolol isolated from *Matricaria chamomilla* L in single and complex form with β-Cyclodextrin as TetK and NorA efflux pump inhibitors in *Staphylococcus aureus* strains. α-Bisabolol potentiated the action of tetracycline and reduced the MIC of norfloxacin to a clinically relevant concentration. The complexed substance showed synergism; however, the effect of the isolated α-Bisabolol was superior to the complex. These results indicate α-Bisabolol is a potential substance to be used as an efflux pump inhibitor (Cruz et al. 2020).

The continued use of medicinal plants and the empirical knowledge of the communities about them has aroused interest in pharmacological research related to plants. Many botanical families are being studied with the purpose of finding biochemical substances in these plants against bacteria and fungi that are resistant to multiple drugs that are currently available in the market. The combination of traditional knowledge about medicinal plants, together with technology and science has enabled countless advances in research, aimed at the search for new alternatives in the treatment of bacterial and fungal infections. Of the native families of the tropical flora, the family Myrtaceae, Asteraceae, Piperaceae, Lauraceae, Verbenaceae, Lamiaceae, among many others stand out to be very important. These tropical plants solvent extracts, decoctions or crude extracts and essential oils have been researched extensively in relation to their bioactive potentials.

Due to the difficulty in the treatment of multi-resistant bacteria, it is notoriously necessary to find new substances that have antimicrobial actions to be used in the combat of these microorganisms. In this context, photodynamic therapy emerges as a new alternative in this scenario. *Light Emitting Diodes* (LED) has been a very promising resource, although infrequent in clinical practice, but it has already proven to be very effective in antibacterial action (Gwynne and Gallagher 2018). It is characterized by producing an absolutely safe irradiation power, consumes little energy, extremely long-life span, good power, and low intensity. Macedo da Silva et al. (2020) reported antibiotic-modulating activities of the essential oils of *Eugenia brasiliensis* Lam and *Piper mosenii* C. DC singly or in association with blue LED (Light-emitting diode) light. They concluded that the association of aminoglycosides with the blue LED light and essential oils represents an effective modulatory potential against resistant bacteria such as multi-resistant strains of *Escherichia coli* and *Staphylococcus aureus. Piper aduncum* essential oil was evaluated in a modulatory experiment associated with blue LED light. The combination of volatile oil with antibiotics showed synergistic effect against S. aureus and *E. coli* which was potentiated in the presence of blue LED. The results obtained in this study showed that the essential oil obtained from *Piper aduncum* interferes with the action of antibiotics against bacteria exposed to blue LED (Barbosa et al. 2018).

## 7.3 Tropical Plants with Relevant Antibacterial Activities

Due to increased bacterial resistance to multiple drugs, antimicrobials arise to concern and the search for new alternatives therapeutic plants, with medicinal plants representing an important source to obtain these medicines. The antimicrobial activity of extracts and oils essential of medicinal plants has been proven in several studies conducted in countries that have a diversified flora (Lautié et al. 2020). In South America, with a wide variety of medicinal plants, research showing that these may be sources of antimicrobial substances has been frequently reported in recent years (Spézia et al. 2020). According to Michelin et al. (2005), plant antibiotics have a chemical structure that differs from that of antibiotics derived from microorganisms, and may regulate the intermediate metabolism of pathogens, activating or by blocking reactions and enzymatic synthesis or even changing the structure of membranes. However, since the advent of antibiotics, the use of plant derivatives as antimicrobials has been little explored (Cowan, 1999).

The acquisition of resistance to antimicrobials is a phenomenon genetic, related to alteration of genes contained in micro-organisms, which codify different biochemical mechanisms that impede the action of drugs, These mechanisms of action can be interference with cell wall synthesis; inhibiting protein synthesis; interfering with nucleic acid synthesis; decreasing permeability to the antimicrobial agent; and destructing the structure of the cell membrane (Tenover 2006). Bacterial resistance can arise by acquisition of mutations or by acquisition of genetic material from other bacteria. Genes encoding proteins involved in resistance mechanisms can be located on the chromosome or on extra-chromosomal elements, such as plasmids and transposons, which move easily from one strain to another, from one species to another, or even from one genus to another (Sultan et al. 2017). Despite the availability of new antibiotics, bacterial resistance occurs at an increasingly rapid rate in the different Gram positive and Gram and represents a major therapeutic challenge (Hayes et al. 2020). In recent decades the focus given to the control of bacterial infections, may have contributed to the emergence of bacteria multi-drug resistant, mainly methicillin-resistant Staphylococcus, Penicillin, and erythromycin-resistant *Pneumumococcus*, *Enterococcus* resistant to vancomycin and also *Pseudomonas aeruginosa* and *Klebsiella pneumoniae* resistant to β-lactams and carbapenens (Ventola 2015).

Considering the increase of multi-resistant bacteria due to the use of antimicrobial agents, The World Health Organization (WHO) has recognized the importance of the potential therapeutic power of plants, and that the associated use of medicinal plants and/or their by-products with antimicrobial drugs can inhibit or intensify the therapeutic effect of conventional medicines. In this context, we can observe some advancements with the products from some species (Table 7.1), which are already used in the medical clinical routine as antibacterial herbal medicines.

## 7.4 Synergisms of Phytochemicals and Conventional Antimicrobial Drugs

Research into natural products with modulating properties has been widely disseminated. Many of these plant products have considerable synergistic effects, positively altering the effect of antibiotics and providing an important alternative to combat increased microbial resistance. The synergistic effects that are usually obtained by associating natural products with antibiotics are related to the increase in the influx of the drug, which alters the permeability of the cell membrane, favoring the penetration of antibiotics and potentiating their effect. Thus, studies on the association of natural products and synthetic drugs are still promising in an attempt to discover new chemical classes with antibiotic potential (Cheesman et al. 2017; Khameneh et al. 2019).

Increasing microbial resistance to existing drugs is a problem that is a serious health issue, and therefore there is a pressing need to look for new classes of antibacterial substances, especially from natural sources. A therapeutic alternative for the treatment of micro-organisms resistant to antibiotics is the use of plant extracts. There are many advantages in using antimicrobial compounds from medicinal plants, as fewer side effects, better patient tolerance, more economical, better acceptance due to long history of use, and being renewable because it is available in nature (Parekh and Chanda 2007). Unlike synthetic drugs, antimicrobials of plant origin are not associated with side effects and have a large therapeutic potential for many infectious diseases (Chanda et al. 2010; Habbal et al. 2011). Numerous studies have proven the antimicrobial activity in vitro plant. However, the problem of drug resistance continues to grow. Thus, the need of the moment is

**Table 7.1** Some of the plant products with antimicrobial activity

| Common name | Scientific name | Compound | Active against | Dosage form |
|---|---|---|---|---|
| Aloés | *Aloe vera* (L.) *Burm.f* | Alloin | Bacteria | Capsule 60 mg |
| Barberry | *Berberis vulgaris* | Berberine | Bacteria, protozoa | Soft gel 1000 mg |
| Black pepper | *Piper nigrum* | Piperine | Fungi, lactobacillus, micrococcus | |
| Calêndula | *Calendula officinalis* | Triterpenoids | Bacteria, fungi | Tinture |
| Canela da índia | *Cinnamomum zeylanicum* | Eugenol | Bacteria, fungi, | extracts 250 mg and 500 mg |
| Burdock | *Arctium lappa* | Tannins | Bacteria, fungi, viruses | Capsule 475 mg |
| Caraway | *Carum carvi* | Carvone, Limonene | Bacteria, fungi, viruses | Capsule 1000 mg |
| Cascara sagrada | *Rhamnus purshiana* | Tannins | Bacteria, fungi, viruses | Capsule 425, 450 mg |
| Chamomile | *Matricaria chamomilla* | Anthemicacid | M. tuberculosis, S. typhimurium, S. aureus | |
| Clove | *Syzygium aromaticum* | Eugenol | General | Capsule 500 mg |
| Cranberry | *Vaccinium spp.* | Fructose | Bacteria | Capsule 500 mg |
| Croton | *Crotonspp*L | proanthocyanidinsand/oralkaloids | General | Teaandpills |
| Erva de Carpinteiro (mil folhas) | *Achillea millefolium* L | Lactones | Bacteria | extracts |
| Eucalyptus | *Eucalyptus globulus* | Tannin | Bacteria, viruses | Inhalerand tablet |
| Garlic | *Alliumsativum* | Allicin, ajoene | General | Tablet |
| Ginger | *Zingiber officinale* | Gingerol | Bacteria | Capsule 100 mg |
| Goldenseal | *Hydrastis canadensis* | Berberine, hydrastine | Bacteria, Giardiaduodenale | Solution, 500 mg per dosage |
| Green tea | *Camelliasinensis* | Catechin | General | |
| Grumixama | *Eugenia brasiliensis* | Tannin | General | An infusion of 10 g of leaves or bark in 300 ml water |
| Licorice | *Glycyrrhiza glabra* | Glabrol | S. aureus, M. tuberculosis | Capsule 450 mg |
| Mentha | *Mentha L* | Menthol | General | Infusionandtinture |

(continued)

**Table 7.1** (continued)

| Common name | Scientific name | Compound | Active against | Dosage form |
|---|---|---|---|---|
| Oak | *Quercus rubra* | Tannins | | Capsule 500, 650 mg |
| | *Allium cepa* | Quercetin | | |
| Onion | *Allium cepa* | Allicin | Bacteria, Candida | |
| Oregon grape | *Mahonia aquifolia* | Berberine | Plasmodium | Capsule 500 mg |
| Pata de Vaca | *Bauhinia L* | Ellagicacid | Bacteria | Infusion/decoction: 2–3 cups of tea a day, preferably after the meals |
| Romã | *Punica granatumLinn* | peletierina, isopeletierina, methylpeletierina | Fungi—Candida | Infusion, decoction |
| Salvia | *V. curassavica* | α-Humuleno | Bacteria Fungi | Tinture |
| Senna St. John's wort | *Hypericum perforatum* | Hypericin, others | General | Table 450 mg |
| Thyme | *Thymusvulgaris* | Caffeicacid | Viruses, bacteria, fungi | Capsule 450 mg |
| Turmeric | *Curcuma longa* | Curcumin, Turmericoil | Bacteria, protozoa | |

Adapted from Khameneh et al. 2019

to develop useful or new antimicrobial agents' ways to treat the resistant microorganism (Negi and Dave 2010). In that case, a new form of therapy would be the use of the combination of synergistic antimicrobial therapy between antimicrobial agents known and bioactive plant extracts. The combination therapy between plant extracts and antibiotics can expand the antimicrobial spectrum, prevent the emergence of mutant resistance, and minimize toxicity. Sometimes the use of a single antibiotic does not produce the inhibitory effects desired effectiveness, but a combination of drugs often has an effect synergistic that surpasses your individual performance. The synergistic effect can be due to the formation of the right complex that becomes more effective in inhibiting a particular species of microorganism, either by inhibition of the synthesis of the cell or causing its lysis or death (Chanda and Rakholiya 2011). Thus, the most economically viable combination of natural products associated with available antibiotics shows as an alternative, since the synergistic effect between the two can provide an increased antibacterial activity against sensitive and resistant microorganisms. Therefore, the potentiated effect of these associations can serve as a new infection treatment strategy, enabling drug use antimicrobial when it alone is not effective on certain bacterial strains. The association of extracts from various plants with ampicillin, chloramphenicol, and tetracycline against sensitive bacteria (*S. aureus, Salmonella choleraesuis, P. aeruginosa, Bacillus subtilis, Proteus spp*) and resistant bacteria isolated from hospital environment (*K. pneumoniae, Shigella spp, Proteus spp, P. aeruginosa, Enterobacter aerogenes, E. coli and S. aureus*) showed that in some cases it occurred synergism enabling already ineffective antibiotics to act on these bacteria (Liu et al. 2017).

## 7.5 Antibiotic Resistance—Combination Therapy (Antibiotic + Phytochemical/Plant Extract)

It is estimated that the herbal medicine market moves up to US$ 300 billion around the world and the traditional pharmaceutical market is growing at an annual rate of 3% to 4% worldwide, while that of herbal medicines rises from 6 to 7% (Calixto 2019). On the other hand, the World Health Organization (WHO) recognizes the importance of the therapeutic potential of plants, but does warnings about improper use and preparation, and recommends caution if they consider the lack of knowledge about possible side effects with a joint administration of prescribed drugs. In any case, what occurs in most societies today is a complementarity between allopathy and the use of medicinal plants. In recent years and in most developing countries, people have resorted to self-medication, which has often led to increased microbial resistance to the drug. However, the trend in drug discovery is gradually returning to natural products because of the constant development of microbial resistance to synthetic drugs. The large-scale use of antibiotic therapy in recent decades is largely a result of the aging process, which influences the increase of infections associated with chronic age-related diseases such as metabolic, cardiovascular, and neoplastic diseases. These conditions are often associated with oxidative stress and act directly on the global pharmaceutical burden. Resistance to antimicrobials represents a growing challenge for modern science. Especially in a hospital environment, the use of powerful, broad-spectrum antibiotics, inadequate or sub-therapeutic treatment, as well as the lack or deficiency of preventive measures contribute to the development of resistance to these drugs. To this end, strategies that prolong the effectiveness of currently available antibiotics, encourage their rational use and reduce or avoid the spread of multi-drug-resistant microorganisms (MDRs) are sought. Understanding the mechanisms of antibiotic resistance is a fundamental step toward the discovery of effective therapeutic measures (Rice 2018; Yelin and Kishony 2018).

Extracts from *Piper* species have been reported to be effective against Gram-positive and Gram-negative bacteria and could be a potential lead to the discovery of new antimicrobial drugs (Mgbeahuruike et al. 2017; Salehi et al. 2019). Antimicrobial photodynamic therapy has grown considerably in recent decades as an effective and inexpensive alternative. *Piper aduncum* is popularly known as monkey pepper with antibacterial activity. This study aimed to evaluate the antibacterial and modulatory activity of *Piper aduncum* essential oil (EOPad) associated with blue LED light, which was exposed to the LED for 20 min. The Germacrene A was identified as the main component. The OEPad presented MIC ≥ 1024 μg/mL against S. aureus and E. coli. The combination of EOPad with antibiotics showed synergistic effect against S. aureus and *E. coli* which was potentiated in the presence of blue LED. The results obtained in this study showed that the essential oil obtained from *Piper aduncum* interferes with the action of antibiotics against bacteria exposed to blue LED (Barbosa et al. 2018).

The antimicrobial potential of leaf of *Piper caldense* essential oil was investigated, and its potential was investigated. The minimum inhibitory concentration (MIC) and minimum fungicidal concentration (MFC) was determined. Besides the essential oil was also tested as a modulator for several antibiotics, and its effect on the morphology of *Candida albicans* (CA) strains was also investigated. The essential oil modulated the activity of fluconazole against CA URM 4387 strain, which was demonstrated by the lower $IC_{50}$ obtained, 2.7 μg/mL, whereas fluconazole itself presented an IC50 of 7.76 μg/ml. No modulating effect was observed in the MFC bioassays. The effect on fungal morphology was observed for both CA INCQS 40,006 and URM 4387 strains. The results demonstrated that the oil has potential as an adjuvant in antimicrobial formulations (Bezerra et al. 2020).

A study was carried out with essential oil from the fresh leaves of *Hyptis martiusii* to investigate the modulating activity in association with different antibiotics against the bacteria *Staphylococcus aureus, Escherichia coli,* and *Pseudomonas aeruginosa*, and to evaluate the cytotoxic activity of the species. The research has shown that the essential oil of *Hyptis martiusii Benth* (OEHM) leaves presents synergism only in association with gentamicin antibiotics and imipenem against the bacteria *Pseudomonas aeruginosa* and *Escherichia coli.* However, it presents antagonism in association with amikacin, gentamicin, and imipenem against the three bacteria studied. Apart from ciprofloxacin, no relevant results were demonstrated. In relation to the cytotoxic activity, the mean lethal concentration (lc50) exposed a value of 263.12 μg/ml. The results reveled that *H. martiusii* presents synergistic cytotoxic activity against the evaluated bacteria (Figueiredo et al. 2018).

A study carried out by Ferreira et al. (2019) identified the chemical composition of the *Senna spectabilis* species, and they analyzed the antimicrobial potential in vitro and its modulating effect on antibiotic activity. The antibiotics used in the experiment were aminoglycosides (amikacin and gentamicin), lincosamides (clindamycin), cephalosporins (cephalexin), and azithromycin. The *Staphylococcus aureus, Escherichia coli, Pseudomonas aeruginosa,* and *Salmonella enterica* strains were used in this approach. The results have shown that the antimicrobial action of the extract of *S. spectabilis* is only regular; however, in modulation, the extract in combination with the tested antibiotics showed a synergistic effect against most of the tested bacteria, potentiating the effect of the antibiotics used.

Ferreira and Costa evaluated the antimicrobial activity of ethanolic extracts of *Banisteriopsis anisandra* (Malpighiaceae) and drug interactions with drugs widely used in the field of oral health. They obtained the leaf extract and performed serial dilutions, which were tested against *Candida albicans* (ATCC 10,231),

*Streptococcus mutans* (ATCC 25,175), *Streptococcus salivarius* (ATCC 7073), and *Staphylococcus aureus* (ATCC 6538). As positive control, antibiotics ampicillin (10 μg/ml), amoxicillin (10 μg/ml), and penicillin (30 μg/ml) were used for bacterial strains; and the antifungal ketoconazole (2 mg/ml), for yeast. There was a positive drug interaction of the extracts with antibiotics and ketoconazole in concentrations of 1: 2 for the extract. The extract of *B. anisandra* has an antimicrobial potential and drug interaction with drugs related to oral health. Studies on the pharmacological properties of the species in question are rare and the results are the first reports on its antimicrobial activity in association with antibiotics and antifungals.

The positive results found not only in these reported studies, but also in countless other studies, with different species increasingly show the importance of the search for new metabolites that can help and reinforce drugs that no longer play the role of eliminating pathogenic microorganisms.

## 7.6 Essential Oils with Antimicrobial Activity from Tropical Medicinal Plants

In nature, EOs play an important role in providing plant protection against pathogenic bacteria, viruses, and fungi and preventing the attack by insect pests. In addition, EOs can attract or repel insects when present in pollen and seeds. To protect chemical compounds' ecological equilibrium, the use of EOs in pharmaceutical, food, bactericidal, and fungicidal is becoming more prevalent in recent times (Swamy et al. 2016). EOs yielding medicinal and aromatic plants are normally native to warm countries, where they represent an important traditional pharmacopeia.

Apiaceae, Lamiaceae, Myrtaceae, Poaceae, and Rutaceae families are of particular importance for medicinal applications. For example, some of the EOs, like anise, caraway, black caraway, clove, oregano, cumin, coriander, sage, basil, dill, lemon balm, peppermint, thyme, and tea oils, already have widespread medicinal applications. Some of the essential oil containing plant families, like Liliaceae, Fabaceae, Pinaceae, Piperaceae, Cupressaceae, Asteraceae and Hypericaceae, also exhibit a considerable medicinal potential.

The Apocynaceae family is considered to be dicotyledonous characterized usually by the presence of latex, consisting of about 155 genus and 2000 species, with distribution in Tropical and Subtropical regions (Lorenzi 1998). This family can be considered one of the most important plant sources of chemical constituents with therapeutic utility in modern medicine. The most important genres in this family are *Alstonia, Aspidosperma, Vinca, Tabernaemontana, Mandevilla, Hancornia, Nerium, Strophantus, Catharanthus, Allamanda, Thevetia, Wrightia, Plumeria, Himatanthus* and *Rauvolfia*. The Apocynaceae family is chemically characterized by structures alkaloids mainly monoterpenic. Several indolic alkaloids have already been isolated from this family, especially from the Aspidosperma and Geissospermum genus (Ekalu et al. 2019). In addition to representatives with numerous medicinal properties, the family Apocynaceae is an important source of economic resources. Rubber is derived from the latex of several species, markedly inferior in quality to that of the extracted from *Hevea brasiliensis* (Willd. ex A. Juss.) Müll. Arg. the rubber tree. At African countries and indigenous peoples of South America, toxic species are used to poison arrows in animal hunting and fishing. Other species provide excellent quality wood for the manufacture of furniture such as case of the genus Aspidosperma, whose most common representative is *A. peroba Allemão* ex Saldanha, known as "peroba-rosa" (Baratto et al. 2010).

The therapeutic potential of the *Himatanthus drasticus* (Mart.) Plumel, Apocynaceae was assayed. This specie presents triterpene-rich fraction which the anti-inflammatory activity is mediated by NF-alpha, iNOS, COX-2 and NF-kB (Almeida et al. 2017). The essential oil of *Mikania cordifolia* and its major constituent limonene was investigated alone or associated against *Pseudomonas aeruginosa, Escherichia coli,* and *Staphylococcus aureus*. The antibiotic-modulating activity was determined in combination with conventional antibacterial drugs. The results demonstrated no relevant antibacterial activity; however, a modulatory effect was observed against *P. aeruginosa*, presenting synergistic effects when associated with gentamicin and norfloxacin. In addition, the oil reduced the MIC of norfloxacin against *E. coli* as well as reduced the MIC of gentamicin against *S. aureus*. The best effect of limonene was obtained against S. aureus. It is possible to conclude that the *Mikania cordifolia* essential oil and the isolated compound limonene modulate the action of antibiotics against MDR bacteria (Araújo et al. 2020). The essential oil bacterial pathogens action is associate to capacity to destabilize the cellular machinery, leading to breakdown of membrane wall, disrupting many cellular activities including energy production and membrane transport. The lipidic components of essential oils induce membrane rupture leading to leakage of cellular components and loss of ions (Fig. 7.2) (Swamy et al. 2016; Tariq et al. 2019; Basavegowda et al. 2020).

## 7.7 Plant-Derived Natural Products with Antifungal Activity

The resistance of microorganisms to therapeutic agents currently used has caused economic and social impact. Fungal resistance has been somewhat neglected when compared to the great repercussion that bacterial resistance has reached. One is strictly related to the other, since the intensive use of antibacterial antibiotic therapy reduces competitiveness in human biota and favors the growth of yeasts of *Candida spp*. which in due course change from diners to pathogens. In addition, other factors contribute to increased resistance, such as the significant increase in the number of immunosuppressed patients and the prolonged use of surgical devices, considered resistance drives.

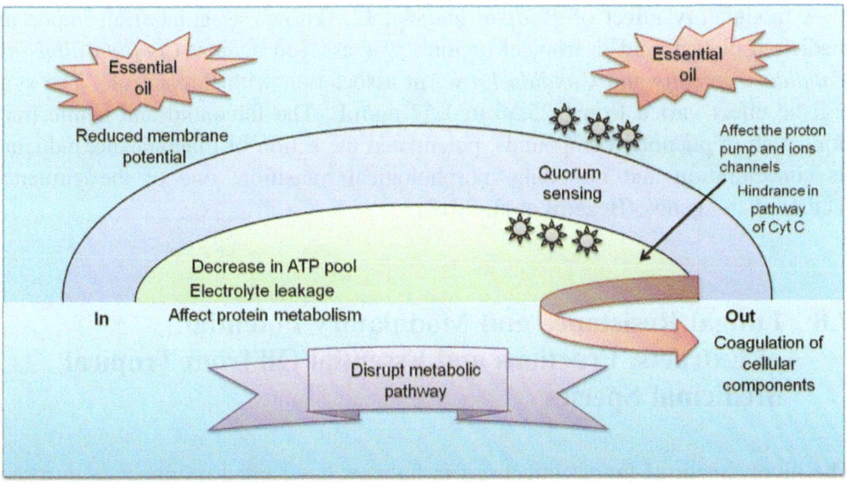


**Fig. 7.2** Antimicrobial mechanisms of essential oils on microbes (Adapted from Swamy et al. 2016)

Infections caused by *Candida spp.* have shown a high level of incidence, where *Candida albicans* and *Candida tropicalis* stand out, with notoriety for the former, as agents capable of causing invasion. Their virulence factors provide adaptability, since they allow the fulfillment of important criteria for survival, such as having tolerance to high temperatures and invasive potential, promoting lysis and absorption of human tissues and resisting immunological defenses. The current context requires searches for bioactive substances with antifungal potential. New therapeutic agents may be contained in plants and consequently, their extracts and essential oils may serve as subsidies for in-depth studies and research that contemplate and target the formulation of new drugs. Research has been conducted to promote the combination of commercial drugs, for which resistance phenomena have already been verified, with natural products in the form of extracts, fractions, essential oils or isolated constituents, in an attempt to circumvent the resistance of *Candida spp.*

Recently, a medicinal plant *Mesosphaerum suaveolens* aqueous extracts of leaves (AELMs) and aerial parts (AEAPMs) was evaluated against strains of the genus *Candida*. The modulatory effect was observed with the drug fluconazole. The AELMs have shown good results since modulated fluconazole activity decreased fluconazole's IC50 from 7.8 μg/mL to an IC50 of 4.7 μg/mL (CA LM 77 strain) and from 28.8 μg/mL to 18.26 μg/mL (CA INCQS 40,006 strain) for the *C. albicans* strains. For the *C. tropicalis* LM 23 strain, the AEPMs obtained an IC50 of 25 μg/mL and the AELMs an IC50 of 359.9 μg/mL. The AEAPMs as well as the AELMs presented clinically relevant activities for *C. tropicalis* strains (Costa et al. 2020).

A modulatory effect of *Psidium guajava* L., (known as goiaba) an important medicinal plant found in tropical regions, was assayed against *Candida albicans*, *Candida tropicalis* and *Candida krusei* in association with fluconazole. The synergistic effect varied from 925.56 to 1.57 μg/mL. The flavonoid and tannic fractions, rich in phenolic compounds, potentiated the action of Fluconazole, reducing its concentration and impeding morphological transition, one of the virulence factors of the genus (Bezerra et al. 2018).

## 7.8 Fungal Resistance and Modulatory Potential of Extracts, Fractions and Essential Oil from Tropical Medicinal Species

The involvement of fungal infections in humans has gained prominence in recent years, and among the most common fungi infections, those of the genus Candida stand out, being among the species causing opportunistic infections. This is justified by the fact that they are very well adapted to the human body, being able to colonize it without producing signs of disease under conditions of physiological normality. However, they can become pathogenic in immunosuppressed patients, generating infections that are difficult to diagnose and are associated with high mortality and morbidity. As for clinical manifestations, we can divide them into cutaneous, systemic, and allergic. The treatment of these pathologies is performed with synthetic drugs with systemic action such as amphotericin B, fluconazole, itraconazole, voriconazole and echinocandins, azoles being the most used, due to their low cost and low toxicity.

The availability of antifungal agents for the treatment of systemic infections is somewhat reduced, becoming a problem with multi-drug resistant yeasts. In addition, one of the stalemates today is resistance to antifungals and difficulties in effective treatment. As a way to ease this resistance and difficulty in treatment, medicinal plants are strongly used by the population for the treatment of infections, and this fact has aroused interest from several research centers for the study of this activity, with the aim of isolating new molecules with antimicrobial activity.

Natural plant products have been evaluated not only for their direct antimicrobial activity, but as a modification agent of resistance to antifungals, which can be a synergistic or antagonistic effect. Thus, the associated use of medicinal plants or their derived compounds may interfere, inhibiting or intensifying the therapeutic effect of conventional antimicrobials. In this context, the synergistic effect of *Piper mikanianum* essential oil species rich in safrole was investigated against *Candida* strains. An interesting result was observed where the oil was combined with gentamicin against the multi-drug resistant *E. coli strain* and also associated with fluconazole against fungal strains. The oil presented a fungistatic effect. The *P. mikanianum* essential oil has shown an inhibitory effect of one of the main virulence factors of the Candida genus, morphological transition, which has been previously shown to be responsible for causing invasive infections in human tissues (Carneiro et al. 2020). The mechanism can be associate with membrane potential across cell wall and disrupt ATP assembly, leading to cell wall damage. Essential oils can also disintegrate mitochondrial membrane interfering with the electron transport system (ETS) pathway (Tariq et al. 2019). Several species of Asteraceae are producers of essential oil of commercial importance. The genus *Baccharis* is represented by more than 500 species, distributed predominantly in South America. In Brazil, 120 species are described, distributed in greater concentration in the South Region of the country (Giuliano 2001). This genus produces many secondary metabolites, and in general, the compounds that stand out the most are flavonoids and terpenoids, especially monoterpenes, sesquiterpenes, diterpenes and triterpenes (Trombin-Souza et al. 2017).

Recent studies have shown the biological activities of the *Baccharis* genus, such as antioxidant (Zuccolotto et al. 2019), antiparasitic (Budel et al. 2018), cytotoxic activity (Fukuda et al. 2006), as well as acts in inflammatory processes (Florão et al.2012). A study carried out with *Baccharis reticulata* demonstrated an antibiotic-modulating activity against both Gram-positive and Gram-negative bacteria. The best result was observed with the volatile oil in association with norfloxacin and gentamicin against the multiresistant strain of *S. aureus*. In addition, the oil exhibited a synergistic effect for Gram-positive and Gram-negative bacteria (Freitas et al. 2020). Evidence of in vitro synergism with MDR strains provides a higher probability of successful in vivo therapy. Thus, these associations can be useful in selecting the most promising combinations of antimicrobials for practical therapy of severe bacterial infections.

*Licania rigida* Benth (known as "oiticica"), belongs to the *Chrisobalanaceae* family and it is distributed in tropical and subtropical regions. A study was carried out with leaf ethanolic extract to evaluate the virulence factor associated with the ability to form biofilms on biotic and abiotic surfaces in association with fluconazole since its action on biofilms is described. The study demonstrated that biofilms formed by *Candida sp.* isolated in acrylic resin discs reduced biofilm formation. It is associated with the reduction of hydrophobicity mechanisms of the cell surface, which reduce the aggregative potential and the formation of the biofilm (Freitas et al. 2019).

These outcomes of the use of medicinal plants by different populations, a fact that passes through generations, have highlighted species that present active principles of recognized clinical relevance, which in some cases have been transformed into modern medicines. Successful therapeutic evidence observed from the daily life of traditional or local communities has inspired pharmacological research, and this forms the basis of the ethnobiological approach, which is guided by popular practices and knowledge, consolidated through time, culture and history of several people, contributing to the discovery of agents with therapeutic potential or even a new drug, more specifically.

## 7.9 Conclusions and Future Prospects

The valorization of biological and cultural diversity, as well as the knowledge acquired by several traditional practices, becomes a necessary tool to know and classify the use of natural resources. In this context, empirical knowledge not only has great relevance in the discovery of new medicinal plants, suppliers of bioactive chemical substances originated mainly from the secondary cellular metabolism of the plant, but also are focused on the planning of actions related to the sustainability of natural resources managed. It is known that in tropical regions some social groups have vast traditional knowledge about the different forms of exploitation and management of natural resources, mainly about plant species. Several ethnic groups have created mechanisms and strategies for disease treatment and prevention. The accumulation of knowledge and practices form empirical medical systems often based on the use of natural plant resources.

The study of these mechanisms provides ethnopharmacology with a supposed efficacy of plants that are selected depending on the concept of disease and health by these groups, and this identification also helps in the improvement of other approaches that use medicinal plants as the object of study. Obtaining new alternatives for the development of new products of natural origin, it can be characterized as a new chemically defined compound with determined pharmacological activity (pharmacotherapeutic) or a simple traditional preparation from parts of a medicinal species (phytotherapeutic) that has evaluated its efficacy.

The discovery of compounds with biological activity may be more effective from active extracts of perennial plants. This is because these plants invest effectively in lifelong chemical defenses and have higher levels of compounds. One of the main advantages of using natural products in the search for new drugs, compared to synthetic compounds, is the high probability of biological activity. Secondary metabolites are synthesized and optimized through the evolution of specific biological interactions between the producer and the environment. In this sense, the chemical diversity of these substances becomes unequalled.

Thus, it is fundamental to understand how medicinal plants behave in the environment, how these resources are used and how this information can contribute to sustainable use strategies. From this conception of traditional knowledge, it is possible to design strategies that lead to alternatives that respect the need for conservation, together with the traditions of the people who use these resources and the scientific knowledge, being able to provide information for a better time of collection and a better period in which it produces high yields of essential substances for medicinal treatment and subsequent pharmacological uses.

## References

Almeida SCX, Monteiro AB, Costa GM, Viana GSB (2017) *Himatanthus drasticus*: a chemical and pharmacological review of this medicinal species, commonly found in the Brazilian northeastern region. Rev Bras Farmacognosia 27(6):788–793. https://doi.org/10.1016/j.bjp.2017.10.002

Araújo ACJ, Freitas PR, Barbosa CRS, Muniz DF, Rocha JE, Silva ACA, Tintino CDMO, Ribeiro-Filho J, Da Silva LE, Confortin C, Amaral W, Deschamps C, Barbosa-Filho JM, Lima NTR, Tintino SR, Coutinho HDM (2020) GC-MS- FID characterization and antibacterial activity of the *Mikania cordifolia* essential oil and limonene against MDR strains. Food Chem Toxicol 136:111023. https://doi.org/10.1016/j.fct.2019.111023

Anand U, Jacobo-Herrera N, Altemimi A, Lakhssassi N (2019) A comprehensive review on medicinal plants as antimicrobial therapeutics: potential avenues of biocompatible drug discovery. Metabolites 9(11):258

Arumugam G, Swamy MK, Sinniah UR (2016) Plectranthus amboinicus (Lour.) Spreng: botani-cal, phytochemical, pharmacological and nutritional significance. Molecules 21:369

Barbosa CRS, Da Silva ACA, Diodatto JS, Coutinho HDM (2018) Atividade antibacteriana e efeito modulador do óleo essencial de Piper aduncum em associação com luz de led azul. Paper presented at the XXI Semana de Iniciação Científica da URCA, Universidade Regional do Cariri, 5–9 nov de 2018

Baratto LC, Hohlemwerger SVA, Guedes MLS, Duarte MR, Santos, CAM (2010) *Himatanthus lancifolius* (Müll. Arg.) Woodson, Apocynaceae: estudo farmacobotânico de uma planta medicinal da Farmacopeia brasileira 1ª edição. Revista Brasileira de Farmacognosia 20 (5):651–658. https://doi.org/10.1590/S0102-95X2010005000015

Basavegowda N, Patra JK, Baek KH (2020) Essential oils and mono/bi/tri-metallic nanocomposites as alternative sources of antimicrobial agents to combat multidrug-resistant pathogenic microorganisms: an overview. Molecules 27; 25(5). pii: E1058. https://doi.org/10.3390/molecules25051058

Beutler JA (2009) Natural products as a foundation for drug discovery. Curr Protocols Pharmacol 46(1):9–11

Bezerra CF, Rocha JE, Silva MKDN, de Freitas TS, de Sousa AK, Dos Santos ATL, da Cruz RP, Ferreira MH, da Silva JCP, Machado AJT, Carneiro JNP, Sales DL, Coutinho HDM, Ribeiro PRV, de Brito ES, Morais-Braga MFB (2018) Analysis by UPLC-MS-QTOF and antifungal activity of guava (*Psidium guajava L.*). Food Chem Toxicol 119:122–132. https://doi.org/10.1016/j.fct.2018.05.021

Bezerra JWA, Rodrigues FC, da Cruz RP, Silva LE, do Amaral W, Rebelo RA, Begnini IM, Bezerra CF, Iriti M, Varoni EM, Coutinho HDM, Morais-Braga MFB, (2020) Antibiotic potential and chemical composition of the essential oil of piper caldense C. DC. (Piperaceae). Appl Sci 10(2):631. https://doi.org/10.3390/app10020631

Blair JM, Webber MA, Baylay AJ, Ogbolu DO, Piddock LJ (2015) Molecular mechanisms of antibiotic resistance. Nat Rev Microbiol 13(1):42–51

Brooks TM, Mittermeier RA, Mittermeier CG, Da Fonseca GA, Rylands AB, Konstant WR, Flick P, Pilgrim J, Oldfield S, Magin G, Hilton-Taylor C (2002) Habitat loss and extinction in the hotspots of biodiversity. Conserv Biol 16(4):909–923

Brown D (2015) Antibiotic resistance breakers: can repurposed drugs fill the antibiotic discovery void? Nat Rev Drug Discov 14(12):821–832

Budel JM, Wang M, Raman V, Zhao J, Khan SI, Rehman JU, Techen N, Tekwani B, Monteiro LM, Heiden G, Takeda IJM, Farago PV, Khan IA (2018) Essential oils of five Baccharis species: investigations on the chemical composition and biological activities. Molecules 23(10):2620. https://doi.org/10.3390/molecules23102620

Calixto JB (2019) The role of natural products in modern drug discovery. An Acad Bras 552 Cienc. 91(3), e20190105. https://doi.org/10.1590/0001-3765201920190105

Carneiro JNP, da Cruz RP, Campina FF, Costa MDS, Dos Santos ATL, Sales DL, Bezerra CF, da Silva LE, de Araujo JP, do Amaral W, Rebelo RA, Begnini IM, de Lima LF, Coutinho HDM, Morais-Braga MFB (2020) GC/MS analysis and antimicrobial activity of the Piper mikanianum (Kunth) Steud. essential oil. Food Chem Toxicol 135:110987. https://doi.org/10.1016/j.fct.2019.110987

Chanda S, Dudhatra S, Kaneria M (2010) Antioxidative and antibacterial effects of seeds and fruit rind of nutraceutical plants belonging to the Fabaceae family. Food Funct 1:308–315

Chanda S, Rakholiya K (2011) Combination therapy: Synergism between natural plant extracts and antibiotics against infectious diseases. Science against microbial pathogens: communicating current research and technological advances. Microbiol Book Series 1(13):520–529

Cheesman MJ, Ilanko A, Blonk B, Cock IE (2017) Developing new antimicrobial therapies: are synergistic combinations of plant extracts/compounds with conventional antibiotics the solution? Pharmacogn Rev 11(22):57–72. https://doi.org/10.4103/phrev.phrev_21_17

Cowan MM (1999) Plant products as antimicrobial agents. Clin Microbiol Rev 12(4):564–582

Costa AR, Bezerra JWA, Cruz RP, de Freitas MA, da Silva VB, Neto JC, dos Santos ATL, Braga MFBM, da Silva LA, Rocha MI, Kamdem JP, Iriti M, Vitalin, S, Duarte AE, Barros LM (2020) In vitro antibiotic and modulatory activity of *Mesosphaerum suaveolens* (L.) Kuntze against Candida strains. Antibiotics 9:46

Cruz RP, Freitas FT, Costa S, Santos L, Thassya A, Campina FF, Pereira RLS, Bezerra JWA, Quintans-Júnior LJ, Araújo AAS, Júnior JPS, Iriti M, Varoni EM, Menezes IRA, Coutinho HDM, Morais-Braga MFB (2020) Effect of α-Bisabolol and Its β-Cyclodextrin complex as TetK and NorA efflux pump inhibitors in Staphylococcus aureus strains. Antibiotics 9:1–8

Chazdon RL (2019) Towards more effective integration of tropical forest restoration and conservation. Biotropica 51(4):463–472

De E, Basle A, Jaquinod M, Saint N, Mallea M, Molle G et al (2001) A new mechanism of antibiotic resistance in Enterobacteriaceae induced by a structural modification of the major porin. Mol Microbiol 41(1):189–198

Ekalu A, Ayo RGO, Habila JD, Hamisu I (2019) A mini-review on the phytochemistry and biological activities of selected Apocynaceae plants. J Herbmed Pharmacol 8(4):269–273. https://doi.org/10.15171/jhp.2019.39

Fazly Bazzaz BS, Khameneh B, Zahedian Ostad MR, Hosseinzadeh H (2018) In vitro evaluation of antibacterial activity of verbascoside, lemon verbena extract and caffeine in combination with gentamicin against drug-resistant Staphylococcus aureus and Escherichia coli clinical isolates. Avicenna J Phytomed 8(3):246–253

Ferreira JA, Sousa Júnior DL, Marquesa AEF, Mendesa RC (2019) Phytochemical Prospection and analysis of the antimicrobial and in vitro modulator activity of the hydroalcoolic extract of the leaves of Senna spectabilis. Ensaios e Ciênc 23(3):262–267. https://dx.doi.org/10.17921/1415-6938.2019v23n3p262-267

Figueiredo FRSDN, Primo AJB, Monteiro AB, Tintino RS, Delmondes GA, Sales VS, Rodrigues CKS, Felipe CFB, Coutinho HDM, Kerntopf MR (2018) Avaliação da atividade moduladora e citotóxica do óleo essencial das folhas de *Hyptis martiusii* Benth. Revista Ciencias de la Salud, 16(1):49–58. https://dx.doi.org/10.12804/revistas.urosario.edu.co/revsalud/a.6489

Florão A, Budel JM, Duarte MR, Marcondes A, Rodrigues RAF, Rodrigues MVN, Santos CAM, Weffort-Santos AM (2012) Essential oils from Baccharis species (Asteraceae) have anti-inflammatory effects for human cells. J Essent Oil Res 24:561–570

Freitas MA, Alves AIS, Andrade JC, Andrade MCL, Dos Santos ATL, de Oliveira TF, Dos Santos F, Buonafina MDS, Coutinho HDM, de Menezes IRA, Morais-Braga MFB, Neves RP (2019) Evaluation of the antifungal activity of the licania rigida leaf ethanolic extract against biofilms formed by Candida sp. isolates in acrylic resin discs. Antibiotics (Basel, Switzerland) 8(4):250. https://doi.org/10.3390/antibiotics8040250

Freitas PR, de Araújo ACJ, Barbosa CRS, Muniz DF, Silva ACA, Rocha JE, Tintino CDMO, Filho JR, Silva LE, Confortin C, Amaral W, Deschamps C, Filho JMB, Lima NTR, Tintino SR, Coutinho HDM (2020) GC-MS-FID and potentiation of the antibiotic activity of the essential oil of Baccharis reticulata (ruiz & pav.) pers. and α-pinene. Ind Crops Prod 145:112106. https://doi.org/10.1016/j.indcrop.2020.112106

Fukuda M, Ohkoshi E, Makino M, Fujimoto Y (2006) Studies on the constituents of the leaves of Baccharis dracunculifolia (Asteraceae) and their cytotoxic activity. Chem Pharm Bull (Tokyo) 54(10):1465–1468

Giuliano DA (2001) Clasificación infragenérica de las especies argentinas de Baccharis (Asteraceae, Astereae). Darwiniana 39(1–2):131–154

Gwynne PJ, Gallagher MP (2018) Light as a broad-spectrum antimicrobial. Front Microbiol 9:119. https://doi.org/10.3389/fmicb.2018.00119

Habbal O, Hasson SS, El-Hag AH, Al-Mahrooqi Z, Al-Hashmi N, Al-Bimani Z, Al- Balushi MS, Al-jabri AA (2011) Antibacterial activity of Lawsonia inermis Linn (henna) against *Pseudomonas aeruginosa*. Asian Pac J Trop Biomed 1:173–176. https://doi.org/10.1016/S2221-1691(11)60021-X

Hayes K, O'Halloran F, Cotter L (2020) A review of antibiotic resistance in group B Streptococcus: the story so far. Crit Rev Microbiol 49:1–17. https://doi.org/10.1080/1040841X.2020.175862

Inui T, Wang Y, Deng S, Smith DC, Franzblau SG, Pauli GF (2007) Counter-current chroma-tography based analysis of synergy in an anti-tuberculosis ethnobotanical. J Chromatogr A 1151:211–215

Khameneh B, Iranshahy M, Soheili V, Bazzaz BSF (2019) Review on plant antimicrobials: a mechanistic viewpoint. Antimicrob Resist Infect Control 8:118. https://doi.org/10.1186/s13756-019-0559-6

Khameneh B, Diab R, Ghazvini K (2016) Fazly Bazzaz BS (2016) Breakthroughs in bacterial resistance mechanisms and the potential ways to combat them. Microb Pathog 95:32–42

Kirubakari B, Sangeetha T, Vijayarathna S, Chen Y, Kanwar JR, Leow CH, Shin LN, Swamy MK, Subramaniam S, Sasidharan S (2019) Antibacterial and Antifungal Agents of Higher Plants. Natural Bio-active Compounds. Springer, Singapore, pp 493–508

Lautié E, Russo O, Ducrot P, Boutin JA (2020) Unraveling plant natural chemical diversity for drug discovery purposes. Front Pharmacol 11:397. https://doi.org/10.3389/fphar.2020.00397

Levy SB (1992) Active efflux mechanisms for antimicrobial resistance. Antimicrob Agents Chemother 36(4):695–703

Liu Q, Meng X, Li Y, Zhao CN, Tang GY, Li HB (2017) Antibacterial and antifungal activities of spices. Int J Mol Sci 18(6):1283. https://doi.org/10.3390/ijms18061283

Lorenzi, H (1998) Árvores brasileiras: manual de identificação e cultivo de plantas arbóreas do Brasil. 2 ed. Nova Odessa: Plantarum, 2v. ISBN do v.2 – 8586714070

Maxwell A (1997) DNA gyrase as a drug target. Trends Microbiol 5(3):102–109

McBain AJ, Bartolo RG, Catrenich CE, Charbonneau D, Ledder RG, Gilbert P (2003) Effects of a chlorhexidine gluconate-containing mouthwash on the vitality and antimicrobial susceptibility of in vitro oral bacterial ecosystems. Appl Environ Microbiol 69(8):4770–4776

Mgbeahuruike EE, Yrjönen T, Vuorela H, Holm Y (2017) Bioactive compounds from medicinal plants: focus on piper species. South African J Botany 112:54–69

Michelin DC, Moreschi P, Lima AC, Nascimento GGF, Paganelli MO, Chaud MV (2005) Avaliação da atividade antimicrobiana de extratos vegetais. Rev Bras Farmacogn 15:316–320

Mohanty SK, Swamy MK, Sinniah UR, Anuradha M (2017) Leptadenia reticulata (Retz.) Wight & Arn. (Jivanti): Botanical, agronomical, phytochemical, pharmacological, and biotechnological aspects. Molecules 22:1019. https://doi.org/10.3390/molecules22061019

Munita JM, Arias CA (2016) Mechanisms of antibiotic resistance virulence mechanisms of bacterial pathogens. Microbiol Spectr 481–511.

Negi BS, Dave BP (2010) In vitro antimicrobial activity of acacia catechu and its phytochemical analysis. Indian J Microbiol 50(4):369–374. https://doi.org/10.1007/s12088-011-0061-1

Olsen I (2015) New promising beta-lactamase inhibitors for clinical use. Eur J Clin Microbiol Infect Dis 34(7):1303–1308

Pages JM, James CE, Winterhalter M (2008) The porin and the permeating antibiotic: a selective diffusion barrier in gram-negative bacteria. Nat Rev Microbiol 6(12):893–903

Pandey AK, Kumar S (2013) Perspective on plant products as antimicrobials agents: a review. Pharmacologia 4:469–480

Parekh J, Chanda S (2007) Antibacterial and phytochemical studies on twelve species of Indian medicinal plants. African J Biomed Res 10:175–181

Paulsen IT, Brown MH, Skurray RA (1996) Proton-dependent multidrug efflux systems. Microbiol Rev 60(4):575–608

Peterson E, Kaur P (2018) Antibiotic resistance mechanisms in bacteria: relationships between resistance determinants of antibiotic producers, environmental bacteria, and clinical pathogens. Front Microbiol 9:2928

Rana R, Sharma R, Kumar A (2018) Repurposing of existing statin drugs for treatment of microbial infections: how much promising? Infect Disord Drug Targets

Rice LB (2018) Antimicrobial Stewardship and Antimicrobial Resistance. Med Clin North Am 102(5):805–818. https://doi.org/10.1016/j.mcna.2018.04.004

Rudramurthy GR, Swamy MK, Sinniah UR, Ghasemzadeh A (2016) Nanoparticles: alternatives against drug-resistant pathogenic microbes. Molecules 21:836

Salehi B, Zakaria ZA, Gyawali R, Ibrahim SA, Rajkovic J, Shinwari ZK, Khan T, Sharifi-Rad J, Ozleyen A, Turkdonmez E, Valussi M, Tumer TB, Monzote Fidalgo L, Martorell M, Setzer WN (2019) Piper species: a comprehensive review on their phytochemistry, biological activities and applications. Molecules 7;24(7). pii: E1364. doi: https://doi.org/10.3390/molecules24071364

Schneider T, Sahl HG (2010) An oldie but a goodie—cell wall biosynthesis as antibiotic target pathway. Int J Med Microbiol 300(2–3):161–169

Shin J, Prabhakaran VS, Kim KS (2018) The multi-faceted potential of plant-derived metabolites as antimicrobial agents against multidrug-resistant pathogens. Microb Pathog 1(116):209–214

Shriram V, Khare T, Bhagwat R, Shukla R, Kumar V (2018) Inhibiting bacterial drug efflux pumps via phyto-therapeutics to combat threatening antimicrobial resistance. Front Microbiol 2018;9:2990. Published 2018 Dec 10. https://doi.org/10.3389/fmicb.2018.02990

Sobrinho TG, Paolucci LN, Muscardi DC, Maradini AC, Silva EA, Solar RR, Schoereder JH. Biodiversity and ecosystem functioning in tropical habitats—case studies and future perspectives in Atlantic rainforest and Cerrado landscapes. In Biodiversity in ecosystems-linking structure and function 2014. Shovonlal Roy. Intech.

Spézia FP, Siebert D, Tenfen AC, Cordova CMM, Alberton MD, Guedes A (2020) Avaliação da a atividade antibacteriana de plantas medicinais de uso popular: Alternanthera brasiliana (penicilina), Plantago major (tansagem), Arctostaphylos uva-ursi (uva-ursi) e Phyllanthus niruri (quebra-pedra). Rev Pan-Amaz Saude 11: e202000127. Available in: https://scielo.iec.gov.br/scielo.php?script=sci_arttext&pid=S2176-62232020000100010&lng=pt. Epub 10-Mar-2020. http://dx.doi.org/https://doi.org/10.5123/s2176-6223202000127.

Sultan I, Rahman S, Jan AT, Siddiqui MT, Mondal AH, Haq QMR (2017). Antibiotics, resistome and resistance mechanisms: a bacterial perspective. Front Microbiol 21(9):2066. https://doi.org/10.3389/fmicb.2018.02066

Swamy MK, Arumugam G, Kaur R, Ghasemzadeh A, Yusoff MM, Sinniah UR (2017) GC-MS based metabolite profiling, antioxidant and antimicrobial properties of different solvent extracts of Malaysian Plectranthus amboinicus leaves. Evid-Based Complement Alternat Med 2017:1517683. https://doi.org/10.1155/2017/1517683

Swamy MK, Rudramurthy GR (2016) Antimicrobial agents: current status and future challenges. Austin Pharmacol Pharm 1:1004

Swamy MK, Sinniah UR, Akhtar MS (2016) Antimicrobial properties of plant essential oils against human pathogens and their mode of action: an updated review. Evid-Based Complement Alternat Med 2016:21. https://doi.org/10.1155/2016/3012462

Tariq S, Wani S, Rasool W, Shafi K, Bhat MA, Prabhakar A, Shalla AH (2019) Rather MA (2019) A comprehensive review of the antibacterial, antifungal and antiviral potential of essential oils and their chemical constituents against drug-resistant microbial pathogens. Microb Pathog 134:103580

Tenover FC (2006) Mechanisms of antimicrobial resistance in bacteria. Am J Infect Control 34(5 Supplement):S3–S10

Trombin-Souza M, Amaral W, Pascoalino J, De Oliveira R, Bizzo H, Deschamps C (2017) Chemical composition of the essential oils of Baccharis species from southern Brazil: a comparative study using multivariate statistical analysis. J Essent Oil Res 29(5):1–7. https://doi.org/10.1080/10412905.2017.1322007

Walsh C (2003) Opinion—anti-infectives: where will new antibiotics come from? Nat Rev Microbiol 1(1):65

Ventola CL. (2015) The antibiotic resistance crisis: part 1: causes and threats. 40(4):277–83

Vila J, Marti S, Sanchez-Cespedes J (2007) Porins, efflux pumps and multidrug resistance in Acinetobacter baumannii. J Antimicrob Chemother 59(6):1210–1215

Yelin I, Kishony R (2018) Antibiotic Resistance. Cell 172(5):1136–1136. https://doi.org/10.1016/j.cell.2018.02.018

Zuccolotto T, Bressan J, Lourenco AVF, Bruginski E, Veiga A, Marinho JVN, Raeski PA, Heiden G, Salvador MJ, Murakami FS, Budel JM, Campos FR (2019) Chemical, Antioxidant, and antimicrobial evaluation of essential oils and an anatomical study of the aerial parts from Baccharis species (Asteraceae). Chem Biodivers 16(4):e1800547. https://doi.org/10.1002/cbdv.201800547

# Chapter 8
# Capillary Electrophoresis: A New Evolutionary Platform of Plant Secondary Metabolites

**Dilipkumar Pal and Souvik Mukherjee**

**Abstract** Capillary electrophoresis (c.e) is a modern $S_{pt}$ technique other than any other chromatographic technique for the detection of plant secondary metabolites (Psm). There are two types of c.e such as capillary zone electrophoresis (c.z.e) and micellar electro kinetic (M.e.k.c.c). All kinds of secondary metabolites ($S_m$) (flavonoids, cardiac glycosides, aglycones, steroids, diterpene, saponin, etc.) are not possible to isolate with the help of HPLC and Gas chromatography. But such type of metabolites may possibly be isolated with the help of c.e smoothly and easily. However, in c.e charged molecules are transferred to the opposite charged molecules in the presence of the electrical field. Very low solvent, low price silica columns, and small amount of samples are needed to run the c.e. There are various characteristics such as voltage ($V_{tg}$), temperature, electrolyte concentration, $B_F$ pH, micelle concentration, and organic modifiers may influence the $S_{pt}$ of different $S_m$. In this book chapter we will describe the different parameters of c.e like methodology, detector sample analysis, and combination of other hypnated technique for the detection of metabolites.

**Keywords** Capillary zone electrophoresis · Micellar electro kinetic capillary chromatography · Phenolics · Alkaloids · Terpenoids

### Abbreviation

| | |
|---|---|
| $A_{NT}$ | Acetonitrile |
| Alkd | Alkaloid |
| $A_{ACT}$ | Ammonium acetate |
| BRT | Borate |
| $B_F$ | Buffer |
| BFS | Buffer solution |
| C.e | Capillary electrophoresis |
| C.w | Capillary wall |

D. Pal (✉) · S. Mukherjee
Department of Pharmaceutical Sciences, Guru Ghasidas Viswavidyalaya
(A Central University), Bilaspur, Chhattisgarh 495 009, India
e-mail: drdilip2003@yahoo.co.in



D. Pal and A. K. Nayak (eds.), *Bioactive Natural Products for Pharmaceutical Applications*, Advanced Structured Materials 140,
https://doi.org/10.1007/978-3-030-54027-2_8

| | |
|---|---|
| C.z.e | Capillary zone electrophoresis |
| C.n | Coumarin |
| Esp | Electrolyte solution parameter |
| ESOP | Electro-osmotic pressure |
| E.c | Eukaryotic cell |
| Fvd | Flavonoid |
| F.s.c.c | Fused silica capillary column |
| I.p | Instrumental parameters |
| M.m | Metabolism |
| $M_td$ | Method |
| M.e.k.c.c | Micellar electro kinetic |
| M.C.L | Micelle |
| NI | Negative ion |
| Ps | Plant species |
| PI | Positive Ion |
| P.c | Prokaryotic cell |
| Qn | Quinone |
| $S_{pt}$ | Separation |
| Tps | Terpenoids |
| $Vt_g$ | Voltage |

## 8.1 Introduction

Metabolism (M.m) is a magical strategy for the formation of various cellular function. It is also observed that M.m is created either the beginning of eukaryotic cell (e.c) and prokaryotic cell (p.c) (Harstad et al. 2016). If M.m does not occur, all the functions related to natural cellular works of a normal cell would be lost. Energy is required for conducting the formation of the cell. This energy is derived from food, cellular function is switched on when energy is derived from food (Han et al. 2017). There are two types of M.m such as catabolism (compound breaking) and anabolism (compound making). The metabolite products (m.ps) of e.c are protein, carbohydrate, lipid, and nucleic acid which are also known as a primary metabolite (Zhang et al. 2017). Besides, m.ps of p.c (plant, bacteria, fungus, etc.) are alkaloid, glycoside, terpenoids, saponin, etc. However, there are different $M_td$ (various spectroscopic technique and chromatographic technique) in regards knowing for what types of P are created in e.c (Junger et al. 2019). By the application of external electrical field, if a sample contained positive (PI) and negative ion (NI), then if these PIS are transferred into the negative end and NIS are positive end then this phenomenon is called as electrophoresis (Fig. 8.1). There are various $M_td$ involved in this technique (Fig. 8.2). So, Capillary electrophoresis (c.e) is a new hyphenated separation ($S_{pt}$) technique (Fig. 8.3). It was first invented by Herten in 1967, thereafter, it was recognized through Jorgensen and Lukas in 1981 (Lu et al. 2018). Based on these techniques there are created two processes such as c.z.e and Micellar electro kinetic capillary chromatography (m.e.k.c.c). At the first time these technologies used for the detection of amino acid, oligonucleotides, nucleotides, DNA, protein, etc., but at present, it is used for the detection of Psm (Voeten et al. 2018).

### 8.1.1 Principle of the Technique

This c.e method ($M_td$) is completely different from any other $S_{pt}$ technique. Suppose we have various particles containing colloidal solution. Now, if these particles are dynamic by an electrical field, then positive charges are accumulated in cathode and negative charge in anode (Aiken and Hue 1993; Singh et al. 2017). Because, during the particles dynamic movement, an electrical potential is generated into the solution. It is the main principle of c.e. However, managing this technique requires a lot of things such as (a) capillary tube, (b) siphon tube (c) UV detector. 50–100 μm diameter and 1 m length containing fused silica capillary column (F.s.c.c) is used this instrument. F.s.c.c is cover by polyamide coat (Rahman 2018). The sample is loaded in the end part of anode. Though there are a various technique for sample loading such as (a) pressure differential (b) Siphoning (c) Split flow injection (d) Electrophoretic injection. μg/ml and Nano/ gm. (Zhang et al. 2020). Concentration containing sample is used for better $S_{pt}$. At the first stage of $S_{pt}$ electrical $Vt_g$ (maximum 50 kV $Vt_g$ but 30 $Vt_g$ is needed) is applied as a result solution containing whole particles are accumulated in cathode by water electro-osmotic pressure (ESOP). ESOP is the main driving force in this technique (Issaadi et al. 2017). The water flow depends on the principle of (a) electro potential of capillary and (b) nature of capillary. In neutral or alkaline PH medium the ESOP is maximum. The silanol group of a capillary wall (C.w) is being polar because of the highest ESOP (Ribeiro et al. 2018). The surface of the C.w is surrounded by the negative charge. When these counter ions are attached with solution then a positive

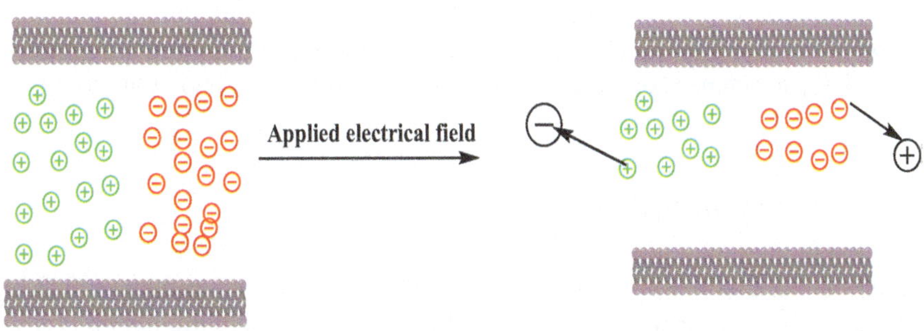


**Fig. 8.1** Gel electrophoresis

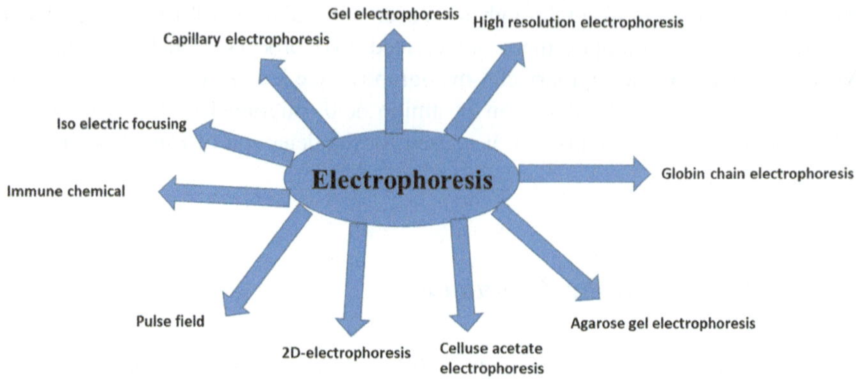


**Fig. 8.2** Various techniques for electrophoresis

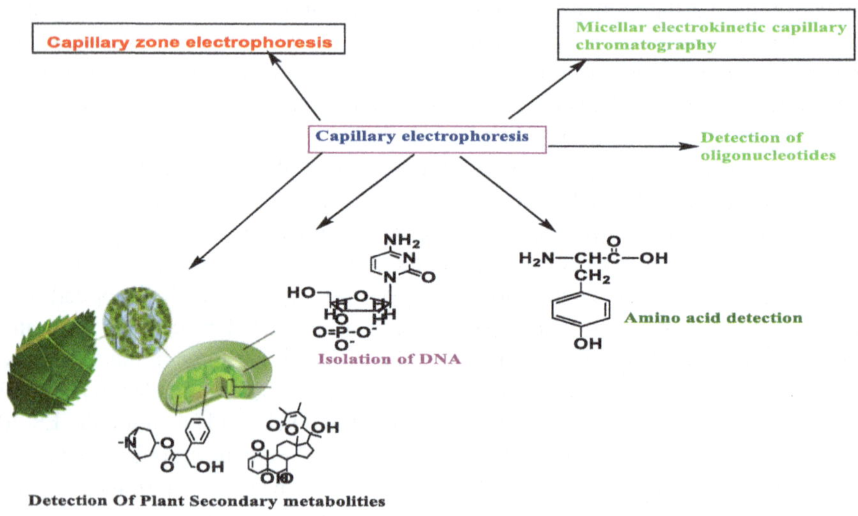


**Fig. 8.3** Brief summary of capillary electrophoresis

charge is formed into the (C.w). C.w has formed a doubled layered wave due to applied $Vt_g$ attachment of water molecule with various ion (Zhao et al. 2016).

## 8.1.2 *Nature of $S_{pt}$*

Based on the C.e principle, there are two types of other $S_{pt}$ technique such as capillary zone electrophoresis (C.z.e) and Meccc. Now, we will discuss these types of $S_{pt}$ technique (Lima et al. 2019).

(a) C.z.e: It is also known as free solution capillary electrophoresis. It depends on charge mass ration of the molecule. This $S_{pt}$ is started when PH value is above 4. If the mass of the element is small and higher negative value then migration time is also increased. Complex sugar molecule and positive, negative neutral molecules are also separated by this technique (Wu et al. 2018).

(b) Meccc: Neutral and hydrophobic molecule are separated in this analytical $M_{td}$. A unique incident occurs while it is processing. Micelle (M.C.L) is activated in the opposite direction of the electrical osmotic pressure when micelle is used (Miyagi et al. 2017). When micelle M.C.L is attached with anionic surfactant then M.C.L is accumulated in anode and *E* of is transfer into the cathode. Thereafter, cationic surfactant is attached with M.C.L then *E* of is entered into the anode and M.C.L in cathode (El et al. 2017).

## 8.2 Factors Effecting $S_{pt}$

There are various factor effect the $S_{pt}$ process. These factors are divided in two section such as (1) Instrumental parameters (I.p) and Electrolyte solution parameter (Esp.). I.p are four types (a) $Vt_g$ (b) polarity (c) Temperature (d) Capillary wall and I.ps are (a) type and concentration of buffer ($B_F$) (b) PH of the $B_F$ (c) Organic modifiers (d) Additives (Gomes et al. 2018).

### *8.2.1 I.P*

(a) Voltage: For the application of the $Vt_g$ in this instrument, Jules law must be followed. The disunion time is inversely proportional to an applied $Vt_g$. When the $Vt_g$ is increased then more amount of heat is produced and increased the c.w temperature (Dresler et al. 2017a, 2017b). As a result resolution of the $S_{pt}$ process is also reduced by the help of viscous $B_F$ of c.w.

(b) Polarity: There are two electrodes such as (a) Cathode (outlet) and (b) Anode (Inlet). ESOP is situated far from the outlet. When the charge of the molecules are highest than the ESOP then molecules are easily accumulated in outlet (Kelley et al. 2019).

(c) Temperature: This parameter is also affected by the $S_{pt}$ process. If increased the C.w temperature then changed the conformation of the molecule (Cong et al. 2017).

(d) Wall of the capillary: During capillary length and internal diameter analysis, the capacitance influences the $S_{pt}$ efficiency and load. Depending on the volume of the capillary and the injected sample, the detection system also affects the detection limit. Limits the adsorption efficiency of the sample material on the capillary wall; Therefore, $M_{td}$ s to avoid these interactions should be considered in the development of segregation $M_{td}$ s (Yang et al. 2019).

### 8.2.2 *Esp*

(a) Type and conc of buffer: $B_F$ suitable for c.e have appropriate $B_F$ capacity at pH range. Choice and low mobility to slow down the current generation (Sharma et al. 2018).
(b) PH of the $B_F$: It can affect the $S_{pt}$ by altering the charge of the analytes or additives ESOP (Huang et al. 2019).
(c) Organic modifiers: We need for operating this process to reduce the either velocity of ESOP or adjustment. For maintaining this system various organic modifier (Methanol and Acetonitrile [ $A_{NT}$]) are added the $B_F$ solution to increase the solubility (Zhao et al. 2018).
(d) Additive: To separate the optical isomers, the chiral selective $S_{pt}$ buffer is additional. Mostly Chiral selectors usually use cyclodextrin, though in some cases crown ethers, bound polysaccharides, or possibly proteins can be used (Dresler et al. 2017a, 2017b), (Rama and Vinutha 2019).

## 8.3 Instrumentation

The main components for running the C.e technique such as (a) To keep the sample in a vial (b) Assemble vial (c) Buffer solution (BFS) (d) Capillary tube (e) $Vt_g$ (Electricity or current) (f) Detector (g) Computer system (Sharma et al. 2017). Firstly, the BFS is made by processing in which type of compound is detected based on chemical nature and PH range. Thereafter, the BFS solution is poured into the vial, assembling vial and capillary (Jones et al. 2018). Then the tip of the capillary is made entered into the vial. The sample goes towards the capillary side by the help of capillary force and siphoning force and returned into the reverse way. It is pertinent to mention here, that main cause for going towards the capillary side an electrical field is generated between the vial and assembling vial (Mendoza and Silva 2018). As a result, positive and negative ions of the samples are accumulated into the middle part of the capillary by an electrical force. At the end samples being isolated into the end part of capillary and assemble into a vial (Memon et al. 2017). Thereafter, detectors identify the chemicals which are existing with the sample through the computer system. The capillary wall is covered with polyamide or Teflon so that smoothly detect the sample (Dresler et al. 2018). The sample is detected by the help of Beers and lamberts law. There are various detector used namely (a) UV (b) UV-Visible (c) 0.01attomole containing lase induced fluorescence detector (d) potentiometric detector (d) Conduct metric detector. But UV or UV-Visible detector is mostly used (Ngoc et al. 2019).

## 8.4 Difference Between C.e and Others $S_{pt}$ Technique

Some differences are observed between C.e techniques with any other $S_{pt}$ $M_{td}$. Especially seen in High-performance Liquid Chromatography (HPLC). To operate the C.e some hindrances are facing such as (a) Uncleanliness of C.e.w (b) Diameter of C.e (c) Reaction between silica-containing C.e.w and basic $B_F$, (Spisso et al. 2018).

To develop the $M_{td}$ of C.e scientists are frequently trying with best efforts. To clean the C.e.w water, sodium hydroxide and any other solvents are used. But the cleaning position does not exist during more times if such solvents are used (Xiao et al. 2018). Let us we discussed about the C.e techniques with others $S_{pt}$ technique. To keep power a compound can be separated by using C.e by using 1–30 min time, high amount of $Vt_g$ and using minimal sample. C.e can separate 87,000–120,000 sample plate within minimum time. On the other hand, HPLC can separate the 2600–18,000 plate (Nimse and Pal 2015). Very less amount sample and minimum samples are separated in HPLC technique. To process the HPLC a very good knowledgeable person is required, but besides in the case of C.e a good person is not required and even glass room is not needed (Pal et al. 2006).

## 8.5 Application

Alkaloid (Alkd) is a nitrogenous basic element. It is also known as PSM. Seens Alkd is a positive charge in nature and less value of PH, It is easily separate by C.e (Pal et al. 2006). Various alkaloidal compounds are separated by C.e (Pal et al. 2005). For example quaternary alkaloidal compound (Fig. 8.4), (Table 8.1).

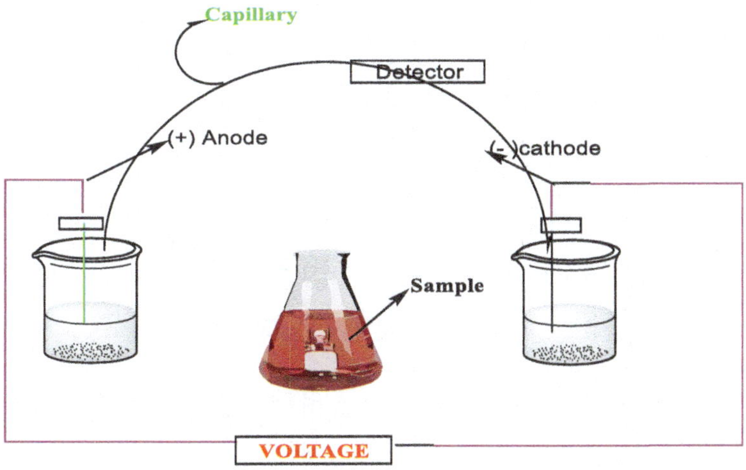


**Fig. 8.4** Schematic representation of capillary electrophoresis

**Table 8.1** Application of C.e in Alkd

| Name of The alkaloid | Structure | Solvent | Diameter of column | $Vt_g$ | Detector | References |
|---|---|---|---|---|---|---|
| Purine | 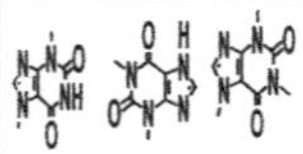 | 20 mM borate, pH 9.2 | 50 cm x 50 um | 10 kV | 254 nm | (Mukherjee et al. 2018) |
| Ephedra | 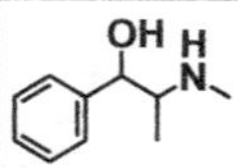 | 20 mM isoleucine, 5 mM barium hydroxide (pH 10.0 with ammonia) methanol, 10 mM phosphate (1:4); 10 mM borate (pH 9.0), 100 mM SDS | 60 cm × 75 pm (52.5 cm 27 °C 60cm × 75pm (52.5 25 ° C 48 cm × 50 pm (26 cm 86) 30 °C | 20–28 | 185 nm | (Chahel et al. 2019) |
| Imidazole | 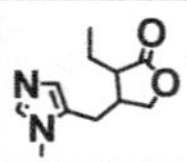 | 100 mM phosphate PH 6.9, 10 mM B-cyclodextrin | 50 cm × 50 pm (40 um to detector) | 8 | 217 nm | (Badria and Aboelmaaty 2019) |
| Oxindole | 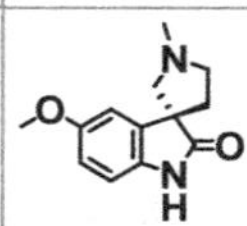 | 20 mM phosphate (PH 5.6) | 57 cm × 75 pm (50 cm to detector) 25 °C | 10 | 254 | (Kolrep et al. 2019) |

Methodology:

## 8.5.1 *Quaternary Alkaloidal Compounds*

At first, 100 mM sodium acetate is made into a 100 ml volumetric flask and make up to 100 ml. Then withdraw 85 ml solution from this 100 ml solution as a result BSF (For better resolution) is made (Pal and Verma 2013). Then 15 ml methanol is added into the $B_F$ for sharp peak. In this $M_{td}$ 100 cm * 100 µm column is used at 28 °C. 25 kV $Vt_g$ is applied in this technique (Sannigrahi et al. 2009). The $S_{pt}$ power of this Alkd is completely depended on which substituted are attached with aromatic ring, So, the serial of separating power is Methylene dioxy < two methoxy group > four methoxy group < one aliphatic hydroxy group < two aromatic hydroxy group (Gupta et al. 2003).

### 8.5.1.1 Opium

Particularly this compound (Fig. 8.5) is separated by the help of M.e.k.c.c. But c.z.e technique is used mostly (Pal and Mitra 2010). 25 mM citrate $B_F$ and 50 cm * 75um fused silica column are used for C.z.e at PH 4.2 to detected this type of Alkd in UV range 214 nm (Mir et al. 2019). However 20 kV $Vt_g$ is also used. 10 mM borate and 50 mM SDS are used in PH 9.2 in M.e.k.c.c. Here 50 cm * 50um diameter Coolum and 20 kV $Vt_g$ are needed. The UV range of this compound is 214 nm (Pal et al. 2019) (Figs. 8.6, 8.7, 8.8 and 8.9).

## 8.5.2 *Flavonoid*

Three carbon phenolic hydroxyl aromatic ring containing Psm is called flavonoid (Fvd). There are 4000 Fvd is discovered till date. Fvd is available in different kinds of fruit, seed, and vegetables. This Fvd is a most important part of our daily life for the prevention of various diseases (Jorge et al. 2019; Pal et al. 2006; Pal and Saha 2019). To know what type or how quantity is contained in a natural product there are various technique are available (GC, GC–MS, TLC, HPLC) but these are more time limited (Soldatou et al. 2019). As a result there is a new introduced technique for the $S_{pt}$ of Fvd which is known as C.z.e (Alagoz et al. 2020). A specific time of 19 min is required for the detection of Fvd with the help of 229 C.z.e. In this field various column (25, 50, 60, 75, 100, and 200um * 68 cm) and 230 various BFS (10 mM NaH2Po4, 50 mM Na2B4o7) are used for $S_{pt}$ of Fvd in PH range of 9.0–9.6. (Poliwoda et al. 2020). In case of polyphenolic flavonoidal compound 20 mM borate (BRT) $B_F$ 2 mM Beta-cyclodextrin (additive) and 7% methanol are used (Brüggemann et al. 2019). 25 mM phosphate $B_F$ is used for the detection of

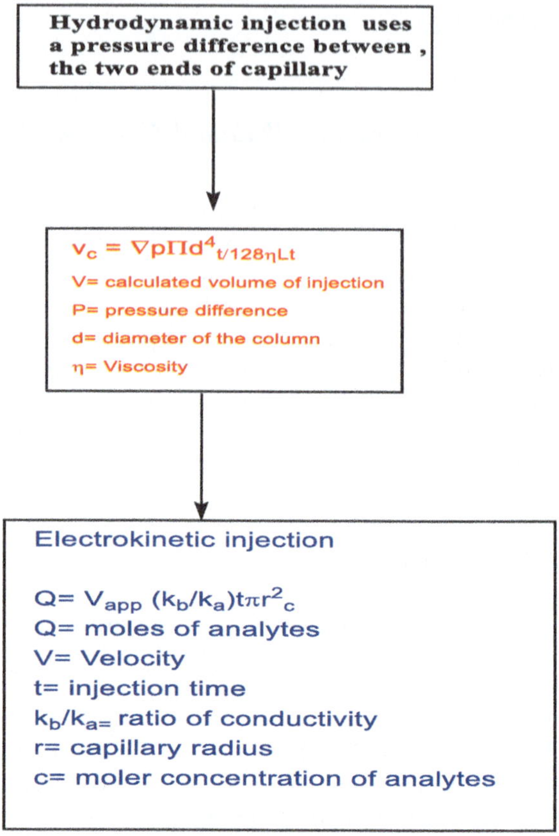


**Fig. 8.5** Equation of injection

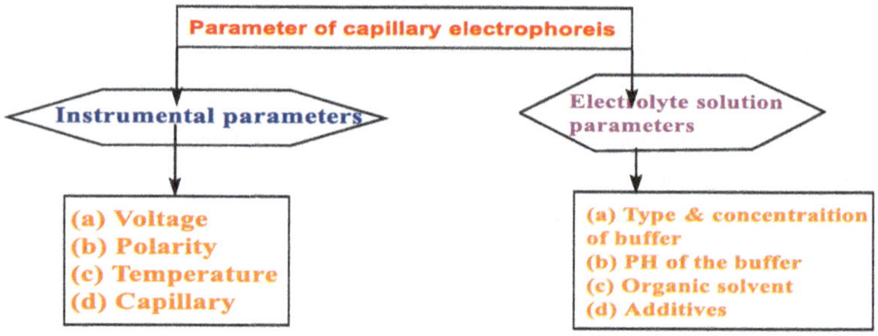


**Fig. 8.6** Parameters of c.e

Epicatechinand Catechin in PH 7.0. 0.20 mm and 300 μm OD containing carbon electrode are used for the detection of which types of Fvd are present in caffeine molecule (Wu et al. 2019). There after here 5 mM ammonium acetate ($A_{ACT}$), $B_F$ and 80 ml water and 20 ml $A_{NT}$ are used in PH 4 and temperature 25 °C (Nawabi et al. 2019).

**Migration velocity v=μep *E, E= μep* v/l**

**Where v= Migration velocity of charged particle in the potential field.**
**μep= Electrophoretic mobility**
**E= Field strength**
**V= Applied voltage**
**L= Length of capillary**

**Electrophoretic mobility** $\mu_{ep}= q/6\Pi\eta n$
Where q= charge on ion
η= viscosity
n= ion radius

**Combining the two effects for migration velocity of an ion (also applies to neutrals) with$\mu_{ep}$=0**
**$v= (\mu ep + \mu e_0).E=(\mu ep + \mu e_0).V/L$**

**Fig. 8.7** Migration current

**Fig. 8.8** Structure of morphine

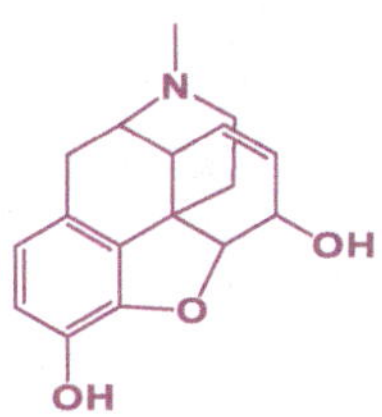


## 8.5.3 *Terpenoids*

C.e is generally used in the field of almost Psm. But in case of terpenoids (Tps), it is not used prevailing system. Yet C.e is used some of the Tps such as mono Tps, di Tps, tri Tps (Shihabi and Oles 1994).

### 8.5.3.1 Mono Tps

Which types of monoterpenes are detected by C.e technique, which has displaced in Fig. 8.10.

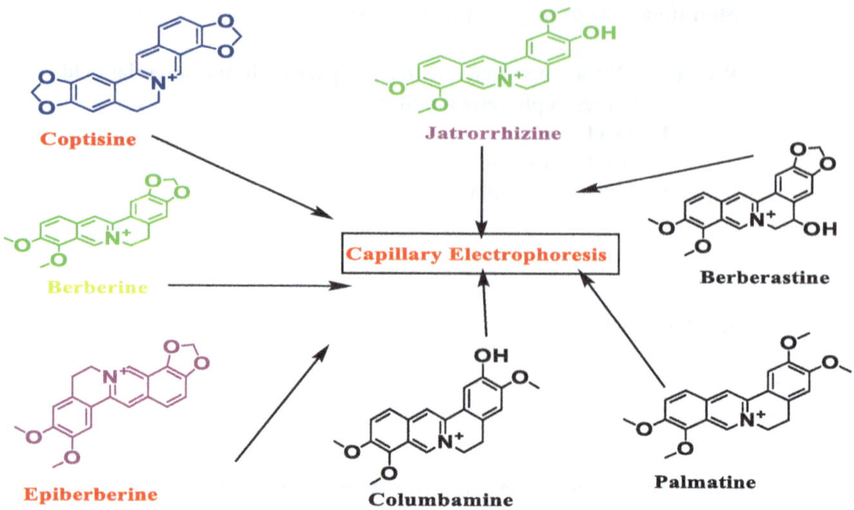


**Fig. 8.9** C.e of quaternary alkaloid compounds

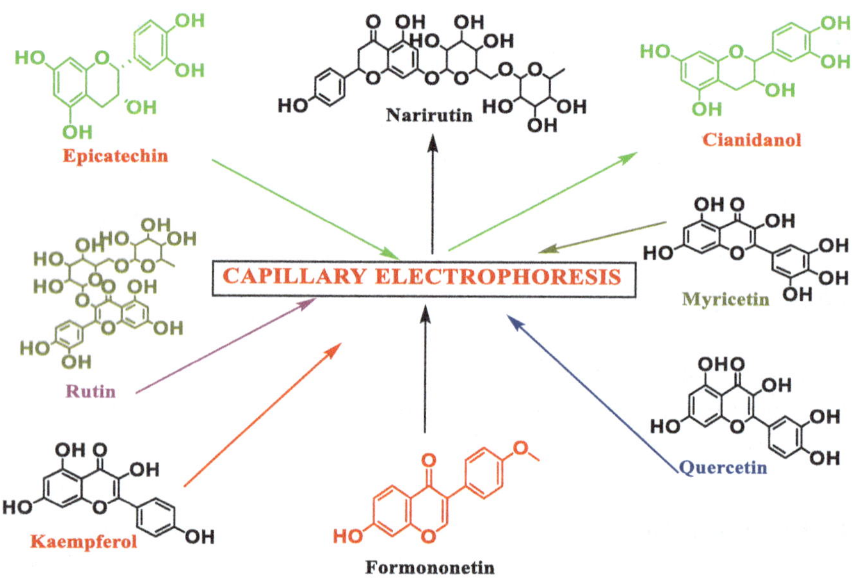


**Fig. 8.10** C.e for flavonoids

#### 8.5.3.2 Di-Tps

It is reminded from the statement of Gibberellin (plant hormone) (Fig. 8.11), that when it was separated by C.e technique where using boro sulfate $B_F$, para/ gamma cyclodextrin at PH 7.5. This isolation technique is depended on such factors as: Hydrogen bond, Vander wall force, and solution power (Weegels et al. 1995), (Arndt et al. 2008).

#### 8.5.3.3 Tri-Tps

Cardiac glycoside (Digitalis) (Fig. 8.12) is a triterpenoid which was formerly invented by the help of C.e techniques, where was used BRT, SDS, cyclodextrin at PH 9.3. But some hindrances have occurred when cyclodextrin used as an additive for isolating this compound (Helander et al. 2005). In this reason, scientists have used a mixture of additives such as cyclodextrin and akylodextrin for the $S_{pt}$ of Tps. 25% ACN + 75 Mm Chlorate and BRT are used for the $S_{pt}$ of saponin at PH 7. Then, SDS + 5% methanol and phosphate $B_F$ are used for isolation of steroidal triterpene at PH 9.4. This isolation is occurred in 7.5 min (Lee and Ong 2000) (Table 8.2).

### 8.5.4 *Coumarin Derivatives*

Coumarin (C.n) is a very important Psm. The existence of various structural containing C.n is observed from the plant to fruit. Cn plays a vital role in preventing various human diseases (Nayak and Pal 2017). Structurally there are 3 types of Cn, which has been shown in Fig. 8.13. In the past situation, there are various chromatographic techniques (HPLC, UPLC, RP–HPLC, TLC, etc.) used for the determination of what types of Cn exist in plant sample (Mandal et al. 2015), (Fiehn et al. 2008). But in the present situation, C.e technique stands a special field of a scientist for Cn $S_{pt}$. Now we will discuss about the how and what types of Cn are separated by C.e in this section Fig. 8.14.

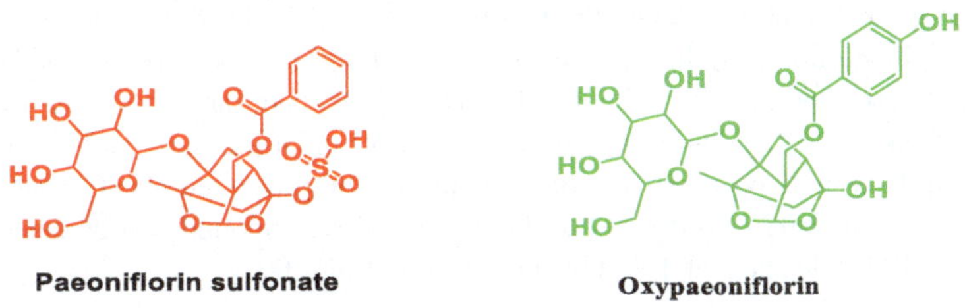


**Fig. 8.11** Monoterpene by capillary electrophoresis

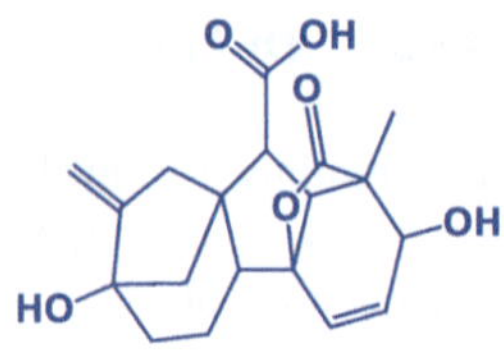


**Fig. 8.12** Structure of Gibbgerellin

**Table 8.2** Application of C.e in terpenoids

| Name | $M_td$ | Solvent | PH | Diameter Of column | $Vt_g$ (kv) | Wave length | References |
|---|---|---|---|---|---|---|---|
| Terpenoids (Monoterpene, Diterpene, Triterpene) | C. z.e | 100 mM borate | 10.5 | 80 cm × 50 um | 20 | 254 | (Troujman and Chottard 1997) |
| Gibberellins | C. z.e | 50 mM phosphate, 100 mM borate, 75 mM Cyclodextrin | 7.54 | 60 cm × 50 um | 15 | 196–210 | (Pal et al. 2019) |
| Digitalis | M. e.k. c.c | 30 mM borate, 7 mM urea or sodium cholate | 9.3 | 80 cm × 50 um | 20 | 225 | (Nayak et al. 2010) |
| Phyto-ecdystereoid | M. e.k. c.c | 40 mM phosphate, 20 mM borate, 20 mM SDS, 5% methanol | 9.4 | 72 cm × 50 um | 20 | 240 | (Pal and Nayak 2012) |
| Saponins | M. e.k. c.c | 20 mM phosphate, 25% CAN, 75 mM cholate | 7.0 | 102 cm × 50 um | 30 | 200 | (Pal2015) |

$M_{td}$ of $S_{pt}$: Firstly, plant species (ps) will be collected from any region of the forest. Then the collected ps will be authenticated by the help of botanist and kept in a hot air oven and shade dryer for removing the moisture (Xiang et al. 2011). Then the dry plant material will be cut into small pieces and powdered by the help of a mixer grinder or ball mill. Then the powder material will be weight and extraction process will be done by the various solvent system (Tolstikov et al. 2003). Now extract materials will be diluted with Me OH or EtOH and transferred into a various conical flask. Then the 24–30 kV electronic $Vt_g$ 50 mbar pressure, and 50um fused silica capillary will be used in extract material for processed the C.e technique (Sugimoto et al. 2012). At the end of the process, the compound will be detected by the help of UV detector at 190–600 nm (Ward et al. 2007).

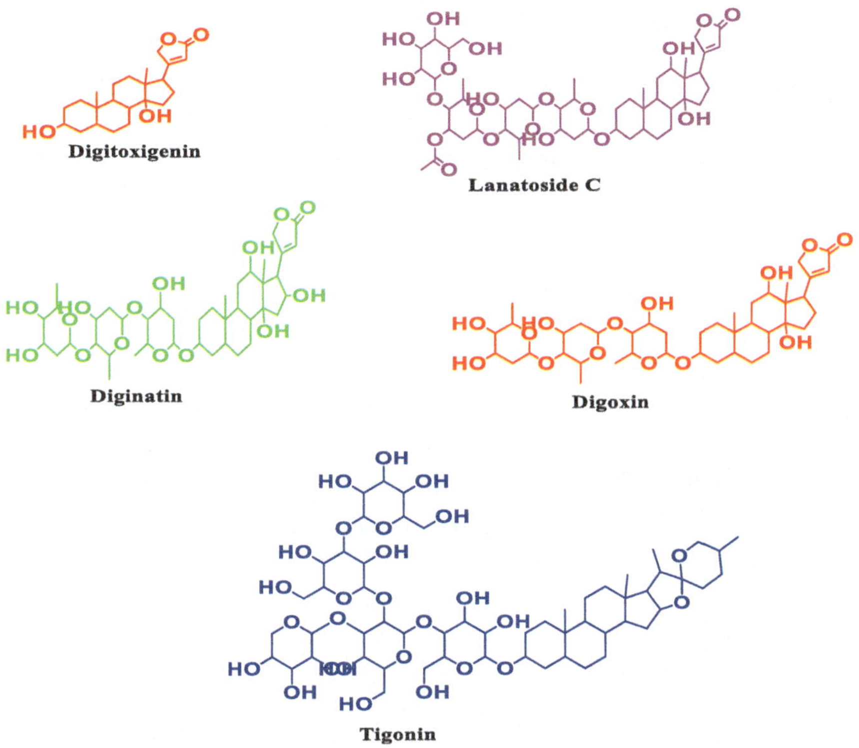


**Fig. 8.13** Structure of digitalis derivatives

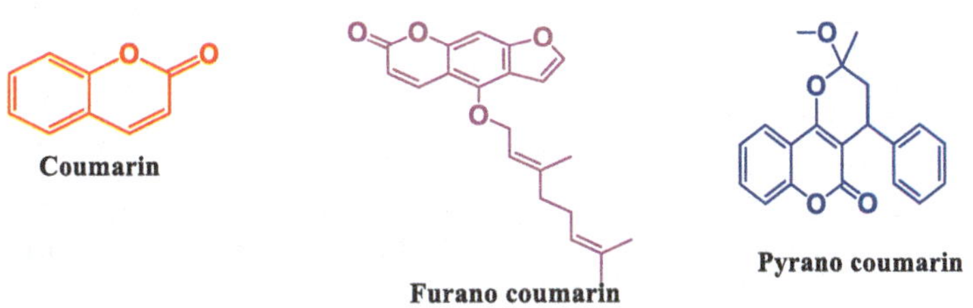


**Fig. 8.14** C.n derivatives

## *8.5.5 Quinones*

Quinone (qn) molecules are especially displayed in the whole plant world. In 1992, the first qn was separated by the help of C.e. Hqn and Aqn were separated by C.e and M.e.k.c.c Fig. 8.15. 1OMm BRT, 75 Mm SDS and 10% methanol, 20 kV $Vt_g$ was used for the detection of Aqn methyl ester at PH 9.5. 50 Mm BRT, 15 kV $Vt_g$ was used for the $S_{pt}$ of Hqn mono sulphate at PH 10 (Lie et al. 2008) (Fig. 8.16).

### 8.5.6 Polyamine

It is a stress resulting metabolite product, which has been shown in Fig. 8.17. These type of compounds are easily separated by C.e technique. 8 mM quinine sulphate, 20% ethanol, quinine sulphate as an electrolyte, and HMC are used for this technique (Kim and Verpoorte 2010).

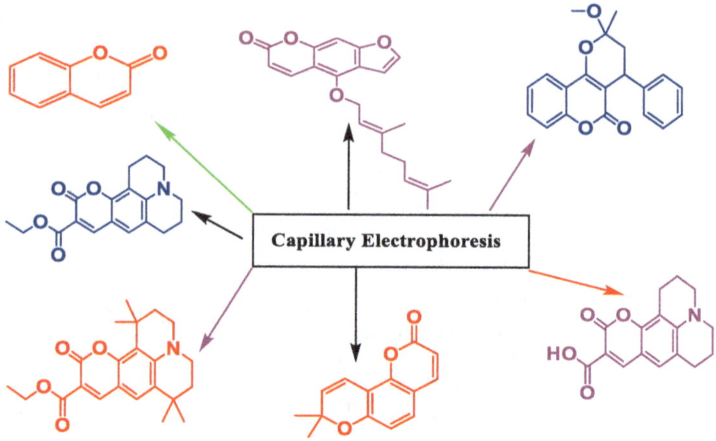


**Fig. 8.15** $S_{pt}$ of C.n.s by C.e

**Fig. 8.16** Structure of quinone derivatives

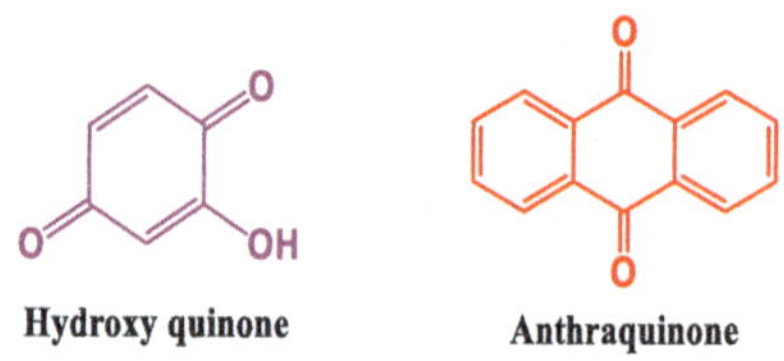


**Fig. 8.17** Structure of various polyamines

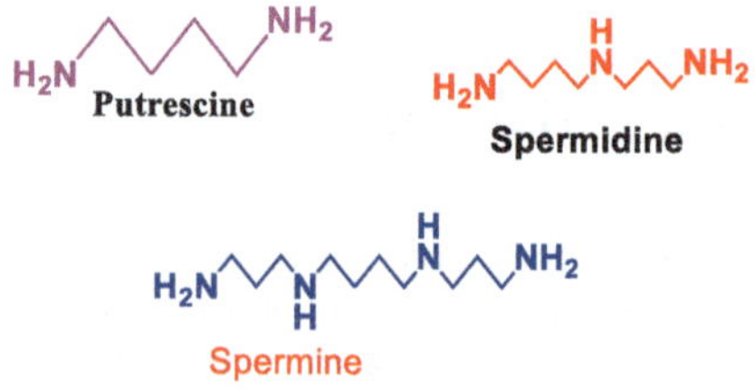

## 8.6 Capillary Electrophoresis Non Aquas ($C.e_N$) for Bioactive Compound

The meaning of C.eN, where does not use any water or moisture containing system. This method is first developed in 1984. This $M_{td}$ is continuously used either beginning of Psm or differentiation of small molecule (Sun et al. 2012). There various $B_F$ systems are used as a result increased the electrical density and power. Even very small pole containing substances are easily soluble in the $B_F$ system. In past, heteroconjugated compounds were not separated in this method but recently, it has been recognized for C.eN (Pomponio et al. 2003). In case of Alkd by the help of C.eN here 75 ml methanol, 25 ml $A_{NT}$, and 50 Mm $A_{ACT}$ are used for preparing the $B_F$. For Di–Tps 250 Mm $A_{ACT}$ is used. For Aqn and C.n there various $B_F$ are used such as 50–60 Mm $S_{CH}$, and 20 Mm $A_{ACT}$ (Huang et al. 2005).

## 8.7 Online C.e Concentration

The detector is frequently used for $S_{pt}$ of the compound in C.e because of the shortest length of C.w, the smallest rate of injection and used picogram amount exam substance (Vanhoenacker et al. 2000). For removing this above problem research invented $M_{td}$ has been progressed, which is known as C.c $_{Ms}$ online CNC $M_{td}$, where the detector is not used frequently (Sun and Yeh 2005). It is also called as sweeping. Here Nano gram amount sample and very complex molecules are detected very easily in this $M_{td}$. Firstly, this $M_{td}$ was applied for $S_{pt}$ Fvd and Riboflavin. This examination described that C.e $M_{td}$ is better than general C.e. Here, 50 Mm borate $B_F$ and 50 Mm phosphate is used in PH 9.5 (Zang et al. 2004).

## 8.8 C.e Versus C.e $_{MS}$

In 1987, a joint $M_{td}$ is introduced for high-velocity compound detecting power, determination of molecular weight in minimal time, and used a small amount of sample. This $M_{td}$ is known as C.e $_{MS}$. As a result, the liquid smoothly flows through C.w. In the past situation, this $M_{td}$ was introduced for qualitative analysis of Alkd (Gao et al. 2004). Firstly, filodren Alkd is produced with the help of $C.e_{MS}$. However, this $M_{td}$ is not limited to Alkd. Now, we will discuss about the difference between C.e and C.e $_{MS}$. In the case of C.e ions are transferred into one end to another end with the help of various factors such as high electrical field, arresting liquid flow, and shape of ions. In the other hand in the case of C.e $_{MS}$, compounds are detected depending on ratio between charge and mass (Bo et al. 2003). Firstly, ions are emitted from C.e then they are transferred into the mass instrument where these ions are formed into small pieces with the effect of electrical and magnetic field. However, the main reason of C.e $M_{td}$ is here compound is not easily detected. In this above point of view, $_{MS}$ is attached with C.e via various technique TOF-MALDI, FAB etc. (Scherz et al. 2007). To know what types of phenolic compounds ($p_c$) have existed in red wine ($r_w$), firstly it was extracted with diethyl ether and analyzed with this $M_{td}$ through ESI ion mode. In spite of the existence of high power of isolation in C.e than HPLC, but C.$e_{MS}$ was stronger because of the best selection and taking better response.13 $p_c$ were analyzed with this Mtd out of 23. With removing the borate $_{ACT}$ here, $A_{ACT}B_F$ was used in this $M_{td}$ in PH 9.3. As a result, it was frequently affected by the hydroxyl group of *rw* is attached with expired liquid of methanol and $B_F$. Adverse position of fingerprint analysis of the natural substance, this instrument shows less response comparatively so, the ion trapping $M_{td}$ is used for Psm determination (Lai and Dabek-Zlotorzynska 1999). Extension of the plant is a very controlled $M_{td}$. Various examination proves that this tragedy is controlled by various substances. These substances are called Phytohormones. It is synthesized in first leaf, bud, flower, and phloem tissue. Auxin ($A_N$), gibberellin, cytokinins are plant hormone ($P_{HR}$). However $A_N$ is an essential $P_{HR.}$ $A_N$ contained indole. So, we will discussed about how this substance is separated by C.e $M_{td}$ (Li et al. 2000), (Ganzera et al. 2003), (Lurie et al. 1998), (Desiderio et al. 2005), (Starkey et al. 2002).

## 8.9 Conclusion

In view of all aspect, we have decided that the C.e technique will open a new door of scientists for Psm detection, especially on the complex molecular structures which are not detected by HPLC. Thereafter, there are various disadvantages in this technique. If we get rid of these disadvantages, as a result, this C.e $M_t$d will open a new research area for the analytical chemistry industry. However, it may be used in the field of drug development from a plant source. In the other hand, various applications have existed in this C.e $M_{td}$ except for bioactive compound. Such as, what types of ions are present in our saliva, to know the characters, and movement of DNA pieces which are derived from polymerase chain reaction as this method act as a gatekeeper for the forensic scientist. DNA is not separated without an electrical field so, C.e method is essential. So that activated liquid is easily passed throughout the C.w so, C.e is needed. One profile is created from a high polymeric genetic marker with the help of C.e. Nowadays finding ink relation is very important for forensic lab because normal ink cannot find out the accurate relation, hence C.e method is essential. In other ways, DNA sequencing is made possible in 540 s with the help of C.e.

## References


Aiken JH, Huie CW (1993) Use of a microemulsion system to incorporate a lipophilic chiral selector in electro kinetic capillary chromatography. Chromatographia 35(8):448–450

Alagoz Y, Dhami N, Mitchell C, Cazzonelli CI (2020) cis/trans carotenoid extraction, purification, detection, quantification, and profiling in plant tissues. Springer, Humana, pp 145–163

Arndt T, van der Meijden BB, Wielders JP (2008) A typical serum transferrin isoform distribution in liver cirrhosis studied by HPLC, capillary electrophoresis and transferrin genotyping. Clin Chim Acta 394(1):42–46

Badria FA, Aboelmaaty WS (2019) Plant histochemistry: A versatile and indispensable tool in localization of gene expression, enzymes, cytokines, secondary metabolites and detection of plants infection and pollution. Acta Sci Pharmaceut Sci 3:88–100

Bo T, Yang X, Li KA, Xu L, Liu H (2003) Comparison of micellar electro kinetic chromatography and microemulsion electro kinetic chromatography for separation of pharmacologically active xanthones from Securidaca inappendiculata. J Separat Sci 26(2):133–136

Brüggemann M, Van Pinxteren D, Wang Y, Yu JZ, Herrmann H (2019) Quantification of known and unknown terpenoid organosulfates in PM10 using untargeted LC–HRMS/MS: contrasting summertime rural Germany and the North China Plain. Environ Chem Lett 16(5):333–346

Chahel AA, Zeng S, Yousaf Z, Liao Y, Yang Z, Wei X, Ying W (2019) Plant-specific transcription factor LrTCP4 enhances secondary metabolite biosynthesis in *Lycium ruthenicum* hairy roots. Plant Cell 136(2):323–337

Cong-Cong XU, Bing WA, Yi-Qiong PU, Jian-Sheng TA, Zhang T (2017) Advances in extraction and analysis of phenolic compounds from plant materials. Chin J Nat medicines 15(10):721–731

Desiderio C, De Rossi A, Sinibaldi M (2005) Analysis of flavanone-7-O-glycosides in citrus juices by short-end capillary electro chromatography. J Chromatogr A 1081(1):99–104

Dresler S, Kováčik J, Strzemski M, Sowa I, Wójciak-Kosior M (2018) Methodological aspects of biologically active compounds quantification in the genus Hypercom. J Pharm Biomed 155:82–90

Dresler S, Rutkowska E, Bednarek W, Stanislavski G, Kubrak T, Bogucka-Kocka A, Wójcik M (2017a) Selected secondary metabolites in *Echium vulgare L.* populations from nonmetalliferous and metalliferous areas. Phyto chem 133:4–14

Dresler S, Szymczak G, Wójcik M (2017b) Comparison of some secondary metabolite content in the seventeen species of the Boraginaceous family. Pharm Biol 55(1):691–695

El Fellah S, Duporté G, Sirén H (2017) Steroid hormones, inorganic ions and botrydial in drinking water. Determination with capillary electrophoresis and liquid chromatography-orbitrap high resolution mass spectrometry. Microchem J 133:126–136

Fiehn O, Wohlgemuth G, Scholz M, Kind T, Lee DY, Lu Y, Moon S, Nikolau B (2008) Quality control for plant metabolomics: reporting MSI-compliant studies. Plant J 53(4):691–704

Ganzera M, Stöggl WM, Bonn GK, Khan IA, Stuppner H (2003) Capillary electro chromatography of boswellic acids in Boswellia serrata Roxb. J Separat Sci 26(15):1383–1388

Gao W, Yaowen C, Yin Y, Chen X, Hu Z (2004) Separation and determination of two sesquiterpene lactones in Radix inulae and Liuwei Anxian San by microemulsion electro kinetic chromatography. Biomed Chromatogra 18(10):826–832

Gomes AF, Ganzera M, Schwaiger S, Stuppner H, Halabalaki M, Almeida MP, Leite MF, Amaral JG, David JM (2018) Simultaneous determination of iridoids, phenylpropanoids and flavonoids in Lippia alba extracts by micellar electro kinetic capillary chromatography. Microchem J 138:494–500

Gupta MA, Mazumdar UK, Pal D, Bhattacharya S, Chakrabarty SU (2003) Studies on brain biogenic amines in methanolic extract of *Cuscuta reflexa Roxb* and *Corchorus olitorius Linn* seed treated mice. Acta Pol Pharm 60(3):207–210

Han M, Pearson JT, Wang Y, Winters D, Soto M, Rock DA, Rock BM (2017) Immunoaffinity capture coupled with capillary electrophoresis-mass spectrometry to study therapeutic protein stability in vivo. Anal Biochem 539:118–126

Harstad RK, Johnson AC, Weisenberger MM, Bowser MT (2016) Capillary electrophoresis. Anal Chem 88(1):299–319

Helander A, Wielders JP, Stroet R, Bergstrom JP (2005) Comparison of HPLC and capillary electrophoresis for confirmatory testing of the alcohol misuse marker carbohydrate-deficient transferrin. Clin Chem 51(8):1528–1531

Huang HY, Chiu CW, Chen YC, Yeh JM (2005) Comparison of microemulsion electro kinetic chromatography and micellar electro kinetic chromatography as methods for the analysis of ten benzophenones. Electrophoresis 26(5):895–902

Huang Y, Jansen O, Frédérich M, Mouithys-Mickalad A, Nys G, Servais AC, Crommen J, Jiang Z, Fillet M (2019) Capillary electrophoresis, high-performance liquid chromatography, and thin-layer chromatography analyses of phenolic compounds from rapeseed plants and evaluation of their antioxidant activity. J Sep Sci 42(2):609–618

Issaadi HM, Hunyadi A, Németh K (2017) Capillary electrophoresis study on the base-catalyzed formation of bioactive oxidized metabolites of 20-hydroxyecdysone. J Pharm Biomed 146:188–194

Jones C, Hatier JH, Cao M, Fraser K, Rasmussen S (2018) Metabolomics of plant phosphorus-starvation response. Ann Plant Rev Online 15:217–236

Jorge TF, Tohge T, Wendenburg R, Ramalho JC, Lidon FC, Ribeiro-Barros AI, Fernie AR, António C (2019) Salt-stress secondary metabolite signatures involved in the ability of Casuarina glauca to mitigate oxidative stress. Environ Exp Bot 166:103808–103816

Junger AS, de Jesus FF, Fracassi da Silva JA, Daniel D, de Jesus DP (2019) A simple and fast method for determination of benzocaine and lidocaine in pharmaceutical formulations by capillary electrophoresis with spectrophotometric detection. SSC Plus 2(11):422–427

Kelley ZD, Rogers DT, Littleton JM, Lynn BC (2019) Microfluidic capillary zone electrophoresis mass spectrometry analysis of alkaloids in Lobelia cardinalis transgenic and mutant plant cell cultures. Electrophoresis 40(22):2921–2928

Kim HK, Verpoorte R (2010) Sample preparation for plant metabolomics phytochemical analysis. Int J Plant Chem Biochem Tech 21(1):4–13

Kolrep F, Schmidt R, Mähnert E, Gehling M, Preiss-Weigert A (2019) Development of an HPLC-MS/MS multi-method for the detection of sesquiterpene lactones and polyphenols occurring in Arnica plant materials. Planta Med 85(18):8–12

Lai EP, Dabek-Zlotorzynska E (1999) Separation of theophylline, caffeine and related drugs by normal-phase capillary electro chromatography. Electrophoresis 20(12):2366–2372

Lee BL, Ong CN (2000) Comparative analysis of tea catechins and theaflavins by high-performance liquid chromatography and capillary electrophoresis. J Chromatogr A 881 (1):439–447

Li Y, Liu H, Ji X, Li J (2000) Optimized separation of pharmacologically active anthraquinone in Rhubarb by capillary electro chromatography. Electrophoresis 21(15):3109–3115

Li Y, Shrestha B, Vertes A (2008) Atmospheric pressure infrared MALDI imaging mass spectrometry for plant metabolomics. Anal Chem 80(2):407–420

Lima NM, Silveira RS, Ramos RR, Santos VN, Nascimento MP, Andrade APT, Carli MO, Almeida MV (2019) Bioactivity and dereplication of phenolic compounds in medicinal plants. Int J Pharm Sci Invent 11(3):105–110

Lu G, Crihfield CL, Gattu S, Veltri LM, Holland LA (2018) Capillary electrophoresis separations of glycan's. Chem Rev 118(17):7867–7885

Lurie IS, Meyers RP, Conver TS (1998) Capillary electro chromatography of cannabinoids. Anal Chem 15:3255–32260

Mandal SC, Mandal V, Das AK (2015) Essentials of botanical extraction: principles and applications. Academic press, pp 23–55

Memon AF, Solangi AR, Memon SQ, Mallah A, Memon N, Memon AA (2017) Simultaneous determination of quercetin, rutin, naringin, and naringenin in different fruits by capillary zone electrophoresis. Food Anal Methods 10(1):83–91
Mendoza N, Silva EM (2018) Introduction to phytochemicals: secondary metabolites from plants with active principles for pharmacological importance. Molecules 5:25–30
Mir SA, Mukherjee S, Makar S, Pal G (2019) Cucurbitacins a vibrant triterpenoid: a review on its anticancer property. PharmaTutor 7(2):43–54
Miyagi A, Uchimiya H, Kawai-Yamada M (2017) Synergistic effects of light quality, carbon dioxide and nutrients on metabolite compositions of head lettuce under artificial growth conditions mimicking a plant factory. Food Chem 218:561–568
Mukherjee S, Roy D, Das S (2018) Apicoplast: a brilliant focus for antimalarial drug development. PharmaTutor 6(5):13–22
Nawabi MY, Uddin W, Ullah S, Fang Z, Sardar S (2019) Identification of two hetero-substituent-aromatic isomers of -4-and 7-hydroxy coumarin with the belousov-zhabotinsky oscillator. Int J Electrochem Sci 14:8676–8685
Nayak AK, Pal D (2017) Tamarind seed polysaccharide: an emerging excipient for pharmaceutical use. Indian J Pharm Educ Res 51:136–146
Nayak AK, Pal D, Pany DR, Mohanty B (2010) Evaluation of Spinacia oleracea L. leaves mucilage as an innovative suspending agent. JAPTR 1(3):338–343
Ngoc HN, Löffler S, Nghiem DT, Pham TL, Stuppner H, Ganzera M (2019) Phytochemical study of Rourea minor stems and the analysis of therein contained Bergenin and Catechin derivatives by capillary electrophoresis, Micro Chem J 149:1040–1063
Nimse SB, Pal D (2015) Free radicals, natural antioxidants, and their reaction mechanisms. Rsc Adv 5(35):27986–8006
Pal D (2015) A review on Cyperus rotundus as a tremendous source of pharmacologically active herbal medicine. IJGP 9(4):13–18
Pal D, Ahuja C, Mukherjee S (2019) Celsia coromandeliane Vahl-A new biomarker with immense pharmaceutical applications. J Drug deliv Ther 9(3):1109–1115
Pal D, Mitra S (2010) A preliminary study on the in vitro antioxidant activity of the stems of *Opuntia vulgaris*. JAPTR 2:268–270
Pal D, Nayak AK (2012) Novel tamarind seed polysaccharide-alginate mucoadhesives microspheres for oral gliclazide delivery: in vitro–in vivo evaluation. Drug Deliv 19(3):123–131
Pal D, Saha S (2019) Chondroitin: a natural biomarker with immense biomedical applications. RSC Adv 9(48):28061–28077
Pal D, Sahoo M, Mishra AK (2005) Analgesic and anticonvulsant effects of saponin isolated from the stems of *Opuntia vulgaris* Mill in mice. Eur Bull Drug Res 13:91–97
Pal D, Verma PR (2013) Flavonoids: a powerful and abundant source of antioxidants. Int J Pharm 5(3):95–98
Pal DK, Dutta S (2006) Evaluation of the antioxidant activity of the roots and Rhizomes of *Cyperus rotundus* L. Indian J. Pharm. Sci 68(2):212–220
Pal DK, Mandal M, Senthilkumar GP, Padhiari A (2006) Antibacterial activity of *Cuscuta reflexa* stem and *Corchorus olitorius* seed. Fitoterapia 77(7):589–591
Poliwoda A, Zielińska K, Wieczorek PP (2020) Direct analysis of psilocin and muscimol in urine samples using single drop micro extraction technique in-line with capillary electrophoresis. Molecules 25(7):1566–1570
Pomponio R, Gotti R, Luppi B, Cavrini V (2003) Microemulsion electro kinetic chromatography for the analysis of green tea catechins: effect of the cosurfactant on the separation selectivity. Electrophoresis 24(10):1658–1667
Rahman M (2018) Application of computational methods in isolation of plant secondary metabolites. In Computational Phytochemistry. Elsevier, pp 107–139
Rama PG, Vinutha T (2019) Metabolomics profiling of plants to understand reasons for plant stress resilience to abiotic stress. In: Recent approaches in omics for plant resilience to climate change, Springer, Cham, pp 57–74

Ribeiro RC, de Carvalho MG, de Moraes MD, Rossiello RO, de Oliveira DR, de Amorim RM, Junior EB (2018) Chemical screening of *Urochloa humidicola*: methods for characterizing secondary metabolites and allelopathic activity on forage legumes. Am J Plant Sci 9(6):1260–1278

Sannigrahi S, Mazumdar UK, Pal D, Mishra SL (2009) Hepatoprotective potential of methanol extract of Clerodendrum infortunatum Linn against CCl4 induced hepatotoxicity in rats. Pharmacogn Mag 5(20):394–398

Scherz H, Huck CW, Bonn GK (2007) CEC and EKC of natural compounds. Electrophoresis 28 (11):1645–1657

Sharma N, Pant BD, Mathur J (2018) MEMS or microchip based capillary electrophoresis tool for analysis of Pathogenesis-related (PRs) proteins in Eruca sativa (Miller) cultivars against Fusarium oxysporum. Micro Technol 24(12):4933–4944

Sharma P, Joshi H, Abdin MZ, Kharkwal AC, Varma A (2017) Analytical techniques to assess medicinal plants value addition after microbial associations. Modern tools and techniques to understand microbes. Springer, Cham, pp 259–271

Shihabi ZK, Oles KS (1994) Felbamate measured in serum by two methods: HPLC and capillary electrophoresis. Clin Chem 40(10):1904–1908

Singh B, Kumar A, Malik AK (2017) Flavonoids biosynthesis in plants and its further analysis by capillary electrophoresis. Electrophoresis 38(6):820–832

Soldatou S, Eldjarn GH, Huerta-Uribe A, Rogers S, Duncan KR (2019) Linking biosynthetic and chemical space to accelerate microbial secondary metabolite discovery. FEMS Microbiol Lett 366(13):142–148

Spisso A, Gomez FJ, Fernanda Silva M (2018) Determination of ellagic acid by capillary electrophoresis in Argentinian wines. Electrophoresis 39(13):1621–1627

Starkey JA, Mechref Y, Byun CK, Steinmetz R, Fuqua JS, Pescovitz OH, Novotny MV (2002) Determination of trace isoflavone phytoestrogens in biological materials by capillary electro chromatography. Anal Chem 74(23):5998–6005

Sugimoto M, Kawakami M, Robert M, Soga T, Tomita M (2012) Bioinformatics tools for mass spectroscopy-based metabolomics data processing and analysis. Curr Bio Inform 7(1):96–108

Sun H, Ni B, Zhang A, Wang M, Dong H, Wang X (2012) Metabolomics study on Fuzi and its processed products using ultra-performance liquid-chromatography/electrospray-ionization synapt high-definition mass spectrometry coupled with pattern recognition analysis. Analyst 137(1):170–185

Sun SW, Yeh PC (2005) Analysis of rhubarb anthraquinones and bianthrones by microemulsion electro kinetic chromatography. J Pharm Biomed Anal 36(5):995–1001

Tolstikov VV, Lommen A, Nakanishi K, Tanaka N, Fiehn O (2003) Monolithic silica-based capillary reversed-phase liquid chromatography/electrospray mass spectrometry for plant metabolomics, Anal Chem 75(23):6737–6740

Troujman H, Chottard JC (1997) Comparison between HPLC and capillary electrophoresis for the separation and identification of the platination products of oligonucleotides with cis-[Pt (NH3) 2 (H2O) 2] 2 + and [Pt (NH3) 3 (H2O)] 2+. Anal Biochem 252(1):177–185

Vanhoenacker G, Rong H, De Keukeleire D, Baeyens W, Van Der Weken G, Sandra P (2000) Simultaneous analysis of hop acids and prenylated flavanones by microemulsion electro kinetic chromatography with diode array detection. Biomed Chromatogra 14(1):34–36

Voeten RL, Ventouri IK, Haselberg R, Somsen GW (2018) Capillary electrophoresis: trends and recent advances. Anal Chem 90(3):1464–1481

Ward JL, Baker JM, Beale MH (2007) Recent applications of NMR spectroscopy in plant metabolomics. FEBS J 274(5):1126–1131

Weegels PL, Hamer RJ, Schofield JD (1995) RP–HPLC and capillary electrophoresis of subunits from glutenin isolated by SDS and osborne fractionation. J Cereal Sci 22(3):211–224

Wu C, Sun Y, Wang Y, Duan W, Hu J, Zhou L, Pu Q (2019) 7-(Diethylamino) coumarin-3-carboxylic acid as derivatization reagent for 405 nm laser-induced fluorescence detection: a case study for the analysis of sulfonamides by capillary electrophoresis. Talanta 201:16–22

Wu T, Yu C, Li R, Li J (2018) Minireview: Recent advances in the determination of flavonoids by capillary electrophoresis. Instrum Sci Technol 46(4):364–386

Xiang Z, Wang XQ, Cai XJ, Zeng S (2011) Metabolomics study on quality control and discrimination of three Curcuma species based on gas chromatograph–mass spectrometry. Phyto chem Analysis 22(5):411–418

Xiao MW, Bai XL, Liu YM, Yang L, Liao X (2018) Simultaneous determination of trace Aflatoxin B1 and Ochratoxin A by aptamer-based microchip capillary electrophoresis in food samples. J Chromatogr A 1569:222–228

Yang F, Zhou K, Lu Y, Yoshida H, Yang H (2019) Simultaneous determination of promethazine and its metabolites by improved capillary electrophoresis coupled with electrochemiluminescence. Int J Electrochem Sci 14:9159–9169

Zhang H, Wu ZY, Wang YZ, Zhou DD, Yang FQ, Li DQ (2020) On-line immobilized trypsin microreactor for evaluating inhibitory activity of phenolic acids by capillary electrophoresis and molecular docking. Food Chem 10(3):125–143

Zhang J, Xie J, Liu J, Tian J, Chen X, Hu Z (2004) Microemulsion electro kinetic chromatography with laser-induced fluorescence detection for sensitive determination of ephedrine and pseudoephedrine. Electrophoresis 25(1):74–79

Zhang W, Hankemeier T, Ramautar R (2017) Next-generation capillary electrophoresis–mass spectrometry approaches in metabolomics. Curr Opin Biotech 43:1–7

Zhao J, Li L, Zhao Y, Zhao C, Chen X, Liu P, Zhou H, Zhang J, Hu C, Chen A, Liu G (2018) Metabolic changes in primary, secondary, and lipid metabolism in tobacco leaf in response to topping. Anal Bioanal Chem 410(3):839–851

Zhao J, Zhao Y, Hu C, Zhao C, Zhang J, Li L, Zeng J, Peng X, Lu X, Xu G (2016) Metabolic profiling with gas chromatography-mass spectrometry and capillary electrophoresis-mass spectrometry reveals the carbon-nitrogen status of tobacco leaves across different planting areas. J Proteome Res 15(2):468–476

# Chapter 9
# Camptothecin: Occurrence, Chemistry and Mode of Action

**Mallappa Kumara Swamy, Boregowda Purushotham, and Uma Rani Sinniah**

**Abstract** Camptothecin (CPT), a planar pentacyclic ring system, constituting a pyrrolo(3, 4-$\beta$)-quinoline group along with $\alpha$-hydroxy lactone is observed mainly from the plants that belong to the genus, *Camptotheca*. They tend to exhibit a broad and significant clinical impact including antitumor, antiviral, antibacterial, antipsoriasis, antirheumatic, antispasmodic, neurotropic, antimalarial activities, etc. Due to their unique mode of action, CPT has a broad clinical interest. These agents turn topoisomerase-I (top-I), an enzyme that exerts the torsional stresses of supercoiled DNA into an intracellular poison. The CPTs stabilize the covalent binding of top-I to its DNA substrates and the formation of these complexes tends to the formation of reversible, single-strand nicks producing potentially lethal double-strand DNA breaks resulting in apoptosis. Owing to high market value, CPT-producing plant species have been scientifically explored in recent years. The present chapter offers a comprehensive info on the sources, chemistry, and mode of action of CPT.

**Keywords** Camptothecin · *Nothapodytes nimmoniana* · Anticancer compounds · Cytotoxicity · Mode of action

## 9.1 Introduction

Bioactive natural compounds obtained from a wide-ranging medicinal plants have gained more attention in the last few decades for treating numerous human health issues. Several types of plant-derived compounds like phenolics, alkaloids, terpenoids, polyketides, and others are being already endorsed to have worthy clinical

M. K. Swamy (✉) · B. Purushotham
Department of Biotechnology, East West First Grade College, Bengaluru, India
e-mail: swamy.bio@gmail.com

U. R. Sinniah
Department of Crop Science, Faculty of Agriculture, Universiti Putra Malaysia, Darul Ehsan, 43400 Serdang, Malaysia

 
D. Pal and A. K. Nayak (eds.), *Bioactive Natural Products for Pharmaceutical Applications*, Advanced Structured Materials 140,
https://doi.org/10.1007/978-3-030-54027-2_9

perspective, including anticancer, antimicrobial, anti-inflammatory, antioxidant, antiinsecticidal activities to name a few (Lodh and Swamy 2019). In recent times, cancer is affecting the lives of many people in the society and is one among the highest death instigating menace around the globe (Ravichandra et al. 2018; Akhtar and Swamy 2018, b; Sasidharan and Saudagar 2019). Herbs are being utilized since from the history in different folk medicines in India, China, Indonesia, Malaysia, etc., for treating several human health issues. The occurrence of chemically diverse biologically active phytoconstituents is accounted for the impartment of medicinal values to plant species. Many such biologically active constituents have been isolated and documented with experimental proofs on their pharmacological activities (Lodh and Swamy 2019). Some of the important medicinal plants having bioactive compounds include *Camptotheca accumulata, Catharanthus roseus, Coleus forskohlii, Tinospora cordifolia, Centella asiatica, Gloriosa superba, Oplopanax horridus,* etc. Today, several well-known therapeutically much-admired compounds, such as camptothecin (CPT), taxol, podophyllotoxins, luteolin, rosmarinic acid, vincristine, ginsenosides, homoharringtonine, and many more are being sequestered and quantified from several herbs. Also, plants are the major natural resources for recovering anticancer agents (Lee et al. 2018; Akhtar and Swamy 2018a, b).

The innovation of an important antitumor alkaloid, CPT has prospered the interest among plant scientists, phytochemists, biologists, medical doctors, drug discovery institutions, and pharmaceutical industries. CPT is a highly valued compound because of its unique therapeutical potential especially in treating various cancers (Rahier et al. 2005; Venditto and Simanek 2010; Kacprzak 2013; Prakash et al. 2016; Pu et al. 2019). The presence of CPT was firstly revealed from the *Camptotheca acuminata* Decne. (Nyssaceae) tree species, available in China. Soon after that CPT was also found in other plant sources such as *Nothapodythes nimmoniana, Chonemorpha grandiflora, Ophiorrhiza mungos*; *Ophiorrhiza pumila* and *Ophiorrhiza filistipula,* etc. Also, some of the entophytic fungi such as *Entrophospora infrequens, Fusarium solani,* etc. associated with these plant species were shown to secrete CPT in low amount (Kaushik et al. 2015; Raveendran 2015; Prakash et al. 2016; Pu et al. 2019). However, the yield of CPT is very low in all these natural resources. Globally, the annual production of CPT from plant sources is only about 600 kg, while the demand is assessed to be roughly 3000 kg per annum in the world market. Thus, alternative approaches such as biotechnology or chemosynthesis must be adopted.

Though, this natural compound is effective in exhibiting cytotoxicity, however, numerous critical issues are presented by CPT including less water solubility, stability in addition to unpredictable adverse effects in patients such as vomiting, myelosuppression, diarrhea, etc. (Martino et al. 2016). For overcoming these issues, several attempts are being made to develop new chemical derivatives, which are readily soluble through semisynthetic approaches involving the basic structure modifications (Venditto and Simanek 2010). Over a time, the chemical exploration of CPT has resulted in few major anticancer chemotherapeutic drugs such as irinotecan and topotecan. Presently, irinotecan, topotecan, 9-aminocamptothecin, and 9-nitrocamptothecin are the widely used water-soluble semisynthetic analogues of CPT for treating cancers all over the world. (Deorukhkar et al. 2007; Ramawat and Goyal 2009; Raveendran 2015). Presently, they represent a distinctive class of clinically approved chemo-drugs of modern medicines (Lee 1999; Ramawat and Goyal 2009). The anticancer mechanism of action of these chemo-drugs is through causing cellular damages by apoptosis intermediated by the inhibition of Top-I (Topoisomerase I) enzyme structure and activity (Wright et al. 2015; Sadre et al. 2016). Though many leads are presently in clinical trials against cancers, only topotecan and irinotecan are permitted for clinical uses (Martino et al. 2016; Pu et al. 2019).

This chapter summarizes the occurrence of CPT in natural resources, chemistry, and its mode of action. Overall, this comprehensive review encourages research interest among chemists and biologists to explore more on this novel compound.

## 9.2 Camptothecin Discovery and Its Chemistry

In 1955, to evaluate the anticancer compounds from nature, Cancer Chemotherapy National Service Center (CCNSC) was created by the National Cancer Institute (NCI). It was initially aimed to screen unknown chemical compounds for chemotherapeutic uses. However, later in the year 1960, the screening program was extended to include different natural resources, mainly plant species and their potential anticancer compounds. In this collaborative work, many pharmaceutical companies and research institutions, including The United States Department of Agriculture (USDA) submitted numerous potential plant species and phytocompounds to be screened for their antitumor activities (Hartwell 1970; Perdue et al. 1970; Moore et al. 1970; Raveendran 2015). The NCI screening work was led by a well-known organic chemist, named Jonathan Hartwell, and it also included data from the documents of traditional medical practices of Egypt, Greece, China, and Rome that employed a wide range of medicinal plant species and their preparations against cancer cure. Different plants species extracted initially at Wisconsin Alumni Research Foundation was directed to many other labs for evaluating their potential of killing cancerous cells. Primarily, the anticancer effects of all samples were evaluated in oral epidermoid carcinoma (KB) cell culture. Also, the samples were screened using tumor xenograft mouse models, i.e., CA-755, an adenocarcinoma; S-180, a sarcoma; and L1210, a lymphoid leukemia (Kessel 1971a,b; Raveendran 2015). From these results, the crude extracts showing active antitumor activities were later subjected to different separation techniques to isolate the active compounds at different laboratories, including Research Triangle Institute (RTI), North Carolina. More than 1000 plant compounds were isolated from different plant species solvent extracts, and among them, only *Camptotheca accuminata*, a tree species bark and stem extract disclosed noteworthy anticancer activity. Further screening and isolation techniques have led to the identification of novel compound, which was named as CPT in 1966, and the new sequestration of this anticancer compound from *C. acuminata* extracts was carried out by Keith Palmer and Harold Taylor. CPT, which is an indole alkaloid was effective against both L1210 and p338 leukemic cells (Wall et al. 1996; Raveendran 2015). *C. accuminata* is used in the traditional medical practices to treat leukemia, psoriasis, liver diseases, spleen, and digestive tract (Govindachari and Viswnathan 1972; Efferth et al. 2007). CPT displayed extraordinary anticancer action in the initial clinical trials. However, it showed poor water solubility, which restricted its ready usage in cancer chemotherapy (Kehrer et al. 2001; Li et al. 2006). Thus, medicinal and synthetic chemists have established several blends of CPT, and several CPT-derivatives were developed to increase its anticancer potentials with good results. So far, four CPT analogues are being permitted by the FDA (Food and Drug Administration), USA to treat various cancer types. These include irinotecan, topotecan, belotecan, and fam-trastuzumab deruxtecan (Wall et al. 1996; Samuelsson 2004).

The chemical structure of CPT comprises a planar pentacyclic ring configuration, encompassed with a pyrrolo[3,4-*β*]-quinoline moiety (rings *A*, *B*, and *C*), conjugated pyridone moiety (ring D) and one chiral center (positioned at C20) within the α-hydroxy lactone ring having (S) configuration (E-ring) (Fig. 9.1). The planar arrangement is believed to be the major factor that inhibits topoisomerase enzymes activity (https://en.wikipedia.org/wiki/Camptothecin). Some of the problems associated with parent CPT is its susceptibility to hydrolysis, which is mainly because of the lactone ring structure, as well as considerable toxicity. The α-hydroxy lactone ring having (S) configuration (E-ring) opening under biological conditions leads to form the carboxylate open form of CPT. Later, sodium salt of CPT was proposed and reported with higher solubility than the parent CPT.

Unfortunately, clinical investigations showed reduced efficiency against cancers associated with severe adverse effects, for example, myelotoxicity and hemorrhagic cystitis (Adams and Burke 2005; Kacprzak 2013). These drawbacks were later addressed by many researchers, and eventually, semisynthetic and total synthetic approaches were used to yield many analogues of CPT. Through the structure activity relationship (SAR), a detailed understanding on CPT structure was made and developed few soluble and very active antitumor drugs, namely irinotecan, topotecan, belotecan, and an active metabolite of irinotecan, named as SN-38 (Fig. 9.2). These aspects are described in detail by a number of review articles (Zunino et al. 2002; Sriram et al. 2005; Kacprzak 2013; Li et al. 2017; Amin et al. 2018), and readers may refer them for better understanding.

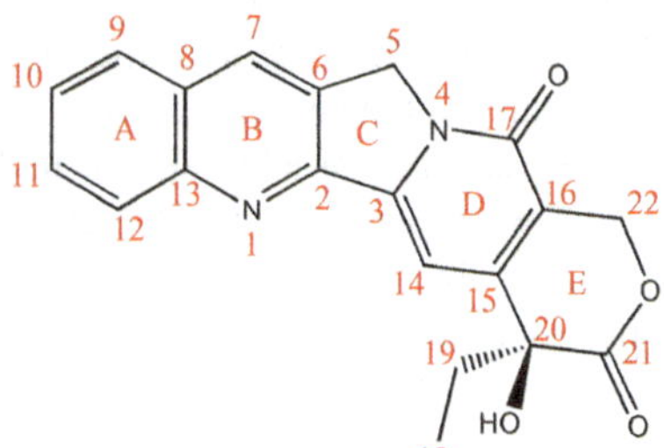

**Fig. 9.1** Structure of camptothecin showing a planar pentacyclic ring structure

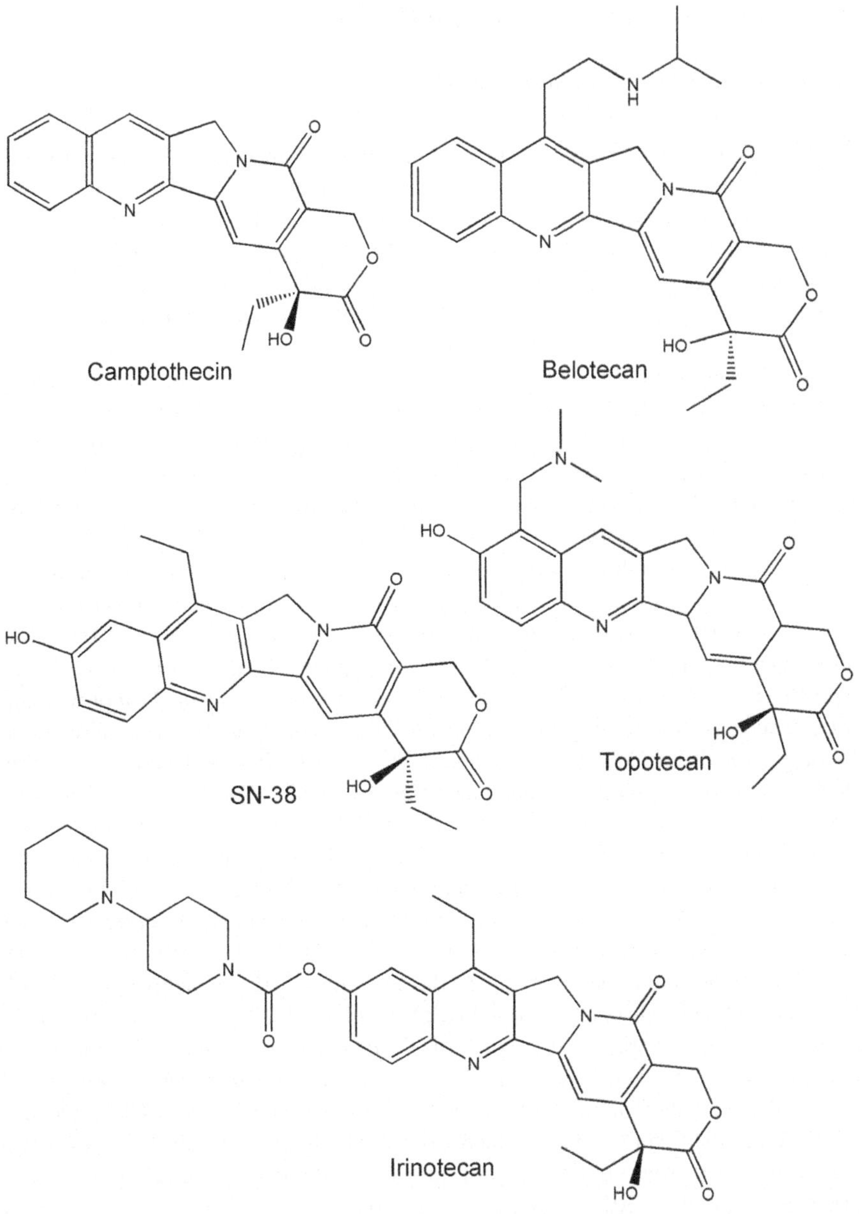


**Fig. 9.2** Camptothecin and its major chemical analogues, approved for cancer chemotherapy

## 9.3 Natural Sources of Camptothecin

### *9.3.1 Camptothecin from Plants*

The botanical classification assists in understanding the presence of phytocompounds in plant species (Larsson 2007). Plant-origin secondary metabolites with similar chemo-structures can be present in dissimilar families of the plant kingdom. Likewise, CPT occurs in different plant species of the distinct orders and families of angiosperms. Owing to high market value, CPT-producing plant species have been scientifically explored in recent years. Further, plant-based CPT biosynthesis still remains as the best and suitable route for production, because of its limited natural sources and productivity. The occurrence of CPT is commonly distributed among angiosperms (Wink 2003; Oberlies and Kroll 2004; Pu et al. 2019). The isolation and characterization of CPT was performed firstly by Wall et al. (1996) from *C. acuminata*. Later, it was re-isolated from a medicinal plant, *O. mungos* L. that belongs to the Rubiaceae family (Tafur et al. 1979). Remarkably, *C. acuminata* and *O. mungos* belongs to totally dissimilar phylogenetic clades, however, they possessed similar biosynthetic competency for CPT (Pu et al. 2019). This discovery revitalized the knowledge of distribution patterns of CPT in plants. Over the next two decades, CPT-producing plant species were identified and isolated from a few other plant species that belong to different plant families, i.e., Icacinaceae, Apocynaceae, and Rubiaceae. These plants include *Mostuea brunonis* Didr. (Loganiaceae), *Tabernaemontana alternifolia* L. (Apocynaceae; *syn. Ervatamia heyneana* (Wall.) T.Cooke); *O. filistipula* Miq. and *O. pumila* Champ. ex Benth. (Rubiaceae); *Nothapodytes nimmoniana* (Grah.) Mabb., *Nothapodytes obtusifolia* (Merr.) R. A. Howard., *Nothapodytes obscura* C.Y. Wu. and *Merrilliodendron megacarpum* (Hemsl.) Sleumer. (Icacinaceae). These plants sources have the potential to serve as alternatives to obtain CPT. Nevertheless, only in two species, namely *C. acuminata* and *N. nimmoniana,* the basic accumulation patterns of CPT is detailed so far (Gunasekera et al. 1981; Arisawa et al. 1981; Arbain et al. 1993; Liu et al. 1998; Ramesha et al. 2008; Bai and Song 2014; Upadhya et al. 2014; Kaushik et al. 2015; Isah and Mujib 2015; Prakash et al. 2016; Pu et al. 2019). CPT occurs in leaves, bark, seeds, fruits of *C. acuminata,* however the maximum quantity can be obtained from young leaves of *C. acuminata*, which is about 5 mg/g DW (dry weight). The matured leaves contain minimum levels of CPT, which is 250% lesser than that can be obtained from the bark, and ten-fold lesser than that can be obtained from young leaves (Lopez et al. 1994; Li et al. 2000; Yan et al. 2003; Kaushik et al. 2015; Prakash et al. 2016).

The increased exploration of plant resources for CPT-producing capability have resulted in the documentation of 34 new plant species that belong to Apocynaceae, Rubiaceae, Gelsemiaceae, Betulaceae, Icacinaceae, Nyssaceae, Meliaceae, and Violaceae. Among them, the majority of plant species belong to Rubiaceae and Icacinaceae family. Importantly, most of these described plant species have their natural distribution in the South-East Asian nations, exclusively in the south part of

China and Western Ghats of India (Pu et al. 2019). The herbaceous species of *Ophiorrhiza* are mainly spread across the Western Ghats of India, particularly at Travancore, Wynaad, Anamalai, and Kemmannugundi. Besides, these species occur in the regions of Assam, Andamans and Nicobars islands, Srilanka, Thailand, Philippines, Malaysia, and China (Namdeo et al. 2012; Kaushisk et al. 2015). The organs of *Ophiorrhiza* species produce low levels of CPT content. Similarly, CPT-yielding *C. grandiflora* (Apocynaceae), which is a latex-bearing shrubby climber has been observed by Kulkarni et al. (2010), and this plant is prevalent in the regions of Kerala and Karnataka in India. The content of CPT in these plants ranges between 0.600 and 1.350%. In *C. acuminata*, CPT content ranges from 0.012 to 0.236%, while in *N. nimmoniana,* it ranges from 0.081 to 0.775%, depending on different organs, and these plant species are the chief sources for CPT extraction in large scale (Liu et al. 1998; Lorence and Nessler 2004; Ramesha et al. 2008; Khan et al. 2013; Isah and Mujib 2015). In *C. acuminata,* the accretion of CPT largely depends on branch, age of the tree, and seasons. Moreover, the increased shade enhances CPT accumulation (Liu et al. 1998). It was established by a study that *N. nimmoniana* inner root bark and stem bark contained with the CPT amount of 0.23 and 0.33%, respectively (Uma et al. 2008; Isah and Mujib 2015).

### 9.3.2 Camptothecin from Endophytes

The plant-derived CPT cannot meet the global demand, and hence constant efforts are being carried out to recognize novel alternative resources of CPT. In this regard, endophytic microbes can be a choice for obtaining CPT (Uzma et al. 2018). There are limited reports on the CPT-yielding endophytic microorganisms that were identified, associated with CPT-producing plants. In a study, an endophytic fungus (*Entrophospora* infrequens) was isolated from *N. foetida,* growing in the coastal regions of India (Puri et al. 2005). The endophytic fungus was cultured in Sabouraud broth, a synthetic liquid media having dextrose (4%) and peptone (1%) under shaking conditions. Cultures, producing CPT was identified using both chromatographic and spectral evaluations. Further, they reported that CPT obtained from fungal cultures with anticancer activity against cancer cells, namely A-549 (a lung cancer), HEP-2 (a liver), and OVCAR-5 (ovarian) cancer cell lines. They testified the sequestration of about 18 μg/mg CPT content from the *E. infrequens* chloroform extract. Likewise, another study by Amna et al. (2006), used a bioreactor for culturing *E. infrequens*, and they found the synthesis of high levels of CPT (4.96 mg/100 g of dry mass) after 48 h. The optimum conditions, i.e., pH of 5.6, temperature of 28 ± 2 °C, agitation rate of 200–230 rpm, and aeration rate of 1 vvm were found optimal for the growth of *E. infrequens* in bioreactor culture, and to produce higher levels of CPT. The above studies clearly suggest the possible use of *E. infrequens* endophyte to produce CPT on a large-scale by employing bioreactors.

Rehman et al. (2008) identified an endophyte (*Neurospora crassa*), and designated it as ZP5SE from *N. foetida* seeds, and was able to produce CPT, when cultured in Sabouraud Dextrose Liquid Medium under the shake flask culturing environments. This endophyte was capable to propagate in Sabouraud broth (100 ml) poured in a 500 ml Erlenmeyer flask, and grown on a shaking incubator, agitated at 220 rpm for a duration of 10 days at a temperature of 28 ± 2 °C. From the growth kinetics study, it was evidenced that this fungal species, showing the exponential phase of growth up to 7 days of culturing. The produced CPT content by this endophyte was analyzed using High Performance Liquid Chromatograhy (HPLC) followed by Liquid chromatography–mass spectrometry (LC/MS) and Tandem mass spectrometry (MS/MS). The obtained CPT was tested against the growth of human lung cancer (A-549) and ovarian cancer (OVCAR-5) cells. The results were positive with effective inhibition of the growth of cancer cells, and the observed data were comparable to the data obtained for the standard drug, CPT.

Another endophyte, *Nodulisporium* sp. was isolated from *N. nimmoniana* inner barks by Rehman et al. (2009). This fungal species possessed thin hyphae, which was found to extend up to 6.4 μm in diameter, and profuse conidiophores that were branched, verticillately. The fungus, when grown on liquid culture (Sabouraud broth) media produced CPT up to 5.5 μg/g DW of mycelia after one week. Further, the authors have explored the possibility of its culturing in an airlift bioreactor with 5 to 18 l volume. They found that mycelium growth initiated within 5 h of inoculation, and the maximum growing was documented on Sixth day of culturing in a bioreactor with the yield of 45 μg/g DW of CPT.

From the plant, *Apodytes dimidiata,* 2 strains of *F. solani* (MTCC 9667 and MTCC 9668) were isolated, and these strains were shown to produce both CPT in their mycelia. Interestingly, MTCC 9668 strain was also shown to produce 10-hydroxycamptothecin, however in low quantities in mycelia. These endophytes, MTCC 9668 and MTCC 9667 after culturing on a liquid broth for 4 days could able to produce 53 and 37 μg CPT/100 g DW of mycelia, respectively (Shweta et al. 2010). Likewise, the produced quantity of 9-methoxycamptothecin and 10-hydroxycamptothecin by MTCC 9668 was observed to be 44.9 and 8.2 μg/100 g DW, correspondingly.

A study carried out by Musavi et al. (2015) identified the occurrence of an endophyte, *F. oxysporum* NFX06 associated with *N. nimmoniana*. They evaluated the production of cell-allied CPT from this endophytic fungus, and proposed its growth kinetics. To determine the consequence of substrate levels, a model was constructed by applying response surface methodology on the basis of central composite design. They used dextrose, magnesium sulfate, and peptone as independent variables to evaluate the production of CPT under sub-merged fermentation conditions. The highest CPT yield noticed from the central composite design was found to be 598 ng/g DW of mycelia. Also, the model-endorsed experimental yield of CPT and the optimal expected yield of CPT from dried mycelia were observed to be 610 ng/g and 628 ng/g DW at the substrate levels of 9.2 g/l peptone, 42.6 g/l dextrose, and 0.26 g/l $MgSO_4$, correspondingly.

Likewise, new endophyte, *Xylaria sp*. occurring in *C. acuminata* was isolated, and it was noticed that this fungus could produce 10-hydroxy-CPT, but failed to produce CPT (Liu et al. 2010). Endophytic microbes that can biosynthesize CPT have been witnessed by many researchers, however, attenuation of CPT synthesis is one of the chief problems in this area of studies.

Later, study by Shweta et al. (2013), described 3 endophytes, namely *Phomopsis* sp., *Fomitopsis* sp., and *Alternaria alternate* that occurred in the shrub species, *Miquelia dentata* Bedd. These fungi produced CPT, 9-methoxy-CPT, and 10-hydroxy-CPT. The obtained yield of CPT from these species was found to be about 42, 73, and 55 μg/g DW, respectively for *Phomopsis* sp., *Fomitopsis* sp., and *A. alternate.* About 161 endophytic fungal species were isolated from *C. acuminata*, and among them, only one fungus, i.e., *Botryosphaeria dothidea* X-4 was shown to be capable of producing 9-methoxy-CPT (Ding et al. 2013).

Likewise, 3 new endophytic fungi (*Aspergillus* sp. LY355, *Aspergillus* sp. LY341, and *Trichoderma atroviride* LY357) were isolated from *C. acuminata.* These species yielded CPT content of 42.9, 7.9, and 197.8 μg per liter of broth culture, respectively. However, repeated sub-culturing of these fungal strains were reported to lead to the diminishing of CPT-yielding ability. However, there was a persistent production of CPT by *Aspergillus* sp. LY357 strain until the eighth generation of sub-culturing (Pu et al. 2013; Kai et al. 2015; Uzma et al. 2018).

An investigation of Soujanya et al. (2017) reported the association of 4 new bacterial strains with the stems, fruits, and leaves of *Pyrenacantha volubilis*, belonging to Icacinanceae plant family. The identified bacteria included *B. amyloliquefaciens* (KY741854), *Bacillus* sp. (KP125956), *Bacillus* sp. (KP125955), and *B. subtilis* (KY741853). All these isolates were reported to produce CPT as revealed by the use of electrospray ionization mass spectrometry (ESI–MS) and nuclear magnetic resonance spectroscopy (NMR) analysis. The crude extract of these bacterial isolates were analyzed for their antitumor activity on colon cancer cells, and the results have shown that crude extracts exhibiting effective cytotoxicity. Also, their study evaluated the part played by plasmid in producing CPT. They proposed the potential role of a plasmid (5 kbp) in bacteria to biosynthesize CPT. However, the exact mechanisms are yet to be revealed. *B. subtilis,* KY741853 strain, producing CPT was attenuated after sub-culturing in the culture medium.

In another study, solid-state fermentation was utilized to culture the endophytic fungus, *Fusarium oxysporum*, and evaluated for its capability to produce CPT (Bhalkar et al. 2016). According to them, soybean meal, among different substrates utilized was observed to be the best medium for producing CPT effectively by *F. oxysporum* under optimized solid-state fermentation conditions. The use of soyabean meal substrate was effective in yielding CPT content of about 128 mg/l dry weight after 3 days of growth in a bioreactor. In addition, the use of whey (the liquid remaining after milk) as a moisture source in solid-state fermentation played an important role in augmenting the yield of CPT. The application of this developed process can be environmentally very effective, and it is evidenced by the fact that after fermentation stage, there was a significant reduction in the levels of biological oxygen demand (BOD), chemical oxygen demand (COD), total soluble solids

(TSS), total dissolved solids (TDS), and total protein content of whey effluent. Hence, this simple solid-state fermentation approach, employing the endophytic fungus and cheaper substrates, such as whey and soybean meal may be useful in producing valued anticancer pro-drug, CPT.

Lately, Clarance et al. (2019), reported the endophyte, *F. solani* strain ATLOY-8 from *Chonemorpha fragrans* (Moon) Alston., and it was capable of producing CPT. The optimized process to yield camptothecin was designed using the conventional approaches and analytical methods. For improved biomass and yield of CPT, the Box–Behnken design matrix ($n = 17$) and one factor at a time method was used to identify and select optimized variables. According to them, the Box–Behnken design matrix method in the basal media with 1% glucose, 5% absolute ethanol, and 0.03% precursors (a mixture of geraniol, tryptophan, and tryptamine) could increase the CPT production up to 1.4 fold increase and biomass up to 1.2 fold in comparison to one factor at a time method.

## 9.4 Mode of Action of Camptothecin

Early researchers have stated the antineoplastic activity of CPT mainly mediated by inhibiting nucleic acids synthesis in both tumor (L1210) and normal cells (HeLa) (Kessel 1971a,b; Horwitz et al. 1971). This inhibitory effect of CPT over macromolecules synthesis was observed to be transient with the removal of the drug. However, its prolonged treatment period may completely suppress nucleic acids synthesis leading to cell death. Also, CPT-induced rapid fragmentation of nucleic acids at concentrations that do not primarily impact on the synthesis of protein (Horwitz et al. 1971). In intact cells, the suppression of RNA synthesis can be fully reversed by removing CPT. Nevertheless, CPT failed to block the activities of purified enzymes (DNA and RNA polymerases). CPT exerts its cytotoxicity by inhibiting topoisomerase enzymes that relieve the topological strains created during the process of chromosomal recombination, replication, and transcription (Champoux 2001; Das et al. 2016). Topoisomerases are of two types (topoisomerase 1 and topoisomerase (2) classified based on their mode of action, i.e., whether they cleave either a single or double DNA strands (Raveendran 2015). Typically, Topoisomerase 1 (Top 1) enzyme relaxes the DNA supercoiling that involves the following steps: (a) cleavage of one of the two strands of DNA after the attachment, (b) relaxation of the strand, and (c) re-annealing of the strand. The incubation of CPT together with isolated DNA results in intact binding and cause fragmentation of DNA. Further, molecular interaction studies using X-ray crystallography has witnessed the establishment of a Topo1-DNA complex with numerous interactions (Redinbo et al. 1998; Liu et al. 2000). This forms the target for many anticancer drugs to induce cell death by trapping this covalent complex formed by Top-I and II enzymes (Zhang et al. 2011). As stated earlier, the obstruction of DNA synthesis and DNA cleavage is because of the suppression of DNA Top 1 enzyme (Hsiang et al. 1985). Top 1 inhibition is mainly mediated by the formation of reversible CPT-induced Top 1-DNA cleavable complex. In particular, CPT attaches to the Top 1-DNA complex and forms the reversible Top 1–CPT–DNA covalent tertiary complexes that inhibit the Top 1-religation reaction (Redinbo et al. 1998; Zhang et al. 2011; Das et al. 2016). Further, it was shown that CPT-induced cytotoxicity is directly related to the CPT-mediated assembly of Top 1-DNA cleavable complexes (Hsiang and Liu 1988; Hsiang et al. 1989; Liu et al. 2000; Pommier 2006). The interaction of CPT with the DNA occurs by intercalating at the cleavage site of the enzyme. Also, it has been suggested that other interactions might occur between CPT and Top 1, as well as CPT and a flipped base of DNA at the +1 site (Liu et al. 2000). However, these molecular interactions at Top 1-CPT-DNA complex are yet to be understood in detail. The cell cytotoxicity effect of CPT is mostly mediated during S-phase (synthetic phase) of the cell cycle (Cliby et al. 2002; Pommier 2006; Raveendran 2015). Based on the understanding of S-phase-specific cytotoxicity of CPT, researchers have proposed a replication fork collision model. According to this model, Top 1-CPT-DNA tertiary complexes are reversible and non-lethal by themselves but, after colliding with the replication forks causes DNA strand break, leading to apoptosis (Hsiang et al. 1989; Liu et al. 2000; Pommier 2006). After the collision of replication fork, the following events are observed: (1) double-strand breakage, (2) replication fork drive arrest, and (3) the establishment of Top 1 induced DNA breaks at the collision site. However, these biomolecular interactions leading to cell toxicity are yet to be understood in detail (Liu et al. 2000; Raveendran 2015). Interfacial inhibitor concept has been critically reviewed for CPT by Pommier (2006) and Pommier (2009). Accordingly, CPT binds at the Top 1-DNA interface and trap cleavage complexes. It is proposed that these topoisomerase inhibitor drugs stack between the DNA base pairs adjoining the cleavage site due to their aromatic nature (Pommier 2009; Koster et al. 2007; Pommier and Marchand 2012). CPT obstructs the rotation of DNA after intercalating at the enzyme–DNA complex through the interaction of $\pi$-$\pi$ electrons between the nucleotide bases flanking the cleavage site and the aromatic structure of the drug (Marchand et al. 2006; Koster et al. 2007). This interference formation inhibits the release of torsional stress and when the replication fork progresses further, it collides with Topo 1-CPT-DNA tertiary complex inducing DNA strand breakage and cell death. CPT, at higher concentrations can also destroy S-phase-independent cells through transcriptionally mediated DNA damage and apoptosis (Morris et al. 1996). Another prominent way of CPT-induced cell toxicity is by arresting RNA synthesis during transcription elongation process (Ljungman et al. 1996; Liu et al. 2000). It is also reported that the Top 1 cleavable complex is attached by ubiquitin/26S proteasome complex that blocks the re-ligation step of the Top 1 reaction (Desai et al. 1997; Liu et al. 2000). CPT, a Top 1-specific poison induces the attachment of SUMO-1 (Small Ubiquitin-like MOdifier) to Top 1 and facilitate its degradation (Mao et al. 2000; Rallabhandi et al. 2002). However, the exact function of SUMOylation of Top 1 in response to CPT is not clear. It is believed that SUMO-1 regulates the cellular localization of Top 1 as it contains nuclear localization signals (Desai et al. 1997, 2001; Mo et al. 2002). On the other hand, Poly(ADP-ribose) polymerase-1 (PARP 1), a chromatin-associated enzyme interacts with various proteins involved in DNA repair mechanisms destabilizes the Top 1-CPT-DNA complex by interacting with the $NH_2$-terminal domain of Top 1 (Park et al. 2005; Das et al. 2014).

## 9.5 Conclusions

CPT and its derivatives are important natural products, and are used to cure different types of cancers. CPT compound also possesses antiprotozoal, antiviral, and insecticidal activities. Since its discovery, numerous studies were made to evaluate its anticancer potential, which can be evidenced by the several hundreds of published research works each year. Though many plant species have been documented to produce CPT, its basic accretion configurations have been recognized in detail only in *C. acuminata* and *N. nimmoniana*. Several endophytic microbes have been identified from CPT-producing plant species, and they have shown to synthesize CPT, however, in small quantities. Semisynthetic CPTs like topotecan and irinotecan exhibit strong anticancer effects, and hence are widely used in cancer therapy. Likewise, several other medicine leads are being established, and some of them are being investigated for anticancer activities. However, many of them are yet to pass through clinical trials. Effective drug delivery, biological stability, and low toxicity of CPTs are some of the areas, which have to be investigated in detail. As research studies are actively involved in these areas, success can be expected in the coming days. Importantly, effective delivery of CPTs via new formulations involving nanotechnology is also being investigated in the present time. Due to higher demand and complexity of chemosynthesis of CPTs have prompted to look into simple and sustainable approaches of its production in large-scale. In this regard, biotechnological approaches, involving plant cell culture, endophytic microbial fermentation processes/techniques are more productive. CPTs production in heterologous systems/organisms and the use of metabolic engineering approaches can be very useful. Hence, future research should focus more on these aspects to fulfill the ever-growing demand for treating different types of cancers throughout the world.

## References

Adams VR, Burke TG (eds) (2005) In: Camptothecins in cancer therapy. Humana Press

Akhtar MS, Swamy MK (2018a) Anticancer plants: properties and application. Springer, Singapore

Akhtar MS, Swamy MK (eds) (2018b) Anticancer plants: properties and application. Springer International, Singapore

Amin SA, Adhikari N, Jha T, Gayen S (2018) A review on camptothecin analogs with promising cytotoxic profile. Anti-Cancer Agents in Med Chem (Formerly Current Medicinal Chemistry-Anti-Cancer Agents)18(13):1796–814

Amna T, Puri SC, Verma V, Sharma JP, Khajuria RK, Musarrat J et al (2006) Bioreactor studies on the endophytic fungus *Entrophospora infrequens* for the production of an anticancer alkaloid camptothecin. Can J Microbiol 52:189–196

Arbain D, Putra DP, Sargent MV (1993) The alkaloids of *Ophiorrhiza filistipula*. Aust J Chem 46 (1993):977–985

Arisawa M, Gunasekera SP, Cordell GA, Farmsworth NA (1981) Plant anticancer agents XXI. constituents of *merrilliodendron megacarpum*. Planta Med 43(1981):404–407

Bai YH, Song QS (2014) Chemical constituents from the roots of Nothapodytes pittosporoides. Nat Prod Res Dev 26:197–201

Bhalkar BN, Bedekar PA, Kshirsagar SD, Govindwar SP (2016) Solid state fermentation of soybean waste and an up-flow column bioreactor for continuous production of camptothecine by an endophytic fungus *Fusarium oxysporum*. RSC Adv 6(61):56527–56536

Champoux JJ (2001) DNA topoisomerases: structure function and mechanism. Annu Rev Biochem 70:369–413

Clarance P, Khusro A, Lalitha J, Sales J, Paul A (2019) Optimization of camptothecin production and biomass yield from endophytic fungus *Fusarium solani* strain ATLOY-8. J Appl Pharm Sci 9(10):035–046

Cliby WA, Lewis KA, Lilly KK, Kaufmann SH (2002) S Phase and G2 arrests induced by topoisomerase I poisons are dependent on ATR kinase function. J Biol Chem 277:1599–1606

Das BB, Shar-yin NH, Murai J, Rehman I, Amé JC, Sengupta S, Das SK, Majumdar P, Zhang H, Biard D, Majumder HK (2014) PARP1–TDP1 coupling for the repair of topoisomerase I–induced DNA damage. Nucleic Acids Res p.gku088.

Das SK, Rehman I, Ghosh A, Sengupta S, Majumdar P, Jana B, Das BB (2016) Poly (ADP-ribose) polymers regulate DNA topoisomerase I (Top1) nuclear dynamics and camptothecin sensitivity in living cells. Nucleic Acids Res p.gkw665.

David A, Michael RZ (2010) Cancer: an old disease, a new disease or something in between?. Nat Rev Cancer 10(10):728–733

De LV, St-Pierre B (2000) The cell and development biology of alkaloid biosynthesis. Trends in Plant Sci 5:168–173

Deorukhkar A, Krishnan S, Sethi G, Aggarwal BB (2007) Back to basics: how natural products can provide the basis for new therapeutics. Expert Opin Investig Drugs 16(11):1753–1773

Desai SD, Li TK, Rodriguez-Bauman A, Rubin EH, Liu LF (2001) Ubiquitin/26S proteasome-mediated degradation of topoisomerase I as a resistance mechanism to camptothecin in tumor cells. Can Res 61(15):5926–5932

Desai SD, Liu LF, Vazquez-Abad D, D'Arpa P (1997) Ubiquitin-dependent destruction of topoisomerase I is stimulated by the antitumor drug camptothecin. J Biol Chem 272 (39):24159–24164

Ding X, Liu K, Deng B, Chen W, Li W, Liu F (2013) Isolation and characterization of endophytic fungi from Camptotheca acuminata. World Journal of Microbiology and Biotechnology. 2013 Oct 1, 29(10):1831–8

Efferth T, Fu YJ, Zu YG, Schwarz G, Konkimalla VS, Wink M (2007) Molecular target-guided tumor therapy with natural products derived from traditional Chinese medicine. Curr Med Chem 14(19):2024–2032

Govindachari TR, Viswnathan N (1972) the stem bark of *Mappia foetida*, a tree native to India, has proved to be another source significant for the isolation of camptothecin. Phytochemistry 11(12):3529–3531

Gunasekera SP, Badawi MM, Cordell GA, Farnsworth NR, Chitnis M (1981) Plant anticancer agents X. Isolation of camptothecin and 9-methoxycamptothecin from *Ervatamia heyneana*. J Nat Prod 42 (1979) 475–477

Hartwell JL (1970) Plants used against cancer a Survey. Lloydia 33:97–194

Horwitz SB, Chang CK, Grollman AP (1971) Studies on Camptothecin. I. effects on nucleic acid and protein synthesis. Mol Pharmacol 7:632–644

Hsiang YH, Liu LF (1988) Identification of mammalian DNA topoisomerase I as an intracellular target of the anticancer drug camptothecin. Cancer Res 48:1722–1726

Hsiang YH, Hertzberg R, Hecht S, Liu LF (1985) Camptothecin induces protein-linked DNA breaks via mammalian DNA topoisomerase I. J Bio Chem. 1985 Nov 25, 260(27):14873–8

Hsiang YH, Liu LF, Wall ME, Wani MC, Nicholas AW (1989) DNA topoisomerase I-mediated DNA cleavage and cytotoxicity of camptothecin analogues. Cancer Res 49:4385–4389

Isah T, Mujib A (2015) Camptothecin from *Nothapodytes nimmoniana*: review on biotechnology applications. Acta Physiol Plant 37(6):106

Kacprzak KM (2013) Chemistry and biology of camptothecin and its derivatives. In: Natural products. Springer, Berlin, Heidelberg, pp 643–682

Kai G, Wu C, Gen L, Zhang L, Cui L, Ni X (2015) Biosynthesis and biotechnological production of anti-cancer drug Camptothecin. Phytochemistry Reviews. 2015 Jun 1;14(3):525–39

Kaushik PS, Swamy MK, Balasubramanya S, Anuradha M. Rapid plant regeneration, analysis of genetic fidelity and camptothecin content of micropropagated plants of *Ophiorrhiza mungos* Linn.—a potent anticancer plant. J Crop Sci Biotechnol 18(1):1–8

Kehrer DFS, Soepenberg O, Loos WJ, Verweil J, Sparreboom A (2001) Modulation of camptothecin analogues in the treatment o cancer: a review. Anticancer Drugs 12:89–106

Kessel D (1971a) Some determinants of camptothecin responsiveness in leukemia L1210 cells. Cancer Res 31:883–887

Kessel D (1971b) Effects of camptothecin on RNA synthesis in leukemia L1210 cells. Biochim Biophys Acta 246:225–232

Khan N, Tamboli ET, Sharma VK, Kumar S (2013) Phytochemical and pharmacolo-gical aspects of *Nothapodytes nimmoniana*, an overview. Herba Pol 59:53–66

Koster DA, Palle K, Bot ES, Bjornsti MA, Dekker NH (2007) Antitumour drugs impede DNA uncoiling by topoisomerase I. Nature 448:213–217

Kulkarni AV, Patwardhan AA, Lele U, Malpathak NP (2010) Production of camptothecin in cultures of *Chonemorpha grandiflora*. Pharm Res 2:296–299

Larsson S (2007) The new chemosystematics: phylogeny and phytochemistry. Phytochemistry 68:2904–2908

Lee KH (1999) Anticancer drug design based on plant-derived natural products1. J Biomed Sci 6 (4):236–250

Lee KW, Ching SM, Hoo FK, Ramachandran V, Swamy MK (2018) Traditional medicinal plants and their therapeutic potential against major cancer types. In: Akhtar MS, Swamy MK (eds) Anticancer plants: natural products and biotechnological implications, vol 2. Springer International, Singapore, pp 383–410

Li F, Jiang T, Li Q, Ling X (2017) Camptothecin (CPT) and its derivatives are known to target topoisomerase I (Top1) as their mechanism of action: did we miss something in CPT analogue molecular targets for treating human disease such as cancer? Am J Cancer Res 7(12):2350

Li QY, Zu YG, Shi RZ, Yao LP (2006) Review camptothecin: current perspectives. Curr Med Chem 13:2021–2039

Li Y, Li CJ, Yu D, Pardeep AB (2000) Potent induction of apoptosis by β- lapachone in human multiple myeloma cell lines and patient cells. Mol Med 6:1008–1015

Liu Z, Carpenter SB, Bourgeois WJ, Yu Y, Constantin RJ, Falcon MJ et al (1998) Variations in the secondary metabolite camptothecin in relation to tissue age and season in *Camptotheca acuminata*. Tree Physiol 18:265–270

Liu K, Ding X, Deng B, Chen W (2010) 10-Hydroxycamptothecin produced by a new endophytic Xylaria sp. M20, from Camptotheca acuminata. Biotechnol Lett 32: 689–693

Liu LF, Desai SD, Li TK, Mao Y, Sun MEI, Sim SP (2000) Mechanism of action of camptothecin. Ann N Y Acad Sci 922(1):1–10

Ljungman M, Hanawalt PC (1996) The anti-cancer drug camptothecin inhibits elongation but stimulates initiation of RNA polymerase II transcription. Carcinogenesis 17(1):31–36

Lodh S, Swamy MK (2019) Phytochemical aspects of medicinal plants of northeast India to improve the gynaecological disorders: an update. In: Natural bio-active compounds. Springer, Singapore, pp 353–367

Lopez MM, Nessler CL, McKnight TD (1994) Sites of the accumulation of the anti tumor alkaloid camptothecin in *Camptotheca acuminata*. Planta Med 60:558–560

Lorence A, Nessler CL (2004) Camptothecin, over four decades of surprising findings. Phytochemistry 65(2004):2735–2749
Mao Y, Sun M, Desai SD, Liu LF (2000) SUMO-1 conjugation to topoisomerase I: a possible repair response to topoisomerase-mediated DNA damage. Proc Natl Acad Sci 97(8):4046–4051
Marchand C, Antony S, Kohn KW, Cushman M, Ioanoviciu A et al (2006) A novel norindenoisoquinoline structure reveals a common interfacial inhibitor paradigm for ternary trapping of the topoisomerase I-DNA covalent complex. Mol Cancer Ther 5:287–295
Martino E, Della Volpe S, Terribile E, Benetti E, Sakaj M, Centamore A, Sala A, Collina S (2016) The long story of camptothecin: from traditional medicine to drugs. Bioorgan Med Chem Lett
Mo YY, Yu Y, Shen Z, Beck WT (2002) Nucleolar delocalization of human topoisomerase I in response to topotecan correlates with sumoylation of the protein. J Biol Chem 277(4):2958–2964
Moore BG, Schwartz HS, Hodo H (1970) Inhibition of Macromolecule Synthesis in LI210 Ascites Tumor Cells. J Cell Biol 47(144A).
Morris EJ, Geller HM (1996) Induction of neuronal apoptosis by camptothecin, an inhibitor of DNA topoisomerase-I: evidence for cell cycle-independent toxicity. J Cell Biol 134(3):757–770
Musavi SF, Dhavale A, Balakrishnan RM (2015) Optimization and kinetic modeling of cell-associated camptothecin production from an endophytic *Fusarium oxysporum* NFX06. Prep Biochem Biotechnol 45:158–172
Namdeo AG, Priya T, Bhosale BB (2012) Micropropagation and production of camptothecin form in vitro plants of *Ophiorrhiza mungos*. Asian Pac J Trop Biomed 2(2):S662–S666
Oberlies NH, Kroll DJ (2004) Camptothecin and taxol: historic achievements in natural products research. J Nat Prod 67(2):129–135
Park SY, Cheng YC (2005) Poly (ADP-ribose) polymerase-1 could facilitate the religation of topoisomerase I-linked DNA inhibited by camptothecin. Can Res 65(9):3894–3902
Perdue RE Jr, Abbott BJ, Hartwell JL (1970) Screening plants for antitumor activity II: a comparison of two methods of sampling herbaceous plants. Lloydia 33:1–6
Pommier Y (2006) Topoisomerase I inhibitors: camptothecins and beyond. Nat Rev Cancer 6:789–802
Pommier Y, Marchand C (2012) Interfacial inhibitors: targeting macromolecular complexes. Nat Rev Drug Discov 11:25–36
Pommier Y (2009) DNA topoisomerase I inhibitors: chemistry biology and interfacial inhibition. Chem Rev 109:2894–2902
Prakash L, Middha SK, Mohanty SK et al (2016) 3 Biotech 6:171. https://doi.org/10.1007/s13205-016-0490-y
Pu X, Zhang CR, Zhu L, Li QL, Huang QM, Zhang L, Luo YG (2019) Possible clues for camptothecin biosynthesis from the metabolites in camptothecin-producing plants. Fitoterapia 134:113–128
Pu X, Qu X, Chen F, Bao J, Zhang G, Luo Y (2013) Camptothecin-producing endophytic fungus Trichoderma atroviride LY357: isolation, identification, and fermentation conditions optimization for camptothecin production. Applied microbiology and biotechnology. 2013 Nov 1;97(21):9365–75
Puri SC, Verma V, Amna T, Qazi GN, Spiteller M (2005) An endophytic fungus from *Nothapodytes foetida* that produces camptothecin. J Nat Prod 68:1717–1719
Rahier NJ, Thomas CJ, Hecht SM (2005) Camptothecin and its analogs. In: Cragg GM, Kingston DGI, Newman DJ (eds) Anticancer agents from natural products. Brunner-Routledge Psychology Press, Taylor and Francis Group, Boca Raton, Chapter 2. 5
Rallabhandi P, Hashimoto K, Mo YY, Beck WT, Moitra PK, D'Arpa P (2002) Sumoylation of topoisomerase I is involved in its partitioning between nucleoli and nucleoplasm and its clearing from nucleoli in response to camptothecin. J Biol Chem 277(42):40020–40026
Ramawat KG, Goyal S. Natural products in cancer chemoprevention and chemotherapy. In: Herbal drugs: ethnomedicine to modern medicine. Springer, Berlin, Heidelberg, pp 153–171

Ramesha BT, Amna T, Ravikanth G, Gunaga RP, Vasudeva R, Ganeshaiah KN, Shaanker RU, Khajuria RK, Puri SC, Qazi GN (2008) Prospecting for camptothecines from *Nothapodytes nimmoniana* in the Western Ghats, south India: identification of high-yielding sources of campthecin and new families of camptothecines. J Chromatogr Sci 46:362–368

Ravichandra V, Ramesh C, Swamy MK, Purushotham B, Rudramurthy GR (2018) Anticancer plants: chemistry, pharmacology, and potential applications. In: Akhtar MS, Swamy MK (eds) Anticancer plants: properties and application, vol 1. Springer Nature, Singapore, pp 485–515

Raveendran VV (2015) Camptothecin-discovery, clinical perspectives and biotechnology. Nat Prod Chem Res 3(175):2

Redinbo MR, Stewart L, Kuhn P, Champoux JJ, Hol WG (1998) Crystal structures of human topoisomerase I in covalent and noncovalent complexes with DNA. Science 279:1504–1513

Rehman S, Shawl AS, Kour A, Andrabi R, Sudan P (2008) Sultan P (2008) An endophytic Neurospora sp from *Nothapodytes foetida* producing camptothecin. Appl Biochem Microbiol 44:203–209

Rehman S, Shawl AS, Kour A, Sultan P, Ahmad K, Khajuria R, Qazi GN (2009) Comparative studies and identification of camptothecin produced by an endophyte at shake flask and bioreactor. Nat Prod Res 23:1050–1057

Sadre R, Magallanes-Lundback M, Pradhan S, Salim V, Mesberg A, Jones AD, DellaPenna D. Metabolite diversity in alkaloid biosynthesis: a multi-lane (diastereomer) highway for camptothecin synthesis in *Camptotheca acuminata*. The Plant Cell Online. tpc-00193

Samuelsson G (2004) Drugs of natural origin: a textbook of pharmacognosy, 5th edn. Swedish pharmaceutical press, Stokkholm

Sasidharan S, Saudagar P (2019) Plant metabolites as new leads to drug discovery. In: Natural bio-active compounds, Springer, Singapore, pp 275–295

Shweta S, Zuhlke S, Ramesha BT, Priti V, Kumar PM, Ravikanth G, Spitellar M, Vasudeva R, Shaanker RU (2010) Endophytic fungal strains of *Fusarium solani* from *Apodytes dimidiata* E. Mey. Ex Arn (Icacinaceae) produce camptothecin, 10-hydroxycamptothecin and 9-methoxycamptothecin. Phytochemistry 71:117–22

Shweta S, Gurumurthy BR, Ravikanth G, Ramanan US, Shivanna MB (2013) Endophytic fungi from Miquelia dentata Bedd., produce the anti-cancer alkaloid, camptothecine. Phytomedicine.:337–42

Soujanya KN, Siva R, Kumara PM, Srimany A, Ravikanth G, Mulani FA, Aarthy T, Thulasiram HV, Santhoshkumar TR, Nataraja KN, Shaanker RU (2017) Camptothecin-producing endophytic bacteria from Pyrenacantha volubilis Hook. (Icacinaceae): a possible role of a plasmid in the production of camptothecin. Phytomedicine 36:160–167

Sriram D, Yogeeswari P, Thirumurugan R, Ratan BT (2005) Camptothecin and its analogues: a review on their chemotherapeutic potential. Nat Prod Res 19(4):393–412

Tafur S, Nelson JD, Delong DC, Svoboda GH (1979) Antiviral components of *Ophiorrhiza mungos*. Isolation of camptothecin and 10-methoxycamptothecin. Lloydia 39:261–262

Uma SR, Ramesha BT, Ravikanth G, Gunaga RP, Vasudeva R, Ganeshaiah KN (2008) Chemical profiling of *Nothapodytes nimmoniana* for camptothecin, an important anticancer alkaloid: towards the development of a sustainable production system. In: Ramawat KG, Merillon JM (eds) Bioactive molecules and medicinal plants. Springer, Berlin, Heidelberg, pp 197–213

Upadhya V, Pai SR, Sharma AK, Hegde HV, Kholkute SD, Joshi RK (2014) Compound specific extraction of camptothecin from *Nothapodytes nimmoniana* and Piperine from Piper nigrum using accelerated solvent extractor. J Anal Methods Chem

Uzma F, Mohan CD, Hashem A, Konappa NM, Rangappa S, Kamath PV, Singh BP, Mudili V, Gupta VK, Siddaiah CN, Chowdappa S (2018) Endophytic fungi—alternative sources of cytotoxic compounds: a review. Front Pharm 9:309

Venditto VJ, Simanek EE (2010) Cancer therapies utilizing the camptothecins: a review of the in vivo literature. Mol Pharm 7(2):307–349

Wall ME, Wani MC, Cook CE, Palmer KH, McPhail AT, Sim GA (1966) Plant antitumor agents. The isolation and structure of camptothecin, a novel alkaloidal leukemia and tumor inhibitor from *Camptotheca acuminata*. J Am Chem Soc 88:3888–3890

Wink M (2003) Evolution of secondary metabolites from an ecological and phylogenetic perspective. Phytochemistry 64:3–19

Wright CM, van der Merwe M, DeBrot AH, Bjornsti MA (2015) DNA topoisomerase I domain interactions impact enzyme activity and sensitivity to camptothecin. J Biol Chem 290 (19):12068–12078

Yan XF, Wang Y, Yu T, Zhang YH, Dai SJ (2003) Variation in camptothecin content in *Camptotheca accuminata* leaves. Bot Bull Acad Sin 44:99–105

Zhang ZB, Cheng B, Tse-Dinh YC (2011) Crystal structure of a covalent intermediate in DNA cleavage and rejoining by Escherichia coli DNA topoisomerase I. Proc Natl Acad Sci USA 108:6939–6944

Zunino F, Dallavalle S, Laccabue D, Beretta G, Merlini L, Pratesi G (2002) Current status and perspectives in the development of camptothecins. Curr Pharm Des 8(27):2505–2520

# Chapter 10
# Secondary Metabolites from Plant Sources

**Chandi Charan Kandar**

**Abstract** A Plant can be considered as a storehouse of a huge number of chemicals biosynthesized by many metabolic pathways like photosynthesis, glycolysis, Kreb's cycle, shikimic acid pathway, mevalonic acid pathway, etc. Primary metabolites include sugars, citric acid, Kreb's cycle intermediates, amino acids, protein, nucleic acid, and polysaccharides. Primary metabolites are identical in all living plant cells and they carry out basic life activities like growth, cell division, storage, respiration, and reproduction. On the other hand, the secondary plant metabolites, well-known as phytoconstituents are derived from primary metabolites by the influence of various surroundings stress like light, temperature, and different metals with the help of several metabolic pathways. The formation of secondary metabolites is very much specific to the plant family concern. By using similar primary metabolites, plants of various families produce a large number of different secondary metabolites having various pharmaceutical values. Generally, secondary metabolites of the plant have a great role to defense from herbivorous and pathogens, attract other animals and protect from UV radiation. Moreover, secondary metabolites show a lot of importance in the pharmaceutical application as medicines used for the treatment of various diseases in the folklore medicine as well as traditional medicine. They are also used as flavors in pharmaceutical ingredients, perfumes in pharmaceutical and perfumery industry, insecticides, dyes, polymers used for the preparation new drug delivery systems and therefore, they have a great value to economic concern.

**Keywords** Secondary metabolites · Biosynthesis · Phenolic compounds · Alkaloids · Glycosides · Terpenes · Saponins · Tannins

---

C. C. Kandar (✉)
Department of Pharmaceutical Chemistry, Institute of Pharmacy, Government of West Bengal, Jalpaiguri, West Bengal 735 101, India
e-mail: cckandar@rediffmail.com



D. Pal and A. K. Nayak (eds.), *Bioactive Natural Products for Pharmaceutical Applications*, Advanced Structured Materials 140,
https://doi.org/10.1007/978-3-030-54027-2_10

## 10.1 Introduction

A plant may be regarded as a large biosynthetic laboratory not only for the preparation of primary metabolites but also for a huge number of secondary metabolites having pharmaceutical importance. Primary metabolite synthesized by green plant includes glucose prepared by photosynthesis process, citric acid, phosphoenolpyruvate from glycolysis, acetyl CoA, citric acid, α- keto glutaric acid as Kreb's cycle intermediates, erythrose-4-phosphate from pentose phosphate pathway, amino acids by transaminase enzyme, protein, nucleic acid by de novo synthesis process, and polysaccharides. The basic roles of primary metabolites are general growth and physiological development, involvement in respiratory, storage, and reproductive system. Practically, primary metabolites are identical from lower to higher plant systems (Seigler 1995; Kokate et al. 2005).

Secondary metabolites biosynthetically produced from primary metabolites are considered as chemical adaptation due to environmental stresses such as light, temperature, and different metals but their distribution is mostly limited, generally restricted to a taxonomical group (Fig. 10.1). A particular family generates a similar group of phytoconstituents due to the presence of the definite enzymatic system in those plants. The different biosynthetic processes occurring in plant cells are dependent on enzymes that act as catalysts for such reactions. As it occurs by the control of enzyme activity, it is always directed into a specific pathway resulting formation of a definite phytoconstituent. For example, tropane alkaloids are obtained from the Solanaceae family, volatile oils from the Umbelliferae family. The idea of the secondary metabolite was primarily given by Albrecht Kossel, who got Nobel Prize for physiology or medicine in 1910, for his contribution (Jones et al. 1953). Later, Czapek described the secondary metabolites as ultimate products and also regarded as waste products or secretory substances of plant metabolism but very much essential for all the animals in the World (Bourgaud et al. 2001). These products are synthesized by nitrogen metabolism, i.e., secondary modification like deamination as per the opinion of the scientist. The development of analytical chemistry in the mid of the previous century become easier to recover of more and more of these phytoconstituents, and ultimately, this was the pillar for the development of the well-known discipline of phytochemistry. Secondary metabolites are very much expensive to produce, as well as accumulate in the different plant parts and that is why they are available in much smaller quantities than primary metabolites. Nowadays, the extraction of a few secondary metabolites has become costly due to their less availability (Kokate et al. 2005; Bourgaud et al. 2001).

The isotopically labeled markers are used to elucidate the biosynthetic pathways in plant cells for the manufacturing of numerous plant metabolites. With the help of radioactive carbon ($C^{14}$), hydrogen ($H^{3}$), and in few cases phosphorus ($P^{32}$), sulphur ($S^{36}$), the biosynthetic pathways are established and become easy to understand the different chemical steps. The specific information regarding biosynthetic pathways of alkaloid, proteins, and amino acids was achieved by using labeled nitrogen atom ($N^{14}$) (Kokate et al. 2005).

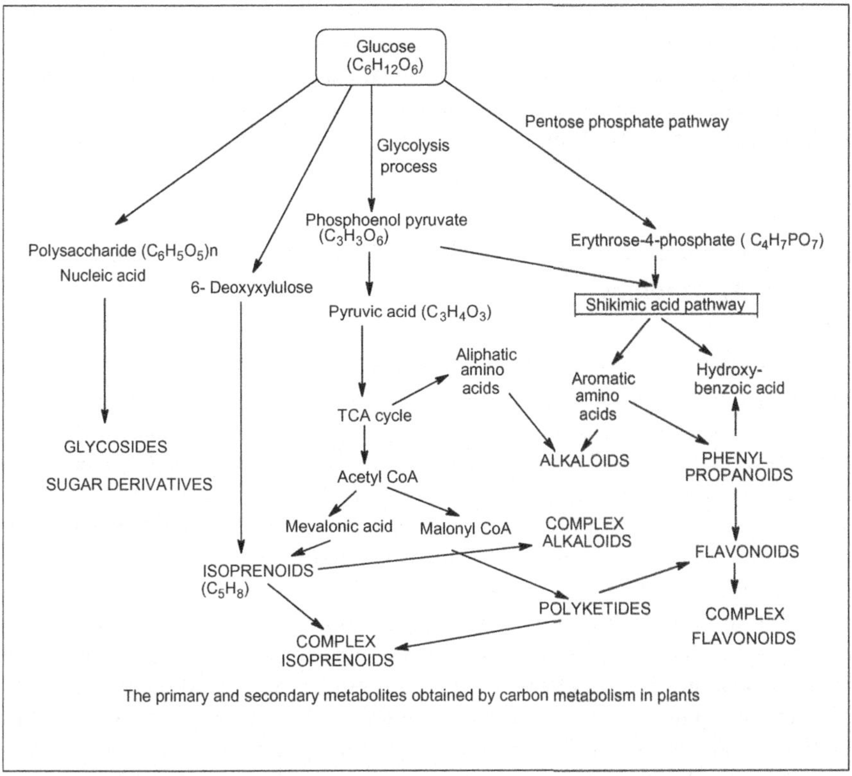


**Fig. 10.1** The primary as well as secondary plant metabolites obtained from metabolism in plants (Kokate et al. 2005)

Secondary metabolites have a great importance to exhibit the pharmacological activities which is established by the application of the extracts of medicinal plants in the primitive peoples and animals to cure or treatment for various illness that is published in different books and journals. Phytoconstituents provide various activities such as analgesic, anti-inflammatory, antidiabetic, antimalarial, antihypertensive, antiprotozoal, antibacterial, antifungal, antiviral activity, etc. It has been observed that some grasses such as alfalfa show biological activity like estrogen that provides an impact on the fertility of animals (Bennets et al. 1946).

Secondary plant metabolites are grouped on the basis of their chemical structures into several classes. This chapter deals with the comparison of secondary metabolites with primary metabolites, description, and structure of various secondary plant metabolites, their biosynthesis, and their pharmacological activities.

Stress compounds are generally produced in plants due to several factors such as injury in the plant and metabolic disturbance. These compounds are the products of either primary metabolism or secondary metabolism in plants. These products are available in the undeveloped flower buds, sub-epidermal glands, and seed kernels of *Gossypium hirsutum* (cotton plant). The synthesis of such compounds is developed by different biological and environmental conditions such as dehydration, chemical treatment, microbial infection, mechanical injury to the plants, and UV irradiation. They are manufactured for pharmaceutical benefits, for example, as oleoresins and gum, e.g., gossypol. Gossypol is useful for male antifertility activity and insecticidal activity (Kaur 2010).

### *10.1.1 Comparison of Secondary Metabolites with Primary Metabolites*

Primary metabolites are available in all types of plants and they perform all essential metabolic reactions by sharing in nutrition, as well as reproduction (Croteau et al. 2000). In some cases, it is difficult to differentiate primary and secondary metabolites of plant origin. Under the class of terpenoids, both primary and secondary metabolites are available. The same phytoconstituent exhibits the role of both primary and secondary metabolites. Actually, secondary metabolites contain a lot of phytoconstituents derived from various plant families in environmental stress conditions that show varieties of activity. Cell pigmentation in seed and flower provided by flavonoids and carotenoids (basically secondary metabolites) shows activities like primary metabolites such as the attraction of pollinators and dispersion of seed and they have, therefore, also involved in plant reproduction (Winkel-Shirley 2001). Plant primary products are mainly glucose, nucleic acids, amino acids, proteins, carbohydrates, fats, and lipids and they are concerned with the structure, physiology, and genetics of the plant, which indicate their significant role in plant development. On the other hand, secondary products are available in very few numbers and also in less concentrations with comparison to primary metabolites. The production of carboxylic acids of the Krebs cycle is under the involvement of primary metabolism. In contrast, secondary metabolites are involved in providing fitness for survival to the plant species. The particular phytoconstituents in a certain species have been used to determine systematically the groups of secondary plant products that are used to classify the plants on the basis of the chemotaxonomic process (Winkel-Shirley 2001).

Plants produce an amazing diversity of low molecular weight compounds. Among the estimated 400,000–500,000 plant species in World, only a small percentage of plants have experimented phytochemically and a small fraction is subjected to biological or pharmacological screening. The capability to produce secondary plant products has been determined through the evolution of different plant progeny when they faced surrounding stresses:

(a) Floral scent volatiles and pigments have developed to attract insect pollinators and thus enhance fertilization that is involved in reproduction.
(b) Preparation of toxic chemicals to safeguard the pathogens and herbivores or to inhibit the development of neighboring plants.
(c) Chemicals found in fruits inhibit spoilage of fruits and provide signals in the form of color, aroma, and flavor to prove the presence of significant materials

such as sugars, vitamins, and flavoring agents. Animals eat fruit and thereby help in the dispersion of seeds.

(d) Some chemicals are involved in cellular functions which are unparalleled to certain plants such as resistance to salt or resistance to drought.

The various classes of secondary plant metabolites include:

(1) Alkaloids
(2) Saponins
(3) Phenolics
(4) Terpenes
(5) Carbohydrates
(6) Lipids
(7) Glycosides.

## 10.2 Alkaloids

Alkaloids are organic heterocyclic compounds having minimum one nitrogen atom in the ring or outside the ring. The term alkaloid is originated from "alkali-like" and hence, they show alkaline character due to the presence of lone pair electrons on the nitrogen atom. They become like the characteristics of some naturally derived complex amines. They include greater than 6000 nitrogenous compounds in about 15% of vascular terrestrial plants and spread in more than 150 plant families. The proper definition of alkaloids may not be described because they exist as a variety of group of compounds (Giweli et al. 2013). The only similarity in these compounds is that all these compounds have nitrogen atom. All alkaloids are not defined by a common definition. A huge number of alkaloids are isolated from different plant sources. Since alkaloids have a vast variation with respect to chemical structure present in these compounds, botanical sources, physiological responses, similar biochemical origin in spite of different taxonomic distribution and pharmacological activities, and therefore, they are classified chemically, taxonomically, pharmacologically and biosynthetic way, respectively (Tadeusz 2015; Nicolaou et al. 2011; Aniszewski 2007).

Chemically alkaloids are subclassified into three groups (Kokate et al. 2005; Clarke1970).

a. True alkaloids (Heterocyclic alkaloids): they are originated from amino acids and contain a heterocyclic ring.
b. Proto alkaloids (non-heterocyclic alkaloids): They are originated from amino acid but the nitrogen atom is not present in the ring structure
c. Pseudo alkaloids: They are not originated from amino acids but from purine or terpenoids. They have a nitrogen-containing heterocyclic ring (Table 10.1). They are classified as follows:

**Table 10.1** Ring structure present in alkaloidal drugs

| Heterocyclic ring | Alkaloidal drugs | Ring structure |
|---|---|---|
| *True alkaloidal drugs (Heterocyclic alkaloids)* | | |
| Indole | Reserpine, Ergotamine, Ergometrine, Physostigmine, Strychnine, Vincristine, Vinblastine | 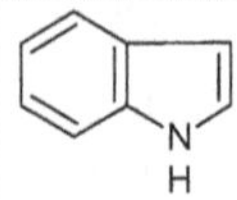 |
| Imidazole | Pilosine, Pilocarpine, | 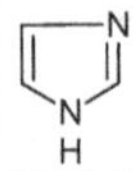 |
| Isoquinoline | Morphine codeine, Papaverine, Narcotine, d-tubocurarine, Emetine, Cephaeline | 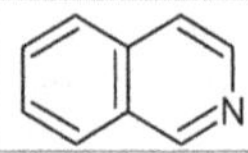 |
| Norlupinane | Lupanine, Sparteine | 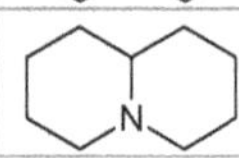 |
| Pyrrole and pyrrolidine | Cocoa, Hygrine | 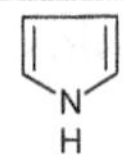 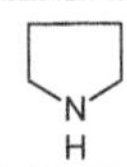 |
| Pyridine and piperidine | Arecoline, Conine, Lobeline | 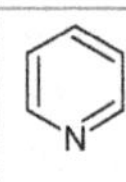 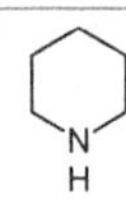 |
| Pyrrolizidine | Senecionine, Symphitine | 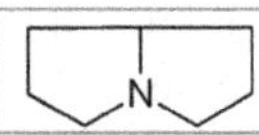 |
| Quinoline | Quinine, Quinidine, Cinchonine, Camptothecin | 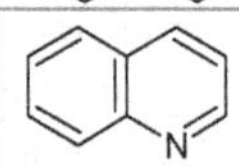 |
| Steroidal | Conessine, Funtumine, Solanidine, Protoveratrine, Withanine | 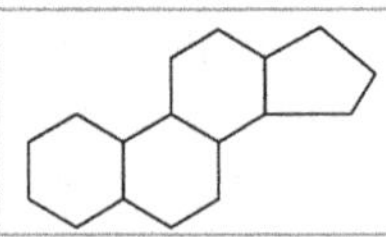 |
| Tropane | Atropine, Hyoscine, Cocaine, Meteloidine | 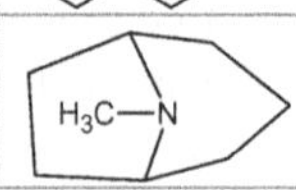 |
| *Proto alkaloidal drugs (non-heterocyclic alkaloids)* | | |
| Alkylamine (Amino alkaloid) | Ephedrine, Colchicine, Pseudoephedrine, Aconite | R-$NH_2$ |
| *Pseudo alkaloidal drugs* | | |
| Diterpenes | Aconitine, Aconine | $C_{10}H_{16}$ |
| Purine | Caffeine, theophylline, theobromine | 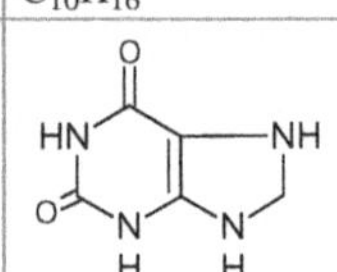 |

Although human beings have been using different parts of plants from the ancient age as drinks like tea, or as medicines like antimalarial, narcotic analgesics, but these alkaloidal substances were not separated and identified from the plant sources up to the beginning of the nineteenth century due to unavailability of modern analytical instruments (Seigler 1995). Alkaloids are not present in the lower plants. Lysergic acid derivatives (LSD) and gliotoxins (alkaloids having sulphur atom) are available in fungi. Other kinds of alkaloids such as ephedra, taxol, and lycopodium present in pteridophytes and gymnosperms are reported for their medicinal values. Based on the taxonomic characteristics, the alkaloids are classified as (a) *Centrospermae* belonging to the family *Chenopodiaceae*, (b) *Magnoliales* belonging to the families *Lauraceae, Magnoliaceae*, (c) *Ranunculales* belonging to the families *Berberidaceae, Menispermaceae, Ranunculaceae*, (d) *Papaverales* belonging to the families *Papaveraceae, Fumariaceae*, (e) *Rosales* belonging to the families *Leguminosae*, subfamily *Papilionaceae*, (f) *Rutales* belonging to the family *Rutaceae*, (g) *Gentiales* belonging to the families *Apocynaceae, Loganiaceae, Rubiaceae*, (h) *Tubiflorae* belonging to the families *Boraginaceae, Convolvulaceae, Solanaceae* (i) *Campanulales* belonging to the family *Campanulaceae* having subfamily *Lobelioideae*, and another family *Compositae* having subfamily *Senecioneae*. But yet, no such scientific proofs are available to show the availability of alkaloids in the plants of various dicot orders such as Salicales, Cucurbitales, Fagales, and Oleales (Evans 2009).

Chemicals found in fruits inhibit spoilage of fruits and provide signals in the form of color, aroma, and flavor to prove the presence of significant materials such as sugars, vitamins, and flavoring agents. Alkaloidal substances exhibit a lot of biological properties such as a) analgesic or painkiller, i.e.,morphine, colchicine; b) local anesthesia, i.e., cocaine; c) respiratory stimulation and relaxation, i.e., arecoline, lobeline; d) vasoconstriction, muscle relaxation, i.e., d-tubocurarine; e) anti-neoplastic, i.e., vincristine, vinblastine; f) hypotensive properties, i.e., reserpine, protoveratrine, and g) cardiac stimulation, i.e., ephedrine. The alkaloids also show another type of biological properties mentioned in the various journals like protection of plants from the herbivorous animals, arrest the growth or killing the bacteria, anticancer property, action against fungi and viruses, causing cancer. Some of the alkaloids are so toxic that they can even cause the death of animals. Several alkaloids like as nicotine and anabasine are useful as insecticides (Seigler 1995; Hoffmann 2003).

Various categories of alkaloids having different taxonomic distribution and pharmacological activities can be brought under the same category. This classification provides the significance of the precursor from which the alkaloids are biosynthesized in the plant. The alkaloids are biosynthetically classified on the basis of their amino acid precursor like lysine, tyrosine, phenylalanine, ornithine, tryptophan, etc. (Kokate et al. 2005).

The chemical structures of some major amino acid precursor are represented below (Fig. 10.2).

The chemical structures of some well-known alkaloids (Fig. 10.3), are as follows (Kokate et al. 2005; Tyler et al. 1988).

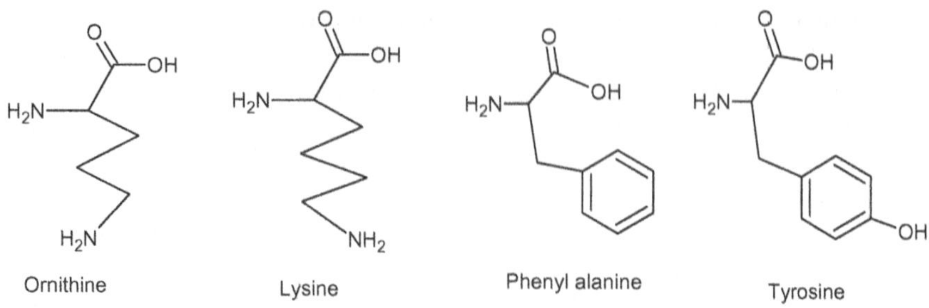


**Fig. 10.2** The structures of amino acid precursors of alkaloidal drugs

The schematic diagram of the biosynthesis of different alkaloids from various amino acids (Figs. 10.4, 10.5, 10.6, 10.7) is depicted below (Kokate et al. 2005; Ali 2019, Evans and Mitch 1982).

## 10.3 Saponins

Plants having saponin glycosides are of economical importance, as well as medicinal value. Saponins are those compounds which contain a polycyclic aglycone moiety known as a sapogenin. The aglycone part is either a steroidal nucleus also called tetracyclic triterpenoid saponins or triterpenoid which contains five rings also popular as pentacyclic triterpenoid saponins (Fig. 10.8). Pentacyclic saponins are further differentiated into (a) $\alpha$-amyrin type (b) $\beta$-amyrin type, and (c) lupeol. The aglycone part is attached to a carbohydrate unit that may be a monosaccharide unit like glucose, mannose, galactose, etc., or sometimes oligosaccharide chain like glucose-gulose-mannose-rhamnose. This glycone part comprises various pentoses and, hexoses sugars, or uronic acids. This hydrophobic-hydrophilic balance made by these types of compounds lower the surface tension between two immiscible liquids and also make lather in their aqueous solutions and hence behave like a soap. Another important property of saponin is the ability to cause hemolytic of red blood corpuscles (RBC) in vitro. The non-sugar part of the saponin compound is known as genin or especially sapogenin due to the ability of the formation of foam. Saponins are found widely among plants and reported more than 500 plants from about 90 different families. These substances are extracted from most of the plant parts like roots bulbs, leaves, stems, flowers, and fruits. The literature survey reveals that saponins are usually found in the roots of various plants like *Chlorophytum borovillianum* known as safed musali, *Dioscorea deltoidea* popular as wild yam, *Eleutherococcus senticosus* called as Siberian ginseng, *Asparagus racemosus* popular as Shatavari, *Gentiana lutea* generally called gentian, *Glycyrrhiza glabra* popular as licorice and *Panax ginseng* called as Korean ginseng (Assa et al. 1973).

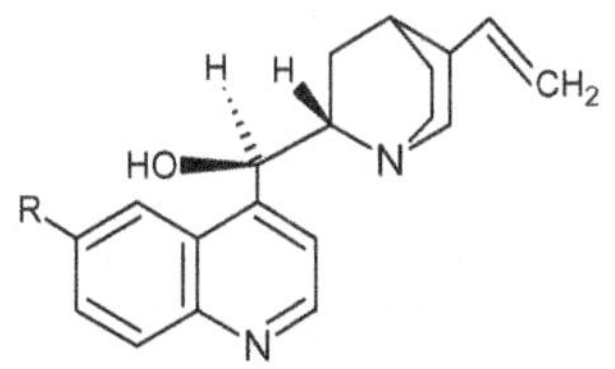

Quinine R= $OCH_3$ ( 8S, 9R)
Cinchonidine R= H

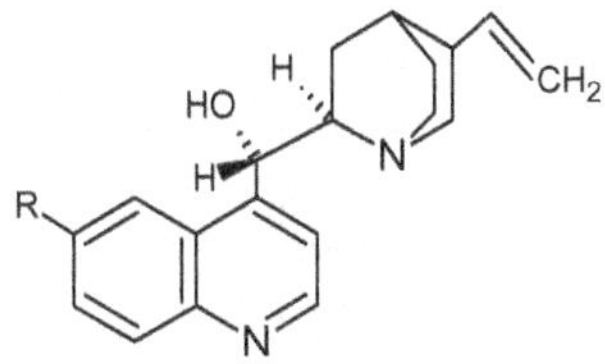

Quinidine R= $OCH_3$ ( 8R, 9S)
Cinchonine R= H

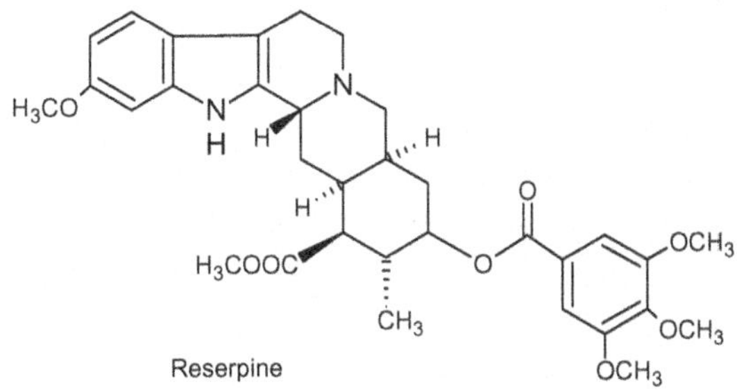

Reserpine

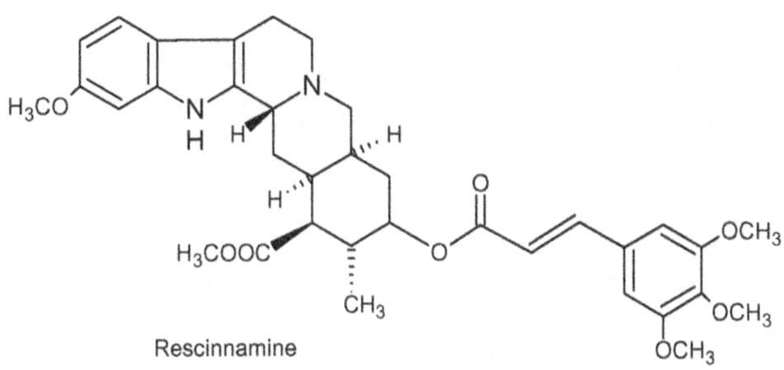

Rescinnamine

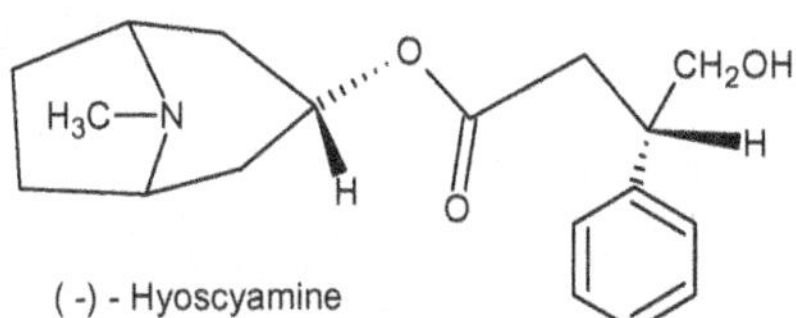

( -) - Hyoscyamine
( $\pm$ ) - Hyoscyamine = Atropine

**Fig. 10.3** The structure of some important alkaloidal drugs

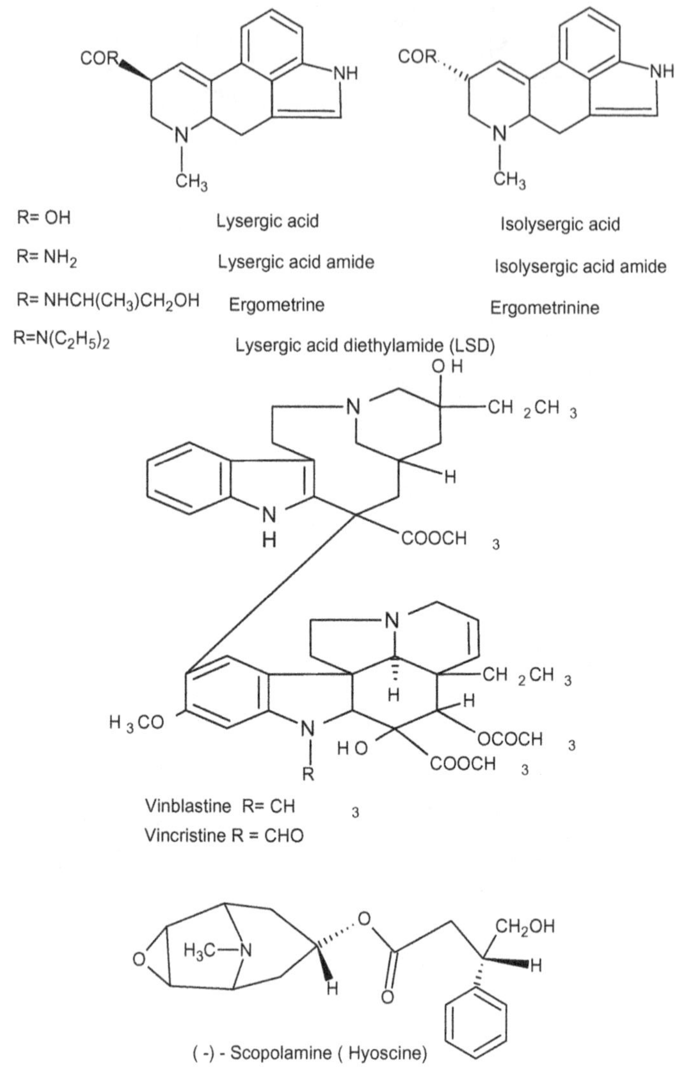


**Fig. 10.3** (continued)

$$\text{Saponin} = \text{Sapogenin} + \text{Sugar}$$

Sapogenin has a steroidal nucleus (neutral saponins) and a pentacyclic triterpenoid nucleus (acid saponins). Sapogenin includes diosgenin, asiatic acid, panaxadiol, glycyrrhetinic acid, carbenoxolone, brahmic acid, and shatavarin I–IV, etc. (Fig. 10.9). Sugars include glucose, xylose, glucuronic acid, arabinose, rhamnose, mannose.

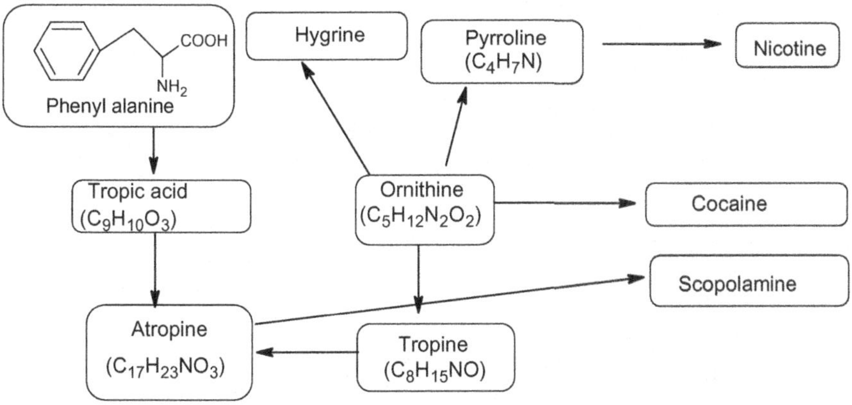


**Fig. 10.4** Schematic diagram of biosynthesis of alkaloids derived from ornithine

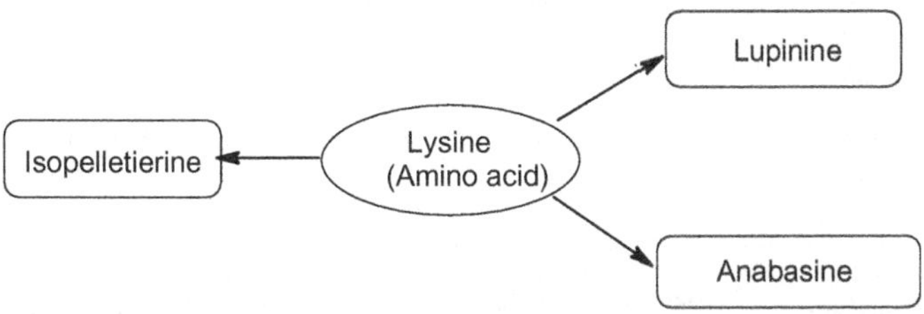


**Fig. 10.5** Schematic diagram of biosynthesis of alkaloids derived from lysine

Saponins have shown a lot of pharmacological activity. Some saponins have hypoglycaemic, antitumor, piscicidal, diuretic, molluscicidal, nervine tonic, adaptogen, spermicidal, galactagogue sedative, stimulant expectorant, and analgesic activities. Glycyrrhizin from *G. glabra* belonging to the family Fabaceae is useful as an expectorant, antitussive agent, and for the treatment of peptic ulcer. Saponin is also useful for the treatment of chronic hepatitis, as well as cirrhosis. The saponins obtained from *Bupleurum falcatum* belonging to the family Apiaceae exhibit potential anti-inflammatory action in the rat paw edema model. In Korean medicine, the saponin derived from the roots of *Phytolacca americana* also shows anti-inflammatory activity. Another saponin, aescin, showing anti-inflammatory activity is obtained from the plant, *Aesculus hippocastanum* popular as horse chestnut. Aescin has been reported 600 times more potent anti-inflammatory action in comparison to another standard molecule, rutin, in rat paw edema model (Guclu-Ustundag et al. 2007).

A list of saponin drugs with their active constituents and uses (Table 10.2), are given below (Kokate et al. 2005, Sparg et al. 2004).

Some chemical structures of important saponins are given below in Fig. 10.9 (Kaur 2010; El Aziz et al. 2019).

The sapogenins are biosynthesized by a mevalonic acid pathway from acetyl CoA to squalene, a 30 carbon-containing compound, with the help of several steps.

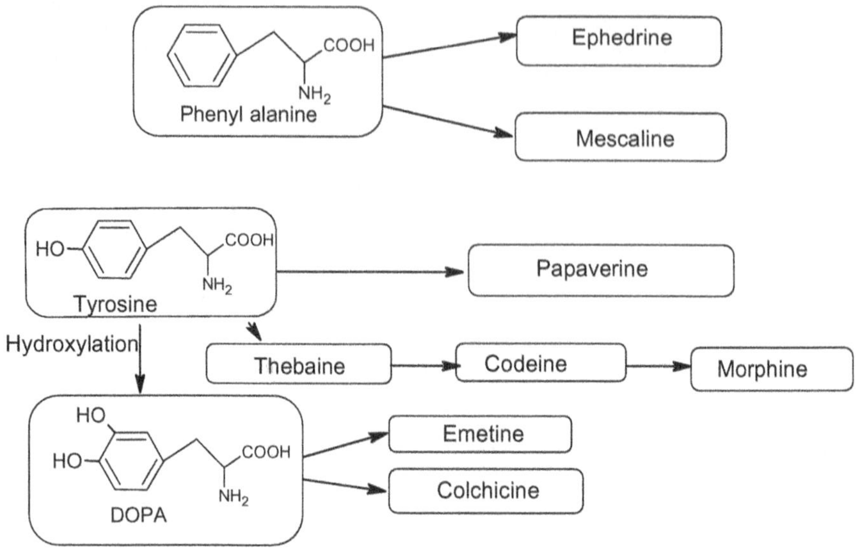


**Fig. 10.6** Schematic diagram of biosynthesis of alkaloids derived from tyrosine and phenyl alanine

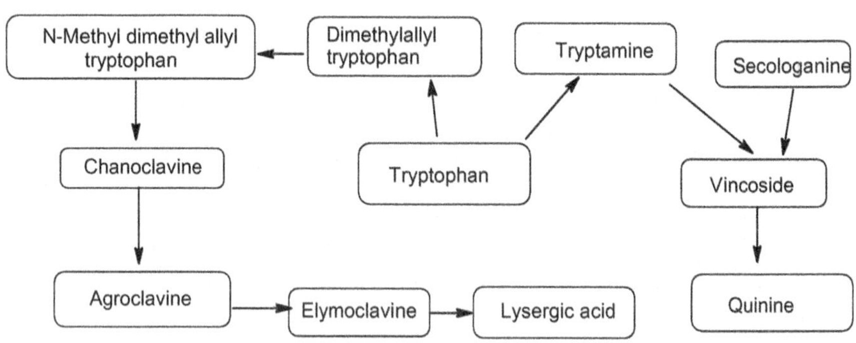


**Fig. 10.7** Schematic diagram of biosynthesis of alkaloids derived from tryptophan

Squalene on further structural modification produces various aglycone namely, diosgenin, yamogenin, tigogenin, etc. (Fig. 10.10).

## 10.4 Phenolic Compounds

Among all the secondary plant products, phenolic compounds have created great importance in plant systems. The color, flavor, and taste of the plant parts, foods, and drinks are due to the presence of phenolic compounds such as flavonoids, coumarins, and tannins, etc. Therefore, phenolic compounds have become the

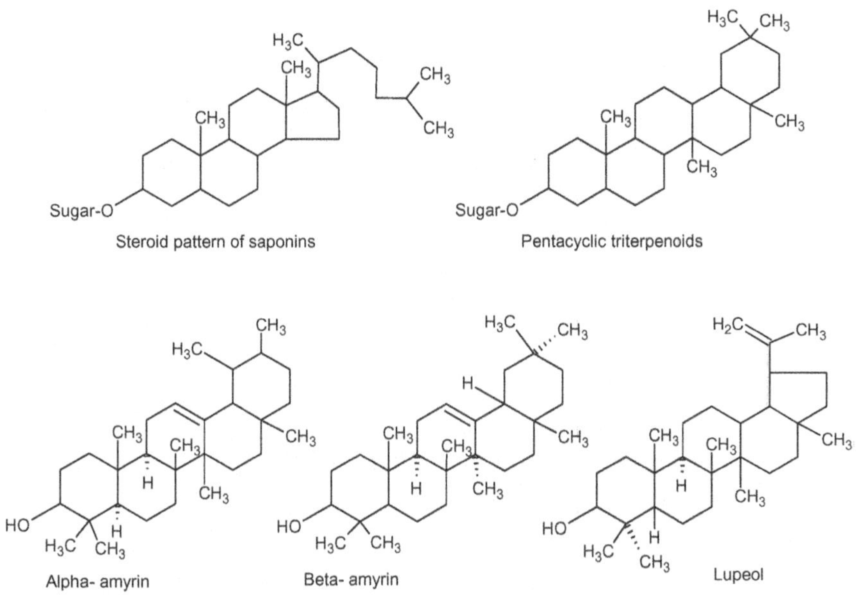


**Fig. 10.8** The structure of the skeleton of sapogenins

largest group of secondary plant products. The common characteristic of such a group is the presence of one or more phenol groups. The compounds are available either as a very simple aromatic structure or high molecular complex structure due to polymerization. Phenolic compounds manufactured by vegetables, leaves, fruits cocoa, tea, and various plants exhibit definite health well-being. Some phenolic compounds exhibit a variety of pharmacological properties like anti-inflammatory properties, e.g., quercetin and hepatoprotective potential, e.g., silybin. Some other compounds show estrogen-like action, e.g., genistein and daidzein are also active against insects, for example, naringenin. Phenolic molecules also act as potent antioxidants and free radical scavengers, especially flavonoid compounds (Golawska et al. 2014).

Other literature revealed that they exhibit anti-carcinogenic and other biological properties along with antioxidant and anti-inflammatory activities (Park et al. 2001). Simple phenolic compounds are antiseptic at low concentration, whereas disinfectant at higher concentration and act against different helminths. Phenol itself shows as an antimicrobial agent (Pengelly 2004). Phenolic compounds are widespread in almost all the plants and involved in various studies in different fields such as biological, chemical, agricultural, and pharmaceutical fields (Dai et al. 2010; Herrmann et al. 1989).

Phenolic compounds may be classified according to their chemical structure or biosynthetic precursor (Fig. 10.11). Based on their structures, phenolic compounds can be classified into (a) Simple phenolics (b) Coumarins (c) Flavonoids (d) Chromones xanthones (e) Tannins (f) Lignans, and (g) Stilbenes.

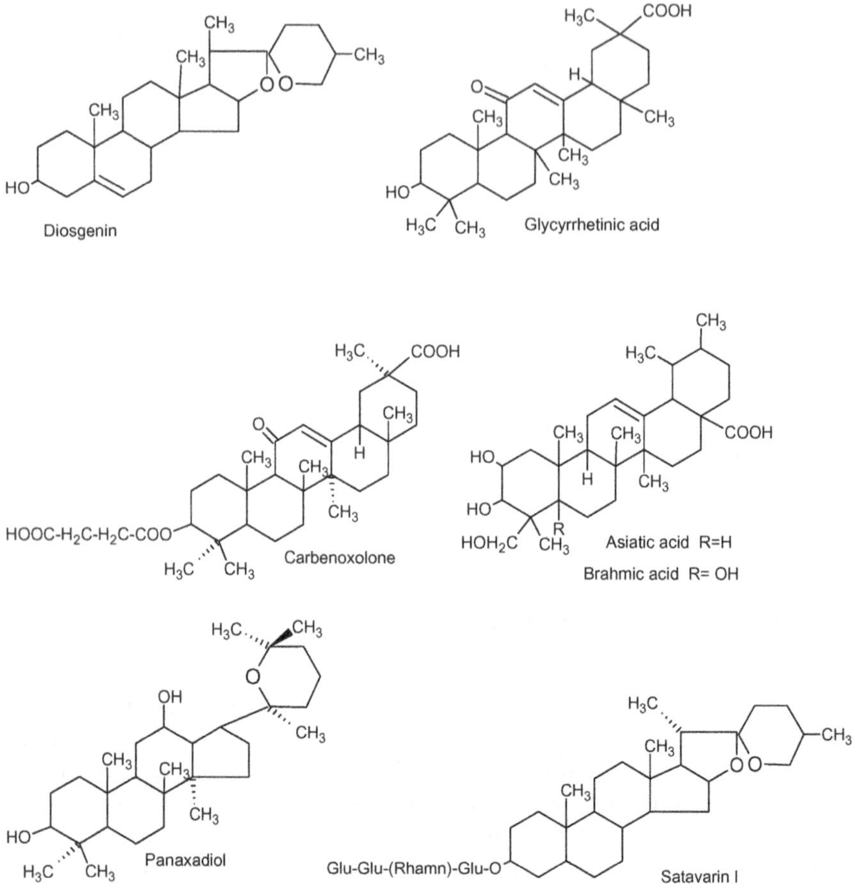


**Fig. 10.9** The chemical structures of important saponins

## 10.4.1 Simple Phenolics

Phenolic substances are universal in plants but the availability of free phenol (also known as carbolic acid) in plants is very rare. Gallic acid (3,4,5-trihydroxy benzoic acid), a phenolic acid is found in the plants as gallotannins. This compound shows astringent property due to its ability to precipitate proteins in cases of internal hemorrhage and also exhibits several pharmacological activities, for examples, anti-inflammatory activity to inhibit inflammation, antiviral, widely used in fungal infection and viral infection, shows cytotoxicity against cancer, some time to treat anaphylactic shock, antibacterial, antimutagenic, choleretic, and bronchodilatory actions. Gallic acid is also used to treat diabetes by inhibiting insulin degradation and enhances the relaxation of smooth muscles (Harborne et al. 1993).

**Table 10.2** The uses and active constituents of saponin obtained from plants

| Sl. No | Drug and synonym | Active constituents | Uses |
|---|---|---|---|
| 1 | Dioscorea (Yam) | Diosgenin, precursor of prostaglandin | Tonic, anti- rheumatoid arthritis, for synthesis of oral contraceptives, cortisone, prednisone, progesterone |
| 2 | Shatavari (satamuli) | Shataverin I and Shataverin II | Galatogogue (increase production of milk), diuretic, tonic, anti-oxytocic, anti-rheumatoid |
| 3 | Brahmi | Bacoside A and Bacoside B | Nerve tonic (increase memory) |
| 4 | Liquorice | Glycyrrhizin, Carbenoxolone, Liquiritin, isoliquiritin | Treatment of peptic ulcer, expectorant and demulcent, antispasmodic, addison's disease |
| 5 | Safed musali | Hicogenin | General tonic, sex stimulant |
| 6 | Ginseng | Panaxadiol, panaxatriol ginsenoside, oleanolic acid | Tonic, sex stimulant, and adaptogen, cosmetics, immunomodulatory |
| 7 | Gokhru | Gitogenin, ruscogenin, chlorogenin | Aphrodisiac, treatment of gout, calculous formation and painful micturation, diuretic |
| 8 | Momordica | Charantin, momordicin | Hypoglycaemic agent |

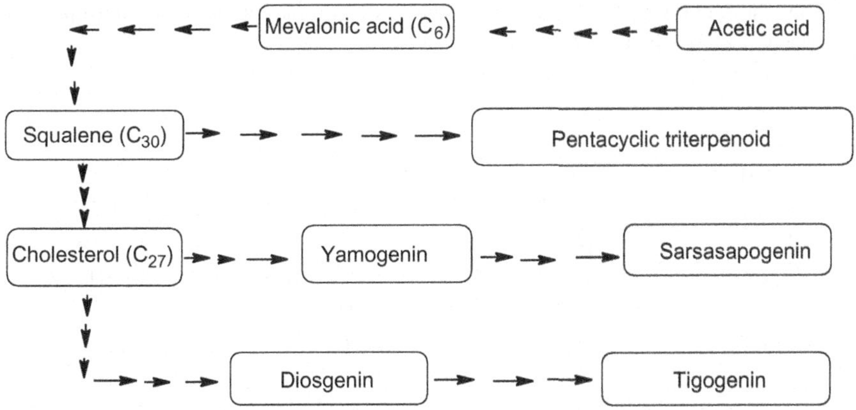


**Fig. 10.10** The Schematic diagram of the biosynthesis of different sapogenins (Kokate et al. 2005, Mugford et al. 2013)

The simple phenolic compounds vary on the basis of the presence of their functional groups, such as hydroxyl (–OH), aldehyde (–CHO), or carboxylic (–COOH) group; these are a phenolic aldehyde, e.g., vanillin, phenolic acids, e.g., salicylic acid, ferulic acid, and caffeic acid and a phenolic phenylpropane, e.g., eugenol. Hydroquinone (4-hydroxy phenol) is abundantly found as simple phenols,

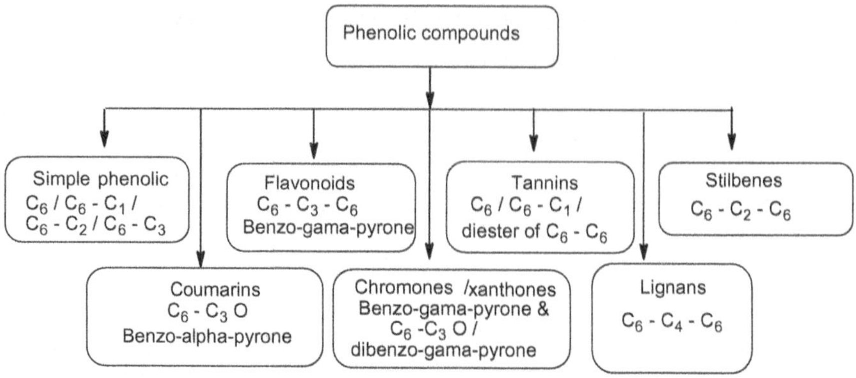


**Fig. 10.11** Classification and carbon skeleton of phenolic compounds of plant sources

in arbutin, a glycoside (Fig. 10.12). The well-known glycoside coniferin and many related substances derived from phenolic cinnamic alcohols are considered as an ingredient for the biosynthesis of lignin (Evans 2009; Hoffmann 2003).

Simple phenolic substances show various pharmacological activities such as the urinary tract infection exhibited by arbutin (Zbigniew et al. 2014), as well as anti-inflammatory action shown by salicylates (Amann and Peskar 2002). It is popular that all phenols at low concentrations act as antimicrobial. In fact, phenol was considered as the first antiseptic used in surgical operation (Pelczar et al. 1988).

The chemical structures of some important simple phenolic compounds are represented below in Fig. 10.12.

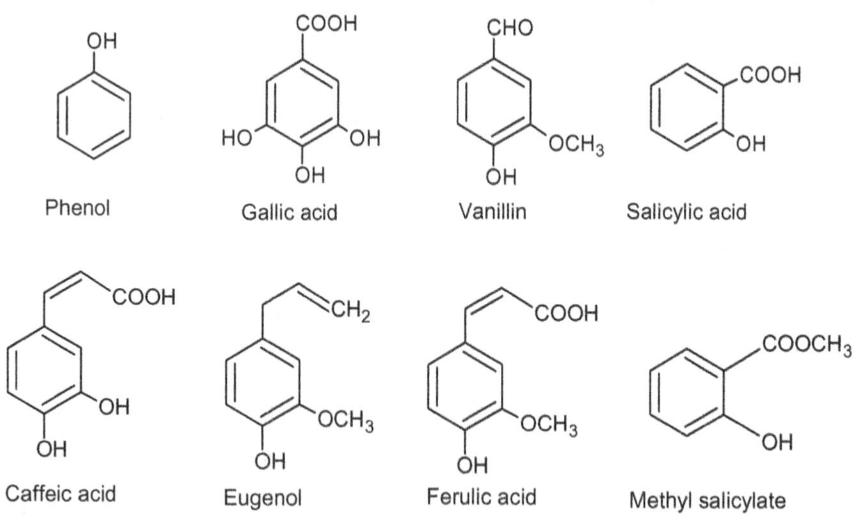


**Fig. 10.12** The chemical structures of simple phenolic compounds derived from plants

## 10.4.2 Coumarins

Coumarins and its derivatives are found in so many plants. Coumarin is ubiquitously available in about 150 species of 30 different families. They are available in the families mainly Solanaceae, Rubiaceae, Leguminosae, Oleaceae, Umbelliferae, Caprifoliaceae, etc. The important source of coumarin compounds is sweet clover belonging to *Melilotus* spp., tonka bean (scientific name: *Dipteryx odorata*), and sweet woodruff (scientific name: *Galium odoratum*) (Hoffmann 2003). They are the derivatives of benzo-α-pyrone, also known as lactone of o-hydroxycinnamic acid, i.e., cyclic ester of phenolic compound (Hoffmann 2003). Aesculetin, aesculin, herniarin, fraxin, umbelliferone, scopolin, and scopoletin are popular coumarin compounds having benzo-α-pyrone nucleus available both in the free state and as glycosides also (Fig. 10.13). Plants having coumarin derivatives are belladonna (*Atropa belladonna*), Datura (*Datura stramonium*) belonging to the family Solanaceae, February daphne (*Daphne mezereum* belonging to the family Thymeliaceae), common rue (*Ruta graveolens* belonging to the family Umbelliferae), and Horse chestnut (*A. hippocastanum* belonging to the family Hippocastanaceae) and certain Rosaceae (Evans 2009).

Furanocoumarin derivatives are also available in bergamot oil, bael fruit, psoralea, etc. These compounds are formed by the fusion of furan ring with coumarin at 6 and 7 positions and 7 and 8 positions, respectively (Bruni et al. 2019). These

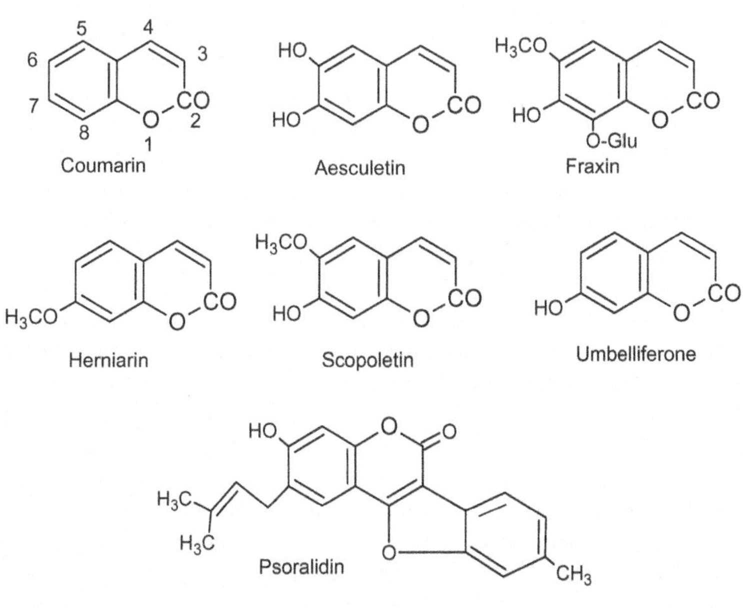


**Fig. 10.13** The chemical structures of some important coumarin drugs of plant sources

compounds show various pharmacological activities like anticoagulant, anticancer, anti-Alzheimer, and anti-inflammatory activities (Xu et al. 2015).

The structures of some important coumarins are shown below in Fig. 10.13.

### *10.4.3 Flavonoids*

Among the normally occurring phenolic compounds, flavonoids have taken a maximum place in the plant kingdom. Out of about 2000 of these known compounds, around 500 compounds are found as a free state (Evans 2009). The most available flavonoids include flavones, flavonols, and anthocyanins. The flavones and their derivative are generally yellow in color (Latin *flavus*, meaning yellow). These compounds are ubiquitously distributed worldwide but they are mainly available in young tissues and in the higher plants. They are found in Pteridophyta (mainly fern and related plants), Gymnosperma (conifer), monocots, and dicots. Among the dicot families, plants belonging to the families Leguminosae, Polygonaceae, Rutaceae, Umbelliferae, and Compositae are a rich source of flavonoids. Researches worldwide have reported the medicinal activities of drugs containing flavonoids. For examples, milk thistle (*Silybum marianum* belonging to the family Compositae), liquorice root (*G. glabra* belonging to the family Leguminosae), Roman chamomile (*Chamaemelum nobile*, Buck–wheat (*Fagopyrum esculentum* belonging to the family Polygonaceae), and ginkgo ( *Ginkgo biloba Linn.* belonging to the family Gingkoaceae). Many herbs having flavonoids have now been incorporated in the *British Pharmacopeia* (B.P.). For example, Birch Leaf (*Betula pendula*), Calendula flower (*Calendula officinalis*), Horsetail (*Equisetum ramosissimum*), Elderflower (*Sambucus nigra*), Motherwort (*Leonurus cardiac*), Lime Flower (*Tilia cordata*), and passionflower (*Passiflora edulis*) are well-known plants. The flavonoid compounds have been shown pharmacological properties like anti-inflammatory, antiallergic, antithrombotic, vasoprotective, antitumor, antifungal, hepatoprotective, and gastric mucosal protective activities (Table 10.3) (Montanher et al. 2007; Serafini et al. 2010). Flavonoids have a 15 carbon skeleton arranged as $C_6$–$C_3$–$C_6$ in which two phenyl rings are linked by a three-carbon chain. A carbonyl group is present at one end of a three-carbon chain. This three-carbon chain may be either open in case of chalcone and dihydrochalcone (simplest naturally occurring flavonoid) or as a part of the heterocyclic ring and the heterocyclic ring fused with benzene ring forms chroman ring. Flavone and flavonol are common flavonoids in which B-ring is attached at 2 position of C-ring, on the other hand, in isoflavone structure, the B-ring is attached to 3 position of C–ring. Flavonoids exist as free molecules or as glycosidic forms (Fig. 10.14). The glycosidic form of flavonoids is generally O-glycosides and few are C-glycosides. Chemically, flavonoids are also known as benzo-γ-pyrone derivatives. Biflavonoids are also available in which two flavonoids molecules are joined through either C–C linkage or C–O linkage. Anthocyanin present in most plants is responsible to provide red, blue, purple color to the flower. Anthocyanin

**Table 10.3** The uses of some flavonoid compounds of plant sources

| Compound or plants | Pharmacological activities |
|---|---|
| Silymarin | Hepatoprotective, chronic inflammatory hepatic disorder |
| Baicalin | Anti- HIV, antihypertensive, anti-inflammatory |
| Genistein | Anticancer, cosmetic |
| Isoflavone of Erythrina sigmoidea | Treatment of fungal infection, antibacterial, anti-venom, antirheumatic, kidney treatment |
| Formononetin | Estrogenic activity, oral contraceptive, antioxidant, treatment of Alzheimer's disease |
| Coumestrol | Estrogenic activity, breast cancer treatment |
| Gingko | Anti-Alzheimer's disease, Raynaud's disease, ant- PAF, acrocyanosis, anti-sepsis |
| Buck wheat | Treatment of retinal hemorrhage |
| Quercetin | Anti-oxidant, estrogenic activity |

exists as flavylium ion in which a positive charge is present on O–atom of chroman ring (Kokateet al. 2005; Wang et al. 2018).

The pharmacological uses of some significant flavonoid compounds are represented in the following Table 10.3 (Wang et al. 2018).

The structures of some important flavonoids are shown below in Fig. 10.14 (Kokate et al. 2005; Panche et al. 2016).

### 10.4.4 Chromones and Xanthones

Chromones and xanthones are also structural derivatives of benzo-γ-pyrone (Figs. 10.15 and 10.16). The biological importance of these compounds is less. Only a few compounds have medicinal importance, for example, eugenin is obtained in the clove plant and khellin is produced from mustard seeds (Bagci et al. 2010). More complex structures of chromone are furanochromones, the active constituents from *Ammi visnaga*.

Xanthones compounds are available mainly in the families of Gentianaceae, as well as Guttiferae, otherwise spread scattered over the plant kingdom in the families of Moraceae, as well as Polygalaceae. Mangostin isolated from mangosteen tree (*Garcinia mangostana*) has been reported for several biological activities like antibacterial, anti-inflammatory, antioxidant, and anti-cancer properties (Jung et al. 2006). *Polygala nyikensis* is mainly used for the treatment of fungal infection by the highlanders of Malawi and bordering countries. Xanthones are also useful as a larvicide. The root of the plant exhibits antifungal activity due to the presence of xanthones (Susana et al. 2011).

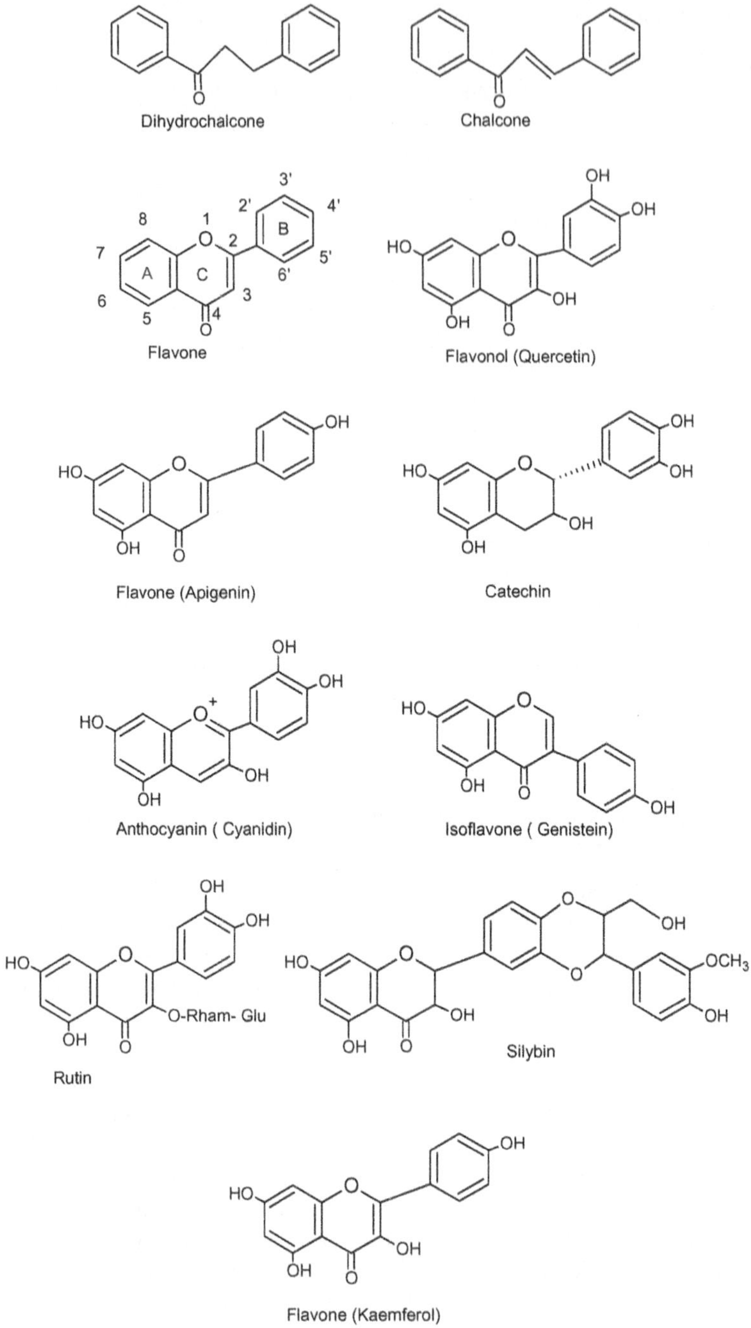


**Fig. 10.14** The structures of some important flavonoids derived from plants

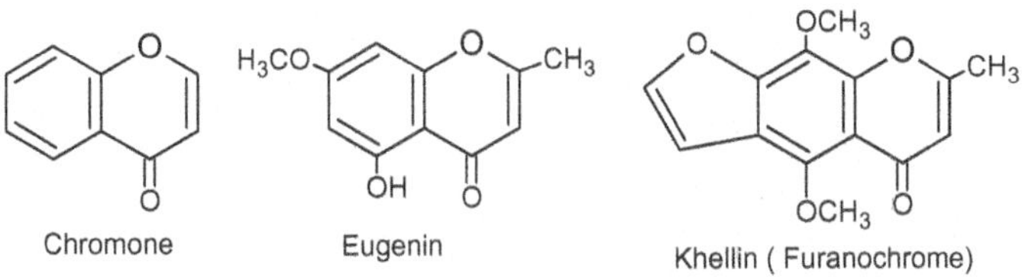


**Fig. 10.15** The structures of chromones obtained from plants

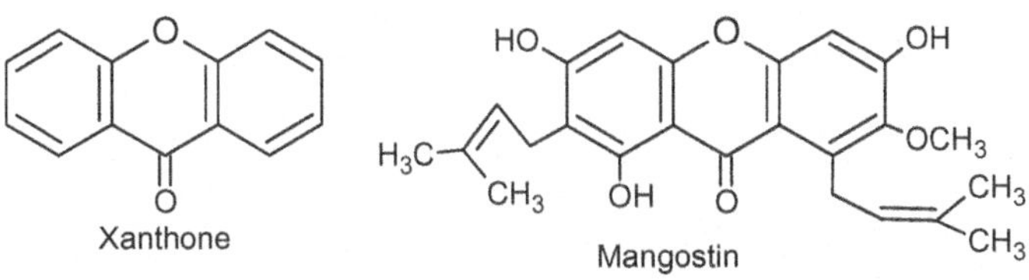


**Fig. 10.16** The structures of xanthones obtained from plants

## *10.4.5 Tannins*

Tannins are the most widely distributed natural compounds found in various families of higher plants. They are non-crystalline substance and make colloid when mixed with water. Tannins are available in vacuoles and cell sap. Chemically they are a mixture of various complex organic compounds containing polyphenols mainly 1,2-dihydroxy benzene (catechol) or 1,2,3-trihydroxy benzene (pyrogallol). Tannins are soluble in dilute alkalies, glycerine, and alcohol due to the formation of hydrogen bonds. Though they are organic compounds, basically they are insoluble in organic solvents except for acetone. Tannins show some characteristic reactions such as precipitation of alkaloids and gelatin, precipitated by copper, lead, and tin salts, as well as concentrated potassium dichromate solution, bluish-black, or brownish-green color with ferric chloride (Sieniawska et al. 2017).

Tannins containing polyphenols have the capability to precipitate protein. Due to its ability to convert raw animal skins into leather, and therefore, they are extensively used in leather technology for decades. Tannins crosslink the proteins of animal hides and therefore hides become more resistant to bacteria, as well as fungus, resulting in the prevention of putrefaction (Hagerman et al. 1981). Based on their chemical characteristic and behavior on dry distillation, tannins are generally divided into two major groups: (a) hydrolyzable tannins (b) condensed tannins or non-hydrolyzable tannins. The hydrolytic product of hydrolyzable tannins is generally gallic acid (3,4,5-trihydroxybenzoic acid) or ellagic acid (di ester of hexahydroxydiphenic acids) which is combined by ester linkage with –OH group of a glucose molecule. They are hydrolyzed quickly by acids or enzymes. Two main types of hydrolyzable tannins are available such as gallotannins and ellagitannins derived from gallic acid and ellagic acid, respectively. Ellagitannins have the

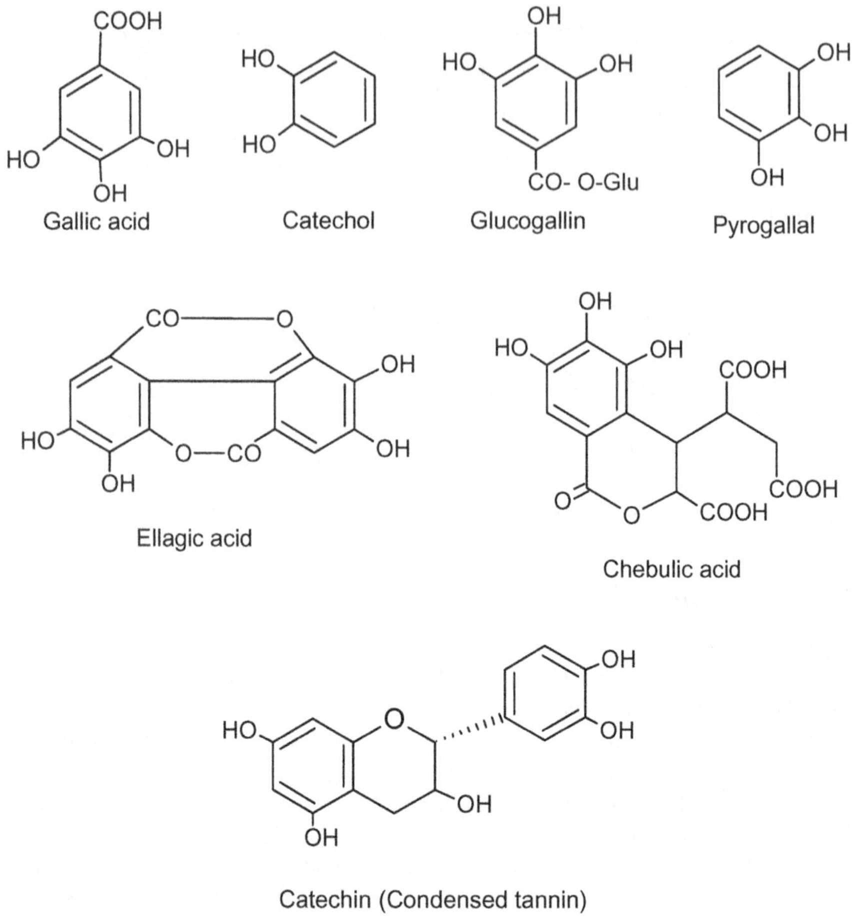


**Fig. 10.17** The structure of some hydrolyzable tannins

medicinal interest, for this reason, the structures of ellagitannins have been elucidated by analytical techniques (Fig. 10.17). The elucidated structures comprise geraniin sequestered from *Geranium robertianum* known as Herb Robert and also *Geranium maculatum* called as American cranesbill (Catarino et al. 2017) and tellimagrandins 1 and 2 sequestered from *Quercus alba* called as Oak bark, *Punica granatum* known as pomegranate and *Filipendula ulmaria* called as Meadowsweet (Yi et al. 2004).

Condensed tannins, or proanthocyanidins, or non-hydrolyzable tannins are very much reverberating to hydrolysis. These substances are oligomeric flavonoids derived from flavones like flavan-3-ol, flavan 3,4-diol, or catechin. The characteristics of such types of tannins depend upon four parameters namely, the type of linkages between flavonoid units; hydroxyl group arrangement, or the enantiomeric effect of the carbon atoms situated at 2, 3, 4 positions of pyran ring and the effect of substituents on the ring structure. Few substances contain both the hydrolyzable and non-hydrolyzable tannins, e.g., bark and leaves of *Hamamelis virginiana and leaves of Camellia sinensis* popular as a tea (Goenka et al. 2013).

The drugs having tannins have a great medicinal value. They are useful for the treatment of diarrhea, to check small hemorrhage in the gastrointestinal tract (g.i.t.), as a mild antiseptic, and as antidotes in poisoning by heavy metals like lead, tin, and alkaloids by precipitating them. It has reported that Epigallocatechin-3-gallate obtained from tea, shows antiangiogenic activity in mice. The juice of *Vaccinium oxycoccos* (cranberry) is used for long times as urinary antiseptic which is also experimentally proven (Jepson et al. 2008).

### 10.4.6 Stilbenes

Stilbenes are another important phytoconstituents broadly distributed worldwide. They are available as heartwood constituents of plant species. They are generally found in the heartwood of trees like as, *Eucalyptus* belonging to the family Myrtaceae, *Madura* belonging to the family Moraceae, and *Pinus* belonging to the family Pinaceae (Seigler 1995). Resveratrol (3,5,4′-trihydroxy-trans-stilbene), a type of natural phenolic compound is widespread in different plants (Fig. 10.18). Resveratrol exhibits estrogen-like property and available in *Picea*, Fabaceae, Myrtaceae, and Vitaceae (Gehm et al. 1997).

### 10.4.7 Lignans

Lignans ($C_{18}$) are 18 carbon-containing compounds that are biosynthetically manufactured by dimerization of two units containing $C_6$–$C_3$ linked at β-carbons of the side chain. Neo-lignans are not linked by β-β' (or 8–8′) -carbon bonds and are derived by the union of the head to tail of two $C_6$–$C_3$ moieties. On the basis of structural types, they subclassified as Aryl tetrahydronaphthalene, arylnaphthalene, dibenzocyclooctadiene, cyclobutane, dibenzylbutane, dibenzyl-butyrolactone, tetrahydrofuran, and furofuran (Fig. 10.19). The plants having medicinally

**Fig. 10.18** The structure of the stilbene derivative

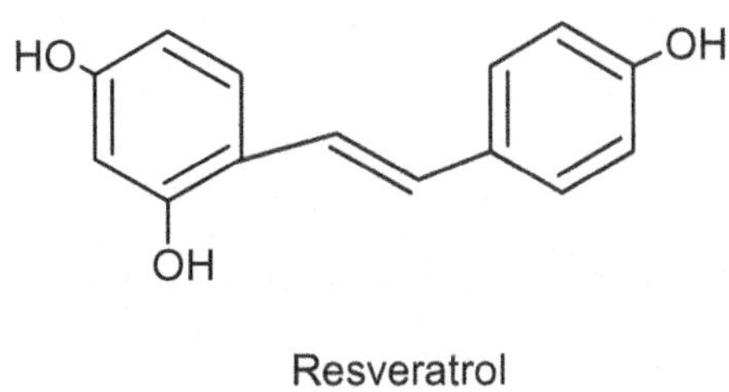

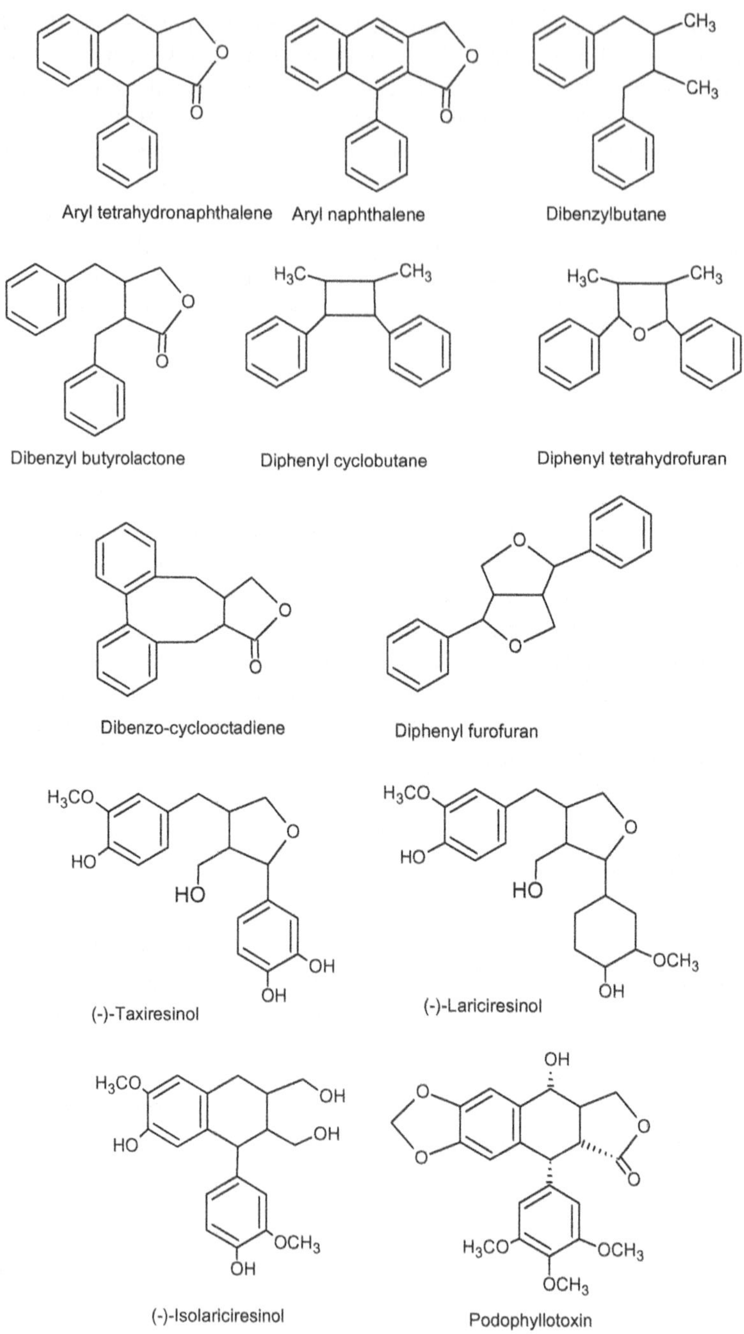


**Fig. 10.19** The structure of lignan derivatives derived from plant sources

important lignans include *Taxus baccata linn., Podophyllum hexandrum, Schisandra chinensis, Linum usitatissimum,* and *S. marianum.* It is reported that taxiresinol and lariciresinol are known as flavonoid lignins whereas isolariciresinol and demethyl-isolariciresinol called as dibenzyl butane lignans, all these lignans show anti-ulcerogenic activity. Taxiresinol, and demethyl-isolariciresinol exhibit antifungal property. Podophyllotoxin is reported as an anticancer agent. Schisandrin A, B, C show hepatoprotective activity. They are also used as antioxidants, tonifier, anti-fatigue anti-inflammatory agents. Silymarin is useful for liver protection and shows lipid-lowering activity. Lignan obtained from linseed is reported to reduce colon and mammary cancer (Cunha et al. 2012).

Various structural nuclei of lignans are given below (Cunha et al. 2012).

The biosynthesis of different phenolic compounds from shikimic acid is shown in the schematic diagram (Fig. 10.20).

## 10.5 Terpenes and Terpenoids

Terpenes and terpenoids comprise one of the largest group of secondary plant metabolites. The term "terpene" is originated from the word "turpentine", meaning resin. Terpene represents hydrocarbon $(C_5H_8)_n$ and terpenoids comprise hydrocarbon and their oxygenated products (Perveen 2018). They are also known as volatile oils or ethereal oils. Most of these compounds are employed in the application of the pharmaceutical industry as pharmaceutical dosage forms or flavoring agents and as perfumes in cosmetic and perfumery industries. Many of these compounds are used as insect repellants, fungicides, as well as waterproofing substances. They mediate electron transport processes in respiration and photosynthesis. They are also used for the treatment of various diseases (Perveen 2018; Lalonde 2005).

Terpenoids are soluble in ether, alcohol, and lipid solvents and basically insoluble in water. They are generally lighter than water and high refractive index. They are mostly optically active either dextrorotatory or levorotatory. These compounds are obtained from the duct, cell, trichomes, and lysigenous or schizogenous glands. They are generally found in the families of Umbelliferae, Labiatae, Myrtaceae, Zingiberaceae, Lauraceae, Rutaceae, and Piperaceae (Perveen 2018; Lalonde 2005). The structure of isoprene $(C_5H_8)$ unit is given below

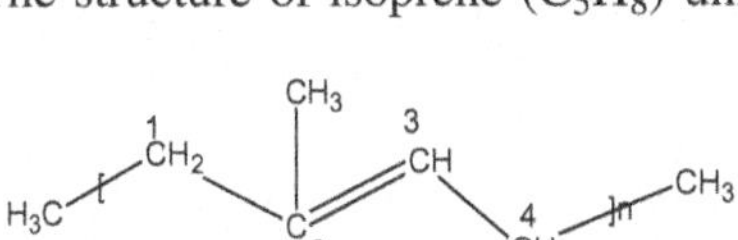


Isoprene unit ( 2-Methyl 1, 3-butadiene]

Based on the number of isoprene $(C_5H_8)$ units, terpenes are divided into the following classes (Table 10.4).

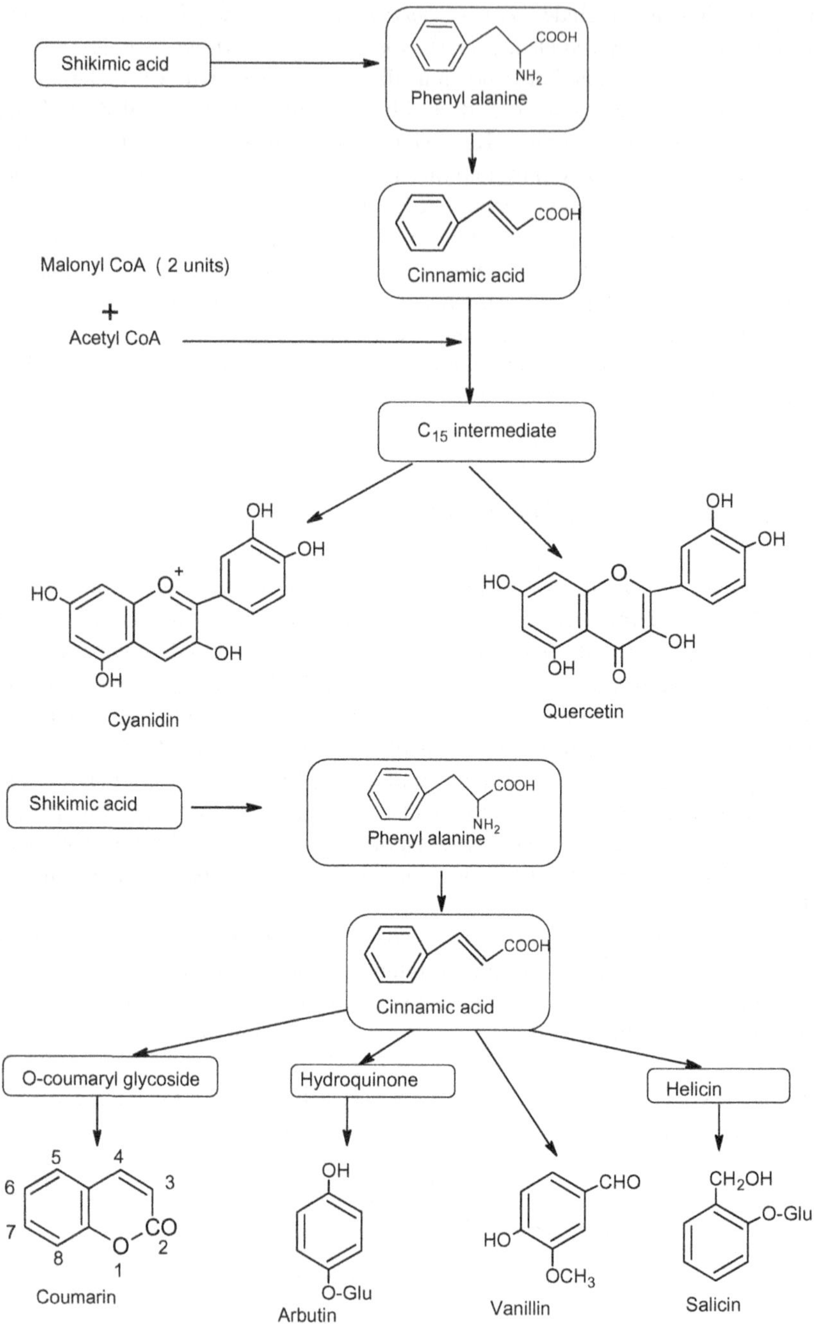


**Fig. 10.20** Biosynthesis of various phenolic compounds (Santos-Sanchez et al. 2019)

**Table 10.4** The formula and examples of different classes of terpenoids of plant sources

| Number of isoprene units | Class ( formula) | Examples of compounds/drugs |
|---|---|---|
| 1 | Hemiterpene ($C_5H_8$) | Isoprene, prenol, isovaleric acid, tiglic acid |
| 2 | Monoterpenes ($C_{10}H_{16}$) | Limonene, eucalyptol, pinene |
| 3 | Sesquiterpenes ($C_{15}H_{24}$) | Abscisic acid ( ABA) |
| 4 | Diterpenes ($C_{20}H_{32}$) | Gibberellin |
| 5 | Sesterterpenes ($C_{25}H_{40}$) | Dysidiolide |
| 6 | Triterpenes ($C_{30}H_{48}$) | Brassinosteroids, squalene, lanosterol |
| 8 | Tetraterpenes ($C_{40}H_{64}$) | Carotenoids, lycopen |
| > 8 | Polyterpenes ($C_{>40}H_{>64}$) | Ubuquinones, rubber, cytokonines, vitamine E |

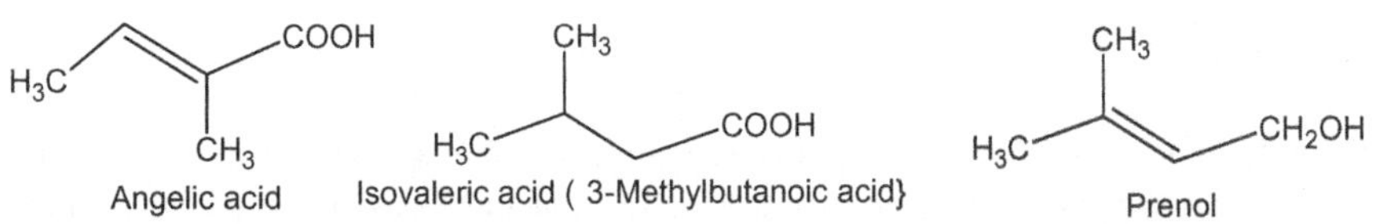


**Fig. 10.21** The structure of hemiterpenes derived from plant sources

## *10.5.1 Hemiterpenes*

Hemiterpenes are the simplest among the terpenoids. They consist of only one isoprene unit. Isoprene itself is considered as hemiterpene, but oxygen-containing isoprene compounds like angelic acid extracted from *Angelica archangelica*, isovaleric acid isolated from *Vaccinium myrtillus and prenol obtained from citrous fruit, tomato, cranberry* are hemiterpenoids (Fig. 10.21) (Seigler 1995; Perveen 2018).

## *10.5.2 Monoterpenes*

Monoterpenes are made up of two isoprene units and 10 carbon backbone structure having molecular formula $C_{10}H_{16}$. They are important constituents of essential oils or volatile oils of plant origin. They are found in certain plant families, such as Lamiaceae, Rutaceae, Apiaceae, and Pinaceae. Few compounds, like geraniol, are almost universal and are available in minute quantity in the volatile excretions of the plants. They are divided into three groups (a) acyclic monoterpenes, e.g., citral, geraniol, citronellal, Linalool (b) monocyclic monoterpenes, e.g., menthol, menthone, limonene, carvone, (c) bicyclic monoterpenes further grouped as (i) Chass I or (6 + 3) membered ring systems: thujane type – α-thujene, sabinene and carane type – carone, car-3-ene, car-2-ene (ii) Class II or (6 + 4) membered ring systems or pinane type ring system: α-pinene, β-pinene, myrtenol, pinocarvone (iii) Class III or (6 + 5) membered ring systems: camphor, camphene, borneol, isoborneol, fenchone (Fig. 10.22). Monoterpenes show a lot of pharmacological applications (Table 10.5). The substances like camphor, as well as menthol are useful as counterirritants, painkillers, and also act against skin infection like itching. They are also used as anthelmintics. A group of monoterpene glycosides has reported showing a vasodilation effect on coronary vessels, as well as the femoral vascular bed (Kaur 2010, Lalonde 2005, Bergman 2019). The pharmaceutical uses of monoterpenes are listed below in Table 10.5.

The chemical structures of different classes of monoterpenoids are provided below in Fig. 10.22.

### *10.5.3 Sesquiterpenes*

Sesquiterpenes composed of three isoprene units combined together in a head to tail fashion and the molecular formula is $C_{15}H_{24}$. On the basis of biogenetic source, about 200 various structural types of sesquiterpenes and many such compounds are reported. Sesquiterpenes can be classified into four different classes in accordance with their structure: (a) acyclic sesquiterpenes (e.g., α-farnesene, β-farnesene, nerolidol), (b) monocyclic sesquiterpenes further subclassified as per basic skeleton (i) Bisabolane type, e.g., zingiberene, β-bisabolene, (ii) Elemane type, e.g., abscisin II, elemol (iii) Humulene type, e.g., humulene (iv) Germacrane type, e.g., germacrone aristolactone, (c) bicyclic sesquiterpenes are further subclassified into (i) Cadinane type, e.g., α-cadinol, δ-cadinolii) Eudesmane type, e.g., β- eudesmol, santonin (iii) Perhydroazulene type, e.g., guaicol, aromadendrene (d) tricyclic sesquiterpenes are further subclassified into (i) Cedrene type, e.g., cedrol, cedrene (ii) Longifolene type, e.g., longifolene (Fig. 10.23) (Seigler 1995; Lalonde 2005; Hikino 1985).

A group of sesquiterpene exhibit antimalarial, antibacterial, anthelmintic, antifungal, anti-inflammatory, antirheumatic, adaptogen, flavoring agent, counterirritant, and antiprotozoal activities (Table 10.6). The list of sesquiterpenes and their uses are mentioned below in Table 10.6 (Kaur 2010; Lalonde 2005).

The chemical structure of sesquiterpenes is represented below in Fig. 10.23.

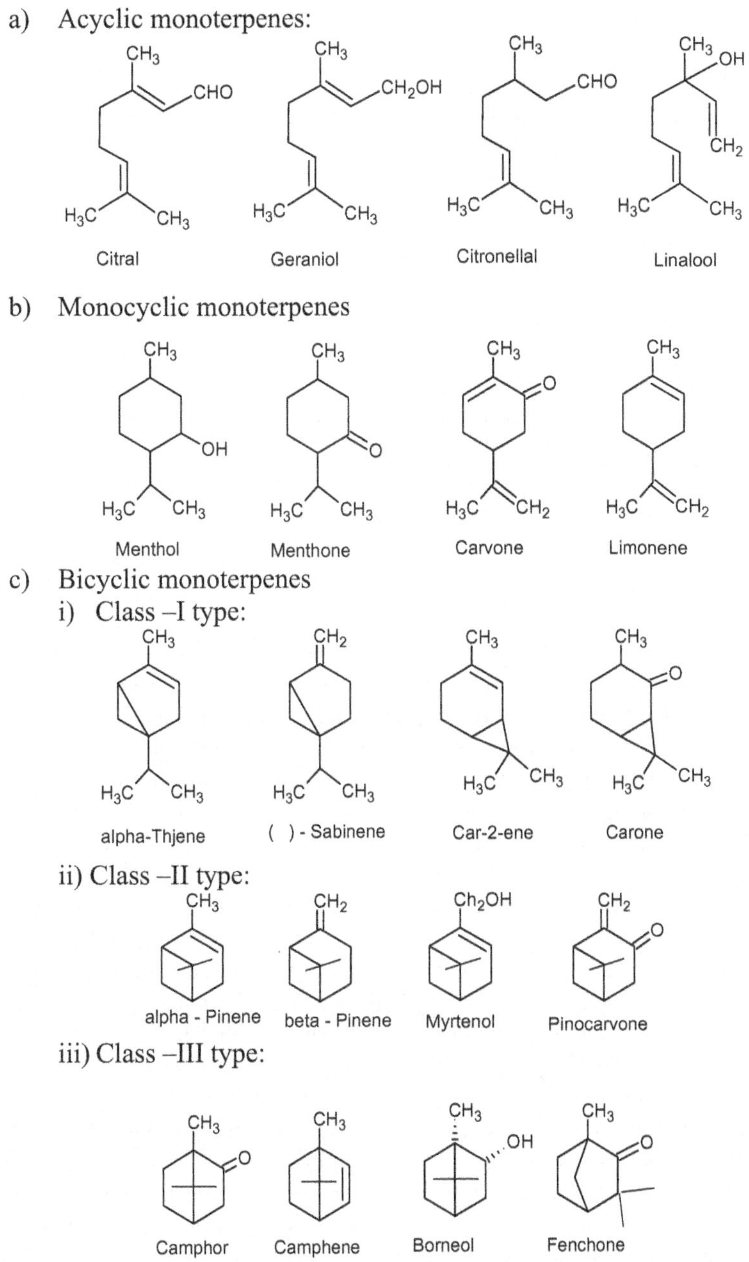


**Fig. 10.22** The structures of different classes of monoterpenes of plant sources

**Table 10.5** The pharmaceutical uses of some important monoterpenes of plant sources

| Sl. No | Name of the drugs | Uses |
|---|---|---|
| 1 | Menthol | Antipruritic, counterirritant, antiseptic, carminative, flavoring agent, and stimulant |
| 2 | Thymol | Antibacterial, and antifungal |
| 3 | Limonene | Flavoring agent, stimulant, stomachic, and carminative |
| 4 | Camphor | Rubifacient, counterirritant, antiseptic, antipruritic, carminative, masking agent in perfumery industry, and insect repellant |
| 5 | Eugenol | Analgesic in dental products, flavoring agent, stimulant, antiseptic, and condiment in cooking |
| 6 | Methyl salicylate | Counter-irritant, antirheumatic, antiseptic, and flavoring agent |
| 7 | Cineole | Analgesic in nasal inhaler and spray, antiseptic in mouthwash, diaphoretic, and expectorant |
| 8 | Terpinol | Stimulant and expectorant |

### 10.5.4 Diterpenes

Diterpenes consist of four isoprene units combined together in a head to tail fashion and the molecular formula is $C_{20}H_{32}$. Diterpenes can be divided into different classes on basis of the number of rings present in the compounds such as (a) acyclic diterpenes, e.g., phytol (b) monocyclic diterpenes, e.g., retinol (vitamin A) (c) tricyclic diterpenes, e.g., abietic acid (d) tetracyclic diterpenes, e.g., gibberellins (Fig. 10.24). Diterpenes contain several rings having various sizes. They may be 6-membered ring structures and also they may contain fused 5- and 7-membered ring structures. Many diterpenes have also additional ring systems. The additional ring is present in the side substitutions as esters or epoxides [14]. The phytoconstituents containing diterpenoids have great importance for medicinal values. Phylloquinone, also known as Vitamin K1 having an antihemorrhagic property is a diterpene that was first discovered in plants in 1929. Vitamin A1, chemically called retinol is also another diterpenoid that is prepared from a tetraterpenoid, carotene found in carrot. The bitter principles of *Jateorhiza palmata* (calumba root) are under to furanoditerpenes. The diterpenes obtained from *Teucrium chamaedrys* (wall germander) and *T. scorodonia* (wood sage) belonging to the family Labiatae, both the products show diaphoretics and antirheumatics activity (Papanov et al. 1980). Like all other classes of terpenes, diterpenes have been reported to exhibit a wide variety of pharmacological activities like as analgesic, antibacterial, antifungal, anti-inflammatory, antineoplastic, antiprotozoal activities and for the treatment of asthma, glaucoma, as well as congestive cardiomyopathy(Winkel-Shirley.2001). Few diterpenes isolated from *Kalmia latifolia* belonging to the family Ericaceae exhibit antifeedant activity. The gibberellins, plant hormones, first isolated from a fungus belonging to genus *Gibberella* and also available in higher plants, are

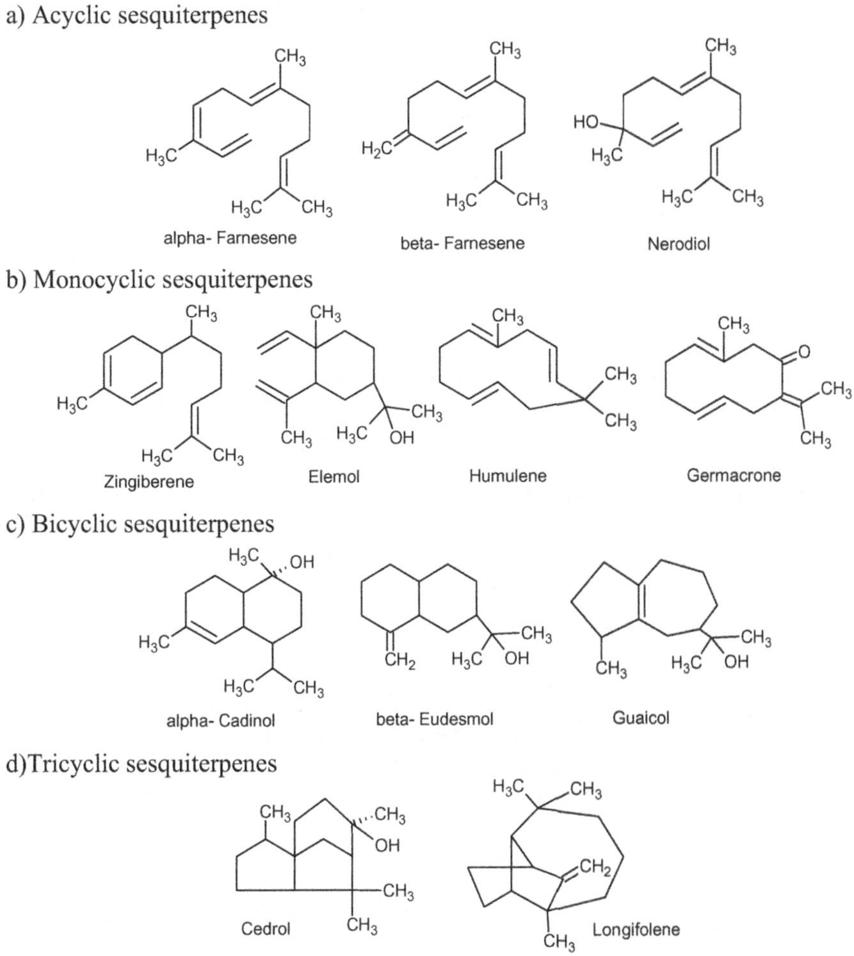


**Fig. 10.23** The structure of various classes of sesquiterpenes derived from plant sources

tetracyclic diterpenoid used for the growth of seedlings in the agricultural field (Evans 2009). The chemical structures of sesquiterpenes are provided below in Fig. 10.24.

## *10.5.5 Sesterterpenes*

The meaning of the term "sester" is half to three, i.e., two and a half. Sesterterpenes having five isoprene units containing 25 carbon atoms are rarely available. Geranyl farnesol, a sesterterpenoid, extracted from seeds of the plant, *Camellia sasanqua*

**Table 10.6** The pharmaceutical uses of some important sesquiterpenes

| Sl.No | Sesquiterpenes | Pharmaceutical uses |
|---|---|---|
| 1 | Artimimisinin | Antimalarial, antiprotozoal |
| 2 | Santonin | Anthelmintic |
| 3 | Gossypol | Male antifertility activity, insecticidal, used in menorrhoea |
| 4 | Zingiberene | Antibacterial, antiulcer. antifungal |
| 5 | Helenalin | Cardiotonic |
| 6 | Farnesene | Antibacterial, anti-inflammatory |
| 7 | Guaiol | Anti-rheumatic, food additives |
| 8 | Cedrol | adaptogen |
| 9 | Vetivone | Treatment of burn, sore and stimulant and diaphoretic |
| 10 | Caryophyllene | Antiseptic, counterirritant, carminative |

called sasanqua and another plant, *Camellia japonica* called as camellia both belonging to the family Theaceae exhibited cytotoxic activity in mouse leukemic M1 cells(Akihisa et al. 1999; Ishikura et al. 1984). Another sesterterpene, dysidiolide was isolated from the Caribbean sponge. Ophiobolin, a cytotoxic agent, is produced through various diverse cyclization (Au et al. 2000).

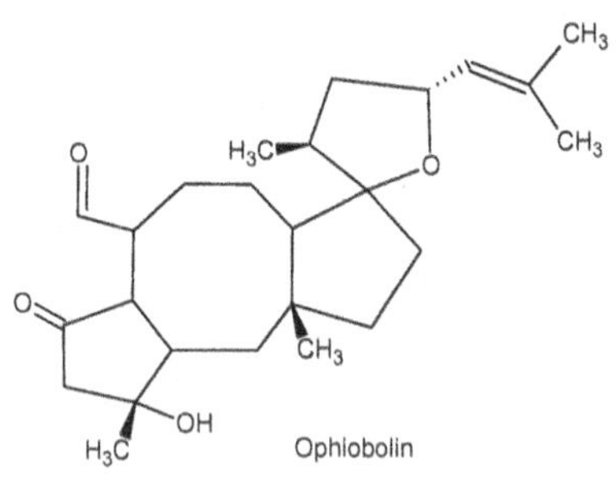


Ophiobolin

## 10.5.6 *Triterpenes*

Triterpenes having six isoprene units containing 30 carbon atoms are found abundantly in nature. Triterpenes may be linear, tetracyclic, or pentacyclic. The linear triterpene, squalene, derived from the coupling of two molecules of farnesyl pyrophosphate by mevalonic acid pathway is found in the animal source, e.g., shark liver oil and plant sources, e.g., olive oil, arachis oil. Triterpenes are mostly lipid substances of all plants and about 4000 triterpenoids have been identified and extracted. These compounds are required to prepare animal steroids, as well as plant steroids. Both triterpenes and steroids (tetracyclic triterpenoids) exist as a free

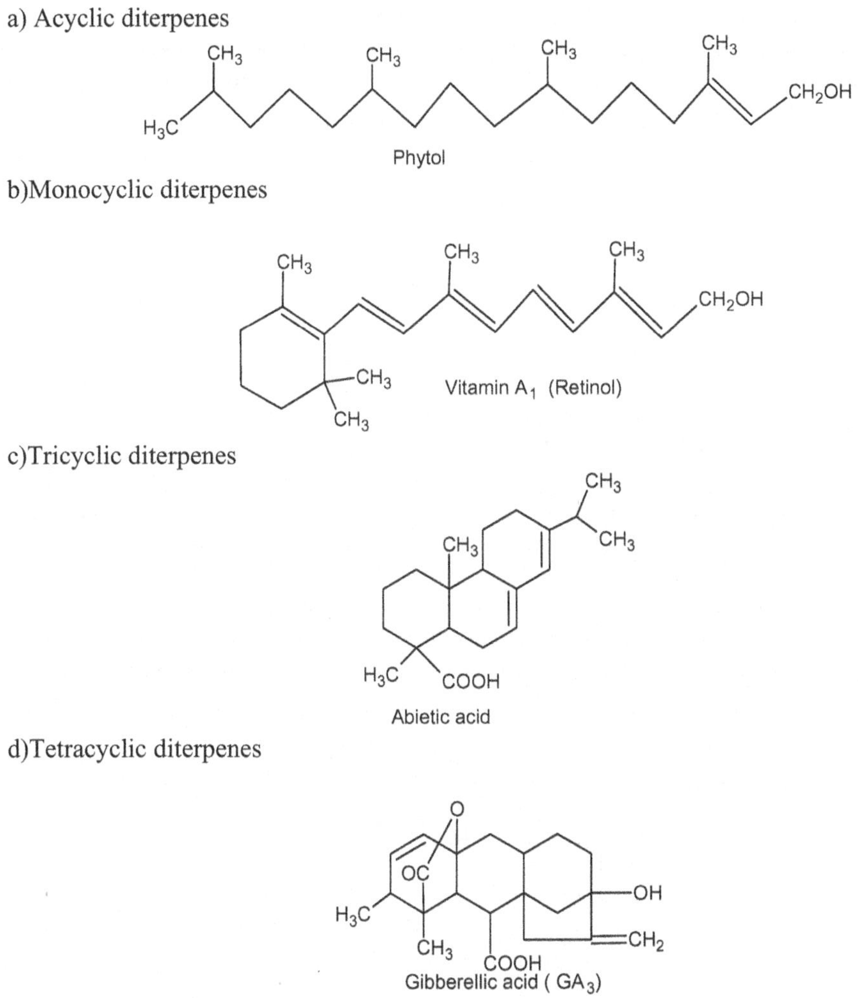


**Fig. 10.24** The structures of lignan derivatives derived from plant sources

molecule, as glycosides form or in other combinations (Seigler 1995). The oleo-gum-resin of *Boswellia carterii* contains two types of triterpenoids namely, one is β-Boswellic acids (ursane-type triterpene) and anther one is α-boswellic acids (oleanane-type triterpene). These compounds exhibit anti-inflammatory, as well as antirheumatic properties. Limonoids are tetracyclic triterpenoids but triterpenoid saponins are pentacyclic triterpenoids. An example of medicinally important triterpenoid is cardiac glycosides or steroid glycosides used as cardiotonic by increasing the force of systolic contraction in congestive cardiac failure (CCF). Another important class of compounds, quassinoids is isolated from *Quassia amara*. Quassinoids are compounds having less than 30 carbons and are synthesized by the degradation of triterpenoids along with the rearrangement of degradable products. Quassia has great biological importance and is used as an insecticide, as a bitter tonic and as anthelmintic for the expulsion of threadworms from the intestine (Culioli et al. 2003).

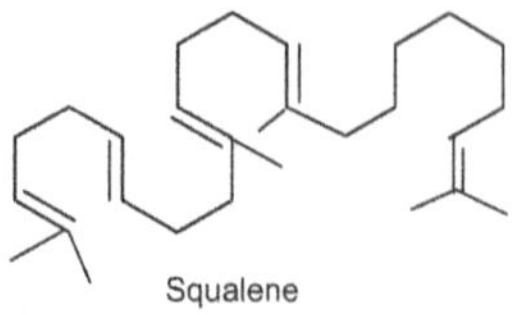

Squalene

### 10.5.7 Sesquarterpenes

Sesquarterpenes contain seven basic units, i.e., isoprene ($C_5H_8$) having molecular formula $C_{35}H_{56}$. They are actually diterpene-sesquiterpene derivatives. Plagiospirolide A, B, C, D are obtained from the liverwort of *Plagiochila moritziana*. Ferrugicadiol and ferrugieudesmol are obtained from the bark of *Calocedrus macrolepis*. C-35 terpene having pentacyclic structure is obtained biosynthetically from *Bacillus subtilis* via cyclization with the help of an enzyme.

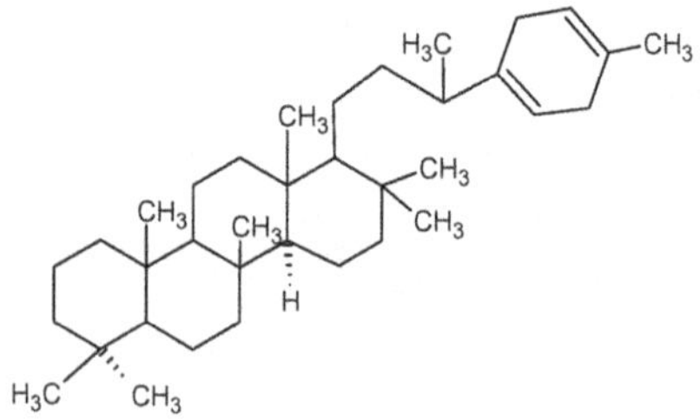


Sesquarterpene obtained from Bacillus subtilis

### 10.5.8 Tetraterpenes

Tetraterpenes are 40 carbon ($C_{40}$) containing compounds. In this class, the medicinally important compounds are well-known carotenoids. They are colored substances usually red, yellow, or orange in color. They are used as food colorants over synthetic color due to their non-toxic and stable nature. The coloring agents are used as pharmaceutical aids. The examples of carotenoids are β-carotene, capsanthin, capsorubin, lycopene, zeaxanthin, lutein, etc. (Lalonde. 2005).

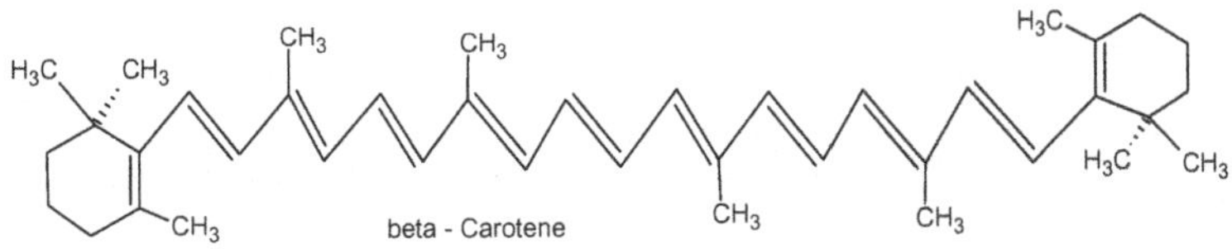

beta - Carotene

### *10.5.9 Polyterpenes*

Polyterpenes consist of more than eight isoprene units. They are available in the plants belonging to the family Euphorbiaceae, Moraceae, Apocynaceae, and Asclepiadaceae.

Pure rubber chemically polymer of cis form of isoprene is obtained from Hevea brasiliensis belonging to the family Euphorbiaceae. Gutta-percha is chemically a polymer of trans form of isoprene (Lalonde 2005).

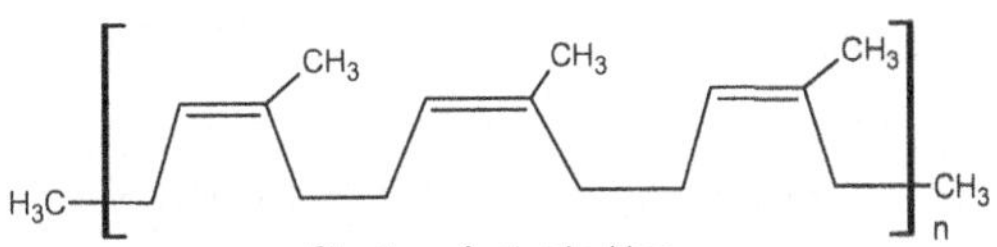

Structure of natural rubber

Terpenoids are a large group of secondary metabolites having various classes of compounds. They are biosynthesized via mevalonic acid pathway from simple compound, acetyl CoA (Fig. 10.25).

## 10.6 Carbohydrates

Carbohydrates are ubiquitously found in living beings on the Earth. Carbohydrates are sugars, starches, and fibers that are available in grains, vegetables, and milk products. Carbohydrate, e.g., glucose is the first product manufactured by photosynthesis from atmospheric carbon dioxide ($CO_2$) and water from the soil in the presence of sunlight with the help of chlorophyll, a green pigment. Glucose is the beginning material for the preparation of all phytoconstituents and also, for the preparation of all animal biochemicals. More carbohydrates occur in nature. The abundance of carbohydrates is maximum among all other kinds of natural products. Cellulose, a polymer of glucose combined by β-glycosidic linkage is the principal structural component of plants and has taken top-ranked abundance as a single

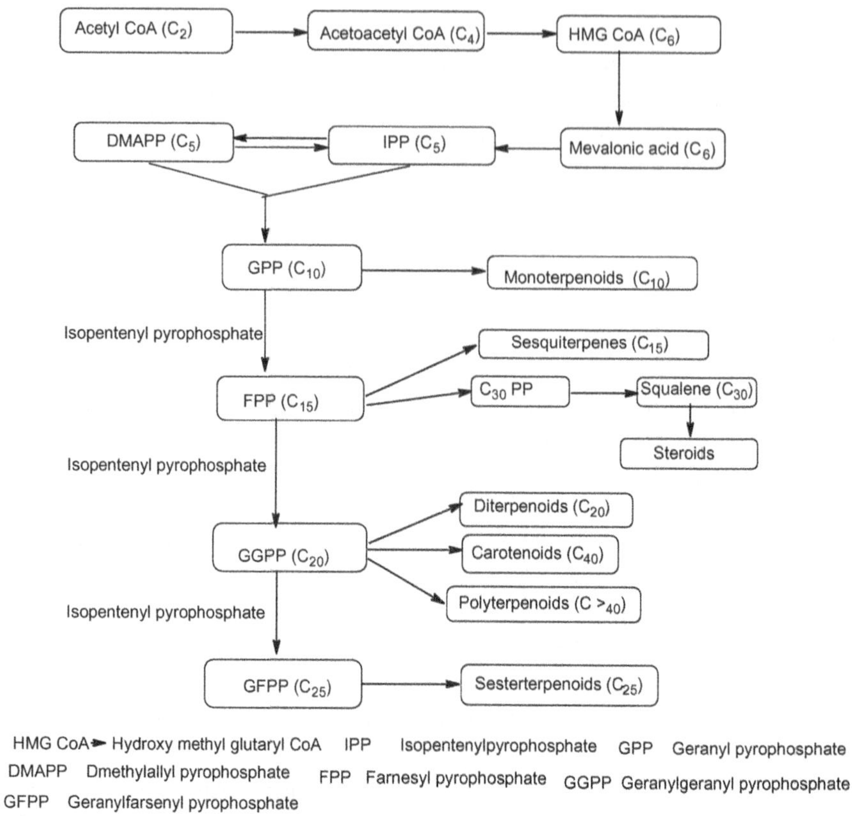


**Fig. 10.25** Flow chart of the biosynthetic pathway of terpenes (Kokate et al. 2005; Lalonde 2005)

organic substance. Although carbohydrates are primary metabolites, they become part of the huge number of secondary metabolites through glycosidation linkages. Glycosides consist of a sugar part known as glycone. The glycone sugars are mainly glucose, galactose, rhamnose, mannose, digitoxose, cymose, etc. Mucilages and gums are the polymeric product of simple sugars and uronic acids (Asif et al. 2011).

Carbohydrates can be defined as carbon, hydrogen, and oxygen-containing polyhydroxy aldehydes or ketones that on hydrolysis produce simple sugars. They are classified into monosaccharides, disaccharides, oligosaccharides (trisaccharides, tetrasaccharides, etc.), and polysaccharides. Carbohydrates are classified based on their presence of saccharide units as follows:

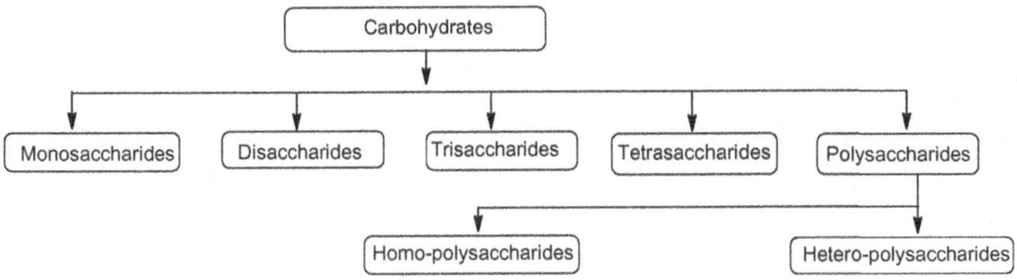


Depending upon reducing property, they are divided into two groups such as

(a) reducing sugars, e.g., all monosaccharides (e.g., glucose, fructose, galactose, mannose, gulose, sorbose, etc.) and disaccharides (e.g., maltose, lactose) except sucrose
(b) non-reducing sugars, e.g., oligosaccharides, polysaccharides.

Depending upon the hydrolytic product, they are grouped as pentosan (e.g., xylan) and hexosan (e.g., starch, cellulose, and inulin).

Cellulose consists of glucose units combined by $\beta$-1,4-glycosidic linkages which are hydrolyzable by cellulase enzyme available in herbivorous animals or cattle class.

Monosaccharides have three-carbon atoms to nine-carbon atoms and accordingly, they are known as triose, tetrose, pentose, hexose, heptose, etc. The simplest carbohydrate is glyceraldehydes containing three-carbon atoms. But those monosaccharides having five carbon atoms like pentoses, $C_5H_{10}O_5$, and six carbon atoms like hexoses, $C_6H_{12}O_6$ ( Fig. 10.26), are collected in plants in the largest amount (Kokateet al. 2005, Morrison et al. 2004).

Depending upon the presence of functional groups (aldehyde or ketone), they are divided into aldose sugar and ketose sugar. For example, glucose is aldohexose, whereas fructose is ketohexose.

Gum, mucilage, pectin, guaran, tragacanthin, and alginic acid are the important polysaccharide derivative having pharmaceutical values. They are used as emulsifying agents, binding agents, suspending agents, thickening agents, adsorbent, laxatives, and pharmaceutical aids. Gums are pathological products of polyuronides, on hydrolysis they produce the mixtures of uronic acids and sugars. Whereas mucilages are the physiological product mainly of sulphuric acid esters.

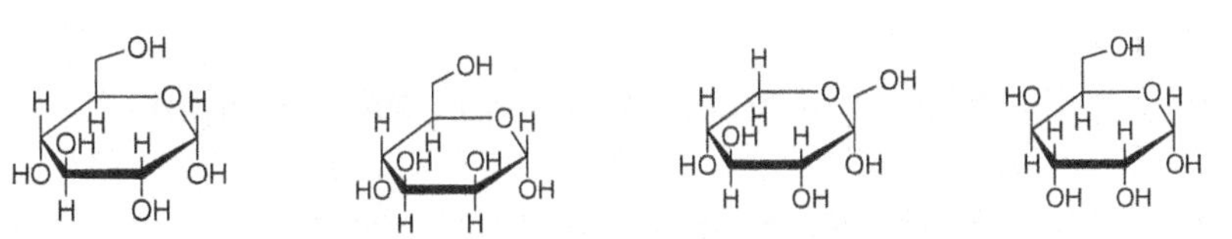

alpha-D-Glucopyranose alpha-D-Mannopyranose alpha-D-Sorbopyranose alpha-D-Gulopyranose

**Fig. 10.26** The structure of various hexose monosaccharides

On decomposition, cellulose produces gum and mucilage. Pectin, methoxy ester of aldobionic acid and pectic acid, is obtained from the inner portion of the skin of citrus fruits, pectins are polyuronides and made of pectic substances such as pectinic acid, protopectin, and calcium pectate. Another viscous sticky material known as mucilage obtained from maximum plants protect the plant by thickening the membranes in plants. It helps to store water and food and also helps in the germination of seed. It is chemically a polar glycoprotein and an exo-polysaccharide. Mucilage mainly acts as a demulcent. The important sources of mucilage are Cactus (and other succulents) and *L. usitatissimum* (flax seeds). The extract of the mucilaginous root of the marshmallow plant exhibits a demulcent effect and hence used as a cough suppressant in respiratory tract infection. When mucilage comes in direct contact with the mucous membrane surface, it acts as an emollient (Kokate et al. 2005; Khowala et al. 2008).

The water-soluble portion approximately 85% of guar gum obtained from the powder of the endosperm of *Cyamopsis tetragonolobus* is called guaran. On hydrolysis, it gives 65% of galactose and 35% of mannose. Honey, a saturated solution of sugar, deposited in the honeycomb by *Apis mellifera* (bees) is used as a demulcent, antiseptic to burns and wounds, and also as a sweetening agent. Tragacanth, obtained by incision from branches of *Astragalus gummifer* belonging to the family Leguminosae, has pharmaceutical importance as a suspending agent, the binding agent in tablets, emollient in cosmetics. Chitin is also a polysaccharide derivative having acetyl and amino groups and are found in skeletal of invertebrates. Chitin is employed in wound healing products and is used in water treatment plants. Agar, also known as vegetable gelatin, obtained from *Gelidium amansii* has two polysaccharides such as agarose and agaropectin. Agarose provides gel strength of the agar, whereas agaropectin of agar provides viscosity when added to the solution. Carrageenan, a sulphated polysaccharide, extracted from seaweed (carrageen) is used as an emulsifying agent and gelling agent. Carrageenan is useful as an experimental tool to produce inflammation in the rat paw edema model of the anti-inflammatory test. Inulin is a polysaccharide which is obtained from the bulbs of dahlia, *Inula helenium*. Inulin is chemically 35–50, 1, 2-linked fructofuranose units, terminated by one glucose unit. Dextrin, an incomplete hydrolytic product of starch with dilute mineral acids, has pharmaceutical value as tablet excipients. Whereas dextran is another polysaccharide that is generated by growing bacteria on sucrose solution. Dextran is a chemical polymer of D-glucose linked by $\alpha$ (1,6) glycosidic linkage (Kokate et al. 2005; Anbalahan 2017).

## 10.7 Lipids (Fixed Oil, Fats, Waxes and Phospholipids)

Lipids consist of a large number of natural products such as fixed oils (triglycerides), waxes (ester of long-chain fatty acid and long-chain alcohol), volatile oils, steroids (having cyclopentanoperhydrophenanthrene ring), fat-soluble vitamins, important ingredients of foods like vit. A, vit. D, vit. E, vit. K, phospholipids, and

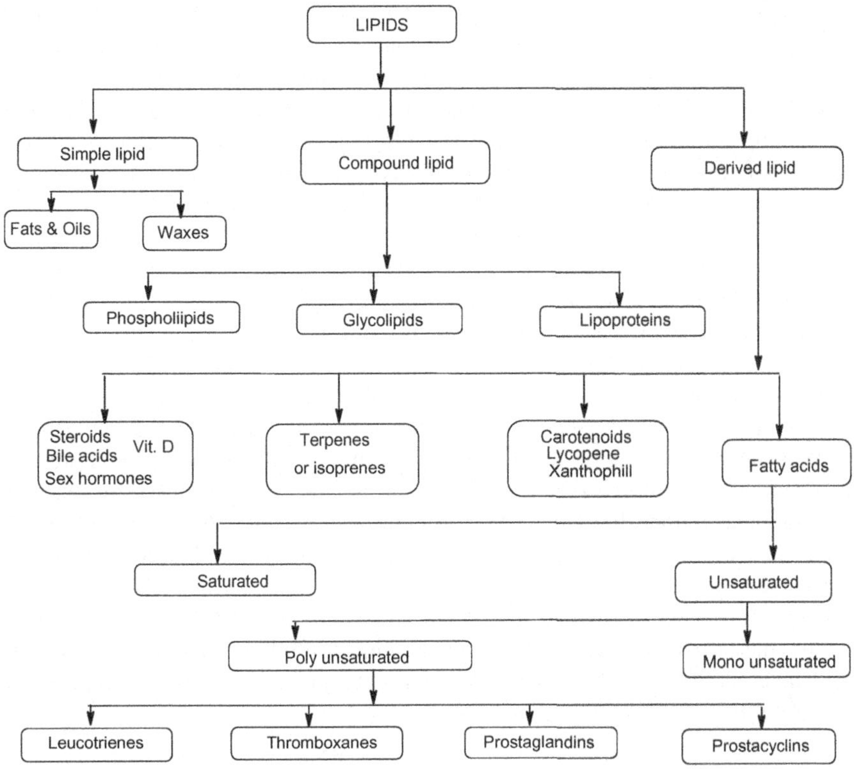


**Fig. 10.27** The classification of lipids

other compounds (Fig. 10.27). They exhibit various biological properties such as the main structural components of all biological membranes which contain lipid-protein lipid layer, as energy reservoirs that provide 9.3 kcal. per mole of fat, as fuel for cellular functions along with other food components such as vitamins and as coordinating compounds, hormones (Fahy et al. 2009; Subramaniam et al. 2011). Lipids are regarded as primary plant metabolites, but lipids have been reported to exhibit several pharmacological actions, and therefore, they considered as phytoconstituents. Lipids are classified as follows on the basis of their structure (Fig. 10.27).

## *10.7.1 Fixed Oils*

Fixed oils obtained from plant sources are available in the seed of the plants. Fats and oil are esters of glycerol and long-chain fatty acids. Fatty acids can be divided into two groups such as saturated fatty acids in which long-chain aliphatic acid contains no double bond and unsaturated fatty acids which contains one or more double bonds. High molecular saturated long-chain fatty acids include palmitic, stearic, arachidic lignoceric acids, whereas unsaturated long-chain fatty acids are oleic acids, linoleic, linolenic, and erucic acids. Fixed oils are liquid in nature at room temperature and have a comparatively high percentage of glycerides containing polyunsaturated fatty acids such as glycerin oleate, whereas fats are solid in nature at room temperature and contain glycerides of saturated fatty acids such as glycerin stearate (Fahyet al. 2009). Fixed oils show a lot of pharmaceutical importance. Arachis oil obtained by expression of the seed kernel of *Arachis hypogea* is used as a solvent for intramuscular injection, for the preparation of liniment and soap. Castor oil obtained by cold expression of seeds of *Ricinus communis* contains ricinoleic acid which exerts laxative property due to its irritant action. Chaulmoogra oil is another fixed oil obtained by cold compression of ripe seeds of *Hydrocarpus anthelmintic*. Chaulmoogra oil is used for the treatment of leprosy due to its strong bactericidal activity. Linseed oil obtained by compression of seeds of *L. usitatissimum*, family Linaceae, is recommended for external preparation as lotions, liniments. Polyunsaturated fatty acids (PUFA) present in few fixed oils can cause the reduction of excretion of lipid peroxidation products and hence they are recommended as good antioxidant potential and also act against inflammation agents. Polyunsaturated fatty acids are generally used to reduce the risk of atherosclerosis disease and for the treatment of other cardiovascular diseases (Kokate et al. 2005; Wallis 2005). The hydrolytic products of glyceryl stearo oleo linolein, a fat, is shown as follows:

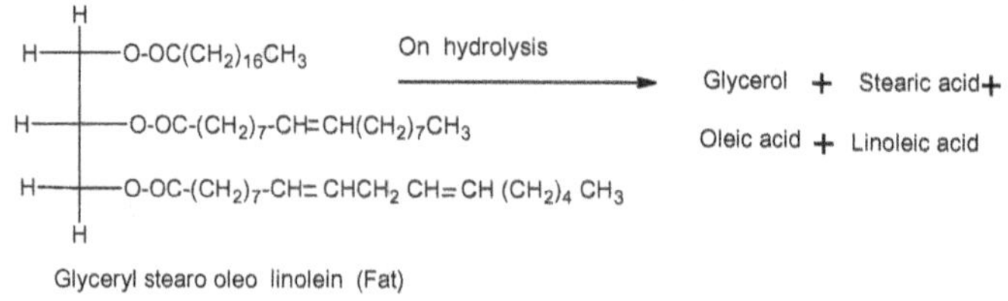


## 10.7.2 Waxes

Waxes are fusible, oily, viscous solid substances with a waxy luster. They are ester of long-chain monohydric alcohols with long-chain fatty acids. High molecular weight monohydric alcohols include cetyl alcohol, cholesterol, and melissyl alcohol. Natural waxes consist of unsaturated bonds and several kinds of functional groups like as primary and secondary alcohol, aldehydes, ketones, aromatic compounds, and ester of fatty acids, whereas synthetic waxes contain aliphatic hydrocarbons having no functional groups. Waxes are comparatively more resistant to saponification than oils and fats. They are available from plant and animal sources. The vegetable waxes include bayberry wax, seasal wax, carnauba wax, etc., whereas animal waxes include beeswax, wool, spermaceti wax. The vegetable waxes contain more amount of the mixture of unesterified hydrocarbons over esterified hydrocarbons. The composition of wax depends upon the species of the plants, as well as the geographical location of the plants (Baker 1982). Jojoba wax, a mixture of liquid wax obtained from seeds of *Simmondsia chinensis*, consists of ester of C-18, C-20, C-22, and C-24 carbon chain monounsaturated acids with alcohols (Wilhelm et al. 2012). Jojoba wax shows various pharmacological activities such as anti-aging, wound healing activities, as well as anti-inflammatory property, and it is also used in several skin diseases due to wound healing property. Jojoba wax has pharmaceutical applications in the preparation of topical medicaments to increase the absorption of drug molecules, i.e., to be a good carrier in dermatological products. It is also used in cosmetics products such as sunscreens and moisturizers. Carnauba wax, the hardest wax, is the exudates of leaves of the Brazilian palm tree, *Copernicia prunifera*, and *Copernicia cerifera* belonging to the family Palmae. Carnauba wax consists of carnaubic acid, cerotic acid, and melissyl cerotate. It is useful for the preparation of cosmetic products and deodorant sticks. It is applied as automobile wax and high-quality shoe polish. It is employed for the coating of tablets (Kokate et al. 2005; Wallis 2005).

### 10.7.3 Phospholipids

Lecithin, an important phospholipid, is obtained from soyabean oil and also available in vegetable seeds and corn. Lecithin consists of glycerin, phosphoric acid, choline, and fatty acids. Generally, a saturated fatty acid is present at α-position of lecithin, and an unsaturated fatty acid is present at β-position of it. Lecithin is combined with protein to form lipoproteins of cells and plasma. It has great importance to transport and utilization of fats, and therefore, it prevents the accumulation of fat in the liver, i.e., fatty liver. It decreases the surface tension of lung alveoli and so becomes easier to expel the liquid cough. It is industrially useful as a lubricant for the textile and petroleum industry (Kokate et al. 2005; Wallis 2005).

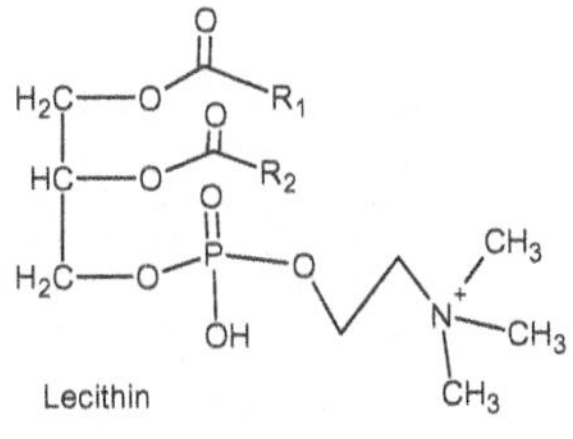


Lecithin

### *10.7.4 Fatty Acids*

Fatty acids are also found in plant origin. Evening primrose oil, obtained from dried seeds of *Oenothera biennis*, contains 9% unsaturated fatty acid called γ-linoleic acid. It is useful as a prostaglandin precursor and to control the severity of the premenstrual syndrome and for the treatment of eczema (Kokate et al. 2005; Wallis 2005). The sources and structures of saturated fatty acids, as well as unsaturated fatty acids derived from plants, are depicted as follows (Tables 10.7 and 10.8):

## 10.8 Glycosides

Glycosides are widely available in plant kingdom throughout the world and become the largest group of secondary plant metabolites. Glycosides contain a sugar part called as glycone and a non-sugar part known as aglycone. Chemically glycosides are acetal which is formed by condensation between the hydroxyl group of glycone (exists as hemiacetal) and another hydroxyl group of aglycone part. The sugar part attached to glycosides is either a monosaccharide such as glucose, mannose, rhamnose, etc., or deoxy sugar such as cymarose, digitoxose, etc. One or more number of monosaccharides molecules are attached to the aglycone moiety. Aglycone moiety includes a wide variety of chemical classes such as anthraquinone, steroid, coumarin, chromone, cyanogenic, flavonoid, phenolic, saponins, aldehyde, etc., which is discussed above of this chapter. On acid or enzymatic

**Table 10.7** The sources and structures of saturated fatty acids derived from plants

| Saturated fatty acid (IUPAC name) | Natural source | Chemical structure |
|---|---|---|
| Butyric acid (Butanoic acid) | Butter fat, Cow's milk | $CH_3(CH_2)_2COOH$ |
| Caproic acid (Hexanoic acid) | Palm kernel oil, butter | $CH_3(CH_2)_4COOH$ |
| Caprylic acid (Octanoic acid) | Coconut oil, palm oil | $CH_3(CH_2)_6COOH$ |
| Capric acid (Decanoic acid) | Palm oil, coconut oil, milk | $CH_3(CH_2)_8COOH$ |
| Lauric acid Dodecanoic acid) | Coconut oil, palm oil | $CH_3(CH_2)_{10}COOH$ |
| Myristic acid (Tetradecanoic acid) | Palm oil, milk fat | $CH_3(CH_2)_{12}COOH$ |
| Palmitic acid (Hexadecanoic acid) | Sesame oil, Arachis oil | $CH_3(CH_2)_{14}COOH$ |
| Stearic acid (Octadecanoic acid | Arachis oil, Sesame oil | $CH_3(CH_2)_{16}COOH$ |
| Arachidic acid (Eicosanoic acid) | Peanut oil, Mustard oil | $CH_3(CH_2)_{18}COOH$ |
| Behenic acid (Docosanoic acid) | Peanut oil, Rapeseed oil | $CH_3(CH_2)_{20}COOH$ |
| Lignoceric acid (Tetracosanoic acid) | Peanut oil | $CH_3(CH_2)_{22}COOH$ |
| Cerotic acid (Hexacosanoic acid) | Bees wax, Carnauba wax | $CH_3(CH_2)_{24}COOH$ |
| Montanic acid (Octacosanoic acid) | Fruit skin, bees wax | $CH_3(CH_2)_{26}COOH$ |
| Melissic acid (Triacontanoic acid) | Cotton wax | $CH_3(CH_2)_{28}COOH$ |

**Table 10.8** The sources and structures of unsaturated fatty acids derived from plants

| Unsaturated fatty acid | Natural source | Chemical structure |
|---|---|---|
| Palmitoleic acid | Cotton seed oil | $CH_3(CH_2)_5CH = CH(CH_2)_7COOH$ |
| Oleic acid | Corn oil, Safflower oil | $CH_3(CH_2)_7CH = CH(CH_2)_7COOH$ |
| Linoleic acid | Sun flower and sesame oil | $CH_3(CH_2)_4CH = CH\,CH_2\,CH = CH(CH_2)_7COOH$ |
| Erucic acid | Rapeseed oil | $CH_3(CH_2)_7CH = CH(CH_2)_{11}COOH$ |
| Ricinoleic acid | Castor oil | $CH_3(CH_2)_5\,CH(OH)\,CH_2\,CH = CH(CH_2)_7COOH$ |
| Chaulmoogric acid | Chaulmoogra oil | 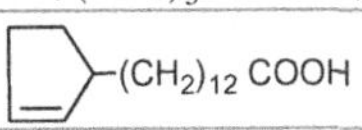 |

hydrolysis, glycosides provide glycone or sugar part and aglycone or non-sugar part (Kaur 2010, Cheeke 2001).

Glycone part is responsible to show the solubility character of the glycosides whereas the aglycone part shows the therapeutic activity. Glycosides exhibit.

Several varieties of pharmacological activities such as purgative, antidepressant, cardiotonic, nervine tonic, demulcent, sedative, counterirritant, rubefacient, hepatoprotective, diuretic, antifungal, etc.

On the basis of the chemical nature of the aglycone part, glycosides are divided as follows:

1. Anthraquinone glycosides, e.g., aloe leaves, rhubarb
2. Steroidal glycosides or cardiac glycosides, e.g., digitalis leaves, squill
3. Coumarin glycosides, e.g. cantharides, psoralea
4. Chromone glycosides, e.g., hypericum
5. Cyanogenic glycosides, e.g., bitter almond, wild cherry bark
6. Flavonoid glycosides, e.g., silymarin, ginkgo
7. Phenolic glycosides, e.g., bearberry
8. Saponin glycosides, e.g., Ginseng, dioscorea, gokhru, licorice
9. Aldehyde glycosides, e.g., vanilla pods
10. Steviol glycosides, e.g., stevia
11. Iridoid glycosides, e.g., nux vomica seed
12. Thioglycosides, e.g., black mustard
13. Steroidal glycol-alkaloids, e.g., solanum.

The biological sources, chemical structure and pharmacological activities of various kinds of glycosides (Table 10.9), are given below (Kokate et al. 2005; Hossain et al. 2019).

**Table 10.9** The biological sources, structure, and uses of various kinds of glycosides

| Type of glycosides | Biological source | Chemical Structure | Therapeutic uses |
|---|---|---|---|
| 1. Anthraquinone glycosides e.g., Sennoside A and B | Dried leaflets of *Cassia angustifolia* | 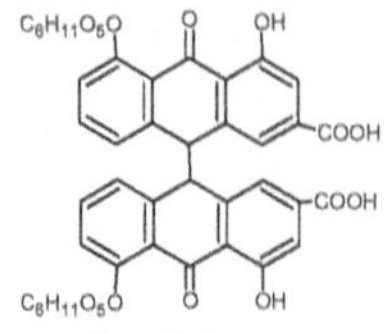 Sennoside A | Laxatives, purgative |
| 2. Steroidal glycosides or cardiac glycosides e.g., digoxin | Dried leaves of *Digitalis purpurea* | 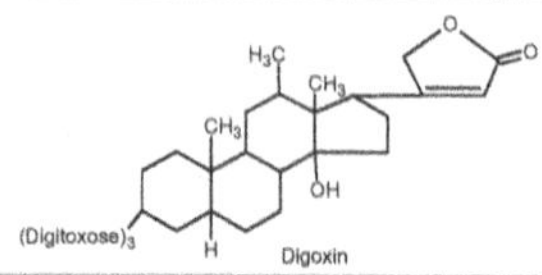 Digoxin | Cardiotonic, cardiac stimulant |
| 3.Coumarin glycosides e.g., psoralidin | Ripe fruits of *Psoralea corylifolia* | 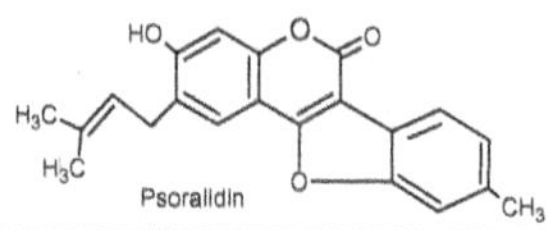 Psoralidin | Anthelmintic, laxative, treatment of leucoderma, leprosy |
| 4. Chromone glycosides e.g., hyperimone | Whole plant of *Hypericum erectum* | 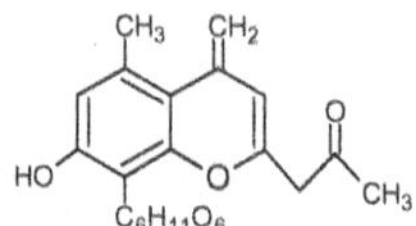 Hyperimone | Anti- H pyroli activity |
| 5. Cyanogenic glycosides e.g., amygdalin | Ripe seeds of *Prunus amygdalus* | 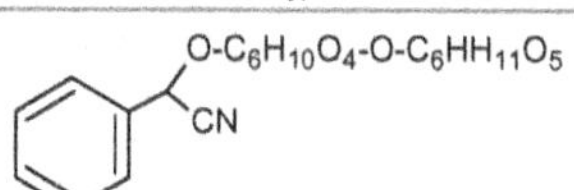 Amygdalin | Sedative, demulcent skin lotion |
| 6. Flavonoid glycosides e.g., silybin | Ripe seeds of *Silybum marianum* | 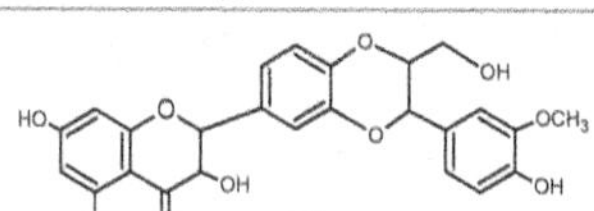 Silybin | Hepatoprotective, antioxidant |
| 7.Phenolic glycosides e.g., arbutin | Dried leaves of *Arctostaphylous uva-ursi* | 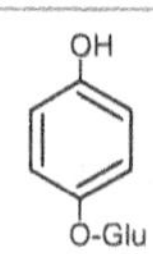 Arbutin | Diuretic, astringent |
| 8. Saponin glycosides e.g., diosgenin | Dried tuber of *Dioscorea deltoidea* | 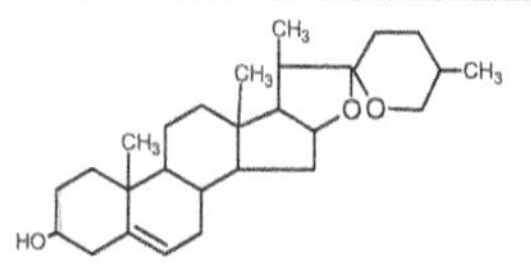 Diosgenin | Precursor of steroids, treatment of rheumatic arthritis |
| 9.Steroidal glycol-alkaloids e.g., solasodine | Dried berries of *Solanum khasianum* | 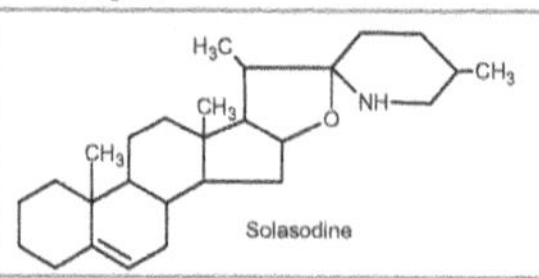 Solasodine | Sex hormone, oral contraceptive |

(continued)

**Table 10.9** (continued)

| Type of glycosides | Biological source | Chemical Structure | Therapeutic uses |
|---|---|---|---|
| 10. Aldehyde glycosides e.g., vanillin | Unripe fruits of *Vanilla planifolia* | 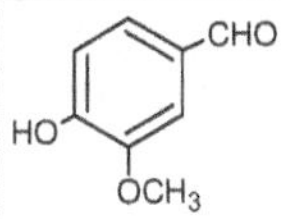 Vanillin | Flavoring agent |
| 11. Steviol glycosides e.g., stevioside | Leaves of *Stevia rebaudiana* | 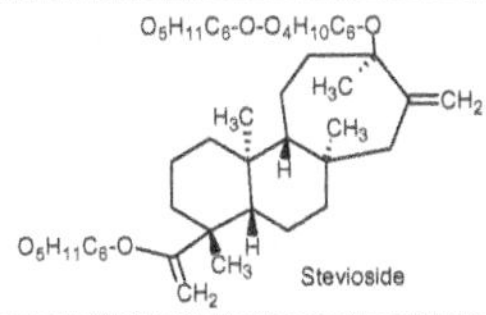 Stevioside | Sweetening agent |
| 12. Iridoid glycosides e.g., loganin | Seeds of *Strychnos nuxvomica* | 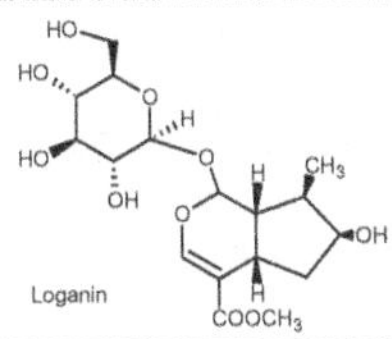 Loganin | Neuroprotective and anti-inflammatory properties |
| 13. Isothiocyanate glycosides e.g., sinigrin | Ripe seeds of Brassica juncea | 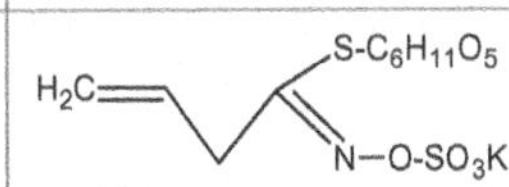 Sinigrin | Emetic, counterirritant, rubifacient |
| 14, Alcohol glycoside e.g., salicin | Bark of *Salix babylonica* (Willow) | 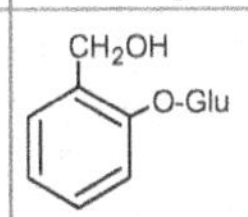 Salicin | Analgesic, antipyretic anti-inflammatory |

## 10.9 Conclusion

As the plant can be considered as a storehouse of the various kinds of secondary metabolites and the production of secondary substances of plant origin fully depends upon the selectivity of the species of a family, i.e., the plants belonging to a particular family provide most specific types phytoconstituents. A huge number of different classes of plants and various types of phytoconstituents are found worldwide but only very few of them are identified, extracted, isolated, elucidated for the chemical structure, and evaluated for their pharmacological activity.

To overcome this, the knowledge regarding various fields of science such as Botany, Organic Chemistry, Biochemistry, Analytical Chemistry, and Pharmacology is needed. A plant extract contains several phytoconstituents showing various biological activities due to complexity of the secondary metabolites in which a component may synergic or counteract the activity, and therefore, specific research or study for gathering more information of the phytoconstituents to develop a new drug or a new combination of drugs having less toxic and more potent is highly essential. Since the chapter summarized the description and structure of various secondary plant metabolites, their biosynthesis and their pharmacological activities may be considered as an informative tool for the development and research purpose of phytoconstituents. Secondary metabolites are available in much smaller quantities than primary metabolites and are very much expensive to produce or extract them. In spite of different difficulties, the herbalist and scientists are giving much effort to discover new molecules for relief and cure of various diseases.

## References

Akihisa T, Koike K, Kimura Y, Sashida N, Matsumoto T, Ukiya M, Nikaido T (1999) Acyclic and incompletely cyclized triterpene alcohols in the seed oils of theaceae and gramineae. Lipids 34 (11):1151–1175

Ali M (2019) Pharmacognosy and phytochemistry, 2nd edn. CBS Publishers and Distribution, New Delhi, pp 1–514

Amann R, Peskar BA (2002) Anti-inflammatory effects of aspirin and sodium salicylate. Eur J Pharmacol 447(1):1–9

Anbalahan N (2017) Pharmacological activity of mucilage isolated from medicinal plants. Int J Appl Pure Sci Agri 3(1): 98–113. e-ISSN: 2394-5532, p-ISSN: 2394-823X

Aniszewski T (2007) Alkaloids-secrets of life: alkaloid chemistry, biological significance, applications and pharmacological role, Elsevier, pp 1–334. https://doi.org/10.1016/B978-0-444-52736-3.X5000-4.

Asif HM, Akram M, Saeed TM, Khan I, Akhtar N, Ur Rehman R, Ali Shah SM, Ahmed K, Shaheen G (2011) Review paper carbohydrates. Int Res J Biochem Bioinformatics 1(1):1–5

Assa Y, Shany S, Gestetner B, Tencer Y, Birk Y, Bondi A (1973) Interaction of alfalfa saponins with components of the erythrocyte membrane in hemolysis. Biochem Biophys Acta 307 (1):83–91

Au TK, Chick PC, Leung PC (2000) The biology of ophiobolins. Life Sci 67:733–742

Bagci E, Yazgin A, Hayta S, Cakilcioglu U (2010) Composition of the essential oil of *Teucrium chamaedrys* L. (Lamiaceae) from Turkey, J Med Plants Res 4(23): 2587–2589

Baker EA (1982) Chemistry and morphology of plant epicuticular waxes. In: Cutler DF, Alvin KL (eds) The plant cuticle, Price C.E Academic press, London, UK. ISBN 0-12-19990-3

Bennets HW, Underwood EJ, Shier FL (1946) A specific breeding problem of sheep in subterranean clover pastures in Western Australia. Aust Vet J 22(1):2–12

Bergman ME, Davis B, Phillips A (2019) Review: medically useful plant terpenoids—biosynthesis occurrence and mechanism of action. Molecules 24:3961–3983. https://doi.org/10.3390/molecules24213961

Bourgaud F, Gravot A, Milesi S, Gontier E (2001) Production of plant secondary metabolites: a historical perspective. Plant Sci 161:839–851

Bruni R et al (2019) Botanical sources chemistry analysis and biological activity of furanocoumarins of pharmaceutical interest. Molecules 24:2163–2187

Catarino MD, Silva AMS, Cruz MT, Cardoso SM (2017) Antioxidant and anti-inflammatory activities of *Geranium robertianum* L. decoctions. Food and Function 8(9): 3355–3365. https://doi.org/10.1039/c7fo00881c

Cheeke PR (2001) Glycosides: naturally occurring, Wiley online library. https://10.1038/npg.els.0000692

Clarke EGC (1970) The forensic chemistry of alkaloids. In: Manske HF (ed) The Alkaloids, vol XII. Academic Press, New York, pp 514–590

Croteau R, Kutchan TM, Lewis NG (2000) Natural products (Secondary metabolites). In: Buchanan BB, Gruissem W, Jones RL (eds) Biochemistry and molecular biology of plants. Courier companies Inc., USA, pp 1250–1318

Culioli G, Mathe C, Archier P, Vieillescazes C (2003) A lupane triterpene from frankincense (*Boswellia* sp., Burseraceae). Phytochemistry 62:537–541. https://doi.org/10.1016/S0031-9422(02)00538-1

Cunha WR, Andrade e Silva ML, Veneziani RCL, Ambrosio SR, Bastos JK (2012) Lignans: chemical and biological properties. In: Phytochemicals—a global perspective of their role in nutrition and health, pp 213–234. 105772/28471

Dai J, Mumper RJ (2010) Plant phenolics: extraction, analysis and their antioxidant and anticancer properties, Molecules 15(10):7313–7352

El Aziz MMA, Ashour AS, Melad ASG (2019) A review on saponins from medicinal plants: chemistry isolation and determination. J Nanomed Res 8(1):6–12

Evans D, Mitch C (1982) Studies directed towards the total synthesis of morphine alkaloids. Tetrahedron Lett 23(3):285–288

Evans WC (2009) Pharmacognosy, 16th edn. Edinburgh, London, New York Philadelphia, St Louis Sydney Toronto, pp 1–616

Fahy E, Subramaniam S, Murphy RC, Nishijima M, Raetz CR, Shimizu T, Spener F, van Meer G, Wakelam MJ, Dennis EA (2009) Update of the LIPID MAPS comprehensive classification system for lipids. J Lipid Res 50(Suppl):S9–S14. https://doi.org/10.1194/jlr.R800095-JLR200

Gehm BD, McAndrews JM, Chien P, Jameson JL (1997) Resveratrol a polyphenolic compound found in grapes and wine is an agonist for the estrogen receptor. Proc Natl Acad Sci USA 94(25):14138–14143

Giweli AA, Dzamic AM, Sokovic M, Ristic M, Janac kovic P, Marin P (2013) The chemical composition, antimicrobial and antioxidant activities of the essential oil of *Salvia fruticosa* Growing Wild in Libya. Archives Bio Sci 1(65):321–329

Goenka P, Sarawgi A, Karun V, Nigam AG, Dutta S, Marwah N (2013) *Camellia sinensis* (tea): implications and role in preventing dental decay. Pharmacognosy Rev 7(14):152–156. https://doi.org/10.4103/0973-7847.120515

Golawska S, Sprawka I, Lukasik I, Golawski A (2013) Are naringenin and quercetin useful chemicals in pest-management strategies. J Pest Sci 87(1):173–180

Guclu-Ustundag O, Mazza G (2007) Saponins: properties, applications and processing, critical reviews in food science and nutrition 47:231–258. ISSN: 1040–8398. DOI: https://doi.org/10.1080/10408390600698197

Hagerman AE, Butler LG (1981) The specificity of proanthocyanidin-protein interactions. J Bio Chem 10(256):4494–4497

Harborne JB, Baxter H (1993) Phytochemical dictionary: a handbook of bioactive compounds from plants. Taylor and Francis, London, Washington, DC

Herrmann K, Nagel CW (1989) Occurrence and content of hydroxycinnamic and hydroxybenzoic acid compounds in foods. Crit Rev Food Sci Nutr 28(4):315–347

Hikino H(1985) Recent research on oriental medicinal plants. In: Wanger H, Hiniko H, Farnsworth NR (eds) Economic and medicinal plant research, vol 1. Academic Press, London, pp 53–85

Hoffmann D (2003) Medical herbalism: The science and practice of herbal medicine. Healing Arts Press, One Park Street, Rochester, Vermont (978-089281749-8)

Hossain MT, Asadujjaman M, Manik MIN, Matin MA, Chowdhury RZ, Rashid MH (2019) Review: a study on the pharmacological effects and mechanism of action of alkaloids glycosides and saponins. Pharmaceutical and Chem J 6(2):112–122

Ishikura H, Mochizuki T, Izumi Y, Usui T, Sawada H, Uchino H (1984) Differentiation of mouse leukemic M1 cells induced by polyprenoids. Leuk Res 8(5):843–852

Jepson RG, Craig JC (2008) Cranberries for preventing urinary tract infections, cochrane database of systematic reviews 1:CD001321, pub 4. https://doi.org/10.1002/14651858.CD001321.

Jones ME, Kossel A (1953) A biographical sketch. Yale J Bio Med 26(1):80–97

Jung H, Su B, Keller W, Mehta R, Kinghorn A (2006) Antioxidant and xanthones from pericarp of *Garcinia mangostana* (Mangosteen). J Agri Food Chem 54(6):2077–2082

Kaur J (2010) Chemistry of natural products, S. Vikas and company (publishing house), 1st ed. Jalandhar, India, pp 1–452. ISBN: 978-81-909385-3-2

Khowala S, Verma D, Banik SP (2008) Carbohydrates—biomolecules: introduction, structure, and function, National Science Digital Library, pp 1–93. https://www.researchgate.net/publication/200787272

Kokate CK, Purohit AP, Gokhale SB (2005) Pharmacognosy. Nirali Prakashan, India, pp 1–618

Lalonde RT (2005) Terpenes and terpenoids. Van Nostrand's Encyclopedia of Chemistry, Wiley Online Library. https://doi.org/10.1002/0471740039.vec2473

Montanher AB, Zucolotto SM, Schenkel EP, Frode TS (2007) Evidence of anti-inflammatory effects of *Passiflora edulis* in an inflammation model. J Ethnopharmacol 109(2):281–288

Morrison RT, Boyd RN (2004) Organic chemistry, 6th edn. Prentice Hall of India Pvt. Ltd., New Delhi, pp 1–1278

Mugford ST, Osbourn A (2013) Saponin synthesis and function. In: Bach TJ, Rohmer M (Eds) Isoprenoid synthesis in plants and microorganisms: new concepts and experimental approaches, Springer Science and Business media, New York

Nicolaou KC, Jason S, Chen EJC ( 2011) In: Classics in total synthesis. further targets, strategies, methods III. Wiley-VCH, Weinheim, pp 1–770. ISBN:978-3-527-32957-1

Panche AN, Diwan AD, Chandra SR (2016) Flavonoids: an overview. . J Nutrional Sci 5:1–15

Papanov GY, Malakov PY (1980) Furanoid diterpenes in the bitter fraction of *Teucrium chamaedrys* L. Naturforsch 35b:764–766

Park ES, Moon WS, Song MJ, Kim MN, Chung KH, Yoon JS (2001) Antimicrobial activity of phenol and benzoic acid derivatives. Int Biodeterior Biodegradation 47(4):209–214

Pelczar MJ, ChanECS KNR (1988) The control of microorganisms by physical agents. Microbiology. McGraw-Hill International, New York, pp 469–509

Pengelly A (2004) In: The constituents of medicinal plants: an introduction to the chemistry and therapeutics of herbal medicine, CABI Publishing, pp 1–184

Perveen S (2018) Introductory chapter: terpenes and terpenoids. In: Terpenes and terpenoids, Intechopen, pp 1–13. DOI:https://doi.org/10.5772/intechopen.79683.

Santos-Sanchez NF, Coronado RS, Carlos BH, Canongo CV(2019) Shikimic acid pathway in biosynthesis of phenolic compounds. In: Plant physiological aspects of phenolic compounds, Intechopen, pp 1–15. https://doi.org/10.5772/intechopen.83815

Seigler DS (1995) Plant Secondary Metabolism. Springer Science Business Media, New York. ISBN: 978-1-4613-7228-8; ISBN: 978-1-4615-4913-0 (eBook). https://doi.org/10.1007/978-1-4615-4913-0

Serafini M, Peluso I, Raguzzini A (2010) Flavonoids as anti-inflammatory agents. Proc Nutr Soc 69(3):273–278. https://doi.org/10.1017/S002966511000162X

Sieniawska E, Baj T (2017) Tannins—an overview. In: Badal S, Delgoda R (eds) Pharmacognosy fundamental, application and strategies. Science Direct, pp 199–232. https://doi.org/10.1016/B978-0-12-802104-0.00010-X

Sparg SG, Staden JV (2004) Biological activities and distribution of plant saponins. J Ethnopharmacology 94(2–3):219–243

Subramaniam S, Fahy E, Gupta S, Sud M, Byrnes RW, Cotter D, Dinasarapu AR, Maurya MR (2011) Bioinformatics and systems biology of the lipidome. Chem Rev 111(10):6452–6490. https://doi.org/10.1021/cr200295k

Susana J, Beatriz GM, Fabiana CM, Maria AD, Moacir GP (2011) Antifungal activity of five species of *Polygala*. Brazilian J Microbiol 42(3):1065–1075. https://doi.org/10.1590/S1517-83822011000300027

Tadeusz A (2015) Alkaloids: chemistry, biology, ecology, and applications, Elsevier, 2nd ed. Amsterdam, Netherlands, pp 1–97. ISBN: 13: 978–0444594334

Tyler VE, Brady .R, Robbers JE (1988) Pharmacognosy, 9th Ed. Lea and Febiger, Philadelphia

Wallis TE (2005) Textbook of pharmacognosy, 5th edn. CBS Publishers and Distributors Pvt. Ltd., New Delhi

Wang TY, Li Q, Bi KS (2018) Bioactive flavonoids in medicinal plants: structure, activity and biological fate. Asian J Pharmaceutical Sci 13(1):12–23

Wilhelm R, Hermann M (2012) "Esters, organic" Ullmann's encyclopedia of industrial chemistry. Wiley-VCH, Verlag GmbH and Co, KGaA, Weinheim. https://doi.org/10.1002/14356007.a09_565.pub2

Winkel-Shirley B (2001) Flavonoid biosynthesis. A colorful model for genetics, biochemistry, cell biology and biotechnology. Plant Physiol 126:485-493

Xu L, Wu Y, Zhao X, Zhang W (2015) The study on biological and pharmacological activity of coumarins. In: Asia-Pacific energy equipment engineering research conference, pp 135–138

Yi Z, Wang Z, Li H, Liu M (2004) Inhibitory effect of tellimagrandin I on chemically induced differentiation of human leukemia K562 cells. Toxicol Lett 147(2):109–119

Zbigniew S, Beata Z, Kamil J, Roman F, Barbara K, Andrzej D (2014) Antimicrobial and antiradical activity of extracts obtained from leaves of three species of the genus *Pyrus*. Microbial Drug Resist 20(4):337–343. https://doi.org/10.1089/mdr.2013.0155

# Chapter 11
# Pharmaceutical and Therapeutic Applications of Fenugreek Gum

**Purusottam Mishra, Amit Kumar Srivastava, Tara Chand Yadav, Vikas Pruthi, and Ramasare Prasad**

**Abstract** Plant-derived gums play an essential role in the biopharmaceutical field due to their advantageous physicochemical properties, nutritional value, ease of availability and biodegradability. Fenugreek gum extracted from the seeds consists of galactomannan with galactose to mannose ratio as one where higher number of galactose enhances the water solubility of fenugreek gum. Exciting contemporary research shows enormous possibilities of fenugreek gum in biopharmaceutical landscape as a natural excipient and therapeutic agent. Since last decade, fenugreek gum has been explored as retarding, mucoadhesive and disintegrating agent for various biopharmaceutical applications. Moreover, seed extract of fenugreek gum has been extensively studied as an antidiabetic, anti-carcinogenic, anti-inflammatory and hepatoprotective agent. This review aims to highlight the applications of fenugreek gum as a natural pharmaceutical excipient and its extensive role in drug delivery systems. Further, physicochemical properties, toxicological aspects and therapeutic efficacy of fenugreek gum has been also discussed.

**Keywords** Antidiabetic · Exicipient · Fenugreek gum · Galactomannan · Mucilage

## 11.1 Introduction

Since the last few decades, we have witnessed a substantial development in the field of drug delivery using a diverse range of excipients (Raghuwanshi et al. 2017). These excipients have been extensively used in the conventional dosage forms for their multiple pharmaceutical purposes such as glidant, binder, sweetner and thickening agent (Rowe et al. 2009), which may alter the physicochemical properties

P. Mishra · A. K. Srivastava · T. C. Yadav · V. Pruthi · R. Prasad (✉)
Department of Biotechnology, Indian Institute of Technology Roorkee, Roorkee 247667, Uttarakhand, India
e-mail: ramasare.prasad@bt.iitr.ac.in



D. Pal and A. K. Nayak (eds.), *Bioactive Natural Products for Pharmaceutical Applications*, Advanced Structured Materials 140,
https://doi.org/10.1007/978-3-030-54027-2_11

of the final formulation. Besides this, they facilitate the regulation of pharmacodynamic and pharmacokinetic properties of the drug in the physiological system. As per the conventional hypothesis, excipients used in the pharmaceutical preparations do not have a prime role in the treatment of disease and are thought to be inert as well. Till date several polymers have been used as excipients for the development of polymer-based drug delivery system with the aim of targeted delivery of the active therapeutics to the specific tissues (Ulbrich et al. 2016). Polymers can play a vital role in disease management as well as they can also modulate the behaviour of the drug in the physiochemical system. Although significant research has been done to investigate the mechanistic behaviour of these polymers when used alone or as a polymer-drug conjugate for the purpose of drug delivery. However, evaluation of the physicochemical properties of the polymer prior to pharmaceutical application is a task of priority (Singh and Pai 2015). A non-toxic, biocompatible and biodegradable polymer is always a need for a biologically safe and effective pharmaceutical dosage form. In recent years, demand of natural polymers has taken a quantum leap in pharmaceutical, food and cosmetic industries than its synthetic counterpart (Zia et al. 2017). Synthetic polymers are used for drug delivery and development of biomedical devices and implants. Though synthetic polymers exhibit high chemical, physical and mechanical stability as well as flexibility to bind with diverse range of therapeutics, bio-incompatibility and cellular toxicity illustrated by them are always a concern for drug delivery purpose (Nair and Laurencin 2007). Natural polymers such as polysaccharides and proteins have been extensively used to develop numerous biomedicines as they possess mighty biocompatible, biodegradable and therapeutic properties. Moreover, natural polymers are more also easily assessable and lower in cost compared to their synthetic counterparts (Li et al. 2015). These natural polymers based drug delivery systems possess several advantages such as high drug pay load, presence of diverse surface functional groups for drug binding and low toxicity (O'Elzoghby et al. 2016).

### *11.1.1 Natural Gums*

Gums are complex hydrocolloid biopolymers composed of polysaccharide consisting of one or more monosaccharides or their derivatives linked with chemical linkages (Prajapati et al. 2013a, b). Natural gums can be classified on the basis of their origin, e.g. marine origin (alginic acid, agar and carrageenans), animal origin (chitin), microbial origin (gellan gum, xanthum gum and lentinan) and plant origin (gum tragacanth, guar gum and cellulose) (Prajapati et al. 2013a, b; Choudhary and Pawar 2014). Natural gums produced by higher plants are consequence of injury or can be obtained from exudates of different parts of the plant. Easy availability, non-toxicity, cost effectiveness and biocompatibility are some of the advantages of natural gums which makes them a better alternative than their synthetic counterparts

(Prajapati et al. 2013a, b). Complicated macromolecular architecture and water solubility offers natural gums adhesiveness and cohesiveness which render natural gums a biopolymer of choice to be used as stabilizing agent, gelling agent, binder, disintegrant and suspending agent in the pharmaceutical preparations (Prajapati et al. 2013a, b). Gums tend to work efficiently at a higher concentration in the drug delivery systems and high swellability possesses challenges. Apart from these above described versatile properties of natural gums, certain drawbacks associated with them are environmental and microbial contamination during procurement and storage, loss of viscosity on long-term storage and high degree of hydration capacity. Chemical modification and copolymer grafting are the two major methods which are followed to neutralize the above-mentioned issues with natural gums (Prajapati et al. 2013a, b).

Various plant-derived gums such as guar gum, locust bean gum, acacia gum and gum karaya are used as natural excipient and therapeutic agent for application in diverse drug delivery systems (Castro et al. 2018; Prajapati et al. 2014; Singh 2018). Fenugreek gum hosts enormous biopharmaceutical possibilities as an excipient and therapeutic agent due its unique physiochemical properties such as low cytotoxicity, water solubility and biocompatibility like other natural plant-derived gums. This article attempts to discuss various physiochemical properties and biosafety issues of fenugreek gum which play a critical role in selection of a polymer as an excipient to develop a formulation. The latest developments in drug discovery using fenugreek gum are surveyed to illustrate areas of research advancing the frontiers of biomedicine.

### 11.1.2 Fenugreek Gum

Fenugreek (*Trigonella foenum-graecum*) belongs to family *Leguminosae (Fabaceae)* and known as one of the oldest medicinal plant cultivated around the globe (Goyal et al. 2016; Ouzir et al. 2016). Both leaves and seeds of fenugreek have been traditionally used as flavuor enhancers in foods and medicinal products. Seeds of fenugreek contain a gummy substance, i.e. mucilage, and consists of galactomannan which is a polysaccharide (Sindhu et al. 2012). Apparently, galactomannan of various gums differs in the physiochemical properties as the ratio of mannose to galactose varies. The higher number of galactose in fenugreek gum reduces this ratio to approximately one as a result, the water dissolving capacity increases (Iurian et al. 2017). Due to the high water solubility of fenugreek gum (Doyle et al. 2009; Kamble et al. 2013) numerous properties such as disintegrating ability (Kumar et al. 2009), suspending capacity (Nayak et al. 2013a, b), mucoadhesiveness (Nayak et al. 2013a, b; Nayak and Pal 2014) have been investigated extensively.

### *11.1.3 Chemical Composition of Fenugreek Seed*

Fenugreek seeds possess a variety of phytochemicals present therein including alkaloids, polyphenols, saponins, carbohydrates, proteins, fibres, amino acids and lipids, (El Nasri and El Tinay 2007; Kakani and Anwer 2012). Potassium and manganese are known to be the major metals present in seeds of fenugreek (Kakani and Anwer 2012). Lipid composition of fenugreek seeds comprise natural lipids (85%), phospholipid (10%) and glycolipid (5%) while its fatty acid profile is dominated by unsaturated fatty acid such as oleic acid and linoleic acid (Kakani and Anwer 2012). Diosgenin, gitogenin, yamogenin, rigogenin, tigogenin and neorigogenin are the major saponins present in fenugreek seeds (Kakani and Anwer 2012; Dawidar et al. 1973). Sesquiterpenes, n-alkanes and non-alactone are the reasons behind the aroma of the fenugreek seeds (Kakani and Anwer 2012). Gentianine, carpaine and trigonelline are the major alkaloids present in the endosperm of fenugreek seeds (Garg 2016). Besides all these phytochemicals, galactomannan extracted from the seeds of fenugreek is also counted as a major bioactive component. The detailed chemical constituents of fenugreek seeds are listed in Table 11.1.

**Table 11.1** Chemical composition of Fenugreek seeds

| Class | Chemical constituent | References |
|---|---|---|
| Saponin | Diosgenin, trigogenin, gitongenin, fenugrin B, yamogenin, rigogenin and neorigogeninetc | Kakani and Anwer (2012), Dawidar et al. (1973) |
| Alkaloids | Trigonelline, gentianine and carpaine | Garg (2016) |
| Polyphenols | Rhaponticin and isovitexin | He et al. (2015) |
| Volatile oils and fixed oil | olfactometry diacetyl, 1-Octene-3-one, sotolon, acetic acid; 3-Isobutyl-2-methoxypyrazine, butanoic acid, isovaleric acid, 3-isopropyl-2-methoxypyrazine, caproic acid, eugenol, 3-Amino-4,5-dimethyl-3, linalool, (Z)-1,5-Octadiene-3-one and 4-dihydro-2(5H)-Furanone | Blank et al. (1997) |
| Amino acid | Glutamic acid, aspartic acid, phenylalanine, leucine, lysine, valine, glycine, isoleucine, serine, proline, alanine, threonine, tyrosine, histidine, phosphoethanolamine and methionine | Wani and Kumar (2018) |
| Metals | Manganese, potassium, phosphorus, copper, calcium, sodium, iron and zinc | Kakani and Anwer (2012) |

## 11.2 Galactomannan: The Chief Constituent of Fenugreek Gum

### 11.2.1 Structural Properties

Galactomannan is the chief constituent of fenugreek seeds composed of β-(1,4)-D-mannose and α-(1,6)-D-galactose subunits (Cerqueira et al. 2011). It is extracted from fenugreek seeds and have an approximately equal number of mannose and galactose residues which makes it readily soluble in water. Generally, the existence of 180-190 number of monosaccharides (both galactose and mannose) provide a molecular weight of approximately 30 KDa to this natural polymer (Prajapati et al. 2013a, b). The backbone of galactomannan consists of 90–95 residues of β-1,4 linked mannopyranosyl units and each backbone monomer comprise an α-1,6 linked galactopyranosyl unit (Prajapati et al. 2013a, b). X-ray diffraction studies show the orthorhombic lattice of hydrated fenugreek gum with a, b and c value of 0.912 nm, 3.335 nm and 1.035 nm, respectively (Song et al. 1989). Gas chromatography-mass spectroscopy (GC-MS) studies of partially methylated alditol acetates gum extracted from fenugreek showed it to be a polysaccharide consists of 1,4 linked mannose having partially substituted galactose at C6 position (Brummer et al. 2003). $^{1}$H NMR and $^{13}$C NMR studies elucidated the structural characteristics of galactomannan and showed that it consists of β-(1,4)-linked D-mannopyranose and α-(1,6)-linked D-galactopyranose. Although the ratio of mannose to galactose residues varies from species to species, it was observed that the polymer backbone consists of mannose residues and about 83% of backbone is substituted by galactose at C6 position of mannose. Coupling constant analysis revealed that D-mannose and D-galactose possess β and α linkage, respectively (Jiang et al. 2007) (Fig. 11.1).

### 11.2.2 Physicochemical Properties

Galactomannans extracted from the fenugreek seeds are neutral polysaccharides with highly substituted galactose residues on mannose backbone (Jiang et al. 2007; Cui 2001; Hefnawy and Ramadan 2011). Extremely branched structure and high molecular weight of fenugreek gum are responsible for its lower viscous nature among other industrially used gums such as locust bean gum and guar gum (Brummer et al. 2003, Jiang et al. 2007; Cui 2001). Fenugreek gum solution appears to be more transparent than the guar gum solution due to the presence of less water insoluble constituents (Jiang et al. 2007). The higher degree of galactose substitution on the mannose backbone has been claimed as the prime reason for this behaviour (Brummer et al. 2003). Viscosity of aqueous fenugreek gum solution depends upon shear stress, temperature and concentration (Wei et al. 2015). Increase in the fenugreek gum concentration enhances the viscosity of solution

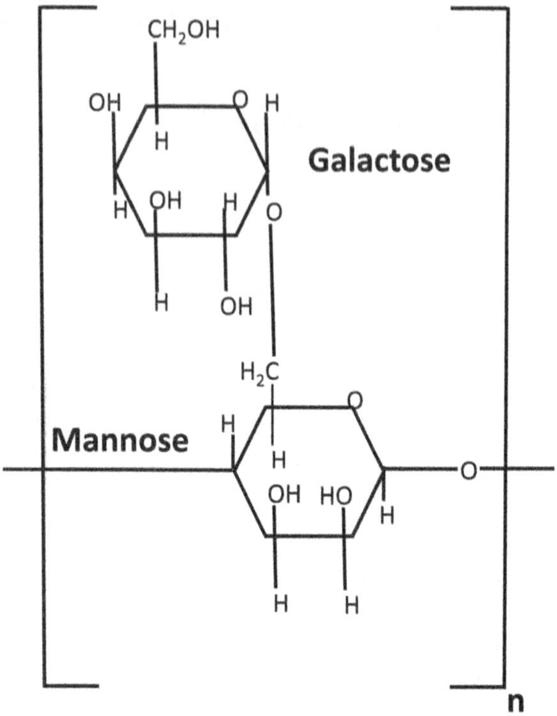


**Fig. 11.1** Structure of galactomannan derived from fenugreek gum

while increasing shear stress and temperature which cause a decline in the viscosity of solution (Wei et al. 2015). Wei et al. observed a drop in viscosity of aqueous fenugreek gum solution in steady shear flow conditions (Wei et al. 2015). Disentanglement of the polymer chain and distortion of the supramolecular structure of gum which might be responsible for this physical change (Singh et al. 2009; Jian et al. 2014). Fenugreek gum, when dissolved in water, undergoes sol–gel transition with an increase in gum concentration (Wei et al. 2015). Fenugreek gum at a concentration of 0.05% (w/v) evinces sol–gel transition whereas gel-like properties have been observed above 1% (w/v) concentration (Wei et al. 2015). Thermogram analysis showed that biopolymer extracted from the fenugreek seeds has a melting range of 66−139 °C and degradation temperature of about 296.45 °C (Rashid et al. 2018). Previous study reported that galactomannan extracted from fenugreek seeds can be effectively recovered from aqueous solution using less amount of isopropanol than ethanol due to its lower dielectric potential (Rashid et al. 2018). It was also observed that precipitation of galactomannan-based gums in polar organic solvents depends on their molecular weight and high density of hydrogen bonding due to large number of galactose substitution (Jian et al. 2014). Viscosity of fenugreek gum rises as hydration time extends to a certain point (12 h) and after that it starts decreasing (Jiang et al. 2007); whereas intrinsic viscosity of fenugreek gum solution decreases with an increase in alkaline environment (Doyle et al. 2009). It was proposed that there could be two reasons for this effect, (a) increment in negative charge along the polymer chains eradicate the hyper entanglement and (b) association between charged chain continues due to enhance ionic strength. However, viscosity of the fenugreek solution is recovered by neutralizing the solution with the addition of salt (Doyle et al. 2009). Previously, flow behaviour of fenugreek gum as a function of concentration and shear rate was examined and it was observed that the shear thinning behaviour was present in all concentration ranges demonstrating pseudoplastic behaviour (Wei et al. 2015). At low shear rates and concentration (<1%, w/v) the apparent viscosity displayed narrow range newtonian plateau as concentration of fenugreek gum increased, while newtonian plateau disappeared at 1% (w/v) concentration (Wei et al. 2015). Pseudoplastic behaviour of fenugreek gum explains the existence of interlacing molecular interaction between polymer chains (Wei et al. 2015). Galactomannan exhibits coiled conformation at small shear rates and high apparent viscosity may induce the interlocking between these coiled complexed macromolecules (Singh et al. 2009). Deformation in interlocking conformation of complexes due to shear rate induction leads to decline in solution viscosity (Wei et al. 2015; Jian et al. 2014). Shear strain also changes the orientation of galactomannan to an aligned fashion, which may reduce the viscosity of solution (Adeli et al. 2015). Fenugreek gum solution at high concentration display elastic behaviour as the thixotropic property enhances with an increment in polymer concentration (Wei et al. 2015).

Numerous physical properties of fenugreek gum make it as an ideal excipient to be used in pharmaceutical industries such as viscosity of fenugreek gum changes with time, which affects the release profiles of loaded therapeutic agents (Jiang et al. 2007; Lee et al. 2009). Viscosity of fenugreek gum also varies with pH, which gives it the pharmacological efficacy to be applied in vaginal and ophthalmic drug delivery (Doyle et al. 2009; Lee et al. 2009).

Thixotropic property of polymer has a vital impact on the therapeutic efficacy of the formulations by affecting their retention time at the administered site and enhancing the bioavailability. Thixotropic behaviour of fenugreek gum at higher concentration might enhance its ability to be used as a smart excipient for numerous formulations such as ointment, hydrogel, emulsions and suspensions through different routes like oral, topical, ophthalmic and mucosal (Lee et al. 2009).

Smart sun screen formulations must have the pseudoplastic behaviour so that they can form coherent protective layer over screen to eradicate the hazardous effect of UV light (Lee et al. 2009). Pseudoplastic behaviour of fenugreek gum at all concentration range makes it a cost effective candidate to develop sunscreen lotions (Wei et al. 2015). These rheological properties make fenugreek gum as flexible and cost effective excipient for various pharmaceutical formulations.

### *11.2.3 Biosafety and Toxicological Studies*

Utilization of natural polymers for biomedical applications is continuously increasing owing to their minimal cost, biocompatibility and minimal side effects. It is of utmost importance to evaluate the probable toxic effects of these polymers before their preclinical and clinical applications. Previous toxicological studies affirm that seed extract/seed powder of fenugreek is non-toxic to experimental animals at a single oral dose of 5 g/kg. Opdyke et al., showed that acute oral $LD_{50}$ of fenugreek was more than 5 g/kg in rats (Opdyke 1978) while the $LD_{50}$ for acute dermal toxicity in rabbits was found to be 2 g/kg. Repeated dose toxicity studies for 90 days concluded that debitterized fenugreek was safe to weaning rats and no significant toxic effects were noticed in behavioural pattern, food intake and growth of animals (Narasimhamurthy et al. 1999). Trigonelline, an alkaloid present in the seeds of fenugreek, was also found to be not altering the weights of mice thymus, kidney, liver, thyroid, adrenals, uterus or ovaries when fed for three weeks. Oral uptake of fenugreek seed extract for six months did not cause any clinical, hepatic, renal and hematological irregularities in diabetic human patients (Sharma et al. 1996a, b). On the other hand, in a randomized, double blinded, placebo-paralleled trial type II diabetes melitus patients suffered from stomach discomfort, nausea and diarrhea due to subchronic administration of saponins derived from fenugreek seeds. Patients were provided with 2.1 g of saponins thrice a day for a period of three months (Lu et al. 2008).

Subchronic oral administration of fenugreek seed powder acts as an antifertility agent in rodents by altering the weight of the reproductive tissues, sperm count and morphological irregularities in sperm cells (Al-Ashban et al. 2010; Al-Yahya 2013). Sharma and Bhinda. observed decline in weights of ovaries and uterine due to administration of steroidal extract of fenugreek (100 mg/day/rat for 15 days) to adult female rats (Sharma and Bhinda 2005). Elbetieha et al. (1996) suggested fenugreek seed powder induced abnormalities in fertility of female rats due to its estrogenic activity that distorts endothelial lining and interfaces of fetal development (Elbetieha et al. 1996).

The toxicological properties of fenugreek seed powder were observed by various animal models and human trials. From the above observations, it was clear that fenugreek act as antifertility agent in both male and female animals. But particular molecule and cellular mechanism responsible for antifertility of fenugreek is still in debate. Further investigations are needed to identify the compositional element of fenugreek seed, which acts as toxic agent. Toxico-metabolomics and metabolic flux analysis experiments may be helpful to understand the cellular metabolism and detoxification pathways (Table 11.2).

**Table 11.2** Toxicological aspects of fenugreek seed powder in various animal studies

| Type of toxicity | Acute oral toxicity | Acute dermal toxicity | Subchronic toxicity (90 days exposure) | Chronic toxicity (exposure over 24 weeks) |
|---|---|---|---|---|
| Dosage | $LD_{50} > 5$ g/kg (rat) Opdyke (1978) $LD_{50} > 5$ g/kg (rat) (Narasimhamurthy et al. 1999) $LD_{50} > 2$ g/kg (mice) Narasimhamurthy et al. (1999) | >2 g/kg (rabbit) Opdyke (1978) | NOAEL= 10% (rat) Diet (~2 g/day) Opdyke (1978) NOAEL= 20% (rat)[87] Diet (~4 g/day) Rao et al. (1996) | 25 g/day (diabetic patients) No toxic effects Sharma et al. (1996a, b) |

LD50—Dose that is expected to be lethal in 50% of test subjects
NOAEL—No Observed Adverse Effect Level

## 11.3 Drug Delivery Applications of Fenugreek Gum

### *11.3.1 Ophthalmic Drug Delivery*

Eyes are delicate and sensitive organs of the human body so ophthalmic formulations strictly need to be non-irritant when applied in the ocular area. Cataracts, macular degeneration, glaucoma, diabetic retinopathy, dry eye syndrome and ocular allergies are the major challenges in the ophthalmology arena. The currently available drugs to treat these diseases are associated with four major problems like poor bioavailability, prolong drug release characteristics, higher dosage and repeated administration. Novel polymer-based drug delivery systems such as liposomes, hydrogels and nano/microparticles addressed these drawbacks mentioned above. Polymers of natural origin proved to be the alternative candidate for ophthalmic drug delivery systems due to less toxic nature, ease of availability and low cost.

Pathak et al. developed nanoparticulate system for ocular delivery using chitosan and fenugreek seed mucilage. Haemocompatibility studies of developed chitosan/fenugreek mucilage nanoparticles showed that it was haemocompatible up to 2000 μg/ml. Acute ocular irritation study of nanoparticle-based system at different concentration ranges (50–2000 μg/ml) was examined using Draize's test in New Zealand white female rabbits. Ocular safety studies concluded that the developed ocular delivery system was non-irritant and can be used for human eyes safely (Pathak et al. 2014).

Mucoadhesive nature of fenugreek mucilage may be were going to enhance pre-cornical drug retention which will provide prolonged drug release to treat back eye diseases. Numerous formulations like viscous polymer vehicles and receptor/transporter targeted nano-molecules can be developed using fenugreek mucilage to build smart ophthalmic drug delivery systems. Concepts of micro-dialysis and imaging technology can be applied to modify animal models to develop fenugreek-based proactive ocular drug delivery systems.

### *11.3.2 Gastroretentive Drug Delivery*

Gastric retention has been known as a key approach to increase the efficacy and bioavailability of drugs having a slender absorption frame. Numerous drug delivery strategies such as swellability, high density, floating and mucoadhesive machineries have been fabricated to enhance the absorption and bioavailability of poorly gastroretentive drugs (Hao et al. 2014; Bera et al. 2015a, b; Deshpande et al. 1997).

Floating drug delivery system is one the most used strategy for gastorentententive drug delivery; however, one major disadvantage of this drug delivery system is its dependency on gastric emptying/transit time (Deshpande et al. 1997). Different strategies have been utilized to address this issue such as combining floating and swelling approaches together (Chen et al. 2013). Floating and swelling behaviour of drugs could be upgraded by applying polymer membrane on the gastroretentive matrix (Deshpande et al. 1997). Polysaccharides extracted from the plants are used as coating agents in gastroretentive drug delivery system as they possess key pharmaceutical properties like stability, swelling and regulated drug release rate.

In this context, an efficient drug delivery system consisting of alginate-fenugreek gum gel membrane covered with hydroxy-propyl-methyl-cellulose (HPMC) based matrix containing quetiapine-fumarate (QF) drug for intra-gastric delivery was developed. This formulation possesses improved buoyancy, prolonged drug release and better swelling ability (Bera et al. 2015a, b). The presence of fenugreek-alginate gel membrane on the surface of the tablet blocks many channels of core tablet, which slow down the drug release rate (Bera et al. 2015a, b). Presence of an air compartment between the coating layer and core tablet might be the reason for the superior floating capability of the alginate-fenugreek gum-based formulation (Bera et al. 2015a, b). Drawbacks such as weight gain and hyperglycemia associated with QF therapy can be avoided using fenugreek gum (Bera et al. 2016). This study not only proved that fenugreek gum as a potential natural polymer to be used in targeted drug delivery system but also proposed an alternative approach to regulate the disadvantages associated with QF therapy.

Though there are significant developments in gastro retentive drug delivery systems formulations with meal independent gastric retention capacity, floating behaviour is yet to be achieved. Fenugreek-based bioadhesive systems, super porous hydrogels, floating systems, high density-based formulations and expandable systems may be acting as game changers in gastro retentive drug delivery arena. Complex in vitro experiments can be designed to evaluate the efficiency of fenugreek-based gastro retentive drug delivery systems by varying the density/viscosity of gastric contents and providing a different degree of contractions in the presence of food. The impact of pharmacokinetic property on gastric retention ability can be observed using sophisticated in vivo imaging techniques.

### 11.3.3 Colon Drug Delivery

Orally administered colon-targeted drugs are expected to protect the therapeutic agent release and degradation in the stomach as well as in the small intestine and able to deliver it in colon. However, the acidic environment of upper gastrointestinal tract induces the degradation of drugs. This obstacle has been addressed by designing drug delivery system that facilitates the release of therapeutic agent in neutral pH. But the difference in the pH of intestine (7.4) and colon (6.8) region adversely affected the ability of pH-dependent drug delivery system to deliver the drug to the colonic region. In order to overcome this pH associated issue, drug encapsulated with polysaccharides offers advantages due to their stability in divergent pH conditions (Park et al. 2010). Natural polysaccharides found to be alternative candidates for the fabrication of colon specific drug delivery systems as colonic bacteria able to degrade them effectively (McConnell et al. 2008). Inter polymer complexes composed of chitosan and carboxymethyl fenugreek gum with tamoxifen was fabricated by Randhawa et al. Drug coated with polymeric system were able to protect the release of bioactive compound in stomach as well as in small intestine, and in the presence of rat cecal content 91% of tamoxifen released was observed from in vitro and in vivo experiments (Randhawa et al. 2012).

Easy availability, low cost, cytocompatibility and mucoadhesive nature of fenugreek gum make it as a gold standard natural polymer-based excipient for colon specific drug delivery. But research involving fenugreek gum based excipients for colon drug delivery is still in nascent stage, and therefore pressure controlled colon specific capsules and osmotic controlled drug delivery can be developed using fenugreek gum. These colon specific drug delivery systems can be evaluated by mimicking various pH conditions of gastric fluid, jejunum, small intestine and ileum. The efficacy of drug delivery system could be observed by incubating it with buffer medium containing ezypectinase and dextranase. Numerous animal models involving dog, guinea pigs and rats can be designed to evaluate the activity of formulation in complex cellular conditions. The absorption and distributional behaviour of fenugreek-based drug delivery system can be evaluated using gamma scintigraphy and high frequency techniques.

### 11.3.4 Vaginal Drug Delivery

Self-cleansing property of vaginal tract has always been a problem for the establishment of drug in the area of infection. Dynamic vaginal delivery system is expected to deliver the therapeutic agents to the site of infection or injury for a longer time span (Pavelić et al. 2001). Vaginal films comprising of polymers loaded with bioactive agents proved to be the best candidate as it offers stability, ease of administration and longer retention (Garg et al. 2005). Bioadhesiveness and biocompatibility of natural polymers draw interest of researchers for their application in vaginal drug delivery system. Polymeric films composed of aminated fenugreek gum and glycerol, loaded with therapeutic agent nystatin were developed for the treatment of vaginal candidiasis. These bioactive films were able to deliver therapeutic agents in 8 h, however, these films cured vaginal candidiasis in rats which were evident from histopathological studies (Bassi and Kaur 2015a, b). Enhanced bioadhesiveness of these films might be due to the interaction between amine groups of aminated fenugreek gum and negatively charged mucin chains through hydrogen bonding or electrostatic interaction (Kaur et al. 2013). In another report Bassi and Kaur (2015a, b) prepared bioengineered films using carboxymethyl fenugreek gum and glycerol functionalized with nystatin which have delivered 100% drug in 5 h. This drug delivery system is found to be compatible with vaginal mucosa which was inferred from the in vivo studies. Derivatives of fenugreek gum (aminated, carboxymethylated) make them as pivotal macromolecule to be explored in the field of targeted drug delivery by enhancing its bioadhesive and drug delivery properties (Bassi and Kaur 2015a, b). Smart nanofibrous mats can also be fabricated using fenugreek-based biopolymer for numerous biomedical applications (Yadav et al. 2019). Mucoadhesive nature of fenugreek gum can be applied to fabricate various potent formulations such as nanoparticle/emulsions, vaginal inserts and hydrogels to treat bacterial vaginosis, vulvo vaginal candidiasis, trichomoniasis, urinary tract infections and aerobic vaginitis.

### *11.3.5 Aerogels*

Aerogel drug delivery system has become the prime area of research in biomedical and pharmaceutical terrain due to its porous structure and large surface area. Aerogels composed of plant-based polysaccharide are biocompatible in nature, and therefore it can be applied as drug carriers. It was shown that aerogel made up of laccase oxidized fenugreek galactomannan act as a hydrolytic glycosidase agent when loaded with lysozyme. Developed drug loaded aerogels were capable of uptaking organic and inorganic solvents 20 times of its own weight. Hydrolytic activity and clemency of lysozyme was noticed when biomaterial was kept on agar plate containing *M. lysodeikticus* cells (Rossi et al. 2016). Chemical composition of galactomannan plays a vital role in enhancement of overall efficiency of aerogel-based drug delivery system. This study compared the aerogels obtained by enzymatic lyophilization of galactomannans of fenugreek, guar and sesbania loaded with therapeutic agents like lysozyme, nicin and polymyxin B. Fenugreek-based aerogels expressed better mechanical properties than guar and sesbania. The highest amount of galactose on the mannose backbone in fenugreek-based galactomannan which could be the possible reason for the better mechanical strength (Campia et al. 2017). Excellent biocompatibility, biodegradability, easy availability, low cost and particularly high porosity with open pore structure make fenugreek-based galactomannan as a potential candidate to synthesize robust aerogel delivery systems for biopharmaceutical industries (Fig. 11.2).

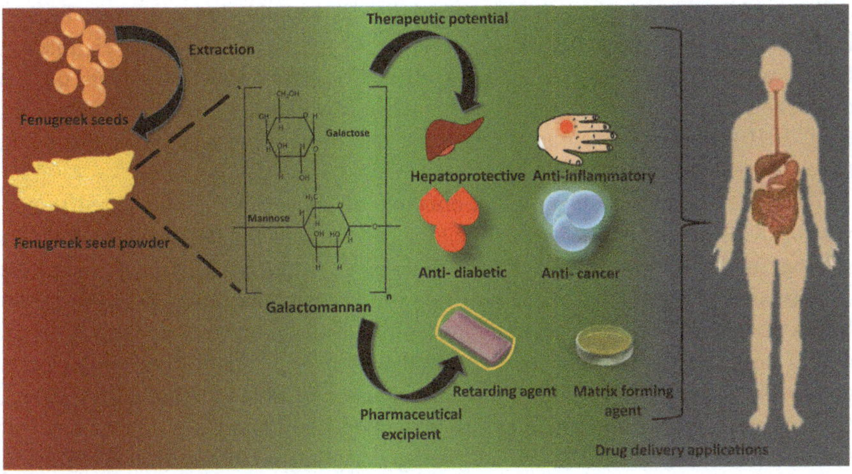


**Fig. 11.2** Numerous applications of fenugreek gum

## 11.4 Fenugreek Gum as a Pharmaceutical Excipient

### *11.4.1 Retarding Agent*

The release of therapeutic agent from the oral delivery system depends on the physicochemical properties of both polymer and drug used in the formulation. The drug release rate can be manipulated by using a combination of hydrophilic or hydrophobic polymers and varying their concentrations (Gade and Murthy 2014). Features of fenugreek gum as a retarding polymer was evaluated by Sav et al. 2013. Both fenugreek gum and its hydrophobic derivative octenylsuccinate anhydride fenugreek gum was used to prepare extended release tablets of metoprolol succinate and carbamazepine as hydrophilic and hydrophobic drug, respectively. The development of extended release tablet formulations displays similar drug release kinetics like marketed formulations, viz., Seloken® XL 50 mg and Tegretal® 200 CR (Sav et al. 2013). It is imperative to mention that controlled drug release can be accomplished not only by chemical alteration of fenugreek gum but also by using various copolymers to enhance unchanged drug concentration in the systemic circulation of poorly bioavailable drugs.

### *11.4.2 Super-disintegrating Agent*

Oral route of administration plays a pivotal role in drug delivery due to simple administration, less patient complication and efficient dosing (Bhowmik et al. 2009). Tablets and capsules are common dosage form to be administered through oral route but physiological changes in paediatrics and elderly cause complications in swallowing the tablets. This kind of obstacle can be addressed by engineering fast dissolving tablets (Karthikeyan et al. 2012). Fast dissolving tablets quickly disintegrate within few seconds by saliva when it is placed in oral cavity (Puttewar et al. 2010). Super-disintegrants induce the rate of dissolution by breaking up the tablets into small particles.

Fenugreek gum was subjected as effective disintegrants and compared to conventional disintegrants like croscarmellose and sodium starch glycolate. Fast dissolving tablets were loaded with diclofenac sodium fabricated by direct compression method using fenugreek gum of different concentration (1–6%, w/w). It was observed that an increase in concentration of fenugreek gum enhanced the rate of drug release and reduced the disintegration time. Application of fenugreek gum in the formulation caused significant reduction of inflammation and diclofenac concentration in the formulation. This reduction of inflammation may be due to synergistic activity of diclofenac and fenugreek gum in curbing the inflammatory mediators like leucocytes migration, cytokinins production and prostaglandin synthesis. Therefore, use of fenugreek gum as a superdisintegrant could minimize the dose related side effects of diclofenac (Kumar et al. 2014).

### *11.4.3 Mucoadhesive and Bioadhesive Agent*

Moderate water solubility, short half life and slow dissolution rate are the major pitfalls associated with conventional antidiabetic drugs such as Metformin HCl (MeHCl) and Glimpiride (Ahmed et al. 2016). Therefore, designing of smart drug delivery system which can overcome these pitfalls is a prime area of research in the pharmaceutical domain. Polymeric beads composed of natural or synthetic polymer loaded with therapeutic agents offer great success as efficient drug carriers (Nayak et al. 2013a, b). Easy availability, low cost, biocompatibility, biodegradability and environment friendly nature of plant-derived polymer and mucilage attracted researchers to use them as bead forming agent for therapeutic applications (Avachat et al. 2011). It was observed that beads consisting calcium pectinate and fenugreek seed mucilage embodied with MeHCl of size range 1.47–2.08 mm shows pH-dependent mucoadhesive and swelling properties. Formulated beads exhibited significant excellent hypoglycemic effect in diabetic rats after oral administration (Nayak et al. 2013a, b).

Nayak et al., developed sodium alginate and fenugreek seed mucilage based mucoadhesive beads using MeHCl as a therapeutic agent (Nayak et al. 2013a, b). In vivo studies revealed mucoadhesiveness and prolonged drug delivery capacity of the developed mucoadhesive beads. In another report, Nayak and Pal used ionotropic gelation technique to fabricate mucoadhesive beads composed of gellan gum and fenugreek seed mucilage functionalized with MeHCl. Release of MeHCl from the drug followed zero-order model with super case-II mechanism. Excellent hypoglycemic effect was noticed when fenugreek-based beads were administered orally to alloxan-induced diabetic rats. Mucoadhesive nature of these system could be due to the OH groups in both of the biopolymers which lead to form hydrogen bonding with mucous biological membranes. Excellent ability to induce chemical interaction with mucous membrane, fenugreek seed mucilage based formulations might enhance bioavailability of drugs, higher gastric retention and effective contact between mucous membrane and the therapeutic agent resulting in number of re-administration of the drug (Nayak and Pal 2014). Glimepiride-loaded mucoadhesive composite beads are made up of carboxymethyl fenugreek galactomannan, gellan gum and calcium silicate. Enhanced drug entrapment efficacy, constant drug release and hypoglycemic activity of glimepiride beads were confirmed from in vivo and in vitro studies (Bera et al. 2018).

Momin et al. 2015 developed a bilayer tablet of venlafaxine. Sustained and bioadhesive layers of tablet were formed using combination of fenugreek seed mucilage (FNM) and xanthan gum, carbopol, hydroxy propyl methyl cellulose (HPMC). The batch containing FNM:HPMC (80:20) displayed substantial bioadhesive force and tablet adhesion retention time as $2.4 \pm 0.028$ g and $24 \pm 2$ h, respectively. Optimized formulation obtained from this study showed enhanced drug release profile than marketed formulation Venlor XR. It was observed that enhanced concentration of fenugreek seed mucilage favoured bioadhesion and bioadhesive rention time of the formulations (Momin et al. 2015).

From the above studies it is clear that there were no significant alterations in drug release rate when fenugreek mucilage/gum is added with any other natural/ synthetic polymers. This property promotes the ability of fenugreek mucilage/gum to be used as a potential biopharmaceutical excipient for oral route of administration.

### *11.4.4 Matrix Forming Agent*

Application of polysaccharides extracted from fenugreek seed in matrix formulation containing propranolol hydrochloride was evaluated by Nokhodchi et al. (2008). In this investigation biopolymer extracted from the seeds of fenugreek compared with Methocel® hypromellose K4M, a conventionally applied matrix forming agent. An increase in fenugreek seed mucilage concentration in matrix formulation retarded propranolol diffusion from the matrix. Addition of lactose to the matrix formulation enhanced the pore size and reduce tortuoisity as a result water diffuses into the tablet by increasing drug release rate. Fenugreek mucilage observed to be a better retardant at concentration 66% (w/w) than hypromellose of equivalent content.

Iurian et al., utilizes freeze drying method to develop fenugreek seed mucilage based oral lyophilisates containing meloxicam as a model drug. It was observed that rise in mucilage concentration retards disintegration of therapeutic agent from the matrix. Longer disintegration time and higher crushing strength was noticed in the case of fenugreek-based tablets than gelatin based tablets produced under same conditions (Iurian et al. 2017). Enhancement in disintegration time with increase in fenugreek seed mucilage concentration in the formulation might have resulted due to the rise in viscosity of liquid medium which ultimately slow down the water absorption into the matrix. The natural polymer extracted from fenugreek seed can be further used to develop wound healing lyophilized products and mucoadhesive films by incorporating synthetic or semi synthetic polymers.

### 11.4.5 Bioavailability Enhancer

Limited intestinal absorption is one of the major problems to deliver bioactive agents efficiently. Krishnakumar et al., developed curcumin impregnated particles of galactomannan derived from fenugreek in the order of 150 ± 20 μm by using ultrasound mediated gel phase technique. Enhancement in bioavailability was observed in formulated particles at pH 1.2 and 6.8 from in vitro studies. Further, prolonged drug release activity was observed in Wistar rats (20 times higher) and human volunteers (15.8 times higher) in case of new formulation compared to raw curcumin (Krishnakumar et al. 2012).

In another study ultra-performance liquid chromatography coupled with triple quadruple electrospray ionization tandem mass spectrometry was used to evaluate blood brain barrier permeability, pharmacokinetics and bio-distribution of above-mentioned drug delivery system and free curcuminoids in Wistar rats.

These formulations of curcumin with fenugreek fibre were administered to the rats with value of 200 mg/kg body weight. Post administration of tissue samples obtained from rats demonstrated 25 times enhancement in bioavailability of fibre-based curcumin than the native. Effective blood brain barrier permeability also noticed from the tissue distribution of free curcuminoids at brain, heart, liver and kidney than standard curcumin (Krishnakumar et al. 2015). A dynamic fenugreek galactomannan-based formulation can be developed in near future by performing various investigations such as drug diffusion, transport, distribution and impact among various tissues.

### 11.4.6 Microencapsulation of Probiotic

Survival in gastrointestinal acidic and bile habitat is the prime criterion for an ideal probiotic. Establishment of an effective drug delivery system which successfully delivers probiotic by oral route and protects it from acidic environment is necessary. Encapsulation of probiotic bacteria offers a solution for these problems. Various polymer such as gelatin, alginate, starch, etc., has been used as matrix forming layer for the probiotic systems (Sohail et al. 2011). Haghshenas et al. designed a smart probiotic system containing *Lactobacillus plantarum* 15HN bacteria encapsulated with alginate–fenugreek–psyllium polymeric blends (Haghshenas et al. 2015a, b). It was observed that alginate combined with fenugreek or psyllium offers great potential as encapsulation matrix for probiotics. These gel formulations were able to protect bacterial cells not only at low pH and high bile salt conditions but also they support the growth of bacteria in gastrointestinal environment. In another report Haghshenas et al. observed that *Enterococcus durans* 39C encapsulated with algenic-psyllium blend with fenugreek gum possess excellent encapsulation efficacy, cell viability as well as enhanced release rates (Haghshenas et al. 2015a, b). Therefore, fenugreek gum can be used as encapsulating agent due to its potential to act as a prebiotic in the formulations. Low toxicity and affordable cost present fenugreek-based polymer as better alternative than synthetic polymers currently being used in probiotic formulations (Table 11.3).

## 11.5 Therapeutic Applications of Fenugreek Gum

### *11.5.1 Antidiabetic Property*

Diabetes mellitus is a metabolic disorder affecting millions of human being day by day. Chronic diabetes tends to offer retinopathy, nephropathy, neuropathy and cardiovascular diseases (Pradeep and Srinivasan 2017; Zarvandi et al. 2017). Currently used hypoglycemic and hypolipidemic agents for the treatment of diabetes offer several side effects such as lactic acidosis, hypoglycemia, abdominal discomfort, peripheral edema, myopathy and heaptic toxicity (Sorrentino 2012). Exploration of drugs of natural origin for the treatment of diabetes is an active area of research and several studies proved their effectiveness as an alternative treatment option than commonly available synthetic drugs. Fenugreek is a plant of traditional and ethnopharmacological relevance used by the tribal people for the treatment of diabetes and other diseases.

Antidiabetic potential of fenugreek seeds has been scientifically validated by several research groups. Different bioactive compounds have been isolated from the fenugreek seed and evaluated for their therapeutic application. Galactomannan extracted from the fenugreek seed also showed its potential for the treatment of type II diabetes which was evident by the clinical studies conducted on experimental animals and human volunteers. Sharma et al. clinically evaluated hypoglycemic effect of fenugreek seed powder in non-insulin dependent diabetic patients. Patients were fed with two equal doses (12.5 g each) of fenugreek seed powder in diet (during lunch and dinner) daily for 24 weeks. Reduced levels of fasting blood glucose, insulin and glycosylated hemoglobin along with enhanced glucose tolerance were observed in the patients fed with fenugreek seed powder. This study showed the potential role of fenugreek seed powder for the treatment of type II diabetes (Sharma et al. 1996a, b).

Raju et al. evaluated the therapeutic efficacy of fenugreek seed powder in type I diabetes. In this study, 5% fenugreek seed powder was orally administered with diet in alloxan-induced diabetic Wistar rats daily for 21 days. Altered activities of

**Table 11.3** Applications of Fenugreek seed powder/mucilage in the drug delivery

| Active constituent | Application | Formulation type | Drug delivery type | Model drug | Reference |
|---|---|---|---|---|---|
| Fenugreek seed/mucilage | Polymer to develop nanoparticles | Nanoparticles | Opthalmic | – | Pathak et al. (2014) |
| alginate–fenugreek gum | Coating gel to matrix | Tablet | Gastroretentive | Quetiapine fumarate | Bera et al. (2016) |
| Chitosan and carboxymethyl fenugreek gum | Protecting agents | Tablet | Colon | Tamoxifen | Randhawa et al. (2012) |
| Amiated fenugreek gum | Bioadhesive agent | Vaginal film | Vaginal | Nystatin | Bassi and Kaur (2015a, b) |
| Carboxy-methylated fenugreek gum | Bioadhesive agent | Vaginal film | Vaginal | Nystatin | Bassi and Kaur (2015a, b) |
| Fenugreek gum | Delivery system | Aerogel | – | Lysozyme | Rossi et al. (2016) |
| Fenugreek gum | Delivery system | Aerogel | – | Polymyxin-B, Nisin, Muraminidase lysozyme | Campia et al. (2017) |
| Octenyl succincate anhydride derivative of fenugreek gum | Retarding agent | Tablet | Oral | Metoprolol succinate | Sav et al. (2013) |
| Octenyl succincate anhydride derivative of fenugreek gum | Retarding agent | Tablet | | Carbamazepine | Sav et al. (2013) |
| Fenugreek gum | Super-disintegrating agent | Tablet | Oral | Diclofenac sodium | Kumar et al. (2014) |
| Fenugreek seed mucilage and calcium pectinate | Mucoadhesive agent | Beads | Oral | MeHCl | Nayak et al. (2013a, b) |

(continued)

**Table 11.3** (continued)

| Active constituent | Application | Formulation type | Drug delivery type | Model drug | Reference |
|---|---|---|---|---|---|
| Sodium alginate and fenugreek seed mucilage | Mucoadhesive agent | Beads | Oral | MeHCl | Nayak et al. (2013a, b) |
| Gellan gum and funugreek seed mucilage | Mucoadhesive agent | Beads | Oral | MeHCl | Nayak and Pal (2014) |
| Carboxymethyl fenugreek galactomannan | Mucoadhesive agent | Beads | Oral | Glimepiride | Bera et al. (2018) |
| Fenugreek seed mucilage and hydroxy propyl methyl cellulose (HPMC) | Bioadhesive and sustained release layer | Tablet | Oral | Venlafaxine hydroclhoride | Momin et al. (2015) |
| Fenugreek seed mucilage | Matrix forming agent | Tablet | Oral | Propranolol hydrochloride | Nokhodchi et al. (2008) |
| Fenugreek seed mucilage | Matrix forming agent | Oral lyophilisates | Oral | Meloxicam | Iurian et al. (2017) |

various enzymes such as gluconeogenic, lipogenic and glycolytic were observed in both liver and kidney of diabetic rats. However, treated animals showed reduced fasting blood glucose level as well as eradication of the abnormalities associated with activities of enzymes (Raju et al. 2001).

Zia et al. 2001 examined hypoglycemic effect of aqueous and methanolic seed extract of fenugreek in normal mice. More prominent blood glucose lowering effect was observed in the case of aqueous extract as compared to methanolic extract. This experiment revealed that the hypoglycemic agent of fenugreek seed extract is polar in nature (Zia et al. 2001). In another study, it was shown that fenugreek seed mucilage is able to lower fasting blood glucose level in streptozotocin-induced diabetic rats (Kumar et al. 2005).

Liver pyruvate kinase (PK), phosphoenolpyruvate carboxykinase (PEPCK) and glucose transporter (GLUT 4) play a pivotal role in glucose homeostasis and abnormal expression of these proteins was noticed in diabetes (Board et al. 1990; Mathupala et al. 1997). Mohammad et al. 2006 explored the ability of fenugreek seed powder in recovering the altered expression of PK, PEPCK, GLUT 4 (in skeletal muscle), and lowering the blood glucose level in alloxan-induced diabetes in rats. Eidi et al. noticed that ethanolic extract of fenugreek seeds was able to lower serum glucose, total cholesterol, urea, uric acid, triacylglycerol, aspartate amino transferase, alanine amino transferase, and enhanced serum insulin level when orally administered to streptozotocin-induced diabetic rats (Eidi et al. 2007). Srichamroen et al. showed that galactomannan is able to reduce the intestinal glucose uptake in genetically determined lean and obese rats and concluded that an increase in viscosity with rise in galactomannan could be the possible reason for declining uptake of glucose (Srichamroen et al. 2009).

Previous studies showed that an increase in oxidative stress during hyperglycemia is a primary reason behind the cardiovascular mortality in diabetic patients. Upregulation of angiotensin converting enzyme (ACE) and receptor $AT_1$ of renin-angiotensin system (RAS) is the reason behind cardiovascular disease (Murça et al. 2012). ACE induce gene expression of extracellular matrix (ECM) components, growth factors and infiltration of inflammatory cells. Due to increment in ECM components (collagen, fibronectin) cardiac stiffness enhances and diastolic dysfunction arises (Ban and Twigg 2008; McKarns and Schwartz 2005). Dietary fenugreek seeds and onion powder have been used as therapeutic agents to treat the cardiovascular damage caused due to diabetes (Pradeep and Srinivasan 2018). It was observed that fenugreek seed and onion powder were able to block renin-angiotensin system (RAS) by downregulation of ECM components in diabetic rats. These observations provided substantial evidences for fenugreek seed powder as a potential antidiabetic agent and offers advantages over marketed drugs in terms of reduced side effects.

Combination of fenugreek seed powder and sodium orthovanadate is used to treat diabetes and long-term difficulties associated with the diabetes like structural abnormalities in sciatic nerves. It was shown that the combination (sodium orthovanadate + fenugreek seed powder) was able to restore the altered glucose level and eradicated structural abnormalities in peripheral nerves (Preet et al. 2005). Several studies reported the defensive role of fenugreek seed against the diabetic nephropathy. Fenugreek seed extract was able to reduce the thickening of glomerular base membrane by inhibiting the accumulation of oxidized DNA in the kidney (Xue et al. 2011). Fenugreek seed powder proved as an effective therapeutic agent to treat both type I and type II diabetes. But fenugreek seed powder based prediabetes treatment is still in its infancy. More studies involving animal models and human volunteers are required to evaluate the efficiency of fenugreek seed powder as a bioactive agent to treat prediabetes.

### *11.5.2 Hypolipidemic Potential and Role in Fat Accumulation*

Vijayakumar et al. observed hypolipidemic potential of fenugreek seed extract in 3T3-L1 and HepG2 cells by quantifying fat accumulation by western blot analysis of lipogenic and adipogenic factors. Fenugreek seed extract decreased the fat accumulation in 3T3-L1 cells by downregulating adipogenic factors like peroxisome proliferators activated-receptors-γ (PPAR-γ), sterol regulatory element-binding protein-1 (SREBP-1), CAAT element-binding proteins-α (c/EBP-α) further, cellular triglycerides and cholesterol concentrations declined in HepG2 cells via reduced expression of (SREBP-1) at mRNA and protein level. Seed extract of fenugreek was also able to upregulate low density lipo-protein receptor (LDLR) expression resulting enhanced LDL uptake (Vijayakumar et al. 2010).

Diosgenin, a steroidal sapogenin (component of fenugreek seed powder) at different concentrations (5–10 mmol/l) inhibit triglyceride accumulation in HepG2 cell lines and expression of lipogenic genes. It inhibited fat accumulation by disrupting the transactivation of liver-X-receptor-alpha (LXR-α), a prime regulator of cholesterol homeostasis in hepatocytes (Uemura et al. 2011).

Kumar et al. fed aqueous extract of fenugreek seed to female Wistar rats for 28 days and it was found to be anti-hyperlipidemic agent as it was able to reduce white adipose tissue weight, body mass index, weight gain, leptin, lipids, serum insulin, blood glucose, lipase, and apolipoprotein-B levels (Kumar et al. 2014). Apart from that rise in antioxidant enzymes (superoxide dismutase, glutathione, catalase) level were noticed and serum aspartate amino transferase (AST), alanine amino transferase (ALT), lactate dehydrogenase levels were upregulated due to feeding of *trigonella* seed extract. Restored activity of liver and uterine WAT lipogenic enzymes to normal level suggests fenugreek seed extract could be used as a bioactive agent to inhibit fat accumulation. Clinical studies can be designed to observe therapeutic nature of diosgenin to treat hyperlipidemia, and diosgenin based drugs can be formulated to address these clinical consequences.

### *11.5.3 Anti-inflammatory Potential*

Rheumatoid arthritis is a chronic inflammatory disease which causes multiple pathophysiological consequences like swelling of joints, cartilage and bone demolition, synovial hyperplasia, vasculogenesis. This disease causes rise in inflammatory enzymes like cyclooxygenase, lipoxygenase, myeloperoxidase and induce free radical formation along with low antioxidant level (Tristano 2009; Hoozemans and O'Banion 2005). Sindhu et al. evaluated antioxidant and anti-inflammatory potential of fenugreek mucilage in adjuvant induced arthritis in rats. Fenugreek mucilage was fed orally to the rats and 73.65% edema inhibition observed with 75 mg/kg mucilage dose. Fenugreek seed mucilage effectively reduced the activities of inflammatory enzymes and upregulated the activities of antioxidant enzymes, vitamin C and reduced glutathione (Sindhu et al. 2012).

Liu et al. investigated lipid peroxidation (LPO) and cyclooxygenase (COX) enzyme inhibitory activity of fenugreek seed extract in different mediums such as water, hexane, methanolic and ethyl acetate by MTT, COX-1, COX-2, LPO enzyme inhibitory assays. These extracts at 250 μg/mL concentration was able to inhibit LPO, COX-1 and COX-2 (Liu et al. 2012). Non-steroidal anti-inflammatory drugs (NSAIDs) are the most preferred drugs to address inflammatory diseases but NSAIDs are associated with certain drawbacks like gastrointestinal ulcerogenicity and renal morbidity (Pincus et al. 1992). Therefore, long-term use of these conventional drugs need a safe alternative like fenugreek seed mucilage as it suppresses inflammatory enzymes activity and upregulate the antioxidants. Fenugreek mucilage based nano-molecules can be engineered to enhance its anti-inflammatory properties and further studies can be performed to gain more insight for molecular mechanism behind its anti-inflammatory properties.

### *11.5.4 Anticancer Potential*

Free radicals generated by inflammatory cells, i.e. neutrophils, eosinophils, macrophages, could be the pathogenic factor for tumour formation in some cases (Rosin et al. 1994). Earlier studies demonstrated, fenugreek seed extract at different concentrations (10–15 μg/mL) are able to inhibit the growth of breast, pancreatic and prostate cancer cell lines but not the healthy cells for 72 h. It was also observed that the crude fenugreek seed extract was able to differentiate between normal and cancerous cells; however, its pure agents like diosgenin and sulforaphane did not differentiate in malignant and normal cells (Shabbeer et al. 2009). In another report, it was informed that not only fenugreek seed extract but also bioactive components, i.e. diosgenin and trigonelline act as anticancer agents (Raju and Rao 2012). From the above-mentioned reports it was observed that there is still a conflict about the particular therapeutic agent of fenugreek seed mucilage. So there is significant demand to evaluate the anticancer potential of fenugreek seed mucilage and its components.

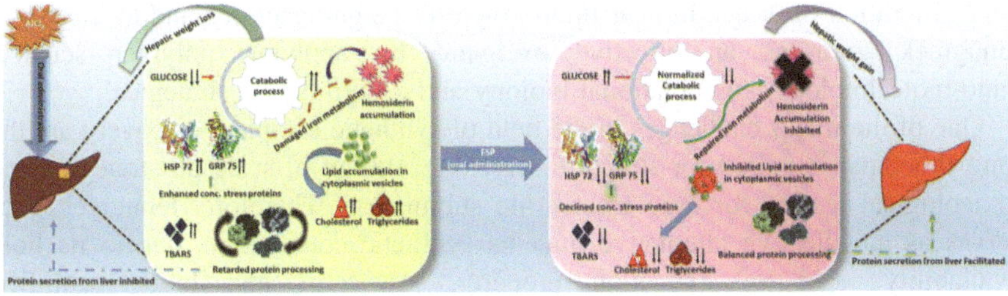


**Fig. 11.3** Hepato-protective mechanism of fenugreek-based seed extract

### *11.5.5 Hepatoprotective Potential*

Hepatoprotective ability of fenugreek seed extract was investigated for ethanol induced hepatic injury and apoptosis rats by Kaviarasan and Anuradha (2007). Upregulation of liver dysfunction markers (alanine aminotransferase (ALT), gamma-glutamyl transferase (GGT), lactate dehydrogenase (LDH), aspartate aminotransferase (AST), alkaline phosphatase (ALP) and bilirubin) in plasma and declined liver glycogen were noticed in rats with chronic ethanol administration (6 g/kg/day) for 60 days. Ethanol administration altered the activities of alcohol metabolizing enzymes. Treatment with polyphenolic extract of fenugreek seed restored the level of liver injury markers to normal levels and this result was compared with silymarin, a hepatoprotective agent (Kaviarasan and Anuradha 2007).

Belaïd-Nouira et al. (2013) observed fenugreek seed powder (5% in diet) efficacy to cure hepatic toxicity due to the oral exposure of $AlCl_3$ (500 mg/kg bw i.g. for one month then 1600 ppm via drinking water) in Wistar rats by preventing the liver weight loss and ameliorating liver dysfunction factors. Regeneration of hepatic cells, management of iron, glucose and lipid metabolism may be the proposed mechanism behind the efficient activity of fenugreek seed extract to cure the hepatic disease (Belaïd-Nouira et al. 2013). Fenugreek seed extract stimulate the liver function and regenerate the liver cells. Thus it can be used as potential pharmaceutical alternative to treat hepatic diseases (Fig. 11.3).

## 11.6 Conclusion and Future Perspectives

In this review, we have outlined numerous applications of fenugreek gum as a potential excipient and therapeutic agent for various biopharmaceutical usage. Chemists and biologists can access fundamentals of rheological, toxicological and bio-macromolecular characteristics of fenugreek gum to design a proactive excipient which can enhance the drug discovery strategies in scalable manner to fabricate cost-effective lifesaving drugs. Indeed, excellent work of industry-academia researches have introduced this biopolymer as one of the prime excipient for the next generation of drugs. In near future, there is a significant demand to transform fenugreek gum as an effective excipient by applying polymer science, nano-biotechnology, computational biology and design driven strategies.

One of the major challenges in the field of synthetic excipient discovery are the long time span required for structure screening, design of scale up strategies and toxicological testing for drug master file submission. Therefore, fenugreek gum serves as an effective alternative than its synthetic counterparts due to its bulk availability, ease of extraction property, low cost and biocompatibility. Mucoadhesive property, rapid disintegration capability, high water solubility and controlled release capacity pose it as a smart excipient to develop tangible pharmaceutical product. Although the application of fenugreek gum in pharmaceutical terrain is in its infancy today, new concepts of biopharmaceutical science can increase the possibilities of fenugreek gum in pharmaceutical industries. Strategies like chemical modification of biopolymer extracted from fenugreek seed mucilage, blending with copolymer and computational biology can be used to tailor the physiochemical property of biopolymer as per the formulation. However, complex in vitro and in vivo models, sophisticated imaging techniques and gamma scintigraphy could be used to analyze the therapeutic potential of fenugreek gum in the circulation. Nanomolecule-based drug delivery systems, liposome-based formulations and polymeric micelles can be fabricated using fenugreek gum with improved solubility, bioavailability, membrane permeation and active targeting.

Recent articles show the huge potential of fenugreek gum as a biopolymer to address formulation associated challenges with current therapeutic regime to treat acute and chronic disease worldwide. As a result, scope of performance, cost effectiveness and accessibility of valuable drugs would increase for growing society. Altogether, this review demonstrates the recent advances related to fenugreek gum aimed at expanding the horizon of drug delivery in the lap of rapidly growing pharmaceutical arena.

**Acknowledgments** PM, AS and TCY are thankful to the Ministry of Human Resource Development, Government of India for Senior Research Fellowship. Authors are thankful to Indian Institute of Technology Roorkee for providing research facilities for this work.

**Conflict of Interest** We confirm that there are no known conflicts of interest associated with this article.

## References

Adeli M, Samavati V (2015) Studies on the steady shear flow behavior and chemical properties of water-soluble polysaccharide from Ziziphus lotus fruit. Int J Biol Macromol 72:580–587

Ahmed OAA, Zidan AS, Khayat M (2016) Mechanistic analysis of Zein nanoparticles/PLGA triblock in situ forming implants for glimepiride. Int J Nanomedicine 11:543–555

Al-Ashban RM, Abou-Shaaban RR, Shah AH (2010) Toxicity studies on Trigonella foenum-graecum L. seeds used in spices and as a traditional remedy for diabetes. Orient Pharm Exp Med 10:66–78

Al-Yahya AA (2013) Reproductive, cytological and biochemical toxicity of fenugreek in male Swiss albino mice. Afr J Pharm Pharmacol 7:2072–2080

Avachat AM, Dash RR, Shrotriya SN (2011) Recent investigations of plant based natural gums, mucilages and resins in novel drug delivery systems. Ind J Pharm Edu Res 45:86–99

Ban CR, Twigg SM (2008) Fibrosis in diabetes complications: pathogenic mechanisms and circulating and urinary markers. Vasc Health Risk Manag 4:575–596

Bassi P, Kaur G (2015a) Bioadhesive vaginal drug delivery of nystatin using a derivatized polymer: development and characterization. Eur J Pharm Biopharm 96:173–184

Bassi P, Kaur G (2015b) Fenugreek gum derivatives with improved bioadhesion and controlled drug release: in vitro and in vivo characterization. J Drug Deliv Sci Tec 29:42–54

Belaïd-Nouira Y, Bakhta H, Haouas Z, Flehi-Slim I, Neffati F, Najjar MF, Cheikh HB (2013) Fenugreek seeds, a hepatoprotector forage crop against chronic AlCl3 toxicity. BMC Vet Res 9:9–22

Bera H, Boddupalli S, Nandikonda S, Kumar S, Nayak AK (2015a) Alginate gel-coated oil-entrapped alginate–tamarind gum–magnesium stearate buoyant beads of risperidone. Int J Biol Macromol 78:102–111

Bera H, Gaini C, Kumar S, Sarkar S, Boddupalli S, Ippagunta SR (2016) HPMC-based gastroretentive dual working matrices coated with Ca+ 2 ion crosslinked alginate-fenugreek gum gel membrane. Mater Sci Eng C Mater Biol Appl 67:170–181

Bera H, Kandukuri SG, Nayak AK, Boddupalli S (2015b) Alginate–sterculia gum gel-coated oil-entrapped alginate beads for gastroretentive risperidone delivery. Carbohydr Polym 120:74–84

Bera H, Mothe S, Maiti S, Vanga S (2018) Carboxymethyl fenugreek galactomannan-gellan gum-calcium silicate composite beads for glimepiride delivery. Int J Biol Macromol 107:604–614

Bhowmik D, Chiranjib B, Krishnakanth P, Chandira RM (2009) Fast dissolving tablet: an overview. J Chem pharm 1:163–177

Blank I, Lin J, Devaud S, Fumeaux R, Fay LB (1997) The principal flavor components of fenugreek (Trigonella foenum-graecum L.), In: Spices. American Chemical Society, pp 12–28

Board M, Humm S, Newsholme EA (1990) Maximum activities of key enzymes of glycolysis, glutaminolysis, pentose phosphate pathway and tricarboxylic acid cycle in normal, neoplastic and suppressed cells. Biochem J 265:503–509

Brummer Y, Cui W, Wang Q (2003) Extraction, purification and physicochemical characterization of fenugreek gum. Food Hydrocoll 17:229–236

Campia P, Ponzini E, Rossi B, Farris S, Silvetti T, Merlini L, Galante YM (2017) Aerogels of enzymatically oxidized galactomannans from leguminous plants: Versatile delivery systems of antimicrobial peptides and enzymes. Carbohydr Polym 158:102–111

Castro PM, Baptista P, Madureira AR, Sarmento B, Pintado ME (2018) Combination of PLGA nanoparticles with mucoadhesive guar-gum films for buccal delivery of antihypertensive peptide. Int J Pharm 547:593–601

Cerqueira MA, Bourbon AI, Pinheiro AC, Martins JT, Souza BWS, Teixeira JA, Vicente AA (2011) Galactomannans use in the development of edible films/coatings for food applications. Trends Food Sci Tech 22:662–671

Chen YC, Ho HO, Lee TY, Sheu MT (2013) Physical characterizations and sustained release profiling of gastroretentive drug delivery systems with improved floating and swelling capabilities. Int J Pharm 441:162–169

Choudhary PD, Pawar HA (2014) Recently investigated natural gums and mucilages as pharmaceutical excipients: an overview. J Pharm 2014:1–9

Cui SW (2001) Cereal non-starch polysaccharides I. (1–3)(1–4)-β-d-Glucans. Polysaccharide gums from agricultural products: processing, structures and functionality. Technomic Publishing Co, Lancaster, pp 103–166

Dawidar AM, Saleh AA, Elmotei SL (1973) Steroid sapogenin constituents of fenugreek seeds. Planta Med 24:367–370

Deshpande AA, Shah NH, Rhodes CT, Malick W (1997) Development of a novel controlled-release system for gastric retention. Pharm Res 14:815–819

Doyle JP, Lyons G, Morris ER (2009) New proposals on "hyperentanglement" of galactomannans: solution viscosity of fenugreek gum under neutral and alkaline conditions. Food Hydrocoll 23:1501–1510

Eidi A, Eidi M, Sokhteh M (2007) Effect of fenugreek (Trigonella foenum-graecum L) seeds on serum parameters in normal and streptozotocin-induced diabetic rats. Nutr Res 27:728–733

El Nasri NA, El Tinay AH (2007) Functional properties of fenugreek (Trigonella foenum graecum) protein concentrate. Food Chem 103:582–589

Elbetieha A, Al-Hamood MH, Al-Kofahi A (1996) Anti-implantation potential of some medicinal plants in female rats. Arch STD HIV Res 10:181–187

Elzoghby OA, Abd-Elwakil MM, Abd-Elsalam K, Elsayed TM, Hashem Y, Mohamed O (2016) Natural polymeric nanoparticles for brain-targeting: implications on drug and gene delivery. Curr Pharm Des 22:3305–3323

Gade R, Murthy TEGK (2014) Effect of hydrophilic and hydrophobic polymers on release kinetics of metoprolol succinate extended release tablets. AJP 5:101–106

Garg RC (2016) Fenugreek: multiple health benefits. In: Nutraceuticals. Academic Press, pp 599–617

Garg S, Vermani K, Garg A, Anderson RA, Rencher WB, Zaneveld LJ (2005) Development and characterization of bioadhesive vaginal films of sodium polystyrene sulfonate (PSS), a novel contraceptive antimicrobial agent. Phar Res 22:584–595

Goyal S, Gupta N, Chatterjee S (2016) Investigating therapeutic potential of Trigonella foenum-graecum L. as our defense mechanism against several human diseases. J Toxicol 2016:1–10

Haghshenas B, Abdullah N, Nami Y, Radiah D, Rosli R, Khosroushahi Yari (2015a) A Microencapsulation of probiotic bacteria L actobacillus plantarum 15 HN using alginate-psyllium-fenugreek polymeric blends. J Appl Microbiol 118:1048–1057

Haghshenas B, Nami Y, Haghshenas M, Barzegari A, Sharifi S, Radiah D, Abdullah N (2015b) Effect of addition of inulin and fenugreek on the survival of microencapsulated Enterococcus durans 39C in alginate-psyllium polymeric blends in simulated digestive system and yogurt. Asian J Pharm Sci 10:350–361

Hao S, Wang Y, Wang B (2014) Sinking-magnetic microparticles prepared by the electrospray method for enhanced gastric antimicrobial delivery. Mol Pharm 11:1640–1650

He Y, Ding C, Wang X, Wang H, Suo Y (2015) Using response surface methodology to optimize countercurrent chromatographic separation of polyphenol compounds from fenugreek (Trigonella foenum-graecum L.) seeds. J Liq Chromatogr Relat Technol 38:29–35

Hefnawy HTM, Ramadan MF (2011) Physicochemical characteristics of soy protein isolate and fenugreek gum dispersed systems. Int J Food Sci Tech 48:371–377

Hoozemans JJ, O'Banion MK (2005) The role of COX-1 and COX-2 in Alzheimer's disease pathology and the therapeutic potentials of non-steroidal anti-inflammatory drugs. CNS Neurol Disord-Dr 4:307–315

Iurian S, Dinte E, Iuga C, Bogdan C, Spiridon I, Barbu-Tudoran L, Leucuţa SE (2017) The pharmaceutical applications of a biopolymer isolated from Trigonella foenum-graecum seeds: focus on the freeze-dried matrix forming capacity. Saudi Pharm J 25:1217–1225

Jian HL, Lin XJ, Zhang WA, Zhang WM, Sun DF, Jiang JX (2014) Characterization of fractional precipitation behavior of galactomannan gums with ethanol and isopropanol. Food Hydrocoll 40:115–121

Jiang JX, Zhu LW, Zhang WM, Sun RC (2007) Characterization of galactomannan gum from fenugreek (Trigonella foenum-graecum) seeds and its rheological properties. Int J Polym Mater 56:1145–1154

Kakani RK, Anwer MM (2012) Fenugreek. In Handbook of herbs and spices. Woodhead Publishing, pp 286–298

Kamble H, Kandhare AD, Bodhankar S, Mohan V, Thakurdesai P (2013) Effect of low molecular weight galactomannans from fenugreek seeds on animal models of diabetes mellitus. Biomed Aging Path 3:145–151

Karthikeyan M, Umarul MA, Megha M, Shadeer HP (2012) Formulation of diclofenac tablets for rapid pain relief. Asian Pac J Trop Biomed 2:308–311

Kaur G, Mahajan M, Bassi P (2013) Derivatized polysaccharides: preparation, characterization, and application as bioadhesive polymer for drug delivery. Int J Polym Mater Po 62:475–481

Kaviarasan S, Anuradha CV (2007) Fenugreek (Trigonella foenum graecum) seed polyphenols protect liver from alcohol toxicity: a role on hepatic detoxification system and apoptosis. Pharmazie 62:299–304

Konapure SA, Chaudhari PS, Oswal RJ, Kshirsagar SS, Antre RV, Chorage TV (2011) Mouth dissolving tablets an innovative technology. Int J Appl Biol Pharm 2:496–503

Krishnakumar IM, Maliakel A, Gopakumar G, Kumar D, Maliakel B, Kuttan R (2015) Improved blood–brain-barrier permeability and tissue distribution following the oral administration of a food-grade formulation of curcumin with fenugreek fibre. J Funct Foods 14:215–225

Krishnakumar IM, Ravi A, Kumar D, Kuttan R, Maliakel B (2012) An enhanced bioavailable formulation of curcumin using fenugreek-derived soluble dietary fibre. J Funct Foods 4:348–357

Kumar GS, Shetty AK, Sambaiah K, Salimath PV (2005) Antidiabetic property of fenugreek seed mucilage and spent turmeric in streptozotocin-induced diabetic rats. Nutr Res 25:1021–1028

Kumar L, Verma S, Vaidya B, Gupta V (2017) Bioadhesive polymers for targeted drug delivery. In: Nanotechnology-based approaches for targeting and delivery of drugs and genes. Academic Press, pp 322–362

Kumar MU, Babu MK (2014) Design and evaluation of fast dissolving tablets containing diclofenac sodium using fenugreek gum as a natural superdisintegrant. Asian Pac J Trop Biomed 4:329–334

Kumar P, Bhandari U, Jamadagni S (2014) Fenugreek seed extract inhibit fat accumulation and ameliorates dyslipidemia in high fat diet-induced obese rats. Biomed Res Int 2014:1–11

Kumar R, Patil S, Patil MB, Patil SR, Paschapur MS (2009) Isolation and evaluation of disintegrant properties of fenugreek seed mucilage. Int J PharmTech Res 1:982–996

Lee CH, Moturi V, Lee Y (2009) Thixotropic property in pharmaceutical formulations. J Control Release 136:88–98

Li Y, Maciel D, Rodrigues J, Shi X, Tomas H (2015) Biodegradable polymer nanogels for drug/nucleic acid delivery. Chem Rev 115:8564–8608

Liu Y, Kakani R, Nair MG (2012) Compounds in functional food fenugreek spice exhibit anti-inflammatory and antioxidant activities. Food Chem 131:1187–1192

Lu FR, Shen L, Qin Y, Gao L, Li H, Dai Y (2008) Clinical observation on trigonella foenum-graecum L. total saponins in combination with sulfonylureas in the treatment of type 2 diabetes mellitus. Chin J Integr Med 14:56–60

Mathupala SP, Rempel A, Pedersen PL (1997) Aberrant glycolytic metabolism of cancer cells: a remarkable coordination of genetic, transcriptional, post-translational, and mutational events that lead to a critical role for type II hexokinase. J Bioenerg Biomembr 29:339–343

McConnell EL, Fadda HM, Basit AW (2008) Gut instincts: explorations in intestinal physiology and drug delivery. Int J Pharm 364:213–226

McKarns SC, Schwartz RH (2005) Distinct effects of TGF-β1 on CD4+ and CD8+ T cell survival, division, and IL-2 production: a role for T cell intrinsic Smad3. J Immunol 174:2071–2083

Mohammad S, Taha A, Akhtar K, Bamezai RNK, Baquer NZ (2006) In vivo effect of Trigonella foenum graecum on the expression of pyruvate kinase, phosphoenolpyruvate carboxykinase, and distribution of glucose transporter (GLUT4) in alloxan-diabetic rats. Can J Physiol Pharmacol 84:647–654

Momin MM, Kane S, Abhang P (2015) Formulation and evaluation of bilayer tablet for bimodal release of venlafaxine hydrochloride. Front Pharmacol 6:144

Murça TM, Moraes PL, Capuruço CA, Santos SH, Melo MB, Santos RA, Ferreira AJ (2012) Oral administration of an angiotensin-converting enzyme 2 activator ameliorates diabetes-induced cardiac dysfunction. Regul Pept 177:107–115

Nair LS, Laurencin CT (2007) Biodegradable polymers as biomaterials. Prog Polym Sci 32:762–798

Narasimhamurthy K, Viswanatha S, Ramesh BS (1999) Acute and subchronic toxicity assessment of debitterized fenugreek powder in the mouse and rat. Food Chem Toxicol 37:831–838

Nayak AK, Pal D (2014) Trigonella foenum-graecum L. seed mucilage-gellan mucoadhesive beads for controlled release of metformin HCl. Carbohydr Polym 107:31–40

Nayak AK, Pal D, Das S (2013a) Calcium pectinate-fenugreek seed mucilage mucoadhesive beads for controlled delivery of metformin HCl. Carbohydr Polym 96:349–357

Nayak AK, Pal D, Pradhan J, Ghorai T (2012) The potential of Trigonella foenum-graecum L. seed mucilage as suspending agent. Ind J Pharm Educ Res 46:312–317

Nayak AK, Pal D, Pradhan J, Hasnain MS (2013b) Fenugreek seed mucilage-alginate mucoadhesive beads of metformin HCl: Design, optimization and evaluation. Int J Biol Macromol 54:144–154

Nokhodchi A, Nazemiyeh H, Khodaparast A, Sorkh-Shahan T, Valizadeh H, Ford JL (2008) An in vitro evaluation of fenugreek mucilage as a potential excipient for oral controlled-release matrix tablet. Drug Dev Ind Pharm 34:323–329

Opdyke DL (1978) Fenugreek absolute. Food Chem Toxicol 16:13–22

Ouzir M, El Bairi K, Amzazi S (2016) Toxicological properties of fenugreek (Trigonella foenum graecum). Food Chem Toxicol 96:145–154

Park JH, Saravanakumar G, Kim K, Kwon IC (2010) Targeted delivery of low molecular drugs using chitosan and its derivatives. Adv Drug Deliv Rev 62:28–41

Pathak D, Kumar P, Kuppusamy G, Gupta A, Kamble B, Wadhwani A (2014) Physicochemical characterization and toxicological evaluation of plant-based anionic polymers and their nanoparticulated system for ocular delivery. Nanotoxicology 8:843–855

Pavelić Ž, Škalko-Basnet N, Schubert R (2001) Liposomal gels for vaginal drug delivery. Int J Pharm 219:139–149

Pincus T, Marcum SB, Callahan LF (1992) Longterm drug therapy for rheumatoid arthritis in seven rheumatology private practices: II Second line drugs and prednisone. J Rheumatol 19:1885–1894

Pradeep SR, Srinivasan K (2017) Amelioration of hyperglycemia and associated metabolic abnormalities by a combination of fenugreek (Trigonella foenum-graecum) seeds and onion (Allium cepa) in experimental diabetes. J Basic Clin Physiol Pharmacol 28:493–505

Pradeep SR, Srinivasan K (2018) Alleviation of cardiac damage by dietary fenugreek (Trigonella foenum-graecum) seeds is potentiated by onion (Allium cepa) in experimental diabetic rats via blocking renin-angiotensin system. Cardiovasc Toxicol 18:221–231

Prajapati VD, Jani GK, Moradiya NG, Randeria NP, Maheriya PM, Nagar BJ (2014) Locust bean gum in the development of sustained release mucoadhesive macromolecules of aceclofenac. Carbohydr Polym 113:138–148

Prajapati VD, Jani GK, Moradiya NG, Randeria NP (2013a) Pharmaceutical applications of various natural gums, mucilages and their modified forms. Carbohydr Polym 92:1685–1699

Prajapati VD, Jani GK, Moradiya NG, Randeria NP, Nagar BJ, Naikwadi NN, Variya BC (2013b) Galactomannan: a versatile biodegradable seed polysaccharide. Int J Biol Macromol 60:83–92

Preet A, Gupta BL, Siddiqui MR, Yadava PK, Baquer NZ (2005) Restoration of ultrastructural and biochemical changes in alloxan-induced diabetic rat sciatic nerve on treatment with Na 3 VO 4 and Trigonella—a promising antidiabetic agent. Mol Cell Biochem 278:21–31

Puttewar TY, Kshirsagar MD, Chandewar AV, Chikhale RV (2010) Formulation and evaluation of orodispersible tablet of taste masked doxylamine succinate using ion exchange resin. J King Saud Univ Sci 22:229–240

Raghuwanshi N, Kumari P, Srivastava AK, Vashisth P, Yadav TC, Prasad R, Pruthi V (2017) Synergistic effects of Woodfordia fruticosa gold nanoparticles in preventing microbial adhesion and accelerating wound healing in Wistar albino rats in vivo. Mater Sci Eng C Mater Biol Appl 80:252–262

Raju J, Gupta D, Rao AR, Yadava PK, Baquer NZ (2001) Trigonella foenum graecum (fenugreek) seed powder improves glucose homeostasis in alloxan diabetic rat tissues by reversing the altered glycolytic, gluconeogenic and lipogenic enzymes. Mol Cell Biochem 224:45–51

Raju J, Rao CV (2012) Diosgenin, a steroid saponin constituent of yams and fenugreek: emerging evidence for applications in medicine. In: Bioactive compounds in phytomedicine. InTech, pp 125–143

Randhawa R, Bassi P, Kaur G (2012) In vitro, in vivo evaluation of inter polymer complexes between carboxymethyl fenugreek gum and chitosan or carboxymethyl guar gum and chitosan for colon delivery of tamoxifen. Asian Pac J Trop Dis 2:202–207

Rao PU, Sesikeran B, Rao PS, Naidu AN, Rao VV, Ramachandran EP (1996) Short term nutritional and safety evaluation of fenugreek. Nutr Res 16:1495–1505

Rashid F, Hussain S, Ahmed Z (2018) Extraction purification and characterization of galactomannan from fenugreek for industrial utilization. Carbohydr Polym 180:88–95

Roberts KT, Cui SW, Wu Y, Williams SA, Wang C, Graham T (2014) Physicochemical evaluation of fenugreek gum and extrusion modified fenugreek gum and effects on starch degradation in bread. Bioact Carbohydr Diet Fibre 4:176–183

Rosin MP, Anwar WA, Ward AJ (1994) Inflammation, chromosomal instability, and cancer: the schistosomiasis model. Cancer Res 54:1929–1933

Rossi B, Campia P, Merlini L, Brasca M, Pastori N, Farris S, Galante YM (2016) An aerogel obtained from chemo-enzymatically oxidized fenugreek galactomannans as a versatile delivery system. Carbohydr Polym 144:353–361

Rowe RC, Sheskey P, Quinn M (2009) Handbook of pharmaceutical excipients, In: Libros Digitales-Pharmaceutical Press, pp 278–298

Sav AK, Fule RA, Ali MT, Amin P (2013) Synthesis and evaluation of octenyl succinate anhydride derivative of fenugreek gum as extended release polymer. J Pharmaceut Invest 43:417–429

Shabbeer S, Sobolewski M, Anchoori RK, Kachhap S, Hidalgo M, Jimeno A, Khan SR (2009) Fenugreek: a naturally occurring edible spice as an anticancer agent. Cancer Biol Ther 8:272–278

Sharma JD, Bhinda A (2005) Antifertility activity of steroidal extract of Trigonella foenum-graecum (seeds) in female rats. Asian J Exp Sci 19:115–120

Sharma RD, Sarkar A, Hazara DK, Mishra B, Singh JB, Sharma SK, Maheshwari PK (1996a) Use of fenuqreek seed powder in the management of non-insulin dependent diabetes mellitus. Nutr Res 16:1331–1339

Sharma RD, Sarkar A, Hazra DK, Misra B, Singh JB, Maheshwari BB (1996b) Toxicological evaluation of fenugreek seeds: a long term feeding experiment in diabetic patients. Phytother Res 10:519–520

Sindhu G, Ratheesh M, Shyni GL, Nambisan B, Helen A (2012) Anti-inflammatory and antioxidative effects of mucilage of Trigonella foenum graecum (Fenugreek) on adjuvant induced arthritic rats. Int Immunopharmacol 12:205–211

Singh B (2018) Gamma radiation synthesis and characterization of gentamicin loaded polysaccharide gum based hydrogel wound dressings. J Drug Deliv Sci Technol 47:200–208

Singh G, Pai RS (2015) Trans-resveratrol self-nano-emulsifying drug delivery system (SNEDDS) with enhanced bioavailability potential: optimization, pharmacokinetics and in situ single pass intestinal perfusion (SPIP) studies. Drug Deliv 22:522–530

Singh V, Sethi R, Tiwari A (2009) Structure elucidation and properties of a non-ionic galactomannan derived from the Cassia pleurocarpa seeds. Int J Biol Macromol 44:9–13

Sohail A, Turner MS, Coombes A, Bostrom T, Bhandari B (2011) Survivability of probiotics encapsulated in alginate gel microbeads using a novel impinging aerosols method. Int J Food Microbiol 145:162–168

Song BK, Winter WT, Taravel FR (1989) Crystallography of highly substituted galactomannans: fenugreek and lucerne gums. Macromolecules 22:2641–2644

Sorrentino MJ (2012) An update on statin alternatives and adjuncts. Clin Lipidol 7:721–730

Srichamroen A, Thomson AB, Field CJ, Basu TK (2009) In vitro intestinal glucose uptake is inhibited by galactomannan from Canadian fenugreek seed (Trigonella foenum graecum L) in genetically lean and obese rats. Nutr Res 29:49–54

Tristano AG (2009) Tyrosine kinases as targets in rheumatoid arthritis. Int Immunopharmacol 9:1–9

Uemura T, Goto T, Kang MS, Mizoguchi N, Hirai S, Lee JY, Tsuge N (2011) Diosgenin, the main aglycon of fenugreek, inhibits LXRα activity in HepG2 cells and decreases plasma and hepatic triglycerides in obese diabetic mice. J Nutr 141:17–23

Ulbrich K, Hola K, Subr V, Bakandritsos A, Tucek J, Zboril R (2016) Targeted drug delivery with polymers and magnetic nanoparticles: covalent and noncovalent approaches, release control, and clinical studies. Chem Rev 116:5338–5431

Vijayakumar MV, Pandey V, Mishra GC, Bhat MK (2010) Hypolipidemic effect of fenugreek seeds is mediated through inhibition of fat accumulation and upregulation of LDL receptor. Obesity 18:667–674

Wani SA, Kumar P (2018) Fenugreek: a review on its nutraceutical properties and utilization in various food products. J Saud Soc Agri Sci 17:97–106

Wei Y, Lin Y, Xie R, Xu Y, Yao J, Zhang J (2015) The flow behavior, thixotropy and dynamical viscoelasticity of fenugreek gum. J Food Eng 166:21–28

Xue W, Lei J, Li X, Zhang R (2011) Trigonella foenum graecum seed extract protects kidney function and morphology in diabetic rats via its antioxidant activity. Nutr Res 31:555–562

Yadav TC, Srivastava AK, Mishra P, Singh D, Raghuwanshi N, Singh NK, Pruthi V (2019) Electrospinning: an efficient biopolymer-based micro-and nanofibers fabrication technique. In: Next generation biomanufacturing technologies. American Chemical Society, pp 209–241

Zarvandi M, Rakhshandeh H, Abazari M, Shafiee-Nick R, Ghorbani A (2017) Safety and efficacy of a polyherbal formulation for the management of dyslipidemia and hyperglycemia in patients with advanced-stage of type-2 diabetes. Biomed Pharmacother 89:69–75

Zia KM, Tabasum S, Nasif M, Sultan N, Aslam N, Noreen A, Zuber M (2017) A review on synthesis, properties and applications of natural polymer based carrageenan blends and composites. Int J Biol Macromol 96:282–301

Zia T, Hasnain SN, Hasan SK (2001) Evaluation of the oral hypoglycaemic effect of Trigonella foenum-graecum L. (methi) in normal mice. J Ethnopharmacol 75:191–195

# Chapter 12
# Antimicrobial Application Potential of Phytoconstituents from Turmeric and Garlic

**Shiv Kumar Prajapati, Gaurav Mishra, Akanksha Malaiya, Ankit Jain, Nishi Mody, and Ashok M. Raichur**

**Abstract** In recent years, natural phytochemicals are gaining much attention for their antimicrobial potential. Garlic and turmeric are most widely used from the natural source, and their constituents directly or indirectly provide various health benefits, especially due to antimicrobial potential. Though the conventional antimicrobial compounds are effective against various pathogens, up till now there is a necessity of effective agents against MDR pathogens. Phytochemicals have been used for their antimicrobial potential from ancient times. These phytochemicals can work by multiple mechanisms, such as by inhibiting target modifying and drug degrading enzymes or as efflux pump inhibitors. The use of natural phytoconstituents (e.g., curcumin from turmeric and allicin from garlic) from these two medicinal plants may be an alternative strategy and can overcome the side effects associated with antibiotics or other allopathic means of treatment. A wide range of indications has revealed the therapeutic efficacy of these compounds on bacterial,

S. K. Prajapati · A. Malaiya
Department of Pharmaceutical Sciences, Bhagyoday Tirth Pharmacy College, Sagar 470002, MP, India
e-mail: shivprajapati1992@gmail.com

A. Malaiya
e-mail: jain.akanksha2107@gmail.com

G. Mishra
Faculty of Ayurveda, Department of Medicinal Chemistry, Institute of Medical Sciences, Banaras Hindu University, Varanasi 221005, UP, India
e-mail: vatsgaurav880@gmail.com

A. Jain (✉) · A. M. Raichur
Department of Materials Engineering, Indian Institute of Science, Bangalore 560012, Karnataka, India
e-mail: ankitjainsagar@gmail.com

A. M. Raichur
e-mail: amr@iisc.ac.in

N. Mody
Department of Pharmaceutical Sciences, Dr. Harisingh Gour Central University, Sagar 470003, MP, India
e-mail: nishimody@gmail.com

 
D. Pal and A. K. Nayak (eds.), *Bioactive Natural Products for Pharmaceutical Applications*, Advanced Structured Materials 140,
https://doi.org/10.1007/978-3-030-54027-2_12

viral, and fungal infections. To improve safety and efficacy, these phytoconstituents have been delivered using nanoformulations such as liposomes, hydrogels, and nanoparticles for the treatment against different bacteria, viruses, fungi, and parasites infections. This chapter is attempted to discuss phytochemistry, antimicrobial mechanisms, and application potential of phytoconstituents from turmeric and garlic.

**Keywords** Phytoconstituents · Antimicrobial activity · Turmeric · Garlic · Curcumin · Allicin · Liposomes · Nanoparticles · Helicobacter pylori infection · Staphylococcus aureus infection

## 12.1 Introduction

The demand for plant-based antimicrobial agents has been increased day by day because of the adverse effect of synthetic antimicrobial agents (Venkatesan and Karrunakaran 2010). Approximately 250,000 to 50,000 plant varieties are available on the earth in which only 10% are of therapeutic benefit till now (Prakashet al. 2020). The traditional system of medicines such as Ayurvedic, Unani, Siddha, Homeopathic, Naturopathy, and Aromatherapy plays an important role to maintain health, hygiene, proper sanitation, and also increases the longevity of life without affecting their sensory properties. India is known as the "Kingdom of spices" because it is one of the biggest producers, exporters, and consumers of herbal medicines (Ansari 2020). Garlic, turmeric, coriander, black pepper, cardamom, onion, anise, fennel, mustard, asafoetida, ginger, rauwolfia, basil, nutmeg, ajowan, chirata, long pepper, gokharu, and pippala, etc., have been directly or indirectly used for the therapeutic benefit because of the presence of phytochemicals. These species and herbs not only reveal antimicrobial properties but also used in cancer, diabetes, inflammation, cardiovascular diseases, hypercholesterolemia, arthritis, as carminative, and antioxidants, etc. (Arora and Kaur 1999). WHO says antimicrobial resistance is one of the major health threats in the twenty-first century against some pathogens (bacteria, pathogen, fungi, and virus) and these pathogens are ever-increasing the infections. Antimicrobial resistance emerges when microbes transform when they are subjected to antimicrobial drugs (such as antibiotics, antifungals, antivirals, etc.). Such type of microorganism which develops resistance are generally termed as "superbugs". Therefore, detection of natural, resourceful, harmless antimicrobial agents, such as phytoconstituents, has noticed since the past decades. Among the above-mentioned herbal drugs, both the turmeric and garlic reveal maximum antimicrobial efficacy against various strains of microbes (Fig. 12.1).

One of the great challenges is to develop herbal drugs for clinical efficacy due to poor bioavailability of phytochemicals within the body and various herbal drug constituents get smashed in the stomach due to high pH before they reached to the blood (Jain et al. 2019). It can be attributed to their low absorption and high metabolism rate and rapid elimination from the body because of their inability to reach the

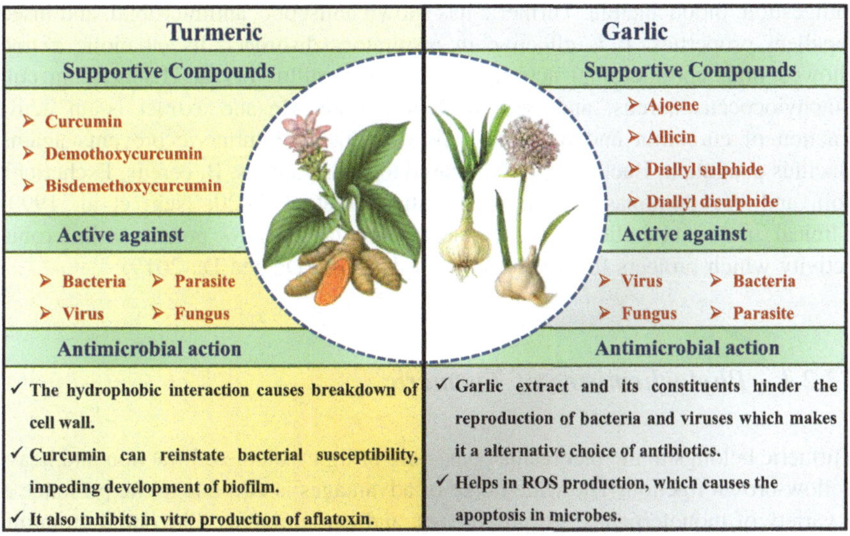


**Fig. 12.1** Antimicrobial application potential of turmeric and garlic phytoconstituents

target site. Nanocarriers have the potential to carry a significant amount of herbal drug to their target site. These carriers also cross all barriers such as acidic pH of stomach and liver metabolites because of the small size of nanocarriers, these enhance permeability and retention effects which cause very lesser side effects (Jain and Jain 2014, 2017). Nanocarriers can easily be altered according to need, and to enhance the blood circulation so that their bioavailability, stability, and solubility of lipophilic compounds can be augmented. Nanocarriers such as liposomes, micelles, nanogels, niosomes, solid-lipid nanoparticles (NPs), carbon nanotubes, polymeric NPs, dendrimers and cyclodextrins are one of the substitutes that deliver the therapeutic concentration of phytoconstituents of garlic and turmeric (Jain et al. 2013, 2020; Jain and Jain 2013, 2016b; Paramanya et al. 2020; Saraf et al. 2020; Subramani et al. 2017; Prajapati and Jain 2020).

## 12.2 Turmeric

Turmeric or Haridra, i.e., Curcuma Longa is a tropical herb and belongs to the family Zingiberaceae family (Zorofchian Moghadamtousi et al. 2014). It acts like a magic bullet in ayurvedic and homeopathic medicine uses externally as well as internally both (Negi et al. 1999). Several studies have reported that Curcuma longa rhizome is broad-spectrum antimicrobial activity including antifungal, antiviral, antibacterial, and antimalarial activities as well as in swellings, stopping pain, cleaning wounds, helping coloration of the skin and also called “Charaka Samhita” as protect skin from itching and also protect from leprosy, urinary disorder, indigestion, blood anemia. Turmeric has shown antiseptic, antimicrobial, and insect repellent properties. It is effective in respiratory disorders. Its ethanolic extract showed high antimicrobial activity against Aspergillus oryzae, Escherichia coli, Staphylococcus aureus, and Saccharomyces sake. Hexane extract is an active fraction of curcumin and it is the yellow pigment of turmeric prevents against Bacillus coagulant, Bacillus subtilis, Staphylococcus aureus, B. cereus, Escherichia coli, and Pseudomonas aeruginosa (Kesharwaniet al. 2020; Negi et al. 1999). Clinical trial data indicate that turmeric and garlic show potent antimicrobial activity which protects from very harmful diseases (De and De 2019).

### *12.2.1 Phytochemistry of Turmeric*

Turmeric belongs to the oleoresins consisting of light volatile oil fraction and heavy yellow-brown fraction. The wide range of advantages is also due to the presence of a variety of monoterpenes, sesquiterpenes, and curcuminoids. Flavonoid curcumin is a chief phytoconstituent of turmeric some volatile oils are also present. Turmeric is a chief source of phenolic compounds, for example, curcuminoid. Curcumin is a phenolic diketone also known as diferuloylmethane (Nisar et al. 2015). The curcuminoids comprise varying concentrations of, i.e., 77%, 17%, and 3% of curcumin I, II, and III, respectively (Fig. 12.2) (Lee et al. 2013). Some volatile oil such as turmerone, zingiberone, and atlantone are also present. Curcuminoids have been recognized by the US Food and Drug Administration (FDA) has been suggested the safer doses of curcuminoids between 4 and 8 mg per day (Hewlings and Kalman 2017). Apart from all these some other phenolic diketones such as demothoxycurcumin and bisdemethoxycurcumin are also available in the rhizomes. The yellow color of curcumin is due to the presence of phenolic diketones (Basnet and Skalko-Basnet 2011). Curcumin I was considered to have minimum stability, while curcumin III was proved to be the most stable among all curcuminoids. Not only the above-mentioned phytochemicals it also consists of some essential oils, for instance, a-phellandrene, a-pinene, caryophyllene, C8-aldehyde, geraniol methyl heptanone, linalool, myrcene, pinene, p-cymene, sabinene, and 1,8-cineole. Curcumin potentially shows antimicrobial activity, anticancer, and antioxidant activity and extensively used in neurodegenerative ailment, inflammatory, and cardiac symptoms (Pandey and Dalvi 2019). The nutritional composition of turmeric is described in Table 12.1.

### *12.2.2 Antimicrobial Mechanism of Curcuminoids*

The antimicrobial activity depends on curcuminoids depend on the mode of their delivery (Shlar et al. 2017). Generally, the antimicrobial action depends on the hydrogen bonding and hydrophobic interaction of various phenolic compounds to

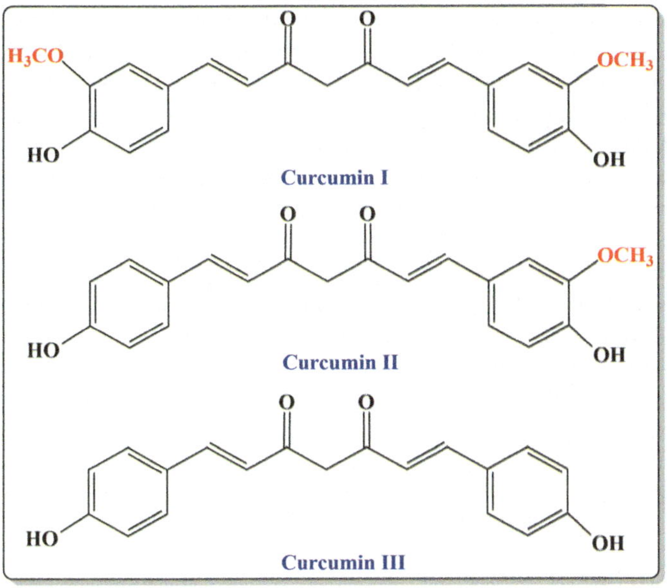


**Fig. 12.2** Chief Constituents of Turmeric

**Table 12.1** Nutritional composition of turmeric (per 100 g)

| Constituent name | Proportion | Constituent name | Proportion |
|---|---|---|---|
| Water (g) | 6 | Iron (g) | 47.5 |
| Calories (Kcal) | 390 | Sodium (mg) | 30 |
| Protein(g) | 8.5 | Phosphorous (mg) | 260 |
| Carbohydrates (g) | 69.9 | Riboflavin (mg) | 0.19 |
| Ash (g) | 6.8 | Ascorbic acid (mg) | 50 |
| Calcium (g) | 0.2 | Niacin (mg) | 4.8 |
| Fat (g) | 8.9 | Thiamine (mg) | 0.09 |
| Potassium (mg) | 2000 | | |

the membrane proteins, which cause cell membrane distraction and breakdown of cell wall and impairment of the electron transport chain. The antibacterial efficacy of aqueous extracts is probably because of the anionic elements present such as chlorides, nitrate, sulphates, and thiocyanate. The alcoholic extracts display improved antimicrobial activity might be due to much solubility than the aqueous extract solution. Curcumin potentially shows antimicrobial efficacy against a wide range of gram-positive and gram-negative bacteria and it is generally much sensitive for gram-positive species than gram-negatives (Betts and Wareham 2014; Kaur et al. 2010). This may be due to the specific cell wall structure of gram-negative bacteria. The lipopolysaccharides of the superficial layer of cell wall represent the furthest penetrability barrier for a diversity of antimicrobials which are the main cause of the remarkably leisurely influx of lipophilic molecule in gram-negative bacteria. Instead, porin proteins entrenched in the superficial sheath signify the main channels for entry of molecules inside the cells of gram-negative bacteria. Curcumin can reinstate bacterial susceptibility, impeding the development of biofilm, and make microorganisms more sensitive to antibiotics. Accordingly, it has been noticed that its antibacterial activity potentiated once it is allied with other antibiotics, such as cefixime, tetracyclines, and vancomycin (Alves et al. 2018). The antimicrobial potential of curcumin can be significantly improved by acquaintance with light. The broad absorption range (300–500 nm) with a maximum absorption band at 425 nm makes curcumin a potent photosensitizer, leading to the enhanced antimicrobial potency (Nikaido 2003). Curcumin also inhibits in vitro production of aflatoxin, it is produced by mold Aspergillus parasiticus, which may grow and contaminate poorly preserved foods and it is a potent biological agent causing injury to the liver.

## 12.3 Garlic

Garlic, i.e., Allium sativum Linn is a bulbous herb of the Alliaceae family and has been used traditionally as a spice for thousand years and extensively used for its medicinal properties such as antibacterial, antiviral, and antifungal properties. The garlic plant is a bulb growing upward to 25-70 cm with hermaphrodite flowers (Mikaili et al. 2013). Not only the root bulb of the garlic but leaves and flowers for medicinal properties for very long. Throughout the time of pre-antibiotics, garlic bulbs assisted as one of the chief treatment approaches for a wide-ranging illness due to its broad-spectrum effects (Majewski 2014; Petrovska and Cekovska 2010). Garlic was also used as a supplement nutrient, as a treatment for weakness and skin infections. The first indication of garlic for its antimicrobial properties was reported in France during the plague of Marseilles in 1721 when four men resorted to consuming a mixture of macerated garlic and wine tincture known as "vinaigre des quatre voleurs" to protect themselves from getting the disease that was afflicting those around them (Harris et al. 2001). Further, around five years later the garlic was utilized at a mass rate by the Egyptians to treat the abnormal growth and also with the circulation problems. Then, North Americans started utilizing this for the treatment of symptoms like flu. Therefore, these shreds of evidence show that it is a prime notification toward the accurate research in the field of science before the development of modern universities and they also develop the attitude to the scientist around the globe. Garlic has a minor, faint smell up until it has been unpeeled. After it is peeled, cut up, or meshed, it instantly activates to feast a strong smell, glycosides. The distillation of garlic with water vapor yields etheric oil with its distinctive sharp smell. The examination of the chemical content of that garlic oil confirmed the presence of few aliphatic unsaturated sulfur compounds (Petrovska and Cekovska 2010). Even in dilution of allicin 1: 85,000 to 1: 250,000, allicin showed strong antibacterial activity against gram-positive and gram-negative bacteria (Dervendži and Karanfilov 1992). Another compound called alliin, with needle-shaped crystals without smell was isolated. Alliin has no antibacterial action but by adding the enzyme alliinase from fresh garlic, allicin having strong antibacterial action is produced (Petrovska and Cekovska 2010). The wide spectrum of pharmacological effects of A. sativum has been reported with the minimum toxicity. It is also used as a spice and food additive Health benefits of garlic are well known and studied to cure ailments caused by infections. The active constituents of garlic showed potential antimicrobial activity due to sulfur compound present in their chemical structure that has been proved against some microorganisms such as Escherichia, Helicobacter, Klebsiella, Mycobacterium, Porphyromonas, Pseudomonas, Salmonella, Streptococcus species (Cellini et al. 1996; Uchida et al. 1975), Candida, Cryptococcus, and Rhodotorula, etc. (Yousuf et al. 2011). Some fungal infections caused by fungi such as Trichophyton and Aspergillus have also been cured by the use of garlic constituents (Saif et al. 2020). Calcium hydroxide show less antibacterial activity compared to garlic. There is also a report that is utilized for the management of the parasite infection and in diarrhea.

### *12.3.1 Phytochemistry of Garlic*

The bulbs of garlic are known to contain hundreds of phytoconstituents including 0.1 to 0.36% volatile oil and these are generally responsible for their pharmacological effects. Garlic includes as ammonium of 33 sulfur compounds, for instance, alliin, allicin (thiosulfinates), ajoenes (E-ajoene and Z-ajoene), allyl propyl, diallyl disulfide, diallyl trisulphide), sallylcysteine, and vinyldithiines, etc. (Motteshard 2008). After the chopping of the garlic bulb and breaking down the parenchyma, the cysteine sulfoxide, i.e., alliin in the presence of alliinase enzyme produces allicin, the chief odoriferous organosulfur compounds are alliin, S-allyl cysteine, S-ethylcysteine, and S-methylcysteine (Al-Snafi 2013; El-Saber Batiha et al. 2020; Zeng et al. 2017). During the storage of garlic bulbs at cool temperatures, alliin accumulates naturally. Allicin can be transformed into stable lipid-soluble allylsulfides such as ajoene. Ajoene from garlic is obtained by the conversion of alliin into allicin by the alliinase-induced cleavage and then in the presence of a polar solvent such as water or alcohol, allicin forms ajoene. Ajoene is chemically more stable than allicin (Viswanathan et al. 2014). Garlic bulbs upon steam distillation yield oil dimethyl mono to hexasulfides (6% of oil), allyl methyl sulfides (37% of oil), and diallylsulphides (57% of oil), which are reported for excellent antibacterial properties (Abubakar 2009; Tripathi 2009). On average, a garlic bulb contains up to 0.9%—glutamylcysteines and up to 1.8% alliin. The pungent offensive odor and its health benefits are generally due to the presence of sulfur compounds. Chemical structures of chief constituents are illustrated in Fig. 12.3. Besides these compounds, garlic also contains 17 amino acids such as leucine, methionine, S-propyl-L-cysteine, S-propenyl-L-cysteine, S-allyl cysteine sulphoxide (alliin),

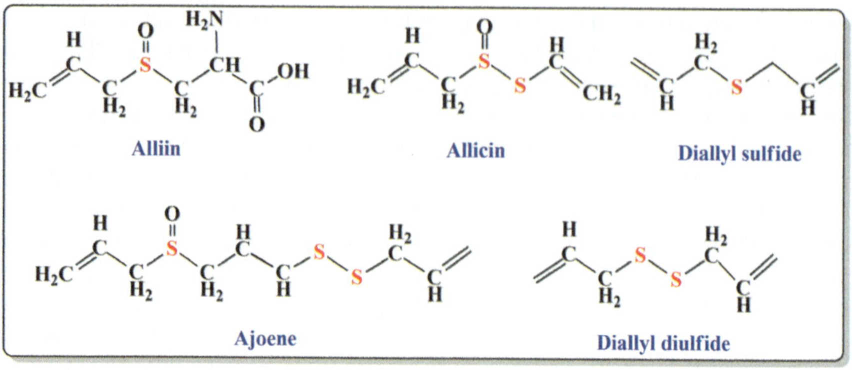


**Fig. 12.3** Chemical structures of phytoconstituents from garlic

S-ethyl cysteine sulphoxide and S-butyl cysteine sulphoxide and their glycosides, arginine, and others (Shah 2009). Some minerals such as selenium and enzymes such as myrosinase, peroxidases, and alliinase are also present. The nutritional composition of garlic is described in Table 12.2.

### 12.3.2 *Antimicrobial Mechanism of Phytochemicals of Garlic*

Among the various phytoconstituents of garlic, allicin shows greater antimicrobial activity than others and it is due to the interaction with thiol-containing enzymes (Mikaili et al. 2013). Allicin elicits antibacterial activity for wide-ranging of gram-negative and gram-positive bacteria's, antiparasitic effect on human intestinal protozoan parasites (Giardia lamblia and Entamoeba histolytica), antiviral activity

**Table 12.2** Nutritional composition of garlic (per 100 g)

| Constituent name | Proportion | Constituent name | Proportion |
|---|---|---|---|
| Water (g) | 59 | Magnesium (mg) | 25 |
| Calories (Kcal) | 149 | Sodium (mg) | 17 |
| Lipids (g) | 0.5 | Vitamin B6 (mg) | 1235 |
| Carbohydrates (g) | 3307 | Vitamin C (mg) | 31 |
| Fiber (g) | 2.1 | Glutamic acid (g) | 0.805 |
| Manganese (mg) | 1672 | Arginine (g) | 0.634 |
| Potassium (mg) | 401 | Aspartic acid (g) | 0.489 |
| Sulfur (mg) | 70 | Leucine (g) | 0.308 |
| Calcium (mg) | 181 | Lysine (g) | 0.273 |
| Phosphorous (mg) | 153 | | |

and antifungal activity for Candida albicans (Chavan et al. 2016). The garlic and its constituents have been described to impede various strains of microbes such as Aeromonas, Aerobacter, Bacillus, Clostridium, Citrobacter, Citrella, Escherichia, Enterobacter, Klebsiella, Leuconostoc, Lactobacillus, Micrococcus, Mycobacterium, Proteus, Providencia, Pseudomonas, Salmonella, Serratia, Shigella, Staphylococcus, Streptococcus, and Vibrio (Sivam 2001). The constituents of garlic can be blended with antibiotics to synergize the antimicrobial effect. Consequently, the level of toxins reduces and prevents their production due to the bactericidal activity of garlic (Karim et al. 2011). The allicin reacts with the multiple enzymes having thiol groups such as thioredoxin reductase, alcohol dehydrogenase, and RNA polymerase, which can affect the metabolism of cysteine proteinase engaged in the virulence of Entamoeba histolytica (Ankri and Mirelman 1999) and in amoeba parasite, allicin hinder the cysteine proteinases by interacting with thioredoxin reductases and alcohol dehydrogenases. A study reported that allicin showed antimicrobial activity owing to the inhibition of RNA synthesis. Though, the protein syntheses and DNA are also inhibited partially, signifying that RNA is the chief target of allicin. The primary aim of allicin to target RNA then partially inhibits DNA, protein synthesis, and membrane damage. The ROS production leads to the DNA and protein damage followed by microbial cell apoptosis (Fig. 12.4). Allicin is a dose-related biocide that can stimulus vital metabolism of cysteine proteinase, and thus, kill all eukaryotic cells due to the presence of thiol groups in all living cells. Ajoene exerts antiparasitic activity by preventing the human glutathione reductase and Trypanosoma cruzi trypanothione reductase.

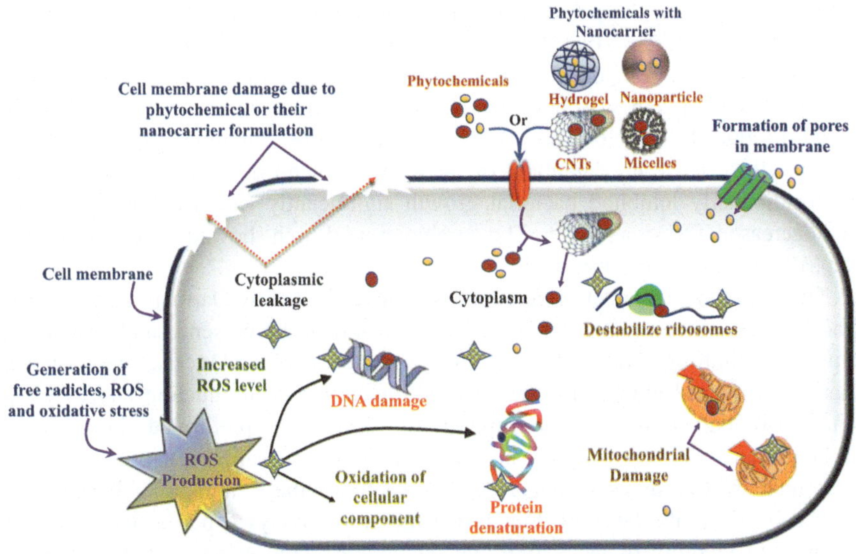


**Fig. 12.4** Proposed mechanism of action by turmeric and garlic extract

## 12.4 Applications

### *12.4.1 Antimicrobial Applications of Phytoconstituents from Turmeric*

Mariselvam et al. (2012) prepared the natural dye from turmeric to evaluate their potential for antimicrobial activity and it is evaluated using agar well diffusion method. The in vitro studies revealed a greater inhibitory effect against Escherichia coli and Vibrio cholera that the curcumin dye with a zone of inhibition 7 mm to 15 mm and 10 mm to 15 mm, respectively (Mariselvam et al. 2012).

A toxicoproteomics approach was followed by Shlar et al. (2017a, b) to evaluate the antibacterial mechanism of action of curcumin on Escherichia coli. The inclusion complex of curcumin was developed with the β-cyclodextrin under the light and dark conditions. When Escherichia coli treated under light, the light-induced curcumin toxicity was conquered by maladaptive responses. The ROS encouraged by the treatment of curcumin over the light dominated the cellular adaptive mechanisms interrupted the metabolism of iron, deregulated the biosynthesis of iron-sulfur mass, and ultimately resulted in cell death. The curcumin activity was potentiated in dark by detoxification of free radicals and modulation of cellular redox status (Shlar et al. 2017). Gopal et al. (2016) prepared the water-soluble extract of curcumin and compared their antimicrobial activity with the ethanolic extract. The entrance of nano curcumin particle within the microbial cell (oral microflora) was markedly improved and was confirmed by the CLSM study. This might be the result of their nanosized, which increased the interaction with cells and led to damage of oral microflora cells. The bioactivity was critically affected by its solubility in water (Gopal et al. 2016). Osteomyelitis is generally caused by Staphylococcus aureus. For the treatment purpose of osteomyelitis, Zhou et al. (2017) gave erythromycin in combination with curcumin. The monotherapy through erythromycin was not found to be efficient to inhibit bacterial growth, while it reduced the TNF-α and IL-6. Contrary to this, curcumin slightly reduced the growth of bacteria. Treatment of rat with the blend of erythromycin and curcumin distinctly inhibited bacterial growth significantly lessened bone infection, and decreased TNF-α and IL-6. The combination of both showed better proficiency for MRSA induced osteomyelitis (Zhou et al. 2017).

The antivirulence effect of curcumin was investigated by Darmani et al. (2020). The efficacy was estimated on Helicobacter pylori in the presence and absence of blue light-emitting diodes (LEDs). In the presence of LEDs, the viability was markedly reduced along with the decrease in the urease production and motility. This also enhanced the interruption of developed preformed biofilms of Helicobacter pylori (Darmani et al. 2020). The in vitro antiparacytic efficacy was examined by Cervantes-Valencia et al. (2019) for the treatment of besnoitiosis caused by Besnoitia besnoiti. Functional inhibition assays exposed that curcumin diminished viability of tachyzoite and encouraged fatal effects in up to 57% of tachyzoites (IC50 in 5.93 μM). Curcumin treatments only inhibited helical gliding

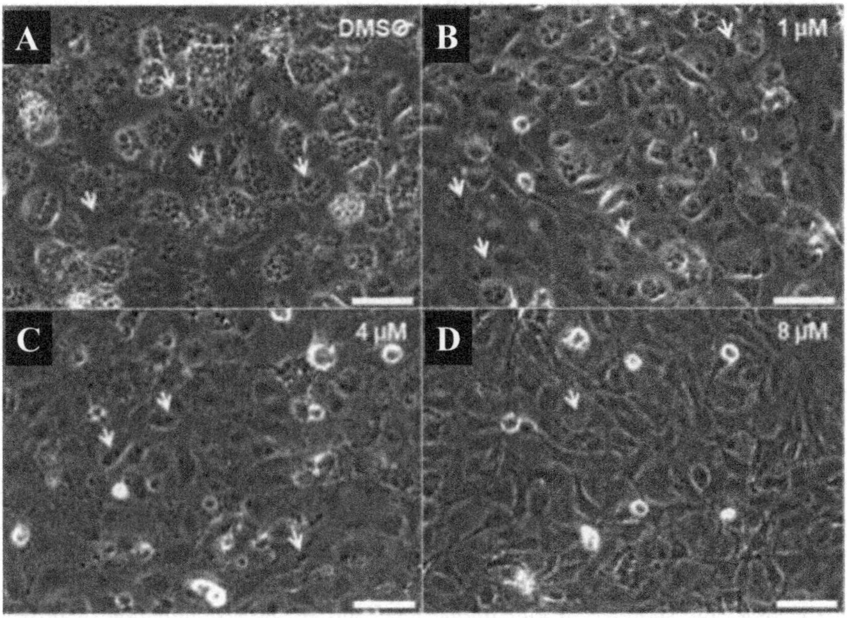


**Fig. 12.5 a–d** Effects of curcumin and parasite proliferation in bovine endothelial host cells (Cervantes-Valencia et al. 2019)

and twirling activities at the same time longitudinal gliding motility was not markedly affected. Pre-treatments by curcumin of tachyzoites caused dose-dependent lessening in host cell invasion (Fig. 12.5a–d) (Cervantes-Valencia et al. 2019). The antifungal efficacy was tested against 23 strains of fungi by Martins et al. (2009). The antifungal exposure was assessed by broth microdilution assay. Paracoccidioides brasiliensis isolates were found to be most susceptible to curcumin while the growth of Aspergillus isolates was not affected. Curcumin showed 2.5-fold more efficacy than fluconazole at inhibiting the adhesion of Candida albicans or Candida parapsilosis to buccal epithelial cells (Martins et al. 2009).

### *12.4.2 Antimicrobial Applications of Phytoconstituents of Garlic*

The main activity of garlic constituent is allicin activity and there are many pieces of evidence of allicin for inhibition of different organisms like gram-positive, gram-negative Escherichia coli and also antibiotic-resistant (Ross et al. 2001). Staphylococcus aureus, Pseudomonas aeruginosa (Kuda et al. 2004), Streptococcus mutans, Streptococcus faecalis, Streptococcus pyogenes, Salmonella enterica,

Klebsiella aerogenes (Cutler and Wilson 2004), Vibrio, Mycobacteria, Proteus vulgaris, and Enterococcus faecalis (Wallock-Richards et al. 2014). There are many forms of extracts of garlic that were reported for antimicrobial activity with inhibition of different pathogenic bacteria. There is another study revealed that the ethanolic extract has better inhibition against S. typhi and Escherichia coli then aqueous extract (Mikaili et al. 2013). Sharma et al. (1977) reported the antibacterial property of garlic water extract against both gram-positive and negative of the gastrointestinal tract of chicks indicating its effectiveness and they also reported the inhibition of some flora which are resistant to few antibiotics. Allicin had shown a very suggestive response toward methicillin-resistant Staphylococcus aureus (Sharma and Kumar 1977). Allicin's which is produced from Alliin showed its antimicrobial property due to the chemical enzymatic reaction having thiol, viz, alcohol dehydrogenase, RNA polymerase, and thioredoxin reductase by oxidizing protein cysteine or glutathione residues under physiological conditions (Gruhlke et al. 2011). Further due to the presence of thiol within allicin group all living cells of eukaryotic cells are killed by increasing the rapid metabolism of cysteine proteinase. Broadly we can say that it also shows the antiviral, antifungal, and antiparasitic activity (Weber et al. 1992). Deeply we can say that the inhibition within the bacteria by S-allylmercapto modification of thiol-containing protein and this leads to the shortage of glutathione level further the accumulation of protein and scratch to the enzyme inactivation (Getti and Poole 2019; Müller et al. 2016). There is another study revealed that allicin in gaseous form showed antimicrobial potential effectively toward lung pathogens (Reiter et al. 2017). Further in case of skin disease caused by methicillin-resistant Staphylococcus aureus showed improved recovery of the skin with allicin there is one drawback with allicin is its stability so it may not be used for inhalation purpose (Freeman and Kodera 1995; Sharifi-Rad et al. 2014). The next compound is a combination of two Ajoenes (Z-ajoene and E- ajoene) which was obtained by the maceration process of garlic having active sulfur compound transformed from three allicin molecule (Block 1985). There is another scientific group of Yoshida and his co-workers have reported the activity against gram-positive and gram-negative bacteria of ajoenes and they evaluated the minimum inhibitory concentration and they find that Z-ajoene has greater activity than E-ajoenes (Yoshida et al. 1998). Another group of Ohata and his team reported that antimicrobial property of three strains of H- pylori and they found that both forms of ajoenes are having an equal inhibitory activity (Ohta et al. 1999). Ajoenes are having activity for fungi, i.e., Aspergillus niger and Candida albicans (Maluf et al. 2008). Further, there is another component which was obtained by the steam distillation of garlic oil is DASn having very low potency toward gram-positive and resistant microbes the potency depends upon the number of sulfur atoms within the compound. It was seen that more than five sulfur atoms showed more potency than the lower number of atoms (Tsao and Yin 2001). Some sulfur free garlic compounds show antimicrobial activity. Matsuura et al. (1989) and Matsuura et al. (1988) isolated new furostanols termed proto-eruboside-B and satiboside-B from a crude glycoside fraction of garlic. They also found that these saponins get transformed into spinostanol by endogenous

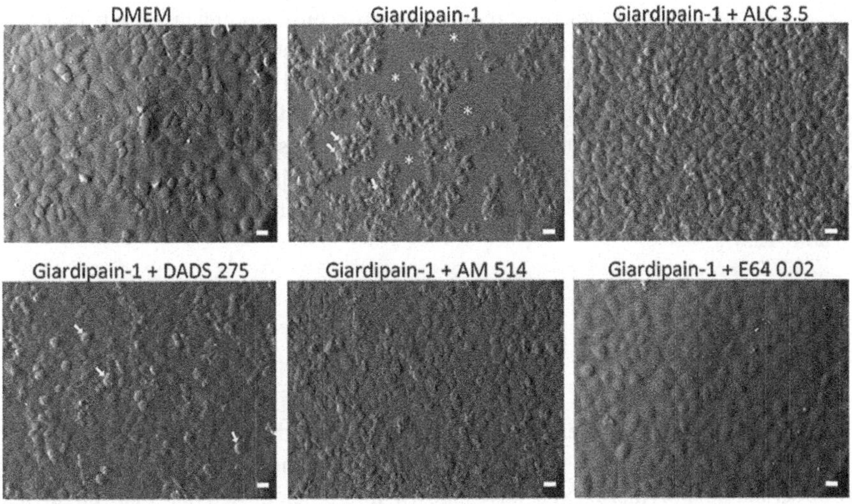


**Fig. 12.6** Effect of garlic TACs on the cytolytic damage caused by giardipain-1 in giardiasis (Argüello-García et al. 2018)

β-glucosidase during processing period. (Matsuura et al. 1989; Matsuura et al. 1988). Argüello-García et al. (2018) studied the efficacy of thioallyl compounds of garlic to measure their potential against Giardia duodenalis Trophozoites. Allicin presented the maximum antigiardial efficacy. Allicin had a cytolytic mechanism in trophozoites with successive loss of diesterase enzyme and oxidoreductase activities. These findings were similar to the fresh aqueous extract (AGE). The partial cell cycle arrest was observed at the G2 phase without any oxidative stress. The intragastric administration of AGE or allicin lessened the number of parasites and removed trophozoites in the experimental animals of giardiasis. Giardipain-1 (a protein) caused a robust demolition of monolayers. While the preincubation of giardipain-1 with allicin (ALC) and allyl mercaptane (AM) prominently diminished such outcomes while diallyl disulphide (DADS) lessened the occurrence of cells presenting apoptotic impairment. In general, the garlic thioallyl compounds showed proteolytic activity of giardipain-1 (Fig. 12.6) (Argüello-García et al. 2018).

### *12.4.3 Nanoformulation Based Applications*

The phytoconstituents are gaining much attention but some of their drawbacks such as poor solubility, stability, and inauspicious bioavailability and lack of targeting ability have constrained their clinical applications. On the other side, nanotechnology has been proved to reduce such types of shortcomings by the use of nanocarriers such as NPs, liposomes, carbon nanotubes, quantum dot, and hydrogel, etc. These nanocarriers not only reduce the dose but also enhance their targeting potential including improvement of the solubility and stability (Bishnoi et al. 2014, 2020; Fahimirad and Hatami 2019; Jain et al. 2016, 2018, 2019, 2020; Jain and Jain 2016a, c, 2016; Kumari et al. 2018; Prajapati et al. 2019; Saraf et al. 2020; Verma et al. 2020). Some potential applications are discussed here and tabulated in Table 12.3. Many research groups were conducted with clinical trials to evaluate their actual performance. The clinical trials were taken from the clinicaltrials.gov (Table 12.4).

### *12.4.4 Nanoparticles*

The NPs due to their smaller size, biocompatibility, and their easy modification has shown its potential for antimicrobial, drug, gene, and vaccine delivery. Biodegradable polymeric NPs have shown microbicidal activity (antifungal and antibacterial efficacy, etc.) by reducing osmotic stability when interacted with microbial cell and cytoplasm membrane (Landriscina et al. 2015; Prajapati et al. 2020). In this series, Bhawana et al. (2011) prepared the curcumin NPs with very small size, freely soluble in water, and have very good antibacterial activity. Therefore due to smaller size, it was penetrated within the cell wall and inhibited more as compared to the pure. The TEM study revealed that the nano curcumin invaded the cell by disrupting the cell wall and destroyed the bacterial cell (Bhawana et al. 2011). Adahoun et al. (2017) prepared curcumin NPs by wet milling apparatus and the prepared formulation was analyzed in vitro for antimicrobial activity on four different bacterial strains two gram-positive (Micrococcus luteus ATCC 9341, Staphylococcus aureus ATCC 29213), two gram-negative (Escherichia coli ATCC 25922, Pseudomonas aeruginosa ATCC 27853) the result showed that the NPs-containing curcumin was effective and safe toward a different type of bacterial strain (Adahoun et al. 2017). dos Santos et al. (2016) reported the Polyethylene glycol containing gold NPs loaded with curcumin and cell viability of different strains of bacteria like Staphylococcus aureus, Staphylococcus epidermidis, Pseudomonas aeruginosa, Escherichia coli, Citrobacter freundii, and Klebsiella pneumoniae was studied using broth dilution assay and the result was enchanting that the percentage of inhibition was more than 80 percent with 32 mM of the prepared formulation (dos Santos et al. 2016). In another study, Sharifi-Rad et al. (2014) used allicin from garlic extract in combination with gold NPs and applied it against the methicillin-resistant Staphylococcus aureus spp. in the rat model. The results proved that there was a synergistic effect of allicin with silver NPs and the inhibition was more in combination as compared to the group treated separately (Sharifi-Rad et al. 2014). Further, El-Refai et al. (2018) prepared different metallic NPs like silver, copper, iron, and zinc and these are investigated against the different microbes such as Staphylococcus aureus, Klebsiella pneumoniae, Candida albicans, Bacillus subtilis, Erwinia. carotovora, and Proteus vulgaris and it was observed that silver and zinc NPs with garlic extract were strongest bactericidal with inhibition

**Table 12.3** Application of turmeric and garlic phytoconstituents

| Active constituent | Nanocarrier | Disease/microbe(s) | Remark | References |
|---|---|---|---|---|
| Curcumin | Silica NPs | *Pseudomonas aeruginosa* and *Staphylococcus aureus* | Bacteria in planktonic condition and bacterial biofilm production and improved wound healing properties | Mirzahosseinipour et al. (2020) |
| Curcumin | Solid lipid NPs | *Escherichia coli* and *Staphylococcus aureus* | The antimicrobial efficacy was improved and showed a synergistic effect when given with other antibacterial compounds | Jourghanian et al. (2016) |
| Curcumin | Hydrogel | *Listeria innocua* | Hydrogel formulation significantly reduced the bacteria level from 4 to 1 log CFU/mL | Tosati et al. (2018) |
| Curcumin | Silver NPs | *Escherichia coli Pseudomonas aeruginosa and Staphylococcus aureus* | MIC showed relatively more efficacy against Gram-negative bacteria than gram-positive | Alves et al. (2018) |
| Curcumin | NPs | *Trichophyton rubrum0* | Antimicrobial photodynamic inhibition by curcumin completely inhibited fungal growth by apoptosis via ROS and RNS | Baltazar et al. (2015) |
| Diallyl sulfide | Niosomes | *Candida albicans* | Niosome formulation markedly decreased fungal load and mortality when compared with the free diallyl sulfide | Alam et al. (2009) |
| Diallyl sulfide | Liposomes | *Candida albicans* | Liposomal formulation significantly reduced residual fungal load when compared with diallyl sulfide | Maroof et al. (2010) |
| Garlic extract | Copper NPs | *Staphylococcus aureus, Escherichia coli* | Extract stabilized the copper NPs. The developed NPs showed much antimicrobial efficacy toward gram-negative bacteria (*Escherichia coli*) | Amatya and Joshi (2020) |
| Diallyl sulfide | Nanorod | *Staphylococcus aureus* | The combination treatment boosted the activity even at a low concentration and markedly inhibited MRSA biofilm | Rauf et al. (2018) |
| Allicin | – | *Trichophyton rubrum* | MIC50 and MIC90 ranged from 0.78–12.5 μg/ml for allicin. Allicin showed better antifungal activity than ketoconazole | Aala et al. (2012) |

(continued)

**Table 12.3** (continued)

| Active constituent | Nanocarrier | Disease/microbe(s) | Remark | References |
|---|---|---|---|---|
| Allicin | – | *Trichophyton rubrum* | Pure allicin extract was more efficient than garlic extract to impede the h0yphae cell growth | Aala et al. (2014) |
| Allicin | – | Antibacterial and antifungal activity | The various thiosulfonates analogous of allicin were tested. The more volatile compounds exhibited substantial antimicrobial properties | Leontiev et al. (2018) |
| Ajoene | – | Antimycobacterial activity | When the ajoene was treated with macrophage. Ajoene stimulated the JNK followed by the initiation of ROS production and accumulation resulted in stimulation of ER stress and autophagy | Choi et al. (2018) |
| Allicin, s-allyl cysteine, | – | Various gram-positive species | Both allicin and s-allyl cysteine showed better antimicrobial activity to treat ear canal and middle ear infections against a wide range of bacteria | Uzun et al. (2019) |
| Curcumin | – | Antibacterial activity against *Escherichia coli* and *Staphylococcus aureus* | A combination of curcumin with tetracycline showed a synergistic effect by rupture of the bacterial cell wall | Sukandar et al. (2018) |

**Table 12.4** Clinical trials of phytoconstituents of turmeric and garlic

| Clinical trial id | Disease | Drug | Phase | Status | Period | Sponsored by |
|---|---|---|---|---|---|---|
| *Turmeric* | | | | | | |
| NCT02018328 | *Helicobacter pylori* infection | Curcumin | Not applicable | Unknown | Jan 2014–Dec 2015 | Gingold Belfer Rachel |
| NCT04012424 | Acute Pulpitis | Curcumin | Not applicable | Not yet required | Aug 2019–Nov 2020 | Cairo University |
| NCT03790605 | Periodontitis | Curcumin | Phase III | Recruiting | Sep 2019–Dec 2020 | KLE Society's Institute of Dental Sciences |
| NCT01035580 | Uterine Cervical Dysplasia | Curcumin | Phase I | Completed | Jan 2010–Jan 2012 | Emory University |
| NCT03746158 | Metabolites Gut Microbiota | Curcumin | Not applicable | Not yet required | Dec 2018–Apr 2019 | University of Massachusetts, Amherst |
| NCT03141918 | HIV infection | Curcumin | Not applicable | Completed | Sept 2017–Dec 2017 | Universidade Federal do Rio Grande do Norte |
| NCT03016039 | Wound infection in female | Curcumin | Not applicable | Unknown | Jan 2017–Oct 2017 | Meir Medical Centre |
| NCT04071106 | Psoriasis | Turmeric extract | Phase II | Not yet required | 19-Sep | Zagazig University |
| | | Turmeric extract in olive oil | Phase III | | | |
| NCT02152475 | Oral disinfection | Curcumin mouthwash | Phase I | Completed | 13-May | University of Sao Paulo |
| NCT02442453 | Chronic Periodontitis | Curenext gel | Phase IV | Completed | 14-Jan | Tatyasaheb Kore Dental College |
| NCT04032132 | Periodontitis | Curcumin paste | Phase IV | Completed | 16-Aug | Ain Shams University |
| *Garlic* | | | | | | |
| NCT03795636 | Intracanal irrigant | Garlic | Not applicable | Completed | Oct 2016–Aug 2018 | Minia University |
| NCT01198223 | Chronic oral candidiasis | Garlic, nystatin | Phase II | Completed | July 2010–Oct 2010 | Qazvin University of Medical Sciences |
| NCT03492723 | Periodontitis | Garlic | Not applicable | Recruiting | Apr 2018–Apr 2020 | Hadassah Medical Organization |

zone 12.6 mm of killing Proteus vulgaris strain (El-Refai et al. 2018). In the next series of modified silver NPs, Vimala et al. (2011) prepared curcumin fabricated chitosan polyvinyl alcohol silver NPs and reviewed its antimicrobial activity on some wound born bacteria and fungi, i.e., Escherichia coli, pseudomonas,

micrococcus, Staphylococcus, and Candida albicans. They found that curcumin loaded chitosan silver nanocomposite having very remarkable potential to inhibit the growth of bacterial colonies as compared to other blank NPs and pure curcumin and this rapid inhibition was noticed for 7 h (Vimala et al. 2011).

#### 12.4.4.1 Nanohydrogels

The need for hydrogel is due to the targeted delivery of the compound and many more like preserving the activity improvement in the stability taste masking of the compound and enhances patient compliance. So that delivery of curcumin with hydrogel was also needed to enhance the biocompatibility. Milk protein is a macronutrient. Bourbon et al. (2016) reported lactoferrin encapsulated glyco-macropeptide nanohydrogels loaded with curcumin and caffeine to find its antimicrobial property against Staphylococcus aureus and Escherichia coli by the disc agar diffusion test and they found that the lactoferrin encapsulated curcumin showed a significant increase in antimicrobial activity ($p < 0.05$). The prepared hydrogel containing curcumin and caffeine was found to release them based on pH, the first release at pH 2 was relaxation dependent and at pH 7 it followed Fick's diffusion (Bourbon et al. 2016). In another study, Ravindra et al. (2012) slightly modified curcumin loaded silver hydrogel nanocomposite by redox co-polymerization method, and the antibacterial activity on nutrient agar medium was noted excellent due to presence of curcumin which extremely inhibited Escherichia coli in comparison to the plain hydrogel without curcumin (Ravindra et al. 2012). Archana et al. (2015) prepared curcumin loaded nano-cubosomal hydrogel for topical delivery which was self-assembled with size 75.2 nm. They found the antibacterial activity against Escherichiacoli and reported that the zone of inhibition of cubosomes was significantly high, i.e., $16.20 \pm 4.26$ mm than pure curcumin $11.36 \pm 1.14$ mm at 24 h (Archana et al. 2015).

#### 12.4.4.2 Quantum Dots

Quantum dots (QDs) reveals size-dependent optical properties and could be the nanomaterial of choice for the antimicrobial application. Under irradiation, QDs produce free radicals, of which the quality and the type are regulated by their core materials. It is well understood that the extra number of free radicals are destructive to microbes (Ipe et al. 2005; Lu et al. 2008). The crucial role for antimicrobial activity of QDs is considered to be reactive oxygen species (ROS) (Fig. 12.4). QDs are usually targeted to the cell wall and cell membrane due to the availability of phosphatidylglycerol and lipoteichoic acid (Manna et al. 2019). There are many reports found that there is biofilm formed around the different strain of bacteria, so to inhibit on a vast scale the QDs with the curcumin was prepared by Singh et al. (2017) and it was having very good dispersion rate in water and its antibacterial activity in contrast to the formation of biofilm on gram-positive and negative bacteria revealed that there was complete inhibition of Escherichia coli, but in case of Pseudomonas aeruginosa, it was completely resistant side by side to the other film formed by the bacteria like Staphylococcus epidermis, the destruction was 50% with the concentration 0.25 mg/ml (Singh et al. 2017).

#### 12.4.4.3 Carbon Nanotubes

Carbon nanotubes (CNTs) have been extensively used for their application in the delivery of antimicrobials. Length of CNTs potentially affect their antimicrobial potential, the shorter length of CNTs shows more bactericidal performance. The open end such type of the CNTs interacts with microorganisms and disrupts the cell membrane. Antimicrobial activity can be improved by functionalizing with chemical groups or by coating with metallic NPs (Al-Jumaili et al. 2017; Prajapati et al. 2020). The antimicrobial application potential of CNTs generally hinges on numerous operational circumstances such as temperature, pH, retention time, and solvent (Kassem et al. 2019). Curcumin was delivered by another bio-nano composite system that was prepared and modified by Chegeni et al. (2020). They had firstly modified the calcium alginate single-walled CNTs which was modified with the surface by glucose. This carrier was proved to be best in carbon nanotube carrier due to its maximum therapeutic activity and minimum undesirable effect or toxicity. The antimicrobial pattern of this modified CNTs loaded with curcumin was evaluated against Escherichia coli and Bacillus cereus. Though, inhibition zone diameter showed that the curcumin loaded CNTs were more effective in comparison to the curcumin alone (Chegeni et al. 2020).

#### 12.4.4.4 Polymeric Micelles

It is a new class of vehicles to deliver poorly soluble drugs which can be modified according to the need (Khan et al. 2018). Therefore, Margaritova et al. (2019) prepared the modified copolymeric micelle based on Pluronic (P123 and F127) having alkylphosphacoline with curcumin to investigate the synergistic antibacterial effect against Staphylococcus aureus strain and it was seen that 1:1 ratio of P123/F127 was enhanced dramatically (Margaritova et al. 2019). In another study, Huang et al. (2017) prepared silver decorated polymeric micelle tethered with curcumin encapsulated into poly(ε-caprolactone) through hydrophobic interaction and it was tested against Pseudomonas aeruginosa (gram-negative) and Staphylococcus aureus (gram-positive) and the antibacterial activity was very good and it was observed that there was a synergistic effect with the inhibition of bacteria when curcumin was incorporated with the silver decorated micelles in comparison to the micelles without curcumin (Huang et al. 2017). Mixed polymeric micelles were developed by Akbar et al. (2018) to compare the antimicrobial efficacy between micellar formulation and curcumin alone. The found results proposed that curcumin loaded micelles had significant inhibitory activity toward bacteria and fungi compared to pure curcumin. It was found that the MICs of curcumin against 10 strains of Staphylococcus aureus varied between 125 and 250 μg/mL (Akbar et al. 2018).

#### 12.4.4.5 Microemulsions

Because of their exceptional properties of easy production, solubilization, long-lasting stability, and biocompatibility, microemulsion have ever-increasing as a looming antimicrobials delivery system, either as a carrier for topical applications or oral delivery for poorly water-soluble active constituent (Shukla et al. 2018). Bactericidal microemulsion using myristic acid was developed for the delivery of curcumin by Liu et al. (2012). Myristic acid is a fatty acid that shows antibacterial potential. Curcumin exposed superior inhibitory capacity toward Staphylococcus epidermidis when compared to myristic acid and there were 43 folds reduction in the IC50 concentration for curcumin was observed compared to azelaic acid. Curcumin loaded microemulsion at the concentration of 0.86 μg/mL inhibited bacterial growth by 50% when it was compared with dimethyl sulfoxide solution of curcumin. The finding revealed the potential of formulation as an antimicrobial and for acne vulgaris (Liu and Huang 2012). The comparative activity against Escherichia coli and Staphylococcus aureus was evaluated using garlic oil and their microemulsion formulation. The garlic oil itself had not shown antibacterial efficacy for Escherichia coli (gram-negative) while it showed significant results for Staphylococcus aureus (gram-positive). The microemulsion having the garlic oil (150 μg/mL) showed marked antibacterial activity (Zheng et al. 2013).

## 12.5 Conclusion

The research community is attracting toward the phytochemicals due to their diversity of applications in various diseases as an antimicrobial, anticancer, anti-inflammatory, antimalarial, and antioxidant agent, etc. Turmeric and garlic both are full of many therapeutic abilities. These phytochemicals have proved their potential as antimicrobial not only in the era of nanotechnology but are being used traditionally for very long without causing cytotoxicity. The nanocarriers have been proved to improve their antimicrobial performance by enhancing their stability, bioavailability, and targeting ability. These nanocarriers can easily be modified for the desired purpose. The phytoconstituents of garlic and turmeric are also given with other allopathic medicines, which showed the potential to improve their efficacy. Looking at the present, some more research studies must be conducted to accelerate their therapeutic efficacy, targeting ability so that their side effects and adverse effects can be diminished.

## References


Aala F, Yusuf UK, Jamal F, Rezaie S (2012) Antimicrobial effects of allicin and ketoconazole on trichophyton rubrum under in vitro condition. Braz J Microbiol 43(2):786–792

Aala F, Yusuf UK, Nulit R, Rezaie S (2014) Inhibitory effect of allicin and garlic extracts on growth of cultured hyphae. Iranian J Basic Med Sci 17(3):150

Abubakar EmM (2009) Efficacy of crude extracts of garlic (Allium sativum Linn.) against nosocomial escherichia coli, staphylococcus aureus, streptococcus pneumoniea and Pseudomonas aeruginosa. J Med Plants Res 3(4):179–185

Adahoun MAA, Al-Akhras MAH, Jaafar MS, Bououdina M (2017) Enhanced anti-cancer and antimicrobial activities of curcumin nanoparticles. Artif Cells Nanomed Biotech 45(1):98–107

Akbar MU, Zia KM, Nazir A, Iqbal J, Ejaz SA, Akash MSH (2018) Pluronic-based mixed polymeric micelles enhance the therapeutic potential of curcumin. AAPS Pharm Sci Tech 19(6):2719–2739

Al-Jumaili A, Alancherry S, Bazaka K, Jacob MV (2017) Review on the antimicrobial properties of carbon nanostructures. Materials 10(9):1066

Al-Snafi AE (2013) Pharmacological effects of allium species grown in Iraq. An overview. Int J Pharm Health Care Res 1(4):132–147

Alam M, Dwivedi V, Khan AA, Mohammad O (2009) Efficacy of niosomal formulation of diallyl sulfide against experimental candidiasis in Swiss albino mice. Nanomed Nanotech Biol Med 4(7):713–724

Alves TF, Chaud MV, Grotto D, Jozala AF, Pandit R, Rai M, Dos Santos CA (2018) Association of silver nanoparticles and curcumin solid dispersion: antimicrobial and antioxidant properties. AAPS Pharm Sci Tech 19(1):225–231

Amatya SP, Joshi LP (2020) Bio-synthesis of copper nanoparticles (CuNPs) using garlic extract to investigate antibacterial activity. BIBECHANA 17:13–19

Ankri S, Mirelman D (1999) Antimicrobial properties of allicin from garlic. Microbes Infect 1(2): 125–129

Ansari MS (2020) Role of traditional knowledge digital library (TKDL) in preservation and protection of indigenous medicinal knowledge of India herbal medicine in India. Springer, pp 609–620

Archana A, Vijayasri K, Madhurim M, Kumar C (2015) Curcumin loaded nano cubosomal hydrogel: preparation, in vitro characterization and antibacterial activity. Chem Sci Trans 4:75–80

Argüello-García R, de la Vega-Arnaud M, Loredo-Rodríguez IJ, Mejía-Corona AM, Melgarejo-Trejo E, Espinoza-Contreras EA, Ortega-Pierres MG (2018) Activity of thioallyl compounds from garlic against giardia duodenalis trophozoites and in experimental giardiasis. Front Cell Infect Microbiol 8:353

Arora DS, Kaur J (1999) Antimicrobial activity of spices. Int J Antimicrob Agents 12(3):257–262

Baltazar LM, Krausz AE, Souza ACO, Adler BL, Landriscina A, Musaev T… Friedman AJ (2015) Trichophyton rubrum is inhibited by free and nanoparticle encapsulated curcumin by induction of nitrosative stress after photodynamic activation. PloS One 10(3)

Basnet P, Skalko-Basnet N (2011) Curcumin: an anti-inflammatory molecule from a curry spice on the path to cancer treatment. Molecules 16(6):4567–4598

Betts JW, Wareham DW (2014) In vitro activity of curcumin in combination with epigallocatechin gallate (EGCG) versus multidrug-resistant *Acinetobacter baumannii*. BMC Microbiol 14(1): 172

Bhawana BR, Buttar HS, Jain V, Jain N (2011) Curcumin nanoparticles: preparation, characterization, and antimicrobial study. J Agric Food Chem 59(5):2056–2061

Bishnoi M, Jain A, Hurkat P, Jain SK (2014) Aceclofenac-loaded chondroitin sulfate conjugated SLNs for effective management of osteoarthritis. J Drug Target 22(9):805–812. https://doi.org/10.3109/1061186X.2014.928714

Bishnoi M, Jain A, Singla Y, Shrivastava B (2020) Sublingual delivery of chondroitin sulfate conjugated tapentadol loaded nanovesicles for the treatment of osteoarthritis. J Lipos Res 1–15. https://doi.org/10.1080/08982104.2020.1730400

Block E (1985) The chemistry of garlic and onions. Sci Am 252(3):114–121

Bourbon AI, Cerqueira MA, Vicente AA (2016) Encapsulation and controlled release of bioactive compounds in lactoferrin-glycomacropeptide nanohydrogels: Curcumin and caffeine as model compounds. J Food Eng 180:110–119

Cellini L, Di Campli E, Masulli M, Di Bartolomeo S, Allocati N (1996) Inhibition of Helicobacter pylori by garlic extract (Allium sativum). FEMS Immunol Med Microbiol 13(4):273–277

Cervantes-Valencia ME, Hermosilla C, Alcalá-Canto Y, Tapia G, Taubert A, Silva LM (2019) Antiparasitic efficacy of curcumin against Besnoitia besnoiti tachyzoites in vitro. Front Veter Sci 5:333

Chavan RD, Shinde P, Girkar K, Madage R, Chowdhary A (2016) Assessment of anti-influenza activity and hemagglutination inhibition of Plumbago indica and Allium sativum extracts. Pharma Res 8(2):105

Chegeni M, Rozbahani ZS, Ghasemian M, Mehri M (2020) Synthesis and application of the calcium alginate/SWCNT-Gl as a bio-nanocomposite for the curcumin delivery. Int J Biol Macromol

Choi J-A, Cho S-N, Lim Y-J, Lee J, Go D, Kim S-H, Song C-H (2018) Enhancement of the antimycobacterial activity of macrophages by ajoene. Innate Immun 24(1):79–88

Cutler R, Wilson P (2004) Antibacterial activity of a new, stable, aqueous extract of allicin against methicillin-resistant Staphylococcus aureus. Br J Biomed Sci 61(2):71–74

Darmani H, Smadi EA, Bataineh SM (2020) Blue light emitting diodes enhance the antivirulence effects of Curcumin against Helicobacter pylori. J Med Microbiol 69(4): 617–624

De AK, De M (2019) Functional and therapeutic applications of some important spices the role of functional food security in global health. Elsevier, pp 499–510

Dervendži V, Karanfilov V (1992) Sovremeno lekuvanje so lekoviti bilki: Tabernakul

dos Santos Courrol D, Teixeira BH, Pereira CBP, Franzolin MR, Courrol LC (2016) Pegylated curcumin with gold nanoparticles: antimicrobial agent evaluation. J Biomed Eng Biosci (JBEB) 3(1):43–47

El-Refai AA, Ghoniem GA, El-Khateeb AY, Hassaan MM (2018) Eco-friendly synthesis of metal nanoparticles using ginger and garlic extracts as biocompatible novel antioxidant and antimicrobial agents. J Nanostruct Chem 8(1):71–81

El-Saber Batiha G, Magdy Beshbishy A, Wasef G, Elewa LYH, Al-Sagan A, El-Hack A, Prasad Devkota H (2020) Chemical constituents and pharmacological activities of garlic (Allium sativum L.): a review. Nutrients 12(3):872

Fahimirad, S., & Hatami, M. (2019). Nanocarrier-Based Antimicrobial Phytochemicals Advances in Phytonanotechnology (pp. 299–314): Elsevier

Freeman F, Kodera Y (1995) Garlic chemistry: stability of S-(2-propenyl)-2-propene-1-sulfinothioate (allicin) in blood, solvents, and simulated physiological fluids. J Agric Food Chem 43(9):2332–2338

Getti G, Poole P (2019) Allicin causes fragmentation of the peptidoglycan coat in *Staphylococcus aureus* by effecting synthesis and aiding hydrolysis: a determination by MALDI-TOF mass spectrometry on whole cells. J Med Microbiol 68(4):667–677

Gopal J, Muthu M, Chun S-C (2016) Water soluble nanocurcumin extracted from turmeric challenging the microflora from human oral cavity. Food Chem 211:903–909

Gruhlke M, Nwachwukwu I, Arbach M, Anwar A, Noll U, Slusarenko A (2011) Allicin from garlic, effective in controlling several plant diseases, is a reactive sulfur species (RSS) that pushes cells into apoptosis. Paper presented at the Modern fungicides and antifungal compounds VI. 16th International Reinhardsbrunn Symposium, Friedrichroda, Germany, 25–29 April 2010

Harris J, Cottrell S, Plummer S, Lloyd D (2001) Antimicrobial properties of *Allium sativum* (garlic). Appl Microbiol Biotechnol 57(3):282–286

Hewlings SJ, Kalman DS (2017) Curcumin: a review of its' effects on human health. Foods 6(10):92

Huang F, Gao Y, Zhang Y, Cheng T, Ou H, Yang L, Liu J (2017) Silver-decorated polymeric micelles combined with curcumin for enhanced antibacterial activity. ACS Appl Mater Interf 9(20):16880–16889

Ipe BI, Lehnig M, Niemeyer CM (2005) On the generation of free radical species from quantum dots. Small 1(7):706–709

Jain A, Gulbake A, Shilpi S, Jain A, Hurkat P, Jain SK (2013) A new horizon in modifications of chitosan: syntheses and applications. Crit Rev Ther Drug Carrier Syst 30(2):91–181

Jain A, Hurkat P, Jain SK (2019a) Development of liposomes using formulation by design: Basics to recent advances. Chem Phys Lipid 224:104764. https://doi.org/10.1016/j.chemphyslip.2019.03.017

Jain A, Jain A, Jain A, Jain A (2016) Quasi emulsion spherical crystallization technique based environmentally responsive Tulsion® (pH dependent) microspheres for colon specific delivery. J Appl Biomed 14(2):147–155. https://doi.org/10.1016/j.jab.2015.11.001

Jain A, Jain SK (2013) Engineered chitosan: a potential tool in biomedical applications. Int J Biotech Bioeng Res 4(3):1–4

Jain A, Jain SK (2016a) In vitro release kinetics model fitting of liposomes: an insight. Chem Phys Lipid 201:28–40. https://doi.org/10.1016/j.chemphyslip.2016.10.005

Jain A, Jain SK (2016b) Liposomes in Cancer Therapy. In: Jimenez C (ed) Nanocarrier systems for drug delivery. Nova Science Publishers, New York, USA, pp 1–42

Jain A, Jain SK (2016c) Multipronged, strategic delivery of paclitaxel-topotecan using engineered liposomes to ovarian cancer. Drug Dev Ind Pharm 42(1):136–149. https://doi.org/10.3109/03639045.2015.1036066

Jain A, Jain SK (2017) Application potential of engineered liposomes in tumor targeting (Chap. 9). In: Grumezescu A (ed) Multifunctional systems for combined delivery, biosensing and diagnostics. Elsevier—Health Sciences Division, pp 171–192

Jain A, Jain SK (2014) Brain targeting using surface functionalized nanocarriers in human solid tumors. In: Singh B, Jain NK, Katare OP (eds) Drug nanocarriers. Studium Press, Houston LLC, USA: Series Nanobiomedicine, pp 203–255

Jain A, Kumari R, Tiwari A, Verma A, Tripathi A, Shrivastava A, Jain SK (2018) Nanocarrier based advances in drug delivery to tumor: an overview. Curr Drug Targets 19(13):1498–1518. https://doi.org/10.2174/1389450119666180131105822

Jain A, Prajapati S, Kumari A, Mody N, Bajpai M (2020) Engineered nanosponges as versatile biodegradable carriers: an insight. J Drug Del Sci Tech. https://doi.org/10.1016/j.jddst.2020.101643

Jain A, Tiwari A, Verma A, Saraf S, Jain S (2019b) Combination cancer therapy using multifunctional liposomes. Crit Rev Ther Drug Carrier Syst. https://doi.org/10.1615/CritRevTherDrugCarrierSyst.2019026358

Jain SK, Jain A (2016) Ligand mediated drug targeted liposomes liposomal delivery systems: advances and challenges, vol 2, pp 145. Future Medicine Ltd Unitec House 2 Albert Place London N3 1QB UK

Jourghanian P, Ghaffari S, Ardjmand M, Haghighat S, Mohammadnejad M (2016) Sustained release curcumin loaded solid lipid nanoparticles. Adv Pharm Bull 6(1):17

Karim A, Sohail MN, Munir S, Sattar S (2011) Pharmacology and phytochemistry of Pakistani herbs and herbal drugs used for treatment of diabetes. Int J Pharm 7(4):419–439

Kassem A, Ayoub GM, Malaeb L (2019) Antibacterial activity of chitosan nano-composites and carbon nanotubes: a review. Sci Total Environ 668:566–576

Kaur S, Modi NH, Panda D, Roy N (2010) Probing the binding site of curcumin in Escherichia coli and Bacillus subtilis FtsZ–a structural insight to unveil antibacterial activity of curcumin. Eur J Med Chem 45(9):4209–4214

Kesharwani P, Jain A, Srivastava AK, Keshari MK (2020) Systematic development and characterization of curcumin loaded nanogel for topical application. In: Drug development and industrial pharmacy, pp 1–36. https://doi.org/10.1080/03639045.2020.1793998

Khan I, Gothwal A, Mishra G, Gupta U (2018) Polymeric micelles. In: Jafar MA, Sheardown MH, Al-Ahmed A (eds) Functional biopolymers, pp 1–29. Cham, Springer International Publishing

Kuda T, Iwai A, Yano T (2004) Effect of red pepper Capsicum annuum var. conoides and garlic Allium sativum on plasma lipid levels and cecal microflora in mice fed beef tallow. Food Chem Toxicol 42(10):1695–1700

Kumari A, Jain A, Hurkat P, Tiwari A, Jain SK (2018) Eudragit S100 coated microsponges for colon targeting of prednisolone. Drug Dev Ind Pharm 44(6):902–913. https://doi.org/10.1080/03639045.2017.1420079

Landriscina A, Rosen J, Friedman AJ (2015) Biodegradable chitosan nanoparticles in drug delivery for infectious disease. Nanomed Nanotech Biol Med 10(10):1609–1619

Lee W-H, Loo C-Y, Bebawy M, Luk F, Mason RS, Rohanizadeh R (2013) Curcumin and its derivatives: their application in neuropharmacology and neuroscience in the 21st century. Curr Neuropharm 11(4):338–378

Leontiev R, Hohaus N, Jacob C, Gruhlke MC, Slusarenko AJ (2018) A comparison of the antibacterial and antifungal activities of thiosulfinate analogues of allicin. Sci Rep 8(1):1–19

Liu C-H, Huang H-Y (2012) Antimicrobial activity of curcumin-loaded myristic acid microemulsions against *Staphylococcus epidermidis*. Chem Pharm Bull 60(9):1118–1124

Lu Z, Li CM, Bao H, Qiao Y, Toh Y, Yang X (2008) Mechanism of antimicrobial activity of CdTe quantum dots. Langmuir 24(10):5445–5452

Majewski, M. (2014). Allium sativum: facts and myths regarding human health. Roczniki Państwowego Zakładu Higieny, 65(1)

Maluf M, Takahachi G, Svidzinski T, Xander P, Apitz-Castro R, Bersani-Amado CA, Cuman RK (2008) Antifungal activity of ajoene on experimental murine paracoccidioidomycosis. Revista Iberoamericana De Micologia 25:163–166

Manna S, Ghosh M, Chakraborty R, Ghosh S, Mandal SM (2019) A review on quantum dots: synthesis to in-silico analysis as next generation antibacterial agents. Curr Drug Targets 20(3):255–262

Margaritova Zaharieva M, Dimitrov Kroumov A, Dimitrova L, Tsvetkova I, Trochopoulos A, Mihaylov Konstantinov S, Miladinov Najdenski H (2019) Micellar curcumin improves the antibacterial activity of the alkylphosphocholines erufosine and miltefosine against pathogenic Staphyloccocus aureus strains. Biotechnol Biotechnol Equip 33(1):38–53

Mariselvam R, Singh AJAR, Kalirajan K (2012) Anti-microbial activity of turmeric natural dye against different bacterial strains. J Appl Pharm Sci 2(6):21–21

Maroof A, Farazuddin M, Owais M (2010) Potential use of liposomal diallyl sulfide in the treatment of experimental murine candidiasis. Biosci Rep 30(4):223–231. https://doi.org/10.1042/bsr20090068

Martins C, Da Silva D, Neres A, Magalhaes T, Watanabe G, Modolo L, De Resende M (2009) Curcumin as a promising antifungal of clinical interest. J Antimicrob Chemother 63(2):337–339

Matsuura H, Ushiroguchi T, Itakura Y, Fuwa T (1989) Further studies on steroidal glycosides from bulbs, roots and leaves of Allium sativum L. Chem Pharm Bull 37(10):2741–2743

Matsuura H, Ushiroguchi T, Itakura Y, Hayashi N, Fuwa T (1988) A furostanol glycoside from garlic, bulbs of Allium sativum L. Chem Pharm Bull 36(9):3659–3663

Mikaili P, Maadirad S, Moloudizargari M, Aghajanshakeri S, Sarahroodi S (2013) Therapeutic uses and pharmacological properties of garlic, shallot, and their biologically active compounds. Iran J Basic Med Sci 16(10):1031

Mirzahosseinipour M, Khorsandi K, Hosseinzadeh R, Ghazaeian M, Shahidi FK (2020) Antimicrobial photodynamic and wound healing activity of curcumin encapsulated in silica nanoparticles. Photodiagn Photodyn Ther 29:101639

Motteshard D (2008) The benefits of the use of garlic in herbal preparations, chemical constituents of garlic

Müller A, Eller J, Albrecht F, Prochnow P, Kuhlmann K, Bandow JE, Leichert LIO (2016) Allicin induces thiol stress in bacteria through S-allylmercapto modification of protein cysteines. J Biol Chem 291(22):11477–11490

Negi P, Jayaprakasha G, Jagan Mohan Rao L, Sakariah K (1999) Antibacterial activity of turmeric oil: a byproduct from curcumin manufacture. J Agric Food Chem 47(10):4297–4300

Nikaido H (2003) Molecular basis of bacterial outer membrane permeability revisited. Microbiol Mol Biol Rev 67(4):593–656

Nisar T, Iqbal M, Raza A, Safdar M, Iftikhar F, Waheed M (2015) Turmeric: a promising spice for phytochemical and antimicrobial activities. Am Eur J Agric Environ Sci 15(7):1278–1288

Ohta R, Yamada N, Kaneko H, Ishikawa K, Fukuda H, Fujino T, Suzuki A (1999) In vitro inhibition of the growth of Helicobacter pylori by oil-macerated garlic constituents. Antimicrob Agents Chemother 43(7):1811–1812

Pandey KU, Dalvi SV (2019) Understanding stability relationships among three curcumin polymorphs. Adv Powder Technol 30(2):266–276

Paramanya A, Sharma S, Bagdat RB, Ali A (2020) Recent practices of medicinal and aromatic plants in nanotechnology. Nanomater Agric Forestry Appl 435–467

Petrovska BB, Cekovska S (2010) Extracts from the history and medical properties of garlic. Pharm Rev 4(7):106

Prajapati SK, Jain A (2020) Dendrimers for Advanced Drug Delivery. In: Nayak A., Hasnain M. (eds) Advanced Biopolymeric Systems for Drug Delivery. Advances in material research and technology. Springer, Cham. pp 339–360

Prajapati SK, Jain A, Shrivastava C, Jain AK (2019) Hyaluronic acid conjugated multi-walled carbon nanotubes for colon cancer targeting. Int J Biol Macromol 123:691–703. https://doi.org/10.1016/j.ijbiomac.2018.11.116

Prajapati SK, Malaiya A, Kesharwani P, Soni D, Jain A (2020) Biomedical applications and toxicities of carbon nanotubes. Drug Chem Toxicol 1–16. https://doi.org/10.1080/01480545.2019.1709492

Prakash, B., Kumar, A., Singh, P. P., & Songachan, L. (2020). Antimicrobial and antioxidant properties of phytochemicals: Current status and future perspective Functional and Preservative Properties of Phytochemicals (pp. 1–45): Elsevier

Rauf MA, Zubair S, Ateeq H, Dabeer K, Pachauri S, Ajmal M, Owais M (2018) Synergistic effect of diallyl sulfide with zinc oxide nanorods: a novel and effective approach for treatment of acute dermatitis in model animals. Frontiers in microbiology 9:586

Ravindra S, Mulaba-Bafubiandi AF, Rajinikanth V, Varaprasad K, Reddy NN, Raju KM (2012) Development and characterization of curcumin loaded silver nanoparticle hydrogels for antibacterial and drug delivery applications. J Inorg Organomet Polym Mater 22(6):1254–1262

Reiter J, Levina N, Van der Linden M, Gruhlke M, Martin C, Slusarenko AJ (2017) Diallylthiosulfinate (Allicin), a volatile antimicrobial from garlic (Allium sativum), kills human lung pathogenic bacteria, including MDR strains, as a vapor. Molecules 22(10):1711

Ross Z, O'Gara EA, Hill DJ, Sleightholme H, Maslin DJ (2001) Antimicrobial properties of garlic oil against human enteric bacteria: evaluation of methodologies and comparisons with garlic oil sulfides and garlic powder. Appl Environ Microbiol 67(1):475–480

Saif S, Hanif MA, Rehman R, Riaz M (2020) Garlic medicinal plants of South Asia, Elsevier, pp 301–315

Saraf S, Jain A, Tiwari A, Verma A, Panda PK, Jain SK (2020) Advances in liposomal drug delivery to cancer: an overview. J Drug Del Sci Tech 101549

Shah BN (2009) Textbook of pharmacognosy and phytochemistry. Elsevier India

Sharifi-Rad J, Hoseini Alfatemi S, Sharifi-Rad M, Iriti M (2014) Antimicrobial synergic effect of Allicin and silver nanoparticles on skin infection caused by methicillin resistant Staphylococcus aureus spp. Annals of medical and health sciences research 4(6):863–868

Sharma V, Kumar A (1977) Antibacterial property of Allium sativum Linn.: in vivo and in vitro studies

Sharma VD, Sethi MS, Arun Kumar, Rarotra JR (1977) Antibacterial property of Allium sativum Linn.: in vivo & in vitro studies. Indian J Exper Bio 15(6):466–468
Shlar I, Droby S, Choudhary R, Rodov V (2017) The mode of antimicrobial action of curcumin depends on the delivery system: monolithic nanoparticles versus supramolecular inclusion complex. RSC Adv 7(67):42559–42569
Shlar I, Droby S, Rodov V (2017b) Modes of antibacterial action of curcumin under dark and light conditions: a toxicoproteomics approach. J Prot 160:8–20
Shukla T, Upmanyu N, Agrawal M, Saraf S, Saraf S, Alexander A (2018) Biomedical applications of microemulsion through dermal and transdermal route. Biomed Pharm 108:1477–1494
Singh AK, Prakash P, Singh R, Nandy N, Firdaus Z, Bansal M, Mishra B (2017) Curcumin quantum dots mediated degradation of bacterial biofilms. Front Microbiol 8:1517
Sivam GP (2001) Protection against helicobacter pylori and other bacterial infections by garlic. J Nutr 131(3):1106S–1108S. https://doi.org/10.1093/jn/131.3.1106S
Subramani PA, Panati K, Lebaka VR, Reddy DD, Narala VR (2017) Nanostructures for curcumin delivery: possibilities and challenges nano-and microscale drug delivery systems. Elsevier, pp 393–418
Sukandar EY, Kurniati N F, Puspatriani,K, Adityas HP (2018) Antibacterial Activity of Curcumin in Combination with Tetracycline Against Staphylococcus aureus by Disruption of Cell Wall. Res J of Med Plant 12(1):1–8
Tosati JV, de Oliveira EF, Oliveira JV, Nitin N, Monteiro AR (2018) Light-activated antimicrobial activity of turmeric residue edible coatings against cross-contamination of Listeria innocua on sausages. Food Contr 84:177–185
Tripathi K (2009) A review–garlic, the spice of life-(Part–I). Asian Journal of Research in Chemistry 2(1):8–13
Tsao S-M, Yin M-C (2001) In vitro activity of garlic oil and four diallyl sulphides against antibiotic-resistant Pseudomonas aeruginosa and Klebsiella pneumoniae. J Antimicrob Chemother 47(5):665–670
Uchida Y, Takahashi T, Sato N (1975) The characteristics of the antibacterial activity of garlic (author's transl). Japan J Antibiot 28(4):638–642
Uzun L, Dal Tuba M, Yürek M, Açıkgöz ZC, Durmaz R (2019) Antimicrobial activity of garlic derivatives on common causative microorganisms of the external ear canal and chronic middle ear infections. Turkish Arch Otorhinolaryngology 57(4):161
Venkatesan D, Karrunakaran C (2010) Antimicrobial activity of selected Indian medicinal plants. J Phytol 2(2):44–48
Verma A, Tiwari A, Panda PK, Saraf S, Jain A, Raikwar S, Jain SK (2020) Liposomes for advanced drug delivery advanced biopolymeric systems for drug delivery, pp. 317–338. https://doi.org/10.1007/978-3-030-46923-8_12
Vimala K, Yallapu MM, Varaprasad K, Reddy NN, Ravindra S, Naidu NS, Raju KM (2011) Fabrication of curcumin encapsulated chitosan-PVA silver nanocomposite films for improved antimicrobial activity. J Biomater Nanobiotech 2(01):55
Viswanathan V, Phadatare A, Mukne A (2014) Antimycobacterial and antibacterial activity of Allium sativum bulbs. Ind J Pharm Sci 76(3):256
Wallock-Richards D, Doherty CJ, Doherty L, Clarke DJ, Place M, Govan JR, Campopiano DJ (2014) Garlic revisited: antimicrobial activity of allicin-containing garlic extracts against Burkholderia cepacia complex. PloS One 9(12)
Weber ND, Andersen DO, North JA, Murray BK, Lawson LD, Hughes BG (1992) In vitro virucidal effects of Allium sativum (garlic) extract and compounds. Planta Med 58(05): 417–423
Yoshida H, Iwata N, Katsuzaki H, NAGANAwA R, Ishikawa K, Fukuda H, Suzuki A (1998) Antimicrobial activity of a compound isolated from an oil-macerated garlic extract. Biosci Biotech Biochem 62(5):1014–1017
Yousuf S, Ahmad A, Khan A, Manzoor N, Khan LA (2011) Effect of garlic-derived allyl sulphides on morphogenesis and hydrolytic enzyme secretion in Candida albicans. Med Mycol 49(4): 444–448

Zeng Y, Li Y, Yang J, Pu X, Du J, Yang X, Yang S (2017) Therapeutic role of functional components in alliums for preventive chronic disease in human being. Evid. Based Complementary Altern. Med. 2017: 13.

Zheng HM, Li HB, Wang DW, Liu D (2013) Preparation methods for monodispersed garlic oil microspheres in water using the microemulsion technique and their potential as antimicrobials. J Food Sci 78(8):N1301–N1306

Zhou Z, Pan C, Lu Y, Gao Y, Liu W, Yin P, Yu X (2017) Combination of erythromycin and curcumin alleviates staphylococcus aureus induced osteomyelitis in rats. Front Cell Infect Microbiol 7(379). https://doi.org/10.3389/fcimb.2017.00379

Zorofchian Moghadamtousi S, Abdul Kadir H, Hassandarvish P, Tajik H, Abubakar S, Zandi K (2014) A review on antibacterial, antiviral, and antifungal activity of curcumin. BioMed Res Int. 2014:186864.

# Chapter 13
# Carvacrol (*Origanum vulgare*): Therapeutic Properties and Molecular Mechanisms

**Arijit Mondal, Sankhadip Bose, Kamalika Mazumder, and Ritu Khanra**

**Abstract** Carvacrol (cymophenol) has been reported to be the major component of aromatic plant essential oils belonging to the Labiatae family, such as *Origanum vulgare* and *Thymus vulgaris*. It is a phenolic moiety containing monoterpenoid compound, chemically known as 5-isopropyl-2-methyl phenol. Carvacrol is used in food products both as a flavoring component and as preservative. Current research is being directed to establish a potent compound carvacrol with diverse pharmacological activities, such as antioxidant, antifungal, antimicrobial, anti-inflammatory, anticancer, hepatoprotective, anti-spasmodic, anti-parasitic, and insecticidal activities. There are various derivatives of carvacrol, which possess antimicrobial action against microbial pathogens. The carvacrol phenolic group does have a good antimicrobial and anti-oxidative function. Its hydrophobic nature is owing to the existence of benzene ring and methyl and isopropyl substituents helped the moiety to bind with guanine present in DNA. This review is based upon an evaluation of the existing data or knowledge regarding the extraction of carvacrol, chemical composition of the potent derivatives synthesized from carvacrol, their pharmacological and toxicological effects. This chapter documents the above-mentioned activities and strived to critically assess the molecular pathway involved in the action of carvacrol.

**Keywords** Carvacrol · Essential oil · Antimicrobial · Anticancer · Toxicity

A. Mondal (✉)
Bengal College of Pharmaceutical Technology, Dubrajpur 731123, India
e-mail: juarijitmondal@gmail.com

S. Bose
Bengal School of Technology, Chuchura, Hooghly 712102, India
e-mail: sankha.bose@gmail.com

K. Mazumder
BCDA College of Pharmacy and Technology, Kolkata 700127, India
e-mail: kamalika@bcdapt.com

R. Khanra
JIS University, Kolkata 700109, India
e-mail: ritukhanra@yahoo.co.in

 
D. Pal and A. K. Nayak (eds.), *Bioactive Natural Products for Pharmaceutical Applications*, Advanced Structured Materials 140,
https://doi.org/10.1007/978-3-030-54027-2_13

## 13.1 Introduction

Essential oils were applied in traditional remedies. It relates to essential oils as ethereal/volatile oils having an aromatic oily texture, collected from the various sections of herbs and then used as food flavors. The convenient form of separating essential oils is the inexpensive and easy-to-use hydro-distillation methodology. Terpenes, terpenoids, aromatic, and aliphatic analogs are the vital components of the essential oils (Marinelli et al. 2018). Carvacrol [2-methyl-5-(1-methyl ethyl)-phenol] mainly comes from many essential oils of the family Labiatae/Lamiaceae (Kirimer et al. 1995). It is a phenolic moiety containing monoterpenoid having a molecular formula $C_{10}H_{14}O$. It is also found in the essential oils of oregano *(Origanum vulgare, Origanum ehrenbergii, Origanum syriacum)*, sage apple *(Salvia fruticosa), Calamintha origanifolia, thyme (Thymus vulgaris), Lepidium flavum, Citrus aurantium*, and many more. Friedel-Crafts alkylation reaction was employed to generate carvacrol using o-cresol and isopropanol as reactants and utilizing mesoporous superacidic UDCaT-5 as catalyst (Yadav and Kamble 2009). Carvacrol is used primarily in food products for its flavoring and preservatives nature, and likewise had a significant application in the pharmaceutical industry (Can Baser 2008). Carvacrol's anticancer activity has gained recognition in recent times (Almanaitytė et al. 2020). It possesses other pharmacological actions such as antioxidant (Han et al. 2017), antimicrobial (Liolios et al. 2009), antibacterial (Menniti et al. 2010), anti-inflammatory (Landa et al. 2009), antiviral activities (Pilau et al. 2011), and hepatoprotective properties (Aristatile et al. 2009). The European Commission (2012) and the Food and Drug Administration (FDA 2017) have classified carvacrol as a safe substance from a toxicological perspective and its application is highly attractive as an additive in foods.

There are only a few previous reports that present an in-depth overview of this important field of research. Many of the preceding publications focus exclusively on the compilation of the various pharmacological activities of the secondary metabolite carvacrol for study on natural products (Mohammad et al. 2017; Sharifi-Rad et al. 2018; Bayir et al. 2019). This review looks at the pharmacology of carvacrol, and its analogs concerning their significant potential anticancer and antimicrobial functionality to become a clinical drug. The data procured covers published preclinical study, research papers, and review of molecule carvacrol isolated from a diverse group of plants. This study would investigate the antimicrobial and anticancer function of the metabolite carvacrol and the derivatives synthesized from carvacrol, emphasizing their chemical structure and discovering the action processes that underlie it, for the discovery of potent metabolites of therapeutic potential.

## 13.2 Literature Search Methodology

In vitro analysis, in vivo experimentation, and human clinical studies which explored the antibacterial and antiproliferative potency of phytoconstituent carvacrol, and its analogs by inhibiting various pathways were screened using authentic databases, such as ScienceDirect, PubMed, Google Scholar, and Springer reference. Relevant full articles published until April 2020 in peer-reviewed journals have been included. Conference abstracts and unpublished findings have not been included. Only papers published in the English language have been considered and included in this review. The keywords used for literature search included carvacrol, antimicrobial, cancer, tumor, proliferation, phytochemicals, terpenoids, in vivo, in vitro, and clinical studies.

## 13.3 Extraction and Isolation of Carvacrol

Like for most naturally occurring substances, the isolation and characterization of carvacrol are based mainly on chromatographic approaches. Carvacrol is an aromatic chemical compound bearing a complex composition, isolated from the plant's raw material either by steam distillation, dry distillation, or a relevant automatic process without heating. Carvacrol is extracted from an aqueous environment by a physical hydro-distillation process which does not adversely impact their composition utilizing Clevenger equipment fitted with a microwave apparatus (Dhifi et al. 2016, Cáceres et al. 2020). Obeying to this information, many scientists processed the isolation of the constituent carvacrol from the stems and leaves of many plants, such as *Origanum ehrenbergii*, *Origanum syriacum*, *Salvia fruticosa*, and *Calamintha origanifolia.* To immerse the whole plant in a flask, a specific amount of water is added and boiled for 3 h. Carvacrol is driven with water vapor into a condensed form owing to its volatile nature and later recollected which was purified with anhydrous sodium bicarbonate. It is then stored in the dark at 4°C in a sealed glass container for further utilization (Galehassadi et al. 2014, Cáceres et al. 2020). Its quantification can be analyzed by high-performance liquid chromatography (HPLC). Nonetheless, HPLC approach for carvacrol isolation has now become commonly adopted. It is very sensitive and has the potentiality to identify extremely small concentrations of a substance. Carvacrol is the major constituents of *Origanum vulgare* with a concentration of 83.7% (Laothaweerungsawat et al. 2020). Another documented report revealed that carvacrol content is 48% in *Origanum vulgare* (Bahmani et al. 2019)

Another method widely considered for carvacrol analysis is gas chromatography. While this approach is not as common as HPLC, both a qualitative and a quantitative study of carvacrol were carried out. Excellent performance but with small productivity and a basic process were the key reasons for its adoption. It was identified and quantified using the gas chromatography-mass spectroscopic

(GC-MS) analysis (Cáceres et al. 2020). Current techniques for the structural elucidation of carvacrol are spectroscopic instruments including high-resolution mass spectrometry (MS) and nuclear magnetic resonance (NMR). These approaches may be used either individually or in combination with chromatographic processes (Cáceres et al. 2020).

## 13.4 Biosynthesis of Carvacrol

Carvacrol was biosynthesized via the mevalonate pathway in various plants. At first, glucose is cleaved to phosphoenolpyruvate and then it is decarboxylated and acetylated to acetyl coenzyme A (acetyl CoA) and transformed to mevalonic acid. The mevalonic acid is then converted to gamma-terpinene, which after aromatization produced *p*-cymene. The hydroxylation *p*-cymene produced carvacrol which is shown in Fig. 13.1 (Friedman 2014).

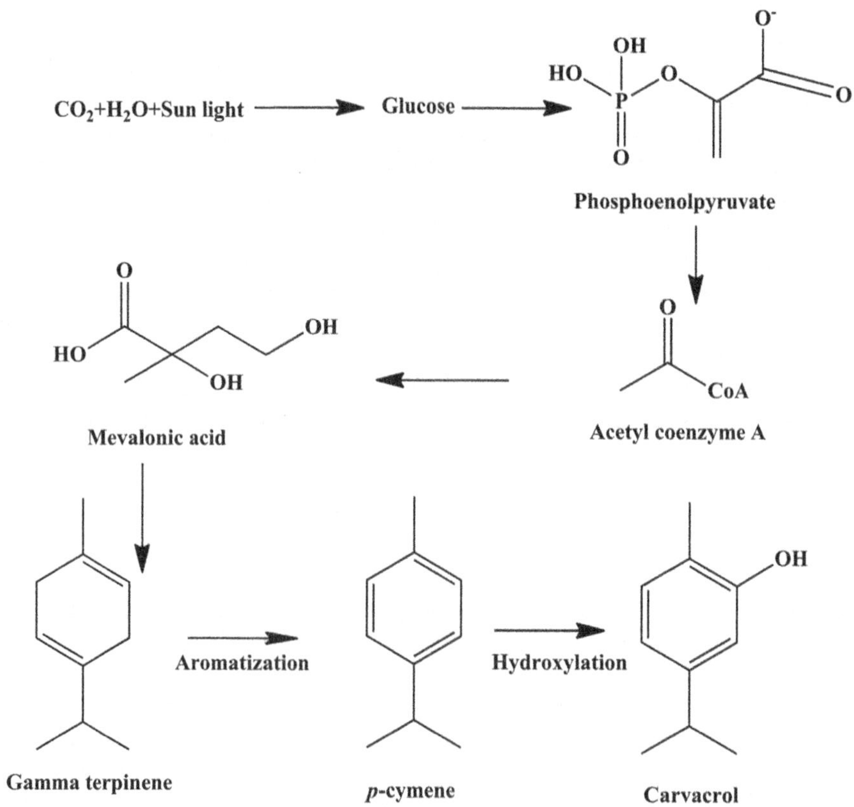


**Fig. 13.1** Biosynthesis of carvacrol through the mevalonate pathway

Another literature has shown an increased biosynthesis of carvacrol in shoot cultures of the Iranian plant *Satureja khuzistanica* by inhibiting the mevalonate pathway (Ramak et al. 2013). The methylerythritol-4-phosphate (MEP) pathway provides isopentenyl diphosphate (IPP) which was the precursor for the biosynthesis of 90% carvacrol. Figure 13.2 showed that carvacrol was biosynthesized from isopentenyl diphosphate (IPP) and dimethylallyl diphosphate (DMADP) which were derived via the MEP pathway located in plastids (Rohmer et al. 1993; Dudareva et al. 2005) of oregano and thyme. IPP is reversibly converted to DMADP by isopentenyl diphosphate isomerase (IDI). The irreversible conversion of 1-deoxy-D-xylulose-5-phosphate (DXP) into 2-C-methyl-D-erythritol-4- phosphate (MEP) was catalyzed by 1-deoxy-D-xylulose-5-phosphate reductoisomerase enzyme (DXR) (Takahashi et al. 1998). The condensation of IPP and DMADP produced geranyl diphosphate (GDP) and that was catalyzed by the geranyl diphosphate synthase (GDS) enzyme (Lichtenthaler 1999; Burke et al. 1999; Majdi et al. 2017). The cyclization of GDP by gamma-terpinene synthase produced gamma-terpinene. The enzymes such as CYP71D178, CYP71D180, and CYP71D181 of cytochrome P450 (CYP) monooxygenases converted γ-terpinene to yield carvacrol (Crocoll et al. 2010; Crocoll 2011).

## 13.5 Physical Properties

It is found in liquid form having molecular weight 150.22. Its boiling point is about 237– 238 °C; density is 0.967 and melting point 0 °C. Its $\lambda_{max}$ value showed 277.5 nm in 95% ethanol in the UV spectrophotometer. Its pKa value is 10.9. It is insoluble in water, soluble in 95% ethanol. Carvacrol has a pleasant pungent taste and odor, sweet and spicy like marjoram (Ramak et al. 2013; Friedman 2014).

## 13.6 Metabolism and Excretion of Carvacrol

Carvacrol's ADME process in the human body system is very rapid and it follows two types of pathways. The major metabolic pathway of carvacrol depends on the phase II metabolism related to the esterification of phenolic group of carvacrol by glucuronidation and sulfation, but when used in low dose, another minor route of carvacrol metabolism was also observed. In the minor pathway transformation of the terminal methyl groups to primary alcohols took place (Alagawany 2015; Dong et al. 2012).

Austgulen et al. 1987 also showed the pattern of metabolism of carvacrol (Fig. 13.3) in male albino rats. After administration by oral route at a dose of 1 mM/kg, the major portion of metabolites was excreted through urine where seven metabolites of carvacrol were identified, such as 2,3-dihydroxy-*p*-cymene, 2-(3-hydroxy-4-methyl phenyl) propan-2-ol, 2-(3-hydroxy-4-methyl phenyl)

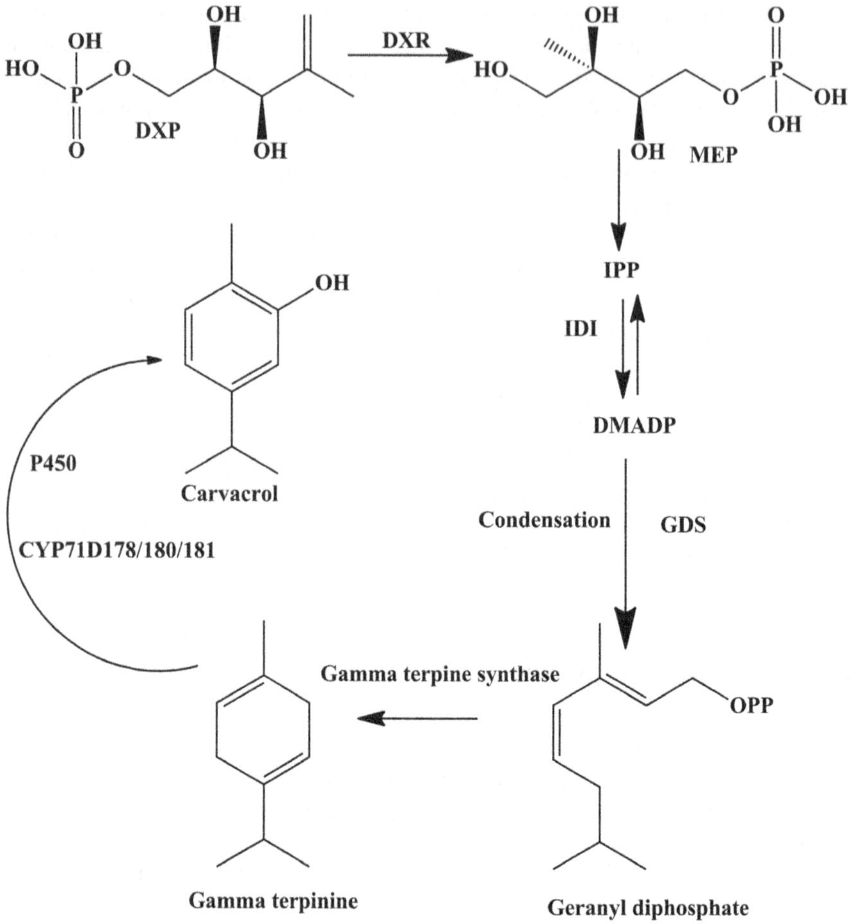


**Fig. 13.2** Biosynthesis of carvacrol through the MEP pathway

propan-1-ol, 2-hydroxy methyl-5-(1-methyl ethyl) phenol, 2-(4-hydroxy methyl-3-hydroxy phenyl) propan-1-ol, 2-(3-hydroxy-4-methyl phenyl) propionic acid, and 2-hydroxy-4-(1-methyl ethyl) benzoic acid (Austgulen et al. 1987). Another experiment was done by Michiels et al. (2008), utilizing gastric fermentation simulation technique on pig. Carvacrol was not degraded in jejunum simulations; rather 29% degradation occurs in the cecum. When piglets were administered with a dose of 13 mg/kg body weight, it was completely absorbed in the stomach and the proximal section of the small intestine where it showed a half-life of 2.05 h in the whole digestive tract (Michiels et al. 2008).

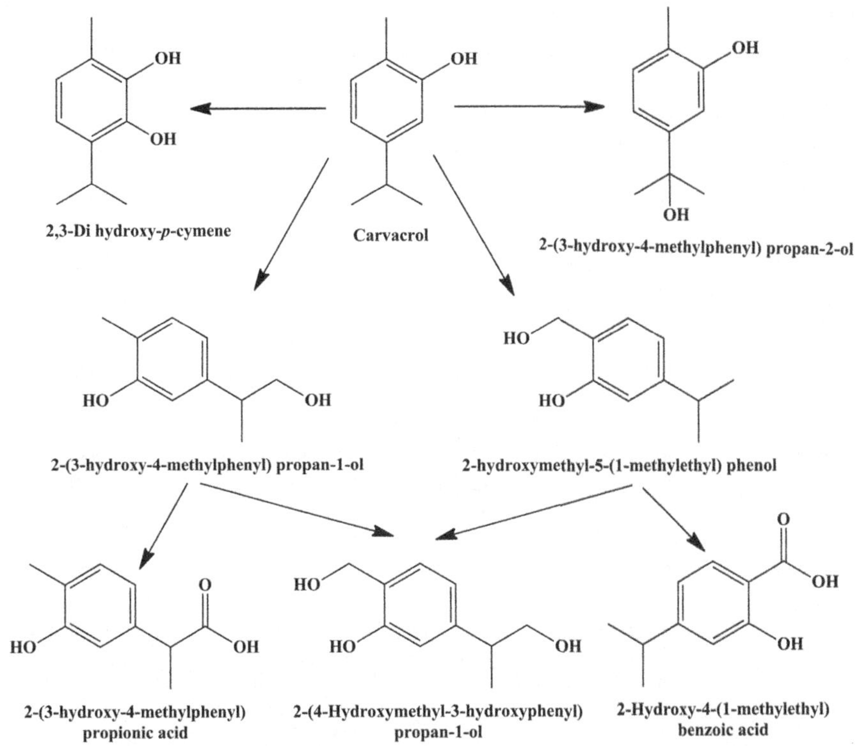


**Fig. 13.3** Metabolic pathway of carvacrol

## 13.7 Acute Toxicity of Carvacrol

Data on carvacrol toxicology is limited. Various toxicology studies have been done in different animal models. Andre et al. (2016) investigated the acute toxicity study in carvacrol administered through esophageal gavage route in Swiss albino mice. The result revealed that depression occurs within 10 min of carvacrol administration, coma, and death of animals within 1 h to 3 days. Administration of the essential oil carvacrol in rabbits, rats, and dogs exhibited the non-toxic nature, although the detailed information was not documented (Budavari 1989; Livingston 1921; Andersen 2006; Ghorani et al. 2019). But some reports showed the toxic behavior of the test drug carvacrol in Osborne-Mendel rats and mice, where depression, ataxia, decreased spontaneous motor activity, somnolence, coma, and death were observed (Jenner et al. 1964; Viana et al. 1981) Detail reports of animal acute toxicity studies have been enlisted in Table 13.1. However, the likely toxicity impacts of carvacrol on humans have not been reported till date.

**Table 13.1** Toxicological study of cavacrol

| Species | Toxicity study | $LD_{50}$/Doses | Toxic signs and death time (DT) | Reference |
|---|---|---|---|---|
| Osborne-Mendel rats | Acute Oral Toxicity | 810 mg/kg body weight | Depression, coma; DT: 1hr-3 days | Jenner et al. (1964) |
| Rabbits | Acute Oral Toxicity | 100 mg/kg body weight | No detail information | Budavari (1989) |
| Swiss albino mice | Acute toxicity through esophageal gavage route | 919 mg/kg body weight | Low toxicity, other effects not specified | Andre et al. (2016) |
| Rats | Acute toxicity | >0.5 g/kg < 5 g/kg body weight | Toxicity not specified; DT: 24 h at 5 g/kg dose | Andersen, (2006) |
| Rabbits | Acute toxicity | Dose: 0.5 g/kg-2 g/kg body weight | Toxicity not specified DT:18 h to 35 days | Livingston (1921) |
| Dog | Acute Parental Toxicity (Intravenously) | 0.31 g/kg body weight | Not specified | Ghorani et al. (2019) |
| Mice | Acute Parental Toxicity (Intraperitoneally) | 73.3 mg/kg body weight | Ataxia, decreased spontaneous motoractivity, somnolence,death; DT-5 days | Viana et al. (1981) |

## 13.8 Antioxidant Activity

Antioxidants are scavenging the ROS, including free radicals, and thereby protecting cells from cellular stress. They also slow down the prostaglandin synthesis, stimulate the drug-metabolizing enzymes, and exhibit other biological activities such as DNA defense, enzyme-induced hepatotoxicity, and cancer imitation inhibition. The explanation of antioxidant activity was the availability of hydroxyl group (OH) linked to the aromatic ring (Aeschbach et al. 1994). During the carvacrol molecule's reaction with free radicals, it donates hydrogen atoms to an unpaired electron and generates another radical that is produced by a molecule resonance structure. Carvacrol can interfere with the cell phospholipid membrane or low-density lipoprotein and minimize the synthesis of lipid peroxidation and nitric oxide, contributing to oxidative degradation of cell membranes (Karimian et al. 2011; Teissedre and Waterhouse 2000). Nitric oxide is created by the spontaneous decomposition of sodium nitroprusside, which has been effectively scavenged by carvacrol (Aeschbach et al. 1994). Differences in vitro and in vivo studies have confirmed that carvacrol has strong antioxidant properties. As an antioxidant, carvacrol can lower stress by enhancing the antioxidant activity of enzymes like superoxide dismutases (SOD), catalase (CAT), and glutathione peroxidase

(GPx), and remove free radicals like peroxide, $H_2O_2$, superoxide, and nitric oxide (Samarghandian et al. 2016). In diabetic rats, carvacrol reduced oxidative stress through the improvement of the activity of antioxidant enzymes (Deng et al. 2013). In vitro analysis showed that the amount of intracellular ROS production of mouse V79 fibroblast cells depends on the dose-dependent manner of carvacrol. The production of ROS decreases with the increasing concentration of carvacrol (Ündeğer et al. 2009a). Some anti-oxidant activities are also performed on in vivo studies. Hydrogen peroxide resistance–induced damage to DNA in hepatic and testicular tissue was gradually decreasing in rats when carvacrol was administered in drinking water (at 30 and 60 mg/kg for 7 days or 15 and 30 mg/kg for 14 days) (Slamenova et al. 2013). Other studies have also shown that carvacrol has a protective role against lipid peroxidation and induces enhanced endogenous antioxidant defensive mechanisms in hepatocellular carcinogenesis caused by N-nitrosodiethylamine and antioxidant function against hepatotoxicity in rats. Carvacrol has enhanced the reduced function of antioxidant enzymes and improved the lipid peroxidation in ethanol-exposed rat hippocampus (Wang et al. 2017; Mehrjerdi et al. 2020).

## 13.9 Antimicrobial Effect

The efficacy of the carvacrol depends on their phenolic group as well as hydroxyl group which is attached with the phenolic structure and increased membrane permeability by working on the micro-organism at low concentration. The inhibitory activity of phenols may be clarified by associations with micro-organisms' cell membrane and is also associated with the compounds' hydrophobicity. Most hydrophobic molecules are usually harmful and the cytoplasmic membrane is often the main source of harmful activity. Besides, lipophilic products showed a strong affinity for cell membranes, and their insertions induced changes in the physico-chemical properties of membranes and increased depletion of intercellular ATP (Ultee et al. 2002; COX et al. 2007). The interactions between antimicrobial compounds and cell membranes are defined to affect both the lipid order and the stability of the bilayer, resulting in a decrease in membrane integrity and an increase in the passive flux of protons through the membrane. This effect is documented particularly for compounds with a log P greater than 3. By consensus, the most effective antimicrobial compound was carvacrol, which has a log P of 3.52 (Ben Arfa et al. 2006). Carvacrol, by partitioning the phospholipid fatty acid chains, created ion channels across the membrane, eventually allowing ions to exit cytoplasm (Ultee et al. 2002). Carvacrol has showed bacteriostatic and bactericidal activity on food pathogens such as *Vibrio cholerae, Campylobacter jejuni, E. Coli, Listeria monocytogenes, S. aureus, Staphylococcus epidermidis, Lactobacillus sakei, P. aeruginosa, Pseudomonas putida, Streptococcus mutans, and Bacillus subtilis* (Friedman et al. 2002; Lambert et al. 2001; Mathela et al. n.d.; Rattanachaikunsopon and Phumkhachorn 2010; Ravishankar et al. 2010). Also, the presence of a free hydroxyl carvacrol group and its influence on the delocalized electron system by growing the gradient across the cytoplasmic membrane is essential for its antibacterial action (Knowles and Roller 2001). Carvacrol not only inhibited bacterial cell growth but also induces cell death through increased depolarization of cell membrane which leads to elevated rapid oxidizing burst in *E. coli.* Cellular DNA, protein, and $K^+$ were released from cells, which have been triggered by carvacrol. On treatment with carvacrol, there is a decrease in the protein expression levels of inflammation, such as TNF-α and COX-2 (Khan et al. 2020)]. Phosphate ion leakage took place in *Staphylococcus aureus* and *P. aeruginosa* cells, which were treated with carvacrol containing oregano essential oil (Friedman et al. 2004; Lambert et al. 2001). Carvacrol also showed antifungal activity that may act through the membrane and cell wall disruption. An analysis of the related performance of carvacrol on different micro-organism strains was presented in Table 13.2.

**Table 13.2** Antimicrobial and antifungal activity of carvacrol using different techniques

| Sl no | Organism | Antimicrobial activity | Reference |
|---|---|---|---|
| 1 | *Aeromonas salmonicida* subsp. *Salmonicida*<br>*ATCC 14174*<br>*CAE 235*<br>*CAE 452*<br>*CAE 258* | Minimum Inhibitory Concentration (MIC)-344 μg/ml<br>MIC-688 μg/ml<br>MIC-344 μg/ml<br>MIC-344 μg/ml | Hayatgheib et al. 2020 |
| 2 | *Staphylococcus aureus ATCC 6538*<br>*Staphylococcus epidermidis ATCC 34984*<br>*Listeria monocytogenes ATCC 13932*<br>*Bacillus subtilis ATCC 6633*<br>*Pseudomonas aeruginosa ATCC9027* | MIC- 12.5% v/v<br>MIC- 25% v/v<br>MIC- 25% v/v<br>MIC- 12.5% v/v<br>MIC- 25% v/v | (Marino et al. 2020) |
| 3. | *Salmonella enterica* | At 1% concentration, no survival of the microorginism | Ravishankar et al. (2010) |
| 4 | *Pseudomonas fluorescens*<br>*Escherichia coli*<br>*Staphylococcus aureus*<br>*Bacillus subtilis*<br>*Lactobacillus plantarum*<br>*Saccharomyces cerevisiae* | MIC-1 g/l<br>MIC-0.25 g/l<br>MIC-0.25 g/l<br>MIC-0.25 g/l<br>MIC->3 g/l<br>MIC-0.25 g/l | Ben Arfa et al. (2006) |
| 5 | *Escherichia coli*<br>*Salmonella typhimurium*<br>*Staphylococcus aureus*<br>*Listeria monocytogenes* | MIC-0.225 μL/ml<br>MIC-0.225 μL/ml<br>MIC-0.175– 0.450 μL/ml<br>MIC-0.375–5 μL/ml | Burt, (2004) |
| 7 | *Malassezia furfur*<br>*Trichophyton rubrum*<br>*Trichosporon beigelii* | Disk diffusion assay: 15 mm<br>Complete inhibition<br>Complete inhibition | (Adam et al. 1998) |

(continued)

**Table 13.2** (continued)

| Sl no | Organism | Antimicrobial activity | Reference |
|---|---|---|---|
| 8 | *Bottytis cinerea*<br>*Monllinla fmcticokz* | MIC100 μg/mL<br>MIC100 μg/mL | Tsao and Zhou, (2000) |
| 9 | *Aspergillus niger PTCC 5154*<br>*Aspergillus fumigatus PTCC 5009*<br>*Aspergillus flavus PTCC 5004*<br>*Aspergillus ochraceus PTCC 5017*<br>*Alternaria alternate PTCC 5224*<br>*Botrytis cinerea ATCC 12481*<br>*Cladosporium spp. PTCC 5202, Penicillium citrinum PTCC 5304*<br>*Penicillium chrysogenum PTCC 5271*<br>*Fusarium oxysporum PTCC 5115*<br>*Rhizopus oryzae PTCC 5174* | MIC-50 μg/mL,MIF -75 μg/mL<br>MIC-100 μg/mL,MIF -125 μg/mL<br>MIC-100 μg/mL,MIF -125 μg/mL<br>MIC-100 μg/mL,MIF -125 μg/mL<br>MIC-350 μg/mL,MIF -400 μg/mL<br>MIC-300 μg/mL,MIF -350 μg/mL<br>MIC-100 μg/mL,MIF -125 μg/mL MIC-150 μg/mL,MIF - 175 μg/mL<br>MIC-125 μg/mL,MIF -150 μg/mL<br>MIC-125 μg/mL,MIF -150 μg/mL<br>MIC-200 μg/mL,MIF -250 μg/mL | Abbaszadeh et al. (2014) |
| 10 | *Candida albicans* | ↓ Colony forming units (CFU) 94.46% | Chami et al. (2004) |
| 11 | *Staphylococcus pseudintermedius*<br>*Pseudomonas aeruginosa*<br>*Proteus mirabilis*<br>*Malassezia pachydermatis* | MIC- 146–292 μg/ml<br>MIC- 585–1120 μg/ml<br>MIC- 146–292 μg/ml<br>MIC- 585 μg/ml | Sim et al. (2019) |
| 12 | *Vibrio parahemolyticus,*<br>*Shewanella putrefaciens,*<br>*Staphylococcus aureus*<br>*Pseudomonas fluorescens* | MIC-0.5 mg/ml<br>MIC-0.5 mg/ml<br>MIC-0.125 mg/ml<br>MIC-0.5 mg/ml | (Fang et al. 2019) |
| 13 | *Bacillus subtilis* | MIC-0.125% v/v | Chraibi et al. 2020 |
| 14 | *Escherichia coli* | MIC-150 μg/ml<br>Minimum bactericidal concentration (MBC)-300 μg/ml | Khan et al. 2020 |

## 13.10 Anticancer Effect of Carvacrol and the Related Mechanism of Actions

Carvacrol extracted from various medicinal plants showed potent cytotoxic activities against various carcinoma cell lines by inducing apoptosis through the influence of numerous proteins associated with the apoptotic pathway, MAPK pathway, and PI3K/Akt pathway as shown in Table 13.3. It inhibited the proliferation of two

**Table 13.3** Pharmacological mechanisms of carvacrol involved in anticancer activities

| S. No. | Cell lines | Effect and mechanism | $IC_{50}$ (μM/l)/dose | Reference |
|---|---|---|---|---|
| 1 | Human colon cancer cell lines: HCT116, LoVo | ↓Proliferation and migration, cell cycle arrest at G2/M phase, ↓Bcl-2, ↑Bax, ↓MMP-2, ↓MMP-9, ↑apoptosis, ↑ MAPK, ↓p-ERK, ↑p-JNKand ↑PI3K/Akt signaling pathway, ↓p-Akt. | HCT116: 544.4<br>LoVo: 530.2 | Fan et al. (2015) |
| 2 | Ovarian epithelial adenocarcinoma cell line (SKOV-3) | Induction of cytotoxicity, ↑apoptosis, | 322.50 | Elbe et al. (2020) |
| 3 | Human colorectal carcinoma: HCT-116<br>Human colorectal carcinoma: HT-29 | antiproliferative effect | HCT-116: 92<br>HT-29: 42 | Pakdemirli et al. (2020) |
| 4 | Prostate cancer cell line:PC3 | Antiproliferative effect, reduce cell viability | 25-200 μg/ml dose | Trindade et al. (2019) |
| 5 | Choriocarcinoma cell line: (JAR and JEG3 cells) | Antiproliferative effect, ↑apoptosis, cell cycle arrest at sub G1 phase, ↑Bax, ↑Bak, ↑cytochrome c, ⊥MMP, ↓cytosolic calcium level, ↑ MAPK, ↓ p-ERK1/2, ↑PI3K/Akt signaling pathway, ↑p-JNK, phosphor-P38, and P90RSK proteins, ↓phosphorylation of PDK1, AKT, P70S6K, and GSK3β | Up to 300 μM doses | Lim et al. (2019) |
| 6 | Human gastric adenocarcinoma cell line AGS cells and normal human fibroblast WA-1, *invivo* analysis in Wister rats | Dose-dependent cytotoxicity, ↑apoptosis, genotoxicity and ROS generation, ↑Bax, ↑caspase-3, ↑caspase-9, ↓ Bcl-2, | WS-1: 138.1 ± 8.7 μM<br>AGS:82.57 ± 5.5 μM | Günes-Bayir et al. (2018a) |
| 7 | Human cervical cancer HeLa cells | ↑Cytotoxicity, ↑apoptosis, ↑ MAPK, ↑phosphor-ERK1/2, ↑caspase-9, ↑PARP cleavage, ↓Cyclin D1, ↑p21, autophagy, ↓p-Akt, ↓phospho-mTOR | HeLa:556 ± 39 μM | Potočnjak et al. (2018) |

(continued)

**Table 13.3** (continued)

| S. No. | Cell lines | Effect and mechanism | $IC_{50}$ (μM/l)/dose | Reference |
|---|---|---|---|---|
| 8 | Human prostate cancer DU145 | Antiproliferative agent, ↑apoptosis, ↑caspase-3, ↓ ΔΨm, cell cycle arrest at G0/G1 phase | DU145:84.39 μM | Khan et al. (2017) |
| 9 | Human gastric adenocarcinoma AGS cells | ⊥proliferation, ↑apoptosis,, ↓GSH levels,, ↑Bax, ↑caspase-3, ↑caspase-9, ↓ Bcl-2, | AGS:82.57 ± 5.58 μM | Günes-Bayir et al. (2017) |
| 10 | Prostate cancer cell lines DU145<br>Prostate cancer cell lines: PC3 | Antiproliferative effect, ↓MMP-2, ↓p-Akt, and ↓p-ERK1/2, ⊥migration, ⊥invasion | DU145: 498.3 ± 12.2<br>PC3: 430.6 ± 21.9 | Luo et al. (2016) |
| 11 | Human T lymphocyte Cells: Jurkat | Inflection of T cell activity by ↓ IL-2, ↓ IFN-γ, ↑AP-1 and ↑NFAT2 | 50 μg/ml (Dose) | Gholijani et al. (2015) |
| 12 | Human acute promyelocytic leukemia cell line (HL-60), and (immortalized human T cell lymphocyte cell line (Jurkat) | Induced apoptosis, ↑caspase-3, ↑Bax, ↓ Bcl-2, ↓ ΔΨm | HL-60: 100 μM<br>Jurkat: 50 μM | Bhakkiyalakshmi et al. (2016) |
| 13 | Human oral squamous cell carcinoma: Tca-8113 | ⊥Proliferation, ↑apoptosis, cell cycle arrest at G1/S phase, ↓cyclin D1, ↓CDK4, ↑Bax/Bcl-2 ratio | 0-80 μM (Doses) | Dai et al. (2016) |
| 14 | Human hepatomacell lines: HepG2 | ⊥Proliferation, ↑apoptosis, cell cycle arrest at G1/S phase | HepG2: 425 μM | Melušová et al. (2014) |
| 15 | Murine melanoma cell line:B16(F10) | ⊥tumer growth | B16F10: 120 ± 15 μM | He et al. (1997) |
| 16 | N2a neuroblastoma cells | ↑cytotoxicity, ⊥proliferation, anticancer activity | >200 mg | Aydin et al. (2014) |
| 17 | MDA-MB 231 | ⊥proliferation, induced morphological changes, ↑apoptosis | MDA-MB 231: 100 | Arunasree (2010) |
| 18 | Human hepatocellular carcinoma cell line HepG2 | Induced anti-proliferation and ↑apoptosis, ↑caspase-3, ↓ Bcl-2, ↑cleavage of PARP, ↓ p-ERK1/2, ↑phosphorylation of p38 | HepG2: 0.4 mM/ml | Yin et al. (2012) |
| 19 | Human non-small cell lung cancer cell line: A549 | Anticancerogenic activity | 100–1000 μM | Koparal and Zeytinoglu (2003) |

(continued)

**Table 13.3** (continued)

| S. No. | Cell lines | Effect and mechanism | $IC_{50}$ (µM/l)/dose | Reference |
|---|---|---|---|---|
| 20 | Human cervical cancer cell lines, HeLa and SiHa | ↑Cytotoxicity, ↑apoptosis, ↑DNA fragmentation | HeLa : 50 ± 5.95 mg/l SiHa: 50 ± 3.89 mg/l | Mehdi et al. (2011 |
| 21 | Human HepG2 cells | Carvacrol protected cells against DNA-damaging effects of $H_2O_2$ was unambiguous | 200 µM dose | Slamenova et al. (2013) |
| 22 | Human breast cancer cell line: MCF-7 | ↑ Cytotoxicity, ↑apoptosis, ↑caspase-3, ↑caspase-9, ↑caspase-6, ↑p53, ↑Bax | MCF-7:244.7 ± 0.71 µM | Al-Fatlawi and Ahmad (2014) |
| 23 | Pparental and drug-resistant human lung cancer cell lines (H1299) | ↑ Cytotoxicity and ↑ membrane and DNA damage | H1299: 380 and 244 µM | Ozkan and Erdogan (2012) |
| 24 | Human colon carcinoma cell line Caco-2 | ↑Oxidative stress at higher concentration | Caco-2: 1832 ± 0.11 µM | Llana-Ruiz-Cabello et al. (2015) |
| 25 | V79 Chinese hamster lung fibroblast cells | DNA damage at higher concentration | Not specified | Ündeğer et al. (2009) |

human colon cancer cells, such as HCT116 and LoVo, by inducing apoptosis and triggering cell cycle arrest at the G2/M phase. It reduced the expression levels of matrix metalloprotease-2 (MMP-2), MMP-9, and cyclin B1. It also downregulated the expression levels of Bcl-2 and upregulated the expression levels of Bax. Apart from this, it decreased the expression levels of phosphorylated protein kinase B (p-Akt) associated with the activation of the PI3K/Akt signaling pathway. In association with the activation of the MAPK signaling pathway, there is a decrease in the expression levels of phosphorylated extracellular signal-regulated kinases (p-ERK) and an increase in the expression levels of phosphorylated c-Jun N-terminal kinase (p-JNK) (Fan et al. 2015). The mitochondrial membrane potential was impaired by carvacrol treatment in choriocarcinoma cell lines, such as JAR and JEG3 cells. The antiproliferative activity was associated with the induction of apoptosis and cell cycle arrest at the sub G1 phase. There was a decrease in cytosolic calcium level in JAR and JEG3 cells. The MAPK and PI3K–AKT signaling pathway was affected similarly in choriocarcinoma cells via regulation of the phosphorylation of various proteins involved in the pathway (Lim et al. 2019). Carvacrol inhibited the proliferation of human gastric adenocarcinoma cell line (AGS cells) and normal human fibroblast (WA-1) by the apoptotic pathway in a dose-dependent manner but the cytotoxicity towards cancer cells was more in comparison to normal cells (Günes-Bayir et al. 2018a, b). Apart from the apoptotic pathway, through autophagy mechanism, the carvacrol exhibited cytotoxic activity against human cervical cancer cells (HeLa). The mTOR, phospho-mTOR, and phospho-Akt protein expression levels were reduced in cancer treated cells (Potočnjak et al. 2018). It exhibited its antiproliferative activity against prostate cancer DU145 cells by inducing apoptosis and activation of caspase 3. Its induced cell cycle arrest at the G0/G1 phase and loss of mitochondrial membrane potential occurred in the cancer cells (Khan et al. 2017). Similar observation was reported by Bhakkiyalakshmi et al. (2016) in which carvacrol showed cytotoxic activity against human acute promyelocytic leukemia cell line (HL-60), and (immortalized human T cell lymphocyte cell line (Jurkat). Carvacrol inhibited the PI3K/Akt and MAPK signaling pathways in prostate cancer cell lines DU145 and PC3, by preventing the proliferation, migration, and invasion of the cancer cells. It reduced the MMP-2, p-Akt, and p-ERK1/2 protein expression levels in the prostate cancer cells (Luo et al. 2016). Carvacrol reduced the production of interleukin-2 (IL-2) and interferon (IFN-$\gamma$) in the human Jurkat T cells through the inhibition of activator protein (AP)-1 and nuclear factors of activated T cells (NFAT2) (Gholijani et al. 2015). It inhibited the proliferation of human oral squamous cell carcinoma (Tca-8113) as well as metastasis and invasion of the cells by inducing apoptosis. But it triggers a cell cycle arrest at the G1/S stage. It reduces the expression level of cyclin D1 and cyclin-dependent kinase (CDK4) (Dai et al. 2016). A similar mechanism of action of carvacrol was reported in human hepatoma cells (HepG2) (Melušová et al. 2014). It reduced the tumor growth of murine melanoma cells B16(F10) (He et al. 1997) and displayed potent cytotoxicity at a higher dose against neuroblastoma cells (N2a cells) (Aydin et al. 2014). Mitogen-activated protein kinase pathway is altered by carvacrol on treating the HepG2 cell line apart from the apoptotic pathway (Yin et al. 2012). So overall carvacrol demonstrated potent anticancer activity against a variety of carcinoma cell lines by inhibiting the proliferation of the cells by apoptotic pathway triggered by activation of caspases, increased DNA fragmentation (Koparal and Zeytinoglu 2003; Mehdi et al. 2011; Al-Fatlawi and Ahmad 2014; Ozkan and Erdogan 2012). It also exhibited protective activity against $H_2O_2$ induced DNA damage (Slamenova et al. 2013; Llana-Ruiz-Cabello et al. 2015; Ündeğer et al. 2009). The efficacy of carvacrol can further be improved by preparing a carvacrol β-cyclodextrin inclusion complex, which exhibited higher antiproliferative activity against prostate cancer cells (PC3), in comparison to free carvacrol (Trindade et al. 2019).

## 13.11 Carvacrol Derivatives with Pharmacological Activities

The structure-activity relationships of carvacrol indicated that the presence of OH group and delocalized electrons cloud were responsible for the pharmacological activities of carvacrol (**1**) (Fig. 13.4) (Ultee et al. 2002). Removal or replacement of the hydroxyl group with methyl ether produces lesser active compound *p*-cymene (**2**) and methyl ether derivative of carvacrol (**3**) due to the lower tendency to donate protons than the hydroxyl group (Ben Arfa et al. 2006; Marinelli et al. 2018). Carvacryl acetate (**4**) is an analog of carvacrol synthesized by refluxing carvacrol, sodium acetate, and acetic anhydride. It was then subsequently purified and characterized [Ben Arfa et al. 2006]. Carvacrol was 28 times more potent than its structural isomer (positional isomer) thymol (**5**) (2-isopropyl-5-methyl phenol) against gram-positive and gram-negative bacteria like *Mycobacterium tuberculosis*, *Escherichia coli*, *Staphylococcus aureus*, and *Candida albicans*. Menthol (**6**) (2-isopropyl-5-methyl cyclohexanol), eugenol (**7**) (4-allyl-2-methoxyphenyl), vanillin (**8**) (4-hydroxy-3-methoxy benzaldehyde) and 3-isopropyl-5-methyl phenol (**9**) were equivalent potent as compared to thymol. Replacement of isopropyl group of menthol with vinyl group produced neo isopulegol derivative (**10**) [2-methyl-5-(prop-1-en-2-yl) cyclohexanol], which was twice more potent antimycobacterial agent than menthol. 4-chloro-5-isopropyl-2-methyl phenol (**11**) is a 4-substituted chloro derivative of carvacrol synthesized from carvacrol exhibited similar antimycobacterial activity as that of 2-methyl-5-(prop-1-en-2-yl) cyclohexanol (**10**) (Alokam et al. 2013).

The aim of the SAR study is to develop carvacrol derivatives with potent biological activity. Patil et al. (2010) suggested carvacryl derivatives were obtained by the attachments of acetanilide or α-chloro acetanilide group with carvacrol via ether bonds. Comparative antimicrobial studies between carvacryl ether derivatives and carvacrol against *Bacillus subtilis*, *Escherichia coli*, *Proteus vulgaris*, and *Staphylococcus aureus*. The phenyl derivative of carvacryl ether (**12**) provided a little antimicrobial activity against *Bacillus subtulis*, and inactive against other

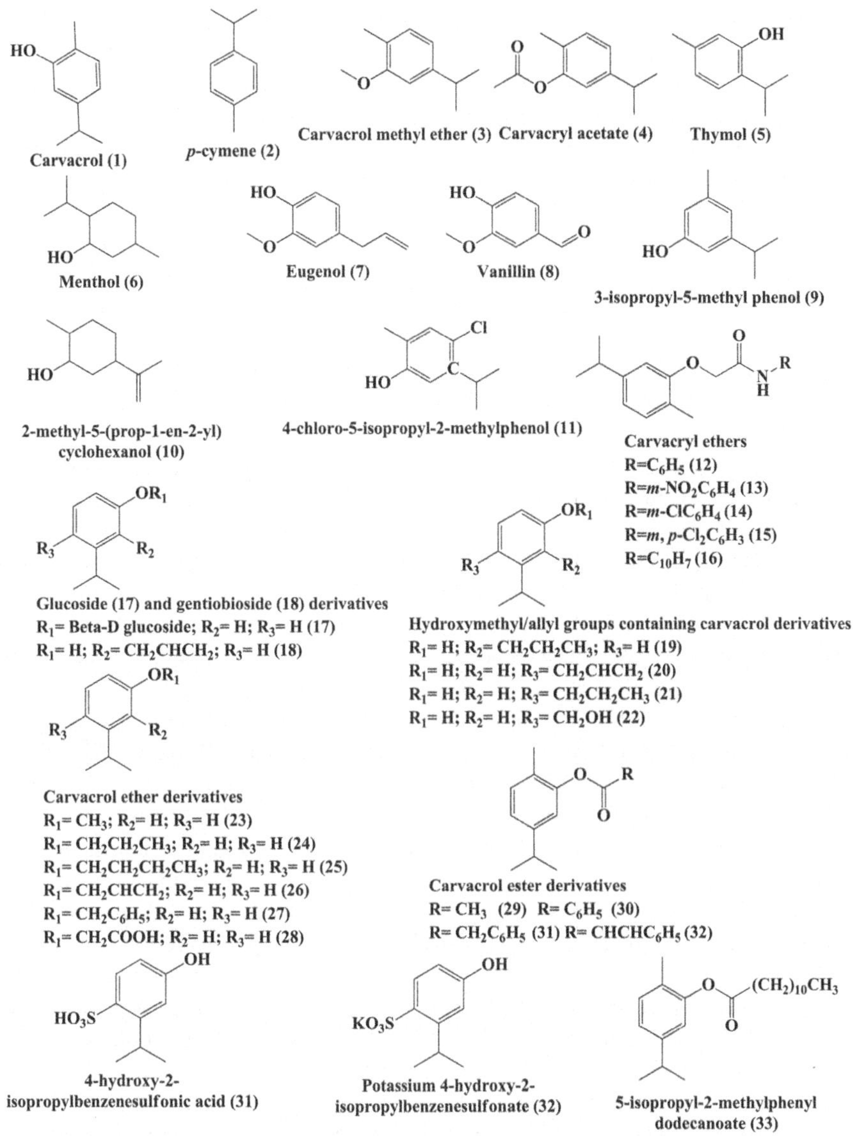


**Fig. 13.4** Chemical structures of some carvacrol derivatives (1-33)

bacterial species. The phenyl *m*-nitro substituted derivative (**13**) showed a mild antimicrobial activity against *Escherichia coli* and remained inactive against other experimental bacteria. The *m*-chloro aryl (**14**) and *m, p*-dichloro aryl (**15**) derivatives showed the most potent activity against all bacteria except *Escherichia coli*; whereas, the naphthyl-substituted derivative of carvacryl ether (**16**) exhibited more potent antibacterial activity than carvacrol against *Proteus vulgaris, and Bacillus* *subtilis*. The parent compound carvacrol also exerted no activity against *Escherichia coli* (Patil et al. 2010). The enzymatic conversions (glycosylation) of carvacrol produced β-glucosides (B7) and β-gentiobiosides (B8) derivatives ameliorated the physicochemical properties and the pharmacokinetic profile of carvacrol (Shimoda et al. 2006). The flavor, fragrance, and antimicrobial properties were improved by introducing hydroxymethyl and allyl groups in carvacrol (**19-22**) (Lupo et al. 2000) but the bioactivity against all the tested Gram-negative bacteria was decreased and antimicrobial activity on Gram-positive bacteria strains were increased in respect to carvacrol. Ortho allyl carvacrol derivative showed the most significant activity against *Propionibacterium acnes* but resistant against *Staphylococcus aureus*. Introduction of saturated and unsaturated aliphatic side chains and aromatic units produced ethers (**23-28**) and esters (**29-32**) derivatives that possessed more potent antimicrobial response than carvacrol against *Bacillus megaterium*, *Bradyrhizobium japonicum*, *Bacillus polymyxa*, and *Bacillus substilits* (Nikumbh et al. 2003). Among ethers, the carboxymethyl derivative (**28**) and among the esters, the phenylacetate derivative (**30**) were the most active antibacterial compounds. Development of 4-hydroxy-2-isopropyl benzene sulfonic acid (**31**), potassium 4-hydroxy-2-isopropyl benzene sulfonate (**32**), and 5-isopropyl-2-methyl phenyl dodecanoate (**33**) derivatives of carvacrol (Fig. 13.4) were synthesized with better physicochemical profiles and antimicrobial activity (Bassanetti et al. 2017). The parent drug and synthesized compounds were tested against *Clostridium perfringens, Salmonella typhimurium, Salmonella enteritidis*, and *Escherichia coli* strains. These three analogs along with carvacrol exhibited potent cytotoxicity against human colorectal carcinoma cell lines (HT-29) by inhibiting the proliferation of the cancer cells (Bassanetti et al. 2017). Cacciatore et al. (2015) synthesized a series of alkyl cysteine carvacrol derivatives by using the co-drug strategy. The hydroxyl group of carvacrol was conjugated with the carboxyl portion of sulfur-containing amino acids via an ester bond. The in vitro studies of all the derivatives showed low water solubility, good stability in an acidic environment, in human and rat plasma, and a good permeability verified by PAMPA-GI (parallel artificial gastrointestinal membrane permeability assay). (R)-5-isopropyl-2-methyl phenyl 2-acetamido-3-(allylthio) propanoate (**34**) was found to be the most active antibacterial compound (Fig. 13.5). The carvacrol sulfonamide derivatives (**35-43**) (Fig. 13.5) were synthesized by inserting selected amines at carvacrol aromatic ring previously treated with chlorosulfonic acid (De Oliveira et al. 2016). The obtained sulfonamides were superior to carvacrol against selected *Staphylococcus aureus* strains and able to reduce bacterial resistance. P-toluene (**35**), p-fluorobenzene (**37**), and 4-amino-N-(thiazol- 2-yl) benzene sulfonamide (42) groups containing carvacrol derivatives revealed antimicrobial activity at lower concentration with lower MIC values (from 3.90 to 62.50 ppm) among the tested compounds. The compound 4-hydroxy-2-isopropyl-5-methyl-N-(p-tolyl) benzene sulfonamide (**35**) had a synergistic effect with erythromycin, and compound N-(4-fluorophenyl)-4-hydroxy-2-isopropyl-5-methyl benzene sulfonamide (**37**) showed a synergistic response when incorporated with ampicillin and tetracycline, respectively (Marinelli et al. 2018).

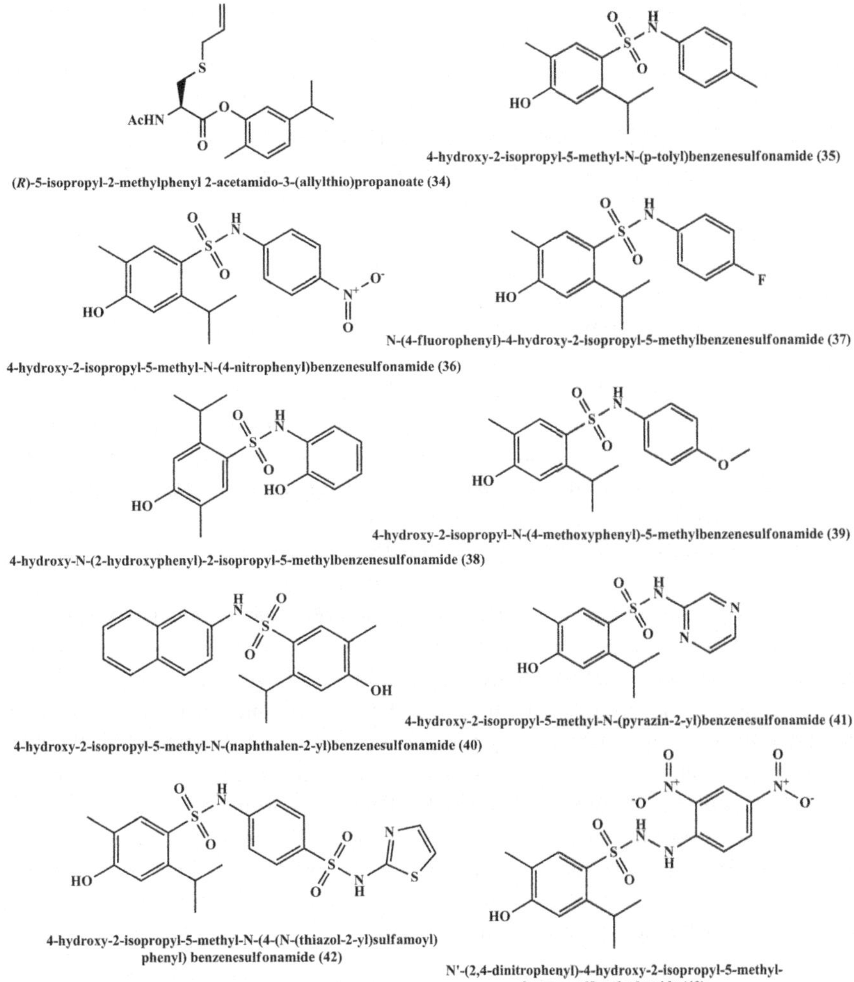


**Fig. 13.5** Chemical structures of some carvacrol derivatives (34-43)

More scientific evaluation of the carvacrol derivatives was required before using in agriculture, food, and medicine as most of the reported documents revealed in vitro results.

## 13.12 Conclusion and Future Perspectives

Cancer is a dangerous health risk that is present worldwide. The morbidity and the mortality rates connected with cancer are alarming, despite the existence of multiple treatment modalities for patients suffering from cancer. Natural substances are important components that can be used for the discovery and the progression of new anticancer medications. Since natural components have no or negligible adverse effects, the utilization of these substances to develop novel anticancer agents provide a promising choice for chemoprevention and novel cancer treatment. The current therapeutic intervention for the management of cancer poses several limitations due to the use of mono-targeted synthetic agents, elevated cost, low effectiveness, and dangerous adverse actions. Consequently, it is necessary to develop novel and innovative medications for the prevention and treatment of cancer. Carvacrol is known to prevent the development of many cancers as well as suppress the growth of cancer cells.

Various signaling pathways were implicated in the proliferation, differentiation, apoptosis evasion, and survival of neoplastic cells. So these pathway activation results in cancer-promoting mechanisms. This review indicates the capacity of phytochemical carvacrol to act as multi-targeted agents to impede cancer cell growth. Carvacrol's cytotoxic actions toward specific cancer by inhibition of cell proliferation, and its antimitotic behavior induced apoptosis and inhibition of movement, invasion, or metastasis of cancer cells.

Further research studies need to be directed in vivo to disclose the long-term impacts of carvacrol usage, the cause of the various pathway inhibition, the impact of carvacrol to prevent cancer in high-risk populations, and its effects when used in combination with existing chemotherapy and when used in conjunction with diversified phytoconstituents. An impressive amount of research findings presented here establishes the promise of this phytochemical as anticancer agents and the necessity for the implementation of this phytochemical into human clinical trials.

Chemopreventive carvacrol responsible for the inhibition of various pathways has been discussed throughout this article. Thus, a coordinated effort to discover new targets to boost the capability to restore the normal signaling process will result in the future chemopreventive and anticancer drugs. The findings of the studies presented in this review may enable researchers to create novel and effective cancer prevention and therapeutic approaches.

## References

Abbaszadeh S, Sharifzadeh A, Shokri H, Khosravi AR, Abbaszadeh A (2014) Antifungal efficacy of thymol, carvacrol, eugenol and menthol as alternative agents to control the growth of food-relevant fungi. J. de. Mycologie Médicale. 24:e51–e56

Adam K, Sivropoulou A, Kokkini S, Lanaras T, Arsenakis M (1998) Antifungal activities of *Origanum vulgare* sub sp. *hirtum, Menthaspicata, Lavandula angustifolia* and *Salvia fruticosa* essential oils against human pathogenic fungi. J Agric Food Chem 46:1739–1745

Aeschbach R, Löliger J, Scott BC, Murcia A, Butler J, Halliwell B, Aruoma OI (1994) Antioxidant actions of thymol, carvacrol, 6-gingerol, zingerone and hydroxytyrosol. Food Chem Toxicol 32:31–36

Alagawany M (2015) Biological effects and modes of action of carvacrol in animal and poultry production and health - A review. Adv Anim Vet Sci 3:73–84

Al-Fatlawi AA, Ahmad RA (2014) Cytotoxicity and pro-apoptotic activity of carvacrol on human breast cancer cell line MCF-7, World. J Pharm Sci 2:1218–1223

Almanaitytė M, Jurevičius J, Mačianskienė R (2020) Effect of carvacrol, TRP channels modulator, on cardiac electrical activity. Biomed Res Int 6:6456805

Alokam R, Jeankumar VU, Sridevi JP, Matikonda SS, Peddi S, Alvala M, Yogeeswari P, Sriram D (2014) Identification and structure-activity relationship study of carvacrol derivatives as Mycobacterium tuberculosis chorismate mutase inhibitors. J Enzyme Inhib Med Chem 29 (4):547–554

Andersen A (2006) Final report on the safety assessment of sodium *p*-chlorom-cresol, *p*-chloro-*m*-cresol, chlorothymol, mixed cresols, *m*-cresol, *o*-cresol, *p*-cresol, isopropyl cresols, thymol, *o*-cymen-5-ol, and carvacrol. Int J Toxicol. 25(Suppl 1):29–127

Andre WPP, Ribeiro WLC, Cavalcante GS, dos Santos JML, Macedo ITF, de Paula HCB, de Freitas RM, de Morais SM, de Melo JV, Bevilaqua CML (2016) Comparative efficacy and toxic effects of carvacryl acetate and carvacrol on sheep gastrointestinal nematodes and mice. Vet Parasitol 218:52–58

Aristatile B, Al-Numair KS, Veeramani C, Pugalendi KV (2009) Effect of carvacrol on hepatic marker enzymes and antioxidant status in d-galactosamine-induced hepatotoxicity in rats. Fundam Clin Pharmacol 23:757–765

Arunasree KM (2010) Anti-proliferative effects of carvacrol on a human metastatic breast cancer cell line, MDA-MB 231. Phytomedicine 17:581–588

Austgulen LT, Solheim E, Scheline RR (1987) Metabolism in rats of *p*-cymene derivatives: Carvacrol and thymol. Pharmacol Toxicol 61:98–102

Aydin E, Türkez H, Keleş MS (2014) The effect of carvacrol on healthy neurons and N2a cancer cells: some biochemical, anticancerogenicity and genotoxicity studies. Cytotechnology 66:149–157

Bahmani M, Taherikalani M, Khaksarian M, Soroush S, Ashrafi B, Heydari R (2019) Phytochemical profiles and antibacterial activities of hydroalcoholic extracts of *Origanum vulgare* and *Hypericum perforatum* and carvacrol and hypericin as a promising Anti-*Staphylococcus aureus*. Int J Pharm 19:923–932

Bassanetti I, Carcelli M, Buschini A, Montalbano S, Leonardi G, Pelagatti P, Tosi G, Massi P, Fiorentini L, Rogolino D (2017) Investigation of antibacterial activity of new classes of essential oils derivatives. Food Control 73:606–612

Bayir AG, Kiziltan HS, Kocyigit A (2019) Chapter 1-Plant family, carvacrol, and putative protection in gastric cancer, Diatary interventions in Gastrointestinal diseases, Foods, Nutrients, and Dietary Supplements, 3–18

Ben Arfa A, Combes S, Preziosi-Belloy L, Gontard N, Chalier P (2006) Antimicrobial activity of carvacrol related to its chemical structure. Lett Appl Microbiol 43:149–154

Bhakkiyalakshmi E, Suganya N, Sireesh D, Krishnamurthi K, Saravana DS, Rajaguru P, Ramkumar KM (2016) Carvacrol induces mitochondria-mediated apoptosis in HL-60 promyelocytic and Jurkat T lymphoma cells. Eur J Pharmacol 772:92–98

Budavari S (ed.1989) The Merck Index. An Encyclopedia of Chemicals, Drugs, and Biologicals, 11th ed. Rahway, NJ: Merck & Co

Burke CC, Wildung MR, Croteau R (1999) Geranyl diphosphate synthase: Cloning, expression, and characterization of this prenyltransferase as a heterodimer. Proc Natl Acad Sci USA 9:13062–13067

Burt S (2004) Essential oils: Their antibacterial properties and potential applications in foods -A review. Int J Food Microbiol 94:223–253

Cacciatore I, Di Giulio M, Fornasari E, Stefano AD, Cerasa LS, Marinelli L, Turkez H, Campli ED, Bartolomeo SD, Robuffo I, Cellini L (2015) Carvacrol codrugs: a new approach in the antimicrobial plan. PLoS ONE 10(4):e0120937

Cáceres M, Hidalgo W, Stashenko E, Torres R, Ortiz C (2020) Essential oils of aromatic plants with antibacterial, anti-biofilm and anti-quorum sensing activities against pathogenic bacteria. Antibiotics 9:147

Can BK (2008) Biological and pharmacological activities of carvacrol and carvacrol bearing essential oils. Curr Pharm Des 14:3106–3120

Chami N, Chami F, Bennis S, Trouillas J, Remmal A (2004) Antifungal treatment with carvacrol and eugenol of oral candidiasis in immune-suppressed rats. Braz. J. Infect. Dis. 8:217–226

Chraibi M, Farah A, Elamin O, Iraqui HM, Fikri-Benbrahim K (2020) Characterization, antioxidant, antimycobacterial, antimicrobial effcts of Moroccan rosemary essential oil, and its synergistic antimicrobial potential with carvacrol. J Adv Pharm Technol Res. 11(1):25–29

Cox SD, Markham JL (2007) Susceptibility and intrinsic tolerance of *Pseudomonas aeruginosa* to selected plant volatile compounds. J Appl Microbiol 103:930–936

Crocoll C, Asbach J, Novak J, Gershenzon J, Degenhardt J (2010) Terpene synthases of oregano (*Origanum vulgare* L.) and their roles in the pathway and regulation of terpene biosynthesis. Plant Mol Biol 73:587–603

Crocoll C (2011) Biosynthesis of the phenolic monoterpenes, thymol and carvacrol, by terpene synthases and cytochrome P450s in oregano and thyme, PhD Dissertation. Germany

Dai W, Sun C, Huang S, Zhou Q (2016) Carvacrol suppresses proliferation and invasion in human oral squamous cell carcinoma. OncoTargets and Therapy. 9:2297–2304

Deng W, Lu H, Teng J (2013) Carvacrol attenuates diabetes-associated cognitive deficits in rats. J Mol Neurosci 51(3):813–9

De Oliveira AS, Llanes LC, Brighente IMC, Nunes RJ, Yunes RA, Junior NM, Baumgart AMK, Aust AN, Cruz AB (2016) New sulfonamides derived from carvacrol: compounds with high antibacterial activity against resistant *Staphylococcus aureus* strains. J Biosci Med. 4:105–114

Dhifi W, Bellili S, Jazi S, Bahloul N, Mnif W (2016) Essential oils' chemical characterization and investigation of some biological activities: A critical review. Medicines (Basel). 3(4):25

Dong RH, Fang ZZ, Zhu LL, Ge GB, Yang L, Liu ZY (2012) Identification of UDP-glucuronosyltransferase isoforms involved in hepatic and intestinal glucuronidation of phytochemical carvacrol. Xenobiotica 42:1009–1016

Dudareva N, Andersson S, Orlova I, Gatto N, Reichelt M, Rhodes D, Boland W, Gershenzon J (2005) The nonmevalonate pathway supports both monoterpene and sesquiterpene formation in snapdragon flowers. Proc Natl Acad Sci USA 102:933–938

Elbe H, Yigitturk G, Cavusoglu T, Baygar T, OzgulOnal M, Ozturk F (2020) Comparison of ultrastructural changes and the anticarcinogenic effects of thymol and carvacrol on ovarian cancer cells: which is more effective? Ultrastruct Pathol.1–10

European Commission (2012) Commission Implimenting Regulation (EU) No 872/2012 of 1 October 2012 adopting the list of flavouring substances provided for by regulation (EC) No 2232/96 of the European parliament and of the council, introducing it in Annex I to Regulation (EC) No 1334. Off J Eur Union L 267:161

European Directorate for the Quality of Medicines & HealthCare of the Council (2016) *Guidance on Essential Oils in Cosmetic Products*. Springer, New York, NY

Fan K, Li X, Cao Y, Qi H, Li L, Zhang Q, Sun H (2015) Carvacrol inhibits proliferation and induces apoptosis in human colon cancer cells. Anticancer Drugs 26:813–823

Fang S, Zhou Q, Hu Y, Liu F, Mei J, Xie J (2019) Antimicrobial carvacrol incorporated in flaxseed gum-sodium alginate active films to improve the quality attributes of Chinese sea bass (*Lateolabrax maculatus*) during cold storage. Molecules 24:3–17

FDA (2017) CFR-Code of federal regulations title 21 3: 25–26

Friedman M (2014) Chemistry and multibeneficial bioactivities of carvacrol (4-Isopropyl-2 methylphenol), a component of essential oils produced by aromatic plants and spices. J Agric Food Chem 62(31):7652–7670

Friedman M, Buick R, Elliott CT (2004) Antibacterial activities of naturally occurring compounds against antibiotic resistant *Bacillus cereus* vegetative cells and spores, *Escherichia coli*, and *Staphylococcus aureus*. J Food Prot 67:1774–1778

Friedman M, Henika PR, Mandrell RE (2002) Bactericidal activities of plant essential oils and some of their isolated constituents against *Campylobacter jejuni*, *Escherichia coli*, *Listeria monocytogenes*, and *Salmonella enterica*. J Food Prot 65:1545–1560

Galehassadi M, Rezaii E, Najavand S, Mahkam M, Mohammadzadeh Gheshlaghi N (2014) Isolation of carvacol from *Origanum vulgare*, synthesis of some organosilicon derivatives, and investigating of its antioxidant, antibacterial activities. Standard Scientific Research and Essays 2(9):438–450

Gholijani N, Gharagozloo M, Kalantar F, Ramezani A, Amirghofran Z (2015) Modulation of cytokine production and transcription factors activities in human Jurkat T cells by thymol and carvacrol. Adv. Pharm. Bull. 5:653–660

Ghorani V, Boskabady M, Boskabady MH (2019) Effect of carvacrol on pulmonary function tests, and total and differential white blood cell counts in healthy volunteers: A randomized clinical trial. Avicenna J Phytomed. 9(2):134–142

Günes-Bayir A, Kiziltan HS, Kocyigit A, Güler EM, Karataş E, Toprak A (2017) Effects of natural phenolic compound carvacrol on the human gastric adenocarcinoma (AGS) cells *in vitro*. Anticancer Drugs 28:522–530

Günes-Bayir A, Kocyigit A, Güler EM, Bilgin MG, Ergün İS, Dadak A (2018a) Effects of carvacrol on human fibroblast (WS-1) and gastric adenocarcinoma (AGS) cells *in vitro* and on Wistar rats *in vivo*. Mol Cell Biochem 448:237249

Günes-Bayir A, Kocyigit A, Güler EM (2018b) *In vitro* effects of two major phenolic compounds from the family *Lamiaceae* plants on the human gastric carcinoma cells. Toxicol Ind Health 34:525–539

Han F, Ma GQ, Yang M, Yan L, Xiong W, Shu JC, Zhao ZD, Xu HL (2017) Chemical composition and antioxidant activities of essential oils from different parts of the oregano. J Zhejiang Univ Sci B. 18(1):79–84

Hayatgheib N, Fournel C, Calvez S, Pouliquen H, Moreau E (2020) *In vitro* antimicrobial effect of various commercial essential oils and their chemical constituents on *Aeromonas salmonicida* subsp. Salmonicida. J Appli Microbio 129:137–145

He L, Mo H, Hadisusilo S, Qureshi AA, Elson CE (1997) Isoprenoids suppress the growth of murine B16 melanomas *in vitro* and *in vivo*. J Nutr 127:668–674

Jamroz D, Wertelecki T, Houszka M, Kamel C (2006) Influence of diet type on the inclusion of plant origin active substances on morphological and histochemical characteristics of the stomach and jejunum walls in chicken. J Anim Physiol Anim Nutr 90:255–268

Jenner PM, Hagan EC, Taylor JM, Cook EL, Fitzhugh OG (1964) Food flavourings and compounds of related structure I. Acute oral toxicity, Food and Cosm Toxico. 2:327–343

Jennings W, Shibamoto T (1980) Qualitative analysis of flavor and fragrance volatiles by glass capillary gas chromatography. Academic Press, Cambridge, MA

Karimian P, Kavoosi G, Saharkhiz MJ (2011) Antioxidant, nitric oxide scavenging and malondialdehyde scavenging activities of essential oils from different chemotypes of *Zataria multiflora*, Nat. Prod. Res. 1–4

Khan F, Khan I, Farooqui A, Ansari IA (2017) Carvacrol induces reactive oxygen species (ROS)-mediated apoptosis along with cell cycle arrest at $G_0$ /$G_1$ in human prostate cancer cells. Nutr Cancer 69:1075–1087

Khan I, Bahuguna A, Shukla S, Aziz F, Chauhan AK, Ansari MB, Bajpai VK, Huh YS, Kang SC (2020) Antimicrobial potential of the food-grade additive carvacrol against uropathogenic *E. coli* based on membrane depolarization, reactive oxygen species generation, and molecular docking analysis, Microb Pathog. 142:104046

Kirimer N, Baer KHC, Tiimen G (1995) Carvacrol-rich plants in turkey. Chem. Nat. Comp. 31:49–54

Knowles J, Roller S (2001) Efficacy of chitosan, carvacrol, and a hydrogen peroxide based biocide against food borne microorganisms in suspension and adhered to stainless steel. J Food Prot 64:1542–1548

Koparal AT, Zeytinoglu M (2003) Effects of carvacrol on a human non-small cell lung cancer (NSCLC) cell line, A549. Cytotechnology 43:149–154

Lambert RJW, Skandamis PN, Coote PJ, Nychas GJE (2001) A study of the minimum inhibitory concentration and mode of action of oregano essential oil, thymol and carvacrol. J Appl Microbiol 91:453–462

Landa P, Kokoska L, Pribylova M, Vanek T, Marsik P (2009) *In vitro* anti-inflammatory activity of carvacrol: Inhibitory effect on COX-2 catalyzed prostaglandin E2 biosynthesisb. Arch. Pharm. Res. 32:75–78

Laothaweerungsawat N, Neimkhum W, Anuchapreeda S, Sirithunyalug J, Chaiyana W (2019) Transdermal delivery enhancement of carvacrol from *Origanum vulgare* L. essential oil by microemulsion, Mini Rev Med Chem. 19(11): 923–932

Lichtenthaler HK (1999) The 1-deoxy-D-xylulose-5-phosphate pathway of isoprenoid biosynthesis in plants. Annu Rev Plant Physiol Plant Mol Biol 50:47–65

Lim W, Ham J, Bazer FW, Song G (2019) Carvacrol induces mitochondria-mediated apoptosis via disruption of calcium homeostasis in human choriocarcinoma cells. J Cell Physiol 234:1803–1815

Liolios CC, Gortzi O, Lalas S, Tsaknis J, Chinou I (2009) Liposomal incorporation of carvacrol and thymol isolated from the essential oil of *Origanum dictamnus* L. and *in vitro* antimicrobial activity. Food Chem 112:77–83

Livingston AE (1921) The comparative toxicity of thymol and carvacrol (isothymol). J. Pharmacol. 17:261–275

Llana-Ruiz-Cabello M, Gutiérrez-Praena D, Puerto M, Pichardo S, Jos Á, Cameán AM (2015) *In vitro* pro-oxidant/antioxidant role of carvacrol, thymol and their mixture in the intestinal Caco-2 cell line. Toxicol In Vitro 29:647–656

Luo Y, Wu JY, Lu MH, Shi Z, Na N, Di JM (2016) Carvacrol alleviates prostate cancer cell proliferation, migration, and invasion through regulation of PI3K/Akt and MAPK signaling pathways, Oxid. Med. Cell. Longev. 1–11

Lupo JAT, Nakatsu T, Caldwell J, Kang RKL, Cilia AT, Van Loveren AG, Villamaria L (2000) Substituted phenols as fragrance, flavor and antimicrobial compounds.US Patent no. US6110888

Majdi M, Mashhady AM, Maroufi A, Crocoll C (2017) Tissue-specific gene-expression patterns of genes associated with thymol/carvacrol biosynthesis in thyme (*Thymus vulgaris* L.) and their differential changes upon treatment with abiotic elicitors. Plant Physiol Biochem 115:152–162

Marinelli L, Di Stefano A, Cacciatore I (2018) Carvacrol and its derivatives as antibacterial agents. Phytochem Rev 17(4):903–921

Marino A, Nostro A, Mandras N, Roana J, Ginestra G, Miceli N, Taviano MF, Gelmini F, Beretta G, Tullio V (2020) Evaluation of antimicrobial activity of the hydrolate of *Coridothymus capitatus* (L.) Reichenb. fil. (Lamiaceae) alone and in combination with antimicrobial agents. BMC Complement. Med. Ther. 20:1–11

Mathela CS, Singh KK, Gupta VK (2010) Synthesis and *in vitro* antibacterial activity of thymol and carvacrol derivatives. Acta. Pol. Pharm. Drug Res. 67:375–380

Mayr E (1961) Cause and Effect in Biology. Science 134:1501–1506

Mehdi SJ, Ahmad A, Irshad M, Manzoor N, Rizvi MMA (2011) Cytotoxic effect of carvacrol on human cervical cancer cells. Bio and Med. 3:307–312

Mehrjerdi ZF, Niknazar S, Yadegari M, Akbari FA, Pirmoradi Z, Khaksari M (2020) Carvacrol reduces hippocampal cell death and improves learning and memory deficits following lead-induced neurotoxicity via antioxidant activity. In press, Naunyn-Schmiedeberg's Archives of Pharmacology

Melušová M, Jantová S, Horváthová E (2014) Carvacrol and rosemary oil at higher concentrations induce apoptosis in human hepatoma HepG2 cells. Interdiscip. Toxicol. 7:189–194

Menniti AM, Gregori R, Neri F (2010) Activity of natural compounds on *Fusarium verticillioides* and fumonisin production in stored maize kernels. Int J Food Microbiol 136:304–309

Michiels J, Missotten J, Dierick N, Fremaut D, Maene P, De Smet S (2008) *In vitro* degradation and *in vivo* passage kinetics of carvacrol, thymol, eugenol and *trans*-cinnamaldehyde along the gastrointestinal tract of piglets. J Sci Food Agric 88:2371–2381

Mohammad YM, Parisa R, Naser A, Aghdam A, Kafil M, Samadi H (2017) Carvacrol and thymol: strong antimicrobial agents against resistant isolates. Rev. Med. Microbiol. 28(2):63–68

Nikumbh VP, Tare VS, Mahulikar PP (2003) Eco-friendly pest management using monoterpenoids-III: antibacterial efficacy of carvacrol derivatives. J Sci Ind Res 62:1086–1089

ÖkeAltuntaş F, DemiRtaş I (2017) Real-time cell analysis of the cytotoxicity of *Origanum acutidens* essential oil on HT-29 and HeLa cell lines. Turk J Pharm Sci 14:29–33

Ozkan A, Erdogan A (2012) A comparative study of the antioxidant/prooxidant effects of carvacrol and thymol at various concentrations on membrane and DNA of parental and drug resistant H1299 cells. Nat Prod Commun 7:1557–1560

Pakdemirli̇ A, Karaca C, Sever T, Daskin E, LeblebiCi̇ A, YiĞiTbaşi T, Başbinar Y (2020) Carvacrol alters soluble factors in HCT-116 and HT-29 cell lines. Turk. J. Med. Sci. 50:271–276

Patil JU, Suryawanshi KC, Patil PB, Chaudhary SR, Pawar NS (2010) Synthesis and antibacterial activity of carvacryl ethers. J Asian Nat Prod Res 12:129–133

Pilau MR, Alves SH, Weiblen R, Arenhart S, Cueto AP, Lovato LT (2011) Antiviral activity of the *Lippia graveolens* (*Mexican oregano*) essential oil and its main compound carvacrol against human and animal viruses. Braz J Microbiol. 42(4):1616–1624

Potočnjak I, Gobin I, Domitrović R (2018) Carvacrol induces cytotoxicity in human cervical cancer cells but causes cisplatin resistance: Involvement of MEK-ERK activation. Phytother. Res. 32:1090–1097

Ramak P, Kazempour Osaloo S, Ebrahimzadeh H, Sharifi M, Behmanesh M (2013) Inhibition of the mevalonate pathway enhances carvacrol biosynthesis and DXR gene expression in shoot cultures of *Satureja khuzistanica* Jamzad. J Plant Physiol 170:1187–1193

Rattanachaikunsopon P, Phumkhachorn P (2010) Assessment of factors influencing antimicrobial activity of carvacrol and cymene against *Vibrio cholerae* in food. J Biosci Bioeng 110:614–619

Ravishankar S, Zhu L, Reyna-Granados J, Law B, Joens L, Friedman M (2010) Carvacrol and cinnamaldehyde inactivate antibiotic-resistant *Salmonella enterica* in buffer and on celery and oysters. J Food Prot 73:234–240

Report Final (2016) The Safety Assessment of Sodium *p*-chloro- *m*-cresol, *p*-chloro- *m*-cresol, chlorothymol, mixed cresols, *m*-Cresol, *o*-cresol, *p*-cresol, isopropyl cresols, thymol, *o*-cymen-5-ol, and carvacrol. Int. J. Toxicol. 25:29–127

Rohmer M, Knani M, Simonin P, Sutter B, Sahm H (1993) Isoprenoid biosynthesis in bacteria: a novel pathway for the early steps leading to isopentenyldiphosphate. Biochem J 295:517–524

Samarghandian S, Farkhondeh T, Samini F, Borji A (2016) Protective effects of carvacrol against oxidative stress induced by chronic stress in rat's brain, liver, and kidney. Biochem Res Int. 2016:1–7

Sharifi-Rad M, Varoni EM, Iriti M, Martorell M, Setzer WN, del Mar Contreras M, Salehi B, Soltani-Nejad A, Rajabi S, Tajbakhsh M, Sharifi-Rad J (2018) Carvacrol and human health: A comprehensive review, Phyto Res. 1–13

Shimoda K, Kondo Y, Nishida T, Hamada H, Nakajima N, Hamada H (2006) Biotransformation of thymol, carvacrol, and eugenol by cultured cells of *Eucalyptus perriniana*. Phytochemistry 67:2256–2261

Sim JXF, Khazandi M, Chan WY, Trott DJ, Deo P (2019) Antimicrobial activity of thyme oil, oregano oil, thymol and carvacrol against sensitive and resistant microbial isolates from dogs with *Otitis externa*. Vet Dermatol 30:524

Slamenova D, Kozics K, Hunakova L, Melusova M, Navarova J, Horvathova E (2013) Comparison of biological processes induced in HepG2 cells by tert-butyl hydroperoxide (t-BHP) and hydroperoxide($H_2O_2$): The influence of carvacrol. Mutat. Res. Genet. Toxicol. Environ. 757:15–22

Takahashi S, Kuzuyama T, Watanabe H, Seto H (1998) A 1-deoxy-D-xylulose 5-phosphate reductoisomerase catalyzing the formation of 2-*C*-methyl-D-erythritol 4-phosphate in an alternative nonmevalonate pathway for terpenoid biosynthesis. Proc Natl Acad Sci USA 95:9879–9884

Teissedre PL, Waterhouse AL (2008) Inhibition of oxidation of human low-density lipoproteins by phenolic substances in different essential oils varieties. J Agric Food Chem 48:3801–3805

Trindade GGG, Thrivikraman G, Menezes PP, França CM, Lima BS, Carvalho YMBG, Souza EPBSS, Duarte MC, Shanmugam S, Quintans-Júnior LJ, Bezerra DP, Bertassoni LE, Araújo AAS (2019) Carvacrol/β-cyclodextrin inclusion complex inhibits cell proliferation and migration of prostate cancer cells, Food Chem. Toxicol. 125:198–209

Tsao R, Zhou T (2000) Antifungal activity of monoterpenoids against postharvest pathogens *Botrytis cinerea* and *Monilinia fructicola*. J Essent Oil Res 12:113–121

Ultee A, Bennik MHJ, Moezelaar R (2002) The phenolic hydroxyl group of carvacrol is essential for action against the food-borne pathogen *Bacillus cereus* the phenolic hydroxyl group of carvacrol is essential for action against the food-borne pathogen *Bacillus cereus*. Appl Environ Microbiol 68:1561–1568

Ündeğer Ü, Başaran A, Degen GH, Başaran N (2009) Antioxidant activities of major thyme ingredients and lack of (oxidative) DNA damage in V79 Chinese hamster lung fibroblast cells at low levels of carvacrol and thymol, Food Chem. Toxicol. 47:2037–2043

Viana GSB, Matos FF, Araujo WL, Matos FJA, Craveiro AA (1981) Essential oil of *Lippia grata*: Pharmacological effects and main constituents. Int J Crude Drug Res 19:1–10

Wang P, Luo Q, Qiao H, Ding H, Cao Y, Yu J, Liu R, Zhang Q, Zhu H, Qu L (2017) The neuroprotective effects of carvacrol on ethanol-induced hippocampal neurons impairment via the antioxidative and antiapoptotic pathways. Oxid Med Cell Longev. 2017:4079425

Yadav GD, Kamble SB (2009) Synthesis of carvacrol by Friedel- Crafts alkylation of o-cresol with isopropanol using superacidic catalyst UDCaT-5. J Chem Technol Biotechnol 84:1499–1508

Yanishlieva NV, Marinova EM, Gordon MH, Raneva VG (1999) Antioxidant activity and mechanism of action of thymol and carvacrol in two lipid systems, Food Chem. 64: 59-66

Yin Q, Yan F, Zu XY, Wu Y, Wu X, Liao M, Deng S, Yin L, Zhuang Y (2012) Anti-proliferative and pro-apoptotic effect of carvacrol on human hepatocellular carcinoma cell line HepG-2. Cytotechnology 64:43–51

# Chapter 14
# Pharmaceutical Application of Bio-actives from *Alstonia* Genus: Current Findings and Future Directions

**Atish T. Paul, Ginson George, Nisha Yadav, Arjun Jeswani, and Prashant S. Auti**

**Abstract** Genus *Alstonia* is comprised of around 155 species throughout the world. Phytochemical screening of *Alstonia* genus has demonstrated the presence of diverse phytochemicals that include at least 800 distinct metabolites. The main classes of metabolites are alkaloids, iridoids, flavonoids, fatty acids, etc. Secondary metabolites from this genus also bring positive results when employed as anti-cancer, anti-spasmodic, antitussive, antiarthritic, antioxidant, etc. *Alstonia boonei* is listed as an antimalarial drug in African pharmacopoeia. In China, a formulation consists of *A. scholaris* leaves—*Dengtaiye* tablets are used for the treatment of cough and fever symptoms. Recently, *A. scholaris* leaves-derived indole alkaloids have been registered as an investigational new botanical drug (No. 2011L01436) and China Food and Drug Administration (CFDA) has approved its phase I/II clinical trials for the treatment of respiratory diseases. Pharmacokinetic and safety analysis of this botanical drug in healthy human subjects have revealed a safe profile under the dose regimen experiment. Apart from that, strictamine, an indole alkaloid isolated from *A. scholaris* exhibited a similar in vitro antiviral activity to that of acyclovir. The present chapter is mainly focused on the critical analysis of various secondary metabolites isolated from *Alstonia* (approximately 15 species). Apart from that, pharmacological, toxicological and intellectual property rights studies have also been included.

**Keywords** Genus *alstonia* · Secondary metabolites · Pharmacological activities · Pharmacokinetic studies · IPR

---

A. T. Paul (✉) · G. George · N. Yadav · A. Jeswani · P. S. Auti
Laboratory of Natural Product Chemistry, Department of Pharmacy, Birla Institute of Technology and Science Pilani (BITS Pilani), Pilani Campus, Pilani 333031, Rajasthan, India
e-mail: atish.paul@pilani.bits-pilani.ac.in



D. Pal and A. K. Nayak (eds.), *Bioactive Natural Products for Pharmaceutical Applications*, Advanced Structured Materials 140,
https://doi.org/10.1007/978-3-030-54027-2_14

## 14.1 Introduction

*Alstonia* is a widespread genus of trees and shrubs in the family of Apocynaceae. This genus is mainly distributed in Central America, Himalayas, China, tropical Africa, Malesia, Philippine, and Australia. In 1810, Robert Brown (1685–1760) described the genus in honor of Charles Alston, a renowned botanist at the University of Edinburgh (1740–1760). *A. scholaris*, *A. spectabilis*, *A. venenata*, and *A. costata* were the four species referred by Robert Brown in the *Alstonia* genus (Sidiyasa 1998). According to the World Checklist of Selected Plant Families (WCSP), approximately 155 species (49 accepted species name) were identified in the *Alstonia* genus (Govaerts and Leeuwenberg 2016). As per The Plant List, 114 scientific plant species rank for the genus *Alstonia*, wherein 43 were accepted as a species name (The Plant List 2013). *Alstonia* genus is documented to possess various health benefits for traditional uses. An ancient Ayurvedic textbook 'Bhavaprakasha' has been well documented the therapeutic significance of *A. scholaris*. According to that saptaparna (vernacular name of *A. scholaris*) can cure ulcer, mitigate vaata and kapha, heal various skin diseases including leprosy, anti-parasitic activity, anti-asthmatic activities, etc. Generally, the *Alstonia* genus is traditionally used for liver and intestinal troubles, heart diseases, respiratory-related symptoms, various skin diseases, fever, and vulnerary conditions (Khyade et al. 2014). In African Pharmacopoeia, *A.booeni* is listed as an antimalarial drug (Olajide et al. 2000). Along with the development of the modern scientific era, investigation of various secondary metabolites from this genus has gained more attention for a diversified pharmacological activity. Among these endeavors, numerous classes of phytochemicals such as alkaloids, iridoids, coumarins, flavonoids, simple phenolics, steroids, and fatty acids have been identified. In addition to that various pharmacological activities of these phytochemicals were also explored. In some of the cases, identified pharmacological activities were further provided a full proof basis for their traditional uses. In China, the monoterpenoid indole alkaloidal component from the *A. scholaris* leaves was formulated as a capsule and currently, it has completed the phase I clinical trial for the respiratory-related ailments (Li et al. 2019). The obtained promising results provided additional value to the traditional uses of these plants for respiratory problems.

The present chapter is mainly focused on the critical analysis of various species of *Alstonia* (approximately 32 species). A comprehensive literature analysis of *Alstonia* genus showed the presence of approximately 800 compounds embracing alkaloids, iridoids, flavonoids, fatty acids, etc., and it exhibited a multifarious range of pharmacological activity such as anti-cancer, anti-spasmodic, antitussive, antiarthritic, and antioxidant. In the present chapter, pharmacological activities of the isolated secondary metabolites are discussed. Further, a higher number of patents have been filed for variety of applications for *Alstonia* that includes pharmaceutical applications also. In many patents, *Alstonia* is a single component in the polyherbal formulation. The *Alstonia* containing polyherbal formulation is beyond the scope of the book chapter. Hence, in this chapter, we have selected the patents, wherein the *Alstonia*

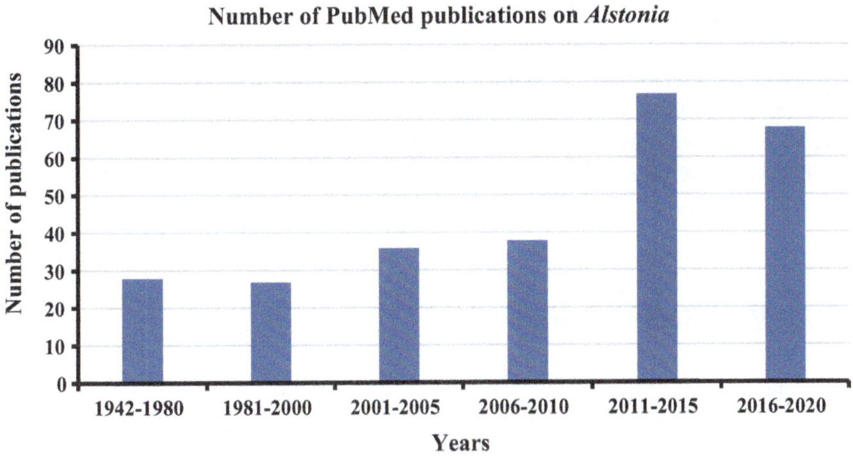


**Fig. 14.1** The number of *Alstonia*-related publications (PubMed)

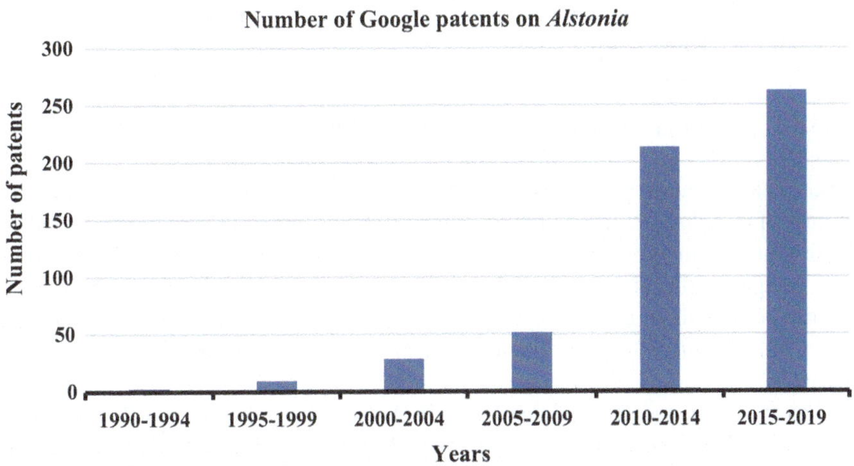


**Fig. 14.2** The number of *Alstonia*-related patents (Google patents)

genus has been claimed for the significant contribution in pharmacological activities. The increase in the amount of the publication and patents (Figs. 14.1 and 14.2) in the last few years evinced that the scientific attention toward the *Alstonia* genus is increasing in a vast manner. Hence, the present book chapter will help to obtain a better understanding of the pharmaceutical application of *Alstonia*-derived phytochemicals.

This chapter mainly deals with the following species namely *A. scholaris*, *A. macrophylla*, *A. boonei*, *A. angustifolia*, *A. venenata*, *A. yunnanesis*. *A. spatulata*, *A. rupestris*, *A. rostrata*, *A. pneumatophora*, *A. penangiana*, *A. mairei*, *A. congensis*,

*A. angustiloba*, *A. actinophylla* (Table 14.1). Other species of *Alstonia genus* such as *A. boulindaensis*, *A. constricta*, *A. coriaceae*, *A. deplanchi*, *A. glaucescenes*, *A. lanceolate*, *A. lanciolifera*, *A. lenormandii*, *A. mulleriana*, *A. odontophora*, *A. plumose*, *A. quaternata*, *A. sphaerocapitata*, *A. undalufolia*, *A. undulata*, *A. villosa*, *A. vitensis* have been reported for the isolation of various phytochemicals but lacks the evaluation of the pharmacological potencies. Hence, these species are beyond the scope of this book chapter and have been avoided for further discussions.

## 14.2 Botanical Description

See Table 14.1.

## 14.3 Phytochemistry and Pharmacological Activities

### *14.3.1 Phytochemistry and Pharmacological Activities of A. Scholaris*

#### 14.3.1.1 Anti-inflammatory and Analgesic Activities

Traditional Chinese medicines have highlighted the importance of the aqueous extract of *A. scholaris* leaves for relieving tracheitis and cold symptoms. Based on these reports, numerous attempts have been tried to identify the anti-inflammatory and analgesic effects of *A. scholaris*. Although the majority of the reports are mainly briefed about the use of the whole extract, a few groups tried to identify the effect of *A. scholaris*-derived phytochemical in exhibiting desired pharmacological activities.

Cai et al. isolated scholarisine I and II (1 and 2, Fig. 14.3) from the alkaloidal fraction of *A. scholaris* leaves and evaluated its in vitro anti-inflammatory activity (Cai et al. 2010a). 1 exhibited a non-selective inhibition toward cyclooxygenase (COX), while 2 resulted in selective inhibition of COX-II with a significant 5-lipoxygenase (5-LOX) inhibitory activity (Table 14.2).

Shang et al. reported the in vitro anti-inflammatory activities (COX-I, COX-II, and 5-LOX inhibition) of 29 alkaloids (3–31, Fig. 14.4) from the ethanolic extract of *A. scholaris* leaves (Shang et al. 2010). Alkaloids fraction resulted in a moderate anti-inflammatory potential (54–64% inhibition). Among the screened compounds, 16-Formyl-5α-methoxystrictamine (12), picralinal (19), and tubotaiwine (29) exhibited a promising anti-inflammatory activity by selectively inhibiting COX-II and 5-LOX (>90 and 70% inhibition, respectively). As represented in Table 14.3, Creabanine (6), isogentialutine (14), scholaricine (24), and strictamine (27) showed the anti-inflammatory activity by selectively inhibiting COX-II enzyme (>90%

**Table 14.1** Botanical description of various *Alstonia* species

| Species | Bark and trunk | Leaves | Flowers or inflorescences | Fruits | Seed |
|---|---|---|---|---|---|
| *A. scholaris* (Jøker 2000; Khyade et al. 2014) | Greyish-brown, lenticellate | Simple, whorled with 4–7 unequal leaves that are narrow elliptic-oblanceolate | Flowers are greenish-white, bisexual, compact, paniculate cyme, pubescent | Linear, 20–50 cm long dehiscent, follicles (two-lobed) | Oblong, hairs at both ends |
| *A. actinophylla* (Flora Malesiana DataPortal; Monachino 1949) | Greyish bark, copious, glabrous branchlets | Whorls of 3–7, elliptic or sublanceolate leaf-blades, prominulous transverse veins | Loosely cymose inflorescence, glabrous corolla and follicles, slender pedicels | Glabrous, a pair of follicles and about 8–20 cm long | Oblong, long ciliated |
| *A. angustifolia* (Monachino 1949) | Bark smooth, grey or brown, flaky or fissured | 3-whorled, stalked leaves have leathery leaf-blades that are narrowly lance-shaped | Terminal inflorescence | Follicles are 50–70 cm long | Oblong, pointed at one end, and 1 cm long |
| *A. angustiloba* (Monachino 1949; NParks Flora & Fauna Web 2019) | Dark grey, rough, fissured, and peeling off in rectangular flakes | Elliptic leaf-blades, broad, and acuminate at the apex | Puberulent inflorescences, ovate to lanceolate calyx tubes, and oblong-ligulate corolla lobes | Dehiscent, develop as a pair of pods and cylindrical follicles. | Numerous dark brown, oblong |
| *A. boonei* (Monachino 1949; Adotey et al. 2012) | Buttresses deep-fluted high and narrow | Whorls at nodes, oblanceolate, apex rounded to acuminate | Flowers are white with long pedicels, tomentose follicles | Pair of slender follicle with brown floss at either end | Seeds bearing a tuft of silky |
| *A. congensis* (Monachino 1949; Lemmens 2005) | Brown, smooth/ rough, lenticellate | Whorls of 4–6, simple and entire, leaf-blades are acuminate at the apex | Puberulent or glabrous inflorescence, glabrous ovary and follicles | Glabrous, a pair linear follicles 20–45 cm long | Brown, oblong, and flattened |

(continued)

**Table 14.1** (continued)

| Species | Bark and trunk | Leaves | Flowers or inflorescences | Fruits | Seed |
|---|---|---|---|---|---|
| *A. macrophylla* (Khyade et al. 2014) | Smooth, blackish brown bark, straight trunk, and a narrow crown | Whorls of 3–4, verticilliate, simple, penni-veined, and glabrous above | Pubescent, slender pedicels, and distinct ovaries | Glabrous, linear, and 20–50 cm long | Oblong, small hairy seeds |
| *A. mairei* (Monachino 1949) | Glabrous, lenticellate branchlets | Whorls of 3–5, oblanceolate and glabrous | Glabrous, not ciliate, cymes longer than leaves, small stigma apiculi | Linear and 5–10 cm long and follicles are distinct | Oblong, cilia at seed apex |
| *A. penangiana* (Sidiyasa 1998) | Grey or pale brown colored smooth barks | Whorls of 3–4, glabrous, obovate | Many-flowered, glabrous, and pubescent | A pair of follicles, glabrous striate | Dark brown, narrowly ovate or elliptic with long cilia |
| *A. pneumatophora* (Monachino 1949; Ferry 2016) | The smooth bark of greyish white color and milky sap | Leaves are simple, sessile or sub-sessile, coriaceous, obovate to spatulate with rounded apex | Cymose terminal inflorescences carrying numerous flowers with funnel-shaped corolla of yellowish-white color | The fruits are dehiscent follicles produced in pair, containing many oblong seeds | Glabrous, blackish brown, oblong, and round ended |
| *A. rostrata* (Monachino 1949) | Pale or yellowish-brown bark without buttresses, granular | Whorls of 3–4, glabrous, oblong or elliptic, undulate and/or sinuate leaf margins | White colored many flowers, glabrous or puberulous, syncarpous ovary | Widest solitary follicles (more than 10 mm), single follicle, glabrous | Numerous, oblong, or elliptical with long cilia |
| *A. venenata* (Monachino 1949) | Greyish-brown bark and bright yellow hard woody root | Whorls of 3–7, elliptic leaf-blades, broad lateral nerves | Ciliate calyx lobes, slender stipes, short and blunt stigma apiculi, sub-umbellar cyme | Fruits are fusiform with five petals, stalked and beaked follicles, tapering both ends | Smooth and flat with tufts |
| *A. yunnanensis* (Monachino 1949) | Shrubs or small trees with smooth or ringed branches | Whorls of 3–6, glabrous; narrowly elliptic or obovate | Many flowers, cymes puberulent, and slender pedicels | A pair of distinct follicles, glabrous, lacks stipes | Oblong or elliptic, round ended, and rough surfaces |

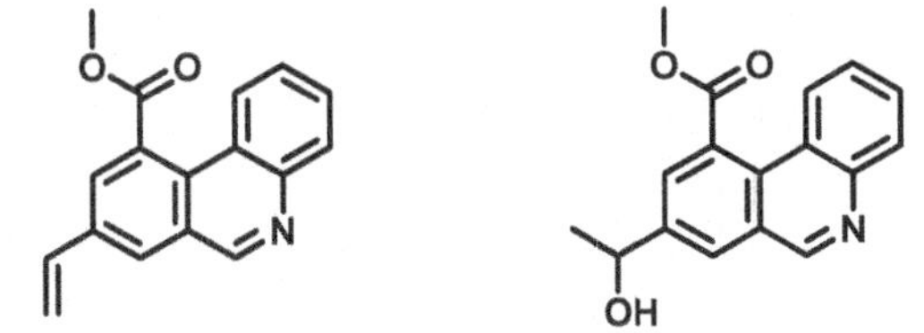

Scholarisine I (1) Scholarisine II (2)

Fig. 14.3 Chemical structures of scholarisine I and II (1 and 2)

Table 14.2 Anti-inflammatory activities of scholarisine I and II (**1 and 2**)

| Compound (at 100 μM) | COX-I inhibition (%) | COX-II inhibition (%) | 5-LOX inhibition (%) |
|---|---|---|---|
| 1 | 83.3 | 95.0 | 4.8 |
| 2 | 53.2 | 96.7 | 80.6 |
| SN-560 | 61.3 | – | – |
| NS-398 | – | 97.1 | – |
| Zileuton | – | – | 83.1 |

inhibition). Corypalmine (5), echitamine (9), and 5-methoxylstrictamine (16) did not exhibit anti-inflammatory properties (inactive). Remaining compounds exhibited a poor to moderate inhibition toward the screened enzymes (10–70% inhibition).

Along with the in vitro anti-inflammatory activity, in vivo anti-inflammatory and analgesic effects of a group of alkaloids (picrinine-20, 24 and vallesamine-30) along with the alkaloids fraction were evaluated in ICR mice. To observe the peripheral and central analgesic effects individually, acetic acid-induced writhing test and hot plate test were used, respectively. Formalin test was used for understanding both central and peripheral analgesic effects. In acetic acid (0.6% acetic acid at 0.1 mL/10 g body weight, *i.p.*)-induced writhing response test, the alkaloids fraction (100 mg/kg) exhibited lesser writhing (20.0) than the remaining fractions, while aspirin resulted in 9.3 writhing responses. Effects of three alkaloids, such as 20 (10 mg/kg), 24 (5 mg/kg), and 30 (8 mg/kg) resulted in the decreased number of writhing response in the ICR rats (20, *i.g.* and *i.p.*—21.7 and 16.8; 24 *i.g.* and *i.p.*—18.7 and 14.9; 30 *i.g.* and *i.p.*—23 and 14.7; aspirin—200 mg/kg, *i.g*—16.0). In the hot plate latent pain response test, the screened alkaloids did not increase the latency time that highlighted the absence of central analgesic effects of these alkaloids. In formalin test, hydrochloric morphine (10 mg/kg, *i.p.*) resulted in an 87.4% inhibition ratio in the licking time (s), while the screened alkaloids exhibited a poor inhibition ratio. However, in the next phase, significant nociception ratios were obtained (24–5 mg/kg, *i.p.*—62.7%; aspirin—200 mg/kg, *i.g.*—67.6%). All these results indicated that the screened alkaloids exhibited an analgesic activity through peripheral mechanisms.

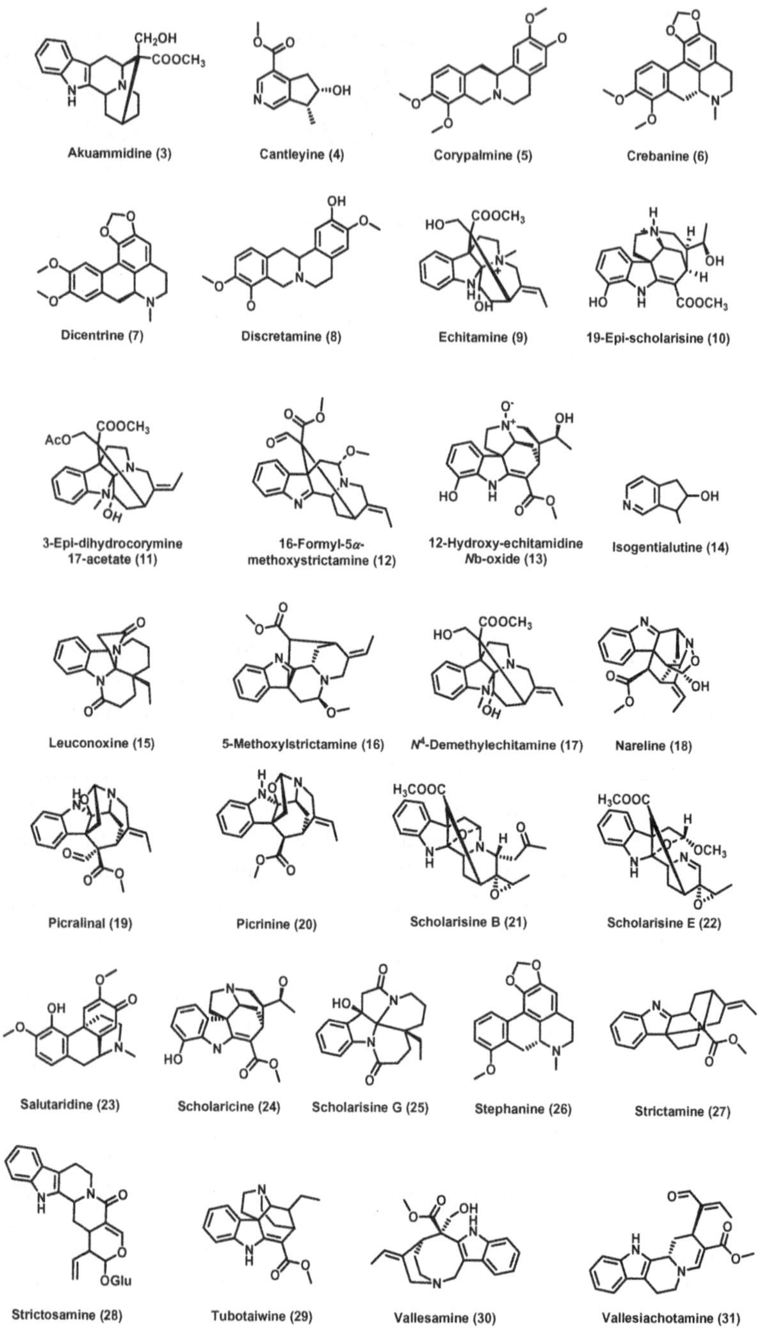


**Fig. 14.4** Chemical structures of various alkaloids from *A. scholaris*

**Table 14.3** Anti-inflammatory activities of various alkaloids from *A. scholaris*

| Compound (at 100 μM) | COX-I Inhibition (%) | COX-II Inhibition (%) | 5-LOX Inhibition (%) |
|---|---|---|---|
| Alkaloids fraction[a] | 54.3 | 63.3 | 57.3 |
| 6 | 53.3 | 90.7 | 38.4 |
| 12 | 45.8 | 95.7 | 79.9 |
| 14 | 43.1 | 95.9 | <0 |
| 19 | 44.7 | 96.4 | 79.5 |
| 24 | 44.7 | 92.0 | 31.0 |
| 25 | 38.5 | 91.1 | 57.3 |
| 27 | 47.0 | 95.6 | <0 |
| 29 | 60.8 | 90.4 | 69.5 |
| SC-560 | 61.3 | – | – |
| NS-398 | – | 97.1 | – |
| Zileuton | – | – | 83.1 |

[a]At 300 μg/mL

Further, anti-inflammatory activities of these alkaloids (20, 24, and 30) were screened by xylene-induced ear edema and the carrageenan-induced air pouch model. In xylene-induced ear edema, early-stage inflammation resulted in edematisation of the ear. Aspirin resulted in 45.7% inhibition in the edematisation. 20 at 10 mg/kg, *i.g.* and *i.p.* resulted in 41.9 and 41.5% inhibition. Although the *i.g.* administration of 24 (5 mg/kg) and 30 (8 mg/kg) did not result in good inhibition, *i. p.* administration exhibited significant inhibition on the edematisation (40.3 and 42.0%, respectively). In the carrageenan-induced air pouch model, various biochemical parameters raised from inflammatory responses were measured. Prostaglandin $E_2$ ($PGE_2$), superoxide dismutase (SOD), nitric oxide (NO), and malondialdehyde (MDA) were measured as the biochemical markers for the inflammation. The screened alkaloids lowered the MDA levels in serum, reduced NO production, and increased levels of SOD in exudate. The results indicated that these alkaloids inhibited the lipid peroxidation and served as a free radical scavenging property by the antioxidant enzymes enhancement. Further, a decreased level of $PGE_2$ in exudates suggested the ability of these alkaloids to interfere with the COX pathways of arachidonic acid metabolism.

Sultana et al. isolated ten triterpenoid derivatives (32–41) from the ethanolic extract of *A. scholaris* aerial parts (Fig. 14.5) and evaluated the anti-inflammatory potential of newly isolated phytochemical, nighascholarene (34), ursane type triterpenoid methyl ester (100 mg/kg), by carrageenan-induced paw edema test in Wistar rats (Sultana et al. 2019). Compound 34 resulted in 44% inhibition in the paw edema, while indomethacin resulted in 85% inhibition (positive control).

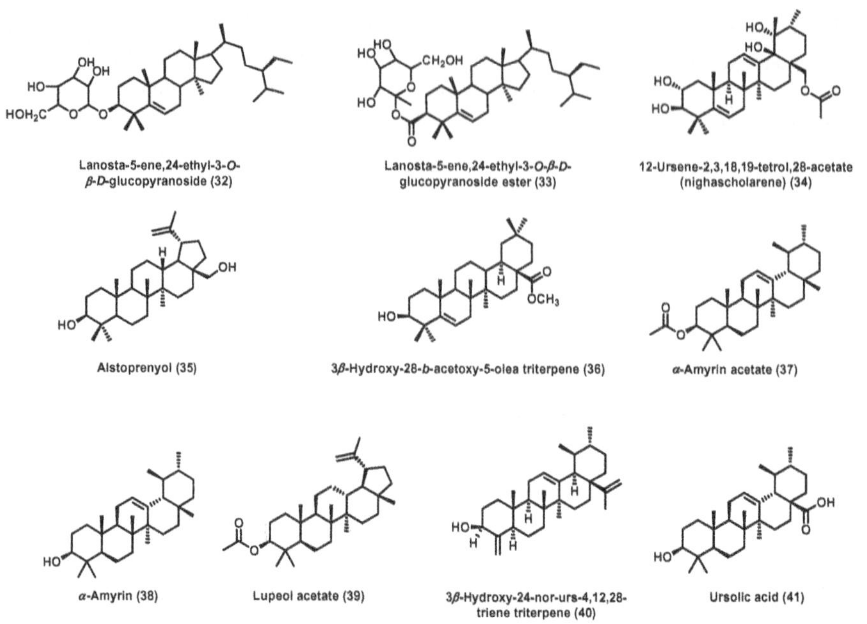


**Fig. 14.5** Chemical structures of various triterpenes from *A. scholaris* (**32–41**)

#### 14.3.1.2 Cytotoxic Activities

Jagetia et al. evaluated the cytotoxic potential of echitamine chloride (42, Fig. 14.6) in cervical carcinoma (HeLa), human liver cancer ($HepG_2$), human myeloid leukemia (HL-60), human oral epidermoid cancer (KB) and human breast cancer (MCF-7) cell lines (in vitro) and in mice bearing Ehrlich ascites carcinoma (EAC) (Jagetia et al. 2005). Dose-dependent cytotoxicity was observed in all the cell lines. Among the various cell lines, KB cells were prone more to cytotoxicity (Table 14.4). To identify the optimum doses in in vivo conditions, a varying concentration of 42 (1, 2, 4, 8,12,16 mg/kg) was screened. A 120 days survival was monitored. A dose-dependent increase in the mean survival time (MST) and the percentage increase in median life span (% IMLS) were observed. The best anti-tumor result was observed for 8 and 12 mg/kg of 42, wherein the 1.3-fold MST increased. Various biochemical parameters (glutathione reduced—GSH, MDA levels) were also evaluated. A time depended reduction in the GSH content was observed with 42 (16 mg/kg) till 3 h (2.2 $\mu$mol/$10^7$ cells), further an elevation was observed till 12 h (2.4 $\mu$mol/$10^7$ cells) after the post-treatment. In contrary, MDA levels were increased till 6 h (4.7 nmol/$10^7$ cells) and reverted to the normal condition after 12 h of pre-treatment (4.2 nmol/$10^7$ cells). From this study, it is evident that an increased lipid peroxidation was observed with the treatment of 42, while GSH concentration was reduced in the tumor cells.

**Fig. 14.6** Chemical structure of echitamine chloride (**42**)

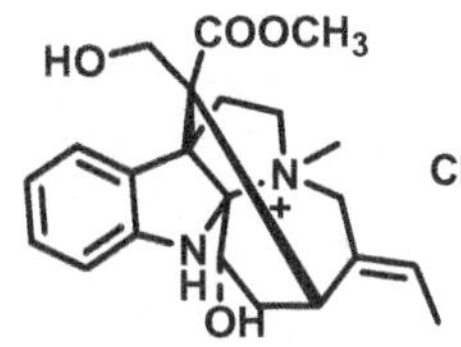


Echitamine Chloride (42)

**Table 14.4** Cytotoxicity effects of echitamine chloride (**42**) against various cell lines

| Cell lines | $IC_{50}$ of 42 (μg/mL) |
|---|---|
| KB | 10.5 |
| HeLa | 30 |
| $HepG_2$ | 50 |
| HL-60 | 51 |
| MCF-7 | 52 |

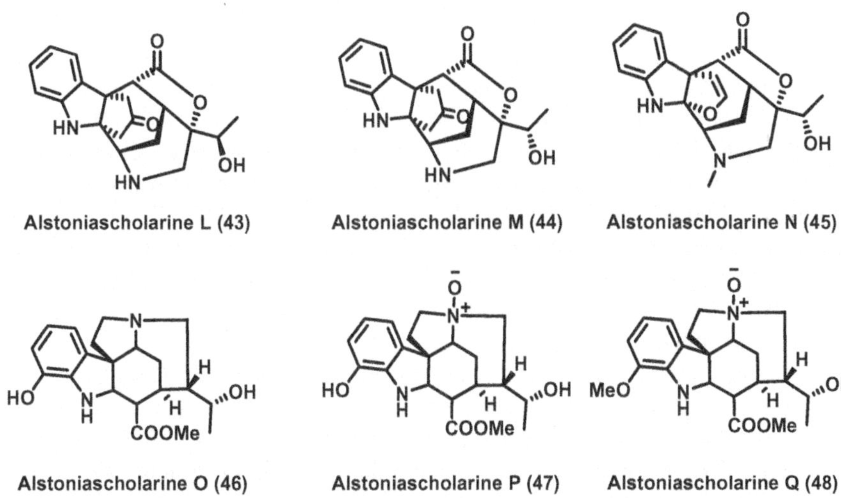


**Fig. 14.7** Chemical structures of various alkaloids from *A. scholaris* (**43–48**)

Qin et al. isolated *Alstonia*scholarines L–Q (43–48, Fig. 14.7) from the methanolic extract of *A. scholaris* inadequately dried leaves (Qin et al. 2015b). The isolated new alkaloids were screened for cytotoxicity assay in various cell lines (human colon adenocarcinoma (SW-480), human hepatocarcinoma (SMMC-7721), HL-60, MCF-7, and human lung epithelial (A-549), by using cisplatin and paclitaxel as a positive control. Further, neurotrophic activities were evaluated in PC12 cells. Unfortunately, the alkaloids neither showed cytotoxic activity nor exhibited neurite outgrowth-promoting activity.

**Fig. 14.8** Chemical structure of echitamidine-*N*-oxide-19-*O*-*β*-*D*-glucopyranoside (**49**)

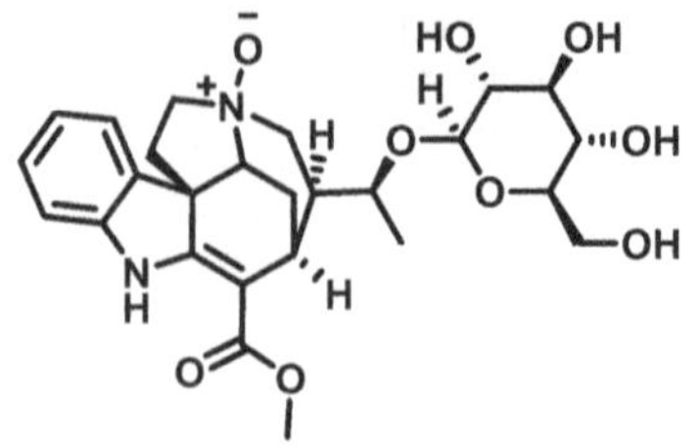

**Echitamidine-*N*-oxide-19-*O*-*β*-*D*-glucopyranoside (49)**

**Table 14.5** Cytotoxicity effects of echitamidine-*N*-oxide-19-*O*-*β*-*D*-glucopyranoside (**49**) against various cell lines

| Cell lines | Cell viability (%) at 100 μg/mL |
|---|---|
| HeLa | 28.77 |
| $HepG_2$ | 36.67 |
| KB | 29.16 |
| MCF-7 | 32.67 |
| U373MG[a] | 29.08 |

[a]U373MG—Human glioblastoma astrocytoma cell line

D. Subba Reddy isolated echitamidine-*N*-oxide-19-*O*-*β*-*D*-glucopyranoside (49, Fig. 14.8) from the methanolic extract of *A. scholaris* stem bark and evaluated the in vitro cytotoxicity in various cell lines (Subba Reddy 2016, 2017). Compound 49 caused concentration-dependent cytotoxicity and maximum effect was observed at 100 μg/mL. Further, it exhibited a moderate potential to KB, MCF-7, and U373MG cell lines (Table 14.5). In another study, the effect of pre-treatment of 49 in the radio sensitizing capacity of KB cells was evaluated. Compound 49 (20 μg/mL) pre-treated KB cells with γ-radiation reduces the clonogenicity of the cells than the individual treatment.

In another study, Wang et al. isolated 13 phytochemicals (38, 41, 50–60, Fig. 14.9) from the hexane fraction of *A. scholaris* leaves and evaluated the anti-proliferation (Wang et al. 2016) effects against non-small-cell lung carcinoma cells (NSCLC) (Wang et al. 2017). Among the isolated compounds (eight triterpenoids (38, 41, 50–55) and five sterols (56–60), terpenoids exhibited a higher potential than the sterols. In the case of isolated sterols (Table 14.6), β -sitosterol (58) resulted in a 20% reduction in the cell viability, while the remaining sterols did not inhibit the viability of A-549 cells. In the case of triterpenoids, ursolic acid (41) and betulinic acid (52) demonstrated a strong anti-NSCLC activity. 41 was reported for the activation of adenosine monophosphate (AMP)-activated protein kinase and inhibition of mammalian target of rapamycin (mTOR) pathways that control protein synthesis. Compound 52 induces the apoptosis in cancer cells via triggering the mitochondrial pathways. Remaining terpenoids exhibited moderate to poor anti-proliferation effects against NSCLC.

Kuok et al. isolated Melosline A (61), Melosline B (62), 1-[2-[2-(carboxymethyl) indole-3-yl] ethyl]-3-ethylpyridinium hydroxide inner salt (63) from the

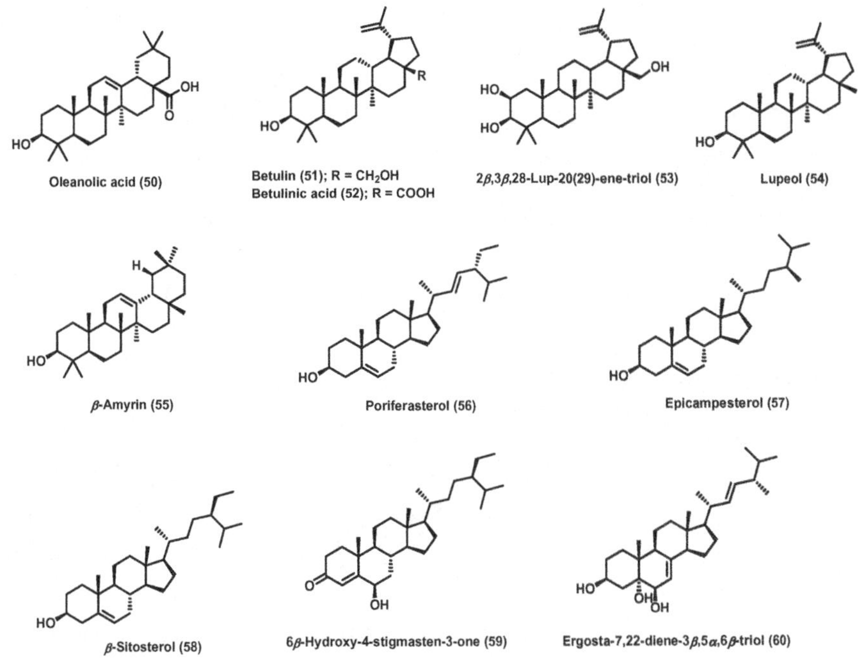


**Fig. 14.9** Chemical structures of various triterpenoids and steroids from *A. scholaris* (**50–60**)

**Table 14.6** Cytotoxicity effects of various triterpenoids from *A. scholaris* against NSCLC

| Compound | $IC_{50}$ values (μM) against A-549 NSCLC cells |
|---|---|
| 41 | 39.8 |
| 52 | 40.1 |
| 53 | 172.6 |
| 51 | 240.5 |
| 50 | >400 |
| 54 | >400 |
| 55 | >400 |
| 38 | >400 |

leaves and twigs of *A. scholaris* (Fig. 14.10) and evaluated their cytotoxicity effects against MCF-7 cell lines (Kuok et al. 2017). 61 exhibited moderate cytotoxicity ($IC_{50}$ > 39.8 μM), while the remaining compounds, 62 and 63 did not exhibit any cytotoxic activity ($IC_{50}$ > 50 μM).

Wang et al. identified the glioma stem cells (GSCs) inhibiting scholarisine Q (64) and scholarisine R (65) (nor-monoterpenoid alkaloids, Fig. 14.11) from the methanolic extract of *A. scholaris* fruits (Wang et al. 2018). As represented in Table 14.7, 64 exhibited a significant effect to that of taxol. Further cell proliferation assay of the tested compounds indicated the inhibition of GSCs proliferation.

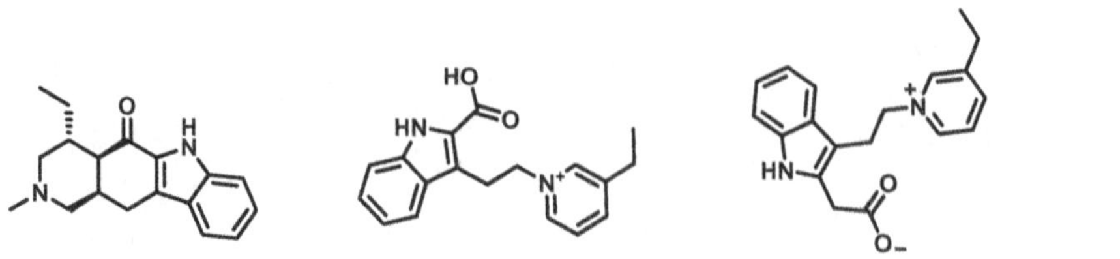

Melosline A (61) Melosline B (62) 1-[2-[2-(carboxymethyl)indole-3-yl]ethyl]-3-ethylpyridinium hydroxide (63)

**Fig. 14.10** Chemical structures of various alkaloids from *A. scholaris* (**61–63**)

**Fig. 14.11** Chemical structures of Scholarisine Q and R (**64** and **65**)

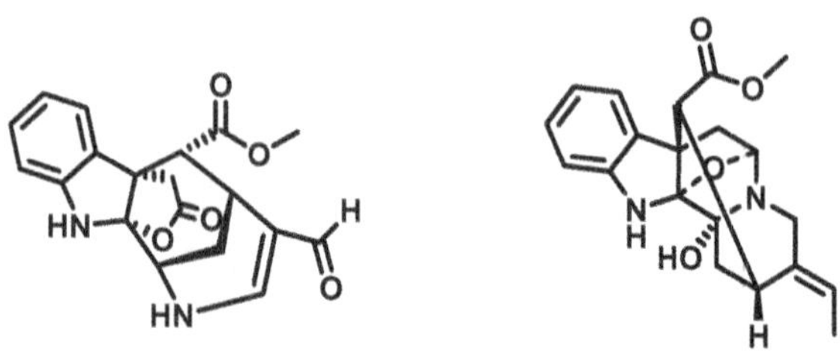

Scholarisine Q (64) Scholarisine R (65)

**Table 14.7** Cytotoxic effects of Scholarisine Q, R (**64** and **65**), taxol against GSCs

| Compound cell lines | $IC_{50}$ values (μg/mL) | | |
|---|---|---|---|
| | 64 | 65 | Taxol |
| GSC-3# | 17.9 | 21.7 | 13.6 |
| GSC-12# | 15.5 | 19.9 | 10.7 |
| GSC-18# | 20.6 | 20.6 | 8.9 |

Moreover, these compounds induced the GSCs apoptosis by increasing the tumor necrosis factor-alpha (TNF-α) expression and cleavage of caspase-3.

Wei et al. reported a novel bis-indole scaffold containing *Alstonia*sidines A and B (66 and 67, Fig. 14.12) from the methanolic extract of *A. scholaris* leaves and evaluated its selective anti-tumor effects in GSCs (Wei et al. 2018). At 25 μM concentration, both the tested compounds inhibited the growth of GSCs (GSC-12# and GSC-18#). Cell viability assay indicated that the compounds exhibited significant cytotoxicity against GSCs (Table 14.8). The isolated compounds were more selective toward the cancer cell types (in normal cell lines $IC_{50} > 90$ μM). In order to identify the apoptotic mechanism, several apoptotic regulators were measured by using real-time polymerase chain reaction (RT-PCR) assay. A significant increase in the Interleukin 1 (IL-1) and TNF-α level (extrinsic pathway regulators) was observed while calpain-1, caspase-12 (endoplasmic reticulum stress apoptotic regulators), Bcl-2, Bax, Bid, Apaf-1 (intrinsic pathway regulators) did not exhibit any changes. Further, they suggested that the anti-tumor effects mainly occurred due to the activation of the extrinsic apoptotic pathways.

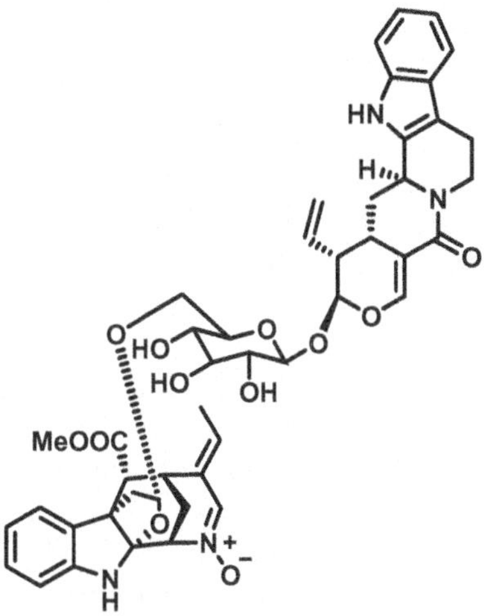

Alstoniasidine A (66)

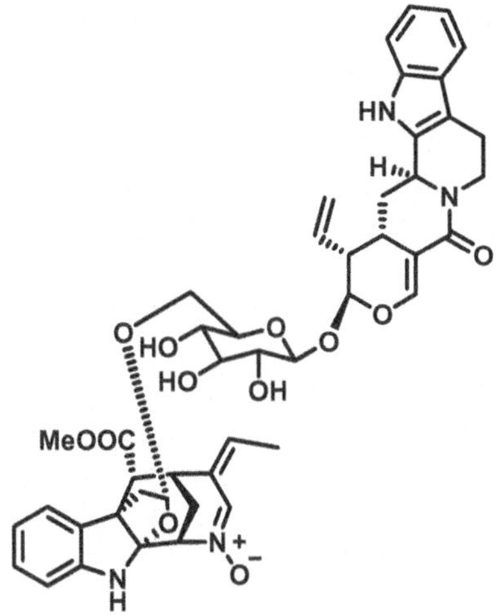

Alstoniasidine B (67)

**Fig. 14.12** Chemical structures of alstoniasidine A and B (**66** and **67**)

**Table 14.8** Cytotoxic effects of alstoniasidine A, B (**66** and **67**) and taxol against GSC cells

| Compound Cell lines | $IC_{50}$ values (μg/mL) | | |
|---|---|---|---|
| | 66 | 67 | Taxol |
| GSC-12# | 30.9 | 20.1 | 12.5 |
| GSC-18# | 19.2 | 16.1 | 10.4 |

**Fig. 14.13** Chemical structure of alstobrogaline (**68**)

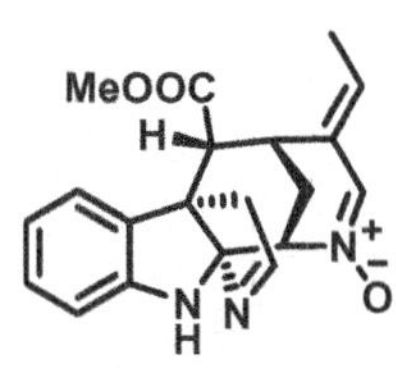

Alstobrogaline (68)

Krishnan et al. isolated alstobrogaline (68, Fig. 14.13) from *A. scholaris* leaves and evaluated for the in vitro cytotoxicity in various human breast cancer cell lines (Krishnan et al. 2019). 68 exhibited significant activity in human breast adenocarcinoma (DA-MB-231) and MCF-7 cells ($IC_{50}$ = 25.3 and 24.1 μM, respectively), while a poor activity was observed in MDA-MB-468, human breast cancer (SK-BR-3), and T47D cells ($IC_{50}$ > 30 μM).

### 14.3.1.3 Antibacterial and Antifungal Effects

Maurya et al. isolated loganetin (69, Fig. 14.14), an iridoid from ethyl acetate extract of *A. scholaris* stem barks by fast centrifugal partition chromatography (FCPC) and evaluated for its antibacterial potential in nalidixic acid-resistant with MTCC No. 1652 (NAREC), nalidixic acid sensitive strain CA8000 (NASEC), and *Escherichia coli* strains (Maurya et al. 2014). Individually, 69 exhibited poor activity against the selected strains (Minimum Inhibitory concentration-MIC value-500 μg/mL). However, a combination of 69 (10 μg/mL) with nalidixic acid exhibited a potential antibacterial activity. Individually nalidixic acid exhibited a MIC of 6.3 and 100 μg/mL for NAREC and NASEC, while in combination, it exhibited a MIC of 1.6 and 12.5 μg/mL with a 4- and 8-fold dose reduction of nalidixic acid concentration, respectively.

Liu et al. isolated 16 compounds (3, 16, 18–19, 24, 27, 30–31, 70–77, Fig. 14.15) from the methanolic extract of *A. scholaris* leaves (Liu et al. 2015) and evaluated the antibacterial effects against *Staphylococcus aureus* (ATCC 25922), *Pseudomonas aeruginosa* (ATCC 27853), *Enterococcus faecalis* (ATCC 10541),

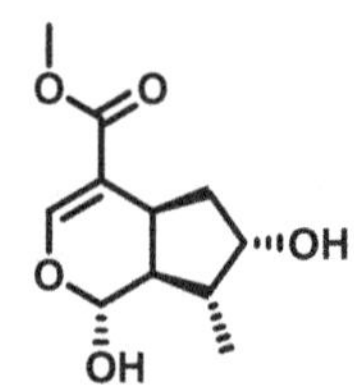

Loganetin (69)

**Fig. 14.14** Chemical structure of loganetin (**69**)

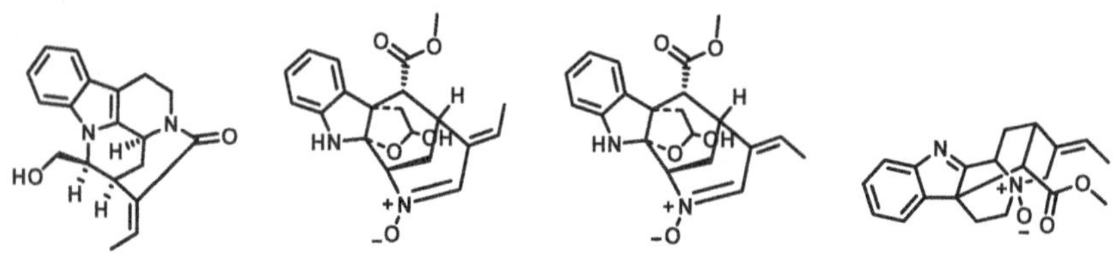

Normavacurine-21-one (70) 5-Hydroxy-19, 20-*E*-alschomine (71) 5-Hydroxy-19, 20-*Z*-alschomine (72) Strictamine $N_4$ oxide (73)

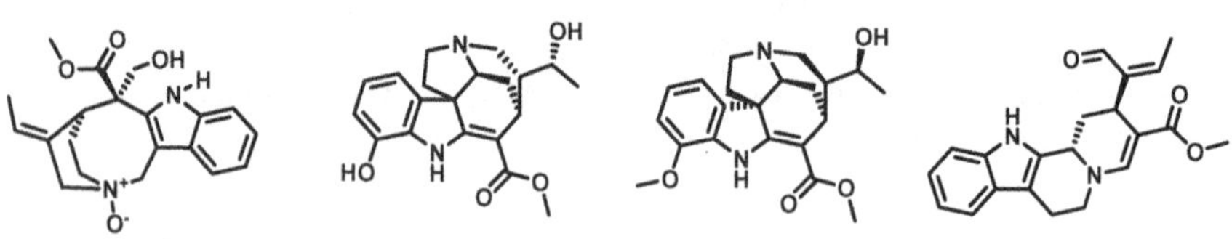

Vallesamine $N^4$-oxide (74) 19-Epischolaricine (75) 12-Methoxyechitamidine (76) Isovallesiachotamine (77)

**Fig. 14.15** Chemical structures of various alkaloids from *A. scholaris* (**70–77**)

**Fig. 14.16** Chemical structure of 3,5,7-Trihydroxyflavone (**78**)

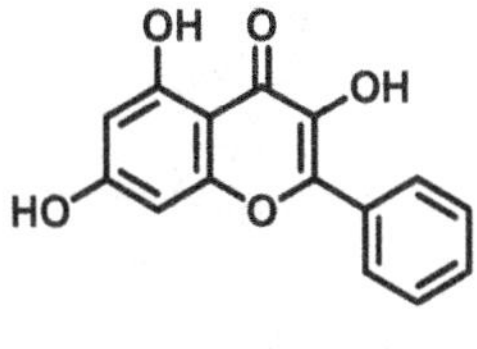


3, 5, 7-Trihydroxyflavone (78)

*E. coli* (ATCC 11775), and *Klebsiella pneumoniae* (ATCC 13883). Cefotaxime was used as a positive control. Normavacurine-21-one (70), 27 and vallesamine $N^4$-oxide (74) exhibited significant antibacterial activity against *E. faecalis* (MIC values—0.8 for each compound; cefotaxime MIC—0.2 µg/mL), 74, 18 and 5-hydroxy-19, 20-*Z*-alschomine (72), and against *P. aeruginosa* (MIC values—0. 8 each, cefotaxime MIC—0.78 µg/mL), while 18 showed moderate activity (MIC value of 1.6 µg/mL) against *K. pneumoniae* (cefotaxime MIC—0.8 µg/mL). The remaining compounds exhibited moderate to poor antibacterial effects (6.25–100 µg/mL).

M. Abinaya et al. isolated 3,5,7-Trihydroxyflavone (78, Fig. 14.16) from the methanolic extract of *A. scholaris* leaves and evaluated the biofilm formation inhibition and quorum sensing activity against *P. aeruginosa* (Abinaya and Gayathri 2019). Among the various concentrations of 78, 6.2 and 12.5 µg/mL concentration resulted in 50 and 55% of the reduction, while 1 mg/mL solution resulted in an 85% reduction in the biofilm formation. Further, the effects of various pathogens (*S. aureus*, *E. coli*, *P. vulgaris*, *B. subtilis*, and *P. aeruginosa*) in the pyocyanin production from *P. aeruginosa* were analyzed by a well diffusion method. The pyocyanin production was more sensitive to *S. aureus* and B. *subtilis* than the remaining species (zone of inhibition of 18.3 and 16.5 mm at 300 µg/mL).

Qin et al. isolated 11 new alkaloids (*Alstonia*scholarines A-K (79–89, Fig. 14.17) along with polyneuridinic acid from the aqueous fraction of *A. scholaris* leaves (Qin et al. 2015a). Antibacterial effects of these alkaloids were screened in various bacterial strains, namely as *S. aureus* (ATCC 25922), *P. aeruginosa* (ATCC27853), *E. faecalis* (ATCC 10541), *E. coli* (ATCC 8739), *P. smaitii* (ATCC 29916), and *K. pneumoniae* (ATCC 13883). Gentamycin was used as a positive control. Antifungal activity was evaluated in fungal strains, namely as *Epidermophyton floccosum* (CBS 566.94), *Microsporum gypseum* (CBS118893), and *Trichophytom mentagrophytes* (ATCC 4439). Griseofulvin was used as a positive control. In the antibacterial study, Alstoniascholarine F (84) and *Alstonia*scholarine J (88) exhibited a potential activity toward *P. aeruginosa* (MIC = 3.1 µg/mL each) while gentamycin exhibited a MIC of 0.78 µg/mL. Further, Alstoniascholarine C (81), 84, and 88 exhibited a potential activity toward *P. smaitii* (MIC = 3.1, 6.3, and 1.6 µg/mL, respectively) while gentamycin exhibited a MIC of 0.20 µg/mL. In the remaining antibacterial strains, the isolated alkaloids exhibited poor activity (>25 µg/mL). Alstoniascholarine D (82), Alstoniascholarine G (85), 88 exhibited moderate activity against *E. floccosum*

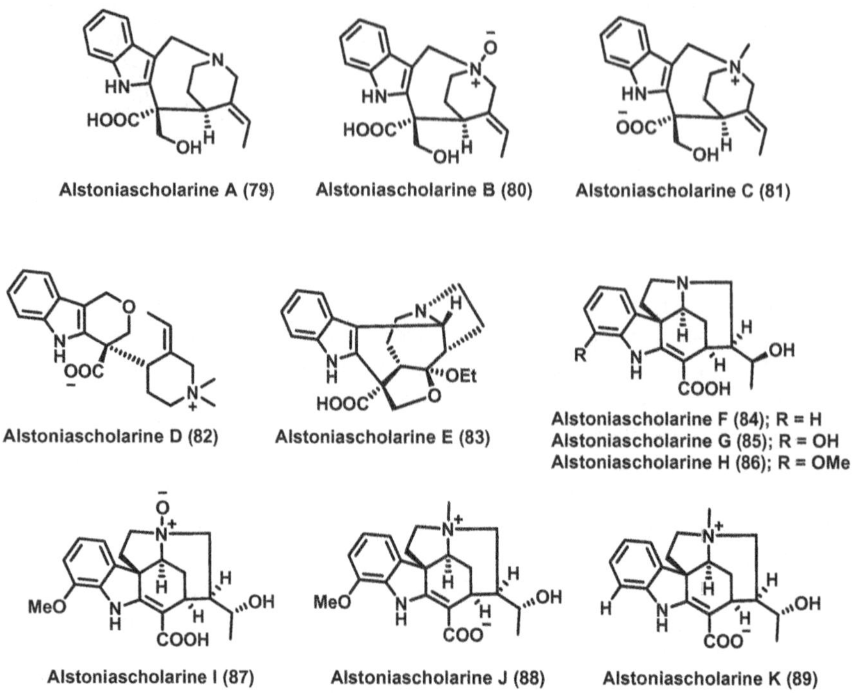


**Fig. 14.17** Chemical structures of Alstoniascholarines A-K (**79–89**)

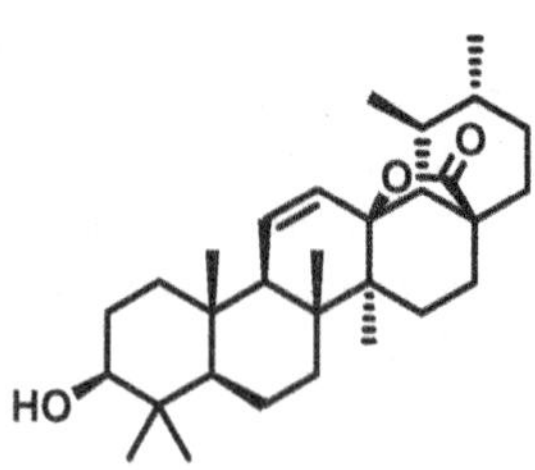

3-Hydroxy-11-ursen-28,13-olide (90)

**Fig. 14.18** Chemical structure of 3-hydroxy-11-ursen-28,13-olide (**90**)

(MIC = 31.6 μg/mL each). The remaining antifungal strains were not susceptible to the reported alkaloids (MIC > 62.5 μg/mL). The MIC values of griseofulvin were in the range of 3.9–7.8 μg/mL.

Wang et al. isolated various pentacyclic triterpenoids (Fig. 14.18) from the methanolic extract of *A. scholaris* leaves and evaluated its antibacterial efficacy (Wang et al. 2016) against *Bacillus cereus* (ATCC 9139), *E. faecalis* (ATCC 29212), *Listeria monocytogenes* (ATCC 7644), methicillin-sensitive *S. aureus* (MSSA, ATCC 29213), methicillin-resistant *S. aureus* (MRSA, ATCC 43300),

**Table 14.9** Antibacterial effects of various triterpenoids from *A. scholaris*

| Compound | Minimum Inhibitory concentrations (μg/mL) | | |
|---|---|---|---|
| | *E. faecalis* | *L. monocytogenes* | *B. cereus* |
| 50 | 4 | 8 | 16 |
| 41 | 1 | 2 | 8 |
| Ampicillin | 2 | 1 | 128 |
| Tetracycline | 4 | 2 | 4 |

*E. coli* (ATCC 35150), *P. aeruginosa* (ATCC 27853), *Salmonella enterica* (ATCC 13311). All the isolated compounds (3-hydroxy-11-ursen-28,13-olide (90), 41, 50–52 and 54) did not exhibit any antibacterial effects in gram-negative bacteria, while in gram-positive bacteria, 50 and 41 exhibited a moderate potential (Table 14.9). Further, a synergistic antibacterial effect was observed against *B. cereus* and *S. aureus* by 41 in combination with ampicillin and tetracycline.

#### 14.3.1.4 Nuclear Factor-κB (NF-κB) Inhibitory Activity and β2 Adrenoreceptor (AR) Activation

Hou et al. identified NF-κB inhibitors and β2-AR agonists from an alkaloidal extract of *A. scholaris* via microfractionation bioactivity-based ultra-performance liquid chromatography/mass spectrometry (Hou et al. 2012b). In the study, cytotoxicity of the compounds was determined via the quantities of lactate dehydrogenase (LDH). For the identification of NF-κB inhibition and β2 AR activation, luciferase reporter assay was used. Preliminary UPLC-MS analysis identified 17 peaks for NF-κB inhibition, while 11 peaks were identified for β2 AR agonist activity. According to the relative percent content, nine alkaloids, namely, 3, 18, 19, 20, 24, 27, 19(*Z*)-vallesamine (91), (*Z*)-alstoscholarine (92), and (*E*)-alstoscholarine (93), were screened for individual screening (Fig. 14.19). Compared to control, these compounds didn't exhibit a significant level of LDH in the serum (10 or 100 μmol/L). The obtained results indicated that these compounds did not show any kind of toxicity to the cells. Further, NF-κB inhibition activity of alkaloids in

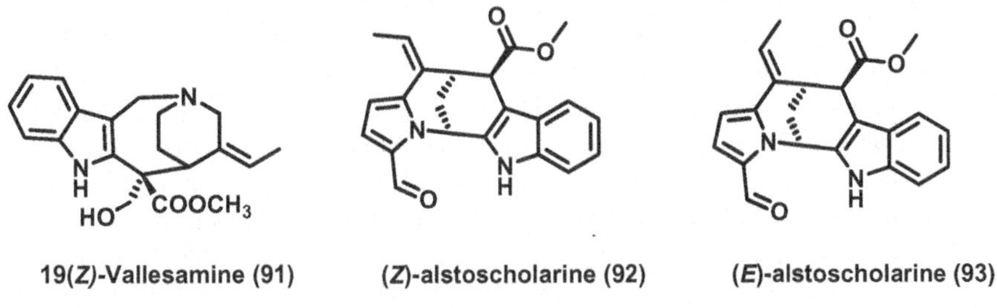


**Fig. 14.19** Chemical structures of various alkaloids from *A. scholaris* (**91–93**)

Fig. 14.20 Chemical structure of 19,20-(*E*)-Vallesamine (**94**)

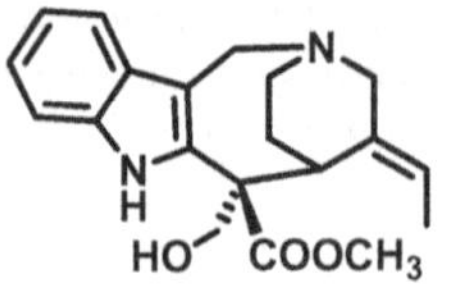

**19,20-(*E*)-vallesamine (94)**

the non-toxic concentrations was determined via luciferase reporter assay in TNF-α induced BEAS-2B cells (5 ng/mL) as compared to dexamethasone (positive control). Although all the tested compounds displayed a good inhibition on TNF-α induced NF-κB production, 24, 92, and 93 showed significant effects at even low concentrations (RFU ratio < 10 at 100 μm/L). Apart from that the cytokine release test (IL-6 and IL-8 expression) was also performed in TNF-α induced BEAS-2B cells (5 ng/mL). Cells treated with 3, 19, 20, 27, 92, and 93 only resulted in decreased expression of IL-6 (<500 pg/mL), while all the alkaloids demonstrated suppression of IL-8 release (<500 pg/mL). Dexamethasone (10 μmol/L) inhibited IL-6 and IL-8 (200 and 300 pg/mL, respectively).

While in another study, Hou et al. identified several new $\beta_2$ adrenergic receptor agonists (3, 13, 20, 24, 27, 92–94) from an alkaloidal extract of *A. scholaris* leaves via bioactivity-based LC/MS analysis (Hou et al. 2012a). $\beta_2$ AR agonist activities were confirmed by performing in vivo relaxant tests on guinea pig tracheal muscles. Compounds 3, 13, 20, 24, 92, 93, and 19,20-(*E*)-vallesamine (94, Fig. 14.20) were identified as the initial hit for $\beta_2$ AR activation. Further, chromatographic separation resulted in the pure form of these phytochemicals (90–95% purity) and $\beta_2$ AR activities of these components (0.5 μg/mL), as well as the positive control (salbutamol, 0.01 μmol/L), were determined. Salbutamol exhibited the highest activity (RFU ratio-2.5). Compared with the control, 3, 92–94 exhibited a significant activity (RFU ratio < 2.0) while 20 and 24 exhibited a poor $\beta_2$ agonist activity (RFU ratio < 1). To confirm the in vitro activity, spasmolytic activity tests were performed by using isolated guinea pig trachea (in vivo) and the obtained activity at 5 μg/mL follows—3 ($EC_{50}$ = 243.9 μmol/L), 92 ($EC_{50}$ = 137.5 μmol/L), 93 ($EC_{50}$ = 74.8 μmol/L), and salbutamol ($EC_{50}$ = 285.7 μmol/L). Remaining alkaloids exhibited a poor spasmolytic activity ($EC_{50}$ < 70 μmol/L). The comparable spasmolytic activity of 3 and 92 highlights the probable role of these alkaloids as $\beta_2$ AR agonists.

Yang et al. isolated various monoterpenoid indole alkaloids (3, 16, 18–20, 24, 30, 75, 95) from the ethanolic extract of *A. scholaris* leaves and determined its NF-κB inhibitory activity by using NF-κB luciferase assay in $HepG_2$-NFκB-Luc cells (Yang et al. 2018). Ammonium pyrrolidine dithiocarbamate (PDTC) was used as a positive control. Scholarisine S (95, Fig. 14.21), 3, 19, 20, and 75 exhibited significant NF-κB inhibitory activity (relative NF-κB luciferase activity < 0.8 folds at 25 μM). The remaining compounds exhibited more than 0.8 folds relative luciferase activity at 25 μM. Further, 19, 20, and 75 inhibited TNF-α induced NF-κB activations.

**Fig. 14.21** Chemical structure of Scholarisine S (**95**)

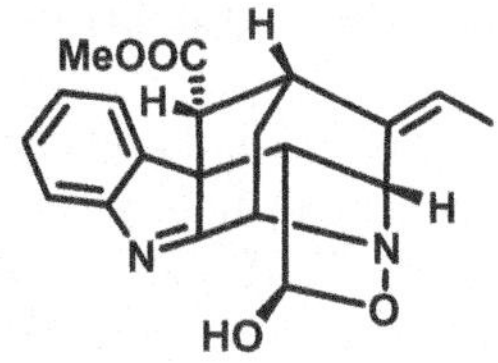


Scholarisine-S (95)

**Fig. 14.22** Chemical structure of 17-nor-excelsinidine (**96**)

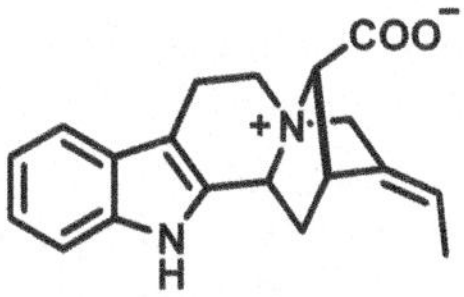


17-nor-excelsinidine (96)

#### 14.3.1.5 Antiviral Activity

Zhang et al. isolated 17-nor-excelsinidine (96, Fig. 14.22) and 27 from the methanolic extract of *A. scholaris* leaves and evaluated its anti-HSV and anti-adenovirus (ADV) activities (Zhang et al. 2014b). HSV-transfected Vero cell line was used for the evaluation of anti-HSV activity while adenovirus-transfected Hep-2 cell line was used for anti-adenovirus activity. Compound 27 exhibited a higher activity ($CC_{50\text{-}5.0}$ and 3.3; $EC_{50\text{-}0.36}$ and 0.28 μg/mL, respectively, for HSV and ADV) than the 96 ($CC_{50\text{-}6.9}$ and 3.3; $EC_{50\text{-}1.09}$ and 0.94 μg/mL, respectively, for HSV and ADV). Acyclovir (positive control) exhibited $CC_{50\text{-}302.0}$ and 302.6; $EC_{50\text{-}0.38}$ and 1.9 μg/mL, respectively, for HSV and ADV. Although 27 exhibited a higher activity than the positive control, the latter is having higher selectivity (Acyclovir—790.8 and 153.0; 27−13.9 and 11.8 for HSV and ADV).

Nguyen et al. identified four cystine knot α-amylase inhibitors (Alstotides 1–4) from the ethanolic extract of *A. scholaris* leaves (Nguyen et al. 2015). These are the types of peptides rich in cysteine and proline molecules with high resistance to heat and protease enzymes. Further, in cytotoxicity assays (Vero cells), Alstotide-1 (AS1) and Alstotide-3 (AS3) did not exhibit a significant activity till 100 μM. Hence, by taking the concentration range of 0–100 μM, antiviral activities of alstotides were evaluated in IBV, DENV2, and RSV A. IBV is a γ-coronavirus and the causative agent of infectious bronchitis. AS1 and AS3 exhibited inhibition of plaque formation in a dose-dependent manner. Furthermore, AS1 exhibited a moderate inhibition toward DENV2 ($EC_{50}$ = approx. 90 μM). However, it didn't exhibit any activity toward the RSV A up to 100 μM concentration.

#### 14.3.1.6 Antiallergic and Antitussive Effects

Shang et al. evaluated antitussive, anti-asthmatic, and expectorant activities of various fractions/phytochemicals (20, 24, and 30) from *A. scholaris* leaves (Shang et al. 2010). The antitussive activity was evaluated in various animal models (Ammonia or sulfur dioxide-induced mice coughing and citric acid-induced guinea pigs coughing) while histamine-induced bronchoconstriction in guinea pig was used for the evaluation of the anti-asthmatic activity. Codeine phosphate was used as a positive control in the antitussive screening, while aminophylline is in bronchoconstriction and ammonium chloride is used in case of expectorant activity. In ammonia liquor-induced cough in mice, among the various fractions evaluated, the alkaloid rich fraction exhibited a significant reduction in the coughing frequency (30.7 and 39.3% for 50 and 100 mg/kg of alkaloid rich fraction), while these activities were comparatively lesser than the codeine phosphate (55.4% at 30 mg/kg). In sulfur dioxide-induced cough in mice, alkaloid rich fraction (100 mg/kg) exhibited a 62.2% increment in the cough latent period, whereas codeine phosphate (30 mg/kg) resulted in 83.2%. A similar kind of results was obtained in citric acid-induced cough in guinea pigs wherein a 153.8 and 419.0% increment in the latent period of cough with the alkaloidal fraction (100 mg/kg), and codeine phosphate (30 mg/kg) groups, with a cough inhibition frequency of 62.4, and 91.9%, respectively. In the anti-asthmatic effects in guinea pigs, alkaloid rich fraction (79.9% at 200 mg/kg) exhibited an increased tumble and delitescence of convulsion with that of aminophylline group (90.1% at 100 mg/kg). In expectorant effect analysis, ammonium chloride (1500 mg/kg) and the alkaloids fraction (at 60 and 120 mg/kg) could markedly enhance tracheal phenol red output (112.7 and 152.1, 138.6% increase, respectively).

The preliminary screening highlighted the potential role of alkaloid rich fraction in exhibiting antitussive and anti-asthmatic effects. Hence, the study was further proceeded with the alkaloids fraction and main alkaloids (20, 24, and 30). In ammonia-induced cough, *i.g.* administration of 20 (10 mg/kg) and 24 (5 mg/kg) resulted in 38.9 and 34.1% increment in the latent period of cough in mice, and these are comparable with the codeine phosphate (30 mg/kg, *i.g.*, resulted in 47.6% increment). The same trend was observed within the case of anti-asthmatic effect screening. At a dose of 10 mg/kg, the 20 group increased the tumble and delitescence of convulsion of guinea pigs by 63.0% while aminophylline group resulted in 96.3% of the parameters.

Qin et al. evaluated the antitussive effects of 43–48 (Qin et al. 2015b). In ammonia-induced cough (mice), the total alkaloids from the inadequately dried leaves exhibited a lesser % inhibition (38.0 and 27.2% at 20 and 10 mg/kg) than the dried leaves (58.0 and 45.6% at 20 and 10 mg/kg). Codeine phosphate (30 mg/kg) was used as a positive control that resulted in 77.6% inhibition.

Zhao et al. identified various alkaloids (20, 24, 75, and 30) from the ethanolic extract of *A. scholaris* leaves and evaluated the effects in airways allergic conditions (Zhao et al. 2017). Ovalbumin-induced airways allergic inflammatory models in SD rats were used in the study, and dexamethasone (2 mg/kg, once daily) was used as a positive control. Various parameters such as cellular infiltration, bronchoalveolar lavage fluid (BALF), IL-4, and IL-10 expression in BALF were analyzed. Along with the various doses of total alkaloids, the following doses of the phytochemicals were also screened. Intra-gastric administration of the following doses of alkaloids were performed once a daily [20 (5 mg/kg); 24 (3 mg/kg); 30 (3 mg/kg); 75 (3 mg/kg)]. The alkaloidal treated group resulted in a decreased amount of total leukocytes and eosinophils % in BALF. All the screened alkaloids exhibited a potential activity (WBC content—<0.5 × $10^9$/mL; eosinophil %—4%). In the disease control group, IgE and eotaxin in serum were expressed at a level of 2.5 μg/mL and 6 pg/mL. However, treatment of the alkaloids resulted in a decreased amount of IgE and eotaxin (<2.0 μg/mL and 4 pg/mL, respectively). Furthermore, the treatment of alkaloids also resulted in the reduction of inflammatory markers such as IL-4 and IL-10 in BALF.

In another study, Yun-Li Zhao et al. evaluated the effect of 20, 24, 30, and 75 in post-infectious cough in mice (Zhao et al. 2018). The animal model was standardized by instilling lipopolysaccharide (LPS—at 80 μg/50 μL/mouse) at tracheal region followed by 30 min exposure of cigarette smoke till 30 days. After the administration of alkaloids (above-mentioned dose, except **73** (*i.g.*), 1 mg/kg), the symptoms of cough in mice were attenuated. Total white blood cells (WBC), neutrophils (NEU) amounts in BALF and MDA, C-reactive protein (CRP), IL-6 in serum were significantly reduced.

#### 14.3.1.7 Metabolic Enzymes (α-Glucoside, α-Amylase, and Pancreatic Lipase) Inhibitory Effects

NJong-Anurakkun et al. reported the α-glucoside inhibitory potential of *A. scholaris* leaves-derived phytochemicals (97–99, Fig. 14.23) Jong et al. (2007). Quercetin 3-*O*-*β*-*D*-xylopyranosyl(1″ → 2″)-*β*-*D* galactopyranoside (97), (+)-lyoniresinol 3-*O*-*β*-*D*-glucopyranoside (98), and (−)-lyoniresinol 3-*O*-*β*-*D*-glucopyranoside (99) were isolated from the rat intestine acetone powder. The screened compounds exhibited a moderate potential toward α-glucoside (Table 14.10).

Nguyen et al. studied the α-amylase inhibitory effect of AS1 to AS3 on α-amylase from human salivary, *T. molitor* larvae (TMA), and fungus (*Aspergillus oryzae*). AS1 to AS3 exhibited a significant activity toward the α-amylase of TMA (1.9–5.2 μM). However, it did not exhibit any activity toward human and fungal α-amylases up to 100 μM concentration (Nguyen et al. 2015).

George et al. evaluated the pancreatic lipase inhibitory potential of *A. scholaris* stem barks (George et al. 2019). Bioassay-guided fractionation of the methanolic extract resulted in the identification of echitamine (**9**) as a potential pancreatic lipase inhibitory leads ($IC_{50}$ = 10.92 μM).

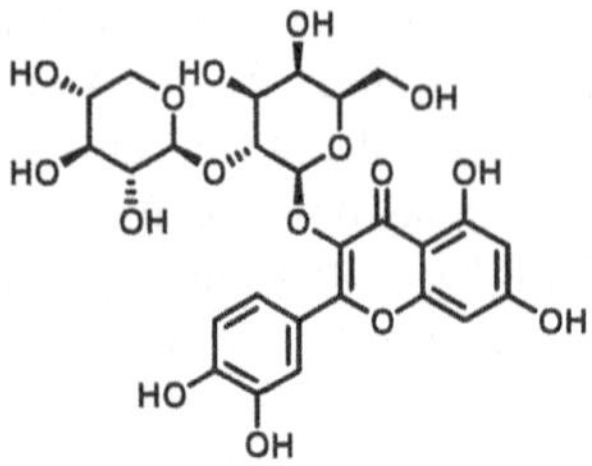

Quercetin 3-*O*-*β*-*D*-xylopyranosyl(1‴→2″)- *β*-*D* galactopyranoside (97)

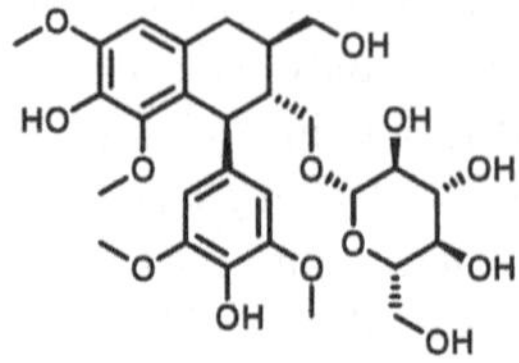

(+)-Lyoniresinol 3-*O*-*β*-*D*-glucopyranoside (98)

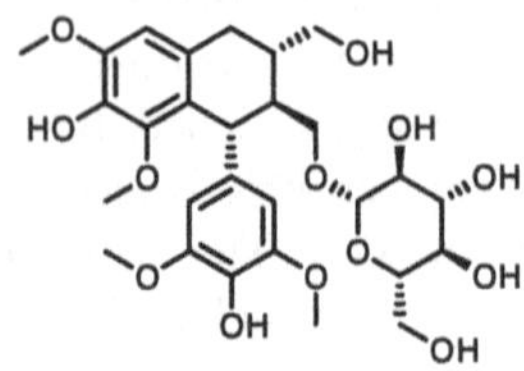

(-)-Lyoniresinol 3-*O*-*β*-*D*-glucopyranoside (99)

**Fig. 14.23** Chemical structures of various lignans from *A. scholaris* (**97–99**)

**Table 14.10** α-glucoside inhibitory potential of various lignans from *A. scholaris* (**97–99**)

| Compound | $IC_{50}$ values(mM) | |
|---|---|---|
| | Sucrase | Maltase |
| 97 | 17.2 | 1.96 |
| 98 | 1.95 | 1.43 |
| 99 | >10 | >10 |

### 14.3.1.8 Anti-tubercular Activities

Macabeo et al. isolated various alkaloids (94, 100–104, Fig. 14.24) from the methanolic extract of *A. scholaris* leaves. Further, the anti-tubercular activity of the isolated compounds was evaluated by microplate alamar blue assay (MABA) in *Mycobacterium tuberculosis* H37Rv by using rifampicin as a positive control (Macabeo et al. 2005, 2008). Among the various compounds screened, 20*S*-tubotaiwine (103) only exhibited a potential anti-tubercular activity (MIC—>100 μg/mL). The remaining compounds exhibited a poor activity toward the *M. tuberculosis* (Table 14.11).

### 14.3.1.9 Antimalarial Activities

Salim et al. isolated seven indole alkaloids (49, 105–110, Fig. 14.25) from the ethanolic extract of *A. scholaris* bark (Salim et al. 2004). Isolated compounds are 49, Akuammiginone (105), echitaminic acid (106), echitamidine-*N*-oxide (107), $N^b$-demethyl alstogustine *N*-oxide (108), akuammicine *N*-oxide (109), and $N^b$-demethyl alstogustine (110). Further, in vitro antimalarial activity was evaluated by hypoxanthine incorporation techniques. Compounds 109 and 110 had an $IC_{50}$ values of 63.2 and 6.75 μg/mL, respectively, against *Plasmodium falciparum* (K1, multidrug-resistance (MDR) strain).

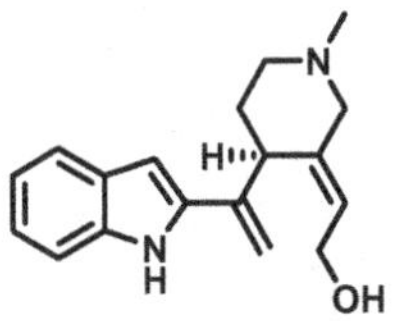

(+)-manilamine (100)

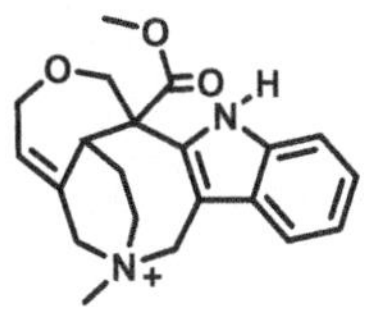

$N^4$-methyl angustilobine B (101)

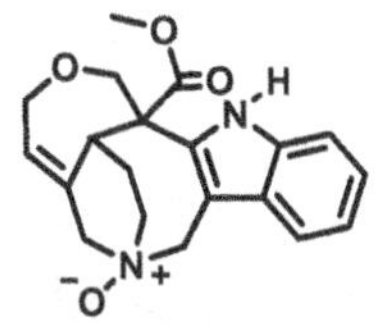

Angustilobine B $N^4$-oxide (102)

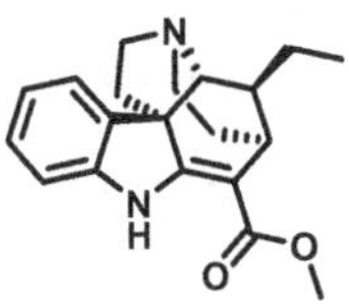

20(*S*)-tubotaiwine (103)

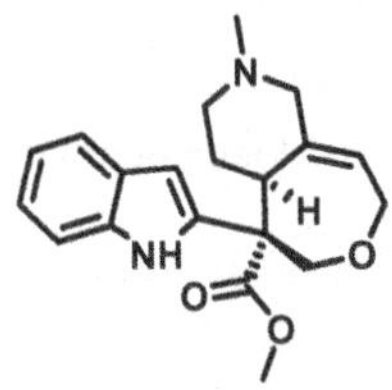

6,7-secoangustilobine (104)

**Fig. 14.24** Chemical structures of various alkaloids from *A. scholaris* (**100–104**)

**Table 14.11** Anti-tubercular activities of various alkaloids from *A. scholaris*

| Compound | MIC (μg/mL) |
|---|---|
| 94 | >128 |
| 100 | >128 |
| 102 | >128 |
| 101 | >128 |
| 103 | >100 |
| 104 | >128 |
| Rifampicin | 98% inhibition at 0.125 μg/mL |

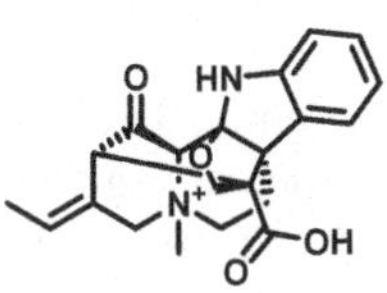

Akuammiginone (105)

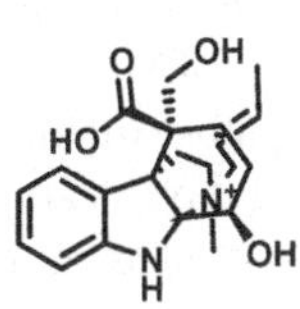

Echitaminic acid (106)

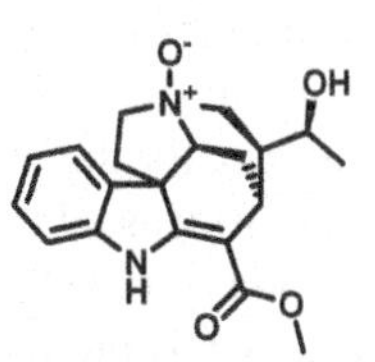

Echitamidine *N*-oxide (107)

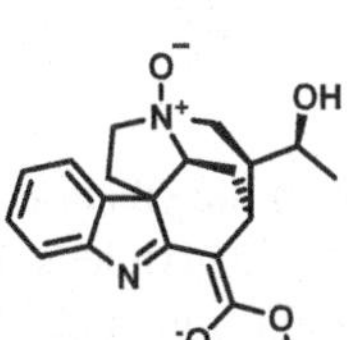

$N^b$-Demethylalstogustine *N*-oxide (108)

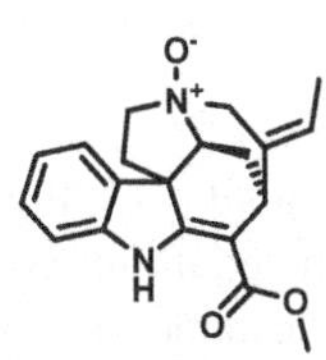

Akuammicine *N*-oxide (109)

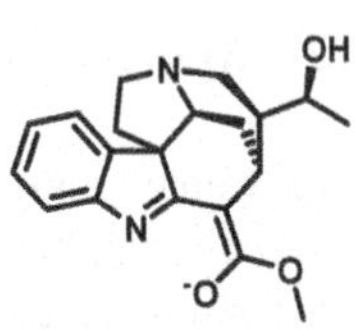

$N^b$-Demethylalstogustine (110)

**Fig. 14.25** Chemical structures of various alkaloids from *A. scholaris* (**105–110**)

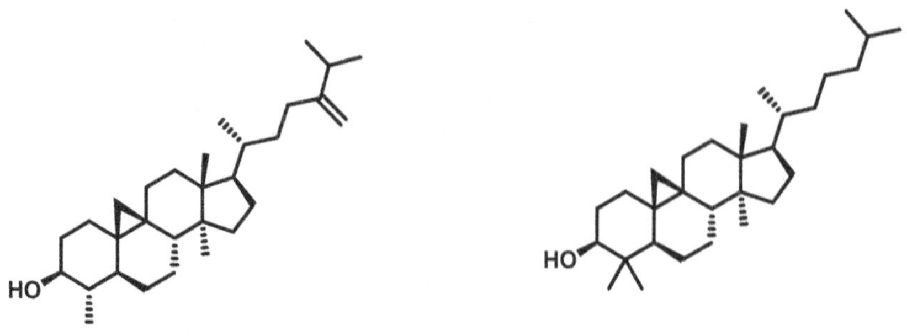


Cycloeucalenol (111)

Cycloartanol (112)

**Fig. 14.26** Chemical structures of cycloeucalenol and cycloartenol (**111** and **112**)

#### 14.3.1.10 Hypoglycemic Activities

Ragasa et al. reported the isolation of a mixture of cycloeucalenol, cycloartanol (111, 112, Fig. 14.26), 39, 51, and 54 from dichloromethane extract of *A. scholaris* leaves (Ragasa et al. 2015). Further, the isolated mixture was screened for hypoglycemic assay in male albino mice (*Mus musculus* L.) at various concentration (25, 50, and 100 mg/kg) by using glimepiride solosa (16.7 μg/kg) as a positive control. Supplementation of 25 mg/kg resulted in a high impact on the hypoglycemic properties. In 0.5 h, the mixture exhibited a 52.61% blood glucose reduction while the glimepiride resulted in 39.07% glucose reduction.

#### 14.3.1.11 Hepatomodular Effects

Singh et al. isolated 38 from the ethanolic extract of *A. scholaris* stem bark and evaluated the hepatomodular effect against $CCl_4$-induced hepatic oxidative stress in Wistar albino rats (Singh et al. 2015). $CCl_4$ (0.2 mL/kg, twice a week, *i.p.*) was administered for the induction of hepato-oxidative stress. Concurrently, 38 (20 mg/kg/day, oral) was administered for 30 days. Various biomarkers such as γ-glutamyl transpeptidase (GGT), aspartate, and alanine transaminases (AST, ALT) were evaluated. $CCl_4$ produced a significant hepatic oxidative stress, while the treatment of 38 resulted in the recovery of the oxidative stress.

#### 14.3.1.12 Anticataract Activity

Soni et al. isolated a novel isoflavonoid (113, Fig. 14.27) from the ethanolic extract of *A. scholaris* stem bark and evaluated its anticataract activity in goat lens (in vitro) and fructose-induced experimental cataract (in vivo) condition (Soni et al. 2019). In the in vitro experiments, cataract was induced in goat lenses by incubating with a

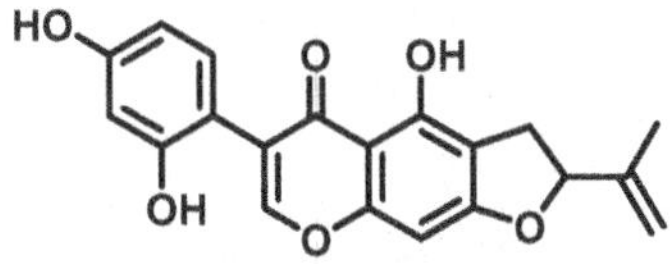


**Fig. 14.27** Chemical structure of chromone analogue from *A. scholaris* (**113**)

higher concentration of glucose (55 mmol/L). Compound 113 (50 μg/mL) resulted in a slight degree of opacity as compared to the toxic control group. During the in vivo experiments (albino male Sprague–Dawley rats), animals were administered with a fructose solution. The high fructose-rich diet will lead to a series of mechanisms in the SD rats. Fructose will lead to the elevation in the blood pressure and blood glucose levels that in turn results in metabolic alterations. Apart from that fructose will also upregulate the angiotensin-II and increase the sympathetic nervous systems that results in hypertensive conditions.

Further, the accumulation of high fructose increases the lenticular opacity, subsequently leading to increased osmotic pressure and oxidative stress in the lens results in cataractogenesis. In the case of systolic and diastolic blood pressure, dose-dependent activity was observed with the administration of 113. Moreover, blood glucose levels were reduced. A significant reduction in the lens opacity was observed with 113 treatment for 8 weeks. Further, various biochemical parameters were restored.

#### 14.3.1.13 Anti-fertility Effects

Gupta et al. isolated **37** from the ethanolic extract of *A. scholaris* stem bark and evaluated the anti-fertility effects (10 mg/rat/day for 60 days) in male albino rats (Gupta et al. 2008). The administration of the **37** did not exhibit a significant weight loss. However, a decreased weight in the reproductive organs (testes, seminal vesicle, ventral prostate, and epididymides) was observed. Apart from that sperm density and motility were also reduced. The seminiferous tubular diameter, Sertoli cells cross-sectional surface area, and counts were also decreased significantly.

### *14.3.2 Phytochemistry and Pharmacological Activities of* A. Macrophylla

#### 14.3.2.1 Cytotoxicity Effects

Keawpradub et al. screened the cytotoxicity effects of various alkaloids (114–116, Fig. 14.28) isolated from the methanolic extract of *A. macrophylla* root bark

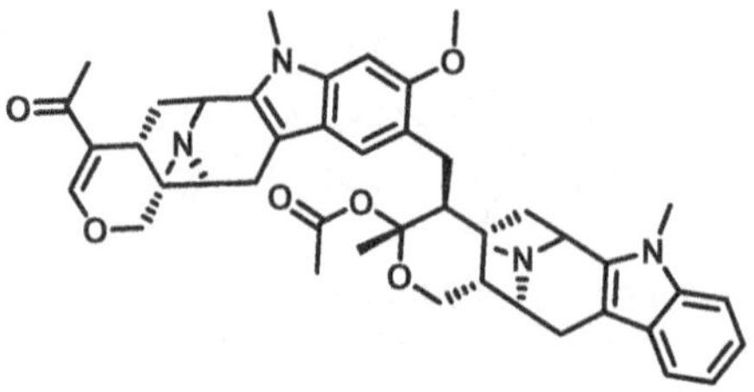

**O-Acetyl-macralstonine (114)**

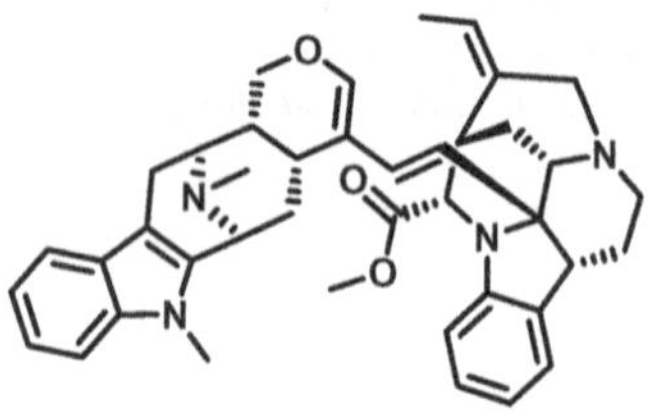

**Macrocarpamine (115)**

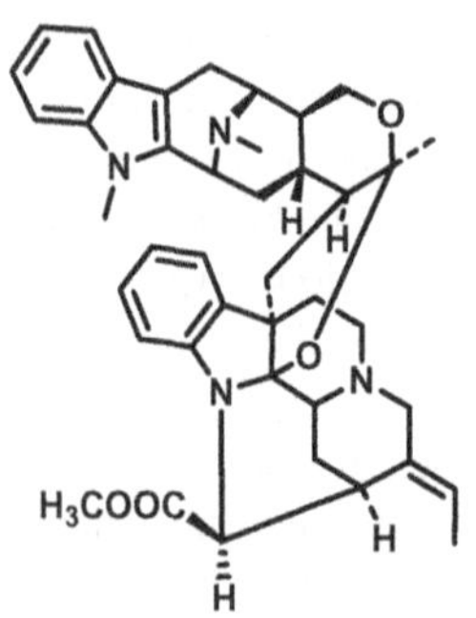

**Villalstonine (116)**

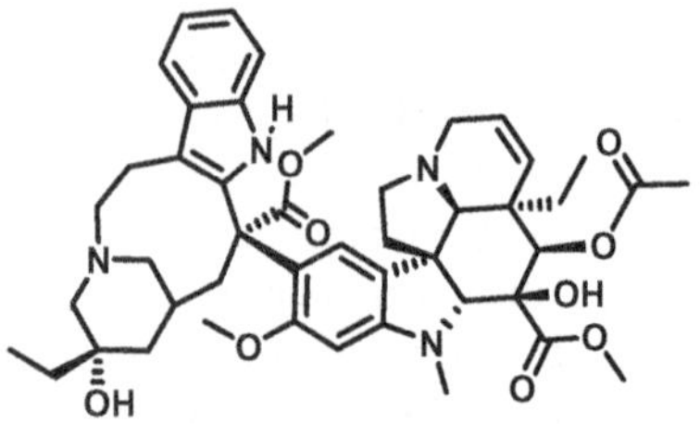

**Vinblastine Sulphate (117)**

**Fig. 14.28** Chemical structures of various alkaloids from *A. macrophylla* (**114–116**)

(Keawpradub et al. 1998). Among the tested alkaloids, only *O*-Acetyl macralstonine (114) displayed selectivity for cancer cell lines (Table 14.12).

In another report, Keawpradub et al. evaluated the cytotoxicity of 14 indole alkaloids (114–116,118–128, Fig. 14.29) from the methanolic extract of *A. macrophylla* root bark in MOR-P, COR-L23 cell lines using SRB assay (Keawpradub and PJ Houghton 1997). Except, macralstonine (124) and villalstonine $N^b$-oxide (127) (89.4 and 62.2 μM), all the tested indole alkaloids exhibited a strong cytotoxicity activity in MOR-P (2.3–19.6 μM) than the monomeric indoles (61.4 to >100 μM), while the same trend was observed in COR-L23 cell lines (2.92–20.2 μM for bisindole alkaloids; 57.2 to >100 μM for monomeric indoles). Bisindole alkaloids such as 114, 115, and 116 exhibited the potent activity (6.3, 4.6, and 2.3 μM in MOR-P cell line; 4.1, 5.3, and 2.9 μM in COR-L23 cell lines, respectively). Further, cytotoxicity effects of these alkaloids were evaluated in various cell lines such as StM 1a, Caki-2, MCF-7, and LS174T. These alkaloids exhibited an $IC_{50}$ in the range of 2–7 μM. However, the results from the normal cell lines (breast fibroblasts) exhibited $IC_{50}$ values of 8.1–21.9 μM highlighting that these alkaloids lack the selectivity toward the cancer cell cytotoxicity.

Changwichit et al. isolated secoiridoid (Fig. 14.30) from the methanolic extract of *A. macrophylla* stems and evaluated its cytotoxicity effects (Changwichit et al. 2011). At a concentration of 10 μg/mL of naresuanoside (129) resulted in a 22% inhibition in the growth of human androgen-sensitive prostate cancer cell (LNCaP),

**Table 14.12** Cytotoxic effects of various alkaloids from *A. macrophylla* and vinblastine sulfate against various cell lines

| Compound Cell lines | $IC_{50}$ values (μM for the compounds and nM for vinblastine) | | | |
|---|---|---|---|---|
| | 114 | 115 | 116 | Vinblastine sulfate |
| MOR-P | 6.3 | 4.6 | 2.3 | 3.1 |
| COR-L23 | 4.1 | 5.3 | 2.9 | 0.9 |
| BF | 21.9 | 8.1 | 8.5 | >100 |
| StMl la | 3.3 | 2.9 | 2.4 | 1.7 |
| Caki-2 | 4.6 | 7.3 | 2.9 | 1.9 |
| MCF-7 | 2.4 | 1.9 | 3.4 | 0.5 |
| Ls 174T | 2.4 | 1.8 | 1.9 | 0.8 |

MOR-P-adenocarcinoma, COR-L23 human lung cancer cell lines; BF-normal human breast fibroblast cell line; StMl la-melanoma; Caki-2-renal cell carcinoma; Ls174T-colon adenocarcinoma

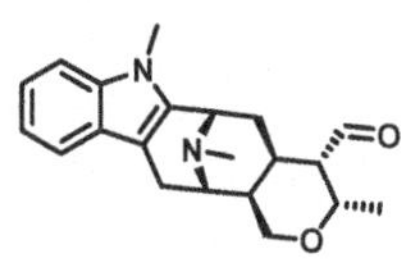

Talcarpine (118)

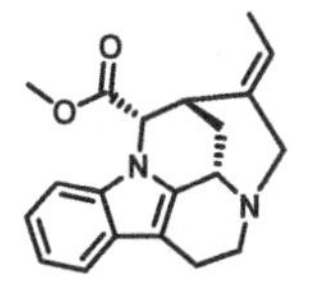

Pleiocarpamine (119)

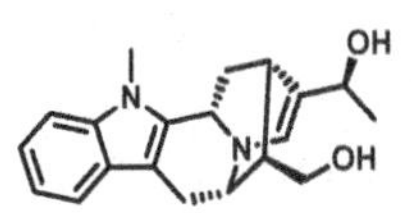

Alstoumerine (120)

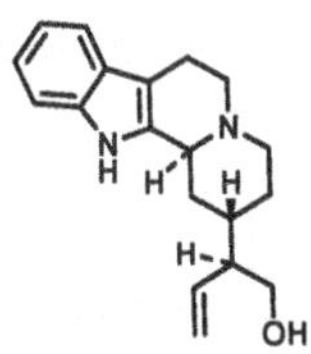

20-*Epi*-antirhine (121)

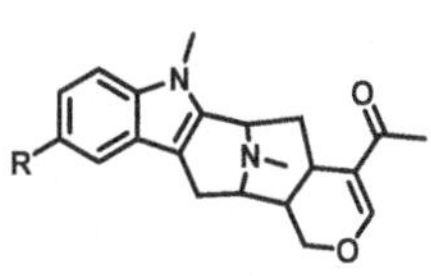

Alstonerine (122), R = H
Alstophylline (123), R = OMe

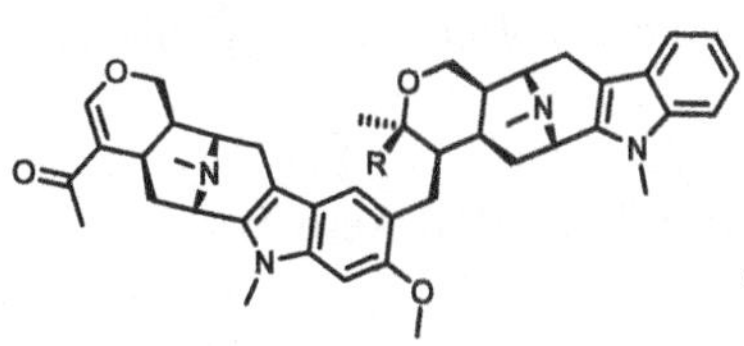

Macralstonine (124); R = OH
*O*-Methyl macralstonine (125); R = OMe

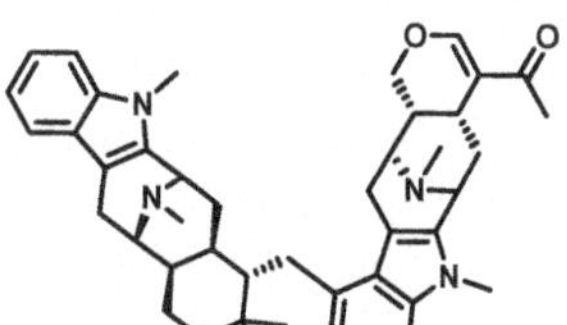

Alstomacrophylline (126)

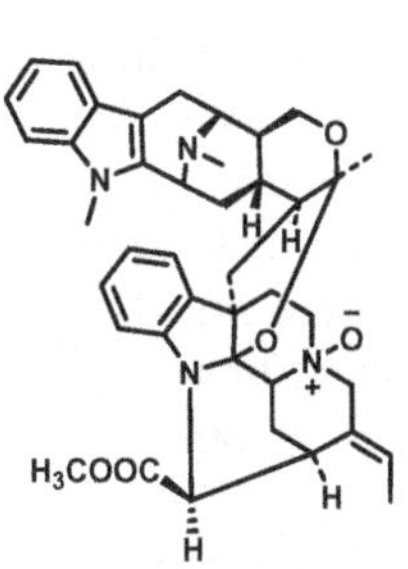

Villalstonine $N^b$-oxide (127)

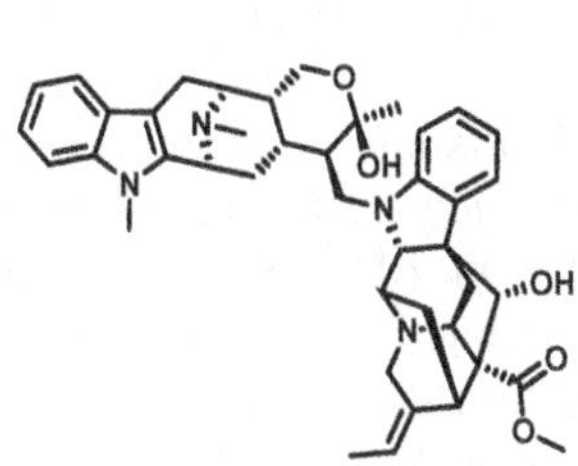

Alstomacroline (128)

**Fig. 14.29** Chemical structures of various alkaloids from *A. macrophylla* (**118–128**)

**Fig. 14.30** Chemical structure of naresuanoside (**129**)

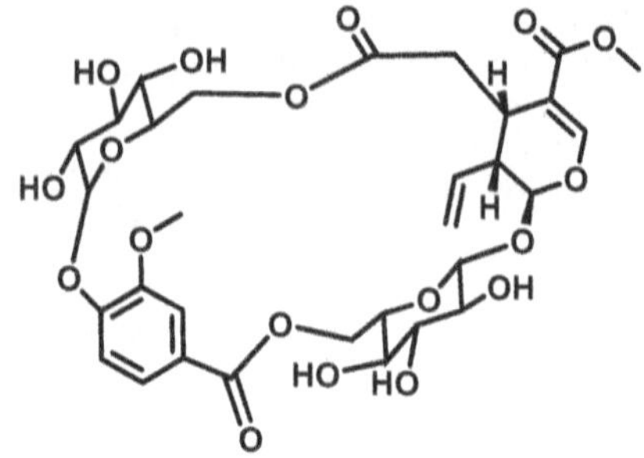

**Naresuanoside (129)**

while in a concentration range of 0.1 ng/mL–10 μg/mL did not exhibit any kind of inhibition in the cell growth of human foreskin fibroblast cells (HF cells, normal cells).

Lim et al. isolated 17 alkaloids (130–146) from the ethanolic extract of *A. macrophylla* stem bark and leaf (Fig. 14.31). Further, they evaluated the MDR reversing capacity in KB/S (vincristine-sensitive) and KB/VJ300 (vincristine-sensitive and vincristine-resistant KB cell lines) (Lim et al. 2014). All the compounds exhibited poor cytotoxicity in the cell lines ($IC_{50}$ > 25 μg/mL). However, 11-methoxyvincorine (141), 11-demethoxyquaternine (144), 19,20-Z-affinisine (136) were reported to reverse MDR in vincristine-resistant KB (VJ300) cells ($IC_{50}$ = 4.60, 6.60 and 7.47 μg/mL, respectively).

#### 14.3.2.2 Antiplasmodial Activity

Keawpradub et al. isolated 13 alkaloids (114–116, 118–128) from the methanolic extract of *A. macrophylla* leaves, stem bark, root bark and evaluated the antiplasmodial activity against the K1 strain of *P. falciparum* by using chloroquine diphosphate as a positive control (Keawpradub et al. 1999). Bisindole alkaloids such as 114, 115, 116, and 128 ($IC_{50}$ values = 0.53, 0.36, 0.27, and 1.12 μM) exhibited a comparable activity with the chloroquine diphosphate ($IC_{50}$ value = 0.20 μM). Further, the effects of these alkaloids were evaluated in chloroquine-resistant (K1) and chloroquine-sensitive strains (T9–96) of *P. falciparum*. These bisindoles were found to be significantly less active against T9–96 (0.94–39 μM; chloroquine diphosphate—0.02 μM). In K1 cells, they exhibited a significant activity (0.36–1.12 μM; chloroquine diphosphate—0.20 μM).

Hirasawa et al. isolated alstiphyllanines A-D (147–150, Fig. 14.32) from the methanolic extract of *A. macrophylla* leaves and evaluated the antiplasmodial activity (against *P. falciparum,* 3D7) (Hirasawa et al. 2009). Screened alkaloids exhibited a moderate in vitro antiplasmodial activity ($IC_{50}$ of 6.85, 0.34, 6.20, and 2.75 μg/mL, respectively, for 147–150). In vitro cytotoxicity evaluation resulted in

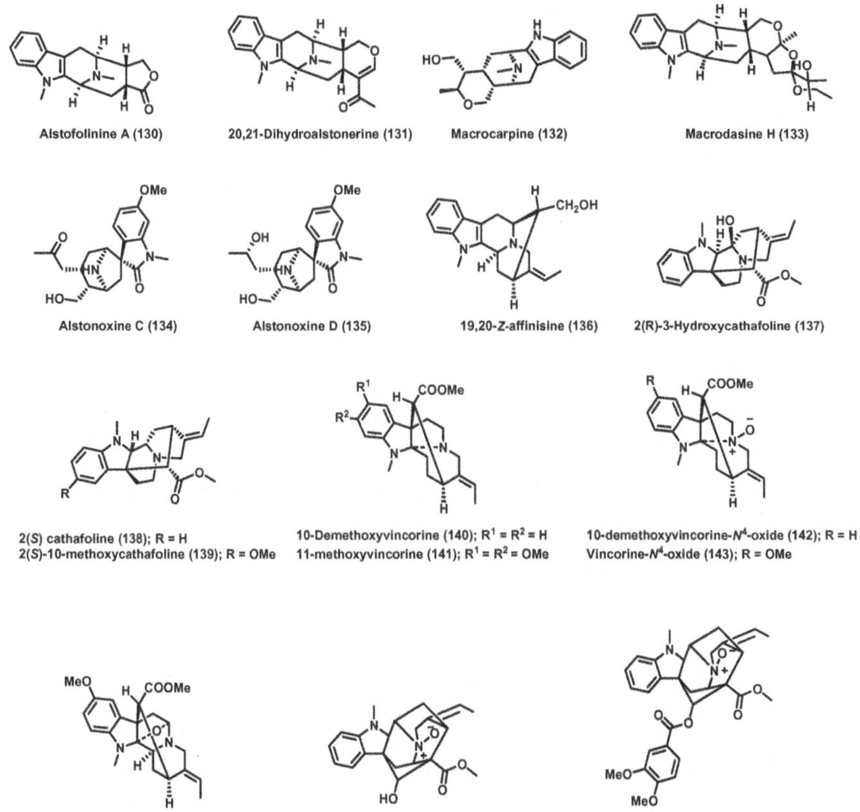


**Fig. 14.31** Chemical structures of various alkaloids from *A. macrophylla* (**130–146**)

the conclusion that these compounds are not cytotoxic in nature ($IC_{50}$ > 25 μg/mL). 148–150 comprised of picraline-type skeletons (arise through a corynanthe-type skeleton) while 147 consists of an ajmaline-type skeleton.

Cheenpracha et al. isolated 11 alkaloids (118, 151–160) from the methanolic extract of *A. macrophylla* bark (Fig. 14.33) and evaluated the antiplasmodial activity (*P. falciparum* K1, MDR) and cytotoxicity activities in KB cell lines (Cheenpracha et al. 2013). Alstonisine (158) exhibited moderate antiplasmodial activity ($IC_{50}$ of 7.6 μM), while dihydroartemisinin was used as a positive control ($IC_{50}$ = 1.41 nM). Remaining screened compounds did not exhibit any antiplasmodial activity. In cytotoxicity assays, all the screened compounds were not cytotoxic to the KB cell line ($IC_{50}$ = >100 μM). Doxorubicin ($IC_{50}$ = 0.55 μM) and ellipticine ($IC_{50}$ = 4.87 μM) were used as reference cytotoxic substance and positive control.

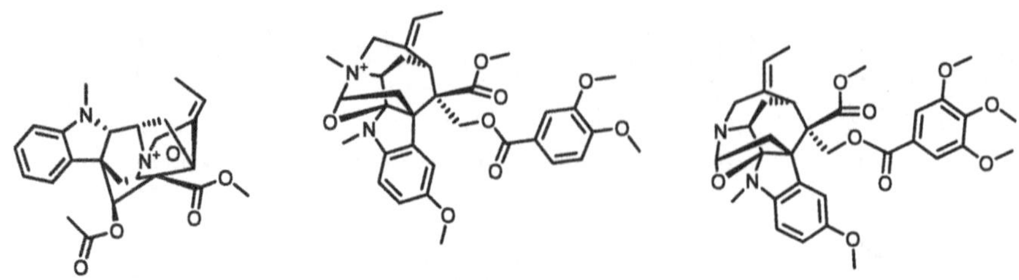

Alstiphyllanine A (147) Alstiphyllanine B (148); R = H Alstiphyllanine C (149); R = OMe Alstiphyllanine D (150)

**Fig. 14.32** Chemical structures of alstiphyllanine A-D from *A. macrophylla* (**147–150**)

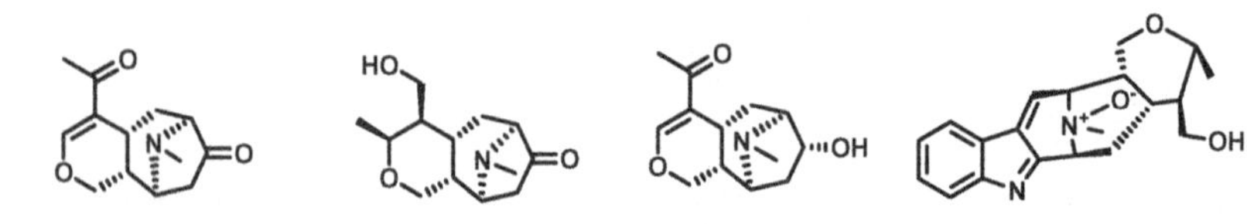

Alstoniaphylline A (151) Alstoniaphylline B (152) Angustimaline (153) Alstoniaphylline C (154)

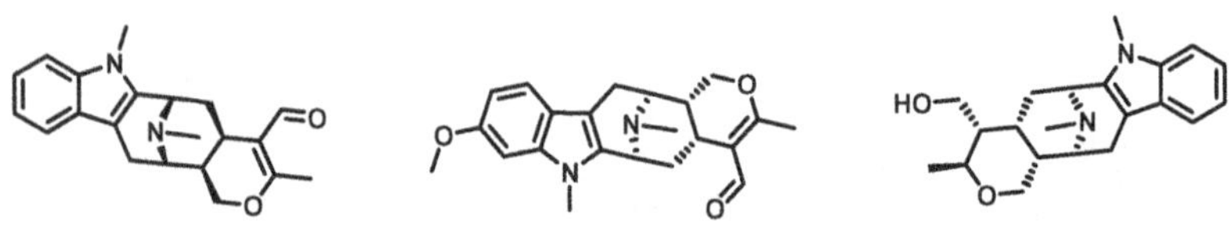

Alstonerinal (155) Alstophyllal (156) Macrocarpine B (157)

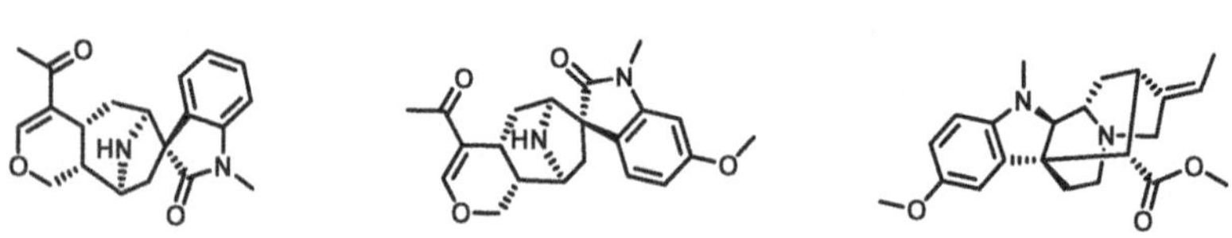

Alstonisine (158) *N*b-Demethylalstophylline Oxindole (159) 10-Methoxy cathafoline (160)

**Fig. 14.33** Chemical structures of various alkaloids from *A. macrophylla* (**151–160**)

#### 14.3.2.3 Vasorelaxant Activities

Hirasawa et al. evaluated the vasorelaxant effects of various *A. macrophylla* leaves-derived alkaloids (147–150) (Hirasawa et al. 2009). The screened molecules exhibited a slow relaxation activity against phenyl epinephrine ($3 \times 10^{-7}$ M)-induced thoracic rat aortic rings with endothelium. Compound 147 exhibited a 70% vasodilation, while 35, 40, and 42% vasorelaxation was observed by the treatment of 148–150.

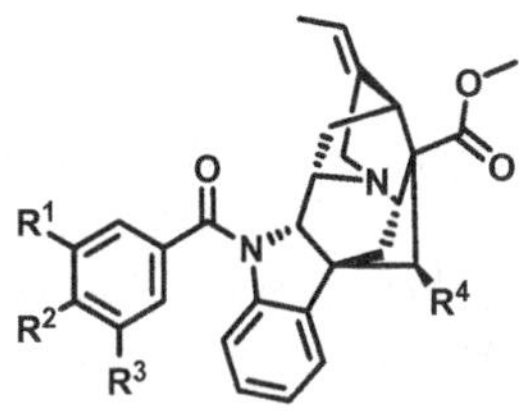

Alstiphyllanine I (161); $R^1$ = H, $R^2$ = $R^3$ = OMe, $R^4$ = $OCOCH_3$
Alstiphyllanine J (162); $R^1$ = $R^2$ = $R^3$ = OMe, $R^4$ = $OCOCH_3$
Alstiphyllanine K (163); $R^1$ = $R^2$ = $R^3$ = H, $R^4$ = $OCOCH_3$
Alstiphyllanine L (164); $R^1$ = H, $R^2$ = $R^3$ = OMe, $R^4$ = OH
Alstiphyllanine N (165); $R^1$ = $R^2$ = $R^3$ = H, $R^4$ = OH
Alstiphyllanine O (166); $R^1$ = $R^2$ = $R^3$ = OMe, $R^4$ = OH

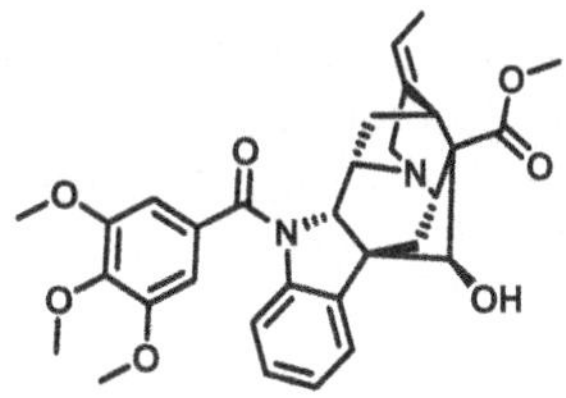

Alstiphyllanine M (167)

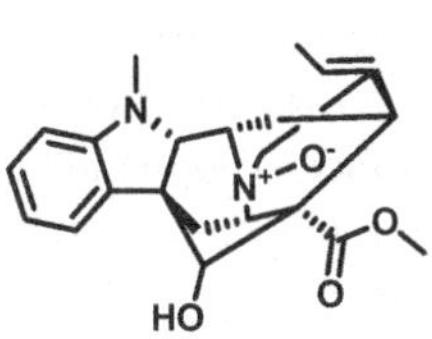

Alstiphyllanine H (168)

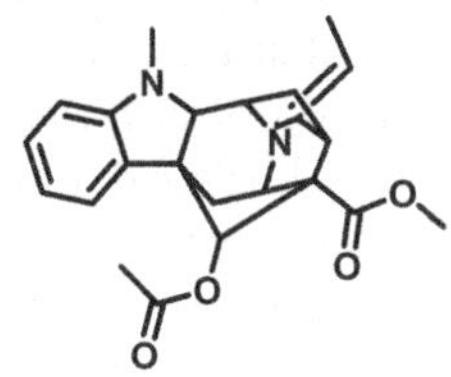

Vincamedine (169)

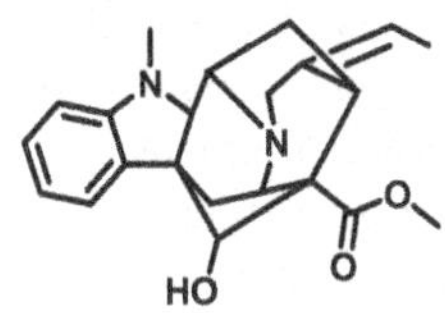

Vincamajine (170)

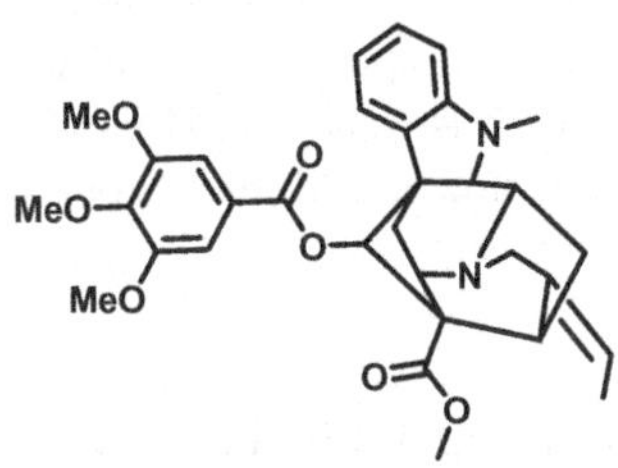

Vincamajine-17-*O*-3',4',5'-trimethoxybenzoate (171)

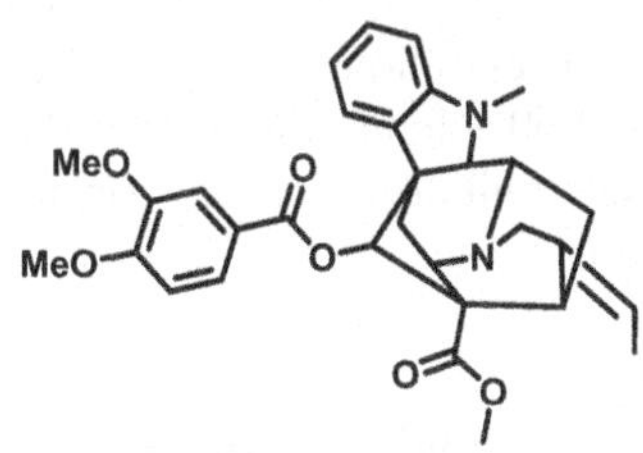

Vincamajine-17-*O*-veratrate (172)

**Fig. 14.34** Chemical structures of various alkaloids from *A. macrophylla* (**161–172**)

Arai et al. reported the ex vivo (rat aorta) vasorelaxant activities of *A. macrophylla*-derived alkaloids (161–172, Fig. 14.34) (Arai et al. 2012). Various mechanisms were proposed for the vasoconstriction. Administration of 0.3 μM of phenylephrine to the thoracic aortic rings with endothelium resulted in vasoconstriction. Vincamedine (169) resulted in a potential activity within 5–15 min, while alstiphyllanine H (168) exhibited poor activity. Further, to understand the involvement of endothelial cells, vasorelaxant activity was tested using endothelium-denuded aorta (-EC rings). In the presence of a nitric oxide synthase (NOS) inhibitor, $N^G$-monomethyl-*L*-arginine (L-NMMA, 100 μM), the relaxation was attenuated in -EC rings, clearly indicating that the vasorelaxation was partially mediated by the NO release from endothelial cells. Further, it was confirmed that the vasorelaxant effect of 169 did not involve a $K^+$ channel.

#### 14.3.2.4 Neuroleptic Activity

Chatterjee and Dey evaluated the neuroleptic activity of 116 (from *A. macrophylla*) by calculating the brain serotonin level in mice (Chatterjee and Dey 1964). Intraperitoneal administration of 116 (20 mg/kg) has shown an effect in the central nervous system. Brain serotonin content was identified as 0.64 and 1.64 μg/g after 15, 30 min of villalstonine administration.

#### 14.3.2.5 Contraceptive Effects

Chattopadhyay et al. screened the effects of methanolic extract and its subsequent n-butanol fractions of *A. macrophylla* leaves on the forward motility (FM) of mammalian (goat and human) spermatozoa (Chattopadhyay et al. 2005). n-butanol fractions exhibited a higher potential than the mother methanolic extract. Further, fractionation resulted in three fractions, such as 58 (fraction A), 41 (fraction B) and β-sitosterol glucoside (173, Fig. 14.35) and a mixture of minor compounds (fraction C). Although a semi-purified fraction was utilized for the pharmacological screening, the majority of activity raised was attributed to the presence of 41 only. A dose-dependent activity was exhibited by 41. In spectrophotometric motility assay of goat cauda sperm, control cells showed 40% vigorous FM. However, a reduced sperm motility was observed with the treatment of 41 at 100 and 200 μg/mL concentrations (25 and 5%, respectively). In human sperm, 41 (100 μg/mL) caused 90% inhibition on the FM. From these results, it is clear that 41 exhibited a higher inhibitory activity toward the human spermatozoa FM than goat cauda sperm.

#### 14.3.2.6 Sodium-Glucose Cotransporters (SGLT) Inhibiting Potential

Arai et al. isolated 20 alkaloids (19, 20, 122, 147–150, 168–172, 174–181) from the methanolic extract of *A. macrophylla* leaves (Fig. 14.36) and evaluated the inhibitory potential of sodium-glucose cotransporters (Arai et al. 2010). The in vitro

**Fig. 14.35** Chemical structure of β-sitosterol glucoside (**173**)

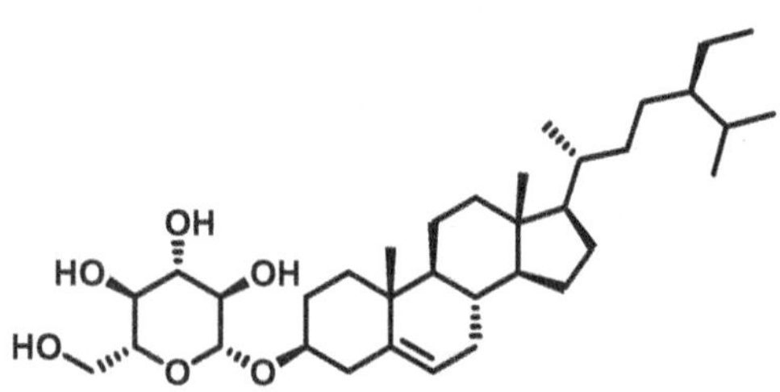


*β*-Sitosterol glucoside (173)

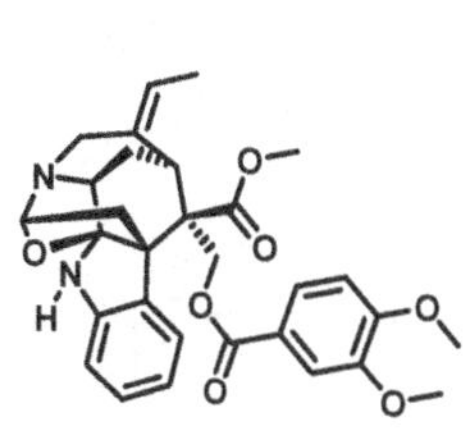

Alstiphyllanine E (174)

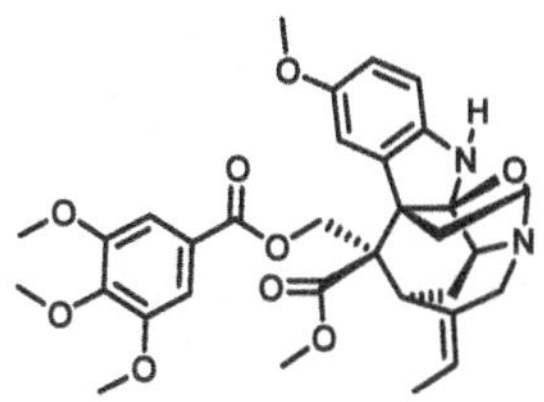

Alstiphyllanine F (175)

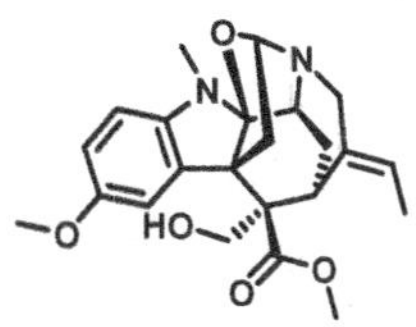

Alstiphyllanine G (176)

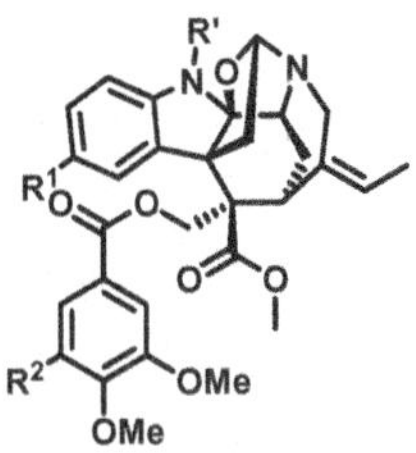

10-Methoxy-*N*(1)-methylburnamine-17-*O*-veratrate (177); $R^1$ = OMe, $R^2$ = H, R' = Me
Burnamine-17-*O*-30,40,50-trimethoxybenzoate (178); $R^1$ = H = R' = H, $R^2$ = OMe

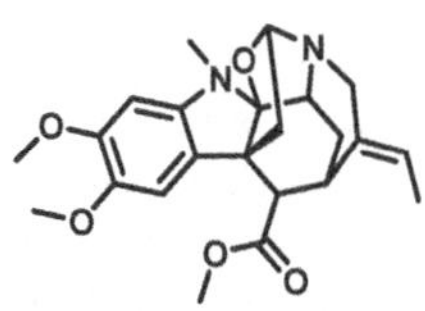

Quaternine (179)

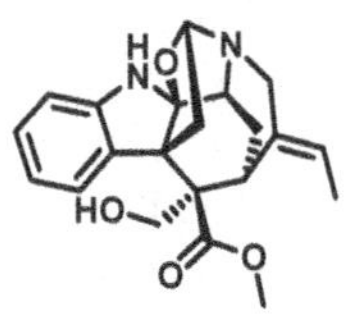

*O*-Deacetylpicraline (180)

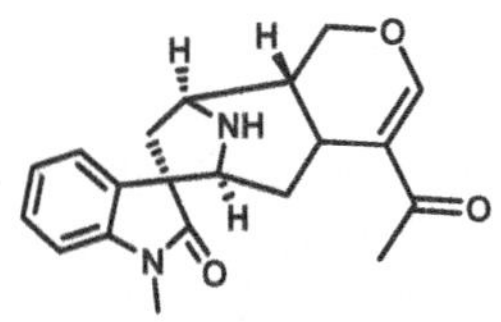

Alstonal (181)

**Fig. 14.36** Chemical structures of various alkaloids from *A. macrophylla* (**174–181**)

**Table 14.13** SGLT inhibitory activities of various alkaloids from *A. macrophylla*

| Compound | % inhibition[a] | |
|---|---|---|
| | SGLT1 | SGLT2 |
| 150 | 89.9 | 101.4 |
| 174 | 60.3 | 85.9 |
| 175 | 65.2 | 103.8 |
| 177 | 95.8 | 102.6 |
| 178 | 53.0 | 87.3 |

[a]At 50 μM

SGLT inhibitory potential was assessed by monitoring inhibition of uptake of methyl-α-*D*-glucopyranoside in cultured cells expressing SGLT1 or SGLT2. As represented in Table 14.13, 10-methoxy-$N^1$-methylburnamine-17-*O*-veratrate (177) resulted in a significant SGLT inhibition, followed by 150 and 175 (95.8, 89.9, and 65.2 for SGLT1 inhibition; 102.6, 101.4, and 103.8 for SGLT2, respectively). The remaining compounds exhibited a moderate to poor SGLT inhibition activity (0–30% inhibition).

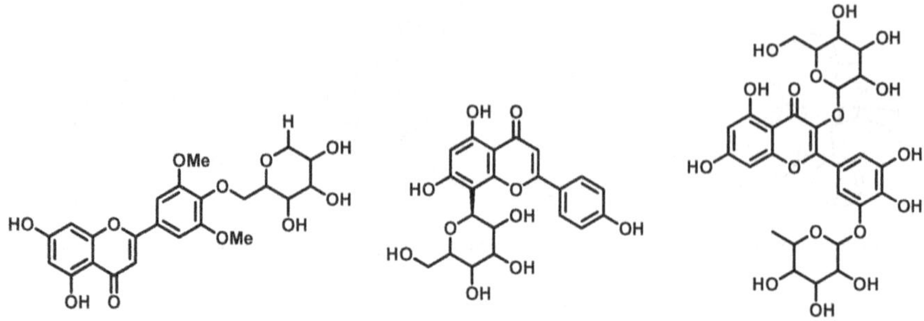

Tricin-4'-*O*-*β*-*L*-arabinoside (182) Vitexin (183) Myricetin-3'-rhamnoside-3-*O*-galactoside (184)

**Fig. 14.37** Chemical structures of flavonoid analogues from *A. macrophylla* (**182–184**)

#### 14.3.2.7 Antifungal and Antibacterial Effects

Parveen et al. isolated three flavonoids (182–184, Fig. 14.37) from the methanolic extract of *A. macrophylla* leaves and evaluated the antifungal and antibacterial effects of triacin-4′-*O*-*β*-*L*-arabinoside (182) by using agar well diffusion methods (Parveen et al. 2010). Antibacterial effects were screened in *S. aureus* (IAO-SA-22) and *E. coli* (K-12) by using chloramphenicol as a positive control. Antifungal activities were performed in *S. typhimurium* (MTCC-98) and *C. albicans* (IAO-109) by using nystatin as a positive control. The compound exhibited a maximum zone of inhibition with *S. typhimurium* (20 mm), *E. coli* (16 mm), followed by *S. aureus* and *C. albicans* (13 and 12 mm, respectively).

#### 14.3.2.8 Cholinesterase Inhibition Activity

Changwichit et al. isolated 129 from the methanolic extract of *A. macrophylla* stems and evaluated the cholinesterase inhibition activity. Compound 129 exhibited a mild acetylcholinesterase (AChE) inhibitory activity on electric eel AChE ($IC_{50}$ = 64.02 μM), human recombinant AChE ($IC_{50}$ = 88.93 μM), and horse BChE ($IC_{50}$ = 110.25 μM). Galantamine was used as a positive control that exhibited an $IC_{50}$ of 0.15, 0.10 and 1.30 in electric eel AChE, human recombinant AChE, and horse BChE, respectively (Changwichit et al. 2011).

### *14.3.3 Phytochemistry and Pharmacological Activities of* A. Angustifolia

#### 14.3.3.1 Anti-protozoal and Antiplasmodial Activities

Wright et al. isolated ten alkaloids (115, 119, 122–124, 170, 185–188) from the methanolic extract of *A. angustifolia* roots (Fig. 14.38) and evaluated the

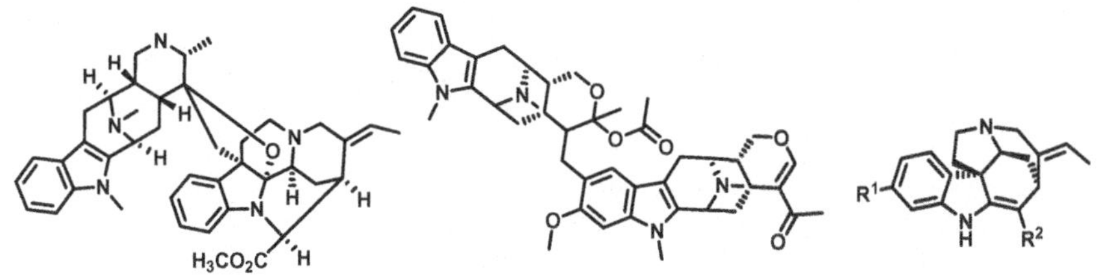


**Fig. 14.38** Chemical structures of various alkaloids reported from *A. angustifolia* (**185–188**)

anti-protozoal (*Entamoeba histolytica*) and antiplasmodial (*P. falciparum*) activities (Wright et al. 1992). Emetine hydrochloride and chloroquine diphosphate were used as positive controls for anti-protozoal and antiplasmodial activities, respectively. Among the screened alkaloids, 115, villastonine (185), and macralstonine acetate (186) exhibited a significant anti-protozoal activity ($ED_{50}$ = 8.12, 11.8, and 15.51 μM, respectively). Compound 124 and 170 were found to be inactive in the anti-protozoal screening. Emetine hydrochloride exhibited an $ED_{50}$ of 2.04 μM. In the case of anti-spasmodial activity, chloroquine diphosphate exhibited an $ED_{50}$ of 0.168 μM. Compound 185 exhibited potential antiplasmodial activity ($ED_{50}$ = 2.92 μM), followed by 186 and 115 ($ED_{50}$ = 3.43 and 9.36 μM, respectively). Compound 124 was found to be inactive against the *P. falciparum*. The remaining compounds exhibited $ED_{50}$ values in the range of 20.50–138 μM.

#### 14.3.3.2 Cytotoxicity Studies

Tan et al. isolated 20 alkaloids (189–208, Fig. 14.39) from the ethanolic extract of *A. angustifolia* barks, leaves and evaluated the reversing ability toward the MDR in vincristine-resistant KB cells (Tan et al. 2014). The screened alkaloids did not exhibit any kind of cytotoxicity against the vincristine-sensitive and vincristine-resistant (KB/VJ300) cells ($IC_{50}$ > 25 μg/mL). However, in the presence of 0.12 μM vincristine, 122 exhibited a strong activity in reversing MDR in drug-resistant KB/VJ300 cells ($IC_{50}$ = 10 μM). Apart from that alkaloids alstolactone A (191), *O*-acetyl talpinine (195), and 119 showed moderate to weak activity in the presence of 0.12 μM wherein, $IC_{50}$ values range between 40 and 60 μM.

Pan et al. isolated ten alkaloids (116, 122, 132, 155, and 209–214) from the methanolic extract of *A. angustifolia* stem bark (Pan et al. 2014) and screened the cytotoxic effects against HT-29 human colon cancer cell line (Fig. 14.40). Paclitaxel (positive control) exhibited an $ED_{50}$ value of 0.006 μM, while the villalstonidine E (214), 116 and 122 exhibited an $ED_{50}$ values of 6.5, 8.0, and 8.6 μM, respectively. The remaining compounds were considered as inactive in nature (>20 μM).

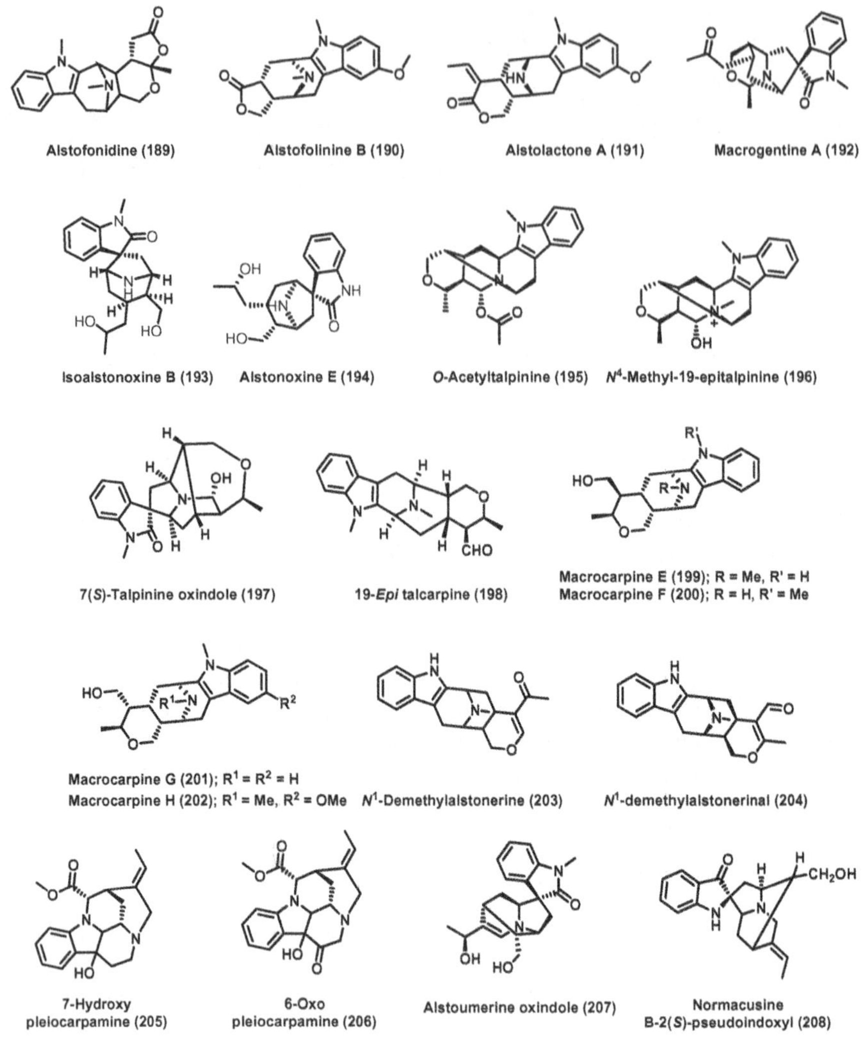


**Fig. 14.39** Chemical structures of various alkaloids from *A. angustifolia* (**189–208**)

#### 14.3.3.3 NF-κB (P65) Inhibitory Activity and Antileishmanial Activity

NF-κB (p65) inhibitory activity of the isolated compounds (116, 122, 132, 155, and 209–214) was performed in HeLa cells by using rocaglamide as a positive control (Pan et al. 2014). Except for $N^4$-methyl talpinine (209; $IC_{50}$ value = 1.2 μM), the remaining compounds were identified as inactive to the NF-κB inhibition. Rocaglamide exhibited an $IC_{50}$ of 0.08 μM. In the antileishmanial activity screening, $N^4$-methyl-$N^4$,21-seco talpinine (210), 214, and villalstonine $N^4$-oxide

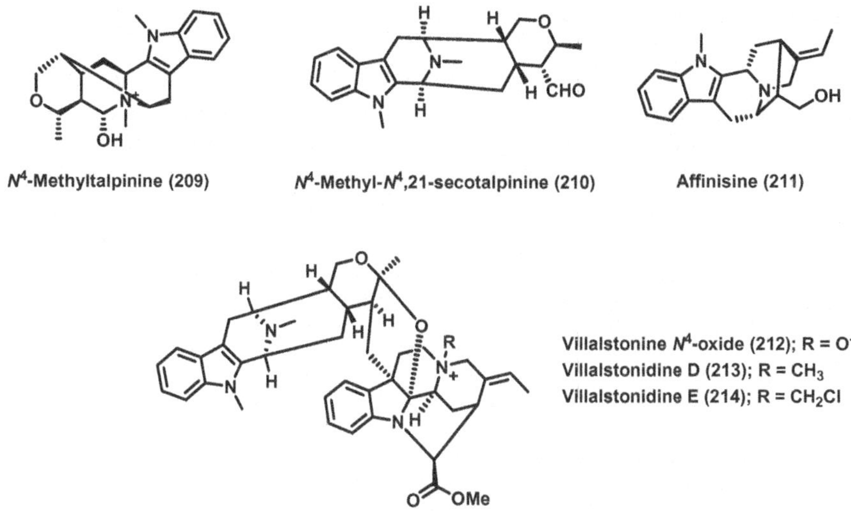


**Fig. 14.40** Chemical structures of various alkaloids from *A. angustifolia* (**209–214**)

(212) exhibited a moderate activity ($IC_{50}$ values = 57.8,78.0, and 80.3 μM) while the remaining compounds exhibited poor $IC_{50}$ values (>100 μM).

## *14.3.4 Phytochemistry and Pharmacological Activities of* A. Boonei

### 14.3.4.1 Behavioral, Acute Toxicity Effects, Diuretic Study, Cardiovascular Activity Studies, and Neuromuscular Activity

Ojewole reported various pharmacological activities of 9, isolated from the stem bark of *A. boonei.* Behavioral and acute toxicity effects (in albino mice), diuretic study (albino rats), cardiovascular activity studies (normotensive albino rats and cats), and neuromuscular activity (cats) were performed in whole animals (Ojewole 2008). Effects of 9 in various muscles were screened by using an isolated muscle, such as isolated atrial muscles and vascular smooth muscles. Guinea pig isolated spontaneously beating atria, electrically driven left atria represented the atrial muscles, whereas, rat isolated portal vein that represented the isolated vascular smooth muscles. The $LD_{50}$ of 9 was found to be 70.5 mg/kg *i.p.* and it did not produce any kind of behavioral changes in rat till 1/7th of $LD_{50}$ values. Two doses of 9 (25 and 50 mg/kg/oral) were administered to the albino rats for understanding the diuretic effects. Hydrochlorothiazide was used as a positive control (157.7% in urine output at 4 h). At 4 h, 9 exhibited the maximum diuretic effect (87.7% and 118.9% increased urinary output for 25 and 50 mg/kg of 9). A dose-dependent reduction in systemic arterial blood pressure and heart rate were observed in normotensive albino rats (for blood pressure-8.5–77.2% for 5–180 min; for heart rate-6–63% at a concentration range of 0.05–10.0 mg/kg *i.v* of 9) and cats (at a concentration range of 0.025–10.0 mg/kg *i.v* of 9). In vivo neuromuscular activity screening revealed that 9 (5–20 mg/kg, *i.p.*) depressed (23–81%) the twitches of the cat anterior *soleus* or *tibialis* muscle preparation induced by indirect electrical stimulation. At a concentration range of 10–100 μg/mL of 9, spontaneously beating atria isolated from the guinea pig reduced the amplitude and rate of contractions. However, in the electrically driven left atria, the force of contraction was also reduced. In the isolated smooth muscles, a depression in the amplitude of the spontaneous myogenic contractions was observed with the administration of 9 (10–100 μg/mL).

#### 14.3.4.2 Antiarthritic Effect

Okai and Carroll isolated **39** from the petroleum ether extract of *A. boonei* root barks and evaluated its antiarthritic effect in complete Freund's adjuvant (CFA)-induced arthritic Wistar rats (Kweifio Okai and Carroll 1993). At the treatment periods, 66 mg/kg body weight of 39 was administered orally in every 48 h from 32 to 40 days of post-adjuvant. Ankle and paw diameters remained unchanged among the treatment and control groups. Among the various biochemical parameters, alkaline phosphatase (U/L), spleen weight (g%) were altered during the arthritic conditions (control-265 U/L and 0.19 g%, respectively; arthritic rats-219 U/L and 0.22 g%, respectively). However, treatment with 39 returned the increase in spleen weight (0.20 g%) and the reduction in serum alkaline phosphatase (272 U/L).

In another study, Okai et al. investigated the antiarthritic effects of α-amyrin palmitate (215, Fig. 14.41) in complete Freund's adjuvant (CFA)-induced arthritic Wistar rats (Kweifio Okai et al. 1995). However, in the study, acute and chronic arthritic conditions were identified and evaluated for the effects of 215. During the acute phase, the ipsilateral ankle swelling was rapidly increased from 60% (11 days after CFA induction) to 94% (19 days after CFA induction). In the chronic phase, the percent swelling was similar from 32 to 50 days. Treatment of 215 prevented the percentage swelling rate by 34% in acute arthritic conditions. Higher blood granulocyte and serum hyaluronate were observed in the acute conditions of arthritis with the control groups (1292–9827 × $10^6$/L and 181–287 μg/L). In chronic conditions also these biochemical parameters were increased from the control groups (Blood granulocytes from 668 to 2287 × $10^6$/L and Serum hyaluronate from 100 to 277 μg/L). However, 215 treatment results in declined biochemical parameters (For acute condition: blood granulocytes-5681 × $10^6$/L and Serum hyaluronate-149 μg/L; for chronic conditions: blood granulocytes—1813 × $10^6$/L and Serum hyaluronate—147 μg/L).

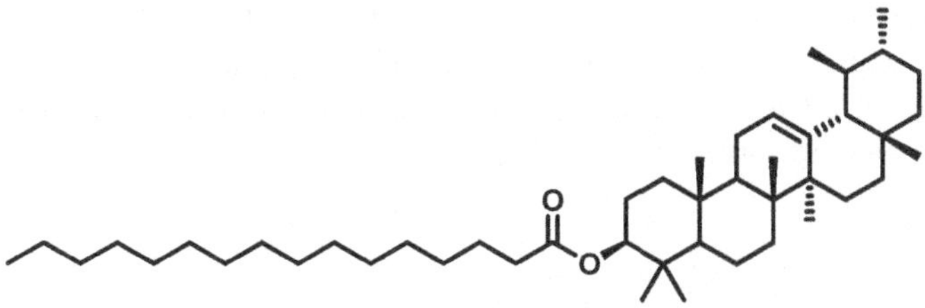

α-amyrin palmitate (215)

Fig. 14.41 Chemical structure of α-amyrin palmitate (**215**)

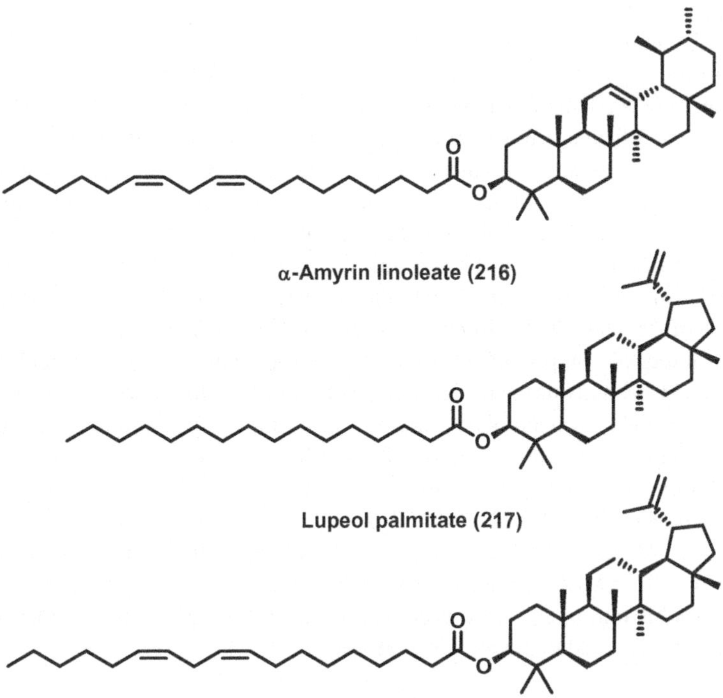


Fig. 14.42 Chemical structures of various triterpenoids *A. boonei* (**216–218**)

#### 14.3.4.3 Serine Protease Inhibitory Potential

Rajic et al. reported the serine protease (trypsin and chymotrypsin) inhibition potential of triterpenoids (38, 39, 215–218) from the barks and roots of *A. boonei* (Fig. 14.42) (Rajic et al. 2000). In the case of trypsin inhibition, lupeol palmitate (217) exhibited a higher potential ($IC_{50}$ = 6.0 μM, with a $K_i$ value of 10.4 μM) than the remaining triterpenoids ($IC_{50}$ = 10–41 μM). However, for chymotrypsin inhibition, α-amyrin linoleate (216) exhibited the potential activities of 16 μM with a $K_i$ value of 27.5 μM. Further, these compounds have inhibited the enzymes through non-competitive inhibition kinetics.

#### 14.3.4.4 Anti-inflammatory Activity

Okoye et al. isolated 37 and 55 from the methanolic extract of *A. boonei* stem bark (Okoye et al. 2014). The anti-inflammatory activity was evaluated by using egg albumen-induced paw edema and xylene-induced ear edema models in rats. Apart from this, gastric ulcerogenic, in vivo leukocyte migration, and RBC membrane stabilization tests were also investigated. Prior (30 min) to the sub-plantar injection of the egg albumen (0.1 mL), 37 (50 and 100 mg/kg) was administered orally. Aspirin (100 mg/kg) was used as a positive control. Aspirin recorded significant reduction in paw edema from the 2nd h, while the screened compound (100 mg/kg) showed a significant activity from the 5th h. Ulcerogenic effect of 37 (50 and 100 mg/kg) was identified by using an ulcer index value. Indomethacin (40 mg/kg) was administered as a positive control. Indomethacin evoked significant irritation of the gastric mucosa compared with the control, while the test compound did not evoke significant irritation. Heat-induced hemolysis and hypotonicity induced hemolysis methods were used for the assessment of membrane stabilization potential of 55 and 37. Diclofenac sodium (100 μg/mL) was used as a positive control. Screened compounds (50 and 100 μg/mL) exhibited higher inhibition on heat-induced haemolysis than diclofenac sodium (40% inhibition). Conversely, in hypotonicity induced haemolysis, diclofenac sodium (50% inhibition) exhibited a higher potential than the compounds. Further, to evaluate the acute topical inflammation, the effects of 37 and 55 in xylene-induced ear inflammation were determined. Compound 37 exhibited a higher % inhibition (>45% at 50 and 100 μg/mL) while 55 exhibited a comparable activity to that of indomethacin (45% inhibition 100 μg/mL). Furthermore, a significant inhibition of leukocyte migration and neutrophil infiltration were produced by 37 (100 mg/kg) in comparison with positive control (indomethacin—50 mg/kg).

#### 14.3.4.5 Antibacterial and Antioxidant Activities

Okoye et al. isolated eight flavonoid glycosides (219–226, Fig. 14.43) from the methanolic extract of *A. boonei* leaves and evaluated the antioxidant activity (DPPH free radical scavenging model) and antimicrobial activity (Agar well diffusion technique) (Okoye and Okoye 2016). In DPPH scavenging assay, kaempferol derivatives did not exhibit any kind of activities, while the quercetin derivatives exhibited dose-dependent activities ($IC_{50}$) ranging from 36.0 to 66.0 μg/mL. Quercetin-3-*O*-[α-L-rhamnopyranosyl (1 → 4)-*β*-*D*-glucopyranoside] 223) and Quercetin-3-*O*-robinobioside (220) exhibited a higher DPPH scavenging potential

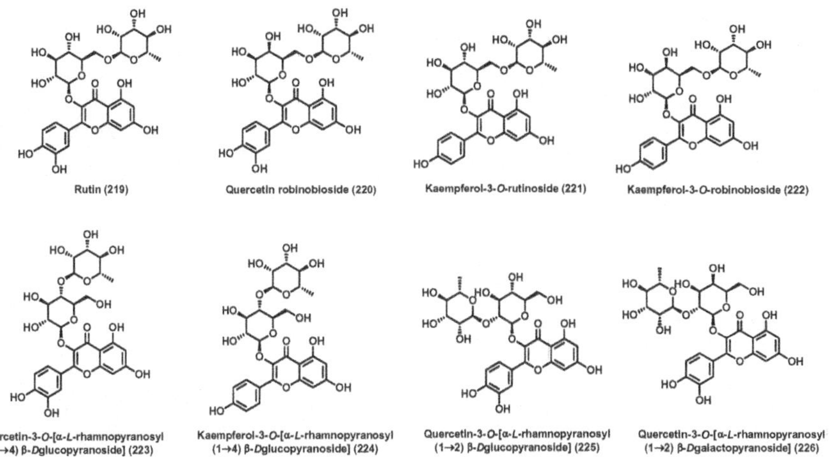


**Fig. 14.43** Chemical structures of various flavonoid glycosides from *A. boonei* (**219–226**)

($IC_{50}$ values = 36.0 and 48.0 μg/mL) than the positive control, Vit. C ($IC_{50}$ of 49.0 μg/mL). Antimicrobial activity was performed against *E. coli.* Quercetin-3-*O*-[α-L-rhamnopyranosyl (1 → 2)-*β*-*D*-glucopyranoside] 225) and Quercetin-3-*O*-[α-L-rhamnopyranosyl (1 → 2)-*β*-*D*-glucopyranoside] (226) exhibited a potential antibacterial activity with a MIC values of 1.77 μg/mL and 1.92 μg/mL, respectively.

#### 14.3.4.6 Cytotoxic Activities

Olaoye S. et al. isolated a cytotoxic indole alkaloid, Alstiboonine (227, Fig. 14.44), from aqueous methanolic extract of *A. boonei* stem bark and confirmed its cytotoxic activities through brine shrimp lethality assay (Balogun et al. 2016). The titled compound exhibited a potent $LD_{50}$ value (39.72 μg/mL) and was claimed as a cytotoxic phytochemical.

### *14.3.5 Phytochemistry and Pharmacological Activities of* A. Venenata

#### 14.3.5.1 Hypertensive and Neuroleptic Properties

P. K Dey evaluated various pharmacological activities of alstovenine (228, Fig. 14.45), an alkaloid from *A. venenata* (Dey 1965). The hypertensive activity of 228 was screened in chloralosed cat, and the desired pharmacological activity was observed with a concentration of 25–30 μg/kg of body weight. Further, a 100–

**Fig. 14.44** Chemical structure of alstiboonine (**227**)

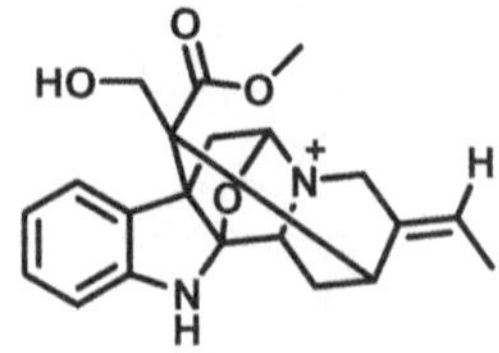


Alstiboonine (227)

**Fig. 14.45** Chemical structure of alstovenine (**228**)

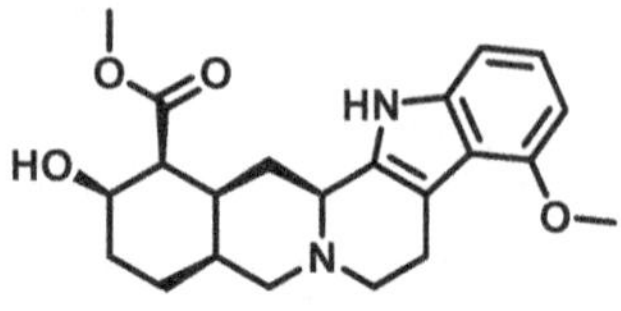


Alstovenine (228)

200 μg bath of 228 did not exhibit any kind of activity in uterine and intestinal contractions. Higher nervous activity was screened in the rabbit, by injecting 30 μg/kg of 228. The treated animal did not exhibit any kind of immediate effect. After 30 min, it became quiet, showed a stupor attitude. Neuroleptic properties of 228 were identified with the help of spontaneous motility tests in mice. Intraperitoneal administration of 500 μg/kg of the drug resulted in somnolence and a state of apprehension after 30–40 min.

### 14.3.5.2 Psychopharmacological Effects

Bhatt acharyaa et al. evaluated the psychopharmacological effects of *A. venenata*-derived alkaloids (228, venenatine (229) and echitovenidine (230), Fig. 14.46) by various methods (Bhattacharya et al. 1975). Compound 228 (1 mg/kg, *i.p.*) produced signs of central excitation such as piloerection, irritability, enhanced locomotor activity, increased startle response, and exophthalmos. A reduction in motor

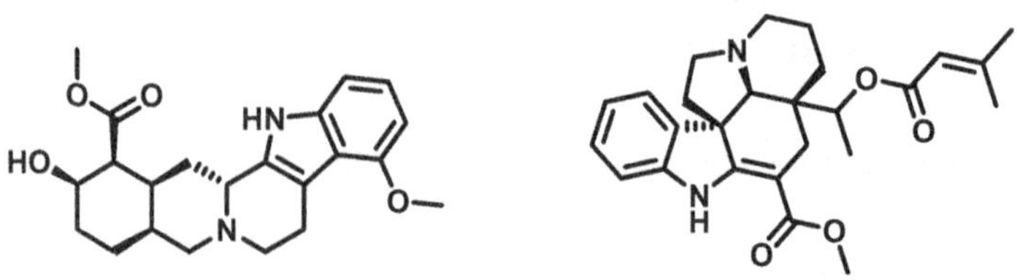


Venenatine (229) Echitovenidine (230)

**Fig. 14.46** Chemical structures of venenatine and echitovenidine (**229** and **230**)

activity and sedation was resulted by the administration of 229 (50 mg/kg, *i.p.*). However, at 100 mg/kg (higher dose), a sedative effect was observed in animals. In the case of 230, a central stimulation in rats and mice was observed. However, in the later phase, a central depression had occurred. All the tested compounds (100 mg/kg, *i.p.*) potentiated the hexobarbitone hypnosis. Both 228 and 230 antagonized reserpine induced sedation, ptosis at 1 and 2.5 mg/kg, *i.p.* respectively. However, 229 (100 mg/kg, *i.p.*) synergistically enhance the ptosis and sedation. Further, 228 and 230 (1 and 2.5 mg/kg, *i.p.*, respectively) potentiated the lethal effects of amphetamine (10 mg/kg, *i.p.*) in aggregated mice. Furthermore, these compounds potentiated the behavioral effects of DOPA and 5-HTP. In contrary, 229 did not exhibit any kind of activity. The analgesic activity of a subanalgesic dose of morphine (2 mg/kg, *i.p.*) was potentiated by 228 (a latent period of tail-flick response from 9.2 to 22.3 s.) and 230 (a latent period of tail-flick response from 10.3 to 25.3 s). While 229 had antagonizing effects on the analgesic activity (a latent period of tail-flick response from 28.6 to 16.2 s.). Diphenylhydantoin (2.5 mg/kg, *i.p.*) did not exhibit any anti-convulsant activity. However, pre-treatment with 228 and 230 (1 and 2.5 mg/kg, *i.p.*, respectively) resulted in 60% of anti-convulsant activity. Compound 229 antagonized the anti-convulsant activity (effective dose—25 mg/kg, *i.p.* of diphenylhydantoin exhibited a 100% inhibition, while pre-treatment resulted in a 40% anti-convulsant activity). Compounds 228 and 230 potentiated the tryptamine induced convulsions (tryptamine control group—10% animals having convulsion, whereas 228 group—70% animal; 230 group—60% animals). 229 did not exhibit any kind of tryptamine-related activities. A transient depressor response along with a stimulation of respiration was produced by 228 in anaesthetized dog's blood pressure, while 230 exhibited the same kind of activity without respiratory stimulation. 229 produced a moderate depressor response. $LD_{50}$ of the compounds was found to be 8.7, 126, and 176 mg/kg, respectively for 228, 230, and 229. These all results indicated the potential of 228 and 230 as an inhibitor of monoamine oxidase.

#### 14.3.5.3 Antifungal Activities

S. K. Singh et al. evaluated the antifungal activities of $\Delta^3$-Alstovenine (231, Fig. 14.47), a quaternary alkaloid isolated from the water-soluble base fraction of the bark of *A. venenata* (Singh et al. 1999). A series of concentrations (0, 250, 500,

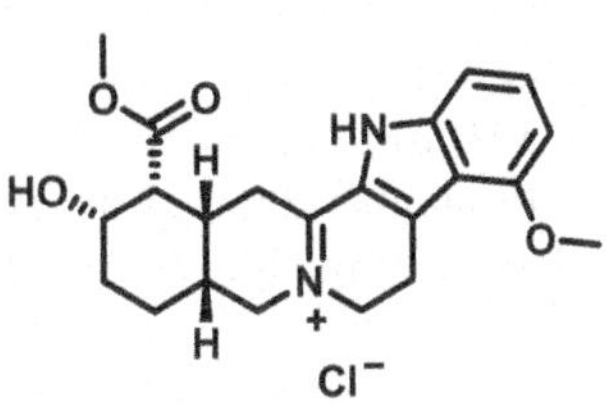


$\Delta^3$-Alstovenine (231)

**Fig. 14.47** Chemical structure of $\Delta^3$-Alstovenine (**231**)

700, and 1000 mg/L) were evaluated against 17 fungal strains. At 250 mg/L concentration, spore germination of *Cercospora* sp. was completely inhibited while at a concentration of 750 or 1000 mg/L, *H. sativum, Helminthosporium maydis* and *Erysiphe polygoni* were found to be sensitive to 231 (100% inhibition of spore germination). *Alternaria* species and *Fusarium udum* were found to be resistant to the tested compounds. Further, 231 was active against facultative and biotrophic pathogens.

In another study, U. P. Singh et al. evaluated the antifungal activity of 229 against ten species of fungi (Singh et al. 2000). Since 229 is insoluble in water, an acetate form was prepared by dissolving the alkaloid in a molar equivalent of 1 mol/L aqueous acetic acid. A series of concentrations (0.5, 1.0, 1.5, 2.0 mg/L) were evaluated for antifungal activities. An aqueous solution of 229 as its acetate showed antifungal activity against all the plant pathogenic and saprophytic fungi. *A. brassicicola* and *U. cynodontis* were most sensitive to 229 (higher inhibitory potential at 0.5 mg/L concentration). At the higher concentrations (2.0 mg/L), *Fusarium udum, Aspergillus flavus, Alternaria brassicicola*, and *Ustilago cynodontis* showed maximum sensitivity (seed germination is less than 10%). Further, the effect of 229 on the germination and development of *E. pisi* conidia on excised pea leaves were also evaluated. Results indicated that 229 had a marked effect in the reduction of germination of *E. pisi* conidia and pre-inoculation treatment with 229 showed greater efficacy than post-inoculation treatment.

### *14.3.6 Phytochemistry and Pharmacological Activities of* A. Yunnanensis

#### 14.3.6.1 Cytotoxic and Anti-inflammatory Activity

Feng et al. isolated various alkaloids (20, 232–255, Fig. 14.48) from the alkaloid extract of *A. yunnanensis* whole plants and cytotoxic and anti-inflammatory potential were evaluated (Feng et al. 2009). Alstoyunines C (234), E (236), and F (237) displayed selective inhibition of COX-II (>75%), while NS-398 (positive control) exhibited a 97.09% COX-II inhibition. All the screened alkaloids exhibited poor activity against COX-I. In the 5-LOX inhibition assay, 236 exhibited 77.72% inhibition, while positive control (zileuton) exhibited 83.05% inhibition. Cytotoxicity evaluations were performed in HL-60, A-549, SMMC-7721, pancreatic cancer (PANC-1), and SK-BR-3cells. Except 237 remaining compounds exhibited a poor activity in all the cell lines ($IC_{50} > 40$ μM). 237 exhibited an $IC_{50}$ of 3.89 μM against HL-60.

In another study, Cao et al. isolated 8 (240, 256–262) monoterpenoid indole alkaloids (Fig. 14.49) from the ethanolic extract of *A. yunnanensis* whole plants and

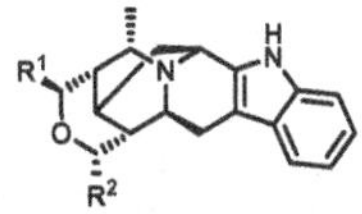

Alstoyunine A (232); R¹ = OMe, R² = OH
Alstoyunine B (233); R¹ = OH, R² = OMe

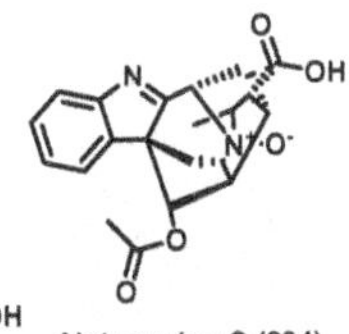

Alstoyunine C (234)

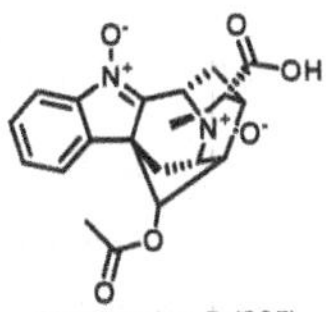

Alstoyunine D (235)

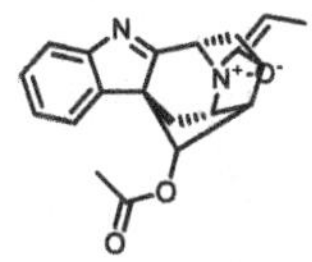

Alstoyunine E (236)

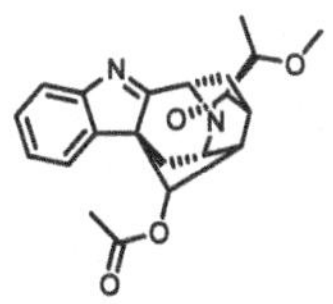

Alstoyunine F (237)

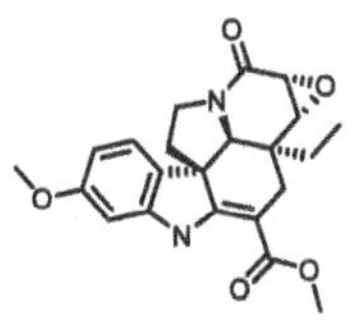

Alstoyunine G (238)

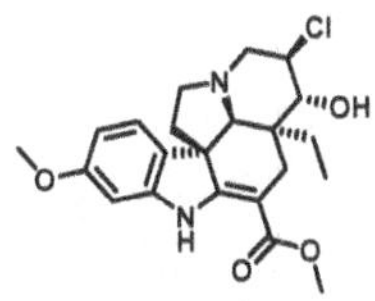

Alstoyunine H (239)

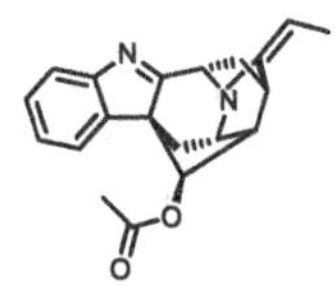

Vinorine (240)

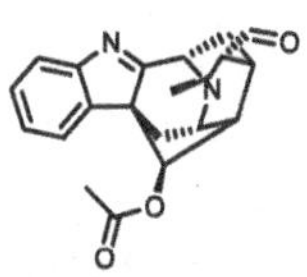

Perakine (241)

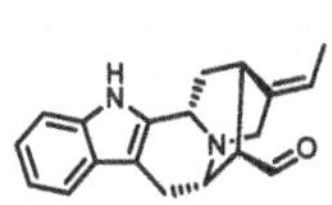

Vellosimine (242)

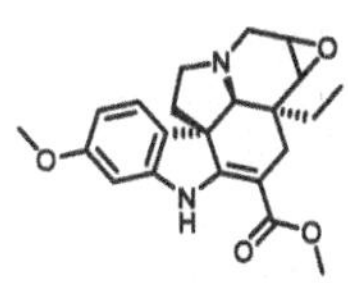

Lochnerinine (243)

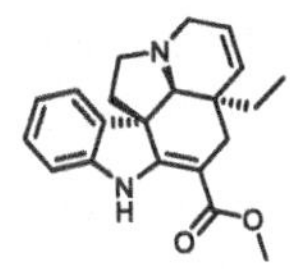

Tabersonine (244)

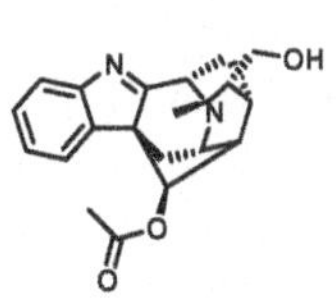

Raucaffrinoline (245)

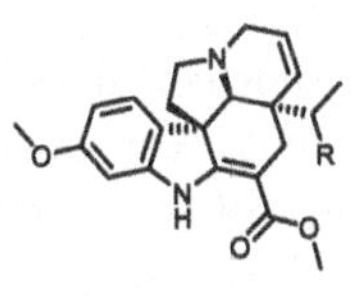

11-Methoxy tabersonine (246); R = H
19-acetoxy-11-methoxytabersonine (247); R = OAc

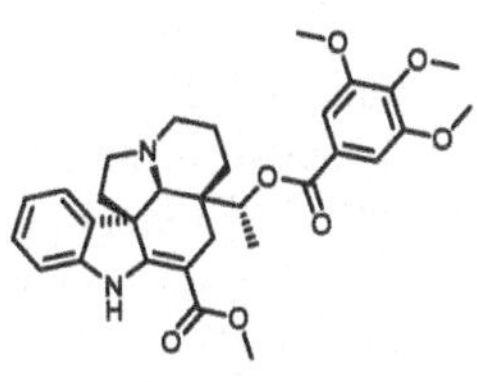

(-)-Echitoveniline (248)

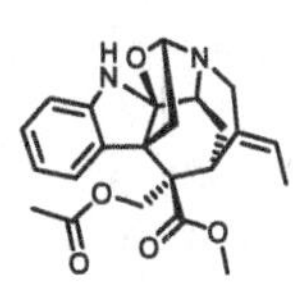

Picraline (249)

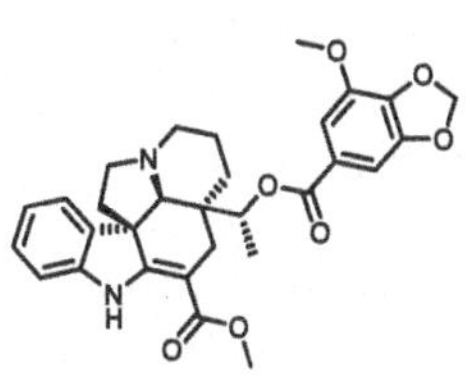

Echitoserpidine (250)

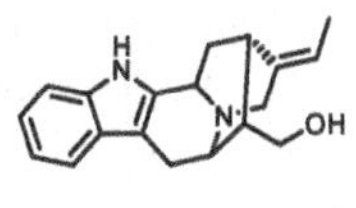

Vellosiminol (251)

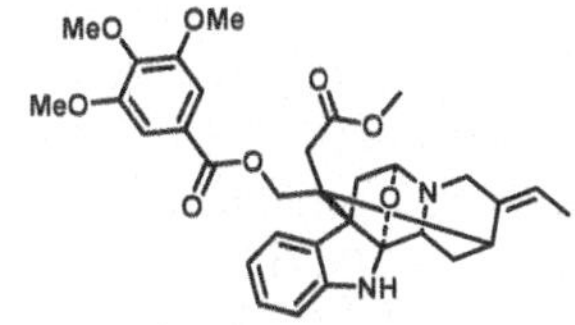

19(*Z*)-burnamine-17-*O*-3',4',5'-trimethoxybenzoate (252)

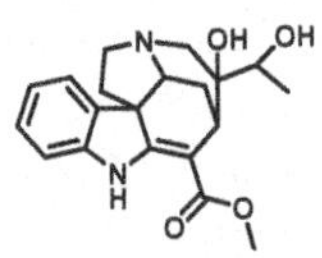

Compactinervine (253)

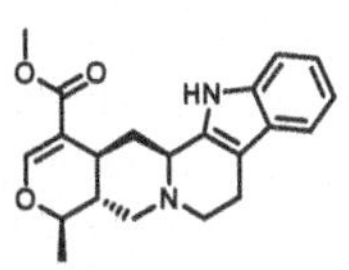

19-*Epi*-ajmalicine (254)

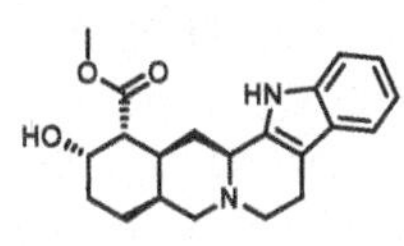

Alloyohimbine (255)

**Fig. 14.48** Chemical structures of various alkaloids from *A. yunnanensis* (**232**–**255**)

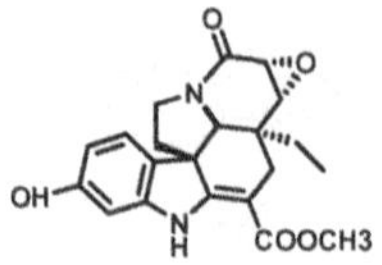

11-Hydroxy-6,7-epoxy-8-oxo-vincadifformine (256)

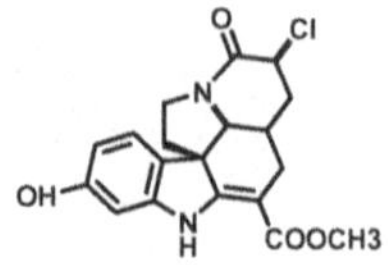

14-Chloro-15-hydroxyvincadifformine (257)

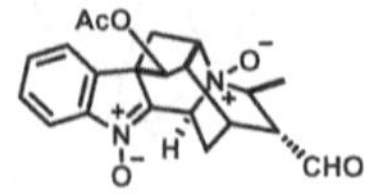

Perakine $N^4$-dioxide (258)

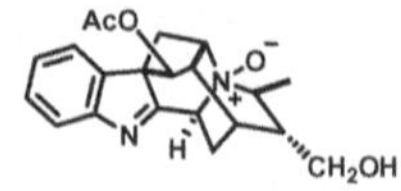

Raucaffrinoline $N^4$-oxide (259)

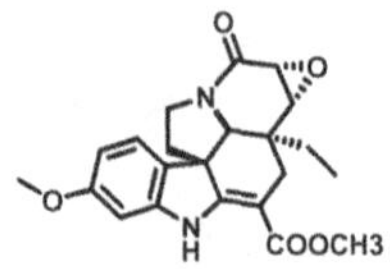

11-Methoxy-6,7-epoxy-8-oxovincadifformine (260)

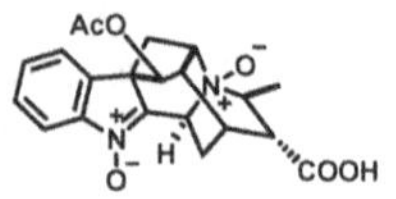

Vinorine $N^1$,$N^4$-dioxide (261)

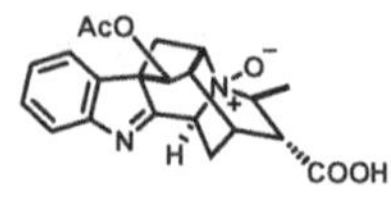

Vinorine $N^4$-oxide (262)

**Fig. 14.49** Chemical structures of various alkaloids from *A. yunnanensis* (**256–262**)

evaluated its cytotoxic and anti-inflammatory potentials (Cao et al. 2012). For the cytotoxicity evaluation, following cell lines were used astrocytoma (CCF-STTG1), glioma (CHG-5, SHG-44, U251), MCF-7 cells, human skin cancer (SK-MEL-2), and meningioma (BEN-MEN-1). Except for meningioma cell lines, perakine $N^4$-oxide (258), raucaffrinoline $N^4$-oxide (259), and vinorine $N^4$-oxide (262) exhibited cytotoxicity against all the tested tumor cell lines ($IC_{50}$ values ranging from 9.2 to 65.5 μM). The remaining compounds were identified as non-toxic in nature ($IC_{50}$ value > 50 μM). Adriamycin (positive control) exhibited strong cytotoxicity wherein the $IC_{50}$ values ranges from 14.1 to 37.6 nM. In in vitro anti-inflammatory screening, similar results were obtained, wherein 258, 262, and 259 exhibited selective COX-II inhibition (94.77, 94.05, and 88.09%, respectively), NS-398 was used as a positive control that exhibited 97.13% of COX-II inhibition. All screened compounds exhibited poor COX-I inhibition. These all results indicate that $N^4$-oxide is required for cytotoxicity and anti-inflammatory activity.

Li et al. isolated six alkaloids (263–268, Fig. 14.50) from the ethanolic extract of *A. yunnanensis* aerial parts and evaluated the in vitro cytotoxicity in eight tumor cell lines (Li et al. 2017). The selected cell lines are human osteosarcoma (SOSP-9607, Mg-63, Saos-2, M663), human gastric carcinoma (BGC-823), $HepG_2$, HL-60 and MCF-7 . Adriamycin was used as a positive control. Among the tested compounds, alstiyunnanenine D (266), E (267) and alstonia scholaine I (268) exhibited the potent cytotoxic activity in osteosarcoma cell lines ($IC_{50}$ = 3.2–5.8 μM). Except Saos-2 and M663 cell lines, alstiyunnanenine A (263) exhibited poor activity against all the selected cell lines ($IC_{50}$ > 35 μM). Alstiyunnanenine B (264) and C

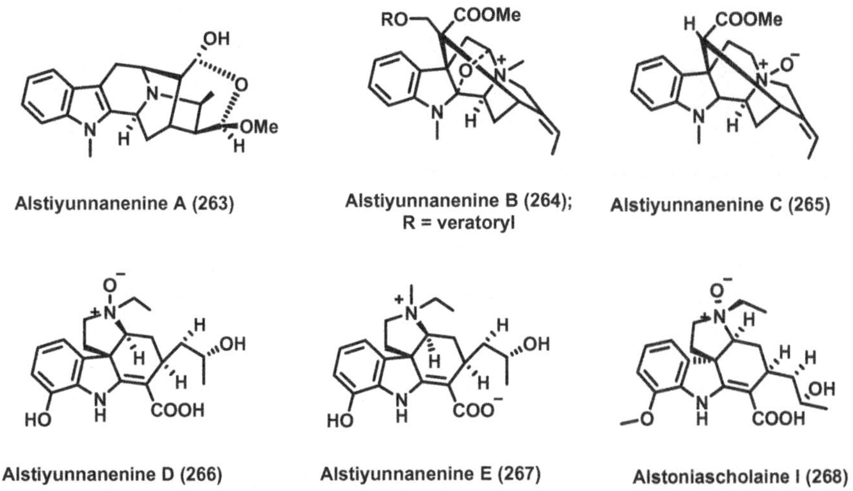


**Fig. 14.50** Chemical structures of various alkaloids from *A. yunnanensis* (**263–268**)

(265) exhibited moderate cytotoxicity against BGC-823, $HepG_2$, HL-60, and MCF-7 cells, wherein $IC_{50}$ values were ranging in 16.9–24.2 μM. *Adriamycin* $IC_{50}$ values were observed in the range of 0.01–0.04 μM.

## 14.3.7 Phytochemistry and Pharmacological Activities of A. Spatulata

### 14.3.7.1 Cytotoxicity Activity

Tan et al. isolated 23 (15, 17, 20, 29, 269–287) alkaloids from ethanolic extract of *A. spatulata* leaves and stembark (Fig. 14.51) (Tan et al. 2010). Further, they evaluated the cytotoxicity effects of these alkaloids against KB/S (vincristine-sensitive) and KB/VJ300 (vincristine-resistant) cell lines. In KB/S cell lines, 15-hydroxyangustilobine A (269) and angustilobine B (281) exhibited a potential activity ($IC_{50}$ = 5.26 and 6.99 μg/mL, respectively), while the remaining screened compounds were found to be inactive ($IC_{50}$ > 25 μg/mL). 281 exhibited significant cytotoxicity ($IC_{50}$ = 8.12 μg/mL). 24 and 269 exhibited moderate cytotoxicity ($IC_{50}$ = 13.35 and 14.39 μg/mL, respectively). However, except alstolucine C (278), nor-6,7-secoangustilobine A (283), and undulifoline (284), all the tested compounds were found to reverse MDR in vincristine-resistant KB (VJ300) cells ($IC_{50}$ = 0.59–19.22 μg/mL).

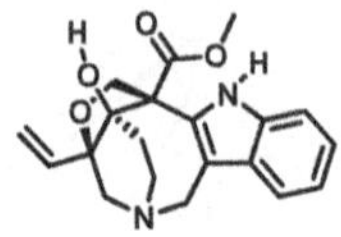

15-Hydroxy angustilobine A (269)

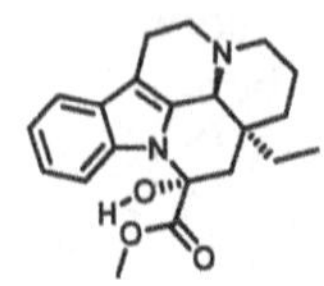

16-Epivincamine (270)

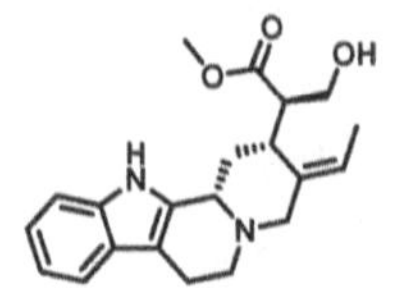

16*R*,19*E*-Isositsirikine (271)

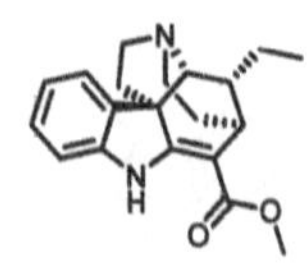

20(*R*)-Tubotaiwine (272)

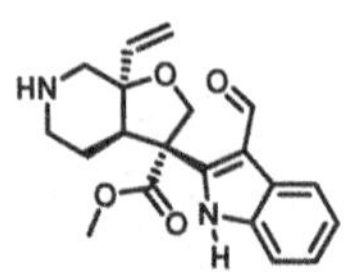

4,6-Secoangustilobinal A (273)

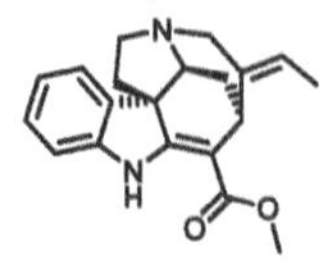

Akuammicine (274)

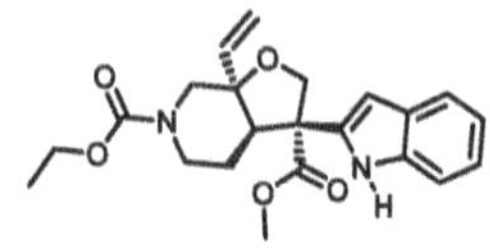

Alstolobine A (275)

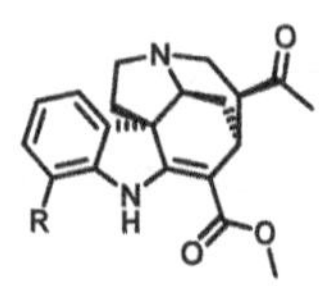

Alstolucine B (276); R = H
Alstolucine D (277); R = OH

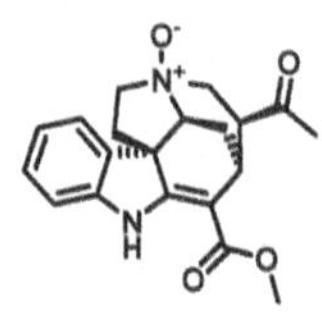

(-)-Alstolucine C (278)

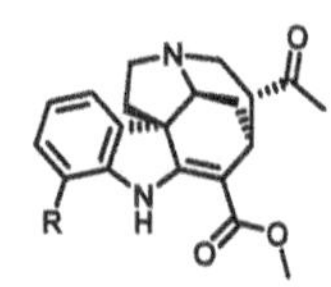

Alstolucine E (279); R = OH
(-)-Alstolucine F (280); R = H

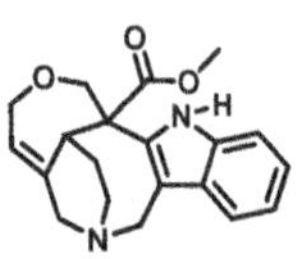

Angustilobine B (281)

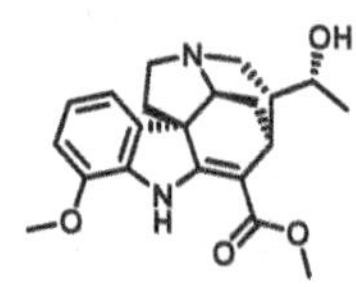

$N^4$-Demethyl-12-methoxyalstogustine (282)

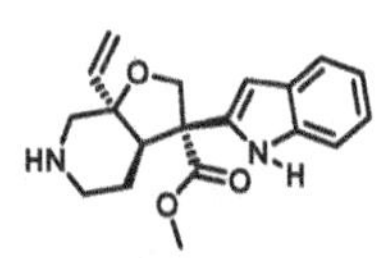

Nor-6,7-Seco angustilobine A (283)

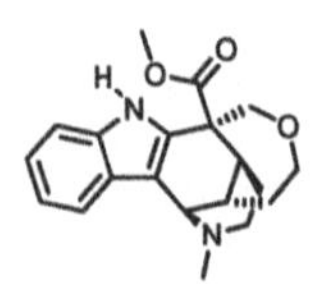

Undulifoline (284)

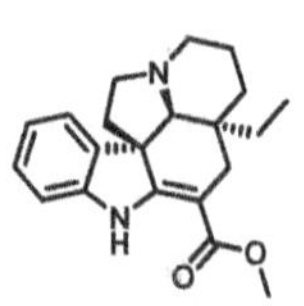

Vincadifformine (285)

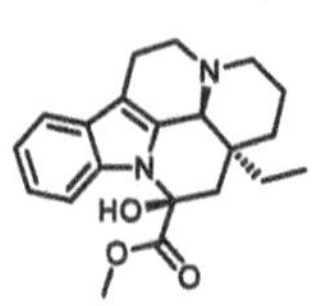

Vincamine (286)

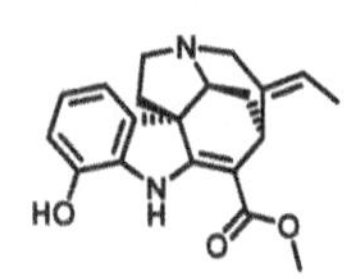

Vinervine (287)

**Fig. 14.51** Chemical structures of various alkaloids from *A. spatulata* (**269**–**287**)

### 14.3.8 Phytochemistry and Pharmacological Activities of A. Rupestris

#### 14.3.8.1 Cytotoxicity Activity

Wang et al. isolated 10 alkaloids (288–297) from the ethanolic extract of *A. rupestris* leaves (Fig. 14.52) and evaluated its cytotoxicity potentials (Wang et al. 2013). Cytotoxicity effects were evaluated in various cell lines, such as A-549, HepG2, SMMC-7721, MCF-7, HL-60, BGC-823, and SW-480. Doxorubicin was used as a positive control. (*E*)-16-formyl-5α-methoxystrictamine (296), scholarisin I (288), and scholarisin VI (293) possessed significant cytotoxicity ($IC_{50}$ values < 30 μM) against all the tested tumor cell lines. Scholarisin IV (291) and scholarisin V (292) were found to be non-cytotoxic ($IC_{50}$ > 80 μM). Structural analysis indicated that these compounds lack a linkage between C-5 and N-4, hence it is speculated that these bonds are essential for cytotoxicity. The remaining compounds demonstrated moderate cytotoxicity ($IC_{50}$ values = 30–80 μM).

Zhang et al. isolated 6 monoterpenoid indole alkaloids (256, 261, 298–301) from the ethanolic extract of *A rupestris* aerial parts (Fig. 14.53) and evaluated its cytotoxic potential (Zhang et al. 2014a). Cytotoxic activity was evaluated in seven head and neck squamous cell carcinomas (CAL-27, Detroit-562, Hep-2, SCL-1, SCC-PKU, TCA-83, and UMSCC-1). Doxorubicin was used as positive control ($IC_{50}$ values in the range of 14.7–35.4 nM). Among the screened compounds, 11-hydroxy-6,7-epoxy-8-oxovincadifformine (256) 6,7-epoxy-8-oxo-vincadifformine (298), and 11-acetyl-6,7-epoxy-8-oxo-vincadifformine (299) exhibited significant cytotoxicity

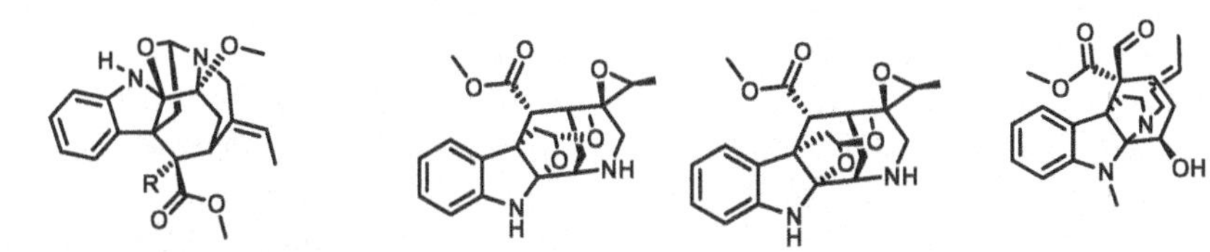

Scholarisin I (288); R = CHO
Scholarisin II (289); R = $CH_2OH$
Scholarisin III (290); R = $OCOCH_3$

Scholarisin IV (291)

Scholarisin V (292)

Scholarisin VI (293)

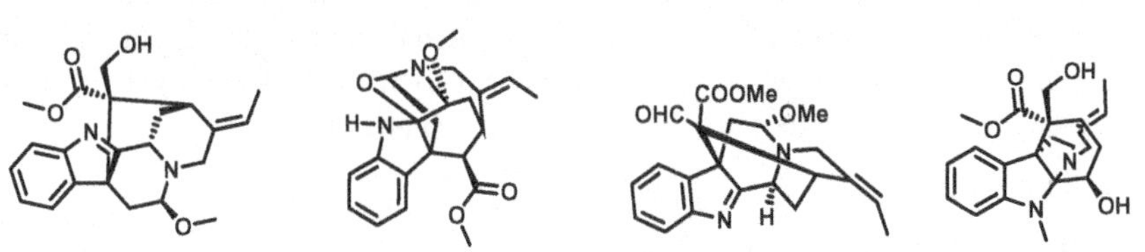

Scholarisin VII (294)

(3*R*,5*S*,7*R*,15*R*,16*R*,19*E*)-Scholarisine F (295)

(*E*)-16-Formyl-5*α*-methoxystrictamine (296)

3-*Epi*-dihydrocorymine (297)

**Fig. 14.52** Chemical structures of various alkaloids from *A. rupestris* (**288–297**)

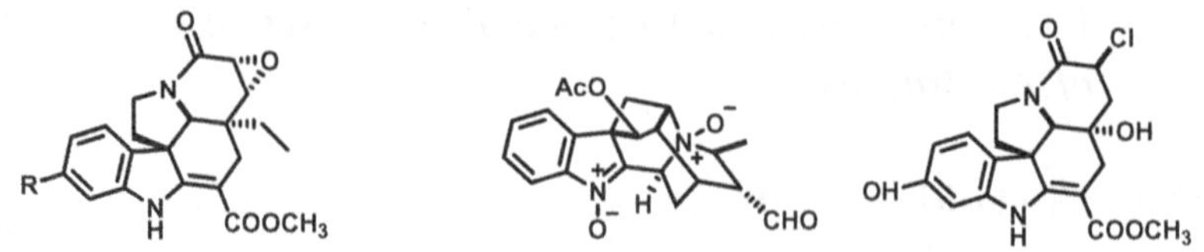


6,7-Epoxy-8-oxo-vincadifformine (298); R = H
11-Acetyl-6,7-epoxy-8-oxo-vincadifformine (299); R = OAc

Perakine $N^1$,$N^4$-dioxide (300)

11-Hydroxy-14-chloro-15-hydroxy-17 vincadifformine (301)

**Fig. 14.53** Chemical structures of various alkaloids from *A. rupestris* (**298–301**)

against all tested cell lines with $IC_{50}$ values less than 20 μM, while remaining compounds were found to be inactive.

#### 14.3.8.2 Antibacterial and Antifungal Activities

Wang et al. further screened antifungal activity of (288–297) against five species of fungi [*Alternaria alternata* (TX-8025), *Colletotrichum nicotianae* (SACC-1922), *Gibberella pulicaris* (KZN 4207), *Gonatopyricularia amomi* (MB-9671), *Phytophthora capsici* (KACC-40157)] (Wang et al. 2013). Compounds 288, 289, 290, and 295 exhibited antifungal activity against *C. nicotianae* and *G. pulicaris* wherein MIC values were ranging from 0.69 to 1.7 and 0.64 to 1.55 mM. Nystatin (positive control) exhibited a MIC value of 0.006 mM.

Zhang et al. further evaluated the in vitro antimicrobial activities (disk diffusion method) of 256, 261, 298–301 against two species of bacteria [*Mycobacterium tuberculosis* (ATCC-25177/H37Ra) and *S. aureus* (ATCC-25923)] by using rifampicin (5 μM/mL) as positive control. Antifungal activity was evaluated against *Alternaria alternata* (TX-8025), *Colletotrichum nicotianae* (SACC-1922), *Gibberella pulicaris* (KZN4207), *Gonatopyricularia amomi* (MB-9671), and *Phytophthora capsica* (KACC-40157)] by using nystatin (10 μM/mL) as a positive control. 298, 256, and 299 exhibited significant activity against *A. alternata* and *P. capsica* with MIC values of 0.66–0.99 mM, 0.87–1.10 mM, and 1.53–1.64 mM, respectively. Further, a moderate antibacterial activity were observed against *S. aureus* (zone of inhibition = 15.72, 16.33 and 14.91 mM). Perakine $N^1$, $N^4$-dioxide (300) and vinorine $N^1$, $N^4$-dioxide (261) demonstrated strong antibacterial activities against *S. aureus* (MIC = 0.49 and 0.83 mM), while 11-hydroxy-14-chloro-15-hydroxy-vincadifformine (301) had no activity (Zhang et al. 2014a).

#### 14.3.8.3 Anti-inflammatory Activity

Wang et al. screened the in vitro anti-inflammatory properties of *A. rupestris*-derived alkaloids (288–297) against COX-I and COX-II (Wang et al. 2013).

Among that 288, 293, and 296 showed selective inhibition of COX-II (96.4, 95.5, and 92.0%) comparable with the positive control, NS-398 (97.1%). Remaining compounds exhibited a poor COX-II inhibition (<50% inhibition). The screened compounds exhibited a poor activity toward COX-I (<45% inhibition).

## 14.3.9 *Phytochemistry and Pharmacological Activities of* A. Rostrata

### 14.3.9.1 Cytotoxic Activities

Cai et al. isolated alstrostine A (302) and B (303) from the methanolic extract of *A. rostrata* leaves (Fig. 14.54) and evaluated the cytotoxic effects in cancer cell lines (MCF-7, SMMC-7721, HL-60, SW-480, and A-549) (Cai et al. 2011). These compounds did not exhibit any kind of cytotoxic activity ($IC_{50} > 40$ μM).

Zhong et al. isolated 21 monoterpenoid indole alkaloids (17, 30, 63, 76, 108, 249, 270, 282, 304–316) from the methanolic extract of *A. rostrata* trunk (Fig. 14.55) and evaluated its cytotoxicity and acetylcholinesterase inhibition activity (Zhong et al. 2017). At 20 μM concentration, screened compounds exhibited poor cytotoxicity against HeLa, SGC-7901 gastric cancer, and A-549 cell lines. Further, these compounds did not exhibit acetylcholinesterase inhibition activity.

In another study, Yuan et al. isolated 6 alkaloids (17, 76, 110, 317–319) from the methanolic extract of *A. rostrata* twigs (Fig. 14.56) and evaluated its cytotoxic effects against five human cancer cell lines, HL-60, SMMC-7721, A-549, MCF-7, and SW-480 cells (Yuan et al. 2018). However, at a concentration of 40 μM, these compounds were found to be inactive in nature.

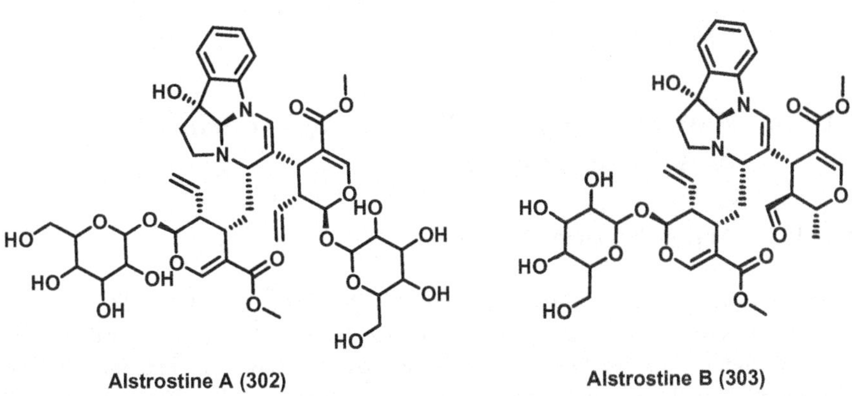


**Fig. 14.54** Chemical structures of alstrostine A and B (**302** and **303**)

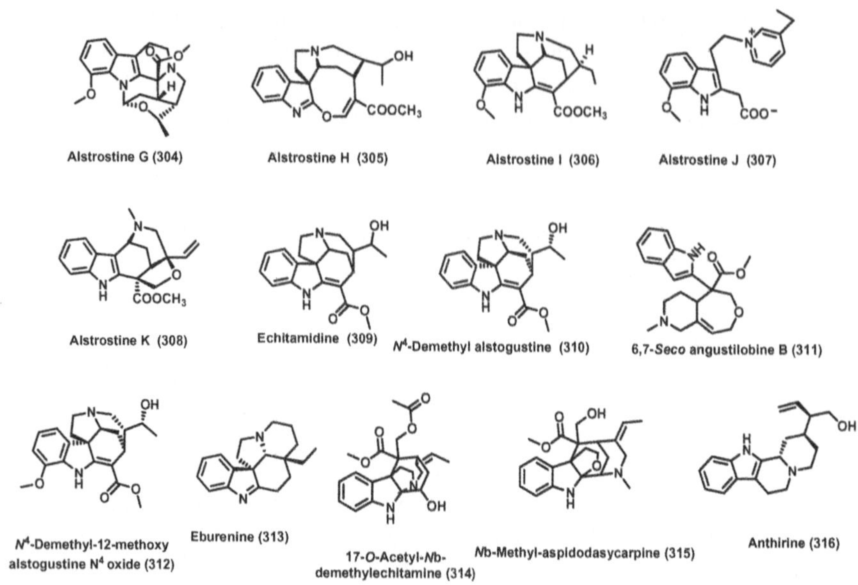


**Fig. 14.55** Chemical structures of alkaloids from *A. rostrata* (**304–316**)

**Fig. 14.56** Chemical structures of various alkaloids from *A. rostrata* (**317–319**)

## *14.3.10 Phytochemistry and Pharmacological Activities of* A. Pneumatophora

### 14.3.10.1 Anti-melanogenic Properties

Koyama et al. isolated various alkaloids (3, 24, 286, 309, 320–327) from the methanolic extract of *A. pneumatophora* leaves (Fig. 14.57) and evaluated its anti-melanogenic properties (Koyama et al. 2010a). A melanin inducer (3-isobutyl-1-methylxanthine) stimulated B16 melanoma cells were cultured with various concentration of (6.3–100 μM) isolated alkaloids for four days. A dose-dependent cell viability ($IC_{50}$ = 58.3, and 68.9 μM, respectively) and a less

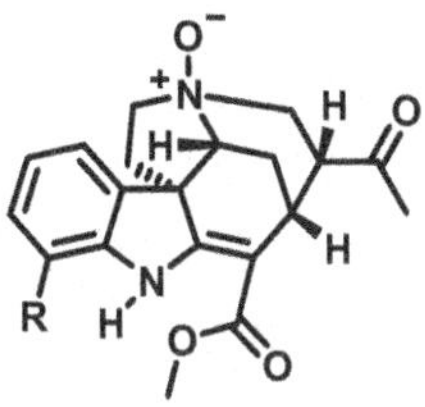


Alpneumine A (320); R = OH
Alpneumine C (321); R = H

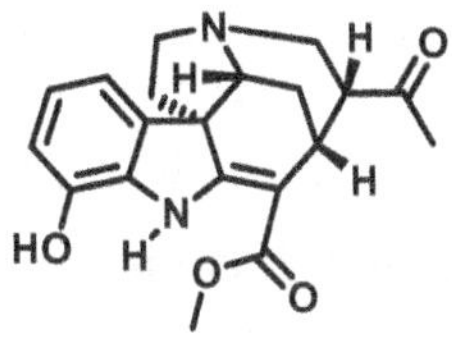


Alpneumine B (322)

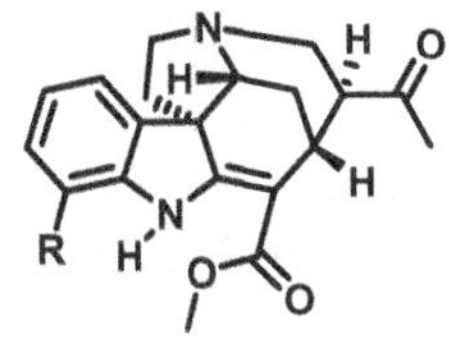


Alpneumine D (323); R = OH
Alpneumine E (324); R = H

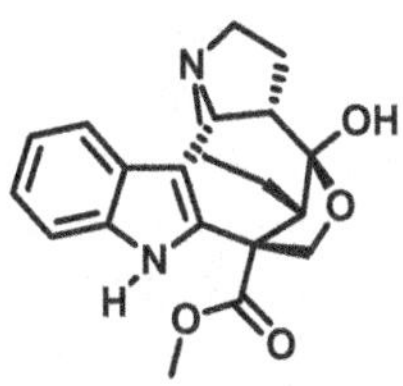


Alpneumine F (325)

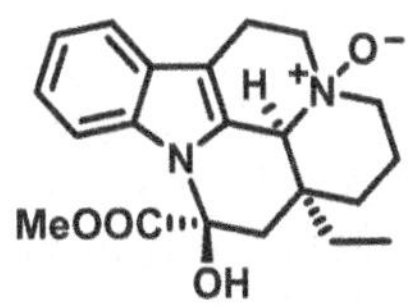


Alpneumine G (326)

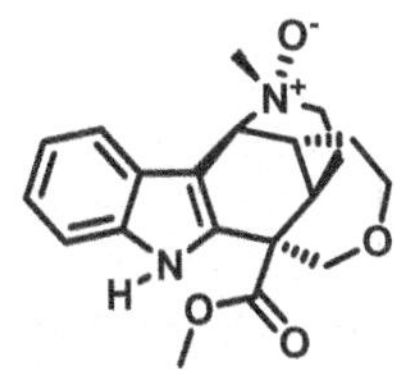


Alpneumine H (327)

**Fig. 14.57** Chemical structures of various alkaloids from *A. pneumatophora* (**320–327**)

than 50% melanin content were observed with the treatment of alpneumine G (326) and vincamine (286). Melanogenesis was moderately inhibited ($IC_{50}$ = 71.4 μM) by alpneumine E (324). Tyrosinase inhibition potential of these alkaloids wasalso examined in mushroom tyrosinase assay (in vitro). However, it exhibited a poor activity ($IC_{50}$ > 100 μM). Furthermore, 326 (50 and 75 μM) in western blotting analysis resulted in a decreased expression of tyrosinase protein.

#### 14.3.10.2 Nitric Oxide Production Inhibition

Koyama et al. isolated alsmaphorazines A (328) and B (329) from the methanolic extract of *A. pneumatophora* leaves (Fig. 14.58) and evaluated the inhibitory potential on the NO production in LPS-stimulated J774.1 (Koyama et al. 2010b).

**Fig. 14.58** Chemical structures of alsmaphorazine A and B (**328** and **329**)

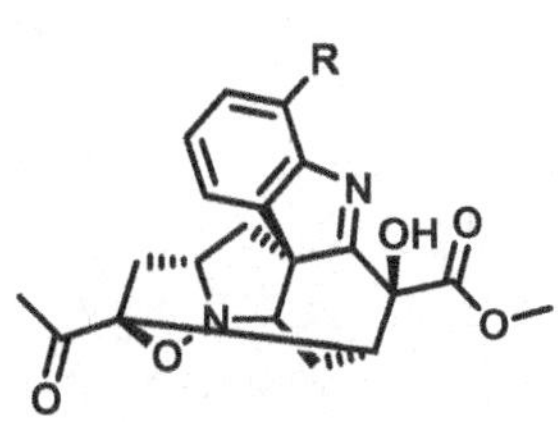


Alsmaphorazine A (328); R = OH
Alsmaphorazine B (329); R = H

Compound 328 exhibited a dose-dependent inhibition ($IC_{50}$ = 49.2 μM), while 329 did not exhibit any inhibition in NO production at 50 μM.

## 14.3.11 Phytochemistry and Pharmacological Activities of A. Penangiana

### 14.3.11.1 Cytotoxic Activities

Yeap et al. isolated ten alkaloids (330–339) from the ethanolic extract of *A. penangiana* leaves (Fig. 14.59) and assessed their cytotoxic effects against various cell lines (KB/S: vincristine-sensitive KB; KB/VJ300: vincristine-resistant KB; PC-3 and LNCaP: MCF-7 and MDA-MB-231: human breast adenocarcinoma; HT-29 and HCT 116: human colorectal carcinoma; A-549) (Yeap et al. 2018). Compound 333 and 334 showed pronounced in vitro growth inhibitory activity against the screened cell lines ($IC_{50}$ values = 0.3 to 8.3 μM). A potent cytotoxic effect was observed in HT-29 cell lines ($IC_{50}$ = 0.7 and 0.3 μM for 333 and 334, respectively).

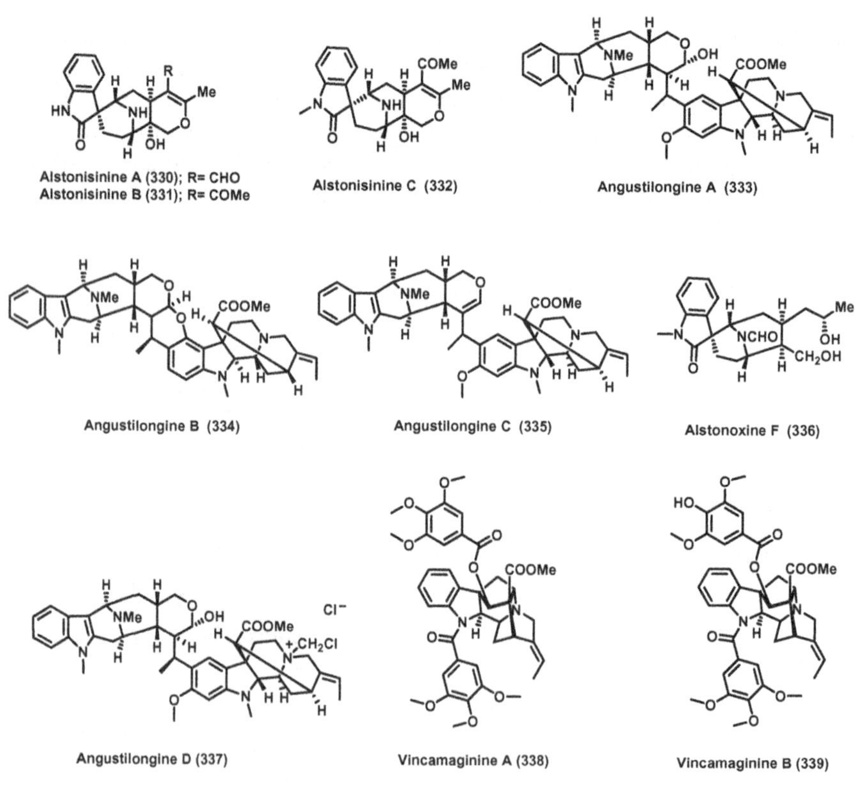


**Fig. 14.59** Chemical structures of various alkaloids from *A. penangiana* (**330–339**)

## 14.3.12 *Phytochemistry and Pharmacological Activities of* A. Mairei

### 14.3.12.1 Cytotoxic Activities

Cai et al. isolated 22 indole alkaloids (340–355, 250, 244, 241, 230, 20, 19) from the ethanolic extract of *A. mairei* leaves and twigs (Fig. 14.60) and evaluated its cytotoxic potential of various monoterpenoid from in A-549, PANC-1, HL-60, SMMC-7721, and SK-BR-3 cells. However, it exhibited poor cytotoxicity ($IC_{50}$ > 40.0 μM) (Cai et al. 2010b).

Yan et al. isolated four alkaloids (320, 356–358) from the ethanolic extract of *A. mairei* leaves (Fig. 14.61) and evaluated its in vitro cytotoxic activities against four osteosarcoma cell lines (Saos-2, Mg-63, U2-OS, and SOSP-9607 cells) (Yan et al. 2017). Alstomairine B (357) and alstomairine C (358) exhibited higher cytotoxic activity ($IC_{50}$ < 15.0 μM) against all tested tumor cell lines. $IC_{50}$ of doxorubicin (positive control) was in the range of 0.02–0.03 μM.

## 14.3.13 *Phytochemistry and Pharmacological Activities of* A. Congensis

### 14.3.13.1 Antiplasmodial Activity

Kanyanga et al. evaluated the antiplasmodial activity of various phytochemicals (9, 55, 311, 359) from *A. congensis* root bark (Cimanga et al. 2019). Two strains of *P. falciparum* (K1 and NF54 A19A) were used for in vitro screening. Boonein (359, Fig. 14.62) was inactive against both the strain (IC50 > 64 μM) while 55 and 311 exhibited a moderate activity (IC50 = 20.57 & and 5.05 and 21.26 & 40.70 μM, respectively for K1 and NF54 A19A strains). Quinine was used as a positive control that exhibited an IC50 of 0.25 and 0.46 μM, respectively, for K1 and NF54 A19A strains. Further, these compounds did not exhibit any kind of cytotoxicity against MRC-5 cells ($CC_{50}$ > 64 μM).

## 14.3.14 *Phytochemistry and Pharmacological Activities of* A. Angustiloba

### 14.3.14.1 Vasorelaxant Activities

Koyama et al. isolated alstilobanines A-E (360–364) from the methanolic extract of *A. angustiloba* leaves (Fig. 14.63) and evaluated its vasorelaxant activities in isolated rat aorta (Koyama et al. 2008). Phenylephrine (PE, $3 \times 10^{-7}$ M) was

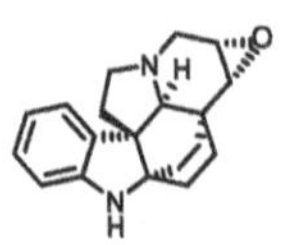
(14-α,15-α)-14,15-Epoxyaspidofractinine (340)

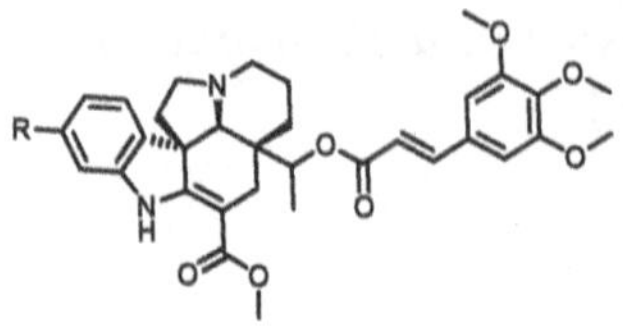
Maireine A (341); R = OMe
Maireine B (342); R =H

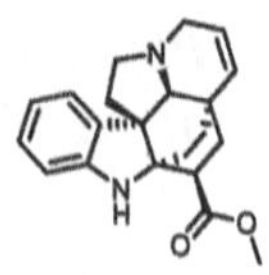
Venalstonine (343)

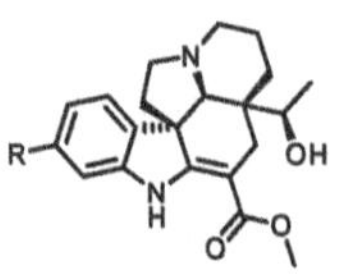
(-)-Minovincinine (344); R = H
(-)-11-Methoxy minovincinine (345); R = OMe

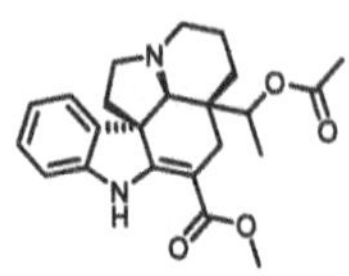
(-)-Echitovenine (346)

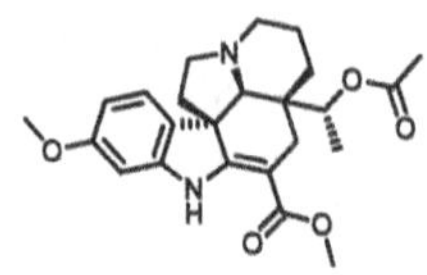
Echitovenaldine (347)

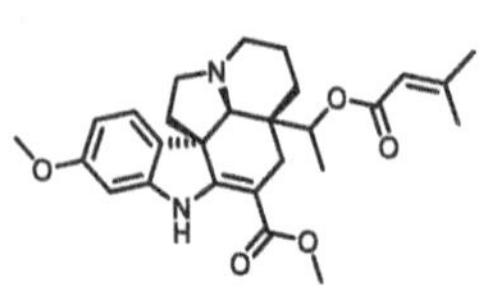
11-Methoxy echitovenidine (348)

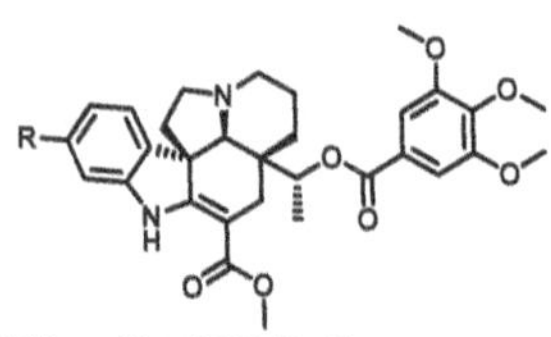
Echitoveniline (349); R = H
11-Methoxy echitoveniline (350); R = OMe

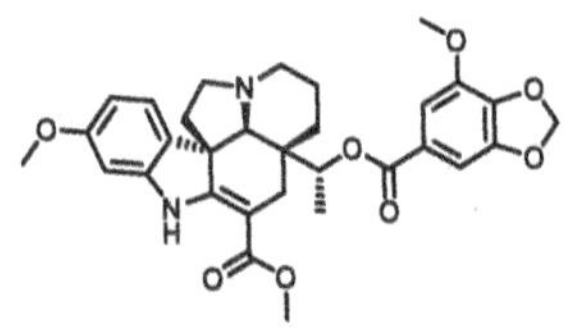
11-Methoxy echitoserpidine (351)

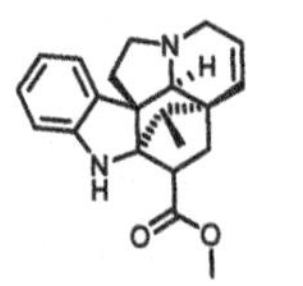
(19*S*)-Mindolinine (352)

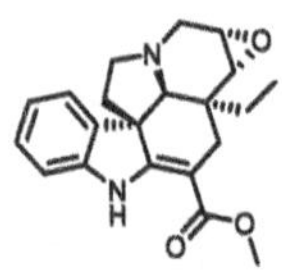
(-)-Lochnericine (353)

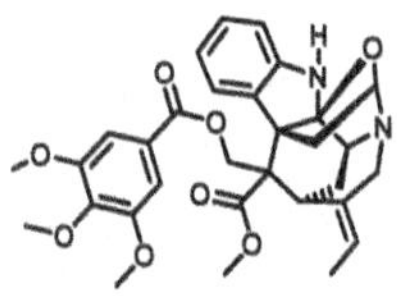
Deacetyl picraline 3,4, 5-trimethoxybenzoate (354)

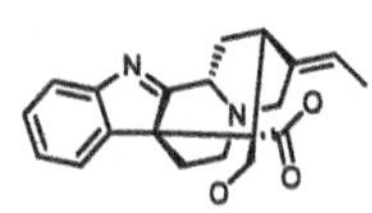
Rhazimol (355)

**Fig. 14.60** Chemical structures of various alkaloids from *A. mairei* (**340–355**)

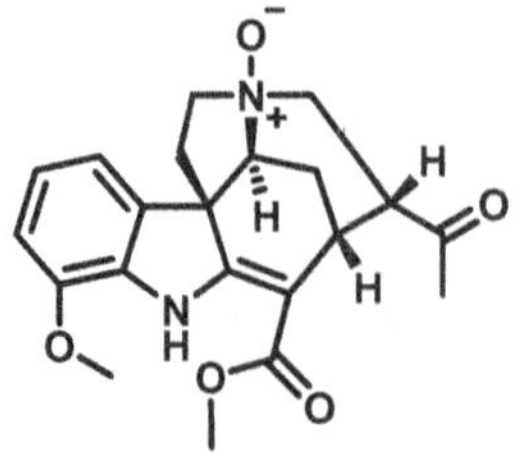
**Alstomairine A (356)**

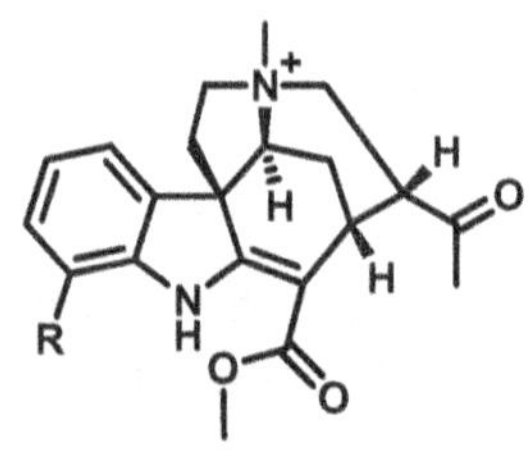
**Alstomairine B (357); R = OMe**
**Alstomairine C (358); R = OH**

**Fig. 14.61** Chemical structures of alstomairine A–C (**356–358**)

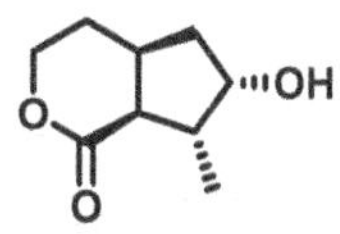

**Boonein (359)**

**Fig. 14.62** Chemical structure of boonein (**359**)

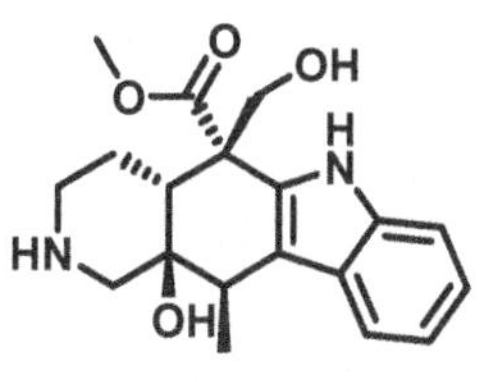

**Alstilobanine A (360)**

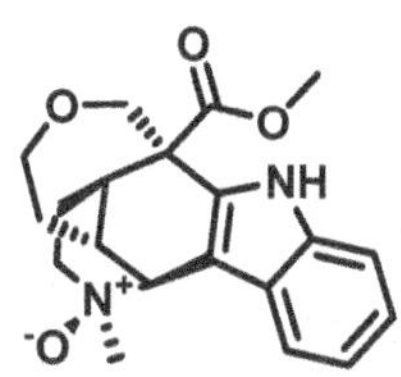

**Alstilobanine B (361)**

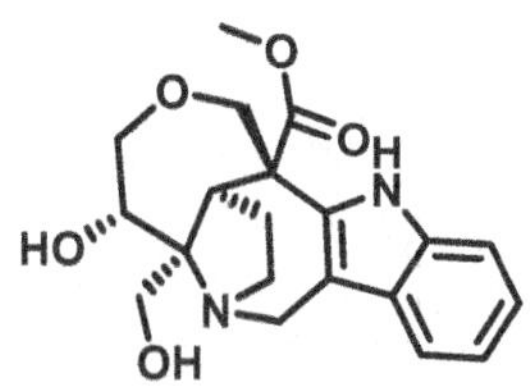

**Angustilobine C (362)**

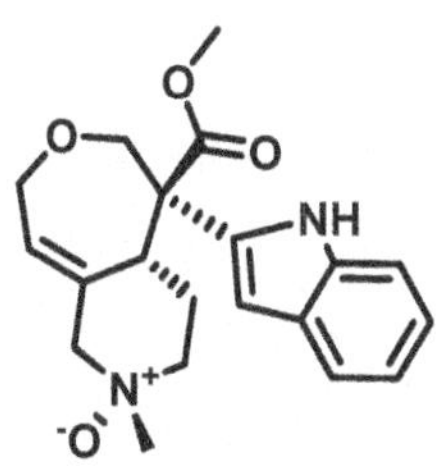

**Alstilobanine D (363)**

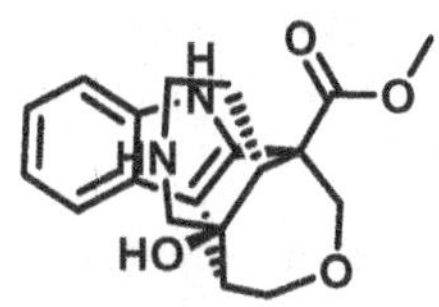

**Alstilobanine E (364)**

**Fig. 14.63** Chemical structures of various alkaloids alstilobanines A–E (**360–364**)

administered for the induction of the contraction and 360 showed more potential relaxation activity (44.3%), while 364 exhibited the least activity (5%).

#### 14.3.14.2 Cytotoxicity Activities

Ku et al. isolated various alkaloids (4,17, 94, 103, 271, 276, 281, 284, 286, 311, 314, 362, 365–372) from ethanolic extracts of *A. angustiloba* leaf and stem bark (Fig. 14.64). Further, they evaluated its cytotoxicity effects in vincristine-sensitive KB and vincristine-resistant (KB/VJ300) cells (Ku et al. 2011). Angustilobine C (362) showed moderate cytotoxicity toward vincristine-sensitive KB ($IC_{50}$ = 7.76 μg/mL) and KB/VJ300 cells ($IC_{50}$ = 7.33 μg/mL). Andransinine (366) did not exhibit appreciable cytotoxicity against both the cell lines ($IC_{50}$ > 25 μg/mL), however, it reversed MDR in KB/VJ300 cells ($IC_{50}$ = 1.61 μg/mL in the presence of 0.1 μg/mL of vincristine).

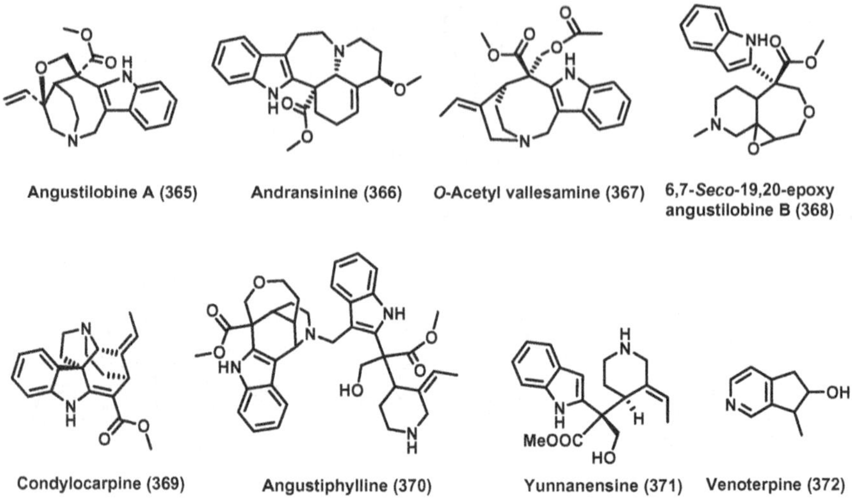


**Fig. 14.64** Chemical structures of various alkaloids from *A. angustiloba* (**365–372**)

## *14.3.15 Phytochemistry and Pharmacological Activities of* A. Actinophylla

### 14.3.15.1 Carboxypeptidase U (CPU) Inhibitor Activity

Carroll et al. reported the bioassay-guided isolation for a carboxypeptidase U (CPU) inhibitors from the aqueous extract of *A. actinophylla* leaves (Carroll et al. 2005). The study resulted in the identification of actinophyllic acid (373, Fig. 14.65) as a CPU inhibitor with an $IC_{50}$ value of 0.84 μM.

## 14.4 Pharmacokinetics and Metabolite Identification

There are few studies that explore the pharmacokinetic properties of the *Alstonia*-derived phytochemicals. These all reported studies are mainly repeated for 20, 24, 30, and 75 (Fig. 14.66).

Cao et al. have characterized the chemical constituents and rat metabolites from the *A. scholaris* leaves alkaloid extract (Cao et al. 2016). After the administration of 10 mg/kg b.w per oral of alkaloid fraction, the metabolites were identified by using an LC/SRM-MS method. Among the 35 metabolites identified, 12 were picrinine-type 11 scholaricine-type, 9 vallesamine-type, and 3 tubotaiwine-type of alkaloids. The developed LC/SRM-MS method was able to separate the R and S confirmation of various components at $MS^3$ spectra (for scholarsine, echitamidine, etc.). A total of 25 biotransformed metabolites were characterized in the

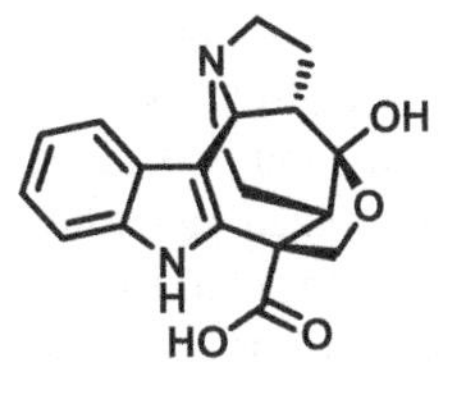

**Actinophyllic Acid (373)**

**Fig. 14.65** Chemical structure of actinophyllic acid (**373**)

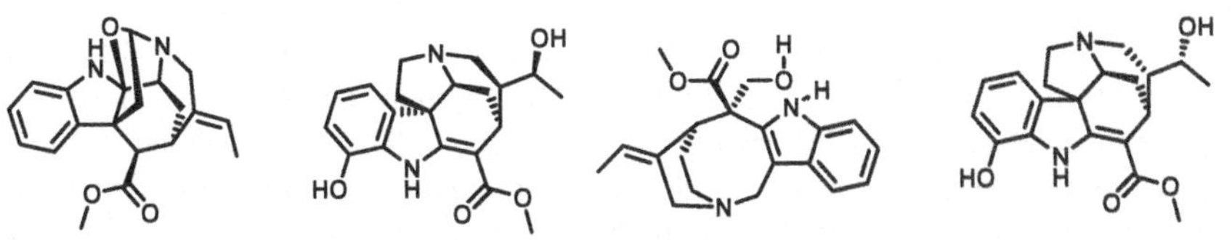

**Picrinine (20)** **Scholaricine (24)** **Vallesamine (30)** **19-Epi scholaricine (75)**

**Fig. 14.66** Chemical structures of various alkaloids from *A. scholaris* (**20**, **24**, **30**, **75**)

experimental animals. Metabolic studies of 24 and 75 highlighted the rapid absorption of these compounds into circulation. Glucuronidation, sulfation, and *N*-oxidation were found to be the metabolic reaction involved in the metabolism of 24, wherein 4 metabolites were characterized in urine, one metabolite in plasma and feces was characterized. In case of 30, 6 metabolites were identified in urine, 4 in plasma, and 4 in fecal samples. Hydroxylation, glucuronidation, N-oxidation, and demethylation were identified as the metabolic pathways. A high amount of 20 in feces and lower amounts in plasma and urine highlighted the low absorption of these compounds. 11 metabolites were identified in urine, 2 in plasma, and 9 in fecal samples. Compound 20 was mainly metabolized into phase I products. Further, these results conclude that the absorption of these molecules is directly related to the polarity of the molecules (Sch/Epi > Val > Pic). Thus, a total of 40 compounds were detected in urine, 33 in plasma, and 38 in feces.

Yun-Li Zhao et al. have studied the pharmacokinetic effects of *A. scholaris* total alkaloids (TA) in SD rat plasma by a validated LC-MS/MS method (Zhao et al. 2018). After the oral administration of various doses (10, 25, and 50 mg/kg) of total alkaloids, the plasma concentration of 4 bioactive alkaloids (24, 75, 30, and 20) fitted an open two-compartment model. $T_{max}$ of 24 and 75 occurred within 10 min while 30 and 20 occurred in 15 min. These compounds were eliminated from plasma after 24 h of the administration. A TA dosage proportional increase in the $C_{max}$ and $AUC_{last}$ was obtained for the four indole alkaloids. Further, a short elimination half-life ($T_{1/2\alpha}$) was also identified. In the study, 20 and 30 were rapidly acting while the remaining alkaloids were exhibited faster activity in in vivo conditions. Thus, it has been suggested that three times a day administration of total alkaloids will produce a best therapeutic effect in humans.

Rui Li et al. registered a mixture of indole alkaloids from the leaf of *A. scholaris* as an investigational new botanical drug (No. 2011L01436) for the treatment of various respiratory-related ailments and was approved for phase I/II clinical trials by CFDA (Li et al. 2019). A randomized, open-labeled, clinical trial was performed in healthy Chinese volunteers (40) for evaluating the safety and pharmacokinetics of capsule of alkaloids from the leaf of *A. scholaris* (CALAS) at different doses (20, 40, 80, 120 mg per oral). 24, 75, 30, and 20 are the major compounds reported for the desired pharmacological activities (respiratory diseases). A dose depended effect were observed in $C_{max}$, AUC $_{0\text{-t}}$, and $AUC_{0\text{-}\infty}$. Among the alkaloids, 30 was detected at higher concentration in blood ($C_{max}$: 12.909–60.889 ng/ml), followed by 24 ($C_{max}$: 2.162–21.025 ng/ml), 19-epischoalricine ($C_{max}$: 1.172–9.391 ng/ml) and least by 20 ($C_{max}$: 1.295–8.674 ng/ml). Further, 30 and 20 had a half-life up to 3–4 h, while 24 and 75 were up to 10 h. 24 exhibited a non-linear nature in a multi-dose linear relationship, while the remaining alkaloids exhibited a linear characteristic. CALAS was safe and well-tolerated without any serious adverse events during the whole trial periods. From these promised results, it can be deduced that 40–80 mg/time t.i.d capsule is required per day to maintain an effective blood concentration, and the effectiveness of these results has to be confirmed by performing further the phase II clinical trials by using these dosage regiments.

## 14.5 Intellectual Property Rights (IPR) Values of *Alstonia* Genus

IPR refers to the general term for the assignment of property rights by an intergovernmental organization in forms of patents, copyrights, and trademarks. These property rights provide an exclusive right to an inventor or assignee to exercise a monopoly on the use of the item for a specified period. Many commercial and academic institutions protect their research information through IPR which makes these as early sources of cutting-edge research. A huge number (>300) of patents has been filed that comes under the *Alstonia* genus. However, the majority of these patents come with *Alstonia* as a single component in polyherbal formulations. These kinds of patents are beyond the scope of this book chapter; hence they were excluded. In the present chapter, Table 14.14 indicates various patents with *Alstonia* as one of the major components. The majority of the patents have to explore the traditional knowledge of *Alstonia* genus-related activities. For instance, a higher number of patents can be observed with respiratory-related ailments and skin-related problems. Apart from that various formulations were also patented.

**Table 14.14** List of patents published on *Alstonia* genus

| S. No. | Composition/ formulation | Use/Indication | Patent ID/Title | References |
|---|---|---|---|---|
| 1 | *Alstonia*-leaf buccal tablet | Antiasthmatic-antitussive effect | CN109985078A A kind of *Alstonia*-leaf buccal tablet and preparation method | Zhi and Xuemei (2019) |
| 2 | *Alstonia*-leaf and de-enzyming tea containing bag | Enriches the drinking health-care efficacy of traditional tea | CN101983574A *Alstonia*-leaf tea bag and processing method thereof | Mingzhi et al. (2010) |
| 3 | *Alstonia*-leaf total flavones dispersible tablet (not less than 62.5 mg/tablet) | phlegm-dispelling functions and relieve inflammations such as throat and tonsillotomes. | CN107693499A A kind of preparation method of *Alstonia*-leaf total flavones dispersible tablet | Xiaolin et al. (2017) |
| 4 | Extract of *Alstonia* species or the extract of Luckynut *Thevetia* Seed species | Treatment of nervous disorders such as Alzheimer disease, Huntington chorea, or apoplexy | CN107412775A With the method for *Alstonia* species or the extract for treating nervous disorders of Luckynut Thevetia Seed species | Addington and Newman (2017) |
| 5 | *A. mairei* extract | To treat liver, lung and colon cancer | CN101347475B Use of *Alstonia mairei* extract in preparing anti-tumor medicament | Yalin et al. (2008) |
| 6 | *Alstonia*-leaf dispersible tablet | For cough-relieving, eliminating the phlegm, anti-inflammatory, spasmolysis, it is antipyretic the effect of, for chronic bronchitis, etc. | CN106727787A A kind of *Alstonia*-leaf dispersible tablet and preparation method thereof | Yuguo et al. (2015) |
| 7 | *Alstonia*-leaf traditional Chinese herbal composite | Anti-cancer (lung) | CN102631396B Anti-lung cancer *Alstonia*-leaf traditional Chinese herbal composite, method for preparing same and application thereof in preparing anti-lung cancer medicine | Yan et al. (2012) |

(continued)

**Table 14.14** (continued)

| S. No. | Composition/ formulation | Use/Indication | Patent ID/Title | References |
|---|---|---|---|---|
| 8 | Sugar-free *Alstonia*-leaf granules | Relief cough, eliminate phlegm, and can diminish inflammation | CN102600220A Sugar-free *Alstonia*-leaf granules and preparation method thereof | Jun (2011) |
| 9 | Extract of *A. scholaris* | Treating cold, fever, and respiratory disease | CN101084951B Medicine for treating diseases concerned with respiratory and preparation method and application thereof | Xiaodong et al. (2006) |
| 10 | Mixture of extracts consists of *Alstonia* leaf, *Momordica grosvenori*, and American Ginseng | Cough-relieving compound preparation | CN106890209A A kind of cough-relieving compound preparation and preparation method thereof | Guangrong et al. (2016a) |
| 11 | Mixture of extract consists of *A.* scholaris leaf *Resina Draconis* and borneol | For producing analgesia, anti-inflammatory, eliminating the phlegm, medicine of the curative effect such as relieving asthma | CN106853148A Compound longxuejie preparation and preparation method thereof | Guangrong et al. (2016b) |
| 12 | A composition consists of picrinine, vallesamine, scholaricine, and 19-epischolaricine | For treating a respiratory disease | US20180344792A1 Pharmaceutical composition for treating respiratory disease | Xiaodong et al. (2015) |
| 13 | A mixture extracts consists of *Alstonia*, Astragalus Root, *Osmanthi Fragrantis* extract, and sulfadimethoxine | For treating livestock and bird's respiratory disease | CN107569538A A kind of pharmaceutical preparation for preventing and treating livestock and bird's respiratory disease and preparation method thereof | Min et al. (2017) |

(continued)

**Table 14.14** (continued)

| S. No. | Composition/ formulation | Use/Indication | Patent ID/Title | References |
|---|---|---|---|---|
| 14 | Mixture extract consists of *A. scholaris, Berberis aristata, Tinospora cordifolia, Andrographis paniculata*, and *Hedychium spicatum* | For the treatment of fever and related symptoms | US20060141069A1 Synergistic antipyretic formulation | Pushpangadan et al. (2004) |
| 15 | Effective amount of a mixture of flavopereirine and alstonine | Preventing prostate cancer and/or alleviating the symptoms of BPH (Benign Prostatic Hyperplasia) or PIN (prostatic intraepithelial neoplasia | CN101027070A Flavopereirine and alstonine combinations in the treatment and prevention of prostate cancer | Hall and Belensky (2005) |
| 16 | Hydrolipid matrix consists of leaf and bark of *A. macrophylla* | Vaginal contraceptive | IN1128/DEL/2005 Herbal vaginal contraceptive an effective birth control measure (topical) | Mandal and Chattopadhyay (2006a) |
| 17 | Echitamidine-*N*-oxide-19-*O*-β-*D*-glucopyranoside isolated from *A. scholaris* | Anti-cancer and anti-metastatic | IN201741044816 Anti-cancer and anti-metastatic activity of EOG (echitamidine-*N*-oxide-19-*O*- β-*D*-glucopyranoside) an isolated compound obtained from ethanolic extract of *Alstonia scholaris* | Subba Reddy (2017) |
| 18 | Ursolic acid from the *A. macrophylla* | Antibacterial, antifungal, anti-inflammatory, and antihistaminic properties | IN1121/DEL/2005 Herbal antimicrobial, anti-inflammatory, and antihistaminic formulation (topical/oral) | Mandal and Chattopadhyay (2016b) |

## 14.6 Conclusion and Future Perspective

*Alstonia* genus consists of many related species, and traditional usage of these species highlighted the potential of such plants in exhibiting various pharmacological activities. Although many reviews are available on *Alstonia* species, the present chapter is mainly focused on the pharmaceutical applications of the secondary metabolites from *Alstonia* genus. A preliminary literature review indicated that more than 800 phytochemicals have been identified/isolated from the *Alstonia* species, and these studies were reported mainly from China, India, Malaysia, South Africa, etc., wherein a strong ethnomedical usage of these species have took place. Further, in African Pharmacopoeia, *A. boonei* was listed as an antimalarial drug. The majority of reported studies is mainly relying on the crude extract; however, the present chapter dealt with the pharmacological application of the isolated phytochemicals from the *Alstonia* genus. To the best of our knowledge, around 32 species of *Alstonia* have been undergone the isolation process. However, 15 species-derived phytochemicals are only evaluated for the various pharmacological activities. Among the vast isolated/identified phytochemicals, a few numbers (approx. 400) have been evaluated for a limited number of pharmacological activities. In that scenario also, all the species-derived phytochemicals did not evaluate all the reported activities. Especially, the search for potent pharmacologically active phytochemicals from the *A. scholaris* is increasing in a dramatic manner, remaining species are getting a negligible focus for their diverse activity. Evaluation of a species correlated pharmacological activities might be helpful for the development of a proper structural activity relationship of these compounds with respect to the individual pharmacological activity as well as the species. Moreover, it will also help to explore the structural similarities and their required modification for the desired pharmacological activity. Monoterpenoid indole alkaloids are the major components explored from the *Alstonia* genus and were exhaustively evaluated for various pharmacological activities such as anti-cancer, antiviral, anti-spasmodic, antitussive, antiarthritic, and antioxidant. Apart from that it also comprises iridoids, flavonoids, steroids, fatty acids, etc.

Among the isolated phytochemicals, a varying degree of pharmacological activities was observed. For instance, 2, 12, and 19 selectively inhibited COX-II. Further, 67 against GSC-12# and GSC-18# exhibited a significant cytotoxicity. Apart from that 27 exhibited a good in vitro antiviral activity similar to that of acyclovir but lacked the selectivity. All these data have suggested that a proper synthetic modification, either via chemical modification or by selecting the active pharmacophore, might also result in the development of various novel drugs.

Although *Alstonia*-derived phytochemicals exhibited numerous activity, the pharmacokinetic effects of these compounds are unexplored. A few studies are only reported for the pharmacokinetics effects that mainly include 24, 75, 30, and 20. The lack of pharmacokinetic studies might be taken care of by the scientific world and will be helpful for obtaining a further exploration of the *Alstonia* genus.

In a nutshell, diverse pharmaceutical application of *Alstonia* genus-derived phytochemicals indicates that a proper evaluation might result in the development of various drug/drug candidates for diverse pharmaceutical applications.

## References


Abinaya M, Gayathri M (2019) Inhibition of biofilm formation, quorum sensing activity and molecular docking study of isolated 3, 5, 7-Trihydroxyflavone from *Alstonia scholaris* leaf against *P. aeruginosa*. Bioorg Chem 87:291–301

Addington OC, Newman RA (2017) With the method for *Alstonia* species or the extract for treating nervous disorders of *Luckynut Thevetia* Seed species, CN107412775A

Adotey JPK, Adukpo GE, Opoku Boahen Y, Armah FA (2012) A review of the ethnobotany and pharmacological importance of *Alstonia boonei* De Wild (Apocynaceae). ISRN Pharmacol 2012:1–9

Arai H, Hirasawa Y, Rahman A, Kusumawati I (2010) Alstiphyllanines E-H, picraline and ajmaline-type alkaloids from Alstonia macrophylla inhibiting sodium glucose cotransporter. Bioorg Med Chem 18:2152–2158

Arai H, Zaima K, Mitsuta E et al (2012) Alstiphyllanines I-O, ajmaline type alkaloids from Alstonia macrophylla showing vasorelaxant activity. Bioorg Med Chem 20:3454–3459

Balogun OS, Ajayi OS, Agberotimi BJ (2016) A cytotoxic indole alkaloid from *Alstonia boonei*. J Biol Act Prod from Nat 6:347–351

Bhattacharya S, Ray A, Dutta S (1975) Psychopharmacological investigations of the 4-methoxy indole alkaloids of *Alstonia venenata*. Planta Med 27:164–170

Cai X, Shang J, Feng T, Luo XD (2010a) Novel alkaloids from *Alstonia scholaris*. Zeitschrift für Naturforsch B 65:1164–1168

Cai XH, Zeng CX, Feng T et al (2010b) Monoterpenoid indole alkaloids from *Alstonia mairei*. Helv Chim Acta 93:2037–2044

Cai XH, Bao MF, Zhang Y et al (2011) A new type of monoterpenoid indole alkaloid precursor from *Alstonia rostrata*. Org Lett 13:3568–3571

Cao P, Liang Y, Gao X et al (2012) Monoterpenoid indole alkaloids from *Alstonia yunnanensis* and their cytotoxic and anti-inflammatory activities. Molecules 17:13631–13641

Cao J, Shen H, Wang Q et al (2016) Characterization of chemical constituents and rats metabolites of an alkaloidal extract of *Alstonia scholaris* leaves by liquid chromatography coupled with mass spectrometry. J Chromatogr B 1026:43–45

Carroll AR, Hyde E, Smith J et al (2005) Actinophyllic acid, a potent indole alkaloid inhibitor of the coupled enzyme assay carboxypeptidase U/hippuricase from the leaves of *Alstonia actinophylla* (Apocynaceae). J Org Chem 70:1096–1099

Changwichit K, Khorana N, Suwanborirux K (2011) Bisindole alkaloids and secoiridoids from *Alstonia macrophylla* Wall. ex G. Don. Fitoterapia 82:798–804

Chatterjee C, Dey PK (1964) Effect of few neuroleptic phytochemical agents on the serotonin content in mouse brain. Naturwissenschaften 51:466

Chattopadhyay D, Dungdung S, AB M (2005) A potent sperm motility-inhibiting activity of bioflavonoids from an ethnomedicine of Onge, *Alstonia macrophylla* Wall ex A. DC, leaf extract. Contraception 71:372–378

Cheenpracha S, Ritthiwigrom T, Laphookhieo S (2013) Alstoniaphyllines A-C, unusual nitrogenous derivatives from the bark of *Alstonia macrophylla*. J Nat Prod 76:723–726

Cimanga RK, Nsaka SL, Tshodi ME et al (2019) In vitro and in vivo antiplasmodial activity of extracts and isolated constituents of *Alstonia congensis* root bark. J Ethnopharmacol 242:111736

Dey PK (1965) Pharmacological properties of a new alkaloid: Alstovenine. Naturwissenschaften 52:187

Feng T, Cai XH, Zhao PJ et al (2009) Monoterpenoid indole alkaloids from the bark of *Alstonia scholaris*. Planta Med 75:1537–1541

Ferry SJW 2009 onwards Plants of Southeast Asia. http://www.asianplant.net/Apocynaceae/Alstonia_pneumatophora.htm. Accessed 25 Apr 2020

Flora Malesiana DataPortal.Flora Malesiana DataPortal. *Alstonia actinophylla*. https://portal.cybertaxonomy.org/flora-malesiana/node/14642#?c=&m=&s=&cv=&manifest=http%3A%2F%2Fapi.cybertaxonomy.org%2Fflora-malesiana%2Fiiif%2Ftaxon%2F784e82b8-b41a-4656-90a6-fd15430e7906%2Fmanifest%3Ftype%3DImageFile%26relationships%3D%26relationshipsIn. Accessed 25 Apr 2020

Flora Malesiana DataPortal. Alstonia actinophylla. http://portal.cybertaxonomy.org/flora-malesiana/node/14642#?c=&m=&s=&cv=&manifest=http%3A%2F%2Fapi.cybertaxonomy.org%2Fflora-malesiana%2Fiiif%2Ftaxon%2F784e82b8-b41a-4656-90a6-fd15430e7906%2Fmanifest%3Ftype%3DImageFile%26relationships%3D%26relationshipsIn. Accessed 25 Apr 2020

NParks Flora & Fauna Web (2019) Alstonia angustiloba Miq. https://www.nparks.gov.sg/florafaunaweb/flora/2/7/2704. Accessed 25 Apr 2020

George G, Dileep PS, Paul AT (2019) Development and validation of a new HPTLC-HRMS method for the quantification of a potent pancreatic lipase inhibitory lead Echitamine from *Alstonia scholaris*. Nat Prod Res 1–5

Govaerts R, Leeuwenberg A (2016) World checklist of apocynaceae. Facilitated by the Royal Botanic Gardens, Kew. In: Publ. Internet; http//wcsp.science.kew.org/ Retrieved

Guangrong Z, Li J, Chengxun S (2016a) A kind of cough-relieving compound preparation and preparation method thereof, CN106890209A

Guangrong Z, Li J, Chengxun S (2016b) Compound longxuejie preparation and preparation method thereof, CN106853148A

Gupta RS, Bhatnager AK, Joshi YC et al (2008) Suppression of fertility in male albino rats following α-Amyrin acetate administration. Pharm Biol 42:98–104

Hirasawa Y, Arai H, Zaima K et al (2009) Alstiphyllanines A-D, indole alkaloids from *Alstonia macrophylla*. J Nat Prod 72:304–307

Hou Y, Cao X, Dong L et al (2012a) Bioactivity-based liquid chromatography-coupled electrospray ionization tandem ion trap/time of flight mass spectrometry for $\beta_2$AR agonist identification in alkaloidal extract of *Alstonia scholaris*. J Chromatogr A 1227:203–209

Hou Y, Cao X, Wang L et al (2012b) Microfractionation bioactivity-based ultra performance liquid chromatography/quadrupole time-of-flight mass spectrometry for the identification of nuclear factor-κB inhibitors and $\beta_2$ adrenergic receptor agonists in an alkaloidal extract of the folk herb *A. schoalris*. J Chromatogr B 908:98–104

Hall JL, Belensky S (2005) Flavopereirine and alstonine combinations in the treatment and prevention of prostate cancer. CN101027070A

Jagetia GC, Baliga MS, Venkatesh P et al (2005) Evaluation of the cytotoxic effect of the monoterpene indole alkaloid echitamine in-vitro and in tumour-bearing mice. J Pharm Pharmacol 57:1213–1219

Jøker D (2000) SEED LEAFLET-*Alstonia scholaris* (L.) R.Br 9

Jong AN, Bhandari MR, Kawabata J (2007) α-Glucosidase inhibitors from Devil tree (*Alstonia scholaris*). Food Chem 103:1319–1323

Jun L (2011) Sugar-free alstonia-leaf granules and preparation method thereof. CN102600220A

Keawpradub N, Houghton PJ, Burke P, Eno Amooguaye EA (1998) The cytotoxicity of alkaloids from *Alstonia macrophylla* in some cancer cell lines. J Pharm Pharmacol 50:229

Keawpradub N, Kirby GC, Steele JCP, Houghton PJ (1999) Antiplasmodial activity of extracts and alkaloids of three *alstonia* species from Thailand. Planta Med 65:690–694

Keawpradub N, Houghton PJ (1997) Indole alkaloids from *Alstonia macrophylla*. Phytochemistry 46:757–762

Khyade MS, Kasote DM, Vaikos NP (2014) *Alstonia scholaris* (L.) R. Br. and *Alstonia macrophylla* Wall. ex G. Don: a comparative review on traditional uses, phytochemistry and pharmacology. J Ethnopharmacol 153:1–18

Koyama K, Hirasawa Y, Zaima K et al (2008) Alstilobanines A-E, new indole alkaloids from *Alstonia angustiloba*. Bioorganic Med Chem 16:6483–6488

Koyama K, Hirasawa Y, Hosoya T, Hoe T (2010a) Alpneumines A-H, new anti-melanogenic indole alkaloids from Alstonia pneumatophora. Bioorg Med Chem 18:4415–4421

Koyama K, Hirasawa Y, Nugroho AE et al (2010b) Alsmaphorazines A and B, novel indole alkaloids from *Alstonia pneumatophora*. Org Lett 12:4188–4191

Krishnan P, Mai CW, Yong KT et al (2019) Alstobrogaline, an unusual pentacyclic monoterpenoid indole alkaloid with aldimine and aldimine-*N*-oxide moieties from *Alstonia scholaris*. Tetrahedron Lett 60:789–791

Ku WF, Tan SJ, Low YY et al (2011) Angustilobine and andranginine type indole alkaloids and an uleine-secovallesamine bisindole alkaloid from *Alstonia angustiloba*. Phytochemistry 72:2212–2218

Kuok CF, Zhang J, Kuok CF et al (2017) Meloslines A and B, two novel indole alkaloids from *Alstonia scholaris*. Tetrahedron Lett 58:2740–2742

Kweifio Okai G, Bird D, Field B et al (1995) Antiinflammatory activity of a Ghanaian antiarthritic herbal preparation: III. J Ethnopharmacol 46:7–15

Kweifio Okai G, Carroll AR (1993) Antiarthritic effect of lupeol acetate. Phyther Res 7:213–215

Lemmens RHMJ (2005) Alstonia congensis (PROTA)-Pl@ntUse English. Wageningen, Netherlands

Li C, Chen S, Sun C et al (2017) Cytotoxic monoterpenoid indole alkaloids from *Alstonia yunnanensis* Diels. Fitoterapia 117:79–83

Li R, Zi MJ, Gou ZP et al (2019) Pharmacokinetics and safety evaluation in healthy Chinese volunteers of alkaloids from leaf of *Alstonia scholaris*: a multiple doses phase I clinical trial. Phytomedicine 61:152828

Lim SH, Low YY, Sinniah SK et al (2014) Macroline, akuammiline, sarpagine, and ajmaline alkaloids from *Alstonia macrophylla*. Phytochemistry 98:204–215

Liu L, Chen YY, Qin XJ et al (2015) Antibacterial monoterpenoid indole alkaloids from *Alstonia scholaris* cultivated in temperate zone. Fitoterapia 105:160–164

Macabeo APG, Krohn K, Gehle D et al (2005) Indole alkaloids from the leaves of Philippine *Alstonia scholaris*. Phytochemistry 66:1158–1162

Macabeo APG, Krohn K, Gehle D et al (2008) Activity of the extracts and indole alkaloids from *Alstonia scholaris* against *Mycobacterium tuberculosis H37Rv*. Philipp Agric Sci 91:348–351

Mandal AB, Chattopadhyay D (2006a) Herbal vaginal contraceptive an effective birth control measure (topical). IN1128/DEL/2005

Mandal AB, Chattopadhyay D (2016b) Herbal antimicrobial, anti-inflammatory and antihistaminic formulation (topical/oral), IN1121/DEL/2005

Maurya A, Dwivedi GR, Darokar MP, Srivastava SK (2014) Preparative isolation of bioenhancer loganetin from *sf Alstonia scholaris* by fast centrifugal partition chromatography. Sep Sci Technol 49:773–777

Min X, Jing H, Liping X et al (2017) A kind of pharmaceutical preparation for preventing and treating livestock and birds respiratory disease and preparation method thereof. CN107569538A

Mingzhi L, Lifei X, Zhifen D, et al (2010) Alstonia-leaf tea bag and processing method thereof, CN101983574A

Monachino J (1949) A revision of the genus *Alstonia* (Apocynaceae). Pacific Sci III:133–182

Nguyen PQT, Ooi JSG, Nguyen NTK et al (2015) Antiviral cystine knot α-amylase inhibitors from *Alstonia scholaris*. J Biol Chem 290:31138–31150

Ojewole JAO (2008) Studies on the pharmacology of echitamine, an alkaloid from the stem bark of *Alstonia boonei* L. (Apocynaceae). Int J Crude Drug Res 22:121–143

Okoye NN, Ajaghaku DL, Okeke HN et al (2014) β-Amyrin and α-Amyrin acetate isolated from the stem bark of *Alstonia boonei* display profound anti-inflammatory activity. Pharm Biol 52:1478–1486

Okoye NN, Okoye COB (2016) Anti-oxidant and antimicrobial flavonoid glycosides from *Alstonia boonei* De wild leaves. J Pharm Res Int 10:1–9

Olajide OA, Awe SO, Makinde JM et al (2000) Studies on the anti-inflammatory, antipyretic and analgesic properties of *Alstonia boonei* stem bark. J Ethnopharmacol 71:179–186

Pan L, Terrazas C, Acuña UM, et al (2014) Bioactive indole alkaloids isolated from *Alstonia angustifolia,* Phytochem Lett 10:liv–lix

Parveen M, Khanam Z, Ali A, Ahmad SM (2010) A novel antimicrobial flavonoidic glycoside from the leaves of *Alstonia macrophylla* Wall ex A. DC (Apocynaceae). Chinese Chem Lett 21:593–595

Pushpangadan P, Rawat A, Rao C et al (2004) Synergistic antipyretic formulation. US20060141069A1

Qin XJ, Zhao YL, Lunga PK et al (2015a) Indole alkaloids with antibacterial activity from aqueous fraction of *Alstonia scholaris.* Tetrahedron 71:4372–4378

Qin XJ, Zhao YL, Song CW et al (2015b) Monoterpenoid Indole Alkaloids from inadequately dried leaves of *Alstonia scholaris.* Nat Products Bioprospect 5:185–193

Ragasa CY, Lim KF, Shen CC, Raga DD (2015) Hypoglycemic potential of triterpenes from *Alstonia scholaris.* Pharm Chem J 49:30–33

Rajic A, Kweifio Okai G, Macrides T et al (2000) Inhibition of serine proteases by anti-inflammatory triterpenoids. Planta Med 66:206–210

Salim AA, Garson MJ, Craik DJ (2004) New indole alkaloids from the bark of *Alstonia scholaris.* J Nat Prod 67:1591–1594

Shang JH, Cai XH, Feng T et al (2010) Pharmacological evaluation of *Alstonia scholaris*: anti-inflammatory and analgesic effects. J Ethnopharmacol 129:174–181

Sidiyasa K (1998) Taxonomy, phylogeny, and wood anatomy of alstonia (apocynaceae). Blumea Suppl 11:1–230

Singh D, Arya P, Sharma A, Dobhal M (2015) Modulatory potential of α-amyrin against hepatic oxidative stress through antioxidant status in wistar albino rats. J Ethnopharmacol 161:186–193

Singh SK, Sarma BK, Srivastava JS et al (1999) Antifungal activity of Δ3-alstovenine, a plant alkaloid isolated from *Alstonia venenata.* Folia Microbiol (Praha) 44:510–512

Singh UP, Sarma BK, Mishra PK, Ray AB (2000) Antifungal activity of venenatine, an indole alkaloid isolated from *Alstonia venenata.* Folia Microbiol (Praha) 45:173–176

Soni P, Choudhary R, Bodakhe SH (2019) Effects of a novel isoflavonoid from the stem bark of *Alstonia scholaris* against fructose-induced experimental cataract. J Integr Med 17:374–382

Subba Reddy D (2016) The antineoplastic activity of Echitamidine-*N*-Oxide-*β*-*D*-glucopyranoside isolated from stem bark extract of *Alstonia scholaris* in KB cells. World J Pharm Res 5:1425–1450

Subba Reddy D (2017) Anti-cancer and anti-metastatic activity of EOG (Echitamidine-*N*-oxide-19-*O*-*β*-*D*-glucopyranoside) an isolated compound obtained from ethanolic extract of *Alstonia scholaris,* IN201741044816

Sultana N, Qazi M, Kamal M (2019) New anti-inflammatory triterpene esters and glycosides from *Alstonia scholaris.* Antiinflamm Antiallergy Agents Med Chem 19:1–17

Tan SJ, Lim JL, Low YY et al (2014) Oxidized derivatives of macroline, sarpagine, and pleiocarpamine alkaloids from *Alstonia angustifolia.* J Nat Prod 77:2068–2080

Tan SJ, Low YY, Choo YM et al (2010) Strychnan and secoangustilobine A type alkaloids from *Alstonia spatulata.* Revision of the C-20 configuration of scholaricine. J Nat Prod 73:1891–1897

The Plant List (2013) The Plant List. http://www.theplantlist.org/tpl1.1/search?q=Alstonia. Accessed 21 Apr 2020

Wang W, Cheng MH, Wang XH (2013) Monoterpenoid indole alkaloids from *Alstonia rupestris* with cytotoxic, anti-Inflammatory and antifungal activities. Molecules 18:7309–7322

Wang CM, Chen HT, Wu ZY et al (2016) Antibacterial and synergistic activity of pentacyclic triterpenoids isolated from *Alstonia scholaris*. Molecules 21:139

Wang CM, Tsai SJ, Jhan YL et al (2017) Anti-proliferative activity of triterpenoids and sterols isolated from *Alstonia scholaris* against non-small-cell lung carcinoma cells. Molecules 22:2119

Wang B, Dai Z, Yang XW et al (2018) Novel nor-monoterpenoid indole alkaloids inhibiting glioma stem cells from fruits of Alstonia scholaris 48:170–178

Wei X, Dai Z, Yang J et al (2018) Unprecedented sugar bridged bisindoles selective inhibiting glioma stem cells. Bioorg Med Chem 26:1776–1783

Wright CW, Allen D, Cai Y et al (1992) In vitro antiamoebic and antiplasmodial activities of alkaloids isolated from *Alstonia angustifolia* roots. Phyther Res 6:121–124

Xiaodong L, Jianhua S, Xiaojiang H, Jingkun W (2006) Medicine for treating diseases concerned with respiratory and preparation method and application thereof, CN101084951B

Xiaodong L, Yunli Z, Ianhua S et al (2015) Pharmaceutical composition for treating respiratory disease, US20180344792A1

Xiaolin W, Fangli Z, Jianfei X et al (2017) A kind of preparation method of alstonia-leaf total flavones dispersible tablet, CN107693499A

Yalin T, Li T, Wei D et al (2008) Use of *Alstonia mairei* extract in preparing anti-tumor medicament, CN101347475B

Yan C, Xuan L, Zhenhai Z, Meng D (2012) Anti-lung cancer *alstonia*-leaf traditional Chinese herbal composite, method for preparing same and application thereof in preparing anti-lung cancer medicine, CN102631396B

Yan TL, Han DX, Hu J et al (2017) Monoterpenoid indole alkaloids from *Alstonia mairei* and their cytotoxicity. J Asian Nat Prod Res 19:550–556

Yang J, Fu J, Liu X et al (2018) Monoterpenoid indole alkaloids from the leaves of *Alstonia scholaris* and their NF-κB inhibitory activity. Fitoterapia 124:73–79

Yeap JSY, Navanesan S, Sim KS et al (2018) Ajmaline, oxindole, and cytotoxic macroline-akuammiline bisindole alkaloids from *Alstonia penangiana*. J Nat Prod 81:1266–1277

Yuan YX, Guo F, He HP et al (2018) Two new monoterpenoid indole alkaloids from *Alstonia rostrata*. Nat Prod Res 32:844–848

Yuguo M, Xin Z, Cuiwen D et al (2015) A kind of alstonia-leaf dispersible tablet and preparation method thereof, CN106727787A

Zhang L, Hua Z, Song Y, Feng C (2014a) Monoterpenoid indole alkaloids from *Alstonia rupestris* with cytotoxic, antibacterial and antifungal activities. Fitoterapia 97:142–147

Zhang L, Zhang C, Zhang D et al (2014b) An unusual indole alkaloid with anti-adenovirus and anti-HSV activities from *Alstonia scholaris*. Tetrahedron Lett 55:1815–1817

Zhao Y, Cao J, Shang J et al (2017) Airways antiallergic effect and pharmacokinetics of alkaloids from *Alstonia scholaris*. Phytomedicine 27:63–72

Zhao YL, Yang ZF, Shang JH et al (2018) Effects of indole alkaloids from leaf of *Alstonia scholaris* on post-infectious cough in mice. J Ethnopharmacol 218:69–75

Zhi Z, Xuemei G (2019) A kind of *alstonia*-leaf buccal tablet and preparation method, CN109985078A

Zhong X, Bao M, Zeng C et al (2017) Polycyclic monoterpenoid indole alkaloids from *Alstonia rostrata* and their reticulate derivation. Phytochem Lett 20:77–83

# Chapter 15
# Role of Natural Bio-active Compounds as Antidiabetic Agents

**Sandra N. Jimenez-Garcia, Lina Garcia-Mier, Moises A. Vazquez-Cruz, Xochitl S. Ramirez-Gomez, Ramon G. Guevara-Gonzalez, Juan Fernando Garcia-Trejo, and Ana Angélica Feregrino-Perez**

**Abstract** Diabetes Mellitus (DM) is a metabolic disease caused by secretion and/or poor use of insulin, which produces an excess of glucose in the blood. DMs are responsible for 1.5 million deaths per year and according to the World Health Organization (OMS) it is estimated that there are 422 million people with DM worldwide, growing dramatically in low- and middle-income countries. A poor public healthcare system, a low purchasing power, and a lack of human resources are just some of the conditions that cause the diabetic population to look for supplements or adjunctive alternatives for this disease. On the other hand, the use of plants to cure and/or prevent diseases has prevailed over time. In this sense, there are various plants in popular use that have been attributed antidiabetic properties since they contain bioactive compounds of secondary metabolites such as saponins, polyphenols, terpenes, among others, which act by inhibiting enzymes involved in the adsorption of carbohydrates, improving secretion and insulin sensitivity, have antioxidant and hepatoprotective properties, among other mechanisms of

S. N. Jimenez-Garcia · X. S. Ramirez-Gomez
División de Ciencias de la Salud e Ingeniería, Universidad de Guanajuato, Ejido de Santa María del Refugio Celaya, Campus Celaya-Salvatierra, C.A. Enfermedades no Transmisibles, Av. Ing. Javier Barros Sierra no. 201 Esq, C.P. 38140 Guanajuato, Baja California, Mexico

L. Garcia-Mier
Departamento de Ciencias de la Salud, Universidad del Valle de México, Campus Querétaro, Blvd, Juriquilla no. 1000 a, Delegación Santa Rosa Jáuregui, C.P. 76230 Santiago de Querétaro, Querétaro, Mexico

MoisesA. Vazquez-Cruz
Departamento de Investigación y Desarrollo, KOPPERT MEXICO, Circuito el Marques Nte. 82, Parque Industrial El Marqués, C.P. 76246 Santiago de Querétaro, Mexico

RamonG. Guevara-Gonzalez · J. F. Garcia-Trejo · A. A. Feregrino-Perez (✉)
División de Estudios de Posgrado Ingeniería de Biosistemas, Facultad de Ingeniería, Universidad Autónoma de Querétaro, C.U. Cerro de las Campanas S/N, Colonia Las Campanas, C.P. 76010 Santiago de Querétaro, Querétaro, Mexico
e-mail: feregrino.angge@hotmail.com

 
D. Pal and A. K. Nayak (eds.), *Bioactive Natural Products for Pharmaceutical Applications*, Advanced Structured Materials 140,
https://doi.org/10.1007/978-3-030-54027-2_15

action, thus helping in prevention and/or as adjunctive therapy in the treatment of blood glucose levels regulation and delaying some of the complications derived of the DM.

**Keywords** Diabetes mellitus · Bioactive compounds · Antidiabetic · Antioxidant

## 15.1 Introduction

Diabetes mellitus (DM) is a serious, chronic, and complex metabolic disorder of multiple aetiologies with profound consequences, both acute and chronic (Sudha et al. 2011; Soumya and Srilatha 2011). Also known only as diabetes, DM and its complications affect people both in the developing and developed countries, leading to a major socioeconomic challenge. The WHO projects that diabetes will be the seventh leading cause of death in 2030 (Kumar et al. 2011b). In 2011, the global prevalence of diabetes for adults of 20–69 years was 7.7%, this was predicted to rise to 9% by 2030 (Wou et al. 2019). More of 415 million people worldwide was affected to diabetes in 2017, and this amount may be increased to 642 million by 2040 due to change in lifestyle, eating habits, and growth of the urban area (Ogurtsova et al. 2017). Diabetes is distinguished by chronic hyperglycemia with disturbances in the macromolecules' metabolism as a result of impairments in insulin secretion, insulin action, or both. There are various types of diabetes of which type 1 DM (T1DM) and type-2 DM (T2DM). The T1DM is also known as insulin-dependent diabetes. It is primarily due to pancreatic islet beta-cell destruction and is characterized by deficient insulin production in the body. On the other hand, T2DM, also known as non-insulin-dependent diabetes, results from the body's inactive use of insulin and hyperglycemia (Baeyens et al. 2018) and accounts for the vast majority of people with diabetes around the world.

The increase in the morbidity and mortality of diabetes is related to whit macrovascular (stroke, peripheral vascular disease, and among other) and microvascular (nephropathy, retinopathy, and neuropathy) complications (Harding et al. 2019). Therefore, avoiding hyperglycemic states without abusing hypoglycemic states in both type 1 and type-2 diabetes helps to retards the development of microvascular complications, without being a determining factor in this (Baeyens et al. 2018; Furman et al. 2020). To achieve the regulation of glucose levels, are administered mainly insulin secretagogues, insulin sensitizers, carbohydrate-degrading enzyme inhibitors, peptide analogs, dipeptidyl peptidase-4 inhibitors, and glucagon-like peptide-1, however, they all have limited efficacy and tolerance, as well as side effects (Rotenstein et al. 2012).

In this sense, medicinal plants have been a good alternative in the search for compounds that help control glucose levels in both developed and developing countries. It is considered that more of 1200 plants and various secondary metabolites such as alkaloids, carotenoids, flavonoids, glycoside molecules, polyphenols, terpenoids, and tannins have been used to treat Diabetes (Sachithanandam et al. 2019).

Other experts indicated than only 800 plants are used in the management of Diabetes and only 30% of plants used in folk medicines for Diabetes have been biologically assayed (Chinsembu 2019). Therefore, the potential to discover new antidiabetic drugs from plants is still untapped.

## 15.2 Mechanisms of Action of Antidiabetic Substances

Hypoglycemic drugs exert antidiabetic effects through different mechanisms, namely, sulfonylureas (reduce blood sugar, mostly by elevating insulin release from islets of Langerhans), biguanides (improve the body's sensitivity towards insulin and reduce the amount of sugar absorbed by the intestine), α-glucosidase inhibitors (reduce the absorption rate of carbohydrates in the body), thiazolidinediones (improve muscle and adipose tissue sensitivity to insulin and reduce liver glucose production. Also augment β-cell function by lowering free fatty acid levels that ultimately lead to β-cell death), and non-sulfonylureas secretagogues (increase the secretion of insulin from active β-cells). These drugs alongside with insulin are the main way for controlling diabetes; however, they have different side effects. The antidiabetic action of plant materials has been accredited to the mixture of phytochemicals (alkaloids, phenolic acids, flavonoids, glycosides, saponins, polysaccharides, stilbenes, and tannin) or to a single component of plant extracts through various mechanisms such as regulation of glucose and lipid metabolism, insulin secretion, stimulating β cells, NF-κB signaling pathway, inhibition of gluconeogenic enzymes (pyruvate carboxylase, PEP carboxykinase, fructose-1, 6-bisphosphatase, and glucose 6-phosphatase), and reactive oxygen species (ROS) protective action (Salehi et al. 2019).

He et al. (2019) summarize naturally antidiabetic drugs and identified their molecular targets. α-Glucosidase and α-amylase are crucial enzymes that hydrolyze complex dietary carbohydrates to glucose or monosaccharides in the digestive system. Glycosidase inhibitors decrease the digestion process and postprandial blood glucose levels through the blockage of various amylases and glycosidases. The inactivation of DPP4 prolong the degradation rate of glucagon-like peptide-1 (GLP-1) and glucose-dependent insulinotropic polypeptide (GIP) normalizing insulin release. GLP-1 and GIP can modulate β-cell differentiation, insulin biosynthesis and release, gastric emptying, and execute some other functions. Plasma levels of GLP-1 and GIP promote insulin secretion, reduce glucagon secretion, and in this way, control blood glucose levels. The stimulation of PPAR γ (a family of nuclear receptors and ligands activated by transcription factors that modulate adipocyte differentiation, lipid metabolism, inflammatory processes, and glucose homeostasis) serves as an insulin sensitizer in adipose tissue and muscle throughout the improvement of glucose and lipid metabolism and inflammatory processes. The antidiabetic effect of alkaloids glycosides, polyphenols, carotenoids, terpenoids, flavonoids, anthocyanins, tocopherols, peptidoglycans, steroids, saponins, xanthones, and polysaccharides may also be related to these other mechanisms: blockage of PTP1B to enhance leptin and insulin sensitivity to exhibit synergistic effects on glucose and lipidic metabolism and modulation of GLUT4 translocation via activating PI3K/AKT pathway in the liver and muscle. Finally, the amelioration of glucose and lipidic metabolism and insulin level through the functional enzymes, insulin-related oxidant stress, and inflammatory factors are also implicated in the antidiabetic action of some natural bioactive compounds.

## 15.3 α-Amilase Inhibitors

Since ancient times, plants have been considered as a source of compounds for the treatment of various diseases. Currently, given the availability of these compounds, their low cost, ease of preparation and/or minimal side effects have increased their use throughout the country world (Sachithanandam et al. 2019), as well as the interest in discovering new antidiabetic drugs.

Several studies have focused on plants that have been attributed to antidiabetic activity by inhibiting carbohydrate-degrading enzymes (Trinh et al. 2020) also called glucosidases. Glucosidases are a group of digestive enzymes that break down the dietary carbohydrates into simple monosaccharides. One of these glucosidases is α-amylase. The α-amylase is one of the major secretory products of the pancreas and salivary glands, catalyze the hydrolysis of starch via a double displacement mechanism involving the formation and hydrolysis of a covalent-glycosyl enzyme intermediate by using active site carboxylic acids for it (Sales et al. 2012). Glucosidases inhibitor enzymes reduce the rate of carbohydrate digestion and delay the carbohydrate absorption from the digestive tract; therefore, they have the potential to prevent the development of diabetes by lowering the after-meal glucose levels (Miao et al. 2015). α-Amylase inhibitors are also called starch blockers since they prevent or slow down the absorption of starch into the body mainly by blocking the hydrolysis of 1,4-glycosidic linkages of starch and other oligosaccharides into maltose, maltotriose, and other simple sugars.

Plants have a wide diversity of compounds including polyphenols than are good inhibitors of enzymes responsible for carbohydrate digestion. The inhibition of α-amylase by a polyphenol is highly related to its molecular structure, as the inhibition results from binding interactions between polyphenols and the enzyme (Sun et al. 2019). Hydroxyl groups, galloyl substituents, and conjugated systems are some of the important characteristics of polyphenols for effective inhibition of the enzyme (Table 15.1).

There are other compounds such as oligosaccharide inhibitors of the trestatin family, proteinaceous inhibitors isolated from plant tissues. Acarbose is obtained from this type of compound. Acarbose is a compound used in the clinical treatment of diabetes mellitus, it is a competitive α-Amylase inhibitor. Terpenoids (oleanane, ursane, and lupane) have been linked to α-Amylase inhibition although their mechanism is not yet known (Sales et al. 2012).

**Table 15.1** Indicates different examples of polyphenols with inhibitory activity of α-amylase

| | Compounds | Inhibition α-amylase ($IC_{50}$) | References |
|---|---|---|---|
| Flavonoids | Quercetagetin | 10.2 uM | Lo Piparo et al. (2008) |
| | Scutellarein | 9.64 uM | Lo Piparo et al. (2008) |
| | Kamferol, Apigenin, Naringenin, Daidzein, Catechin | 0.5–6.0 mM | Barrett et al. (2013) |
| | Luteolin | 0.01 mg/ml | Komaki et al. (2003) |
| Phenolic acids | Caffeic acid | 0.4 mM<br>3.49 mg/ml | Narita and Inouye (2011), Sun et al. (2016a, b) |
| | Quinic acid | 26.5 mM | Narita and Inouye (2011) |
| | Chlorogenic acids (Caffeoylquinic, Feruloylquinic, Dicaffeoylquinic, Dihydrocaffeic)<br>Chlorogenic acid | 0.02–2.55 mM<br>1.96 mg/ml | Narita and Inouye (2011)<br>Sun et al. (2016a, b) |
| | Cinnamic acid | >6.0 mM | Narita and Inouye (2011) |
| | Coumaric acid | 4.51–4.86 mM | Narita and Inouye (2011) |
| | Ferulic acid | 4.27–5.45 mM | Narita and Inouye (2011) |
| Tannic acid | Tannic acid | 0.301 mg/ml | Sun et al. (2016a, b) |
| Galloyl moiety | Epigallocatechin | 2.07 mg/ml | Miao et al. (2015) |
| | Catechin | 3.552 mg/ml | Sun et al. (2016b) |
| | Epicatechin | 9.32 mg/ml | Sun et al. (2016b) |
| | Theaflavin | 0.412 mg/ml | Sun et al. (2016b) |

On the other hand, worldwide the use of plants for the treatment of diabetes is increasing, some of them have been attributed an inhibitory effect on α-Amylase. Table 15.2 indicates some parts of plants that have been used to regulate glycemic indexes. According to the review made by Seetaloo et al. (2018), it can be observed that there is a wide diversity of plants, plant parts, and types of extracts used to maintain the hypoglycemic state. Additionally, Chinsembu (2019) in his review of various plants used in the world for the control of diabetes, where alpha-amylase inhibition is included, concludes that plants are a prelude in a more realistic and more inclusive to control diabetes a world level.

**Table 15.2** Plants whit α-Amylase inhibitory activity

| Scientific name | Part used | % Inhibition (IC50) | References |
|---|---|---|---|
| *Amaranthus spinosus* | Leaves | 2.4 mg/ml | Sudha et al. (2011) |
| *Antidrographis lineata* Wallich | Leaves | 46.02 ug/ml | Rahmatullah et al. (2012), Kumar et al. (2011a) |
| *Artocarpus heterophyllus* | Leaves | 70.58 ug/ml | Nair et al. (2013), Mahomoodally et al. (2016) |
| *Bixa orellana* | Leaves | 50 ug/ml | Ponnusamy et al. (2011), Ezuruike and Prieto (2014) |
| *Calendula officinalis* | Leaves | 72.60 ug/ml | Kumar et al. (2011b), Ezuruike and Prieto (2014) |
| *Camellia sinensis* | Leaves | 0.52 to 1.54 mg/ml | Olennikov and Kashchenko (2014) |
| *Cichorium intybus* | Leaves | 18.3 mg/ml | Dalar and Konczak (2014), Street et al. (2013) |
| *Cinnamomum verum* | Leaves | 40 ug/ml | Ponnusamy et al. (2011), Telli et al. (2016) |
| *Cinnamomum zeylanicum* | Bark | 40–46 ug/ml | Ali-Shtayeh et al. (2012), Salehi et al. (2013) |
| *Cecropia obtusifolia* | Leaves | 14 ug/ml | Andrade-Cetto et al. (2008) |
| *Cyperus esculentus* | Tuber | 5.19 mg/ml | Khan and Yadava (2010), Sabiu et al. (2017) |
| *Ficus natalensis* | Leaves | 17.85 ug/ml | Olaokun et al. (2013), Keter and Mutiso (2012) |
| *Kandelia candel* | Bark | 206 ug/ml | Trinh et al. (2020), Sathe et al. (2014) |
| *Malva neglecta* | Fruit | 15.2 mg/ml | (Yaseen et al. 2015; Türker and Dalar 2013) |
| *Malmea depressa* | Roots | 21 ug/ml | Andrade-Cetto et al. (2008) |
| *Momordica dioica* | Fruit | 8 ug/ml | Yaseen et al. (2015), Rao and Mohan (2017) |
| *Orthosiphon stamineus* | Leaves | 36.70 mg/ml | Mohamed et al. (2012), Ching et al. (2013), Ameer et al. (2012) |
| *Parkia roxburghii* | Pod | 7.09–37.2 ug/ml | Sheikh et al. (2015) |
| *Phoenix dactylifera* | Leaves, stem, whole plant | 10.43 to 985.45 ug/ml | Mahomoodally et al. (2016) |
| *Pipper betel* Blanco | Leaves | 84.63 ug/ml | Mahomoodally et al. (2016), Nair et al. (2013) |
| *Rubus fruticosus* | Leave | 53.7 ug/ml | Salehi et al. (2013), Shaheen et al. (2017) |
| *Rumex crispus* | Root | 113.3–199.1 ug/ml | Minh et al. (2019) |

(continued)

**Table 15.2** (continued)

| Scientific name | Part used | % Inhibition (IC50) | References |
|---|---|---|---|
| *Senna urattensis* | Leave | 123.95 ug/ml | Thilagam et al. (2013) |
| *Stevia rebaudiana* | Leave | 198.40 ug/ml | Mootoosamy and Mahomoodally (2014), Ruiz-Ruiz et al. (2015) |
| *Trigonella foenum-graecum* | Leave | 1.92 mg/ml | Nickavar and Yousefian (2011) |
| *Urtica dioica* | Leave | 1.89 mg/ml | Nickavar and Yousefian (2011), Barkaoui et al. (2017) |
| *Vaccinium arctostaphylos* | Fruit | 57 ug/ml | Salehi et al. (2013), Nowbandegani et al. (2015) |
| *Zea mays* | Flower | 5.89 mg/ml | Ezuruike and Prieto (2014), Sabiu et al. (2017) |

## 15.4 α-Glucosidase Inhibition

Among the therapeutic options in the treatment of type-2 diabetes mellitus (T2DM) α-glucosidase enzyme inhibition is an effective one. One strategy that has been used for the treatment of type-2 diabetes is the inhibition of the activity of alpha-glucosidases using synthetic drugs. α-glucosidase inhibitors are known to exist for clinical use (e.g.,voglibose, acarbose, and miglitol); nevertheless, these inhibitors are habitually connected with gastrointestinal side effects as well as long synthetic routes. For this reason, the development of inhibitors from natural products could be an alternative option for the management of postprandial hyperglycemic condition in T2DM (Dhameja and Gupta 2019). Several studies have been directed to identify α-glucosidases inhibitors from natural sources, and many candidates have resulted to be secondary metabolites (alkaloids, flavonoids, phenols, and terpenoids) (Assefa et al. 2020; Choudhury et al. 2018; Di Stefano et al. 2018).

α-glucosidase, located in the brush border of the small intestine, catalyzes the liberation of α-glucose from the non-reducing end of the substrate. By the inhibition of α-glucosidase in the intestine, the process of carbohydrate digestion spreads to the lower part of small intestine delaying the absorption rate of glucose into the blood. Several phytochemicals have shown antidiabetic activity by lowering the glucose absorption throughout the inhibition of hydrolyzing enzymes α-glucosidase (Ota and Ulrih 2017).

The World Health Organization has listed a plethora of plants for medical purposes (21,000). According to the review made by Amjad et al. (2019) around 1200 plants are being used worldwide for the control of diabetes mellitus; from this, only 30% were chemically and pharmacologically investigated. In this review, it mentions a list of commonly used herbs in India for diabetes treatment. In the same way, Bagetta et al. (2020) reviewed the bioactive substances contained in some foods typical of the Mediterranean area (olive oil, onion, liquorice, rosemary, oregano, hazelnut, pistachio, apple, red wine, hot pepper, Citrus sp. fruits, saffron, and garlic) focusing on their impact on insulin resistance and T2DM, endothelial dysfunctions, inflammatory response, oxidative stress, and dyslipidaemic and hypercholesterolemic effects. A broad range of vegetables displayed differente ranges of α-glucosidade inhibition activity (pepper, tomato, eggplant, potato, onion, bitter melon, pumpkin, shallot, red cabbage, kiwi, cinnamon, mustard, Chinese rhubarb, soybeans, among other). Assefa et al. (2020) addressed the percentage of α-glucosidade inhibition according to different parts of plants and plant extracts. Also, it is aborded the chemical compounds implicated in such inhibition such as luteolin-7-O-glucoside a flavonoid isolated from pepper leaves and lactucaxanthin a carotenoid extracted from lettuce. Cyanidin-3-rutinoside as well as different glucosides derived from cyanidin has been found to exert α-glucosidase inhibitory activity. Different anthocyanins are postulated as bioactive components against diabetes (Gowd et al. 2017). Many α-glucosidase inhibitors belonging to plants such as flavonoids, alkaloids, terpenoids, anthocyanins, glycosides, phenolic compounds, and so on are aborded in the review made by Kumar et al. (2011b) with their inhibitory potency along with $IC_{50}$ values.

Plants commonly used to help manage blood glucose include bitter melon, fenugreek gurmar, ivy gourd, nopal, ginseng, Russian tarragon, cinnamon, psyllium, and garlic. The review made by Ota and Ulrih (2017) states that many of them prevent glucose absorption by inhibiting intestinal α-amylase and α-glucosidase throughout the action of several flavonoids (e.g., quercetin and kaempferol). In this same sense is the review by Bi et al. (2017) about the used of spices in the management of diabetes mellitus. A wide range of plant species from Africa, Central America, Mexico, South Asia, and Iran implicate in the decrease of blood sugar levels by inhibiting the enzymes α-amylase and α-glucosidase are summarized by Chinsembu (2019). Here it is stated that the mechanisms of action are mediated by phytochemical agents such as saponins, polyphenols, ellagitannins, triterpenes, and elements (Mg, P, Ca, K, Mn, Cu, Zn, S, Cr, Co, Ni, and V). Sesquiterpene lactones that could be used to treat diabetes because of α-glucosidase inhibition are summarized by Chen et al. (2019b). Reviews from plants all over the word state as an antidiabetic mechanisms the inhibition of α-glucosidase (Samarakoon et al. 2019; Hamza et al. 2019; Rafe 2017; Rezaei et al. 2015; Esquivel-Gutiérrez et al. 2012; Salehi et al. 2019).

It is well known the richens of secondary metabolites of *Moringa oleifera* (quercetin, kaempferol, myricetin, ferulic acid, gallic acid, rutin, caffeic acid, oleic acid, α-tocopherol, β-sitosterol, stigmasterol, campesterol and $\Delta^5$-avenasterol, among others), hence their implications in ameliorating different health diseases have also been evaluated. Hexane root extract exhibited the high α-glucosidase inhibition of $88.045 \pm 0.765\%$ at 1.00 mg/ml extract concentration ($IC_{50}$ value of $0.382 \pm 0.006$ mg/ml). This value is higher than that of acarbose (Magaji et al. 2020). Due to significant amount of saponins, alkaloids, flavonoids, phenols, terpenoids, and steroids found in *Carica papaya* seeds extracts, inhibitory effects of on α-glucosidase enzymes have been evaluated. The hexane extract ($IC_{50} = 75.78$ mg/ml) displayed the most potent and prominent effect compared to other extracts

(Agada et al. 2020). In the same sense, hexane extract of the roots of *Paramignya trimera* (Rutaceae), a woody shrub endemic from the south of Vietnam, exhibited α-glucosidase inhibitory activity presumably related to new compound found in there, comprising two coumarins, pyranocoumarins A and B, and an acridone alkaloid, paramiacridone (Trinh et al. 2020). Methanolic, aqueous, petroleum, and hexane extracts from seeds, barks, and leaves of *S. mahogany* have shown antidiabetics activities by different mechanisms, among them, inhibition of glucosidase. The compounds implicated in such activity are phenolics (flavonoids (swietemacrophyllanin, catechin, and epicatechin) and tannins), triterpenoids, tetranortriterpenoid (limonoid: mahonin, secomahoganin, swietmanins, swiemahogins, swietenine, and swietenolide), saponins, and alkaloids (Ervina 2020). Alkaloids isolated from the mushroom *Hericium erinaceus* exhibited $IC_{50}$ values for the inhibition of α-glucosidase in the range of 12.7–23.3 μM (Wang et al. 2015). A vast array of plants displays antidiabetic activity throughout α-glucosidase inhibition; Table 15.3 shows some plants that have been reported with α-glucosidase inhibition activity.

**Table 15.3** Overview of plants whit α-glucosidase inhibitory activity

| Plant/vegetable specie | Part used | % Inhibition (IC50) | References |
|---|---|---|---|
| *Moringa oleifera* | Root | 0.382 ± 0.006 mg/ml | Magaji et al. (2020) |
| *Carica papaya* | Seeds | 75.78 mg/ml | Agada et al. (2020) |
| *Paramignya trimera* | Root | Paramiacridone: 62.5 μM Paramicoumarin B: 65.5 μM | Trinh et al. (2020) |
| *Ziziphus oxyphylla* | Leaves | Nummularine-R: 212.1 ± 1.6 μM, Nummularin-C: 215.1 ± 1.2 μM, Hemsine-A: 394.0 ± 2.4 μM | Choudhary et al. (2011) |
| *Curcuma longa* | Rhizomes | 1.32–0.38 μg/ml | Lekshmi et al. (2012) |
| *Callistephus chinensis* | Flower | 8.14 μg/ml | Zhang et al. (2013) |
| *Thymus praecox* | Aerial parts | – | Cam et al. (2019) |
| *Cinnamomum zeylanicum* | Bark | 670 mg/ml | Shihabudeen et al. (2011) |
| *Ficus benghalensis* | Bark | – | Deepa et al. (2018) |
| *Swietenia mahagoni* | Seed | – | Wresdiyati et al. (2015) |

(continued)

**Table 15.3** (continued)

| Plant/vegetable specie | Part used | % Inhibition (IC50) | References |
|---|---|---|---|
| *Clitoria ternatea* L. | Petals | – | Escher et al. (2020) |
| *Phaseolus vulgaris* L. | Fruit | 19.52 ± 1.10 μg/ml | Tan and Chang (2017) |
| *Glycine max* L. | Fruit | 13.81 ± 0.08 μg/ml | Tan and Chang (2017) |
| Red cabbage | Leaves | 280.03 ± 18.02 μg/ml | Tan and Chang (2017) |
| Broccoli | Flower | 260.05 ± 10.20 μg/ml | Tan and Chang (2017) |
| Blueberry | Fruit | 31.22 ± 1.10 μg/ml | Tan and Chang (2017) |
| Blackberry | Fruit | 42.05 ± 1.81 μg/ml | Tan and Chang (2017) |
| *Salacia chinensis* | Heartwood | 5.01 ± 1.51 μg/ml | Thengyai et al. (2019) |
| *Vitex glabrata* | Bark | 11.22 ± 1.70 μg/ml | Thengyai et al. (2019) |
| *Senna siamea* | Buds Heartwood | 14.12 ± 1.59 μg/ml | Thengyai et al. (2019) |
| *Terminalia catappa* | Leaves | 15.84 ± 1.34 μg/ml | Thengyai et al. (2019) |
| *Phyllanthus amarus* | Whole plant | 25.11 ± 1.44 μg/ml | Thengyai et al. (2019) |
| *Erythroxylum laurifolium* | – | – | Picot et al. (2014) |
| *Elaeodendron orientale* | – | – | Picot et al. (2014) |
| *Antidesma madagascariensis* | – | – | Picot et al. (2014) |
| *Stillingia lineata* | – | – | Picot et al. (2014) |
| *Polyscias fruticosa* | Leaves. 3-*O*-[-D-glucopyranosyl-(1 → 4)–D-glucuronopyranosyl] oleanolic acid 28-*O*–D-glucopyranosyl Ester | 440.5 ± 12.7 μg/ml | Hanh et al. (2016) |

(continued)

**Table 15.3** (continued)

| Plant/vegetable specie | Part used | % Inhibition (IC50) | References |
|---|---|---|---|
| *Pheidole punctulata* | Leaves | 0.874 mg/dl | Kidane et al. (2018) |
| *Meriandra bengalensis* | Leaves | 0.599 mg/dl | Kidane et al. (2018) |
| *Agaricus blazei* | Fruit body | 357.23 ± 20.11 | Stojkovic et al. (2019) |
| *Coprinus comatus* | Fruit body | 322.74 ± 1.21 | Stojkovic et al. (2019) |
| *Canna indica* L. | Rhizome | 2.35 μg/ml, 27.1 μg/ml. | Ayusman et al. (2020) |
| *Canarium tramdenum* | Bark | 18.93 μg/ml | Quan et al. (2019) |
| *Vitis aestivalis* | Peel | 0.384 mg/ml | Zhang et al. (2011) |

– Not specified

## 15.5 Activation of Glucose Transporters

Glucose is a key fuel for most cells and a vital substrate for many biochemical reactions. Since glucose is essential for every cell of the body, so are the glucose transporters. There are three main of glucose transporters: sodium–glucose linked transporters (SGLTs), facilitated diffusion glucose transporters (GLUT), this one can be divided into subclasses (Class I, II, and III), and SWEET (this one was recently discovered. SWEET1 is the one found in humans and it is suggested to provide glucose for lactose synthesis in the mammary gland). About GLUT transporters, Class I facilitative glucose transporters are represented by GLUT1 to GLUT4. Class II of glucose transporters has four members, namely, GLUT5, GLUT7, GLUT9, and GLUT11. There are five known Class III glucose facilitative transporters, namely, GLUT6, GLUT8, GLUT10, GLUT12, and GLUT13 (HMIT). GLUT isoforms vary in their tissue specificity and affinity for glucose (Navale and Paranjape 2016; Szablewski 2019).

Peripheral resistance to insulin is a prominent feature of both insulin-dependent and non-insulin-dependent diabetes. Skeletal muscle is the principal site in charge of the decreased insulin-induced glucose consumption in diabetic subjects. The rate-limiting step for glucose utilization in muscle is the glucose transporters. This cellular process malfunctions in human and animal diabetes. Glucose transport crosswise the muscle cell plasma membrane is facilitated by glucose transporter proteins. The isoforms GLUT1 and GLUT4 are expressed in muscle. Insulin increases glucose transport in muscle by the recruitment of the GLUT4 transporter (but not GLUT1) from an intracellular pool to the plasma membrane. GLUT4 is involved in the pathophysiology of T2DM. Defects in GLUT4 expression or translocation to the peripheral cell plasma membrane in T2DM patients hinders the entrance of glucose into the cell for energy production. In addition to suitable drugs, a proper diet and/or exercise can be executed to target the increase in GLUT4 expression. Glucose transporters have major implications for the control of the delivery of glucose to mammalian cells; for this reason, they will become more and more prominent for the management as diabetes (Alam et al. 2016).

In the next lines are going to be aborded phytochemicals as source materials for the modulation of glucose transporters and therefore implicated in the development of new antidiabetic drugs or therapies for the management of diabetes.

The potential of flavonoids such as anthocyanins and myricetin against diabetes is primarily through the modulation effects on glucose transporter by enhancing GLUT2 expression in pancreatic β cells and increasing expression and encouraging translocation of GLUT4 via PI3K/AKT, CAP/Cb1/TC10, and AMPK pathway (Hajiaghaalipour et al. 2015). Resveratrol is a compound derived from the phenylpropanoid pathway found in Japanese knotweed, peanuts, berries, grape skins, red wine, and roots of rhubarb; studies have reported blood-glucose-lowering effects of resveratrol in animal models, among the propose mechanisms is the enhancement of GLUT4 translocation (Ota and Ulrih 2017). Quercetin influences glucose and lipid metabolism, probably because of the increase of the glucose uptake by means of mitogen-activated protein kinases (MAPKs) insulin-dependent mechanism, resulting in the translocation of GLUT4. Rutin also has shown stimulation of glucose transport into muscle through activating the synthesis and translocation of the transporter GLUT4 (Bagetta et al. 2020). Isoflavones improve glucose uptake in white adipose tissue by means of increasing the expression of GLUT4 whereas improves insulin resistance in the skeletal muscles by means of upregulating GLUT 4 translocation (Duru et al. 2018). Epigallocatechin gallate, a bioactive polyphenol in green tea, promotes GLUT4 translocation to improve adipose insulin resistance (Xu et al. 2018), this same mechanism is activated by ginsenosides (Shao et al. 2019) and by para-nitro derivative of caffeic acid phenethyl ester (Li et al. 2019). Dinda et al. (2020) in their comprehensive review about role of dietary flavonoids in prevention of obesity and diabetes states that flavonoids improve GLUT4 expression and translocation to plasma membrane. In the same sense is the review made by Unuofin and Lebelo (2020). Abscisic acid, a terpenoid, in in vitro condition has been reported to adjust GLUT4-mediated glucose uptake. Plumbagin, a quinone, contributes to glucose homeostasis through GLUT4 translocation (Xu et al. 2018). *P*-coumaric acid has potentially beneficial effects in improving or treating metabolic disorders by modulating modulates glucose and lipid metabolism via GLUT2 (Amalan et al. 2016).

Regarding SGLT inhibitors, alkaloid compounds isolated from the leaves of *A. macrophylla* have shown good SGLT1 and SGLT2 inhibition. Two stilbene trimers, gneyulin A and B from the dried bark of *G. gnemonoides*, displayed inhibition for SGLT1 and SGLT2. Also, inhibition for these transporters was found for *Schisandra chinens.* This plant contains polyphenols, lignans, and triterpenoids whose pharmacological effects have been evaluated (Choi 2016). SGLT2 inhibitors inhibit renal glucose reabsorption and promote glucosuria, thereby leading to

**Table 15.4** Overview of plants with antidiabetic effect through effect in glucose transporters

| Plant/vegetable specie | Effect in glucose transporters | References |
|---|---|---|
| Saffron | GLUT4 | Bagetta et al. (2020) |
| Onion | GLUT4 | Akash et al. (2014), Bagetta et al. (2020) |
| *Opuntia ficus-indica* | GLUT1 | Dinda et al. (2020) |
| *Brachylaena elliptica* | GLUT2 | Sagbo et al. (2018) |
| Cinnamon | GLUT4 | Şanlier and Gencer (2020) |
| Ginger | GLUT4 | Şanlier and Gencer (2020) |
| *Astragalus membranaceus* | GLUT4 | Venkatakrishnan et al. (2019) |
| *Gymnema sylvestre* | GLUT2, GLUT4 | Venkatakrishnan et al. (2019) |
| Fenugreek | GLUT4 | Venkatakrishnan et al. (2019) |
| *Nigella sativa* | GLUT4 | Benhaddou-Andaloussi et al. (2011) |
| Black bean | GLUT2, SGLT1 | Mojica et al. (2017) |
| *Centratherum anthelminticum* | GLUT2, GLUT4 | Arya et al. (2012) |
| *Thymus praecox* | SGLT-1, SGLT-2, GLUT2 | Cam et al. (2019) |
| Curcumin | GLUT4 | Chauhan et al. (2018) |
| *Acer nikoense* | SGLT | Choi (2016) |
| *Sophora flavescens* | SGLT | Choi (2016) |

reductions in plasma glucose concentrations, SGLT1 are similar, but their mode of action is in the intestines. Phlorizin from the bark of *P. communis* was the first compound to be known to exert SGLT1 and SGLT2 inhibition. This flavonoid along with some other flavonoids and some alkaloids (10-methoxy-N(1)-methylburnamine-17-O-veratrate and allstiphyllanine D) with SGLT inhibitor capacity are reviewed by Blaschek (2017). Table 15.4 presents an overview of plant with antidiabetic effect through the modulation of glucose transporters.

## 15.6 Activation of Insulin Secretion

DM has become an important health problem with its high morbidity and mortality since it is considered a chronic-degenerative disease with an increase in its prevalence. Etiologically, DM suggests that alterations in genetic and environmental factors, as well as increased insulin resistance (Fang et al. 2016). Insulin resistance in DM is mainly caused by the modification of insulin signaling making the body insulin sensitive (Wang et al. 2019). The result of these modifications is the reduction of glucose absorption from myocytes, hepatocytes, adipocytes, and the elevation of blood glucose levels, that is, the development of insulin resistance and hyperglycemia (Perry et al. 2014). For many years of research, insulin resistance mechanisms remain a focal point of study and the ability to maintain glycemic homeostasis improves the progression and severity of diabetes mellitus, for this reason, drugs that include oral antidiabetic agents are used, insulin and others like incretin to control blood homeostasis (Chen et al. 2019a). Medicinal plants have been used for many years to treat various diseases. The use of these medicinal plants alone or in combination with other medicinal plants is currently known as an alternative therapy for the treatment of chronic-degenerative diseases such as DM using phytochemical compounds and the bioactivity of medicinal plants (Engin et al. 2018). Plants can have many phytochemical compounds with an effect on various metabolic pathways. On the other hand, the bioactive compounds of medicinal plants and the antidiabetic, anti-hyperglycemic, and hypoglycemic potential have been reviewed in various investigations in different models in the long or short term, as well as these can be used in conjunction with pharmacological drugs (Martín and Ramos 2017). Table 15.5 shows the compounds with antidiabetic potential such as bitter gourd, ginger, turmeric, black cumin, cinnamon, garlic, green tea, among others. The most common compounds found in these medicinal plants are anthocyanins, quercetin, cinnamaldehyde, curcumin, catechins, S-methyl cysteine sulfoxide, and oleoresin A among others (Beidokhti and Jäger 2017). These phytochemical compounds have been shown to improve insulin sensitivity, stimulate insulin secretion, increase peripheral glucose uptake, inhibit hepatic glycogenolysis, exert antioxidant effects, inhibit hepatic glycogenolysis, and potentiate endogenous incretins. Glycogenolysis and the enhancement of endogenous incretins are some pharmacological mechanisms of these herbs. Currently, there is biological knowledge about the specifications and the process of action in the treatment of diabetes of medicinal plants. Polyphenols are the compounds that we find mostly in fruits, vegetables, cereals, and most foods. Polyphenol amounts are specific to each food. The products with the highest amounts of polyphenols are spices, dried herbs, cocoa, purple or red berries and seeds, which can even contain 200–300 mg per 100 g. Epicatechin and catechin monomers or oligomers of epicatechin and/or catechin (procyanidins) and flavan-3-ols have various beneficial effects such as possible effects to improve insulin resistance (Ghorbani et al. 2019). Rowley et al. (2017) demonstrated in a clinical analysis that catechins improve the cellular redox state. Thus, migration of Nrf2 (Nuclear factor erythroid 2-related factor-2) to the nucleus is stimulated, key genes for mitochondrial respiration are regulated, glucose-stimulated insulin secretion is increased, and beta-cell function is improved as a result (Fu et al. 2016). On the other hand, they also beneficially modulate the concentrations of transforming growth factor beta1 (TGF-beta1) and TNF(Tumor necrosis factor-alpha) in blood mononuclear cells, regulate NF-kappaB (Nuclear factorkappa B), and IL-1beta (interleukin-1-beta), thus considerably increasing the stimulated Akt phosphorylation (Fu et al. 2016; Matzinger et al. 2018). Insulin resistance is a cause, at least in part, of endothelial dysfunction. Therefore, flavonoids improve endothelial dysfunction, stimulate the elevation of NO (Nitric oxide) bioavailability, therefore, protect the vascular endothelium and decrease risk factors for cardiovascular disease (Bhattacharya et al. 2019). Thus, the use of in vivo models with rats and different types of extracts, such as Aloe Vera extracts, reduces blood glucose levels, $\alpha$-glucosidase activity (Chang et al. 2013). In other investigations, the use of phenolic extracts of avocado that reduces serum insulin levels and in a clinical model use of omega fatty acids was able to maintain glycemic control (Thenmozhi et al. 2012). Also, in an in vivo model using the *musa paradisiaca* flowers, and green fruits it was possible to control plasma glucose and Hba1c (glycosylated hemoglobin) (Dikshit et al. 2012). Also, in a medicinal shrub *berberis integerrima* conducted in vivo and clinical studies in which it produced a decrease in blood glucose and a decrease in serum glucose, as well as Hbca1 levels, respectively (Moazezi and Qujeq 2014). Likewise, for cinnamon, clinical studies showed that the polyphenols it contains were applied to women with polycystic ovary syndrome, and during the period of application there is insulin resistance and hyperinsulinemia (Singab et al. 2014). Other in vivo studies showed that black tea and black cumin in water and hexane extracts, respectively, are associated with a lower risk of DM, showing hypoglycemic, anti-hyperglycemic, and oral antidiabetic activity in rats, and a sensitizing action in the black tea (Abeywickrama et al. 2011; Ahmad et al. 2013). Based on this, treatment with *Nigella sativa* (black cumin extract) caused a decrease in serum glucose, an increase in decreased serum insulin levels and a partial proliferation/regeneration of pancreatic $\beta$ cells (Mohebbati and Abbasnezhad 2020). On the other hand, garlic, ginger, and shallot were applied in models with diabetic rats, and when isolating in S-alyl cysteine sulfoxide from garlic it stimulated in vitro insulin secretion of $\beta$ cells, besides that the ginger extracts showed increased adipocytes of mouse 3T3-L1 preadipocytes and thus insulin sensitivity was improved (Jafri et al. 2011; Moradabadi et al. 2013). Besides, the hypoglycemic properties of the shallot were shown by the quercetin content in the extracts applied in diabetic models producing a hypoglycemic effect and also the extract decreased plasma glucose and HbA1c levels and increased insulin level (Mehdi et al. 2012). There is increasing evidence showing that flavonoids such as eupatilin, fisetin, genistein, naringenin, proanthocyanidins, puerarin, quercetin, and rutin increase plasma insulin tested in vivo models (Akash et al. 2014; Gowd et al. 2017). The increase in plasma insulin can be attributed to the preservation of beta-cell survival, and the stimulation of insulin release. These beneficial effects are mediated by influencing the activation and amplification pathways of insulin secretion. Thus, the accumulation of bioactive compounds inhibits the electron transport chain in the mitochondria and alters the function of the KATP channel (Şanlier and Gencer 2020). Other research indicates that flavonoids can improve glucose-induced insulin secretion by influencing the Amplification pathways including PLC/PKC and cAMP/PKA signaling cascades as demonstrated in the pancreatic islets of rats by applying various doses of routine, also, genistein, and kaemferol have an effect on cAMP/PKA signaling cascades (Fu et al. 2016). Chen et al. (2019a) demonstrated that quercetin participates in ERK1/2 phosphorylation by enhancing insulin release.

**Table 15.5** Antidiabetic activity of dietary plants in vivo, in vitro and clinical trials

| Plant/vegetable specie | Extract/compounds | Effects | References |
|---|---|---|---|
| Aloe vera | Aloeresin A/no quantity | Decrease fasting blood glucose and improve insulin levels in plasma | Chang et al. (2013) |
| Avocado | Monounsaturated fatty acids of avocado/300 mg/kg | Shown significant recovery in the level of serum insulin, glycosylated hemoglobin and activities of carbohydrate metabolic enzymes | Thenmozhi et al. (2012) |
| Banana (*Musa paradisiaca fruit*) | Flower and lyophilized stem juice/50 mg/kg | Reduced significantly fasting and postprandial plasma glucose and HbA1c and increased serum insulin | Dikshit et al. (2012) |
| Berberis integerrima | Anthocyanin/1 mg, twice daily | Showed a significant decrease in serum glucose and HbA1c levels | Moazezi and Qujeq (2014) |
| Bitter melon (*Momordica charantia*) | Triterpene, phenolic compounds/10 mg/kg significantly | The increased glucose utilization by the liver, stimulation of insulin release from pancreatic β-cell, increasing number of β-cell, and suppression of the key gluconeogenic enzymes glucose-6-phosphatase and fructose-1,6-bisphosphatase | Joseph and Jini (2013) |
| Black cumin or black seed (*Nigella sativa*) | Oil/1 ml/kg | Significantly reduced the glucose levels | Ahmad et al. (2013) |
| Black tea (*Camellia sinensis* L.) | Flavonoids and caffeine/480 mg/ml | Hypoglycemic, anti-hyperglycemic, and antidiabetic actions. | Abeywickrama et al. (2011) |
| Blueberry (*Vaccinium angustifolium Aiton*) | Anthocyanin/22.5 g of powdered dry blueberries/twice daily | Increased insulin sensitivity and decreased glucose concentrations | Stull et al. (2010) |
| Caraway, coriander, and fennel | Flavonoids (rutin, quercitin, and kaempferol)/Caraway (10 mg/kg), coriander (40 mg/kg), and fennel (30 mg/kg) | Improving kidney and pancreas cell dysfunction | Shaffie et al. (2010) |
| Cinnamon | Bark extract (cinnamaldehyde)/200 mg/kg | A significant decrease in blood glucose, α-glycosidase activity, and increase in serum insulin levels and HDL-cholesterol levels | Singab et al. (2014) |

(continued)

**Table 15.5** (continued)

| Plant/vegetable specie | Extract/compounds | Effects | References |
|---|---|---|---|
| Coffee (*Coffea arabica* L.) | Quinides and chlorogenic acids/ 120 mg/kg | Improvement in glucose tolerance and insulin sensitivity and a lower risk of DM | Chang et al. (2013), Tunnicliffe and Shearer (2008) |
| Garlic (*Allium sativum* L.) | Oil/150 mg/kg | Significantly lowered the levels of blood glucose, and improved the levels of plasma insulin, increased insulin secretion and insulin sensitivity | Moradabadi et al. (2013) |
| Ginger (*Zingiber officinale Roscoe*) | Gingerol/500 mg/kg | Showed significant decreased blood glucose | Jafri et al. (2011) |
| Grape (*Vitis vinifera* L.) | Condensed tannins and flavonoinds/ 100 mg/kg | Showed significantly reduces the postprandial plasma glucose | Sapwarobol et al. (2012) |
| Guava (*Psidium guajava* L.) | Leaf Tea/200 mg/kg | Showed significantly reduces the postprandial plasma glucose | Deguchi and Miyazaki (2010) |
| Jackfruit (*Artocarpus heterophyllus* Lam.) | Condensed tannins and flavonoinds/ 400 mg/kg | Significant decrease in blood glucose | Shahin et al. (2012) |
| Loquat (*Eriobotrya japonica* Lindl.) | Total triterpene acid fraction at 300 mg/ kg day and Total sesquiterpenes at 30 g/kg day | Caused significant hypoglycemic effects on normal, alloxan and STZ-induced diabetic mice | Li et al. (2007) |
| Lychee (*Litchi chinensis Sonn*) | Oligonol and polyphenol/10 mg/ kg | Attenuated diabetes-induced hepatic damage via regulation of oxidative stress and lipid metabolism | Noh et al. (2011) |
| Mango (*Mangifera indica* Linn. (Anacardiaceae)) | Aqueous extract of leaves/250, 500 and 1000 mg/kg | Was effective in maintaining the long-term hypoglycemic effect, as well as, significantly increased the sensitivity of diabetic animals to insulin and the plasma insulin level | Villas Boas et al. (2020) |
| Nopal (*Opuntia ficus-indica* var. saboten) | Aqueous extract (water-soluble dietary fiber, vitamin c, and polyphenols/ flavonoids)/250, 500, and 1000 mg/kg | Play role in intestinal glucose absorption, glucose uptake to peripheral tissues, and improvement of insulin action | Leem et al. (2016) |

(continued)

**Table 15.5** (continued)

| Plant/vegetable specie | Extract/compounds | Effects | References |
|---|---|---|---|
| Onion (*Allium cepa* L.) | S-methyl cysteine sulphoxide/200 mg/kg | Significantly controlled the blood glucose and altered the activities of liver hexokinase and glucose 6-phosphatase towards normal | Ogunmodede et al. (2012), Ozougwu (2011) |
| Papaya (*Carica papaya* L.) | Phenolics, and β-cryptoxanthin/ 100 mg/kg | Reduced blood glucose levels and possessed antidiabetic and anti-hyperglycemic activities | Venkateshwarlu et al. (2013) |
| Pomegranate (*Punica granatum* L.) | Oleanolic, ursolic, and gallic acids/ 300 mg/kg | Reduced blood glucose levels, and stimulation, regeneration, and increased number of β-cell, by protecting pancreatic tissue and subsequent release of insulin | Li et al. (2008) |
| Saffron (*Crocus sativus* L.) | Carotenoids/ 600 mg/kg | significantly decreased blood glucose levels and increased serum insulin and improved liver and kidney functions | Elgazar et al. (2013) |
| Shallot (*Allium ascalonicum* L.) | Flavonoids (quercetin)/100 mg/ kg | Lowered the plasma glucose and HbA1c levels and enhanced insulin level | Mehdi et al. (2012) |
| Soybean (*Glycine max* L.) | Isoflavones/100 g/ dia | Produced effective antidiabetic activity and blood glucose regulators | Nanri et al. (2010) |
| Turmeric (*Curcuma longa* L.) | Curcuminoids, (vanillin, and ferulic acid/200 mg/kg | Increased the production of insulin and actions likely via regulation of insulin resistance, β-cell function, and gut absorption | Chang et al. (2013), Pawar et al. (2014) |
| | Curcumol/40 mg/kg | Is 9.4% more potent than turmeric extract (100 mg/ kg) in subchronic management of diabetes | Raafat and Omar (2016) |
| Watermelon (*Citrullus lanatus* (*Thunb*)) | L-citrulline/94.5% watermelon juice | Increase serum arginine concentration and has a significant hypoglycemic, hypolipidemic effect and significantly | El-Razek and Sadeek (2011) |

## 15.7 Future Perspective and Conclusion

DM is a metabolic syndrome that is difficult to define due to the conditions it has and the alternatives used to control or decrease it. The use of phytochemical compounds plays important roles in DM control based on the state of homeostasis inducing antioxidant enzymes by large amounts of reactive oxidation species by controlling Nrf2. And this provides us with tools to apply antioxidant therapies to treat DM. Therefore, precision medicine targeting a certain type of antioxidant in specific cells in a limited period of time can be helpful in treating the disease. Lifestyle changes and better control of diet are more effective than medicines to prevent the disease. On the other hand, clinical trials have positively described changes in biomarkers in diabetes and other diseases when using polyphenolic compounds, but there is still a long way to go in establishing the dose and type of phenolic compound responsible for the preventive potential and/or therapeutic effects against these chronic diseases. Accordingly, all of the medicinal plants described in this review suggest that the content of phytochemical compounds they contain may be used as a strategy to improve glucose/insulin homeostasis through increased insulin sensitivity, leading to preservation. of function of pancreatic islets in a prediabetic condition. The low availability of the isolated compounds also leads us to investigate other alternatives to improve bioavailability such as the use of microencapsulates.

## References

Abeywickrama K, Ratnasooriya W, Amarakoon A (2011) Oral hypoglycaemic, antihyperglycaemic and antidiabetic activities of Sri Lankan Broken Orange Pekoe Fannings (BOPF) grade black tea (Camellia sinensis L.) in rats. J Ethnopharmacol 135(2):278–286

Agada R, Usman WA, Shehu S, Thagariki D (2020) In vitro and in vivo inhibitory effects of Carica papaya seed on α-amylase and α-glucosidase enzymes. Heliyon 6(3):e03618

Ahmad A, Husain A, Mujeeb M, Khan SA, Najmi AK, Siddique NA, Damanhouri ZA, Anwar F (2013) A review on therapeutic potential of Nigella sativa: a miracle herb. Asian Pac J Trop Biomed 3(5):337–352

Akash MSH, Rehman K, Chen S (2014) Spice plant Allium cepa: dietary supplement for treatment of type 2 diabetes mellitus. Nutrition 30(10):1128–1137

Alam F, Islam A, Khalil I, Hua Gan S (2016) Metabolic control of type 2 diabetes by targeting the GLUT4 glucose transporter: intervention approaches. Curr Pharm Des 22(20):3034–3049

Ali-Shtayeh MS, Jamous RM, Jamous RM (2012) Complementary and alternative medicine use amongst Palestinian diabetic patients. Complementary Ther Clin Pract 18(1):16–21

Amalan V, Vijayakumar N, Indumathi D, Ramakrishnan A (2016) Antidiabetic and antihyperlipidemic activity of p-coumaric acid in diabetic rats, role of pancreatic GLUT 2: In vivo approach. Biomed Pharmacother 84:230–236

Ameer OZ, Salman IM, Asmawi MZ, Ibraheem ZO, Yam MF (2012) Orthosiphon stamineus: traditional uses, phytochemistry, pharmacology, and toxicology. J Med Food 15(8):678–690

Amjad S, Jafri A, Sharma A, Serajuddin M (2019) A novel strategy of nanotized herbal drugs and their delivery in the treatment of diabetes: present status and future prospects. J Herb Med 100279

Andrade-Cetto A, Becerra-Jiménez J, Cárdenas-Vázquez R (2008) Alfa-glucosidase-inhibiting activity of some Mexican plants used in the treatment of type 2 diabetes. J Ethnopharmacol 116 (1):27–32

Arya A, Looi CY, Cheah SC, Mustafa MR, Mohd MA (2012) Anti-diabetic effects of Centratherum anthelminticum seeds methanolic fraction on pancreatic cells, β-TC6 and its alleviating role in type 2 diabetic rats. J Ethnopharmacol 144(1):22–32

Assefa ST, Yang E-Y, Chae S-Y, Song M, Lee J, Cho M-C, Jang S (2020) Alpha glucosidase inhibitory activities of plants with focus on common vegetables. Plants 9(1):2

Ayusman S, Duraivadivel P, Gowtham H, Sharma S, Hariprasad P (2020) Bioactive constituents, vitamin analysis, antioxidant capacity and α-glucosidase inhibition of Canna indica L. rhizome extracts. Food Biosci 35:100544

Baeyens L, Lemper M, Staels W, De Groef S, De Leu N, Heremans Y, German MS, Heimberg H (2018) (Re) generating human beta cells: status, pitfalls, and perspectives. Physiol Rev 98 (3):1143–1167

Bagetta D, Maruca A, Lupia A, Mesiti F, Catalano R, Romeo I, Moraca F, Ambrosio FA, Costa G, Artese A (2020) Mediterranean products as promising source of multi-target agents in the treatment of metabolic syndrome. Eur J Med Chem 186:111903

Barkaoui M, Katiri A, Boubaker H, Msanda F (2017) Ethnobotanical survey of medicinal plants used in the traditional treatment of diabetes in Chtouka Ait Baha and Tiznit (Western Anti-Atlas), Morocco. J Ethnopharmacol 198:338–350

Barrett A, Ndou T, Hughey CA, Straut C, Howell A, Dai Z, Kaletunc G (2013) Inhibition of α-amylase and glucoamylase by tannins extracted from cocoa, pomegranates, cranberries, and grapes. J Agric Food Chem 61(7):1477–1486

Beidokhti MN, Jäger AK (2017) Review of antidiabetic fruits, vegetables, beverages, oils and spices commonly consumed in the diet. J Ethnopharmacol 201:26–41

Benhaddou-Andaloussi A, Martineau L, Vuong T, Meddah B, Madiraju P, Settaf A, Haddad PS (2011) The in vivo antidiabetic activity of Nigella sativa is mediated through activation of the AMPK pathway and increased muscle Glut4 content. Evid-Based Complementary Altern Med

Bhattacharya S, Maji U, Khan GA, Das R, Sinha AK, Ghosh C, Maiti S (2019) Antidiabetic role of a novel protein from garlic via NO in expression of Glut-4/insulin in liver of alloxan induced diabetic mice. Biomed Pharmacother 111:1302–1314

Bi X, Lim J, Henry CJ (2017) Spices in the management of diabetes mellitus. Food Chem 217:281–293

Blaschek W (2017) Natural products as lead compounds for sodium glucose cotransporter (SGLT) inhibitors. Planta Med 83(12/13):985–993

Cam ME, Hazar-Yavuz AN, Yildiz S, Ertas B, Adakul BA, Taskin T, Alan S, Kabasakal L (2019) The methanolic extract of Thymus praecox subsp. skorpilii var. skorpilii restores glucose homeostasis, ameliorates insulin resistance and improves pancreatic β-cell function on streptozotocin/nicotinamide-induced type 2 diabetic rats. J Ethnopharmacol 231:29–38

Chang CL, Lin Y, Bartolome AP, Chen Y-C, Chiu S-C, Yang W-C (2013) Herbal therapies for type 2 diabetes mellitus: chemistry, biology, and potential application of selected plants and compounds. Evid-Based Complementary Altern Med

Chauhan P, Tamrakar AK, Mahajan S, Prasad G (2018) Chitosan encapsulated nanocurcumin induces GLUT-4 translocation and exhibits enhanced anti-hyperglycemic function. Life Sci 213:226–235

Chen L, Gnanaraj C, Arulselvan P, El-Seedi H, Teng H (2019a) A review on advanced microencapsulation technology to enhance bioavailability of phenolic compounds: based on its activity in the treatment of Type 2 diabetes. Trends Food Sci Technol 85:149–162

Chen L, Lu X, El-Seedi H, Teng H (2019b) Recent advances in the development of sesquiterpenoids in the treatment of type 2 diabetes. Trends Food Sci Tech

Ching SM, Zakaria ZA, Paimin F, Jalalian M (2013) Complementary alternative medicine use among patients with type 2 diabetes mellitus in the primary care setting: a cross-sectional study in Malaysia. BMC Complementary Alt Med 13(1):1–7

Chinsembu KC (2019) Diabetes mellitus and nature's pharmacy of putative antidiabetic plants. J Herb Med 15:100230

Choi C-I (2016) Sodium-glucose cotransporter 2 (SGLT2) inhibitors from natural products: discovery of next-generation antihyperglycemic agents. Molecules 21(9):1136

Choudhary MI, Adhikari A, Rasheed S, Marasini BP, Hussain N, Kaleem WA (2011) Cyclopeptide alkaloids of Ziziphus oxyphylla Edgw as novel inhibitors of α-glucosidase enzyme and protein glycation. Phytochem Lett 4(4):404–406

Choudhury H, Pandey M, Hua CK, Mun CS, Jing JK, Kong L, Ern LY, Ashraf NA, Kit SW, Yee TS (2018) An update on natural compounds in the remedy of diabetes mellitus: a systematic review. J Tradit Complementary Med 8(3):361–376

Dalar A, Konczak I (2014) Cichorium intybus from Eastern Anatolia: Phenolic composition, antioxidant and enzyme inhibitory activities. Ind Crops Prod 60:79–85

Deepa P, Sowndhararajan K, Kim S, Park SJ (2018) A role of Ficus species in the management of diabetes mellitus: a review. J Ethnopharmacol 215:210–232

Deguchi Y, Miyazaki K (2010) Anti-hyperglycemic and anti-hyperlipidemic effects of guava leaf extract. Nutr Metab 7(1):9

Dhameja M, Gupta P (2019) Synthetic heterocyclic candidates as promising α-glucosidase inhibitors: an overview. Eur J Med Chem

Di Stefano E, Oliviero T, Udenigwe CC (2018) Functional significance and structure–activity relationship of food-derived α-glucosidase inhibitors. Current Opin Food Sci 20:7–12

Dikshit P, Shukla K, Tyagi MK, Garg P, Gambhir JK, Shukla R (2012) Antidiabetic and antihyperlipidemic effects of the stem of Musa sapientum Linn. in streptozotocin-induced diabetic rats. J Diabet 4(4):378–385

Dinda B, Dinda M, Roy A, Dinda S (2020) Dietary plant flavonoids in prevention of obesity and diabetes. Adv Protein Chem Struct Biol 120:159

Duru K, Kovaleva E, Danilova I, Van der Bijl P, Belousova A (2018) The potential beneficial role of isoflavones in type 2 diabetes mellitus. Nutr Res 59:1–15

Elgazar AF, Rezq AA, Bukhari HM (2013) Anti-hyperglycemic effect of saffron extract in alloxan-induced diabetic rats. Eur J Biol Sci 5(1):14–22

El-Razek FHA, Sadeek EA (2011) Dietary supplementation with watermelon (Citrullus ianatus) juice enhances arginine availability and modifies hyperglycemia, hyperlipidemia, and oxidative stress in diabetic rats. Aust J Basic Appl Sci 5(6):1284–1295

Engin AB, Tsatsakis AM, Tsoukalas D, Engin A (2018) Do flavanols-rich natural products relieve obesity-related insulin resistance? Food Chem Toxicol 112:157–167

Ervina M (2020) The recent use of Swietenia mahagoni (L.) Jacq. as antidiabetes type 2 phytomedicine: a systematic review. Heliyon 6(3):e03536

Escher GB, Marques MB, do Carmo MAV, Azevedo L, Furtado MM, Sant'Ana AS, da Silva MC, Genovese MI, Wen M, Zhang L (2020) Clitoria ternatea L. petal bioactive compounds display antioxidant, antihemolytic and antihypertensive effects, inhibit α-amylase and α-glucosidase activities and reduce human LDL cholesterol and DNA induced oxidation. Food Res Int 128:108763

Esquivel-Gutiérrez E, Noriega-Cisneros R, Bello-González M, Saavedra-Molina A, Salgado-Garciglia R (2012) Plantas utilizadas en la medicina tradicional mexicana con propiedades antidiabéticas y antihipertensivas. Biológicas 14(1):45–52

Ezuruike UF, Prieto JM (2014) The use of plants in the traditional management of diabetes in Nigeria: pharmacological and toxicological considerations. J Ethnopharmacol 155(2):857–924

Fang P, Shi M, Zhu Y, Bo P, Zhang Z (2016) Type 2 diabetes mellitus as a disorder of galanin resistance. Exp Gerontol 73:72–77

Fu J, Hou Y, Xue P, Wang H, Xu Y, Qu W, Zhang Q, Pi J (2016) Nrf2 in Type 2 diabetes and diabetic complications: Yin and Yang. Curr Opin Toxicol 1:9–19

Furman BL, Candasamy M, Bhattamisra SK, Veettil SK (2020) Reduction of blood glucose by plant extracts and their use in the treatment of diabetes mellitus; discrepancies in effectiveness between animal and human studies. J Ethnopharmacol 247:112264

Ghorbani A, Rashidi R, Shafiee-Nick R (2019) Flavonoids for preserving pancreatic beta cell survival and function: a mechanistic review. Biomed Pharmacother 111:947–957

Gowd V, Jia Z, Chen W (2017) Anthocyanins as promising molecules and dietary bioactive components against diabetes—a review of recent advances. Trends Food Sci Technol 68:1–13

Hajiaghaalipour F, Khalilpourfarshbafi M, Arya A (2015) Modulation of glucose transporter protein by dietary flavonoids in type 2 diabetes mellitus. Int J Biol Sci 11(5):508

Hamza N, Berke B, Umar A, Cheze C, Gin H, Moore N (2019) A review of Algerian medicinal plants used in the treatment of diabetes. J Ethnopharmacol 111841

Hanh TTH, Dang NH, Dat NT (2016) α-Amylase and α-glucosidase inhibitory saponins from Polyscias fruticosa leaves. J Chem 2016

Harding JL, Pavkov ME, Magliano DJ, Shaw JE, Gregg EW (2019) Global trends in diabetes complications: a review of current evidence. Diabetologia 62(1):3–16

He J-H, Chen L-X, Li H (2019) Progress in the discovery of naturally occurring anti-diabetic drugs and in the identification of their molecular targets. Fitoterapia

Jafri SA, Abass S, Qasim M (2011) Hypoglycemic effect of ginger (Zingiber officinale) in alloxan induced diabetic rats (Rattus norvagicus). Pak Vet J 31(2):160–162

Joseph B, Jini D (2013) Antidiabetic effects of Momordica charantia (bitter melon) and its medicinal potency. Asian Pac J Trop Disease 3(2):93–102

Keter LK, Mutiso PC (2012) Ethnobotanical studies of medicinal plants used by traditional health practitioners in the management of diabetes in Lower Eastern Province, Kenya. J Ethnopharmacol 139(1):74–80

Khan MH, Yadava P (2010) Antidiabetic plants used in Thoubal district of Manipur, Northeast India

Kidane Y, Bokrezion T, Mebrahtu J, Mehari M, Gebreab YB, Fessehaye N, Achila OO (2018) In vitro inhibition of-amylase and-glucosidase by extracts from Psiadia punctulata and Meriandra bengalensis. Evid-Based Complementary Altern Med

Komaki E, Yamaguchi S, Maru I, Kinoshita M, Kakehi K, Ohta Y, Tsukada Y (2003) Identification of anti-α-amylase components from olive leaf extracts. Food Sci Technol Res 9 (1):35–39

Kumar R, Pate DK, Prasad SK, Sairam K, Hemalatha S (2011a) Antidiabetic activity of alcoholic leaves extract of Alangium lamarckii Thwaites on streptozotocin–nicotinamide induced type 2 diabetic rats. Asian Pacific J Trop Med 4(11):904–909

Kumar S, Narwal S, Kumar V, Prakash O (2011b) α-glucosidase inhibitors from plants: a natural approach to treat diabetes. Pharmacogn Rev 5(9):19

Leem K-H, Kim M-G, Hahm Y-T, Kim HK (2016) Hypoglycemic effect of Opuntia ficus-indica var. saboten is due to enhanced peripheral glucose uptake through activation of AMPK/p 38 MAPK pathway. Nutrients 8(12):800

Lekshmi P, Arimboor R, Indulekha P, Nirmala Menon A (2012) Turmeric (Curcuma longa L.) volatile oil inhibits key enzymes linked to type 2 diabetes. Int J Food Sci Nutr 63(7):832–834

Li W-L, Wu J-L, Ren B-R, Chen J, Lu C-G (2007) Pharmacological studies on anti-hyperglycemic effect of folium eriobotryae. Am J Chinese Med 35(04):705–711

Li Y, Qi Y, Huang TH, Yamahara J, Roufogalis BD (2008) Pomegranate flower: a unique traditional antidiabetic medicine with dual PPAR-α/-γ activator properties. Diabetes Obes Metab 10(1):10–17

Li S, Huang Q, Zhang L, Qiao X, Zhang Y, Tang F, Li Z (2019) Effect of CAPE-pNO2 against type 2 diabetes mellitus via the AMPK/GLUT4/GSK3β/PPARα pathway in HFD/STZ-induced diabetic mice. Eur J Pharmacol 853:1–10

Lo Piparo E, Scheib H, Frei N, Williamson G, Grigorov M, Chou CJ (2008) Flavonoids for controlling starch digestion: structural requirements for inhibiting human α-amylase. J Med Chem 51(12):3555–3561

Magaji U, Sacan O, Yanardag R (2020) Alpha amylase, alpha glucosidase and glycation inhibitory activity of Moringa oleifera extracts. South Afr J Botany 128:225–230

Mahomoodally MF, Mootoosamy A, Wambugu S (2016) Traditional therapies used to manage diabetes and related complications in Mauritius: a comparative ethnoreligious study. Evid-Based Complementary Altern Med

Martín MÁ, Ramos S (2017) Health beneficial effects of cocoa phenolic compounds: a mini-review. Current Opin Food Sci 14:20–25

Matzinger M, Fischhuber K, Heiss EH (2018) Activation of Nrf2 signaling by natural products-can it alleviate diabetes? Biotechnol Adv 36(6):1738–1767

Mehdi M, Seyed-Mostafa H-z, Javad H, Hashemi B, Seyed-Mohammad H-z (2012) The effects of Persian shallot extract on the levels of some blood biochemical parameters in streptozotocin-induced diabetic rats. Afr J Agric Res 7(22):3308–3313

Miao M, Jiang B, Jiang H, Zhang T, Li X (2015) Interaction mechanism between green tea extract and human α-amylase for reducing starch digestion. Food Chem 186:20–25

Minh TN, Van TM, Andriana Y, Hau DV, Duyen DH, Guzman-Gelani Cd (2019) Antioxidant, Xanthine Oxidase, α-Amylase and α-glucosidase inhibitory activities of bioactive compounds from rumex Crispus L. Root. Molecules 24(21):3899

Moazezi Z, Qujeq D (2014) Berberis fruit extract and biochemical parameters in patients with type II diabetes. Jundishapur J Nat Pharm Prod 9(2)

Mohamed EAH, Siddiqui MJA, Ang LF, Sadikun A, Chan SH, Tan SC, Asmawi MZ, Yam MF (2012) Potent α-glucosidase and α-amylase inhibitory activities of standardized 50% ethanolic extracts and sinensetin from Orthosiphon stamineus Benth as anti-diabetic mechanism. BMC Complementary Alt Med 12(1):176

Mohebbati R, Abbasnezhad A (2020) Effects of Nigella sativa on endothelial dysfunction in diabetes mellitus: a review. J Ethnopharmacol 112585

Mojica L, de Mejia EG, Granados-Silvestre MÁ, Menjivar M (2017) Evaluation of the hypoglycemic potential of a black bean hydrolyzed protein isolate and its pure peptides using in silico, in vitro and in vivo approaches. J Funct Foods 31:274–286

Mootoosamy A, Mahomoodally MF (2014) Ethnomedicinal application of native remedies used against diabetes and related complications in Mauritius. J Ethnopharmacol 151(1):413–444

Moradabadi L, Kouhsari SM, Sani MF (2013) Hypoglycemic effects of three medicinal plants in experimental diabetes: inhibition of rat intestinal α-glucosidase and enhanced pancreatic insulin and cardiac glut-4 mrnas expression. Iranian J Pharm Res: IJPR 12(3):387

Nair SS, Kavrekar V, Mishra A (2013) In vitro studies on alpha amylase and alpha glucosidase inhibitory activities of selected plant extracts. European J Exp Biol 3(1):128–132

Nanri A, Mizoue T, Takahashi Y, Kirii K, Inoue M, Noda M, Tsugane S (2010) Soy product and isoflavone intakes are associated with a lower risk of type 2 diabetes in overweight Japanese women. J Nutr 140(3):580–586

Narita Y, Inouye K (2011) Inhibitory effects of chlorogenic acids from green coffee beans and cinnamate derivatives on the activity of porcine pancreas α-amylase isozyme I. Food Chem 127(4):1532–1539

Navale AM, Paranjape AN (2016) Glucose transporters: physiological and pathological roles. Biophys Rev 8(1):5–9

Nickavar B, Yousefian N (2011) Evaluation of α-amylase inhibitory activities of selected antidiabetic medicinal plants. J für Verbraucherschutz und Lebensmittelsicherheit 6(2):191–195

Noh JS, Park CH, Yokozawa T (2011) Treatment with oligonol, a low-molecular polyphenol derived from lychee fruit, attenuates diabetes-induced hepatic damage through regulation of oxidative stress and lipid metabolism. Br J Nutr 106(7):1013–1022

Nowbandegani AS, Kiumarcy S, Rahmani F, Dokouhaki M, Khademian S, Zarshenas MM, Faridi P (2015) Ethnopharmacological knowledge of Shiraz and Fasa in Fars region of Iran for diabetes mellitus. J Ethnopharmacol 172:281–287

Ogunmodede O, Saalu L, Ogunlade B, Akunna G, Oyewopo A (2012) An evaluation of the hypoglycemic, antioxidant and hepatoprotective potentials of onion (Allium cepa L.) on alloxan-induced diabetic rabbits. Int J Pharmacol 8(1):21–29

Ogurtsova K, da Rocha Fernandes J, Huang Y, Linnenkamp U, Guariguata L, Cho NH, Cavan D, Shaw J, Makaroff L (2017) IDF Diabetes Atlas: Global estimates for the prevalence of diabetes for 2015 and 2040. Diabetes Res Clin Pract 128:40–50

Olaokun OO, McGaw LJ, Eloff JN, Naidoo V (2013) Evaluation of the inhibition of carbohydrate hydrolysing enzymes, antioxidant activity and polyphenolic content of extracts of ten African Ficus species (Moraceae) used traditionally to treat diabetes. BMC Complementary Alt Med 13 (1):94

Olennikov DN, Kashchenko NI (2014) Componential profile and amylase inhibiting activity of phenolic compounds from Calendula officinalis L. leaves. The Scientific World Journal 2014

Ota A, Ulrih NP (2017) An overview of herbal products and secondary metabolites used for management of type two diabetes. Front Pharmacol 8:436

Ozougwu JC (2011) Anti-diabetic effects of Allium cepa (onions) aqueous extracts on alloxan-induced diabetic Rattus novergicus. J Med Plants Res 5(7):1134–1139

Pawar H, Karde M, Mundle N, Jadhav P, Mehra K (2014) Phytochemical evaluation and curcumin content determination of turmeric rhizomes collected from Bhandara District of Maharashtra (India). Med Chem 4(8):588–591

Perry RJ, Samuel VT, Petersen KF, Shulman GI (2014) The role of hepatic lipids in hepatic insulin resistance and type 2 diabetes. Nature 510(7503):84–91

Picot C, Subratty AH, Mahomoodally MF (2014) Inhibitory potential of five traditionally used native antidiabetic medicinal plants on α-amylase, α-glucosidase, glucose entrapment, and amylolysis kinetics in vitro. Adv Pharmacol Sci

Ponnusamy S, Ravindran R, Zinjarde S, Bhargava S, Ravi Kumar A (2011) Evaluation of traditional Indian antidiabetic medicinal plants for human pancreatic amylase inhibitory effect in vitro. Evid-Based Complementary Altern Med

Quan NV, Xuan TD, Tran H-D, Thuy NTD, Trang LT, Huong CT, Andriana Y, Tuyen PT (2019) Antioxidant, α-amylase and α-glucosidase inhibitory activities and potential constituents of Canarium tramdenum bark. Molecules 24(3):605

Raafat KM, Omar AG (2016) Phytotherapeutic activity of curcumol: Isolation, GC–MS identification, and assessing potentials against acute and subchronic hyperglycemia, tactile allodynia, and hyperalgesia. Pharmaceutical Biol 54(8):1334–1344

Rafe MR (2017) A review of five traditionally used anti-diabetic plants of Bangladesh and their pharmacological activities. Asian Pac J Trop Med 10(10):933–939

Rahmatullah M, Azam NK, Khatun Z, Seraj S, Islam F, Rahman A, Jahan S, Aziz S (2012) Medicinal plants used for treatment of diabetes by the Marakh sect of the Garo tribe living in Mymensingh district, Bangladesh. Afr J Tradit Complement Altern Med 9(3):380–385

Rao PS, Mohan GK (2017) In vitro alpha-amylase inhibition and in vivo antioxidant potential of Momordica dioica seeds in streptozotocin-induced oxidative stress in diabetic rats. Saudi J Biol Sci 24(6):1262–1267

Rezaei A, Farzadfard A, Amirahmadi A, Alemi M, Khademi M (2015) Diabetes mellitus and its management with medicinal plants: a perspective based on Iranian research. J Ethnopharmacol 175:567–616

Rotenstein LS, Ran N, Shivers JP, Yarchoan M, Close KL (2012) Opportunities and challenges for biosimilars: what's on the horizon in the global insulin market? Clin Diabetes 30(4):138–150

Rowley TJ IV, Bitner BF, Ray JD, Lathen DR, Smithson AT, Dallon BW, Plowman CJ, Bikman BT, Hansen JM, Dorenkott MR (2017) Monomeric cocoa catechins enhance β-cell function by increasing mitochondrial respiration. J Nutr Biochem 49:30–41

Ruiz-Ruiz J, Moguel-Ordoñez Y, Matus-Basto A, Segura-Campos M (2015) Antidiabetic and antioxidant activity of Stevia rebaudiana extracts (Var. Morita) and their incorporation into a potential functional bread. J Food Sci Technol 52(12):7894–7903

Sabiu S, Ajani EO, Sunmonu TO, Ashafa AOT (2017) Kinetics of modulatory role of Cyperus esculentus L. on the specific activity of key carbohydrate metabolizing enzymes. African J Tradit Complement Altern Med 14(4):46–53

Sachithanandam V, Lalitha P, Parthiban A, Mageswaran T, Manmadhan K, Sridhar R (2019) A review on antidiabetic properties of indian mangrove plants with reference to island ecosystem. Evid-Based Complementary Altern Med

Sagbo IJ, van de Venter M, Koekemoer T, Bradley G (2018) In vitro antidiabetic activity and mechanism of action of Brachylaena elliptica (Thunb.) DC. Evid-Based Complementary Altern Med

Salehi P, Asghari B, Esmaeili MA, Dehghan H, Ghazi I (2013) α-Glucosidase and α-amylase inhibitory effect and antioxidant activity of ten plant extracts traditionally used in Iran for diabetes. J Med Plants Res 7(6):257–266

Salehi B, Ata A, V Anil Kumar N, Sharopov F, Ramírez-Alarcón K, Ruiz-Ortega A, Abdulmajid Ayatollahi S, Valere Tsouh Fokou P, Kobarfard F, Amiruddin Zakaria Z (2019) Antidiabetic potential of medicinal plants and their active components. Biomolecules 9(10):551

Sales PM, Souza PM, Simeoni LA, Magalhães PO, Silveira D (2012) α-Amylase inhibitors: a review of raw material and isolated compounds from plant source. J Pharm Pharmaceutical Sci 15(1):141–183

Samarakoon DNAW, Uluwaduge DI, Siriwardhene MA (2019) Mechanisms of action of Sri Lankan herbal medicines used in the treatment of diabetes: a review. J Integr Med

Şanlier N, Gencer F (2020) Role of spices in the treatment of diabetes mellitus: a minireview. Trends Food Sci Technol

Sapwarobol S, Adisakwattana S, Changpeng S, Ratanawachirin W, Tanruttanawong K, Boonyarit W (2012) Postprandial blood glucose response to grape seed extract in healthy participants: a pilot study. Pharmacogn Mag 8(31):192

Sathe SS, Lavate RA, Patil SB (2014) Ethnobotanical and medicinal aspects of mangroves from southrn Kokan (Maharashtra). Int J Emerg Trends Pharmaceutical Sci 3(4):11–17

Shaffie NM, Morsy FA, Ali AG, Sharaf HA (2010) Effect of caraway, coriander and fennel on the structure of kidney and islets of Langerhan in alloxan-induced diabetic rats: histological and histochemical study. J Am Sci 6(9):405–418

Shaheen R, Asgher M, Hussain F, Bhatti HN (2017) Immobilized lignin peroxidase from Ganoderma lucidum IBL-05 with improved dye decolorization and cytotoxicity reduction properties. Int J Biol Macromol 103:57–64

Shahin N, Alam S, Ali M (2012) Pharmacognostical standardisation and antidiabetic activity of Artocarpus heterophyllus leaves lam. Int J Drug Dev Res 4(1):346–352

Shao J-W, Jiang J-L, Zou J-J, Yang M-Y, Chen F-M, Zhang Y-J, Jia L (2019) Therapeutic potential of ginsenosides on diabetes: from hypoglycemic mechanism to clinical trials. J Funct Foods 103630

Sheikh Y, Maibam BC, Biswas D, Laisharm S, Deb L, Talukdar NC, Borah JC (2015) Anti-diabetic potential of selected ethno-medicinal plants of north east India. J Ethnopharmacol 171:37–41

Shihabudeen HMS, Priscilla DH, Thirumurugan K (2011) Cinnamon extract inhibits α-glucosidase activity and dampens postprandial glucose excursion in diabetic rats. Nutr Metabol 8(1):46

Singab AN, Youssef FS, Ashour ML (2014) Medicinal plants with potential antidiabetic activity and their assessment. Med Aromat Plants 3(151):2167–0412. 1000151

Soumya D, Srilatha B (2011) Late stage complications of diabetes and insulin resistance. J Diabetes Metab 2(9):1000167

Stojkovic D, Smiljkovic M, Ciric A, Glamoclija J, Van Griensven L, Ferreira IC, Sokovic M (2019) An insight into antidiabetic properties of six medicinal and edible mushrooms: inhibition of α-amylase and α-glucosidase linked to type-2 diabetes. South Afr J Bot 120:100–103

Street RA, Sidana J, Prinsloo G (2013) Cichorium intybus: traditional uses, phytochemistry, pharmacology, and toxicology. Evid-Based Complementary Altern Med

Stull AJ, Cash KC, Johnson WD, Champagne CM, Cefalu WT (2010) Bioactives in blueberries improve insulin sensitivity in obese, insulin-resistant men and women. J Nutr 140(10):1764–1768

Sudha P, Zinjarde SS, Bhargava SY, Kumar AR (2011) Potent α-amylase inhibitory activity of Indian Ayurvedic medicinal plants. BMC Complementary Alt Med 11(1):5

Sun L, Chen W, Meng Y, Yang X, Yuan L, Guo Y (2016a) Interactions between polyphenols in thinned young apples and porcine pancreatic α-amylase: inhibition, detailed kinetics and fluorescence quenching. Food Chem 208:51–60

Sun L, Warren FJ, Netzel G, Gidley MJ (2016b) 3 or 3′-Galloyl substitution plays an important role in association of catechins and theaflavins with porcine pancreatic α-amylase: The kinetics of inhibition of α-amylase by tea polyphenols. J Funct Foods 26:144–156

Sun L, Warren FJ, Gidley MJ (2019) Natural products for glycaemic control: polyphenols as inhibitors of alpha-amylase. Trends Food Sci Technol

Szablewski L (2019) Introductory chapter: glucose transporters. In: Blood Glucose Levels. IntechOpen

Tan Y, Chang SK (2017) Digestive enzyme inhibition activity of the phenolic substances in selected fruits, vegetables and tea as compared to black legumes. J Funct Foods 38:644–655

Telli A, Esnault M-A, Khelil AOEH (2016) An ethnopharmacological survey of plants used in traditional diabetes treatment in south-eastern Algeria (Ouargla province). J Arid Environ 127:82–92

Thengyai S, Thiantongin P, Sontimuang C, Ovatlarnporn C, Puttarak P (2019) α-Glucosidase and α-amylase inhibitory activities of medicinal plants in Thai antidiabetic recipes and bioactive compounds from Vitex glabrata R. Br. stem bark. J Herb Med 100302

Thenmozhi A, Shanmugasundaram C, Rao UM (2012) Biochemical evaluation of anti-diabetic phytomolecule through bioactivity guided solvent fractionation and subfractionation from hydromethanolic (2: 3) extract of Alligator pear Fruit in streptozotocin induced diabetic rats. J Appl Pharmaceutical Sci 2(1):61

Thilagam E, Parimaladevi B, Kumarappan C, Mandal SC (2013) α-Glucosidase and α-amylase inhibitory activity of Senna surattensis. J Acupunct Meridian Stud 6(1):24–30

Trinh DH, Tran PT, Trinh BT, Nguyen HT, Nguyen HD, Ha LD, Nguyen L-HD (2020) Coumarins and acridone alkaloids with α-glucosidase inhibitory and antioxidant activity from the roots of Paramignya trimera. Phytochem Lett 35:94–98

Tunnicliffe JM, Shearer J (2008) Coffee, glucose homeostasis, and insulin resistance: physiological mechanisms and mediators. Appl Physiol Nutr Metab 33(6):1290–1300

Türker M, Dalar A (2013) In vitro antioxidant and enzyme inhibitory properties and phenolic composition of M. neglecta Wallr. (Malvaceae) fruit: A traditional medicinal fruit from Eastern Anatolia. Ind Crops Prod 51:376–380

Unuofin JO, Lebelo SL (2020) Antioxidant effects and mechanisms of medicinal plants and their bioactive compounds for the prevention and treatment of type 2 diabetes: an updated review. Oxid Med Cell Longevity

Venkatakrishnan K, Chiu H-F, Wang C-K (2019) Popular functional foods and herbs for the management of type-2-diabetes mellitus: a comprehensive review with special reference to clinical trials and its proposed mechanism. J Funct Foods 57:425–438

Venkateshwarlu E, Dileep P, Rakesh K, Sandhya P (2013) Evaluation of antidiabetic activity of Carica papaya seeds on streptozotocin-induced type-II diabetic rats. J Adv Sci Res 4(2):38–41

Villas Boas GR, Rodrigues Lemos JM, de Oliveira MW, Dos Santos RC, Stefanello da Silveira AP, Barbieri Bacha F, Ito CNA, Bortolotte Cornelius E, Brioli Lima F, Sachilarid Rodrigues AM (2020) Aqueous extract from Mangifera indica Linn.(Anacardiaceae) leaves exerts long-term hypoglycemic effect, increases insulin sensitivity and plasma insulin levels on diabetic Wistar rats. Plos One 15(1):e0227105

Wang K, Bao L, Ma K, Liu N, Huang Y, Ren J, Wang W, Liu H (2015) Eight new alkaloids with PTP1B and α-glucosidase inhibitory activities from the medicinal mushroom Hericium erinaceus. Tetrahedron 71(51):9557–9563

Wang N, Zhu F, Shen M, Qiu L, Tang M, Xia H, Chen L, Yuan Y, Ma S, Chen K (2019) Network pharmacology-based analysis on bioactive anti-diabetic compounds in Potentilla discolor bunge. J Ethnopharmacol 241:111905

Wou C, Unwin N, Huang Y, Roglic G (2019) Implications of the growing burden of diabetes for premature cardiovascular disease mortality and the attainment of the Sustainable Development Goal target 3.4. Cardiovasc Diagn Ther 9(2):140

Wresdiyati T, Sa'Diah S, Winarto A, Febriyani V (2015) Alpha-glucosidase inhibition and hypoglycemic activities of Sweitenia mahagoni seed extract. HAYATI J Biosci 22(2):73–78

Xu L, Li Y, Dai Y, Peng J (2018) Natural products for the treatment of type 2 diabetes mellitus: pharmacology and mechanisms. Pharmacol Res 130:451–465

Yaseen G, Ahmad M, Zafar M, Sultana S, Kayani S, Cetto AA, Shaheen S (2015) Traditional management of diabetes in Pakistan: ethnobotanical investigation from traditional health practitioners. J Ethnopharmacol 174:91–117

Zhang L, Hogan S, Li J, Sun S, Canning C, Zheng SJ, Zhou K (2011) Grape skin extract inhibits mammalian intestinal α-glucosidase activity and suppresses postprandial glycemic response in streptozocin-treated mice. Food Chem 126(2):466–471

Zhang X, Liu Z, Bi X, Liu J, Li W, Zhao Y (2013) Flavonoids and its derivatives from Callistephus chinensis flowers and their inhibitory activities against alpha-glucosidase. EXCLI J 12:956

# Chapter 16
# An Overview of the Bioactivities of Gedunin

**Yong Sze Ong, Kooi Yeong Khaw, Loh Teng-Hern Tan, Peng-Nian Yew, Kai-Boon Tan, Wei Hsum Yap, Siah Ying Tang, Liang Ee Low, Learn-Han Lee, and Bey-Hing Goh**

**Abstract** Gedunin is a naturally occurring pentacyclic triterpenoid secondary metabolite presents in Meliaceae family. Since the first isolation and characterization of this compound at 1960s, the chemical synthesis and bioactive properties of this compound have been continuing researched. The plant has been historically used as the folk medicine to treat malaria and infectious diseases. Evidence-based pharmacology study showed that gedunin and its derivatives exerted enormous potential in the treatment of cancer, neurodegenerative and infectious diseases. Considering this year marks 60 years since the isolation of gedunin, this review presents an update of the bioactivity of gedunin and its derivatives with future prospect of this bioactive compound.

**Keywords** Gedunin · Bioactivities · Anti-cancer · Triterpenoid · Pharmacology

---

Y. S. Ong · K. Y. Khaw · K.-B. Tan · L. E. Low · B.-H. Goh (✉)
Biofunctional Molecule Exploratory (BMEX) Research Group, School of Pharmacy,
Monash University Malaysia, Bandar Sunway, 47500 Selangor Darul Ehsan, Malaysia
e-mail: goh.bey.hing@monash.edu

L. T.-H. Tan · L.-H. Lee
Novel Bacteria and Drug Discovery (NBDD) Research Group, Microbiome and Bioresource Research Strength, Jeffrey Cheah School of Medicine and Health Sciences, Monash University Malaysia, Bandar Sunway, 47500 Selangor Darul Ehsan, Malaysia

P.-N. Yew
Faculty of Applied Sciences (FOAS), Department of Bioscience, Tunku Abdul Rahman University College (TARUC), Jalan Genting Kelang, 53300 Kuala Lumpur, Setapak, Malaysia

K.-B. Tan
The University of Edinburgh, Centre for Discovery Brain Sciences, Hugh Robson Building, 15 George Square, Edinburgh EH8 9XD, Scotland

W. H. Yap
School of Biosciences, Taylor's University, Subang Jaya, 47500 Selangor Darul Ehsan, Malaysia



D. Pal and A. K. Nayak (eds.), *Bioactive Natural Products for Pharmaceutical Applications*, Advanced Structured Materials 140,
https://doi.org/10.1007/978-3-030-54027-2_16

## 16.1 Introduction

Ever since humankind begins, a close connection between humans and mother earth has been partly attributed to the discovery of cures for illnesses from nature. The use of medicinal plants to treat various illnesses back then was totally based on experience, due to the fact that there was no evidence-based knowledge behind the cause and pathogenesis of illnesses. Over the time, medicinal plants play an important role in the discovery and development of the drugs due to the advantages of readily available and cost effectiveness (Atanasov et al. 2015). As a result of nutritional needs and the adaption to the environmental challenges, plants produce chemical compounds which then become the valuable source for novel therapeutic agents (Khaw et al. 2017). Examples of clinically available plant-derived therapeutic compounds are vincristine from *Catharanthus roseus*, capsaicin from *Capsicum spp.*, paclitaxel from *Taxus brevifolia* and camptothecin from *Camptotheca acuminate.*

Plant-derived compounds have contributed to the drug development as they served as the lead compounds for some of the chemically-modified therapeutics such as acetylsalicylic acid from salicylic acid, warfarin from dicoumarol and topotecan from campthothecin. Public gives attention to the plant-derived compounds as proven by the award of Nobel Prize in Physiology or Medicine in 2015 for the discovery of artemisinin which was isolated from *Artemisia annua* for the treatment of parasitic diseases (Efferth et al. 2015). The therapeutic effects of

S. Y. Tang
Chemical Engineering Discipline, School of Engineering, Monash University Malaysia, Bandar Sunway, Selangor Darul Ehsan, Malaysia

Advanced Engineering Platform, Monash University Malaysia, Bandar Sunway, Selangor Darul Ehsan, Malaysia

L. E. Low
Institute of Pharmaceutics, College of Pharmaceutical Sciences, Zhejiang University, Hangzhou, Zhejiang 310058, P.R. China

Key Laboratory of Biomedical Engineering of the Ministry of Education, College of Biomedical Engineering & Instrument Science, Zhejiang University, Hangzhou, Zhejiang 310027, P.R. China

L.-H. Lee
Health and Well-Being Cluster, Global Asia in the 21st Century (GA21) Platform, Monash University Malaysia, Bandar Sunway, 47500 Selangor Darul Ehsan, Malaysia

B.-H. Goh
College of Pharmaceutical Sciences, Zhejiang University, Hangzhou, Zhejiang 310058, P.R. China

medicinal plants on an array of chronic diseases including cancer, diabetes, and neurological disorders are still being extensively studied, partly due to the necessity arisen from decreasing efficacy of the contemporary drugs.

Meliaceae, the fast-growing flowering plant family, consists of medicinal evergreens members that could withstand dry season in the tropical and subtropical areas. Neem tree (*Azadirachta indica*) is one of the most well-known species among members of the Meliaceae family for its wide health-promoting activities. The scientific name of *Azadirachta indica* (*Azadi* = free, *diracht* = tree, *indica* = of India, in Persian) literally denotes neem as a 'free tree of India', describing that the tree is intrinsically free from insect and illnesses (Kumar and Navaratnam 2013). Over the centuries, neem has been traditionally used by folklores for various purposes: The aqueous extract of neem leaf is used by Indians and Burmese to treat malaria (Badam et al. 1987) while the neem leaf is added in toothpaste by German and Indians for anti-septic purpose (Uko et al. 1995).

Over the decades, much attentions have been paid to discover the bioactive compounds of the neem tree, in which more than 300 structurally complex bioactive compounds have been purified and structurally characterized (Subapriya and Nagini 2005). It has been reported that the vast biological activities of the neem are contributed to some of the active constituents such as azadirachtin, gedunin, nimbolinin, nimbin and quercetin (Alzohairy 2016). Over the years, tremendous amount of attentions has been drawn to one of the bioactive compounds known as gedunin. This limonoid type of tetranortriterpenoid compound consists of four *trans*-fused six-membered rings with an oxirane annelated to the fourth ring. It has been reported that gedunin possesses a vast array of biological activities such as anti-cancer, anti-parasitic, anti-inflammatory, antimicrobial and insect growth inhibitory effects.

In addition to neem, gedunin and its natural occurring derivatives can be purified from the other Meliacae family including cannonball mangrove (*Xylocarpus granatum*), African mahogany (*khaya grandifoliola*) and Spanish cedar (*Cedrela odorata*) (Bickii et al. 2000; Lakshmi et al. 2010; Brandt et al. 2008). The current review summarizes the recent findings on the bioactivities of gedunin and its derivatives, and also evaluate the potential and significance of gedunin and its derivatives as a potential therapeutic agent.

## 16.2 Bioactivities of Gedunin

### *16.2.1 Anti-Cancer Properties of Gedunin*

More than 60% of the clinically-approved anti-cancer drugs are derived from plants and microorganisms. Owing to the genetic elasticity of cancer cells and the consequent emergence of chemoresistance, considerable attention has been drawn to natural bioactive compounds with anti-cancer properties. Gedunin has been studied extensively for its in vitro anti-cancer properties against a variety of cancers such as colon (Uddin et al. 2007), ovarian (Johnson et al. 2014; Patwardhan et al. 2013), mammary (Brandt et al. 2008; Kikuchi et al. 2011), pancreatic (Boopalan et al. 2013; Subramani et al. 2017), lung, stomach cancers, melanoma and leukemic (Kikuchi et al. 2011) (Table 16.1). The anti-cancer activities of neem compounds including gedunin in gynecological cancers (breast, cervical and ovarian) was recently reviewed by Moga et al. (2018). A number of literatures reported gedunin showed notable anti-cancer mechanisms such as induction of autophagy, apoptosis, mitotic arrest and inhibition of metastasis (Fig. 16.1). The compound exhibited a promising in vitro efficacy with $IC_{50}$ values in a range of 3.22–30 μM; in which the efficacy was more prominent in breast cancer cell lines (MCF-7 and SkBr3) and leukemia cell line (HL60).

In 2006, a research group led by Lamb et al. (2006) discovered that the anti-proliferative property of gedunin was modulated by the 90-kDa heat shock protein (Hsp90) through connectivity map. Hsp90 is a molecular chaperone that is responsible for the stability and function of a wide variety of client proteins for cells growths and survivals. In cancerous cells, these client proteins are frequently mutated or/and overexpressed, henceforth it is being actively pursued as individual therapeutic target (Garg et al. 2016). Hsp90 is essential for normal cellular homeostasis. It is also known to play an important role in several pathological conditions. Hsp90 and its co-chaperones (which help to regulate the function of the Hsp90 protein-folding machine) are extensively studied in tumorigenesis. Hsp90 inhibitors bind and inactivate Hsp90 N-terminal ATP-binding site, causing proteasomal degradation of Hsp90-dependent client protein (Patwardhan et al. 2013). Unlike most of the drug inhibitor of Hsp90, gedunin interacted with Hsp90 without the involvement of competitive inhibitor of ATP, suggesting that a novel and unique mechanism of Hsp90 modulation (Hieronymus et al. 2006).

Through Hsp90 inhibition, gedunin has been further investigated for the downstream mechanisms of action responsible for its anti-cancer effect. Research revealed that gedunin inhibited cancerous cell proliferation by inducing mitotic arrest between metaphase and anaphase. The significant overexpressed of checkpoint kinase-1 (CHK1) and polo-like kinase-1 (PLK1) in the cancerous cells has led to the mitotic halt. In a study performed with pancreatic cancer cells, gedunin has shown to exhibit anti-proliferative effect by suppressing mTOR and 4EBP, which are the key phosphoproteins responsible for protein translation.

Induction of apoptosis in cancerous cells has been the main strategy in cancer treatment (Ong et al. 2018). Several studies reported that treatment with gedunin has induced apoptosis and autophagy in various cancer cells through modulation of cell death executors such as LC3B, WIPI-1, VMP-1, pJNK, caspases and PARP. Gedunin demonstrated its potential as effective agent against pancreatic cancer which currently lacks of effective treatment. In vitro study by Subramani and colleagues showed that gedunin at 25 μM has induced high percentage of apoptotic cell death in three pancreatic cancer cell lines HPAC (46.5%), MIAPaCa-2 (37.6%) and PANC-1 (44.7%), but it was relatively safe against non-cancerous pancreatic cell (hTERT-HPNE). In addition, gedunin induced apoptosis through intrinsic and

**Table 16.1** Bioactivities of gedunin or derivatives tested with respective experimental model

| Bioactivities | Gedunin/Derivatives | Experimental model | Efficacy | References |
|---|---|---|---|---|
| Anti-cancer | Gedunin | Caco2 colon cancer cell line | 16.8 μM[a] | Uddin et al. (2007) |
| | | MCF-7 breast cancer cell lines | 8.84 ± 0.03 μM[a] | Brandt et al. (2008) |
| | | SkBr3 breast cancer cell lines | 3.22 ± 0.06 μM[a] | Brandt et al. (2008) |
| | | Pancreatic cancer cell lines | Unidentified | Boopalan et al. (2013) |
| | | ID8, ID8TaxR, A2870, C30 and CP70 ovarian cancer cell lines | 0–30 μM[b] | Johnson et al. (2014) |
| | | A2870 and ID8 chemoresistant ovarian cancer cell lines | 2.5 μM (Synergizes with cisplatin and paclitaxel)[b] | Johnson et al. (2014) |
| | | Cervical cancer cell line HeLa-$PR_B$ and Breast carcinoma cell lines MDA-MB-231, MDA-MB-453, Hs578T, T47D and MCF7 | 0–40 μM[b] | Patwardhan et al. (2013) |
| | | Leukemia cell line HL60 | 5.9 μM[a] | Kikuchi et al. (2011) |
| | | Lung cancer cell line A549 | >20 μM[a] | Kikuchi et al.(2011) |
| | | Stomach cancer cell line AZ521 | 16.9 μM[a] | Kikuchi et al. (2011) |
| | | Breast cancer cell line SK-BR-3 | 8.3 μM[a] | Kikuchi et al. (2011) |
| | | Melanoma cell line CRL1579 | >20 μM[a] | Kikuchi et al. (2011) |
| | | Caco2 colon cancer cell line | Unidentified | Uddin et al. (2007) |

(continued)

Table 16.1 (continued)

| Bioactivities | Gedunin/Derivatives | Experimental model | Efficacy | References |
|---|---|---|---|---|
| | | Pancrease cell line | 5.0 μM | |
| | 1α-hydroxy-1,2-dihydrogedunin | Leukemia cell line HL60 | >20 μM[a] | Kikuchi et al. (2011) |
| | 7-deacetylgedunin | Lung cancer cell line A549 | >20 μM[a] | Kikuchi et al. (2011) |
| | | Stomach cancer cell line AZ521 | 16.9 μM[a] | Kikuchi et al. (2011) |
| | | Breast cancer cell line SK-BR-3 | 12 μM[a] | Kikuchi et al. (2011) |
| | | Melanoma cell line CRL1579 | >20 μM[a] | Akihisa et al. (2009), Kikuchi et al. (2011) |
| | | In vitro epstein-bar virus early antigen (EBV-EA) activation assay | 488 mol ratio/32 pmol TPA[a] | Akihisa et al. (2009) |
| | | Leukemia cell line HL60 | 2.9 μM[a] | Kikuchi et al. (2011) |
| | 7-deacetyl-7-benzoylgedunin | Lung cancer cell line A549 | >20 μM[a] | Kikuchi et al. (2011) |
| | | Stomach cancer cell line AZ521 | >20 μM[a] | Kikuchi et al. (2011) |
| | | Breast cancer cell line SK-BR-3 | >20 μM[a] | Kikuchi et al. (2011) |
| | | Melanoma cell line CRL1579 | >20 μM[a] | Kikuchi et al. (2011) |
| | | Neuroblastoma cell line SH-SY5Y | 38.95 ± 3.225 | Lu et al. (2009) |

(continued)

**Table 16.1** (continued)

| Bioactivities | Gedunin/Derivatives | Experimental model | Efficacy | References |
|---|---|---|---|---|
| Cryoprotective | Gedunin | Experimental rat (in vivo) | 56.86 μg/mL[a] | Lakshmi et al. (2010) |
| | | Primary neuronal and astrocyte cells | 0–24 μM[b] | Smirnova et al. (2011) |
| | | Experimental rat (in vivo) | 66.54 μg/mL[a] | Lakshmi et al. (2010) |
| | Photogedunin | Primary rat hippocampal neurons | 500 nM[b] | Jang et al. (2010) |
| | Deoxygedunin | Neuroblastoma cell line SK-N-SH | 24.03[c] | Zhang et al. (2009) |
| | | Neuroblastoma cell line SK-N-SH | 12.77[c] | Zhang et al. (2009) |
| | Deacetylgedunin | Neuroblastoma cell line SK-N-SH | 6.02[c] | Zhang et al. (2009) |
| | Deacetoxy-7-oxogedunin | In vivo articular inflammation experimental model | Pretreatment: 0.005–5 mg/kg[b]<br>Post-treatment: 0.05 mg/kg[b] | Conte et al. (2015) |
| Anti-neurological | Gedunin | Inhibit NF-*κB activation* | 3.5–7.5 μM | Tom et al. (2019) |
| | | Neuro-2a cells | 20 μM | Yang et al. (2019) |
| Anti-inflammation | Gedunin | TPA-induced ear edema inflammation in Mice | 0.16[d] | Akihisa et al. (2009) |
| | | Inhibited NLRP3 expression in vivo | 0.5 mg/kg | Borges et al. (2017) |

(continued)

**Table 16.1** (continued)

| Bioactivities | Gedunin/Derivatives | Experimental model | Efficacy | References |
|---|---|---|---|---|
| | 7-deacetylgedunin | In vitro parasitized human $O^+$ erythrocyte with *Plasmodium falciparum* | 1 μM[a] | Khalid et al. (1986) |
| | Salannin<br>7-deacetylgedunin | Inhibit expression levels of nitric oxide and COX proteins | 30–50 μM | Akihisa et al. (2017) |
| Anti-parasitic | Gedunin | In vitro parasitized human $O^+$ erythrocyte with *Plasmodium falciparum* | 1.25 μg/mL[a] | Bickii et al. (2000) |
| | | In vivo *Plamodium berghei*-infected experimental model | 50 mg $kg^{-1}$ $day^{-1b}$ to achieve 44.6% parasitaemia clearance<br>50 mg $kg^{-1}$ $day^{-1b}$ synergize with 25 mg $kg^{-1}$ $day^{-1}$ dillapiol to achieve 70% parasitaemia clearance | Omar et al. (2003b) |
| | | In vivo *Plamodium berghei*-infected experimental model | 50 mg $kg^{-1}$ $day^{-1b}$ to achieve 44% parasitaemia clearance<br>50 mg $kg^{-1}$ $day^{-1b}$ synergize with 25 mg $kg^{-1}$ $day^{-1}$ dillapiol to achieve 75% parasitaemia clearance | Omar et al. (2003a) |
| | | In vitro parasitized human $O^+$ erythrocyte with *Babesia bovis, Babesia bigemina, Babesia caballi and Theleria equi* | *Babesia bovis*: 21.72 μM[a]<br>*Babesia bigemina*: 15.25 μM[a]<br>*Babesia caballi*: 22.1 μM[a]<br>*Theleria equi*: 33.21 μM[a] | Azirwan et al. (2019) |

(continued)

**Table 16.1** (continued)

| Bioactivities | Gedunin/Derivatives | Experimental model | Efficacy | References |
|---|---|---|---|---|
| | | In vitro antimalarial microdilution assay against *Plasmodium falciparum* chloroquine-sensitive clone W2 and chloroquine-resistance clone D6 | D6: 306 ng/mL[a]<br>W2: 315 ng/mL[a] | Omar et al. (2003a) |
| | | In vitro human erythrocyte infected with *Plasmodium falciparum* chloroquine-sensitive clone D10 and chloroquine-resistance clone W2 | D10: $1.66 \pm 0.37$[a]<br>W2: $1.31 \pm 0.42$[a] | Chianese et al. (2010) |
| | | In vitro MTT assay for female *Brugia malayi* adult worm | 0.24 μM[a] | Misra et al. (2011) |
| | | In vitro MTT assay for female *Brugia malayi* microfilaria | 2.03 μM[a] | MacKinnon et al. (1997), Misra et al. (2011) |
| | | In vivo experimental model intra-peritonealy inoculated with *Brugia malayi* adult worm | $-80.0 \pm 10.0\%$[b] changes in worm recovery | |
| | | In vitro antimalarial microdilution assay against *Plasmodium falciparum* chloroquine-sensitive clone W2 and chloroquine-resistance clone D6 | D6: 39 ng/mL[a]<br>W2: 20 ng/mL[a] | |
| | | In vitro antimalarial microdilution assay against *Plasmodium falciparum* chloroquine-sensitive clone W2 and chloroquine-resistance clone D6 | D6: >10,000 ng/mL[a]<br>W2: 840 ng/mL[a] | MacKinnon et al. (1997), Chianese et al. (2010) |
| | 1,2-Dihydrogegunin | In vitro antimalarial microdilution assay against *Plasmodium falciparum* chloroquine-sensitive clone W2 and chloroquine-resistance clone D6 | D6: 2580 ng/mL[a]<br>W2: 980 ng/mL[a] | |
| | 1,2-Epoxygedunin | | D6: 4210 ng/mL[a]<br>W2: 2440 ng/mL[a] | |
| | 1,2-Dihydro-3β-gedunol | In vitro human erythrocyte infected with *Plasmodium falciparum* | D6: > 10,000 ng/mL[a]<br>W2: > 10,000 ng/mL[a] | |

(continued)

**Table 16.1** (continued)

| Bioactivities | Gedunin/Derivatives | Experimental model | Efficacy | References |
|---|---|---|---|---|
| | 7-Ketogedunin | chloroquine-sensitive clone D10 and chloroquine-resistance clone W2 | D6: 2500 ng/mL[a]<br>W2: 900 ng/mL[a] | |
| | Tetrahydrogedunin | | D6: 133 ng/mL[a]<br>W2: 39 ng/mL[a] | |
| | 21-Acetylgedunin | | D6: 832 ng/mL[a]<br>W2: 156 ng/mL[a] | |
| | 23-Acetylgedunin | | D6: 10,000 ng/mL[a]<br>W2: 2130 ng/mL[a] | |
| | Hexahydrogedunin | | D6: 2610 ng/mL[a]<br>W2: 1280 ng/mL[a] | |
| | 7-Deacetylgedunin | | D10: 5.14 ± 1.23[a]<br>W2: 3.29 ± 0.59[a] | |
| | | In vivo *Plamodium berghei*-infected experimental model | 50 mg $kg^{-1}$ $day^{-1b}$ to achieve 67.5% parasitaemia clearance<br>50 mg $kg^{-1}$ $day^{-1b}$ synergize with<br>25 mg $kg^{-1}$ $day^{-1}$ dillapiol to achieve 80.7% parasitaemia clearance | Omar et al. (2003b) |
| | 7-Methoxygedunin | In vivo *Plamodium berghei*-infected experimental model<br>In vitro antimalarial microdilution assay against *Plasmodium falciparum* chloroquine-sensitive clone W2 and chloroquine-resistance clone D6 | 50 mg $kg^{-1}$ $day^{-1b}$ to achieve 67% parasitaemia clearance<br>50 mg $kg^{-1}$ $day^{-1b}$ synergize with<br>25 mg $kg^{-1}$ $day^{-1}$ dillapiol | Omar et al. (2003a) |

(continued)

**Table 16.1** (continued)

| Bioactivities | Gedunin/Derivatives | Experimental model | Efficacy | References |
|---|---|---|---|---|
| | | | to achieve 80% parasitaemia clearance | |
| | | | D6: 783 ng/mL[a]<br>W2: 749 ng/mL[a] | Omar et al. (2003a) |
| | | In vitro antimalarial microdilution assay against *Plasmodium falciparum* | D6: 556 ng/mL[a]<br>W2: 473 ng/mL[a] | |
| | 7-Benzyloxygedunin | chloroquine-sensitive clone W2 and chloroquine-resistance clone D6 | D6: 1828 ng/mL[a]<br>W2: 1998 ng/mL[a] | |
| | 7-Hydroxygedunin | In vitro MTT assay for female *Brugia malayi* adult worm | 0.21 μg/mL[a] | |
| | Photogedunin | In vitro MTT assay for female *Brugia malayi* microfilaria | 2.23 μg/mL[a] | Misra et al. (2011), Okhale et al. (2012) |
| | | In vivo experimental model intra-peritoneally inoculated with *Brugia malayi* adult worm | −70.0 ± 10.0%[b] changes in worm recovery | |
| | | Minimum inhibitory concentration (MIC) against: *Bacillus subtilis, Staphylococcus aureus, Klebsiella pneumonia, Escherichia coli* | Bs: NA<br>Sa: NA<br>Kp: NA<br>Ec: NA | |
| Antimicrobial | Gedunin | Minimum inhibitory concentration (MIC) against: *Bacillus subtilis, Staphylococcus aureus, Klebsiella pneumonia, Escherichia coli* | Bs: NA<br>Sa: 2000 μg/mL[b]<br>Kp: 1000 μg/mL[b]<br>Ec: 2000 μg/mL[b] | Okhale et al. (2013), Cespedes et al. (2016) |
| | 7-deacetoxy-7α-hydroxygedunin potassium salt | Minimum inhibitory concentration (MIC) against: *Bacillus subtilis, Staphylococcus aureus, Klebsiella pneumonia, Escherichia coli* | Bs: 2000 μg/mL[b]<br>Sa: NA[b]<br>Kp: 1000 μg/mL[b]<br>Ec: 2000 μg/mL[b] | |
| | 7-deacetoxy-7α-hydroxygedunin | In vitro bioassay with fall armyworm larvae | 39.0 ppm[b] with 48.3% larvae survival | |

(continued)

**Table 16.1** (continued)

| Bioactivities | Gedunin/Derivatives | Experimental model | Efficacy | References |
|---|---|---|---|---|
| Insect growth inhibitory | Gedunin | In vitro bioassay with the gram pod borer, *Helicoverpa armigera* (Hubner) and the Asian armyworm larvae, *Spodoptera litura* (Fabricius) | *H. armigera*: 50.8 ppm[a]<br>*S. litura*: 40.4 ppm[a] | Koul et al. (2003) |
| | | In vitro bioassay with *Anopheles stephensi*'s larvae, pupae and adult | 0.1 ppm[b]<br>Larval mortality: 55–67%<br>Pupal mortality: $61.6 \pm 5.9\%$<br>Adult mortality: $61.0 \pm 5.9\%$ | Nathan et al. (2005) |
| | | In vitro bioassay with fall armyworm larvae | 8.0 ppm[b] with 17% larvae survival | Céspedes et al. (2000) |
| | Photogedunin acetate | In vitro bioassay with fall armyworm larvae | 10.0 ppm[b] with 50% larvae survival | Céspedes et al. (2000) |
| | Photogedunin epimeric | In vitro bioassay with the gram pod borer, *Helicoverpa armigera* (Hubner) and the Asian armyworm larvae, *Spodoptera litura* (Fabricius) | *H. armigera*: 24.2 ppm[a]<br>*S. litura*: 21.5 ppm[a] | Koul et al. (2003) |
| | 6β-hydroxygedunin | In vitro bioassay with *Anopheles stephensi*'s larvae, pupae and adult | 0.1 ppm[b]<br>Larval mortality: 78–94%<br>Pupal mortality: $84.3 \pm 7.6\%$<br>Adult mortality: $87.5 \pm 8.5\%$ | Nathan et al. (2005) |

[a]50% inhibitory dose; [b]Dose with highest reported efficacy; [c]Half maximal effective concentration; [d]Effective dose for 50% of sample population

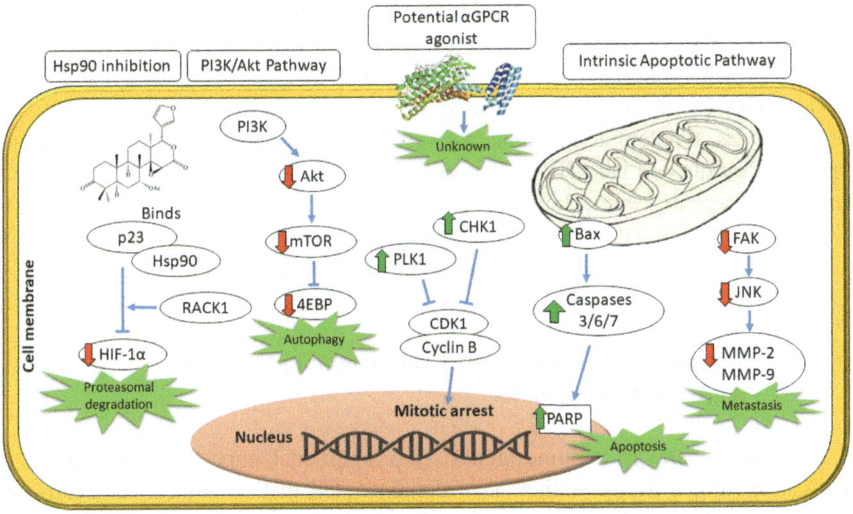


**Fig. 16.1** Graphical summary of the anti-cancer mechanisms of gedunin, which includes the degradation of proteasome via binding to p23, activation of autophagy via modulation of PI3K/Akt pathway, arrest cell cycle by upregulation of CHK1 and PLK1, induction of apoptosis via intrinsic pathway and inhibition of metastasis via downregulation of FAK, JNK, MMP-2 and MMP-9

extrinsic pathways by modulation of Bax, caspase 3 and PARP in pancreatic cancer cells PANC-1 (Subramani et al. 2017).

Gedunin showed anti–metastatic effect through inhibition of sonic hedgehog signaling. In the study performed by Subramani and colleagues, the migratory characteristics of HPAC, MIAPaCa-2 and PANC-1 were significantly inhibited (45–86%) by 15 μM of gedunin in a scratch assay (Subramani et al. 2017). Additionally, 5 μM gedunin efficiently inhibited rhShh–induced invasion of pancreatic cancer cells by downregulation of proteins involved in Hedgehog/Gli signaling pathway such as PTCH1, PTCH2, Gli1, SUFU and Shh proteins. Another study by Li et al. (2018) demonstrated that gedunin inhibited cell migration and invasion at concentration of 15 μM in glioma primary brain tumor cell U-251 MG by reducing the expression of FAK, MMP-2, MMP-9 and uPA after 48 h of incubation. Even though gedunin has been well-studied in vitro, its mechanisms of action have not been studied extensively under in vivo conditions, leaving a huge gap towards the realization of its clinical translation as an anti-cancer agent.

Apart from gedunin, some studies have been carried out to investigate the anti-cancer activities of its derivatives. The isoform, 7-deacetyl-7-benzoylgedunin, exhibited lower anti-cancer efficacy with higher $IC_{50}$ (more than 20 μM) as compared to gedunin across the variety of cancerous cell lines. In 2008, Brandt's research team has synthesized a series of gedunin derivatives and found out that none of those were more effective than gedunin against breast cancer cells (MCF-7 and SkBr3) (Brandt et al. 2008). In fact, there is no literature available to explain why its derivatives cannot outperform gedunin.

A recent study reported that gedunin derivatives, 3-α-acetoxydihydrodeoxygedunin (3-α-DOG) were partial agonist towards adhesion G protein coupled receptors (aGPCR) GPR56/ADGRG1 (Stoveken et al. 2018). aGPCRs is broadly distributed in various tumor cells and has become prominent as a potential therapeutic target in various neurological diseases and cancers (Paavola and Hall 2012). The activating mechanism of most aGPCRs have only been recently characterized and limited modulatory compound are being defined. Hence, 3-α-DOG and possibly other gedunin derivatives as partial agonist to aGPCRs could indicate an invaluable area for further research.

#### 16.2.1.1 Combinatorial Treatment Using Gedunin Enhances Anti-cancer Properties

The complex pathway of tumorigenesis has significant impact on the efficacy of anti-cancer drugs. Many studies have demonstrated that combinatorial treatments using multiple agent have proven to improve the clinical therapeutic efficacy of anti-cancer drugs as compared to single agent regimen. Specifically, gedunin in neem has been shown to synergize with other anti-cancer drugs apart from its anti-cancer properties as a single agent. (Sharma et al. 2014; Gupta et al. 2017; Bodduluru et al. 2014; Kamath et al. 2009).

Combinatory treatment with gedunin and epalrestat has shown to attenuate the Aldose Reductase (AR) enzyme in the SCC131 oral cancer cell line (Tanagala et al. 2018). Inhibition of diabetic associated enzymes Aldose Reductase (AR) are correlated to the inhibition of SCC131 cell proliferation, apoptosis evasion and angiogenesis through enervating downstream PI3K/Akt/mTOR/ERK/NF-κB signalling axis. It is reported that the combinatory treatment resulted in diversion of cell death in favor of apoptosis in the treated cell lines. Further, pro-invasive and proangiogenic proteins were downregulated in treated groups, hence inhibiting cancer cell migration. In all events, combined treatment of gedunin and epalrestat was reported to be more effective than single agents. In a similar study, Kishore and the team (2016) suggested that the abrogation of AR enzyme, PI3K/Akt and NF-κB pathways by gedunin has downregulated the expression of miR-21, hypoxia inducible factor-1 alpha (HIF-1α) and the pro-angiogenic factors vascular endothelial growth factor in a hamster model of oral carcinogenesis, which inhibited angiogenesis (Kishore et al. 2016).

Interestingly, gedunin acted synergistically with the chemotherapy agents cisplatin (C30, CP75) and paclitaxel (ID8TaxR) at doses as low as 2.5 μM, which have no effect on the chemoresistant cancer cell lines when treated alone (Johnson et al. 2014). In a similar study, the anti-cancer effect of neem-derived gedunin alone and in combination with cisplatin was also tested against SKOV3, OVCAR4, and OVCAR8 ovarian cancer cell lines. Gedunin reduced cell proliferation by 20% when treated alone. When treat synergistically with cisplatin, further 47% decrease in cell proliferation was reported (Kamath et al. 2009).

In another study, Sharma et al. (2014) proved the synergistic effect sub-lethal dose of ethanolic neem leaf extract and cisplatin in reducing the viability of breast (MCF-7) and cervical (HeLa) cancer cells compared to individual compound alone. It was reported that 1 μM of cisplatin used in combination with 50–100 μg/mL neem leaf extract resulted in a significant (71–82%) reduction of MCF-7 cells as compared to the compounds alone (85–93.1%). The calculated Combinational indices (CI) in the studies were reported to be <1, indicating a synergistic interaction between the two drugs at their respective treatment dosage (Sharma et al. 2014).

### *16.2.2 Anti-neurological Disorders and Cryoprotective Effects of Gedunin*

Neurodegeneration is a progressive deterioration of neuronal structures and functions resulting to cognitive disability and dementia primarily affecting aging population (Ramanan and Saykin 2013). Alzheimer's disease, Parkinson's disease, Amyotrophic lateral sclerosis and Hungtington disease are the few commonly known neurodegenerative disorders. These diseases remain uncured despite extensive basic research and drug discovery to ameliorate the conditions. Some research showed that gedunin and its derivatives were able to protect the neurons against β-amyloid (Aβ) and other stressors.

It is widely known that the accumulation of Aβ oligomers or fibrils contributes to neurodegenerative pathology in Alzheimer's disease (Murphy and LeVine 2010). Gedunin significantly suppressed the cell death in neuronal cells subjected to Aβ cytotoxicity (Lu et al. 2009). Gedunin derivatives, deacetylgedunin, deacetoxy-7-oxogedunin and deoxygedunin are the HSF1/Hsp70 activators and thus protecting neuronal cells from cell death induced by MG-132 cytotoxicity (Zhang et al. 2009). Recent study by Tom and his colleagues demonstrated that gedunin inhibited $A\beta_{1\text{-}42}$-induced NF-κB activation, the expression of nitric oxide and IL-1β level (Tom et al. 2019).

Brain-derived neurotrophic factors (BDNF) are a family of neurotrophins that plays an essential role in the maintenance of central nervous system (Thoenen 1991). BDNF binds to transmembrane TrkB receptor and activates downstream signaling cascades that are responsible for a variety of neuronal protections against ischemic injury, glutamate toxicity, and programmed cell death (Leeds et al. 2005; Lindholm et al. 1993; Schabitz et al. 2000). Smirnova et al. (2011) reported that gedunin was able to rescue neuronal cells from the cell death caused by glutamate and its analog by acting as a direct activator of NF-E2-related factor 2 (Nrf2), a transcriptional regulator of cellular detoxification and antioxidant defense. Deoxygedunin has been reported as an agonist of BDNF by eliciting the strongest stimulatory effect on TrkB signaling cascades and executing neuronal protection against cell death induced by oxygen–glucose deprivation (OGD) and glutamate. In a similar study, deoxygedunin appeared to have anti-depression and learning enhancement effects (Jang et al. 2010).

Huntington disease is a neurological disease characterized by abnormal movements, cognitive dysfunction, and psychiatric disease (Frank 2014). Till date, there are no treatment to reverse the progression of this disease. A study by Yang and the team showed that gedunin disaggregated mutant Huntingtin protein in Neuro-2a cells, endogenous mutant Huntingtin progression and intranuclear inclusions in Hungtinton disease patients derived cells (Yang et al. 2019).

### *16.2.3 Anti-inflammatory Effects of Gedunin*

Inflammation is an adaptive immune response that underlies myriads of physiological and pathological conditions (Medzhitov 2008). Dysregulated inflammation plays a crucial role in the development of many cellular and systemic disorders such as cardiovascular diseases, cancer, neurodegenerative disorders, metabolic syndromes and autoimmune disease. For this reason, researchers have begun to understand the underlying mechanism of inflammation in order to control or cure the aforementioned diseases.

Ferraris et al. (2012) showed that pretreatment of gedunin reduced the level of chemotactic mediators such as CCL2, CCL3, CCL5, CCL11, IL-5 and the lipid mediator $LTB_4$ thus impairing the recruitment of eosinophil and T lymphocyte to the site of inflammation of experimental mice. Subsequent study by Ferraris et al. (2012) revealed that treatment with gedunin and derivatives in zymosan-induced inflammation experimental model has significantly improved knee edema progression, reduced neutrophil accumulation and production of inflammation mediators such as CXCL8/IL-8, IL-1β, IL-6, TNF-α, LBT4 and PGE2 (Conte et al. 2015).

Penido et al. (2005) studied on the anti-inflammatory and analgesic effects of bioactive compounds from the seeds of *Carapa guianensis* Aublet, which consist of gedunin and 6-acetoxygedunin, 7-deacetocy-7-oxogedunin. Orally given tetranortriterpenoids (TNTP) appeared to be able to reduce ovalbumin- and zymosan-induced edema inflammation and protein extravasation to the site of inflammation in mice. It is noteworthy that such efficacy of TNTP at 25–100 mg/kg was comparable to that of commercially available anti-inflammatory drugs such as dexamethasone and promethazine. Besides, the influx of leukocyte and the increase in inflammation mediators such as TNF-α, IL-1β, CXCL8/IL-8 and prostaglandin $E_2$ induced by various noxious stimuli were markedly observed after the treatment with TNTP (Penido et al. 2006, 2005). Gedunin derivatives, 7-deacetylgedunin was reported to reduce ear edema inflammation promoted by 12-O-tetradecanoylphorbol-13-acetate (TPA) in experimental mice (Akihisa et al. 2009). A continuous investigation by Akihisa et al. (2017) showed that Salannin and 7-deacetylgedunin reduced the expression levels of the inducible nitric oxide synthase and cyclooxygenase proteins in a dose dependent manner (Akihisa et al. 2017). Another study showed that gedunin inhibited NLRP3 expression and inpaired production of TNF-α, IL-6 and nitric oxide in-vivo (Borges et al. 2017).

### 16.2.4 Anti-parasitic Effects of Gedunin and Derivatives

Malaria is a vector-borne infectious disease caused by *Plasmodium* spp. parasitic protozoan that belongs to the Apicomplexa phylum. Malaria is one of the major public health problems that negatively impact on social and economy. In 2010, 1.2 million people worldwide were dead due to malaria (Murray et al. 2012). The situation continues to deteriorate due to the fact that no preventive vaccination against *Plasmodium* is available. The control of the disease thus relies heavily on the treatment with antimalarial medications such as quinine, chloroquine, sulfadoxine-pyrimethamine and amodiaquine over the decades (Sinha et al. 2014). Due to massive spreads of antimalarial resistance, alternative therapeutic strategies are urgently needed.

Gedunin was reported to possess antimalarial activities against *Plasmodium falciparum,* which accounted for most of the malaria cases worldwide (Khalid et al. 1986). The $IC_{50}$ of gedunin was about 1 μM upon 48 h of treatment, which was comparable to the established antimalarial compound quinine. MacKinnon et al. (1997) tested the antimalarial activities of gedunin and 9 derivatives isolated from neem tree against *P. falciparum.* However, only gedunin exhibited the highest antimalarial activity with the $IC_{50}$ of 39 ng/mL against chloroquine-sensitive *P. falciparum* clone, which was even higher than chloroquine and quinine (MacKinnon et al. 1997). On top of that, synergistic study between gedunin and chloroquine showed that gedunin extracted from the bark and the seed of *Khaya grandifoliola* was able to exert synergistic effect when treated together with chloroquine (Bickii et al. 2000). Moreover, gedunin and derivative 7-methoxygedunin were shown to have synergistic effect when acting together with dillapiol, which is a commonly known antimalarial compound extracted from dill weed, in the mice infected with *Plasmodium berghei* (Omar et al. 2003b). When acting alone, 7-methoxygedunin was able to inhibit the parasite level in infected mice by 67% at 50 mg/kg, while 80% of inhibition was achieved when 25 mg/kg dillapiole was administered together with the same dose of 7-methoxygedunin (OMAR et al. 2003a).

On the other hand, gedunin and photogedunin have been reported for their antifilarial activity against parasitic filarial nematodes *Brugia malayi.* In vitro study showed that gedunin and photogedunin were active against adult worm ($IC_{50}$ of 0.24 and 0.21 μg/mL) comparable to the antiparasitic drug ivermectin ($IC_{50}$ of 1.61 μg/mL). In addition, in vivo study showed that gedunin and photogedunin significantly suppressed the survival of the parasite in jird transplanted model (Misra et al. 2011). Gedunin inhibited the growth of *Babesia bovis, Babesia bigemina, Babesia caballi and Theleria equi* which are the causative agents of bovine and equine piroplasmosis (Azirwan et al. 2019). Taken together, it is plausible to mention that gedunin possesses strong antiparasitic activities that is noteworthy for further investigations. Research by Yerbanga et al. (2014) proved that the ethanol extract of neem leaves with gedunin as a major component exhibited in vitro anti-malarial effect through transmission blocking activity in *P. falciparum*. Yet, no further study has been done to investigate the underlying mechanisms of action (Yerbanga et al. 2014).

### *16.2.5 Antimicrobial Effects of Gedunin and Derivatives*

The occurrence of bacterial resistance has increased over the years in several countries where by common antibiotics such as aminoglycosides, cephalosporins and penicillins are no longer effective against pathogens (Chaves et al. 2015; Davies and Davies 2010). Such resistance causes negative impacts on public health with increasing mortality and morbidity and increase in economic burden with rising cost for antimicrobial treatment. Therefore, myriads of natural products derived from animal, plant and microorganism have been deemed as an alternative to antibiotics.

Okhale et al. (2012) revealed that gedunin was inactive against *Bacillus subtilis, Staphylococcus aureus, Escherichia coli* and *Klebsiella pneumoniae* whereas the semisynthetic derivative 7-deacetoxy-7α-hydroxygedunin potassium salt possessed greater antimicrobial activity against all tested bacterial species except *B. subtilis*. The subsequent study by the group showed that another derivative, 7-deacetoxy-7α-hydroxygedunin exhibited antimicrobial activity against *B. subtilis, K. pneumonia* and *Escherichia coli* but not *S. aureus* (Okhale et al. 2012, 2013). Although various studies have been carried out to investigate the bioactivities of gedunin, the antimicrobial property of gedunin is overlooked by the research community.

### *16.2.6 Insect Growth Inhibition Effects of Gedunin and Derivatives*

Due to insect resistance and ecological impacts caused by commercialized insecticides, more efforts have been devoted to investigate the potential of natural products on the insect's growth inhibitory effect. One of the functions of the plant-based phytochemicals as the defense system is to dissuades herbivores and insecticides from consuming them. Over the years, studies have been carried out to investigate the inhibitory effect of gedunin and its derivatives against insect growth (Cespedes et al. 2016). Gedunin and photogedunin from *Cedrela dugessi* and *Cedrela salvadorensis* were tested against the growth larvae of fall armyworm. The results showed that these compounds were able to inhibit the growth of larvae in a concentration dependent manner (Céspedes et al. 2000). Another study has reported the larval growth inhibitory effects of gedunin derivative, 6β-hydroxygedunin, against the gram pod borer, *Helicoverpa armigera* (Hubner) and the Asian armyworm, *Spodoptera litura* (Fabricius). The 6β-hydroxygedunin inhibited the growth of the larvae by 50% at a concentration 21.5 μg/mL. Interestingly, the efficacy of 6β-hydroxygedunin was two-fold greater than gedunin. In addition, gedunin and deacetylgedunin were tested against *Anopheles stephensi* which is the major vector of malaria. Nathan et al. (2005) reported that deacetylgedunin possessed high inhibitory activity against the growth of larvae, pupae and adult as well as the production of egg of *A. stephensi*. However, the mechanisms of the insecticidal, repellent and growth inhibitory of these compounds are not well understood.

## 16.3 Commercial Potential of Gedunin

Considering the multiple therapeutic effects of gedunin that have been reported in the literature, it is not surprised that patents have been registered for gedunin and its derivatives. Thus far, Vinson-Hieronymus et al. (2011) has patented gedunin and its derivatives (publication number: US 2011/0,263,693 A1), together with celastrol to inhibit Hsp90. As mentioned earlier, Hsp90 is a multitasking molecule that governs the maturation of an array of diverse cellular functions. It is also a specific client protein that acts as signal transducers implicated in the molecular mechanisms of normal cellular biology, disease and evolutionary processes (Kaplan and Li 2012). Thus, by inhibiting Hsp90, pathological processes like cancer and inflammation can thus be suppressed (Kaplan and Li 2012; Chatterjee et al. 2007). On the other hand, gedunin has been patented as a substrate of polymeric wound care material (publication number: US 2013/0,209,534 A1) for its antimicrobial properties and the ability to aid in wound healing. The neuroprotective effects of gedunin and a number of limonoids have also granted the inventor Steiner and colleagues a patent (publication number: US 2010/0,056,617 A1) disclosing the method of preparation and the utilization of these compounds.

## 16.4 Conclusion and Gaps of Knowledge

The tremendous robust evidences to date have shown that gedunin and its derivative possess multiple therapeutic effects including anti-cancer, anti-inflammatory, anti-parasitic, antimicrobial and insect growth inhibition. However, most of the studies are limited to in vitro phase with or without the comparison to commercialized drugs. We thus look forward to have extensive animal model-based evidences in order to understand how gedunin and its derivatives work in in vivo systems. This will then fill the gaps of the pharmacological understandings of gedunin and its derivative such as bioavailability, pharmacokinetics and pharmacodynamics. Nonetheless, toxicology screening is yet to be carried out to ensure the safety of gedunin and its derivatives, especially for semisynthetic derivatives of gedunin, which involve inorganic components and chemicals in the production. To entirely elucidate the molecular mechanism action of gedunin, more in-depth and holistic studies have yet to be carried out.

**Acknowledgements** This work is financially supported by Monash Global Asia in the 21st Century (GA21) research grant (GA-HW-19-L01, GA-HW-19-L06 and GA-HW-19-S02); University grants from the University of Malaya (PG136-2016A and PG135-2016A), External Industry Grants from Biotek Abadi Sdn Bhd (vote no. GBA-81811A) and Fundamental Research Grant Scheme (FRGS/1/2019/WAB09/MUSM/02/1 and FRGS/1/2019/SKK08/MUSM/02/7).

**Author Contributions** The writing was performed by K-YK, Y-SO, LT-HT, P-NY, K-BT. The manuscript was further critically reviewed and improved by W-SY, S-YT, K-YK, Y-SO, LT-HT, L-EL, L-HL and B-HG. While B-HG, LT-HT, L-HL provided vital guidance and insight to the work. The project was conceptualized by B-HG.

## References

Akihisa T, Nishimoto Y, Ogihara E, Matsumoto M, Zhang J, Abe M (2017) Nitric oxide production-inhibitory activity of limonoids from Azadirachta indica and Melia azedarach. Chem Biodiv 14(6).https://doi.org/10.1002/cbdv.201600468

Akihisa T, Noto T, Takahashi A, Fujita Y, Banno N, Tokuda H, Koike K, Suzuki T, Yasukawa K, Kimura Y (2009) Melanogenesis inhibitory, anti-inflammatory, and chemopreventive effects of limonoids from the seeds of Azadirachta indicia A. Juss. (neem). J Oleo Sci 58(11):581–594. https://doi.org/10.5650/jos.58.581

Alzohairy MA (2016) Therapeutics role of Azadirachta indica (Neem) and their active constituents in diseases prevention and treatment. Evidence Complement Alter Med ECAM 2016:7382506–7382506. https://doi.org/10.1155/2016/7382506

Atanasov AG, Waltenberger B, Pferschy-Wenzig E-M, Linder T, Wawrosch C, Uhrin P, Temml V, Wang L, Schwaiger S, Heiss EH, Rollinger JM, Schuster D, Breuss JM, Bochkov V, Mihovilovic MD, Kopp B, Bauer R, Dirsch VM, Stuppner H (2015) Discovery and resupply of pharmacologically active plant-derived natural products: a review. Biotechnol Adv 33 (8):1582–1614. https://doi.org/10.1016/j.biotechadv.2015.08.001

Azirwan G, Thillaiampalam S, Muncharee T, Yokoyama N, Igarashi I (2019) In vitro inhibitory effect of gedunin on Babesia and Theileria parasites. J Protozool Res 23(1–2):1–6

Badam L, Deolankar RP, Kulkarni MM, Nagsampgi BA, Wagh UV (1987) In vitro antimalarial activity of neem (Azadirachta indica A. Juss) leaf and seed extracts. Indian J Malariol 24 (2):111–117

Bickii J, Njifutie N, Foyere JA, Basco LK, Ringwald P (2000) In vitro antimalarial activity of limonoids from Khaya grandifoliola C.D.C. (Meliaceae). J Ethnopharmacol 69(1):27–33. https://doi.org/10.1016/s0378-8741(99)00117-8

Bodduluru LN, Kasala ER, Thota N, Barua CC, Sistla R (2014) Chemopreventive and therapeutic effects of nimbolide in cancer: the underlying mechanisms. Toxicol In vitro Int J Publ Assoc BIBRA 28(5):1026–1035. https://doi.org/10.1016/j.tiv.2014.04.011

Boopalan T, Chaudhary A, Murali S, Arumugam A, Lopez R, Nandy S, Gutierrez C, Parada J, Agullo P, Lakshmanaswamy R (2013) Abstract 2264: anti-cancer effect of gedunin against pancreatic cancer cell lines. Can Res 73(8 Supplement):2264–2264. https://doi.org/10.1158/1538-7445.am2013-2264

Borges PV, Moret KH, Raghavendra NM, Maramaldo Costa TE, Monteiro AP, Carneiro AB, Pacheco P, Temerozo JR, Bou-Habib DC, das Gracas Henriques M, Penido C (2017)

Protective effect of gedunin on TLR-mediated inflammation by modulation of inflammasome activation and cytokine production: Evidence of a multitarget compound. Pharmacol Res 115:65–77. https://doi.org/10.1016/j.phrs.2016.09.015

Brandt GEL, Schmidt MD, Prisinzano TE, Blagg BSJ (2008) Gedunin, a novel Hsp90 inhibitor: semisynthesis of derivatives and preliminary structure−activity relationships. J Med Chem 51 (20):6495–6502. https://doi.org/10.1021/jm8007486

Céspedes CL, Calderón JS, Lina L, Aranda E (2000) Growth inhibitory effects on fall armyworm Spodoptera frugiperda of some limonoids isolated from Cedrela spp.(Meliaceae). J Agric Food Chem 48(5):1903–1908

Cespedes CL, Lina-Garcia L, Kubo I, Salazar JR, Ariza-Castolo A, Alarcon J, Aqueveque P, Werner E, Seigler DS (2016) Calceolaria integrifolia sl complex, reduces feeding and growth of Acanthoscelides obtectus, and Epilachna varivestis. A new source of bioactive compounds against dry bean pests. Ind Crops Prod 89:257–267

Chatterjee A, Dimitropoulou C, Drakopanayiotakis F, Antonova G, Snead C, Cannon J, Venema RC, Catravas JD (2007) Heat shock protein 90 inhibitors prolong survival, attenuate inflammation, and reduce lung injury in murine sepsis. Am J Respir Crit Care Med 176 (7):667–675. https://doi.org/10.1164/rccm.200702-291OC

Chaves TP, Clementino EL, Felismino DC, Alves RR, Vasconcellos A, Coutinho HD, Medeiros AC (2015) Antibiotic resistance modulation by natural products obtained from Nasutitermes corniger (Motschulsky, 1855) and its nest. Saudi J Biol Sci 22(4):404–408. https://doi.org/10.1016/j.sjbs.2014.12.005

Chianese G, Yerbanga SR, Lucantoni L, Habluetzel A, Basilico N, Taramelli D, Fattorusso E, and Taglialatela-Scafati O (2010) J Nat Prod 73(8):1448–1452. https://doi.org/10.1021/np100325q

Conte FP, Ferraris FK, Costa TEMM, Pacheco P, Seito LN, Verri WA Jr, Cunha FQ, Penido C, Henriques MG (2015) Effect of gedunin on acute articular inflammation and hypernociception in mice. Molecules (Basel, Switzerland) 20(2):2636–2657. https://doi.org/10.3390/molecules20022636

Davies J, Davies D (2010) Origins and evolution of antibiotic resistance. Microbiol Mol Biol Rev MMBR 74(3):417–433. https://doi.org/10.1128/mmbr.00016-10

Efferth T, Zacchino S, Georgiev MI, Liu L, Wagner H, Panossian A (2015) Nobel Prize for artemisinin brings phytotherapy into the spotlight. Phytomed Int J Phytotherapy Phytopharmacol 22(13):A1-3. https://doi.org/10.1016/j.phymed.2015.10.003

Ferraris FK, Moret KH, Figueiredo AB, Penido C, Henriques M (2012) Gedunin, a natural tetranortriterpenoid, modulates T lymphocyte responses and ameliorates allergic inflammation. Int Immunopharmacol 14(1):82–93. https://doi.org/10.1016/j.intimp.2012.06.002

Frank S (2014) Treatment of Huntington''s disease. Neurotherapeutics 11(1):153–160. https://doi.org/10.1007/s13311-013-0244-z

Garg G, Khandelwal A, Blagg BS (2016) Anti-cancer inhibitors of Hsp90 function: beyond the usual suspects. Adv Cancer Res 129:51–88. https://doi.org/10.1016/bs.acr.2015.12.001

Gupta SC, Prasad S, Tyagi AK, Kunnumakkara AB, Aggarwal BB (2017) Neem (Azadirachta indica): an Indian traditional panacea with modern molecular basis. Phytomed Int J Phytotherapy Phytopharmacol 34:14–20. https://doi.org/10.1016/j.phymed.2017.07.001

Hieronymus H, Lamb J, Ross KN, Peng XP, Clement C, Rodina A, Nieto M, Du J, Stegmaier K, Raj SM, Maloney KN, Clardy J, Hahn WC, Chiosis G, Golub TR (2006) Gene expression signature-based chemical genomic prediction identifies a novel class of HSP90 pathway modulators. Cancer Cell 10(4):321–330. https://doi.org/10.1016/j.ccr.2006.09.005

Jang S-W, Liu X, Chan CB, France SA, Sayeed I, Tang W, Lin X, Xiao G, Andero R, Chang Q, Ressler KJ, Ye K (2010) Deoxygedunin, a natural product with potent neurotrophic activity in mice. PLoS ONE 5(7):e11528. https://doi.org/10.1371/journal.pone.0011528

Johnson J, Venugopal A, Kwatra D, Roby K, Godwin A, Anant S (2014) Abstract 4553: Gedunin, a novel HSP-90 inhibitor, synergizes with cisplatin and paclitaxel to inhibit growth of chemoresistant ovarian cancer cell lines. Can Res 74(19 Supplement):4553–4553. https://doi.org/10.1158/1538-7445.am2014-4553

Kamath SG, Chen N, Xiong Y, Wenham R, Apte S, Humphrey M, Cragun J, Lancaster JM (2009) Gedunin, a novel natural substance, inhibits ovarian cancer cell proliferation. Int J Gynecol Cancer Offic J Int Gynecol Cancer Soc 19(9):1564–1569. https://doi.org/10.1111/IGC.0b013e3181a83135

Kaplan KB, Li R (2012) A prescription for 'stress'—the role of Hsp90 in genome stability and cellular adaptation. Trends Cell Biol 22(11):576–583. https://doi.org/10.1016/j.tcb.2012.08.006

Khalid SA, Farouk A, Geary TG, Jensen JB (1986) Potential antimalarial candidates from African plants: an in vitro approach using Plasmodium falciparum. J Ethnopharmacol 15(2):201–209. https://doi.org/10.1016/0378-8741(86)90156-X

Khaw KY, Parat MO, Shaw PN, Falconer JR (2017) Solvent supercritical fluid technologies to extract bioactive compounds from natural sources: a review. Molecules 22(7). https://doi.org/10.3390/molecules22071186

Kikuchi T, Ishii K, Noto T, Takahashi A, Tabata K, Suzuki T, Akihisa T (2011) Cytotoxic and apoptosis-inducing activities of limonoids from the seeds of Azadirachta indica (neem). J Nat Prod 74(4):866–870. https://doi.org/10.1021/np100783k

Kishore TK, Ganugula R, Gade DR, Reddy GB, Nagini S (2016) Gedunin abrogates aldose reductase, PI3K/Akt/mToR, and NF-kappaB signaling pathways to inhibit angiogenesis in a hamster model of oral carcinogenesis. Tumour Biol J Int Soc Oncodevelop Biol Med 37(2):2083–2093. https://doi.org/10.1007/s13277-015-4003-0

Koul O, Multani JS, Singh G, Daniewski WM, and Berlozecki S (2003) J Agric Food Chem 51(10):2937–2942. https://doi.org/10.1021/jf021049m

Kumar VS, Navaratnam V (2013) Neem (Azadirachta indica): prehistory to contemporary medicinal uses to humankind. Asian Pacific J Tropic Biomed 3(7):505–514

Lakshmi V, Singh N, Shrivastva S, Mishra SK, Dharmani P, Mishra V, Palit G (2010) Gedunin and photogedunin of Xylocarpus granatum show significant anti-secretory effects and protect the gastric mucosa of peptic ulcer in rats. Phytomed Int J Phytotherapy Phytopharmacol 17(8–9):569–574. https://doi.org/10.1016/j.phymed.2009.10.016

Lamb J, Crawford ED, Peck D, Modell JW, Blat IC, Wrobel MJ, Lerner J, Brunet JP, Subramanian A, Ross KN, Reich M, Hieronymus H, Wei G, Armstrong SA, Haggarty SJ, Clemons PA, Wei R, Carr SA, Lander ES, Golub TR (2006) The connectivity map: using gene-expression signatures to connect small molecules, genes, and disease. Science (New York, NY) 313(5795):1929–1935. https://doi.org/10.1126/science.1132939

Leeds P, Leng Y, Chalecka-Franaszek E, Chuang DM (2005) Neurotrophins protect against cytosine arabinoside-induced apoptosis of immature rat cerebellar neurons. Neurochem Int 46(1):61–72. https://doi.org/10.1016/j.neuint.2004.07.001

Li H, Wu J-G, Zhang H-W, Wang W, Zhang Y-X, Zhang J-n (2018) Inhibition of proliferation, invasion and migration in U-251 mg glioblastoma cell line by gedunin. Int J Pharmacol 14(4):522–527

Lindholm D, Dechant G, Heisenberg CP, Thoenen H (1993) Brain-derived neurotrophic factor is a survival factor for cultured rat cerebellar granule neurons and protects them against glutamate-induced neurotoxicity. Eur J Neurosci 5(11):1455–1464. https://doi.org/10.1111/j.1460-9568.1993.tb00213.x

Lu Y, Ansar S, Michaelis ML, Blagg BS (2009) Neuroprotective activity and evaluation of Hsp90 inhibitors in an immortalized neuronal cell line. Bioorg Med Chem 17(4):1709–1715. https://doi.org/10.1016/j.bmc.2008.12.047

MacKinnon S, Durst T, Arnason JT, Angerhofer C, Pezzuto J, Sanchez-Vindas PE, Poveda LJ, Gbeassor M (1997) Antimalarial activity of tropical Meliaceae extracts and gedunin derivatives. J Nat Prod 60(4):336–341. https://doi.org/10.1021/np9605394

Medzhitov R (2008) Origin and physiological roles of inflammation. Nature 454(7203):428–435. https://doi.org/10.1038/nature07201

Misra S, Verma M, Mishra SK, Srivastava S, Lakshmi V, Misra-Bhattacharya S (2011) Gedunin and photogedunin of Xylocarpus granatum possess antifilarial activity against human lymphatic filarial parasite Brugia malayi in experimental rodent host. Parasitol Res 109 (5):1351–1360. https://doi.org/10.1007/s00436-011-2380-x

Moga MA, Balan A, Anastasiu CV, Dimienescu OG, Neculoiu CD, Gavris C (2018) An overview on the anti-cancer activity of azadirachta indica (neem) in gynecological cancers. Int J Mol Sci 19(12).https://doi.org/10.3390/ijms19123898

Murphy MP, LeVine H 3rd (2010) Alzheimer's disease and the amyloid-beta peptide. J Alzheimers Dis 19(1):311–323. https://doi.org/10.3233/JAD-2010-1221

Murray CJ, Rosenfeld LC, Lim SS, Andrews KG, Foreman KJ, Haring D, Fullman N, Naghavi M, Lozano R, Lopez AD (2012) Global malaria mortality between 1980 and 2010: a systematic analysis. Lancet 379(9814):413–431

Nathan SS, Kalaivani K, Murugan K (2005) Effects of neem limonoids on the malaria vector Anopheles stephensi Liston (Diptera: Culicidae). Acta Trop 96(1):47–55

Okhale S, Amupitan J, Ayo R, Oladosu P, Okogun J (2013) Synthesis and antibacterial activity of 7-deacetoxy-7α-hydroxygedunin. African J Pure Appl Chem 7(4):157–163

Okhale S, Amupitan J, Ndukwe I, Oladosu P, Okogun J (2012) Synthetic modification of gedunin and comparative antibacterial activity of gedunin and 7-deacetoxy-7α-hydroxygedunin potassium salt. African J Pure Appl Chem 6(14):183–189

Omar S, Godard K, Ingham A, Hussain H, Wongpanich V, Pezzuto J, Durst T, Eklu C, Gbeassor M, Sanchez-Vindas P, Poveda L, Philogene BJR, Arnason JT (2003a) Antimalarial activities of gedunin and 7-methoxygedunin and synergistic activity with dillapiol. Annals Appl Biol 143 2):135–141. https://doi.org/10.1111/j.1744-7348.2003.tb00279.x

Omar S, Zhang J, MacKinnon S, Leaman D, Durst T, Philogene BJ, Arnason JT, Sanchez-Vindas PE, Poveda L, Tamez PA, Pezzuto JM (2003) Traditionally-used antimalarials from the Meliaceae. Curr Top Med Chem 3(2):133–139. https://doi.org/10.2174/1568026033392499

Ong YS, Saiful Yazan L, Ng WK, Abdullah R, Mustapha NM, Sapuan S, Foo JB, Tor YS, How CW, Abd Rahman N (2018) Thymoquinone loaded in nanostructured lipid carrier showed enhanced anti-cancer activity in 4T1 tumor-bearing mice. Nanomedicine 13(13):1567–1582

Paavola KJ, Hall RA (2012) Adhesion G protein-coupled receptors: signaling, pharmacology, and mechanisms of activation. Mol Pharmacol 82(5):777–783. https://doi.org/10.1124/mol.112.080309

Patwardhan CA, Fauq A, Peterson LB, Miller C, Blagg BS, Chadli A (2013) Gedunin inactivates the co-chaperone p23 protein causing cancer cell death by apoptosis. J Biol Chem 288 (10):7313–7325. https://doi.org/10.1074/jbc.M112.427328

Penido C, Conte FP, Chagas MS, Rodrigues CA, Pereira JF, Henriques MG (2006) Antiinflammatory effects of natural tetranortriterpenoids isolated from Carapa guianensis Aublet on zymosan-induced arthritis in mice. Inflam Res Offic J Eur Histamine Res Soc 55 (11):457–464. https://doi.org/10.1007/s00011-006-5161-8

Penido C, Costa KA, Pennaforte RJ, Costa MF, Pereira JF, Siani AC, Henriques MG (2005) Anti-allergic effects of natural tetranortriterpenoids isolated from Carapa guianensis Aublet on allergen-induced vascular permeability and hyperalgesia. Inflam Res Offic J Eur Histamine Res Soc 54(7):295–303. https://doi.org/10.1007/s00011-005-1357-6

Ramanan VK, Saykin AJ (2013) Pathways to neurodegeneration: mechanistic insights from GWAS in Alzheimer's disease, Parkinson's disease, and related disorders. Am J Neurodegen Disease 2(3):145–175

Schabitz WR, Sommer C, Zoder W, Kiessling M, Schwaninger M, Schwab S (2000) Intravenous brain-derived neurotrophic factor reduces infarct size and counterregulates Bax and Bcl-2 expression after temporary focal cerebral ischemia. Stroke 31(9):2212–2217. https://doi.org/10.1161/01.str.31.9.2212

Sharma C, Vas AJ, Goala P, Gheewala TM, Rizvi TA, Hussain A (2014) Ethanolic Neem (Azadirachta indica) leaf extract prevents growth of MCF-7 and HeLa cells and potentiates the therapeutic index of cisplatin. J Oncol 2014:10. https://doi.org/10.1155/2014/321754

Sinha S, Medhi B, Sehgal R (2014) Challenges of drug-resistant malaria. Parasite 21:61–61. https://doi.org/10.1051/parasite/2014059

Smirnova NA, Haskew-Layton RE, Basso M, Hushpulian DM, Payappilly JB, Speer RE, Ahn YH, Rakhman I, Cole PA, Pinto JT, Ratan RR, Gazaryan IG (2011) Development of Neh2-luciferase reporter and its application for high throughput screening and real-time monitoring of Nrf2 activators. Chem Biol 18(6):752–765. https://doi.org/10.1016/j.chembiol.2011.03.013

Stoveken HM, Larsen SD, Smrcka AV, Tall GG (2018) Gedunin- and Khivorin-derivatives are small-molecule partial agonists for adhesion G protein-coupled receptors GPR56/ADGRG1 and GPR114/ADGRG5. Mol Pharmacol 93(5):477–488. https://doi.org/10.1124/mol.117.111476

Subapriya R, Nagini S (2005) Medicinal properties of neem leaves: a review. Curr Med Chem Anticancer Agents 5(2):149–146. https://doi.org/10.2174/1568011053174828

Subramani R, Gonzalez E, Nandy SB, Arumugam A, Camacho F, Medel J, Alabi D, Lakshmanaswamy R (2017) Gedunin inhibits pancreatic cancer by altering sonic hedgehog signaling pathway. Oncotarget 8(7):10891–10904. https://doi.org/10.18632/oncotarget.8055

Tanagala KKK, Baba AB, Kowshik J, Reddy GB, Nagini S (2018) Gedunin, a neem limonoid in combination with epalrestat inhibits cancer hallmarks by attenuating aldose reductase-driven oncogenic signaling in SCC131 oral cancer cells. Anticancer Agents Med Chem 18(14):2042–2052. https://doi.org/10.2174/1871520618666180731093433

Thoenen H (1991) The changing scene of neurotrophic factors. Trends Neurosci 14(5):165–170. https://doi.org/10.1016/0166-2236(91)90097-e

Tom S, Rane A, Katewa AS, Chamoli M, Matsumoto RR, Andersen JK, Chinta SJ (2019) Gedunin inhibits oligomeric abeta1-42-induced microglia activation via modulation of Nrf2-NF-kappaB signaling. Mol Neurobiol 56(11):7851–7862. https://doi.org/10.1007/s12035-019-1636-9

Uddin SJ, Nahar L, Shilpi JA, Shoeb M, Borkowski T, Gibbons S, Middleton M, Byres M, Sarker SD (2007) Gedunin, a limonoid from Xylocarpus granatum, inhibits the growth of $CaCo_2$ colon cancer cell line in vitro. Phytotherapy Res PTR 21(8):757–761. https://doi.org/10.1002/ptr.2159

Uko O, Edafiogho I, Ataja A (1995) Possible laxative acivity of azadirachta indica (neem). West African J Pharmacol Drug Res 11(1)

Vinson-Hieronymus H, Golub TR, Lamb J, Stegmaier K (2011) Celastrol, gedunin, and derivatives thereof as HSP90 inhibitors. Google Patents

Yang W, Xie J, Qiang Q, Li L, Lin X, Ren Y, Ren W, Liu Q, Zhou G, Wei W, Saiyin H, Ma L (2019) Gedunin degrades aggregates of mutant huntingtin protein and intranuclear inclusions via the proteasomal pathway in neurons and fibroblasts from patients with huntington's disease. Neurosci Bulletin 35(6):1024–1034. https://doi.org/10.1007/s12264-019-00421-5

Yerbanga RS, Lucantoni L, Ouédraogo RK, Da DF, Yao FA, Yaméogo KB, Churcher TS, Lupidi G, Taglialatela-Scafati O, Gouagna LC, Cohuet A, Christophides GK, Ouédraogo JB, Habluetzel A (2014) Transmission blocking activity of Azadirachta indica and Guiera senegalensis extracts on the sporogonic development of Plasmodium falciparum field isolates in Anopheles coluzzii mosquitoes. Parasites Vect 7(1):185. https://doi.org/10.1186/1756-3305-7-185

Zhang B, Au Q, Yoon IS, Tremblay MH, Yip G, Zhou Y, Barber JR, Ng SC (2009) Identification of small-molecule HSF1 amplifiers by high content screening in protection of cells from stress induced injury. Biochem Biophys Res Commun 390(3):925–930. https://doi.org/10.1016/j.bbrc.2009.10.079

# Chapter 17
# Biological Activities of Marine Products and Nutritional Importance

**Dilipkumar Pal and Khushboo Raj**

**Abstract** Bioactive compounds, also known as phyto-nutrients are those compounds which enhance or we can say that promote good health. They are found in a limited amount in plants, animals, marine, and other natural food sources which help in the prevention of many diseases e.g. cancers, cardiac disorders, diabetes, etc. It is well known to us that nearly half of the worldwide biodiversity is constituted by different types of marine species and as a result, oceans, sea are enriched with valuable natural bioactive compounds such as proteins, peptides, amino acids, fatty acids, sterols, oligosaccharides, vitamins, and minerals, etc. These agents also help to enhance the nutritional as well as the therapeutic value of the food products. Every year nearly thousands of new compounds are isolated from a marine organism which further help in the discovery of new leads for the development of new drugs to treat or diagnose human diseases like cancer, viral diseases, inflammation, etc. For example, thyrsiferol, which is isolated from marine red algae (genus: Laurencia) on the experimental studies shows potent anti-viral and anti-tumour activities. Another example is marine sponge (genus: Insignia) which is a good source of terpenoids consisting of tetronic acids which act as anti-inflammatory agent, analgesic, and antibiotics. There are numerous examples of such marine products that are enriched with nutritional values and show many potent biological activities. In this article, we are going to discuss different marine products one by one with their biological and nutritional importances and also their role in the development of new drugs in the treatment of various human diseases.

**Keywords** Bioactive compounds · Cancers · Cardiac disorders · Diabetes · Marine organism · Thyrsiferol · Anti-viral

---

D. Pal (✉) · K. Raj
Department of Pharmaceutical Sciences, Guru Ghasidas Viswavidyalaya (A Central University), Bilaspur, Chhattisgarh 49551, India
e-mail: drdilip2003@yahoo.co.in



D. Pal and A. K. Nayak (eds.), *Bioactive Natural Products for Pharmaceutical Applications*, Advanced Structured Materials 140,
https://doi.org/10.1007/978-3-030-54027-2_17

## 17.1 Introduction

Ocean or Marine life denotes the life of various organism which inhabits in the saltwater of the sea or near the coastal subculture and due to its remarkable biodiversity, it has become a huge stock of natural resources for many biologically active compounds (Bellisle et al. 1998). More than 200,000 marine species are already reported and 2 million are yet to be documented (Fiszman and Salvador 2003). Marine species vary in size including small phytoplankton (0.02 mm) to large cetaceans e.g. whales, dolphins, blue whale, etc. It also includes various microorganisms like bacteria, algae, fungi, viruses, etc. (Plaza et al. 2008). Today marine organism is well known as a potent source of the bioactive compound because of its complex habitat and exposure to extreme condition. Thus, these potent bioactive compounds which act as a good source of nutrition also exhibit potent biological activities. Peptides, polysaccharides which are isolated from various marine sources produce anticancer, anticoagulant activities. A marine sponge, (genus: Insignia) for example is a good source of terpenoids and these terpenoids consists of tetronic acids which act as anti-inflammatory, analgesic and antibiotic agents (Diplock et al. 1999).

## 17.2 Marine Organisms: Source of Nutrition

### *17.2.1 Proteins*

Proteins are considered as one of the most essential nutrients for our body system. It consists of one or more than one long chain of amino acid residues (the building block of protein) which is coded by the genetic (DNA) code along with a no. of other amino acids (Kadam and Prabhasankar 2005). There are twenty different types of amino acids which are classified into two categories (i) EAA (Essential amino acid) and (ii) NEAA (Non-Essential amino acid). EAAs are nine in numbers and they cannot be produced by the body itself, thus they are taken from foods or outer sources. On another hand, NEAAs are eleven in numbers and they can be produced by the body itself (Amado et al. 2013). In Table 17.1 name of twenty amino acids with their 3 letters abbreviation is mentioned (Kim et al. 2010).

The primary function of protein includes catalytic reactions, DNA replication, production of antibodies, etc. (Sheraji et al. 2013). Seafood is well known for its outstanding source of protein which includes all essential amino acids in an adequate amount for humans (Freitas et al. 2012). On various chemical analysis, it was found that marine food consists of 9–24% of protein which is well digestible to human beings. For example, Halibut fish consists of 19% proteins with all EAA in an adequate amount (Ar-6%, His-1.66%, Cys-6.16%, Trp-1.64%, and Cys 1.45%) (Damodaran et al. 1997; Lum et al. 2013).

**Table 17.1** Name of essential amino acid and non-essential amino acid

| S. No. | Essential amino acids (eaas) | Non-essential amino acids (neaas) |
|---|---|---|
| 1 | Histidine (His) | Alanine (Ala) |
| 2 | Isoleucine (Ile) | Arginine (Arg) |
| 3 | Leucine (Leu) | Asparagines (Asn) |
| 4 | Lysine (Lys) | Aspartic acid (Asp) |
| 5 | Methionine (Met) | Cysteine (Cys) |
| 6 | Phenylalanine (Phe) | Glutamic acid (Glu) |
| 7 | Threonine (Thr) | Glutamine (Gln) |
| 8 | Tryptophan (Trp) | Glycine (Gly) |
| 9 | Valine (Val) | Proline (Pro) |
| 10 | | Serine (Ser) |
| 11 | | Tyrosine (Tyr) |

Also, the other protein sources are marine mammals, algae, bacteria, etc. Seal liver is recently documented as an enriching source of protein as same as the amount of it present in beef liver (Asha et al. 2014). Different species of Whales reported as an excellent source of protein ranges from (17–24%) (Lordan et al. 2011). Another example of protein source is Laminaria (alga) and their different species having a total content of 14% of crude protein. The EAA which is found in Laminaria includes aspartic acid, glutanic acid, glycine, alonine, valive, leucine, isoleucine, and also some serine, threonine, proline, phenylalanine, lysine and a small amount of arginine (Rasmussen and Morrissey 2007; Cho et al. 2008). Spirulina, which is a biomass of cyanobacteria is reported as rich in high protein content which ranges from 60 to 70% with an ample amount of EAA like leucin tryptophan, methionine, phenylalanine, lysine, threonine, isoleucine, and valine (Kim et al. 2010). It also helps in repairing damages, lowering total cholesterol (CDC), raising good cholesterol (HDC) and acts as an anti-inflammatory agent (Bocanegra et al. 2009). In Table 17.2 protein percentage in a marine organism is mentioned.

### 17.2.2 *Lipids and Fatty Acids*

Omega-3 or polyunsaturated fatty acids (PUFA) play a vital role to keep the human body healthy and free from various diseases like CVS, retinal diseases, diabetics, etc. On various chemical and biological analyses, it is found that a regular consumption of seafood like fishes, crabs, etc. can help to lower the risk of cardiac

**Table 17.2** Different types of marine species with their protein percentage

| Species | Protein content (%) | Ar (%) | His (%) | Lys (%) | Trp (%) | Cys (%) | References |
|---|---|---|---|---|---|---|---|
| Fishes amiuriscatus (catfish) | ~28 | | | | ~0.97 | | Zheng et al. (2013) |
| Oncorhynchus tschawytscha (salmon) | ~16 | ~5.02 | ~1.41 | ~6.27 | ~1.20 | ~1.27 | Jun et al. (2004) |
| O.nerka (salmon) | ~11 | – | – | – | – | ~1.25 | Rocha et al. (2015) |
| O.keta (salmon) | ~5 | ~5.55 | ~1.30 | ~5.69 | ~1.33 | – | |
| O.gorbuscha (salmon) | ~6 | – | – | – | ~1.09 | ~1.15 | |
| O.kisutch (salmon) | ~8 | ~5.68 | ~1.87 | ~6.57 | ~1.94 | ~1.39 | |
| Germaalalunga (tuna) | | – | – | – | ~1.18 | – | Kanazawa (2001) |
| Melanogrammusaeglefinus (Haddock) | 18 | 5.70 | 1.17 | 6.41 | 0.85 | 1.16 | |
| Algae *Spirulina platensis* | 60 | 7.3 | 2.2 | 4.8 | 0.3 | 0.9 | Karuppasamy et al. (2013) |
| *Porphyratenra* | 20 | 16.4 | 1.4 | 4.5 | 1.3 | – | Ge et al. (2006) |
| Ulva | 13.6 | 3.7 | 0.7 | 0.0 | 0.6 | – | |
| Sargassum | 9–20 | 4.0 | 1.9 | 4.5 | 1.8 | – | |
| Shellfish crabs | 17 | 7.6 | 1.5 | 6.4 | 1.1 | | Mayer et al. (2010) |
| Oysters | 7 | 5.7 | 1.8 | 5.2 | – | – | Paul et al. (2011) |
| Clams | 9 | 5.3 | 1.5 | 5.4 | 1.2 | – | |
| Shrimp | 25 | 6.6 | 1.8 | 8.3 | 1.2 | | |

related disorders (CHD,CAD) and other diseases also because seafood has a very low content of saturated fatty acid and higher content of omega-3 fatty acids. The main reason of a high amount of PUFA in fishes is because of their daily consumption of algae where the synthesis of long-chain PUFA takes place (Aneiros and Garateix 2004).

Eicosapentenoic (EPA) acid and docosahexenoic (DHA) acid are the examples of Omega-3 fatty acid that is mainly found in the brain, cerebral cortex, skin, retinal part of the human body system. They can be synthesized in the human body from their precursor alpha-linolenic acid (ALA), but the rate of conversion is very low because of the insufficient amount of the enzyme Δ6-desaturase (Dharmaraj et al. 2009). That's why PUFA is taken in diet from marine food (fish oil, algae oil, etc.) or other food resources. In Fig. 17.1 the synthesis process of Eicosapentenoic acid and docosahexenoic acid are illustrated. There are so many examples of marine sources available which enrich PUFA, EPA, DHA, etc. For example, salmon, swordfish and halibut are excellent sources of PUFA (EPA, DPA, DHA) with respective average levels of 1434, 3625, 3358, 2654 mg-PUFAs/100 g fw (Ma et al. 2008).

Also, C. Vulgaris, the microalgae is enriched with oleic acid, palmitic acid and linolenic acid. Haematococcus, a green microalgae includes short-chain fatty acids and exhibits antimicrobial activity (Sijtsma et al. 2004). Figure 17.2 exemplify various marine sources with the fatty acid profiles (LøvstadHoldt and Kraan 2011).

### *17.2.3 Sterols*

These are another class of lipids that are also found in marine food. On experimental studies, it has been noticed that the sterols which are extracted from macro and microalgae, fish and other marine invertebrates (like corals, molluscs, sponges) are capable of lowering the level of LDC (low- density lipoprotein) and also show anti-inflammation bioactivity (Wall et al. 2010). *Sargasseum ringgoldianum* (brown algae) has been reported to have various types of sterols in which saringosterol is associated with anti-tubercular activity. *Pelvetia siliquosa* (marine algae) has also been reported to have fucosterol and anti-oxidant activity (Pal 2013; Nimse and Pal 2015). Phytosterols (C28 and C29 sterols) are essential precursors of some vitamins. For instance, ergosterol is a precursor of vitamin D2 and cortisone. Clionasterol which is found in *Spirulina* is associated with increasing PAF in vascular endothelial cells. Fucosterol, which is isolated from *P. Siliquosa* exhibits potent anti-diabetic activity at a dose level of 100 and 300 mg/kg. It also decreases the glycogen degradation of mouse liver by 23–29% (Wang et al. 2006). Various sterols with their sources are mentioned In Table 17.3 (Özyurt et al. 2013).

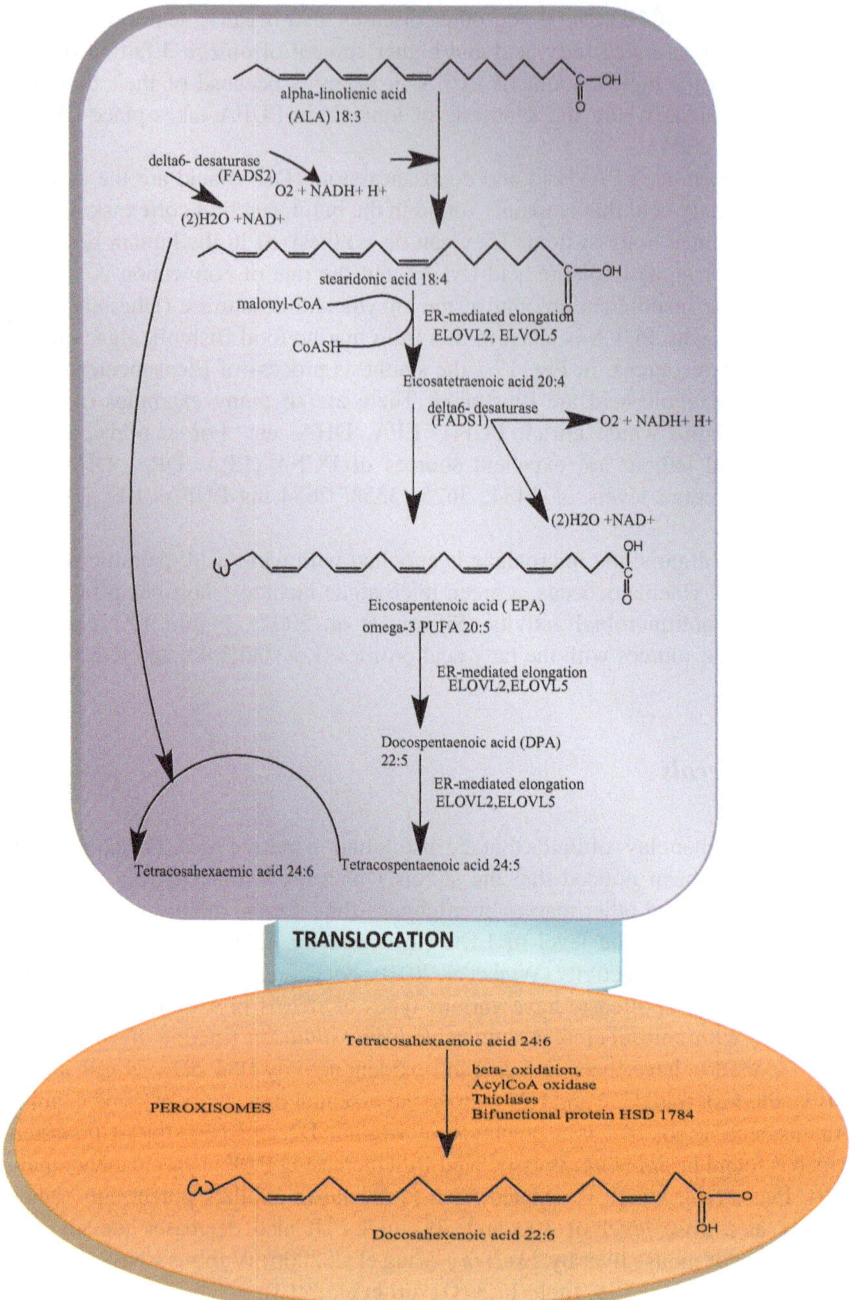


**Fig. 17.1** Synthesis process of eicosapentenoic acid and docosahexenoic acid

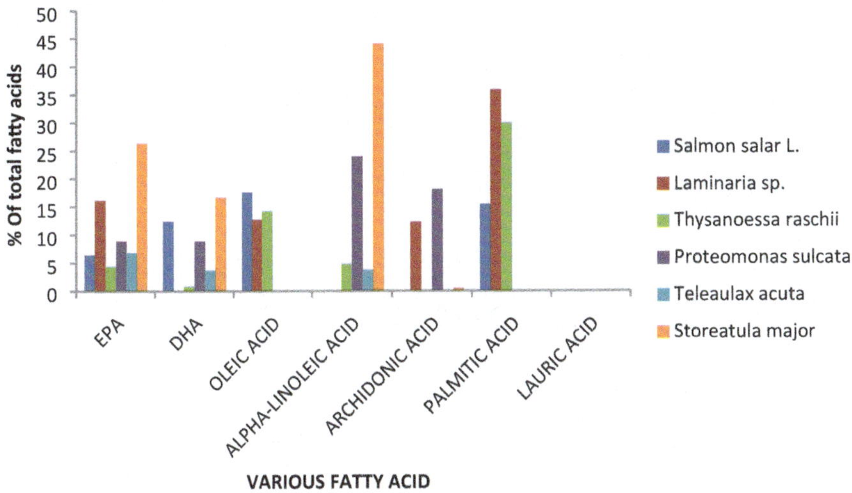


Fig. 17.2 Marine sources with fatty acid profile

**Table 17.3** Structures of sterols with their sources and uses

| Sterols | Structure | Sources | Use |
|---|---|---|---|
| Ergostérol | 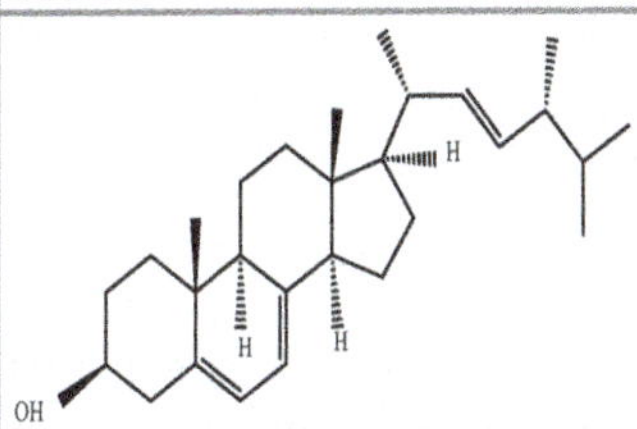 | Chlamydomonas reinhardtii | In breast cancer treatment |
| Clionasterol | 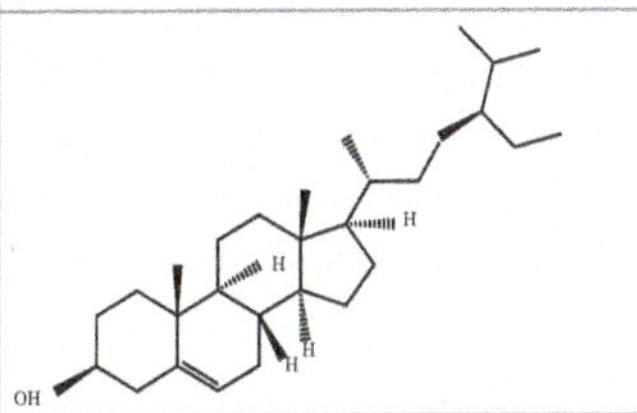 | Spirulina | PAF activity |
| Fucosterol | 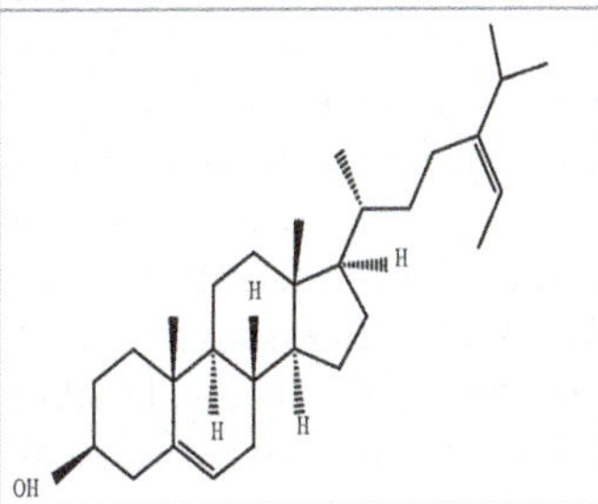 | Turbinariaornata | Anti-cancer activity |

(continued)

**Table 17.3** (continued)

| Sterols | Structure | Sources | Use |
|---|---|---|---|
| Saringasterol | 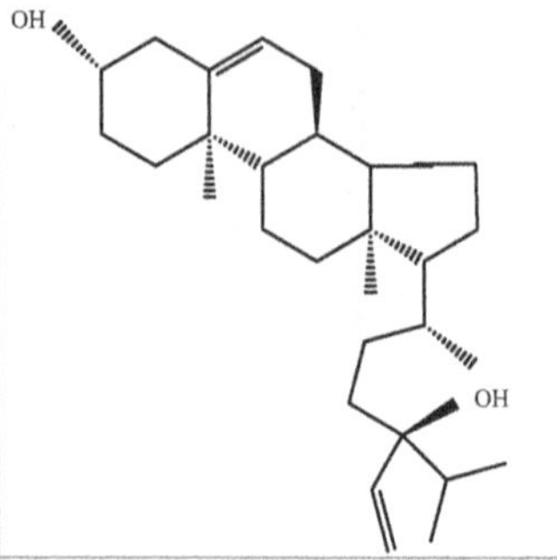 | Lessonia nigrescenes | (anti-tubercular activity) |
| Chondrillasterol | 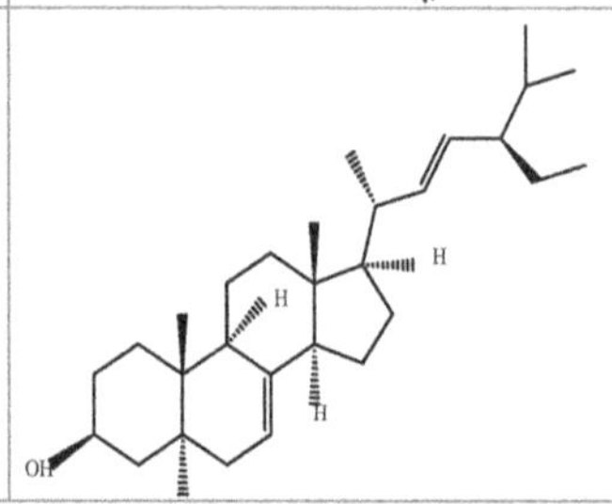 | Brown algae (scenedesmusobliquus) | Anti-bacterial activity |

## 17.2.4 Carbohydrates

### 17.2.4.1 Polysaccharides

Polysaccharides (glycan) are a long chain of polymeric carbohydrates composed of monosaccharide (simple sugar) linked by glycosidic bonds. They are found in high content in marine seaweed and algae. Some of the examples of polysaccharides include agar, alginate, carvageen, isolated from macroalgae whereas other polysaccharides like chitin, chitosan, isolated from cuticles of various crustaceans, crabs and shrimps (Spolaore et al. 2006). Red algae-like gelidium, arcicilaria, hypnea, and gigartina are the main source of agar and they are used extensively in food, pharmaceutical, cosmetic, paper and textiles industries for various purposes like food gums, emulsifying agent, etc. (Prasad et al. 2006 and Calder 2009). Various polysaccharides are not digestible for human GIT because of the absence of suitable degradation enzymes and thus they are regarded as dietary fibre. The total dietary fiber content of seaweeds (25–75% of the dry weight of marine algae) is higher than the fiber content of most fruits and vegetables. Dietary fibres are categorized into two classes, one is soluble and another is insoluble. Marine algae are a reservoir of different types of carbohydrates. For example, Phaeophyta (brown algae) includes alginates, fucans, laminarans, etc. and all these are soluble fibres.

Other examples are xylans, floridean starch, sulphated galactans and mannans, ionic polysaccharides which are obtained from Rhodophyta (red algae) and green algae (Lloret 2010) respectively.

It is well proven now that human consumption of algal fibre is very beneficial and can resolve many health-related problems with its many biological activities like antitumor, anticoagulant, antiviral, and many others.

Fucidan which is a sulfated polysaccharide and made up of primarily L-fucose can be found from the various sources of brown seaweeds like *Saccharina japonica*, *Fucus vesiculosus*, *Undaria pinnatifida*, and *Hizikia fusiformis* and also from marine invertebrates such as sea cucumber. Experimental analysis shows that, fucoidans can be widely used in cosmetic products. When it is used topically it forms a protective layer, enhances skin hydration and also helps in prevention and treatment of skin photo aging. (Zheng et al. 2013 and Charissoux et al. 2013).

Carrageenan is also an example of sulfated polysaccharides which is found in Rhodophyta (red algae). It is composed of D-galactose units. Carrageenan is used in cosmetic industries as skin lotions, toothpaste binders, shaving foams, etc. It also exhibits antioxidant, antiaging, and anti-cancer properties (Li et al. 2008).

Also, laminaran is an example of marine polysaccharides that produces anti-coagulant activity when structural modification like sulfatin, reduction, or oxidation occurs (Kanekiyo et al. 2007; Heo et al. 2002; Je et al. 2014).

#### 17.2.4.2 Oligosaccharides

They are another class of carbohydrates that contains 3–10 sugar units. Marine oligosaccharides are produced either in algae by natural process or by hydrolysis of derived polysaccharides.

Oligosaccharides derived from seaweed have been proven as a potent antifungal, antibacterial precursor. They also show defence action by enhancing pathogens protection. In addition marine algae also includes oligosaccharides that are used as prebiotic as egxylo/func-oligosaccharides and non-digestible oligomers (cannot pass through GIT) so they can be used to enhance the growth of beneficial probiotic bacteria (Rodríguez-Meizoso et al. 2010; Le et al. 2009). Bifidobacteria is an example of probiotics which live in intestines and stomach and helps in digestion and also provide protection from pathogens. K-Carrageenan isolated from *Kappaphycus alvarezii* is very much beneficial for the prevention of colon carcinogenesis and also exerts cholesterol-lowering, anticoagulant, immunomodulatory, antiviral and antioxidant properties. On the other hand, alginate oligosaccharides exerts antioxidant activities along with inhibitory action on neuro-inflammation and enhances the microglial phagocytosis, thus promises a potent nutraceutical agent against neurodegenerative diseases e.g. Alzheimer's disease, Parkinson's diseases etc. Marine oligosaccharides have also been used in food products, cosmetics, biomedicine etc. (Zaporozhets et al. 2014).

### *17.2.5 Antioxidants*

Today food industry has been more focused towards the development of antioxidants from the natural sources as it is a safer option in comparison to many synthetic commercial anti- oxidants. Antioxidants are those agents which prevents the oxidation of lipids due to which there is no spoilage of food. Lipid oxidation is caused by reactive oxygen species (ROS) e.g. hydrogen peroxide, free radicals, etc. and its leads to the deterioration of nutritional values of lipid content of food. Thus to reduce lipid peroxidation. Antioxidants like propyl gallate, butylated hydroxytoluene, butylated hydroxytoluene, tert-butylhydroquinone, etc. are used (Wijesekara et al. 2011; Pal and Nandi 2005; Soni et al. 2008) to prevent oxidation. The main problem with synthetic antioxidants is that they are not safe: may include serious side effects. That's why researchers are now more focused on naturally derived anti-oxidants with proper utilisations. Apart from deterioration of food products, these ROS are also responsible for causing various diseases such as cancer, neurodegenerative disease, inflammatory diseases, etc. (Rinaudo et al. 2006, Pal 2013; Nimse and Pal 2015).

The reaction of ROS with biomolecules like proteins, membrane lipids and DNA result in cellular or tissue level injuries. Equilibrium between endogenous anti-oxidant systems and oxidant formation protect cellular biomolecules, however a disturbance in this balance can lead to oxidative stress. Therefore, anti-oxidants play a vital role in maintaining the cellular redox state and protecting the body against damage caused by ROS (Mata et al. 2010, Pal 2013; Nimse and Pal 2015) (Fig. 17.3).

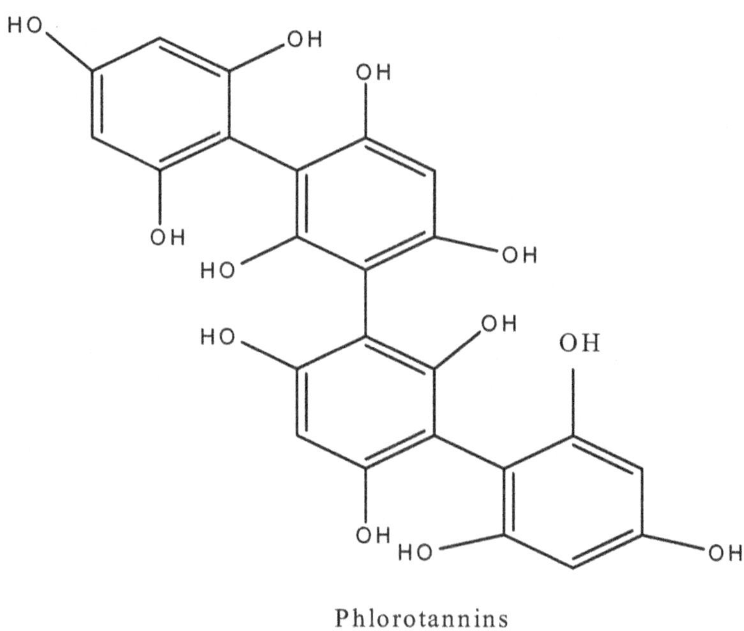


**Fig. 17.3** Structure of phlorotannins

#### 17.2.5.1 Polyphenolic Compounds

Marine polyphenolic compounds have been extracted from micro and macroalgae and they are characterised in ten different categories according to their structures. These ten classes include- phenolic acid, hydroxycinnamic acids, simple phenols, coumarines, xanthones, naphthoquinones, flavonoids, stibenes, anthraquinones and lignins (Fig. 17.4) (Vo et al. 2011).

Phlorotannins that is one of the classes of polyphenol compounds produced by brown algae families as a secondary metabolite is very effective in bacterial infection. It also shows UV- protective and anti-proliferative effects (Fig. 17.3) (Patel and Goyal 2011).

#### 17.2.5.2 Photosynthetic Pigments

These are the pigments that capture solar energy and use it for photosynthesis. In marine life generally micro and microalgae, plant, fungus have these types of photosynthetic pigment. Carotenoids and chlorophyll are such examples of photosynthetic pigment that are present in macroalgae (Ishihara et al. 2010). Carotenoids pigment exhibit anti-oxidant property and also act as pro-vitamin A, which is further converted into vitamin A. Carotenoids also provide profusion against cancer, cardio-vascular disease and AMD (age-related macular degeneration) (Mussatto and Mancilha 2007).

Beta-carotene and astaxanthins, found in microalgae along with carotenoids also act as a antioxidants, anti-cancer agents and helps in treatment of arthritis disease (Bin et al. 2013).

### *17.2.6 Vitamins and Minerals*

They are considered as necessary nutrients which performs many vital roles to keep human body healthy. They provide strength to bones, boost the immune system, heal wounds and also repair cellular damages. Seafoods are considered as a good source of vitamins and minerals. *Palmarialongat*, a macroalgae is an excellent source of vit. B12 and green seaweeds like *Laminaria digitata, U. Pinnatifida* is a great source of Vit. C and E. Also *U. Pinnatifida and sargassum sp.* contain an adequate amount of major minerals like Na, Ca and Mg as well as trace minerals (Fe, Zn, Mn and Cu) (Kuroiwa et al. 2009). Vitamin D is very important for the growth and development of the bones. It makes the bone stronger and prevents various bone related diseases like rickets (in infants and children) and Osteoporosis (in adults). Thus fatty fishes like salmon, pilchard etc. are good source of vit. D (Ngo et al. 2010). In Table 17.4, examples of different microalgae with different number of vitamins are mentioned. Figure 17.5 describes mineral content of different algae (Shahidi and Zhong 2010; Miyashita et al. 2011).

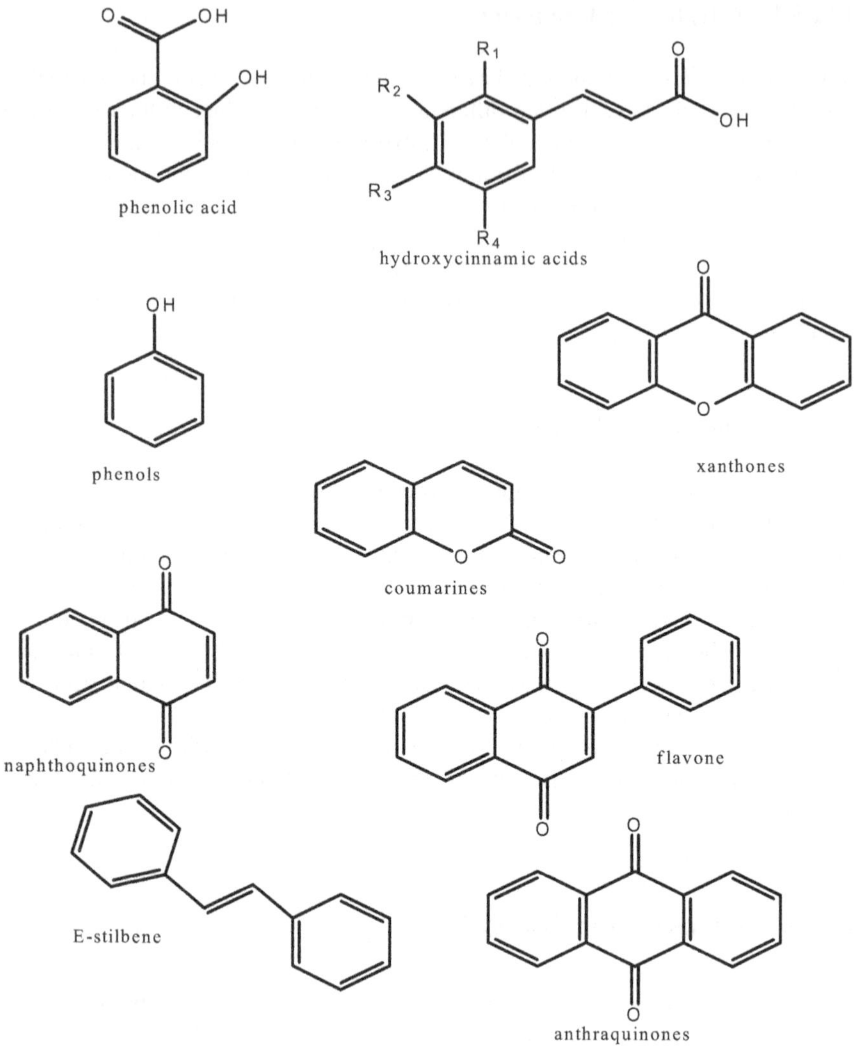


**Fig. 17.4** Structure of various polyphenolic derivatives

## 17.3 Marine Sources for Pharmacological Effect

### *17.3.1 Anti-cancer*

Cancer is a frightful human disease and has become a major burden to human worldwide. According to a recent report in 2015, 90.5 million people have been suffering from cancer and about 14.1 million new cases occurring a year with 8.8 million death. According to an estimate of 2021, 21 million new cases of cancer are

**Table 17.4** Different microalgae with different amount of vitamins

| | Microalga (mg/kg) | | | | | | |
|---|---|---|---|---|---|---|---|
| Vitamins (mg/kg) | *Spirulina platensis* | *Chlorella pyrenoidosa* | *Senedesmusquadricauda* | *T. suecica* | *L galbana* | *D. tertiolecta* | *C. stigmatophora* |
| A | 840 | 480 | 554 | 263 850 | 76 500 | 82 500 | 49,380 |
| Thiamin (B1) | 44 | 10 | 11.5 | 32.3 | 14.0 | 29.0 | 14.6 |
| Riboflavin (Bz) | 37 | 36 | 27 | 19.1 | 30.0 | 31.2 | 19.6 |
| Pyridoxin (B6) | 3 | 23 | – | 2.8 | 1.8 | 2.2 | 1.9 |
| Pantothenic acid (B5) | – | – | – | 37.7 | 9.1 | 13.2 | 21.4 |
| Biotin (H) (B8) | 0.3 | 0.15 | – | 0.8 | 1.0 | 0.9 | 1.1 |
| Ascorbic acid (C) | 80 | | 396 | 191.0 | I 19.0 | 163.2 | 100.2 |
| Tocopherol (E) | 120 | | | 421.8 | 58.2 | 116.3 | 669.0 |

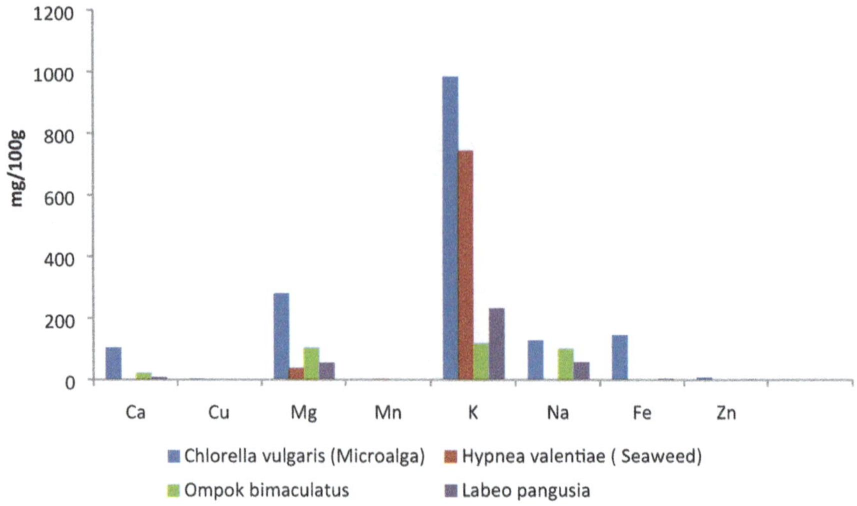


**Fig. 17.5** Mineral content of different algae

expected with 13 million deaths of cancer patients (Miyashita 2013). However, to fight with this dreadful disease many chemotherapeutic drugs are currently available in the market but due to their serious side effects they become risky to use for the cancer patients. Thus, today researches are more focused on natural products and their derivatives to treat the cancer patients and 60% of drugs from natural origin have already been documented for cancer treatment (Saha and Pal 2016; Saha et al. 2017; Sannigrahi et al. 2012). As we know oceans cover more than 70% of earth's surface and an exultingly marine source has emerged as a new hope for cancer treatment. The chemical adaptations generally take the form of so-called "secondary metabolites" and involve such well-known chemical classes as terpenoids, alkaloids, polypeptides, peptides, shikimic acid derivatives, sugars, steroids and multitude of mixed biogenesis metabolites (Prokschet et al. 2002). Marine bioactive compounds are self-sufficient to treat cancer by imitating insensitivity to anti-growth signals evasions of apoptosis, limitless replication, presenting metastasis and tissue invasion.

There are many examples of marine compounds like *Parmariapatmate* (seaweed) is effective against cancer disease. It inhibits cancer cell proliferation. It has been already discussed that seaweed is enriched with protein polysaccharides PUFA, vitamins, and minerals. It is also used for different pharmaceutical purposes, especially for cancer treatment (Anitha et al. 2010). Among the three types of seaweeds, phaeophyta (brown seed) has the highest phytochemical content such as terpenes, carotenoids and phenolic compounds. These phytochemicals exhibit a potent anticancer activity.

Since past decades marine sponges have been considered as a huge source of bioactive natural chemical substances. Poriferas is the oldest one with excellent living fossils and approx. 8000 documented species and perhaps twice as many un-described species. Sponges are capable of surviving in every type of marine environment; either it is polar seas or temperate and tropical waters. They vary in shape, size, and colours. From many years marine compounds derived from sponges are under doing preclinical and clinical trials for anticancer activity and the compounds are discodermolide, hemiasterlins A and B, modified halichondrin B, KRN-70000, Alipkinidine (alkaloid), fascaphysins (alkaloid), isohomohalichondrin B, Halichondrin B, Laulimalide/Fijianolide, 5-methoxyamphimedine (alkaloid) and Variolin (alkaloid) respectively. In recent years, 43 new marine originated anti-angiogenic agents have been investigated with unknown molecular mechanisms (Halim et al. 2011; Migliore and Coppede 2009).

Ara-C (1-beta-D-Arabnofuranosylcytosine or Cytarabine) is the first documented marine derived anti-cancer agent. It is a nucleoside derivative of spongothymidine and spongouridine, isolated from *Tectibethrya crypta*. Thyrsiferol, which is isolated from marine red algae (genus: *Laurencia*), on the experimental studies shows a potent anti-viral and anti-tumour activity. Renieramycins is isolated from a marine sponge *Reniera sp.* which persuades apoptosis by following p53—a dependent pathway and enable to inhibit progression and metastasis of lung cancer. Monanchocidin is a polycyclic guanidine alkaloid isolated from *Monanchora pulchra* (a marine sponge) that encourages cell death in human monocytic leukemia, human cervical cancer and mouse epidermal cells (Perdicaris et al. 2013). Smenospongine, a sesquiterpene aminoquinone, extracted from the sponge *Smenospongia* sp. is associated with anti-proliferetive and antiangiogenic activities. Spongistatin 1, a macrocyclic lactone polyether, isolated from the marine sponge *Spongia* sp. is associated with the inhibition of mitosis process, inclusion of microtubule assembly and induction of cytotoxicity in cancer cells. In addition, scientists purify a lectin from the marine sponge *Cinachyrellaapion* (CaL) and find that it exerts hemolytic, cytotoxic and anti-proliferative activities. A marine sestertespene, isolated from the sponge *Hyrtios sp.* acts as an anticancer agent by inhibiting chronic myclogenousrecekmia cells by regulating cell cycle, apoptosis, mitogen activated protein kinase (MAPKs) pathway and nuclear factor kappa B (NF-Ke) signalling cascade (Thakur and Müller 2004).

Marine algae include phytoplankton, seaweeds, sea anemones, etc. Seaweeds are further categorized into four classes: (i) Chlorophyta (green algae) (ii) Phaeophyta (brown algae) (iii) Rhodophyta (red algae) and (iv) Cyanobacteria (blue-green algae). There are 6000 species of rhodophyta, 2000 species of phaeophyta and 1200 species of chlorophyta. Seaweeds are used in a variety of fields such as food industry, fertilizer industry, pharmaceutical industry, cosmetic industry, etc. Seaweeds have been proved as promising sources of bioactive compounds that exert many biological activities including anti-cancer action. In many works of literature, it has mentioned that daily intake of dietary carotenoids lowers the risk of cancer. Dimethylsulfonioacetate, derived from Chlorophyta can bring improvement in cancer cells and neural degeneration caused by brain cancer (Solanki et al. 2008; Nguemfo et al. 2004; Zubia et al. 2007).

Microbes are the single-celled marine phytoplankton which constitutes 90% of the ocean's total biomass. They are tiny microscopic organisms. Bacteria, fungi, and plankton along with the viruses originate from them. They are in billion inhabiting in seawater. Many marine microbes constitute biochemical diversity along with a reservoir of novel drugs. They are considered as essential sources for drug discovery and development. Marine bacteria are associated with the development of novel drugs like antibiotics and other pharmaceuticals. Meroterpenoids are the class of secondary metabolites in which the terpenoid moieties are linked to molecules from different biosynthetic pathways (Rahman and McFadden 2011). They include quinones with prenylated naphthoquinones and reduced hydroquinone analogues and are mainly derived from marine fungi and actinomycetes. Meroterpenoids exert anti-cancer activity. Polyketide synthases belong to a class of enzymes associated with the biosynthesis of secondary metabolites and show potent anti-cancer activity. Actinomycetes come under the systematic groups of secondary metabolite producers. They exert various biological activities including anti-cancer property (Huheihel et al. 2002).

Arisostatin A and arisostatin B, glycosylated polyketides are isolated from the *Micromonospora sp.*. They come under the class of tetrocarcin–type cytotoxic compounds and especially Arisostatin A exhibit potent anti-cancer activity because of its cytotoxic effect on human cancer cells. It also activates caspase 3, a key effector protease responsible for apoptosis induction (Wang et al. 2013).

Toluquinol, isolated from marine fungus interferes with one of the factors causing cancer by impairing the unlimited replicative potential, characteristic features of tumour cells. It suppresses the proliferation of the promyelocytic leukemia cell line, fibrosarcoma cell, and colon adenocarcinoma cell. Diketopiperazines, isolated from marine-derived fungi, have gained a lot of attention from the researchers' attention because of their varying chemical structure and biological activities. Halimide ((-)-phenylahistin), originated from *Aspergillus ustus,* is a fungal prenylated Diketopiperazines showed anticancer activity by arresting the cell cycle of cancer (P388) in its G2/M phase (Schumacher et al. 2010).

Tunicates or urochordates, comes under subphylum Tunicata or Urochordata. They have been shown as a primitive model organism to study immunodefense since the innate immune system has been hypothesized as an important functional component that may partially explain the lack of metastatic tumors in invertebrates (Janakiram et al. 2010). Marine-derived compounds such as didemnin B, Aplidine, and ecteinascidin have reached clinical trials for antitumor activity. Didemnin B, isolated from *Trididemnum solidum*, is a cyclic depsipeptide and today it is recognized as a first marine-originated compound to enter phases I and II clinical trials. The phase II studies suggest the complete or partial remissions with non-Hodgkins lymphoma, but due to cardiotoxicity as a side effect, scientists chose didemnin B for further study. The closely related dehydrodidemnin B (DDB, Aplidine) is isolated from *Aplidium albicans* in 1988. Spectroscopic studies assign a structural formula in which a pyruvyl group in dehydrodidemnin B is replaced by the lactyl group in Didemnin B and syntheses of dehydrodidemnin B have been achieved. It is also found that aplidine is more active than Didemnin B and lacks Didemnin B's cardiotoxicity (Rabelo et al. 2012).

Ecteinascidins, which are considered as the second family of tunicates, originated from the extracts of *Ecteinascidia turbinate.* They exert anti-cancer activity, this property of ecteinascidins is first described in 1969 (Menna et al. 2013). Apart from sponge, algae, tunicate, microbes' other marine organisms like sea cucumber, sear hare, molluscs and bryozoans also have an anticancer function including microtubule-interfering action, DNA-interaction, phosphatase inhibitions, etc. (Smithet al. 2010). Alkaloids pyridoacridines isolated from various marine sources have been reported to possess significant cytotoxicity against cultured cells, and the family as a whole seems to be of great interest as a source of new lead structures for the development of a future generation of therapeutic agents. Sea cucumbers are those marine animals that are considered as an essential human food resource, and their extracts have been used for OTC dietary health supplements. Triterpene glycosides from sea cucumbers show a wide range of biological effects, such as antifungal, antitumor, hemolytic, cytostatic, pro-apoptotic, and immunomodulatory activities (Thiansilakul et al. 2007). Frondoside A and Cucumariosides exhibit anti-cancer effects on both in vitro and in vivo models. The dolastatins are extracted from the Indian Ocean sea hare, *Dolabella auricularia.* Subsequently, several dolastatins and related molecules are isolated from filamentous marine cyanobacteria, which are the natural diet of the sea hares. The dolastatins are the most active molecules in inhibiting cancer cell growth (Yam et al. 2001).

### 17.3.2 Anti-Cardiovascular Effect

Omega-3 or (PUFA) is very much beneficial to our health. It lowers the risk of cardiovascular disease. Many studies and reports suggest that fish consumption has a direct relation in reducing the risk of congenital heart defect, myocardial infarction, etc. Many research reports suggest that omega-3-fatty acids help to lower the blood pressure (systolic and diastolic) in people with hypertension problems. Also, it suggests that fish oil supplements are very much beneficial for arrhythmia patients (Xia et al. 2011).

A study on Mediterranean people suggests that people who intake a high amount of marine food have a low risk of congenial heart disease because Omega-3-fatty acids reduce the risk factors associated with triglyceride concentrations, blood pressure, platelet aggregation and heart arrhythmias. Also the report shows that on daily consumptions of fish can lower the severe symptoms of depressions in adults and asthma and respiratory disorders in children. Also, a protein that is derived from marine (macroalgae) is better ACF-1 inhibition because of fewer side effects and also, they act as a potent drug beneficial for hypertension patients (Mozaffarian and Wu 2011).

### 17.3.3 Anti-coagulant Activity

Literature survey suggest that marine organisms also exert anti-coagulant activity. They show anti-coagulant effect either by inhibiting thrombin or by activating anti-thrombin (III) or by increasing the clotting time both in the intrinsic and extrinsic pathways (Mauro et al. 2012). In a recent study, marine-derived sulphated glycans have emerged with a potent coagulant and thrombosis property. These glycans are categorized into two class (i) GAGs such as fucosylated chondroitin sulfates (isolated from sea cucumber) and (ii) GAG mimetics like sulphated galactans and sulphated fucans. Moreover, sulphate content plays a key role in anti-coagulant activity because the presence and amount of distribution of sulphate decide the process of coagulation or platelet aggregation (Fan et al. 2011). However, in some cases of fucoidan and fucans, the outcome of anti-coagulant activity is related to some factors such as (i) the content of sulfate or disulfate or fucose (ii) the higher molecular weight that usually induced a stronger anticoagulant activity and (iii) the molecule presents a linear backbone. Laminaran is also an example of a marine polysaccharide which after structural modification like sulfation, reduction, or oxidation exerts anticoagulant activity (Mauro et al. 2012).

Heparin is a sulfated polysaccharide that is mainly present in mammalian tissues, one of the most common anti-coagulant drugs that has been used for the last 15 years as a commercial anti-coagulant in thromboembolic disorders. However, it shows several side effects like thrombocytopenia and is not able to inhibit thrombin bound to fibrin, and shows deficiencies and unwanted bleeding. More than 300 marine organisms have been documented as heparin alternatives and around 301 new anti-thrombotic and anti-coagulant derivatives have been documented as well. These molecules vary from polysaccharides to protein structures and are originated from a variety of marine sources (Sharon and Lis 2004).

In vivo studies suggest that S-galactofucan, isolated from Spatoglossum schroederi (brown seaweed) exerts potent antithrombotic activity. Spirulan extracted from *Arthrospira platensis* interferes with the blood coagulation-fibrinolytic system and exhibited anti-thrombogenic properties. The degree of sulfation of chitosan is an important point. Highly sulfated chitosans induce an increase of thrombin, activated partial thromboplastin time and thrombin time (Desai 2004).

### 17.3.4 Anti-obesity

Obesity is a complex disease that includes the deposition of excessive fat in the body's tissue. It is considered as chronic metabolic disorders that occur due to an imbalance between energy intake and energy outlay. Obesity is a matter of concern because it can lead to many health-related problems such as B.P., cardiac related disorders, sleep apnoea, type-2 diabetes mellitus, high cholesterol levels; etc. Obesity also harms people on the social platform. Sometimes obese people may face multiple forms of prejudice and discrimination due to their over-weight, thus obesity has become a medical as well as a social problem. Many studies showed that fish oil is capable to reduce body weight (Mauro et al. 2012). A study on 324 men and women aged between 20 and 40 years from different countries (Spain, Iceland, and Ireland) is conducted. They are given sunflower oil and fish oil as their diet. Various measurements are noted down which include: total cholesterol, high-density lipoprotein, low-density lipoprotein, cholesterol, triacylglycerol, and anthropometric measurements. It is found that the weight loss diet with fish oil has high triacylglycerol, low total cholesterol, and high-density lipoprotein. Other studies are also done by comparing the body mass index between fish consuming people and meat consuming people. The study is conducted on men and women aged 20–97 years. A study report shows that people consuming fish has a lower body mass index in comparison to people consuming meat (Maeda et al. 2009).

Eicosapentenoic (EPA) acid and docosahexenoic (DHA) acid also play a major role to prevent obesity. They mainly inhibit lipid synthesis enzymes like fatty acid synthase and stearoyl-CoA desaturase-1. EPA and DHA prevent the entry of fatty acid to adipocytes for lipogenesis (Aminin et al. 2010). Polyunsaturated fatty acids are also very effective against obesity as they suppress the factors which are involved in adipocyte differentiation and fat deposition. Animal studies show that if EPA and DHA intake increases then, it can protect the body from obesity, and can also reduce fat when they are already obese. A study on mice is performed in which they are received sunflower and fish oil diet. In this study, plasma lipid level, hepatic triglycerides, cholesterol, hepatic mRNA of expression of lipogenic and lipidolytic genes are measured. The results which is found suggest that a group of mice who are fed with fish oil has lower body weight in comparison with those who are fed with sun flower oil (Imhoff et al. 2011).

Marine algae also exert anti-obesity activity e.g. *U. pinnatifida, Hijikia fusiformis*, and *Sargassum fulvellum*, which are part of the Asian diet but are also consumed in many other places (Venegas-Calerón et al. 2010). Fucoxanthin has a unique structure involving an allene bond and a-monoepoxide which is associated with anti-obesity activity as well as anti-diabetic activity. Also a nutrigenomic report suggest that fucoxanthin induces uncoupling protein 1 (UCP1) expression in mitochondria of white adipose tissue resulting in oxidation of fatty acids and heat generation in white adipose tissue. Uncoupling protein 1 is known as co-factor of anti-obesity enzyme because its expression can result in the total organism's energy expenditure and its dysfunction can cause obesity. Fucoxanthin induces a clear UCP1 signal and its mRNA is detected using Western and Northern blot analyses of abdominal WAT). Further experiments with obese rats and mice show that the fucoxanthin-fed groups always has improved insulin resistance and decreased blood glucose In another study, purified Fx and macroalgae lipids containing fucoxanthin are compared, in which the latter induces a higher expression of UCP1 when Fucoxanthin is present with the other components, especially lipids. Furthermore, a synergistic action of omega-3 fatty acids is noticed on the anti-obesity effect of fucoxanthin (Tsukui et al. 2007).

Fucoidan, as we all know, can reduce lipid accumulation by stimulating lipolysis which decreases weight gain. Researchers have studied the lipolitic activity of fucoidan by examining the protein level of the hormone-sensitive lipase (HSL) and the phosphorylated HSL (p-HSL) using Western blots. HSL is one of the most important targets of lipolytic regulation (Menna et al. 2013). The subsequent phosphorylation and activation of HSL result in an increase in the hydrolysis of stored triacylglycerols into mono-acylglycerols and free fatty acids. The level of HSL and p-HST has increased compared to the controls after fucoidan treatment of differentiated 3T3-L1 adipocytes (Venegas-Calerón et al. 2010).

### *17.3.5 Bone Growth and Healing*

Osteoporosis is the most common bone disease in which the density of bone decrease. It mainly occurs when the body reabsorbs more bone tissue and produces less to replace it. According to IOF (International Osteoporosis foundation) every 1 in 3 women over age 50 will experience osteoporotic fractures, as will 1 in 5 men aged over 50. Overall, 61% osteoporotic fractures occur in women with a female to male ratio of 1:6. Marine organisms have been proved itself as a potent source of osteogenic bioactive. Fucoidan isolated from *Apostichopus japonicas* (a sea cucumber) exerts anti-resorptive and osteogenic activity (Granito et al. 2017; Al et al. 2013).

Kim et al. (2014) demonstrate the osteogenic effect with brown algal extracts added to bone marrow macrophage cultures. This includes inhibition of RANKL-dependent MAPKs and downregulation of c-Fos and NFATc1 transcription factors. Other examples of algal extract which show osteogenic activity is *Sargassum horneri* (brown algae) which stimulate osteoblast genesis and inhibit osteoclast genesis in vitro in preosteoblastic and monocytic cell lines. Similarly in vitro and in vivo work is performed on rat femoral tissue to study the ability of *S. horneri* extracts and it is found that it increases the bone calcium content and inhibits bone resorption.

Nacre which is generally available in powder form and also known as the mother of pearl has now become the body of research. It is a lustrous aragonitic inner layer that is found in molluscan shells and belongs to mussels and abalone taxa (Singh et al. 2011). Similar to the bone, nacre also consists of both inorganic and organic contents with an organic shell matrix comprises of proteins and other nutrition. It is used as a template for calcium carbonate mineralisation. Various studies on nacre have been conducted since the early 1990s, with initial in vitro work showing its capacity to stimulate the mineralization of human osteoblasts and the ability of nacre to aid bone reconstruction in human maxillary defects. Here, nacre powder is mixed with the blood of patients and injected into the defect site of eight middle-aged female patients. The results show no evidence of toxic effect and demonstrate enhanced mineralisation and good bio-dissolution of nacre within the area of injection (Akakabe et al. 2014). The significance of this discovery is not realised until a subsequent commentary is published, emphasising the remarkable ability of a raw and unrefined natural product to promote bone growth. Since this early work, there has been a surge of research effort, including in vitro and in vivo studies, as well as those specifically focusing on the proteins and mechanisms involved in enhancing cellular activity, making nacre an excellent case study of bioactive research (Kose et al. 2016).

A good example of the in vitro work which is conducted with used water soluble matrix (WSM) that is extracted from the oyster *Pinctada fucata*. This study demonstrates both the ability of nacre to enhance osteoblast differentiation (increased Col-I, osteocalcin, and ALP expression) and its ability to scavenge free radicals, suggesting an antioxidant potential that may also support bone regeneration. WSM has also been shown to increase bone mineral density (BMD) in an ovariectomized mouse model of osteoporosis, in part attributed to increased Runx2 and Fos-related antigen-1 expression as a result of JNK pathway stimulation in osteoblasts. Furthermore, the extract suppresses actin ring formation and RANKL-induced upregulation of c-Fos and NFATc1 in osteoclasts. Other in vivo work shows nacre implanted into rat femurs supported new bone formation, implant/bone fusion, and increased expression of numerous markers indicative of increased BMU action (Uchiyama et al. 2003).

Aquamin derived from *Lithothamnioncorallioides* (red alage) includes calcium, magnesium and approx 72 other trace minerals. *L. corallioides* is famous for its uniqueness that comes under the few algal species who produces a calcareous skeleton. *L. corallioides* generally found on muddy/sandy substrates (at less than 20 m in depth) with assemblage of unattached algae (Yamaguchi and Matsumoto 2012).

Other examples include *Haliotis discus*, which is associated with osteoblastic activity. *Zoanthus sp.* Includes Norzoanthamine, an alkaloid, responsible for anabolic effects on bones and increases the formation of a collagen-hydroxyapatite composite. Another natural marine compound with osteogenic potential is Phorbaketal A, derived from the marine sponge *Phorbas sp.* This bioactive is shown to stimulate osteoblast differentiation in mesenchymal stem cells, predominantly through activation of the extracellular signal-regulated kinase (ERK) pathway (Kose et al. 2016).

### *17.3.6 Anti-inflammatory Activities*

Marine food or seafood also exerts anti-inflammatory effects because of the inhibition of certain inflammation mediators like cytokines, prostaglandins, leukotrienes, etc. by polyunsaturated fatty acids (PUFAs) especially omega-3. Many studies show that enhancing the ratio of omega-3 to omega-6 fatty acids (which have pro-inflammatory and immunoactivity properties) in diet, can result in a decrease in inflammation because eicosanoids derived from omega-3 have anti-inflammatory activity (Sonani et al. 2014). Many works of literature and experimental studies suggest that daily intake of fish or fish oil is very much effective on inflammation. It is also found that fish oil can be taken as an alternative for NSAID (Non-steroidal anti-inflammatory drugs) to avoid the side effects of NSAIDs like gastric ulcer, bleeding, etc. (Ellis 2001).

Marine seaweeds also exert anti-inflammatory activities. For example, fucoxanthin derived from brown algae helps to decrease the production of prostaglandin (E2) (PGE2) and inhibits COX-2 protein expressions. The sulfoglycolipidic fraction (SF) derived from *Porphyridium cruentum* includes a high concentration of palmitic acids (26.1%), archidonic acid (6.8%), EPA (16.6%) and omega -9 fatty acids (10.5%) and thus it exhibits anti-inflammatory activity (Park et al. 2011). In addition, heterofucan isolated from brown algae *Dictyota menstrualis* inhibits the leukocyte migration and reduces the level of pro-inflammatory cytokines. Moreover, Fucoidan isolated from seaweed *E. Cava* decreases COX-2, nitric acid, and prostaglandins (Gunnarsdottir et al. 2008).

### *17.3.7 Neuroprotective Agents*

Neurodegenerative diseases refer to the death of a certain part of the brain due to some illness which generally occurs in the older age group. These are generally incurable which results in progressively degeneration of the nervous system (CNS as well as ANS) and neurons Ferrari et al. (2008), Gordaliza (2010). The most common neurodegenerative diseases are Parkinson's disease and Alzheimer's disease. Many synthetic compounds are available in the market for the treatment of neurodegenerative diseases but they all involve certain side effects such as anxiety, nervousness, drowsiness (Pal et al. 2008, 2009; Pal and Mazumder 2014; Gupta et al. 2003), mouth dryness or tiredness, etc. (Shibata et al. 2008; Davies-Coleman 2012). For this reason, many scientists and researchers have focused on the development of novel naturally derived drugs with low side effects.

There are various causes for neurodegenerative disease and oxidative stress is one of them. Oxidative stress occurs due to an imbalance between pro-oxidant and anti-oxidant homeostasis that further results in the occurrence of toxic reactive oxygen species. Our CNS is very much sensitive to oxidative stress and that results in lipid peroxidation, DNA, and protein damage and ultimately becomes the reason for neuronal death. Thus, antioxidants become a major saviour for the prevention of neurodegenerative diseases. Lim et al. demonstrate that *Neorhodomela aculeate*, which is also known as *Rhodomela confervoides*, is able to scavenge DPPH with an IC50 = 90 μg/mL It at a concentration of 20 μg/mL completely suppresses $H_2O_2$ induced lipid peroxidation in rat brain homogenate (Maeda et al. 2007). In addition, lowering the risk of neurodegenerative disease and high DHA blood levels is interlinked. Experimental studies on animal models suggest that DHA depletion can result in amyloid protein accumulation and hyperphosphorylation of tau and can become the reason for Alzheimer's disease (Santiago-Santos et al. 2004). Many works of literature and reports suggest that an adequate intake of fish and marine algae can help to maintain the DHA blood level and can help in the prevention of Alzheimer's disease. Houghton et al. Suggest in his report that

Crinum *jagus* and *Crinum glaucum,* two Nigerian Crinum species consist of cholinesterase (ChE) inhibitory activity that is very much beneficial for the treatment of Alzheimer disease. There are many examples of algal species that are endowed with cholinesterase (ChE) inhibitory activity and can be used for the treatment of neurodegenerative diseases for e.g. (Dalisay DS et al. 2009) *Dictyotahumifusa, Hypneavalentiae, Padina gymnospora, Ulva reticulate*, etc.

## 17.4 Summary

In Table 17.5 various marine sources along with their bioactive ingredients and biological activity is mentioned (Ustyuzhanina et al. 2013; Theodore and Kristinsson 2007; VonPost-Skagegard et al. 2006).

**Table 17.5** A summary of marine sources with their bioactive ingredients and biological activities

| Sources | Bioactive ingredients | Biological activities |
|---|---|---|
| Fish oil | Omega-3 fatty acid | Antitumoral and Antimetastasis activities |
| Ascidian | Omega-3 PUFA | Anticancer |
| Crustaceans (shrimp, crab, crayfish) | chitin and chitosan | Anticancer, Antimicrobial Anti-inflammatory Hypocholesterolemic activities |
| Marine sponge *Monanchora pulchra* | Polycyclic guanidine alkaloid | Anticancer activities |
| Sponge *Smenospongia*sp. | Smenospongine, a sesquiterpene aminoquinone | Antiproliferetive and Antiangiogenic activities |
| *Micromonospora*sp. | arisostatin A and arisostatin B | Cytotoxic effects |
| *Aspergillus ustus* | Halimide ((-)-phenylahistin) | Anticancer effects |
| *Cladophora rupestris* | Crude extract | Osteogenic effects |
| *Laurencia undulata* | Floridoside | Osteogenic effects |
| *Haliotis laevigata* | Perlucin protein | Osteogenic effects |
| *Lithothamnioncorallioides* | Aquamin | Osteogenic effects |
| *Nannochloropsisoculata* | Peptide | Osteogenic activity |
| *Symploca* sp. | Largazole (depsipeptide) | Osteogenic effects |
| *Phorbas* sp. | Phorbaketal A | Neuroprotective activity |
| *B. triquetrum* | ferulic acid | Neuroprotective activity |
| *L. japonica* | fucoidan | Neuroprotective actvity |
| *Eisenia bicyclis* | Phlorotannins | Neuroprotective activity |
| *Ecklonia stolonifera* | Sterols, phlorotannins | AChE, Neuroprotective actvity |

(continued)

**Table 17.5** (continued)

| Sources | Bioactive ingredients | Biological activities |
|---|---|---|
| *Nemalionhelminthoides* | Mannans | Antiviral immunomodulatory actvity |
| *Codium fragile* | Galactan | Antiviral immunostimulating activity |
| *Arthrospira platensis* | Spirulan | Antiviral anticoagulant effects |
| Cuttlefish (*Sepiellamaindroni, Euprymnaberryi*) | Polysaccharide | Antimutagenic effects<br>Antimicrobial effects |
| Krill (*Euphausiasuperba*) | Omega-3 PUFA (DHA and EPA | Anti-cardiovascular diseases effects<br>Cholesterol reducing effects |
| Bonito | Omega-3 PUFA (DHA and EPA) | Anti-cardiovascular effects |
| Herring | Omega-3 PUFA (DHA and EPA) | Anticardiovascular effects<br>Anti-obesity actvity<br>Antitumor effects |
| Cuttelfish | Peptides | Antihypertensive activity<br>Antioxidant actvity |
| *Chrysophrys (pagrus) major* | Chrysophsins | Antimicrobial actvity |
| Microalgae *Chlorella zofingiensis* | Astaxanthin | AGE formation inhibition |
| Microalgae *Chlorella pyrenoidosa* | – | Antioxidant potential, α-amylase and α-glucosidase inhibition |
| Fungus *Cosmospora* sp. | Aquastatin A | PTP1B inhibition |
| Actinomycetes *Streptomyces* sp. | Pyrostatins A and B | N-acetyl-glucosaminidase inhibition |

## 17.5 Conclusion

Marine sources are reservoir of essential bioactive ingredients which are beneficial for human body system. They can be used in wide ranges of fields such as, cosmetics, pharmaceuticals, food industry, etc. Marine foods have ability to prevent and cure many diseases like CVS disorders, bone related disease, cancer, etc. The marine organisms offer an enormous resource for drug discovery and development, and are considered as the largest remaining reservoir for natural molecules. They may be used as functional ingredients in the food industry. In this regards, efforts should be made to create awareness about the beneficial effects of marine food since their consumption and utilisation in day to day life will help prevention of many chronic diseases.

## References


Akakabe M, Kumagai K, Tsuda M, Konishi Y, Tominaga A, Tsuda M, Fukushi E, Kawabata J (2014) Amphirionin-5, a novel linear polyketide from a cultured marine dinoflagellate Amphidinium species with a potent cell proliferation-promoting activity. Tetrahedron Lett 55:3491–3494

Amado IR, Vázquez JA, González MP, Murado MA (2013) Production of antihypertensive and antioxidant activities by enzymatic hydrolysis of protein concentrates recovered by ultrafiltration from cuttlefish processing wastewaters. Biochem Eng J 76:43–54

Aminin D, Chaykina E, Agafonova I, Avilov S, Kalinin V, Stonik V (2010) Antitumor activity of the immunomodulatory lead Cumaside. Int Immunopharmacol 10(6):648–654

Aneiros A, Garateix A (2004) Bioactive peptides from marine sources: pharmacological properties and isolation procedures. J Chromatogr B 803:41–53

Anitha A, Maya S, Deepa N, Chennazhi KP, Nair SV, Tamura H, Jayakumar R (2010) Efficient water soluble Ocarboxymethylchitosan nanocarrier for the delivery of curcumin to cancer cells. CarbohydrPolym 83:452–461

Asha KK, Anandan R, Mathew S, Lakshmanan PT (2014) Biochemical profile of oyster Crassostrea madrasensis and its nutritional attributes. Egypt J Aquat Res 40:35–41

Al S, Ismail SH, Manap A, Mustafa S, Yusof R, Hassan FA (2013) Prebiotics as functional foods: a review. J Func Foods 5:1542–1553

BellisleF DAT, Hornstra G, Koletzko B, Roberfroid M, Salminen S (1998) Functional food science in Europe. Br J Nutr 8:1–193

Bin S, Zhi-ping W, Xin-ying L, Jin-xin Y, Jin-jin L, Jing-mei W, Li-fang M, Zi-yuan C (2013) Breeding of a Chlorella strain with high yield of polysaccharide and its effect on growth and immunoregulation of Litopenaeusvannamei. J Nucl Agric Sci 27:168–172

Bocanegra A, Bastida S, Benedí J, Ródenas S, Sánchez-Muniz FJ (2009) Characteristics and nutritional and cardiovascular-health properties of seaweeds. J Med Food 12:236–258

Calder PC (2009) Polyunsaturated fatty acids and inflammatory processes: new twists in an old tale. Biochimie 91:791–795

Charissoux JL, Vergnenegre G, Pelissier M, Fabre T, Mansat P (2013) Epidemiology of distal humerus fractures in the elderly. OrthopTraumatolSurgRes 99:765–769

Cho SS, Lee HK, Yu CY, Kim MJ, Seong ES, Ghimire BK, Son EH, Choung MG, Lim JD (2008) Isolation and characterization of bioactive peptides from Hwangtae (yellowish dried Alaska pollack) protein hydrolysate. Food Sci Nutr 13:196–203

Dalisay DS, Morinaka BI, SkepperCK, Molinski TF(2009) A tetrachloro polyketide hexahydro-1H-isoindolone, muironolide A, from the marine sponge *Phorbas sp.* natural products at the nanomole scale. J Am Chem Soc 131(22):7552

Damodaran S (1997) Food proteins, an overview. In: Srinivasan D, Paraf A (eds) Food proteins and their applications. CRC Press, New York, pp 1–6

Desai UR (2004) New antithrombin-based anticoagulants. Med Res Rev 24:151–181

Davies-Coleman MT (2012) Cytotoxic and antioxidant marine prenylated quinines and hydroquinones. Nat Prod Rep 29(5):513–535

Dharmaraj S, Ashokkumar B, Dhevendaran K (2009) Food-grade pigments from streptomyces sp. Isolated from the marine sponge callyspongiadiffusa. Food Res Int 42:487–492

Diplock AT, AggetPJ AM, Bornet F, Fern EB, Roberfroid MB (1999) Scientific concepts of functional foods in Europe: consensus document. Br J Nutr 81:S1–S27

Ellis AE (2001) Innate host defense mechanisms of fish against viruses and bacteria. Dev Comp Immunol 25:827–839

Fan L, Jiang L, Xu Y, Zhou Y, Shen Y, Xie W (2011) Synthesis and anticoagulant activity of sodium alginate sulphates. CarbohydrPolym 83:11797–11803

Ferrari D, Cysneiros RM, Scorza CA, Arida RM, Cavalheiro EA, de Almeida ACG, Scorza FA (2008) Neuroprotective activity of omega-3 fatty acids against epilepsy-induced hippocampal damage: quantification with immunohistochemical for calcium-binding proteins. Epilepsy Behav 13:36–42

Fiszman SM, Salvador A (2003) Recent developments in coating batters. Trends Food Sci Technol 14:399

Freitas AC, Rodrigues D, Rocha-Santos TA, Gomes AM, Duarte AC (2012) Marine biotechnology advances towards applications in new functional foods. Biotechnol Adv 30:1506–1515

Ge B, Qin S, Han L, Lin F, Yuhong Ren Y (2006) Antioxidant properties of recombinant allophycocyanin expressed in Escherichia coli. J PhotochemPhotobiol B 84:175–180

Gordaliza M (2010) Cytotoxic terpene quinones from marine sponges. Mar Drugs 8(12):2849–2870

Granito RN, Custódio MR, Rennó ACM (2017) Natural marine sponges for bone tissue engineering: the state of art and future perspectives. J Biomed Mater Res Part B Appl Biomater 105:1717–1727

Gunnarsdottir I, Tomasson H, Kiely M, Martinéz JA, Bandarra NM, Morais MG, Thorsdottir I (2008) Inclusion of fish or fish oil in weight-loss diets for young adults: effects on blood lipids. Int J Obes 32:1105–12

Gupta M, Mazumder UK, Pal D, Bhattacharya S and Chakrabarty S (2003) Studies on brain biogenic amines in methanolic extract of *Cuscuta reflexa* Roxb stem and *Corchorus olitorius* Linn. seed treated mice. Acta-Polon Pharm-Drug Res 60(3):207–210

Halim H, Chunhacha P, Suwanborirux K, Chanvorachote P (2011) Anticancer and antimetastatic activities of renieramycin M, a marine tetrahydroisoquinoline alkaloid, in human nonsmall cell lung cancer cells. Anticancer Res 31(1):193–201

Heo SJ, Hwang JY, Choi JI, Han JS, Kim HJ, Jeon YJ (2002) Diphlorethohydroxycarmalol isolated from Ishigeokamurae, a brown algae, a potent [alpha]-glucosidase and [alpha]-amylase inhibitor, alleviates postprandial hyperglycemia in diabetic mice. Eur J Pharmacol 615:252–256

Huheihel M, Ishanu V, Tal J, Arad S (2002) Activity of Porphyridium sp. polysaccharide against herpes simplex viruses in vitro and in vivo. J BiochemBiophys Meth 50:189–200

Imhoff JF, Labes A, Wiese J (2011) Bio-mining the microbial treasures of the ocean: new natural products. Biotechnol Adv 29(5):468–482

Ishihara K, Oyamada C, Sato Y, Suzuki T, Kaneniwa M, Kunitake H, Muraoka T (2010) Prebiotic effect of glycerol galactoside isolated from color-faded nori in rats. Fish Sci 76:1015–1021

Janakiram NB, Mohammed A, Zhang Y, Choi C-I, Woodward C, Collin P, Steele VE, Rao, CV (2010) Chemopreventive effects of frondanol A5, a cucumariafrondosa extract, against rat colon carcinogenesis and inhibition of human colon cancer cell growth. Cancer Prev Res 3 (1):82–91

Je JY, Cho YS, Ahn CB (2014) Purification and anti-inflammatory action of tripeptide from salmon pectoral fin byproduct protein hydrolysate. Food Chem 168:151–156

Jun SY, Park PJ, Jung WK, Kim SK (2004) Purification and characterization of an antioxidative peptide from enzymatic hydrolysate of yellowfin sole (Limanda aspera) frame protein. Eur Food Res Technol 219:20–26

Kadam SU, Prabhasankar P (2010) Marine foods as functional ingredients in bakery and pasta products. Food Res Int 43:1975–1980

Kanazawa A (2001) Sterols in marine invertebrates. Fisheries Sci 67:997–1007

Kanekiyo K, Hayashi K, Takenaka H, Lee JB, Hayashi T (2007) Anti-herpes simplex virus target of an acidic polysaccharide, nostoflan, from the edible blue-green alga Nostocflagelliforme. Biol Pharm Bull 30:1573–1575

Karuppasamy PK, Sakthi Priyadarshini RS, Ramamoorthy N, Sujatha R, Ganga S, Jayalakshmi T, Santhanam P (2013) Comparison of proximate, amino and fatty acid composition of Penaeus monodon (Fabricius, 1798), Fenneropenaeus indicus (Milne Edwards H, 1837) and Aristeusvirilis (Bate, 1881) of Nagapattinam landing centre, Tamil Nadu. J Mar Biol Ass India 55:5–10

Kim SK, Wijesekara I (2010) Development and biological activities of marine-derived bioactive peptides: a review. J Funct Foods 1–9

Kose O, Arabaci T, Yemenoglu H, Kara A, Ozkanlar S, Kayis S, Duymus Z (2016) Influences of fucoxanthin on alveolar bone resorption in induced periodontitis in rat molars. Mar Drugs 14:70

Kuroiwa T, Izuta H, Nabetani H, Nakajima M, Sato S, Mukataka S, Ichikawa S (2009) Selective and stable production of physiologically active chitosan oligosaccharides using an enzymatic membrane bioreactor. Process Biochem 44:283–287

Le QT, Li Y, Qian ZJ, Kim MM, Kim SK (2009) Inhibitory effects of polyphenols isolated from marine alga *Ecklonia cava* on histamine release. Process Biochem 44:168–217

Li JJ, Huang CJ, Xie D (2008) Anti-obesity effects of conjugated linoleic acid, docosahexaenoic acid, and eicosapentaenoic acid. Mol Nutr Food Res 52:631–645

Lloret J (2010) Human health benefits supplied by Mediterranean marine biodiversity. Mar Pollut Bull 60:1640–1646

Lordan S, Ross RP, Stanton C (2011) Marinebioactives as functional food ingredients: potential to reduce the incidence of chronic diseases. Mar Drugs 9:1056–1100

Lum KK, Kim J, Lei XG (2013) Dual potential of microalgae as a sustainable biofuel feedstock and animal feed. J Anim Sci Biotechnol 4:53

LøvstadHoldt S, Kraan S (2011) Bioactive compounds in seaweed: functional food applications and legislation. J ApplPhycol 23:543–597

Ma LZ, Zhen RY, Zhang JR, Shi H (2008) Study on processing technique of enzyme hydrolysis of catfish collagen peptide with immunite activity. Sci Technol Food Ind 4:152–155

Maeda H, Hosokawa M, Sashima T, Murakami-Funayama K, Miyashita K (2009) Anti-obesity and anti-diabetic effects of fucoxanthin on diet-induced obesity conditions in a murine model. Mol Med Rep 2:897–902

Maeda H, Hosokawa M, Sashima T, Murakami-Funayama K, Miyashita K (2007) Effect of medium-chain triacylglycerols on anti-obesity effect of fucoxanthin. J Oleo Sci 56:615–621

Mata TM, Martins AA, Caetano NS (2010) Microalgae for biodiesel production and other applications: a review. Renew Sust Energ Rev 14:217–232

Mauro SG, Paulo P, Mourão AS (2012) Challenges for heparin production: artificial synthesis or alternative natural sources? Glycobiol Insights 3:1–6

Mayer AM, Glaser KB, Cuevas C, Jacobs RS, Kem W, Little RD, McIntosh JM, Newman DJ, Potts BC, Shuster DE (2010) The odyssey of marine pharmaceuticals: a current pipeline perspective. Trends Pharmacol Sci 31:255–265

Menna M, Imperatore C, Aiello D (2013) A Meroterpenes from marine invertebrates: structures, occurrence, and ecological implications. Mar Drugs 11(5):1602–1643

Migliore L, Coppedè F (2009) Environmental-induced oxidative stress in neurodegenerative disorders and aging. Mutat Res Gen Tox En 674:73–84

Miyashita K, Nishikawa S, Beppu F, Tsukui T, Abe M, Hosokawa M (2011) The allenic carotenoid fucoxanthin, a novel marine nutraceutical from brown seaweeds. J Sci Food Agric 91:1166–1174

Miyashita K (2013) Anti-obesity therapy by food component: unique activity of marine carotenoid, fucoxanthin. Obes Control Ther 1:4

Mozaffarian D, Wu JHY (2011) Omega-3 fatty acids and cardiovascular disease effects on risk factors, molecular pathways, and clinical events. J Am Coll Cardiol 58:2047–2067

Mussatto SI, Mancilha IM (2007) Non-digestible oligosaccharides: a review. CarbohydrPolym 68:587–597

Ngo DN, Kim MM, Qian ZJ, Jung WK, Lee SH, Kim SK (2010) Free radical scavenging activities of low molecular weight chitin oligosaccharides lead to an antioxidant effect in live cells. J Food Biochem 34:161–177

Nguemfo EL, Dimo T, Azebaze AGB, Asongalem EA, Alaoui K, Dongmo AB (2007) Anti-inflammatory and anti-nociceptive activities of the stem bar extracts from Allanblackiamonticola STANER L.C. (Guttiferae). J Ethnopharmacol 114:417–424

Nimse SB, Pal D (2015) Free radicals, natural antioxidants and their reaction mechanisms. RSC Adv 5(35):27986–28006

Özyurt G, Kuley E, Etyemez M, Özoğul F (2013) Comparative seasonal sterol profiles in edible parts of Mediterranean fish and shellfish species. Int J Food Sci Nutr 64:476–483
Pal D, Nandi M (2005) CNS activities of *Celsia coromandeliane* Vahl in mice. Acta Polon Pharm Drug Res 62(5):355–361
Pal D, Kumar M, Chakrabarty P, Kumar S (2008) Evaluation of the antioxidant activity of aerial parts of *Cynodon dacvtylon*. Asian J Chem 20(3):2479–2481
Pal D, Sannigrahi S, Mazumder UK (2009) Analgesic and anticonvulsant effects of isolated saponin from the leaves of *Clerodendrum infortunatum* Linn. in mice. Indian J Exp Biol 47:743–747
Pal D (2013) Comparative analysis of in vitro antioxidant activity of two selected plants with a reference to antidiabetic profile. Asian J Chem 25(4):2165–2169
Pal D, Mazumder UK (2014) Isolation of compound and studies on CNS depressant activities of *Mikania scandens* Wild with special emphasis to brain biogenic amines in mice. Indian J Exp Biol 52:1186–1194
Park MK, Jung U, Roh C (2011) Fucoidan from marine brown algae inhibits lipid accumulation. Mar Drugs 9:1359–1367
Patel S, Goyal A (2011) Functional oligosaccharides: production, properties and applications. World J Microbiol Biotechnol 27:1119–1128
Paul VJ, Ritson-Williams RK, Sharp (2011) Marine chemical ecology in benthic environments. Nat Prod Rep 28:345–388
Perdicaris S, Vlachogianni T, Valavanidis A (2013) Bioactive natural substances from marine sponges: new developments and prospects for future pharmaceuticals. Nat Prod Chem Res 1:1–8
Plaza M, Cifuentes A, Ibañez E (2008) In the search of new functional food ingredients from algae. Trends Food Sci Technol 19:31–39
Prasad K, Goswami AM, Meena R, Ramavat BK, Ghosh PK, Siddhanta AK (2006) Superior quality agar from red alga *Gelidiellaacerosa* (Rhodophyta, Gelidiales) from Gujarat coast of India: an evaluation. Indian J Mar Sci 35:268–274
Proksch P, Edrada R, Ebel R (2002) Drugs from the seas-current status and microbiological implications. Appl Microbiol Biotech 59(2–3):125–134
Rabelo L, Monteiro N, Serquiz R, Santos P, Oliveira R, Oliveira A, Rocha H, Morais AH, Uchoa A, Santos E (2012) A lactose-binding lectin from the marine sponge *Cinachyrellaapion* (Cal) induces cell death in human cervical adenocarcinoma cells. Mar Drugs 10(4):727–743
Rahman MM, McFadden G (2011) Modulation of NF-κB signalling by microbial pathogens. Nat Rev Microbiol 9:291–306
Rasmussen RS, Morrissey MT (2007) Marine biotechnology for production of food ingredients. Adv Food Nutr Res 52:237–292
Rinaudo M (2006) Chitin and chitosan: properties and applications. Prog Polym Sci 31:603–632
Rocha J, Calado R, Leal M (2015) Marine bioactive compounds from cnidarians. In: Springer handbook of marine biotechnology, pp 823–849
Rodríguez-Meizoso I, Jaime L, Santoyo S, Señoráns F, Cifuentes A, Ibáñez E (2010) Subcritical water extraction and characterization of bioactive compounds from Haematococcuspluvialis microalga. J Pharm Biomed Anal 51:456–463
Saha S, Pal D (2016) Molecular docking, synthesis, in vitro HDAC inhibition activity and antiproliferative activity against breast cancer cell line of hydroxyacetamide derivatives. Trop J Pharm Res 15(7):1401–1411
Saha S, Pal D, Kumar S (2017) Hydroxyacetamide derivatives: cytotoxicity, genotoxicity, antioxidative and metal chelating studies. Indian J Exp Biol 55(2):831–837
Sannigrahi S, Mazumder UK, Pal D, Mishra SL (2012) Terpenoids of methanol extract of *C. infortunatum* exhibit anticancer activity against Ehrlich's ascites carcinoma (EAC) in mice. Pharm Biol 50(3):304–309
Santiago-Santos MC, Ponce-Noyola T, Olvera-Ramírez R, Ortega-López J, Cañizares-Villanueva RO (2004) Extraction and purification of phycocyanin from Calothrix sp. Process Biochem 39:2047–2052

Schumacher M, Cerella C, Eifes S, Chateauvieux S, Morceau F, Jaspars M, Dicato M, Diederich M (2010) Heteronemin, a sponge ansesterterpene, inhibits TNFα-induced NF-κB activation through proteasome inhibition and induces apoptotic cell death. Biochem Pharmacol 79(4):610–622

Schumacher M, Kelkel M, Dicato M, Diederich M (2011) Gold from the sea: marine compounds as inhibitors of the hallmarks of cancer. Biotechnol Adv 29(5):531–547

Shahidi F, Zhong Y (2010) Novel antioxidants in food quality preservation and health promotion. Eur J Lipid Sci Technol 112:930–940

Sheraji Al, Ismail SH, Manap A, Mustafa, S, Yusof R, Hassan FA (2013) Prebiotics as functional foods: A review, J Funct Foods5: 1542–53

Sharon N, Lis H (2004) History of lectins: from hemagglutinins to biological recognition molecules. Glycobiology 14:53R-62R

Shibata T, Ishimaru K, Kawaguchi S, Yoshikawa H, Hama Y (2008) Antioxidant activities of phlorotannins isolated from Japanese Laminariaceae. J Appl Phycol 20:705–711

Sijtsma L, de Swaaf ME (2004) Biotechnological production and applications of the omega-3 polyunsaturated fatty acid docosahexaenoic acid. Appl Microbiol Biotechnol 64:146–153

Singh A, Nigam PS, Murphy JD (2011) Mechanism and challenges in commercialisation of algal biofuels. Bioresour Technol 102:26–34

Smith VJ, Desbois AP, Dyrynda EA (2010) Conventional and unconventional antimicrobials from fish, marine invertebrates and micro-algae. Mar Drugs 8:1213–1262

Solanki R, Khanna M, Lal R (2008) Bioactive compounds from marine actinomycetes. Indian J Microbiol 48(4):410–431

Sonani RR, Singh NK, Kumar J, ThakarD MD (2014) Concurrent purification and antioxidant activity of phycobiliproteins from *Lyngbya sp.* A09DM: an antioxidant and anti-aging potential of phycoerythrin in Caenorhabditiselegans. Process Biochem 49:1757–1766

Soni B, Trivedi U, Madamwar D (2008) A novel method of single step hydrophobic interaction chromatography for the purification of phycocyanin from Phormidiumfragile and its characterization for antioxidant property. Bioresour Technol 99(881):188–194

Spolaore P, Joannis-Cassan C, Duran E, Isambert A (2006) Commercial applications of microalgae—review. J Biosci Bioeng 101:87–96

Thakur NL, Müller WE (2004) Biotechnological potential of marine sponges. Curr Sci 86 (11):1506–1512

Theodore AE, Kristinsson HG (2007) Angiotensin-converting enzyme inhibition of fish protein hydrolysates prepared from alkaline-aided channel catfish protein isolate. J Sci Food Agric 87:2353–2357

Thiansilakul Y, Benjakul S, Shahidi F (2007) Antioxidative activity of protein hydrolysate from round scad muscle using alcalase and flavourzyme. J Food Biochem 31:266–287

Tsukui T, Konno K, Hosokawa M, Maeda H, Sashima T, Miyashita K (2007) Fucoxanthin and fucoxanthinol enhance the amount of docosahexaenoic acid in the liver of KKAy obese/diabetic mice. J Agr Food Chem 55:5025–5029

Uchiyama S, Yamaguchi M (2003) Preventive effect of marine Alga Sargassum Horneri extract on bone loss in streptozotocin-diabetic rats in vivo. J Health Sci 49:149–155

Ustyuzhanina NE, Ushakova NA, Zyuzina KA, Bilan MI, Elizarova AL, Somonova OV, Madzhuga AV, Krylov VB, Preobrazhenskaya ME, Usov AI, Kiselevskiy MV, Nifantiev NE (2013) Influence of fucoidans on hemostatic system. Marine Drugs 12 11(7):2444–2458

Venegas-Calerón M, Sayanova O, Napier JA (2010) An alternative to fish oils: metabolic engineering of oil-seed crops to produce omega-3 long chain polyunsaturated fatty acids. Prog Lipid Res 49(2):108–119

Vo TS, Ngo DH, Ta QV, Kim SK (2011) Marine organisms as a therapeutic source against Herpes simplex virus infection. Eur J Pharm Sci 44:11–20

VonPost-Skagegard M, VessbyB KB (2006) Glucose and insulin responses in healthy women after intake of composite meals containing cod-, milk-, and soya protein. Eur J Clin Nutr 60:949–954

Wall R, Ross RP, Fitzgerald GF, Stanton C (2010) Fatty acids from fish: the anti-inflammatory potential of long-chain omega-3 fatty acids. Nutr Rev 68:280–289

Wang Y, Han F, Hu B, Li JB, Yu WG (2006) In vivo prebiotic properties of alginate oligosaccharides prepared through enzymatic hydrolysis of alginate. Nutr Res 26:597–603

Wang Y-Q, Miao Z-H (2013) Marine-derived angiogenesis inhibitors for cancer therapy. Mar Drugs 11(3):903–933

Wijesekara I, Pangestuti R, Kim SK (2011) Biological activities and potential health benefits of sulfated polysaccharides derived from marine algae. Carbohydr Polym 84:14–21

Xia W, Liu P, Zhang J, Chen J (2011) Biological activities of chitosan and chitooligosaccharides. Food Hydrocolloids 25:170–179

Yam D, Peled A, Shinitzky M (2001) Suppression of tumor growth and metastasis by dietary fish oil combined with vitamins E and C and cisplatin. Cancer Chemother Pharmacol 47:34–40

Yamaguchi M, Matsumoto T (2012) Marine algae Sargassum Horneri bioactive factor stimulates osteoblastogenesis and suppresses osteoclastogenesis in vitro. OA Biotechnol 1:3

Zaporozhets TS, Besednova NN, Kuznetsova TA, Zvyagintseva TN, Makarenkova ID, KryzhanovskySP MVG (2014) The prebiotic potential of polysaccharides and extracts of seaweeds. Russ J Mar Biol 40:1–9

Zheng JS, Hu XJ, Zhao YM, Yang J, Li D (2013) Intake of fish and marine n-3 polyunsaturated fatty acids and risk of breast cancer: meta-analysis of data from 21 independent prospective cohort studies. BMJ 346:1–10

Zubia M, Robledo D, Freile-Pelegrin Y (2007) Antioxidant activities in tropical marine macroalgae from the Yucatan peninsula Mexico. J Appl Phycol 19:449–458

# Chapter 18
# Cardiac Tissue Engineering: A Role for Natural Biomaterials

**Pallavi Pushp and Mukesh Kumar Gupta**

**Abstract** Cardiovascular diseases (CVDs) are the major cause of death all over the world, being responsible for 7.4 million deaths in a year. In the past two decades, advances in biomaterials, stem cell biology, and engineering have allowed the generation of tissue constructs that can imitate the complex structure of the heart and, upon transplantation into animal models, improved the cardiac function. Consequently, stem cell-based cardiac tissue engineering (CTE) has emerged as a potential therapeutic strategy for the treatment of CVDs. The choice of appropriate biomaterials that can provide differentiating and paracrine milieu to mimic the extracellular matrix (ECM) is the key to the proper functioning of cardiac tissue construct. Plant- or animal-derived natural polymers such as alginate, chitosan, collagen, fibrin, gelatin, and glycosaminoglycan are biocompatible and facilitate cell adhesion, proliferation, and differentiation for development of cardiac tissue constructs. On the other hand, natural biomaterials such as collagen, fibronectin, laminin, nephronectin, etc. are often used to biofunctionalize the synthetic biomaterials-derived scaffolds to enhance cell adhesion and cell-to-cell communication. This book chapter discusses various approaches for CTE and the role of different natural biomaterials and their importance in the biofunctionalization of synthetic biomaterials for developing cardiac tissue constructs.

**Keywords** Bioactive polymers · Biomaterials · Cardiac tissue engineering · Hybrid polymers · Scaffolds · Stem cells

P. Pushp
Department of Biotechnology, Institute of Engineering and Technology (IET), Bundelkhand University, Jhansi, Uttar Pradesh 284 128, India

P. Pushp · M. K. Gupta (✉)
Department of Biotechnology and Medical Engineering, National Institute of Technology, Rourkela, Odisha 769 008, India
e-mail: guptam@nitrkl.ac.in



D. Pal and A. K. Nayak (eds.), *Bioactive Natural Products for Pharmaceutical Applications*, Advanced Structured Materials 140,
https://doi.org/10.1007/978-3-030-54027-2_18

## 18.1 Introduction

Cardiovascular diseases (CVDs) such as myocardial infarction (MI), atherosclerosis, cardiomyopathies, heart valve diseases, and long-QT syndromes (LQTS) are the major causes of death worldwide. They are generally associated with apoptotic and necrotic changes in the cardiac tissue and loss of functional cardiomyocytes (CMs). Since the self-renewal or regenerative potential of CMs is extremely limited, the lost CMs are challenging to be compensated by therapeutic approaches. Thus, despite significant research in diagnosis and treatment, CVDs remain major therapeutic challenges in the medical field. In recent years, cardiac tissue engineering (CTE) has been emerging as a potential therapeutic strategy for the treatment of CVDs. These CTE approaches involve isolation and expansion of autologous patient cells, including stem cells such as mesenchymal stem cells (MSCs) or induced pluripotent stem cells (iPSCs) and their differentiation into cardiac tissue on an appropriate extracellular matrix (ECM) for in vitro incubation followed by in vivo transplantation (Fig. 18.1).

In initial studies, 'cell-based therapy' by cardiomyoplasty were attempted wherein embryonic, fetal and neonatal CMs, skeletal myoblasts, bone marrow stem cells, or resident cardiac progenitors were directly injected into the injured heart. Several clinical trials on cardiomyoplasty reported functional improvements in CVDs but the functional benefits were transient. The cardiomyoplasty procedure also suffer from poor availability of cells, low cell retention and poor survival of

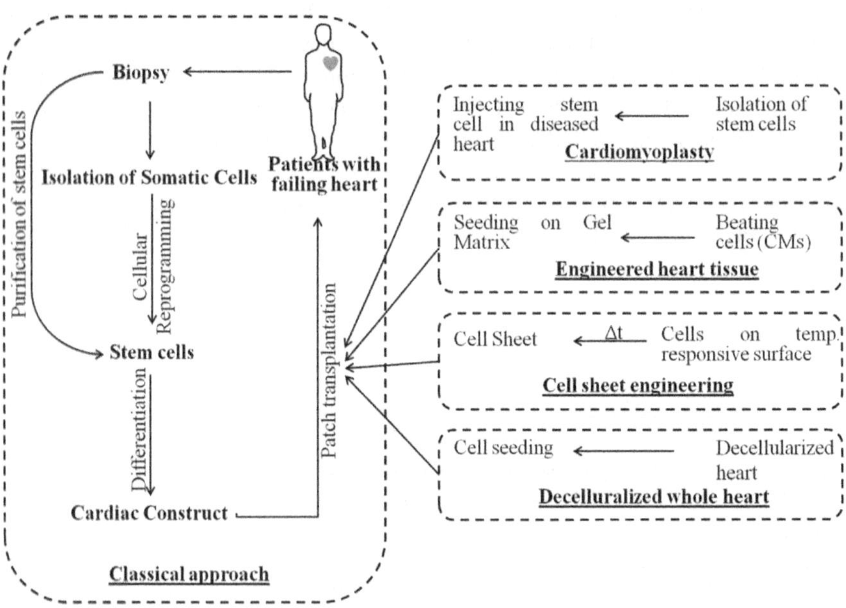


**Fig. 18.1** Approaches for cardiac tissue engineering

implanted cells, loss of intracellular communication between CMs and ECM, acute inflammation and injury caused by injection procedure and blockage of the microcirculatory network by the injected cells. Some studies even reported negative effect of cardiomyoplasty, resulting in arrhythmias due to inappropriate electrical integration. Thus, in vitro developed cardiac tissue constructs, seeded with stem cells or their progenitors, were explored as alternative options. Since then, the field of CTE exponential advanced with advancements in the design and development of ECM-mimicking scaffolds, efficient perfusion bioreactors and, newer sources of stem cells, and advanced analytical techniques. This book chapter summarises various approaches for CTE and the role of natural biomaterials in fabricating the scaffolds for seeding of stem cells.

## 18.2 Approaches for CTE Using Natural Biomaterials, Cell-Sheet and Decellularized Tissues

In 'classical' CTE, cardiac tissue construct is created by seeding desired cells into biomaterial-derived scaffolds. The scaffolds are initially fabricated, without cells, using natural or synthetic or hybrid biomaterials and, the cells are then seeded into the scaffold. The scaffolds are expected to allow neovascularisation, neuronal innervations and electromechanical cell coupling with host myocardium following their transplantation into the host. In order to achieve these characteristics, several modifications such as incorporation of growth factors and pro-angiogenic factors, co-culturing of CMs with other cell types etc. are often required to further improve the survival of the tissue construct (Fig. 18.1). The scaffolds are traditionally fabricated using natural biomaterials such as alginate, chitosan, collagen, fibrin, gelatin, and glycosaminoglycan, fibronectin, laminin, nephronectin. Many natural polymers have low elastic modulus and poor electrical conductivity required for the CTE and necessitates doping with synthetic biomaterials or nanomaterials such as carbon nanofibers to develop a conductive scaffold (Martins et al. 2014). Natural biomaterials such as gelatine, collagen, fibronectin, laminin, stromal-derived factor-1, nephronectin, and bioactive glass etc. are also used to bio-functionalize the synthetic biomaterials-derived scaffolds to enhance cell adhesion, cell-to-cell communication and cellular interaction with scaffold (Oliver et al. 2019; Yang et al. 2018).

In a newer development, engineered heart tissue (EHT) were created by seeding CMs into collagen I and Matrigel™ (an extracellular matrix compound derived from Engelbreth-Holm-Swarm mouse sarcoma) matrix followed by mechanical stretching into circular form (Yildirim et al. 2007). Guo et al. generated embryonic stem cell (ESC)-derived myocardium in mouse wherein CMs were seeded into circular moulds with collagen I and Matrigel™ to produce EHT, which were subsequently subjected to unidirectional cyclic stretch at 10% strain and 2 Hz (Guo et al. 2006). Transplantation of the EHT improved ventricular function in MI model but formed transmural thrombus in a heterotropic heart transplant model (Zimmermann et al. 2006).

Scaffold-free cell sheet/cell patch engineering is a yet another attractive approach for producing multi-layered cardiac patch. This technique was first developed in Japan by Kushida and co-workers. In this technology, a temperature-responsive culture surface is first fabricated by layering a temperature-responsive polymer (e.g. poly (N-isopropylacrylamide)) onto polystyrene petri plates. Monolayers of CMs are then grown on such cell culture plates and, upon confluency, are recovered as an intact cell sheet by lowering the temperature to 20 °C. Scaffold-free multi-layered cell sheets are then generated as 3D tissue by stacking of these monolayer sheets using a silicon mould (Haraguchi et al. 2014). Scaffold-free vascularized CMs patches have also been developed by Stevens et al. (2009) by utilizing a rotating orbital shaker. The 3D myocardial tissue, developed from these stacked CMs sheets, resembled native heart tissue and showed synchronous beats (Haraguchi et al. 2011). In a clinical trial conducted in Japan, transplantation of autologous myoblast cell sheets was shown to improve the cardiac dysfunction in heart failure cases (Matsuura et al. 2014). However, the stacked cell sheets are generally limited in thickness up to 3 layers ($\sim$80 μm) due to the lack of vascularization and oxygen transport to the centre of the tissue. Furthermore, cell sheet tissue lack native orientation and spatial placement of cardiac cells. The later can however, be overcome by photolithography-based microtexturing of culture dishes before growing the monolayers (Isenberg et al. 2008).

Decellularized whole heart tissue has also been used as scaffold for CTE. Cadaveric hearts can be decellularized by using detergents to obtain whole heart ECM scaffolds with well preserved natural 3D structure of ECM, vascular architecture, mechanical anisotropy, competent acellular valves and intact chamber geometry. The decellularized whole heart can then be re-seeded with CMs and endothelial cells (ECs) and, cultured in a perfusion bioreactor to generate a spontaneously beating 'whole heart', which can be used as allograft or xenograft (Ozlu et al. 2019). Such strategy has been successful in both human and animal. It is also possible to develop personalized bioartificial hear by seeding of patient's own cells (Kang et al. 2020). Unfortunately, re-seeding of cells into the decelluralized heart in a uniform manner is quite difficult and requires specialized multi-stimulation bioreactor to provide coordinated mechanical and electrical stimulation and nutrient supply (Wang et al. 2013a). To circumvent non-uniform re-seeding some studies have suggested the sectioning of myocardium into 2 mm thick portions before decellularization and using them as cardiac patches (Wang et al. 2010). In other studies, decellularized matrix of heart or other organ has been incorporated into scaffolds for CTE.

## 18.3 Natural Biomaterials Used in 'Classical' CTE

Cells within a tissue are surrounded by ECMs that are composed of biomolecules and fibers, synthesized by cells themselves, blood vessels and innervating neurons. The ECMs provide micro-niche whose physico-chemical microenvironment allows the cells to interact through cell signalling molecules and perform their physiological function. Thus, a tissue engineered cardiac construct must be designed to the mimic ECM of the heart. Ideally, a scaffold for human cardiac tissue must have a thickness of $\sim$0.5 cm with adequate porosity for oxygen transport and to allow a cell density of $\sim 10^5$ cells/cm$^3$ of the construct. The construct should also have the ability to generate a force of 2–4 mN/mm$^2$ during contraction and to support electrical signals propagation at $\sim$25 cm/s. It should also allow neovascularisation, neuronal innervations and electromechanical cell coupling with the host myocardium upon transplantation into the host. An ideal biomaterial must be biocompatible and bio-mimetic yet biodegraddable in concomitant with the development of new ECM after transplantation. It should also be amenable to laboratory procedures for fabricating the scaffolds of desired physical, chemical and biological properties and should be scalable of large-scale clinical applications. Some of the commercially available biomaterials for CTE are shown in Table 18.1.

Scaffolds for CTE have been developed using plant- or animal-derived natural polymers such as alginate (Choe et al. 2019), chitosan (He et al. 2018), collagen (Hosoyama et al. 2018), fibrin (Bagheri-Hosseinabadi et al. 2018), gelatin (Majidi et al. 2018), hyaluronic acid (HA) (Hadisi et al. 2020), Poly (3-hydroxyoctanoate) (Bagdadi et al. 2016) and glycosaminoglycan (Flanagan et al. 2006). Natural polymer-derived scaffolds are generally biocompatible and biodegradable and provide better cellular ECM-like environments to facilitate cell adhesion, proliferation and differentiation (Yi et al. 2017). Some polymers may be easily degraded by body's natural enzymes or have relatively weak mechanical properties, which can be controlled by combing with synthetic polymers. The rate of degradation can also be controlled by chemical cross-linking (Wu et al. 2014) of the biomaterials. Some natural polymers have also been shown to have immunogenicity or may act as a potential reservoir of infectious agents (Jawad et al. 2008). The later can however, be controlled by use of genetically engineered and 'immunologically virgin' animals reared under gnotobiotic conditions (Hwang et al. 2015).

### *18.3.1 Collagen I*

Collage I is a predominant ECM protein of heart and is FDA approved for therapeutic use in skin and bone. It contains peptide sequences that interact with integrins to promote cell attachment. It can also modify the substrate stiffness and thereby alter the cellular functions such as growth, proliferation and differentiation. Thus, it has been used extensively in CTE as hydrogel, sponge-like scaffold or as coating material for enhancing hydrophilicity of synthetic biomaterials. Electrospun of sub-micron diameter fibrils of collagen has also been possible, which could be rapidly and densely infiltrated by the cells and helped in the formation of functional blood vessels (Telemeco et al. 2005). Hamdi et al. (2009) observed significant increase in angiogenesis at the site of MI upon transplantation of myoblast-seeded collagen scaffold. Similar results were also observed with bone marrow-derived

Table 18.1 Commercially available biomaterials for cardiac tissue engineering

| Sl. No | Product | Company | Material | Processing | Application |
|---|---|---|---|---|---|
| 1 | Restore ® | DePuy Orthopaedics, Inc | Small Intestine Submucosa | Biological | Soft tissue |
| 2 | PEEK-OPTIMA® | Invibio | Polyether ether ketone | Polymer | Heart valve |
| 3 | ENDURAGen® | Stryker | Collagen | Polymer | Vascularization |
| 4 | SurgiMend® | TEI Biosciences | Acellular Collagen | Biological | Soft tissue and vascularization |
| 5 | Cymetra® | LifeCell | Tissue | Biological | Revascularization and cell population |
| 6 | Ultramax® | Atrium Medical Corporation | Polyethylene terephthalate | Polymer | Vascular grafts |
| 7 | CardioPass® | CardioTech | Polyurethane | Polymer | Synthetic coronary artery |
| 8 | HYAFF-11® | Fidia Advanced Biopolymer | Hyaluronic acid | Biological | Myocytes |

$CD133^+$ cells loaded on collagen patch (Joshi et al. 2018). Zhang et al. (2013) microfabricated chitosan-collagen scaffold with micropores (200 μm) by using an array of parallel channel designed for CTE. The parallel channels enhanced oxygen transport, established cell to cell contact and improved cell alignment and elongation. Another way to improve number of stem cell growth on collagen matrix is to conjugate it with anti-Sca-1 monoclonal antibody (Shi et al. 2011).

Collagen has also been used in combination with other natural polymers such as glycosaminoglycan, gelatin or Matrigel™ to have a synergistic effect (Kleinman and Martin 2005). Transplantation of MSCs in collagen-glycosaminoglycan composite to the infarct region induced neovascularisation at the site of the infarct and increased MSCs in both the scaffold and the heart wall (Qu et al. 2018). The structural integrity and bio-stability of collagen can be improved by cross-linking of with gluteraldehyde (GA), dimethylaminopropyl-ethlcabodiimide (EDC) (Ahn et al. 2013) glyoxal, β-glycerophosphate (Wang et al. 2013b), malic acid derivatives (Saito et al. 2008), dendrimers (Duan and Sheardown 2005), diamine and diaminohexane (McKegney et al. 2001), formaldehyde, genipin, and carbodiimide hydrochloride (EDAC) (Madhavan et al. 2010).

### *18.3.2 Gelatin*

Gelatin is generally formed by thermal denaturation or partial hydrolysis of animal-derived collagen I. Thus, it has been used in CTE to promote cellular attachment and their integration into the infracted heart. It can also be blended with other natural polymers such as gelatine. A blend of alginate/gelatin in 20:80 has been successfully used for fabrication of scaffolds for CTE (Rosellini et al. 2009). In another study, Ravichandran et al. (2013a) have electrospun nanofibrous tissue construct by using a blend of gelatin, hemoglobin and fibrinogen and seeding them with 5-azacytidine-treated mesenchymal stem cells (MSCs). The developed cardiac tissue construct was shown to provide superior biological and functional effects and high oxygen carrying potential for CTE. Combination of chitosan-agarose-gelatin (Bhat and Kumar 2012) and fibrinogen-gelatin (Balasubramanian et al. 2013) composite nanofibers was also shown to enhance the proliferation of CMs and expression of functional cardiac proteins such as α-actinin, troponin I, connexin-43 and myosin heavy chain.

The gelatine has also been combined with synthetic biomaterials for growing aligned CM with improved contractility. Composite of gelatine with polyglycerol sebacate (PGS) could be used to develop an hybrid scaffold which allowed aligned cell proliferation and resulted into well-defined anisotropy with enhanced contractibility of the CMs (Kharaziha et al. 2013). Aligned patterned 3D bioprinted gelatin hydrogel with synchronized beating of CMs, differentiated from MSCs, has also been reported (Tijore et al. 2018).

### 18.3.3 Fibrin

Fibrin is a naturally occurring blood clotting protein and is FDA approved for use as a surgical sealant. The fibrin network has a nanometric fibrous structure, which upon treatment with thrombin can create fibrin gel. Upon introduction into the heart, the fibrin gels can get replaced by natural ECM and therefore, it serve as a temporary scaffold in CTE (Grassl et al. 2003). Fibrin glue (Christman et al. 2004), and hydrogel scaffolds (Kaiser et al. 2019) have been explored as possible cardiac grafts. Thomson et al. described a high-density microtemplated fibrin scaffold seeded with a tri-cell mixture of CM, Epithelial cells (ECs), and fibroblasts to mimic native cardiac tissue in structure and cellular composition (Thomson et al. 2013).

### 18.3.4 Alginate

Alginate of algal origin has also been used as natural biomaterials for CTE. Sodium alginate has been extensively used as hydrogel for encapsulation of cells and delivery of growth factors for CTE (Sun et al. 2010). The physico-chemical properties such as gelation time, uniformity, mechanical strength, tensile properties, and mechanics of alginate gel formation can be controlled or altered by using different cation sources such as $CaCl_2$, $CaSO_4$, and $CaCO_3$ (Kuo and Ma 2001), temperature (Drury et al. 2004) and molecular weight distribution (Augst et al. 2006). Alginates containing higher amount of glucuronic acid (G-type) formed stiffer gels due to the presence of diaxial links and had higher elastic and Young's moduli than mannuronic acid (M-type) alginate gels and were more adaptable for cardiac implantation (Ceccaldi et al. 2012). Fetal CMs, seeded onto alginate hydrogels, stimulated neovascularization and healing of the infracted myocardium in a rat model of MI (Leor et al. 2000). In addition to hydrogel system, alginate has also been used to develop fibrous scaffold by electrospinning and porous scaffold by freeze-drying (Lee et al. 2009). Macroporous scaffolds with interconnected pores showed efficient cell seeding and maintained high metabolic activity of the cardiac cells (Dar et al. 2002).

In earlier studies, it was difficult to electrospun the alginates as continuous and uniform nanofibers due to lack of chain entanglements caused by rigid and extended chain conformation. Electrospinning of alginates could successfully be done by using glycerol as a co-solvent with water in different volume ratio (Nie et al. 2008) and combining with water–soluble synthetic polymers such as polyethylene oxide (PEO) and poly (vinyl alcohol) (PVA). Alginate/PEO nanofibers showed good uniformity, integrity, cell adherence, cell viability and cellular compatibility (Narayan and Miqin 2007). The cell adherence of alginate scaffolds could be increased by incorporation of natural peptides such as fibronectin, laminin, nephronectin or immobilization of RGD (arginine-glycine-aspartic acid) peptides (Shachar et al. 2011).

Several researchers have also used alginate to create hybrid biomaterials for CTE. Bai et al. (2011) made alginate/collagen composite microbeads for EHT. The 3D porous composite scaffold formed from different ratio of alginate and chitosan by freeze drying method showed significant increase in ejection fraction, improved neo-vascularisation, attenuated fibrosis and proliferation of CMs (Ceccaldi et al. 2013). Mollar et al. developed hydrogel composites from methacrylated hyaluronic acid, alginate and gelatin (Moller et al. 2011). The composite of alginate/elastin and PEG showed large pore size (approx 60–75 μm) as compared to alginate and elastin composite at 20 °C (Chandy et al. 2003). The larger pore size membranes retained less water content and high degradation profile and hence, were suitable for CTE. Sapir et al. also created a functional cardiac patch from rat cardiac cells seeded on macroporous alginate scaffold incorporated with magnetically responsive nanoparticles (MNPs) (Sapir et al. 2014b).

### *18.3.5 Chitosan*

Chitosan, a natural cationic polysaccharide and linear polymer obtained from chitin through deacetylation process and linked via β, 1–4 linkage, has also been found suitable as for CTE (Hussain et al. 2013). The low mechanical strength of chitosan scaffolds can be overcome by formation of composite materials and creation of nanofibers (Albanna et al. 2012). Chitosan hydrogel, incorporated with glutathione, suppressed reactive oxygen species (ROS) and showed high survival of CMs (Li et al. 2013). Liu et al. (2012) reported that chitosan hydrogel can scavenge ROS molecules and improve the environment for cell engraftment, survival and homing for ischemic heart.

Chitosan is occasionally used as hybrid biomaterials along with other natural or synthetic biomaterials. For example, copolymer of chitosan and poly(ε-caprolactone) (PCL) provided better cell adhesion and mechanical strength for development of cardiac tissue construct than Chitosan or PCL alone (Chatzinikolaidou et al. 2014). The multilayer scaffold prepared from combination of chitosan, gelatin and PCL was also found to be suitable for CTE (Pok et al. 2013). Similarly, micro-patterned aligned collagen-chitosan hydrogel enhanced the success rate of beating CMs, oxygen transport and provided better cell connections with construct (Zhang et al. 2013). Baei et al. (2016) developed a thermo-sensitive conductive hydrogel with a highly porous network of interconnected pores by incorporating nanoscale electro-conductive gold nanoparticles into chitosan hydrogels.

### *18.3.6 Fibroin/Silk Fibroin*

Fiboin present as natural form in silkworms, moths, spiders, silverfish, beetles etc. is also a good source of natural biomaterial for CTE. Silk from mulberry *Bombyx mori* as well as non-mulberry *Antheraea mylitta* has been used for CTE (Patra et al. 2012). It is composed of heavy and light chain of core proteins called fibroins, which is coated by another protein called sericin—a glue, which help to bind silk fibroins together due to its sticky nature. It is a natural biopolymer with excellent mechanical strength, extensibility, elasticity and strain hardening. The strength-to-density ratio of silk is much higher than that of steel. The mechanical properties of silk, which include extreme elasticity (5–17 GPa), toughness, tensile strength (500–900 MPa), and compression resistance, depends on the hydrophobic region composed of repetitive amino acid sequences. Silk also has small diameter and occur as aligned fibers which increases the surface area to volume ratio. Properties such as slow degradation, high tensile strength, good oxygen and water permeability support better cell adhesion and growth, which made it one of the demanding biopolymer for CTE. In fact, scaffolds prepared from silk fibroin not only induced alignment of the cells but also helped in synthesis of titin, a protein critical to sarcomere assembly that provide elasticity to the muscles (Cutts et al. 2015).

The main limitation in the use of silk is its hypersensitivity reactions and adverse immunological effects due to non-mammalian origin. The immune response can be improved by removing sericin through a process called degumming (Altman et al. 2003). After degumming, the silk used to dissolve in appropriate solvent such as formic acid, calcium nitrate, calcium chloride, ionic liquids, hexafluoroisopropanol, lithium bromide (Kluge et al. 2008) to form scaffold. Silk fibrous has been used as films, hydrogel, 3-D macro or micro capsules, porous scaffolds, and electro-spun and wet-spun fibers. The porous silk-based matrix can be formed by salt leaching, gas foaming and freeze drying whereas fibrous scaffolds can be formed by electrospinning or microfluidic approaches. The porosity and mechanical properties of matrix formed can be controlled by concentration of silk fibroin and particle size of salt. Several authors have reported efficient seeding of CM into silk fibroin scaffolds to form contractile patches (Patra et al. 2012). Rahimi et al. developed a thermo-sensitive conductive hydrogel with a highly porous network of interconnected pores by incorporating nanoscale electro-conductive gold nanoparticles (Au-NPs) into chitosan hydrogels (Rahimi et al. 2014). Silk fibroin exhibited similar properties of cell attachment as fibronectin, a component of the natural matrix for CM, probably due to its RGD domains.

Composite of silk fibroin with chitosan or hyaluronic acid has also been used to generate cardiac patch from seeded MSCs, which promoted cell growth and cardiomyogenic differentiation (Yang et al. 2010) and, upon transplantation into rat heart model of MI, improved cardiac repair (Chi et al. 2013). Silk fibroin has also been used as a coating material for surface modification of synthetic biopolymers. Liang et al. immobilized silk fibroin on poly (ethylene terephthalate) (PET) film via plasma pre-treatment followed by dip coating, which gave high surface roughness and promoted MSCs culture (Liang et al. 2013).

### 18.3.7 Decellularized ECM

Decellularized ECM of heart (Singelyn et al. 2009), and other organs such as small intestinal submucosa (SIS) (Crapo and Wang 2010), and urinary bladder (Robinson et al. 2005) has also been incorporated into scaffolds for CTE. Decellularization of tissues by detergents such as SDS can remove cells without affecting the biochemical composition and mechanical properties of the ECM in terms of elasticity, strength and durability (Eitan et al. 2010). The viscoelasticity and strength of the decelluralized cardiac ECM was reported to be similar to those of native tissue, although its elasticity and apparent viscosity are higher (Bronshtein et al. 2013). Seeding and culturing of decelluralized cardiac tissue with MSCs partially restored the mechanical properties lost after decellularization (Bronshtein et al. 2013). Thus, decellularized ECM of heart are being extensively explored as a natural biomaterial for CTE and variable factors such efficient methods of ECM removal, homogeneous re-cellularization strategies and tailoring of ECM for enhancing bioactivity (Kc et al. 2019).

Decellularized ECM has also been used in combination with other natural and synthetic biomaterials. Stoppel et al. developed a silk-based scaffold containing decellularized ECM of heart (Stoppel et al. 2015). These silk-ECM composite scaffolds had tunable architectures, degradation rates, and mechanical properties and their subcutaneous implantation in rats resulted in remarkable endogenous cell infiltration and vascularization after 4 weeks in vivo. In vitro, silk-cECM scaffolds maintained both the atrial CMs and the human ESC-derived CMs (Stoppel et al. 2015).

Interestingly, xenogenic ECMs were also shown to be beneficial in promoting human CM proliferation (Sarig et al. 2015). Crapo and Wang (2010) combined rat CMs with pig SIS gel and cultured on porous elastomeric PCS scaffolds. They observed that cardiac tissue engineered from pig SIS gel had a more physiological rate of contraction than those produced on Matrigel™. However, the composition of ECM varies from batch to batch and chances of immunogenicity and zoonosis with xenogeneic sources of ECM cannot be ruled out.

### 18.3.8 Other Natural Biomaterials

Other natural biomaterials such as glycosaminoglycans, hyaluronic acid, Poly (3-hydroxyoctanoate) etc. have also been used in CTE. Glycosaminoglycans are the most abundant polysaccharide present in the body in the form of long unbranched disaccharide units, where units contain either of two such as N-acetylgalactosamine or N-acetylglucosamine and a uronic acid such as glucuronate or iduronate. Hyaluronic acid is a macromolecule made of linear glycosaminoglycans and has been reported to improve neovascularization of tissues and normalize the left ventricular function in infracted rat hearts (Ventura et al. 2007). Poly (3-hydroxyoctanoate), Poly

(3-hydroxyoctanoate) is a medium chain length polyhydroxyalkanoate produced by bacterial fermentation. Poly (3-hydroxyoctanoate) polymer was found to support cell viability, proliferation and adhesion and resembled cardiac tissues in terms of their mechanical properties (Bagdadi et al. 2016). In another study, Mussel-inspired conductive Ti2C cryogel was shown to promote the functional maturation of CMs (Ye et al. 2020). The cryogel was developed as 3D vessels-shape framework and when seeded with CMs, they remarkable aligned sarcomere with primitive intercalated disc between the CMs and showed synchronous beating. Transplanted of the Ti2C cryogel into MI rat model, could improve cardiac function (Ye et al. 2020).

More recently, Walker et al. (2019) engineered an adhesive and electroconductive cardiac patch by electrospinning of gelatin methacryloyl (GelMA) and conjugation of a choline-based bio-ionic liquid (Bio-IL). These patches strongly adhered to the myocardium by forming ionic bonds and did not require suturing. Culture of primary CMs and fibroblasts could form contractile tissue constructs and established gap junction with superior mechanical and electroconductivity. List of natural biopolymers and other commercially available biomaterials used in CTE is shown in Tables 18.1 and 18.2.

**Table 18.2** Overview of natural biomaterials used for cardiac tissue engineering

| Polymer | Co-polymer/ Crosslinking | Fabrication method | Cell type seeded | References |
|---|---|---|---|---|
| Collagen | Chitosan, Glycosaminoglycan, gelatin or Matrigel/ GA, EDC or EDAC | Electrospinning | MSCs, $CD133^{+}$ cells | Shi et al. (2011), Telemeco et al. (2005), Xiang et al. (2006) |
| Gelatin | Alginate, Chitosan-agarose, Fibrinogen, PGS | Electrospinning | MSCs | Balasubramanian et al. (2013), Bhat and Kumar (2012), Ravichandran et al. (2013a) |
| Fibrin | Thrombin | Hydrogel | CMs, EC, fibroblast | Black et al. (2009), Thomson et al. (2013) |
| Alginate | PEO, PVA, Elastin, and PEG | Electrospinning, Hydrogel | Fetal CMs | Moller et al. (2011), Nie et al. (2008) |
| Chitosan | PCL, Gelatin | Hydrogel | CMs | Chatzinikolaidou et al. (2014), Chiu et al. (2012) |
| Fibroin | Chitosan/HA | Films, Hydrogel, | MSCs, CMs | Rahimi et al. (2014), Yang et al. (2009) |
| Decellularized ECM | Silk | Porous aligned scaffold | MSCs, CMs | Bronshtein et al. (2013), Stoppel et al. (2015) |

## 18.4 Enhancing the Elasticity and Electrical Conductivity of Natural Biomaterials for CTE

Polymeric scaffolds for CTE must exhibit elasticity and high electrical conductivity to mimic the flexibility, electrical conductance, and contractility of native cardiac tissues (Ye et al. 2020). Many of biomaterials currently used in CTE lack of electrical conductivity and appropriate mechanical properties. This can be overcome by use of hybrid biomaterials combining two or more natural and synthetic biomaterials or combining with conductive nanomaterials. For examples, natural polymers such as chitosan have low elastic modulus and poor electrical conductivity, which could be improved by doping with carbon nanofibers to develop a highly conductive composite scaffold (Martins et al. 2014). Chitosan/carbon composite scaffolds had an elastic modulus of 28.1 $\pm$ 3.3 kPa, similar to that measured for rat myocardium, and excellent electrical properties, with a conductivity of 0.25 $\pm$ 0.09 S/m (Martins et al. 2014). Similarly, incorporation of carbon nanohorns into collagen was reported to promote the proliferation of rat CMs and inhibit the proliferation of cardiac fibroblasts (Wu et al. 2016).

Metallic nanoparticles such as Au-NPs have also been combined with natural polymers to develop thermosensitive and/or electro-conductive scaffolds (Min et al. 2018). Baei et al. (2016) developed a thermosensitive conductive hydrogel with a highly porous network of interconnected pores by incorporating nanoscale electro-conductive Au-NPs into chitosan hydrogels. AuNP-collagen matrix with localized nanoscale stiffness in the substrate activated the $\beta$1-integrin signalling, which mediates the activation of integrin-linked kinase and its downstream signal kinase for cardiac cell differentiation (Li et al. 2016). Sapir et al. (2014a) created a functional cardiac patch by incorporating magnetically responsive nanoparticles into macroporous alginate scaffold and applying external magnetic stimulation. In recent years, incorporation of reduced graphene oxide (rGO) into biomaterials has also attracted great attention. Shin et al. engineered cardiac constructs using rGO-incorporated gelatin methacryloyl (GelMA) hybrid hydrogels (Shin et al. 2016). The incorporation of rGO into the GelMA matrix significantly enhanced electrical conductivity and mechanical properties of material and exhibited better viability, proliferation, and maturation of CMs compared to those cultured on GelMA hydrogels. Similarly, Jiang et al. (2019) developed electroconductive porous scaffold of chitosan for CTE by blending with GO. Incorporation of GO into chitosan resulted in an electro-conductivity of 0.134 S/m, which is in the range of conductivity required for native heart tissue. Combination of natural biomaterials with synthetic biomaterials has also been tried. Bhaarathy et al. used copolymer of poly (l-lactic acid)-co-poly ($\varepsilon$-caprolactone) (PLACL), silk fibroin and Aloe Vera for fabricating biocomposite nanofibrous scaffolds for CTE (Bhaarathy et al. 2014).

## 18.5 Enhancing the Mechanical Properties of Natural Polymers for CTE

While natural biomaterials such as gelatine, chitosan etc. provide good hydrophilicity to support cell attachment, they are difficult to handle during surgical reconstruction of heart defects. It can be overcome by use of sandwiched scaffold wherein biomaterial of high tensile strength such as PCL is sandwiched in natural biomaterials such as gelatine-chitosan (Pok et al. 2013). The PCL core could provide surgical handling, suturability and high initial tensile strength, while the gelatin-chitosan scaffold allowed for cell attachment, with pore size and mechanical properties conducive to CMs migration and function. Similarly, Ravichandran et al. fabricated PGS/fibrinogen, core/shell fibers with core as elastomeric PGS that provided suitable mechanical properties comparable to that of native tissue and shell as fibrinogen to promote cell-biomaterial interactions (Ravichandran et al. 2013b). The PCL has high mechanical strength with slow degradation rate, therefore incorporation of hybrid composite with polydioxanone fibers by electrospinning improves mechanical properties compared to PCL alone for vascular smooth muscles generation (Pan et al. 2017). Hence, hybrid materials with tunable mechanical properties have attracted researcher's attention in the field of tissue engineering. The hybrid material of polylactide with aniline tetramer shows excellent ductility and enhanced modulus for the differentiation of myoblast to myotubes (Xie et al. 2015). In addition, crosslinking of hybrid materials with genipin represented lower elastic modulus suitable for soft tissue engineering (Garnica-Palafox and Sanchez-Arevalo 2016). The first reported hybrid scaffolds with the help of 3-D stereolithographic printing of chitosan and polyethylene glycol diacrylate were found to have $\sim$400 kPa elastic modulus with interconnected pores suitable for the repair of soft tissues (Morris et al. 2017).

## 18.6 Natural Biomaterials as Coating Materials

Natural biomaterials such as gelatine, collagen, Matrigel™, fibronectin, laminin, bioactive glass etc. have also been used commonly to bio-functionalize the synthetic biomaterials-derived scaffolds to enhance cell adhesion, cell-to-cell communication and cellular interaction with scaffold (Oliver et al. 2019). Synthetic RGD has also been used to mimic biochemical signals for cell attachment (Lambshead et al. 2018). Most cell types attach to the biomaterials in an RGD domain-dependent manner and therefore, synthetic RGD peptides were effecting interaction of CMs, osteocytes and other cells with biomaterials (Hoyos-Nogues et al. 2019). CMs are also responsive to synthetic fragments of fibronectin, vitronectin, and stromal-derived factor-1 (Liao et al. 2018). Thus, inclusions of these synthetic sequences into biomaterial formulation are expected to improve cell adhesion and play an important role in cell retention during dynamic cell seeding of scaffolds.

## 18.7 Types of Scaffolds for CTE from Natural Biomaterials

Generally, scaffolds are designed to be fibrous or porous in nature. Fibrous scaffolds mimic the nanoscale structure of the native ECM and their topology and porosity can be designed to modify the cell behaviour. They contain a large surface-to-volume ratio, high permeability and interconnecting pore structure, which allow efficient infiltration and proliferation of cells. On the other hand, porous scaffolds support host cell infiltration and vascularization, and provide a large surface area for cell adhesion, differentiation and proliferation, thus making them advantageous for growing functional 3D tissues.

### *18.7.1 Fibrous Scaffolds*

Fibrous scaffolds are generally created by electrospinning of polymer solutions. The process of electrospinning involves dispensing of a biopolymer out of a syringe needle and its exposure through a charged voltage gradient between the needle and a collector. The diameter, density, and alignment of electrospun fibers can be controlled by optimizing the viscosity and flow-rate of the polymer solution, the voltage gradient, and the topography of the collector (Zhou et al. 2019). Natural and hybrid biomaterials have been used for electrospinning (Pushp et al. 2017). In addition, composite materials comprised of layers of different polymers such as PLGA/gelatine can also be created by electrospinning for developing cardiac patches (Prabhakaran et al. 2011). Bio-functionalized fibrous scaffolds can also be electrospun by adding functional materials to the polymers before or after electrospinning. Kai et al. incorporated Polypyrrole (PPy) to PCL/gelatine solution before electrospinning to impart electroconductivity to the nanofibers.

Various modifications for creating specialized nanofibrous scaffolds have also been reported. Ravichandran et al. (2013b) fabricated core/shell fibers by co-axial electrospinning, with PGS as core material and fibrinogen as shell material. Among various modifications, the creation of aligned nanofibrous scaffolds is considered as the most important for CTE as they structurally mimic oriented ECM of the heart (Parrag et al. 2012). Gelatin-coated aligned PCL nanofibrous scaffold greatly promoted cell attachment and alignment because of biological components and ordered topography of scaffolds (Kai et al. 2011). In our study, we observed that PCL aligned nanofibers, coated with collagen I, promoted cell attachment and proliferation of MSCs cells. The MSCs attached, penetrated, and spread well on the surface of aligned nanofibrous scaffold than random nanofiber (Pushp et al. 2019). Similar results were also observed with iPSCs (Fig. 18.2). However, the biggest limitation of electrospinning is the inability to form thick scaffolds due to low fiber packaging density and entrapment of cells (Venkataraman et al. 2018).

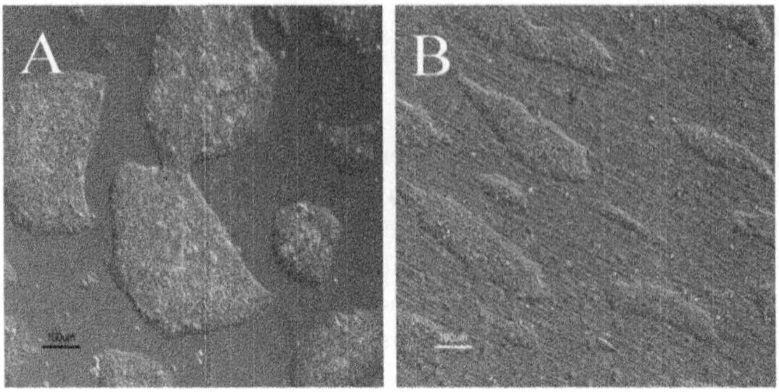


**Fig. 18.2** Growth of human induced pluripotent stem cells (iPSCs) on Matrigel coated (**A**) plates and aligned PCL nanofibers (**B**)

### *18.7.2 Porous Scaffolds/sponges*

Porous scaffolds or sponges can provide true 3D scaffolds with desired thickness. Highly porous scaffolds with interconnected pores can be created by lyophilizing the biomaterial solution (Petrauskaite et al. 2016), salt extraction (Caspi et al. 2007), selective laser sintering (Mota et al. 2015) or microfabrication (Engelmayr et al. 2008). Porous scaffolds for cardiac grafts have been most commonly created using collagen (Copes et al. 2019). Collagen composite with fibrin resulted into compressive stiffness of 18 and 21 kPa comparable to native myocardium for iPSCs-derived CMs whereas the beating CMs were appeared only on pure collagen groups (Kaiser et al. 2019). A major limitation of porous scaffolds is difficulty in uniform seeding of cells into the scaffold (Alaribe et al. 2016) during the cell culture. Cell seeding generally results in higher cell densities towards the surface and lower towards core of the scaffolds (Nikolova and Chavali 2019).

### *18.7.3 Hydrogels*

In another approach, cells can be suspended into non-porous, non-fibrous hydrogels, which are then cross-linked for complete encapsulation of the cells into construct. Hydrogels are cross-linked networks of polymers with high water content between polymer chains. They not only acts as scaffolds to provide physical support but also permit the suspended cells to migrate and assemble into contractile tissues to form spheroid-like micro-tissues (Fang and Eglen 2017). Hydrogels have been created using natural biomaterials such as fibrin (Myu Mai Ja et al. 2018), collagen (Hasan et al. 2015), and chitosan (Lu et al. 2009) etc. Composite hydrogels have also been created by combining natural and synthetic biomaterials such as PEGylated fibrin (Zhang et al. 2008). Cross-linking of polymers for hydrogel formation can be done by various methods such as temperature modulation, enzymatic activation, alteration of pH, or ionic cross-linkage. Unlike porous or fibrous scaffolds, the shape, stiffness and thickness of cell-encapsulated hydrogels can be easily controlled and therefore, they can be directly injected into the heart tissue in vivo as in situ construct (Ye and Black 2011). A major limitation in the use of hydrogels for CTE is its inherent heterogeneous nature which can influence mechanical properties of the construct and may result in graft failure due to improper mechanical and electrical integration into recipient myocardium.

### *18.7.4 Bio-Fabricated/Micro-Fabricated Scaffolds*

In recent years, need for microfabrication of scaffolds has been felt to mimic the natural structure of the heart. Mammalian heart has unique and intricate structural organization such as parallel aligned myocytes and myofibers which varies along the depth of the ventricular wall. These topological cues and structural anisotropy of heart are important for CTE and therefore, strategies such as nanopatterning of biomaterials, stamping of ECM proteins on scaffolds, 3D tissue printing, textile-templated scaffolds etc. have been attempted as novel means to mimic native ECM and enhance functionality of cardiac tissue construct (Zhang et al. 2019). Thomson et al. microtemplated fibrin scaffolds with uniform architecture of parallel microchannels (60 μm), surrounded by an interconnected microporous network of pores (27 μm) and mechanical stiffness (70–90 kPa) comparable to native cardiac tissues (Thomson et al. 2013). Tri-culture of CMs, EC and fibroblasts on microtemplated fibrin scaffolds mimicked native cardiac tissue in structure and cellular composition.

Natural biomaterials have also been used as bioink to develop scaffolds of pre-defined and organized architecture by 3D tissue printing or prototyping. The 3D bioprinting allows the deposition of cells, biomaterials and signalling molecules into precisely organized geometrics and therefore, helps in biofabrication of tissue constructs similar to those found in the native heart (Alonzo et al. 2019). A number of bioprinting technology including extrusion, inkjet, laser-assisted and stereolithography with natural as well as synthetic bioinks have been developed and applied to the CTE. Gaetani et al. (2012) successfully used tissue printing technology to fabricate scaffold for culturing human cardiac progenitor cells. In yet another methodology, using knitted conventional textiles made of cotton or polyester yarns as template targets. Hong et al. (2020) used digital lighting processing printer for 3D printing of silk fibroin natural polymer with gycidyl-methacrylate as bioink. Silk fibroin was also combined with Collagen I for 3D printing and development of cardiac patches after seeding with MSC (Sanz-Fraile et al. 2020). Mixture of different cells with hydrogels have been printed for functional vascularized patches with the help 3D printing that meets immunological and anatomical properties of native heart tissue (Noor et al. 2019).

## 18.8 Conclusion and Future Direction

Physico-chemical properties such as biodegradability, biocompatibility, porosity, cell adhesion property, cytocompactibility etc. make natural biomaterial an excellent choice for fabrication of scaffolds for CTE. However, natural biomaterials may require to be tuned up for mechanical integrity and electrical coupling for mitigating native cardiac tissue. Such tuning requires generation of hybrid or composite biomaterials with nanomaterials or synthetic polymers. Conversely, natural polymers are commonly used for biofunctionalization of synthetic biomaterials and improving their hydrophilicity, cell adhesion property and biocompatibility. Newer scaffold fabrication techniques such as 3D printing, prototyping and microfabrication technology are also being developed which allows generation of aligned and anisotropic heart tissue.

Since the first description of engineered heart tissue nearly two decades ago, various research groups have also developed methods of cultivating contractile 3D cardiac grafts. These cardiac grafts displayed functional and morphological properties of native tissue and upon grafting, they integrated into native tissue to improve the cardiac function. However, despite progressive improvements, several critical challenges such as identification of the best cell source and biofabrication technique for generation of 3D vascularised tissue constructs exist. Furthermore, automation and standardization are also needed for translation of the technology from laboratory animals to human.

## References

Ahn JI, Kuffova L, Merrett K, Mitra D, Forrester JV, Li F, Griffith M (2013) Crosslinked collagen hydrogels as corneal implants: effects of sterically bulky vs. non-bulky carbodiimides as crosslinkers. Acta Biomater 9:7796–7805

Alaribe FN, Manoto SL, Motaung SCKM (2016) Scaffolds from biomaterials: advantages and limitations in bone and tissue engineering. Biologia 71:353–366

Albanna MZ, Bou-Akl TH, Walters HL 3rd, Matthew HW (2012) Improving the mechanical properties of chitosan-based heart valve scaffolds using chitosan fibers. J Mech Behav Biomed Mater 5:171–180

Alonzo M, AnilKumar S, Roman B, Tasnim N, Joddar B (2019) 3D Bioprinting of cardiac tissue and cardiac stem cell therapy. Transl Res 211:64–83

Altman GH, Diaz F, Jakuba C, Calabro T, Horan RL, Chen J, Lu H, Richmond J, Kaplan DL (2003) Silk-based biomaterials. Biomaterials 24:401–416

Augst AD, Kong HJ, Mooney DJ (2006) Alginate hydrogels as biomaterials. Macromol Biosci 6:623–633

Baei P, Jalili-Firoozinezhad S, Rajabi-Zeleti S, Tafazzoli-Shadpour M, Baharvand H, Aghdami N (2016) Electrically conductive gold nanoparticle-chitosan thermosensitive hydrogels for cardiac tissue engineering. Mater Sci Eng C Mater Biol Appl 63:131–141

Bagdadi AV, Safari M, Dubey P, Basnett P, Sofokleous P, Humphrey E, Locke I, Edirisinghe M, Terracciano C, Boccaccini AR, Knowles JC, Harding SE, Roy I (2016) Poly (3-hydroxyoctanoate), a promising new material for cardiac tissue engineering. J Tissue Eng Regen Med

Bagheri-Hosseinabadi Z, Mesbah-Namin SA, Salehinejad P, Seyedi F (2018) Fibrin scaffold could promote survival of the human adipose-derived stem cells during differentiation into cardiomyocyte-like cells. Cell Tissue Res 372:571–589

Bai XP, Zheng HX, Fang R, Wang TR, Hou XL, Li Y, Chen XB, Tian WM (2011) Fabrication of engineered heart tissue grafts from alginate/collagen barium composite microbeads. Biomed Mater 6:045002

Balasubramanian P, Prabhakaran MP, Kai D, Ramakrishna S (2013) Human cardiomyocyte interaction with electrospun fibrinogen/gelatin nanofibers for myocardial regeneration. J Biomater Sci Polym Ed 24:1660–1675

Bhaarathy V, Venugopal J, Gandhimathi C, Ponpandian N, Mangalaraj D, Ramakrishna S (2014) Biologically improved nanofibrous scaffolds for cardiac tissue engineering. Mater Sci Eng C Mater Biol Appl 44:268–277

Bhat S, Kumar A (2012) Cell proliferation on three-dimensional chitosan-agarose-gelatin cryogel scaffolds for tissue engineering applications. J Biosci Bioeng 114:663–670

Black LD 3rd, Meyers JD, Weinbaum JS, Shvelidze YA, Tranquillo RT (2009) Cell-induced alignment augments twitch force in fibrin gel-based engineered myocardium via gap junction modification. Tissue Eng Part A 15:3099–3108

Bronshtein T, Au-Yeung GC, Sarig U, Nguyen EB, Mhaisalkar PS, Boey FY, Venkatraman SS, Machluf M (2013) A mathematical model for analyzing the elasticity, viscosity, and failure of soft tissue: comparison of native and decellularized porcine cardiac extracellular matrix for tissue engineering. Tissue Eng Part C Methods 19:620–630

Caspi O, Lesman A, Basevitch Y, Gepstein A, Arbel G, Habib IH, Gepstein L, Levenberg S (2007) Tissue engineering of vascularized cardiac muscle from human embryonic stem cells. Circ Res 100:263–272

Ceccaldi C, Bushkalova R, Alfarano C, Lairez O, Calise D, Bourin P, Frugier C, Rouzaud-Laborde C, Cussac D, Parini A, Sallerin B, Fullana SG (2013) Evaluation of polyelectrolyte complex-based scaffolds for mesenchymal stem cell therapy in cardiac ischemia treatment. Acta Biomater

Ceccaldi C, Fullana SG, Alfarano C, Lairez O, Calise D, Cussac D, Parini A, Sallerin B (2012) Alginate scaffolds for mesenchymal stem cell cardiac therapy: influence of alginate composition. Cell Transplant 21:1969–1984

Chandy T, Rao GH, Wilson RF, Das GS (2003) The development of porous alginate/elastin/PEG composite matrix for cardiovascular engineering. J Biomater Appl 17:287–301

Chatzinikolaidou M, Kaliva M, Batsali A, Pontikoglou C, Vamvakaki M (2014) Wharton's jelly Mesenchymal stem cell response on chitosan-graft-poly (epsilon-caprolactone) copolymer for myocardium tissue engineering. Curr Pharm Des 20:2030–2039

Chi NH, Yang MC, Chung TW, Chou NK, Wang SS (2013) Cardiac repair using chitosan-hyaluronan/silk fibroin patches in a rat heart model with myocardial infarction. Carbohydr Polym 92:591–597

Chiu LL, Reis LA, Momen A, Radisic M (2012) Controlled release of thymosin beta4 from injected collagen-chitosan hydrogels promotes angiogenesis and prevents tissue loss after myocardial infarction. Regen Med 7:523–533

Choe G, Park J, Jo H, Kim YS, Ahn Y, Lee JY (2019) Studies on the effects of microencapsulated human mesenchymal stem cells in RGD-modified alginate on cardiomyocytes under oxidative stress conditions using in vitro biomimetic co-culture system. Int J Biol Macromol 123:512–520

Christman KL, Fok HH, Sievers RE, Fang Q, Lee RJ (2004) Fibrin glue alone and skeletal myoblasts in a fibrin scaffold preserve cardiac function after myocardial infarction. Tissue Eng 10:403–409

Copes F, Pien N, Van Vlierberghe S, Boccafoschi F, Mantovani D (2019) Collagen-based tissue engineering strategies for vascular medicine. Front Bioeng Biotechnol 7:166

Crapo PM, Wang Y (2010) Small intestinal submucosa gel as a potential scaffolding material for cardiac tissue engineering. Acta Biomater 6:2091–2096

Cutts J, Nikkhah M, Brafman DA (2015) Biomaterial approaches for stem cell-based myocardial tissue engineering. Biomark Insights 10:77–90

Dar A, Shachar M, Leor J, Cohen S (2002) Optimization of cardiac cell seeding and distribution in 3D porous alginate scaffolds. Biotechnol Bioeng 80:305–312

Drury JL, Dennis RG, Mooney DJ (2004) The tensile properties of alginate hydrogels. Biomaterials 25:3187–3199

Duan X, Sheardown H (2005) Crosslinking of collagen with dendrimers. J Biomed Mater Res A 75:510–518

Eitan Y, Sarig U, Dahan N, Machluf M (2010) Acellular cardiac extracellular matrix as a scaffold for tissue engineering: in vitro cell support, remodeling, and biocompatibility. Tissue Eng Part C Methods 16:671–683

Engelmayr GC Jr, Cheng M, Bettinger CJ, Borenstein JT, Langer R, Freed LE (2008) Accordion-like honeycombs for tissue engineering of cardiac anisotropy. Nat Mater 7:1003–1010

Fang Y, Eglen RM (2017) Three-dimensional cell cultures in drug discovery and development. SLAS Discov 22:456–472

Flanagan TC, Wilkins B, Black A, Jockenhoevel S, Smith TJ, Pandit AS (2006) A collagen-glycosaminoglycan co-culture model for heart valve tissue engineering applications. Biomaterials 27:2233–2246

Gaetani R, Doevendans PA, Metz CH, Alblas J, Messina E, Giacomello A, Sluijter JP (2012) Cardiac tissue engineering using tissue printing technology and human cardiac progenitor cells. Biomaterials 33:1782–1790

Garnica-Palafox IM, Sanchez-Arevalo FM (2016) Influence of natural and synthetic crosslinking reagents on the structural and mechanical properties of chitosan-based hybrid hydrogels. Carbohydr Polym 151:1073–1081

Grassl ED, Oegema TR, Tranquillo RT (2003) A fibrin-based arterial media equivalent. J Biomed Mater Res A 66:550–561

Guo XM, Zhao YS, Chang HX, Wang CY, E LL, Zhang XA, Duan CM, Dong LZ, Jiang H, Li J, Song Y, Yang XJ (2006) Creation of engineered cardiac tissue in vitro from mouse embryonic stem cells. Circulation 113:2229–2237

Hadisi Z, Farokhi M, Bakhsheshi-Rad HR, Jahanshahi M, Hasanpour S, Pagan E, Dolatshahi-Pirouz A, Zhang YS, Kundu SC, Akbari M (2020) Hyaluronic acid (HA)-based silk fibroin/zinc oxide core-shell electrospun dressing for burn wound management. Macromol Biosci 20:e1900328

Hamdi H, Furuta A, Bellamy V, Bel A, Puymirat E, Peyrard S, Agbulut O, Menasche P (2009) Cell delivery: intramyocardial injections or epicardial deposition? head-to-head comparison. Ann Thorac Surg 87:1196–1203

Haraguchi Y, Shimizu T, Matsuura K, Sekine H, Tanaka N, Tadakuma K, Yamato M, Kaneko M, Okano T (2014) Cell sheet technology for cardiac tissue engineering. Methods Mol Biol 1181:139–155

Haraguchi Y, Shimizu T, Yamato M, Okano T (2011) Regenerative therapies using cell sheet-based tissue engineering for cardiac disease. Cardiol Res Pract 2011:845170

Hasan A, Khattab A, Islam MA, Hweij KA, Zeitouny J, Waters R, Sayegh M, Hossain MM, Paul A (2015) Injectable hydrogels for cardiac tissue repair after myocardial infarction. Adv Sci (Weinh) 2:1500122

He S, Song H, Wu J, Li SH, Weisel RD, Sung HW, Li J, Li RK (2018) Preservation of conductive propagation after surgical repair of cardiac defects with a bio-engineered conductive patch. J Heart Lung Transplant 37:912–924

Hong H, Seo YB, Kim DY, Lee JS, Lee YJ, Lee H, Ajiteru O, Sultan MT, Lee OJ, Kim SH, Park CH (2020) Digital light processing 3D printed silk fibroin hydrogel for cartilage tissue engineering. Biomaterials 232:119679

Hosoyama K, Ahumada M, McTiernan CD, Davis DR, Variola F, Ruel M, Liang W, Suuronen EJ, Alarcon EI (2018) Nanoengineered electroconductive collagen-based cardiac patch for infarcted myocardium repair. ACS Appl Mater Interf 10:44668–44677

Hoyos-Nogues M, Falgueras-Batlle E, Ginebra MP, Manero JM, Gil J, Mas-Moruno C (2019) A dual molecular biointerface combining RGD and KRSR sequences improves osteoblastic functions by synergizing integrin and cell-membrane proteoglycan binding. Int J Mol Sci 20
Hussain A, Collins G, Yip D, Cho CH (2013) Functional 3-D cardiac co-culture model using bioactive chitosan nanofiber scaffolds. Biotechnol Bioeng 110:637–647
Hwang JH, Gupta MK, Park CK, Kim YB, Lee HT (2015) Establishment of major histocompatibility complex homozygous gnotobiotic miniature swine colony for xenotransplantation. Anim Sci J 86:468–475
Isenberg BC, Tsuda Y, Williams C, Shimizu T, Yamato M, Okano T, Wong JY (2008) A thermoresponsive, microtextured substrate for cell sheet engineering with defined structural organization. Biomaterials 29:2565–2572
Jawad H, Lyon AR, Harding SE, Ali NN, Boccaccini AR (2008) Myocardial tissue engineering. Br Med Bull 87:31–47
Jiang L, Chen D, Wang Z, Zhang Z, Xia Y, Xue H, Liu Y (2019) Preparation of an electrically conductive graphene oxide/chitosan scaffold for cardiac tissue engineering. Appl Biochem Biotechnol 188:952–964
Joshi J, Brennan D, Beachley V, Kothapalli CR (2018) Cardiomyogenic differentiation of human bone marrow-derived mesenchymal stem cell spheroids within electrospun collagen nanofiber mats. J Biomed Mater Res A 106:3303–3312
Kai D, Prabhakaran MP, Jin G, Ramakrishna S (2011) Guided orientation of cardiomyocytes on electrospun aligned nanofibers for cardiac tissue engineering. J Biomed Mater Res B Appl Biomater 98:379–386
Kaiser NJ, Kant RJ, Minor AJ, Coulombe KLK (2019) Optimizing blended collagen-fibrin hydrogels for cardiac tissue engineering with human iPSC-derived cardiomyocytes. ACS Biomater Sci Eng 5:887–899
Kang K, Chuai JB, Xie BD, Li JZ, Qu H, Wu H, Fang SH, Cui JJ, Xiu LL, Han JC, Cao TH, Leng XP, Tian H, Li RK, Jiang SL (2020) Mesenchymal stromal cells from patients with cyanotic congenital heart disease are optimal candidate for cardiac tissue engineering. Biomaterials 230:119574
Kc P, Hong Y, Zhang G (2019) Cardiac tissue-derived extracellular matrix scaffolds for myocardial repair: advantages and challenges. Regen Biomater 6:185–199
Kharaziha M, Nikkhah M, Shin SR, Annabi N, Masoumi N, Gaharwar AK, Camci-Unal G, Khademhosseini A (2013) PGS: Gelatin nanofibrous scaffolds with tunable mechanical and structural properties for engineering cardiac tissues. Biomaterials 34:6355–6366
Kleinman HK, Martin GR (2005) Matrigel: basement membrane matrix with biological activity. Semin Cancer Biol 15:378–386
Kluge JA, Rabotyagova O, Leisk GG, Kaplan DL (2008) Spider silks and their applications. Trends Biotechnol 26:244–251
Kuo CK, Ma PX (2001) Ionically crosslinked alginate hydrogels as scaffolds for tissue engineering: part 1 structure, gelation rate and mechanical properties. Biomaterials 22:511–521
Lambshead JW, Meagher L, Goodwin J, Labonne T, Ng E, Elefanty A, Stanley E, O'Brien CM, Laslett AL (2018) Long-Term maintenance of human pluripotent stem cells on cRGDfK-presenting synthetic surfaces. Sci Rep 8:701
Lee KY, Jeong L, Kang YO, Lee SJ, Park WH (2009) Electrospinning of polysaccharides for regenerative medicine. Adv Drug Deliv Rev 61:1020–1032
Leor J, Aboulafia-Etzion S, Dar A, Shapiro L, Barbash IM, Battler A, Granot Y, Cohen S (2000) Bioengineered cardiac grafts: a new approach to repair the infarcted myocardium? Circulation 102:III56–61
Li J, Shu Y, Hao T, Wang Y, Qian Y, Duan C, Sun H, Lin Q, Wang C (2013) A chitosan-glutathione based injectable hydrogel for suppression of oxidative stress damage in cardiomyocytes. Biomaterials 34:9071–9081

Li Y, Shi X, Tian L, Sun H, Wu Y, Li X, Li J, Wei Y, Han X, Zhang J, Jia X, Bai R, Jing L, Ding P, Liu H, Han D (2016) AuNP-Collagen matrix with localized stiffness for cardiac-tissue engineering: enhancing the assembly of intercalated discs by beta1-integrin-mediated signaling. Adv Mater 28:10230–10235

Liang M, Yao J, Chen X, Huang L, Shao Z (2013) Silk fibroin immobilization on poly(ethylene terephthalate) films: comparison of two surface modification methods and their effect on mesenchymal stem cells culture. Mater Sci Eng C Mater Biol Appl 33:1409–1416

Liao YX, Zhang ZP, Zhao J, Liu JP (2018) Effects of fibronectin 1 on cell proliferation, senescence and apoptosis of human glioma cells through the PI3K/AKT signaling pathway. Cell Physiol Biochem 48:1382–1396

Liu Z, Wang H, Wang Y, Lin Q, Yao A, Cao F, Li D, Zhou J, Duan C, Du Z, Wang C (2012) The influence of chitosan hydrogel on stem cell engraftment, survival and homing in the ischemic myocardial microenvironment. Biomaterials 33:3093–3106

Lu WN, Lu SH, Wang HB, Li DX, Duan CM, Liu ZQ, Hao T, He WJ, Xu B, Fu Q, Song YC, Xie XH, Wang CY (2009) Functional improvement of infarcted heart by co-injection of embryonic stem cells with temperature-responsive chitosan hydrogel. Tissue Eng Part A 15:1437–1447

Madhavan K, Belchenko D, Motta A, Tan W (2010) Evaluation of composition and crosslinking effects on collagen-based composite constructs. Acta Biomater 6:1413–1422

Majidi SS, Slemming-Adamsen P, Hanif M, Zhang Z, Wang Z, Chen M (2018) Wet electrospun alginate/gelatin hydrogel nanofibers for 3D cell culture. Int J Biol Macromol 118:1648–1654

Martins AM, Eng G, Caridade SG, Mano JF, Reis RL, Vunjak-Novakovic G (2014) Electrically conductive chitosan/carbon scaffolds for cardiac tissue engineering. Biomacromol 15:635–643

Matsuura K, Utoh R, Nagase K, Okano T (2014) Cell sheet approach for tissue engineering and regenerative medicine. J Control Release 190:228–239

McKegney M, Taggart I, Grant MH (2001) The influence of crosslinking agents and diamines on the pore size, morphology and the biological stability of collagen sponges and their effect on cell penetration through the sponge matrix. J Mater Sci Mater Med 12:833–844

Min JH, Patel M, Koh WG (2018) Incorporation of conductive materials into hydrogels for tissue engineering applications. Polymers (Basel) 10

Moller L, Krause A, Dahlmann J, Gruh I, Kirschning A, Drager G (2011) Preparation and evaluation of hydrogel-composites from methacrylated hyaluronic acid, alginate, and gelatin for tissue engineering. Int J Artif Organs 34:93–102

Morris VB, Nimbalkar S, Younesi M, McClellan P, Akkus O (2017) Mechanical properties, cytocompatibility and manufacturability of chitosan: PEGDA hybrid-gel scaffolds by stereolithography. Ann Biomed Eng 45:286–296

Mota C, Puppi D, Chiellini F, Chiellini E (2015) Additive manufacturing techniques for the production of tissue engineering constructs. J Tissue Eng Regen Med 9:174–190

Myu Mai Ja KP, Lim KP, Chen A, Ting S, Li SQ, Tee N, Ramachandra C, Mehta A, Wong P, Oh S, Shim W (2018) Construction of a vascularized hydrogel for cardiac tissue formation in a porcine model. J Tissue Eng Regen Med 12:e2029–e2038

Narayan B, Miqin Z (2007) Controlled synthesis and structural stability of alginate-based nanofibers. Nanotechnology 18:455601

Nie H, He A, Zheng J, Xu S, Li J, Han CC (2008) Effects of chain conformation and entanglement on the electrospinning of pure alginate. Biomacromol 9:1362–1365

Nikolova MP, Chavali MS (2019) Recent advances in biomaterials for 3D scaffolds: a review. Bioact Mater 4:271–292

Noor N, Shapira A, Edri R, Gal I, Wertheim L, Dvir T (2019) 3D printing of personalized thick and perfusable cardiac patches and hearts. Adv Sci (Weinh) 6:1900344

Oliver JN, Su Y, Lu X, Kuo PH, Du J, Zhu D, Hoyos-Nogues MA-O, Falgueras-Batlle E, Ginebra MA-O, Manero JM, Gil J, Mas-Moruno CA-O, Yang X, Li Y, He W, Huang Q, Zhang R, Feng Q, Liao YX, Fau-Zhang Z-P, Zhang ZP, Fau-Zhao J, Zhao J, Fau-Liu J-P, Liu JP, Lambshead JW, Meagher L, Goodwin J, Labonne T, Ng E, Elefanty A, Stanley E, O'Brien CM, Laslett ALA-O, Kapp TG, Rechenmacher F, Neubauer S, Maltsev OV, Cavalcanti-Adam

EA, Zarka R, Reuning U, Notni J, Wester HJ, Mas-Moruno C, Spatz J, Geiger B, Kessler H (2019) Bioactive glass coatings on metallic implants for biomedical applications; A dual molecular biointerface combining RGD and KRSR sequences improves osteoblastic functions by synergizing integrin and cell-membrane proteoglycan binding. LID—E1429 [pii] LID. https://doi.org/10.3390/ijms20061429; Hydroxyapatite/collagen coating on PLGA electrospun fibers for osteogenic differentiation of bone marrow mesenchymal stem cells; Effects of fibronectin 1 on cell proliferation, senescence and apoptosis of human glioma cells through the PI3K/AKT signaling pathway; Long-term maintenance of human pluripotent stem cells on cRGDfK-presenting synthetic surfaces; A comprehensive evaluation of the activity and selectivity profile of ligands for RGD-binding integrins

Ozlu B, Ergin M, Budak S, Tunali S, Yildirim N, Erisken C (2019) A bioartificial rat heart tissue: perfusion decellularization and characterization. Int J Artif Organs 42:757–764

Pan Y, Zhou X, Wei Y, Zhang Q, Wang T, Zhu M, Li W, Huang R, Liu R, Chen J, Fan G, Wang K, Kong D, Zhao Q (2017) Small-diameter hybrid vascular grafts composed of polycaprolactone and polydioxanone fibers. Sci Rep 7:3615

Parrag IC, Zandstra PW, Woodhouse KA (2012) Fiber alignment and coculture with fibroblasts improves the differentiated phenotype of murine embryonic stem cell-derived cardiomyocytes for cardiac tissue engineering. Biotechnol Bioeng 109:813–822

Patra C, Talukdar S, Novoyatleva T, Velagala SR, Muhlfeld C, Kundu B, Kundu SC, Engel FB (2012) Silk protein fibroin from Antheraea mylitta for cardiac tissue engineering. Biomaterials 33:2673–2680

Petrauskaite O, Juodzbalys G, Viskelis P, Liesiene J (2016) Control of the porous structure of cellulose-based tissue engineering scaffolds by means of iyophilization. Cellul Chem Technol 50:23–30

Pok S, Myers JD, Madihally SV, Jacot JG (2013) A multilayered scaffold of a chitosan and gelatin hydrogel supported by a PCL core for cardiac tissue engineering. Acta Biomater 9:5630–5642

Prabhakaran MP, Kai D, Ghasemi-Mobarakeh L, Ramakrishna S (2011) Electrospun biocomposite nanofibrous patch for cardiac tissue engineering. Biomed Mater 6:055001

Pushp P, Ferreira FC, Cabral JMS, Gupta MK (2017) Improved survival of cardiac cells on surface modified electrospun nanofibers. Polym Sci Ser A 59:515–523

Pushp P, Sahoo B, Ferreira FC, Sampaio Cabral JM, Fernandes-Platzgummer A, Gupta MK (2019) Functional comparison of beating cardiomyocytes differentiated from umbilical cord-derived mesenchymal/stromal stem cells and human foreskin-derived induced pluripotent stem cells

Qu H, Xie BD, Wu J, Lv B, Chuai JB, Li JZ, Cai J, Wu H, Jiang SL, Leng XP, Kang K (2018) Improved left ventricular aneurysm repair with cell- and cytokine-seeded collagen patches. Stem Cells Int 2018:4717802

Rahimi M, Mohseni-Kouchesfehani H, Zarnani AH, Mobini S, Nikoo S, Kazemnejad S (2014) Evaluation of menstrual blood stem cells seeded in biocompatible Bombyx mori silk fibroin scaffold for cardiac tissue engineering. J Biomater Appl 29:199–208

Ravichandran R, Seitz V, Reddy Venugopal J, Sridhar R, Sundarrajan S, Mukherjee S, Wintermantel E, Ramakrishna S (2013) Mimicking native extracellular matrix with phytic acid-crosslinked protein nanofibers for cardiac tissue engineering. Macromol Biosci 13:366–375

Ravichandran R, Venugopal JR, Sundarrajan S, Mukherjee S, Sridhar R, Ramakrishna S (2013) Expression of cardiac proteins in neonatal cardiomyocytes on PGS/fibrinogen core/shell substrate for Cardiac tissue engineering. Int J Cardiol 167:1461–1468

Robinson KA, Li J, Mathison M, Redkar A, Cui J, Chronos NA, Matheny RG, Badylak SF (2005) Extracellular matrix scaffold for cardiac repair. Circulation 112:I135-143

Rosellini E, Cristallini C, Barbani N, Vozzi G, Giusti P (2009) Preparation and characterization of alginate/gelatin blend films for cardiac tissue engineering. J Biomed Mater Res A 91:447–453

Saito H, Murabayashi S, Mitamura Y, Taguchi T (2008) Characterization of alkali-treated collagen gels prepared by different crosslinkers. J Mater Sci Mater Med 19:1297–1305

Sanz-Fraile H, Amoros S, Mendizabal I, Galvez-Monton C, Prat-Vidal C, Bayes-Genis A, Navajas D, Farre R, Otero J (2020) Silk-reinforced collagen hydrogels with raised multiscale stiffness for mesenchymal cells 3D culture. Tissue Eng Part A 26:358–370

Sapir Y, Polyak B, Cohen S (2014) Cardiac tissue engineering in magnetically actuated scaffolds. Nanotechnology 25:014009

Sapir Y, Ruvinov E, Polyak B, Cohen S (2014) Magnetically actuated alginate scaffold: a novel platform for promoting tissue organization and vascularization. Methods Mol Biol 1181:83–95

Sarig U, Nguyen EB, Wang Y, Ting S, Bronshtein T, Sarig H, Dahan N, Gvirtz M, Reuveny S, Oh SK, Scheper T, Boey YC, Venkatraman SS, Machluf M (2015) Pushing the envelope in tissue engineering: ex vivo production of thick vascularized cardiac extracellular matrix constructs. Tissue Eng Part A 21:1507–1519

Shachar M, Tsur-Gang O, Dvir T, Leor J, Cohen S (2011) The effect of immobilized RGD peptide in alginate scaffolds on cardiac tissue engineering. Acta Biomater 7:152–162

Shi C, Li Q, Zhao Y, Chen W, Chen B, Xiao Z, Lin H, Nie L, Wang D, Dai J (2011) Stem-cell-capturing collagen scaffold promotes cardiac tissue regeneration. Biomaterials 32:2508–2515

Shin SR, Zihlmann C, Akbari M, Assawes P, Cheung L, Zhang K, Manoharan V, Zhang YS, Yuksekkaya M, Wan KT, Nikkhah M, Dokmeci MR, Tang XS, Khademhosseini A (2016) Reduced graphene oxide-GelMA hybrid hydrogels as scaffolds for cardiac tissue engineering. Small 12:3677–3689

Singelyn JM, DeQuach JA, Seif-Naraghi SB, Littlefield RB, Schup-Magoffin PJ, Christman KL (2009) Naturally derived myocardial matrix as an injectable scaffold for cardiac tissue engineering. Biomaterials 30:5409–5416

Stevens KR, Pabon L, Muskheli V, Murry CE (2009) Scaffold-free human cardiac tissue patch created from embryonic stem cells. Tissue Eng Part A 15:1211–1222

Stoppel WL, Hu D, Domian IJ, Kaplan DL, Black LD 3rd (2015) Anisotropic silk biomaterials containing cardiac extracellular matrix for cardiac tissue engineering. Biomed Mater 10:034105

Sun Q, Silva EA, Wang A, Fritton JC, Mooney DJ, Schaffler MB, Grossman PM, Rajagopalan S (2010) Sustained release of multiple growth factors from injectable polymeric system as a novel therapeutic approach towards angiogenesis. Pharm Res 27:264–271

Telemeco TA, Ayres C, Bowlin GL, Wnek GE, Boland ED, Cohen N, Baumgarten CM, Mathews J, Simpson DG (2005) Regulation of cellular infiltration into tissue engineering scaffolds composed of submicron diameter fibrils produced by electrospinning. Acta Biomater 1:377–385

Thomson KS, Korte FS, Giachelli CM, Ratner BD, Regnier M, Scatena M (2013) Prevascularized microtemplated fibrin scaffolds for cardiac tissue engineering applications. Tissue Eng Part A 19:967–977

Tijore A, Irvine SA, Sarig U, Mhaisalkar P, Baisane V, Venkatraman S (2018) Contact guidance for cardiac tissue engineering using 3D bioprinted gelatin patterned hydrogel. Biofabrication 10:025003

Venkataraman M, Mishra R, Militky J, Xiong X, Marek J, Yao J, Zhu G (2018) Electrospun nanofibrous membranes embedded with aerogel for advanced thermal and transport properties. Polym Adv Technol 29:2583–2592

Ventura C, Cantoni S, Bianchi F, Lionetti V, Cavallini C, Scarlata I, Foroni L, Maioli M, Bonsi L, Alviano F, Fossati V, Bagnara GP, Pasquinelli G, Recchia FA, Perbellini A (2007) Hyaluronan mixed esters of butyric and retinoic Acid drive cardiac and endothelial fate in term placenta human mesenchymal stem cells and enhance cardiac repair in infarcted rat hearts. J Biol Chem 282:14243–14252

Walker BW, Lara RP, Yu CH, Sani ES, Kimball W, Joyce S, Annabi N (2019) Engineering a naturally-derived adhesive and conductive cardiopatch. Biomaterials 207:89–101

Wang B, Borazjani A, Tahai M, Curry AL, Simionescu DT, Guan J, To F, Elder SH, Liao J (2010) Fabrication of cardiac patch with decellularized porcine myocardial scaffold and bone marrow mononuclear cells. J Biomed Mater Res A 94:1100–1110

Wang B, Wang G, To F, Butler JR, Claude A, McLaughlin RM, Williams LN, de Jongh Curry AL, Liao J (2013) Myocardial scaffold-based cardiac tissue engineering: application of coordinated mechanical and electrical stimulations. Langmuir 29:11109–11117
Wang L, Rao RR, Stegemann JP (2013) Delivery of mesenchymal stem cells in chitosan/collagen microbeads for orthopedic tissue repair. Cells Tissues Organs 197:333–343
Wu C, Yu S, Lin S (2014) Graphene modified electrospun poly (vinyl alcohol) nanofibrous membranes for glucose oxidase immobilization. Express Polym Lett 8
Wu Y, Shi X, Li Y, Tian L, Bai R, Wei Y, Han D, Liu H, Xu J (2016) Carbon nanohorns promote maturation of neonatal rat ventricular myocytes and inhibit proliferation of cardiac fibroblasts: a promising scaffold for cardiac tissue engineering. Nanoscale Res Lett 11:284
Xiang Z, Liao R, Kelly MS, Spector M (2006) Collagen-GAG scaffolds grafted onto myocardial infarcts in a rat model: a delivery vehicle for mesenchymal stem cells. Tissue Eng 12:2467–2478
Xie M, Wang L, Guo B, Wang Z, Chen YE, Ma PX (2015) Ductile electroactive biodegradable hyperbranched polylactide copolymers enhancing myoblast differentiation. Biomaterials 71:158–167
Yang MC, Chi NH, Chou NK, Huang YY, Chung TW, Chang YL, Liu HC, Shieh MJ, Wang SS (2010) The influence of rat mesenchymal stem cell CD44 surface markers on cell growth, fibronectin expression, and cardiomyogenic differentiation on silk fibroin—hyaluronic acid cardiac patches. Biomaterials 31:854–862
Yang MC, Wang SS, Chou NK, Chi NH, Huang YY, Chang YL, Shieh MJ, Chung TW (2009) The cardiomyogenic differentiation of rat mesenchymal stem cells on silk fibroin-polysaccharide cardiac patches in vitro. Biomaterials 30:3757–3765
Yang X, Li Y, He W, Huang Q, Zhang R, Feng Q (2018) Hydroxyapatite/collagen coating on PLGA electrospun fibers for osteogenic differentiation of bone marrow mesenchymal stem cells. J Biomed Mater Res A 106:2863–2870
Ye G, Wen Z, Wen F, Song X, Wang L, Li C, He Y, Prakash S, Qiu X (2020) Mussel-inspired conductive Ti(2)C-cryogel promotes functional maturation of cardiomyocytes and enhances repair of myocardial infarction. Theranostics 10:2047–2066
Ye KY, Black LD 3rd (2011) Strategies for tissue engineering cardiac constructs to affect functional repair following myocardial infarction. J Cardiovasc Transl Res 4:575–591
Yi S, Ding F, Gong L, Gu X (2017) Extracellular matrix scaffolds for tissue engineering and regenerative medicine. Curr Stem Cell Res Ther 12:233–246
Yildirim Y, Naito H, Didie M, Karikkineth BC, Biermann D, Eschenhagen T, Zimmermann WH (2007) Development of a biological ventricular assist device: preliminary data from a small animal model. Circulation 116:I16-23
Zhang D, Shadrin IY, Lam J, Xian HQ, Snodgrass HR, Bursac N (2013) Tissue-engineered cardiac patch for advanced functional maturation of human ESC-derived cardiomyocytes. Biomaterials 34:5813–5820
Zhang G, Hu Q, Braunlin EA, Suggs LJ, Zhang J (2008) Enhancing efficacy of stem cell transplantation to the heart with a PEGylated fibrin biomatrix. Tissue Eng Part A 14:1025–1036
Zhang W, Wang Z, Xie C, Wang X, Luo F, Hong M, Zhou R, Ma C, Lin N, Zhang J, Hu X, Chan JKY, Wen F, Wang Y (2019) Scaffold with micro/macro-architecture for myocardial alignment engineering into complex 3D cell patterns. Adv Healthc Mater 8:e1901015
Zhou Y, Hu Z, Du D, Tan GZ (2019) The effects of collector geometry on the internal structure of the 3D nanofiber scaffold fabricated by divergent electrospinning. Int J Adv Manuf Technol 100:3045–3054
Zimmermann WH, Didie M, Doker S, Melnychenko I, Naito H, Rogge C, Tiburcy M, Eschenhagen T (2006) Heart muscle engineering: an update on cardiac muscle replacement therapy. Cardiovasc Res 71:419–429

# Chapter 19
# The Importance of Natural Products in Cosmetics

**Nagarjuna Reddy Desam and Abdul Jabbar Al-Rajab**

**Abstract** People from worldwide have been using plant-based substances (Natural Products) to enhance the appearance since the existence of mankind. In the ancient Egypt, around 3000 BC, there is evidence of using cosmetics, and their usages have been a necessary part in our everyday life in all cultures. Initially, natural products have been used for beauty products; occasionally augment with paints and dyes. Natural products have approached back with present trend cosmetic products which are mainly derived from plant sources. Since from longer time, plant products (Natural Products) are source of food and medicines. A broad range of natural products is used in cosmetics preparations, skin care such as treatment of dryness, treatment of eczema and acne, as well as antioxidant, anti-inflammatory, anti-aging, hair care products such as hair growth imputes, hair color, scalp complaints like dandruff, and skin protection, and also toiletry preparations. Essential oils are major source of plants; essential oils have been used in preparation of perfumes, hair care substances, emollient of the skin. For example, natural products have been used in cosmetic industry avoiding side effects with traditional preparations for herbal beauty such as *Emblica officinalis* (Amla), *Acacica concinna* (Shikakai), and *Callicarpa macrophylla* (Priyangu) have been used strongly in skin care and hair care. Moreover, Indian women are still using natural products such as *Pterocarpus santalinus* L. and *Curcuma longa* (skin care), *Lawsonia inermis* L. (hair color), and natural oils such as coconut, olive, shea butter, jojoba, and essential oils in perfumes for their bodies. The present book chapter represents the importance of natural products in cosmetics.

**Keywords** Natural products · Herbs · Cosmetics · Essential oils

N. R. Desam (✉)
Anantha Lakshmi Institute of Technology & Sciences, Anantapur, Andhra Pradesh 515001, India
e-mail: dndnrchem@gmail.com

A. J. Al-Rajab
Etcetera Publications, Chesterville, ON K0C 10, Canada

 
D. Pal and A. K. Nayak (eds.), *Bioactive Natural Products for Pharmaceutical Applications*, Advanced Structured Materials 140,
https://doi.org/10.1007/978-3-030-54027-2_19

## 19.1 Introduction

Since the ancient times, humans from all advancements have been using infinite substance as a source to improve their beauty, looking younger, enhance their sexiness, and normally protect their health (Freitas et al. 2015). The substances used commonly nowadays for these momentums are generally known as cosmetics and cosmeceuticals (Freitas et al. 2015). Normally, cosmetics do not consist of pharmacologically active substances, at the minimum level, and do not demonstrate the amount of benefit of skin scientifically (Freitas et al. 2015). Moreover, in comparison with the classic cosmetics, they are safe, universally available from supermarkets, medical shops, beauty salons, and online traders. Classic cosmetic products include maquillage, toothpastes, conditioners and shampoos, hair colors, nail enamel, perfumes, and antiperspirant (Joshi and Pawar 2015; Wanjari and Waghmare 2015).

Cosmeceuticals, also known as cosmedics or cosmedicals, are skin care substances with active natural products with the therapeutic or drugs like interest includes cosmetic property (Joshi and Pawar 2015; Wanjari and Waghmare 2015). The skin care substances methodological demonstrate measures that productive affects the skin and are issued by the functional substances (Joshi and Pawar 2015; Wanjari and Waghmare 2015). Cosmetics could summon constructional adjustments in pores and skin and beneficial outcomes in skin situations including blackheads, pimples, hyperpigmentation, and rosacea (Joshi and Pawar 2015; Wanjari and Waghmare 2015). Although cosmeceuticals are not considered as medicines or pharmaceutical substances, so they don't require any prescription to omit, and they could be used frequently without risk and major side effects (Joshi and Pawar 2015; Wanjari and Waghmare 2015). Cosmeceuticals of herbs and artificial substances and their derivatives include antioxidants, antidandruff, vitamins, shampoos, antifungal compound, sunscreen lotions for sun protection factor, anti-acne, anti-wrinkle, anti-aging, toothpastes, and deodorants that contain antiperspirants (Freitas et al. 2015; Joshi and Pawar 2015; Wanjari and Waghmare 2015). Natural products such as haldi, chandan, manjistha, yastimadhu, khas, and nagkheshara are used to gleam complexion; while arusa, amala bavchi, guduchi, and chakmarad are mentioned as kustaharan (Kuno and Matsumoto 2004). Moreover, natural products such as haridra, khadira, vidyanga, amalaki, abhaya, jatisaptaparna, karacira of various promises from Knahshthag and Mahakashiya are mentioned productively for skin disorder. Charak and Sages stated natural products are used to remove toxins from the body, clear the entangled that conduct grow on the skin and also secured from kushtha and boils (Kuno and Matsumoto 2004). For example, natural products such as Eladi Gana contain ela, kusstha, tagar, jatamani, tvak, dhmamaka, potra harenuka, shutki, stuuneyaka, choraka, guggol sarjarasa, agaru, devedaru, and padmakesher.

As per 1989 safety regulations, consumer substance or substance deliberate application should be used on the outermost surface of human body which includes lips, nails, hair system, epidermis, and external medicine, as well as denticle, mucous membrane for wash, fragrances used for protecting from bad smell for the determination of treating and stopping disease (Zuorro and Lavecchia 2010).

Novel natural active chemical constituents are derived from the plant kingdom, earth, and sea. Approved active chemical constituents include vitamins, food fibers, minerals, enzymes, hormones, antioxidants, multitude of naturals, and Chinese herbs. Historically, natural products have been used commercially, and till now, several new natural products consist of natural oil and herbs that are available in the market. In cosmetics, plants are the major source and base. Furthermore, natural products are used as drugs in pharmaceutical industry, but they are not acceptable to be used as cosmetics products. According to the cosmetic safety regulatory act in 1989, *Atropa belladonna* and *Digitalis purpurea* plant materials were forbidden (Zuorro and Lavecchia 2010). Natural product complete extracts and specific extracts have been used in cosmetics. Total natural products and specific extracts are employed which are the major cause of determination of their specific activities. Some selective natural products were applied in different zone of use, for example, *Glycyrrhiza glabra* for skin annoyance; *Ginkgo biloba* as antioxident; *Berberis vulgaris* as skin glowing (Zuorro and Lavecchia 2010). Natural products are major source to make soaps such as *S. officinalis* (soapwort) in Europe; *Y. glauca* (yucca) in southern USA; *S. indica* as soapnut in India; *Phytolacca dodecandra* L., and *Quillaja saponaria* L., in Africa and America, respectively (Zuorro and Lavecchia 2010). This is extremely interesting to know how plants are differing from other plants used which differ from tree bark to berries. The present chapter represents the importance of natural products in cosmetics and cosmeceuticals.

## 19.2 Background

In Africa Homo sapiens, over 100,000 years ago, African Middle Stone Age discovered the red bister including crayons are the application of cosmetic body (Murube 2013). An additional proof in ancient Egypt, about 10,000 BC, humans used scented oil, lotions to clean, skin soothe, and mask their body odor (Murube 2013). Then, about 3,000 BC, the use of cosmetics such as hair and skin care products became more common in Egypt and also expanded to large parts of Asia and Africa (Hetta 2016a, b; Murube 2013). Historically, women and men were applied kohl—as eye makeup which also protects the eyes from dry winds and sun radiation (Hetta 2016a, b; Murube 2013). Castor oil skin lotions and creams were prepared and used for the skin protection. Also, extracts from red algae for abrade lipsticks; sometimes from fish scales to obtain the glittering substance and crunch on *Glycyrrhiza glabra* root sticks to improve their breath. Silent basic natural active ingredients such as almond oil, sesame oil, lily, rosemary, peppermint, rose, aloe, lavender, chamomile marjoram, and thyme oils were used in perfumes (Hetta 2016a, b; Murube 2013).

Around 100 AD, Ancient Rome traditionally used makeup and beauty products. It was their trend and faith that became a necessary part of their daily life

(Blanco-Dávila 2000). Beauty products like body lotions, eyeshadow, liners, talcum powders, nail products, perfumes, and toothpaste products were used by women in Roman daily and also beauty masks to make their faces glow, while Roman men were used it as hair dye. Particularly, women's social status, attractiveness, wealth, clothes, makeup embellish in. In 215 BC, *Lex Oppia* established a law extravagance to control gorgeous and luxurious; women could purchase and wear (Hsieh et al. 2016; Naidoo et al. 2016; Stutesman 2016; Watson 2012). After six years, this law was reversed; but people from Ancient Rome took cosmetics to new heights and limits. Now, in Milan, there are the largest beauty empires in the world.

Prior to twentieth century, about the year 1910, cosmetics were used more in fashion in Europe and USA. In 1910, in Paris, they introduced color makeup. Hence, in 1920s, film industry introduced in Hollywood had an advanced influence on cosmetics (Jones 2011). Theda Bara was a great movie star of the silent movie period, and her makeup artist has been developing the cosmetics production (Jones 2011). In 1920s, the cosmetic substances were highly demanded to inculcate consciousness in American women. The conduct among others to the wing variety which was identified by fearless dark eyes, red nail polish, red lipstick, eyebrow pencil, mascaras artificial hair color, brownish sunscreens (Murnen and Seabrook 2012; Schlessinger 2007). In the entire world, cosmetics were shortsupply between 1939 and 1945 during the Second World War. During this war, important part owed to the fact that alcoholic and petroleum products are the best active ingredients of several cosmetic products.

The late twentieth century, between 1960 and 1970s, women in the western world used the cosmetic products desert to cheapen women to sex objects (Oréal 2015). In 1970s, it was a trend to develop the natural looking products. Coincidently different lipstick colors (like green, red lilac, silver, and pink), non-allergenic cosmetics makeup powders with longer staying have been been developed (Oréal 2015), and also males started to use cosmetics to improve facial features (Ribeiro et al. 2015). They have been developed the most popular cosmetics such as mask for dark circles, age spots, and blemishes on the skin and large pores on the skin (Ribeiro et al. 2015).

Cosmetics and cosmeceuticals industries have a large market worldwide which was estimated over 200 billion US dollars in the year 2015 (Chaudhri and Jain 2009; Ribeiro et al. 2015). This might be qualified to increase in world economy, changing the way of life, increasing utilization and personal skin and hair care of new commercial naturally developed substances. Worldwide in the main retailers, 36.9% of share is from Asia and Pacific regions, and share from America, Europe, Africa, and Middile East holds 30.8%, 29.5%, and 2.9%, respectively (Chaudhri and Jain 2009). In global markets, accounted skin, hair, makeup, fragrances, and hygiene products are 36.3%, 22.9%, 18.2%, 12%, and 10.5%, respectively. These cosmetics and beauty products are usually purchased from online stores, retail shops, supermarkets, drug stores, departmental stores, and brand outlets (Chaudhri and Jain 2009).

## 19.3 From Past to Future

A modern development of cosmetics and cosmeceuticals industry is the natural beauty products (Atanasov et al. 2015). Since ancient times, extracts from various parts of aromatic and medicinal plants are introduced in cosmetics and cosmeceutical products (Barbulova et al. 2015; Kapoor 2005). Around early 1900s, naturally a derived product adequately increases in cosmetics and cosmeceuticals industry (Jones 2011; Kapoor 2005). Currently, organic substances or products are more on demand throughout the world as the utilizing of natural products at low risk of undesirable effects (Carola et al. 2012).

## 19.4 Source of Natural Products

Natural products are generally acquired from aromatic and medicinal plants, mixture of volatile and nonvolatile chemical constituents with strong odor (Bakkali et al. 2008). Natural products are considered as one of the most control plant products in agriculture, as they exhibit antiviral, antibacterial, anti-cancer, antioxidant, antidiabetic, insect repellent, antifungal, anti-inflammatory, and cosmetic properties (Reddy 2019; Said et al. 2016; Swamy et al. 2016). For example, from willow tree bark, *Digitalis lanata* flowers and opium medicinal products derived as aspirin, digoxin, and morphine, respectively, and also extracted from *T. indica* and *S. alata* areal parts and seeds contains polyphenols, flavonoids, and hemicellulose xyloglucan that were used for skin protection from sun radiation. Leaf extracts from *T. serrulata* (Vahl) have been used for treatment for hair growth and hair loss (Freitas et al. 2015). Essential oils extracted from *L. aestuans* are used as facial toners, lotions, lip balms, ointments, creams, scrubs, massage oils, masks, and shampoos for antidandruff and also used for treatment for eczema, acne, chicken fox, insect bites and scarring from burns (Reddy 2019). *M. olefira* leaf and seed oil extracts contain considerable amount of β-carotenes, vitamins A, C, & E, and polyphenols using anti-inflammatory and antioxidant activities also have been used for body oils, scrubs, lotions, balms, creams, hair care, moistures sun protection, and perfumes (General Bureau of statistics 2014). Some of these plants were used for traditional use because monoterpenes, sesquiterpenes and triterpenoids and phenolic compounds are present. For example, *P. amboinics* consist of mono, di, and sesquiterpenes which are used for antioxidant, anti-inflammatory, antimicrobial, skin cleansers, anti-wrinkle, anti-skin cleansers, anti-aging night creams, and cosmetics also used for itchy skin and insect bite (Freitas et al. 2015; Kleynhans et al. 2017). *A. indica* L. oils extracted from fruits and seeds are used for preparing soaps, creams, balms, shampoos, nail polishes and toothpastes (Mans et al. 2017; Nagarjuna et al. 2017). These essential oils could be used as perfumes, flavoring agents, fragrances for different cosmetics and cosmeceuticals (Swamy et al. 2016).

## 19.5 Extraction and Isolation of Natural Products or Essential Oils

Natural products or volatile oils extracted by using different methods could be applied for essential oil extraction, such as steam distillation, hydro distillation, solvent extraction, or continuous extraction such as Soxhlet extraction, liquid–solid extraction, liquid carbon dioxide, microwave extraction. Generally, for essential oil extraction hydrodistillation or steam distillation quintessential is used, resulting in less molecular weight, mixture of chemical constitutes is obtained. For example, Lamiaceae and Citrus families have used steam and hydrodistillation. For extraction of natural products, Soxhlet extraction and liquid–solid extraction quintessential are used, resulting in high molecular weight of the compound.

All these natural products and essential oils are mixtures with the other chemical constituents from the natural sources; the products must be isolated and purified. Isolated purified compounds are acquired in milligrams to grams. After extraction, natural products or essential oils are isolated using these methods by adsorption, precipitation, chromatography, and crystallization methods, etc. After isolation product needs to study chemical and physical properties, chemical structure elucidation, and structural activity of relationship of the product.

## 19.6 Plants-Derived Cosmetics and Cosmeceuticals

A rich and diverse traditional medicinal system has been employed for traditional medicine using different variety of plants worldwide. The uses of these plants in the world thoroughly have been addressed by Raghoenandan (Hindustani), van't Klooster (Maroons), van Adel and Ruysschaert (Creoles), and Tjong Ayong (Javanese) (Raghoenandan 1994; Tjong Ayong 1989; Van Andel and Ruysschaert 2011; Van't Klooster et al. 2016). Around 789 medicinal plants have been published in the Surinamese, among 10% (≈seventy two) of plants were used for the purpose of cosmetics and cosmeceuticals. Among seventy-two plants, three quarter (fifty eight) is used cosmeceuticals for treating scars, warts, pimple, pigment skin, scaly skin, skin vexation, inflammation, pustules, and boils as well as skin fungi and skin parasites. These properties indicated plants have significant cosmaceuticals and medicinal properties. More than twenty-five plant families are belonging to the cosmeceuticals, which are recommended for treating the skin-related problems. More than 20% of medicinal plants have been used for cosmetics and cosmeceuticals which include perfumes, skin care, hair care, refreshment and steam bath and some of the cosmetics and cosmeceutical plants as shown in Table 19.1.

**Table 19.1** Plants used for cosmetics and cosmeceuticals

| Plant Scientific name | Condition |
|---|---|
| *P. guajava* L. | Body odor |
| *T. serrulata (Vahl.)* | Genital steam baths |
| *E. hieracifolia* (L.), *S. alata* (L.) | Flaky skin |
| *M. maripa* L., *M. citrifolia* L., *B. excels* L. | Hair care |
| *M. indica* L., *M. micrantha, S. trilobata, U.camphorata, M. charantia* L., *S. alata* L., *A. indica, A. Juss, C. odorata* L., *C. citrates, Citrus* sp., *Cecropia* sp. | Local skin lesions (e.g., boils and pustules) |
| *B. orellana* L., *Z. mays* L. | Makeup |
| *Sida rhombifolia* L. *Cedrela odorata* L. | Pigment spots |
| *A. vera* L., *C. odorata* L., *A. galanga* L. | Scars |
| *E. foetidum* L., *C. nucifera* L., *E. oleracea, M. maripa, M. citrifolia* L., *L. aestuans* L. | Skin care |
| *A. sativum* L., *L. purpureus* L., *S. alata* L., *S. reticulate* L. | Skin fungi |
| *P. stellis, L. aestuans* L. | Skin rejuvenation |
| *P. amboinicus* Lour. *G. barbadense* L., Orchid spp. | Skin refreshment |
| *Amaranthus* sp., *C. nucifera* L., *C. odorata* L., *E. prostrata* L., *M. micrantha* Kunth, *R. fruticosa* L., *U. camphorata* L., *O. cochinellifera* Steud., *V. guianensis, C. sativus* L., *J. curcas* L., *R. communis* L., *C. guyanensis* Desf., *S. alata* L., *T. indica* L., *B. guianensis* Aubl, *C. alatus* Aubl, *C. ramosa* Aubl, *M. glabra* L. *A. moschatus* Medik., *A.purpurascens* Aubl, *A. indica, A. Juss, C. odorata* L., *A. bilimbi* L., *P. coccinea* Aubl., *O. sativa* L., *Q.amara* L., *C. latifolium* Lam., *S.leucocarpon* Dual., *S. stramoniifolium* Jacq., *W. indica* L., *L. rosea, A. galanga* L. | Pimples |
| *A. curassavica* L., *P. acuminata* Ait., *C. difussa* Burm.f. | Warts |
| *B. pinnatum* (Lam), *G. barbadense* L. | Warts |
| *E. prostrata* L., *M. micrantha* Kunth., *U.camphorata* L., *C. cujete* L., *B. orellana* L., *T. ulei* Vaupel., *C. difussa* Burm.f., *C. urens* L., *M. guianensis* Aubl., *S. alata* L., *L. nepetifolia* L., *O. tenuiflorum* L., *A. moschatus* Medik., *A. indica, A. Juss, C. guianensis* Aubl., *C. odorata* L., *F. schumacheri, P. marginatum* Jacq. | Skin irritation (e.g., rash, dermatitis, and eczema) |
| *U. camphorata* L., *C. guianensis* Aubl. | Skin parasites |

## 19.6.1 Manufacture of Plant-Derived Cosmetics and Cosmeceuticals

Based on the earlier abundant plant, raw material could be processed into cosmetics, cosmeceuticals, and medicinal preparations. As a result in the worldwide increasing number of entrepreneurs such as individuals, small, medium, and large scale gross on medicinal plants sector. Hence, either collecting or cultivating the raw materials homeless they act as retailers or mediator or supplier for processing or preparation of cosmetics or cosmeceuticals intermediates or final products (Playfair et al. 2011). For example, in Suriname Skin Glanz cosmetics Odany Jewa, and Jomi cosmetics are specialized in skin care-related cosmetics and cosmeceuticals from start. These include scrubs, face washes, liver spots, eye bags, day and night creams, ointments, removal of impurities on the skin and facial cleaners, etc. (Grant 2017). However, while using the cosmetics/cosmeceutical materials caution and possible side effects such as photosensitivity and skin allergy are perfectly mentioned in the user instructions (Grant 2017).

Cosmetics and cosmeceuticals are usually produced from different fresh plant parts such as leaves, fruits, barks, seeds as well as unfinished substances such as waxes and vegetable oils from plants as shown in Table 19.2. The acquire plant raw materials are from growers, collectors, and vendors who mainly operate in different places in the world. Hence, plants of these parts grow easily and persistently encountered in the wild. *L. aestuans* is an exception West Indian wood nettles relatively rare (Grant 2017). Extraction of other important constitutes such as oils are complicated process from these plants such as *C. nucifera* (coconut tree), *C. guyanensis, V. paradoxa* (shea butter), *A. chinensis* (jojoba oil), and *C, guianensis*.

**Table 19.2** Plants used for preparation of cosmetics and cosmeceuticals

| Family | Plant Scientific name | Use of cosmetics/ cosmeceuticals |
|---|---|---|
| Asphodelaceae | *A. vera* (L.) | Skin care products |
| Arecaceae | *C. nucifera* L., *E. oleracea* Mart., *M. maripa* | Skin, Hair, and Eye care products |
| Asteraceae | *E. prostata* L. | Skin care products |
| Crassulaceae | *B. pinnatum* | Skin care products |
| Cucurbitaceae | *C. sativus* L. | Skin care products |
| Commelinaceae | *T. serrulata* L. | Hair care products |
| Fabaceae | *C. guyanensis* | Eye and hair care products |
| | *S. alata* | Deep skin cleaners |
| | *T. indica* L. | Skin care products |
| Lecythidaceae | *B. excelsa* | Hair and skin care products |
| Lamiaceae | *L. nepetifolia* | Eye care products |
| | *P. amboinicus* | Skin care products |
| Moringaceae | *M. oleifera* | Skin care products |
| Malvaceae | *H. sabdariffa* | Skin and hair care products |
| Meliaceae | *C. guianensis* | Eye, Hair and Skin care products |
| Orchidaceae | Orchid spp. | Skin care products |
| Poaceae | *C. citratus* | Skin care products |
| Urticaceae | *L. aestuans* | Skin care products |
| Rubiaceae | *M. citrifolia* | Hair and skin care products |

### 19.6.2 Substances of Plant-Based Cosmetics and Cosmeceuticals

The most frequently used plants in cosmetic and cosmeceuticals are shown in Table 19.2. There are some plant leaves especially for facial washes such as *A. indica*, *A. vera*, *M. oleifera*, *C. citrates,* and *Y. indica*. *C. citrates* and *P. amboinicus* leaves are used for facial milks. The leaves and fruit juices of these important plants such as *E. oleracea*, *C. sativus*, *M. citrifolia*, *A. indica*, *C. nucifera*, *T. indica,* and *H. subdariffa* L. are used to synthesis day and night skin lotions. To remove dead skin cells, stimulate blood circulation and facial scrubs *E. oleracea* fruit granules are used mainly and also for deep skin cleaning *E. prostate*, *S. alata,* and *T. indica* essential components of leaf juices are used. *M. pleifera* and *L. aestuans* leaves extracts contain active ingredients for treating eczema, chapped lips, insect bites, and acne, also leaves extracts of *T. serrulata* are used for treating hair growth. Particularly most of the above-mentioned plant extracts are used for skin care creams, lotions, and scrubs. Oils extracted from leaves, fruits, and bark from these plants *B. excels* (Brazil oil), *C. nucifera* (coconut oil), *C. guianensis* (karapa oil), *M. maripa* (maripa oil), and *C. guyanensis* (hoped oil from bark) are used for treating hair growth and hair care products. *A. indica*, *B. pinnatum,* and *L. nepetifolia* leaves extracts are used eye masks, ointments, and treating eye bags and acne. Significant unfinished products such as shea butter, grapeseed oil, jojoba oil, coconut oils are used for preparing hair care, skin care, soaps, moisturizers and sunscreens, hair conditioners, anti-wrinkle formulations, hair conditioners, ointments, baby oils, skin inflammation, and juices extracted from *A. indica* and *P. amboinicus* are commonly used for preparations of facial masks.

Some of the above indicated medicinal plants used for cosmetics and cosmeceuticals and their plants phytochemical composition are widely presented in Table 19.3. Few of the plants are summarized below, for example,

***Aloe vera* (L) Burm.f.**

This plant is mainly used in traditional medicinal treatment of microbial infections, skin conditions, constipation, and diabetic mellitus (Manvitha and Bidya 2014; Mahor and Ali 2016). This indicates that leaves of *A. vera* extracts contain significant therapeutic agents such as polymannans, lectins, anthraquinones, and acetylated mannans (Hamman 2008; Moghaddasi and Verma 2011); that's why the leaf gels are included in soft drinks and dilatory additives for digestion (Eshun and He 2004; Qadir 2009). It is moreover used in sunburns, ointments, shampoos, conditioners, skin lotions, sunscreens, facial tissues, soaps, makeup, moisturizers, and shaving creams which are used for the main applications for smoothening and moisturizing effects of *A. vera* leaf gel (Eshun and He 2004; Hamman 2008; Manvitha and Bidya 2014; Mahor and Ali 2016; Moghaddasi and Verma 2011; Periasamy et al. 2014; Qadir 2009).

**Table 19.3** Plants and its main parts used for preparation of cosmetics and cosmeceuticals and its presumed active chemical constituents

| Plant Scientific name | Main plant part's used | Active chemical constituent |
|---|---|---|
| *A. vera* L. | Gel from fresh leaves | Polysaccharides |
| *A. indica, A. Juss* | Fresh leaves and seed | Limonoids |
| *B. excelsa* | Fresh leaves | Glycosides, saponins |
| *C. guianensis* | Seeds | Limonoids, fatty acids |
| *C. nucifera* L | Fruits and seed | Phenolic compounds, vitamin E, terpenes, saponins |
| *C. guyanensis* | Trunk | Terpenes |
| *C. sativus* L. | Fruits | Water, antioxidants |
| *C. citratus* | Leaves | Essential oils |
| *E. prostata* L. | Leaves | Coumestans, glycosides, amyrins |
| *E. oleracea* Mart | Fruits, seeds | Anthocyanins |
| *H. sabdariffa* L. | Fresh calyces | Anthocyanins, vitamin E, flavonoids |
| *L. aestuans* L. | Leaves | Essential oils |
| *L. nepetifolia* L. | Leaves | Flavonoids, terpenes, essential oils, coumarins |
| *M. maripa* | Seeds | Vitamins A and E, fatty acids |
| *M. citrifolia* L. | Fruits and seeds | Flavonoids, fatty acids |
| *M. olifeira* | Leaves | Ben oil |
| Orchid spp. | Seed | Essential oils |
| *P.boinicus* | Leaves | Essential oils |
| *S. alata* L. | Leaves | Phenolic compounds, essential oils |
| *T. indica* L. | Leaves and seeds | Polyphenols, flavonoids, xyloglucan, fatty acids |

## *Azadirachta indica* A. juss.

In India, *A. indica* leaves, fruits, seeds, bark, and root extracts were used against different types of diseases in Ayurveda over two millennia such as diabetic mellitus, high blood pressure, fever, cold respiratory conditions, and cosmetics. Especially, oils extracted from seeds and fruits have been used for traditional medicine against redness, inflammation of the skin and acne. These results indicate that *A. indica* leaves extract contains nimbinin and Azadirachtin phytochemicals, which are useful for antimicrobial and antiparasitic activities. Neem oil contains oleic acid, palmitic acid, linoleic acid, and stearic acid phytochemicals. Hence, it could be used in synthesis of different cosmetics like nail polishes, toothpastes, shampoos, soaps, and balms (Djibril et al. 2015; Galeane et al. 2017; Hashmat et al. 2012; Mak-Mensah and Firempong 2011).

***Bertholletia excelsa*** **Humb. & Bonpl.**

*B. excelsa* is a native of Amazon forest, essential seeds of the Brazilian nut. These seeds are more abundant of vitamins, digestible minerals, staple diet, and dietary fiber. Brazilian nuts are commercially collected and mixed with different nuts. Oils extracted from these nuts consist of more than 75% of unsaturated fatty acids, mainly selenium, linoleic acid, oleic acid, phytosterols, polyphenolic acids, and vitamin E, particularly polyphenols and vitamin E show antioxidant properties. Hence, these phytochemicals could be used mainly in lotions, creams, shampoos, hair care products, aging skin, flaky skin, skin inflammation, and acne. Hence, the fatty acids are the applications of moisturizing effects (Chunhieng et al. 2004; Chunhieng et al. 2008; John and Shahidi 2010; Kluczkovski et al. 2015; Yang 2009).

***Bryophyllum pinnatum*** **(Lam) Oken.**

Since ancient times, mother-of-thousands *B. pinnatum* (Fig. 19.1) is especially used for the sores, boils, wounds, treat for burns, boils, as well as it has antiseptic, amphiphatic, antimicrobial, anti-inflammatory and conventional astringent properties. Leaves of *B. pinnatum* extracts consist of mainly fumaric acid and saponins (from various parts of plant), which is more essential for cosmetics and cosmeceuticals substances for skin regeneration formulas, hair growth syrups, facial creams, creams for wrinkles, treatment for acne, and anti-aging lotions (Afzal et al. 2012; Akpuaka and Ezem 2011; Amenta et al. 2000; Dey et al. 2012; El-Abdellaoui et al. 2010; Kaur et al. 2014; Nagaratna and Hegde 2015).

**Fig. 19.1** The mother-of-thousands *Bryophyllum pinnatum* (Lam) Oken. (Crassulaceae)

***Carapa guianensis* Audl.**

The bark and seed oil extracts of *C. guianensis* are composed of alkaloids and terpenoids, respectively, which is used for antipyretic and antiallergic, antiparasitic, antimicrobial, anti-inflammatory and also having anti-wound healing medication purpose, respectively. Leaf and fruit extracts were traditional medicine for treatment of itching and intestinal worms. The crab or carap oil is composed of many fatty acids such as oleic acid, linolic acid, and palmitic acid which is used for preparation of skin and hair care products (Campos et al. 2007; Cabral et al. 2013; Henriques and Penido 2014; Letawe et al. 1998; Miranda Júnior et al. 2012; Nayak et al. 2010; Pereira et al. 2014).

***Cocos nucifera* L.**

Parts of *C. nucifera* plant are used for the traditional medicine, for example, in Brazil, Haiti, Papua New Guinea it used for treatment of arthritis from the husk fiber, root extracts are used to treat diarrhea, amenorrhea, and stomach pins, respectively. As per Indonesian and Fijan beliefs, coconut oil prevents hair loss. These are causes for the presence of tannins, phenols, flavonoids, triterpenes, steroids, saponins, alkaloids, and condensed tannins as well as fatty acids and vitamin E in different parts of plant. For these causes, coconut oil is base essential element for hair conditioners, shampoos, shower gel, sunscreens, moisturizers, body butters, nourishing, and emollients, skin infections, anti-aging, prevents dry skin and anti-redness (Da Fonseca et al. 2014; Esquenazi et al. 2002; Gopala et al. 2010; Holdsworth 1992; Lima et al. 2015; Sachs et al. 2002; Weniger et al. 1986).

***Copaifera guyanensis* Desf.**

Copaiba oils extracted from *C. guyanensis* (Fig. 19.2) tree trunk and bark have been many medicinal and cosmetic properties, were already known to American Indians. The trunk of *C. guyanensis* is used for injured animals to heal their wound and also it could be used for the treatment of skin diseases, to stimulate wound healing, infections, inflammations, and malignancies. Copaiba oils consist of biologically active chemical constituents such as diterpenens and sesquiterpenes, kaurenoic acid, copalic acid and caryophyllene. Because of these, they exhibit antibacterial, anti-eczema, anti-inflammatory, antibacterial, antifungal (especially *S. aureus*) and antioxidant properties. Hence, copaiba oils are significantly used in cosmetic industry in scars, stretch masks, shampoos, capillary lotions, soaps, bathing foams, and anti-acne (Da Silva et al. 2012; Leandro et al. 2012; Lucas et al. 2017; Veiga et al. 2007).

***Cucumis sativus* L.**

The plant of *C. sativus* is composed of more than 90% of water; also contains vitamin B, C, & K and β-carotene phenols and flavonoids phytochemicals. *C. sativus* fruits, fruit juices, fruit water, and fruit extracts are essential to be used in cosmetic formulation, the water considerably is essential benefit for the skin such as skin conditioning, eye and facial makeup, neck and face products, bath foams,

**Fig. 19.2** Harvesting copaiba oil from the trunk of *Copaifera guyanensis* Desf. (Fabaceae)

soaps, skin hydrating products, cleaning products, detergents, facial peel products, skin rejuvenation products, body and hand lotions and hair care products (Akhtar et al. 2011; Ibrahim et al. 2010; Kumar et al. 2010; Mukherjee et al. 2013; Murad and Nyc 2016; Rajasree et al. 2016).

***Cymbopogon citrates* (Dc) stapf.**

The essential oils and phytochemicals are extracted from the leaves of *C. citates* (lemongrass), extensively used for the antimicrobial, antipyretics, diuretics, insect repellents, anti-inflammatory, antiprotozoals, antidyspeptics, spasmolytics, antibacterial, antifungal, antidiarrheal, antioxidants, antipyretics, cytoprotective. Oils are extracted from the leaves, phytochemical is composed of flavonoids, phenolic compounds, and essential oil is composed of citral, geraniol, citronella, citronellol, and myrcene. Essential oils are nice smelling, resulting in it could be used as fragrances, perfumes, creams, detergents, and creams. Preparation of ionones for cosmetics and perfumes citral is the main starting molecule. Essential oils are mainly used for the synthesis of skin care products, like creams, lotions, facial cleansers (Ekpenyong et al. 2014, 2015; Ganjewala 2009; Maheshwari et al. 2014; Mosquera et al. 2016; Olorunnisola et al. 2014; Pandey 2017; Shah et al. 2011).

***Eclipta prostate* (L.) L.**

Extracts from different parts of *E. prostate* are used for respiratory disorders, gastrointestinal complaints, fever, skin problems, microbial, parasitic infections, spleen enlargement, liver ailments, as well as hair greying and hair loss. Ethonoilc and petroleum ether extracts of *E. prostrate* leaves are used for hair oils in Ayurveda and hair growth respectively. Extracts are composed of various phytochemicals such as oleanane type glycosides (eclalbasaponins) and triterpenes (oleanolic acid, ursolic acid, & amyrins). These compounds show potential antibacterial, antieczema, anti-inflammatory, and antioxidant properties. Hence, because of these reasons oils and extracts of *E. prostrara* are used for skin nourishing and skin anti-aging agents, as well as hair and hair growth (Baldi et al. 2011; Chan et al. 2014; Datta et al. 2009; Kaur and Chandola 2010; Kumar et al. 2005; Roy et al. 2008; Saraswat et al. 2015; Uddin et al. 2010).

***Euterpe oleracea* Mart.**

Various parts of *E. oleracea* extracts are used worldwide especially in Brazil processed into pulp, which is used for food products, juice in beverages, smoothies, and dietary supplements. Extracts of *E. oleracea* pulps are mainly composed of polyphenolic anthocyanin, palmitic acid, and oleic acids. Resultant extract shows potential antioxidant properties. Phytochemicals such as anthocyanins and phenolic compounds are significantly used for the anti-aging lotions, skin regenerating creams, treatment for skin damage, sunscreens, and anti-inflammatory products as well as fatty acids are used as cosmetics as hair conditioners, skin moisturizers, soaps, and shampoos (Bobbio et al. 2000; Daher et al. 2014; De Santana et al. 2017; Hogan et al. 2010; Kang et al. 2010; Portinho et al. 2012; Schauss et al. 2006; Yamaguchi et al. 2015).

***Hibiscus sabdariffa* L.**

*H. sabdariffa* L. (Fig. 19.3) is extracted from calyces used in traditional folk medicine system for a wide range of kidney disease, cough and bronchitis, gastrointestinal problems microbial infections and cosmetics. These applications indicate that the plant extracts contain vitamin E, flavonoids, phenolic substances as well as antioxidant anthocyanins, antibacterial and anti-inflammatory effects. Hence, *H. sabdaritha* L extracts are considerably used for preparations of hair and skin care products as well as flowers of *H.sabdaritha* L. crude polysaccharides acts as stimulatory effects (Abou et al. 2011; Brunold et al. 2004; Da-Costa-Rocha et al. 2014; Ismail et al. 2008; Liu et al. 2002; Mohamed et al. 2007; Mounnissamy et al. 2002).

***Laportea aestuans* (L.) *Chew.***

Extracts from *L. aestuans* are potentially used for traditional medicine system which includes eye infections, parasitic infections, and microbial as well as gonorrhea and syphilis and also exhibit antioxidant and antimicrobial activity. Essential oils extracts from the leaves of *L. aestuans* are composed of methyl salicylate and

**Fig. 19.3** The roselle *Hibiscus sabdariffa* L. (Malvaceae)

chrysene -2-ol derivatives. Hence, essential oils are used for face masks, massage oils, lotions, creams, scrubs, facial tones, antidandruff shampoos and conditioners, lip blames as well as used for treatment for insect bites, chicken fox, acne, eczema, and blemishes (Chukwuma et al. 2015; Essiett et al. 2011; Lans 2007; Okereke et al. 2017; Oloyede 2016; Oloyede and Ayanbadejo 2014).

***Lenotis nepetifolia* (L.)**

*L. nepetifolia* (Fig. 19.4) plants parts (sepals, leaves, flowers, and roots) are used in traditional medicinal system. Hence, extracts are composed of active ingredients such as terpenes, terpenoids, quinines, alkaloids, Saponins, and coumarins as well as essential oils. Because of these active chemical constituents, they exhibit antibacterial, antioxidant, insecticidal, radical scavenging anti-inflammatory, and cosmetic properties. *L. neperifolia* has been used for skin allergies, calm agitation, counteract muscle spasms, treatment for bronchial asthma, heal burns, pain, fever, cold, anti-malaria, and arthritis as well as it used for cosmetics such as skin regenerating agents, skin rashes, skin infections, and skin rejuvenating. Essential oils from this plant shown pleasant fragrance, it could be used for preparation of perfumes, because of the diterpenens and coumarins (Imran et al. 2012; Niteshwar and Kumari 2012; Oyedeji and Afolayan 2005; Oyedeji et al. 1999; Pedro et al. 1991; Udaya et al. 2013).

**Fig. 19.4** The lion's ear *Leonotis nepetifolia* (L) R. Br. (Lamiaceae)

***Maximiliana marpia* (Correa) Drude.**

*M. maripa* edible oils are extracted from fruit, seeds, and pulp, composed of fatty acids, phytosterols, tocotrienols, glycolipids, tocopherols, and α-carotene. These chemical constituents exhibit antimicrobial, antioxidant, and moistening properties. For these reasons, *M. maripa* edible oils are used traditionally in skin ageing, skin care products, smoothen scared skin, massage oil, moisturizers, rejuvenation, soaps, shampoos, treating acne, and hair conditioners (Balslev et al. 2008; Bereau et al. 2001; Dos Santos et al. 2015a, b; Dos Santos et al. 2015a, b; Fernández et al. 2016; Pereira et al. 2013).

***Morinda citrifolia* L.**

The extracts from *M. citrifolia* are potentially used for the treatment of wounds, sprains, sores and diabetic mellitus, AIDS, high blood pressure, malignant neoplasm's and also used in skin injury, infection by ultraviolet radiation, and seed oil is used as anti-inflammatory for acne. Hence, these extracts are used in preparation of skin and hair care cosmetics and cosmeceuticals like shampoos and conditioners, body and foot lotions, deodorants, ointments, hand and facial soaps, eye creams, face masks, moisturizers, day and night creams as well as for treatment for acne.

(Assi et al. 2017; Kakad et al. 2015; Krishnaiah et al. 2012; Levand and Larson 1979; Palu et al. 2012; Potterat and Hamburger 2007; West and Sabin 2012).

***Moringa oleifera*** **Lam.**

*M. oleifera* leaf and seed oils are used in traditional medicinal system, *M. oleifera* seed oil is known as ben oil, ben oil is composed of fatty acids mainly behenic acid. Because of this active substance, it could be used for moisturizers, massage oils, and aromatherapy. Leaf extracts of *M. oleifera* are composed of significant amount of vitamins A, C, & E, polyphenols and β-carotene which might own antioxidant and anti-inflammatory properties. Hence, for these active ingredients of leaf and seed oil extracts might be used in sunscreens, scrubs, body oils, creams, lotions, balms, hair care products, moisturizers as well as seed oils are used in preparation of perfumes. *M. olifeira* extracts relieve spasms, cardiac stimulant, diabetes mellitus, cardiac stimulant, antimicrobial, antiparasitic, and other conditions could be used (Ali et al. 2013; Ashraf and Gilani 2007; Kale and Megha 2011; Ogbunugafor et al. 2011; Ojiako and Okeke 2013; Taher et al. 2017; Warra 2012; Warra 2014).

**Orchidaceae.**

Generally, Orchidaceae (Fig. 19.5) plants and its parts show potentially various traditional medicinal systems. Especially these family plants are used in Chinese traditional medicine for treatment of cancer, diabetes mellitus, hypertension, and urinary tract infections. For example, in this family, *V. planifolia* is used for the preparation of perfumes, deodorants, and aromatherapy products. Some of these family plant extracts are used in cosmetics and cosmeceuticals. Some of them are composed of flavonoids and phenolic compound, exhibit anti-inflammatory,

**Fig. 19.5** The tiger orchid *Oncidium jonesianum* (Orchidaceae)

anti-aging, and antioxidant property. Some plants from this family are water bound or hydrophilic chemical constituents. These constituents increase corneum hydration and have been used for skin moisturizers and emollients. Flowers of this family are the source of perfumes, fragrances, skin care products, hair care products, and bathing products, as well as flower bouquets, are useful to identity scent and perfumes (Bulpitt et al. 2007; Hadi et al. 2015; Hossain 2011;Menon and Nayeem 2013; Minh et al. 2016; Paul et al. 2013; Ribeiro et al. 2015; Sadler et al. 2011).

***Plectranthus amboinicus* Lour.**

Extracts from *P. amboinicus* have been used for the treatment of fever, infections, and genitourinary disorders, gastrointestinal and respiratory diseases. Essential oils extracted from this species are used for scent as well as in the treatment of skin diseases like wounds, burns, sores, and also used for insect bites, allergies, antiseptic dressing for wounds and parasitic infections. Plant extracts are composed of phenolic compounds, monoterpenes, sesquiterpenes, and diterpenens. These active constituents show significant antioxidant, anti-inflammatory, and antimicrobial properties. These properties indicate that essential oils could be used in soaps, skin cleansers, anti-wrinkle, anti-ageing, day and night creams, ointments, skin itchy, moisturizers and skin rejuvenating (Bhatt and Negi 2012; Chifundera 2001; Erny et al. 2014; Harsha et al. 2003; Lukhoba et al. 2006; Rabe and Van Staden 1998; Roshan et al. 2010).

***Senna alata* (L.) Roxb.**

*S. alata* L. (Fig. 19.6) extracts of natural products and essential oils from various parts such as roots, bark, fruits, seeds, and leaves are composed of bioactive substances such as phenolic substances, alkaloids, terpenes, anthraquinones, tannins, and phytosterols. Few of these chemical constituents exhibit laxative and purgative properties. Essential oils are used for treatment of ringworms infections, fungal infections, as well as scabies. Bioactive chemical substances from this plant show antifungal, antioxidant, and anti-inflammatory activity. Essential oils reduce skin damage from ultraviolet irradiation. Hence, *S. alata* leaf extracts are considered as cosmetics. Leaf extracts are used in skin care products like skin-repairing agents, anti-ageing, sunscreens, and soaps (Adelowo and Oladeji 2017; Ehiowemwenguan et al. 2014; Meenupriya et al. 2014; Moriyama et al. 2003; Oladeji et al. 2016; Oresajo et al. 2010; Sule et al. 2011).

***Tamaringus Indica* L.**

Preparation of *T. indica* extracts is composed of flavonoids, polyphenols, and linoleic acid and oleic acid. These phytochemicals exhibit antioxidant, antibacterial, antifungal, antiviral, antiparasitic, and anti-inflammatory properties. Hence, plant extracts could be used in traditional medicinal system like malaria, fever, parasitic infections, and respiratory problems. Flavonoids and polyphenols are present in *T. indica* leaves extract, so it exhibits wound healing properties and hemicelluloses

**Fig. 19.6** The candle bush *Senna alata* (L.) Roxb. (Fabaceae)

xyloglucan substances extracted from seeds of *T. indica* used for skin damage from UV radiation. Hence, extracts from *T. indica* are used for preparation of skin care products such as moisturizers, face masks, skin rashes, anti-aging night creams, ointments, sunscreens, soaps, body lotions facemasks, facial toners and lip balms (Al-Fatimi et al. 2007; Attah et al. 2015; Escalona-Arranz et al. 2010; Havinga et al. 2010; Kuru 2014; Luzia and Jorge 2011; Mesfin et al. 2012; Naik et al. 2017; Strickland et al. 2004).

***Tripogandra serrulata*** **(vahl) Handlos.**

*T. serruulata* extracts are prepared from various parts of the plant; it is native of Caribbean and Southern American countries. These extracts are traditional medicine used for kidney disorders, uterus cleaning, treating traumas, wounds, and oviducts. Moreover, leaves are used to treat hair growth and hair loss treatment. For these reasons, plant extracts are used for the preparation of hair care products. Despite, the shortage of comprehensive data on the phytochemical information about the plant extracts (Caballero-George and Gupta 2011; De Filipps et al. 2017; Funasaki et al. 2016; May 1982; Pereira and Bartolo 2016; Valadeau et al. 2010; Sedoc 1992; Venugopalan et al. 2011).

## 19.7 Applications of Natural Products in Cosmetics

Some of the natural products from the plants have been used for skin and hair problems. For instance, different parts of these plants *Rubia cordifolia*, Linn. *Callicarpa macrophylla* Vahl, *Acacia concinna* DC, *emblica officinalis* Gaertn, and *Curcuma longa* have been used directly for face and hair problems. Ethnic and community group's historically natural products have been used for the treatment of different hair conditions and skin diseases. Natural products intimate potential properties like antioxidants, antimicrobial, anti-inflammatory, antimelanogenesis, antihyluronidase, and antityrosinase in cosmetics. Some of natural products have been used in different applications shown below.

### *19.7.1 Natural Products as Skin Care Agents*

**Dry skin treatment.**

Stratum corneum binding of water indissoluble could be compromised and ineffective. For this reason, it is beneficial to decrease the transepidermal water loss by applying occlusive films. Hence, there is no cause why mineral oil or petrolatum should not be used. Hence, vegetable oils have more benefits for dry skin treatment.

Castor oil formed from (*Ricinum communis*) castor seeds, consists of more than 50% of fixed oil, more viscous, pure when colorless and has slight odor. Castor oil is composed of ricinoleic acid and unsaturated fatty acids. Hence, castor oil is used to skin smoothing, protect skin from harsh climate as well as used to prepare soaps. Ricinolic acid and its analogs are used to improve the skin moistening and smoothing qualities as well as enhance the skin conditions such as acne and rough skin. Hydrogenated castor oil or its esters oils has been used for solubilizers for toiletry as well as skin and hair care cosmetic formulations and are useful condition and cleaning of the skin and hair (Sato 2002). Cocoa butter is used for the smoothing of sunburn and windburn. Cocoa butter is obtained from *T. cacao*. Cocoa butter composed of triglycerides consists of palmitic acid, stearic acid, and oleic acid as well as monounsaturated fatty acid. Hence, it is widely used in moisturizer and also used in cosmetic preparations and it has reported as potential antioxidant (Sato 2002). Sunflower oil is composed of polyunsaturated fats, triglycerides, or linoleic acid as well as essential fatty acids. These constituents are used for maintaining good skin. Linoleic acid reduces the water loss transepidermal and eliminates the scaly lesions normally with patients. Hydrated sunflower oil and oleic acid contents are used in natural and functional raw materials in cosmetic preparations. Olive oil is composed of triglycerides, tocopherols, squalene, carotenoids, sterols, chlorophylla, polyphenols, squalene, volatile oils, flavored compounds, and fatty acids. These constituents are useful for treatment of dry skin, as well as used in lip balm, soap, hand lotions, shampoos and conditioners and massage oils. Extracts of leaves and fruits of olive tree show anti-inflammatory and oxygen scavenging effects (Tehara and Hachimaki 2002). Olive oil shows potential free radical scavenging effects, due to rich in polyphenols; it is applied for contact dermatitis, skin damage, atopic dermatitis, eczema, xerosis, seborrhea, psoriasis, rosacea thermal and radiation burns as well as other skin aging and inflammations.

**Eczema.**

Eczema is a skin condition specified by scaling, itching, redness, and swelling. For treating eczema, turmeric has been used potentially. *C. longa* L; turmeric is processed rhizome portion, which is used for traditional medicine. It is usually boiled yellow powder. Curcumin is a major chemical component in turmeric, which has potential biological activities such as anti-HIV, antiparasitic, antibacterial, anticarcinogenic, antioxidant, anti-inflammatory as well as wound healing powder, applied to septic and aseptic wounds and inhibition of lipid peroxidation. Curcumin is also used to check and prevention or treatment for psoriasis, skin damage from sun radiation, wounds, burns, acne, and premature aging (Phan et al. 2001).

**Acne, spots, and pimples.**

Skin condition damage causes whiteheads, blackheads, inflammation, sweat glands, and hair follicles. Some natural product extracts are traditionally used for the treatment of acne, spots, and pimples. Extracts from *A. vulgaris*, *A. absinthum,* and *A. campestris* are used for rapid healing wounds, skin ulcers as well as eczema, herpes, and purulent scabies (Aniya et al. 2000). Extracts from *O. gratissimum* and *O. basilicum* essential oils are used for treatment of acne, pimples and spots as well as it shows antibacterial treatment for acne, antiseptic and antimicrobial activities (Orafidiya et al. 2002). *P. sativum* extracts are used for the treatment of acne, due to peas composed of fats, salts, proteins, lecithins, and carbohydrates. For example, crushed peas are used for face masks, acne, and wrinkled skin (Orafidiya et al. 2002). *C. pepo* (pumpkin) extracts are composed of fatty acids; it has been used for traditional medicine, and it shows potential anti-inflammatory properties due to fatty acids are mainly composed of palmitic, stearic, and oleic acid. The roots, leaves, and seeds of these extracts are used for pimples, blackheads, sores, and herpes lesion (Orafidiya et al. 2002). *A. cepa* (red onion) has been used for traditional medicinal system, it could be used externally for boils, blackheads, and abscesses to draw out of the infection as well as reduce the inflammation and improve the healing. Red onions are composed of high amount of flavonoids; hence, it shows potential anti-inflammatory and antiallergic properties as well as onion juice shows antimicrobial and antifungal effects. In Africa, onion juice is applied for scalds, infection, and burns, especially in East Africa onion skin has been used for body sores and facial (Aburjai and Natsheh 2003).

**Anti-aging skin treatment.**

Human Skin aging caused by UV radiation from the sun is one major environmental factor. Skin aging is due to exceeded degenerative and regenerative changes and epidermises are analyzed by wrinkling and thinning together in the aspect of

**Table 19.4** Plants used in skin cosmetics and toiletries as cosmeceuticals

| Plant Scientific name | Active chemical constituent | Cosmetic use |
|---|---|---|
| *A. Catechu* | Catechin | Antioxidant |
| *A. Vera* | Aloin | Antidermatitis |
| *A. Recutita* | Chammomile | Antiphologistic |
| *A. Sativum* | Alliin and Allicin | Antioxidant |
| *B. Seeds* | Rutin | Anti-wrinkle |
| *C. Sativus* | Crocetin | Protective |
| *C. Longa* | Curcumin | Antibacterial |
| *C. Asiatica* | Centella | Skin Firming/Conditioning |
| *C. Limonus* | Hesperedin | Fungal Infection of Skin |
| *G. Glabra* | Glycyrrhizin | Skin Whitner |
| *G. Tea* | Chammomile | Photoprotective |
| *C. Murula* | Lupenol | Anti-aging |
| *R. Officinalis* | Rosemary | Anti-aging |
| *E. Officinale* | Ascorbic Acid, Tannins | Protective |
| *G. Biloba* | Ginki | Skin Tonic |
| *P. Corlifolia* | Psorolin | Skin Staining & Pigmenting Agen |
| *T. Viridis* | Gallic Acid, Catechin and Rutin | Antioxidant |
| *V. Vinifera* | Carotene | Eczema |
| *D. Carota* | Beta Corotene | UV Protection |
| *L. Esculantum* | Tamotine and Tamotidine | Potent Bacteriostatic |
| *H. Virginiana* | Gallic Acid | Cooling Agent |

lines, groove, crack, and wrinkle, especially in lines of facial expressions. This is the reason for rapidly evident morphological alterations. There are many so-called anti-aging that is traditionally applied material nothing more than moisturizers. Ginseng (*P. ginseng*) is a traditional medicinal system for treatment of anti-ageing, for more than 2000 years in Korea. Compared to other countries, Korean Ginseng is chemically and physically different. Extract of this plant is more active to enhance the skin metabolism, provide soften and moisture as well as enhance the skin whiteness and reduce keratinization. The anti-aging effect leads due to the increase of blood circulation and increase of skin nutrition and cell proliferation (Aburjai and Natsheh 2003). Moreover, natural extracts or oils are a major source of phytosterols, and tocopherol chemical constituents, which has been used for antioxidant and bioactivity skin formulations. Generally, wheat germ oil, corn oil, and seaweed extracts help to maintain skin elasticity and moisture. The details of plants and their benefits in cosmetic benefits are shown in (Table 19.4).

**Free-radical scavenging effects.**

Natural product extracts are major source of plants, hence plant extracts shows potential free-radical activity, due to polyphenols its derivatives, tannins, and flavonoids (Ashawat et al. 2000; Pietta 2000). *C. sinesis* yields Black and Green Tea. Black and green tea acquire leaves fermentation and after harvest leaves are immediately steamed and dried, respectively. Tea is composed of more than 500 chemical constituents such as amino acids, tannins, flavonoids, vitamins, caffeine and polysaccharides, which are similarly found in lemons. Black and Green tea contain almost same amount of vitamin B6, vitamin E, and vitamin K, but 90% of vitamin C destroyed during the fermentation process. Flavonoids which is present in tea proven potential properties such as antibacterial, antiviral, antioxidant, anti-inflammatory, antiallergic and while tannins show antiseptic and antioxidant properties. Root extracts from tea are composed of Saponins; it shows potential anti-inflammatory and antioxidant properties. Green tea consists of polyphenols which are major active constituents; catechins are major important polyphenols, while flavonoids such as phenolic acids and flavones. Recent days Green tea is now subjected key attention, it is proven as potential antioxidant property for its capability to repair UV photo-damage and phototoxicity. Green tea and its extracts are used for dry skin treatment and thus it shows potential anti-inflammatory and anticarcinogenic effects of skin disorders (Katiyar and Elmets 2001). While composition of polyphenols is less abundance in black tea when compared with green tea; but, it is still considered as a good source of antioxidant properties. Applying black and green tea oral extracts decrease the photochemical damages to the skin (Katiyar and Elmets 2001). Green and black tea extracts are remaining essential in preventing the early sign of UV radiation phototoxic effects (Katiyar and Elmets 2001). *V. vinifera* L. (grape seed) is composed of different types of polyphenolic proanthocyanidins. These polyphenols show potential antioxidant activity when compared with Vitamin C and Vitamin E and it shows tyrosinase-inhibiting activity, which has shown potential skin-lightening and anti-aging cosmetics (Lee et al. 2001).

**Anti-inflammatory effects**.

In many diseases, inflammation is usual response. It's bearing and controlling in the treatment of these pathologies. There are many natural products that show anti-inflammatory properties. *T. pretense* L. (Red clover) shows potential anti-inflammation for multiple skin conditions like acne, rash, psoriasis, and eczema. Red clover consists of isoflavones, which is used for UV radiation protection, reduces the inflammatory aedema reaction hypersensitivity induces simulated UV radiation. *M. recutia* L. (German chamomile) and *A. nobilis* Linn. (Roman chamomile) extracts similar chemical constitutes with almost same ratio. Extracts from these plants traditionally are used in the form lotions, ointments, and inhalations. Essential oil extracts and its isolated substances from these plants are used for treatment and prevention of different skin disorders. Chamomile essential oil extracts consist of flavonoids like apigenin and glycosides, which shows potential antipruritic, anti-inflammatory, and antierythema effect (Aburjai and Natsheh 2003). Extracts from chamomile consist of chamazulene and bisabolol shows anti-inflammatory activity. *T. foenum* (Fenugreek) is fragrant herb, whose seed is used for traditional medicinal system for all ages. Europe, Greek, Romans,

Egyptians, and among othera are used for medicinal and delicious purpose. Seeds of this extract show potential antioxidant and anti-inflammation and emollient properties. Extracts are traditionally used for treatment of skin inflammations, mouth ulcers, and chapped lips. Worldwide *G. glabra* L. (Licorice root) extracts used in traditional medicinal system. Root extracts of these plants composed of 5–10% of glycyrrhizin, a sweet taste and in general, it is less soluble in blood compared with Saponins. Hence, Licorice root extracts consist of glycyrrhetic acid, which shows potential anti-inflammatory effects as well as used for skin problems like irritations and acne (Aburjai and Natsheh 2003).

**Miscellaneous**.

Cucumber (*C. sativa* Linn.) is palliative, which cools and heals the irritated skin by sun radiation or cutaneous eruption. Cucumber and lemon extracts are used in cosmetics and cosmedicals products that are used for the treatment of hyperpigmentation. Both extracts are not interfering with each other and provide skin lighting. Many scientific reports indicate the presence of antioxidant enzymes and superoxide's activity with cucumber fruit and highest in the skin (Aburjai and Natsheh 2003).

**Enumeration**.

*E. officinalis* Gaertn; commonly known as Amla, dry fruits of this plant are used for traditional medicinal system and cosmetics. Fruit extracts are composed of phyllembic acid as well as isolated other chemical constituents such as ellagic acid, corilagin, terchebin, trigalloylglucose from fruits. These extracts are used for treatment of cooling, diuretic, laxative, and it is rich source of vitamin C. Vitamin C is five times rich when compared with orange juice; for this reason, it is used for hair dyes. The fruit powder and extracts are used as hair care products such as shampoos and used as medicine for hair roots. *Rubia coedifolia* Linn. its common name Manjit, from this plant dry roots and stems are used for traditional medicine, especially used in cosmetics. Roots and stem extracts are composed of purpurin, munjistin, Alzarin, and glucosides as well as anthraquinone derivatives. These chemical constituents are used for antiseptic, antidysenteric deobstruent and root ionic. Decoctions of leaves and stems extracts are used for vermifuge. Especially extracts from stem are used for rhinosinal infections due to the presence of septilin drug as well as roots extracts are used as coloring medicinal oils. Extract could be applied directly in skins care such as removal of dark spots on the face, treatment for acne, skin regeneration, and anti-ageing.

*Acacia concunna* DC. Is commonly known as Ritha and Shikakai in India. The extracts from this plant are used especially as hair care products; treatment for hair growth, hair splitting, dandruff, and hair fall. It is used to keep original hair color. These applications due to plant extracts are composed of saponin, mixture of acacinin-A and B as well as carbohydrates are composed of fructose, glucose, and xylose. Sometimes seed extracts could be used for fish poison.

According to the ethnic or community groups understanding, the above-mentioned plants are successfully used for skin and face problems such as dark shadows, wrinkles on the face, acne, and pimples. Skikakai and amla could be used for the hair problems such as hair color, scalp care, hair falling, and dandruff. Particularly, Indians used cosmetics and cosmedicals for skin and hair care as shown in (Table 19.5). Besides some other medicinal plants having soap properties which can be used for skin and hair care products are shown in (Table 19.6) (Sharma et al. 2003).

### *19.7.2 Natural Products as Hair Care Agents*

Natural products are used for stimulating hair growth, hair color, and dyes as well as scalp problems like dandruff.

**Hair growth stimulants.**

Nowadays, natural products and its derivatives are used for hair growth and hair tonic products and precaution of alopecia, due to increasing blood circulation,

**Table 19.5** Plants used for cosmetics for skin and hair care

| Plant Scientific name | Family | Part used | Cosmetic use |
|---|---|---|---|
| *A. concinna* DC | Mimosaceae | Pods | Soaps, hair splitting, hair falling, and dandruff |
| *A. barbadensis* | Liliaceae | Leaves | Hair falling, dandruff, and sun burn |
| *A. racemous* | Liliaceae | Roots | Used in cure of wrinkle on face |
| *A. indica* | Meliaceae | whole plant | Skin, hair, and scalp care |
| *B. orellana* | Bixaceae | Seed pulp | Seeds used for color, mascaras, & lipcare |
| *C. macrophyllaia* | Verbenaceae | Fruits | The fruits are blended in creams to treat acne |
| *C. longa* | Zingiberaceae | Rhizome | Improves the color of the skin |
| *C. amada* | Zingiberaceae | Rhizome | A good face pack |
| *E. prostrate* | Asteraceae | Whole plant | Used in keep the hairs in original color |
| *E. officinalis* | Euphorbiaceae | Fruits | Commonly used in hair care |
| *A. moschatus* | Malvaceae | Seeds | Used to provide musk like fragrance to cosmetics |
| *H. abelmoschus* | Malvaceae | Seeds | Used to provide musk like fragrance to cosmetics |
| *L. inermis* | Lythraceae | Leaves | Used to color for hairs |
| *O. dillenii* | Cactaceae | Fruits | Used in lipcare |
| *R. cardifolia* | Rubiaceae | Whole plant | Application on skin and lip care and treat for acne and pimples |

**Table 19.6** Selected plants in India used in soaps as cosmetics

| Plant Scientific name | Family | Common name |
|---|---|---|
| *A. concinna* DC | Mimosaceae | Shikakai |
| *A. millefolium* Linn | Asteraceae | Gandana |
| *A. nobilis* Linn | Asteraceae | Roman Chamomile |
| *L. inermis* Linn | Lythraceae | Henna |
| *M. chamomilla* Linn | Asteraceae | Babuna |
| *M. fragrans* Houtt | Myristicaceae | Jayaphal |
| *N. jatamansi* DC | Valerianaceae | Jatamansi |
| *P. emblica* Linn | Euphorbiaceae | Amla |
| *S. mukorossi* Gaertn | Sapindaceae | Reetha (soap nut) |
| *S. saponaria* Lour | Sapindaceae | Reetha (soap nut) |
| *S. trifoliatus* Linn | Sapindaceae | Reetha (soap nut) |
| *V. officianalis* Linn | Valerianaceae | Tagger |

activation of dermal papilla, increased nutrition through increased blood flow and antitestosterone action is not yet clear (Mans and Grant 2017). Grape seeds and *G. biloba* leaf extracts are composed of proanthocyanidins. This has promoted proliferation of apoptosis hair follicle cells. Thus, these extracts are suggesting potential hair tonic (Aburjai and Natsheh 2003). But other plants claimed potential hair growth such as aloe, henna, rosemary, and sage. But it required further clinical trials to use traditional medicine. *A. barbadensis* and *A. vera* gels are used traditionally for hair growth (alopecia) and hair loss. The plant extracts consists of aloenin that is a major chemical constituent, which has major responsible for hair growth without any skin irritation and side effects (Aburjai and Natsheh 2003). Henna (*L. alba* L.) has been used for hair dyes. Since ancient time, Egyptians are used to hair loss. More than 500 species of sage is used for traditional medicine as general medicine. Generally, it is used for healing purpose. *S. officinalis* L. is also known as garden sage. The extracts of these sages are used as lotions to improve the skin and hair growth as well as conditioner. Combination of sage and rosemary used to maintain the dark wavy hair and stimulate hair growth. *S. officinalis* L active chemical constituents such as tannins, Saponins as well as camphor and borneol are accountable for the effect on hair (Aburjai and Natsheh 2003). Rosemary (*R. officinalis* Linn), is a known aromatic traditional medicinal plant. It is used in folk medicine to stimulate hair growth. It is delicious, medicinal, and cosmetic properties. Rosemary consists of coffeic acid and its derivatives like rosmarinic acid which shows potential antioxidant property.

**Dandruff treatment**.

In the recent years, dandruff became a major problem due to fungal infections, microbial, and environmental conditions. The dandruff has been knowledgeable, because of scaling and flaking of the scalp. Dandruff sufferers get damaged scalp as well as decreased lipid level, ceramides, cholesterol and fatty acids. Hence, epidermal water level reduces on the scalp which causes dandruff (Harding et al. 2002). Rosemary and sage extracts are potentially used for dandruff. The extracts are used for the treatment of greasy hair; hair loss, and skin as well as extract sage massaged on the scalp could control dandruff, loss of hair, or hair falling. Thyme essential oils also used to inhibit the dandruff, scalp rub to avert hair loss as well as rosemary and thyme encourage hair health. Garlic extracts are used traditional medicine for dandruff and as a vegetable, it includes potential medicinal properties such as antifungal, antibacterial, antioxidant, antiseptic, tonic, and anti-inflammatory (Agiga and Seki 2000). Garlic could not be used directly; it causes burning sensation, contact dermatitis, and allergic reactions for some people. Walnut leaves extracts are used in external applications for hair loss, itching, eczema, acne, peeling, and dandruff as well as skin disorder treatment such as itching, emollient, abrasions, frostbite, treatment for sun burns, and nappy rashes as well as dandruff and scalp problems (Aburjai and Natsheh 2003).

**Hair coloring**.

For hair coloring, vegetable oils are usually recommended, because of low allergenic power. For hair color, natural products do not have great advance, sinceplant extract dyes or natural dyes are unstable in solution form, liable to oxidation, pH color shift, fading and discoloration and single dye may not give right color for hair. But only walnut and henna are suitable for hair color by mixing of leaves of other plants (Aburjai and Natsheh 2003). *L. inermis* is known as henna. It is used for hair color, feet, and hand color as well as used for certain skin disorders. Henna consists of lawsone chemical constituent, which is responsible for developing red color in henna. The chemical constituent is isolated from leaves as brown powder. Lawsone has strong binding capacity to the hair, due to the chemical reaction between thiol groups with keratin. German chamomile leaves extracts consist of apigenin flavonoids, which could be used for dull golden yellow color as well as flowers extracts are used for hair rinse. Turmeric also is used for hair color to convert yellow to deep orange color which is due to Curcumin pigment present, which is responsible for yellow color of herb. *H. sabdariffa* L. extracts show red color due to extracts composed of red color anthocyanidins such as delphinidin could be used. But red color depends on pH of the solution.

### *19.7.3 Essential Oils Used as Cosmetics*

Essential oils are the complex and composition of mixtures. Aromatic and medicinal plants and its oils have been used in perfumes, cosmetics, medicines, and culinary applications. Essential oils could be used for bath and skin massage, as scent (inhaled directly or diffused). Since ancient time, essential oils are used for pain relief, tension alleviation, skin care, fatigue, produce sense of relaxation and revitalize the entire body which is potential benefit for cosmetics. Essential oils and

**Table 19.7** Plant essential oils are used in cosmetics

| Plant | Family | Essential oils extracted from parts |
|---|---|---|
| *A. squarrosus* L. | Gramineae | Roots |
| *V. zizanioides* L. | Gramineae | Roots |
| *C. aurantiacum* L. | Rutaceae | Flowers |
| *E. dives* Schauer | Myrtaceae | Leaves |
| *J. officinale* Linn | Oleaceae | Flowers |
| *J. virginiana* Linn | Pinaceae | Wood |
| *M. piperita* Linn | Labiateae | Aerial parts |
| *M. champaca* Linn | Magnoliaceae | Flowers |
| *M. elenhi* Linn | Sapotaceace | Flowers |
| *P. fascicularis* Lamk. Syn | Pandanaceae | Male inflorescence |
| *R. damascene* Mill | Rosaceae | Flowers |
| *S. album* Linn | Santalaceae | Wood |
| *S. lappa* Clarke | Compositae | Roots |
| *S. robusta* Roth | Dipterocarpaceae | Stem |
| *S. aromaticum* Linn | Myrtaceae | Flower bud |
| *V. odorata* Linn | Violaceae | Flowers |

its derivatives are used for skin emollient, hair conditioners, pleasant aroma, and skin elasticity. Essential oils are composed of terpenoids, phenylpropanoids, and fatty acids. Terpenoids belong to mainly mono, di, and sesquiterpenes. These largest terpenoids more than 30,000 groups are synthesized by isopentenyl diphosphate and other chemical constituents are present such as small chain alcohols and aldehydes, it is formed by conversion of fatty acids and phospholipids. Essential oils show different medicinal and cosmetic properties (Reddy 2019) as well as some essential oils producing plants are used in cosmetics shown (Table 19.7).

**Cosmetic benefits**.

Essential oils could be used in perfumes, skin lotions, hair care products and pleasant smell and glow. Essential oils show potential antibacterial activity, hence it could be used in preservative system in cosmetics. The essential oil efficiency depends on both concentration as well as the microbial strain. Essential oils could be used in cosmetics, for instance, menthol, mint, camphor, and eucalyptus oils as cooling agents, as well as menthol analogs such as methoxypropanediol, menthyl hydroxybutyrate, menthyl glucoside, menthoxy furan, and menthy lactate as a refreshing the skin (Aburjai and Natsheh 2003).

**Perfumery**.

Perfume word is derived from Latin. Essential oils are major applications for the perfumes. In ancient times, perfumes are obtained by simple method to produce perfume, soak flower petals in fat known as pomade (Barel et al. 2001). Plant flower fragrance has rapidly converted volatile constituents into high impact commercial commodity. Cosmetics, perfumes, and therapeutic fragrances are extracted from flowers of some aroma plants such as narcissus, tuberose, gardenia, and rose. Perfumes are mainly used in two purposes, one it could be used as cosmetics and toiletries precuts such as hair products, personal care products, bath products, deodorants, and fine fragrance and other types used as household products include laundry products, surface cleaners, room fresheners, washing liquids, and disinfections. Essential oils are major ingredients for preparation of perfumes. Essential oils are composed of hundreds of suitable ingredients, which are added to fragrance composition to improve the smell. Still successful fragrances contain significant quantities of naturally occurring essential oils (Barel et al. 2001).

**Hair care**.

Essential oils are used for hair care products to remove negative smell from perm lotion; it helps for hair conditioning and shampoos, improvement of hair texture and also longer lasting pleasant smell. Rosemary and chamomile essential oils are added straight to mild shampoo. Rosemary oil can help hair condition and potential hair growth as well as tea tree could help controlling dandruff. Lavender could be used to repel lice and fleas. Essential oils mixed with hair care products; it would be improved shine and conditioning effects.

**Aromatic skin care**.

Skin problems are generally the surface manifestation of deeper conditions such as environmental pollution, toxins from blood, hormonal imbalance, nervous and emotional difficulties. Essential oils are potential use for such problems. Essential oils are soluble in oils and alcohols and important scent to water. Hence, essential oils provide ideal chemical constituents for cosmetics and specific disease. Essential oils are made up of very small molecules, and interact through the skin, it means of lipophilic fractions reacting with the lipid parts of the cell membrane. For instance, citrus oils such as lemons, grapefruits, oranges, and tangerines are composed of mixture of alcohol, acids, esters, carbohydrates, ketones, and aldehydes. These chemical constituents are rich in fragrance and flavor. Hence, essential oils are used to important taste to different products, aroma, and taste. Essential oils are used personal perfumes widely, where they are used pure and frequently combined with synthetic substances. Lavender oils distillated from mainly *Lavandula* have been used therapeutically and cosmetic agents. These oils are extremely used for the preparation of soaps, perfumes, and skin lotions. Lavender oils show potential antibacterial, antioxidant, antifungal properties. Essential oils from black cumin are composed of thymoquinone, carvacrol, 4-terpineol, and t-anethole shows antioxidant, antibacterial, antifungal and anticarcinogenic, analgesic and anti-inflammatory properties. Hence, these oils could be used for skin care (Aburjai and Natsheh 2003).

## 19.8 New Trends in Cosmetics (Plant Origin of By-Products)

In recent days, fruits, vegetables, and foods (agronomical) industries increase the amount of waste, which produce different types of expendable by-products; rich in valuable components such as pharmaceuticals, food, and cosmetics. There are many cosmetic active constituents which are extracted from meat, fish, and dairy products. Furthermore, in agronomical disposable wastes are used in cosmetic field. Hence, these products are commercially less expensive, bio-feasible, more effective, and environmental friendly. In consequences to plant-derived extracts from wastes are adopted in cosmetic industry. Moreover, the generated wastes or products from organic farming are absolutely more valuable source and safe to use in cosmetics. Since decades, fruits and vegetables and agricultural products are considerable more significant in our everyday diet and their indubitable nutritional values have been comprehensively studied (Barbulova et al. 2014). Since decades, human kinds used fruits and vegetables for fragrance, perfumes, flavoring, preservatives, cosmetics, and pharmaceuticals. Vegetables, fruits, dairy, and agricultural products have enormous health benefits such as heart diseases, heart stokes, respiratory disorders as well as variety of cancers. Fruits and vegetables are composed of flavonoids exhibit that shows potential antioxidant activity; grapes and apples present active substances such as flavonoids, catechins, epicatechins, and procyanidins, which are used for heart diseases; some vegetables such as red tomatoes and orange are composed of carotenoids, β-carotenes is present, which are effective for neutralizing the free radicals; in cherries, red grapes, and berries anthocyanidins are present, which are used for brain function; citrus fruits are more dominant for flavonoids. Because of the above reasons, fruits and vegetable consumptions considerably increased in last few decades and also the amount of wastes and residues increased substantially. Without observing waste during food processing, every year, agro-food industries generate 800,000 tons wastes around worldwide (Ayala-Zavala et al. 2010; Tuck et al. 2012).

Food and agriculture industries are generating 10–60% (Table 19.8) of solid waste and raw materials. It was demonstrated that in few cases main products are less valuable than waste products (Liu et al. 2012). It was indicated in percentages, which is preferred to call by-products. Industry by-products are capable to be recycled as valuable products. These by-products are potentially made by fruits, skin, stems, seeds, leaves unusable pulp, and wastewaters generally through way. In few cases, wastes represent more than 40% of total plant food (like mango, papaya, pineapple, citrus fruits, artichoke, and asparagus) (Liu et al. 2012). Food by-products know to minerals, sugars, and organic acids, bioactive constituents as flavonoids, carotenoids, and polyphenols parallel to their essential counterparts. Because of this phytochemical composition, finding natural chemical constituents are as an alternate for synthetic substances. The by-products in industrial fields are used in food, pharmaceutics, nutraceutics, and cosmetics.

**Table 19.8** Percentages of by-products generated from fruits and vegetables processing industries

| Plant | Edible part (%) | % of by-products |
|---|---|---|
| Agave | 60 | 40 (rind and pith) |
| Apple | 89 | 11 (pulp and seed core) |
| Artichoke | 40 | 60 (outer bracts, receptacles, and stems) |
| Asparagus | 50–60 | 40–50 (spear) |
| Banana | 70 | 30 (peel) |
| Cactus cladodes | 80 | 20 (spines, glochids, and peel) |
| Carrot | 60–70 | 30–40 (pomace) |
| Mandarin | 84 | 16 (peels) |
| Mango | 58 | 42 (seeds, peels, unusable pulp) |
| Citrus fruits | 44 | 66 (peel) |
| Papaya | 53 | 47 (seeds, peels, unusable pulp) |
| Passion fruit | 25 | 75 (rind and seeds) |
| Pineapple | 48 | 52 (core, peels, top, pulp) |
| Potato | 60–85 | 15–40 (peel) |
| Tomato | 93–97 | 3–7 (peel and seeds) |

### *19.8.1 By-Products from Citrus Fruits*

*Rutaceae* (citrus family plants) also known as *agrumes* are one of the largest fruit crops worldwide. Citrus fruits are rich in vitamin C (citric acid) and vitamin B (like thiamin, niacin, pyridoxine, riboflavin, pantothemic acid, and folate), as well as flavonoids, limonoids, and carotenoids chemical constituents are present. These ingredients are the major source of dietary fibers, lowered circulating cholesterol, and gastrointestinal diseases. Citrus fruits contribute to human diet and exhibit good antioxidant properties. From citrus family, some of them are eaten fresh such as grapes, oranges, and tangerins, as well as others around 85% of industrial processed consumption such as lemon and orange juices. Phytochemicals of these natural sources are used for human color for skin, hair color, and eyes, as well as skin protection from ultraviolet radiation. Ethanol extracts waste by-products from the citrus family shown potential antimicrobial, antioxidant, and anti-inflammatory properties, because these extracts are major source of essential oils and flavonoids. Essential oils are used particularly as preservatives against spoilage, pharmaceuticals, and cosmetics. For example, *C. unshiu* peel essential oils composed of limonene (80.5%), γ-terpinene (6.80%), and cymene (4.02%) are major compounds. These essential oils exhibit potential antibacterial activity, anti-inflammatory, antimicrobial, and antioxidant properties. Essential oils and extracts of citrus family by-products are used as hair and skin care products.

### *19.8.2 By-Products from Tomato and Olive*

Tomato and olive are the most important food products worldwide. Tomato and olive oil food processed products are mainly located in the Mediterranean areas and shipped worldwide.

The annual production of tomato in the world is around 160 million tons, about 40 million tons of which are processed such as tomato peel, tomato paste, unpeeled, and chopped tomatoes (Tomato Processing industries 2014). High amount of important tomato by-products (peels, pulps, and seed) generated due to tomato processed industrially are rich in lycopene which has antioxidant properties, and used for disease prevention. Tomatoes are highly demanded for food, pharmaceuticals, and cosmetic industries. Lycopene is extracted from peel approximately five times more than pulp. Extraction of lycopene commercially is expensive due to its difficult extraction process. Tomatoes by-products' high moisture content and sensitivity to microbial spoilage make the storage and processing of this material problematic (Papaioannou and Karabelas 2012). Hence, lycopene ensuring to be used in food and cosmetic industry (Papaioannou and Karabelas 2012).

Mediterranean diet olive oil has most important to improve health status in terms of heart disease. It is already know that olive oils are principle ingredients which are used to variety of cosmetic products such as face and body creams, soaps, lotions, shampoos, body massage oil, and hair oils. Olive oils by-products are composed of phenolic compounds and monounsaturated fatty acids. Olive oil and by-products show potential biological activity in epithelial, endothelial, platelets, neurons, neoplastic cells, and immune system. By-products are composed of phenolic compounds such as hydroxytyrosol, oleuropein, and other derivates show in vivo and in vitro antioxidant property. A lot of research efforts were published especially about oleuropein and its pharmacological activities which include anti-inflammatory, anti-atherogenic, anti-cancer, antimicrobial, antiviral, antioxidant, hypoglycemic, and hypolipidemic effect (Omar 2010). Olive oil and its by-products already known to skin and hair care products.

### *19.8.3 By-Products Processing from Coffee*

Coffee is among the most crops generating large amount of by-products in the coffee processing industry. Globally, coffee (*C. arabica* and *C. robusta*) is usually cultivated in South Asia, India, Latin America, and Africa. By-products of coffee are produced from the coffee pulp and husk processing, it has less applications such as prepare compost, fertilizer, and livestock feed (Murthy and Naidu 2012). While collection and selection of coffee beans some of coffee berries are unused green beans, they have been mechanically damaged and sufficiently ripe and big to pass the next processing step. The green beans have valuable antioxidants, which not yet been broken by the roasting process (Murthy and Naidu 2012).

Otherwise, the unroasted beans are extracted with water, carbon dioxide, and ethanol; these extracts are rich in bioactive antioxidant constituents. The removal of ethanol from the extract results in jelly like extract which is used to improve the skin tone, natural skin cell renewal, reinforced the epidermal barrier, and decreasing inflammatory process.

## 19.9 Future Prospects and Conclusions

This chapter presents the selected plants extracts and their by-products used in the preparation of cosmetic products worldwide. Globally, more than 5,000 plants are used in cosmetics and cosmeceuticals. Currently to develop cosmetics and cosmeceuticals, natural products are untapped reservoir.

This clench identifying and developing many plants-derived natural beauty products. For occasion *B. orellana* L (Fig. 19.7) seed extracts are used for orange-red wax or lipstick plant, Amazon base native people has been used for body and facial embellishment, as well as used in lipsticks, powders, nail polishes, eye shadows, and cream blushes. Baobab (*Adansonia digitatal* L) is generally known as monkey tree or upside-down tree. It is native of Madagascar and southwest of Africa. This plant seed extracts have been used for normal and dry skin anti-irritating, antioxidant, and anti-sensitizing as well as hair and nail conditioning purpose (Hetta 2016a, b). Mafura (*Trichilia emetica* L.) common name Natal mahogany is growing in Zimbabwe, Sudan, South Africa, and Uganda. Its plant extracts show potential antioxidant, and anti-inflammatory effects, preparing

**Fig. 19.7** Seeds of the annatto *Bixa orellana* L. (Bixaceae)

natural soaps, making candles, as well as used for hair and skin care products traditionally (Hetta 2016a, b). *Ximenia american* L. (common name Sour plum), it has been distributed in Africa. Extracts of this plant are used in traditional medicinal system like skin problems, angina, fever, toothache, and others, it is also used traditionally in cosmetics such as lipsticks, lubricant, soaps, vegetable butters, emollient, conditioner, body massage oil, hair oil and conditioner and skin softener (Hetta 2016a, b). The early shoots of *C. aurantium* L. (sour orange) chewed for fresh breath in elderly age women's in Creole as well as essential oils from this plant might be used in soaps, skin care products, deodorants, and mouthwashes. Essential oils from seeds of *D. odorata* are extensively used in perfumes industry due to major amounts of coumarins with pleasant odors and also used in preparation of skin and hair care products.

Moreover, an important pathology to take advantage of these opportunities to develop active chemical constituents, mechanism and its characterization must be necessary for cosmetics and cosmaseuticals. For example, *A. vera* is used for wound healing, but it needs high level of supporting evidence and its promoting activity, *A. vera* analog products are used as topical agents for treatment of skin lesion (Pereira and Bártolo 2016). *E. olearacea* products are used worldwide in most foods, cosmetics, and cosmeceuticals; but, health and scientific evidence are still insufficient (Yamaguchi et al. 2015). *M. citrifolia* fruits are used in general health disease, power and energy drinks, fruit juices as well as leaves formulated as capsules or pills high in demand. But most of assert of noni plant products are not carried by hard scientific evidence (Assi et al. 2017).

Over few decades, enormous understanding of cosmetic skin formulations and skin biology was developed. Different types of commercial cosmetics are available for skin care products such as skin whitening, skin protection lotions, skin creams, and anti-ageing. Several natural products are serving cosmetic industry in addition to their medicinal benefits to the skin. More research efforts are required for delivery of natural products cosmaceuticals ingredients and report of their activities in cosmetic field could lead the development in the next decade.

## References


Abou AAA, Abu SFM, Abou AEA (2011) Physicochemical properties of natural pigments (anthocyanin) extracted from Roselle calyces (*Hibiscus subdariffa*). J Am Sci 7:445–456

Aburjai T, Natsheh FM (2003) Plants used in cosmetics, phytotheraphy research 17:987–1000

Adelowo F, Oladeji O (2017) An overview of the phytochemical analysis of bioactive compounds in *Senna alata*. Am J Chem Biochem Eng 2:7–14

Afzal M, Kazmi I, Khan R, Singh R, Chauhan M, Bisht T, Anwar F (2012) Bryophyllum pinnatum: a review. Int J Res Biol Sci 2:143–149

Agiga T, Seki T (2000) Flavor components of garlic (*Allium sativum* L.) and their multiple functions. Aroma Res 1:16–27

Akhtar N, Mehmood A, Khan BA, Mahmood T, Khan MHS, Saeed T (2011) Exploring cucumber extract for skin rejuvenation. Afr J Biotechnol 10:1206–1216

Akpuaka MU, Ezem SN (2011) Preliminary phytochemical screening of some Nigerian dermatological plants. J Basic Phys Res 2:1–5

Al-Fatimi M, Wurster M, Schroder G, Lindequist U (2007) Antioxidant, antimicrobial and cytotoxic activities of selected medicinal plants from Yemen. J Ethnopharmacol 111:657–666

Ali A, Akhtar N, Khan MS, Rasoo F, Iqbal FM, Khan MT, Din MU, Elahi E (2013) Moisturizing effect of cream containing *Moringa oleifera* (sohajana) leaf extract by biophysical techniques: in vivo evaluation. J Med Plants Res 7:386–391

Amenta R, Camarda L, Di Stefano V, Lentini F, Venza F (2000) Traditional medicine as a source of new therapeutic agents against psoriasis. Fitoterapia 71:S13-20

Aniya Y, Shimabukuro M, Shimoji M, Kohatsu M, Gyamfi MA, Miyagi C, Kunii D, Takayama F, Egashira T (2000) Antioxidant and hepatoprotective actions of the medicinal herb *Artemisia campestris* from the Okinawa islands. Biol Pharm Bull 23:309–312

Ashawat MS, Banchhor M, Saraf S, Saraf S (2000) Herbal cosmetics: trends in skin care formulations. Pharmacognosy Rev 3:72–79

Ashraf F, Gilani SR (2007) Fatty acids in *Moringa oleifera* oil. J Chem Soc Pakistan 29:343–345

Assi RA, Darwis Y, Abdulbaqi IM, Khan AA, Vuanghao L, laghan MH (2017) *Morinda citrifolia* (noni): a comprehensive review on its industrial uses, pharmacological activities, and clinical trials. Arab J Chem 10:691–707

Atanasov AG, Waltenberger B, Pferschy WEM, Linder T, Wawrosch C et al (2015) Discovery and resupply of pharmacologically active plant-derived natural products: a review. Biotechnol Adv 33:1582–1614

Attah MO, Ishaya HB, Chiroma MS, Amaza DS, Balogun SU, Jacks TW (2015) Effect of *Tamarindus indica* (Linn) on the rate of wound healing in adult rabbits. IOSR J Dent Med Sci 14:80–84

Ayala-Zavala J, Rosas-Domínguez C, Vega-Vega V, González-Aguilar GA (2010) Antioxidant enrichment and antimicrobial protection of fresh-cut fruits using their own byproducts: looking for integral exploitation. J. Food Sci 75:R175–R181

Bakkali F, Averbeck S, Averbeck D, Idaomar M (2008) Biological effects of essential oils-a review. Food Chem Toxicol 46:446–475

Baldi A, Gupta R, Panwar MS (2011) Evaluation of in vitro antioxidant activity of *Eclipta alba*. Int J Pharm Biol Arch 2:767–771

Balslev H, Grandez C, Zambrana P, Møller AL, Hansen SL (2008) Palmas (Arecaceae) utiles en los alrededores de Iquitos. Amazonia Peruana. Rev Per Biol 15:121–132

Barbulova A, Colucci G, Apone F (2015) New trends in cosmetics: by-products of plant origin and their potential use as cosmetic active ingredients. Cosmetics 2:82–92

Barbulova A, Apone F, Colucci G (2014) Plant cell cultures as source of cosmetic active ingredients. Cosmetics 2:94–104

Barel AO, Paye M, Maibach HI (2001) Handbook of cosmetic science and technology, 1st edn. London, Marcel Dekker

Bereau D, Benjelloun-Mlayah B, Delmas M (2001) *Maximiliana maripa* Drude mesocarp and kernel oils: fatty acid and total tocopherol compositions. J Am Oil Chem Soc 78:213–214

Bhatt P, Negi PS (2012) Antioxidant and antibacterial activities in the leaf extracts of Indian borage (*Plectranthus ambionicus*). Food Nutr Sci 3:146–152

Blanco-Dávila F (2000) Beauty and the body: the origins of cosmetics. Plast Reconstr Surg 105:1196–1204

Bobbio FO, Druzian JI, Abrao PA, Bobbio PA, Fadelli S (2000) Identification and quantification of anthocyanins from the açai fruit (*Euterpe oleracea*) Mart. Ciênc Tecnol Aliment 20:388–390

Brunold C, Deters A, Knoepfel-Sidler F, Hafner J, Müller B, Hansel A (2004) Polysaccharides from *Hibiscus sabdariffa* flowers stimulate proliferation and differentiation of human keratinocytes. Planta Med 7:370–373

Bulpitt CJ, Li Y, Bulpitt PF, Wang J (2007) The use of orchids in Chinese medicine. J R Soc Med 100:558–563

Caballero-George C, Gupta MP (2011) A quarter century of pharmacognostic research on Panamanian flora: a review. Planta Med 77:1189–1202

Cabral EC, Da Cruz GF, Simas RC, Sanvido GB, Gonçalves LDV et al (2013) Typification and quality control of the andiroba (*Carapa guianensis*) oil via mass spectrometry fingerprinting. Anal Methods 5:1385–1391

Campos V, Pompei R, Ferreira C (2007) Clinical study of a cosmetic cream to reduce dark circles under the eyes. J Am Acad Dermatol 56:AB97

Carola A, Tito A, Bimonte M, Mustilli A, Cucchiara M, Monoli I, Hill J, Apone F, Colucci G (2012) Liposoluble extracts of *Vitis vinifera* grape marc and cell cultures with synergistic anti-ageing effects. HPC Today 7:42–46

Chan CF, Huang WY, Guo HY, Wang BR (2014) Potent antioxidative and UVB protective effect of water extract of *Eclipta prostrata* L. Sci World J 759039

Chaudhri SK, Jain NK (2009) History of cosmetics. Asian J Pharm 3:164–167

Chifundera K (2001) Contribution to the inventory of medicinal plants from the Bushi area, South Kivu Province, Democratic Republic of Congo. Fitoterapia 72:351–368

Chukwuma EC, Soladoye MO, Feyisola RT (2015) Traditional medicine and the future of medicinal plants in Nigeria. J Med Plants Stud 3:23–29

Chunhieng T, Hafidi A, Pioch D, Brochier J, Montet D (2008) Detailed study of Brazil nut (*Bertholletia excelsa*) oil microcompounds: phospholipids, tocopherols and sterols. J Braz Chem Soc 19:1374–1380

Chunhieng T, Pétritis K, Elfakir C, Brochier J, Goli T, Montet D (2004) Study of selenium distribution in the protein fractions of the Brazil nut, *Bertholletia excelsa*. J Agric Food Chem 52:4318–4322

Da Fonseca AM, Marques DD, Lemos TLG, Aguiar GR, Bizerra AMC (2014) Fatty chemical composition and antioxidant activities of coconut oils (*Cocos nucifera* L.). J Med Plants Res 8:1081–1085

Da Silva AG, Puziol PdF, Leitão RN, Gomes TR, Scherer R et al (2012) Application of the essential oil from copaiba (*Copaifera langsdorffii* Desf.) for acne vulgaris: a double- blind, placebo-controlled clinical trial. Altern Med Rev 17:69–75

Da-Costa-Rocha I, Bonnlaender B, Sievers H, Pischel I, Heinrich M (2014) *Hibiscus sabdariffa* L. - a phytochemical and pharmacological review. Food Chem 165:424–443

Daher CC, Fontes IS, Rodrigues RdO, Damasceno GAdB, Soares DdS, Aragão CFS, Gomes APB, Ferrari M (2014) Development of O/W emulsions containing *Euterpe oleracea* extract and evaluation of photo protective efficacy. Braz J Pharm Sci 50:1–10

Datta K, Singh AT, Mukherjee A, Bhat B, Ramesh B et al (2009) *Eclipta alba* extract with potential for hair growth promoting activity. J Ethnopharmacol 124:450–456

De Filipps RA, Maina SL, Crepin J (2017) Medicinal plants of the Guianas. Department of Botany, National Museum of Natural History, Smithsonian Museum, Washington DC (USA)

De Santana AC, De Santana AL, De Santana AL (2017) Açaí pulp demand in the retail market of Belém, state of Para. Rev Bras Frutic 39:1–10

Dey A, Gupta B, De JN (2012) Traditional phytotherapy against skin diseases and in wound healing of the tribes of Purulia district, West Bengal India. J Med Plants Res 6:4825–4835

Djibril D, Mamadou F, Gérard V, Geuye MDC, Oumar S et al (2015) Physical characteristics, chemical composition and distribution of constituents of the neem seeds (*Azadirachta indica* A. Juss) collected in Senegal. Res J Chem Sci 5:52–58

Dos Santos MFG, Alves RE, Roca M (2015) Carotenoid composition in oils obtained from Amazonian palms fruits. Grasas Aceites 66:086–093

Dos Santos Md FG, Mamede RVS, Rufino Md SM, de Brito ES, Alves RE (2015) Amazonian native palm fruits as sources of antioxidant bioactive compounds. Antioxidants 4:591–602

Ehiowemwenguan G, Inetianbor JE, Yakubu JM (2014) Antimicrobial qualities of *Senna alata*. IOSR J Pharm Biol Sci 9:47–52

Ekpenyong CE, Akpan E, Nyoh A (2015) Ethnopharmacology, phytochemistry, and biological activities of *Cymbopogon citratus* (DC.) Stapf extracts. Chin J Nat Med 13:321–337

Ekpenyong CE, Akpan EE, Daniel NE (2014) Phytochemical constituents, therapeutic applications and toxicological profile of *Cymbopogon citratus* Stapf (DC) leaf extract. J Pharmacogn Phytochem 3:133–141

El-Abdellaoui S, Destandau E, Toribio A, Elfakir C, Lafosse M, Renimel I, André P, Cancellieri P, Landemarre L (2010) Bioactive molecules in *Kalanchoe pinnata* leaves: extraction, purification, and identification. Anal Bioanal Chem 398:1329–1338

Erny SMN, Razali M, Mirfat AHS, Mohd SMA (2014) Antimicrobial activity and bioactive evaluation of *Plectranthus amboinicus* essential oil. Am J Res Commun 2:121–127

Escalona-Arranz JC, Péres-Roses R, Urdaneta-Laffita I, Camacho-Pozo MI, Rodriguez-Amado J et al (2010) Antimicrobial activity of extracts from *Tamarindus indica* L. leaves. Pharmacogn Mag 6:242–247

Eshun K, He Q (2004) Aloe vera: a valuable ingredient for the food, pharmaceutical and cosmetic industries—a review. Crit Rev Food Sci Nutr 44:91–96

Esquenazi D, Wigg MD, Miranda MM, Rodrigues HM, Tostes JB et al (2002) Antimicrobial and antiviral activities of polyphenolics from *Cocos nucifera* Linn. (Palmae) husk fiber extract. Res Microbiol 153:647–652

Essiett UA, Edet NI, Bala DN (2011) Phytochemical and physicochemical analysis of the leaves of *Laportea aestuans* (Linn.) Chew and *Laportea ovalifolia* (Schumach.) Chew (male and female). Asian J Plant Sci Res 1:35–42

Fernández IM, Mozombite DMS, Santos RC, Melo Filho AA, Ribeiro RE et al (2016) Oil in inajá pulp (*Maximiliana maripa)*: fatty acid profile and anticholinesterase activity. Orbital: Electron J Chem 8:80–83

Freitas A, Moldao-Martins M, Costa H, Albuquerque T, Valentea A, Sanches-Silvaa A (2015) Effect of UV-C radiation on bioactive compounds of pineapple (*Ananas. comosus* L. Merr.) by-products. J Sci Food Agric 95:44–52

Funasaki M, Barroso HS, Fernandes VLA, Menezes IS (2016) Amazon rainforest cosmetics: chemical approach for quality control. Quim Nova 39:194–209

Galeane MC, Martins CHG, Massuco J, Bauab TM, Sacramento LVS (2017) Phytochemical screening of *Azadirachta indica* A. Juss for antimicrobial activity. Afr J Microbiol Res 11:117–122

Ganjewala D (2009) Cymbopogon essential oils: chemical compositions and bioactivities. Int J Essen Oil Ther 3:56–65

General Bureau of Statistics (2014) Suriname in numbers 303–2014-04. Basic indicators, General Bureau of Statistics, Paramaribo (Suriname)

Gopala KAG, Raj G, Bhatnagar AS, Kumar PPK, Chandrashekar P (2010) Coconut oil: chemistry, production and its applications - a review. Indian Coconut 73:15–27

Grant A (2017) Planten die gebruikt worden voor gezichtsbehandelingen. Eenonderzoek onder de natuurproducten in Suriname, Instituut voor de Opleiding van Leraren, Paramaribo (Suriname)

Hadi H, Razali SN, Awadh AI (2015) A comprehensive review of the cosmeceutical benefits of Vanda species (Orchidaceae). Nat Prod Commun 10:1483–1488

Hamman JH (2008) Composition and applications of *Aloe vera* leaf gel. Molecules 13:1599–1616

Harding CR, Moore AE, Rogers JS, Meldrum H, Scott AE, McGlone FP (2002) Dandruff: a condition characterized by decreased levels of intercellular lipids in scalp stratum corneum and impaired barrier function. Arch Dermatol Res 294:221–230

Harsha VH, Hebbar SS, Shripathi V, Hegde GR (2003) Ethnomedicobotany of Uttara Kannada District in Karnataka, India–plants in treatment of skin diseases. J Ethnopharmacol 84:37–40

Hashmat I, Azad H, Ahmed A (2012) Neem (*Azadirachta indica* A. Juss)—a nature's drugstore: an overview. Int Res J Biol Sci 1:76–79

Havinga RM, Hartl A, Putscher J, Prehsler S, Buchmann C et al (2010) *Tamarindus indica* L. (Fabaceae): patterns of use in traditional African medicine. J Ethnopharmacol 127:573–588

Henriques Md, Penido C (2014) The therapeutic properties of *Carapa guianensis*. Curr Pharm Des 20:850–856

Herndon CN, Uiterloo M, Uremaru A, Plotkin MJ, Emanuels-Smith G et al (2009) Disease concepts and treatment by tribal healers of an Amazonian forest culture. J. Ethnobiol Ethnomed 5:1–22

Hetta MH (2016a) Phytocosmetics in Africa. Int J Phytocosm Nat Ingred 3:1–7

Hetta MH (2016b) Phytocosmetics in Africa. Int J Phytocosmetics Natural Ingredients 3:1–7

Hogan S, Chung H, Zhang L, Li J, Lee Y et al (2010) Antiproliferative and antioxidant properties of anthocyanin-rich extract from açai. Food Chem 118:208–214
Holdsworth D (1992) Medicinal plants of the Gazelle peninsula, New Britain Island, Papua New Guinea. Part I Int J Pharmacogn 30:185–190
Hossain MM (2011) Therapeutic orchids: traditional uses and recent advances—an overview. Fitoterapia 82:102–140
Hsieh S, Maranda EL, Salih T, Nguyen A, Jimenez J (2016) The quest for the ultimate skin-lightening agent. JAMA Dermatol 152:372
Ibrahim T, El-Hefnawy H, El-Hela A (2010) Antioxidant potential and phenolic acid content of certain cucurbitaceous plants cultivated in Egypt. Nat Prod Res 24:1537–1545
Imran S, Suradkar SS, Koche DK (2012) Phytochemical analysis of *Leonotis nepetifolia* (L.) R. Br., a wild medicinal plant of Lamiaceae. Biosci Discov 3:197–199
Ismail A, Ikram EHK, Nazri HSM (2008) Roselle (*Hibiscus sabdariffa* L.) seeds nutritional composition protein quality and health benefits. Food 2:1–16
John JA, Shahidi F (2010) Phenolic compounds and antioxidant activity of Brazil nut (*Bertholletia excelsa*). J Funct Foods 2:196–209
Jones G (2011) Globalization and beauty: a historical and firm perspective. Euramerica 41:885–916
Joshi LS, Pawar HA (2015) Herbal cosmetics and cosmeceuticals: an overview. Nat Prod Chem Res 3:1–10
Kakad SL, Pise SS, Dhembare AJ (2015) Evaluation of phytochemical, antibacterial, antifungal activities of leaf extracts of *Morinda citrifolia* (Linn). Der Pharmacia Sinica 6:9–12
Kale S, Megha G (2011) Formulation and in vitro evaluation for sun protection factor of *Moringa oleifera* Lam (family moringaceae) oil sunscreen cream. Int J Pharm Pharm Sci 3:371–375
Kang J, Li Z, Wu T, Jensen GS, Schauss AG et al (2010) Anti-oxidant capacities of flavonoid compounds isolated from açai pulp (*Euterpe oleracea* Mart.). Food Chem 122:610–617
Kapoor VP (2005) Herbal cosmetics for skin and hair care. Nat Prod Radiance 4:306–314
Katiyar SK, Elmets CA (2001) Green tea polyphenolic antioxidants and skin photoprotection. Int J Oncol 18:1307–1313
Kaur M, Chandola HM (2010) Role of rasayana in cure and prevention of recurrence of vicharchika (eczema). Ayu 31:33–39
Kaur N, Bains R, Niazi J (2014) A review on *Bryophyllam pinnatum* - A medicinal herb. J Med Pharm Innov 1:13–19
Kleynhans R, Singh S, Kishore N, Lall N (2017) Investigation towards propagation and cosmeceutical application of *Athrixia phylicoides* DC. South African J Botany 112:319–321
Kluczkovski AM, Martins M, Mundim SM, Simões RH, Nascimento KS et al (2015) Properties of Brazil nuts: a review. Afr J Biotechnol 14:642–648
Krishnaiah D, Nithyanandam R, Sarbatly R (2012) Phytochemical constituents and activities of *Morinda citrifolia* L. In: Rao V (ed) Phytochemicals—A Global Perspective of Their Role in Nutrition and Health. InTech, Rijeka, Croatia, pp 127–150
Kumar D, Kumar S, Singh J, Narender R et al (2010) Free radical scavenging and analgesic activities of *Cucumis sativus* L. fruit extract. J Young Pharm 2:365–368
Kumar SS, Sivakumar T, Chandrasekar MJ, Suresh B (2005) Evaluation of anti-inflammatory activity of *Eclipta alba* in rats. Anc Sci Life 24:112–118
Kuno N, Matsumoto M (2004) Skin Beautifying agent, Antiaging agent for skin, Whitening Agent and External Agent for the skin. US Patent, 6682763
Kuru P (2014) *Tamarindus indica* and its health related effects. Asian Pac J Trop Biomed 4:676–681
Lans C (2007) Comparison of plants used for skin and stomach problems in Trinidad and Tobago with Asian ethnomedicine. J Ethnobiol Ethnomed 3:3
Leandro LM, Vargas Fde S, Barbosa PC, Neves JK, da Silva JA et al (2012) Chemistry and biological activities of terpenoids from copaiba (*Copaifera* spp.) oleoresins. Molecules 17:3866–3889

Lee YH, Lee SJ, Jan JK, Park SY, Lee HK (2001) Grape seed extracts as tyrosinase inhibitors and their uses as drugs, cosmetics, and food. Jpn Kokai Tokkyo Koho. 10:JP2000159681

Letawe C, Boone M, Pierard GE (1998) Digital image analysis of the effect of topically applied linoleic acid on acne microcomedones. Clin Exp Dermat 23:56–58

Levand O, Larson HO (1979) Some chemical constituents of *Morinda citrifolia*. Planta Med 36:186–187

Lima EBC, Sousa CNS, Meneses LN, Ximenes NC, Santos MA Jr et al (2015) *Cocos nucifera* (L.) (Arecaceae): a phytochemical and pharmacological review. Braz J Med Biol Res 48:953–964

Liu CL, Wang JM, Chu CY, Cheng MT, Tseng TH (2002) In vivo protective effect of protocatechuic acid on tert-butyl hydroperoxide-induced rat hepatotoxicity. Food Chem Toxicol 40:635–641

Liu YQ, Heying E, Tanumihardjo SA (2012) History, global distribution, and nutritional importance of citrus fruits. Compr Rev Food Sci Food Saf 11:530–545

Lucas Fd A, Kandrotas AL, Neto EN, de Siqueira CE, André GS et al (2017) Bromerschenkel I, Perri SHV. Copaiba oil in experimental wound healing in horses. Ciencia Rural. 47: e20151292

Lukhoba CW, Simmonds MS, Paton AJ (2006) Plectranthus: a review of ethnobotanical uses. J Ethnopharmacol 103:1–24

Luzia DMM, Jorge N (2011) Antioxidant activity, fatty acid profile and tocopherols of *Tamarindus indica* L. seeds. Ciênc Tecnol Aliment 31:497–501

Maheshwari RK, Mohan L, Malhotra J, Updhuay B, Rani B (2014) Invigorating efficacy of *Cucumis sativas* for healthcare. Int J Chem Pharm Sci 2:737–744

Mahor G, Ali SA (2016) Recent update on the medicinal properties and use of *Aloe vera* in the treatment of various ailments. Biosci Biotech Res Comm 9:273–288

Mak-Mensah EE, Firempong CK (2011) Chemical characteristics of toilet soap prepared from neem (*Azadirachta indica* A. Juss) seed oil. Asian J Plant Sci Res 1:1–7

Mans DRA, Ganga D, Karto J (2017) Meeting of the minds: traditional herbal medicine in multiethnic Suriname pp 111–132. In: El-Shemy H (ed) Aromatic and medicinal plants—back to nature. Croatia, InTech, Rijeka, pp 111–132

Mans RAD, Grant A (2017) A thing of beauty is a joy forever. Plant–Plant based preparations for facial care in Suriname. Clin Med Investigations 2:1–16

Manvitha K, Bidya B (2014) *Aloe vera*: a wonder plant its history, cultivation and medicinal uses. J Pharmacogn Phytochem 2:85–88

Matkovic Z, Zivkovic V, Korica M, Plavec D, Pecanic S (2010) Efficacy and safety of *Astragalus membranaceus* in the treatment of patients with seasonal allergic rhinitis. Phytother Res 24:175–181

May AF (1982) Surinamese folk medicine. A collection of Surinamese medicinal herbs, De Walburg Pers, Amsterdam (The Netherlands)

Mazimba O (2015) *Leonotis leonurus*: a herbal medicine review. J Pharmacogn Phytochem 3:74–82

Meenupriya J, Vinisha AS, Priya P (2014) *Cassia alata* and *Cassia auriculata*—review of their bioactive potential. World J Pharm Sci 2:1760–1769

Menon S, Nayeem N (2013) *Vanilla planifolia*: a review of a plant commonly used as flavouring agent. Int J Pharm Sci Rev Res 20:225–228

Mesfin A, Giday M, Animut A, Teklehaymanot T (2012) Ethnobotanical study of antimalarial plants in Shinile District, Somali Region, Ethiopia, and in vivo valuation of selected ones against *Plasmodium berghei*. J Ethnopharmacol 139:221–227

Minh TN, Khang DT, Tuyen PT, Minh LT, Anh LH et al (2016) Phenolic compounds and antioxidant activity of Phalaenops is orchid hybrids. Antioxidants 5:1–10

Miranda Júnior RN, Dolabela MF, da Silva MN, Póvoa MM, Maia JG (2012) Antiplasmodial activity of the andiroba (*Carapa guianensis* Aubl., Meliaceae) oil and its limonoid-rich fraction. J Ethnopharmacol 142:679–683

Moghaddasi SM, Verma SK (2011) *Aloe vera* their chemicals composition and applications: a review. Int J Biol Med Res 2:466–471

Mohamed R, Fernández J, Pineda M, Aguilar M (2007) Roselle (*Hibiscus sabdariffa*) seed oil is a rich source of gamma-tocopherol. J Food Sci 72:S207-211

Moriyama H, Iizuka T, Nagai M, Miyataka H, Satoh T (2003) Anti-inflammatory activity of heat-treated *Cassia alata* leaf extract and its flavonoid glycoside. Yakugaku Zasshi 123:607–611

Mosquera TT, Noriega RP, Cornejo JC, Pardo ML (2016) Biological activity of *Cymbopogon citratus* (DC) Stapf and its potential cosmetic activities. Int J Phytocosm Nat Ingred 3:07

Mounnissamy VM, Kavimani S, Gunasegaran R (2002) Antibacterial activity of gossypetin isolated from *Hibiscus sabdariffa*. the Antiseptic 99:81–82

Mukherjee PK, Nema NK, Maity N, Sarkar BK (2013) Phytochemical and therapeutic potential of cucumber. Fitoterapia 84:227–236

Murad H, Nyc MA (2016) Evaluating the potential benefits of cucumbers for improved health and skin care. J Aging Res Clin Practice 5:139–141

Murnen SK, Seabrook R (2012) Feminist perspectives on body image and physical appearance. In: Cash TF (ed) Encyclopedia of body image and human appearance, vol 1. Academic Press, San Diego, USA, pp 438–443

Murthy S, Naidu M (2012) Sustainable management of coffee industry by-products and value addition—a review. Resour Conserv Recycl 66:45–58

Murube J (2013) Ocular cosmetics in ancient times. Ocul Surf 11:2–7

Nagaratna A, Hegde PL (2015) A comprehensive review on Parnabeeja [*Bryophyllum pinnatum* (Lam.) Oken]. J Med Plants Stud 3:166–171

Nagarjuna RD, Al-Rajab AJ, Sharma M, Mary Moses M, Ramachandra RG, Mohammed A (2017) Chemical composition, antibacterial and antifungal activities of Saudi Arabian *Mentha longifolia* L. essential oil. J Costal Life Med 5:441–446

Naidoo L, Khoza N, Dlova NC (2016) A fairer face, a fairer tomorrow? A review of skin lighteners. Cosmetics 3:1–10

Naik TI, Shrikanth P, Mundugaru R, Bairy ST (2017) Wound healing activity of *Tamarindus indica* Linn. seed and cork ash Ayu Med Sci 2:129–135

Nayak B, Kanhai J, Milne D, Swanston W, Mayers S et al (2010) Investigation of the wound healing activity of *Carapa guianensis* L. (Meliaceae) bark extract in rats using excision, incision, and dead space wound models. J Med Food 13:1141–1146

Niteshwar PR, Kumari H (2012) Ethnomedicinal claims of *Leonotis nepetifolia* (L). R. Br. A Review. Int J Res Ayurveda Pharm 6:783–785

Ogbunugafor HA, Eneh FU, Ozumba AN, Igwo-Ezikpe MN, Okpuzor J et al (2011) Physico-chemical and antioxidant properties of *Moringa oleifera* seed oil. Pakistan J Nutr 10:409–414

Ojiako EN, Okeke CC (2013) Determination of antioxidant of *Moringa oleifera* seed oil and its use in the production of a body cream. Asian J Plant Sci Res 3:1–4

Okereke SC, Chukwudoruo CS, Nwaokezie CO (2017) Antioxidant enzyme activities and chemopreventive potentials of *Laportea aestuans* on urinary inflammatory markers using albino rats. J Med Plants Stud 5:288–291

Oladeji O, Adelowo F, Odelade K, Aremu S, Adisa M (2016) *Senna alata*: phytochemical components, characterization and medicinal benefits. CIBTech J Pharm Sci 5:40–51

Olorunnisola SK, Asiyanbi HT, Hammed AM, Simsek S (2014) Biological properties of lemongrass: an overview. Int Food Res J 21:455–462

Oloyede GK, Ayanbadejo OE (2014) Phytochemical, toxicity, antimicrobial and antioxidant screening of extracts obtained from *Laportea aestuans* (Gaud). J Med Sci 14:51–59

Oloyede GK (2016) Toxicity, antimicrobial and antioxidant activities of methyl salicylate dominated essential oils of *Laportea aestuans* (Gaud). Arabian J Chem 9:S840–S845

Omar SH (2010) Oleuropein in olive and its pharmacological effects. Sci Pharm 78:133–154

Orafidiya LO, Agbani EO, Oyedele AO, Babalola OO, Onayemi O (2002) Preliminary clinical tests on topical preparations of *Ocimum gratissimum* Linn leaf essential oil for the treatment of acne vulgaris. Clin Drug Invest 22:313–319

Oréal L (2015) Finance. Annual Reports

Oresajo C, Yatskayer M, Galdi A, Foltis P, Pillai S (2010) Complementary effects of antioxidants and sunscreens in reducing UV-induced skin damage as demonstrated by skin biomarker expression. J Cosmet Laser Ther 12:157–162

Oyedeji OA, Afolayan AJ (2005) Comparative study of the essential oil composition and antimicrobial activity of *Leonotis leonurus* and *L. ocymifolia* in the Eastern Cape. South Africa S Afr J Bot 71:114–116

Oyedeji OA, Ekundayo O, Koenig WA (1999) Composition of the essential oil from the leaves of *Leonotis nepetaefolia* (L.). J Essent Oil Res 11:716–718

Palu AK, West BJ, Jensen CJ (2012) Noni seed oil topical safety, efficacy, and potential mechanisms of action. J Cosm Dermatol Sci Appl 2:74–78

Pandey B (2017) Screening of antibacterial activity of essential oil of lemmon grass (*Cymbopogon citratus*). Indian J Sci Res 13:29–31

Papaioannou E, Karabelas A (2012) Lycopene recovery from tomato peel under mild conditions assisted by enzymatic pre-treatment and non-ionic surfactants. Acta Biochim Polonica 59:71–74

Paul P, Chowdhury A, Nath D, Bhattacharjee MK (2013) Antimicrobial efficacy of orchid extracts as potential inhibitors of antibiotic resistant strains of Escherichia coli. Asian J Pharm Clin Res 6:108–111

Pedro LG, Barroso JG, Marques NT, Ascensao L, Paid MS et al (1991) Composition of the essential oil from sepals of *Leonotis leonurus* R. Br J Essent Oil Res 3:451–453

Pereira AS, Alves HP, Souse CM, Costa GL (2013) Prospeção sobre o conhecimento de espécies amazônicas – inajá (*Maximiliana maripa* Aubl.) e bácaba (*Oenocarpus bacaba* Mart.). Revista Geintec 3:110–122

Pereira RF, Bártolo PJ (2016) Traditional therapies for skin wound healing. Adv Wound Care (New Rochelle) 5:208–229

Pereira TB, Rocha E, Silva LF, Amorim RC, Melo MR et al (2014) In vitro and in vivo anti-malarial activity of limonoids isolated from the residual seed biomass from *Carapa guianensis* (andiroba) oil production. Malar J 13:1–5

Periasamy G, Kassa S, Sintayehu B, Mebrahtom Glibanos M, Geremedhin G et al (2014) Cosmetic use of *Aloe vera*—a review. World J Pharm Pharm Sci 3:342–458

Phan TT, See P, Lee ST, Chan SY (2001) Protective effects of curcumin against oxidative damage on skin cells in vitro: Its implication for wound healing. J Trauma: Injury, Infect Crit Care 51:927–931

Pietta P (2000) Flavonoids as antioxidant. J Nat Prod 63:155–161

Playfair M, Esseboom M, Zschuschen E (2011) Marktverkenning van potentiële non-timber forest products en niet-industriële houtproducten voor de Marchallkreek community. CELOS, Paramaribo (Suriname)

Portinho JA, Zimmermann LM, Bruck MR (2012) Beneficial effects of açai. Int J Nutrol 5:15–20

Potterat O, Hamburger M (2007) Morinda citrifolia (noni) fruit–phytochemistry, pharmacology, safety. Planta Med 73:191–199

Qadir MI (2009) Medicinal and cosmetological importance of *Aloe vera*. Int J NatTher 2:21–26

Rabe T, van Staden J (1998) Screening of Plectranthus species for antibacterial activity. S Afr J Bot. 64:62–65

Raghoenandan UPD (1994) An ethnobotanical investigation among Hindustanis in Suriname. Anton de Kom Universiteit van Suriname, Paramaribo (Suriname)

Rajasree RS, Sibi P, Francis F, William H (2016) Phytochemicals of Cucurbitaceae family – A review. Int J Pharmacogn Phytochem Res 8:113–123

Reddy DN (2019) Essential oils extracted from medicinal plants and their applications. In: Akhtar M, Swamy M, Sinniah U (eds) Natural bio-active compounds. Springer, Singapore

Ribeiro AS, Estanqueiro M, Oliveira MB, Lobo JMS (2015) Main benefits and applicability of plant extracts in skin care products. Cosmetics 2:48–65

Roshan P, Naveen M, Manjul PS, Gulzar A, Anita S et al (2010) *Plectranthus amboinicus* (Lour) Spreng: an overview. Pharm Res 4:1–15

Roy RK, Thakur M, Dixit VK (2008) Hair growth promoting activity of *Eclipta alba* in male albino rats. Arch Dermatol Res 300:357–364

Sachs M, von Eichel J, Asskali F (2002) Wound management with coconut oil in Indonesian folk medicine. Chirurg 73:387–392

Sadler JJ, Smith JM, Zettler LW, Alborn HT (2011) Richardson, L.W., Fragrance composition of *Dendrophylax lindenii* (Orchidaceae) using a novel technique applied in situ. Eur J Environ Sci 1:137–141

Said ZB, Haddadi-Guemghar H, Boulekbache-Makhlouf L, Rigou P, Remini H, Adjaoud A, Khoudja NK, Madani K (2016) Essential oils composition, antibacterial and antioxidant activities of hydro distillated extract of *Eucalyptus globulus* fruits. Ind Crop Prod 89:167–175

Santos AO, Ueda-Nakamura T, Dias FBP, Veiga VF Jr, Pinto AC et al (2008) Antimicrobial activity of Brazilian copaiba oils obtained from different species of the *Copaifera genus*. Mem Inst Oswaldo Cruz 103:277–281

Saraswat VP, Verma S, Musale SV, Jaiswal ML (2015) A review on traditional and folklore uses, phyto-chemistry and pharmacology of *Eclipta alba* (L) Hassk. Int Ayurvedic Med J 3:2462–2469

Sato N (2002) Topical cleansing cosmetics. Jpn Kokai Tokkyo Koho. 7:JP2002145761

Schauss AG, Wu X, Prior R, Ou B, Patel D et al (2006) Phytochemical and nutrient Composition of the freeze-dried amazonian palm berry *Euterpe oleracea* Mart. (açai). J Agric Food Chem 54:8598–8603

Schlessinger J (2007) Skin care for men and its marketing. Dermatol Ther 20:452–456

Sedoc NO (1992) Afro-Surinamese natural remedies: over two hundred commonly used medicinal herbs. Vaco Press, Paramaribo (Suriname)

Shah G, Shri R, Panchal V, Sharma N, Singh B et al (2011) Scientific basis for the therapeutic use of *Cymbopogon citratus*, stapf (lemon grass). J Adv Pharm Technol Res 2:3–8

Sharma L, Agarwal G, Kumar A (2003) Medicinal plants for skin and hair care. Indian J Traditional Knowl 2:62–68

Strickland FM, Kuchel JM, Halliday GM (2004) Natural products as aids for protecting the skin's immune system against UV damage. Cutis 74:24–28

Stutesman D (2016) The silent screen. In: McLean A (ed) Costume, makeup, and hair. Rutgers University Press, New Brunswick, pp 21–46

Sule WF, Okonko IO, Omo-Ogun S, Nwanze JC, Ojezele MO et al (2011) Phytochemical properties and in vitro antifungal activity of *Senna alata* Linn. Crude stem bark extract. J Med Plants Res 5:176–183

Swamy MK, Sinniah UR, Akhtar MS (2016) Antimicrobial properties of plant essential oils against human pathogens and their mode of action: An updated review. Evidence-Based Compl Altern Med 3012462

Taher MA, Nyeem MAB, Ahammed MM, Hossain MM, Islam MN (2017) *Moringa oleifera* (shajna): the wonderful indigenous medicinal plant. Asian J Med Biol Res 3:20–30

Tehara T, Hachimaki H (2002) Antiallergic cosmetic or topical compositions containing olive extracts. Jpn Kokai Tokkyo Koho 11:JP2002332238

Tomato Processing Industries (2014) Available online: http://www.tomatonews.com/resources.html. Accessed on 2 Dec 2014

Tranggono RIS (2000) The influence of SPA products treatment for skin-care. J Appl Cosmetol 18:177–184

Tuck C, Perez E, Horvath I, Sheldon R, Poliakoff M (2012) Valorization of biomass: Deriving more value from waste. Science 337:695–699

Udaya PNK, Bhuvaneswari S, Balamurugan A, Vaishnavi S, Sugunan S et al (2013) Studies on antibacterial, antioxidant, larvicidal, pesticidal activities and phytochemistry of *Leonotis nepetifolia* (Linn) R.Br. Int J Res Pharm Sci 4:303–309

Uddin N, Rahman A, Ahmed NU, Rana S, Akter R et al (2010) Antioxidant, cytotoxic and antimicrobial properties of *Eclipta alba* ethanol extract. Int J Biol Med Res 1:341–346

Valadeau C, Castillo JA, Sauvain M, Lores AF, Bourdy G (2010) The rainbow hurts my skin: medicinal concepts and plants uses among the Yanesha (Amuesha), an Amazonian Peruvian ethnic group. J Ethnopharmacol 127:175–192

Van Andel TR, Ruysschaert S (2011) Medicinal and ritual plants of Suriname. KIT Publishers, Amsterdam (The Netherlands)

Van't Klooster C, van Andel T, Reis R (2016) Patterns in medicinal plant knowledge and use in a Maroon village in Suriname. J Ethnopharmacol 189:319–330

Veiga VF, Rosas EC, Carvalho MV, Henriques MGMO, Pinto AC (2007) Chemical composition and anti-inflammatory activity of copaiba oils from *Copaifera cearensis* Huber ex Ducke, *Copaifera reticulata* Ducke and *Copaifera multijuga* Hayne—a comparative study. J Ethnopharmacol 112:248–254

Venugopalan A, Giridhar P, Ravishankar GA (2011) Food, ethanobotanical and diversified applications of *Bixa orellana* L.: a scope for its improvement through biotechnological mediation. Ind J Fund Appl Life Sci 1:9–31

Wanjari N, Waghmare J (2015) A review on latest trend of cosmetics-cosmeceuticals. Int J Pharma Res Rev 45:45–51

Warra AA (2014) A review of *Moringa oleofera* Lam seed oil prospects in personal care formulations. Res Rev J Pharm Nanotechnol 2:31–34

Warra AA (2012) Production of soap from an Indigenous *Moringa oleifera* Lam Seed Oil. J Raw Mater Res 7:23–30

Watson P (2012) Cosmetics, Greece and Rome. The Encyclopedia of Ancient History

Weniger B, Rouzier M, Daguilh R, Henrys D, Henrys JH et al (1986) Traditional medicine in the Central Plateau of Haiti. 2. Ethnopharmacologic Inventory. J Ethnopharmacol 17:13–30

West BJ, Sabin RA (2012) Efficacy of a *Morinda citrifolia*-based skin care regimen. Curr Res J Biol Sci 4:310–314

Yamaguchi KK, Pereira LF, Lamarão CV, Lima ES, da Veiga-Junior VF (2015) Amazon acai: chemistry and biological activities: a review. Food Chem 179:137–151

Yang J (2009) Brazil nuts and associated health benefits: a review. LWT Food Sci Technol 42:1573–1580

Zuorro A, Lavecchia R (2010) Mild enzymatic method for the extraction of lycopene from tomatopaste. Biotechnol Biotechnol Equip 24:1854–1857

# Chapter 20
# Encapsulation of Bioactive Compound and Its Therapeutic Potential

**Lalduhsanga Pachuau, Laldinchhana, Probin Kumar Roy, James H. Zothantluanga, Supratim Ray, and Sanjib Das**

**Abstract** Micro- or nanoencapsulation has become one of the most attractive approaches to enhance the stability and bioavailability of bioactive compounds isolated from natural products. In addition, such encapsulation also provides an opportunity to modulate the release of bioactive compounds by using functional polymers that can be fine-tuned according to the need of the body. Over the last 10 years, increasing research works have been dedicated toward the encapsulation of bioactive compounds for various purposes. Numerous techniques, embracing micro- and nanoplatforms including spray drying, freeze drying, micro- and multiple emulsification, electrospinning, and coacervation, have been utilized to achieve the encapsulations. Such encapsulations have been found to improve the physicochemical properties of the bioactive compound, provide stability and enhanced bioavailability, controlled release of the compound, enhancing bioactivity, and masking of flavor or taste, along with several other benefits. Wide ranges of materials including lipids, synthetic, and natural polymers have been utilized, and the type and amount of the wall formers have been found to influence the performance and functionality of these preparations.

**Keywords** Encapsulation · Bioavailability · Bioactive compounds · Hydrogels · Electrodynamic processes

---

L. Pachuau (✉) · S. Ray · S. Das
Department of Pharmaceutical Sciences, Assam University, Silchar 788011, India
e-mail: aduhapc@gmail.com

Laldinchhana · P. K. Roy
Department of Pharmacy, Regional Institute of Paramedical and Nursing Sciences, Aizawl 796017, India

J. H. Zothantluanga
Department of Pharmaceutical Sciences, Dibrugarh University, Dibrugarh 786004, India

 
D. Pal and A. K. Nayak (eds.), *Bioactive Natural Products for Pharmaceutical Applications*, Advanced Structured Materials 140,
https://doi.org/10.1007/978-3-030-54027-2_20

## 20.1 Introduction

Phytochemicals in plants that produce therapeutic activities are collectively known as bioactive compounds. These bioactive compounds are predominantly plant secondary metabolites and nutrients such as vitamins and minerals that may induce pharmacological activity at high doses are usually excluded from the definition of bioactive compounds (Bernhoft 2010). Abundant bioactive phytochemicals are found in vegetables, fruits, seeds, nuts, legumes, leaves, and other parts of plants (Al Juhaimi et al. 2018; Pachuau et al. 2019; Sagar et al. 2018; Septembre-Malaterre et al. 2018; Xiao and Bai 2019). Phenolic compounds such as flavonoids, courmarins, phenolic acids, xanthones, and ellagitannins along with alkaloids, phytosterols, carotenoids, anthocyanins, and tocopherols are some of the commonly found bioactive compounds from plant sources that are responsible for their bioactivities (Fig. 20.1) (Altemini et al. 2017; Da Silva et al. 2016; Xiao and Bai 2019). In addition to their bioactivities, phytochemicals such as phenolics in vegetables and fruits may also contribute to their stability against oxidation, taste, color, flavor, and odor (Naczk and Shahidi 2006).

Studies have reported myriads of pharmacological activities including antimicrobial, anti-hyperglycemic, anti-inflammatory, antioxidant, anti-proliferative, and anti-diabetic effects of bioactive phytochemicals isolated and characterized from various plant species (Martins et al. 2016; Ramesh Kumar et al. 2018; Rodriguez-Garcia et al. 2017). Diverse preparations made from these wide varieties of plant species have been used in the treatment of different diseases since ancient times as documented in Eber's Papyrus, Traditional Chinese Medicines (TCM), and Ayurvedic formulations (Atanasov et al. 2015; Khan et al. 2019; Yang and Yue 2012). These traditional practices still remain valuable sources of information for modern drug discovery and development programs. In fact, data from Food and

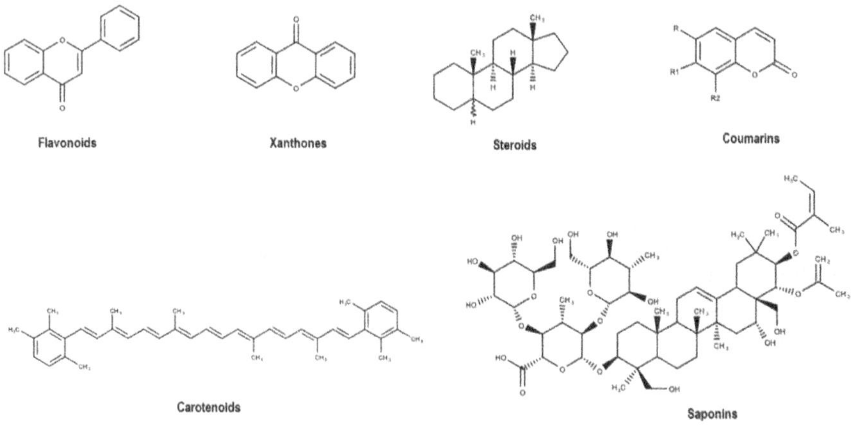


**Fig. 20.1** Structures of some phytochemicals

Drug Administration (FDA) revealed that about 40% of the approved molecules are either natural compounds or inspired by them and, from these, 74% are indicated for anticancer therapy (Seca and Pinto 2018). Still, it has been reported that about 95% of the world's biodiversity are yet to be systemically investigated for their pharmacological activity and only 15% have been evaluated phytochemically (Atanasov et al. 2015; David et al. 2015). The current global scenario and the urgent need for effective therapy to treat chronic as well as infectious diseases including cancer and HIV implied that natural products will continue to play a pivotal role in drug discovery and development processes (Cragg and Newman 2013).

Although natural products are considered to be safe and their bioactive phytochemicals exhibit excellent pharmacological activities in vitro, they often failed to translate this into clinical or therapeutic effects in vivo. The major problems associated with bioactive phytochemicals include their low oral absorption and rapid systemic clearance which ultimately lowers their bioavailability and therapeutic efficacy in vivo (Kumar et al. 2010; Rein et al. 2012). For instance, oral administration of curcumin at 500 mg/kg in Sprague-Dawley rats produced maximum plasma concentration ($C_{max}$) of $0.06 \pm 0.01$ μg/ml after $41.7 \pm 5.4$ min (Yang et al. 2007), while in healthy human volunteers, curcumin was detected in the serum only when it was administered orally at a dose higher than 8 g (Lao et al. 2006). Similarly, other potent bioactive compounds such as quercetin (Almeida et al. 2018; Kasikci and Bagdatlioglu 2016), paclitaxel (Malingre et al. 2001; Tiwari and Amiji 2006), and ellagic acid (Bala et al. 2006; Lei et al. 2003) also produced extremely low oral bioavailability eventually resulting in suboptimal oral therapy in vivo. Predominant factors that led to the poor oral absorption of these natural bioactive compounds are their eminently low aqueous solubility, poor dispersibility, and bioaccessibility which are the prerequisites for absorption through the gastrointestinal (GI) tract. For example, the maximum solubility of curcumin in aqueous buffer pH 5.0 is only 11 ng/ml (Ma et al. 2019) and the aqueous solubility of ellagic acid is about 9.7 μg/ml (Alfei et al. 2019). In addition, instability of various bioactive compounds on exposure to environmental factors such as pH, enzymes, light, heat, oxygen, and moisture may also contribute to their degradation in vitro and in vivo (Bohn et al. 2015; Islam Shishir et al. 2018). Moreover, biological factors including permeability of the intestinal epithelium, the presence of efflux transporters such as P-glycoprotein and the induction or inhibition of metabolizing enzymes may also serve as barriers to oral bioavailability of several bioactive compounds (Chu et al. 2008; Ma et al. 2019; Xie et al. 2011). Overcoming these challenges has become one of the most important focuses of the current research on improving the oral bioavailability of bioactive compounds. Techniques such as micronization (Aguiar et al. 2018) or nanonization (Aditya et al. 2019; Borhan et al. 2014) to reduce the particle size, co-administration of phytochemicals with bio-enhancers such as piperine (Gorgani et al. 2017), micro- and nanoencapsulation with different types of coating materials (Dias et al. 2017; Ezhilarasi et al. 2013) are some of the approaches designed to achieve the oral bioavailability enhancement of bioactive compounds.

The process of applying relatively thin coatings around a substance which may be solids, liquids, or even gases is called encapsulation. Based on the size of the coated particles, the process may be microencapsulation or nanoencapsulation. Encapsulation separates the encapsulated materials also known as the core materials from the environment with the help of a coating material, usually a polymer. Thus, encapsulation provides protection of the core material and it may also add certain functionalities such as color for identification, masking of taste, or sustaining the release of the core materials. As a result, micro- or nanoencapsulation has become an indispensable technology in various industries including pharmaceuticals, cosmetics, nutraceuticals, and agriculture.

Encapsulation is one of the techniques that can be applied to enhance oral bioavailability, solubility, dispersibility, and stability of bioactive compounds (Beevers and Huang 2011; Ezhilarasi et al. 2013; Gomez-Estaca et al. 2015; Li et al. 2015; Shaikh et al. 2009). Numerous methods have been applied in the encapsulation of various natural bioactive compounds utilizing wide range of coating materials such as lipids, natural, and synthetic polymers (Anirudhan and Binusree 2016; Islam Shishir et al. 2018; Oehlke et al. 2017; Pinho et al. 2013; Shao et al. 2011; Sharma et al. 2007; Young et al. 2005). The selection of the coating materials is critical as the size, shape, and stability of the final encapsulated product as well as the release characteristics of the core material is largely determined by the type and amount of the coating materials used (Dias et al. 2017).

## 20.2 Rationale for Encapsulation of Bioactive Compounds

Encapsulation of natural bioactive compounds into various nano- or micro-platforms has the potential to resolve several drawbacks that are inherent to these phytochemicals. This may range from improving their physicochemical properties such as solubility and stability, to their pharmacokinetics such as absorption and the overall bioavailability. Some of the advantages offered by encapsulation of bioactive compounds can be summarized as follows:

(a) *Enhanced bioavailability and therapeutic activity*: Improvement in bioavailability of poorly absorbable phytochemicals such as polyphenols has been reported after their encapsulation into various systems such as solid lipid nanoparticles (SLN), micelles, liposomes, and others (Aqil et al. 2013; Murugan et al. 2009; Puligundla et al. 2017; Xie et al. 2011; Yang et al. 2008). This enhancement in bioavailability can be manyfold, as at least ninefold increase in oral bioavailability had been reported for poly(lactic-co-glycolic acid) (PLGA) nanoencapsulation of curcumin (Shaikh et al. 2009), while 942.53% enhancement in relative bioavailability was also reported for curcumin encapsulated in SLNs (Ji et al. 2014).

(b) *Stability*: Most of the bioactive compounds from plants are susceptible to light, heat, and pH changes developing unpleasant flavor or colors (Alborzi et al. 2012; Fang and Bhandari 2010; Prakash et al. 2018; Rezaei et al. 2019; Soukoulis and Bohn 2018). Encapsulation can prevent in vitro or in vivo degradation of bioactive compounds against these factors.

(c) *Taste masking*: Food or therapeutic applications of several phytochemicals have been limited by their unpleasant flavor, astringency, and bitter taste (De Souza et al. 2020). Encapsulation can eliminate this problem as the bioactive compound is not in contact with the taste buds located in the oral cavity. For instance, microencapsulation of quercetin with carnauba wax, shellac, or zein was able to reduce or mask the bitter taste of the compound (Khor et al. 2017), while the flavor of turmeric extract was effectively masked by brown rice flour and β-cyclodextrin-based microcapsules (Laokuldilok et al. 2016).

(d) *Improvement of physicochemical properties*: Micro- or nanoencapsulation offers the opportunity to improve the overall physicochemical properties such as morphology and wettability, presenting bioactive compounds into spherical, uniform size, and free-flowing powders which also facilitate their processing during manufacturing (Bertoni et al. 2019; Dima et al. 2016; Lu et al. 2011). Such improvement in powder flow property reduces the possibility of variation in the quality of the end products (Guajardo-Flores et al. 2015). Dissolution rate of poorly aqueous-soluble phytochemicals such as naringin and quercetin has also been reported to be enhanced by 20–55% (Pai et al. 2015) and 100 times (Barras et al. 2009), respectively, probably due to the dispersion of the bioactive compound in amorphous form within the matrix.

(e) *Controlled or targeted release of bioactive compounds*: Encapsulation enables controlled and targeted release of the encapsulated bioactive molecules, protecting against degradation and first-pass metabolism while enhancing their bioactivity following oral administration (Goiun 2004; Moreno et al. 2018; Sun et al. 2015; Yao et al. 2015) (Table 20.1).

## 20.3 Encapsulation of Bioactive Compounds

### *20.3.1 Microencapsulation*

Microencapsulation is one of the most common methods employed to provide wide range of functionalities including protection of bioactive compounds against environmental factors and to enhance their bioavailability (Table 20.2). It is also possible to encapsulate bioactive oils such as essential oils and is a means of converting these oils into free-flowing powders which facilitate its handling, stability, and bioactivities. Due to their small and uniform particle size, microcapsules are distributed homogenously along the GI tract which promote absorption of their encapsulated compounds and enhance oral bioavailability (Lengyel et al. 2019). Various methods such as spray drying (Boonchu and Utama-ang 2015), spray congealing (Tomsik et al. 2019), coacervation phase separation (Silva et al. 2011; Jain et al. 2015), solvent evaporation (Paulo and Santos 2018; Sawale et al. 2017),

**Table 20.1** List of some bioactive compounds with nanoencapsulation technology for improved characteristics

| Bioactive compound | Encapsulation type/ techniques | Improvement after encapsulation | References |
|---|---|---|---|
| Eugenol | Oil-in-water emulsion and ionic gelation | Thermal stability | Woranuch and Yoksan (2013) |
| Curcumin | Nano-precipitation | Solubility and antioxidant activity | Dutta et al. (2018) |
| | Nanoemulsion | Solubility and stability | Ahmed et al. (2012) |
| | Liposomes | Increased bioavailability compared to suspension | Takahashi et al. (2009) |
| Caffeic acid | Inclusion complex by β-cyclodextrin | Enhanced bioactivity | Pinho et al. (2015) |
| | Liposomes | Enhanced antioxidant properties | Katuwavila et al. (2016) |
| Chlorogenic acid | Nanoparticles by ionic gelation techniques | Sustained release property, retained antioxidant activity, and enhanced bioavailability | Nallamuthu et al. (2015) |
| | Liposomes | Enhanced oral bioavailability, increased antioxidant properties | Feng et al. (2016) |
| Coumaric acid | Nano-sized multi-phase emulsion | Increased stability and bioaccessibility, sustained release | Huang et al. (2019) |
| Resveratrol | Nanoemulsion by high-pressure homogenization | Solubility, stability, and bioavailability | Sessa et al. (2014) |
| Quercetin | Solid lipid nanoparticle | Increased bioavailability | Li et al. (2009) |
| | Microemulsion | Increased bioavailability | Sun et al. (2010) |
| | Nanoliposomes by film-sonication | Enhanced bioactivity, enhanced release | Rodriguez et al. (2019) |
| Capsaicin | Nanoemulsion | Increased bioavailability | Choi et al. (2013) |
| Retinoic acid | Solvent evaporation | Controlled release | Ezpeleta et al. (1996) |
| Ellagic acid | Emulsion–diffusion–evaporation | Sustained release, enhanced bioavailability | Mady and Shaker (2017) |
| Sesamol | Phosphatidyl choline micelles | Enhanced bioavailability and bioactivity | Yashaswini et al. (2017) |
| d-limonene | High-pressure homogenization | Enhanced antimicrobial activity | Donsi et al. (2011) |
| Carvacrol | Liposome/lipid film hydration technique | Sustained released, increased activity | Engel et al. (2017) |

(continued)

**Table 20.1** (continued)

| Bioactive compound | Encapsulation type/ techniques | Improvement after encapsulation | References |
|---|---|---|---|
| Thymol | Liposome/lipid film hydration technique | Sustained released, increased activity | Engel et al. (2017) |
| Naringenin | Liposome/thin film hydration | Enhanced solubility and bioavailability | Wang et al. (2017) |

**Table 20.2** Reasons for microencapsulation of bioactive compounds

| Reasons |
|---|
| To minimize bitterness and astringency of bioactive compounds |
| Encapsulation of bioactive fixed and essential oils |
| Controlling of organoleptic properties such as color, flavor, and odor |
| Protection against degradation due to oxidation, pH, enzymatic reactions, etc. |
| Controlled release and improved therapeutic activity of bioactive compounds |
| Ease of administration |
| Improvement in physicochemical properties |
| Enhance dissolution rate |

and ionic gelation (Borgogna et al. 2010) have been utilized for encapsulating bioactive compounds. Wide range of wall-forming materials from natural and synthetic sources have been used in microencapsulation of bioactive compounds and the nature and quantity of these wall formers determined, to a large extent, the quality and functionality of the resultant microcapsules.

In an ideal delivery system, the bioactive compounds should not be allowed to affect the color or flavor or interact with the sensory organs and also the particles are presented small enough not to interfere with the texture (Champagne and Fustier 2007). Microencapsulation is a technology to achieve this objective in effective delivery of bioactive compounds. An extract of *Gentiana lutea* root containing bitter secoiridoids was coated with ethylcellulose–stearate system into microcapsules and was found to mask the bitterness in the mouth and the release of the bioactive compound in the GI effectively reduces the daily energy intake in human subjects (Mennella et al. 2016). Not only with the bitter taste alone, natural compounds often come with strong odor or flavor, leaving behind a sensation of astringency. Apart from enhancing its stability, the strong flavor and astringency of cinnamon extract, a rich source of proanthocyanidins, were successfully concealed by microcapsulation with gelatin/gum arabic and gelatin/κ-carrageenan systems (De Souza et al. 2020).

Spray drying is one of the most common methods employed for microencapsulation of bioactive compounds. Spray drying was employed to effectively encapsulate both the seed oils and peel extract of pomegranate to provide protection of the phenolics and fatty acid contents which are otherwise highly sensitive to environmental factors during their storage (Bustamante et al. 2017). Procyanidins extracted from grape seed oil was also encapsulated by spray-drying method using gum arabic and maltodextrin as coating materials and in this manner, the stability and shelf-life of the product were enhanced (Zhang et al. 2007). When freeze drying was compared against spray drying for microencapsulation of bioactive compounds extracted from *Hibiscus Calyces* using various encapsulating materials, encapsulation of anthocyanins was higher with freeze drying using gum arabic and yields higher antioxidant activity, while in terms of physical properties, better results were obtained with spray-drying method (Piovesana and Norena 2018).

Another improvement that microcapsules offered is the prospect of enhancing the solubility, and hence the oral bioavailability of bioactive compounds. Spray-congealing microencapsulation of wild garlic (*Allium ursinum* L.) extract was found to enhance the aqueous solubility more than 18 times of the pure extract, and in addition, there was only a minor decrease in the content of the bioactive compounds, allicin, and S-methyl methanethiosulfonate over 3-month period of storage (Tomsik et al. 2019). Microencapsulation also provides the opportunity for sustained or controlled release of bioactive compounds for prolonged therapeutic activity (Jain et al. 2015; Saifullah et al. 2019).

## 20.3.2 Nano-based Encapsulation Platforms for Bioactive Compounds

Various nanotechnology-based platforms have been developed to overcome the low oral bioavailability, systemic toxicity, or stability problems facing bioactive compounds. Enhancing oral bioavailability has been one of the major objectives of these nanomedicinal preparations. The main focus of such enhancement schemes encompasses the three oral bioavailability requirement steps like increasing the aqueous solubility of bioactive compounds, enhancing the lipid partitioning or permeability and reduction of first-pass metabolism (Pachuau 2019). Studies have shown that encapsulation of bioactive phytochemicals in various nanoparticulate systems was successful in achieving these objectives by controlling their release, facilitating their systemic absorption, bypassing pre-systemic metabolism, and enhancing the cellular uptake while also providing their stability (Ahmed et al. 2012; Aqil et al. 2013; Mukherjee et al. 2015).

### 20.3.2.1 Polymeric Nanocapsules

Nanocapsules differ from microcapsules only in their size; however, reduction of particle size to nanometer range leads to remarkable advancement in their physicochemical properties. Nanoencapsulation enhances dissolution of the drug, membrane permeability stabilization of labile drugs, and controlled release of the encapsulated drug (Pandey et al. 2005). As a result, polymeric nanocapsules facilitate both oral and parenteral administration of bioactive compounds. The type and amount of the wall formers in nanocapsules have been shown to influence the size, surface properties, stability, drug-loading capacity, aqueous solubility, and the release profile of the encapsulated compounds (Dos Santosa et al. 2016). Therefore, selection of appropriate wall-forming material is crucial to impart the desired functionality to the nanocapsules. Biocompatible and biodegradable poly(lactic acid) (PLA), poly(glycolic acid) (PGA), and their copolymer poly (lactide-coglycolide) (PLGA) are among the most commonly used polymers for nanoencapsulation (Liu and Feng 2015). Encapsulation of ellagic acid in PLGA nanoparticles was found to increase the intestinal permeability from 66% in aqueous suspension of ellagic acid from to 87% in nanocapsules while also exhibiting its antioxidant effects (Bala et al. 2006). PLGA-based silymarin nanocapsule was also shown to sustain the release, improve bioavailability, and exhibit preferential toxicity toward prostate cancer cells, thus enhancing its overall therapeutic efficacy (Snima et al. 2014). Nanoencapsulation of curcumin with PLGA also delayed the progression of diabetic cataract in animal model (Grama et al. 2013) (Table 20.3).

Poorly aqueous-soluble anticancer phytochemical paclitaxel, isolated from *Taxus brevifolia*, which also suffers from P-glycoprotein-mediated efflux and metabolism by cytochrome P450 metabolic enzymes (Singla et al. 2002), had been alleviated through nanoencapsulation. The improvement in permeability of nanoencapsulated paclitaxel across Caco-2 monolayers was found to be 6.7–7.4 times of the lone paclitaxel and the plasma concentration time curve (AUC) and $C_{max}$ following oral administration also increased up to 5.7 times and 7.3 times, respectively (Iqbal et al. 2011).

Coating with mucoadhesive polymers is another means to improve oral bioavailability of bioactive compounds due to their ability to bring prolonged and intimate contact with GI tract surfaces resulting in better absorption. Chitosan has often been used to coat the polymeric nanocapsules due to its ability to adhere to the negatively charged mucosal surface, thus improving absorption and hence therapeutic activity (Chuah et al. 2014; Ensign et al. 2012; Mazzarino et al. 2012).

Capsaicin is a natural bioactive compound isolated from peppers with pungent odor and quick degradation and nanoencapsulation with natural wall formers gelatin and acacia was able to mask the pungency and instability of capsaicin (Jincheng et al. 2010; Wang et al. 2008). pH-dependent release and thermal stabilization of bioactive compounds have also been achieved using natural-based encapsulating materials (Akbari and Wu 2016; Pinheiro et al. 2015).

#### 20.3.2.2 Solid Lipid Nanoparticles (SLN) and Nanostructured Lipid Carriers (NLC)

Lipid-based nanocarriers like SLNs have become effective alternative delivery systems to conventional colloidal carriers like emulsions, liposomes, and polymeric

**Table 20.3** Various methods for microencapsulation of bioactive compounds

| Methods | Bioactive compounds | Coating materials | Particle size (μm) | Percent encapsulation (%) | Purpose |
|---|---|---|---|---|---|
| Spray drying (Mennella et al. 2016) | *Gentiana lutea* root extract (secoiridoids) | Ethylcellulose and stearate | – | – | Bitter-taste masking |
| Spray-drying (Boonchu and Utama-ang 2015) | Grape (*Vitis vinifera* L.) pomace extract | Maltodextrin and Carboxymethyl-cellulose | – | – | Masking of bitterness and astringency |
| Spray drying (Bustamante et al. 2017) | Pomegranate (*Punica granatum* var. Wonderful) peel extract and seed oil | Capsul C (modified starch) | 4.34–4.67 | 35.1–92.7 | Stabilization |
| Spray drying (Zhang et al. 2007) | Grape seed extract (Procyanidins) | Gum arabic and maltodextrin | 5–30 | 88.84 | Stabilization |
| Spray drying and freeze drying (Piovesana and Norena 2018) | Bioactive compounds from *Hibiscus calyces* extract | Partially hydrolyzed guar gum, poly-dextrose, gum arabic | 5.43–143.08 | 59.84–74.28 | Improvement of physical properties |
| Spray drying (Romo-Hualde et al. 2012) | Bioactive compounds from red pepper (*Capsicum annum* L.) | Gum arabic | 5.46 | 63.60–77.10 | Stabilization |
| Spraydrying (Saenz et al. 2009) | Bioactive compounds from cactus pear (*Opuntia ficus-indica*) | Maltodextrin and inulin | – | 39.41–99.49 | Stabilization |
| Double emulsion solvent evaporation (Paulo and Santos 2018) | Caffeic acid | Ethylcellulose | 1.4–298 | 65.9–92.3 | Controlled release cosmetic formulations |
| Spray congealing (Tomsik et al. 2019) | Wild garlic (*Allium ursinum* L.) extract | Gelucire 50/13 (Stearoyl polyoxyl-32 glycerides) | 100–200 | 99 | Stability and oral bioavailability |
| Complex coacervation (Jain et al. 2015) | β-Carotene | Whey protein isolates and gum acacia | 1–500 | 70 | Controlled release |

micelles (Lin et al. 2017). In SLNs, the bioactive compounds are part of the solid lipid core matrix (Fig. 20.2) which is stabilized by a surfactant or a mixture of surfactants (Weiss et al. 2008). They provide a safe means of effective delivery system for poorly aqueous-soluble bioactive phytochemicals, enhancing their stability while also promoting their permeability and bioavailability. They originate from an O/W-type emulsion where the liquid oil (lipid) is replaced by solid oil that remains solid at body temperature (Akhavan et al. 2018). SLNs are biocompatible, considerably stable with good encapsulation efficiency, and can prolong the release of the encapsulated compounds (Akhavan et al. 2018; Weiss et al. 2008). The surface of SLN can also be functionalized with agents to facilitate their uptake and targeting the release of the drug to the desired site of action (Ganesan et al. 2018). In addition, SLN can also carry both lipophilic and hydrophilic drugs and can be sterilized and produced in large scale (Liu and Feng 2015). Nanostructured lipid carriers (NLC) are new generations of SLNs developed in the 1990s which incorporate a mixture of solid and liquid lipids to overcome the limitation of SLNs in effective drug delivery (Akhavan et al. 2018). NLCs improve the bioactive retention capacity of these nanostructures and facilitate controlled release of the bioactive compounds.

Stabilization to oxidation offered by SLNs seemed to depend on the physicochemical properties of the encapsulated bioactives. Within the SLN, hydrophobic and high melting bioactive compounds like β-carotene are reported to be located within the inner matrix and kept at the α-subcell of the crystal structure thereby protecting it from oxidation, while less hydrophobic compounds like vitamin A arranged on the surfaces of the SLN leading to rapid oxidation (Salminen et al. 2016). However, between the SLNs and NLCs, NLCs were reported to provide better stability to β-carotene than the SLNs (Qian et al. 2013). The tendency of droplets aggregation and to coalescence still exists in SLNs which have the capacity to expel the encapsulated β-carotene to the particle exterior on storage, making it sensitive to degradation than when it is encapsulated in NLCs.

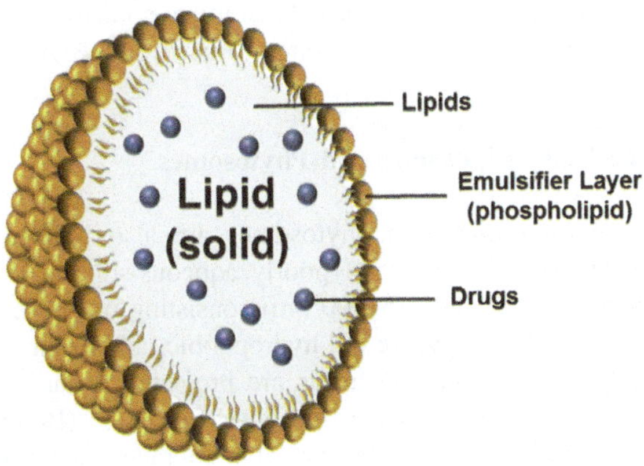


**Fig. 20.2** Structures of solid lipid nanoparticles (SLNs) (Reproduced with permission from *Lin* et al. 2017, Copyright © Food and Drug Administration, Taiwan. Published by Elsevier Taiwan LLC)

Flavonoids like quercetin and resveratrol suffer from low aqueous solubility, rapid metabolism, and photosensitivity which limited their oral bioavailability when taken orally (Li et al. 2009; Pandita et al. 2014). Formulation of quercetin into SLN remarkably enhances its gastrointestinal absorption and pharmacokinetic study in animal model demonstrated that the relative bioavailability of the SLN formulation against the quercetin suspension 571.4% indicating phenomenal increase in its oral bioavailability (Li et al. 2009). Stearic acid-based SLN of resveratrol was also reported to increase the oral bioavailability of the bioactive compound by eightfolds when compared to its suspension. Similarly, oral bioavailability of curcumin was also increased in a dose-dependent manner when formulated into soy lecithin-based SLN in animals. Improvement in oral bioavailability of curcumin-loaded SLN against solubilized curcumin rat plasma was reported to be 155 times at the dose of 1 mg/kg, 59 times at 12.5 mg/kg, 32 times at 25 mg/kg, and 39 times at the dose of 50 mg/kg (Kakkar et al. 2011). Bioavailability enhancement through SLN formulation has also been reported for other bioactive phytochemicals like an alkaloid vinpocetine which exhibit poor aqueous solubility and extensive first-pass metabolism (Luo et al. 2006; Medina 2010). Oral bioavailability from SLN vinpocetine formulation was found to be 4.16–4.17 times that of the suspension (Luo et al. 2006).

Recently, SLN-based sclareol was prepared following hot homogenization technique and evaluated its anti-proliferative activity in vitro against A549 human lung epithelial cancer cells which showed similar cytotoxicity to the plain sclareol but long-term stability and sustained release of sclareol were obtained with the SLN formulation (Hamishehkar et al. 2018).

NLC-based formulations also exhibit promising results in enhancing bioavailability of bioactive compounds. Silymarin was encapsulated in NLC-based formulation using glycerol distearates, oleic acids, lecithin, and Tween-80, which increases the oral bioavailability of silymarin against solid dispersion pellets and commercially available silymarin preparation, Legalon®, was 2.54- and 3.10-folds, respectively (Shangguan et al. 2014). However, when NLC was compared against microemulsion in its capacity to enhance oral bioavailability of luteolin in animal model, microemulsion fared better than the NLCs (Liu et al. 2014). The relative bioavailability of NLC-based and microemulsion-based luteolin formulations to the luteolin suspension were found to be 515.06% and 885.46%, respectively.

#### 20.3.2.3 Liposomes and Phytosomes

Both liposomes and phytosomes are also lipid-based formulations suitable to facilitate the delivery of poorly aqueous-soluble phytochemicals. Liposomes are spherical vesicles (15–100 nm) consisting of phospholipid with a liquid core and variable layers, where the hydrophobic tails of the phospholipids face each other, while the hydrophilic heads are projected toward the inner aqueous core or the outside boundary of the liposome (Fig. 20.3) (Bonechi et al. 2019; Emami et al. 2018). They resemble biological membrane and can accommodate both hydrophilic

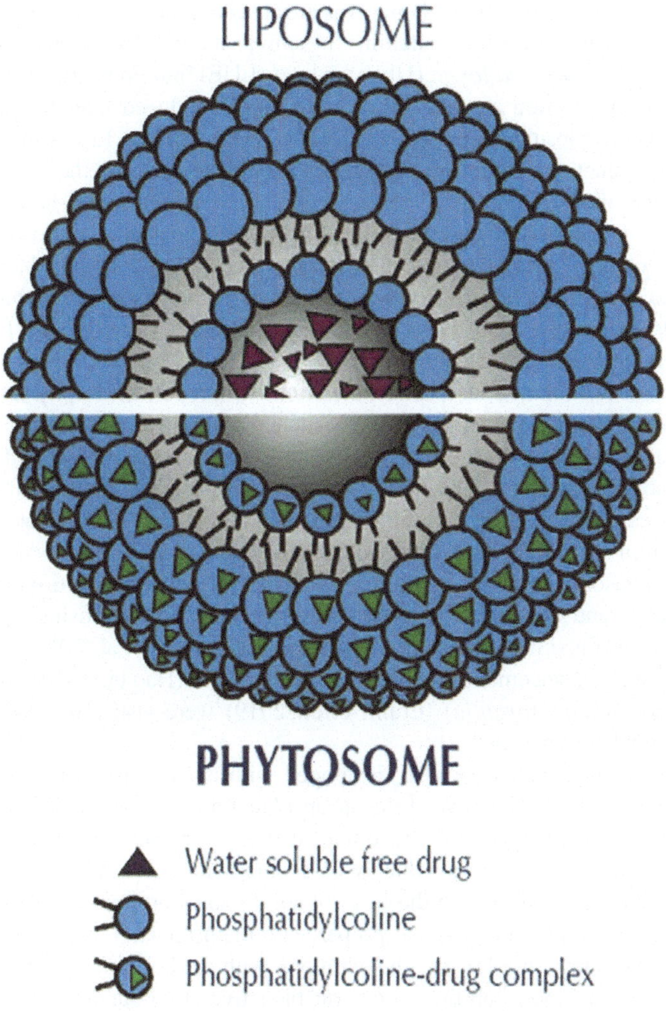


**Fig. 20.3** Major difference between liposome and phytosome® (Reproduced with permission from *Selmaty* et al. 2010, Copyright © Elsevier Science BV 2009)

and hydrophobic drugs within their layers which make them unique carrier for drug delivery. In spite of their advantages, successful delivery of liposomes through oral route has always been a challenge as the question remains whether they would remain stable along the GI tract conditions (Park et al. 2011). However, the effective delivery of phytochemicals through oral route has been demonstrated in animal models (Yi et al. 2013). *Flammulina velutipes* sterols was orally delivered through liposomes in Kunming mice and relative bioavailability of ergosterol and 22,23- dihydroergosterol at 162.9 and 244.2%, respectively, was obtained. The sterols, however, was rapidly eliminated within 4 h. A faster and better absorption of curcumin was also reported when it was encapsulated in leicithin-based liposome exhibiting improved oral bioavailability in animal model (Takahashi et al. 2009). Formulating the *Curcuma longa* (Ukon) extract (LUE) into liposomes was found to deliver better protection against carbon tetrachloride-induced liver injury as compared to the uncapsulated extract (Takahashi et al. 2008). Liposomes are also adaptable to functionalization with targeting ligands to improve their effectiveness in cancer therapy. *Kappaphycus alvarezii* extract containing multi-bioactive compounds was encapsulated into PEGylated-liposome and then functionalized with folic acid to target folate positive breast cancer cells (MCF-7) (Baskararaj et al. 2020). The constructed liposome was highly stable in the physiological buffers with steady drug release and also exhibit cytotoxicity toward MCF-7 in a concentration-dependent manner.

Liposomal formulation also contributes to the improvement in physicochemical properties of natural bioactive compounds such as solubility and stability. Chemical stability of resveratrol was enhanced by liposomes and along with 3-oxo-$C_{12}$-homoserine lactone, about 70% inhibition of tumor growth in murine tumor model was reported when they are administered intravenously (Coimbra et al. 2011). Encapsulation of enriched-phenolic fraction of the extract from *Pistacia vera* L into nanoliposome exhibits multiple bioactivity including antioxidant, anti-inflammatory, and anti-melanogenic activities making it a promising cosmeceutical preparation for skin pigmentation disorders (Oskoueian et al. 2020). Promotion of stability and enhancement in bioactivity of quercetin (Hao et al. 2017) and bitter gourd (*Momordica charantia*) (Erami et al. 2019) were also reported with their nanoliposomal formulations.

Proliposome systems were also developed with an objective of enhancing the stability and oral bioavailability of the encapsulated drugs. Compared to liposomes, proliposomes are highly stable and obtained in dry, free-flowing particles which form liposomal suspension immediately on contact with water (Yan-yu et al. 2006). Presenting the whole system in the form of solid makes proliposome highly stable while the intrinsic pharmacological properties remain intact.

Phytosomes are complex of phospholipid with herbal extracts developed to enhance the oral bioavailability of natural bioactive compounds. It is a patented technology developed in Italy around 1989 and several phytosome products are already available in the market (Pachuau 2019). The complexation of phosphatidylcholine (phospholipid) with polyphenols in phytosomes led to the formation of intermolecular bond between these two and the amphiphilic nature of the phosphatidylcholine promotes the miscibility of the complex with aqueous and lipid solvents which guide the polyphenol to permeate the lipid-based GI tract lumen (Kidd 2009). Delivery of polyphenols in the form of phytosomes has the capacity to enhance the oral bioavailability of polyphenols at least by 2–6 times (Ajazuddin and Saraf 2010; Kidd 2009). The ratio of the bioactive compounds to the phosphatidylcholine has been considered to be important factor in phytosome formulations. This bioactive compound:phospholipid ratio may range from 0.5:1 to 3:1 and the optimum ratio may depend on the kind of phytochemicals being complexed (Pachuau 2019). Frequently, a combination of bioactive phytochemicals has been incorporated into phytosomes to attain their synergistic effects (Rathee and Kamboj 2018; Sharma and Sahu 2016).

#### 20.3.2.4 Self-emulsifying Systems

Both micro- and nano-self-emulsifying systems are capable of enhancing oral bioavailability of poorly aqueous-soluble drugs by enhancing their solubility and dissolution rate. They are composed of a lipid, surfactant, bioactive compound, and a co-surfactant, and following oral administration, they swiftly form microemulsions with particle size less than 100 nm within the GI fluid bringing the drug into solution for enhanced absorption (Cui et al. 2009; Dwivedi et al. 2014). Micro- and nano-self-emulsifying systems have been demonstrated to enhance the oral bioavailability of various phytochemicals including curcumin (Cui et al. 2009), arteether (Dwivedi et al. 2014), silymarin (Li et al. 2010; Wu et al. 2006), curcumin, and thymoquinone (Alwadei et al. 2019), naringenin (Khan et al. 2015) as well as plant extracts (Arun and Maneesh 2016).

#### 20.3.2.5 Niosomes

Niosomes are self-assembled non-ionic surfactant vesicles analogous to phospholipid vesicles of liposomes where hydrophilic drugs are encapsulated and hydrophobic drugs are partitioned into the hydrophobic tails of the non-ionic surfactant (Hu and Rhodes 1999; Uchegbu and Vyas 1998). Niosomes are more stable than liposomes and are more economical to prepare from a wide range of surfactants which make them an attractive alternative to liposomes (Song et al. 2015). Niosomal preparations are reportedly initiated from the cosmetic industry in the 1970s and gradually expanded toward drug delivery applications including phytochemicals (Muzzalupo and Mazzotta 2019). Niosomes of poorly soluble D-limonene were reported to prolong the release of D-limonene and its cytotoxicity toward HepG2, Macf-7, and A549 cancer cells was significantly enhanced (Hajizadeh et al. 2019). Niosome also facilitates the solubilization of Lawsone and compared to free Lawsone solution, the antitumor activity of Lawsone noisome against MCF-7 breast cancer cells was increased significantly (Barani et al. 2018).

Studies have shown that niosomes are also effective carrier for the delivery of bioactive phytochemicals across the skin. Compared to their solutions, more efficient delivery of ellagic acid across the human epidermis and dermis was observed when it was formulated as niosomes and this penetration of the skin was reported to be dependent on the vesicle size (Junyaprasert et al. 2012).

Niosomes may suffer from the issue of aggregation, fusion, and leakage of the encapsulated bioactive compounds. To alleviate this problem, proniosomes have been prepared which are dry and free-flowing type that spontaneously form niosomes on gentle agitation with hot water. Proniosomes were demonstrated to increase the oral bioavailability of poorly soluble alkaloid, vinpocetine by about 4- to 4.9-folds (Song et al. 2015).

#### 20.3.2.6 Hydrogels

Hydrogels are hydrophilic three-dimensional polymer networks which can take up large amount of water. They are biodegradable and the three-dimensional network of the hydrogel disintegrates into non-toxic materials with excellent biocompatibility (Akhtar et al. 2016). Hydrogels are prepared from wide ranges of natural and synthetic polymers and can be presented in the various physical forms such as beads, microparticles, nanoparticles, and films and they have become one of the promising encapsulating systems to provide protection to sensitive phytochemicals and for their controlled delivery (Abaee et al. 2017; Gomez-Mascaraque et al. 2016a). The advantages hydrogels offer in encapsulation of bioactive compounds include the following:

(i) Enhancing stability and bioacessibility of sensitive, poorly aqueous-soluble bioactive molecules (Gomez-Mascaraque et al. 2016a; Han et al. 2020; Zhang et al. 2016).
(ii) pH-dependent and controlled release of bioactive compounds (Bourbon et al. 2016; Lopez Cordoba et al. 2013; Rutz et al. 2013).
(iii) Efficient encapsulation of both hydrophilic and hydrophobic bioactive compounds (Bourbon et al. 2016).
(iv) Enhancing bioactivity (Chan et al. 2010).
(v) Improving bioavailability (McClements 2017).
(vi) High drug-loading capacity due to their large spacing within the polymeric network (Teng et al. 2015).

Various hybrid hydrogel systems such as emulsion hydrogels, organogels and bigels have also been evaluated for encapsulation of bioactive compounds. Encapsulation of bioactive compounds within these networks can improve their stability, solubility and dispersibility thereby improving their overall bioavailability (Mao et al. 2019). Suitable functionalities such as responsiveness to pH, temperature, enzymes etc. can also be imparted to these hybrid hydrogels to provide the desired release characteristics of the formulation (McClements 2017). In emulsion (O/W) gel systems, the continuous aqueous phase gets solidified or gelled to provide structural integrity and flexibility to the whole system. As a result, it has wide application in functional food industry. Bioactive compounds such as curcumin (Geremias-Andrade et al. 2017), quercetin (Chen et al. 2018), α-tocopherol (Freire et al. 2018) have been delivered through emulsion gel systems with improved stability, bioaccessibility and controlled release. In addition, the structural state of emulsion hydrogels also allows them to be used as an efficient fat-replacement in various food products without affecting their sensory characteristics thereby promoting health benefits to consumers (Pintado et al. 2015, 2016).

#### 20.3.2.7 Electrodynamic Processes

In recent years, electrodynamic encapsulation processes such as electrospraying and electrospinning have received increasing attention as an alternative to the more established and commonly employed techniques such as microencapsulation (Alehosseini et al. 2018; Jacobsen et al. 2018). During electrospraying, the high voltage electrostatic force converts the drug-loaded liquid solutions into fine droplets and gets evaporated during their flight toward the ground electrode (Alehosseini et al. 2018). A high degree of molecular interaction in the sprayed liquid produces electrospun fibers, while lower degree of interaction results in electrosprayed particles (Fig. 20.4). In conventional spray-drying process, drugs or bioactive compounds are exposed to high temperature in the range of 170–220 °C to dry and encapsulate the material and this condition of drying may degrade thermolabile drug substances (Jacobsen et al. 2018). However, in electrospraying, exposure to such high temperature has been eliminated which makes it more conducive for encapsulation of heat-sensitive compounds. Electrodynamic encapsulation of bioactive compounds also offers the following advantages (Drosou et al. 2017; Ghorani and Tucker 2015; Wen et al. 2017a; Wen et al. 2017b):

(i) It is a relatively simple, cost-effective, and flexible method for fabrication of nanomaterials.
(ii) It has the potential to scale up to large-scale manufacturing.
(iii) The method is adaptable to wide range of encapsulating materials.

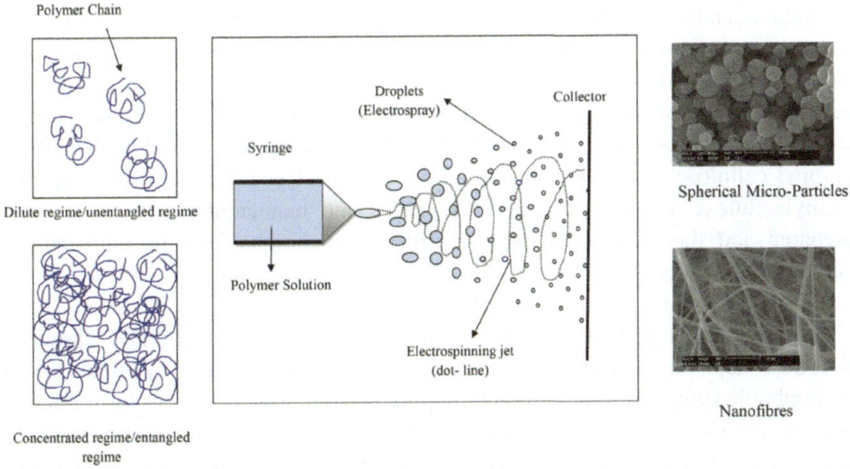


**Fig. 20.4** Left: Schematic of physical representation at the molecular level of entanglement regimes for dilute and concentrated polymer concentration. Middle: Schematic diagram of a basic electrospinning (jet formation) and electrospraying (liquid-droplet atomization) processes. Right: Examples of SEM image of microspheres (electrospraying) and nanofibers (Electrospinning) (Reproduced with permission from Ghorani and Tucker 2015, Copyright © Elsevier Science BV 2015)

(iv) Enhanced stability of bioactive compounds.
(v) High surface-to-volume ratio of the nanoscale materials results in better bioavailability.
(vi) Both hydrophilic and hydrophobic bioactive compounds can be encapsulated.

Electrospraying was employed to successfully encapsulate β-carotene in its solubilized form to enhance its stability and also to improve its bioavailability (Basar et al. 2020). When exposed under UV light, the free-β-carotene degrades within 180 min while only 20% degradation was found with the encapsulated β-carotene. Wide ranges of polymers are adaptable to electrodynamic processing including polysaccharides cellulose and its derivatives, guar gum, pectin, starch, chitosan, alginate, etc. (Wen et al. 2017b). Viscosity, surface tension, and electrical conductivity of the polymeric solution were found to influence the morphology of the electrosprayed capsules or fibers (Gomez-Mascaraque et al. 2016b).

#### 20.3.2.8 Solid Dispersion and Micelles

Solid dispersion technique is one of the most commonly employed methods to improve the solubility and bioavailability of poorly aqueous-soluble drugs. Crystalline drugs often exhibit poor solubility in aqueous-based solvents as the lattice energy must be overcome to bring them into solution (Li et al. 2013). Formulation into amorphous solid dispersion helps solubilization of such crystalline drugs as the encapsulating polymers in the solid dispersions trapped them in the amorphous state which facilitates their dissolution once they come in contact with the aqueous solvents. Various cellulose derivatives have been reported to exhibit strong enhancement of curcumin dissolution while also protecting from chemical degradation and affecting pH-dependent release (Li et al. 2013). Various formulations of curcumin such as nanocrystal solid dispersion prepared with hydroxypropyl cellulose SL (CSD-Cur), amorphous solid dispersion with hydroxypropyl methylcellulose acetate succinate (ASD-Cur), and nanoemulsion (NE-Cur) were prepared and the increase in oral bioavailability compared to the unformulated curcumin was reported to be 12-folds for ASD-Cur, 16-folds for CSD-Cur and, 9-folds for NE-Cur (Onoue et al. 2010).

Self-assembled colloidal dispersion of amphiphilic surfactants, also known as micelles, has been another important technique to enhance the solubility of hydrophobic drugs. Certain amphiphilic polymers are also known to form micelles and are known as polymeric micelles (Husseini and Pitt 2008). Polymeric micelles have been shown to enhance the solubility and bioactivity of bioactive compounds. When Pluronic P123- and Solutol HS15-based polymeric micelles were prepared loaded with naringenin, sustained release of naringerin and significant enhancement in its cytotoxicity against cancer cell line were observed (Zhai et al. 2013). Oral bioavailability of paclitaxel in glycyrrhizic acid micelle was also found to enhance sixfolds as compared to Taxol (Yang et al. 2015).

## 20.4 Conclusion

Enhancement in stability and bioavailability of phytochemicals has been one of the most important challenges in their effective utilization. Thousands of bioactive compounds have been isolated and characterized over the years; however, their therapeutic effectiveness in vivo is often limited by their poor physicochemical properties and modest oral absorption. Several studies have demonstrated that encapsulation into various micro-and nanoplatforms is capable of alleviating such obstacle. Microencapsulation techniques have been proven and commonly employed method for such enhancement. Novel techniques like hydrogels and their variants such as emulsion gels, organo gels, and bigels along with electrodynamic processes like electrospraying and electrospinning techniques are certainly promising approaches to facilitate the effective delivery of these bioactive compounds.

## References

Abaee A, Mohammadian M, Jafari SM (2017) Whey and soy protein-based hydrogels and nano-hydrogels as bioactive delivery systems. Trend Food Sci Technol 70:69–81

Aditya NP, Hamilton IE, Noon J, Norton IT (2019) Microwave assisted nanonization of poorly water-soluble curcumin. ACS Sustain Chem Eng 7:9771–9781

Aguiar GPS, Arcari BD, Chaves LMPC, Magro CD, Boschetto DL, Piato AL et al (2018) Micronization of trans-resveratrol by supercritical fluid: dissolution, solubility and in vitro antioxidant activity. Ind Crop Prod 112:1–5

Ahmed K, Li Y, McClements DJ, Xiao H (2012) Nanoemulsion-and emulsion-based delivery systems for curcumin: encapsulation and release properties. Food Chem 132:799–807

Ajazuddin and Saraf S (2010) Applications of novel drug delivery system for herbal formulations, Fitoterapia 81: 680–689

Akbari A, Wu J (2016) Cruciferin nanoparticles: Preparation, characterization and their potential application in delivery of bioactive compounds. Food Hydrocolloid 54:107–118

Akhavan S, Assadpour E, Katouzian I, Jafari SM (2018) Lipid nano scale cargos for the protection and delivery of food bioactive ingredients and nutraceuticals. Trend Food Sci Technol 74:132–146

Akhtar MF, Hanif M, Ranjha NM (2016) Methods of synthesis of hydrogels. A review, Saudi Pharm J 24:554–559

Al Juhaimi F, Ozcan MM, Ghafoor K, Babiker EE, Hussain S (2018) Comparison of cold-pressing and soxhlet extraction systems for bioactive compounds, antioxidant properties, polyphenols, fatty acids and tocopherols in eight nut oils. J Food Sci Technol 55:3163–3173

Alborzi S, Lim LT, Kakuda Y (2012) Encapsulation of folic acid and its stability in sodium alginate-pectin-poly(ethylene oxide) electrospun fibres. J Microencapsul 30:64–71

Alehosseini A, Ghorani B, Sarabi-Jamab M, Tucker N (2018) Principles of electrospraying: a new approach in protection of bioactive compounds in foods. Crit Rev Food Sci Nutr 58:2346–2363

Alfei S, Turrini F, Catena S, Zunin P, Parodi B, Zuccari G et al (2019) Preparation of ellagic acid micro and nano formulations with amazingly increased water solubility by its entrapment in pectin or non-PAMAM dendrimers suitable for clinical applications. New J Chem 43:2438–2448

Almeida AF, Borge GIA, Piskula M, Tudose A, Tudoreanu L, Valentová K et al (2018) Bioavailability of quercetin in humans with a focus on interindividual variation. Compr Rev Food Sci F 17:714–731

Altemini A, Lakhssassi N, Baharlouei A, Watson DG, Lightfoot DA (2017) Phytochemicals: extraction, isolation, and identification of bioactive compounds from plant extracts. Plants 6:42

Alwadei M, Kazi M, Alanazi FK (2019) Novel oral dosage regimen based on self-nanoemulsifying drug delivery systems for codelivery of phytochemicals—curcumin and thymoquinone. Saudi Pharm J 27:866–876

Anirudhan TS, Binusree J (2016) Dextran based nanosized carrier for the controlled and targeted delivery of curcumin to liver cancer cells. Int J Biol Macromol 88:222–235

Aqil F, Munagala R, Jeyabalan J, Vadhanam MV (2013) Bioavailability of phytochemicals and its enhancement by drug delivery systems. Cancer Lett 334:133–141

Arun K, Maneesh J (2016) Effect of Self Nanoemulsifying Drug Delivery System (SNEDDS) on Intestinal Permeation and Anti-Diabetic Activity of *Berberis aristata* Extract: *In-Vitro* and *Ex-Vivo* Studies. J Nanopharm Drug Deliv 3:51–62

Atanasov AG, Waltenberger B, Pferschy-Wenzig EM, Linder T, Wawrosch C, Uhrin P et al (2015) Discovery and resupply of pharmacologically active plant-derived natural products: a review. Biotechnol Adv 33:1582–1614

Bala I, Bhardwaj V, Hariharan S, Kharade SV, Roy N, Ravi Kumar MNV (2006) Sustained release nanoparticulate formulation containing antioxidant ellagic acid as potential prophylaxis system for oral administration. J Drug Target 14:27–34

Barani M, Mirzaei M, Torkzadeh-Mahani M, Nematollahi MH (2018) Lawsone-loaded Niosome and its antitumor activity in MCF-7 breast cancer cell line: a nano-herbal treatment for cancer. Daru 26:11–17

Barras A, Mezzetti A, Richard A, Lazzaroni S, Roux S, Melnyk P et al (2009) Formulation and characterization of polyphenol-loaded lipid nanocapsules. Int J Pharm 379:270–277

Basar AO, Prieto C, Durand E, Villeneuve P, Sasmazel HT, Lagaron J (2020) Encapsulation of β-carotene by emulsion electrospraying using deep eutectic solvents. Molecules 25:981

Baskararaj S, Panneerselvam T, Govindaraj S, Arunachalam S, Parasuraman P, Pandian SRK et al (2020) Formulation and characterization of folate receptor-targeted PEGylated liposome encapsulating bioactive compounds from *Kappaphycus alvarezii* for cancer therapy. Biotech 10:136

Beevers CS, Huang S (2011) Pharmacological and clinical properties of curcumin. Botanics 1:5–18

Bernhoft A (2010) A brief review on bioactive compounds in plants. Bernhoft A(ed) Bioactive compounds in plants -benefits and risks for man and animals. Norwegian Acad Sci Lett, Oslo, pp 11–17

Bertoni S, Albertini B, Passerini N (2019) Spray Congealing: an emerging technology to prepare solid dispersions with enhanced oral bioavailability of poorly water soluble drugs. Molecules 24:3471

Bohn T, McDougall GJ, Alegría A, Alminger M, Arrigoni E, Aura AM et al (2015) Mind the gap-deficits in our knowledge of aspects impacting the bioavailability of phytochemicals and their metabolites–a position paper focusing on carotenoids and polyphenols. Mol Nutr Food Res 59:1307–1323

Bonechi C, Donati A, Tamasi G, Pardini A, Rostom H, Leone G (2019) Chemical characterization of liposomes containing nutraceutical compounds: Tyrosol, hydroxytyrosol and oleuropein. Biophys Chem 246:25–34

Boonchu T, Utama-ang N (2015) Optimization of extraction and microencapsulation of bioactive compounds from red grape (*Vitis vinifera* L.) pomace. J Food Sci Technol 52:783–792

Borgogna M, Bellich B, Zorzin L, Lapasin R, Cesaro A (2010) Food microencapsulation of bioactive compounds: Rheological and thermal characterisation of non-conventional gelling system. Food Chem 122:416–423

Borhan MZ, Ahmad R, Rusop M, Abdullah S (2014) Effect of nanonization on physicochemical properties of *Centella asiatica* powders. Adv Mat Res 917:106–112

Bourbon AI, Cerqueira MA, Vicente AA (2016) Encapsulation and controlled release of bioactive compounds in lactoferrin-glycomacropeptide nanohydrogels: Curcumin and caffeine as model compounds. J Food Eng 180:110–119

Bustamante A, Hinojosa A, Robert P, Escalona V (2017) Extraction and microencapsulation of bioactive compounds from pomegranate (*Punica granatum* var. Wonderful) residues. Int J Food Sci Technol 52:1452–1462

Champagne CP, Fustier P (2007) Microencapsulation for the improved delivery of bioactive compounds into foods. Curr Opin Biotechnol 18:184–190

Chan E, Yim Z, Phan S, Mansa RF, Ravindra P (2010) Encapsulation of herbal aqueous extract through absorption with ca-alginate hydrogel beads. Food Bioprod Process 88:195–201

Chen X, McClements DJ, Zhu Y, Zou L, Li Z, Liu W et al (2018) Gastrointestinal fate of fluid and gelled nutraceutical emulsions: impact on proteolysis, lipolysis, and quercetin bioaccessibility. J Agric Food Chem 66:9087–9096

Choi AY, Kim CT, Park HY, Kim HO, Lee NR, Lee KE et al (2013) Pharmacokinetic characteristics of capsaicin-loaded nanoemulsions fabricated with alginate and chitosan. J Agric Food Chem 61:2096–2102

Chu Z, Chen JS, Liau CT, Wang HM, Lin YC, Yang MH et al (2008) Oral bioavailability of a novel paclitaxel formulation (Genetaxyl) administered with cyclosporin a in cancer patients. Anticancer Drugs 19:275–281

Chuah LH, Roberts CJ, Billa N, Abdullah S, Rosli R (2014) Cellular uptake and anticancer effects of mucoadhesive curcumin-containing chitosan nanoparticles. Colloid Surface B 116:228–236

Coimbra M, Isacchi B, van Bloois L, Torano JS, Ket A, Wu X et al (2011) Improving solubility and chemical stability of natural compounds for medicinal use by incorporation into liposomes. Int J Pharm 416:433–442

Cragg GM, Newman DJ (2013) Natural products: a continuing source of novel drug leads. Biochim Biophys Acta - General Subjects 1830:3670–3695

Cui J, Yu B, Zhao Y, Zhu W, Li H, Lou H et al (2009) Enhancement of oral absorption of curcumin by self-microemulsifying drug delivery systems. Int J Pharm 371:148–155

Da Silva BV, Barreira JCM, Oliveira MBPP (2016) Natural phytochemicals and probiotics as bioactive ingredients for functional foods: extraction, biochemistry and protected-delivery technologies. Trends Food Sci Technol 50:144–158

David B, Wolfender J, Dias DA (2015) The pharmaceutical industry and natural products: historical status and new trends. Phytochem Rev 14:299–315

De Souza VB, Thomazini M, Chaves IE, Ferro-Furtado R, Favaro-Trindade CS (2020) Microencapsulation by complex coacervation as a tool to protect bioactive compounds and to reduce astringency and strong flavor of vegetable extracts. Food Hydrocoll 98:105244

Dias DR, Botrel DA, De Barros Fernandes VR, Borges SV (2017) Encapsulation as a tool for bioprocessing of functional foods. Curr Opin Food Sci 13:31–37

Dima C, Patraşcu L, Cantaragiu A, Alexe P, Dima S (2016) The kinetics of the swelling process and the release mechanisms of *Coriandrum sativum* L. essential oil from chitosan/alginate/ inulin microcapsules. Food Chem 195:39–48

Donsi F, Annunziata M, Sessa M, Ferrari G (2011) Nanoencapsulation of essential oils to enhance their antimicrobial activity in foods. LWT-Food Sci Technol 44:1908–1914

Dos Santosa PP, Flores SH, Rios AO, Chiste RC (2016) Biodegradable polymers as wall materials to the synthesis of bioactive compound nanocapsules. Trend Food Sci Technol 53:23–33

Drosou CG, Krokida MK, Biliaderis CG (2017) Encapsulation of bioactive compounds through electrospinning/electrospraying and spray drying: a comparative assessment of food-related applications. Drying Technol 35:139–162

Dutta RS, Hauzel L, Roy PK, Kalita P, Devi TB, Deka D et al (2018) Nanoprecipitated ethylcellulose-curcumin particles for controlled release and enhanced antioxidant activity. Curr Nanosci 14:298–306

Dwivedi P, Khatik R, Khandelwal K, Srivastava R, Taneja I, Rama Raju KS et al (2014) Self-nanoemulsifying drug delivery system (SNEDDS) for oral delivery of arteether: pharmacokinetics, toxicity and antimalarial activity in mice. RSC Adv 4:64905–64918

Emami S, Azadmard-Damirchi S, Peighambardoust SH, Valizadeh H, Hesariet J (2018) Liposomes as carrier vehicles for functional compounds in food sector. J Experiment Nanosci 11:737–759

Engel JB, Heckler C, Tondo EC, Daroit DJ, Da Silva Malheiros P (2017) Antimicrobial activity of free and liposome-encapsulated thymol and carvacrol against *Salmonella* and *Staphylococcus aureus* adhered to stainless steel. Int J Food Microbiol 252:18–23

Ensign LM, Cone R, Hanes J (2012) Oral drug delivery with polymeric nanoparticles: The gastrointestinal mucus barriers. Adv Drug Deliv Rev 64:557–570

Erami SR, Amiri ZR, Jafari SM (2019) Nanoliposomal encapsulation of Bitter Gourd (*Momordica charantia*) fruit extract as a rich source of health-promoting bioactive compounds. LWT - Food Sci Technol 116:108581

Ezhilarasi PN, Karthik P, Chhanwal N, Anandharamakrishnan C (2013) Nanoencapsulation techniques for food bioactive components: A review. Food Bioprocess Technol 6:628–647

Ezpeleta I, Irache JM, Stainmesse S, Chabenat C, Gueguen J, Popineau Y et al (1996) Gliadin nanoparticles for the controlled release of all-trans-retinoic acid. Int J Pharm 131:191–200

Fang Z, Bhandari B (2010) Encapsulation of polyphenols: a review. Trends Food Sci Technol 21:510–523

Feng Y, Sun C, Yuan Y, Zhu Y, Wan J, Firempong CK et al (2016) Enhanced oral bioavailability and in vivo antioxidant activity of chlorogenic acid via liposomal formulation. Int J Pharm 501:342–349

Freire M, Cofrades S, Perez-Jimenez J, Gomez-Estaca J, Jimenez-Colmenero F, Bou R (2018) Emulsion gels containing n-3 fatty acids and condensed tannins designed as functional fat replacers. Food Res Int 113:465–473

Ganesan P, Ramalingam P, Karthivashan G, Ko YT, Choi DK (2018) Recent developments in solid lipid nanoparticle and surface-modified solid lipid nanoparticle delivery systems for oral delivery of phyto-bioactive compounds in various chronic diseases. Int J Nanomed 13:1569–1583

Geremias-Andrade IM, Souki NPDBG, Moraes ICF, Pinho SC (2017) Rheological and mechanical characterization of curcumin-loaded emulsion-filled gels produced with whey protein isolate and xanthan gum. LWT Food Sci Technol 86:166–173

Ghorani B, Tucker N (2015) Fundamentals of electrospinning as a novel delivery vehicle for bioactive compounds in food nanotechnology. Food Hydrocoll 51:227–240

Goiun S (2004) Microencapsulation: industrial appraisal of existing technologies and trends. Trend Food Sci Technol 15:330–347

Gomez-Estaca J, Gavara R, Hernandez-Munoz P (2015) Encapsulation of curcumin in electrosprayed gelatin microspheres enhances its bioaccessibility and widens its uses in food applications. Innov Food Sci Emerg Technol 29:302–307

Gomez-Mascaraque LG, Soler C, Lopez-Rubio A (2016a) Stability and bioaccessibility of EGCG within edible micro-hydrogels. Chitosan vs. gelatin, a comparative study. Food Hydrocoll 61:128–138

Gomez-Mascaraque LG, Sanchez G, Lopez-Rubio A (2016b) Impact of molecular weight on the formation of electrosprayed chitosan microcapsules as delivery vehicles for bioactive compounds. Carbohydr Polym 150:121–130

Gorgani L, Mohammadi M, Najafpour GD, Nikzad M (2017) Piperine—The bioactive compound of black pepper: From isolation to medicinal formulations. Compr Rev Food Sci F 16:124–140

Grama CN, Suryanarayana P, Patil MA, Raghu G, Balakrishna N, Ravi Kumar MNV et al (2013) Efficacy of biodegradable curcumin nanoparticles in delaying cataract in diabetic rat model. PLoS ONE 8:e78217

Guajardo-Flores D, Rempel C, Gutierrez-Uribe JA, Serna-Saldivar SO (2015) Influence of excipients and spray drying on the physical and chemical properties of nutraceutical capsules containing phytochemicals from black bean extract. Molecules 20:21626–21635

Hajizadeh MR, Maleki H, Barani M, Fahmidehkar MA, Mahmoodi M, Torkzadeh-Mahani M (2019) *In vitro* cytotoxicity assay of D-limonene niosomes: an efficient nano-carrier for enhancing solubility of plant-extracted agents. Res Pharm Sci 14:448–458

Hamishehkar H, Bahadori MB, Vandghanooni S, Eskandani M, Nakhlband A, Eskandani M (2018) Preparation, characterization and anti-proliferative effects of sclareol loaded solid lipid nanoparticles on A549 human lung epithelial cancer cells. J Drug Deliv Sci Technol 45:272–280

Han J, Zhang Z, Shang W, Yan J, McClements DJ, Xiao H et al (2020) Modulation of physicochemical stability and bioaccessibility of β-carotene using alginate beads and emulsion stabilized by scallop (*Patinopecten yessoensis*) gonad protein isolates. Food Res Int 129:108875

Hao J, Guo B, Yu S, Zhang W, Zhang D, Wang J (2017) Encapsulation of the flavonoid quercetin with chitosan-coated nano-liposomes. LWT - Food Sci Technol 85:37–44

Hu C, Rhodes DG (1999) Proniosomes: a novel drug carrier preparation. Int J Pharm 185:23–35

Huang H, Belwal T, Liu S, Duan Z, Luo Z (2019) Novel multi-phase nano-emulsion preparation for co-loading hydrophilic arbutin and hydrophobic coumaric acid using hydrocolloids. Food hydrocolloid 93:92–101

Husseini GA, Pitt WG (2008) Micelles and nanoparticles for ultrasonic drug and gene delivery. Adv Drug Deliv Rev 60:1137–1152

Iqbal J, Sarti F, Perera G, Bernkop-Schnurch A (2011) Development and in vivo evaluation of an oral drug delivery system for paclitaxel. Biomaterials 32:170–175

Islam Shishir MR, Xie L, Sun C, Zheng X, Chen W (2018) Advances in micro and nano-encapsulation of bioactive compounds using biopolymer and lipid-based transporters. Trends Food Sci Technol 78:34–60

Jacobsen C, Garcia-Moreno PJ, Mendes AC, Mateiu RV, Chronakis IS (2018) Use of electrohydrodynamic processing for encapsulation of sensitive bioactive compounds and applications in food. Annu Rev Food Sci Technol 9:525–549

Jain A, Thakur D, Ghoshal G, Katare OP, Shivhare US (2015) Microencapsulation by complex coacervation using whey protein isolates and gum acacia: an approach to preserve the functionality and controlled release of β-carotene. Food Bioprocess Technol 8:1635–1644

Ji H, Tang J, Li M, Ren J, Zheng N, Wu L (2014) Curcumin-loaded solid lipid nanoparticles with Brij78 and TPGS improved *in vivo* oral bioavailability and *in situ* intestinal absorption of curcumin. Drug Deliv 23:459–470

Jincheng W, Xiaoyu Z, Sihao C (2010) Preparation and properties of nanocapsulated capsaicin by complex coacervation method. Chem Eng Comm 197:919–933

Junyaprasert VB, Singhsa P, Suksiriworapong J, Chantasart D (2012) Physicochemical properties and skin permeation of Span 60/Tween 60 niosomes of ellagic acid. Int J Pharm 423:303–311

Kakkar V, Singh S, Singla D, Kaur IP (2011) Exploring solid lipid nanoparticles to enhance the oral bioavailability of curcumin. Mol Nutr Food Res 55:495–503

Kasikci MB, Bagdatlioglu N (2016) Bioavailability of quercetin. Curr Nutr Food Sci 4:146–151

Katuwavila NP, Perera AD, Karunaratne V, Amaratunga GA, Karunaratne D (2016) Improved delivery of caffeic acid through liposomal encapsulation. J Nanomat 9701870:1–7

Khan AW, Kotta S, Ansari SH, Sharma RK, Ali J (2015) Self-nanoemulsifying drug delivery system (SNEDDS) of the poorly water-soluble grapefruit flavonoid Naringenin: design, characterization, in vitro and in vivo evaluation. Drug Deliv 22:552–561

Khan H, Marya Belwal T, Tariq M, Atanasov AG, Devkota HP (2019) Genus Vanda: A review on traditional uses, bioactive chemical constituents and pharmacological activities. J Ethnopharmacol 229:46–53

Khor CM, Ng WK, Kanaujia P, Chan KP, Dong Y (2017) Hot-melt extrusion microencapsulation of quercetin for taste-masking. J Microencapsul 34:29–37

Kidd PM (2009) Bioavailability and activity of phytosome complexes from botanical polyphenols: the silymarin, curcumin, green tea, and grape seed extracts. Altern Med Rev 14:226–246

Kumar A, Ahuja A, Ali J, Baboota S (2010) Conundrum and therapeutic potential of curcumin in drug delivery. Crit Rev Ther Drug 27:279–312

Lao CD, Ruffin MT, Normolle D, Heath DD, Murray SI, Bailey JM et al (2006) Dose escalation of a curcuminoid formulation. BMC Complement Altern Med 6:10

Laokuldilok N, Thakeow P, Kopermsub P, Utama-ang N (2016) Optimisation of microencapsulation of turmeric extract for masking flavor. Food Chem 194:695–704
Lei F, Xing D, Xiang L, Zhao Y, Wang W, Zhang L (2003) Pharmacokinetic study of ellagic acid in rat after oral administration of pomegranate leaf extract. J Chromatogr B 796:189–194
Lengyel M, Kallai-Szabo N, Antal V, Laki AJ, Antal I (2019) Microparticles, microspheres, and microcapsules for advanced drug delivery. Sci Pharm 87:20
Li H, Zhao X, Ma Y, Zhai G, Li L, Lou H (2009) Enhancement of gastrointestinal absorption of quercetin by solid lipid nanoparticles. J Control Release 133:238–244
Li X, Yuan Q, Huang Y, Zhou Y, Liu Y (2010) Development of silymarin self-microemulsifying drug delivery system with enhanced oral bioavailability. AAPS PharmSciTech 11:672–678
Li B, Konecke S, Wegiel LA, Taylor LS, Edgar KJ (2013) Both solubility and chemical stability of curcumin are enhanced by solid dispersion in cellulose derivative matrices. Carbohydr Polym 98:1108–1116
Li Z, Jiang H, Xu C, Gu L (2015) A review: Using nanoparticles to enhance absorption and bioavailability of phenolic phytochemicals. Food Hydrocoll 43:153–164
Lin C, Chen C, Lin Z, Fang J (2017) Recent advances in oral delivery of drugs and bioactive natural products using solid lipid nanoparticles as the carriers. J Food Drug Anal 25:219–234
Liu Y, Feng N (2015) Nanocarriers for the delivery of active ingredients and fractions extracted from natural products used in traditional Chinese medicine (TCM). Adv Colloid Interfac 221:60–76
Liu Y, Wang L, Zhao Y, He M, Zhang X, Niu M et al (2014) Nanostructured lipid carriers versus microemulsions for delivery of the poorly water-soluble drug luteolin. Int J Pharm 476:169–177
Lopez Cordoba A, Deladino L, Martino M (2013) Effect of starch filler on calcium-alginate hydrogels loaded with yerba mate antioxidants. Carbohydr Polym 95:315–323
Lu Q, Li D, Jiang J (2011) Preparation of a tea polyphenol nanoliposome system and its physicochemical properties. J Agric Food Chem 59:13004–13011
Luo Y, Chen D, Ren L, Zhao X, Qin J (2006) Solid lipid nanoparticles for enhancing vinpocetine's oral bioavailability. J Control Release 114:53–59
Ma Z, Na W, He H, Tang X (2019) Pharmaceutical strategies of improving oral systemic bioavailability of curcumin for clinical application. J Control Release 316:359–380
Mady FM, Shaker MA (2017) Enhanced anticancer activity and oral bioavailability of ellagic acid through encapsulation in biodegradable polymeric nanoparticles. Int J Nanomed 12:7405
Malingre MM, Beijnen JH, Schellens JHM (2001) Oral delivery of taxanes. Invest New Drug 19:155–162
Mao L, Lu Y, Cui M, Miao S, Gao Y (2019) Design of gel structures in water and oil phases for improved delivery of bioactive food ingredients. Crit Rev Food Sci Nutr 20:1–16
Martins N, Barros L, Ferreira ICFR (2016) *In vivo* antioxidant activity of phenolic compounds: Facts and gaps. Trends Food Sci Technol 48:1–12
Mazzarino L, Travelet C, Murillo SO et al (2012) Elaboration of chitosan-coated nanoparticles loaded with curcumin for mucoadhesive applications. J Colloid Interfac Sci 370:58–66
McClements DJ (2017) Recent progress in hydrogel delivery systems for improving nutraceutical bioavailability. Food Hydrocoll 68:238–245
Medina AE (2010) Vinpocetine as a potent antiinflammatory agent. P Natl Acad Sci USA 107:9921–9922
Mennella I, Fogliano V, Ferracane R, Arlorio M, Pattarino F, Vitaglione P (2016) Microencapsulated bitter compounds (from *Gentiana lutea*) reduce daily energy intakes in humans. Br J Nutr 116:1841–1850
Moreno T, Cocero MJ, Rodriguez-Rojo S (2018) Storage stability and simulated gastrointestinal release of spray dried grape marc phenolics. Food Bioprod Process 112:96–107
Mukherjee PK, Harwansh RK, Bhattacharyya S (2015) Bioavailability of herbal products: approach towards improved pharmacokinetics, In: Mukherkee PK (ed) Evidence-based validation of herbal medicines. Elsevier, Amsterdam

Murugan V, Mukherjee K, Maiti K, Mukherjee PK (2009) Enhanced oral bioavailability and antioxidant profile of ellagic acid by phospholipids. J Agric Food Chem 57:4559–4565

Muzzalupo R, Mazzotta E (2019) Do niosomes have a place in the field of drug delivery? Expert Opin Drug Deliv 16:1145–1147

Naczk M, Shahidi F (2006) Phenolics in cereals, fruits and vegetables: occurrence, extraction and analysis. J Pharm Biomed Anal 41:1523–1542

Nallamuthu I, Devi A, Khanum F (2015) Chlorogenic acid loaded chitosan nanoparticles with sustained release property, retained antioxidant activity and enhanced bioavailability. Asian J Pharm Sci 10:203–211

Oehlke K, Behsnilian D, Mayer-Miebach E, Weidler PG, Greiner R (2017) Edible solid lipid nanoparticles (SLN) as carrier system for antioxidants of different lipophilicity. PLoS ONE 12: e0171662

Onoue S, Takahashi H, Kawabata Y, Seto Y, Hatanaka J, Timmermann B, Yamada S (2010) Formulation design and photochemical studies on nanocrystal solid dispersion of curcumin with improved oral bioavailability. J Pharm Sci 99:1871–1881

Oskoueian E, Karimi E, Noura R, Ebrahimi M, Shafaei N, Karimi E (2020) Nanoliposomes encapsulation of enriched phenolic fraction from pistachio hulls and its antioxidant, anti-inflammatory, and anti-melanogenic activities. J Microencapsul 37:1–13

Pachuau L (2019) Nanostructures for improving oral bioavailability of herbal medicines. In: Mazumder B et al (eds) Nanotechnology: Therapeutic, Nutraceutical and Cosmetic Advances. CRC Press, Boca Raton, FL

Pachuau L, Devi CM, Goswami A, Sahu S, Dutta RS (2019) Seed oils as a source of natural bioactive compounds, In: Akhtar MS, Swamy MK, Sinniah UR(eds.) Natural Bio-active Compounds. Springer Nature, Singapore, pp 209–236

Pai DA, Vangala VR, Ng JW, Ng WK, Tan RBH (2015) Resistant maltodextrin as a shell material for encapsulation of naringin: production and physicochemical characterization. J Food Eng 161:68–74

Pandey R, Ahmad Z, Sharma S, Khuller GK (2005) Nano-encapsulation of azole antifungals: potential applications to improve oral drug delivery. Int J Pharm 301:268–276

Pandita D, Kumar S, Poonia N, Lather V (2014) Solid lipid nanoparticles enhance oral bioavailability of resveratrol, a natural polyphenol. Food Res Int 62:1165–1174

Park K, Kwon IC, Park K (2011) Oral protein delivery: current status and future prospect. React Funct Polym 71:280–287

Paulo F, Santos L (2018) Microencapsulation of caffeic acid and its release using a w/o/w double emulsion method: assessment of formulation parameters. Drying Technol 37:950–961

Pinheiro AC, Bourbon AI, Cerqueira MA, Maricato E, Nunes C, Coimbra MA et al (2015) Chitosan/fucoidan multilayer nanocapsules as a vehicle for controlled release of bioactive compounds. Carbohydr Polym 115:1–9

Pinho E, Grootveld M, Soares G, Henriques M (2013) Cyclodextrins as encapsulation agents for plant bioactive compounds. Carbohydr Polym 101:121–135

Pinho E, Soares G, Henriques M (2015) Evaluation of antibacterial activity of caffeic acid encapsulated by β-cyclodextrins. J Microencapsul 32:804–810

Pintado T, Herrero A, Ruiz-Capillas C, Triki M, Carmona P, Jimenez-Colmenero F (2015) Effects of emulsion gels containing bioactive compounds on sensorial, technological, and structural properties of frankfurters. Food Sci Technol Int 22:132–145

Pintado T, Herrero AM, Jimenez-Colmenero F, Ruiz-Capillas C (2016) Emulsion gels as potential fat replacers delivering β-glucan and healthy lipid content for food applications. J Food Sci Technol 53:4336–4347

Piovesana A, Norena CPZ (2018) Microencapsulation of bioactive compounds from *Hibiscus calyces* using different encapsulating materials. Int J Food Eng 14:20170170

Prakash B, Kujur A, Yadav A, Kumar A, Singh PP, Dubey NK (2018) Nanoencapsulation: an efficient technology to boost the antimicrobial potential of plant essential oils in food system. Food Control 89:1–11

Puligundla P, Mok C, Ko S, Liang J, Recharla N (2017) Nanotechnological approaches to enhance the bioavailability and therapeutic efficacy of green tea polyphenols. J Funct Food 34:139–151

Qian C, Decker EA, Xiao H, McClements DJ (2013) Impact of lipid nanoparticle physical state on particle aggregation and β-carotene degradation: Potential limitations of solid lipid nanoparticles. Food Res Int 52:342–349

Ramesh Kumar B, Anupam A, Manchikanti P, Rameshbabu AP, Dasgupta S, Dhara S (2018) Identification and characterization of bioactive phenolic constituents, anti-proliferative, and anti-angiogenic activity of stem extracts of *Basella alba and rubra*. J Food Sci Technol 55:1675–1684

Rathee S, Kamboj A (2018) Optimization and development of antidiabetic phytosomes by the Box-Behnken design. J Liposome Res 28:161–172

Rein MJ, Renouf M, Cruz-Hernandez C, Actis-Goretta L, Thakkar SK, Pinto MS (2012) Bioavailability of bioactive food compounds: a challenging journey to bioefficacy. Br J Clin Pharmacol 75:588–602

Rezaei A, Fathi M, Jafari SM (2019) Nanoencapsulation of hydrophobic and low-soluble food bioactive compounds within different nanocarriers. Food Hydrocolloid 88:146–162

Rodriguez EB, Almeda RA, Vidallon ML, Reyes CT (2019) Enhanced bioactivity and efficient delivery of quercetin through nanoliposomal encapsulation using rice bran phospholipids. J Sci Food Agric 99:1980–1989

Rodriguez-Garcia A, Hosseini S, Martinez-Chapa SO, Cordell GA (2017) Multi-target activities of selected alkaloids and terpenoids. Mini-Rev Org Chem 14:272–279

Romo-Hualde A, Yetano-Cunchillos AI, Gonzalez-Ferrero C, Saiz-Abajo MJ, Gonzalez-Navarro CJ (2012) Supercritical fluid extraction and microencapsulation of bioactive compounds from red pepper (*Capsicum annum* L.) by-products. Food Chem 133:1045–1049

Rutz JK, Zambiazi RC, Borges CD, Krumreich FD, da Luz SR, Hartwig N et al (2013) Microencapsulation of purple Brazilian cherry juice in xanthan, tara gums and xanthan-tara hydrogel matrixes. Carbohydr Polym 98:1256–1265

Saenz C, Tapia S, Chavez J, Robert P (2009) Microencapsulation by spray drying of bioactive compounds from cactus pear (*Opuntia ficus-indica*). Food Chem 114:616–622

Sagar NA, Pareek S, Sharma S, Yahia EM, Lobo MG (2018) Fruit and vegetable waste: Bioactive compounds, their extraction, and possible utilization. Compr Rev Food Sci F 17:512–531

Saifullah M, Islam Shishir MR, Ferdowsi R, Rahman MRT, Vuong QV (2019) Micro and nano encapsulation, retention and controlled release of flavor and aroma compounds: a critical review. Trend Food Sci Technol 86:230–251

Salminen H, Gommel C, Leuenberger BH, Weiss J (2016) Influence of encapsulated functional lipids on crystal structure and chemical stability in solid lipid nanoparticles: towards bioactive-based design of delivery systems. Food Chem 190:928–937

Sawale PD, Patil GR, Abdul Hussain S, Singh AK, Bijoy Singh RR (2017) Release characteristics of polyphenols from microencapsulated *Terminalia arjuna* extract: Effects of simulated gastric fluid, Int J Food Prop 20:3170–3178

Seca AML, Pinto DCGA (2018) Plant secondary metabolites as anticancer agents: successes in clinical trials and therapeutic application. Int J Mol Sci 19:263

Selmaty A, Selmaty M, Rawat MSM, Francheschi F (2010) Supramolecular phospholipids–polyphenolics interactions: The PHYTOSOME® strategy to improve the bioavailability of phytochemicals. Fitoterapia 81:306–314

Septembre-Malaterre A, Remize F, Poucheret P (2018) Fruits and vegetables, as a source of nutritional compounds and phytochemicals: changes in bioactive compounds during lactic fermentation. Food Res Int 104:86–99

Sessa M, Balestrieri ML, Ferrari G, Servillo L, Castaldo D, D'Onofrio N et al (2014) Bioavailability of encapsulated resveratrol into nanoemulsion-based delivery systems. Food Chem 147:42–50

Shaikh J, Ankola DD, Beniwal V, Singh D, Ravi Kumar MNV (2009) Nanoparticle encapsulation improves oral bioavailability of curcumin by at least 9-fold when compared to curcumin administered with piperine as absorption enhancer. Eur J Pharm Sci 37:223–230

Shangguan M, Lu Y, Qi J, Han J, Tian Z, Xie Y et al (2014) Binary lipids-based nanostructured lipid carriers for improved oral bioavailability of silymarin. J Biomater Appl 28: 887–896

Shao J, Zheng D, Jiang Z, Xu H, Hu Y, Li X et al (2011) Curcumin delivery by methoxy polyethylene glycol–poly(caprolactone) nanoparticles inhibits the growth of C6 glioma cells. Acta Biochim Biophys Sin 43:267–274

Sharma S, Sahu AN (2016) Development, characterization, and evaluation of hepatoprotective effect of *Abutilon indicum* and *Piper longum* phytosomes. Pharmacogn Res 8:29–36

Sharma G, Italia JL, Sonaje K, Tikoo K, Ravi Kumar MNV (2007) Biodegradable in situ gelling system for subcutaneous administration of ellagic acid and ellagic acid loaded nanoparticles: evaluation of their antioxidant potential against cyclosporine induced nephrotoxicity in rats. J Control Release 118:27–37

Silva DF, Favaro-Trinade CS, Rocha GA, Thomazini M (2011) Microencapsulation of lycopene by gelatin-pectin complex coacervation. J Food Process Pres 36:185–190

Singla AK, Garg A, Aggarwal D (2002) Paclitaxel and its formulations. Int J Pharm 235:179–192

Snima KS, Arunkumar P, Jayakumar R, Lakshmanan VK (2014) Silymarin encapsulated Poly(D, L-lactic-co-glycolic acid) nanoparticles: A prospective candidate for prostate cancer therapy. J Biomed Nanotechnol 10:559–570

Song S, Tian B, Chen F, Zhang W, Pan Y, Zhang Q et al (2015) Potentials of proniosomes for improving the oral bioavailability of poorly water-soluble drugs. Drug Develop Ind Pharm 41:51–62

Soukoulis C, Bohn T (2018) A comprehensive overview on the micro- and nano-technological encapsulation advances for enhancing the chemical stability and bioavailability of carotenoids. Crit Rev Food Sci Nutr 58:1–36

Sun M, Gao Y, Pei Y, Guo C, Li H, Cao F et al (2010) Development of nanosuspension formulation for oral delivery of quercetin. J Biomed Nanotechnol 6:325–332

Sun Y, Xia Z, Zheng J, Qiu P, Zhang L, McClements DJ, Xiao H (2015) Nanoemulsion-based delivery systems for nutraceuticals: Influence of carrier oil type on bioavailability of pterostilbene. J Funct Food 13:61–70

Takahashi M, Kitamoto D, Imura T, Oku H, Takara K, Wada K (2008) Characterization and bioavailability of liposomes containing a Ukon extract. Biosci Biotechnol Biochem 72:1199–1205

Takahashi M, Uechi S, Takara K, Asikin Y, Wada K (2009) Evaluation of an oral carrier system in rats: bioavailability and antioxidant properties of liposome-encapsulated curcumin. J Agric Food Chem 57:9141–9146

Teng Z, Xu R, Wang Q (2015) Beta-lactoglobulin-based encapsulating systems as emerging bioavailability enhancers for nutraceuticals: a review. RSC Adv 5:35138–35154

Tiwari SB, Amiji MM (2006) Improved oral delivery of paclitaxel following administration in nanoemulsion formulations. J Nanosci Nanotechnol 6:3215–3221

Tomsik A, Saric L, Bertoni S, Protti M, Albertini B, Mercolini L et al (2019) Encapsulations of wild garlic (*Allium ursinum* L.) extract using spray congealing technology. Food Res Int 119:941–950

Uchegbu IF, Vyas SP (1998) Non-ionic surfactant based vesicles (niosomes) in drug delivery. Int J Pharm 172:33–70

Wang JC, Chen SH, Xu ZC (2008) Synthesis and properties research on the nanocapsulated capsaicin by simple coacervation method. J Disper Sci Technol 29:687–695

Wang Y, Wang S, Firempong CK, Zhang H, Wang M, Zhang Y et al (2017) Enhanced solubility and bioavailability of naringenin via liposomal nanoformulation: preparation and in vitro and in vivo evaluations. AAPS PharmSciTech 18:586–594

Weiss J, Decker EA, McClements DJ, Kristbergsson K, Helgason T, Awad T (2008) Solid lipid nanoparticles as delivery systems for bioactive food components. Food Biophys 3:146–154

Wen P, Zong M, Linhardt RJ, Feng K, Wu H (2017a) Electrospinning: a novel nano-encapsulation approach for bioactive compounds. Trend Food Sci Technol 70:56–68

Wen P, Wen Y, Zong MH, Linhardt RJ, Wu H (2017b) Encapsulation of bioactive compound in electrospun fibers and its potential application. J Agric Food Chem 65:9161–9179

Woranuch S, Yoksan R (2013) Eugenol-loaded chitosan nanoparticles: I. Thermal stability improvement of eugenol through encapsulation. Carbohydr Polymer 96:578–585

Wu W, Wang Y, Que L (2006) Enhanced bioavailability of silymarin by self-microemulsifying drug delivery system. Eur J Pharmaceut Biopharm 63:288–294

Xiao J, Bai W (2019) Bioactive phytochemicals. Crit Rev Food Sc 59:827–829

Xie X, Tao Q, Zou Y, Zhang F, Guo M, Wang Y et al (2011) PLGA nanoparticles improve the oral bioavailability of curcumin in rats: characterizations and mechanisms. J Agric Food Chem 59:9280–9289

Yang S, Yue J (2012) Discovery of structurally diverse and bioactive compounds from plant resources in China. Acta Pharmacol Sin 33:1147–1158

Yang KY, Lin LC, Tseng TY, Wang SC, Tsai TH (2007) Oral bioavailability of curcumin in rat and the herbal analysis from Curcuma longa by LC–MS/MS. J Chromatogr B Analyt Technol Biomed Life Sci 853:183–189

Yang CS, Sang S, Lambert JD, Lee MJ (2008) Bioavailability issues in studying the health effects of plant polyphenolic compounds. Mol Nutr Food Res 52:S139–S151

Yang FH, Zhang Q, Liang QY, Wang SQ, Zhao BX, Wang YT et al (2015) Bioavailability enhancement of paclitaxel via a novel oral drug delivery system: paclitaxel-loaded glycyrrhizic acid micelles. Molecules 20:4337–4356

Yan-yu X, Yunmei S, Zhipeng C, Qineng P (2006) Preparation of silymarin proliposome: a new way to increase oral bioavailability of silymarin in beagle dogs. Int J Pharm 319:162–168

Yao M, McClements DJ, Xiao H (2015) Improving oral bioavailability of nutraceuticals by engineered nanoparticle-based delivery systems. Curr Opin Food Sci 2:14–19

Yashaswini PS, Kurrey NK, Singh SA (2017) Encapsulation of sesamol in phosphatidyl choline micelles: Enhanced bioavailability and anti-inflammatory activity. Food Chem 228:330–337

Yi C, Fu M, Cao X, Tong S, Zheng Q, Firempong CK (2013) Enhanced oral bioavailability and tissue distribution of a new potential anticancer agent, Flammulina velutipes sterols, through liposomal encapsulation. J Agric Food Chem 61:5961–5971

Young S, Wong M, Tabata Y, Mikos AG (2005) Gelatin as a delivery vehicle for the controlled release of bioactive molecules. J Control Release 109:256–274

Zhai Y, Guo S, Liu C, Yang C, Dou J, Li L et al (2013) Preparation and in vitro evaluation of apigenin-loaded polymeric micelles. Colloid Surface A 429:24–30

Zhang L, Mou D, Du Y (2007) Procyanidins: extraction and microencapsulation. J Sci Food Agric 87:2192–2197

Zhang Z, Zhang R, McClements DJ (2016) Encapsulation of β-carotene in alginate-based hydrogel beads: Impact on physicochemical stability and bioaccessibility. Food Hydrocoll 61:1–10

# Chapter 21
# Tannins and Polyphenols Extracted from Natural Plants and Their Versatile Application

**Suvadeep Mal and Dilipkumar Pal**

**Abstract** From the beginning of lives on earth, nature is contributing different products to the system constantly and endlessly. Plants synthesize a large number of organic compounds, which are commonly known as primary and secondary metabolites with various applications. Tannins are one of the secondary metabolites solely obtained from the natural or plant sources where it present in the woods, barks, leaves, fruits, cell sap or in vacuoles. Chemically, they are polyphenolic colloidal solutions with complex astringent properties and it has the ability to tan or convert the skin of animals into leather. Depending on the complexity of chemical nature, tannins are classified into two types i.e., hydrolysable tannins and condensed tannins. More than 8000 different tannins of free or bound forms have been detected which can be used in various sector. Despite of its astringent property, tannins and polyphenols can show their identity with different applications with properties like anti-oxidant, anti-inflammatory, anti-microbial, anti-aging, stomachic, cardio-tonic, diuretics, laxatives, hypoglycemic, anti-corrosive or in photography, food, neutraceuticals or cosmeceuticals. In this review, we discuss about different tannins and polyphenols obtained from different sources, their types, about important chemicals and their remarkable applications in different fields of the system.

**Keywords** Secondary metabolite · Colloidal solution · Tannins · Hydrolysable tannin · Condensed tannin · Astringent · Polyphenols · Laxative · Cardio-tonic · Anti-oxidants · Anti-inflammatory · Anti-microbial · Diuretics · Neutraceuticals · Cosmeceuticals

---

S. Mal
Haldia Institute of Pharmacy, Kshudiram Nagar, West Bengal, Haldia-721657, India

D. Pal (✉)
Department of Pharmaceutical Sciences, Guru Ghasidas Viswavidyalaya
(A Central University), Bilaspur, Chhattisgarh 495009, India
e-mail: drdilip2003@yahoo.co.in



D. Pal and A. K. Nayak (eds.), *Bioactive Natural Products for Pharmaceutical Applications*, Advanced Structured Materials 140,
https://doi.org/10.1007/978-3-030-54027-2_21

## 21.1 Introduction

Natural products are the chemicals and compounds that are given to us by nature. The significance of the nature and the natural products are well understood when we come to the ecosystem. For the thousands of years, plants are defending and reigning rank holder which works day and night in their own body for giving us remarkable natural products. The various and different parts of a plant can be a good source for natural compounds. Plants metabolize different organic compounds in their body by complex mechanism comprises physical and chemical episodes of photosynthesis, respiration, degradation. There are two types of metabolism occurs in plant system, like primary and secondary metabolism. Primary metabolism yields primary metabolites which are essential for the growth, development, functioning and the survival of the plant whereas, secondary metabolites play a direct or indirect role in plant's growth and they are not required for a plant to survive. The secondary metabolites of a plant play a key role in plant's defence system, fighting herbivores and pathogens. Also they are found useful in other important functions like giving plant or plant portions a color, maintaining and signaling the primary metabolic pathways etc. (Mazid et al. 2011). Some of the much known important secondary metabolites are terpenes, phenolic metabolites, glycosides and alkaloids.

Tannins and polyphenols are one of the most widely used secondary metabolites obtained from different species of higher altitude plants. It is found from different parts of myrobalan, nutgall, chestnut, rhubarb, bahera, arjuna, Indian goose berry, ashoka bark, black and pale catechu, pterocarpus etc. and they are extracted by maceration. Chemically, they are the mixture of complex organic substances in which polyphenols are present. Tannins are basically phenylpropanoids, condensed to lengthy polymers with molecular weight ranging from 300 to 3000 (Khanbabaee and Van Ree 2001). They are non-nitrogenous and not crystalline in nature, they do possess a bitter taste and they give positive results with Goldbeater's skin test (Nierenstein 1932). Tannins are classified into two types generally, hydrolysable tannins and condensed tannins. Hydrolysable tannins can be hydrolysed easily by enzymes or different acids into gallic acid or ellagic acid, which further process pyrogallol, glucogallin, etc., where condensed tannins do not respond to hydrolysis and being a flavone derivative, they can be related to flavonoid dyes and pigments or phlobatannins or proanthocyanidins (Schofield et al. 2001). It is having astringent property and it helps in precipitation of different proteins, gelatins, glycosides etc. from any solution. The word 'tannin' helps to give a simple knowledge about its property to tan, i.e. ability to convert leather from skin surface (Sieniawska and Baj 2017). Because of this property, tannins and tannin containing drugs can be used in burn treatment. Apart from the astringent property, tannins do possess versatile applications like medicinal uses, industrial uses and biological applications. Tannins are used to treat different diseases and biological conditions like fever, diarrhea, diabetes, eye infections, gall bladder stone, intestinal disorder, constipation etc. (Sieniawska and Baj 2017). Different tannins and tannic acid derivatives obtained from different parts of plants like roots, barks, stems, leaves, galls, fruits, herbs are used as anti-viral, anti-oxidant, anti-inflammatory, anti-microbial, anti-aging, stomachic, cardio-tonic, diuretics, laxatives, anti-hyperglycemic, anti-cancer, anti-coagulant etc. (Smeriglio et al. 2017). Despite of all these beneficial effects on body, tannins also cause side effects like loss of appetite, sickness, nausea, stomach irritation, over urination, or sometimes more serious side effects like hepatotoxicity, sore and throat cancer, and heart failure (Kumar and Singh 1984).

## 21.2 Tannin Occurrence: Plants Containing Tannins

Tannin, being a secondary metabolite, can be found in almost all higher plants. Tannins are found commonly in angiosperms and gymnosperms. Among the plants, tannins are distributed through the different parts of plants like, fruits, roots, barks, woods, leaves, seeds and plant galls. In major cases, tannins are found normally in the plants belonging to different families like combretaceae, fagaceae, euphorbiaceae, leguminosae, rubiaceae, anacardiaceae, aceraceae, actinidiaceae, bixaceae, burseraceae, dipterocarpaceae, ericaceae, myricaceae among dicot plants and najadaceae, typhaceae among monocot plants. Condensed tannins are most abundant polyphenols and it is found in almost all families of plants. Tannins are often found in the xylem and secondary phloem or in the layer between the epidermis and cortex (Ashok and Upadhyaya 2012). The production of condensed tannins occurred inside tannosome, a chlorophyllus organelle, enclosed within tannoplast, attached on the cytosolic face of the endoplasmic reticulum (Brillouet et al. 2013). After synthesis, tannins are stored in vacuoles or in the surface wax of the plants giving protection to the plants from the outer environmental predators. While staying in the vacuoles, tannins present in the active form but it does not show any of its activeness i.e., it does not interfere with plant metabolism or proteins until cell breakdown or cell death occurs and it is released (Cannas 2008). Tannin can also be accumulated in the vacuole of tannin cells, which are the idioblasts of parenchyma cells (Kanzaki et al. 2001). The presence of tannin can be seen generally in the outer part of the bud and in case of leaf tissues, the most common part where tannin is obtained is in the upper epidermis. They protect the leaf tissues against predators as it is distributed in leaf tissues also. In root tissues the most common region is the hypodermis, where they act as a chemical barrier of pathogens to penetration and colonization in roots. In seeds, they are located in between outer integument and aleuronic layer, where they exhibit allopathic and bactericidal properties (Ashok and Upadhyaya 2012). The different tannins obtained from different plant sources with their pharmacological indications are given in Table 21.1.

**Table 21.1** Tannin containing plants, their sources and pharmacological indications

| Source | Plant parts | Tannins obtained | Indications | References |
|---|---|---|---|---|
| *Acacia catechu* (L.) Family: Fabaceae | Bark | Catechin, epicatechin, catechutannic acid, epicatechin-3-O-gallate, and epigallocatechin-3-O-gallate | Cough, diarrhoea, protection against skin ulcer | Shen et al. (2006) |
| *Acacia nilotica* (L.) Family: Fabaceae | Pod | Gallocatechin-gallate methyl gallate, catechin, epicatechin, catechin gallate, galloylglucose | Diarrhoea, diabetes, sore gums, and skin infection | (arim and Azlan (2012) |
| *Agrimonia eupatoria* (L.) Family: Rosaceae | Herb | Catechin; procyanidin $B_3$, agrimoniin | Eye infections, diarrhoea and disorders of gall bladder, liver, and kidneys | Granica et al. (2013) |
| *Camellia sinensis* (L.) Family: Theaceae | Leaves | Catechin, epicatechin gallate, epigallocatechin gallate, epigallocatechin | Aiding digestion, blood purification, boosting immune system, heart disorder, antiviral, hyperglycemia | Shoji and Nakashima (2006) |
| *Diospyros kaki* Family: Ebenaceae | Fruit | Proanthocyanidin catechin, gallocatechin, catechin-3-O-gallate, gallocatechin-3-Ogallate | Antiseptic, hypercholesterolamia, cardiac disorder | Zhang et al. (2011) |
| *Emblica officinalis* Family: Euphorbiaceae | Fruits | Phyllemblin | Laxative, diuretic | Variya et al. (2016) |

(continued)

**Table 21.1** (continued)

| Source | Plant parts | Tannins obtained | Indications | References |
|---|---|---|---|---|
| *Geranium Thunbergii* Family: Geraniaceae | Leaves | Geraniin | Intestinal disorders | Okuda and Ito (2011) |
| *Geum japonicum* (L.) Family: Rosaceae | Herb | O-galloyl-betaglucoside, pedunculagin, tellimagrandin II, 2,6-di-O-galloyl-D-glucose, casuariin, and 5-desgalloylstachyurin | Diuretic, anti-coagulant | Xie et al. (2007), Dong et al. (1998) |
| *Hamamelis virginiana* (L.) Family: Hamamelidaceae | Bark and leaves | Hamamelitannin, pentagalloylglucose, derivatives of epicatechin-(4B-8)-catechins, proanthocyanidins | Swellings, wound healing, burns therapy, anti-inflammatory, and anti-tumors | Hartisch and Kolodziej (1996) |
| *Krameria trianda* (L.) Family: Krameriaceae | Root | Tannic acid, rhataniatannic acid, crameric acid, phlobaphene, phloroglucin, proanthocyanidins | Dysentery, menorrhagia, incontinence of urine, haematuria, sore throat | (Singh) |
| *Mallotus japonicas* (L.) Family: Euphorbiaceae | Bark | 3,4,11-Tri-Ogalloylbergenin | Gastritis, gastric ulcer, diarrhoea, and constipation | Okada et al. (1973) |

(continued)

**Table 21.1** (continued)

| Source | Plant parts | Tannins obtained | Indications | References |
|---|---|---|---|---|
| *Mouriri pusa* Family: Melastomataceae | Leaves | Catechins and condensed tannins | Gastritis and ulcers | Vasconcelos et al. (2010) |
| *Phyllanthus Muellerianus* Family: Phyllanthaceae | Leaves, Stem, bark | Geraniin, corilagin, furosin | Chronic wounds, skin eruptions, anti-tumor | Agyare et al. (2011) |
| *Potentilla erecta* Family: Rosaceae | Roots | Pentadigalloylglucose, pedunculagin, agrimoniin, epigallocatechin, catechins, proanthocyanidins | Inflammations, wounds, bleeding, dysentery, diarrhoea, bowel disease, microbial infections, cancer, antiseptic for the mouth and throat | Tomczyk and Latté (2009) |
| *Potentilla kleiniana* Family: Rosaceae | Aerial parts | Agrimoniin, potentillin | Diarrhoea, bleeding, influenza, parotitis, lymphadenitis, hepatitis, scare, numbness of limbs, dysmenorrhea, ulcer | Okuda et al. (1984) |
| *Pterocarpus marsupium* Family: Leguminosae | Whole plant | Kinotannic acid, k-pyrocatechin | Anti-hyperglycemic, anti-hyperlipidemic, astringent | Dhanabal et al. (2006) |
| *Quercus infectoria* (Oak) Family: Fagaceae | Gall (Turkish gall) | Tannic acid | Astringent, inflammation, local anesthetic, bacterial, fungal and viral infections | Ikram and Nowshad (1977) |

(continued)

**Table 21.1** (continued)

| Source | Plant parts | Tannins obtained | Indications | References |
|---|---|---|---|---|
| *Quercus robur* (L.) Family: Fagaceae | Bark | Grandinin, castalagin, glucogallin | Diarrhoea, oral mucosa or skin inflammation, hemorrhoids | Mämmelä et al. (2000) |
| *Rhus chinensis* (Mill.) Family: Anacardiaceae | Gall (Chinese gall) | Chinese gallotannin | Antiseptic, burn therapy, traumatic injuries, haemorrhoids, and ulcers in the mouth | Djakpo and Yao (2010) |
| *Sanguisorba officinalis* (L.) Family: Rosaceae | Root | Sanguiin H-6 | Dysentery with blood, nosebleeds, burns | Tsukahara et al. (2001) |
| *Saraca indica* Family: Leguminosae | Bark | Catechol | Oxytocic, larvicidal activity | Mathew et al. (2009) |
| *Syzygium cumini* | Bark | Corilagin and related ellagitannins | Sore throat, bronchitis, asthma, dysentery and ulcers | Lanitis et al. (2012) |
| *Terminalia arjuna* (Arjuna) Family: Combretaceae | Stem bark | β-sitosterol | Cardioprotection, angina pectoris, hypercholesterolemia treatment | Bharani et al. (2002) |
| *Terminalia belerica* (Bahera) Family: Combretaceae | Dried mature fruits | Gallic acid, chebulagic acid | Anti-diabetic and anti-oxidant | Hazra et al. (2010) |

(continued)

**Table 21.1** (continued)

| Source | Plant parts | Tannins obtained | Indications | References |
|---|---|---|---|---|
| *Terminalia chebula* (Myrobalan) Family: Combretaceae | Dried mature fruit | 2,4-Chebulyl-β-Dglucopyranose, chebulinic acid, punicalagin, terflavin A, terchebin, tannic acid | Cold-related coughs, ulcer, purgative | Pfundstein et al. (2010) |
| *Uncaria gambier* Family: Rubiaceae | Leave, young shoot | Catechutannic acid, catechin | Anti-oxidant, anthelmintic | Amir et al. (2012) |

## 21.3 Classification of Tannins

Chemically, tannins are high molecular weight compounds, which are soluble in water and can be broadly classified in various types according to their complexity of their chemical nature or according to the structural moieties or according to the behavior on dry distillation or on the action of enzyme tannase. Firstly, on the basis of the chemical nature or better to say, according to the functionality, tannins are classified as "true tannins' and "pseudo tannins". In case of this first divisional group, true tannins are those, which is having high molecular weight and give positive result in Goldbeater's skin test (Rangari 2008), so they can be used as tanning agents. Whereas, the latter is of low molecular weight, do not respond to the Goldbeater's skin test i.e. they cannot be used as the tanning agents. The basic nucleus of true tannins and pseudo tannins are also in contrast like, true tannins are made of gallic acid or egallic acid derivatives but the basic nucleus of pseudo tannins are chlorogenic acid or catechins (Shah 2009). Catechins, and chlorogenic acid from cocoa and *Nux vomica* respectively are two common examples of pseudotannins (Ashok and Upadhyaya 2012). True tannins are further classified as hydrolysable tannins, condensed tannins and complex tannins, according to their behavior in acid hydrolysis. Hydrolysable tannins are those, which upon hydrolysis by acids like HCl or $H_2SO_{4,}$ yield gallic acid or ellagic acid. Since, it is a pyrogallol type of tannin, further dry distillation of gallic acid and ellagic acid gives pyrogallol (Ramakrishnan and Krishnan 1994). On the other hand, condensed tannins are resistant to hydrolysis by acids. Instead of hydrolysis, they depolymerize or decompose into red or brown colored pigment called phlobaphene (Foo and Karchesy 1989). So, the hydrolysable tannins comprise gallotannins and hexahydroxydiphenic acid made polyesters, whereas condensed tannins contain favan-3-ol nuclei (proanthocyanidins) (De Bruyne et al. 1999). Complex tannins are mixtures of both, hydrolysable and condensed tannins. But according to modern studies, structural characteristics leads to categorise tannins into four major groups i.e. gallotannins, ellagitannins, complex tannins and condensed tannins (Khanbabaee and Van Ree 2001).

### *21.3.1 Gallotannins*

Gallotannins are one of the compounds, belonging to the subclass of hydrolysable tannins because it can be hydrolysed by mineral acids and yield gallic acid. It is a polymer, obtained from the condensation of the carboxy group (–COOH) of gallic acid and its polymeric derivatives with the hydroxyl groups (–OH) of a carbohydrate (mainly glucose). In case of partial substitution with galloyl unit (1), the remaining –OH groups of polyol can make substitutions with various other residues or remain un-substituted (Mueller-Harvey 2001).

The gallotannins 2,3,4,6-*tetra-O-g*alloyl-D-*g*lucopyranose (TGG) (2) and 1,2,3,4,6-*penta-O-g*alloyl-β-D-*g*lucopyranose (β-PGG) (3), found in many plant families, are the major intermediates for the biosynthesis of hydrolysable tannins (Gross 1993) (Hagenah and Gross 1993). Most of the galloyl units at the anomeric centre of their D-glucosyl unit have the β configuration at the anomeric centre. But in some cases like 1,4-di-*O*-galloyl-α-D-glucopyranose (4), the anomeric centre of the D-glucopyranose has the α configuration (Khanbabaee and van Ree 2001). Examples of some gallotannins are: tannic acid (Chinese gallotannin) (5), Turkish gallotannin (6), acertannin (7), hamamelitannin (8) etc. (Fig. 21.1).

Tannic acid is a gallotannin with chemical formula given as $C_{76}H_{52}O_{46}$. It contains decagalloyl glucose (5) moiety and it is a commercially available tannin. Tannic acid is found in the nutgalls formed by insects on twigs of certain oak trees (*Quercus infectoria* and other Quercus species). While Chinese gallotannin contains decagalloylglucose residue, Turkish gallotannins tends to have penta-, hexa-, or heptagalloylglucose moiety.

### *21.3.2 Ellagitannins*

Ellagitannins contains at least two galloyl units coupled to each other by C–C bonds, but they do not form any catechin units. Ellagitannins are formed by the first-stage biogenetic oxidative coupling of at least two galloyl units, yielding an axially chiral hexahydroxydiphenoyl moiety or HHDP unit (9) (Khanbabaee and Van Ree 2001). Remarkably, all ellagitannins, with HHDP units bound to the polyol unit (basically, D-glucopyranose) in 2,3- or 4,6- or 1,6-positions, always exert the (*S*)-configuration, while 2,4- or 3,6-coupled HHDP units show the (*R*)-configuration generally (Feldman and Ensel 1994). Ellagitannins differ from gallotannins, in their galloyl groups linkage. Ellagitannins link their galloyl units through C–C bonds, whereas gallotannins follows depside bond linkage (Radebe et al. 2013). Some of the ellagitannins are chebulinic acid (10), corilagin (11), geraniin (12), tellimagrandin II (13), casuarinin (14), potentillin (15) etc. (Fig. 21.2).

### *21.3.3 Condensed Tannins*

These tannins are derivatives of flavonoid (16), catechins (17), flavan-3–4-diol. They do not contain sugar residues. Monomeric catechins and leukoanthocyanidins (flavan-3–4-diol) (18) don't have any tanning properties, but when they oligomerize under acidic condition or enzymatic action, they can produce significant tanning action (Khanbabaee and van Ree 2001) (Ferreira and Slade 2002). Condensed tannins are the oligomeric and polymeric proanthocyanidins formed via linking catechins one another in C-4 to C-6 or C-8 positions.

Proanthocyanidins (19) can be classified according to variation in hydroxylation pattern into several subgroups like propelargonidins (3,4′,5,7–OH) (20), procyanidins (3,3′,4′,5,7–OH) (21), prodelphinidins (3,3′,4′,5,5′,7-OH) (22), proguibourtinidins (3,4′,7-OH) (23), profisetinidins (3,3′,4′,7-OH) (24), prorobinetinidins (3,3′,4′,5′,7-OH) (25), proteracacidins (4′,7,8-OH) (26), promelacacidins (3′,4′,7,8-OH) (27), pro-apigeninidins (4′,5,7-OH), and proluteolinidins (3′,4′5,7-OH) (Fig. 21.3) (Khanbabaee and Van Ree 2001).

## *21.3.4 Complex Tannins*

Complex tannins are formed when, a hydrolysable unit binds with a gallotannins or an ellagitannins through a glycoside linkage. An example of complex tannin is acutissimin A (**27**), which is isolated from the bark Korean chestnut (*Castanea*

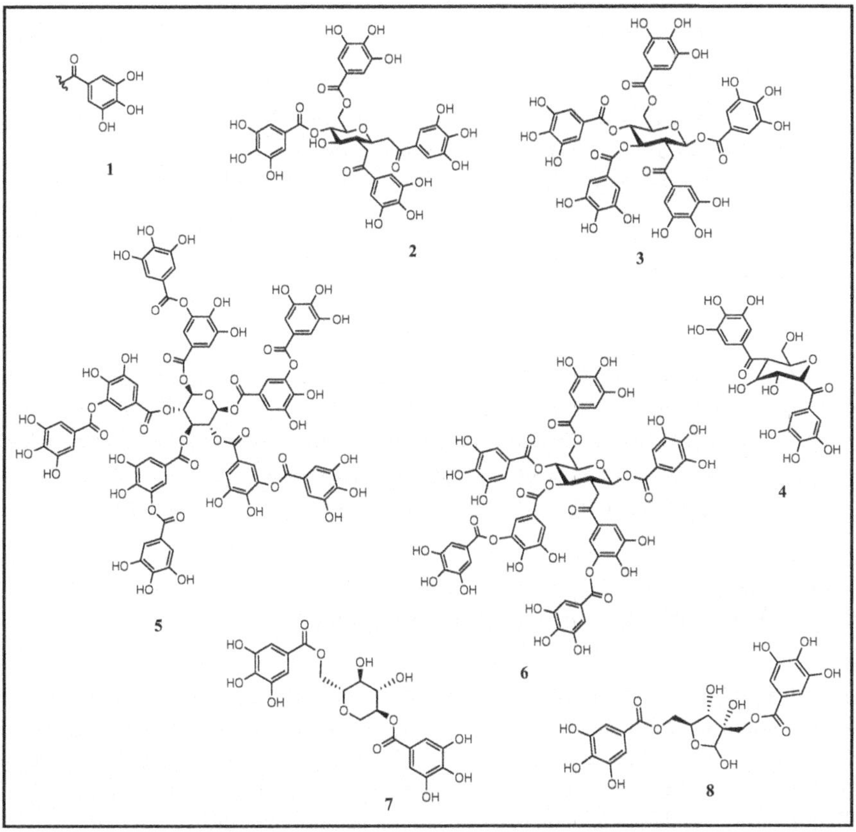

**Fig. 21.1** Structures of some gallotannins

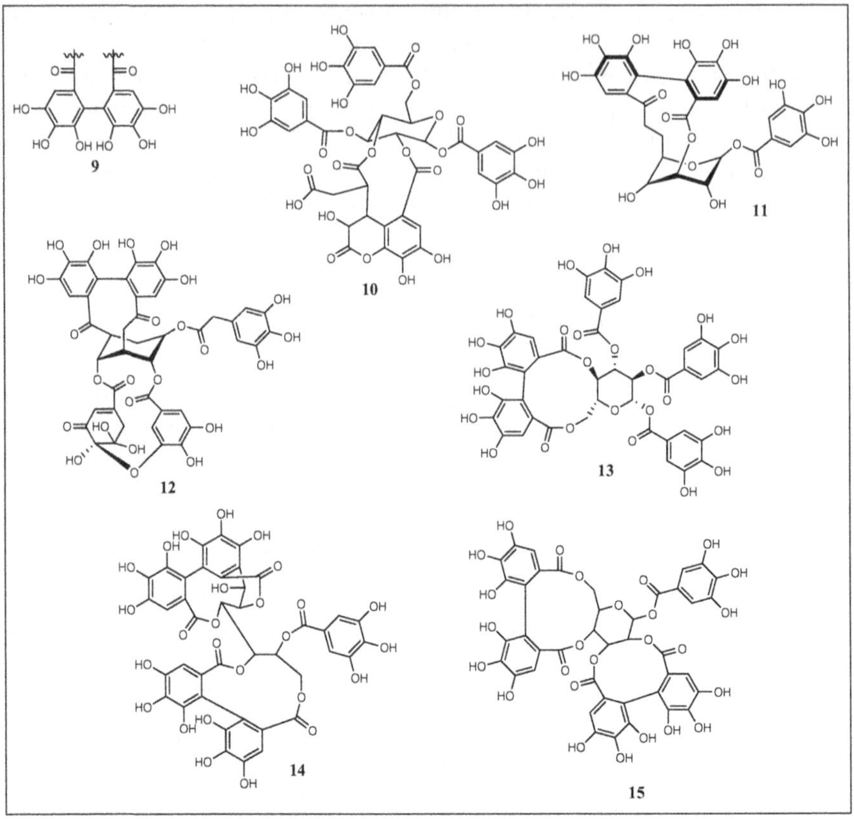

Fig. 21.2 Structures of some ellagitannins

*crenata*) (Okuda and Ito 2011). This is a flavo-galloyl unit bound glucosidically to C-1, with an additional three hydrolysable ester bonds to a D-glucose-derived open-chain polyol. More example of complex tannin is epigallocatechin gallate (28), which is widely present in tea leafs and responsible for variety of medicinal activities or camelliatannin A (29) (Fig. 21.4) obtained from *Camellia japonica*.

## 21.4 Biosynthetic Pathways of Tannins

The biosynthesis of tannins follow two different synthetic pathways for hydrolysable tannins and condensed tannins. The biosynthesis of hydrolysable tannins comprises via the key intermediate formation of 1,2,3,4,6-pentagalloyl glucose (Hagenah and Gross 1993). The starting material or the precursor is quinic acid. Gallotannins or depsidic metabolites can be formed by the galloylation of the

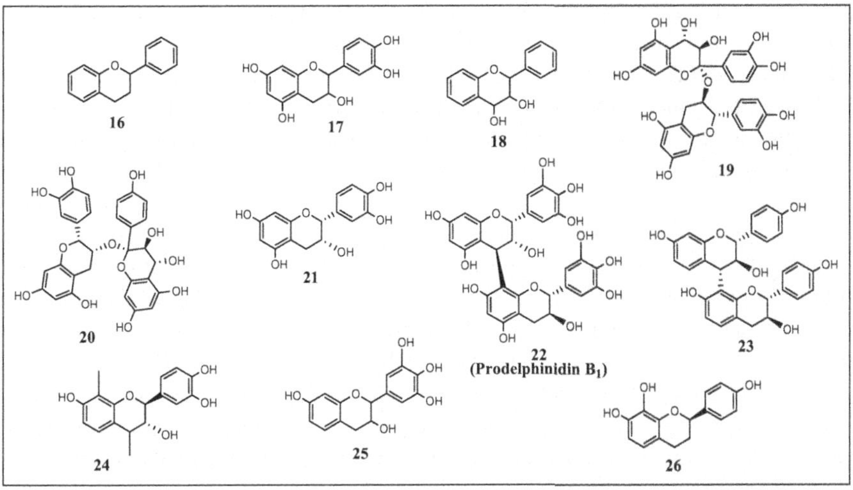


**Fig. 21.3** Structures and basic units of condensed tannins

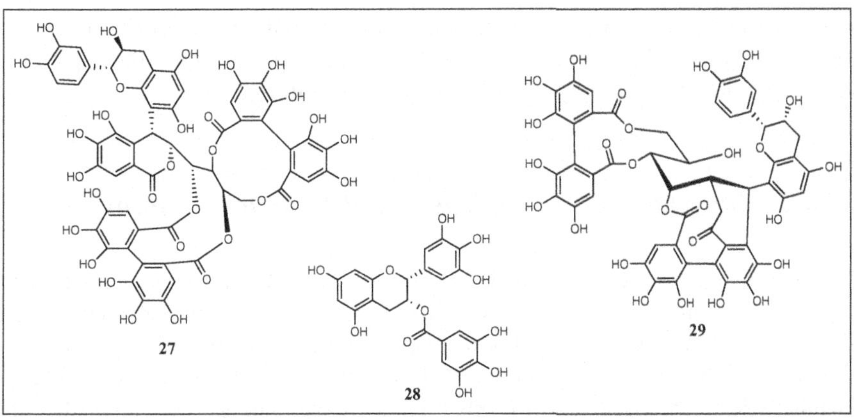


**Fig. 21.4** Structures of some complex tannins

intermediates and ellagitannins are formed due to the oxidation leading to C–C linkage.

## *21.4.1 Biosynthesis of Hydrolysable Tannins*

Hydrolysable tannins biosynthesis starts from gallic acid synthesis from quinic acid. Quinic acid comes from the one of the most important intermediate in glycolysis in plant, phosphoenolpyruvate. Phosphoenolpyruvate via phosphoenolpyruvate translocator (PEPTr), forms quinic acid. Under the catalytic action of gallate 1-β-glycosyl transferase enzyme, and via the formation of UDP from UDP-α-glucose, gallic acid yields 1-O-galloyl-β-D-glucose. From 1-O-galloyl-β-D-glucose, one molecule of D-glucopyranose is cleaved consecutively in each step and it results in galloylation to form 1,6-digalloyl-β-D-glucose, 1,2,6-trigalloyl-β-D-glucose, 1,2,3,6-tetragalloyl-β-D-glucose and finally 1,2,3,4,6-pentagalloyl glucose under the catalysis of enzymes like β-glucogallin-dikisgalloyl glucose-O-galloyl transferase, β-glucogallin-trikisgalloyl glucose-O-galloyl transferase, β-glucogallin-tetrakisgalloyl glucose-O-galloyl transferase respectively. 1,2,3,4,6-pentagalloyl-β-D-glucopyranose or simply 1,2,3,4,6-pentagalloyl glucose is the key intermediate found in many plant families. After the intermediate 1,2,3,4,6-pentagalloyl glucose formation, the galloylation of pentagalloylglucose continues and ester linkage between two galloyl moieties forms gallotannins or depsidic metabolites. The enzymatic catalysis of 1,2,3,4,6-pentagalloyl glucose by different galloyltransferase enzymes forms different linkages to form different gallotannins (Niemetz and Gross et al. 2005).

After the formation of intermediate 1,2,3,4,6-pentagalloyl glucose, oxidation takes place by the action of different oxido-reductase enzymes, leading to C–C linkage formation between suitably orientated galloyl residues of glucogalloyl molecules and it results in hexahydroxydiphenoyl (HHDP) units formation which is the core moiety of ellagitannins (Grundhöfer et al. 2001). The biosynthetic pathway of hydrolysable tannin and complex tannin is given in Fig. 21.5.

### *21.4.2 Biosynthesis of Condensed and Complex Tannins*

The phosphoenolpyruvate obtained from glycolysis in plants, forms quinic acid by the phosphoenolpyruvate translocator (PEPTr). Quinic acid in the next step, yields shikimic acid and chorismate. Shikimic acid is found to be the main precursor of different metabolite formation in plants. In the next step, chorismate gives phenylalanine when asparagine (Asn) forms aspartate (Asp) side by side. After this, the enzyme called phenylalanine ammonia lyase (PAL) helps to convert phenylalanine into cinnamic acid or cinnamate. Hydroxyl group insertion occurs to the para position of cinnamic acid by the help of the enzyme called cinnamic acid hydroxylase (CAH) to form 4-coumaroyl-CoA. P-coumaroyl-CoA gives chalcone, when it is condensed with malonyl CoA. The enzymes involved in this step are 4-coumaroyl-CoA ligase (4CL) and chalcone synthase (CHS). Naringenin chalcone isomerizes under the enzymatic condition produced by chalcone isomerase (CHI) to form naringenin, which is a flavanone. Naringenin then forms a dihydroflavonol called dihydrokaempferol via the attachment of a hydroxyl group in 3$^{rd}$ position by flavanone 3-hydroxylase (F3H). In the next step, another hydroxyl group insertion occurs in dihydrokaempferol and gives dihydroquercetin by flavonoid 3′-hydroxylase (F3′H). Further, reduction takes place by dihydroflavonol reductase (DFR) which results in the formation of leucocyanidin. Leucocyanidin cleaves the

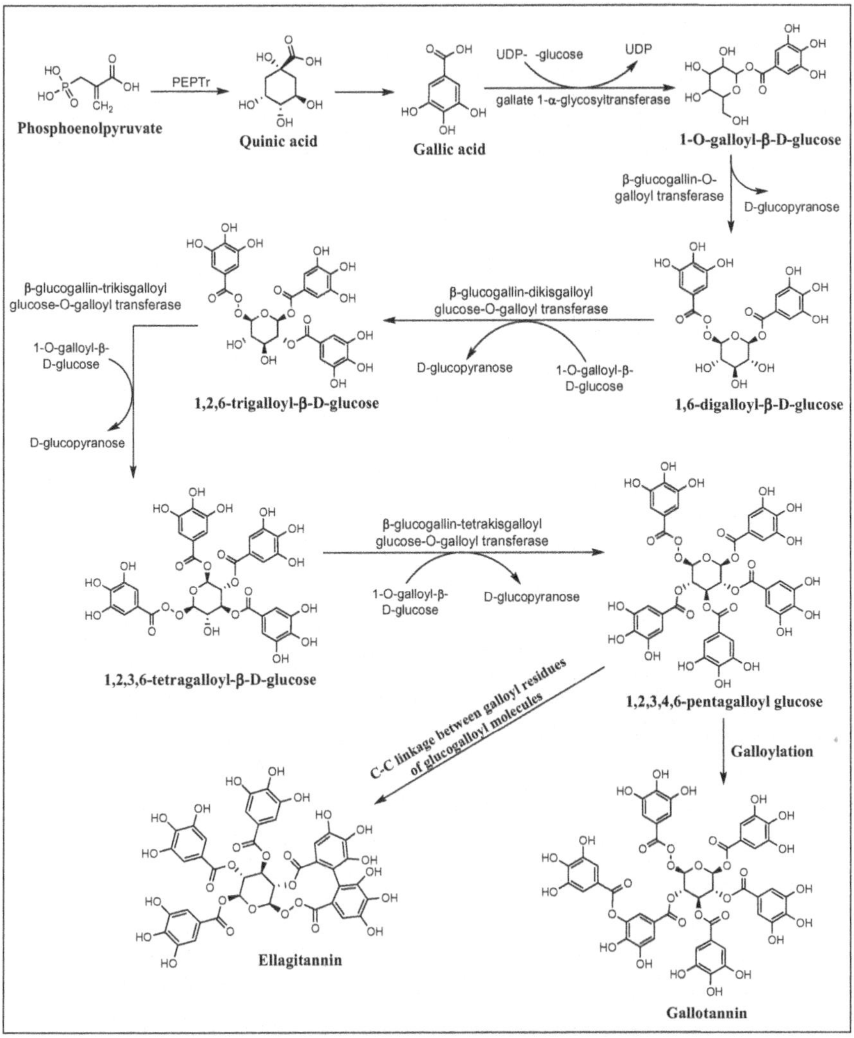


**Fig. 21.5** Biosynthetic pathway of gallotannin and ellagitannin

hydroxyl group in 4th position by the help of leucocyanidin deoxygenase (LDOX) to produce cyanidin or anthocyanidin, which gives (–)-epicatechin by the reduction of ring double bond. Leucocyanidin upon the enzymatic action of leucoantho-cyanidin 4-reductase (LAR), gives (+)-catechin. Leucocyanidin polymerises in the presence of anthocyanidin synthase (ANS) to produce dimer and trimer procyanidin via $C_4$–$C_6$ or $C_4$–$C_8$ linkage. Procyanidin polymer finally gives condensed tannin by different C–C linkage formation. Epicatechin units can go through glycosylation to produce complex tannins (Harding et al. 2014). The biosynthetic pathway of condensed tannin and complex tannin is given in Fig. 21.6.

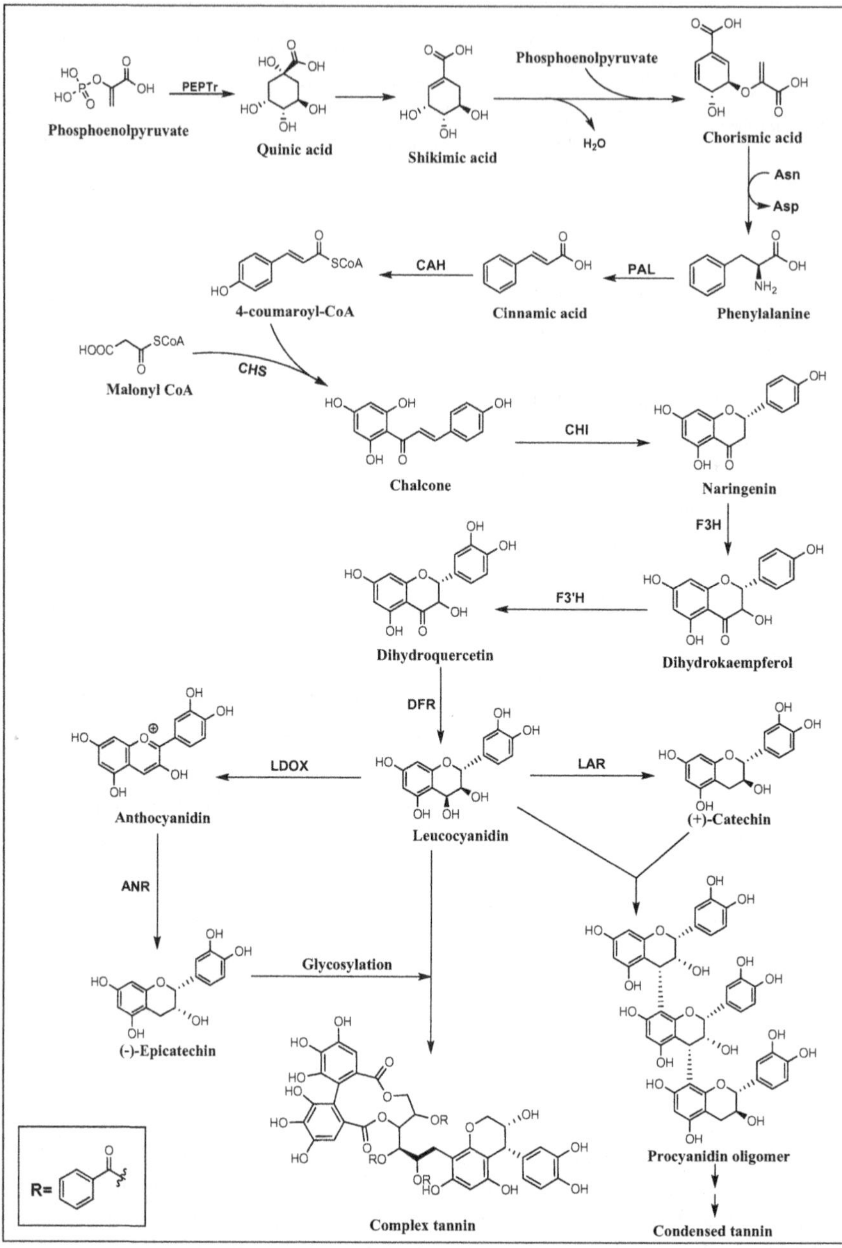


**Fig. 21.6** Biosynthetic pathway of condensed tannins and complex tannins

## 21.5 Extraction Process of Tannins

As tannins show its solubility in water, alcohols, acetone, ethyl acetate etc., the solvents which are commonly used to extract different tannins (better say hydrolysable tannin) from natural sources are: aqueous solutions of methanol, ethanol, or acetone and also ethyl acetate. Nonpolar organic solvents like benzene, n-hexane, petroleum ether, chloroform, dichloromethane etc. are usually used in pre-treatment of sample for the purpose of removing lipids and chlorophyll. These non-polar solvents with low extraction strengths are also beneficial in preventing enzymatic reactions of tannins (Arapitsas 2012). In case of handling tannins with low molecular weight or in the processing of matrices containing large amounts of enzymes (i.e., bark or fruit), extraction is done with methanol treatment, while the use of acetone is preferentially done with tannins of higher mol. Wt. But the temperature should be taken into account as higher temperatures for extended times results in hydrolysis of the galloyl moiety and it may force ellagitannins releasing ellagic acid (Okuda et al. 1989). In addition, extraction with alcohols like ethanol and methanol may produce ethyl or methyl esters of gallic acid, respectively (Mueller-Harvey 2001) (Okuda et al. 1989). Not only the extraction solvents, but also the other factors like temperature, pH, solvent/material ratio etc. contributes in the extraction procedure. Some of the extraction process applied in the extraction of tannins from different plants are shortly described below:

1. **Classic extraction** is done via boiling the solvent. The optimal conditions for extracting Assam green tea polyphenols lies under maintaining sample and water ratio (1:20), with a pH of 4–5, utilizing fresh green tea leaves (Wati et al. 2009).
2. **Pressurized liquid extraction** is another way of extracting tannins, treating with the solvent at an elevated temperature (usually between 50 and 200 °C) and under high pressure (between 10 to 15 Mpa). Herein, the best recoveries were obtained by way of employing this technique for releasing catechin and epicatechin from grape seeds, as well as from non-fermented tea leaves, medium-fermented tea leaves, and fermented tea leaves. Such release comes about within 10 min (Piñeiro et al. 2004).
3. **Supercritical carbon dioxide extraction** is applicable using supercritical carbon dioxide (SC–$CO_2$) with critical point at 31.1 °C and under a pressure of 73.8 bar. But while performing this technique, one condition should be taken care of. Normally, carbon dioxide ($CO_2$) is not a suitable solvent for extracting polyphenols because of its nonpolar nature, hence, to overcome this problem, a polar organic solvent such as a solution of ethanol and water, can be used during extraction or before the extraction (Lang and Wai 2001). Ethanol is used in this technique, as a co-solvent, which was successfully utilized in the extraction of green tea catechins (Gadkari et al. 2015).
4. **Microwave-assisted extraction** is done with solvent that is directly heated by microwaves. The optimal performance of tea polyphenols extraction was obtained under a microwave intensity of 600 W, a microwave radiation time of

3 min, and a one-time microwave radiate action, with tea/water ratio of 1:20 (Li and Jiang 2010).

5. **Ultrasound/sonication-assisted extraction** method demonstrates excellent solvent penetration into cellular materials. In this process, optimal extraction conditions for several hydrolysable tannins came about by way of treatment with 19.7% ethanol, for 26.4 min, at 24 °C (Lee et al. 2013). Of note, for deriving catechins from commercial Chinese tea samples, the novel dynamic ultrasound-assisted extraction method was found to be more effective than static ultrasound-assisted extraction (Gu et al. 2007).
6. **Extraction by lyophilisation** can be done by lyophilizing sample and maceration by acetone and water in 4:1 ratio overnight. After that, extracted sample is shaken on planar shaker for several hours (3 h.) and it is then centrifuged to get the supernatant, which is concentrated and frozen. Then the frozen material is taken and extracted again with methanol (Suvanto et al. 2017).

## 21.6 Physical and Chemical Properties of Tannins

- These are high molecular weight phenolic compounds. These are complex organic, non-nitrogenous and non-crystalline substances.
- Physically, tannins have a colour of reddish brown and dark brown, it possess a state of non-crystalline matter and a taste with bitterness and puckering (Jaiswal et al. 2018). A tannin can also possess light yellow or white amorphous or shiny powder like appearance (Li et al. 2013). Tannins are soluble in water, acetone, glycerol, dilute alkalies and alcohols; sparingly soluble in ethyl acetate, chloroform and insoluble in benzene. Gallotannins are rapidly soluble but ellagitannins are slowly soluble in water in comparison to gallotannin (Mingshu et al. 2006). They are practically insoluble in ether, carbon-disulphide and benzene (Maximilian Nierenstein and Skene 1934).
- Tannins show its individual characteristics when they are treated with mineral acids. Hydrolysable tannin is hydrolyzed into gallic acid or ellagic acid when it gets a treatment with different mineral acids like HCl or $H_2SO_4$. Upon dry distillation, hydrolysable tannin results in the production of pyrogallol and phenolic compounds. Condensed tannins, upon acid treatment in elevated temperature, it polymerizes and is converted into a red insoluble complexes, called phlobaphenes (Pizzi 2019).
- Precipitation and astringency: Tannins form insoluble cross-linked complexes by interacting with protein molecules. They combine with skin and hide to form leather through a process of crosslinking the skin collagen (Khanbabaee and Van Ree 2001) and with gelatin and isinglass to form an insoluble compound. They combine with alkaloids to form tannates, most of which are insoluble in

water. Tannins have the ability to precipitate different compounds like gelatin, alkaloid, protein, glycoside, skin hide and heavy metal.

- Reaction with salts or bases: Tannins yield purple, violet or black precipitates with iron compounds. They are precipitated by number of metallic salts notably potassium dichromate, and lead acetate and sub acetate. Tannins are precipitated by copper acetate solution. Tannins when mixed with potassium ferricyanide in the presence of ammonia, it produce deep red colour (Wei and Xingjin 1989).
- When it is boiled with acetic anhydride, it is converted into a crystalline acetyl derivative, which melts at 137 °C (Ramakrishnan and Krishnan 1994).
- Carcinogenicity: Prolonged use of tannins can cause cancer. An evidence about the carcinogenic effect of *Areca catechu* was observed by typical use of it that resulted in oral and esophageal cancer (Hernandez et al. 2017).
- Anti-oxidizing properties: Tannins can show their anti-oxidant property because of the large number of the hydroxyl (OH) groups present in it (Bagyalakshmi et al. 2019).

## 21.7 Chemical Tests of Tannins

The presence and the amount of tannins can be clarified with the help of different chemical tests of tannins. It can be divided into two groups, like, qualitative tests and quantitative test. By qualitative test we can understand about the presence of different tannins and by the quantitative tests, the amount of the tannin present in the sample can be determined.

### *21.7.1 Qualitative Tests*

1. Goldbeater's skin test: This is the major qualitative test for the confirmation of true tannins based on the astringent property of tannins. Goldbeater's shin is the untanned hide (membrane) prepared from the intestine of ox. The gold-beater's skin test is originally described by Atkinson and Hazleton (Honorá Price 1924). The first stage of performing this test is to make the Gold-beater's skin to soak in 2% HCl and make the skin hide permeable for tannin. After that, the skin is rinsed with distilled water and is dipped in the tannin solution for 5 min. After this step, the skin is washed with water and is treated with 1% ferrous sulphate solution. The brown or black colouration of skin gives the confirmation about the presence of tannins. Pseudo tannins don't give positive result to this test (Baig and Anand).
2. Water test: Tannins create a light yellow to dark brown discoloration in the water. A simple test for tannins involves filling a clear glass with water and letting it sit overnight. If the colour settles to the bottom of the glass, the

discoloration is most likely caused by iron and/or manganese and not tannins, but if the intensity of the colour remains intact, it is most likely caused by tannins (Franzmann et al. 2001).

3. Ferric chloride test: In two millilitres (2 mL) of the hot aqueous solution of tannin extract, few drops of 10% Ferric chloride solution is added. The occurrence of blackish blue colour showed the presence of gallotannins and a green-blackish colour indicated presence of catechol tannins (Auwal et al. 2014) and also the occurrence of greenish brown colour indicates the presence of condensed tannins (Pizzolato 1977).
4. Alkaline reagent test: To 1 ml of the plant extract, a few drops of 10% Sodium hydroxide (NaOH) solution was added. An instant appearance of yellow to red precipitate indicates the presence of tannins (Bharudin et al. 2013).
5. Phenazone test: In a 5 ml aqueous extract of tannin, sodium acid phosphate (0.5 gm.) is added and the mixture is heated, cooled and filtered. A solution of 2% phenazone is added to the filtrate. A bulky coloured precipitate is formed (Alotaibi 2019).
6. Match stick test (Catechin test): Match stick test is specially done for detection of catechins. The principle of this qualitative test based upon simple reaction of catechins, which can produce phloroglucinol in the presence of acid. A match stick (which contains lignin) is dipped in aqueous extract of tannin (catechins), dried and moistened with concentrated hydrochloric acid and is warmed in the flame. Catechins in the presence of HCl produces phloroglucinol which stains the lignin (present in match stick) produces pink or red colour (Thiruchenduran et al. 2017).
7. Gambir-Fluorescein test: It is often performed to identify the pale catechu. The alcoholic extract of pale catechu is mixed with sodium hydroxide solution, and petroleum ether. After the mixing, the whole mixture is shaken and kept aside for few minutes. The petroleum ether layer show green fluorescence due to the presence of fluorescein present in pale catechu. Black catechu gives negative result in this test (Ray et al. 2006).
8. Bromine water test: In an aqueous extract of condensed tannins, the addition of acetic acid followed by bromine water gives a result of buff coloured precipitate. Hydrolysable tannin doesn't response to this test (El Sissi and El Sherbeiny 1967).
9. Lead sub-acetate test: In the ethanolic extract of tannin, addition of acetic acid and lead sub-acetate solution results in gelatinous cream coloured precipitate (Ukoha et al. 2011).
10. Chlorogenic acid test: When an extract of chlorogenic acid containing sample is treated with aqueous ammonia. A green colour is formed on exposure to air (Ashok and Upadhyaya 2012).
11. Vanillin-hydrochloric acid test: This test is performed by taking alcoholic extract of sample, vanillin alcohol (vanillin 1 gm, alcohol 10 ml) and concentrated hydrochloric acid in a ratio of 1:10:10. The presence of tannin is confirmed by a pink or red colour formation due to phloroglucinol (Rangra et al. 2019).

12. Phlobatannin test: The occurrence of phlobatannin can be confirmed by the deposition of red precipitate when the aqueous solution of tannin is mixed with 5 ml of 1% hydrochloric acid (Ezeonu and Ejikeme 2016).

### *21.7.2 Quantitative Tests*

1. Hide-powder method: The aqueous solution of tannin is mixed with previously dried chromated hide powder in vacuum for 24 h. over $CaCl_2$ and the mixture is stirred for 1 h. at ambient temperature. The suspension forms and it is filtered through sintered glass filter. The weight gain of the hide-powder after filtration is expressed as a percentage of the weight of the starting material is referred to the percentage of tannin in the sample (Guangcheng et al. 1991).
2. Stiasny's method: In the aqueous solution of tannin, 1 ml of 10 (M) HCl and 2 ml of 37% formaldehyde are added. The reaction mixture is heated over reflux for 30 min and it is then filtered through sintered glass. The precipitate obtained, is washed with hot water for several times and dried over $CaCl_2$ bed. The percentage yield of tannin is calculated with the starting material (Zhang et al. 2008).
3. Folin–Ciocalteau spectrophotometric method: The Folin-Ciocalteu reagent (FCR) or Folin-Denis reagent, is named after Otto Folin, Vintilă Ciocâlteu, and Willey Glover Denis. This reagent is prepared by dissolving 10 g of sodium tungstate and 2.5 g of sodium molybdate in 70 ml of water, followed by the addition of 5 ml of 85% phosphoric acid and 10 ml of concentrated hydrochloric acid and 15 g of lithium sulphate, 5 ml of water, and 1 drop of bromine (Peterson 1979). Polyphenols in plant extracts react with the reagent to form a phosphotungstic-phosphomolybdenum complex, which gives blue fluorescence under UV visible spectrophotometry (Schofield et al. 2001). The maximum absorption of the chromophore depends on the alkaline solution and the concentration of phenolic compounds. The maximum absorption is generally obtained in a wave length of 760 nm (Blainski et al. 2013).
4. Turbidimetric method: This method is based upon the simple precipitation reaction between $Cu^{2+}$ from copper acetate and tannin at 4.5 pH and the absorbance is being monitored at around 470 nm (McDonald et al. 1996). The precipitate is formed due to the presence of functional groups such as ortho-dihydroxyphenyl and carboxyl groups in tannins, which act as mono or bidentate ligands and forms chelates (Lima et al. 2012).
5. Ferrous tartrate method (Colorimetric method): This method is based on the formation, in a buffered medium (pH 6.8) of a complex between ferrous tartrate and tannins in the tea, which absorbs light at around 560 nm (Iwasa and Torii 1962). The ferrous tartrate method is more selective than the Folin–Ciocalteau method, and is unaffected by the co-existence of reducing agents, such as ascorbic acid (Y T Hung et al. 2010).

6. Iodometric method: This method follows simple titration. Due to the phenolic nature of tannins, they can consume iodine from any alkaline medium. The titration is done by taking standard sodium thiosulphate solution and using starch as indicator, and the excess consumed iodine determined which is then used to determine total phenolic contains in the extract (Al-Alimi et al. 2017).
7. Agglutination method: Erythrocytes have a tendency of agglutination when it comes in contact with tannins but the RBCs do not take part in haemolysis. The agglutination occurs because of the formation of sticky membrane on their surface. The end point is reached when all RBCs present, forms precipitation i.e. solution remains colourless on shaking. The quantity of precipitated RBC determines the amount of tannin present in sample (Pirofsky et al. 1962).

## 21.8 Activities of Tannins

Lots of research provide the knowledge about tannins that they are used biologically or can be used in different therapeutic purposes, in manufacturing purposes etc. From the discovery of this secondary metabolites, tannins are extensively studied and the outcome of those research shows tannin's ability to perform in a versatile way which can show its different activities like anti-oxidant, anti-inflammatory, anti-microbial, anti-aging, stomachic, cardio-tonic, diuretics, laxatives, hypoglycemic, anti-corrosive property etc. or it can be used in photography, food, neutraceuticals or cosmeceuticals and so in different sectors. It is believed that tannins show its applications by two mechanistic way.

1. As un-absorbable, which is usually the complexes with binding properties which may exert local actions like antioxidant, radical scavenging, antimicrobial, antiviral, anti-mutagenic, and anti-nutrient effects, or
2. As absorbable, which is having a benefit of possessing lower molecular weight so that they are easily absorbed, and produce systemic effects in various organs (Serrano et al. 2009).

### *21.8.1 Anti-oxidant Properties*

Among all of the in-vitro studies, the most extensively studied property is its anti-oxidant property. Tannins may show the anti-oxidant property because of its several hydroxyl (OH) groups present in the phenolic moiety. Tannins do have an ability of scavenging free radicals and inhibiting lipid peroxidation depending on their structure and degree of polymerization (Tian et al. 2012). The antioxidant activities of tannins are generally calculated by DPPH radical scavenging method. The absorbance values of the compounds which changes the colour from violet to yellow were measured at 517 nm. The percentage inhibition of DPPH free radical scavenging activity was calculated using the following equation:

$$\%\text{Inhibition} = [(A_{\text{DPPH}} - A_{\text{Sample}})/A_{\text{DPPH}}] \ * \ 100\%$$

Where, $A_{\text{DPPH}}$ = Absorbance of DPPH, $A_{\text{Sample}}$ = Absorbance of sample (Bora et al. 2019; Pal 2013; Nimse and Pal 2015).

Different hydrolysable tannins like geraniin, pentagalloyl glucose, chebulinic acid, geraniin isoterchebin, mallotusinic acid, pedunculagin, pedunculagin, tellimagrandins I and II exert their anti-oxidant property by lipid peroxidation prevention interfering with xanthin-xanthin oxidase complex, adenine diphosphate and ascorbic acid or with DPPH (Okuda 2005). Different types of condensed tannins such as procyanidins $B_1$ and $B_3$ (Ariga et al. 1988) and kaki tannin also helps to inhibit $H_2O_2$ or $Fe^{2+}$ mediated lipid peroxidation (Jerez et al. 2007). Another study using DPPH assay and oxygen radical absorbance capacity (ORAC) assays, showed that Bengal coffee extract is having more potency towards anti-oxidant property than Arabic coffee (Patay et al. 2016). Arsenic mediated liver injury and obese-diabetic rat models are protected by administering tannin rich cocoa extract for 4 weeks (Chandranayagam et al. 2013; Jalil et al. 2008). Different anti-oxidant properties of tannins is listed below in Table 21.2.

**Table 21.2** Anti-oxidant property of different tannins

| Tannin | Characteristics | Assay type |
|---|---|---|
| Geraniin, Pentagalloyl glucose Chebulinic acid, geraniin isoterchebin, mallotusinic acid, pedunculagin, pentagalloylglucose, tellimagrandins I and II | Lipid peroxidation, interferes with xanthin-xanthin oxidase complex, adenine diphosphate and ascorbic acid and scavenging effects on DPPH free radical | In-*vitro* Assay |
| Pedunculagin, epigallocatechin gallate | Lipid peroxidation in rat liver mitochondria stimulated by adenine di-phosphate and ascorbic acid | |
| Kaki-tannin | Significant inhibition of the auto-peroxidation of lipids caused with $H_2O_2$ or $Fe^{2+}$/ Ascorbic acid | |
| Tannin-rich cocoa | Arsenic mediated liver injury protection | In-*vivo* |
| Cocoa extracts | Anti-oxidant effects in the obese-diabetic rats | |

**Table 21.3** Anti-microbial property of different tannins

| Tannin | Characteristics | Assay type |
|---|---|---|
| Gallotannin | Inhibits glucan synthesis. Activity seen on *S. Mutans, S. Salivarius, and A. Viscosus* | In-vivo |
| Catechins and epigallocatechin gallate | Potency against *H. Pylori, E. coli* | In-vitro |
| Tellimagrandin I, rugosin B, corilagin and theasinensin A, procyanidin $B_3$ and $B_4$ | Reduce the MIC of antibiotics in the MRSA treatment | |
| Catechin | Inhibition towards *C. histolyticum* | |

### *21.8.2 Anti-microbial Properties*

The anti-microbial properties of tannins also have been a field of interest for different researchers. Tannins can show their anti-microbial activity in direct way or indirect way. The direct way comprises of the action of tannin in microbial metabolism by affecting oxidative phosphorylation, whereas by inhibiting extracellular microbial enzymes and depriving the substrate required for microbial growth, tannins exert their indirect way of anti-microbial action (Cardona et al. 2013; Mohanta et al. 2007; Gurjar and Pal 2020; Howell 2001). Gallotannins can exert bacteriocidal activity on *Streptococcus mutans, S. Salivarius, and Actinomyces viscosus* by inhibiting the synthesis of water insoluble glucan (Wu-Yuan et al. 1988). Different complex tannins like catechins and epigallocatechin gallate showed in-*vitro* anti-bacterial potency against *Helicobacterium pylori, Escherichia coli* (Díaz-Gómez et al. 2014) (Díaz-Gómez et al. 2013). Different hydrolysable tannins like Tellimagrandin I, rugosin B, corilagin and theasinensin A (complex tannin) and procyanidin $B_3$ and $B_4$ can remarkably reduce the MIC (minimum inhibitory concentration) of antibiotics like oxacillin, penicillin G, aminoglycosides and ampicillin in the treatment of methicillin resistant *Staphylococcus aureus* (MRSA) (Hatano et al. 2005). Catechins are shown to have inhibitory property towards Clostridium histolyticum (Cardona et al. 2013). The potency of different tannins obtained from different sources are listed in Table 21.3.

## 21.9 Anti-viral Properties

Tannins show anti-viral activity via the inhibition of virus absorption (Fukuchi et al. 1989). Different hydrolysable tannins like chebulagic acid and punicalagin can be effective against hepatitis C, dengue, measles and bronchitis (Lin et al. 2013). Potency against *Herpes simplex* can be exerted by some hydrolysable tannin or galloylated condensed tannins by inhibiting its absorption (Fukuchi et al. 1989;

**Table 21.4** Anti-viral property exhibited from different tannins

| Tannin | Characteristics | Assay type |
|---|---|---|
| Chebulagic acid and punicalagin | Effective against human cytomegalovirus, hepatitis c virus, dengue virus, measles virus, and respiratory syncytial virus | In-vitro |
| Hydrolysable tannin | Potency against *Herpes simplex* by inhibiting its absorption | |
| Casuarinin | Potency against *Herpes simplex* by inhibiting its attachment and penetration | |
| Oenothein B, Coriariin A, and Agrimoniin | Activity against Human Immuno deficiency Virus (HIV) | |

Takechi et al. 1985) or by some ellagitanninns like casuarinin, through the inhibition of the attachment and penetration power of the virus (Cheng et al. 2002). Dimeric ellagitannins like oenothein B, coriariin A, and agrimoniin showed activity against HIV (Cheng et al. 2002). The anti-viral properties are listed below (Table 21.4).

### *21.9.1 Cardioprotective Activity*

The mechanism by which tannins show their cardiac activity is believed to be via the inhibition of enzymatic degradation of elastin, induced stabilization of pericardial tissue, stabilization of endothelial dependant vaso-relaxation (Flesch et al. 1998), depressing cardiac papillary muscle contraction (Lee et al. 2010) and the reduction of the calcification of the glutaraldehyde-fixed aortic wall (Sieniawska and Baj 2017). Condensed tannins can also give cardioprotective effect against myocardial injury in rat models (Karthikeyan et al. 2007). Hydrolysable tannins give protection to the heart and show anti-ischemic activity by TNF-$\alpha$ inhibition, scavenging free radicals and reactive oxygen species and endothelial nitric oxide synthase stimulation (Beretta et al. 2009). Purified hydrolysable tannins obtained from *Geum japonicum* relaxes phenylephrine mediated vaso-contraction by nitric oxide (NO)-cGMP pathway and gives hypotensive effects (Xie et al. 2007). The galloyl moiety of gallic acids, digallic acid and 1-desgalloyl regusin F is responsible for the inhibitory effect on propranolol induced negative inotropism (H Lee et al. 2010). Condensed tannins tend to form endothelium derived relaxing factor which relaxes noradrenaline mediated contracted pulmonary artery (Russell and Rohrbach 1989). Proanthocyanins exert long lasting anti-hypertensive effects by regulating nitric oxide mediated endothelium related factors (Magos et al. 2008) and give cardio-protection against myocardial infarction induced by isoproterenol (Karthikeyan et al. 2007). Tannic acid relaxes pre-contraction of human coronary artery in both endothelium dependant and non-dependant manner, by interfering or

**Table 21.5** Cardioprotective, cardiotonic and cardio-active properties of tannins

| Tannin | Characteristics | Assay type |
|---|---|---|
| Tannin from red and white wine | Prevention of elastin degredation, stabilizes tissues on heart, endotheial dependant vaso-relaxation | In-vitro |
| Hydrolysable tannin | Action on cardiac papillary muscle, anti-ischemic activity by TNF-α inhibition, scavenging free radicals and reactive oxygen species and endothelial nitric oxide synthase stimulation | |
| Digallic acid and 1-desgalloyl regusin F | Inhibition o negative inotropic effect | |
| Condensed tannins | Relaxes pulmonary artery, myocardial injury relief | |
| Proanthocyanin | Fights against myocardial infarction and interferes with cGMP level | |

increasing vascular cyclic guanosin monophosphate (cGMP) levels (Flesch et al. 1998). The cardioactive effects of tannins are listed in Table 21.5.

### 21.9.2 Anti-histaminic Property

Depending upon the number and composition of the galloyl groups, HHDP groups and phenolic moiety present, hydrolysable tannins like geraniin, agrimoniin, euphorbin C, oenothein B can significantly inhibit $\kappa O_2$ induced histamine release (Kanoh et al. 2000; Gurjar and Pal 2018, 2020).

### 21.9.3 Cytotoxic and Anticancer Activity

Procyanidins and various monomeric flavanols can exert their cytotoxic activity by the presence of the galloyl moiety present in the 3′ position of C ring (digalloyl procyanidin) (Actis-Goretta et al. 2008). Among different hydrolysable tannins, geraniin and corilagin inhibits TNF-α release (Okuda 2005) whereas hydrolysable tannins obtained from *Eugenia jambolana* like, vescalagin, acutissimin A/B, epia-cutissimin A/B, grandinin/roburin E, hexagalloyl glucose and heptagalloyl glucose etc. inhibits α-amylase which helps in cancer growth (Tong et al. 2014). Different complex tannins like catechins also exhibit cytotoxic activity by the previous mechanism or interfering with cellular signalling pathways (Miao et al. 2014). Proanthocyanidins like Procyanidin $B_1$ and $B_2$ inhibit cell proliferation (Actis-Goretta et al. 2008), effectively suppress the epidermal growth factor receptor phosphoryla-tion, inhibiting the growth of human colon carcinoma cell line (Serrano et al. 2009). Some in-*vivo* activities are also reported. Inhibition of skin-tumor promotion can be seen by different tannins like 1,2,3,4,6-penta-O-galloyl-β-D-glucose, tenophyllanin

A and alienanin B and C, epigallocatechin gallate in the rat models which had been pre-treated with different tumour initiators like 7,12-dimethylbenzo[a]anthracene (DMBA), tetradecanoylphorbol-13-acetate, teleocidin (Okuda 2005). Apple and red wine proanthocyanidins inhibited tumor promotion with azoxymethane induced colon carcinomas in rat models (Serrano et al. 2009). Procyanidins hexamer from cocoa and Japanese quince exerted anti-cancer activity arrests $G_2$/M cell cycle colorectal cancer cells, which can possibly be mediated by the Akt pathway (Choy et al. 2016) and showed pro-apoptotic effects on caco-2 colon cancer cells respectively (Gorlach et al. 2011; Saha and Pal 2016; Sannigrahi et al. 2012; Saha et al. 2017; Rani et al. 2016; Pal et al. 2012).

Ellagic acid inhibits the growth of MCF-7 breast cancer cells, by $G_0/G_1$ cell cycle arrest (Chen et al. 2015). Another research briefly explains about androgen independent prostate cancer cells suppression by cell invasion and motility or the down-regulation of MMPs by ellagitannins (Pitchakarn et al. 2013) and at higher dose, ellagic acid treatment was found to be effective to induce growth inhibition and caspase-dependent apoptosis in PC3 prostate cancer cells in a dose approachable manner (Malik et al. 2011).

Gallic acid from blackberry, raspberry, walnuts, chocolate, wine, green tea and vinegar may inhibits the migration of AGS gastric cancer cells possibly by up-regulating RhoB as well as down-regulating AKT/small GTPase signalling pathways and regulating NF-κB activity (Ho et al. 2013). It also showed ROS-dependent pro-apoptotic effects in different cell lines, such as HCT-15 colon cancer cells (L H Russell et al. 2012) and LNCaP prostate cancer cells (Subramanian et al. 2016). Gallic acid treatment can selectively inhibited growth of liver cancer cells through the mitochondria-mediated apoptotic pathways (Sun et al. 2016). However, gallic acid suppressed the invasion and migration of PC3 prostate cancer cells via the down-regulation of MMP-2 and MMP-9 (Liu et al. 2011) or decreased cell proliferation, invasion and angiogenesis of HeLa and HTB-35 cervical cancer cell lines (Zhao and Hu 2013).

Among anthocyanins, delphinidin exhibits strong anticancer activities by inducing apoptosis and cell cycle arrest in several types of cancer. The cytotoxic effect of delphidine is believed to be due to the suppression of the nuclear factor kappa B (NF-κB) pathway (Bin Hafeez et al. 2008). Another research study proves that peonidin-3-glucoside treatment pointedly down-regulates the matrix metalloproteinase (MMP) and prevents lung cancer cells invasion and metastasis (Liu et al. 2013).

A study showed that epigallocatechin gallate treatment suppressed nicotine-induced migration and invasion of lung cancer cell lines in-vitro as well as in-vivo through inhibition of angiogenesis and epithelial-mesenchymal transition (EMT) (Shi et al. 2015). Another study found that epigallocatechin gallate inhibits survivin (a potent anti-apoptotic protein) and induces apoptosis (Onoda et al. 2011) or prevents β-catenin/Wnt oncogenic signalling pathway to several gastric cancer cell lines (Tanaka et al. 2011). Another study on colon cancer suggested that the Akt, extracellular signal-related kinase (ERK 1 or 2) and alternative p38MAPK signaling pathways were involved in the chemopreventive effects of epigallocatechin gallate

**Table 21.6** In-vitro and in-vivo activities of tannins against cancer

| Tannin | Characteristics | Assay type |
|---|---|---|
| Geraniin and corilagin | Inhibits TNF-$\alpha$ | In vitro |
| Hydrolysable tannins from *Eugenia jambolana*, catechin | Inhibits $\alpha$-amylase | |
| Procyanidin B1 and B2 | Epidermal growth factor receptor phosphorylation, inhibiting the growth of human colon carcinoma cell line | |
| Gallic acid from Blackberry, raspberry, walnuts, chocolate, wine, green tea and vinegar | RhoB regulation and AKT/GTPase signalling down-regulation | |
| Delphinidin | Prevents NF-$\kappa$B pathway | |
| Peonidin-3-glucoside | Prevents the invasion and metastasis of lung cancer cells | |
| Epigallocatechin gallate | Inhibits nicotine-induced lung cancer, inhibits surviving, interferes with Wnt pathway in gastric cancer, Akt, ERK and p38MAPK signaling pathways in colon cancer | |
| Tenophyllanin A and alienanin B and C, Epigallocatechin gallate | Skin tumour promotion by DMBA, tetradecanoylphorbol-13-acetate, teleocidin is inhibited | In vivo |
| Proanthocyanidins from Apple and red wine | Azoxymethane induced colon carcinoma prevention | |
| Procyanidins hexamer from cocoa and Japanese quince | Arrests $G_2$/M cell cycle colorectal cancer cells via Akt pathway and interferes with Caco-2 colon cancer cell | |
| Ellagic acid | $G_0/G_1$ cell cycle arrest in breast cancer cell, androgen independent prostate cancer suppression, inhibition of MMP-9, caspase-dependent apoptosis of prostate cancer cell line | |
| Gallo tannin | ROS-dependent pro-apoptotic effects on colon and prostate cancer cells | |
| Gallic acid | down-regulation of MMP-2 and MMP-9 in prostate and cervical cancer cell line | |
| Epigallocatechin gallate | Prevents angiogenesis and EMT, inhibits androgen action in prostate cancer and estrogen action in breast cancer | |
| Anthocyanidins | Inhibits tumor growth in lung cancer cell line, suppressing the growth of HER2-positive tumor | |

(Cerezo-Guisado et al. 2015). Anti-cancer activity on hormonal regulation is also shown by epigallocatechin as it suppresses estrogen (estradiol, $E_2$) induced breast cancer cell proliferation (Tu et al. 2011) or it antagonizes androgen, leading to suppression of prostate cancer growth both in vitro and in vivo (Siddiqui et al. 2011). Different anti-tumor and anti-cancer activity of annis is as follow listed in Table 21.6.

### *21.9.4 Anti-diabetic Property*

Tannins possess potential anti-diabetic action by several mechanisms (Karan et al. 2013). Firstly, tannins do possess the ability to lower blood glucose levels by delaying intestinal glucose absorption or through the inhibition of α-amylase and as well as by exerting α-glucosidase activity. It has been suggested that the interaction of tannins with human α-amylase depends on the possessed free hydroxyl group that are able to participate in hydrogen bonding (Kandra et al. 2004) and the inhibition of intestinal α-glucosidase activity is given by proanthocyanidin content (McDougall et al. 2005). Tannins can also work on insulin and insulin receptor cells by inducing insulin-like effect on insulin sensitive tissues or by delaying the onset of insulin-dependent diabetes mellitus by regulating the antioxidant environment of pancreatic β-cells (Serrano et al. 2009). Procyanidin found in grape seed can show insulinomimetic action resulting hypoglycaemia (Pinent et al. 2004). 1,2,3,4,6-penta-o-galloyl-β-D-glucopyranose binds to insulin receptor, stimulates glucose transport in adipocytes and maintains blood glucose levels in body (Li et al. 2005). In-vivo activities have been reported in defining diabetes prevention by bark extract of *Syzygium cumini*. Tannins from *Syzygium cumini* have been reported to increase plasma insulin and C-peptide levels with reduction of blood glucose levels of both normal and diabetic rats after 45 days oral administration (Saravanan and Leelavinothan 2006). Tannins with anti-diabetic activity are listed below (Table 21.7).

**Table 21.7** Anti-diabetic property of different tannins

| Tannin | Characteristics | Assay type |
|---|---|---|
| Proanthocyanidin | Inhibits of intestinal α-glucosidase activity | In-vitro |
| Hydrolysable tannin | Controls anti-oxidant environment of pancreatic β-cells | |
| Procyanidin | Insulinomimetic action is shown | |
| 1,2,3,4,6-penta-o-galloyl-β-D-glucopyranose | Stimulates glucose transport in adipocytes | |
| Tannins from *Syzygium cumini* | Up-regulates concentration of plasma insulin and C-peptide levels | In-vivo |

**Table 21.8** Tannins in obesity prevention

| Tannin | Characteristics | Assay type |
|---|---|---|
| Epigallocatechin gallate | Inhibits lipid accumulation | In-vitro |
| Complex tannin | G0/G1 phase or G2/M phase cell cycle arrest in adipocyte differentiation | |
| Tannin from green tea | Adipocyte apoptosis via caspase 3 activation and ERK and Cdk2 pathway regulation | |

## *21.9.5 Anti-obesity Action*

It is believed that polyphenol-rich diets may relate to anti-obesity action as tannins possess the ability to interact with adipose tissues like pre-adipocytes, adipose stem cells, and immune cells directly or indirectly (Sieniawska and Baj 2017) (Table 21.8). It can inhibit lipid accumulation (Moon et al. 2007) and adipogenesis, or can induce apoptosis of adipocytes (Lin et al. 2005). Different complex tannins like (+)/(-) epigallocatechin galllate, epigallocatechin or epicatechin arrest cell cycle of adipocyte cells in adipocyte differentiation at $G_0/G_1$ phase or $G_2/M$ phase and induce apoptosis in mature adipocytes or murine pre- adipocytes (Wang et al. 2014). It also activates caspase 3 and follows ERK and Cdk2 pathway which is responsible for adipocyte cell apoptosis (Wu et al. 2005; Hung et al. 2005).

## *21.9.6 Anti-inflammatory Action*

Several laboratory researches confirm that tannins can inhibit hyaluronidase enzyme and elastase enzyme which is connected with the inflammation factors (Piwowarski et al. 2011). Studies on tannins obtained from *Terminalia chebula* (pentagalloylglucose and trigalloyl glucose) or grape seed proanthocyanidins like geraniin, corilagin, and furosin showed good wound healing property (Zhang et al. 2005). Tannic acid crosslinks fibrous collagen and prevents collagen matrix degradation which inhibits MMP (matrix metalloprotease) activity in-vitro (Zhang et al. 2009). Corilagin showed its potency against lung inflammation by reducing bleomycin-induced lung fibrosis number in apoptotic lung cells and prevented membrane breakdown of lung epithelial cells in-vivo (Wang et al. 2014; Rani et al. 2014). The anti-inflammatory action of tannins are listed in Table 21.9.

**Table 21.9** Anti-inflammatory property of different tannins

| Tannin | Characteristics | Assay type |
|---|---|---|
| Tannin from *Terminalia chebula* | Keratinocyte and fibroblast proliferation mediated wound healing property | In-vitro |
| Grape seed proanthocyanidins | Amelioration of mitochondrial oxidative stress | |
| Tannic acid | Crosslinks fibrous collagen and prevents collagen matrix degradation | |
| Corilagin | Prevents membrane breakdown of lung epithelial cell and fibrosis | In-vivo |

## *21.9.7 Anti-aging Properties*

The extract of *Boerhaavia diffusa* can be served as diuretics, anti-convulsants, opthalmic and anti-aging products (Rajpoot and Mishra 2011). The gall of *Terminalia chebula* was tested for inhibitory activity against tyrosinase as well as the proliferative and MMP-2 inhibition activity on early aging human skin fibroblasts in order to evaluate their in vitro anti-aging activity and it resulted in a potent tannin extract that could be used as anti-aging drug (Manosroi et al. 2010). *Phyllanthus emblica* extracted tannins showed marked anti-aging and skin protection activity (Chaudhuri et al. 2007). The tannins, catechins, epigallocatechin gallate, naringenin obtained from *Cochlospermum vitifolium* also possess anti-aging potency due to its ability to inhibit collagenase, elastase, hyaluronidase enzymes (German-Baez et al. 2017).

## *21.9.8 Other Therapeutic Activities of Tannins*

Tannins from *Mouriri pusa* showed a good in-vivo anti-ulcer activity by giving cyto-protective effects against gastric ulcers in rats (Vasconcelos et al. 2010). Kaki-tannin is seen to have the ability in preventing the rise of total plasma cholesterol, non-HDL cholesterol, triglycerides in type 2 diabetic mice fed high fat diet. It also induces the genes which can metabolize cholesterol (Matsumoto and Yokoyama 2012). Induction of immune response by stimulating interleukin 1 (IL-1) can be seen in case of different ellagitannin oligomers such as oenothein A and B, camellin B, woodfordin C, D, E and F (Okuda and Ito 2011).

Some in-vivo and in-vitro assays had been done to check its effectiveness against alzheimer's disease and epilepsy (convulsion) and tannic acid was seen to inhibit β-secretase activity via interfering with amyloid beta production and also it prevented cognitive impairment (Pal et al. 2008, 2009; Pal and Mazumder 2014; Gupta et al. 2003; Pal and Nandi 2005; Mori et al. 2012). The hydrolysable tannins obtained from *Terminalia chebula* also can act as anti-alzheimer by inhibiting acetylcholinesterase and NMDA receptor (Afshari et al. 2016).

In-vivo anti-diarrhoeal activity was evaluated of tannin (procyanidin) obtained from the bark of *Sclerocarya birrea* as it inhibits intestinal transit rather than to inhibition of net secretion of fluid and electrolytes provoked by the laxative agents (Galvez et al. 1991). The gallotannin-enriched extract isolated from *Galla Rhois* (GEGR) gave laxative effects due to its ability to recover the colon structure, to increase the production of mucin, and downstream signaling pathway of muscarinic acetylcholine receptors via G-protein coupled receptor (Kim et al. 2016). Laxative effect can also be seen by tannin extract of *Mareya micrantha, Aloe ferox, Urginea indica, Fumaria parviflora, Phyllanthus emblica* etc. (Hwang 2018).

## 21.10 Tannins in Industry

Tannins or tannic acid are found to be in charge of the application in ink manufacture, dye industry, plastic resin industry, water purification process, manufacturing of adhesives, surface coatings, and cosmetics etc. (Ramakrishnan and Krishnan 1994). Tannins are mainly considered in processing leather from animal skin in industry. They bind to the protein collagen present in the skin and make crosslinking among them to produce leather, thus preventing the disintegration and decompose of the fibres. A variety of tanning agents are used for tanning but the three main types of tanning are vegetable tanning, chromium III and synthetic tanning (Falcão and Araújo 2018; Covington 1997). The leather possess the properties of better resistance, elasticity and softness due to its better durability. In this way, the leather becomes an extremely flexible material that can be used for different applications from bags to shoes, from belts to jewellery, but even for leather sofas and armchairs, cell phone and tablet cases, clothing and outerwear.

## 21.11 Tannins in Cosmeceuticals

For the effectiveness of cosmeceuticals, the topically applied cosmetic products have to overcome stratum corneum barrier and penetrate into the skin epidermis and dermis layer after the release of active ingredients to give their activities. This whole process solely depends upon different properties such as molecular weight, lipophilicity, vehicle formulation etc. (Löf et al. 2011). It was reported that polyphenols can show slight rheological properties as well as their stability, particularly, the decrease of viscosity, surface active properties (Di Mambro and Fonseca 2005). For example, catechin can decrease the surface tension by emerging itself at the air/water interface. In the o/w emulsion, different tannin units like gallic acid, catechin etc. decreases the surface tension, which results in alteration of droplet sizes of emulsions in some cases and thus improving the dispersion state of emulsion

(Di Mattia et al. 2010). Despite of the ability of tannins to work on the air–water interface, tannins can provide the faster release of active ingredients. The smallest and the most hydrophilic protocatechuic acid exhibited the highest permeation rates, followed by catechins and epigallocatechin gallate (Dal Belo et al. 2009).

Due to their antioxidant properties, polyphenols can improve the oxidative stability and storage stability of emulsions (Di Mattia et al. 2010). In case of tannins with anti-ageing property, the tannin should release from the formulation and work on the dermis and epidermis of the skin (Zillich et al. 2015). Tannins from *Phyllanthus emblica* can be used in formulation of cosmetics as it possess good skin protection activity and the skin lightening lotion with 2% extract showed remarkable in-vivo activity by inhibiting the conversion of dihydroxy indole and dihydroxy indol carboxylic acid to melanin (Chaudhuri et al. 2007). Hydrophilic tannic acid and lipophilic ellagic acid had been tasted for skin protection and it gave anti-aging property by protecting dermal elastin from enzymatic degradation and its deposition into fibroblast (Jimenez et al. 2006).

Different hydrolysable tannins and catechins obtained from chestnut can be used as adhesives, cosmeceuticals, and food processors (Aires et al. 2016). Another research study gave a knowledge about the collagen stabilization ability of vegetable tannin. It was found that pre-treated collagen fibres with acrylic polymer and with tannin extract of *Acacia mollissima* exhibited an increase in hydrothermal stability by 25 °C (Madhan et al. 2002).

## 21.12 Tannins in Neutraceuticals

Tannins can be found in some food bearings in an extensible amount and by taking those in a definite amount can be beneficial for our health. *Camellia sinensis* (tea plant) contain high tannin proportion (Ashok and Upadhyaya 2012) and the tannins found in tea can provide health benefits beyond satisfying traditional nutritional requirement. The tannins obtained from tea can be effective in conditions like hypertension (Hara and Suzuki 1989), diabetes (Tong et al. 2014), and obese (Ra 2002). Chewing gum, containing tea polyphenols has been claimed to be effective against influenza, and said to inhibit dissemination of the virus (Hara and Nakayama 2001). High amounts of tannins (mainly condensed tannins and procyanidins) can be found in wines (more in red wines) and beers. Procyanidins was shown to improve pathological oxidative state in a diabetic condition (Kumari and Jain 2012) or asthma (Schmitz 2001). Pomegranate seed extract associated with ellagitannins can be effective against atherosclerosis. De-caffeinated coffee extract and chlorogenic acid supplement can make a person to lose weight. The stability of flora in the intestine and the anti-fungal activity can be monitored with the help of tannins or condensed tannins extract from acerin fruit, canaigre root or Swedish birch bark (Sieniawska and Baj 2017).

## 21.13 Conclusion

Tannins possess an important place in medicinal chemistry, phytochemistry, as well as in the industrial or manufacturing field. For the many past years, this secondary metabolite has been extensively studied and for their application in different sectors, tannin holds a favourite and remarkable place for the researchers in the way of phytochemical researches. Many more years to come, many more researches to be done and hopefully more applicability of tannins in more different fields yet to be discovered. Giving a little contribution, this book chapter describes about different types of tannins, their occurrence, extraction processes and about their versatile applications and importance in the field of science.

## References


Actis-Goretta L, Romanczyk LJ, Rodriguez CA, Kwik-Uribe C, Keen CL (2008) Cytotoxic effects of digalloyl dimer procyanidins in human cancer cell lines. the J Nutri Biochem 19(12):797–808

Afshari AR, Sadeghnia HR and Mollazadeh H (2016) A review on potential mechanisms of *Terminalia chebula* in Alzheimer's disease. Adv Pharmacol Sci

Agyare C, Lechtenberg M, Deters A, Petereit F and Hensel A (2011) Ellagitannins from *Phyllanthus muellerianus* (Kuntze) Exell.: Geraniin and furosin stimulate cellular activity, differentiation and collagen synthesis of human skin keratinocytes and dermal fibroblasts. Phytomed 18(7):617–624

Aires A, Carvalho R, Saavedra MJ (2016) Valorization of solid wastes from chestnut industry processing: Extraction and optimization of polyphenols, tannins and ellagitannins and its potential for adhesives, cosmetic and pharmaceutical industry. Waste Manage 48:457–464

Al-Alimi K, Razak A A, Saub R and Alabsi A (2017) Tannins acid, ascorbic acid and fluoride from khat chewing plant. Int J Dent Oral Health 3(4)

Alotaibi S (2019) Biophysical properties and finger print of Boswellia Sp. Burseraceae, Saudi J Biol Sci 26(7):1450–1457

Amir M, Mujeeb M, Khan A, Ashraf K, Sharma D and Aqil M (2012) Phytochemical analysis and in vitro antioxidant activity of *Uncaria gambir*. Int J of Green Pharma (IJGP) 6(1)

Arapitsas P (2012) Hydrolyzable tannin analysis in food. Food Chem 135(3):1708–1717

Ariga T, Koshiyama I, Fukushima D (1988) Antioxidative properties of procyanidins B-1 and B-3 from azuki beans in aqueous systems. Agri Biol Chem 52(11):2717–2722

Ashok PK, Upadhyaya K (2012) Tannins are astringent. J Pharmacog and Phytochem 1(3):45–50

Auwal MS, Saka S, Mairiga IA, Sanda KA, Shuaibu A and Ibrahim A (2014) Preliminary phytochemical and elemental analysis of aqueous and fractionated pod extracts of *Acacia nilotica* (*Thorn mimosa*). Paper presented at the Veterinary research forum: an International quarterly Journal

Bagyalakshmi B, Nivedhitha P, Balamurugan A (2019) Studies on phytochemical analysis, antioxidant and antibacterial activity of *Ficus racemosa* L. leaf and fruit extracts against wound pathogens. Vegetos 32(1):58–63

Baig MS, Anand R Tannin containing unani drugs-a review

Beretta G, Rossoni G, Santagati NA, Facino RM (2009) Anti-ischemic activity and endothelium-dependent vasorelaxant effect of hydrolysable tannins from the leaves of *Rhus coriaria* (Sumac) in isolated rabbit heart and thoracic aorta. Planta Med 75(14):1482–1488

Bharani A, Ganguli A, Mathur L, Jamra Y, Raman P (2002) Efficacy of *Terminalia arjuna* in chronic stable angina: a double-blind, placebo-controlled, crossover study comparing *Terminalia arjuna* with isosorbide mononitrate. Indian Heart J 54(2):170–175

Bharudin MA, Zakaria S, Chia CH (2013) Condensed tannins from *acacia mangium* bark: characterization by spot tests and FTIR. Paper presented at the AIP Conference Proceedings

Bin Hafeez B, Asim M, Siddiqui IA, Adhami VM, Murtaza I, Mukhtar H (2008) Delphinidin, a dietary anthocyanidin in pigmented fruits and vegetables: a new weapon to blunt prostate cancer growth. Cell Cycle 7(21):3320–3326

Blainski A, Lopes GC, De Mello JCP (2013) Application and analysis of the Folin Ciocalteu method for the determination of the total phenolic content from *Limonium Brasiliense* L. Mol 18(6):6852–6865

Bora G, Gogoi D, Saikia S, Pareek A, Handique JG (2019) Stannous chloride catalyzed synthesis of Schiff bases from hydroxybenzaldehydes and determination of their antioxidant activity by ABTS and DPPH assay. JC Sci 131(8):68

Brillouet JM, Romieu C, Schoefs B, Solymosi K, Cheynier V, Fulcrand H, Conéjéro G (2013) The tannosome is an organelle forming condensed tannins in the chlorophyllous organs of Tracheophyta. Ann Bot 112(6):1003–1014

Cannas A (2008) Tannins: fascinating but sometimes dangerous molecules. Cornell University, NY, USA. URL https://www.ansci. cornell.edu/plants/toxicagents/tannin.html

Cardona F, Andrés-Lacueva C, Tulipani S, Tinahones FJ, Queipo-Ortuño MI (2013) Benefits of polyphenols on gut microbiota and implications in human health. the J Nutri Biochem 24 (8):1415–1422

Cerezo-Guisado MI, Zur R, Lorenzo MJ, Risco A, Martín-Serrano MA, Alvarez-Barrientos A, Centeno F (2015) Implication of Akt, ERK1/2 and alternative p38MAPK signalling pathways in human colon cancer cell apoptosis induced by green tea EGCG. Food and Chem Toxicol 84:125–132

Chandranayagam C, Veeraraghavan G, Subash A, Vasanthi HR (2013) Restoration of arsenite induced hepato-toxicity by crude tannin rich fraction of *Theobroma cacao* in Sprague Dawley rats. Food Res Int 50(1):46–54

Chaudhuri R, Lascu Z, Puccetti G (2007) Inhibitory effects of *Phyllanthus emblica* tannins on melanin synthesis. Cosm and Toiletries 122(2):73–81

Chen HS, Bai MH, Zhang T, Li GD, Liu M (2015) Ellagic acid induces cell cycle arrest and apoptosis through TGF-β/Smad3 signaling pathway in human breast cancer MCF-7 cells. Int J Oncol 46(4):1730–1738

Cheng HY, Lin CC, Lin TC (2002) Antiherpes simplex virus type 2 activity of casuarinin from the bark of *Terminalia arjuna* Linn. Antiviral Res 55(3):447–455

Choy YY, Fraga M, Mackenzie GG, Waterhouse AL, Cremonini E, Oteiza PI (2016) The PI3K/Akt pathway is involved in procyanidin-mediated suppression of human colorectal cancer cell growth. Mol Carcinogenesis 55(12):2196–2209

Covington AD (1997) Modern tanning chem. Chem Society Rev 26(2):111–126

Dal Belo S, Gaspar L, Campos PM, Marty JP (2009) Skin penetration of epigallocatechin-3-gallate and quercetin from green tea and *Ginkgo biloba* extracts vehiculated in cosmetic formulations. Skin Pharmacol and Phys 22(6):299–304

De Bruyne T, Pieters L, Deelstra H, Vlietinck A (1999) Condensed vegetable tannins: biodiversity in structure and biological activities. Biochem Systematics and Ecol 27(4):445–459

Dhanabal S, Kokate C, Ramanathan M, Kumar E, Suresh B (2006) Hypoglycaemic activity of Pterocarpus marsupium Roxb. Phytotherapy Res: An International Journal Devoted to Pharmacological and Toxicological Evaluation of Natural Product Derivatives 20(1):4–8

Di Mambro VM, Fonseca MJ (2005) Assays of physical stability and antioxidant activity of a topical formulation added with different plant extracts. J Pharm Biomed Analysis 37(2):287–295

Di Mattia CD, Sacchetti G, Mastrocola D, Sarker DK, Pittia P (2010) Surface properties of phenolic compounds and their influence on the dispersion degree and oxidative stability of olive oil O/W emulsions. Food Hydrocolloids 24(6–7):652–658

Díaz-Gómez R, López-Solís R, Obreque-Slier E, Toledo-Araya H (2013) Comparative antibacterial effect of gallic acid and catechin against *Helicobacter pylori*. LWT-Food Sci and Tech 54 (2):331–335

Díaz-Gómez R, Toledo-Araya H, López-Solís R, Obreque-Slier E (2014) Combined effect of gallic acid and catechin against *Escherichia coli*. LWT-Food Sci and Tech 59(2):896–900

Djakpo O, Yao W (2010) Rhus chinensis and Galla Chinensis-folklore to modern evidence. Phytotherapy Res 24(12):1739–1747

Dong H, Chen SX, Kini RM, Xu HX (1998) Effects of Tannins from *Geum japonicum* on the Catalytic Activity of Thrombin and Factor Xa of Blood Coagulation Cascade. J Nat Prod 61 (11):1356–1360

El Sissi H, El Sherbeiny A (1967) Local plants as potential sources of tannins in Egypt, part III. Qualitas Plantarum Et Materiae Vegetabiles 14(1–2):28–36

Ezeonu CS, Ejikeme CM (2016) Qualitative and quantitative determination of phytochemical contents of indigenous Nigerian softwoods. New J Sci

Falcão L, Araújo MEM (2018) Vegetable tannins used in the manufacture of historic leathers. Molecules 23(5):1081

Feldman KS, Ensel SM (1994) Ellagitannin chemistry. Preparative and Mechanistic Studies of the Biomimetic Oxidative Coupling of Galloyl Esters, J ACS 116(8):3357–3366

Ferreira D, Slade D (2002) Oligomeric proanthocyanidins: naturally occurring O-heterocycles. Nat Prod Reports 19(5):517–541

Flesch M, Schwarz A, Böhm M (1998) Effects of red and white wine on endothelium-dependent vasorelaxation of rat aorta and human coronary arteries. a J Physiol-Heart and Circulatory Physiol 275(4):H1183–H1190

Foo LY and Karchesy JJ (1989) Chemical nature of phlobaphenes chemistry and significance of condensed tannins. Springer, pp 109–118

Franzmann PD, Heitz A, Zappia LR, Wajon JE, Xanthis K (2001) The formation of malodorous dimethyl oligosulphides in treated groundwater: the role of biofilms and potential precursors. Water Res 35(7):1730–1738

Fukuchi K, Sakagami H, Okuda T, Hatano T, Tanuma S, Kitajima K, Nonoyama M (1989) Inhibition of *herpes simplex* virus infection by tannins and related compounds. Antiviral Res 11(5–6):285–297

Gadkari PV, Balarman M, Kadimi US (2015) Polyphenols from fresh frozen tea leaves (*Camellia assamica* L.,) by supercritical carbon dioxide extraction with ethanol entrainer-application of response surface methodology. J Food Sci and Tech 52(2):720–730

Galvez J, Zarzuelo A, Crespo M, Utrilla M, Jimenez J, Spiessens C, de Witte P (1991) Antidiarrhoeic activity of *Sclerocarya birrea* bark extract and its active tannin constituent in rats. Phytotherapy Res 5(6):276–278

German-Baez L, Valdez-Flores M, Figueroa-Perez M, Garduno-Felix K, Valdez-Ortiz R, Meza-Ayala K, Valdez-Oritz A (2017) Anti-aging and nutraceutical characterization of plant infusions used in traditional medicine. Pakistan J Nutri 16(4):285–292

Gorlach S, Wagner W, Podsędek A, Szewczyk K, Koziołkiewicz M, Dastych J (2011) Procyanidins from Japanese quince (*Chaenomeles japonica*) fruit induce apoptosis in human colon cancer Caco-2 cells in a degree of polymerization-dependent manner. Nutri and Cancer 63(8):1348–1360

Granica S, Krupa K, Kłębowska A, Kiss AK (2013) Development and validation of HPLC-DAD-CAD–MS3 method for qualitative and quantitative standardization of polyphenols in *Agrimoniae eupatoriae herba* (Ph. Eur). J Pharm and Biomed Analysis 86:112–122

Gross GG (1993) In vitro-studies on the biosynthesis of gallotannins and ellagitannins. Paper presented at the international symposium on natural phenols in plant resistance, p 381

Grundhöfer P, Niemetz R, Schilling G, Gross GG (2001) Biosynthesis and subcellular distribution of hydrolyzable tannins. Phytochem 57(6):915–927

Gu X, Cai J, Zhang Z, Su Q (2007) Dynamic Ultrasound-Assisted Extraction of Catechins and Caffeine in Some Tea Samples. Annali Di Chimica: J of Analytical Environ and Cultural Heritage Chem 97(5–6):321–330

Guangcheng Z, Yunlu L, Yazaki Y (1991) Bark tannin contents of *Acacia mearnsii* provenances and the relationship between the hide-powder and the Stiasny methods of estimation. Australian Fores 54(4):209–211

Gupta M, Mazumder UK, Pal D, Bhattacharya S, Chakrabarty S (2003) Studies on brain biogenic amines in methanolic extract of *Cuscuta reflexa* Roxb stem and *Corchorus olitorius* Linn. seed treated mice. Acta-Polon Pharm Drug Res 60(3):207–210

Gurjar VK, Pal D (2018) Design, synthesis, biological evaluation and In Silico ADMET studies of 1,8-naphthyridine derivatives as a H1 receptor inhibitor. International J Pharm and Biol Arch 9 (4):265–273

Gurjar VK, Pal D (2020) Design, in silico studies, and synthesis of new 1,8naphthyridine-3-carboxylic acid analogues and evaluation of their H1R antagonism effects. RSC Adv 10:13907–13921

Gurjar VK, Pal D, Mazumder A, Mazumder R (2020) Synthesis, biological evaluation and molecular docking studies of novel 1,8 Naphthyridine-3-carboxylic acid derivatives as potential antimicrobial agents (Part -1). Indian J Pharm Sci 82(1):41–53

Hagenah S, Gross GG (1993) Biosynthesis of 1, 2, 3, 6-tetra-O-galloyl-β-d-glucose. Phytochem 32 (3):637–641

Hara Y and Nakayama M (2001) Chewing gum and production of the same, U.S. Patent No. 6,248,346. Washington, DC: U.S. Patent and Trademark Office

Hara Y, Suzuki Y (1989) Method of treating hypertension, US Patent no: 4,840,966. Washington, DC: U.S. Patent and Trademark Office

Harding SA, Xue LJ, Du L, Nyamdari B, Lindroth RL, Sykes R, Tsai CJ (2014) Condensed tannin biosynthesis and polymerization synergistically condition carbon use, defence, sink strength and growth in Populus. Tree Physiol 34(11):1240–1251

Hartisch C, Kolodziej H (1996) Galloylhamameloses and proanthocyanidins from *Hamamelis virginiana*. Phytochem 42(1):191–198

Hatano T, Kusuda M, Inada K, Ogawa T, Shiota S, Tsuchiya T and Yoshida T (2005) Effects of tannins and related polyphenols on methicillin-resistant *Staphylococcus aureus*. Phytochem 66 (17):2047–2055

Hazra B, Sarkar R, Biswas S, Mandal N (2010) Comparative study of the antioxidant and reactive oxygen species scavenging properties in the extracts of the fruits of Terminalia chebula Terminalia Belerica and Emblica Officinalis. BMC Comp Alter Medicine 10 1 20

Hernandez BY, Zhu X, Goodman MT, Gatewood R, Mendiola P, Quinata K, Paulino YC (2017) Betel nut chewing, oral premalignant lesions, and the oral microbiome. PLoS ONE 12(2): e0172196

Ho HH, Chang CS, Ho WC, Liao SY, Lin WL, Wang CJ (2013) Gallic acid inhibits gastric cancer cells metastasis and invasive growth via increased expression of RhoB, downregulation of AKT/small GTPase signals and inhibition of NF-κB activity. Toxicol and Applied Pharmacol 266(1):76–85

HonoráPrice P (1924) The gold-beater's skin test for tannins. Analyst 49(574):25–29

Howell A (2001) In vivo evidence that cranberry proanthocyanidins inhibit adherence of P-fimbriated *E. coli* bacteria to uroepithelial cells. FASEB J 15:A284

Hung P, Wu BT, Chen HC, Chen YH, Chen CL, Wu MH, Liu HC, Lee MJ, Kao YH (2005) Antimitogenic effect of green tea (−)-epigallocatechin gallate on 3T3-L1 preadipocytes depends on the ERK and Cdk2 pathways. Am J Physiol Cell Physiol 288:C1094–C1108

Hung YT, Chen PC, Chen RL, Cheng TJ (2010) Sequential determination of tannin and total amino acid contents in tea for taste assessment by a fluorescent flow-injection analytical system. Food Chem 118(3):876–881

Hwang DY (2018) Therapeutic Role of Natural Products Containing Tannin for Treatment of Constipation. IntechOpen, Constipation

Ikram M, Nowshad F (1977) Constituents of *Quercus infectoria*. Planta Med 31(03):286–287

Iwasa K, Torii H (1962) A colorimetric determination of tea tannin with ferrous tartrate. Chagyo Kenkyu Hokoku (Tea Res J) 1962(19):104–108

Jaiswal H, Singh O, Chauhan A, Sahu M, Dv SP (2018) A review on tannins. Biosci. J 6:16–17

Jalil AMM, Ismail A, Pei CP, Hamid M, Kamaruddin SHS (2008) Effects of cocoa extract on glucometabolism, oxidative stress, and antioxidant enzymes in obese-diabetic (Ob-db) rats. J Agri and Food Chem 56(17):7877–7884

Jerez M, Selga A, Sineiro J, Torres JL, Núñez MJ (2007) A comparison between bark extracts from *Pinus pinaster* and *Pinus radiata*: Antioxidant activity and procyanidin composition. Food Chem 100(2):439–444

Jimenez F, Mitts TF, Liu K, Wang Y, Hinek A (2006) Ellagic and tannic acids protect newly synthesized elastic fibers from premature enzymatic degradation in dermal fibroblast cultures. J Investigative Derma 126(6):1272–1280

Kandra L, Gyémánt G, Zajácz Á, Batta G (2004) Inhibitory effects of tannin on human salivary α-amylase. Biochem and Biophys Res Communications 319(4):1265–1271

Kanoh R, Hatano T, Ito H, Yoshida T, Akagi M (2000) Effects of tannins and related polyphenols on superoxide-induced histamine release from rat peritoneal mast cells. Phytomed 7(4):297–302

Kanzaki S, Yonemori K, Sugiura A, Sato A, Yamada M (2001) Identification of molecular markers linked to the trait of natural astringency loss of Japanese persimmon (*Diospyros kaki*) fruit. JAS for Horticul Sci 126(1):51–55

Karan SK, Mondal A, Mishra SK, Pal D, Rout KK (2013) Antidiabetic effect of Streblus asper in streptazotacin–induced diabetic rats. Pharm Biol 51(3):369–375

Karim AA, Azlan A (2012) Fruit pod extracts as a source of nutraceuticals and pharmaceuticals. Molecules 17(10):11931–11946

Karthikeyan K, Bai BS, Devaraj SN (2007) Cardioprotective effect of grape seed proanthocyanidins on isoproterenol-induced myocardial injury in rats. Int J Cardio 115(3):326–333

Khanbabaee K, van Ree T (2001) Tannins: classification and definition. Nat Prod Rep 18(6):641–649

Kim JE, Go J, Koh EK, Song SH, Sung JE, Lee HA, Hwang DY (2016) Gallotannin-enriched extract isolated from *Galla Rhois* may be a functional candidate with laxative effects for treatment of loperamide-induced constipation of SD rats, PloS one 11(9)

Kumar R, Singh M (1984) Tannins: their adverse role in ruminant nutrition. J Agri and Food Chem 32(3):447–453

Kumari M, Jain S (2012) Tannins: An antinutrient with positive effect to manage diabetes. Res J Rec Sci ISSN 2277:2502

Lang Q, Wai CM (2001) Supercritical fluid extraction in herbal and natural product studies-a practical review. Talanta 53(4):771–782

Lanitis S, Khan M, Sgourakis G, Kontovounisios C, Papaconstandinou T, Karaliotas C (2012) Severe monobacterial necrotizing soft tissue infection by group A Streptococcus: A surgical emergency. Asian Pacific J Tropical Biomed 2(3):250–252

Lee H, Lee JY, Suh MH, Sim SS, Lee MW, Kim CJ (2010) Hydrolysable tannins depress cardiac papillary muscle contraction and propranolol-induced negative inotropism. Fitoterapia 81 (7):820–825

Lee LS, Lee N, Kim YH, Lee CH, Hong SP, Jeon YW, Kim YE (2013) Optimization of ultrasonic extraction of phenolic antioxidants from green tea using response surface methodology. Molecules 18(11):13530–13545

Li DC, Jiang JG (2010) Optimization of the microwave-assisted extraction conditions of tea polyphenols from green tea. Int J Food Sci and Nutri 61(8):837–845

Li X, Pizzi A, Lacoste C, Fierro V and Celzard A (2013) Physical properties of tannin/furanic resin foamed with different blowing agents

Li Y, Kim J, Li J, Liu F, Liu X, Himmeldirk K, Chen X (2005) Natural anti-diabetic compound 1, 2, 3, 4, 6-penta-O-galloyl-D-glucopyranose binds to insulin receptor and activates insulin-mediated glucose transport signaling pathway. Biochem Biophys Res Commun 336 (2):430–437

Lima MB, Andrade SI, Harding DP, Pistonesi MF, Band BS, Araújo MC (2012) Turbidimetric and photometric determination of total tannins in tea using a micro-flow-batch analyzer. Talanta 88:717–723

Lin J, Della-Fera MA, Baile CA (2005) Green tea polyphenol epigallocatechin gallate inhibits adipogenesis and induces apoptosis in 3T3-L1 adipocytes. Obesity Res 13(6):982–990

Lin LT, Chen TY, Lin SC, Chung CY, Lin TC, Wang GH, Richardson CD (2013) Broad-spectrum antiviral activity of chebulagic acid and punicalagin against viruses that use glycosaminoglycans for entry. BMC Microbiol 13(1):187

Liu KC, Huang AC, Wu PP, Lin HY, Chueh FS, Yang JS, Chung JG (2011) Gallic acid suppresses the migration and invasion of PC-3 human prostate cancer cells via inhibition of matrix metalloproteinase-2 and-9 signaling pathways. Oncol Rep 26(1):177–184

Liu W, Xu J, Wu S, Liu Y, Yu X, Chen J, Li X (2013) Selective anti-proliferation of HER2-positive breast cancer cells by anthocyanins identified by high-throughput screening. PloS one 8(12)

Löf D, Schillén K, Nilsson L (2011) Flavonoids: precipitation kinetics and interaction with surfactant micelles. J Food Sci 76(3):N35–N39

Madhan B, Muralidharan C, Jayakumar R (2002) Study on the stabilisation of collagen with vegetable tannins in the presence of acrylic polymer. Biomaterials 23(14):2841–2847

Magos GA, Mateos JC, Páez E, Fernández G, Lobato C, Márquez C, Enríquez RG (2008) Hypotensive and vasorelaxant effects of the procyanidin fraction from *Guazuma ulmifolia* bark in normotensive and hypertensive rats. J Ethnopharmacol 117(1):58–68

Malik A, Afaq S, Shahid M, Akhtar K, Assiri A (2011) Influence of ellagic acid on prostate cancer cell proliferation: A caspase-dependent pathway. Asian Pacific J Tropical Med 4(7):550–555

Mämmelä P, Savolainen H, Lindroos L, Kangas J, Vartiainen T (2000) Analysis of oak tannins by liquid chromatography-electrospray ionisation mass spectrometry. J Chromatography a 891 (1):75–83

Manosroi A, Jantrawut P, Akihisa T, Manosroi W, Manosroi J (2010) In vitro anti-aging activities of *Terminalia chebula* gall extract. Pharma Biol 48(4):469–481

Mathew N, Anitha M, Bala T, Sivakumar S, Narmadha R, Kalyanasundaram M (2009) Larvicidal activity of *Saraca indica*, *Nyctanthes arbortristis*, and *Clitoria ternatea* extracts against three mosquito vector species. Parasitol Res 104(5):1017–1025

Matsumoto K, Si Y (2012) Induction of uncoupling protein-1 and-3 in brown adipose tissue by kaki-tannin in type 2 diabetic NSY/Hos mice. Food Chem Toxicol 50(2):184–190

Mazid M, Khan T, Mohammad F (2011) Role of secondary metabolites in defense mechanisms of plants. Biol and Med 3(2):232–249

McDonald M, Mila I, Scalbert A (1996) Precipitation of metal ions by plant polyphenols: optimal conditions and origin of precipitation. J Agri Food Chem 44(2):599–606

McDougall GJ, Shpiro F, Dobson P, Smith P, Blake A, Stewart D (2005) Different polyphenolic components of soft fruits inhibit α-amylase and α-glucosidase. J Agri Food Chem 53(7):2760–2766

Miao M, Jiang H, Jiang B, Li Y, Cui SW, Zhang T (2014) Structure elucidation of catechins for modulation of starch digestion. LWT-Food Sci Tech 57(1):188–193

Mingshu L, Kai Y, Qiang H, Dongying J (2006) Biodegradation of gallotannins and ellagitannins. J Basic Microbiol 46(1):68–84

Mohanta TK, Patra JK, Rath SK, Pal D, Thatoi HN (2007) Evaluation of antimicrobial activity and phytochemical screening of oils and nuts of *Semicarpus anacardium* L.f. Sci Res Essays 2 (11):486–490

Moon HS, Chung CS, Lee HG, Kim TG, Choi YJ, Cho CS (2007) Inhibitory effect of (−)-Epigallocatechin-3-gallate on lipid accumulation of 3T3-L1 cells. Obesity 15(11):2571–2582

Mori T, Rezai-Zadeh K, Koyama N, Arendash GW, Yamaguchi H, Kakuda N, Town T (2012) Tannic acid is a natural β-secretase inhibitor that prevents cognitive impairment and mitigates Alzheimer-like pathology in transgenic mice. J Biol Chem 287(9):6912–6927

Mueller-Harvey I (2001) Analysis of hydrolysable tannins. Animal Feed Sci Tech 91(1–2):3–20

Niemetz R, Gross GG (2005) Enzymology of gallotannin and ellagitannin biosynthesis. Phytochem 66(17):2001–2011

Nierenstein M (1932) Application of the Gold-Beater's Skin Test to Some Synthetic Tannins. Nature 130(3291):813–813

Nierenstein M, Skene M (1934) Natural organic tannins
Nimse SB, Pal D (2015) Free radicals, natural antioxidants and their reaction mechanisms. RSC Adv 5(35):27986–28006
Özyurt G, Kuley E, Etyemez M, Özoğul F (2013) Comparative seasonal sterol profiles in edible parts of Mediterranean fish and shellfish species. Int J Food Sci Nutr 64:476–483
Pal D, Nandi M (2005) CNS activities of Celsia coromandeliane Vahl in mice. Acta Polon Pharm-Drug Res 62(5):355–361
Pal D, Kumar M, Chakrabarty P, Kumar S (2008) Evaluation of the antioxidant activity of aerial parts of Cynodon dacvtylon. Asian J Chem 20(3):2479–2481
Pal D, Sannigrahi S, Mazumder UK (2009) Analgesic and anticonvulsant effects of isolated saponin from the leaves of Clerodendrum infortunatum Linn. mice. Indian J Exp Biol 47:743–747
Pal D (2013) Comparative analysis of in vitro antioxidant activity of two selected plants with a reference to antidiabetic profile. Asian J Chem 25(4):2165–2169
Pal D, Mazumder UK (2014) Isolation of compound and studies on CNS depressant activities of Mikania scandens Wild with special emphasis to brain biogenic amines in mice. Indian J Exp Biol 52:1186–1194
Pal D, Banerjee S, Ghosh A (2012) Dietary Induced Cancer Prevention: An Expanding Research Arena of Emerging Diet, related to health care system. J Adv Pharm Tech Res 3(1):16–24
Okada T, Suzuki T, Hasobe S, Kisara K (1973) Bergenin. 1. Antiulcerogenic activities of bergenin. Nihon yakurigaku zasshi. Folia pharmacol Japonica 69(2):369
Okuda T (2005) Systematics and health effects of chemically distinct tannins in medicinal plants. Phytochem 66(17):2012–2031
Okuda T, Ito H (2011) Tannins of constant structure in medicinal and food plants-hydrolyzable tannins and polyphenols related to tannins. Molecules 16(3):2191–2217
Okuda T, Yoshida T, Hatano T (1989) New methods of analyzing tannins. J Nat Prod 52(1):1–31
Okuda T, Yoshida T, Kuwahara M, Memon MU, Shingu T (1984) Tannins of rosaceous medicinal plants. I. Structures of potentillin, agrimonic acids A and B, and agrimoniin, a dimeric ellagitannin. Chem Pharm Bull 32(6): 2165–2173
Onoda C, Kuribayashi K, Nirasawa S, Tsuji N, Tanaka M, Kobayashi D, Watanabe N (2011) (-)-Epigallocatechin-3-gallate induces apoptosis in gastric cancer cell lines by down-regulating survivin expression. Int J Oncol 38(5):1403–1408
Patay ÉB, Sali N, Kőszegi T, Csepregi R, Balázs VL, Németh TS, Papp N (2016) Antioxidant potential, tannin and polyphenol contents of seed and pericarp of three Coffea species. Asian Pacific J Tropical Med 9(4):366–371
Peterson GL (1979) Review of the Folin phenol protein quantitation method of Lowry. Rosebrough, Farr Randall, Analytical Biochem 100(2):201–220
Pfundstein B, El Desouky SK, Hull WE, Haubner R, Erben G, Owen RW (2010) Polyphenolic compounds in the fruits of Egyptian medicinal plants (*Terminalia bellerica*, *Terminalia chebula* and *Terminalia horrida*): characterization, quantitation and determination of antioxidant capacities. Phytochem 71(10):1132–1148
Piñeiro Z, Palma M, Barroso CG (2004) Determination of catechins by means of extraction with pressurized liquids. J Chromatography a 1026(1–2):19–23
Pinent M, Blay M, Blade MC, Salvado MJ, Arola L, Ardevol A (2004) Grape seed-derived procyanidins have an antihyperglycemic effect in streptozotocin-induced diabetic rats and insulinomimetic activity in insulin-sensitive cell lines. Endocrinology 145:4985–4990
Pirofsky B, Cordova M, Imel TL (1962) The function of proteolytic enzymes and tannic acid in inducing erythrocyte agglutination. J Immunol 89(6):767–774
Pitchakarn P, Chewonarin T, Ogawa K, Suzuki S, Asamoto M, Takahashi S, Limtrakul P (2013) Ellagic acid inhibits migration and invasion by prostate cancer cell lines. Asian Pacific J Cancer Prevention 14(5):2859–2863
Piwowarski JP, Kiss AK, Kozłowska-Wojciechowska M (2011) Anti-hyaluronidase and anti-elastase activity screening of tannin-rich plant materials used in traditional Polish medicine for external treatment of diseases with inflammatory background. J Ethnopharmacol 137(1):937–941

Pizzi A (2019) Tannins: Prospectives and actual industrial applications. Biomolecules 9(8):344

Pizzolato TD (1977) Staining of *Tilia mucilages* with Mayer's tannic acid-ferric chloride, Bulletin the Torrey Bot club 277–279

Ra A (2002) Tea enhances insulin activity. J Agri Food Chem 50(24):7182–7186

Radebe N, Rode K, Pizzi A, Giovando S, Pasch H (2013) MALDI-TOF-CID for the microstructure elucidation of polymeric hydrolysable tannins. J Appl Polymer Sci 128(1):97–107

Rajpoot K, Mishra R (2011) Boerhaavia diffusa roots (Punarnava mool)-review as rasayan (rejuvenator/antiaging). Int J Res in Pharm Biomed Sci 2(4):1451–1460

Ramakrishnan K, Krishnan M (1994) Tannin-classification, analysis and applications. Ancient Sci Life 13(3–4):232

Rangari VD (2008) Tannin containing drugs

Rangra N, Samanta S, Pradhan K (2019) In vivo antiinflammatory potential of leaf extracts of *Acacia auriculiformis* Benth. Indian J Pharm Sci 81(4):709–719

Rani P, Pal D, Hegde RR (2016) Hashim SR (2016) Acetamides: Chemotherapeutic agents for inflammation associated cancers. J Chemother 28(4):255–265

Rani P, Pal D, Hegde RR, Hashim SR (2016) Leuckart synthesis and pharmacological assessment of novel acetamide derivatives. Anticancer Agent Med Chem 16(7):898–906

Rani P, Pal D, Hegde RR, Hashim SR (2014) Anticancer, anti-inflammatory, analgesic activities of synthesized 2-(Substituted phenoxy) acetamide derivatives. Biomed Int. Article ID 386473:1–9

Ray D, Sharatchandra K and Thokchom I (2006) Antipyretic, antidiarrhoeal, hypoglycaemic and hepatoprotective activities of ethyl acetate extract of *Acacia catechu* Willd. in albino rats. Indian J Pharmacol 38(6):408

Russell JA, Rohrbach MS (1989) Tannin induces endothelium-dependent contraction and relaxation of rabbit pulmonary artery. Am J Respiratory Critical Care Med 139(2):498–503

Russell LH, Mazzio E, Badisa RB, Zhu ZP, Agharahimi M, Oriaku ET, Goodman CB (2012) Autoxidation of gallic acid induces ROS-dependent death in human prostate cancer LNCaP cells. Anticancer Res 32(5):1595–1602

Saha S, Pal D (2016) Molecular Docking, synthesis, in vitro HDAC inhibition activity and antiproliferative activity against breast cancer cell line of hydroxyacetamide derivatives. Trop J Pharm Res 15(7):1401–1411

Saha S, Pal D, Kumar S (2017) Hydroxyacetamide derivatives: cytotoxicity, genotoxicity, antioxidative and metal chelating studies. Indian J Exp Biol 55(2):831–837

Sannigrahi S, Mazumder UK, Pal D, Mishra SL (2012) Terpenoids of methanol extract of *C. infortunatum* exhibit anticancer activity against Ehrlich's ascites carcinoma (EAC) in mice. Pharm Biol 50(3):304–309

Saravanan G and Leelavinothan P. (2006). Effects of *Syzygium Cumini* bark on blood glucose, plasma insulin and C-peptide in streptozotocininduced diabetic rats

SchmitzHH, (2001) The use of procyanidins in the modulation of cytokine gene expression and protein secretion. Patent no: WO 2001045726:A2

Schofield P, Mbugua D, Pell A (2001) Analysis of condensed tannins: a review. Animal Feed Sci and Tech 91(1–2):21–40

Serrano J, Puupponen-Pimiä R, Dauer A, Aura AM, Saura-Calixto F (2009) Tannins: current knowledge of food sources, intake, bioavailability and biological effects. Mol Nutri Food Res 53(S2):S310–S329

Shah BN (2009) Textbook of pharmacognosy and phytochemistry, Elsevier India.

Shen D, Wu Q, Wang M, Yang Y, Lavoie EJ, Simon JE (2006) Determination of the predominant catechins in *Acacia catechu* by liquid chromatography/electrospray ionization- mass spectrometry. J Agri Food Chem 54(9):3219–3224

Shi J, Liu F, Zhang W, Liu X, Lin B, Tang X (2015) Epigallocatechin-3-gallate inhibits nicotine-induced migration and invasion by the suppression of angiogenesis and epithelial-mesenchymal transition in non-small cell lung cancer cells. Oncol Rep 33(6):2972–2980

Shoji Y, Nakashima H (2006) Glucose-lowering effect of powder formulation of African black tea extract in KK-A y/TaJcl diabetic mouse. Archives Pharmacal Res 29(9):786

Siddiqui IA, Asim M, Hafeez BB, Adhami VM, Tarapore RS, Mukhtar H (2011) Green tea polyphenol EGCG blunts androgen receptor function in prostate cancer. FASEB J 25(4):1198–1207

Sieniawska E, Baj T (2017) Tannins. In: Pharmacognosy, Elsevier, pp 199–232

Singh AP. Krameria triandra (R. and P.): Source of Tannin, Ethnobotanical Leaflets 2007(1):11

Smeriglio A, Barreca D, Bellocco E, Trombetta D (2017) Proanthocyanidins and hydrolysable tannins: occurrence, dietary intake and pharmacological effects. Brit J Pharmacol 174 (11):1244–1262

Subramanian AP, Jaganathan SK, Mandal M, Supriyanto E, Muhamad II (2016) Gallic acid induced apoptotic events in HCT-15 colon cancer cells. World J of Gastroenterol 22(15):3952

Sun G, Zhang S, Xie Y, Zhang Z, Zhao W (2016) Gallic acid as a selective anticancer agent that induces apoptosis in SMMC-7721 human hepatocellular carcinoma cells. Oncol Lett 11 (1):150–158

Suvanto J, Nohynek L, Seppänen-Laakso T, Rischer H, Salminen JP, Puupponen-Pimiä R (2017) Variability in the production of tannins and other polyphenols in cell cultures of 12 Nordic plant species. Planta 246(2):227–241

Takechi M, Tanaka Y, Takehara M, Nonaka GI, Nishioka I (1985) Structure and antiherpetic activity among the tannins. Phytochem 24(10):2245–2250

Tanaka T, Ishii T, Mizuno D, Mori T, Yamaji R, Nakamura Y, Akagawa M (2011) (−)-Epigallocatechin-3-gallate suppresses growth of AZ521 human gastric cancer cells by targeting the DEAD-box RNA helicase p68. Free Rad Biol Med 50(10):1324–1335

Thiruchenduran S, Maheswari KU, Prasad T, Rajeswari B, Suneetha W (2017) UV-Vis scanning coupled with PCA as an alternative method for phytochemical screening of natural products-*Costus Igneus* leaf metabolites. J Pharmacog Phytochem 6(1):411–416

Tian Y, Zou B, Cm Li, Yang J, Sf Xu, Hagerman AE (2012) High molecular weight persimmon tannin is a potent antioxidant both ex vivo and in vivo. Food Res Int 45(1):26–30

Tomczyk M, Latté KP (2009) Potentilla-A review of its phytochemical and pharmacological profile. J Ethnopharmacol 122(2):184–204

Tong WY, Wang H, Waisundara VY, Huang D (2014) Inhibiting enzymatic starch digestion by hydrolyzable tannins isolated from *Eugenia jambolana*. LWT-Food Sci Tech 59(1):389–395

Tsukahara K, Moriwaki S, Fujimura T and Takema Y (2001) Inhibitory effect of an extract of *Sanguisorba officinalis* L. on ultraviolet-B-induced photodamage of rat skin. Biolo Pharma Bull 24(9):998–1003

Tu SH, Ku CY, Ho CT, Chen CS, Huang CS, Lee CH, Chang CH (2011) Tea polyphenol (−)-epigallocatechin-3-gallate inhibits nicotine-and estrogen-induced α9-nicotinic acetylcholine receptor upregulation in human breast cancer cells. Mol Nutri Food Res 55(3):455–466

Ukoha PO, Cemaluk EA, Nnamdi OL, Madus EP (2011) Tannins and other phytochemical of the *Samanaea saman* pods and their antimicrobial activities. African J Pure Applied Chem 5 (8):237–244

Variya BC, Bakrania AK, Patel SS (2016) *Emblica officinalis* (Amla): A review for its phytochem, ethnomedicinal uses and medicinal potentials with respect to molecular mechanisms. Pharmacol Res 111:180–200

Vasconcelos P, Andreo M, Vilegas W, Hiruma-Lima CA, Pellizzon C (2010) Effect of *Mouriri pusa* tannins and flavonoids on prevention and treatment against experimental gastric ulcer. J Ethnopharmacol 131(1):146–153

Wang S, Moustaid-Moussa N, Chen L, Mo H, Shastri A, Su R, Shen CL (2014) Novel insights of dietary polyphenols and obesity. J Nutri Biochem 25(1):1–18

Wang Z, Guo QY, Zhang XJ, Li X, Li WT, Ma XT, Ma LJ (2014) Corilagin attenuates aerosol bleomycin-induced experimental lung injury. Int J Mol Sci 15(6):9762–9779

Wati R, Sampanvejsobha S and Punbusayakul N (2009) Effect of extraction conditions on Assam green tea. Agri Sci J 40:47–50

Wei Y and Xingjin Q (1989) Determination of tannin in hops by visible absorptiometry with complex of potassium ferrous ferricyanide. J Shandong Agricultural University 2

Wu-Yuan C, Chen C, Wu R (1988) Gallotannins inhibit growth, water-insoluble glucan synthesis, and aggregation of mutans streptococci. J Den Res 67(1):51–55

Wu BT, Hung PF, Chen HC, Huang RN, Chang HH, Kao YH (2005) The apoptotic effect of green tea (−)-epigallocatechin gallate on 3T3-L1 preadipocytes depends on the Cdk2 pathway. J Agri Food Chem 53(14):5695–5701

Xie YW, Xu HX, Dong H, Fiscus RR, But PP (2007) Role of nitric oxide in the vasorelaxant and hypotensive effects of extracts and purified tannins from *Geum japonicum*. J Ethnopharmacol 109(1):128–133

Zhang H, Zhu Sj, Wang D, Wei Yj and Hu SS (2009) Intramyocardial injection of tannic acid attenuates postinfarction remodeling: a novel approach to stabilize the breaking extracellular matrix. J Thoracic Cardio Surgery 137(1):216–222. e212

Zhang T, Li G, Mo H and Zhi C (2011) Persimmon tannin composition and function. Adv in Biomed Eng 1–2.

Zhang XY, Li WG, Wu YJ, Gao MT (2005) Amelioration of doxorubicin-induced myocardial oxidative stress and immunosuppression by grape seed proanthocyanidins in tumour-bearing mice. J Pharm Pharmacol 57(8):1043–1051

ZHANG Zj, LI Xs and CHENG R (2008) Comparative study on the three techniques of removal tannins from acorns. Food Sci Tech 9

Zhao B, Hu M (2013) Gallic acid reduces cell viability, proliferation, invasion and angiogenesis in human cervical cancer cells. Oncol Lett 6(6):1749–1755

Zillich O, Schweiggert-Weisz U, Eisner P, Kerscher M (2015) Polyphenols as active ingredients for cosmetic products. Int J Cos Sci 37(5):455–464

# Chapter 22
# Piperine: Sources, Properties, Applications, and Biotechnological Production

**Neetu Sachan, Dilipkumar Pal, and Phool Chandra**

**Abstract** From ancient times, phytopharmaceuticals have played an important role in the management of human health. Piperine, an alkaloid with the piperidine nucleus was discovered and isolated by Hans Christian Ørsted, from the fruits of *Piper nigrum*. Piperine forms is slightly water soluble and forms monoclinic needles and possess a strong pungent taste. Piperine contains plentiful established health effects and beneficial therapeutic properties. Cells and enzymes are key elements in biotechnological processes to carry out a wide variety of very specific reactions under judicious conditions to produce piperine and their products. Piperine also serves as bio-enhancers in conjunction with drugs to stimulate drug molecules' activity across different routes by improving the drug's bioavailability across the membrane, raising the drug's effect across conformational interaction, and working as a drug receptor. In recent years, there has been significant interest in the use of piperine to treat many illnesses, its health-beneficial effects, and its work as bio-enhancers. Due to their biological activity, piperine has the potential to be used in health and medicine.

**Keywords** Piperine · Alkaloid · Piper nigrum · Bio-enhancers · Therapeutics · Phytopharmaceuticals

---

N. Sachan (✉) · P. Chandra
School of Pharmaceutical Sciences, IFTM University, Lodhipur Rajput, Delhi Road (NH-24), Moradabad 244 102, Uttar Pradesh, India
e-mail: neetuphool@gmail.com

D. Pal
Institute of Pharmaceutical Sciences, Guru Ghasidas Viswavidyalaya, Bilaspur-495 009, Koni, Chhattisgarh, India

 
D. Pal and A. K. Nayak (eds.), *Bioactive Natural Products for Pharmaceutical Applications*, Advanced Structured Materials 140,
https://doi.org/10.1007/978-3-030-54027-2_22

## 22.1 Introduction

Piperine contains numerous pharmacological activities and many health benefits, particularly against chronic conditions such as insulin resistance reduction, anti-inflammatory activity, and correction in hepatic steatosis (Derosa et al. 2016). Piperine was first isolated in 1820 by the Danish chemist Hans Christian Orstedt; it occurs as a yellow crystalline solid (MW = 285.33 g $mol^{-1}$, m.p. = 128–130 °C) poorly soluble in water and presents weak base properties (Chavarria et al. 2016; Koleva et al. 2012).

## 22.2 Biosynthesis of Piperine

Piperine is a secondary metabolite biosynthetically derived from L-lysine and a cinnamoyl-CoA precursor. Decarboxylation of L-lysine by lysine decarboxylase (LDC) yields cadaverine, which undergoes oxidative deamination by copper amine oxidase (CuAO,) originating 5- aminopentanal. This compound is rapidly cyclized into $\Delta^1$-piperidine Schiff base and subsequently reduced to form piperidine. In parallel, piperonyl-CoA is generated from a cinnamoyl –CoA precursor; this precursor undergoes chain elongation with malonylCoA in a Claisen-like reaction, generating a keto-ester that is reduced by NADPH and then dehydrated to afford piperonyl-CoA. The piperine unit reacts with piperonyl-CoA affording piperine (Fig. 22.1).

## 22.3 Extraction Techniques

The extraction of piperine from the plants can be made with different techniques and that are present in Fig. 22.2 (Raman and Gaikar 2002; Rathod 2014).

## 22.4 Effect on Heart

Piperine produced both positive chronotropic and inotropic responses in the isolated rat atria but not in the ventricular muscles. A tachyphylaxis to piperine occurred rapidly depending on the dose of preincubation. It has been revealed that piperine diminishes the level of substance P in the rat spinal cord, possibly as a result of an extensive discharge of this neuropeptide. In the heart, it has been shown that capsaicin releases CGRP from NANC nerves and the released CGRP shows positive chronotropic and inotropic effects (Franco-Cereceda and Lundberg 1985; Franco-Cereceda et al. 1988; Miyauchi et al. 1987, 1988). Therefore, the lack of

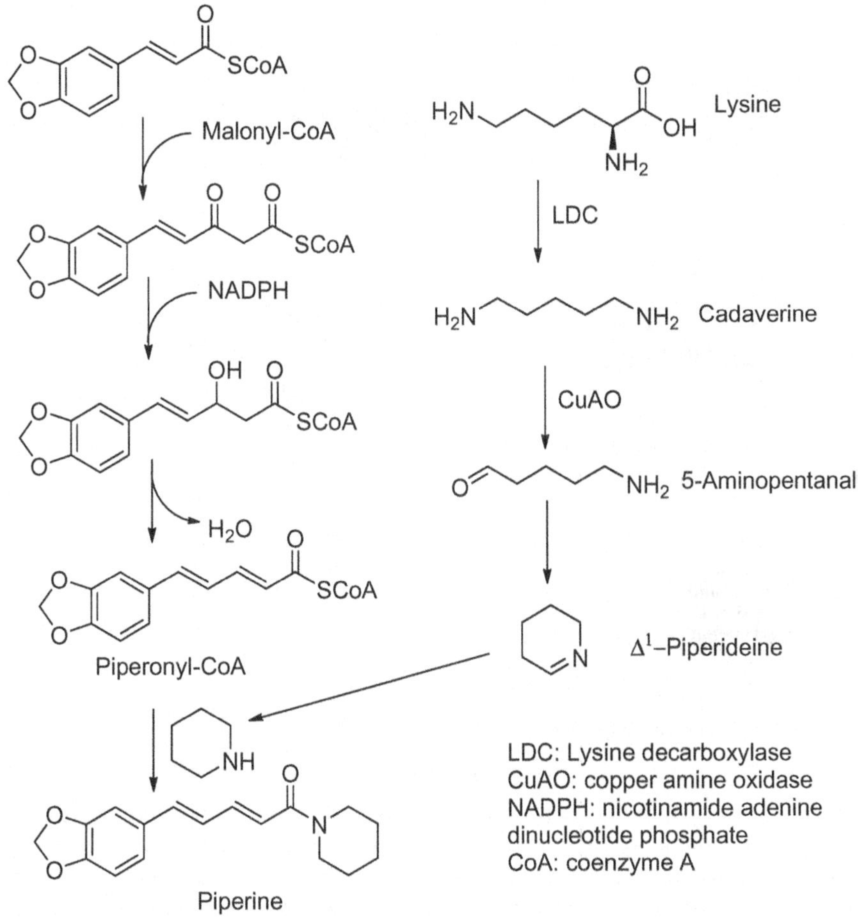


**Fig. 22.1** Biosynthesis of piperine

effects of capsaicin after preincubation with piperine suggests that endogenous CGRP might be depleted during preincubation with piperine. This is supported by the immunohistochemical study, where a marked reduction of CGRP-I was found after the incubation of the tissue with piperine. It has been demonstrated that CGRP produces positive motropic responses and an increase in the formation of cyclic AMP in the atria but not in ventricles of rats and also of guinea pigs. In the present study, CGRP also did not show positive inotropic effects on the ventricular muscle. However, these results are not in accordance with the demonstration that CGRP-binding sites are present in both atrial and ventricular muscle membranes (Sigrist et al. 1986). They showed that the number of receptors is highest in the right atrium followed by the left atrium, and then the right and left ventricles. The exact reasons for the difference in the effects of CGRP between the atria and

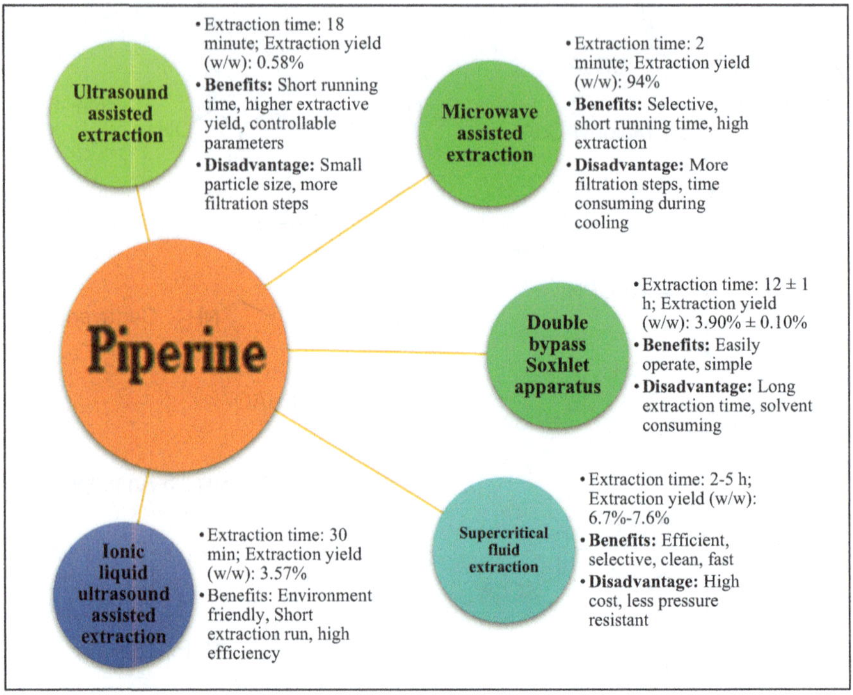


**Fig. 22.2** Different extraction techniques used to extract piperine

ventricles have yet to be determined. The supply of CGRP-containing nerves was considerably poor in the ventricles compared to those in the atria, indicating that the amount of releasable CGRP is small in the ventricles. Even if CGRP is released from sparse CGRP-containing nerves, it does not exert positive inotropic actions on the ventricles. This may be a reason why both piperine and capsaicin failed to produce positive inotropic responses in the ventricles (Miyauchi et al. 1988).

## 22.5 Effect on Pentobarbitone Sleeping Time

Piperine potentiated pentobarbitone sleeping time in dose-dependant manner, with peak effect at 30 min. Blood and brain pentobarbitone levels were higher in piperine-treated animals. Piperine treatment in rats, treated chronically with phenobarbitone, significantly potentiated pentobarbitone sleeping time, as compared to the controls. There was no alteration in barbital sodium sleeping time. It is possible that piperine inhibits liver microsomal enzyme system, and thereby potentiates the pentobarbitone sleeping time (Gupta et al. 2003; Mujumdar et al. 1990; Pal and Nandi 2005; Pal and Mazumder 2014; Sachan et al. 2015; Pal et al. 2009).

## 22.6 Bioavailability of Drugs

Piperine, an essential constituent of black and long peppers, has been mentioned to boost the bioavailability of drugs. The goal of the present studies was to understand piperine 's interaction with intracellular drug biotransforming processes in vitro and in vivo liver tissue. During in vitro findings in rat filtrate, piperine suppressed removal of the hydroxy group from aryl hydrocarbon, N-methyl group from ethylmorphine, O-ethyl group from 7-ethoxycoumarin, and 3-hydroxy-benzo(a) pyrene glucuronidation in graded order. Piperine suppression of certain consequences from 3-methylcholanthrene- and phenobarbital-treated rats in post mitochondrial filtrate was identical to those controls. Piperine suppression of arylhydrocarbon hydroxylase from rats treated with 3-methylcholanthrene was similar to 7,8-benzoflavone. Piperine has induced noncompetitive suppression of hepatic hydroxylase from non-treated and 3-methylcholanthrene-treated animals with a Ki of thirty microM below the significant Km of arylhydrocarbon hydroxylase detected at controls. Additionally, the kinetic studies of ethylmorphine-N-demethylase suppression from monitoring rat hepatic microsomes showed noncompetitive suppression with significant 0.8 mM Km and thirty-five microM Ki. Such studies have shown that piperine is a non-specific drug metabolic inhibitor that displays no distinction between the various cytochrome P-450. Oral piperine administration in rats significantly suppressed the functions of the hepatic arylhydrocarbon hydroxylation and UDP-glucuronyltransferase. Absolute arylhydrocarbon hydroxylation inhibition detected within 1 h was restored to normal range within 6 h. Piperine pretreatment extended in time of sleeping and paralysis in animals at 50% of the SKF-525A dose. These verdicts confirm that piperine as effective drug metabolism inhibitor (Atal et al. 1985).

## 22.7 Effect on Enzymes

In vitro and in vivo modulation of drug-metabolizing enzymes by piperine was investigated in microsomes of rats and guinea pigs. In vitro piperine caused concentration-related inhibition (50% at 100 microM) of arylhydrocarbon hydroxylase (AHH) and 7-ethoxycourmarin deethylase (7ECDE) activities, which were comparable in control and 3-methylcholanthrene (3MC) treated rats. In guinea pig microsomes, however, piperine caused strong inhibition at lower concentrations (35% at 10 microM) and relatively much lesser inhibition with further increase in piperine concentrations. A Dixon plot of the kinetic data of both AHH and 7ECDE indicated noncompetitive inhibition with a Ki of approx. 100 microM. In vivo, piperine given at a dose of 25 mg/kg body wt to rats caused a maximal inhibition at 1 h of both the enzymes, while only AHH returned to normal value within 4 h. Similarly, upon daily treatment of piperine (15 mg/kg body wt) to rats for 7 days, 7ECDE was consistently inhibited, while AHH showed faster recovery. Piperine thus appeared to cause differential inhibition of two forms of cytochrome P450, and thus would accordingly affect the steady-state level of those drugs metabolized by these pulmonary forms of cytochromes P450 (Reen and Singh 1991).

## 22.8 Effects on Antioxidant Pathways in Tissues from Diabetic Rats

Excessive oxidative stress has been implicated in the pathology and complications of diabetes mellitus (Matough et al. 2012), as well as other conditions including cancer, multiple sclerosis, muscular dystrophy, emphysema, Parkinson's disease, Alzheimer's disease, and the aging process (Kehrer and Klotz 2015). Hyperglycemia generates abnormally high levels of free radicals by a mechanism involving autoxidation of glucose, followed by oxidative degeneration and protein glycation (Hunt et al. 1988). In addition, nonenzymatic glycosylation of those enzymes that normally detoxify free radical species may exacerbate oxidative stress in diabetes. Thus, clinical complications in diabetes and other oxidative stress-related diseases may be due partially to the inability of key antioxidant enzymes to function at normal levels (Karan et al. 2012; Pal et al. 2008; Pal 2013; Rauscher et al. 2000).

Complications of diabetes mellitus, which include atherosclerosis, ischemic heart disease, fatty liver, retinopathy, cataract, nephropathy, and neuropathy, are associated with oxidative modifications of various tissue components (Matough et al. 2012; Oberley 1988; Wolff 1993). In diabetes, cell death resulting from inflammation or oxidative damage may release products of protein catabolism, including transition metals. These may serve to perpetuate the generation of reactive oxygen species such as hydroxyl and superoxide radicals. The effects of diabetes on antioxidant defense, consistent with those seen by other researchers, include alterations in detoxifying enzyme activity in every tissue studied, as well as increases in hepatic lipid peroxidation and decreases in the hepatic concentration of GSH, an endogenous free radical scavenger. In diabetic rats, piperine treatment resulted in the reversal of diabetic effects on cardiac lipid peroxidation (the 33% reduction by diabetes was returned to untreated normal values after piperine), on cardiac glutathione reductase activity (the 50% increase by diabetes was reduced to 25% above untreated normal values after piperine), on renal glutathione peroxidase activity (145% increase by diabetes was reversed to 52% above normal after piperine), and on superoxide dismutase activity in kidney (the 90% increase by diabetes was somewhat decreased after piperine). Likewise, piperine reversed the diabetes-induced changes in the concentration of GSSG in brain (the 54% increase by diabetes was decreased almost to untreated normal values after piperine). Piperine treatment appeared to have no effect on hepatic TBA and GSH concentrations or catalase or glutathione peroxidase activities or on cardiac catalase or glutathione peroxidase activities that were altered by diabetes. The 19% increase in glutathione peroxidase activity in the brain of diabetic rats after piperine and the 35% decrease in concentration of hepatic GSSG in piperine-treated diabetic rats as compared to untreated diabetic rats appear unrelated to the effects of diabetes. In the study, the lipophilic piperine was more likely to interact with membranes, influencing lipid peroxidation and GSH transport. Superoxide dismutase and catalase, on the other hand, are part of the defense system in the hydrophilic compartment of the cell, and were less likely to be modulated by piperine treatment.

## 22.9 Effect on the CNS

After exposure to piperine for 96 h, the concentration toxicity relationships of piperine on hippocampal neurons and astrocytes displayed marked differences in terms of cellular vulnerability to piperine. Cultured astrocytes showed only a moderate increase in LDH release without any significant change in morphology at 100 piperine. On the contrary, cultured hippocampal neurons showed marked injuries both biochemically and morphologically. In the time course study, piperine induced delayed neuronal injury in which significant damage was apparent after 48 h and increased gradually up to 96 h of incubation, whereas the same concentrations did not affect astrocytes at any of the times studied (Unchern et al. 1997).

Their data suggest that piperine is selectively cytotoxic to cultured neurons compared with cultured astrocytes. It was notable that MTT reduction of piperine-treated astrocytes increased at all concentrations used in spite of the unchanged protein content. Mitochondrial dehydrogenases in viable cells are mainly responsible for MTT reduction activity. These findings imply that the apparent increase in MTT reduction in piperine-treated astrocytes may be due to enhanced mitochondrial enzymatic activity rather than increased cell proliferation. In addition, the fact that astrocyte cultures were grown to confluence before being used in the experiment does not support piperine-induced astrocyte proliferation. The relationship between increased mitochondrial function and the tolerance of astrocytes to piperine toxicity is still unknown.

Several lines of evidence indicate that astrocytes possess more defences, particularly the glutathione system, against free radicals than neurons. In neurons, the level of glutathione is much lower than in astroglial cells (Bolaños et al. 1995; Makar et al. 1994). This is likely because the activity of $\gamma$-glutamyl cysteine synthetase, a key enzyme in glutathione synthesis, is approximately eightfold higher in astrocytes compared with neurons. Therefore, cultured neurons seem to be more vulnerable to reactive compounds such as peroxynitrite (ONOO–) than cultured astroglial cells. In this connection, it is possible that the differences in sensitivity of cultured neurons and cultured astrocytes to piperine may be explained by their different cellular content of glutathione. Piperine neurotoxicity was prevented by free radical scavengers thereby suggesting the involvement of lipid peroxidation and/or free radicals in its toxic action. The reason for the failure of SOD and catalase to protect hippocampal neurons may be because these agents cannot cross the cell membrane, whereas the effective agents, DO-tocopherol and trolox, are expected to cross the cell membrane readily. These results suggest that intracellular target(s) may be involved in the cytotoxicity of piperine. The protective emcacy of a lipoxygenase inhibitor against piperine-induced neurotoxicity suggests that the lipoxygenase pathway may play a role in piperine-induced free radical generation. In addition to the harmful effects of the free radicals formed, lipoxygenase metabolites by themselves may provoke various kinds of cell damage (Ochi et al. 1992; Sakagami et al. 1989).

It had been reported that SOD and catalase, but not fail to protect cultured pulmonary artery (Kachel et al. 1990) endothelial cells against amidarone-induced injury. However, SOD and catalase prevent the generation of reactive oxygen species and glutamate-induced neurotoxicity in cultured cerebellar granule neurons (Gunasekar et al. 1995). This discrepancy may be due to the different sites of free radical generation in these insulting conditions. SOD and catalase form the defense system in the hydrophilic compartment of the cell whereas u-tocopherol is the key membrane-bound antioxidant (Cotgreave et al. 1988). In this connection, it is also significant that piperine, due to its high lipophilicity, may directly interact with membrane and may facilitate lipid peroxidation resulting in neuronal injury. Although the piperine molecule contains unsaturated double-bonds in the side-chain which may suggest antioxidant properties, recent studies indicate that piperine contains, if any, very (Joe and Lokesh 1994; Reddy and Lokesh 1992) weak antioxidant and free radical scavenging activity. Indeed, piperine in a concentration range similar to that used in our study was shown to increase lipid peroxidation (Johri et al. 1992) in isolated rat jejunum epithelial cells.

## 22.10 Effect on Acute Kidney Injury

Acute kidney injury (AKI) is a serious complication that may be observed in up to 5% of the hospitalized patients (Basile and Yoder 2014). A chief cause of Acute renal injury is renal ischemia–reperfusion (IR). Renal injuries caused by ischemia–reperfusion include oxidative stress, inflammation, and damages of tubular epithelium and vascular endothelium (Abuelo 2007). After ischemia–reperfusion, inflammation initiates as a consequence of cell injury, subsequent molecular products which trigger parenchymal cells of kidney and dendritic cells (DCs), demanding the exudation of chemokines (Chen and Nuñez 2010; Kurts et al. 2013; Rabb et al. 2016). Renal ischemia primes to leukocyte permeation and tissue injury progress through upregulating the adhesive molecules (including ICAM-1) in vascular endothelium, and creating cytokines (proinflammatory), viz., interleukin-6 (IL-6), interleukin-1 (IL-1), and TNF-$\alpha$ in kidney (Kurts et al. 2013; Thurman 2007), Moreover, renal ischemia–reperfusion increases the amalgamation of inducible nitric oxide synthase (iNOS), enhancing the production of NO which bind with reactive oxygen species (ROS) to form peroxynitrite (ONOO), prevents endothelial nitric oxide synthase (eNOS), diminishes NO production in endothelium, and consequences in vasoconstriction (Chatterjee et al. 2003; Goligorsky et al. 2002; Ling et al. 1999). They reported that piperine gavage shields kidney from 30-min ischemia and 24-h reperfusion influenced injuries in rat. Piperine was ingested by gavage with 97% of absorption without any modification. In the research, renal ischemia–reperfusion increased plasma creatinine and urea-nitrogen concentrations. Given the contrary relationship between plasma creatinine concentration and glomerular filtration rate (GFR), it is likely that the rise in the plasma creatinine and urea-nitrogen concentrations is because of the diminution in GFR. It has been shown that different factors are at play in reducing GFR following IR, among which, mention can be made of the back-leak of filtered substances (due to damaged tubular epithelial cells), activation of tubuloglomerular feedback because of increased NaCl (Mohammadi et al. 2019).

## 22.11 Anticonvulsant Mechanisms of Piperine

Experimental findings suggest that piperine considerably slowed down the onset of seizures in the PTZ-induced seizure test, telling a potential involvement of GABA A receptors. In addition, prolonged onset of seizures in picrotoxin and strychnine-initiated piperine-treated seizure models, although deprived of upsetting mortality, also indicates the involvement of GABAergic and glycinergic signaling paths, possibly with a widely recognized downstream pathway. Certainly, piperine's acute effect on cortical and hippocampal GABA estimate the prevalence that piperine intervention has increased GABA levels in these areas and fits into the PTZ experiment with its anticonvulsant influence. Earlier works indicate the involvement of GABA transmission in animal models with anticonvulsant responses of piperine (Zaugg et al. 2010). In addition, da Cruz et al. study (2013) encourage our results because it advocates that piperine enhances basal GABA and glycine amounts in the brain (da Cruz et al. 2013). From the other hand, serotonergic modulation in MES induced convulsions could be a potential mechanism for the protective role (Browning et al. 1983). In fact, our acute study indicated that piperine facilitated the release of serotonin in cortex and hippocampus (Li et al. 2007) which may result in increased seizure threshold and this facilitatory effect on serotonin release (Pei 1983). Interestingly, piperine has provided effective protection in seizures provoked by BAYK-8644 (a glutamate receptor agonist dependent on L-type voltage) (Gasior et al. 1995), in line with the PASS prediction that piperine may affect L-type channels. Piperine modulatory impacts of the $Ca^{2+}$ channel have been reported earlier in in vitro studies utilizing rat hippocampal neurons (Fu et al. 2010). In patch clamp tests, piperine merely poorly blocked L-type channels. Piperine may reduce the inclination of BAYK-8644 to L-type calcium channels allosterically, thus compensating for the difference between both the in vitro and in vivo effects (Bukhari et al. 2013; D'Hooge et al. 1996; Mishra et al. 2015). The dose-dependent anticonvulsant impact of piperine in the convulsions induced by MES mentioned the idea of piperine inhibiting sodium channel activity. In fact, this possibility had been predicted by in-silico PASS prediction, and our in vitro findings show that piperine reduces the peak current of the Na C isoform Nav1.4 canal.

We note also that Nav1.4 converts the muscle tissue isoform of sodium channels, but many of the antagonistic locations are preserved in the sodium channel family as well as we thus anticipate that other types of sodium channels, including neuronal ones, would also be similarly blocked (Fozzard et al. 2005). Therefore, it is possible that piperine in animal models evaluated by inhibiting sodium channel interaction may reflect its anti-convulsive attributes. We further found that the TRPV1 receptor has also, in recent times, reported to play a role in the epileptogenesis method. TRPV1 is a broad—spectrum, high $Ca^{2+}$ permeability cation channel (Kauer and Gibson 2009), and this specific receptors usually promotes the release of glutamate by boosting the excitability of neurons and synaptic C-terminals (Chávez et al. 2010; Gibson et al. 2008; Schöbel et al. 2012; Zsombok et al. 2011). Piperine has been shown to inhibit TRPV1 receptors, and this activity may thus also contribute to the anticonvulsant action of this compound (Chen et al. 2013; McNamara et al. 2005), perhaps in conjunction with sodium channel block.

## 22.12 Immunomodulatory and Antitumor Activity

The dosage of piperine was selected on the basis of cytotoxicity. 1.14 mg/dose/animal for piperine is the lowest concentration with maximum activity. Administration of piperine showed increased number of total WBC count. This indicates piperine can stimulate the hemopoietic system. The differential count shows the drug did not alter the ratio of different WBC types. Bone marrow fills in as the significant wellspring of all blood cells, including lymphocytes. Administration of piperine demonstrated an expansion in bone marrow cellularity and α-esterase positive cells showing its impact on stem cell multiplication. Piperine was found to expand the coursing immunizer titer and counter-acting agent framing cells demonstrating its stimulatory impact on the humoral arm of the immune system. Administration of piperine could likewise altogether repress the development of strong tumor incited by DLA cells and ascites tumor instigated by EAC cells. Immunomodulators may enact cytotoxic effector cells, for example, cytotoxic T lymphocytes, natural killer (NK) lymphocytes, macrophages, and initiated neutrophils (Fidler 1985, 1988). Utilization of chemotherapy in addition to target-explicit immunomodulators hold a sensible guarantee for clinical utility in future (Sunila and Kuttan 2004). Immunomodulatory activity of piperine may be due to the combined action of humoral and cell-mediated immune responses. Hence, the results indicated that piperine could act as a non-toxic immunomodulator which possesses antitumor property also (Sunila and Kuttan 2004).

## 22.13 Larvicidal Effects

The group of scientist studied the consequences of piperine against Anopheles larvae. Larvae were given to eat mixtures of typical larval diet and piperine in changed amounts. Mortality was documented 48 h after the test was placed to the larval containers. Piperine mixtures produced high number of deaths in the *Anopheles gambiae*. The *Anopheles funestus* were noticeably less penetrating to piperine which may reproduce a noticeable change in the nourishing ways of this species to Gambiae complex or a change in nutrition breakdown by alterations in breeding habitation flanked by species. They inferred that Insecticide safe and vulnerable by species demonstrated defenseless to piperine (Samuel et al. 2016).

## 22.14 Inhibits B Lymphocyte Activation and Effector Functions

B cells play a vital role in humoral immune responses that guard against microbial pathogens are yet also connected with the pathogenesis of certain autoimmune and allergic disorders (Gadermaier et al. 2014; Khan et al. 2013). Keeping this in mind a group of scientists started to keep on searching and identifying a novel pharmacological agent that is able to modulate B cell activation and effector functions. The inhibitory effect of the test compound was devoid of TRPV1 as piperine inhibited the multiplication of B cells from TRPV1-deficient mice. In addition, piperine repressed B cell production of interleukin (IL)-6 and IL-10 cytokines, as well as IgM, IgG2b, and IgG3 immunoglobulins (Soutar et al. 2017).

## 22.15 The Anti-tumor Effectiveness and Mechanisms Accompanied with the Combination of Docetaxel-Piperine

Malignant growth of Prostate is the utmost widely recognized non-cutaneous carcinoma in men. Metastatic castrate-resistant prostate cancer (mCRPC) can be treated by docetaxel a chemotherapeutic agent but soon it develops resistance (Guirgis 2015), and that can be overcome by simultaneous treatment with docetaxel and piperine. These effects can be due to the suppression response of piperine to CYPs and P-gp intervention along with alterations in gene expression linked to tumorigenesis and cellular responses (Li et al. 2018).

## 22.16 Allergic Encephalomyelitis

Multiple sclerosis (MS), an inflammatory T cell facilitated autoimmune complication of CNS (Ridderstad Wollberg et al. 2014). In clinical MS patients and Experimental Allergic Encephalomyelitis (EAE) preclinical model, T leukocytes transmigrate from the bloodstream into the CNS to attack myelin sheath encircling nerve fibers (Zozulya and Wiendl 2008). Autoreactive CD4+ and CD8 + T cells were considered as the key driver events in EAE models, indicating important immunopathologic features of MS (Friese and Fugger 2005; Lassmann and Bradl 2017). In highly proliferative cells, such as activated T lymphocytes, increased de novo pyrimidine biosynthesis by Human Dihydroorotate dehydrogenase can enable their superior growth capacity (Quéméneur et al. 2003), thereby inhibiting this enzyme serve the purpose. The scientists' result showed that piperine is a potent, direct DHODH inhibitor with an $IC_{50}$ value of $0.88 \pm 0.04$ μM. Meanwhile, piperine effectively reduced T cell proliferation through DHODH inhibition in vitro, suggesting a potential role of piperine in immune function modulating. Furthermore, piperine treatment significantly reduced immune cell infiltration in central nervous tissue and resultant remission of EAE symptoms. The 13.2–15.8 h terminal half-life ($t_{1/2}$) for elimination of piperine (Ren and Zuo 2019) is largely shorter than A771726. Since DHODH is a main functional target of EAE, we corroborated the efficacy of piperine in MS model by inhibiting DHODH activity, although it remains to be investigated in the future whether other proteins were involved in the therapeutic effect of piperine in EAE (Liu et al. 2020).

## 22.17 Memory Enhancer and Restoration of Myelin Damage

Piperine enhanced remembrance damage in lysolecithin-stimulated demyelination model by mitigating astrocytes initiation, iNOS, and provocative cytokines appearance level by enhancing total antioxidant capability in demyelination situation. Also, Piperine amplified the level of IL10, Foxp3, and antioxidant gene markers. Piperine upgraded the myelin restoration progression by intensifying the amount of MBP and BDNF (Roshanbakhsh et al. 2020).

## 22.18 Effect on Carbamazepine Metabolism

Carbamazepine (CBZ) is a drug of choice to treat different type of seizures. It works for seizure control in the dose range of 4–12 μg/ml in human plasma. Increase in concentration may cause side effects such as diplopia, nystagmus, and aplastic anemia (Bialer et al. 1998). Carbamazepine-10, 11-epoxide (CBZE), a metabolic product of Carbamazepine in liver by CYP3A4 with lesser concern of CYP2C8 (Kerr et al. 1994) and contributes to seizure control and toxicity (Tomson et al. 1990). The impact of piperine on the CBZ metabolic pathways was verified by nurturing 50 μM CBZ with piperine in animal or human hepatic microsome nurture arrangement at various concentration levels. The intensity of CBZ metabolism was measured by the amount of CBZE formation. CBZE formation was shown to have decreased after adding piperine in both the rat and the human nurture arrangement, suggesting that piperine could supress both animal and human CBZ metabolism. The measured piperine inhibition $IC_{50}$ on CBZ breakdown in hepatic microsomes, as shown in Fig. 22.3, was 9.20 ± 1.36 μM and 10.00 ± 1.22 μM individually for alive primates and human. The related $IC_{50}$ values of piperine between alive primates and human hepatic microsomes direct that alive primates hepatic microsome can be exercised to predict such a suppressive activity by piperine against mammalian CBZ metabolism (Ren et al. 2019b).

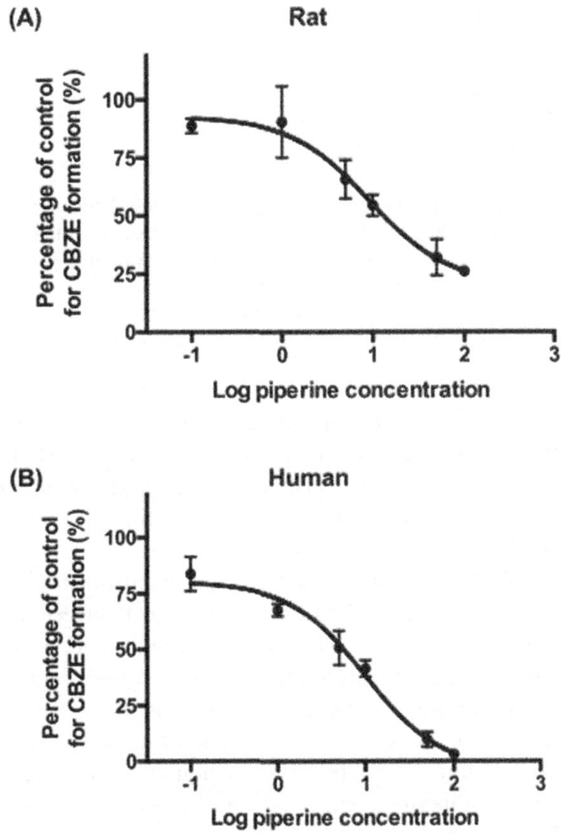


**Fig. 22.3** Sigmoidal dose–response curves of piperine on CBZ metabolism in rat (**A**) and human (**B**) liver microsome incubation system. Each point represents mean ± S.D. in triplicate experiments. With copyright permission (Ren et al. 2019b)

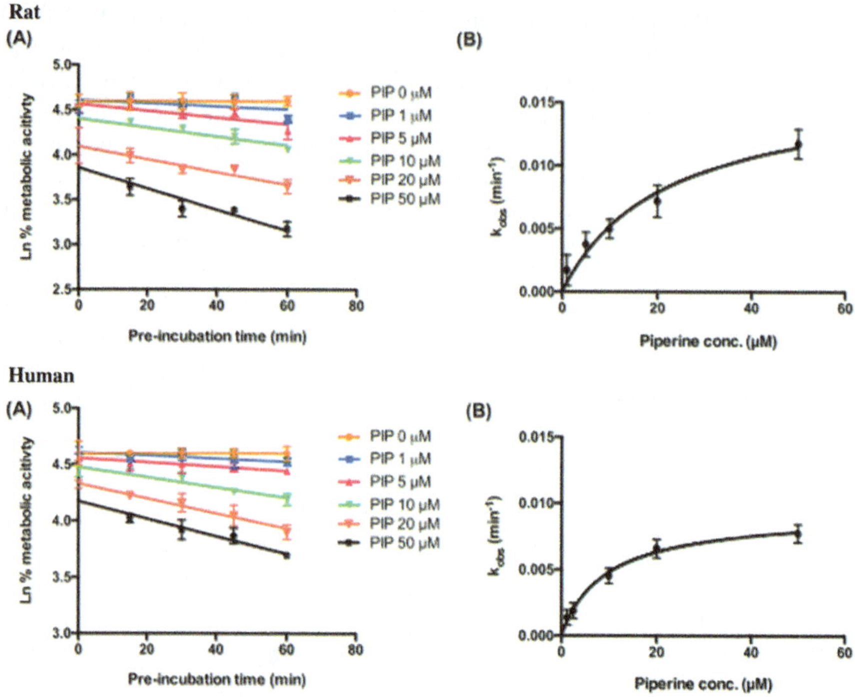


**Fig. 22.4** Representative plots for irreversible inhibition of CBZ metabolism by piperine in rat (Upper) and human (Bottom) microsome incubation system. **A** CBZ metabolic activity remained after preincubation with different concentration of piperine versus preincubation time plot. **B** Observed inactivation rates (kobs) versus piperine concentration plot for inactivation kinetic parameter calculation (kinact and KI). Each point represents mean ± S.D. in triplicate experiments. With copyright permission (Ren et al. 2019b)

Meanwhile, noncompetitive repression and time-dependent repression established comparable kinetic variations in the demonstrative graphs, the time-dependent repression power was also examined by preincubation of hepatic microsome with piperine for dissimilar time and then verified the residual metabolic pursuit on CBZ breakdown (Fig. 22.4) (Ren et al. 2019b).

The consequence of piperine on microsomal action was explored, additionally, using the hepatic microsome from animals ingested with 3.5 mg/kg and 35 mg/kg piperine for 14 days. The metabolic events were examined by CBZE creation using CBZ as enquiry substratum. Associated with regulator group, a substantial inhibition of CBZE formation was found in higher amount of piperine ingested group with comparative CBZE creation of 67.0 ± 18.8% against control, indicating suppression of CBZ metabolic action by chronic piperine administration (Ren et al. 2019b). Piperine's effect on mRNA and protein appearance of genes controlling CBZ breakdown was further examined by means of hepatic illustrations from animals ingested with manifold piperine dose (Fig. 22.5). Against the control

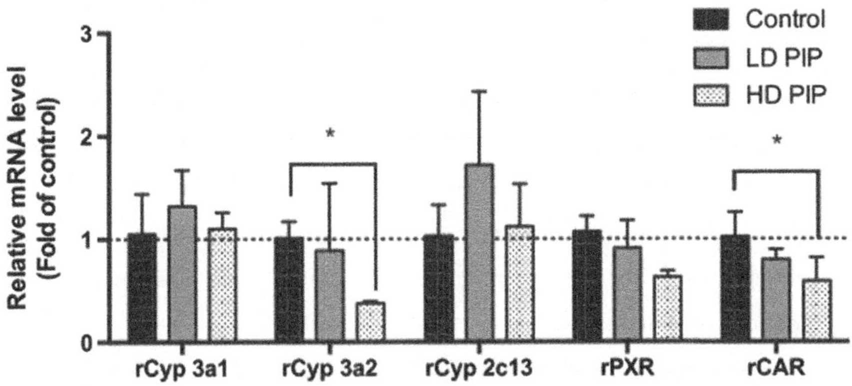


**Fig. 22.5** Comparison of liver mRNA expression levels of genes regulating CBZ metabolism from rats after 14 days treatment of 3.5 mg/kg or 35 mg/kg piperine (LD PIP or HD PIP) or vehicle (Control). (*$p < 0.05$, compared with control group). Each point represents mean ± S.D ($n = 6$). With copyright permission (Ren et al. 2019b)

group, significant decreases were found in the expression level of rCyp3a2 and rCAR mRNA in the elevated dose piperine treatment group (Ren et al. 2019b).

Finally, Ren et al. (2019a) concluded that the study demonstrated time-dependent repression of carbamazepine metabolism through piperine as a CBZ and piperine interference effect. Sustained use of high-dose piperine could not only suppress the metabolic response of the CBZ but also decrease the level of rCyp3a2 mRNA and protein creation and rCAR mRNA illustration. Simultaneous ingestion of CBZ with piperine, particularly after extended usage at a high-dose level, therefore requires further attention.

## 22.19 Bioenhancer Effects

The piperine enhances the bioavailability of different drugs explored in Fig. 22.6 (Atal et al. 1985; Balakrishnan et al. 2001; Bano et al. 1987; Dama et al. 2008; Gupta et al. 1998; Janakiraman and Manavalan 2008; Karan et al. 1988; Kasibhatta and Naidu 2007; Mujumdar et al. 1990; Pooja et al. 2007; Singh et al. 2010, 2005).

## 22.20 Ayurvedic Formulations

The piperine is used in the ayurvedic formulations and they are presented in Fig. 22.7 (Gupta and Jain 2011a, b, c; Patel et al. 2012; Rode et al. 2013; Rout et al. 2007; Sarkar et al. 2011; Shailajan et al. 2011; Singh et al. 2011; Vyas et al. 2011).

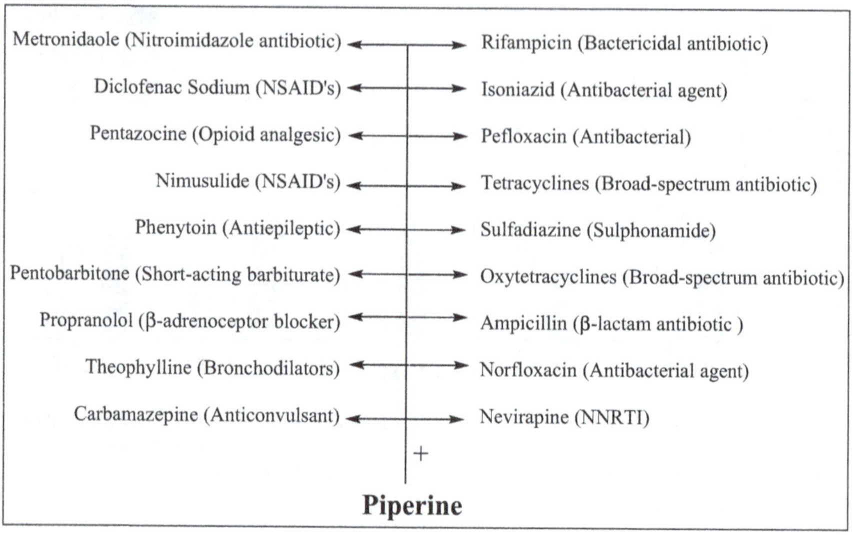


**Fig. 22.6** Bioenhancer effects of piperine on various drugs

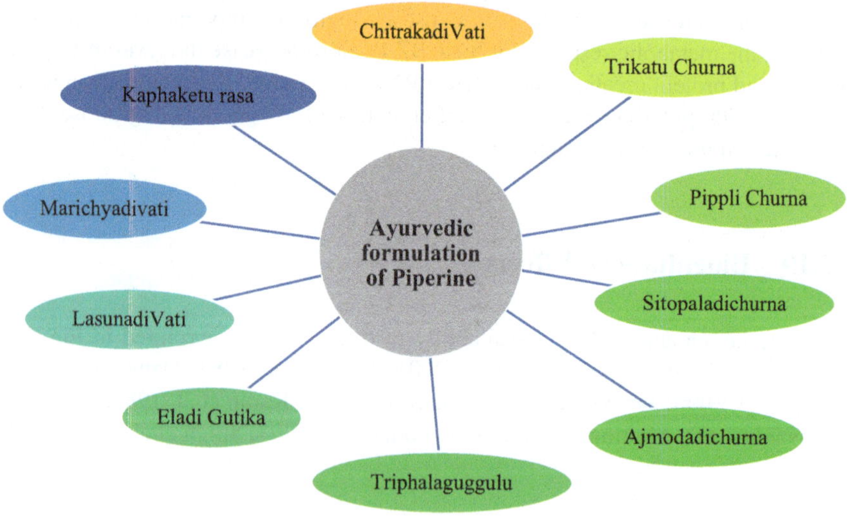


**Fig. 22.7** Marketed ayurvedic formulations made up of piperine

## 22.21 Death of Cerebellar Granule Neurons Induced by Piperine is Distinct from that Induced by Low Potassium Medium

In this study, piperine-induced cell death in cultured cerebellar granule neurons was compared with that induced by K+ withdrawal. Exposures to piperine resulted in concentration-related neuronal death. In contrast to low K+-induced granule neuronal death, piperine-induced death was unaffected by inhibitors of protein synthesis and endonuclease activity. It is well-known that death of cerebellar granule neurons induced by K+ withdrawal displays the requisite morphological and biochemical characteristics of apoptosis including the dependence on de novo protein synthesis and endonuclease activation (Unchern et al. 1998). Therefore, our data suggest an involvement of non-apoptotic mechanism in piperine-induced granule neuronal death. It was notable that a decrease in cellular MTT reduction preceded the onset of neuronal death induced by both piperine and low K+ exposures. This indicated that compromised mitochondrial function may be, at least partly, responsible for the observed neuronal death. It was evident that neurons rapidly lost their mitochondrial membrane potential, energy charge, and subsequently the ability to metabolize MTT before proceeding to necrosis. Due to its high lipophilicity and unsaturated double bonds, it is conceivable that piperine may directly interact with neuronal membranes and initiate lipid peroxidation. This speculation is in accordance with an observation that piperine (25–100 μM) increased lipid peroxidation in isolated epithelial cells of rat jejunum (Johri et al. 1992). The consequences of piperine-induced lipid peroxidation are the disruption of neuronal membrane integrity and the liberation of cytotoxic reactive oxygen species (ROS). ROS have been implicated as mediators of both glutamate-induced cell death (Gunasekar et al. 1995) and low K+ -induce apoptosis of cerebellar granule neurons (Schulz et al. 1996). However, we did not observe any protective effects of lipophilic free radical scavengers against low K+ -induced granule neuronal death. This ineffectiveness may be due to the fact that major intracellular sites of oxygen free radical generation in apoptosis include mitochondria, endoplasmic reticulum, and perhaps nuclear membranes (Korsmeyer et al. 1995); whereas the free radical scavengers used in our study exert their effects mainly on cell membranes. Schulz et al. (1996) observed that vitamin E (>2 mM) partially protected cerebellar granule neurons against apoptosis induced by 24 h K+ withdrawal (Schulz et al. 1996). At very high concentrations, the protective effect of vitamin E may be related to its stabilizing effect on the cell membrane rather than its antioxidant or free radical scavenging property (Diplock 1982).

## 22.22 KV Channel as Therapeutic Target for Prostate Cancer Treatment

The concept of KV channel as therapeutic target for prostate cancer treatment attracts increasing interest, but the lack of effective and selective KV channel modulators has hindered the progression of treatment strategy (Schönherr 2005). Exploring the functional properties of ion channels in cancer progression and metastatic behavior is emerging as a novel approach for the development of effective anticancer treatment. Potassium (K+) channels in the plasma membrane of tumor cells contribute to a wide range of cellular processes including cell cycle progression, cell proliferation, and apoptosis (Ouadid-Ahidouch and Ahidouch 2008). In particular, K+ channels play a crucial role in cell proliferation, as its activation is a prime factor for cell cycle progression through early G1 phase of the cell cycle (Wonderlin and Strobl 1996). Hence, the blockage of K+ channel activity has shown to inhibit cell proliferation in several cancer cell lines including prostate (Ouadid-Ahidouch and Ahidouch 2013). In prostate cancer cells, KV channel is quite prominently expressed and involved in cell proliferation (Prevarskaya et al. 2007). Blockade of KV channel by K+ channel blocker 4-aminopyridine (4-AP) inhibited cell growth in both androgen-sensitive (AT-2) and androgen-insensitive (MAT-LyLu) rat prostate cancer cell lines (Fraser et al. 2000). Similarly, the channel blockers, dequalinium, amiodarone, and glibenclamide have shown to induce apoptosis in PC-3 cells (Abdul and Hoosein 2002). Characterization of Kv channel in LNCaP and PC-3 Cells is presented in Fig. 22.8, Concentration-dependent effect of piperine on IK in LNCaP cells Fig. 22.9 and Concentration-dependent effect of piperine on IK in PC-3 cells Fig. 22.10 (George et al. 2019; Rauscher et al. 2000).

## 22.23 Piperine Impairs the Migration and T Cell-Activating Function of Dendritic Cells

Group of scientist attribute piperine-mediated inhibition of DC migration to impairment of the chemokine receptor switch from CCR5 to CCR7 since CCL21 causes DC migration via stimulation of CCR7 (Yoshida et al. 1998). Reduced in vivo migration of piperine-treated DCs was likely also caused by reduced expression of CCR7. In addition, piperine-treated DCs exhibited decreased LPS-induced expression of CD40 and MHC II molecules, as well as reduced synthesis of the pro-inflammatory cytokines, TNF-α, IL-6, MCP-1. Furthermore, exposure to piperine caused a significant decrease in MHC II high-expressing DC, which is indicative of a reduction in highly mature DCs (Lutz et al. 1999). Taken together, these findings suggest that exposure to piperine caused DCs to retain an immature or semi-mature DC phenotype in spite of TLR4 stimulation with LPS. Interestingly, piperine-treated DCs have increased phagocytic activity (Bae et al. 2012), which is also consistent with an immature DC phenotype. Our findings are in

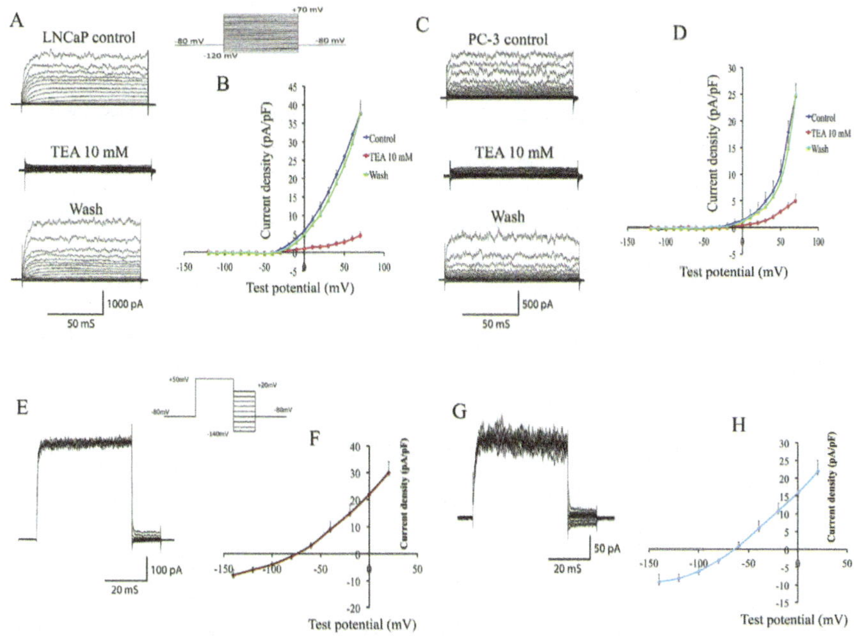


**Fig. 22.8** Characterization of K V channel in LNCaP and PC-3 Cells. Typical recordings of the whole cell outward K + current were elicited by voltage command pulses of increasing step pulses from −140 mV to +70 mV in 10 mV increments. (**A** and **C**) Representative current traces of I recorded in the absence and presence of 10 mM TEA in LNCaP and PC-3 cells. (**B** and **D**) The I-V relationship in the absence and presence of TEA in LNCaP and PC-3 cells. (**E** and **G**) Representative tail current traces recorded in LNCaP and PC-3 cells. (**F** and **H**) Tail current–voltage relationship of LNCaP and PC-3 cells. Data are plotted as mean ± SEM ($n > 7$). With copyright permission (George et al. 2019)

general agreement with two previous studies that showed reduced expression of maturation markers and proinflammatory cytokines following DC exposure to piperine (Bae et al. 2012); however, unlike these earlier reports, we observed reduced synthesis of IL-6, but no effect on CD86 expression by piperine-treated DCs. Reduced production of proinflammatory cytokines has also been reported in piperine-treated B16-F10 melanoma cells, adipocytes, and macrophages (Pradeep and Kuttan 2003, 2004; Woo et al. 2007) indicating that this effect is not cell type-specific. Moreover, normal expression of ICAM-1, MAC-1, and LFA-1 adhesion molecules by piperine-treated DCs shows selective inhibition of DC surface marker expression, which may reflect targeting of specific signal transduction pathways. Indeed, piperine is a known inhibitor of extracellular signal-regulated and c-Jun N-terminal kinases that are activated in mouse DCs as a result of LPS stimulation (Bae et al. 2012).

Piperine-mediated inhibition of DC maturation marker expression induced by LPS, CpG, and poly I:C indicates that the inhibitory effect of piperine on DC maturation caused by pattern recognition receptor stimulation was not restricted to the

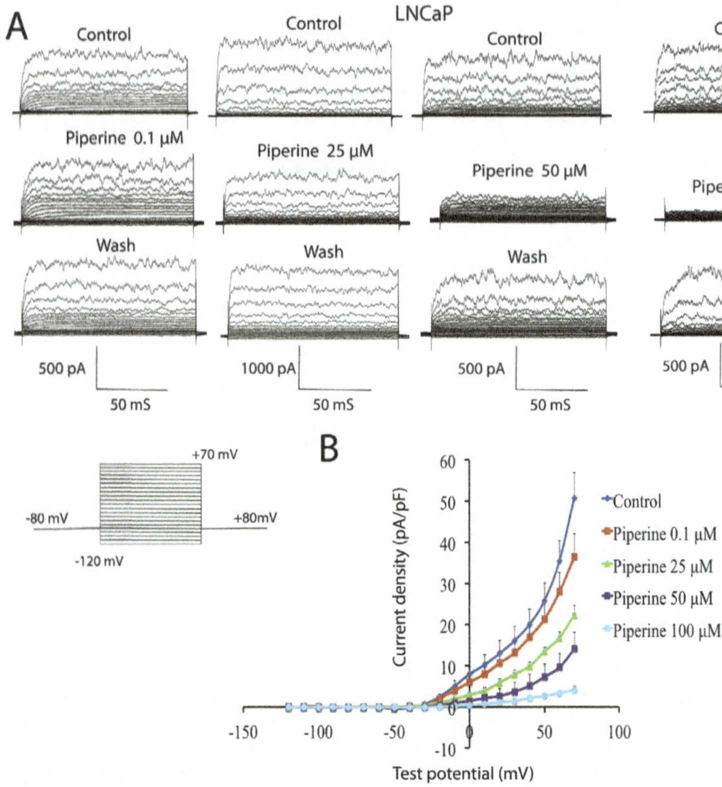


**Fig. 22.9** Concentration-dependent effect of piperine on $I_K$ in LNCaP cells. **A** Representative current traces of I recorded in the absence and presence of piperine at different concentrations (0.1, 25, 50 and 100 μM). **B** The mean I-V of $I_K$ in the absence and presence of piperine at different concentrations. Data are plotted as mean ± SEM ($n > 7$). With copyright permission (George et al. 2019)

LPS/TLR4 axis. Furthermore, this finding suggests that piperine modulates signaling pathways associated with the TLR adaptor molecule Toll/IL-1R domain-containing adaptor inducing interferon (TRIF) and the TLR adaptor protein MyD88, since TLR3 agonists such as poly I:C act via TRIF, TLR9 agonists such as CpG act via MyD88, and TLR4 agonists such as LPS activate both TRIF and MyD88 signaling pathways (Trinchieri and Sher 2007). It, therefore, seems likely that piperine acts downstream of TRIF and MyD88 by targeting NF-κB, which stimulates the production of genes participating in the inflammatory response (Gasparini and Feldmann 2012). Other investigators have also reported piperine-mediated suppression of the NF-κB route. For instance, nuclear aggregation of the NF-κB subunits p65, p50, and cRel is repressed in piperine-medicated B16-F10 melanoma cells (Pradeep and Kuttan 2004), while piperine interferes with NF-κB transcriptional activation in HT-1080 fibrosarcoma cells (Hwang et al. 2011). Piperine also inhibits the degradation of IκBα

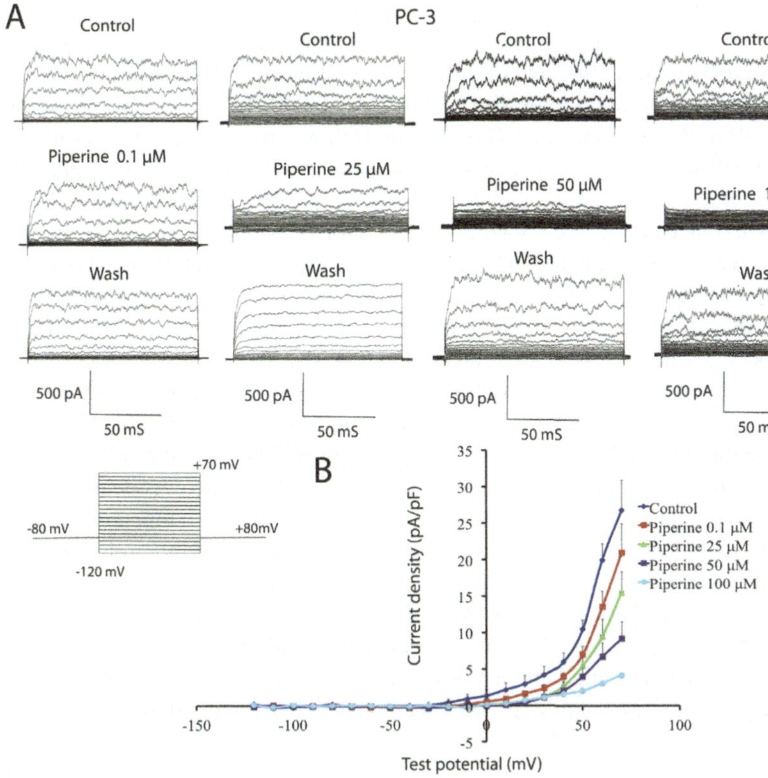


**Fig. 22.10** Concentration-dependent effect of piperine on $I_K$ in PC-3 cells. **A** Representative current traces of IK recorded in the absence and presence of piperine at different concentrations (0.1, 25, 50 and 100 μM). **B** The mean I-V of $I_K$ in the absence and presence of piperine at different concentration. Data are plotted as mean ± SEM ($n > 7$). With copyright permission (George et al. 2019)

in TNFα-stimulated endothelial cells and PMA-stimulated HT-1080 cells (Hwang et al. 2011; Kumar et al. 2007). However, piperine does not affect the IL-1β-induced transcriptional activity of NF-κB in synoviocytes (Bang et al. 2009). Piperine-mediated inhibition of the NF-κB pathway, therefore, appears to be cell type-specific. We report here that piperine-treated DCs had decreased nuclear levels of the NFκB subunit RelB, which is normally abundant in the nuclei of DCs matured by stimulation with pathogen-associated molecular pattern molecules (Neumann et al. 2000). Since RelB is critical for DC maturation and antigen-presenting function (Zanetti et al. 2003) and NF-κB controls CCR7 transcript (Sánchez-Sánchez et al. 2006), and reduced RelB expression in the cores of piperine pickled DCs shows partially for suppressive response for piperine in DC development, immigration, and T cell excitation detected in the experiment. Though, piperine suppresses mitogen-activated protein kinase signaling in numerous cells (Bang et al. 2009; Hwang et al. 2011), including DCs (Bae et al. 2012). We suggest that piperine-mediated inhibition of mitogen-activated protein kinases also contributes to the failure of DCs to migrate in response to CCL21 since CCR7-induced chemotaxis of DCs involves the activation of extracellular signal-regulated kinase and p38 mitogen-activated protein kinase (Riol-Blanco et al. 2005). Antigen-specific stimulation of OVA323–339-specific CD4+ T-cells with piperine-treated DCs resulted in decreased production of IL-2 and IFN-$\gamma$, indicating that T helper cells were directed away from the Th1 phenotype that is typically induced by DCs differentiated with granulocyte–macrophage colony-stimulating factor (Eksioglu et al. 2007). In contrast, there was no reduction in CD4+ T-cell production of IL-4 and IL-17, suggesting that T helper cells had not been alternatively directed to a Th2 or Th17 phenotype. A similar decrease in CD4+ T-cell synthesis of IFN-$\gamma$ with no change in IL-4 production occurs when CD4+ T-cells are stimulated with DCs that were treated with dexamethasone prior to LPS maturation (Matyszak et al. 2000). Interestingly, multiple restipulation of CD4+ T-cells with dexamethasone-treated DCs resulted in the selective induction of T regulatory cells. Although we did not determine whether piperine-treated DCs also induced T regulatory cells, such an outcome seems unlikely given that T regulatory cells require IL-2 for their development and function (Burchill et al. 2007; de la Rosa et al. 2004). We also observed that the in vivo proliferation of OVA323–339-specific CD4+ T-cells was virtually ablated when piperine-treated DCs were used as antigen-presenting cells. This profound inhibitory effect on CD4+ T-cell activation was likely due to reduced migration of piperine-treated DCs to lymph nodes and impaired interactions with CD4+ T-cells as a result of decreased DC expression of CD40 and MHC II molecules (Rodgers et al. 2016).

## 22.24 Piperine-Laden Nanoparticles with Increased Dissolving and Improved Bioavailability for Controlling Epilepsy

The reduction of size is a regular exercise to enhance the drug solubility by raising the surface area to mass quotient, altering particle curvature that institute defects into crystal lattice (Williams et al. 2013). The solubility and dissolution of piperine have been improved by several formulation techniques, such as solid dispersion with hot melt extrusion technology, Tween 80 coated solid lipid nanoparticles and microemulsion with self-emulsifying drug delivery system (Ashour et al. 2016; Elnaggar et al. 2015; Shao et al. 2015). However, none of these formulations have been verified to confirm its anti-epileptic effect. Although nanoprecipitation method has been used to develop a piperine and curcumin co-loaded nanoparticle preparation before, the feasibility for nanoprecipitation with piperine only and its subsequent effect for epilepsy control have never been attempted (Moorthi et al. 2012). In the study, they developed a nanosized formulation of piperine by nanoprecipitation method to enhance its solubility, dissolution, and improve systemic exposure of piperine after oral administration, which is a noninvasive route that is suitable for long-term use during epilepsy control. The anti-epileptic effects of piperine nanoparticles were further evaluated in acute seizure modes in mice (Pal et al. 2009).

The nanosized piperine was articulated by employing nanoprecipitation method (Ren et al. 2019a) and they observed the characteristics of nanoparticles under TEM with high spacial resolution. As shown in Fig. 22.11, piperine nanoparticles appeared as spherical black spots with a smooth surface. Smaller particle size (around 67 nm) was found for piperine nanoparticles under TEM observation compared with DLS, which may be attributed to the shrinkage of nanoparticles upon drying process during TEM preparation. Thus, the outcome of the particle size from DLS would be reflected as a more precise result for piperine nanoparticles particle size.

The release pattern of piperine nanoparticles was constructed to demonstrate its dissolution pattern and compared with that of unformulated piperine (Fig. 22.12).

They studied the anti-epileptic effect of piperine nanoparticles with a mouse model with PTZ-induced acute seizures. Characteristic behavioral patterns such as myoclonic twitch and tonic limb extension were observed after i.p. injection of PTZ at 80 mg/kg in Kunming mice, indicating the successful development of seizure model in mice (Raol and Brooks-Kayal 2012). As shown in Fig. 22.13, unformulated piperine at 15 mg/kg failed to prevent the PTZ-induced seizure since there was no significant difference in average seizure frequency or latency to first seizure compared with the PTZ group. In contrast, no seizure was found during the 30 min observation for piperine nanosuspensions at same dose level (15 mg/kg), which was significantly different from PTZ group for seizure frequency and latency to first seizure ($p < 0.01$). After lowering the dose of piperine nanoparticles to 7.5 mg/kg, effective anti-seizure effect was still demonstrated with reduced seizure frequency

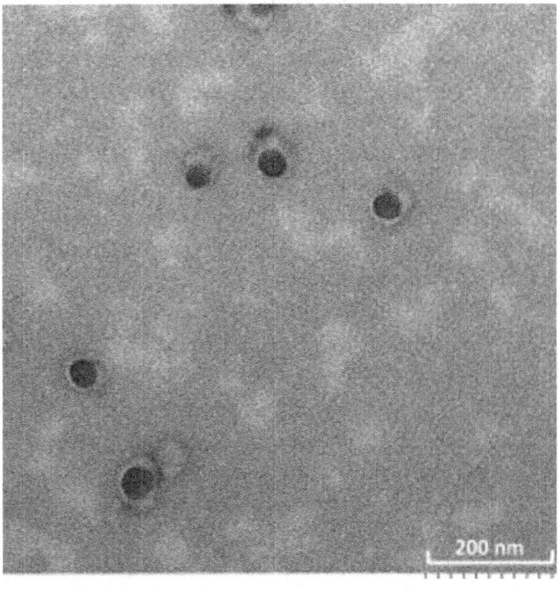


**Fig. 22.11** Micrograph of piperine nanoparticles revealed by transmission electron microscopy (TEM). With copyright permission (Ren et al. 2019a)

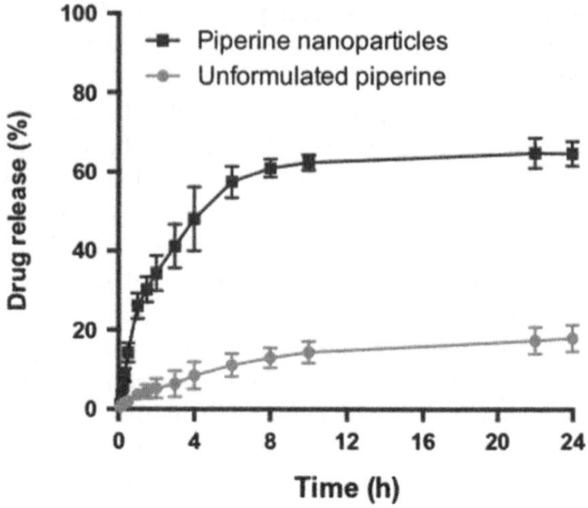


**Fig. 22.12** In vitro release profiles of piperine nanoparticles and unformulated piperine in phosphate buffer at 37 °C for 24 h. Each point represents mean ± S.D ($n = 3$). With copyright permission (Ren et al. 2019a)

and delayed onset of seizure compared with PTZ group ($p < 0.05$). Therefore, they concluded that the nanosuspensions preparation could significantly improve the anti-epileptic effect piperine (Ren et al. 2019a).

## 22.25 In Vitro Cytotoxic and In Silico Activity

The obtained MTT values showed that piperine has a cytotoxic effect because the $IC_{50}$ was recorded as 61.94 ± 0.054 μg/ml (Nimse and Pal 2015; Paarakh et al. 2015; Sannigrahi et al. 2012; Saha and Pal 2016). The tyrosine kinase receptors have multidomain extracellular Ligands for specific Ligand, a signal pass transmembrane hydrophobic helix and tyrosine kinase domain. The receptor tyrosine kinases are not only cell surfaces transmembrane receptors, but are also enzymes having kinase activity (Bari et al. 2012). Angiogenesis in cancer is a significant stage in which new capillaries form to supply a vasculature for nutrient supply and waste material removal. Thus, the kinase inhibitor is a new cancer treatment as such an anti-angiogenic agent. The trend nowadays is to develop herbal drugs and drug candidates as inhibitors.

Low molecular weight entities originated in the extracellular part of the receptor impede tyrosine kinase phosphorylation block signaling (Manley et al. 2002). Since the type I receptor tyrosine kinase is a major regulator of several distinct and diverse cellular pathways. They took Piperine and docked to get the superior conformer. The docking demonstrates that piperine has a $-7.6$ kJ $mol^{-1}$ binding energy with two hydrogen bonds formed. Molecular docking with EGFR tyrosine kinase domain revealed that piperine has suppressive capability and thereby has interactions in active pockets one or the other amino acids. The topology of the active position of EGFR tyrosine kinase was similar in all synthesized molecules, which is

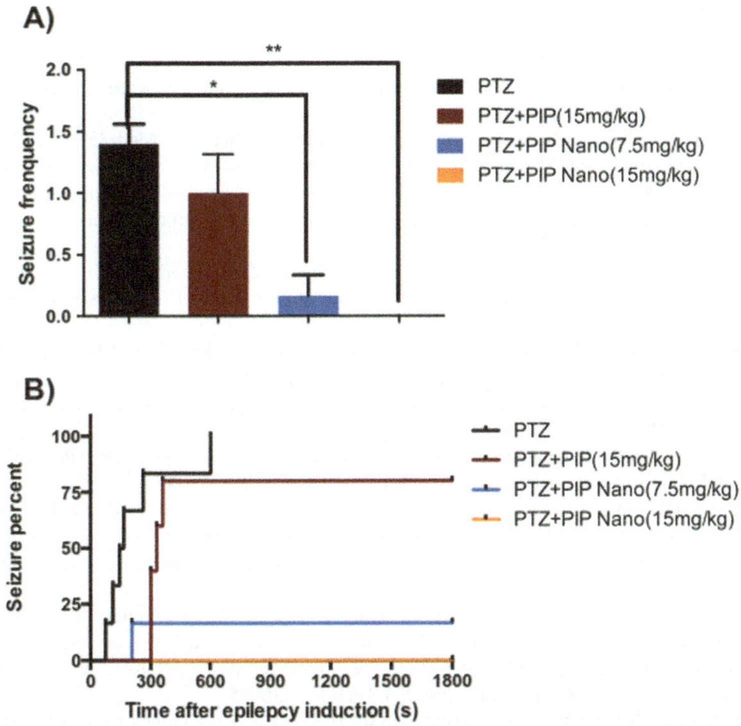


**Fig. 22.13** The seizure frequency (**A**) and latency to first seizure (**B**) of Kunming mice within 30 min after inducing acute seizure by PTZ (80 mg/kg, i.p.) in absence or presence of oral administration of unformulated piperine at 15 mg/kg or piperine nanoparticles at 7.5 or 15 mg/kg 45 min before PTZ injection. Each point represents mean ± S.D ($n = 6$, $*p < 0.05$, $**p < 0.01$ compared with PTZ group). With copyright permission (Ren et al. 2019a)

lined by interacting amino acids as predicted from the ligplot. In in vitro experiments, the molecule emerged to be active against the cell line used in inhibiting the cell growth (Paarakh et al. 2015).

## 22.26 Conclusion

Piperine is a versatile chemical entity and it possesses different anticancer properties as well it has the capability to enhance the bioavailability of important drugs either in its form or in a modified dosage form. Further activity is required to understand its utility in other important metabolic pathways.

## References


Abdul M, Hoosein N (2002) Expression and activity of potassium ion channels in human prostate cancer. Cancer Lett 186(1):99–105

Abuelo JG (2007) Normotensive ischemic acute renal failure. N Engl J Med 357(8):797–805

Ashour EA, Majumdar S, Alsheteli A, Alshehri S, Alsulays B, Feng X, Gryczke A, Kolter K, Langley N, Repka MA (2016) Hot melt extrusion as an approach to improve solubility, permeability and oral absorption of a psychoactive natural product, piperine. J Pharm Pharmacol 68(8):989–998

Atal CK, Dubey RK, Singh J (1985) Biochemical basis of enhanced drug bioavailability by piperine: evidence that piperine is a potent inhibitor of drug metabolism. J Pharmacol Exp Ther 232(1):258–262

Bae GS, Kim JJ, Park KC, Koo BS, Jo IJ, Choi SB, Lee CH, Jung WS, Cho JH, Hong SH, Song HJ, Shin YK, Park SJ (2012) Piperine inhibits lipopolysaccharide-induced maturation of bone-marrow-derived dendritic cells through inhibition of ERK and JNK activation. Phytother Res 26(12):1893–1897

Balakrishnan V, Varma S, Chatterji D (2001) Piperine augments transcription inhibitory activity of rifampicin by severalfold in Mycobacterium smegmatis. Curr Sci 80(10):1302–1305

Bang JS, Oh DH, Choi HM, Sur BJ, Lim SJ, Kim JY, Yang HI, Yoo MC, Hahm DH, Kim KS (2009) Anti-inflammatory and antiarthritic effects of piperine in human interleukin 1beta-stimulated fibroblast-like synoviocytes and in rat arthritis models. Arthritis Res Ther 11(2):R49

Bano G, Amla V, Raina RK, Zutshi U, Chopra CL (1987) The effect of piperine on pharmacokinetics of phenytoin in healthy volunteers. Planta Med 53(6):568–569

Bari SB, Adhikari S, Surana SJ (2012) Tyrosine kinase receptor inhibitors: A new target for anticancer drug development. J Pharma Sci Tech 1(2):36–45

Basile DP, Yoder MC (2014) Renal endothelial dysfunction in acute kidney ischemia reperfusion injury. Cardiovasc Hematol Disord Drug Targets 14(1):3–14

Bialer M, Levy RH, Perucca E (1998) Does carbamazepine have a narrow therapeutic plasma concentration range? Ther Drug Monit 20(1):56–59

Bolaños JP, Heales SJ, Land JM, Clark JB (1995) Effect of peroxynitrite on the mitochondrial respiratory chain: differential susceptibility of neurones and astrocytes in primary culture. J Neurochem 64(5):1965–1972

Browning RA, Smith JK, Brandon MT (1983) Comparison of regional brain 5-HT and 5-HIAA content in flexor and extensor rats. Pharmacol Biochem Behav 18(4):525–528

Bukhari IA, Pivac N, Alhumayyd MS, Mahesar AL, Gilani AH (2013) The analgesic and anticonvulsant effects of piperine in mice. J Physiol Pharmacol 64(6):789–794

Burchill MA, Yang J, Vogtenhuber C, Blazar BR, Farrar MA (2007) IL-2 receptor beta-dependent STAT5 activation is required for the development of Foxp3+ regulatory T cells. J Immunol 178 (1):280–290

Chatterjee PK, Patel NS, Sivarajah A, Kvale EO, Dugo L, Cuzzocrea S, Brown PA, Stewart KN, Mota-Filipe H, Britti D, Yaqoob MM, Thiemermann C (2003) GW274150, a potent and highly selective inhibitor of iNOS, reduces experimental renal ischemia/reperfusion injury. Kidney Int 63(3):853–865

Chavarria D, Silva T, Magalhães e Silva D, Remião F, Borges F, (2016) Lessons from black pepper: piperine and derivatives thereof. Expert Opin Ther Pat 26(2):245–264

Chávez AE, Chiu CQ, Castillo PE (2010) TRPV1 activation by endogenous anandamide triggers postsynaptic long-term depression in dentate gyrus. Nat Neurosci 13(12):1511–1518

Chen CY, Li W, Qu KP, Chen CR (2013) Piperine exerts anti-seizure effects via the TRPV1 receptor in mice. Eur J Pharmacol 714(1–3):288–294

Chen GY, Nuñez G (2010) Sterile inflammation: sensing and reacting to damage. Nat Rev Immunol 10(12):826–837

Cotgreave IA, Moldéus P, Orrenius S (1988) Host biochemical defense mechanisms against prooxidants. Annu Rev Pharmacol Toxicol 28:189–212

D'Hooge R, Pei YQ, Raes A, Lebrun P, van Bogaert PP, de Deyn PP (1996) Anticonvulsant activity of piperine on seizures induced by excitatory amino acid receptor agonists. Arzneimittelforschung 46(6):557–560

da Cruz GMP, Felipe CFB, Scorza FA, da Costa MAC, Tavares AF, Menezes MLF, de Andrade GM, Leal LKAM, Brito GAC, da Graça N-M, Cavalheiro EA, de Barros Viana GS (2013) Piperine decreases pilocarpine-induced convulsions by GABAergic mechanisms. Pharmacol Biochem Behav 104:144–153

Dama MS, Varshneya C, Dardi MS, Katoch VC (2008) Effect of trikatu pretreatment on the pharmacokinetics of pefloxacin administered orally in mountain Gaddi goats. J Vet Sci 9 (1):25–29

de la Rosa M, Rutz S, Dorninger H, Scheffold A (2004) Interleukin-2 is essential for CD4+CD25+ regulatory T cell function. Eur J Immunol 34(9):2480–2488

Derosa G, Maffioli P, Sahebkar A (2016) Piperine and Its Role in Chronic Diseases. Adv Exp Med Biol 928:173–184

Diplock AT (1982) The modulating influence of vitamin E in biological membrane unsaturated phospholipid metabolism. Acta Vitaminol Enzymol 4(4):303–309

Eksioglu EA, Mahmood SS, Chang M, Reddy V (2007) GM-CSF promotes differentiation of human dendritic cells and T lymphocytes toward a predominantly type 1 proinflammatory response. Exp Hematol 35(8):1163–1171

Elnaggar YS, Etman SM, Abdelmonsif DA, Abdallah OY (2015) Novel piperine-loaded Tween-integrated monoolein cubosomes as brain-targeted oral nanomedicine in Alzheimer's disease: pharmaceutical, biological, and toxicological studies. Int J Nanomedicine 10:5459–5473

Fidler IJ (1985) Immunomodulation of macrophages for cancer and antiviral therapy. Wiley , New York

Fidler IJ (1988) Targeting of immunomodulators to mononuclear phagocytes for therapy of cancer. Adv Drug Deliv Rev 2(1):69–106

Fozzard HA, Lee PJ, Lipkind GM (2005) Mechanism of local anesthetic drug action on voltage-gated sodium channels. Curr Pharm Des 11(21):2671–2686

Franco-Cereceda A, Lundberg JM (1985) Calcitonin gene-related peptide (CGRP) and capsaicin-induced stimulation of heart contractile rate and force. Naunyn Schmiedebergs Arch Pharmacol 331(2–3):146–151

Franco-Cereceda A, Lundberg JM, Saria A, Schreibmayer W, Tritthart HA (1988) Calcitonin gene-related peptide: release by capsaicin and prolongation of the action potential in the guinea-pig heart. Acta Physiol Scand 132(2):181–190

Fraser SP, Grimes JA, Djamgoz MB (2000) Effects of voltage-gated ion channel modulators on rat prostatic cancer cell proliferation: comparison of strongly and weakly metastatic cell lines. Prostate 44(1):61–76

Friese MA, Fugger L (2005) Autoreactive CD8+ T cells in multiple sclerosis: a new target for therapy? Brain 128(Pt 8):1747–1763

Fu M, Sun ZH, Zuo HC (2010) Neuroprotective effect of piperine on primarily cultured hippocampal neurons. Biol Pharm Bull 33(4):598–603

Gadermaier E, Levin M, Flicker S, Ohlin M (2014) The human IgE repertoire. Int Arch Allergy Immunol 163(2):77–91

Gasior M, Kleinrok Z, Czuczwar SJ (1995) Influence of BAY k-8644, a calcium channel agonist, on the anticonvulsant activity of conventional anti-epileptics against electroconvulsions in mice. Neuropharmacology 34(4):433–438

Gasparini C, Feldmann M (2012) NF-κB as a target for modulating inflammatory responses. Curr Pharm Des 18(35):5735–5745

George K, Thomas NS, Malathi R (2019) Piperine blocks voltage gated K(+) current and inhibits proliferation in androgen sensitive and insensitive human prostate cancer cell lines. Arch Biochem Biophys 667:36–48

Gibson HE, Edwards JG, Page RS, Van Hook MJ, Kauer JA (2008) TRPV1 channels mediate long-term depression at synapses on hippocampal interneurons. Neuron 57(5):746–759

Goligorsky MS, Brodsky SV, Noiri E (2002) Nitric oxide in acute renal failure: NOS versus NOS. Kidney Int 61(3):855–861

Guirgis HM (2015) The value of anticancer drugs in metastatic castrate-resistant prostate cancer: economic tools for the community oncologist. J Commun Support Oncol 13(10):362–366

Gunasekar PG, Kanthasamy AG, Borowitz JL, Isom GE (1995) NMDA receptor activation produces concurrent generation of nitric oxide and reactive oxygen species: implication for cell death. J Neurochem 65(5):2016–2021

Gupta M, Mazumder UK, Pal D, Bhattacharya S and Chakrabarty S (2003) Studies on brain biogenic amines in methanolic extract of *Cuscuta reflexa* Roxb stem and *Corchorus olitorius* Linn. seed treated mice, Acta-Polon Pharm Drug Res 60(3):207–210

Gupta SK, Velpandian T, Sengupta S, Mathur P, Sapra P (1998) Influence of piperine on nimesulide induced antinociception. Phytother Res 12(4):266–269

Gupta V, Jain UK (2011a) Estimation of piperine by UV-spectrophotometric method in herbal formulation, Pippli Churna. Int J Res Pharmaceut Biomed Sci 2(2):550–553

Gupta V, Jain UK (2011b) Quantitative analysis of piperine in ayurvedic formulation by UV Spectrophotometry. Int J Pharmaceut Sci Res 2(2):58–61

Gupta V, Jain UK (2011c) Status of piperine content in ayurvedic formulation: method standardization by HPTLC. Res J Pharmaceut Biol Chem Sci 2(2):524–532

Hunt JV, Dean RT, Wolff SP (1988) Hydroxyl radical production and autoxidative glycosylation. Glucose autoxidation as the cause of protein damage in the experimental glycation model of diabetes mellitus and ageing. *Biochem J* 256(1): 205–212

Hwang YP, Yun HJ, Kim HG, Han EH, Choi JH, Chung YC, Jeong HG (2011) Suppression of phorbol-12-myristate-13-acetate-induced tumor cell invasion by piperine via the inhibition of PKCα/ERK1/2-dependent matrix metalloproteinase-9 expression. Toxicol Lett 203(1):9–19

Janakiraman K, Manavalan R (2008) Studies on effect of piperine on oral bioavailability of ampicillin and norfloxacin. Afr J Tradit Complement Altern Med 5(3):257–262

Joe B, Lokesh BR (1994) Role of capsaicin, curcumin and dietary n-3 fatty acids in lowering the generation of reactive oxygen species in rat peritoneal macrophages. Biochim Biophys Acta 1224(2):255–263

Johri RK, Thusu N, Khajuria A, Zutshi U (1992) Piperine-mediated changes in the permeability of rat intestinal epithelial cells. The status of gamma-glutamyl transpeptidase activity, uptake of amino acids and lipid peroxidation. Biochem Pharmacol 43(7): 1401–1407

Kachel DL, Moyer TP, Martin WJ 2nd (1990) Amiodarone-induced injury of human pulmonary artery endothelial cells: protection by alpha-tocopherol. J Pharmacol Exp Ther 254(3):1107–1112

Karan RS, Bhargava VK, Garg SK (1988) Effect of trikatu (PIPERINE) on the pharmacokinetic profile of isoniazid in rabbits. Ind J Pharmacol 30(4):254–256

Karan S, Mishra S, Pal D (2012) Isolation of beta sitosterol and evaluation of antidiabetic activity of *Aristolochia indica* in alloxan induced diabetic mice with a special reference to *in vitro* antioxidant activity. J Med Plant Res 6(7):1219–1223

Kasibhatta R, Naidu MU (2007) Influence of piperine on the pharmacokinetics of nevirapine under fasting conditions: a randomised, crossover, placebo-controlled study. Drugs R D 8(6):383–391

Kauer JA, Gibson HE (2009) Hot flash: TRPV channels in the brain. Trends Neurosci 32(4):215–224

Kehrer JP, Klotz LO (2015) Free radicals and related reactive species as mediators of tissue injury and disease: implications for Health. Crit Rev Toxicol 45(9):765–798

Kerr BM, Thummel KE, Wurden CJ, Klein SM, Kroetz DL, Gonzalez FJ, Levy RH (1994) Human liver carbamazepine metabolism. Role of CYP3A4 and CYP2C8 in 10,11-epoxide formation. Biochem Pharmacol 47(11):1969–1979

Khan WN, Wright JA, Kleiman E, Boucher JC, Castro I, Clark ES (2013) B-lymphocyte tolerance and effector function in immunity and autoimmunity. Immunol Res 57(1–3):335–353

Koleva II, van Beek TA, Soffers AE, Dusemund B, Rietjens IM (2012) Alkaloids in the human food chain–natural occurrence and possible adverse effects. Mol Nutr Food Res 56(1):30–52

Korsmeyer SJ, Yin XM, Oltvai ZN, Veis-Novack DJ, Linette GP (1995) Reactive oxygen species and the regulation of cell death by the Bcl-2 gene family. Biochim Biophys Acta 1271(1):63–66

Kumar S, Singhal V, Roshan R, Sharma A, Rembhotkar GW, Ghosh B (2007) Piperine inhibits TNF-alpha induced adhesion of neutrophils to endothelial monolayer through suppression of NF-kappaB and IkappaB kinase activation. Eur J Pharmacol 575(1–3):177–186

Kurts C, Panzer U, Anders HJ, Rees AJ (2013) The immune system and kidney disease: basic concepts and clinical implications. Nat Rev Immunol 13(10):738–753

Lassmann H, Bradl M (2017) Multiple sclerosis: experimental models and reality. Acta Neuropathol 133(2):223–244

Li C, Wang Z, Wang Q, Ka Yan Ho RL, Huang Y, Chow MSS, Kei Lam CW, Zuo Z (2018) Enhanced anti-tumor efficacy and mechanisms associated with docetaxel-piperine combination- in vitro and in vivo investigation using a taxane-resistant prostate cancer model. Oncotarget 9(3):3338–3352

Li S, Wang C, Li W, Koike K, Nikaido T, Wang MW (2007) Antidepressant-like effects of piperine and its derivative, antiepilepsirine. J Asian Nat Prod Res 9(3–5):421–430

Ling H, Edelstein C, Gengaro P, Meng X, Lucia S, Knotek M, Wangsiripaisan A, Shi Y, Schrier R (1999) Attenuation of renal ischemia-reperfusion injury in inducible nitric oxide synthase knockout mice. Am J Physiol 277(3):F383-390

Liu Z, Hu Q, Wang W, Lu S, Wu D, Ze S, He J, Huang Y, Chen W, Xu Y, Lu W, Huang J (2020) Natural product piperine alleviates experimental allergic encephalomyelitis in mice by targeting dihydroorotate dehydrogenase. Biochem Pharmacol 177: 114000

Lutz MB, Kukutsch N, Ogilvie ALJ, Rößner S, Koch F, Romani N, Schuler G (1999) An advanced culture method for generating large quantities of highly pure dendritic cells from mouse bone marrow. J Immunol Methods 223(1):77–92

Makar TK, Nedergaard M, Preuss A, Gelbard AS, Perumal AS, Cooper AJ (1994) Vitamin E, ascorbate, glutathione, glutathione disulfide, and enzymes of glutathione metabolism in cultures of chick astrocytes and neurons: evidence that astrocytes play an important role in antioxidative processes in the brain. J Neurochem 62(1):45–53

Manley PW, Furet P, Bold G, Brüggen J, Mestan J, Meyer T, Schnell CR, Wood J, Haberey M, Huth A, Krüger M, Menrad A, Ottow E, Seidelmann D, Siemeister G, Thierauch KH (2002) Anthranilic acid amides: a novel class of antiangiogenic VEGF receptor kinase inhibitors. J Med Chem 45(26):5687–5693

Matough FA, Budin SB, Hamid ZA, Alwahaibi N, Mohamed J (2012) The role of oxidative stress and antioxidants in diabetic complications. Sultan Qaboos University Med J 12(1):5–18

Matyszak MK, Citterio S, Rescigno M, Ricciardi-Castagnoli P (2000) Differential effects of corticosteroids during different stages of dendritic cell maturation. Eur J Immunol 30(4):1233–1242

McNamara FN, Randall A, Gunthorpe MJ (2005) Effects of piperine, the pungent component of black pepper, at the human vanilloid receptor (TRPV1). Br J Pharmacol 144(6):781–790

Mishra A, Punia JK, Bladen C, Zamponi GW, Goel RK (2015) Anticonvulsant mechanisms of piperine, a piperidine alkaloid. Channels 9(5):317–323

Miyauchi T, Ishikawa T, Sugishita Y, Saito A, Goto K (1987) Effects of capsaicin on nonadrenergic noncholinergic nerves in the guinea pig atria: role of calcitonin gene-related peptide as cardiac neurotransmitter. J Cardiovasc Pharmacol 10(6):675–682

Miyauchi T, Ishikawa T, Sugishita Y, Saito A, Goto K (1988) Effects of piperine on calcitonin gene-related peptide (CGRP)-containing nerves in the isolated rat atria. Neurosci Lett 91 (2):222–227

Mohammadi M, Najafi H, Mohamadi Yarijani Z, Vaezi G, Hojati V (2019) Protective effect of piperine in ischemia-reperfusion induced acute kidney injury through inhibition of inflammation and oxidative stress. J Trad Complem Med

Moorthi C, Krishnan K, Manavalan R, Kathiresan K (2012) Preparation and characterization of curcumin-piperine dual drug loaded nanoparticles. Asian Pac J Trop Biomed 2(11):841–848

Mujumdar AM, Dhuley JN, Deshmukh VK, Raman PH, Thorat SL, Naik SR (1990) Effect of piperine on pentobarbitone induced hypnosis in rats. Indian J Exp Biol 28(5):486–487

Neumann M, Fries H, Scheicher C, Keikavoussi P, Kolb-Mäurer A, Bröcker E, Serfling E, Kämpgen E (2000) Differential expression of Rel/NF-kappaB and octamer factors is a hallmark of the generation and maturation of dendritic cells. Blood 95(1):277–285

Nimse SB, Pal D (2015) Free radicals, natural antioxidants and their reaction mechanisms. RSC Adv 5(35):27986–28006

Oberley LW (1988) Free radicals and diabetes. Free Radic Biol Med 5(2):113–124

Ochi H, Morita I, Murota S (1992) Mechanism for endothelial cell injury induced by 15-hydroperoxyeicosatetraenoic acid, an arachidonate lipoxygenase product. Biochim Biophys Acta 1136(3):247–252

Ouadid-Ahidouch H, Ahidouch A (2008) K+ channel expression in human breast cancer cells: involvement in cell cycle regulation and carcinogenesis. J Membr Biol 221(1):1–6

Ouadid-Ahidouch H, Ahidouch A (2013) K(+) channels and cell cycle progression in tumor cells. Front Physiol 4:220

Pal D, Nandi M (2005) CNS activities of *Celsia coromandeliane* Vahl in mice. Acta Polon Pharm Drug Res 62(5):355–361

Pal D, Kumar M, Chakrabarty P, Kumar S (2008) Evaluation of the antioxidant activity of aerial parts of *Cynodon dacvtylon*. Asian J Chem 20(3):2479–2481

Pal D, Sannigrahi S, Mazumder UK (2009) Analgesic and anticonvulsant effects of isolated saponin from the leaves of *Clerodendrum infortunatum* Linn. in mice. Indian J Exp Biol 47:743–747

Pal D (2013) Comparative analysis of *in vitro* antioxidant activity of two selected plants with a reference to antidiabetic profile. Asian J Chem 25(4):2165–2169

Pal D, Mazumder UK (2014) Isolation of compound and studies on CNS depressant activities of *Mikania scandens* Wild with special emphasis to brain biogenic amines in mice. Indian J Exp Biol 52:1186–1194

Paarakh PM, Sreeram DC, D SS, Ganapathy SP, (2015) In vitro cytotoxic and in silico activity of piperine isolated from Piper nigrum fruits Linn. Silico Pharmacol 3(1):9

Patel SG, Patel JK, Patel K (2012) Comparison and quantification of marker compound of triphala guggulu by using HPTLC method. Am J Pharm Tech Res 2(4):999–1013

Pei YQ (1983) A Review of Pharmacology and Clinical Use of Piperine and Its Derivatives. Epilepsia 24(2):177–182

Pooja S, Agrawal RP, Nyati P, Savita V, Phadnis P (2007) Analgesic activity of Piper nigrum extract per se and its interaction with diclofenac sodium and pentazocine in albino mice. Int J Pharmacol 5(1):30

Pradeep CR, Kuttan G (2003) Effect of piperine on the inhibition of nitric oxide (NO) and TNF-alpha production. Immunopharm Immun 25(3):337–346

Pradeep CR, Kuttan G (2004) Piperine is a potent inhibitor of nuclear factor-kappaB (NF-kappaB), c-Fos, CREB, ATF-2 and proinflammatory cytokine gene expression in B16F–10 melanoma cells. Int Immunopharm 4(14):1795–1803

Prevarskaya N, Skryma R, Bidaux G, Flourakis M, Shuba Y (2007) Ion channels in death and differentiation of prostate cancer cells. Cell Death Differ 14(7):1295–1304

Quéméneur L, Gerland LM, Flacher M, Ffrench M, Revillard JP, Genestier L (2003) Differential control of cell cycle, proliferation, and survival of primary T lymphocytes by purine and pyrimidine nucleotides. J Immunol 170(10):4986–4995

Rabb H, Griffin MD, McKay DB, Swaminathan S, Pickkers P, Rosner MH, Kellum JA, Ronco C (2016) Inflammation in AKI: current understanding, key questions, and knowledge gaps. J Am Soc Nephrol 27(2):371–379

Raman G, Gaikar VG (2002) Microwave-Assisted Extraction of Piperine from *Piper nigrum*. *Indus*. Eng Chem Res 41:2521–2528

Raol YH, Brooks-Kayal AR (2012) Experimental models of seizures and epilepsies. Prog Mol Biol Transl Sci 105:57–82

Rathod SS (2014) Extraction of piperine from Piper longum using ultrasound. Industr Crops Prod 58:259–264

Rauscher FM, Sanders RA, Watkins Iii JB (2000) Effects of piperine on antioxidant pathways in tissues from normal and streptozotocin-induced diabetic rats. J Biochem Mol Toxicol 14 (6):329–334

Reddy AC, Lokesh BR (1992) Studies on spice principles as antioxidants in the inhibition of lipid peroxidation of rat liver microsomes. Mol Cell Biochem 111(1–2):117–124

Reen RK, Singh J (1991) In vitro and in vivo inhibition of pulmonary cytochrome P450 activities by piperine, a major ingredient of piper species. Indian J Exp Biol 29(6):568–573

Ren T, Hu M, Cheng Y, Shek TL, Xiao M, Ho NJ, Zhang C, Leung SSY, Zuo Z (2019a) Piperine-loaded nanoparticles with enhanced dissolution and oral bioavailability for epilepsy control. Eur J Pharm Sci 137:104988

Ren T, Yang M, Xiao M, Zhu J, Xie W, Zuo Z (2019b) Time-dependent inhibition of carbamazepine metabolism by piperine in anti-epileptic treatment. Life Sci 218:314–323

Ren T, Zuo Z (2019) Role of piperine in CNS diseases: pharmacodynamics, pharmacokinetics and drug interactions. Expert Opin Drug Metab Toxicol 15(10):849–867

Ridderstad Wollberg A, Ericsson-Dahlstrand A, Juréus A, Ekerot P, Simon S, Nilsson M, Wiklund SJ, Berg AL, Ferm M, Sunnemark D, Johansson R (2014) Pharmacological inhibition of the chemokine receptor CX3CR1 attenuates disease in a chronic-relapsing rat model for multiple sclerosis. Proc Natl Acad Sci U S a 111(14):5409–5414

Riol-Blanco L, Sánchez-Sánchez N, Torres A, Tejedor A, Narumiya S, Corbí AL, Sánchez-Mateos P, Rodríguez-Fernández JL (2005) The chemokine receptor CCR7 activates in dendritic cells two signaling modules that independently regulate chemotaxis and migratory speed. J Immunol 174(7):4070–4080

Rode S, Bhujbal P, Sandhya P (2013) Quantification of piperine and 18-β glycyrrhetinic acid in herbal formulation "Eladi Gutika." Acta Chromatographica 25(1):135–146

Rodgers G, Doucette CD, Soutar DA, Liwski RS, Hoskin DW (2016) Piperine impairs the migration and T cell-activating function of dendritic cells. Toxicol Lett 242:23–33

Roshanbakhsh H, Elahdadi Salmani M, Dehghan S, Nazari A, Javan M, Pourabdolhossein F (2020) Piperine ameliorated memory impairment and myelin damage in lysolecethin induced hippocampal demyelination. Life Sci 253:117671

Rout K, Rout O, Mishra S (2007) Estimation of piperine in commercial Ayurvedic formulations. J Planar Chromatogr – Modern TLC 20(6): 447–450

Sachan N, Chandra P, Pal D (2015) Assessment of gastroprotective potential of *Delonix regia* (Boj Ex Hook) Raf against ethanol and cold restrain stress-induced ulcer in rats. Trop J Pharm Res 14(6):1063–1070

Saha S, Pal D (2016) Molecular Docking, synthesis, in vitro HDAC inhibition activity and antiproliferative activity against breast cancer cell line of hydroxyacetamide derivatives. Trop J Pharm Res 15(7):1401–1411

Sakagami Y, Mizoguchi Y, Seki S, Kobayashi K, Morisawa S, Yamamoto S (1989) Effect of arachidonate derivatives on cell viability in isolated rat hepatocytes: cytotoxicity of 15-hydroperoxyeicosatetraenoic acid. Gastroenterol Jpn 24(5):583

Samuel M, Oliver SV, Coetzee M, Brooke BD (2016) The larvicidal effects of black pepper (Piper nigrum L.) and piperine against insecticide resistant and susceptible strains of Anopheles malaria vector mosquitoes. Parasites Vectors 9(1): 238

Sánchez-Sánchez N, Riol-Blanco L, Rodríguez-Fernández JL (2006) The multiple personalities of the chemokine receptor CCR7 in dendritic cells. J Immunol 176(9):5153–5159

Sannigrahi S, Mazumder UK, Pal D, Mishra SL (2012) Terpenoids of methanol extract of *C. infortunatum* exhibit anticancer activity against Ehrlich's ascites carcinoma (EAC) in mice. Pharm Biol 50(3): 304–309

Sarkar B, Tiwari S, Patel R, Dubey G (2011) High performance thin layer chromatographic method for quantification of piperine in an ayurvedic formulation. Asian J Pharm Life Sci 1 (2):128–132

Schöbel N, Radtke D, Lübbert M, Gisselmann G, Lehmann R, Cichy A, Schreiner BS, Altmüller J, Spector AC, Spehr J, Hatt H, Wetzel CH (2012) Trigeminal ganglion neurons of mice show intracellular chloride accumulation and chloride-dependent amplification of capsaicin-induced responses. PLoS ONE 7(11):e48005

Schönherr R (2005) Clinical relevance of ion channels for diagnosis and therapy of cancer. J Membr Biol 205(3):175–184

Schulz JB, Weller M, Klockgether T (1996) Potassium deprivation-induced apoptosis of cerebellar granule neurons: a sequential requirement for new mRNA and protein synthesis, ICE-like protease activity, and reactive oxygen species. J Neurosci 16(15):4696–4706

Shailajan S, Sayed N, Joshi H, Tiwary B (2011) Standardization of an ayurvedic formulation: Trikatu Churna using bioanalytical tools. Int J Res Ayurveda Pharm 2(6):1676–1678

Shao B, Cui C, Ji H, Tang J, Wang Z, Liu H, Qin M, Li X, Wu L (2015) Enhanced oral bioavailability of piperine by self-emulsifying drug delivery systems: in vitro, in vivo and in situ intestinal permeability studies. Drug Deliv 22(6):740–747

Sigrist S, Franco-Cereceda A, Muff R, Henke H, Lundberg JM, Fischer JA (1986) Specific receptor and cardiovascular effects of calcitonin gene-related peptide. Endocrinology 119 (1):381–389

Singh A, Pawar VK, Jakhmola V, Parabia MH, Awasthi R, Sharma G (2010) In Vivo assessment of enhanced bioavailability of metronidazole with piperine in rabbits. Res J Pharm Biol Chem Sci 1:273–278

Singh M, Varshneya C, Telang RS, Srivastava AK (2005) Alteration of pharmacokinetics of oxytetracycline following oral administration of Piper longum in hens. J Vet Sci 6(3):197–200

Singh NK, Kumar P, Gupta DK, Singh S, Singh VK (2011) UV-spectrophotometric method development for estimation of piperine in Chitrakadi Vati. Der Pharmacia Lettre 3(3):178–182

Soutar DA, Doucette CD, Liwski RS, Hoskin DW (2017) Piperine, a Pungent Alkaloid from Black Pepper, Inhibits B Lymphocyte Activation and Effector Functions. Phytotherapy Research 31 (3):466–474

Sunila ES, Kuttan G (2004) Immunomodulatory and antitumor activity of Piper longum Linn. and piperine. J Ethnopharm 90(2):339–346

Thurman JM (2007) Triggers of inflammation after renal ischemia/reperfusion. Clin Immunol 123 (1):7–13

Tomson T, Almkvist O, Nilsson BY, Svensson JO, Bertilsson L (1990) Carbamazepine-10, 11-epoxide in epilepsy pilot study. Arch Neurol 47(8):888–892

Trinchieri G, Sher A (2007) Cooperation of Toll-like receptor signals in innate immune defence. Nat Rev Immunol 7(3):179–190

Unchern S, Saito H, Nishiyama N (1997) Selective cytotoxicity of piperine on cultured rat hippocampal neurons in comparison with cultured astrocytes: the possible involvement of lipid peroxidation. Biol Pharm Bull 20(9):958–961

Unchern S, Saito H, Nishiyama N (1998) Death of cerebellar granule neurons induced by piperine is distinct from that induced by low potassium medium. Neurochem Res 23(1):97–102

Vyas A, Jain V, Singh D, Singh M, Shukla SS, Pandey R, Swarnlata S, Saraf S (2011) TLC densitometric method for the estimation of piperine in ayurvedic formulation Trikatu churna. Orient J Chem 27(1):301–304

Williams HD, Trevaskis NL, Charman SA, Shanker RM, Charman WN, Pouton CW, Porter CJH (2013) Strategies to address low drug solubility in discovery and development. Pharmacol Rev 65(1):315

Wolff SP (1993) Diabetes mellitus and free radicals. Free radicals, transition metals and oxidative stress in the aetiology of diabetes mellitus and complications. Br Med Bull 49(3):642–652

Wonderlin WF, Strobl JS (1996) Potassium channels, proliferation and G1 progression. J Membr Biol 154(2):91–107

Woo HM, Kang JH, Kawada T, Yoo H, Sung MK, Yu R (2007) Active spice-derived components can inhibit inflammatory responses of adipose tissue in obesity by suppressing inflammatory actions of macrophages and release of monocyte chemoattractant protein-1 from adipocytes. Life Sci 80(10):926–931

Yoshida R, Nagira M, Kitaura M, Imagawa N, Imai T, Yoshie O (1998) Secondary lymphoid-tissue chemokine is a functional ligand for the CC chemokine receptor CCR7. J Biol Chem 273(12):7118–7122

Zanetti M, Castiglioni P, Schoenberger S, Gerloni M (2003) The role of relB in regulating the adaptive immune response. Ann N Y Acad Sci 987:249–257

Zaugg J, Baburin I, Strommer B, Kim H-J, Hering S, Hamburger M (2010) HPLC-based activity profiling: discovery of piperine as a positive GABAA receptor modulator targeting a benzodiazepine-independent binding site. J Nat Prod 73(2):185–191

Zozulya AL, Wiendl H (2008) The role of regulatory T cells in multiple sclerosis. Nat Clin Pract Neurol 4(7):384–398

Zsombok A, Bhaskaran MD, Gao H, Derbenev AV, Smith BN (2011) Functional plasticity of central TRPV1 receptors in brainstem dorsal vagal complex circuits of streptozotocin-treated hyperglycemic mice. J Neurosci 31(39):14024–14031

# Chapter 23
# Protein and Enzymes Isolated from Plant Sources and Their Utilization in Pharmaceutical Field

**Om Prakash Panda, Sitansu Sekhar Nanda, Dong Kee Yi, Dilipkumar Pal, and Souvik Mukherjee**

**Abstract** For transporting out original $B^{D}{}_{Y}$ functions, minerals, $V^{i}{}_{T}$, carbohydrates, $p^{R}{}_{t}$ and fibres necessitated, received from animal or $P^{l}{}_{T}$ sources or both. $p^{R}{}_{t}$ reckoned as important compounds among all nutrients for the $H^{u}{}_{M}$ $B^{D}{}_{Y}$ because they facilitated in cells to build up and tissues repair in the $B^{D}{}_{Y}$. The $B^{D}{}_{Y}$ used $p^{R}{}_{t}$ for energy production in the shortage of carbohydrates and fats, is essential for the $B_{u}{}^{Ld}$ of $M^{us}{}_{L}$ mass. An active $E^{z}{}_{m}$ is extracted from any living organism. Sources of $E^{z}{}_{m}$ are fungi, yeast, bacteria, animals and $P^{l}{}_{T}$. A very much larger number of $E^{z}{}_{m}$ is found its use in diagnosis and chemical analysis. Non-microbial sources provided a larger proportion of enzyme. $E^{z}{}_{m}$ prevailed from $P^{l}{}_{T}$ sources are bromelain, actinidin, ficin, a-amylase-amylase, papain, $L_{ip}{}^{OX}{}_{se}$. Application of $E^{z}{}_{m}$ finds its way in industries for food and beverage processing, animal feed, detergents biosensors, Pharmaceuticals, wastewater treatment and recent biofuels.

**Keywords** Protein · Nutrient building muscle mass · Enzyme · Chemical analysis · Chemical diagnosis · Bromelanin · Ficin · Lipooxygenase

## Abbreviations

| | |
|---|---|
| $A_{L}{}^{B}{}_{M}$ | Albumins |
| $A_{m}{}^{a}$ | Amino acid |
| $B^{D}{}_{Y}$ | Body |
| $B_{u}{}^{Ld}$ | Building |
| $C_{o}{}^{Nj}{}_{T}$ | Conjugated |
| $D^{I}{}_{se}$ | Diseases |
| $E^{z}{}_{m}$ | Enzymes |

O. P. Panda
Centurion University of Technology and Management, Balasore, Odisha, India

S. S. Nanda (✉) · D. K. Yi
Department of Chemistry, Myongji University, Yongin, South Korea
e-mail: nandasitansusekhar@gmail.com

D. Pal · S. Mukherjee
Department of Pharmaceutical Sciences, Guru Ghasidas Viswavidyalaya (A Central University), Bilaspur, Chhattisgarh 495 009, India



D. Pal and A. K. Nayak (eds.), *Bioactive Natural Products for Pharmaceutical Applications*, Advanced Structured Materials 140,
https://doi.org/10.1007/978-3-030-54027-2_23

| | |
|---|---|
| $F^{b}{}_{R}{}^{s}$ | Fibrous |
| $H_{IS}{}^{td}$ | Histidine |
| $H^{u}{}_{M}$ | Human |
| $I_{SL}{}^{e}$ | Isoleucine |
| $L^{e}$ | Leucine |
| $L_{ip}{}^{OX}{}_{se}$ | Lipooxygenase |
| $M^{us}{}_{L}$ | Muscles |
| $P_{E}{}^{pt}$ | Peptide |
| $P^{l}{}_{T}$ | Plants |
| $p^{R}{}_{t}$ | Proteins |
| $R_{e}{}^{at}$ | Reactions |
| $S_{e}{}^{D}$ | Seeds |
| $S_{ys}{}^{T}$ | System |
| $V^{a}{}_{L}$ | Valine |
| $V^{i}{}_{T}$ | Vitamin |

## 23.1 Introduction

Proteins ($p^{R}{}_{t}$) are the most abundant organic molecules of the living system ($S_{ys}{}^{T}$). The name protein is derived from the Greek word "proteios" meaning holding the first place. In 1838 the Dutch chemist Mulder used the term $p^{R}{}_{t}$ to denote the high molecular weight nitrogen-rich and most abundant substances present in animals and plants ($P^{l}{}_{T}$). The body ($B^{D}{}_{Y}$) regularly needs nutrients for carrying out original $B^{D}{}_{Y}$ functions, among these, $p^{R}{}_{t}$ are the most important compounds because they aid in building ($B_{u}{}^{Ld}$) cells and tissues for a human($H^{u}{}_{M}$) $B^{D}{}_{Y}$. A high $p^{R}{}_{t}$ richen diet is needed for $B_{u}{}^{Ld}$ $B^{D}{}_{Y}$ or muscles ($M^{us}{}_{L}$). If the $B^{D}{}_{Y}$ deficiencies in fat and carbohydrates, it creates energy production by using $p^{R}{}_{t}$ as they are requisite for $M^{us}{}_{L}$ mass $B_{u}{}^{Ld}$ $M^{us}{}_{L}$. It reduces the long polymer chain of $p^{R}{}_{t}$ from the amino acid ($A_{m}{}^{a}$) monomer. The $A_{m}{}^{a}$ monomers is coupled with the peptide ($P_{E}{}^{pt}$) bonds between the amino and carboxyl groups. It is applied in $p^{R}{}_{t}$ in industry and in the $H^{u}{}_{M}$ $B^{D}{}_{Y}$. So, an effort produced to review on enzymes and $p^{R}{}_{t}$ and their pharmaceutical application.

### 23.1.1 *Classification of $p^{R}{}_{t}$*

Based on the structure, composition, shape and solubility of $A_{m}{}^{a}$, $p^{R}{}_{t}$ are assorted into following classes: simple, fibrous ($F^{b}{}_{R}{}^{s}$), conjugated ($C_{o}{}^{Nj}{}_{T}$) and derived $p^{R}{}_{t}$. Simple $p^{R}{}_{t}$ are two types such as (a) Globular $p^{R}{}_{t}$ (Albumins ($A_{L}{}^{B}{}_{M}$), Globulins, Glutelins, Histones, Globins, Protamines, and Prolamins etc.), (b) $F^{b}{}_{R}{}^{s}$ $p^{R}{}_{t}$ (Collagens, Elastin, and Keratins etc.). The example of $C_{o}{}^{Nj}{}_{T}$ $p^{R}{}_{t}$ is Nucleo, Glyco,

Muco, Lipo, Phospho, and Metallo $p^R_t$ etc. However Derived $p^R_t$ are two types such as Primary $p^R_t$ (Coagulated $p^R_t$, Proteans, and Meta $p^R_t$ etc.), and Secondary $p^R_t$ (Proteoses, Peptones, Poly $P_E^{pt}$, $P_E^{pt}$ etc.). The simple $p^R_t$ only comprise of $A_m^a$ and are further classified in to: Globular $p^R_t$: These $p^R_t$ are oval or spherical shaped, soluble in water or in other solvents and digestible (Table 23.1). $F^b{}_R{}^s$ $p^R_t$: These are Fiber like shape, water- insoluble and cannot be digested. $A_L{}^B{}_M$ ous or sclero $p^R_t$ establish the major group of $F^b{}_R{}^s$ $p^R_t$. (Table 23.2) $C_o{}^{Nj}{}_T$ $p^R_t$: It makes these $p^R_t$ up of $A_m^a$ and non- $p^R_t$ moiety which is also known as a prosthetic or $C_o{}^{Nj}{}_T$ group (Table 23.3). Derived $p^R_t$**:** These $p^R_t$ are derived from simple and $C_o{}^{Nj}{}_T$ $p^R_t$ by denaturation or degradation process. These are further divided into primary derived $p^R_t$. It may also know as first hydrolyzed products of $p^R_t$. For example: Coagulated, Proteans, Meta $p^R_t$. Secondary Derived $p^R_t$: These $p^R_t$ are the degraded products of $p^R_t$ or the progressive hydrolytic products of $p^R_t$ undergoing hydrolysis. Examples include proteoses, peptones, poly $P_E^{pt}$ and $P_E^{pt}$ (Fig. 23.1).

**Table 23.1** Examples of Globular $p^R_t$

| $p^R_t$ | Properties | Examples |
|---|---|---|
| ALBMs | Coagulates on heating and soluble in water forming dilute salt solutions | Serum $A_L{}^B{}_M$, Ov $A_L{}^B{}_M$ (egg), Lact $A_L{}^B{}_M$ (milk) |
| Globulins | Soluble in neutral and dilute salt solutions | Serum globulins, Vitelline (egg yolk) |
| Glutelins | Soluble in dilute acids & alkalis and are mostly found in $P^l_T$ | Oryzenin (rice), Glutelin (wheat) |
| Prolamins | Soluble in 70% alcohol | Zein (maize), Gliadin (wheat) |
| Histones | Soluble in dilute acids & water but insoluble in dilute ammonium hydroxide | Thymus histones Histones of codfish sperm |
| Globins | Behaves like histones but is not basic and is not precipitated by ammonium hydroxide | Thymus histones, Histones of codfish sperm |
| Protamines | Strongly basic and resemble histones but smaller and soluble in ammonium hydroxide | Sperm $p^R_t$ |

**Table 23.2** Examples of $F^b{}_R{}^s$ $p^R_t$

| Examples | Properties |
|---|---|
| Collagens | Connective tissue $p^R_t$ but do not have tryptophan ($T_{ryp}{}^N$). On boiling with dilute acids or water, yield gelatin which is soluble and digestible |
| Elastin | $p^R_t$ of Elastic tissues such as arteries and tendons |
| Keratins | Present in an exoskeletal structure like hairs, horns, nails, etc. |

**Table 23.3** Examples of $C_o^{Nj}{}_T$ $p^{R}{}_{t}$

| $p^{R}{}_{t}$ | Properties | Examples |
|---|---|---|
| Nucleo $p^{R}{}_{t}$ | Here prosthetic group is nucleic acid, either RNA or DNA | Nucleohistones, nucleoprotamines |
| Glyco $p^{R}{}_{t}$ | Here prosthetic group is carbohydrate | Mucin (saliva), ovomucoid (egg white) |
| Lipo $p^{R}{}_{t}$ | Prosthetic group is lipid | Serum lipo $p^{R}{}_{t}$, membrane lipo $p^{R}{}_{t}$ |
| Phospho $p^{R}{}_{t}$ | Prosthetic group is phosphoric acid | Vitelline (egg yolk), Casein (milk) |
| Chromo $p^{R}{}_{t}$ | It colors prosthetic group substance | Hemoglobin, cytochromes |
| Metallo $p^{R}{}_{t}$ | $p^{R}{}_{t}$ contain metal ions such as Co, Cu, Mg, Fe, Zn | Carbonic anhydrase (Zn), Ceruloplasmin (Cu) |

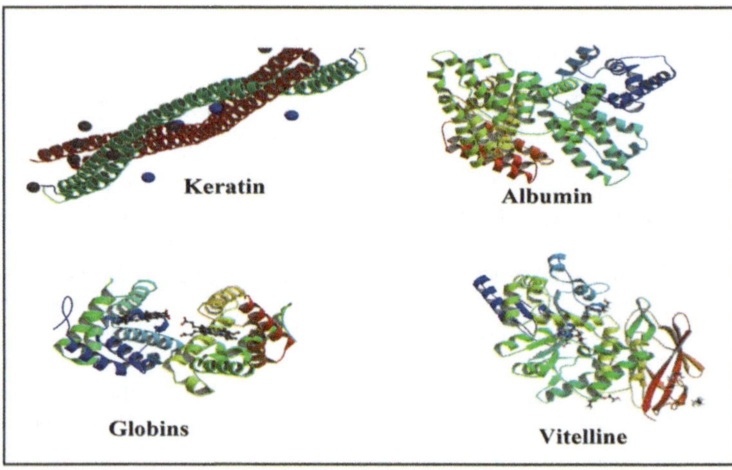


**Fig. 23.1** Structure of various proteins

## *23.1.2 Elemental Composition of Proteins*

$p^{R}{}_{t}$ are predominantly formed by five major elements in following proportion Hydrogen: 6–7.3%,

Carbon: 50–55%, Nitrogen: 13–19%, Oxygen: 19–24%, Sulfur: 0–4%**.** Essential $A_m{}^a$ are the chief constituents of $p^{R}{}_{t}$ found in naturally occurring organic compounds containing $A_m{}^a$ and carboxyl groups, and necessary for and animal growth and nutrition, Hence the source of essential $A_m{}^a$ is rich food consumption as a $H^{u}{}_{M}$ $B^{D}{}_{Y}$ does not produce them. $p^{R}{}_{t}$ or poly $P_E{}^{pt}$ are the substance made by $A_m{}^a$ which are joined by $P_E{}^{pt}$ bonds (Millward 1999). It highlights the roles of the various $A_m{}^a$ in Table 23.4.

**Table 23.4** $A_m{}^a$ and their biological role

| $A_m{}^a$ | Structure | Role |
|---|---|---|
| $I_{SL}{}^e$ | 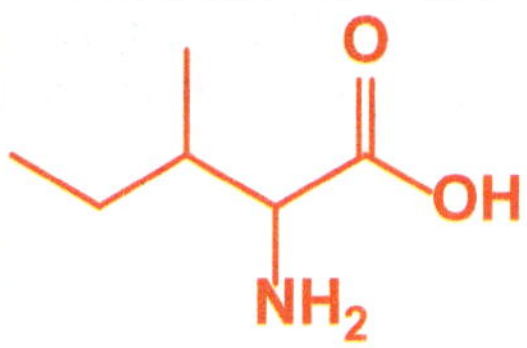 | Forms hemoglobin; prevents $M^{us}{}_L$ wasting in debilitated individuals |
| $L^e$ | 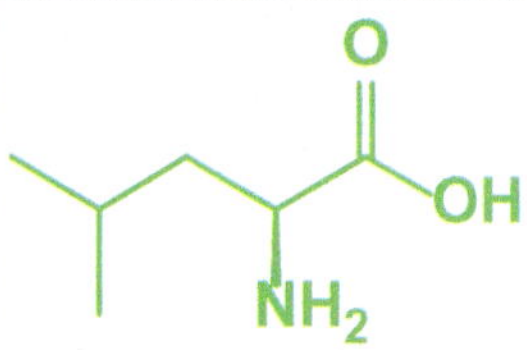 | Promotes healing of skin and broken bones; reduces $M^{us}{}_L$ $p^R{}_t$ breakdown |
| $V^a{}_L$ | 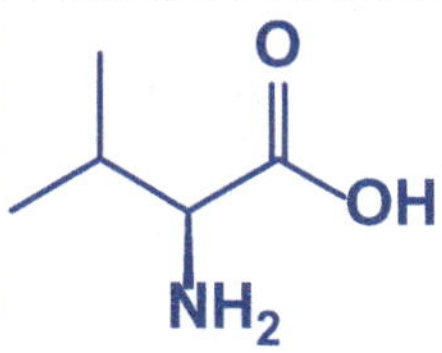 | Influences brain uptake of other neurotransmitter precursors ($T_{ryp}{}^N$, phenylalanine and tyrosine) |
| $H_{IS}{}^{td}$ | 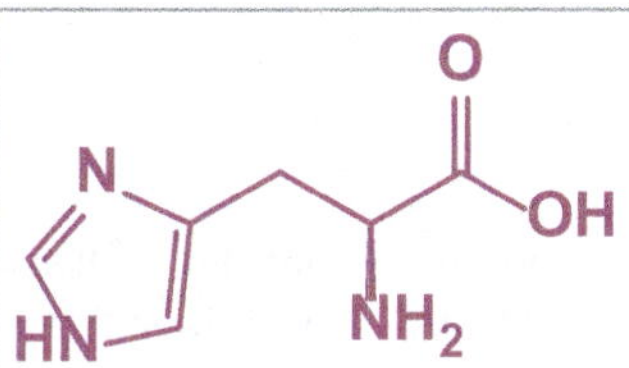 | Produces red and white blood cells; used in the treatment of anemia |
| Lysine | 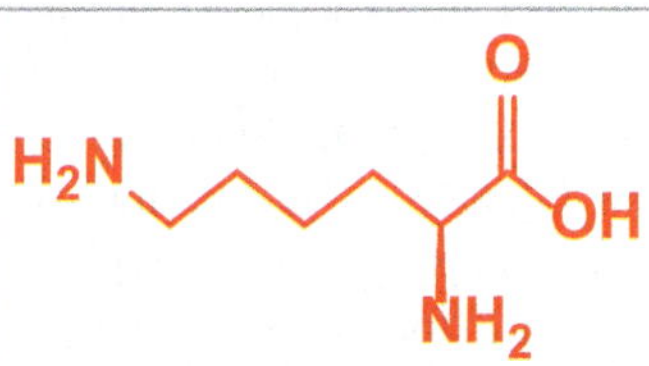 | Inhibits viruses; used in the treatment of herpes simplex; Lysine and vitamin ($V^i{}_T$)-c together form L-carnitine: a biochemical that enables $M^{us}{}_L$ tissue to use oxygen more efficiently, delaying fatigue |
| Methionine | 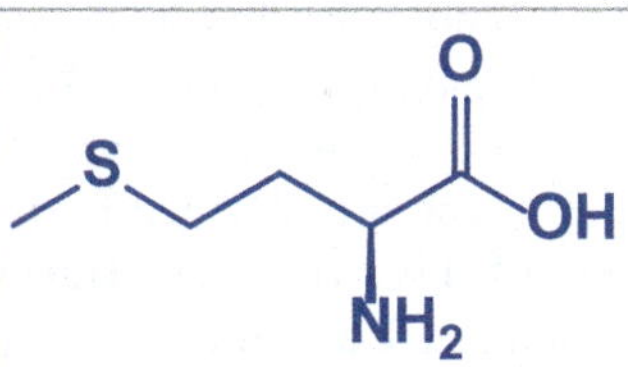 | Increases the antioxidant levels (glutathione) and reduces blood cholesterol level` |
| Phenylalanine | 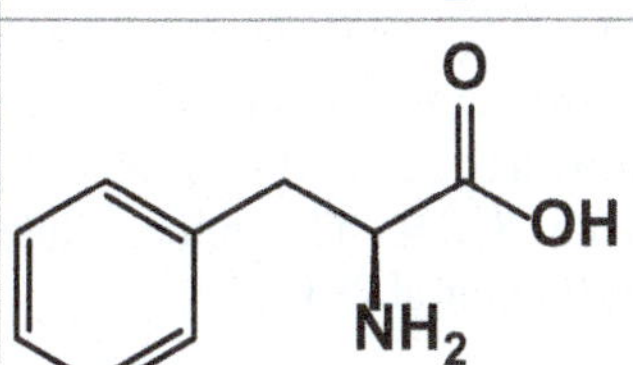 | Produces collagen, it acts as precursor of tyrosine; enhances learning, memory, mood and alertness |

(continued)

**Table 23.4** (continued)

| $A_m^a$ | Structure | Role |
|---|---|---|
| Threonine | 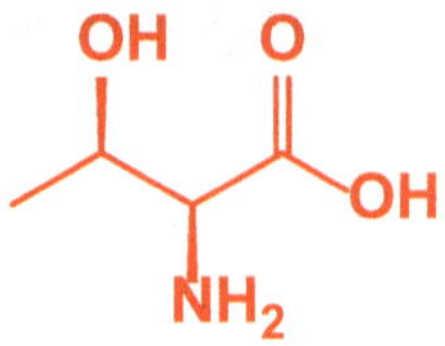 | Prevents fatty build up in the liver; it is also amino detoxifier |
| $T_{ryp}^N$ | 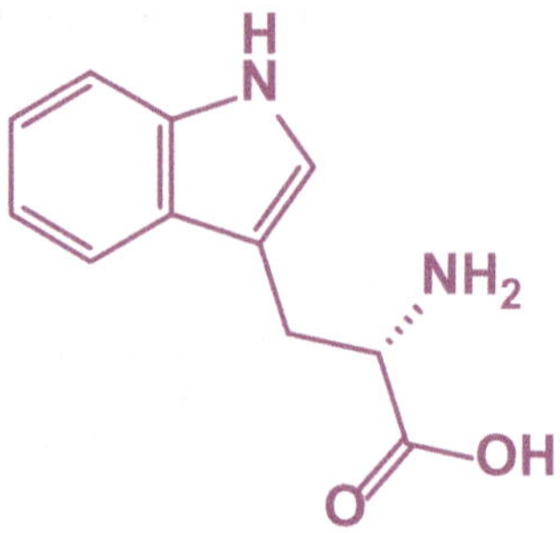 | Prevents fatty build up in the liver, it acts as precursor of key neurotransmitter serotonin; which exerts a calming effect |

### *23.1.3 Classification $A_m^a$*

They are classified on the basis of carbon chains present, nutritional requirement and polarity. Based on carbon chain present $A_m^a$ are of three types: Aliphatic $A_m^a$: These are further classified in to Neutral (Monoamine-Monocarboxylic Acid) for example, Glycine, Alanine, Serine, Threonine, Valine ($V^a_L$), Leucine ($L^e$) and Isoleucine ($I_{SL}^e$). Acidic (Monoamine-Dicarboxylic Acids) for example, Aspartic acid, Asparagine, Glutamic acid and Glutamine. Basic: for example, Arginine, Lysine and Hydroxyl sine. Sulphur containing $A_m^a$ for Example, Cysteine, Cystine (di cysteine) and Methionine, Aromatic $A_m^a$ for example, Phenylalanine, Tyrosine and Thyroxine. Heterocyclic $A_m^a$: for example, Proline, hydroxy Proline, $T_{ryp}^N$, Histidine ($H_{IS}^{td}$). Based on nutritional requirements amino $A_m^a$ are two types: Essential $A_m^a$: they are essentials for the body and these $A_m^a$ are also indispensable $A_m^a$. They are not synthesized in the $B^D_Y$ and are obtained from dietary sources. Examples; $V^a_L$ $I_{SL}^e$, $T_{ryp}^N$, Methionine, $L^e$, Phenylalanine, Threonine and lysine. Exceptionally $H_{IS}^{td}$ and Arginine are indicated as semi-essential $A_m^a$ as a little amount of them are synthesized in the $B^D_Y$. Deficiency of these $A_m^a$ may give adverse effects like retarded growth, weak immunity, early ageing, etc. Non-Essential $A_m^a$: These $A_m^a$ are also termed as dispensable $A_m^a$ and are synthesized in the $B^D_Y$. Examples; Glycine, Serine, Threonine, Glutamate, Cysteine, Glutamine, Proline, etc. (Table 23.5) (Fink et al.2012).

**Table 23.5** Essential and non-essential $A_m{}^a$

| Essential | Non-essential |
|---|---|
| $V^a{}_L$ | Glycine |
| $I_{SL}{}^e$ | Tyrosine |
| $T_{ryp}{}^N$ | Proline |
| Methionine | Cysteine |
| Arginine | Aspartic acid |
| $L^e$ | Alanine |
| Phenylalanine | Serine |
| Threonine | Hydroxyproline |
| Lysine | Cystine |
| $H_{IS}{}^{td}$ | Glutamic acid |

## 23.1.4 *Properties of $p^R{}_t$*

### 23.1.4.1 Solubility

The solubility properties of $p^R{}_t$ is like as follows: Forms Colloidal solutions in water (because of its enormous size), Solubility depends on electrostatic charges: identity net charge depends on number, pH of solvent, location of $A_m{}^a$ and pH of solvent. It depends upon isoelectric point (range 5–8.5): Isoelectric point depends on seven-charge $A_m{}^a$, example aspartate (β-carboxyl group), glutamate (δ-carboxyl group), cysteine (thiol group), $H_{IS}{}^{td}$ (imidazole side chains), tyrosine (phenol group), arginine (guanidinium group) and lysine (ammonium group).

### 23.1.4.2 Molecular Weight

$p^R{}_t$ molecular weight variation depends on the number of $A_m{}^a$ residues. Each $A_m{}^a$ contributes 110 value increases in a $p^R{}_t{}'$ molecular weight e.g., Myoglobin—1700; Insulin—5700; Hemoglobin—64,450. $p^R{}_t$ shape varies as globular (insulin), $F^b{}_R{}^s$ or elongated (fibrinogen), oval ($A_L{}^B{}_M$). For identification of $A_m{}^a$ present in $p^R{}_t$ the below given in Table 23.6 is recommended;

### 23.1.4.3 Chemical Nature of $p^R{}_t$

Most of the $p^R{}_t$ fold up to form a unique 3-D structure and the shape in order to which a $p^R{}_t$ naturally folds is its native conformation. Depending upon the chemical properties of $A_m{}^a$, most of the $p^R{}_t$ fold up accordingly while others fold up in to their native states with the help of molecular chaperones. Four original structure of $p^R{}_t$ have been introduced by the biochemists: (A) Primary, (B) Secondary, (C) Tertiary & (D) Quaternary.

**Table 23.6** Color $R_e^{at}$ of $p^R_t$

| $R_e^{at}$ | Observations | Specific group/$A_m^a$ |
|---|---|---|
| Biuret $R_e^{at}$ | Violet or purple color | Two $P_E^{pt}$ linkages |
| Ninhydrin $R_e^{at}$ | Blue color | $\alpha$-$A_m^a$ |
| Xanthoproteic $R_e^{at}$ | Orange color | Aromatic $A_m^a$ |
| Millions $R_e^{at}$ | Red color | Phenolic group |
| Hopkins-Cole | Violet ring | Indole ring |
| Sakaguchi | Intense red color | Guanidino group |
| Nitroprusside | – | Sulfhydryl groups (Cys) |
| Sulfur test | Black ppt | Sulfhydryl groups (Cys) |
| Pauly's test | – | Imidazole ring (His) |
| Folin–Coicalteau's test | – | Phenolic groups |

Primary Structure

In 1953 Frederick Sanger determined the complete sequence of $A_m^a$ of Insulin by studying the primary structure of $p^R_t$, this sequence can be determined. Actually, each $p^R_t$ has a unique sequence of $A_m^a$ which is determined by the genes contained in DNA. The primary structure of a $p^R_t$ is largely responsible for its function (Kyte and Doolittle 1982). The $A_m^a$ composition of a $p^R_t$ determines its physical and chemical properties. The primary structures formed during the translation (a process of $p^R_t$ biosynthesis) are held together by covalent $P_E^{pt}$ bonds. Depending on the nature of the free group on each terminal of the $P_E^{pt}$ chain, each extremely is referred as the carboxyl terminus (C-terminus) and the amino terminus (N-terminus). Numbering of residues always starts at the N-terminal end (–$NH_2$ group) from where the amino group is involved in a $P_E^{pt}$ bond. The primary structure of $p^R_t$ depends on the gene corresponding to the $p^R_t$ through transcription mRNA is synthesized by a particular sequence of nucleotides. The mRNA is further read by the ribosome in a process known as translation. The sequence of $p^R_t$ is specific for a $p^R_t$ which defines the function and structure of that $p^R_t$. For the determination of $A_m^a$ sequence in a $p^R_t$, they commonly use methods like Edman degradation or tandem mass spectrometry (Bender and Smith 1997).

Secondary Structure

It refers to the conformation of poly $P_E^{pt}$ chain by twisting or folding to a secondary structure. Two types of secondary structures are mainly identified i.e., $\alpha$ helix, & $\beta$ sheet.

(a) α helix

Pauling and Corey propose this in 1951, which is regarded as one milestone in the biochemistry research. In this structure, the poly $P_E^{pt}$ chains fold around the long axis in such a way that the -NH group of each $A_m^a$ bind to the –CO group of the fourth residue by a hydrogen bond, throughout the chain These hydrogen bonds are parallel to the long axis with the side chains protruding outward in a manner that each turn of α helix comprises 3.6 $A_m^a$ residues. Each $A_m^a$ residue is present at a distance of 0.15 nm from one another along the axis. Example of α helical $p^R_t$ structure is hair $p^R_t$ i.e., Keratin (Ikai 1980).

(b) β sheet

The β sheet also known as β -Pleated sheet is a common motif of regular secondary structure of $p^R_t$. Beta sheet consists of beta strands connected laterally by at least two or three backbone hydrogen bonds forming a twisted, pleated sheet. A beta strand is a stretch of poly $P_E^{pt}$ chain typically 3–10 $A_m^a$ long with backbone in an extended conformation. They have implicated the supramolecular association of beta sheets in formation of the $p^R_t$ aggregates and fibrils observed in many $H^u_M$ diseases ($D^I_{se}$), notably the amyloidosis such as Alzheimer's $D^I_{se}$'.

### Tertiary Structure

Tertiary structure represents the folding of $p^R_t$ that also controls the basic functioning of $p^R_t$. Stabilization of tertiary structure is achieved by non-local interactions, through salt bridges, hydrogen bonds, di sulphide bonds and post-translational modifications. Tertiary structure involves three-dimensional folding of the chain; stabilized by the bonding between the distant parts of the sequence. The long poly $P_E^{pt}$ chains are tightly folded in to a compact form because of the following interactions of -R groups of side chains of $A_m^a$.

### Hydrogen Bonding

Along with the hydrogen bonding between $P_E^{pt}$ bonds that give rise to the secondary structure of $p^R_t$, the hydrogen linkage may also be present between the $A_m^a$ side chains.

### Disulphide Bonding

The disulphide bonds cross-link with the poly $P_E^{pt}$ chain. The $p^R_t$ folds to bring two cysteine residues together thus the two -SH side chains oxidize and form a covalent disulphide (S–S) linkage (Mc Ardle et al. 2001).

Electrostatic/Ionic Bonding

The forces of attraction can stabilize the tertiary structure of $p^R{}_t$ between $A_m{}^a$ side chains of opposite charge. For example; an electrostatic bond is present between $-NH_3{}^+$ side chains of Leu and –COO side chains of Asp.

Hydrophobic Bonding

Sometimes $p^R{}_t$ folds so that the hydrophobic side chains of $A_m{}^a$ (e.g.; Gly, Ala, Val, Leu, lie, Pro, Met, Phe and Trp) are suppressed within the $p^R{}_t$, where they can interact to form hydrophobic linkage and thus stabilizes the structure of $p^R{}_t$.

#### 23.1.4.4 Quaternary Structure

Quaternary structure arises when two or more $p^R{}_t$ molecules interact with each other and form a large assembly or complex of $p^R{}_t$. Often, $p^R{}_t$ comprise multiple poly $P_E{}^{pt}$ chains and are mostly referred as $p^R{}_t$ subunits in this context. These $p^R{}_t$ subunits may have different (as in a heterodimer) or same (as in a homodimer) poly $P_E{}^{pt}$ chains. Thus, the quaternary structure is the interaction and arrangement of these $p^R{}_t$ with each other to form a larger aggregate $p^R{}_t$ complex. The $p^R{}_t$ with two or more poly $P_E{}^{pt}$ linked with non-covalent interactions are quaternary $p^R{}_t$, e.g.; Hemoglobin ($\alpha_2\beta_2$) (Fig. 23.2).

## 23.2 $p^R{}_t$ Sources: Animals and $P^l{}_T$

Meat, milk products, milk, egg, fish, and poultry are rich sources of $p^R{}_t$ containing $A_m{}^a$ with a balanced level. Legumes, $P^l{}_T$ food items, and nuts are also a source of the same. $P^l{}_T$ (vegetable) $p^R{}_t$ and animal $p^R{}_t$ are differentiated as: When animals' $p^R{}_t$ consumed in enormous amounts, it leads to high risks of $D^I{}_{se}$ like high blood pressure and heart $D^I{}_{se}$ because of high fat content in animals' $p^R{}_t$. Animal $p^R{}_t$ hollered complete $p^R{}_t$ because they balance it in a combination of all $A_m{}^a$; hence, $P^l{}_T$ (vegetable) $p^R{}_t$ is incomplete $p^R{}_t$; an exception is soya bean $p^R{}_t$. $p^R{}_t$ can find from vegetables, legumes and fruits whereas vegetable and fruits have less content of $p^R{}_t$ than legumes. It distributes $p^R{}_t$ in every part of the plants. $P^l{}_T$ are essential to isolate $p^R{}_t$ and useful for pharmaceutical industry (Tables 23.7 and 23.8).

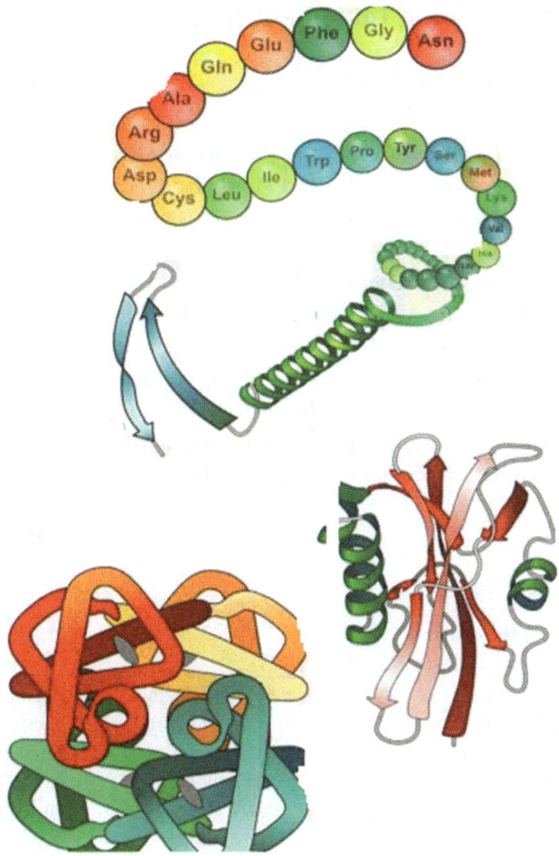


**Fig. 23.2** Structural idea of protein

**Table 23.7** Parts of $P^{l}_{T}$ as a source of $p^{R}_{t}$

| $p^{R}_{t}$ source as part of part of $P^{l}_{T}$ | Examples |
|---|---|
| Legume | Garbanzo beans, kidney beans, lentils, lima beans, navy beans, soybeans, split peas |
| Grain | Barley, brown rice, buckwheat, millet, oatmeal, quinoa, rye, wheat germ, wheat, wild rice |
| Vegetable | Barley, brown rice, buckwheat, millet, oatmeal, quinoa, rye, wheat germ, wheat, wild rice |
| Fruit | Apple, banana, cantaloupe, grape, grapefruit, honeydew, melon, orange, papaya, peach, pear, pineapple, strawberry, yangerine, watermelon |
| Nuts and seeds ($S_{e}^{D}$) | Almonds, cashews, filberts, hemp $S_{e}^{D}$, pea nuts, pumpkin, sesame, sunflower $S_{e}^{D}$, walnuts (black) |

## 23.3 Some Important $p^R_t$, Their Characteristics and Uses

### 23.3.1 Soybean

Soy $p^R_t$ obtained from the $P^l_T$ species Glycine max, family Fabaceae. It composed of conglycinin (140–170 kDa with glycosylated three subunits)–globular $p^R_t$ and Glycinin (340–375 kDa with six AB subunits comprising a basic [B] poly $P_E^{pt}$ and an acidic [A] linked via disulfide bonds). Based on the mole***cular*** weight ***a***nd sedimentation coefficient, it separates into fractions, 2S, 7S, 11S or 15S. The 7S globulin and 11S globulin. With other film forming coating combination, glycinin is known as the emulsifier, gelling agent, and foaming agent. B-conglycinin gets denatured at a temperature of 70–80 degree centigrade and also less stable than glycinin. The $p^R_t$ which are found from Soybean are called Soy $p^R_t$ are used to replace the animal $p^R_t$ in an individual diet. As discussed above the soybean is a legume that contains no cholesterol and is low in saturated fat. Soybeans are the

**Table 23.8** $P^l_T$ and their $p^R_t$

| $P^l_T$ (Scientific name) | Part of $P^l_T$ | $p^R_t$ | Pharmaceutical/Medicinal uses |
|---|---|---|---|
| Soybean (*Glycine max*) | $S_e^D$ | B-Conglycinin (7Sglobulin) | Potential diagnostic marker for severe allergic $R_e^{at}$ to soy |
| Pea (*Plsum sativum*) | $S_e^D$ | Glycinin (11sglobulin) α and β-pisavins | Emulsifying and surfactant properties, RNA N-glycosidase activity to release an Endo's fragment |
| Pea (*Pisum sativum var macrocarpon*) | $S_e^D$ | Plsumin, Legumin, Vicilin, $A_L{}^B{}_M$ | Antifungal, micro particle preparation |
| Peanuts (*Arachis hypogaea*) | $S_e^D$ | Hypogin | Antifungal |
| Rice (*Oryza sativa*) | $S_e^D$ | Glutenin, Globulin, $A_L{}^B{}_M$, Prolamin | Microencapsulation |
| Quinoa (*Chenopodium quinoa*) | $S_e^D$ | $A_L{}^B{}_M$, Globulins | Microencapsulation |
| Peach (*Prunus persica*) | Fruit | Thaumatin like $p^R_t$ | Protection against chilling injury in peach fruit |
| Almonds (*Prunus dulcis*) | $S_e^D$ | Amadin | Essential $A_m{}^a$ |
| Spinach (*Spinacia Oleracea*) | Leaves | Biotinyl $p^R_t$ | $V^i_T$ |
| Wheat (*Triticum aestivium*) | $S_e^D$ | Gluten | Microencapsulation |

only vegetable food that contains all eight essential $A_m{}^a$. These are also excellent sources of fiber, iron, calcium, zinc and B $V^i{}_T$.

### 23.3.2 *Wheat*

Based on solubility, the wheat $p^R{}_t$ fractions are classified as, albumins (water soluble), globulins (dilute salt solutions soluble), gliadins (soluble in 70–90% ethanol, glutenin (insoluble under all the previously mentioned conditions, comprise 34% of total $p^R{}_t$) and comprise 47% of the total $p^R{}_t$). Gliadin (40 kDa) have four distinct fractions, a single chain $P_E{}^{pt}$, and containing intermolecular disulfide bonds. These play an important role in strength, film formation, strength and elasticity. A mixture of $p^R{}_t$, Glutenin, has a molecular weight distribution between 100 and 1000 kDa. The strength of the $p^R{}_t$ matrix determined by the disulfide bonds present in gliadin and glutenin (Fig. 23.3).

### 23.3.3 *Corn Zein*

A class of prolamin $p^R{}_t$, Zein, found in maize (corn). It is a powder from corn gluten meal, one of the best $P^l{}_T$ $p^R{}_t$. Pure Zein is clear, tasteless, water-insoluble, odorless, hard, and edible. High percentages of non-polar $A_m{}^a$, like leucine (20%), glutamine (26%), proline (10%) and also basic and acidic $A_m{}^a$ in low proportions contained by Zein. Two major fractions of Zein are α-Zein, soluble in 95% ethanol and β-Zein soluble in 60% ethanol. Commercially, it is used for coating of tablets and in biodegradable packaging.

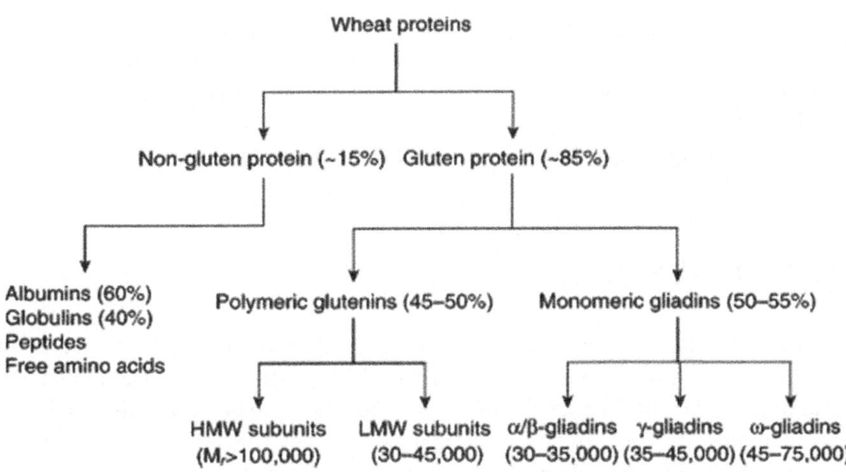


**Fig. 23.3** Composition of wheat protein

### 23.3.4 Pea $p^R{}_t$

It is food with a neutral taste that is used in diary alternative such as cheeses and yogurt. They extract it from the yellow pea, *Pisum sativum* and has a typical legume $A_m{}^a$ profile. Pea $p^R{}_t$ includes four major classes of $p^R{}_t$ such as globulin, albumin, prolamin and glutelin. Mainly globulin (65–80%) and some fractions of albumins, prolamins and Glutelins are extracted from pea $S_e{}^D$. Three different $p^R{}_t$ - legumin, vicilin and convicilin are comprising by globulin. The 11S globulin fraction with a molar mass between 350–400 kDa is represented by legumin, while the 7S globulin fraction with a molar mass of 150 kDa is represented by vicilin and convicilin. Pea $p^R{}_t$ can be found in energy bars, meal-replacement shakes and vegetarian burgers, etc.

### 23.3.5 Rice $p^R{}_t$

Rice is a vegetarian $p^R{}_t$ isolate that is an alternative to the more common soy and whey $p^R{}_t$ isolates. They can treat Brown rice with enzymes ($E^z{}_m$) that will cause carbohydrates to separate from $p^R{}_t$. The resulting powder is then sometimes added to smoothies or flavored or health shakes. Most other forms of $p^R{}_t$ powder, Rice $p^R{}_t$ powder, has a more distinct taste. Pea $p^R{}_t$ is high in lysine, low in cysteine and methionine. Thus, the combination of pea $p^R{}_t$ and rice offer a superior $A_m{}^a$, comparable to egg $p^R{}_t$ or diary, but without the intestinal issues or potential for allergies that some users have with these $p^R{}_t$. Compared with rice bran, the $p^R{}_t$ content in rice grains is slightly lower, varying from 6 to 15%, prepared by alkali extraction, followed by subcritical water treatment and by isoelectric precipitation. After sequential extraction, it has received the following distribution. About 6% albumin, 15% globulin, 3% prolamin and majorly 75% glutenin Albumin from egg white have foaming properties, found similar also from rice $p^R{}_t$; the emulsifying capacities of albumin from bovine serum (BSA) are significantly higher than those of rice $p^R{}_t$; isoelectric point at pH 4, minimum $p^R{}_t$ solubility occurred whereas maximum at pH 10; main $A_m{}^a$ content of rice $p^R{}_t$ is like that of soy and casein $p^R{}_t$; and denaturation temperature of the rice $p^R{}_t$ isolate is about 83.4 °C.

### 23.3.6 Sunflower $p^R{}_t$

Sunflower oil cakes are the source for major constituents of $p^R{}_t$. A high quantity of $p^R{}_t$, defatted sunflower flour, contains around 27% in dry weight. The hulled $S_e{}^D$ contains about 20–40% crude $p^R{}_t$. Four fractions of $p^R{}_t$ are present in the sunflower $p^R{}_t$: Albumins, 17–23% of total $p^R{}_t$; major Globulins, 55–60%; and two minor fractions, prolamins and glutelin, comprising 1–4% and 11–17% of the total $p^R{}_t$

fractions, respectively. It shows two major fractions: 2S albumins and 11S globulins (also named helianthinin). Helianthinin, a globular oligomeric $p^{R}{}_{t}$ with a molecular weight of 300–350 kDa and this $p^{R}{}_{t}$ mainly exists in the 11S hexametric form.

## 23.4 Isolation of $p^{R}{}_{t}$

They can use selective precipitation of $p^{R}{}_{t}$ as: Fractionate a subset of $p^{R}{}_{t}$ from a $p^{R}{}_{t}$ solution. Bulk method used major recovery of the $p^{R}{}_{t}$ from a crude lysate and recover a single $p^{R}{}_{t}$ of interest from a purification step.

### *23.4.1 Selective Precipitation Methods*

#### 23.4.1.1 Salting Out

They apply ammonium sulfate to form co precipitation with $p^{R}{}_{t}$ because the saturation concentration provides high molarity. It causes precipitation of most $p^{R}{}_{t}$, and does not have a large heat of solution. So, the generated heat gets easily dissipated; a saturated solution (4.40 M at 20 °C) of $p^{R}{}_{t}$ has a density of 1.235 g/cm$^3$, does not interfere with the precipitated $p^{R}{}_{t}$ sedimentation by centrifugation. It also has bacteriostatic properties and its concentrated solutions are protecting most $p^{R}{}_{t}$ from denaturation in solution state.

### *23.4.2 Isoionic Precipitation*

#### 23.4.2.1 Column Method

$p^{R}{}_{t}$ are less soluble and showing more precipitation when they are isoionic. The $p^{R}{}_{t}$ are their least hydrated conformation (a phenomenon that closely associated to the condition of $p^{R}{}_{t}$ at their isoelectric point) when they are isoionic salt- free state. Tan ford proposes the procedure applicable for this method. To determine the solubility of many $p^{R}{}_{t}$, it applies two important parameters:( a) Low salt concentration (0 to 0.1 to 0.2 M salt) (b) Solution pH regarding each $p^{R}{}_{t}$ Isoionic point. Column method can determine by adjusting the $p^{R}{}_{t}$ to their Isoionic pH to $p^{R}{}_{t}$ that remain soluble at their Isoionic point to strip away all salts from $p^{R}{}_{t}$ the mixed-bed deionization methods.

#### 23.4.2.2 Dialysis Method

For rendering $p^R{}_t$ become salt free (Isoionic), it is one of the oldest methods, however, two problems arise often with conventional dialysis: Because of presence of appreciable amount of $p^R{}_t$, the salt diffuses outward and swelling of the dialysis bag occurs, these total things for osmotic effects. It is uncertain where the Isoionic point is? If it is workable to deionize by dialysis against buffer. Automatically adjusts a $p^R{}_t$ precisely to its Isoionic pH without prior knowledge of the same by resin deionization method, so this method also known as Dintzis method. It traps the counter ions exchange between salt ions and $p^R{}_t$ through the membrane and outside in the exchanger resins. The precipitated $p^R{}_t$ in the dialysis tubing is recovered by centrifugation of the bag's contents when the exchange is completed. The dialysis technique is time consuming method as compare to flow-through column method.

### *23.4.3 Reversed-Phase High-Performance Liquid Chromatography (RP-HPLC)*

The $P_E{}^{pt}$ and $p^R{}_t$ obtained from synthetically or biologically are used for both preparations and analytical applications can isolate by a versatile technique known as RP-HPLC. Separation of molecule from the mobile phase to immobilized hydrophobic ligands attached to the stationary phase is the function of this technique. The stationary phase containing n-alkyl silica whereas acetonitrile containing an ionic modifier trifluoroacetic (TFA) used in a gradient analysis of RP-HPLC. Pharmaceutically important $p^R{}_t$ like; $P_E{}^{pt}$, globular $p^R{}_t$, and $p^R{}_t$ having molecular weight 10,000 undergo purification by this technique. Because of hydrophobicity of *n*-alkyl silica and acidic buffering $S_{ys}{}^T$ supports results into loss of biological activity of larger poly $P_E{}^{pt.}$ This technique is very limited in large scale of utilisation.

### *23.4.4 Mass Spectrometry of $p^R{}_t$*

Electrospray and matrix-assisted laser desorption ionization (MALDI) techniques are used because it is too difficult to ionize and transfer of large and polar biomolecules in to the gas phase. The major areas where the Mass spectrometry in proteomics are used (a) Identification of $p^R{}_t$, either in large scale proteomic ones or in classical biochemical projects (b) Characterization of macromolecule, recombinant $p^R{}_t$, and quality control of $p^R{}_t$ in biotechnology (c) It applies for detection of molecular weight of the $p^R{}_t$.

## 23.5 Application of $p^R_t$

They use $p^R_t$ as albumin-based nanoparticles, hydrogels, film coater, micro particles, as beads, as composites and also as $p^R_t$-based nano carriers in drug and gene delivery $S_{ys}^T$. Some examples are nanoparticles $S_{ys}^T$ for encapsulation and controlled delivery of bioactive compounds. When $p^R_t$ used as hydrogels, because of $p^R_t$-micelle structure, solubility of curcumin is enhanced, acts as a nano vehicle in the food industry. It also acts as vehicles for bioactive like milk $p^R_t$. Plsumin $p^R_t$ obtained from *Pisum sativum* acts as novel antifungal. Casein-derived four main bioactive $P_E^{pt}$ act on immune, cardiovascular, nervous $S_{ys}^T$, and nutrition $S_{ys}^T$ so applied as a source of bioactive $P_E^{pt}$. It uses vegetable $p^R_t$ in microencapsulation.

## 23.6 Introduction of $E^z_m$

It defines a catalyst as a substance that increases the velocity or rate of a chemical $\mathbf{R_e^{at}}$ without itself undergoing any changes in the overall process. $E^z_m$ are the biocatalyst or $p^R_t$ that catalyze the chemical $\mathbf{R_e^{at}}$. Synthesize $E^z_m$, $p^R_t$ in nature (except RNA acting as ribozyme), colloidal and thermo labile in properties and also specific in their action (Mauguet et al. 2002). French scientist Louis Pasteur in the year of 1850s states that fermentation of sugar in to alcohol by the yeast is catalyzed by ferments. As per suggestion, they cannot separate ferments from the structure of living yeast cells, this hypothesis of a scientist is called vitalism (Cho et al. 2004). It continues for a decade, later in the year 1897, Eduard Buchner finds that yeast extracts can ferment sugar solution in to alcohol and he also proves that chemicals which continues to function when is removed from the cells facilitates that fermentation (Renkema and Vliet 2002). Finally, Frederick gives name such chemical molecules $E^z_m$. James Sumner in 1926 isolates and crystallizes $E^z_m$ urease and he finds that crystals of urease are formed only of $p^R_t$, thus he postulates that chemically all $E^z_m$ are $p^R_t$. Because of the lack of other examples, this discovery remains controversial for years. This hypothesis is widely accepted, when John Northrop and Moses Kunitz crystallize pepsin, trypsin, and other digestive $E^z_m$ and chemically find them in $p^R_t$. Over the last decades of the twentieth century, enormous advances have been made like isolation purification, and structure elucidation of several $E^z_m$ have been done (Shukla and Cheryan 2002).

### 23.6.1 *Nomenclature and Classification of $E^z_m$*

All most all $E^z_m$ contain a $p^R_t$ backbone. In some $E^z_m$ $p^R_t$ are the major component in their structure whereas some $E^z_m$ also have $p^R_t$ and additional non-$p^R_t$ moieties, which may or may not take part in the catalytic activity of the $E^z_m$. We commonly encounter carbohydrates groups which are covalently attached to structural features, which have no catalytic activity but they support $E^z_m$ in stability and solubility (Edsall and Wyman 1958). Other factors which are associated in $E^z_m$ activity are cofactors metal ions and low molecular weight organic molecules i.e. co $E^z_m$. These may be loosely or tightly bound by monovalent or covalent forces, contributing to both $E^z_m$ activity and stability (Morr and Ha 1993). $E^z_m$ are named by adding the suffix "-ase" to substrate e.g.; DNA polymerase $E^z_m$ catalyzes polymerization of nucleotides to form DNA (Gonzalez-Perez and Vereijken 2007). Urease $E^z_m$ catalyzes urea hydrolysis, etc. Based on functions some $E^z_m$ are also named e.g. pepsin a digestive $E^z_m$, derived from Greek word Pepsis' means digestion; lysozyme responsible for cell wall lysis of bacteria (Ordonez et al. 2001). It derives some $E^z_m$ names from their source from where they are obtained, e.g. trypsin is produced by pancreatic tissue after rubbing it with glycerin hence is named after Greek word "tryein" meaning "to wear down". As the time passes away, several new $E^z_m$ are discovered, some $E^z_m$ have multiple names. To overcome these problems, they are classified $E^z_m$ based on the $\mathbf{R_e^{at}}$ they catalyze after the international agreement (Chandi and Sogi 2007). So, it classifies $E^z_m$ according to the report of a Nomenclature committee appointed by the International Union of Biochemistry. As a result, all $E^z_m$ get a formal E.C. ($E^z_m$ Commission) number and some have their trivial names (Hamada 2000). The $E^z_m$ commission (EC) numbers divide $E^z_m$ in order to six major groups according to the $\mathbf{R_e^{at}}$ catalyzed (Table 23.9).

### *23.6.2 Chemical Nature and Properties of $E^z_m$*

As discussed above, all $E^z_m$ are $p^R_t$, and even it is found that it also acts on some RNA molecules. Each $E^z_m$ has its own specific structure and specific conformation, which are essential things for catalytic activity (Garfin 2003). Holo $E^z_m$ which comprises of Apo $E^z_m$ (the $p^R_t$ part) and a co $E^z_m$ (non- $p^R_t$ organic part) is the functional unit of the $E^z_m$. The non-$p^R_t$ moiety, which is covalently bound with the Apo $E^z_m$ termed as prosthetic group, is the integral part of $E^z_m$ structure. Dialysis can separate the co $E^z_m$ from the $E^z_m$, whereas they cannot separate the prosthetic group (Aguilar 2004). If $E^z_m$ is made up of a single poly $P_E^{pt}$, it is referred as monomeric $E^z_m$, if over one poly $P_E^{pt}$ chain present, known as oligomeric $E^z_m$, e.g. lactate dehydrogenase. Multi $E^z_m$ complexes possessing $E^z_m$ have specific sites to catalyze uncommon $R_e^{at}$ in a sequence. Some process can regulate $E^z_m$ activity such as phosphorylation, glycosylation. These processes alter the structure of a $p^R_t$ found in the $E^z_m$. It requires $E^z_m$ in very minute quantity (Aguilar and Hearn 1996). It does not absorb them in the overall $E^z_m$ $R_e^{at}$, rather is received whole after completion of $R_e^{at}$. $E^z_m$ do not alter the equilibrium constant (K) of a $R_e^{at}$ but only increases the velocity of $R_e^{at}$.

**Table 23.9** International classification of $E^z{}_m$

| S. No. | Classes | Type of $R_e{}^{at}$ catalysed |
|---|---|---|
| 1 | Oxidoreductases | Transfer of electrons (Hydride ions or H atom) |
| 2 | Transferases | Group transfer $R_e{}^{at}$ |
| 3 | Hydrolases | Hydrolysis $R_e{}^{at}$ (Transfer of functional groups of water) |
| 4 | Lyases | Addition of groups to double bonds, formation of double bonds by removal of groups |
| 5 | Isomerases | Transfer of groups within molecules to yield isomeric forms |
| 6 | Ligases | Formation of C–C, C–S, C–O, and C–N bonds by condensation $R_e{}^{at}$ coupled to ATP cleavage |

### 23.6.3 Mechanism of $E^z{}_m$ Action

$E^z{}_m$ are the powerful catalyst; the nature of catalysis taking place in the biological $S_{ys}{}^T$ is like that of non-biological catalysis. In any $R_e{}^{at}$ the reactants have to be a triggered state or transition state. We know the energy required by the reactants to carry out $R_e{}^{at}$ as activation energy (Aguilar and Hearn 1996). When the reactant is heated then attain the energy. The catalyst reduces the activation energy ($E^z{}_m$ in the biological $S_{ys}{}^T$) and this causes the $R_e{}^{at}$ to proceed at a lower temperature. As we have discussed that $E^z{}_m$ do not alter the equilibrium constant, it just enhances the rate of $R_e{}^{at}$. The $E^z{}_m$ lowers energy barrier of reactants, making the $R_e{}^{at}$ go faster. The $E^z{}_m$ reduce the activation energy of the reactants in such a way that all the biological $S_{ys}{}^T$ occur at $B^D{}_Y$ temperature (below 40 °C) (Mant and Hodges 1996).

### 23.6.4 Important Industrial $E^z{}_m$ and Their Sources

It may extract $E^z{}_m$ those biologically active from any living organism. It uses a wide range of sources for commercial $E^z{}_m$ production from spinach to snake venom. More than a hundred $E^z{}_m$ being used industrially, it may prevail $E^z{}_m$ from $P^l{}_T$, animals, bacteria, fungi or yeast sources (Aguilar 2004). A very much larger number of $E^z{}_m$ find use in chemical analysis and chemical diagnosis (Lin and Karger 1990). They prefer microbes to $P^l{}_T$ and animals as a source of $E^z{}_m$. The details are remarked in the Tables 23.10, 23.11, 23.12, 23.13 and 23.14.

**Table 23.10** List of $E^z{}_m$ received from animal source

| $E^z{}_m$ | EC number | Source | Industrial use |
|---|---|---|---|
| Catalase | 1.11.1.6 | Liver | Food |
| Chymotrypsin | 3.4.21.1 | Pancreas | Leather |
| Lipase | 3.4.23.4 | Pancreas | Food |
| Rennet | 3.4.23.4 | Abomasum | Cheese |
| Trypsin | 3.4.21.4 | Pancreas | Leather |

**Table 23.11** List of $E^{z}_{m}$ received from $P^{l}_{T}$ source

| $E^{z}_{m}$ | EC number | Source | Industrial use |
|---|---|---|---|
| Actinidin | 3.4.22.14 | Kiwi fruit | Food |
| α-Amylase | 3.2.1.1 | Malted barley | Brewing |
| β-Amylase | 3.2.1.2 | Malted barley | Brewing |
| Bromelain | 3.4.22.4 | Pineapple latex | Brewing |
| β-Glucanase | 3.2.1.6 | Malted barley | Brewing |
| Ficin | 3.4.22.3 | Fig latex | Food |
| Lipoxygenase | 1.13.11.12 | Soybeans | Food |
| Papain | 3.4.22.2 | Pawpaw latex | Meat |

**Table 23.12** List of $E^{z}_{m}$ received from bacterial source

| $E^{z}_{m}$ | EC number | Source | Industrial use |
|---|---|---|---|
| α-Amylase | 3.2.1.1 | *Bacillus* | Starch |
| β-Amylase | 3.2.1.2 | *Bacillus* | Starch |
| Asparaginase | 3.5.1.1 | *Escherichia coli* | Health |
| Glucose isomerase | 5.3.1.5 | *Bacillus* | Fructose syrup |
| Penicillin amidase | 3.5.1.11 | *Bacillus* | Pharmaceuticals |
| Protease | 3.4.21.14 | *Bacillus* | Detergent |
| Pullulanase | 3.2.1.41 | *Klebsiella* | Starch |

**Table 23.13** List of $E^{z}_{m}$ received from fungal source

| $E^{z}_{m}$ | EC number | Source | Industrial use |
|---|---|---|---|
| α-Amylase | 3.2.1.1 | *Aspergillus* | Baking |
| Amino acylase | 3.5.1.14 | *Aspergillus* | Pharmaceutical |
| Glucoamylase | 3.2.1.3 | *Aspergillus* | Starch |
| Catalase | 1.11.1.6 | *Aspergillus* | Food |
| Cellulase | 3.2.1.4 | *Trichoderma* | Waste |
| Dextranase | 3.2.1.11 | *Penicillium* | Food |
| Glucose oxidase | 1.1.3.4 | *Aspergillus* | Food |
| Lactase | 3.2.1.23 | *Aspergillus* | Diary |
| Lipase | 3.1.1.3 | *Rhizopus* | Food |
| Rennet | 3.4.23.6 | *Mucor miehei* | Cheese |
| Pectinase | 3.2.1.15 | *Aspergillus* | Drinks |
| Pectin lyase | 4.2.2.10 | *Aspergillus* | Drinks |
| Protease | 3.4.23.6 | *Aspergillus* | Baking |
| Raffinase | 3.2.1.22 | *Mortierella* | Food |

**Table 23.14** List of $E^z{}_m$ received from yeast source

| $E^z{}_m$ | EC number | Source | Industrial use |
|---|---|---|---|
| Invertase | 3.2.1.26 | *Saccharomyces* | Confectionery |
| Lactase | 3.2.1.23 | *Kluvveromyces* | Dairy |
| Lipase | 3.1.1.3 | *Candida* | Food |
| Raffinase | 3.2.1.22 | *Saccharomyces* | Food |

## 23.6.5 $E^z{}_m$ Derived from $P^J{}_T$ Sources

### 23.6.5.1 Actinidin

Kiwi fruits are the source of actinidin $E^z{}_m$, which is a proteolytic $E^z{}_m$. Actinidin plays a vital role in aiding the digestive process (Mann et al. 2001). Kiwi fruits rather than actinidin also contain other $E^z{}_m$ which have no functions or little functions (Yamada et al.2008). There is also a wide range of $E^z{}_m$ involved in the ripening of kiwi fruit, particularly $E^z{}_m$ involved in polysaccharide and oligosaccharide metabolism and in the development of flavor and aroma compounds (Zengion and Yarnell 2011). Some $E^z{}_m$ influence flavor, texture and nutritional values, during storage, processing and preparation of kiwifruit (Caballero et al. 2003).

### 23.6.5.2 α-Amylase

The major source of alpha-amylase (α-Amylase) is malted barley. It is an oligosaccharide endoglycosidase having the character to cleave an internal glycosidic bond within a poly or oligosaccharide. It requires calcium for producing activity along with certain anions like chloride, phosphate and others (Ott and Lu 1991). α-Amylase can be produced from certain tissues, the forms found in serum are most often from the pancreas and salivary glands. α-Amylase can be found in a variety of $B^D{}_Y$ fluids and some $E^z{}_m$ are also found in urine in healthy individuals. The chief purpose in testing amylase, especially when the symptoms are present, is to diagnose pancreatitis and other primary and secondary pancreatic pathologies (Moridani and Bromberg 2003). We can make this more specific by testing amylase iso $E^z{}_m$ specific to pancreas. It also plays a vital role in diagnosing cancers other than pancreatic cancer like myeloma, ovarian cancer, etc. (Moriyama 2008).

### 23.6.5.3 β-Amylase

The key source of beta-Amylase (β -Amylase) is also malted barley. It is exo $E^z{}_m$ that able to cleave alpha (1, 4) linkage from the non-reducing end of the polymeric chains and release maltose, the disaccharide (Pal et al. 2020). It acts alone and can degrade amylose completely to maltose. This cannot attack the alpha (1, 6) linkages or alpha (1, 4) linkages close to the alpha (1, 6) links. β-Amylase is made up in endosperm in barley and so they do not synthesize it during germination (Calvo-Villas et al. 2007).

#### 23.6.5.4 Bromelain

The fundamental source of this $E^z{}_m$ is pineapple, it is a mixture of $E^z{}_m$ and has proteolytic activity (Pal et al. 2019a, b.b). The major function of bromelain is that it stimulates fibrinolysis by increasing plasmin and also prevent kinin production by inhibiting platelet aggregation because its mechanism of action is anti-inflammatory. So, it is used to treat a variety of pain and inflammation (Pal et al. 2020). When it is applied to reduce pain, it must be dispensed away from food because it acts as a digestive $E^z{}_m$ if consumed with food. It has also wound healing activity and a useful tool to shorten healing time post surgically and to reduce levels of edema, pain and ecchymoses (Mir et al. 2019).

#### 23.6.5.5 β–Glucanase

It is derived from Malted barley. It is also produced by Bacillus amyloliquefaciens. Fungal b-Glucans are also made by fungi of the Aspergillus group. It is also formed as a side activity pectinase preparation (Nayak et al. 2019). The structure of $P^l{}_T$ glucans differs from that of the fungal glucan: the former being composed of beta (1, 3)–(1, 4) structure which are hydrolyzed by the beta (1, 4) glucanase and the latter the fungal glucans presenting a beta (1, 3) backbone being hydrolyzed by the beta (1, 3) glucanase. In wine making, specific glucanase have been developed to hydrolyze Botrytis and yeast glucans in order to improve clarification and filterability of wines (Nayak et al. 2020).

#### 23.6.5.6 Ficin

It is derived from figs latex and a family of proteases known as the cysteine endo peptidases, it is mainly found in alcoholic beverages and acts as chill proofing agent for beer, meat tenderizer, dough conditioner, rennet substitute, processing aid for precooked cereals. It is one of the most commonly used for differentiating a good deal of blood group antigens: e.g. destroy M, N, S, Duffy a & Duffy b and enhance some other antigens (Gurjar and Pal 2020).

#### 23.6.5.7 Lipooxygenase ($L_{ip}{}^{OX}{}_{se}$)

In the year of 1932, Andre et al. find that the bean flavor in soybeans is mainly caused by lipoxygenase (LOX) and in 1947, Theorell et al. first extract $L_{ip}{}^{OX}{}_{se}$ crystals from soybeans. The $p^{R}{}_{t}$ content in soybean is about 40% and in mature $S_{e}{}^{D}$, $L_{ip}{}^{OX}{}_{se}$ accounts for 1–2% of total $p^{R}{}_{t}$ content (Pal et al. 2020). As compared to other $P^{l}{}_{T}$ the $L_{ip}{}^{OX}{}_{se}$ activity is higher in soybeans. Algae, baker's yeast, fungi and cyanogen bacteria also contain $L_{ip}{}^{OX}{}_{se}$. The soybean $S_{e}{}^{D}$ have three iso $E^{z}{}_{m}$ types i.e. LOX-1, LOX-2, LOX-3 and soybean leaves (young), flowers and immature pods have LOX-7 AND LOX-8 isozymes. LOX is an Oxido reductase, a non-heme iron-containing $p^{R}{}_{t}$ that specifically catalyzes polyunsaturated fatty acids with a cis-1, 4-pentadiene producing a hydrogen peroxide derivative having a $C_{o}{}^{Nj}{}_{T}$ double bond by intermolecular oxygenation. Though the active site of $L_{ip}{}^{OX}{}_{se}$ is not fully understood but active site group may contain iron, aromatic $A_{m}{}^{a}$ and methionine residue (Pal et al. 2019a; b).

#### 23.6.5.8 Papain

It is derived from the roots, leaves, latex, and fruits of the $P^{l}{}_{T}$ *Carica papaya* (Papaya). It catalyzes the breakdown of $p^{R}{}_{t}$ by hydrolysis (Pal et al. 2019a, b). It is useful for the analysis of $p^{R}{}_{t}$, in tenderizing meat, in clarifying beer. It is also used in toothpastes and cosmetics and also preparation of various remedies for indigestion, ulcers, fever and swelling (Saha and Pal 2020). Papain can trigger allergic $R_{e}{}^{at}$ in susceptible individuals. Skin $R_{e}{}^{at}$ may occur following contact with fresh latex from papaya. Hypersensitivity $R_{e}{}^{at}$ may be especially pronounced in persons allergic to latex (Sachan et al. 2019).

## 23.7 Pharmaceutical Applications

Many $E^{z}{}_{m}$ find many applications in pharmaceutical industries and various other industries. It uses $E^{z}{}_{m}$ for production of glucose syrups, crystalline glucose, high fructose corn syrups, maltose syrups, and food, detergent, paper and textile industries (Saha et al. 2017). It may be used in detergent industry as additives to remove starch-based stains. It uses textile industry amylases for warp sizing of textile fibers. In paper industry, it uses them for the reduction of starch viscosity for coating of paper. ${}^{E}{}_{zm}$ like proteoses, lipases or xylanases have a significant contribution in food industry. Bromelain is used in treatment of inflammation of soft tissues and edema because of surgery and injury. Papain is used in clarification of beverages and meat tenderizer; it also find its applications in cheese manufacture as a substitute of renin (Nayak et al. 2018). Papain is used as an anti-inflammatory agent; it has shown relieving symptoms of episiotomy. It uses $E^{z}{}_{m}$ like pancreatic as a digestive aid for converting starch in to dextrin and sugar (Soni et al.2017). Renin is used to coagulate milk and hence making the milk easily digestible for weak patients. $^{E}{}_{zm}$ is applied as a contact lens cleaner (Pal et al. 2019a, b). It is used to manipulate DNA in genetic engineering and detect the amount of glucose present in blood. It uses supplement for $E^{z}{}_{m}$ deficiencies. Prolactin $E^{z}{}_{m}$ is used to treat lactose intolerance (Nayak et al. 2016). Collagenase is used for skin ulcers. Asparaginase is used to treat leukemia.

## 23.8 Conclusion

This book chapter highlights the pharmaceutical importance of $p^{R}{}_{t}$ and $E^{z}{}_{m}$ isolated from the $P^{l}{}_{T}$ sources along with their classifications, structure and strategies used for isolation of $P^{l}{}_{T}$ $p^{R}{}_{t}$. It may be concluded that $P^{l}{}_{T}$ $p^{R}{}_{t}$ and $E^{z}{}_{m}$ are very much useful in the pharmaceutical field. The present review would help the scientific community to think about further modifications or development towards phytochemical screening and isolation of $p^{R}{}_{t}$ and $E^{z}{}_{m}$ prevailing in $P^{l}{}_{T}$ sources. The researcher and academicians will also be benefited towards achieving fundamental knowledges regarding enzymes and proteins.

## References


Aguilar MI (2004) HPLC of peptides and proteins. Springer, Totowa, pp 3–8

Aguilar MI, Hearn MT (1996) High resolution reversed phase performance liquid chromatography of peptides and proteins. Methods Enzymol 270:3–26

Bender W, Smith M (1997) Population: food and nutrition. Popul Bull 51:5–25

Caballero B, Trugo LC, Finglas PM (2003) Encyclopedia of food sciences and nutrition. Academic, London, pp 23–55

Calvo-Villas JM, Alvarez I, Carretera E, Espinosa J, Sicilia F (2007) Para neoplastic hyper amylasaemia in association with multiple myeloma. ACTA HAEMATOL-BASEL 117 (4):242–245

Chandi GK, Sogi DS (2007) Functional properties of rice bran protein concentrate. J Food Eng. 79:592–597

Cho SY, Rhee C, Wiss L (2004) Mechanical properties and water vapor permeability of edible films made from fractional soy proteins with ultrafiltration. U-Technol 37:833–839

Edsall JT, Wyman I (1958) Biophysical chemistry. Academic, London, pp 591–662

Fink NH, Mikesky AE, Burgoon LA (2012). Practical application in sports nutrition. Springer, Berin, pp 1–569

Garfin DE (2003) Gel electrophoresis of proteins. Essential Cell Biol 1(7):197–268

Gonzalez-Perez S, Vereijken JM (2007) Sunflower proteins: overview of their physicochemical, structural and functional properties. J Sci Food Agric 87:2173–2191

Gurjar VK, Pal D (2020) Design, in silico studies, and synthesis of new 1, 8-naphthyridine-3-carboxylic acid analogues and evaluation of their H1R antagonism effects. RSC Adv 10(23):13907–13921

Hamada JS (2000) Characterization and functional properties of rice bran proteins modified by commercial exoproteases and endoproteases. J Food Sci 65:305–310

Ikai A (1980) Thermo stability and aliphatic index of globular proteins. J Biochem 88:1895–1898

Kyte J, Doolittle RF (1982) A simple method for displaying the hydropathic character of a protein. J Mol Biol 157:105–132

Lin S, Karger BL (1990) Reversed phase chromatographic behavior of proteins in different unfold states. J Chromatogram 499:89–102

Mann M, Hendrickson RC, Pandey A (2001) Analysis of proteins and proteomes by mass spectrometry. Annu. Rev. Biochem 70(1):437–473

Mant CT, Hodges RS (1996) Analysis of peptides by high performance liquid chromatography. Methods Enzymol 271:3–50

Mauguet MC, Legrand J, Brujes L, Carnelle G, Larre C, Popineau Y (2002) Gliadin Matrices for microencapsulation processes by simple coacervation method. J Microencapsul 19:377–384

Mc Ardle WD, Katch FI, Katch VL (2001) Exercise physiology: energy, nutrition, and human performance. Lippincott Williams and Wilkins, Philadelphia, pp 47–81

Millward DJ (1999) The nutritional value of plant-based diets in relation to human amino acid and protein requirements. Proc Nutr Soc 58:249–260

Mir SA, Mukherjee S, Makar S, Pal G (2019) Cucurbitacins a vibrant triterpenoid: a review on its anticancer property. Pharma Tutor 7(2):43–54

Moridani MY, Bromberg IL (2003) Lipase and pancreatic amylase versus total amylase as biomarkers of pancreatitis: an analytical investigation. Clin Biochem 36(1):31–43

Moriyama T (2008) Sialyl salivary-type amylase associated with ovarian cancer. Clin Chim Acta 391(2):106–111

Morr CV, Ha EY (1993) Whey protein concentrates and isolates: processing and functional properties. Crit Rev Food Sci Nutr 33:431–476

Nayak AK, Bera H, Hasnain MS, Pal D (2018) Synthesis and characterization of graft copolymers of plant polysaccharides. Elsevier, Amsterdam, pp 1–62

Nayak AK, Hasnain MS, Pal D (2019) Natural polymers for pharmaceutical applications, plant-derived polymers. CRC Press, Boca Raton, pp 187–210

Nayak AK, Hasnain MS, Pal K, Banerjee I, Pal D (2020) Gum-based hydrogels in drug delivery. In: Biopolymer-based formulations. Elsevier, Amsterdam, pp 605–645

Nayak AK, Pal D, Santra K (2016) Swelling and drug release behavior of metformin HCl-loaded tamarind seed polysaccharide-alginate beads. Int J Biol Macromol 82:1023–1027

Ordonez C, Asenjo MG, Benitez JL, Gonzalez JL (2001) Obtaining a protein concentrate from integral defatted sunflower flour. Bioresour Technol 78:187–190

Ott JA, Lu CJ (1991) Lipase isoforms and amylase isozymes: assays and applications in the diagnosis of acute pancreatitis. Clin Chem 37(2):361–368

Pal D, Ahuja C, Mukherjee S (2019) *Celsia coromandeliane* Vahl—a new biomarker with immense pharmaceutical applications. J Drug Deliv Ther 9(3):1109–1115

Pal D, Chandra P, Sachan N (2020) Sesame seed in controlling human health and nutrition. Nuts and Seeds in Health and Disease Prevention. Elsevier, Academic Press, London, pp 183–210

Pal D, Nayak AK, Saha S (2019b) Cellulose-based hydrogels: present and future. In: Natural bio-active compounds. Springer, Berin, pp 285–332

Renkema JM, Vliet TV (2002) Heat-induced gel formation of soy proteins at neutral pH. J Agric Food Chem 50:1569–1573

Sachan N, Shrivastav A, Pal D (2019) Evaluation of antidepressant activity of ethanolic extract of *Abies webbiana* and *Berberis aristata* in laboratory animals. J. Drug Deliv Ther 9(1):244–247

Saha S, Pal D (2020) Gymnemic acids: sources, properties, and biotechnological production. Springer, Berlin, pp 177–193

Saha S, Pal D, Kumar S (2017) Antifungal and antibacterial activities of phenyl and ortho-hydroxy phenyl linked imidazolyl triazolo hydroxamic acid derivatives. Inventi Rapid Med Chem 2:42–49

Shukla R, Cheryan MZ (2002) The industrial protein from corn. Ind Crops Prod 13:171–192

Soni S, Verma N, Verma A, Pandit JK, Nayak AK, Dubey J, Hedge RR, Pandit JK, Raj V, Mishra AK, Mishra A (2017) Gelucire based floating emulsion gel beads: a potential carrier for sustained stomach specific drug delivery. FARMACIA 65(1):142–152

Yamada Y, Nishizawa K, Yokoo M, Zhao H, Onishi K, Teraishi M, Utsumi S, Ishimoto M, Yoshikawa M (2008) Anti-hypertensive activity of genetically modified soybean seeds accumulating novokinin. Peptides 29(3):331–337

Zengion AH, Yarnell E (2011) Herbal nutritional supplements for painful conditions. In: Pain procedures in clinical practice. Springer, Berlin, pp 187–204

# Chapter 24
# Advances and Perspectives of Gamma-Aminobutyric Acid as a Bioactive Compound in Food

**Priti Jain and Mangesh S. Ghodke**

**Abstract** GABA is a novel potent bioactive amino acid, non-proteinaceous in nature and comprising of four carbon units. Its role as an inhibitory neurotransmitter of the neuronal cortex is well known and it had proven its effects on central nervous system. Recent studies reveal that this neurotransmitter is also able to act as a hypotensive agent, anti-diabetic and as tranquilizer. Besides this, GABA is also a bioactive compound of food, pharmaceutical, and feed industries. It is present in beans, pulses, milk, green, black, and oolong tea, as well as in fermented foods including kefir, yogurt, and tempeh and other dairy products. It is normally synthesized in the plant from Glutamate and nowadays various advanced strategies are being used for enhanced production of GABA. The reason for this may be accounted for use of GABA as a supplement to treat high blood pressure, stress, and anxiety, and sleep, as well as to stimulate the body's natural growth hormone, often by athletes. This chapter would focus on the production of GABA, its effects on humans as neurotransmitter along with the physiological effects being studied recently. The chapter will also cover various types of food containing GABA and later it deals with the strategies being adopted for large scale production of GABA.

**Keywords** Gamma-amino butyric acid · Glutamate · GAD · Lactobacilli · Lactic acid bacteria · Hypertension · Neurodegeneration · Obesity · Anti-diabetic

## 24.1 Introduction

The human nervous system comprises of more than 40 neurotransmitters. Few have excitatory roles while few have an inhibitory role. Out of them one of the important neurotransmitter is Gamma aminobutyric acid. Gamma aminobutyric acid commonly known as GABA is an amino acid, non-proteinaceous in nature and comprising of four carbon atoms. It is predominating inhibitory neurotransmitter of the

P. Jain (✉) · M. S. Ghodke
R C Patel Institute of Pharmaceutical Education and Research, Shirpur 425405, India
e-mail: pmj.grv@gmail.com

 
D. Pal and A. K. Nayak (eds.), *Bioactive Natural Products for Pharmaceutical Applications*, Advanced Structured Materials 140,
https://doi.org/10.1007/978-3-030-54027-2_24

Central nervous system. This amino acid is widely distributed not only in our physiological system but is also present in plants, animals, and micro-organisms where its role in Krebs cycle is known in plants and microbes while in animals as a neural signal neurotransmitter. Its wide applications in pharmaceutical, medical, nutraceuticals and recently for manufacturing of nylon and other compounds make this simple amino acid "GABA" very important (Dhakal et al. 2012; Dai-Hung and Sang 2019).

GABA binds to GABA receptors which are sub-classified as GABA-A and GABA-B. Another minor class GABA-C has also been reported and it has a major role in retinal signal processing. GABA is biosynthesized from L-glutamic acid by the decarboxylation reaction catalyzed by Glutamate decarboxylase (GAD, EC 4.1.1.15) and pyridoxal phosphate (PLP) that acts as a cofactor. GABA after biosynthesis gets packaged into synaptic vesicles by VGAT (vesicular GABA transporter) (Fig. 24.1), later released in the synaptic cleft and then binds to the GABA receptors on postsynaptic region which is either GABA-A or GABA-B. Though the main concern of this chapter is not to deal with GABA receptors, we are presenting in very brief these receptors for reader's concerns. Both receptors vary in their physiological and pharmacological properties. GABA-A receptors are ligand gated chloride channel receptors located post-synaptically while GABA-B is GPCR's located pre as well as post-synaptic. On one hand, GABA-A receptors are involved in mediating cardiovascular, anti-anxiety and anti-convulsive activities while on the other hand, GABA–B receptors are mainly indulged in depression and analgesic effects (Matsumoto 1989; Diana et al. 2014; Richard and Timothy 1999).

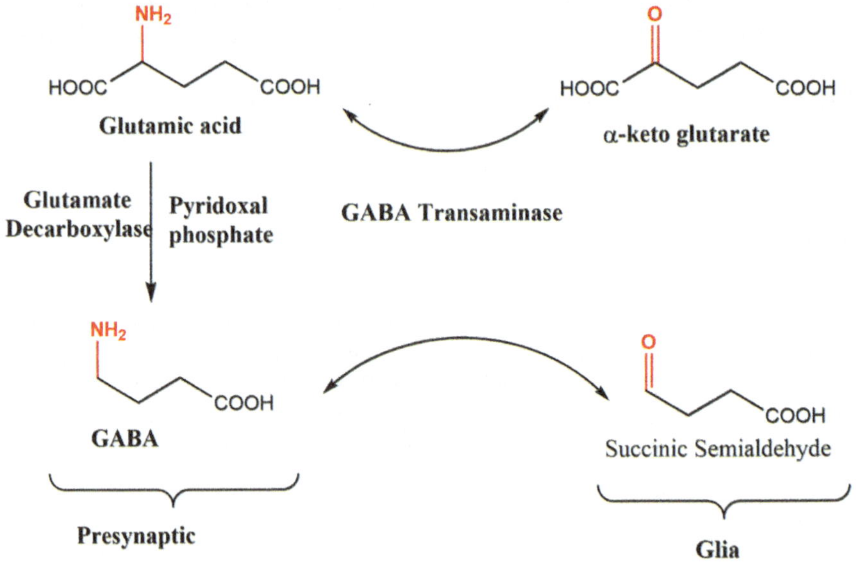


**Fig. 24.1** Biosynthesis of GABA and its fate in Glia

Alongside the binding to GABA receptors, some GABA undergoes reuptake and some reach the glia with the help of GABA transporters. GABA which reaches presynapse may be reused but those GABA entities that move to glia are metabolized to form succinic semialdehyde using GABA-Transaminase (GABA-T or EC 2.6.1.19) (Seigel et al. 1999; Rashmi et al. 2018).

### *24.1.1 Why is GABA Important?*

GABA is considered to be an important inhibitory neurotransmitter because of the pivotal role it plays in mammalian CNS. It has vital roles to play in neuronal development, synaptic activities, counteracting depression and sleep disorders (Ziskind-Conhaim 1998; Wong et al. 2003). Besides documented shreds of evidence for its activities on CNS, its innervation to other peripheral systems is also considered crucial. GABA has proven to be effective anti-hypertensive, anti-cancer, anti-oxidant, anti-obesity, anti-diabetic, tranquilizer, and many other effects are well documented (Sheng et al. 2012; Teresa 2013; Injae et al. 2019; Shimada et al. 2009). Hence, the pharmaceutical applications of GABA give it an edge over other neurotransmitters and its abundance in a variety of food products make it usable by humans for health benefits. The enzyme GAD that catalyzes the conversion of glutamate to GABA is found in many bacterial strains like *Escherchia*, *Aspergillus*, *Lactic acid bacteria*, *Neurospora* etc. (Tavakoli et al. 2015; Diana et al. 2014). This enzyme is also present in various plants such as soyabean, tomatoes, tea, germinated brown rice and also in few insects like flies, housefly and cockroaches (Diana et al. 2014; Nikmaram et al. 2017). Scientists have been struggling hard to utilize these microorganisms for GABA production through various technologies. The levels of GABA may vary from trace quantity to micromolar range depending on various factors. Despite of these low quantities present in plants, the pharmaceutical and food industries and recently other manufacturing industries are paying strong attention in increasing the concentration by various means. Several such techniques include anoxia, enzymatic treatment, germination, old conventional methods, microbial fermentation, use of excess glutamic acid and many more (Yongqi et al. 2018; Benincasa and Falcinelli 2019; Li et al. 2010a, b).

Very recent reports suggest that GABA is not only important from pharmaceutical and nutraceutical front but it is also important in industrial biotechnology due to its need as a precursor for production of polymer called "Nylon 4" (Park et al. 2013).

### *24.1.2 Alternative Synthetic Methods of GABA*

Since biosynthesis of GABA does not always provide sufficient levels, it is important to synthesize it by alternative methods. Hence, various synthetic methods are used for amplifying the GABA production. Few such synthetic methods are given below:

Usually, GABA is synthesized using 4-bromobutyric acid ester with pthalimide or ammonia as nitrogen source. Zhinyong et al. reported the synthesis from γ-butyrolactone with thionyl chloride in the presence of methanol which needs to be aminated with ammonium hydroxide as the Nitrogen source. (Scheme 24.1) (Zhiyong et al. 2013; Mohamed et al. 2019).

Another simple method was reported Nudelman et al. They reacted 2-bromo propionic acid with radio labeled cyano compound, which on reduction with platinum dioxide gave desired product (Scheme 24.2) (Song et al. 2014; Nudelman et al. 2008).

Hua and Cai filed a patent exploring simple synthesis for GABA. They used glutamic acid and converted it to glutaric anhydride. It was then treated with aqueous ammonia solution to produce imide which was then oxidized with sodium hypochlorite to produce the desired product. (Scheme 24.3) (Hua and Cai 1995; Mohamed et al. 2019).

Though many synthetic schemes have been proposed for GABA synthesis but none is suitable for large scale synthesis because of the corrosive nature of the chemicals and health-related hazards. Therefore, it is always preferred to use "**green synthesis**". Since the green synthetic approaches for GABA mainly use microorganisms, these methods are cumbersome because of the low levels present in

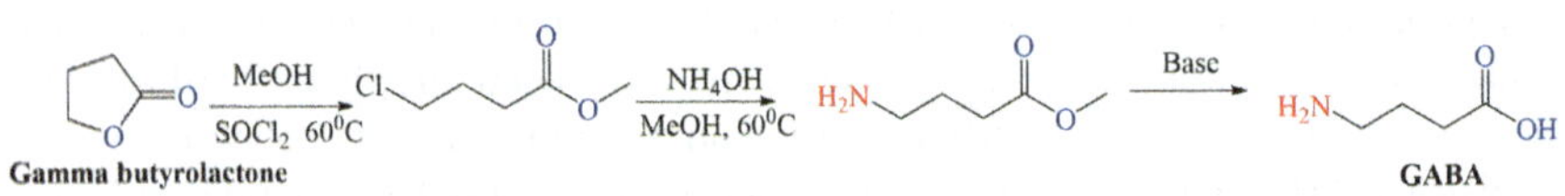


**Scheme 24.1** Synthesis of GABA using Gamma butyrolactone

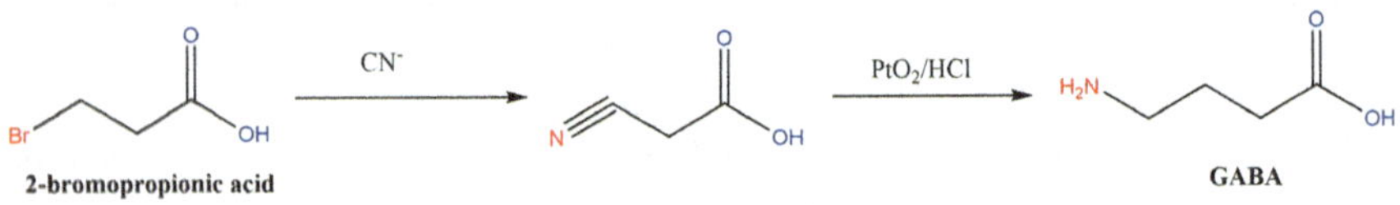


**Scheme 24.2** Synthesis of GABA using 2-bromopropionic acid

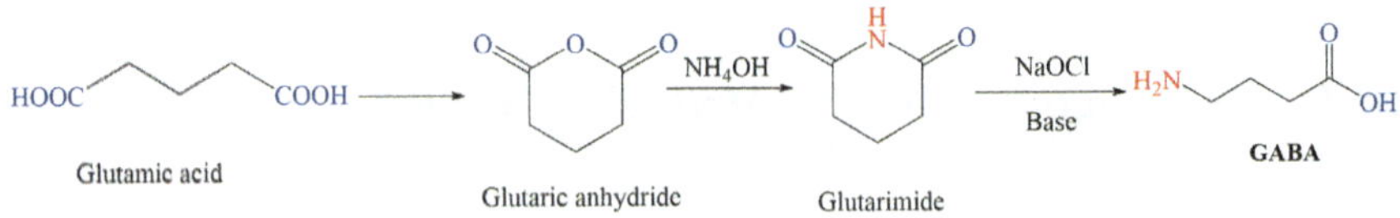


**Scheme 24.3** Synthesis of GABA using Glutamic acid

microbes but are efficient and safer than normal chemical synthetic methods. The greener synthesis will be discussed later in this chapter.

GABA, owing to its large profile of pharmacological activities and beneficial effects in humans, is regarded as a "BIOACTIVE COMPOUND". In this chapter, we will, therefore, discuss in detail the Pharmaceutical applications of GABA, its implication in various diseases, GABA as the bioactive compound in food and microbes and finally, we would discuss the advances in techniques being adopted for increased production of GABA.

## 24.2 Pharmaceutical Properties of GABA

### *24.2.1 Anti-Hypertensive Effect of GABA*

In recent years because of drastic changes in eating habits and decreased physical activity most of the people are associated with increased metabolic syndrome. The metabolic syndrome is a collection of interrelated risk factors that seem to promote atherosclerotic cardiovascular disease. The most commonly known metabolic risk factors are atherogenic dyslipidemia, elevated blood pressure, and elevated plasma glucose (Fig. 24.2) (Akama et al. 2009; Ebizuka et al. 2009).

GABA has been reported to reduce blood pressure in experimental animals and humans. In a study reported by Morio Shimada et al., the anti-hypertensive effect of GABA-rich *Chlorella* was given by oral administration for 12 weeks in subjects

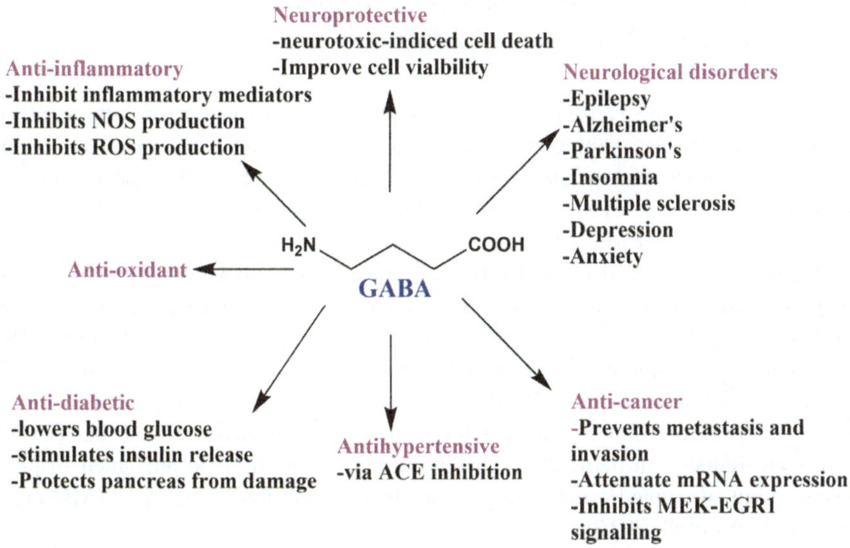


**Fig. 24.2** Different pharmaceutical roles of GABA

with high-normal blood pressure (130–139 mmHg for systolic blood pressure-SBP), 85–89 mmHg for diastolic blood pressure (DBP), and borderline hypertension (140–159 mmHg for SBP and 90–99 mmHg for DBP) in a placebo-controlled, double-blind study. In the GABA group, SBP significantly decreased from week 4 to week 12, and also on week 16 of follow-up compared to the baseline level ($p < 0.05$ or $p < 0.01$). Systolic blood pressure levels were significantly lower on weeks 8, 10, and 12 than those in the placebo group. Systolic blood pressure in the placebo group did not significantly change from the baseline throughout the study period. Diastolic blood pressure in the placebo group was not significant to the baseline throughout the study period. Though, mechanism for hypotensive action of GABA has not been fully explained; it is reasonable to consider that the anti-hypertensive effect of GABA is attributable to its vasodilator action or the inhibition of peripheral sympathetic nerve or a decrease in total vascular resistance, because, GABA poorly passes the blood–brain barrier due to its low lipid solubility. It is known that Angiotensin converting enzyme (ACE) plays essential role in regulating blood pressure. Several food supplements containing GABA are found to elicit antihypertensive effect by inhibiting ACE. Eg. *Lactococcus lactis* DIBCA2 and *Lactobacillus plantarum* PU11 fermented milk, soybean fermented by GABA enriched dairy products, GABA enriched rice grains, white rice, Chingshey purple sweet potato-fermented milk by lactic acid bacteria, GABA rich tomatoes, bread have also been reported to reduce BP in animal studies (Kawakami et al. 2018a, b; Nishimura et al. 2016; Tsai et al. 2013; Lin et al. 2012; Suwanmanon and Hsieh 2014; Aoki et al. 2003).

### 24.2.2 GABA as Neuroprotective Compound and for Neurological Disorders

Injuries and damage caused to nervous system leads to increased inflammatory responses which lead to release of inflammatory mediators. Several reactive oxygen species, cytokines, leukotrienes, nitric oxides are released which trigger neurodegeneration and may serve as a responsible factor for diseases like Alzheimer's, Parkinson's and sclerosis (Cho et al. 2007).

Many studies have been performed to find newer GABA containing food that may be used to treat neurodegeneration or reduce neurodegenerative effects. It is proven that GABA agonist protects neurons from inflammation and injury. The most common symptom observed in Parkinson's disease (PD) are visual hallucinations, which increases as the disease advances, and may range from simple flashes of light or color to complex hallucinations. It is also reported that GABA levels are reduced in dementia with lewy bodies. This may be attributed to poor visual input, disrupted connectivity with other visual areas, reduced GABAergic inhibition (Li et al. 2016).

Scientists harnessed magnetic resonance spectroscopy to govern GABA levels in occipital lobe and used 36 subjects with Parkinson's, 19 with and 17 without complex visual hallucinations, together with 20 healthy controls without hallucinations. Through this study they found lower GABA + /creatine in PD with visual hallucinations (0.091 ± 0.010) versus those without (0.101 ± 0.010) and controls (0.099 ± 0.010) (F2.49 = 4.5; $p$ = 0.016).

It was observed that there were widespread reductions in white matter integrity in the visual hallucinations group but was insignificant after controlling the cognition. Since GABA is reported to be involved in neurodegeneration, the GABA enriched food may directly or indirectly be beneficial in controlling the neurodegeneration, mainly by protecting through the toxins or injury (Li et al. 2016; Okada et al. 2000).

Besides Parkinson's, GABA also has shown to improve the cognitive abilities and suppress neurodegeneration. Eg. GABA enriched rice germ, Gaba-rich Monascus-fermented food, Gaba-enriched fermented Laminaria japonica was examined for the treatment of anxiety, sleeplessness, and depression, cognition (Yamatsu et al. 2016; Reid et al. 2018).

### *24.2.3 GABA as Anti-obesity Agent*

Reports suggest that the worldwide prevalence of obesity is high and constantly increasing. Hence, there is an urgent need for search of newer therapies to treat obesity-related pathologies (NCD Risk Factor Collaboration (NCD-RisC), 2017). Brown adipose tissues (BAT) were characterized as a thermogenic organ and as per the reports adult humans also possess functional BAT, and studies have shown that it is a metabolically active organ with the potential to regulate systemic metabolism. The metabolomic analyses reveal that GABA levels are increased in the interscapular BAT of mice with dietary obesity.

The study results reveal that constitutive activation of GABA/GABA-BR1 signaling causes BAT dysfunction and systemic metabolic derangement in conditions of obesity. The results also provide a substance that, in healthy volunteers, GABA-BR1 is significantly increased in individuals with low UCP1 expression. Obesity is linked to functional decline in BAT levels and metabolic stress reduces Ucp1 expression. It implicates that GABA-BR1 levels are related inversely to UCP1 in humans (Ippei 2018).

### *24.2.4 Antimutagenic and Antimicrobial Activities of γ-Aminobutyric Acid*

The most widely appreciated and consumed beverage in vast quantities worldwide is Tea. GABA tea is a special kind of tea enriched with GABA. Alternative cycles of anaerobic and aerobic conditions help to accumulate higher GABA levels. GABA is proposed to act as a natural relaxant and anti-anxiety compound, and its administration could concurrently enhance immunity under stress conditions. Jeng-Leun Mau et al. reported a study using GABA tea powder as test materials. Ethanolic extract was prepared using the optimal extraction conditions (50% aqueous ethanol and 75 °C). The extraction yield determined was 32.21 ± 0.68%, and phenol content was 492.08 ± 3.16 mg/g. The results presented that 50% ethanolic extract of GABA tea was nontoxic and non-mutagenic towards Salmonella typhimurium (TA98 and TA 100). This extract of TA 98 and TA 100 was highly anti-mutagenic as experimental microbes in the presence of metabolic activators. In addition, the growth of *V. parahaemolyticus*, *S. aureus*, and *B. cereus* was inhibited by the addition of tea extract (Mau et al. 2012).

Further studies revealed that Gaba tea extract exhibited inhibitory activity against *Vibrio parahaemolyticus, Staphylococcus aureus, Bacillus cereus, Salmonella typhimurium, and Escherichia coli* (Mau et al. 2012).

### *24.2.5 GABA as Anti-stress Compound*

Stress is defined as a disruption of the body's homeostasis and their activation is essential for our survival because it helps us to cope with varying internal and external factors. Tea (*Camellia sinensis*) is reported to be the most widely consumed beverage worldwide and it is also reported to reduce physiological stress, anxiety, and induce relaxation. A popular intervention in complementary medicine to remove chronic stress involves dietary food supplemented with GABA which is said to exert the effect peripherally, by acting on the autonomic nervous system ganglia, and central processes. One such commercially available product is GABA-enriched oolong tea (Hinton et al. 2019).

Jelinek et al. reported a study to measure the effects of GABA-fortified tea consumption on heart rate variability (HRV) and stress in 30 participants using a pre-post cohort study design. ECG recordings, frequency domain parameters including total power, high and low-frequency power and heart rate were determined. The control group comprised of subjects with non–fortified tea consumption. Two-way ANOVA was performed and it was concluded that oolong tea consumption led to a significant decrease in the immediate stress. The authors concluded that autonomic imbalance and HRV in people with acute stress is significantly reduced following a cup of GABA fortified oolong tea and highlighted the complex interaction between autonomic nervous system function and mood (Zhenxing et al. 2014). A study carried in 2009 showed that chocolate containing GABA was successful in reducing the stress condition.

### 24.2.6 Gamma-Aminobutyric Acid in Thyroid Dysfunction

Obesity is a worldwide epidemic and a key factor for the development of metabolic syndrome and type 2 diabetes (TD2). Nowadays, high energy diet is an important cause of obesity occurrence for which the endocrine mechanism remains unclear. Recently, many surveys have shown that obesity is closely associated with hypothyroidism (Zhenxing et al. 2014).

Le et al. performed a study where 20-week high-fat diet-fed (HFD) mice were utilized to study redox status and thyroid functions of diet-induced obesity (DIO) and DIO-resistant (DIOR) mice in their 1$^{st}$ study. The study also aimed at determining whether anti-obesity activity of GABA was related to antioxidant effect and if it improved thyroid function by using GABA in drinking water (0.2, 0.12 and 0.06%). It was observed that in DIO mice, TSH (thyroid stimulating hormone) levels increased, free thyroid hormone decreased. On the other hand, DIO-R mice showed normal TSH levels, increased THs and its functions. Down-regulated thyroid and THs functions in DIO mice may be accounted for obesity. GABA could prevent obesity by ameliorating oxidative stress and HFD-disrupted functions of thyroid and THs (Zhenxing et al. 2014).

### 24.2.7 GABA as Renoprotective

Chronic renal failure (CRF) is characterized as a pathophysiological condition of the kidney due to the permanent loss of nephrons, leading to the development of glomerular and tubular lesions. The kidney also contains significant levels of GABA, and specific binding sites for GABA have been confirmed in the kidney. Patients with end-stage renal disease and dialysis encephalopathy showed reduction in GABA levels in many areas of their brains, especially in the cerebral cortex and thalamus (Sumiyo et al. 2006). A study by Sasaki et al., used a remnant kidney model with 5/6 nephrectomized rats to analyze the protective effect of γ-aminobutyric acid against chronic renal failure (CRF). Nephrectomy causes renal dysfunctioning, which was evaluated via several parameters like serum urea nitrogen, creatinine levels and creatinine clearance. However, GABA administration ameliorated renal dysfunction, and a longer administration period of GABA increased its protective effect. In addition, nephrectomized control rats showed an elevation in the fractional excretion of sodium (FENa) with an increase in urinary sodium, while GABA led to a significant decline in FENa. Rats administered with GABA showed improvement in marked levels associated with CRF caused due to nephrectomy. This might be a proof that GABA inhibits disease progression and has a protective role against CRF. The protective role may also be accounted for improvement in the serum lipid profile, with reduced triglyceride and total cholesterol levels (Kim et al. 2004).

Furthermore, nephrectomy also caused renal oxidative stress with a decrease in the activity of antioxidative enzymes and elevation of lipid peroxidation. The administration of GABA attenuated oxidative stress induced by nephrectomy through an increase in superoxide dismutase and catalase, and decrease in lipid peroxidation. The histopathological lesions, including glomerular, tubular and interstitial lesions, under nephrectomy were also improved by GABA with the inhibition of fibronectin expression. This study revealed that GABA regulated blood pressure and lipid profile and hence, attenuated renal dysfunction, and it also ameliorated the oxidative stress induced by nephrectomy, suggesting the promising potential of GABA in protecting against renal failure progression. In addition to this, it is documented that oral administration of GABA and its supplements improved the other comorbidities associated with renal failure (Kim et al. 2004; Talebi et al. 2016).

## 24.3 GABA as Bioactive Compound in Food

As discussed in the previous section, GABA is produced from glutamic acid by enzymatic action of GAD. This reaction occurs in anaerobic conditions at pH below 5. GAD is, therefore, an important enzyme found in many strains of bacteria and fungi (*Lactic acid bacteria, Escherichia, Aspergillus, Streptococcus, Neurospora, Lactobacillus, Streptomyces, Monascus, Enterococcus, Rhizopus, Haematococcus*), plants (tomato, soybean, tea, mulberry leaves, petunia, pulses, germinated rice, pulses, brown rice), dairy products and human brain. Besides bacteria, algae are also a good source of GABA. Section 24.2 discussed the importance of GABA in various physiological functions and this is the reason why the search and development of food containing GABA are given high priority. Using GABA enriched food products represent a natural and economic way to get rid of many ailments.

**Table 24.1** Few GABA containing plants

| Botanical Name | Common name | Amount of GABA | Part of plant |
|---|---|---|---|
| *Zinziber officinale* | Ginger | 0.0114% | Rhizome |
| *Solanum torvum* | Potato | 0.0119% | Plant |
| *Ananas comosos* | Pineapple | 124 ppm | Fruit |
| *Arctium lappa* | Beggars button | 25 ppm | Root |
| *Hypericum perforatum* | Goat weed | 700 ppm | Plant |
| *Lycopersicum esculentum* | Tomato | 220–480 | Fruit |
| *Phoenix dactylifera* | Date palm | 2660–3370 | Fruit |
| *Pisum sativum* | shoot | 153 | Green pea |
| *Rehmannia gltinosa* | Root | 4000–31,000 | Chinese fox-glove |
| *Urtica diocia* | Root | 250 | Common nettle |

https://www.naturalmedicinefacts.info/chemical-detected/12318-1.html

Chemical modifications in existing food products is a method that may be used to enhance the GABA levels and use them as nutraceuticals (Table 24.1).

### *24.3.1 Microorganisms as Sources of GABA/GAD*

Microorganisms like bacteria, fungi, algae, molds, yeast isolated from different foodstuff are a big pool of GABA and GAD. The first instance of GABA is known to be isolated from yeast extracts which were acid-treated and later it was also isolated from red yeast named *Rhodotorula glutinis, Neurospora crassa spore germination, Aspergillus nidulans and A. niger.* In case of bacteria, the specific class called Lactic acid bacteria (LAB) are major group for GABA production. From amongst the several, strains of Lactobacillus and Lactococcus have been isolated from variety of fermented foodstuff like Kimchi, paocai, raspberry juice, etc. The best known GABA producing strains are *Lactobacillus paracasei* PF6, *Lactobacillus delbrueckii* subsp. *Bulgaricus* PR1, *Lactobacillus lactis* PU1, and *Lactobacillus brevis* PM17. It has also been detected in red yeast, *Rhodotorula glutinis* (Kawakami et al. 2018a, b). GABA pool was also seen in *Neurospora crassa* spore germination during early phases (Krishnaswamy and Giri 1953; Schmit and Brody 1975). Besides this, fungi like *Aspergillus nidulans* and *A. niger* contain GABA (Kubicek et al. 1979). As previously stated, the biosynthetic methods of GABA production are beneficial due to simple and mild reactions, high efficiency, simple workup, high yield, and environmental compatibility. Therefore, isolated enzymes like GAD are used for GABA production using different strategies. The other method is to perform microbial fermentation of low-cost feedstock. This is called a "natural method of production" and is preferred more than chemical methods. Many factors govern the amount of GABA produced which in turn depends upon the characteristics of GAD. These are discussed in Sect. 24.4.

### *24.3.2 Plants as a Source of GABA and GABA Enriched Food*

GABA is found naturally in small quantities in many plant sources. The first source where GABA was found is potato tubers. Besides this, natural GABA is found in many fruits, vegetables, and cereals. Vegetables such as spinach, tomatoes, cabbage, broccoli, ginger, asparagus, peas, mushrooms, etc. and fruits like Pineapple, apples, grapes, sweet potato, chestnuts contain GABA. Besides fruits and vegetables, GABA is also found in a number of cereals and pulses as such or they are enriched by fermentation using different LAB. Many cereals, legumes contain GABA producing bacterial strains which are further utilized to generate GABA enriched breads and sourdoughs. Sourdough breads are better quality breads than normal because of the effect of Lactobacillus and other LAB which enrich them better than baker's yeast. *Lactobacillus plantarum* C48 and *Lactococcus lactis* subsp. *lactis* PU1 have been used for fermentation of sourdough of many cereals, pseudocereals and flour obtained from legumes (Coda et al. 2010). Various flours that have been GABA enriched are amaranth, chickpea, buckwheat, quinoa. Bathura sourdough bread has been an example of achieving highest levels of GABA enrichment i.e. 226.22 mg/100 gm (Bhanwar et al. 2013). Oats when fermented with *Aspergillus oryzae* has very high level of GABA i.e. 435.2 microg/g) (Briguglio et al. 2018; Shengbao et al. 2014).

Fermentation is not the only process to improve the GABA content. Alteration of physical conditions needed for plant growth are also important for improving the GABA content in plants. Changing the germination conditions, pH, stimuli and other controlled condition help to improve the GABA contents in many cereals. e.g. 42.9 mg/100 g GABA levels are reported in foxtail millet (Bai and Fan 2008) and 14.3 mg/100 g in germinated waxy hull-less barley (Hyun et al. 2009). The most important and commonest cereal to note is RICE. Exceptional attention has been paid to increase GABA content in different varieties of rice. The highest GABA content is found in germinated brown rice as compared to white rice (10 times less) and brown rice (2 times less) (Patil and Khan 2011). Other cereals rich in GABA are sprouts of brown rice, barley, oats, millets, corn and beans.

Pulses, also known as leguminous crops are also an excellent source of GABA because faba beans and mung beans contain high levels of glutamic acid and applying environmental stress further enhances the levels. Abscisic acid acts as a stress hormone and under hypoxic-NaCl conditions, it also increases GABA levels in beans. Azduki beans, wultari beans also possess high GABA levels after proper treatment. Azduki beans are reported to reduce the risk of cardiac disease and acetaminophen-induced liver damage. Kidney beans when treated with glutamic acid during solid state or liquid state fermentation with *Baccilus subtilis* and *Lactobacillus plantarum* respectively, resulted in high GABA levels (Limón et al. 2014).

Overall, pulses may be categorized in 11 groups: dry beans, dry broad beans, dry peas, chickpeas, lentils, lupins, pigeon peas, black-eyed peas, bambara groundnut, vetch, other pulses. All are associated with health benefits and have a little amount of GABA which may be enriched on treatment (Nikmaram et al. 2017).

### *24.3.3 Dairy Products and Beverages as GABA Sources*

Dairy products like cheese, yogurt, milk are GABA enriched by LAB. *Lactobacillus lactis* is used for GABA enrichment in cheese (manages hypertension), fermented goats milk with LAB has very high GABA amount (28 mg/kg). *L. plantarum* fermented skim milk is reported to reduce systolic and diastolic blood pressure in rats. Many varieties of cheese also have natural presence of GABA with cheese strain like ULAAC-A23 and ULAAC-H13 having highest content (Lacroix et al. 2013). Chen et al. evaluated anti-diabetic effects in GABA enriched yogurt and reported it to lower glucose levels and raise serum insulin concentration. Yogurt-sake, an alcoholic fermented beverage had high levels of GABA than the total amount observed in yogurt and sake alone. *Streptococcus thermophilus* Hp was responsible for high levels while *Streptococcus thermophilus* Lp produced low levels (Ohmori et al. 2018).

Many beverages like white tea, fermented juices, alcoholic beverages also contain GABA and these are being demonstrated for beneficial effects mainly in treatment of hypertension. Maintaining anaerobic conditions have been reported to enhance the GABA in tea leaves by 8.9 folds. A variety of Gaba enriched tea is known as Gabaron tea which contains greater than 150 mg GABA per 100 gm. This helps to reduce blood pressure and improves sleep disorders. Few such commercialized compounds are oolong tea, GABA black tea and GABA green tea (Lacroix et al. 2013).

### *24.3.4 Marine Sources of GABA*

Marine sources have phenomenal biodiversity and this makes them very useful as a source of healthy food. Marine organisms are also a proven source of GABA and GAD. Many Irish marine cyanobacteria are reported to be GABA producers. Few such examples are *Calothrix contarenii* SABC022701, *Chlorogloea microcystoides* SABC022904, *Phormidium africanum* SABC010301, *P. angustissimum* SABC022612 and *P. laminosum* SABC022613 possessing $99*10^2$–$7284 \times 10^2$ nmol $g^{-1}$ on dry-weight biomass (Shiels et al. 2019). Marine *Pseudomonas* is another example to possess this metabolite. pH, stress, osmotic pressure are known to enhance GABA production in marine and freshwater cyanobacteria. Since cyanobacteria are microphototrophs, they have become choice for the development of various bioactive metabolites. *Synechocystis* sp. PCC6803 has been characterized for improved GAD activity and in another case, double engineering was performed (GADox/ΔKgd) to improve GABA production (Kanwal and Incharoensakdi 2019).

As discussed earlier, GABA gets metabolized by GABA shunt pathway which is important to maintain carbon and nitrogen balance intracellularly. This pathway is also an important metabolic pathway found in cyanobacteria.

Masuda et al. have demonstrated the presence of GABA in unicellular marine fungi (Masuda et al. 2008). Microalgae also possess GABA which acts as neurotransmitter, anti-oxidant, and anti-inflammatory (Gupta and Dhan 2018). Halotolerant cyanobacterium *Aphanothece halophytica* were treated in stress conditions and the normal levels of GABA increased to double than initial amounts along with enhanced GAD activity. These results also have proven successful in *Arthrospira platensis* (Boonburapong and Incharoensakdi 2016). Red Microalgae *Rhodosorus marinus* also contains GABA and GABA-Alanine which have proven to exhibit neuro-soothing effect and regulate skin sensitization by decreasing

TRPV1 over-expression in normal human astrocytes under PMA-induced inflammatory conditions (Scandolera et al. 2018). One of the Pseudomonas species of marine origin also produces GABA (Mountfort and Pybus 1992).

## 24.4 Techniques for GABA Enrichment and Advances in GABA Production

Recent consumer awareness and hence need for adopting healthier products with added advantages, called as "Nutraceuticals" have resulted in the market growth of such bioactive or functional components. The natural abundance of bioactive compounds like GABA is usually not very high and does not meet the ever-increasing market needs. Hence, scientists work untiringly to develop newer methods and techniques to enhance the concentration of bioactive compounds by various methods.

Similarly, reports reveal many novel techniques used for optimization and improvement of GABA content in plants, microbes, and dairy products. (Details on these techniques are discussed in this section below).

### *24.4.1 GABA Production by LAB*

Microorganisms contain low GABA content which makes it utmost difficult to extract. Also, the chemical synthetic methods as described previously are hazardous and difficult to adopt for large scale production. In this case, scientists have discovered many species of Lactic acid bacteria (LAB) that can produce GAD or produce high levels of GABA. LAB are gram-positive bacteria found ubiquitously in vegetables, fermented food and LAB are residents and normal flora present in human gastrointestinal tract. Hence, these are safe for human use and are widely used by the food industry. These are used as cell factories for GABA production. LAB generally includes strains of Lactococci and Lactobacilli. Many LAB species like *Lactobacillus brevis* PM17, *Lactobacillus plantarum* C48, *Lactobacillus paracasei* PF6, *Lactobacillus delbrueckii* subsp. *bulgaricus* PR1 and *Lactococcus lactis* PU1 were isolated from cheese and these have been used for GABA production (15–63 mg/Kg) (Diana et al. 2014). It has been reported that GABA produced by LAB possesses higher biological activities compared to natural and synthetically produced form. Besides these, *Lb. buchneri, Lb. helveticus and Streptococcus salivarius* subsp. Thermophilus are also GABA producing and the highest levels have been determined in *Lb. brevis* i.e. 345.83 mM.

Zhong et al. have reported the use of fermented mulberry fruits to isolate *Lactobacillus pentosus* SS6. This was then used as a starter culture for mulberry leaves fermentation and incubated at $30^0$C for 6 h. 10% saccharose, 6% peptone,

1.6% $K_2HPO_4$ and 1% L-sodium glutamate addition with above incubation also helped to increase the GABA levels. Mulberry leaves have multiple pharmaceuticals like anti-diabetic, anti-bacterial, anti-inflammatory, etc. Therefore, many efforts have been needed to increase GABA content in these leaves (Zhong et al. 2019). *Lb. plantarum* DSM19463 has been used to ferment grapes which on GABA enrichment are used as anti-hypertensive and dermatological protecting agents. *Pediococcus acidilactici, P. pentosaceus, E. durans, E. faecalis, E. faecium and Leuconostocs* (L.) are other GABA producing bacteria.

Yogurt, cheese, kimchi sourdough, paocai are few GABA enriched food which have been fermented with LAB. Except for *Lb. brevis* CGMCC 1306 (fresh milk without pasteurization), most other strains may be obtained from sources given below (Table 24.2).

## *24.4.2 GABA Production by Other Microorganisms*

Though maximum GABA production relies on LAB, there are other microbes that also produce GABA. *Aspergillus niger, Aspergillus nidulans, Neurospora crassa* are few filamentous fungi that have GABA pool. *Monascus purpureus* is reported to increase GABA levels in rice (Jannoey et al. 2010) and Rhizopus microspores increases its level in fermented soybeans.

## *24.4.3 Factors Affecting GABA Levels*

GABA producing ability of each LAB strain is different and it can be modulated by optimizing the culture conditions and medium composition. Stress conditions (biotic or abiotic) like drought, wounds, salt level, infection, hypoxia, germination, soaking are few factors that contribute to enhanced GABA levels. Other contributing factors are pH, temperature, carbon source, nitrogen source, glutamate concentration, PLP. All these parameters are further dependant on GAD properties which is essential for glutamate decarboxylation (Li and Cao 2010). The optimal pH needed for GAD activity is specifically governed by the strain of the organism. e.g. *E.coli* needs pH of 3.8, *N. crassa* needs 5.0 and *L. brevis* needs 4.2.

### 24.4.3.1 Effect of pH

It is observed that the conversion of glutamate to GABA by GAD is stoichiometric and it eventually leads to increased pH of the cytosol and neighboring environment. This increase in pH may further hamper the GABA biosynthesis and hence it needs to be maintained to acidic pH. $H_2SO_4$ is used in fermentation broth to maintain the acidic pH using *Lb. brevis* for GABA production. Similarly, maintaining pH 5.0 for

**Table 24.2** List of GABA producing strains and their isolation sources

| LAB strain | Isolation sources | Culture medium | GABA levels | References |
|---|---|---|---|---|
| *Lactobacillus brevisPM17* | Korean fermented vegetable kimchi, Chinese traditional paocai, fresh milk, alcohol distillery lees and black raspberry juice | Sodium acetate buffer | 15 mg/Kg | Siragusa et al. (2007) |
| *Lb. delbrueckii* subsp. *bulgaricus*, *Lb. plantarum* and *Lb. paracasei* | Cheese, Japanese fermented fish | Sodium acetate buffer | 63, 16, 99.9 mg/Kg respectively | Siragusa et al. (2007) |
| *Lb. paracasei* PF6, *Lb. delbrueckii* subsp. *bulgaricus* PR1, *L. lactis* PU1 and *Lb. brevis* PM17 | Pecorino di Filiano, Pecorino del Reatino, Pecorino Umbro, and Pecorino Marchigiano cheeses, respectively, | Sodium acetate buffer | 100 mg/Kg | Siragusa et al. (2007) |
| *Lb. brevis* CGMCC 1306 | Fresh milk without pasteurization | GYP | 4599.2 mg/l | Huang et al. (2007) |
| *Lactobacillus brevis* NCL912 | Paocai | Nutrient medium, MSG | 19.3 mg/L | Li et al. (2010a, b) |
| *Lb paracasei* PF6 | Indian cheese | Sodium acetate buffer, MSg | 100 mg/L | Siragusa et al. (2007) |
| *L. lactis ssp. lactis* 01–4, 01–7, 53–1, and 53–7 | cheese starters | Wheat flour | 258.71 mg/kg | Rizzello et al. (2008), Dhakal et al. (2012) |
| *Lb helveticus* | Koumiss | Skim milk | 165.11 mg/l | Sun et al. (2009) |
| *Lb. paracasei* NFRI 7415 | Fermented crucians | MRS Broth | 31,145.3 mg/l | Komatsuzakiet al. (2005) |

fermentation conditions of *Lb. paracasei* NFRI 7415 produced 302 mM GABA from 500 mM glutamate and fermentation at pH 4.5 for *Streptococcus salivarius subsp. Thermophilus* Y2 also enhanced the production. These examples imply that optimum pH needed for GABA production is highly species-dependent. Maintaining the pH towards acidic conditions is also important because increase in the pH shall increase the activity of GABA degrading or metabolizing enzymes like GABA transaminase to form succinic semi-aldehyde. This enzyme is highly active at pH 8.5 in *Pseudomonas aeruginosa* and transaminases GABA. Succinic semi-aldehyde dehydrogenase also has been reported to be active at pH 8.4 when isolated from Saccharomyeces.

#### 24.4.3.2 Effect of Temperature and Time

Incubation temperature during fermentation is also an important factor that governs GABA production. It affects the thermodynamic equilibrium, biocatalyst activity and stability. It is observed that the growth of *Lb. brevis* NCL912 increases up to temperature of 35 °C and then on the further rise, growth declines. The usual temperature needed by most strains for high GABA production ranges between 25 and 405 °C. Further, the time for which the incubation is allowed is also a determining factor. Table 24.3 represents some cases revealing the importance of time, temperature and pH. The time needed for fermentation is equally important as pH and temperature. *Lb. plantarum* DSM19463 and *Lb. paracasei* NFRI 7415 produced high levels of GABA (4.83 mM and 60 mM) when fermented for 72 hr and 144 h respectively. In addition, the time at which the additives like GABA substrate, MSG (monosodium glutamate), PLP are added also make a big difference in the yield. Adding PLP at 0, 24 and 48 h produced 6272, 6570 and 7333 mg/l GABA respectively after 72 h (Dhakal et al. 2012).

#### 24.4.3.3 Effect of Additives

GABA levels are also influenced by nutrient composition and media additives; PLP and glutamate being the major additives needed during fermentation while carbon and nitrogen source being the other additives. 1.25% glucose is found to be the best carbon source chosen from different pool of carbohydrates like glucose, galactose, fructose, ribose, xylose, arabinose, melibiose, etc. and 0.5% urea as the nitrogen source in general. Glutamate is an important additive which generally increases GABA production. This has been observed in *Lb. paracasei* NFRI 7415 (500 mM glutamate for 144 h yielded 161 mM GABA) and *Lb. brevis* but no effect was observed in the case of *S. salivarius*. Besides glutamate, PLP used as a cofactor can also be given due consideration as it enhances GAD activity. *S. salivarius subsp. Thermophilus* Y2, *Lb. paracasei* NFRI 74,150 and *Lb. plantrum* C48demonstrated GABA levels of 7333 mg/l, 200 mM and 504 mg/kg respectively on PLP addition. Sulfate ion also in some cases play a beneficial role by

**Table 24.3** Various conditions adapted to increase GABA levels

| Strain | Optimum pH | Amount of GABA produced at given pH | Optimum temp | Amount of GABA produced at given temp |
|---|---|---|---|---|
| *Lb. brevis* | 5.0 | Higher than normal | NC | – |
| *Lb. paracasei* NFRI 7415 | 5.0 | 302 mM | NC | – |
| *Streptococcus salivarius* subsp. *thermophilus* Y2 | 4.5 | – | NC | – |
| *Lb. plantarum* DSM19463 | 6.0 | 59micromole/h | NC | – |
| *Lb. paracasei* NFRI 7415 | 5.0 | 210 mM | NC | – |
| *S. salivarius* subsp. *thermophilus* | 4.5 | 7984 mg/l | NC | – |
| *Lactobacillus, Lb. paracasei* PF6, PF8, PF13, *Lb. plantarum* PF14, Lb. sp. strain PF7 and *E. durans* PF15 | 4.6–5.7 | 289 mg/kg - 391 mg/kg | NC | – |
| *Lb. paracasei* | 5 | 210 Mm | NC | – |
| *Lb. lactis* | 7.5–8 | 7.2 g/l | NC | – |
| *Lb. brevis* NCL912 | NC | – | 35C | – |
| *Lb. plantarum* DSM19463 | NC | – | 30–35 | 59microM/h |
| *Lb. brevis* GAD | NC | – | 30 | — |
| *Lb. brevis* CGMCC 1306 | NC | – | 37 | — |
| Lb. *brevis* GABA | 3.5 | – | 30 | 27.6 mg/ml |
| *Lb. buchneri* in MRS broth medium | NC | – | 30 | — |
| Immobilized whole cells of *Lb. brevis* | NC | – | 40 | 92% |
| *Lc. Lactis* | NC | – | 33<br>34 | 310 mg/ml<br>439 mg/ml |
| *Lb. paracasei* NFRI 7415 | NC | – | 37 | 302 mM |
| Black raspberry juice fermented with *Lb. brevis* GABA 100 | 4<br>5.5<br>3.5 | – | 25<br>37<br>30 | 25.4 mg/ml on 15th day<br>26.5 mg/ml on 15th day<br>Highest level on 12th day |

NC indicates the normal conditions used

increasing GAD activity by hydrophobic interactions. Many other substrates like sourdough, tomo koji, erythritiol, skim milk, iso-malto oligosaccharide, pectin are also being utilized for increasing GABA production (Dhakal et al. 2012; Yang et al. 2008; Li et al. 2010a, b).

## 24.4.4 *Advances in GABA Production Techniques*

### 24.4.4.1 Immobilized Cell Technology

As discussed previously in GABA biosynthesis, GAD plays the vital role of decarboxylation of glutamate to synthesize GABA. GAD is also important to maintain the intracellular pH and for oxidative stress tolerance. This enzyme is present in prokaryotes and eukaryotes but those which are obtained from microbial sources are considered prime important for large scale synthesis of GABA. Most of the GAD obtained from bacterial sources is active only at acidic pH which limits their application. Hence, newer technologies of molecular engineering have been adopted to make them activated towards near-neutral pH. To ensue this, the techniques used are immobilized cell technology and gradient controlling fermentation.

Immobilized cell technique leads to the reduction of non-productive growth phase which in turn increases the cell density of immobilized cells and hence enhances the overall product yield. This technique also protects the cells from stress conditions like pH, temperature, salts, and self-destruction (Ying 2007). Immobilization is considered to be an economical and powerful strategy for GABA production. *E coli* GAD has been immobilized by his-tag method which resulted in 223.8 g/l GABA in 100 min and 58% activity could be retained even after ten cycles of consecutive uses (Lee et al. 2013). Choi et al. immobilized *Lb. brevis* GABA 057 to enhance GABA yield as a result of increased glutamate from 2 to 12%. Moreover, these immobilized cells can be used for four such cycles (Choi et al. 2006).

### 24.4.4.2 Gradient–Controlling Fermentation

It is to be noted that high cell densities are required for high GABA yields and all the bacterial strains do not possess the required amount of cell density. For such cases, gradient-controlling fermentation is used. Such an experiment was designed by Yang et al (2008) where they used temperature control and double stage pH and succeeded in achieving high GABA levels from *S. salivarius subsp. thermophilus* Y2 (Li and Cao 2010).

### 24.4.4.3 Molecularly Engineered GAD

It is notable that most GAD's act in acidic pH. Hence, they are being molecularly engineered to extend active pH ranges. Knowledge of crystal structure of various GAD has been helpful to understand the mechanism and active site responsible for the activity. The method has been successfully applied to GAD isoforms present in E coli. Its active site was determined which lies in PLP-binding domain and comprises of Lys276, Glu89, His275, Leu306, and His465 as important amino acids of binding cleft. Exploiting this information, site-specific mutation was performed and resultant mutants had activity in expanded ph ranges (Pennacchietti et al. 2009; Thu et al. 2013; Jun et al. 2014). GAD enzymes for *L. brevis* and *B. megaterium* have been molecularly engineered. The crystal structure of these two is not yet known but taking the clues from structural sequence alignment, the engineering was performed. In the case of L. brevis CGMCC 1306, C-terminal mutant was designed and it was found to have an activity to near neutral pH values. For *B. megaterium*, E294R and H467A, two mutants were designed which succeeded in displaying enhanced catalytic activity at pH 5.0 (Xu and Wei 2017).

In some studies the genes that encode GAD have also been overexpressed either homologously or heterologously in various bacterial strains like *E coli, Corynebacterium glutamicum, Lactobacillus sakei, Lactobacillus plantarum, Bifidobacterium longum*. In the case of *C glutamicum*, the GABA production increased more than double by co-expressing two genes gadB1 nad gadB2, obtained from *Lb brevis* Lb85 (Cui et al. 2020).

#### 24.4.4.4 Coculturing GABA Producing Strains

Co-culturing techniques have been in use for long to enhance the production using more than one strain for culture. This technique has also been used for enhancing GABA production. One such study for GABA has been reported by Barrett et al. where they used human-derived strains. Strains of lactobacilli and bifidobacteria were cultured in MRS broth supplemented with 0·05% (w/v) L-cysteine-hydrochloride (mMRS) under anaerobic conditions at 37 °C. Strains were then subcultured in mMRS broth for 16–24 h prior to inoculation. The study aimed at assessing the ability of human intestinally derived strains of *Lactobacillus and Bifidobacterium* to produce GABA. From a total of 91 intestinally derived bacterias assessed, one Lb strain and four Bifidobacterium strains produced GABA. *Lactobacillus brevis* DPC6108 was found to be the most efficient onverting up to 100% of MSG to GABA. The addition of *Lb brevis* DPC6108 to a faeces-based fermentation significantly improved GABA concentration, evidencing that this biosynthesis could occur in vivo. The result of this study shows that the production of GABA by bifidobacteria exhibited considerable interspecies variation. *Lactobacillus brevis* and *Bifidobacterium dentium* proved to most efficient GABA producing strains among the range of microorganisms tested. The addition of *Lact. brevis* DPC6108 to the culturable gut microbiota increased the GABA concentration in fermented faecal slurry at physiological pH (Barrett et al. 2012).

#### 24.4.4.5 Other Techniques

Various other techniques have also been used to improve the GABA content. Some of them are Batch fermentation, Sourdough fermentation and improving variety of medium culture. Many studies have been reported for improving the culture media and conditions and one such study was done by Alejandra et al. They used wild GABA-producing LAB isolated from artisanal Mexican cheese and evaluated the conditions needed for fermentation in milk. The experiments were performed different conditions and additive concentrations [using two inoculum concentrations (107 and 109 CFU/mL), two incubation temperatures (30 and 37 °C), three glutamate concentrations (1, 3, and 5 g/L), and three pyridoxal 5′-phosphate (PLP) concentrations (0, 100, and 200 μM)] to determine appropriate conditions to enhance the GABA. Results revealed that from a total of 94 LAB strains, fermented milk with two *Lb lactis* strains (L-571 or L-572) presented the highest GABA production (Santos-Espinosa et al. 2020). Batch fermentation has been performed on fungal strains Basidiomycetes and Ascomycetes (Wan et al. 2019) and sourdough fermentation has been utilized on lactic acid bacteria strains, isolated from Andean amaranth (A) and Real Hornillos quinoa (Qr) sourdoughs (Villegas et al. 2016).

## 24.5 Conclusion

The ever increasing search for safe and healthy food products, rose the interest of researchers and customers in GABA enriched products. Literature reveals that GABA, an important inhibitory neurotransmitter plays an important functional role in various diseases and pathologies. The pharmaceutical industries in association with the food industry are therefore focused on the search of novel methods for enhancing the quantity of GABA in food products and also in microorganisms and to inculcate it as a bioactive molecule to control many diseases. There are many roadblocks and challenges that need to be considered because most of the studies have been done on animals to date and their efficacy in humans is yet to be established.

## References


Akama K, Kanetou J, Shimosaki S, Kawakami K, Tsuchikura S, Takaiwa F (2009) Seed-specific expression of truncated OsGAD2 produces GABA-enriched rice grains that influence a decrease in blood pressure in spontaneously hypertensive rats. Transgenic Res 18:865–876

Aoki H, Furuya Y, Endo Y, Fujimoto KE (2003) Effect of gamma-aminobutyric acid-enriched tempeh-like fermented soybean (GABA-Tempeh) on the blood pressure of spontaneously hypertensive rats. Biosci Biotechnol Biochem 67:1806–1808

Bai Q, Fan G (2008) Effects of culture conditions on gama-aminobutyric acid accumulation during germination of foxtail millet (*Setaria italica* L.). Euro Food Res and Tech 228(2):169–175

Barrett E, Ross RP, O'Toole PW, Fitzgerald GF, Stanton C (2012) Gama-aminobutyric acid production by culturable bacteria from the human intestine. J Appl Microbiol 113: 411–417

Benincasa P, Falcinelli B (2019) Stanley Lutts sprouted grains. Nutrients 11(2):421

Bhanwar S, Bamnia M, Ghosh M, Ganguli A (2013) Use of *Lactococcus lactis* to enrich sourdough bread with γ-aminobutyric acid. Interna J of Food Sci and Nutri 64:77–81

Boonburapong SL, Incharoensakdi A (2016) Accumulation of gamma-aminobutyric acid in the halotolerant cyanobacterium *Aphanothece halophytica* under salt and acid stress. Bongkoj J Appl Phycol 28:141–148

Briguglio M, Dell'Osso B, Panzica G (2018) Dietary neurotransmitters: a narrative review on current knowledge. Nutrients 10(5):59

Cho YR, Chang JY, Chang HC (2007) Production of gamma-aminobutyric acid (GABA) by *Lactobacillus buchneri* isolated from kimchi and its neuroprotective effect on neuronal cells. J Microbiol Biotechnol 17:104–109

Choi SI, Lee JW, Park SM, Lee MY, Ji GE, Park MS, Heo TR (2006) Improvement of gamma-aminobutyric acid (GABA) production using cell entrapment of *Lactobacillus brevis* GABA 057. J Microbiol Biotechnol 16:562–568

Coda R, Giuseppe C, Gobbetti M (2010) Use of sourdough fermentation and pseudo-cereals and leguminous flours for the making of a functional bread enriched of γ-aminobutyric acid (GABA). Int J Food Microbiol 137:236–245

Cui Y, Miao K, Niyaphornand S, Qu X (2020) Production of gamma-aminobutyric acid from lactic acid bacteria a systematic review. Int J Mol Sci 21(3):995

Dai-Hung N, Sang T (2019) An updated review on pharmaceutical properties of gamma-aminobutyric Acid. Molecules 24:2678

Dhakal R, Bajpai VK, Kwang-Hyun B (2012) Production of GABA (gama-aminobutyric acid) by microorganisms: a review. Braz J Microbiol 1230–1241

Diana M, Quílez J, Rafecasa M (2014) Gamma-aminobutyric acid as a bioactive compound in foods. J Funct Foods 10:407–420

Ebizuka H, Ihara M, Arita M (2009) Antihypertensive effect of pre-germinated brown rice in spontaneously hypertensive rats. Food Sci Technol Res 15:625–630

Gupta C, Dhan P (2018). Nutraceuticals from microbes of marine sources. https://doi.org/10.5772/intechopen82369

Hinton T, Herbert FJ, Viengkhou V, Graham AJ, Slade M (2019) Effect of GABA-fortified oolong tea on reducing stress in a University Student Cohort. Front Nutr 6:27. https://doi.org/10.3389/fnut.2019.00027

https://www.creative-diagnostics.com/gabaergic-synapse-pathway.htm

https://www.naturalmedicinefacts.info/chemical-detected/12318-1.html

Hua W, Cai C (1995) Process for synthesizing alpha-pyrrolidone acetamide, Patent Number: CN1105357

Huang J, Mei L, Sheng Q, Yao S, Lin D (2007) Purification and characterization of glutamate decarboxylase of lactobacillus brevis CGMCC 1306 isolated from fresh milk. Chin J Chem Eng 15:157161

Hyun J, Chung SuHJ, Hong Y (2009) Effects of steeping and anaerobic treatment on GABA (γ-aminobutyric acid) content in germinated waxy hull-less barley. LWT Food Sci Technol 42 (10):1712–1716

Injae H, Jo K, Kyung CS, Jong K (2019) GABA-stimulated adipose-derived stem cells suppress subcutaneous adipose inflammation in obesity. PNAS 116(24):11936–11945

Ippei S (2018) Signaling in brown adipose tissue promotes systemic metabolic derangement in obesity. Cell Reports 24:2827–2837

Jannoey P, Niamsup H, Lumyong S, Suzuki T, Katayama T, Chairote G (2010) Comparison of gamma-aminobutyric acid. World J Microb Biotech 26(2):257–263

Jun C, Joo JC, Lee JH (2014) Thermostabilization of glutamate decarboxylase B from *Escherichia coli* by structure-guided design of its pH-responsive N-terminal interdomain. J Biotechnol 174:22–28

Kanwal S, Incharoensakdi A (2019) The role of GAD pathway for regulation of GABA accumulation and C/N balance in *Synechocystis* sp. PCC6803. J of App Phyco 31:3503–3514

Kawakami K, Yamada K, Yamada T, Nabika T, Nomura M (2018) Antihypertensive effect of aminobutyric acid-enriched brown rice on spontaneously hypertensive rats. J Nutr Sci Vitaminol 64:56–62

Kawakami K, Yamada K, Yamada T, Nabika T, Nomura M (2018) Antihypertensive effect of aminobutyric acid-enriched brown rice on spontaneously hypertensive rats. J Nutr Sci Vitaminol 64:56–62

Kim HY, Yokozawa T, Nakagawa T, Sasaki S (2004) Protective effect of gamma-aminobutyric acid againstglycerol-induced acute renal failure in rats. Food Chem Toxicol 42:2009–2014

Komatsuzaki N, Shima J, Kawamotoa S, Momosed H, Kimurab T (2005) Production of g-aminobutyric acid (GABA) by *Lactobacillus paracasei* isolated from traditional fermented foods. Food Microbiol 22:497–504

Krishnaswamy PR, Giri KV (1953) The occurrence of 4-aminobutyric acid and glutamic acid decarboxylase in red yeast (*Rhodotorula glutinis*). Curr Sci 22:143–144

Kubicek CP, Hampel W, Rohr M (1979) Manganese deficiency leads to elevated amino acid pools in critic acid accumulating *Aspergillus niger*. Arch Microbiol 123(1):73–79

Lacroix N, St-Gelais D, Champagne CP, Vuillemard JC (2013) Gamma-aminobutyric acid-producing abilities of lactococcal strains isolated from old-style cheese starters. Dairy Sci Technol 93:315–327

Lee S, Ahn J, Kim YG, Jung JK (2013) Gamma-aminobutyric acid production using immobilized glutamate decarboxylase followed by downstream processing with cation exchange chromatography. Int J Mol Sci 14:1728–1739

Li H, Cao Y (2010) Lactic acid bacterial cell factories for gamma-aminobutyric acid. Amino Acids 39:1107–1116

Li H, Qiu T, Gao D, Cao Y (2010) Medium optimization for production of gamma-aminobutyric acid by *Lactobacillus brevis* NCL912. Amino Acids 38:1439–1445

Li H, Qiu T, Huang G, Cao Y (2010) Production of gamma-aminobutyric acid by *Lactobacillus brevis* NCL912 using fed-batch fermentation. Microb Cell Fact 9: Article number 85

Li W, Wei M, Wu J, Rui X, Dong M (2016) Novel fermented chickpea milk with enhanced level of aminobutyricacid and neuroprotective effect on PC12 cells. Peer J 4:2292

Limón RI, Peñas E, Martínez-Villaluenga C, Frias J (2014) Role of elicitation on the healthpromoting properties of kidney bean sprouts. LWT Food Sci Technol 56:328–334

Lin PP, Hsieh YM, Kuo WW, Lin CC, Tsai FJ, Tsai CH, Huang CY, Tsai CC (2012) Inhibition of cardiac hypertrophy by probiotic-fermented purple sweet potato yogurt in spontaneously hypertensive rat hearts, Int J Mol Med 30:1365–137

Masuda KK, Guo X, Uryy T, Hagiwara Watabe S (2008) Isolation of marine yeasts collected from the Pacific Ocean showing a high production of gamma-aminobutyric acid. Biosci. Biotechnol. Biochem 72:3265–3272

Matsumoto R (1989) GABA receptors: are cellular differences reflected in function? Brain Res Rev 14(3):203–225

Mau JL, Chiou SY, Hsu CA, Tsai HL, Lin SD (2012) Antimutagenic and antimicrobial activities aminobutyric acid (Gaba) tea extract. Int Conf Nutr Food Sci 39:178–182

Mohamed K, Frederick D, Shahin R, Kumar M (2019) Gamma-aminobutyric acid. Encycl Food Chem 528–534

Mountfort DO (1992) Regulatory influences on the production of gamma-aminobutyric acid by a marine pseudomonad. Appl Environ Microbiol 58(1):237–242

Nikmaram N, Dar BN, Roohinejad S (2017) Recent advances in $\gamma$-aminobutyric acid (GABA) properties in pulses. J. Sci. Food Agric 97(9):2681–2689

Nishimura M, Yoshida S, Haramoto M, Mizuno H, Fukuda T, Kagami-Katsuyama H, Tanaka A, Ohkawara T, Sato Y, Nishihira JE (2016) Effects of white rice containing enriched gamma-aminobutyric acid on blood pressure. J Tradit Complement Med 6:66–71

Nudelman A, Gil-Ad I, Shpaisman N, Terasenko I, Ron H, Savitsky K, Geffen Y, Weizman A, Rephaeli A (2008) A mutual prodrug ester of GABA and perphenazine exhibits antischizophrenic efficacy with diminished extrapyramidal effects. J Med Chem 51:2858–2862

Ohmori T, Tahara M, Ohshima T (2018) Mechanism of gamma-aminobutyric acid (GABA) production by a lactic Process biochemistry volume acid bacterium in yogurt-sake. Process Biochem 74:27

Okada T, Sugishita T, Murakami T, Murai H, Saikusa T, Horino T, Onoda A, Kajimoto O, Takahashi R, Takahashi TE (2000) Effect of the defatted rice germ enriched with GABA for sleeplessness, depression, autonomic disorder by oral administration. J Jpn Soc Food Sci 47:596–603

Park SJ, Kim EY, Noh W (2013) Synthesis of nylon 4 from gamma-aminobutyrate (GABA) produced by recombinant *Escherichia coli*. Bioprocess Biosyst Eng 36(7):885–892

Patil SB, Khan MdK (2011) Germinated brown rice as a value added rice product. J Food Sci Technol 48(6):661–667

Pennacchietti E, Lammens TM, Capitani G (2009) Mutation of His465 alters the pH-dependent spectroscopic properties of *Escherichia coli* glutamate decarboxylase and broadens the range of its activity toward more alkaline pH. J Biol Chem 284:31587–31596

Rashmi D, Zanan R, John S (2018) γ-aminobutyric acid (GABA): biosynthesis, role commercial production and productions and applications. Stud Nat Prod Chem 57:413–452

Reid SNS, Ryu JK, Kim Y, Jeon BH (2018) The effects of fermented Laminaria japonica on short-term working memory and physical fitness in the elderly, Evid Based Complement Altern Med 8109621

Richard WO, Timothy MD (1999) GABA receptor physiology and pharmacology basic neurochemistry: molecular, cellular and medical aspects, 6th edn. Lippincott-Raven, Philadelphia

Rizzello CG, Cassone A, Cagno RD, Gobbetti M (2008) Synthesis of angiotensin I-converting enzyme (ACE)-inhibitory peptides and γ-aminobutyric acid (GABA) during sourdough fermentation by selected lactic acid bacteria. J Agric Food Chem 56:6936–6943

Santos-Espinosa A, Beltrán-Barrientos LM, Reyes-Díaz R, González-Córdova AF (2020) Gamma-aminobutyric acid (GABA) production in milk fermented by specific wild lactic acid bacteria strains isolated from artisanal Mexican cheeses. Ann Microbiol 70:12

Scandolera A, Hubert J, Humeau A (2018) GABA and GABA-alanine from the red microalgae *Rhodosorus marinus* exhibit a significant neuro-soothing activity through inhibition of neuro-inflammation mediators and positive regulation of TRPV1-related skin sensitization. Mar Drugs 16(3):96

Schmit JC, Brody S (1975) *Neurospora crassa* conidial germination: role of endogenous amino acid pools. J Bacteriol 124:232–242

Sheng DL, Mau J-L, Hsu C-A Bioactive components and antioxidant properties of γ-aminobutyric acid (GABA) tea leaves. LWT Food Sci Tech 64–70

Shengbao C, Fengyi G, Xudong Z (2014) Evaluation of γ- aminobutyric acid, phytate and antioxidant activity of tempeh-like fermented oats (Avena sativa L.) prepared with different filamentous fungi. J Food Sci Technol 51(10):2544–2551

Shiels K, Patrick M, Saha SK (2019) Marine cyanobacteria as potential alternative source for GABA production. Bioresource Tech Rep 8:100342

Shimada M, Hasegawa T, Nishimura C (2009) Hypertension-anti-hypertensive effect of γ-aminobutyric acid (GABA)-rich *Chlorella* on high-normal blood pressure and borderline hypertension in placebo-controlled double blind study. J Clinic Experim Hypertension 31 (4):342–354

Siragusa S, De Angelis M, Di Cagno R et al (2007) Synthesis of gamma-aminobutyric acid by lactic acid bacteria isolated from a variety of Italian cheeses. Appl Environ Microb 73:7283–7290

Song P, Zhang Y, Li J (2014) Design, synthesis, and potent antiepileptic activity with latent nerve rehabilitation of novel γ-aminobutyric acid derivatives. Chem Pharm Bull 62:967–978

Sumiyo S, Yokozawa T, Cho E, Shigeru O, Kim M (2006) Protective role of *g*-aminobutyric acid against chronic renal failure in rats. JPP 58:1515–1525

Sun TS, Zhao SP, Wang HK et al (2009) ACE-inhibitory activity and gamma-aminobutyric acid content of fermented skim milk by *Lactobacillus helveticus* isolated from *Xinjiang koumiss* in China. Eur Food Res Technol 228:607–612

Suwanmanon K, Hsieh PC (2014) Effect of-aminobutyric acid and nattokinase-enriched fermented beans on the blood pressure of spontaneously hypertensive and normotensive Wistar-Kyoto rats. J Food Drug Anal 22:485–491

Talebi N, Nematbakhsh M, Monajemi R (2016) The protective effect of γ-aminobutyric acid on kidney injury induced by renal ischemia-reperfusion in ovariectomized estradiol-treated rats. Int J Prev Med 7:6

Tavakoli Y, Esmaeili A, Rabbani M (2015) Identification and molecular cloning of glutamate decarboxylase gene from *Lactobacillus casei*. Mol Biol Res Commun 4(3):161–165

Teresa CD (2013) Glutamate and GABA in appetite regulation. Front Endocrinol (Lausanne) 4:103

Thu Ho NA, Hou CY, Kim WH (2013) Expanding the active pH range of *Escherichia coli* glutamate decarboxylase by breaking the cooperativeness. J Biosci Bioeng 115:154–158

Tsai CC, Chiu TH, Ho CY, Lin PP, Wu TY (2013) Effects of anti-hypertension and intestinal microflora of spontaneously hypertensive rats fed gamma aminobutyric acid-enriched Chingshey purple sweet potato fermented milk by lactic acid bacteria. Afr J Microbiol Res 7:932–940

Villegas JM, Brown L, Hebert EM (2016) Optimization of batch culture conditions for GABA production by *Lactobacillus brevis* CRL 1942, isolated from quinoa sourdough. LWT Food Sci Tech 67:22–26

Wan AA, Kadir S, Sarina AH, Ilham Z, Hajar-Azhari S, Saari N (2019) Vital parameters for high gamma-aminobutyric acid (GABA) production by an industrial soy sauce koji *Aspergillus oryzae* NSK in submerged-liquid fermentation. Food Sci Biotechnol 28(6):1747–1757

Wong CG, Bottiglieri T, Snead OC (2003) GABA, gamma-hydroxybutyric acid, and neurological disease. Ann Neurol 54(6):S3-12

Xu N, Wei L (2017) Biotechnological advances and perspectives of gamma-aminobutyric acid production. World J Microbiol Biotechnol 33:64

Yamatsu A, Yamashita Y, Pandharipande T, Maru I, Kim M (2016) Effect of oral aminobutyric acid (GABA) administration on sleep and its absorption in humans. Food Sci Biotechnol 25:547–551

Yang SY, Lu FX, Lu ZX, Bie XM, Jiao Y, Sun LJ, Yu B (2008) Production of gamma-aminobutyric acid by *Streptococcus salivarius* subsp. *thermophilus* Y2 under submerged fermentation. Amino Acids 34:473–478

Ying Z (2007) Immobilized cell fermentation for production of chemicals and fuels, in bioprocessing for value-added products from renewable resources. New Tech Appl 373–396

Yongqi Y, Cheng C, Fang W (2018) Effects of the inhibitor of glutamate decarboxylase on the development and GABA accumulation in germinating fava beans under hypoxia-NaCl stress. RSC Adv 20456–20461

Zhenxing X, Shufang X, Guo-Wei Le (2014) Gamma-aminobutyric acid improves oxidative stress and function of the thyroid in high-fat diet fed mice. J Function Foods 76–86

Zhiyong F, Xin W, Xinyi L, Haiyan L, Qixiong L, Cimin L (2013) Preparation method of green feed additive gamma-aminobutyric acid. Patent Number: CN103242183

Zhong Y, Wu S, Chen F, He M, Lin J (2019) Isolation of high γ-aminobutyric acid producing lactic acid bacteria and fermentation in mulberry leaf powders. Exp Ther Med 18(1):147–153

Ziskind-Conhaim L (1998) Physiological functions of GABA-induced depolarizations in the developing rat spinal cord. Perspect Dev Neurobiol 5(2–3):279–287

# Chapter 25
# Medicinal Attribution of Ginsenoside: A Huge Source of Plant Bioactive Compound

**Dilipkumar Pal, Souvik Mukherjee, Satish Balasaheb Nimse, and K. K. Chandra**

**Abstract** Ginsenosides ($G_N{}^D$) are chemically triterpenoid saponin in nature. According to the presence of aglycones, dammarane and oleanane are the two types of $G_N{}^D$. These are mostly observed in species of Panax. The researchers have discovered over one hundred fifty substances from stocks, grasses, shoots, florets, drupes from the ginseng plant. $G_N{}^D$ and their derivatives are the main chemical constituents of the ginseng plant. Recently, $G_N{}^D$ are gaining increasing interests among natural product scientists. $G_N{}^D$ have many significant pharmacological activities, including anti-oxidation, mmunomodulation, and preventive actions in cancer, inflammation, stress, and hypertension, etc. The metabolism of $G_N{}^D$ involves two significant metabolic reactions, including acid hydrolysis and hydrolytic reactions oriented from bacterial origins. After metabolism, $G_N{}^D$ are transformed into a more active $G_N{}^D$ derivatives. The utilization and changes of unblemished $G_N{}^D$, which appears to assume a significant job for their potential wellbeing impacts, are discussed in this chapter.

**Keywords** Triterpene · Ginseng · Saponin · Gut flora · Biosynthesis · Metabolism

## List of Abbreviations

| | |
|---|---|
| $B_{syt}$ | Biosynthesis |
| $C_N{}^r$ | Cancer |
| $C_e{}^L$ | Cell |

D. Pal (✉) · S. Mukherjee
Department of Pharmaceutical Sciences, Guru Ghasidas Viswavidyalaya (A Central University), Bilaspur, Chhattisgarh 49551, India
e-mail: drdilip2003@yahoo.co.in

S. B. Nimse
Institute of Applied Chemistry and Department of Chemistry, Hallym University, Chuncheon 200-702, Korea

K. K. Chandra
Department of Forestry, Wildlife and Environmental Science, Guru Ghasidas Viswavidyalaya (A Central University), Bilaspur, Chhattisgarh 49551, India

 
D. Pal and A. K. Nayak (eds.), *Bioactive Natural Products for Pharmaceutical Applications*, Advanced Structured Materials 140,
https://doi.org/10.1007/978-3-030-54027-2_25

$C_y{}^T{}_k$ Cytokines
$C_A{}^{sp}$ Caspase
$C^h{}_L{}^n$ Cholinergic
$_C{}^{ytx}$ Cytotoxicity
$D_E{}^a{}_S$ Disease
$G^n{}_S$ Ginseng
$G_N{}^D$ Ginsenosides
$I_n{}^F{}_l$ Inflammatory
$I_m{}^M$ Immuno
$L_y{}^M{}_T$ Lymphocytes
$L^S$ Lanosterol
$M_{Lg}{}^T$ Malignant
$M^L{}_g$ Malignancy
$M_a{}^T{}_s$ Metastasis
$N_e{}^{PL}$ Neoplasm
$N_E{}^{ur}$ Neuro
$P_t{}^W$ Pathway
$P_L{}^t$ Plant
$P_r{}^{Lf}$ Proliferative
$P_H{}^{SP}$ Phosphate
$P^{NX}$ Panax
$S_p{}^N$ Saponin
$S_Q{}^L$ Squalene
$T_r{}^I d$ Triterpenoid
$T_U{}^m$ Triterpenoid

## 25.1 Introduction

Ginseng ($G^n{}_S$) is a significant restorative plant ($P_L{}^t$) having a place in family Araliaceae. Ginsenosides ($G_N{}^D$) and Gintonin are the primary concoction constituents of $G^n{}_S$. As indicated by the nearness of synthetic constituents and different land inception, there are four kinds of $G^n{}_S$ seen in different nations of the world, for example, South China, American, Vietnam $G^n{}_S$. It is the root of Panax ($P^{NX}$) family (Bilia and Bergonzi 2019). The base of $G^n{}_S$ has been utilized on antiquated occasions where it gives protection from stress, infection, and fatigue. There are a variety of $G^n{}_S$ based products in the market that are used to advance personal satisfaction. The $G^n{}_S$ products contain a variety of active constituents, including $G_N{}^D$ (Fig. 25.1), polyacetylenes, polyphenolic mixes, and acidic polysaccharides. The $G_N{}^D$ derivatives are steroidal nature and are sometimes called as triterpene saponin ($S_p{}^N$), which is a particular form of oleanane families . $G_N{}^D$ is made from the cytoplasm and plastid region of the $G^n{}_S$ $P_L{}^t$. The oral route is the most favorable way for $G^n{}_S$ administration. It is metabolized by gut flora. Configuration of those compound

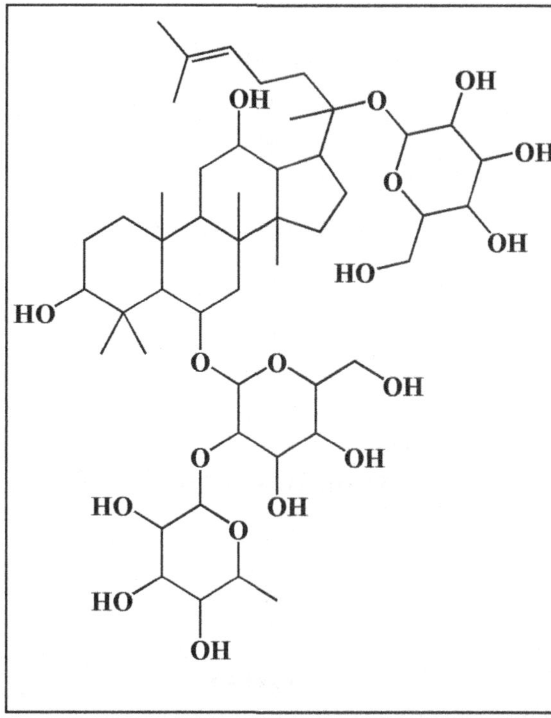


**Fig. 25.1** Structure of ginsenoside

shows that it contains four rings with steroidal moieties (Nimse and Pal 2015). In proto $P^{NX}$ adiols, sugar gatherings are found at the three-position within the carbon frame, whereas carbohydrate congregations append at the six positions of the same frame. $G_N^D$ from the oleanane family are pentacyclic containing a five-membered ring carbon skeleton (Xue et al. 2019).

## 25.2 Biosynthesis of $G_N^D$

Biosynthesis ($B_{syt}$) of $G_N^D$ can be divided into four steps, which are shown in Fig. 25.2. Firstly 2, 3-oxidosqualene ($S_Q^L$) is produced from acetyl-CoA. Then it is cyclized and modified with the help of the glycosylation process. There two carbocations (CT) have existed, namely 3-isopentyl salt and dimethylallyl radical salt, which is measured as the primary precursor for synthesis (Tang et al. 2019). Isopentenyl Di-phosphate ($P_H^{SP}$) pathway (IPP) is originated from the radical acetyl CoA of Mevalonic acid pathway ($P_t^W$). Di $P_H^{SP}$ Mevalonic acid $P_t^W$ is generated from phosphoglyceraldehyde. Here condensation additionally occurs. As a result, geranyl salt is created (Zhang et al. 2019). Geranyl Pyro $P_H^{SP}$ e is condensed with IPP, fashioned farnesyl salt. However, the Farnesyl Pyro $P_H^{SP}$ skeleton comprising

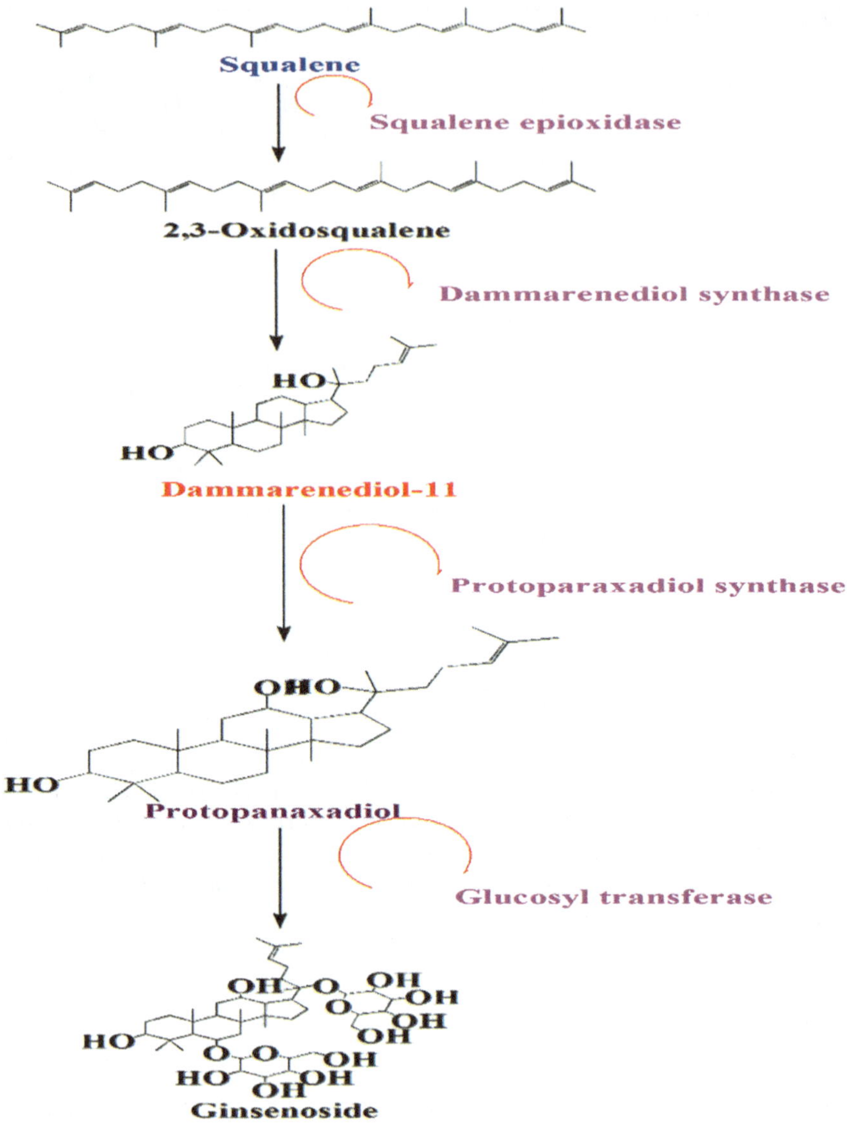


**Fig. 25.2** Biosynthesis of ginsenoside

of thirty carbon structures is referred to as $S_Q^L$ (Zhang et al. 2019). Cyclization of 2, 3-oxido compound is the second step (Rahimi et al. 2019). Oxido $S_Q^L$ is transferred into cyclic derivatives by the assistance of protonation and ring gap feature. During this method, a stereospecific reaction has additionally happened. Here c–c-c [triterpenoid ($T_r^I d$) $S_p^N$] or c-b-c [steroid] conformation is discovered and produces a tetracyclic prostosteryl as result lanosterol ($L^S$) is produced. (Senbagalakshmi et al. 2019). More $L^S$ carbocation is created as a Cyclo fused ring, as found in cycloartenol. Finally, $L^S$ has converted again into $L^S$. When de-methylation occurs, it produces steroid alcohol. Ultimately, p450, amyrin, dammarenediol CT are changed into $G_N^D$ via hydroxylation and glycosylation method (Hong et al. 2019).

## 25.3 Biotransformation of $G_N^D$

Therapeutic herb items or dietary enhancements are regularly taken orally, including those based on $G^n_S$ (Kim et al. 2019a). At the point when most therapeutic herbs/dietary enhancements are taken orally, then their constituents will be contacted with stomachic liquids (stomach acids), and the small floral proteins within the organ are well maintained in the alimentary tract (Liu et al. 2019). There is a bit of the mixture at the starting point within the medicinal herb items/dietary elements, which are used to different alimentary tract, which may be GD secondary metabolites. The metabolic destiny of the segments of dietary enhancements consequently may be beneficial to a superior comprehension of their organic movement and also the medical specialty activities of individual elements (Hu et al. 2019). Moreover, it has been planned that the small floral metabolic movement is influenced by intake programmed modification and physiological components, rather than by varieties within the microscopic organisms. It will be important in association to the medical specialty activities and impacts of bioactive constituents (Dou et al. 2019). The metabolism and retention of $G_N^D$ are contemplated seriously. Thus clarification is needed for the medical specialty activities of $G_N^D$ and their constituents to the clinical adequacy of $G_N^D$. For many examinations that are embraced to clarify the destiny of $G_N^D$ through the alimentary tract utilizing acids, chemicals, and human enteral microscopic organisms, clearly, a massive piece of the perfect $G_N^D$ is used/changed to $G_N^D$ with all the additional upgrading natural impacts contrasted distinguishable in plasma. For $G_N^D$ of the PPD-type, for instance, Rb1, Rb2, and Rc, it has been exhibited in each in vitro and in vivo examinations (He et al. 2019). Then again, it has been incontestable that the $G_N^D$ Rb1, Rb2, and Rc may be decayed to $G_N^D$ Rg3 (GRg3) by mellow corrosive treatment, for instance, abdomen acids, in spite of the very fact that a few of examinations have shown that PPD $G_N^D$ are scarcely disintegrated within the abdomen. In any case, if $G_N^D$ Rg3 is delivered within the abdomen, it is transformed to $G_N^D$ Rh2 or 20(S)-PPD by human enteral microorganisms as exhibited in some in vitro studies (Fu 2019). When the pharmacokinetic profile of $G_N^D$ is concerned in rodents, it is found that once an intra-gastric organization of 10 mg/kg $G_N^D$ Rg3 make a plasma grouping in around forty hours and likewise a 100 mg/ml of $G_N^D$ Rh2 or 20(S)-PPD form the same separately in four hours. $G_N^D$ Rh2 and 20(S)-PPD has been significantly changed physiological response than the unflawed $G_N^D$ Rb1, Rb2 and Rc. $G_N^D$ Rh2 and 20(S)-PPD have, as an example, of additional cytotoxicity($C_y^{tx}$) against neoplasm($N_e^{PL}$) cell($C_e^L$) lines and also the unflawed $G_N^D$ (Kim et al. 2019b). For $G_N^D$ of the PPT-type, for instance, Rg1 and Re, a couple of investigators have exhibited that these mixtures are processed to $G_N^D$ Rh1 and $G_N^D$ F1 and then at long last to 20(S)-PPT following numerous conditions suggesting thereby a stepwise cleavage of the sugar moieties. $G_N^D$ Rg1 on the oral organization may be transformed into $G_N^D$ Rh1 and hydrous subsidiaries of Rh1 in the abdomen. It is the fact that all $G_N^D$ Rg1 is not hydrolyzed within the abdomen, and unflawed $G_N^D$ Rg1 might hit the interior organ, wherever enteral microbes use it to $G_N^D$ F1 and 20(S)-PPT $G_N^D$ Re. Then again, it can be hydrolyzed by stomachic liquids to form $G_N^D$ Rg2, which is then modified within the system to $G_N^D$ Rh1 by the disposal of rhamnose through enteral microorganisms. Unflawed $G_N^D$ Re might likewise hit the interior organ wherever it tends to be used by enteral microbes to $G_N^D$ F1 and 20(S)-PPT through GD Rg1. Apart from enteral microscopic organisms, a couple of nourishment microorganisms have incontestable conditions to deliver specific sorts of $G_N^D$, together with those created by enteral human microbes. This demonstrates that it would be doable to create up a selected bioconversion procedure to induce expressly structured useful things by the acceptable mixture of $G_N^D$ substrate and specific microorganism chemicals from sustenance microorganisms (Darsandhari et al. 2019). Pharmacokinetic examines have exhibited that $G_N^D$, once taken orally, may be distinguished in plasma. Pee tests as unflawed $G_N^D$ or glycosylated corruption things, the first debasement things recognized in urine, and plasma tests on oral admission of PPD and PPT $G_N^D$ are considered as the monoglycosylated $G_N^D$ compound K, $G_N^D$ Rh1 and $G_N^D$ F1. Deglycosylase $G_N^D$ are usually additional promptly eaten into the cardiovascular system considered as more dynamic mixtures than the relating unrotten $G_N^D$. The bioavailability of perfect $G_N^D$ is poorly contrasted, as also found in the case of de-glycosylated $G_N^D$ (Yang et al. 2019). Thusly, the direct physiological impact of unflawed $G_N^D$ in vivo will consequently be talked regarding and want examinations. Compound K and $G_N^D$ F1 are usually distinguished in plasma from seven h once the admission of $G^n{}_S$ and in pee from twelve h after the intake, whereas $G_N^D$ Rh1 is recognizable from one h in plasma and three h in pee once oral. The pharmacodynamics of compound K for which once endovenous organization is established to mice have become incontestable than compound K for the foremost half discharged in biliary tract. In any case, some compound K are esterified with unsaturated fats at C-3 of the aglycone moiety or C10 of the aldohexose moiety within the liver. Thusly, the esterified sorts of compound K are collected longer within the liver than compound K itself. Further, it has been incontestable that esterified compound K inhibits growth more than compound K in vivo (Zuo et al. 2019). These outcomes propose that liver chemicals may be involved within the digestion of $G_N^D$ and in the development of dynamic standards of $G_N^D$ in the body, which is shown in Fig. 25.3 (Lee et al. 2019).

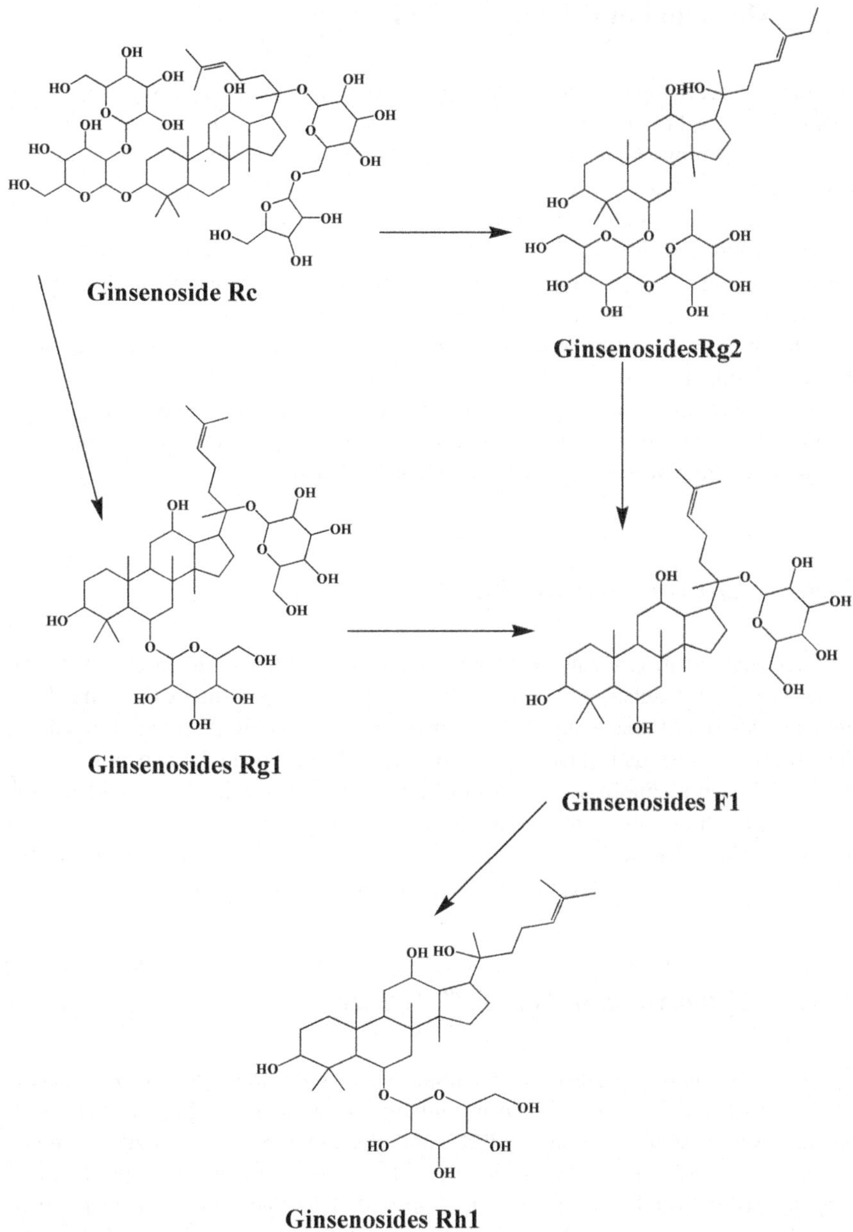


**Fig. 25.3** Biotransformation process of ginsenoside

## 25.4 Medicinal and Nutraceutical Applications

Pharmacological impacts of $G^n{}_S$ are exhibited within the focal sensory system (CNS). Besides, broad diagnosis and epidemiologic examinations have shown that $G^n{}_S$ and $G^n{}_S$ things have potential cancer ($C_N{}^r$) preventive impacts even as consequences for hyperglycemia. The dynamic segments in $G^n{}_S$ comprise primarily of polysaccharides, polyacetylenes, and $G_N{}^D$, of that the $G_N{}^D$ are viewed because of the real dynamic standards of $G^n{}_S$. The $G_N{}^D$ have exhibited a capability to focus on varied types of tissues, making a range of medicine reactions. Since $G_N{}^D$ might produce impacts that do not seem to be constant as one another and single $G_N{}^D$ and in addition, their used things might begin varied activities in a very similar tissue, the overall medicine of $G^n{}_S$ things is quite unpredictable (Nguyen and Nguyen 2019). Within the concomitant, the fascinating medicinal impact of $G_N{}^D$ are obtained, and hence their potential successfulness with advancing impacts are talked concerning with forwarding (Karra et al. 2019).

### 25.4.1 *Anti-carcinogenic Effects*

$G_N{}^D$ are applied to evaluate their anti-carcinogenic impacts in vitro and in vivo using varied approaches. Several numbers of them show a direct $C_y{}^{tx}$ and development restrictive impacts against tumor($T_U{}^m$) $C_e{}^L$, whereas others are looked as if they would repress metastasis ($M_a{}^T{}_s$) and $T_U{}^m$ development (Wang et al. 2018). Results from epidemiologic and companion contemplate with white and red $G^n{}_S$ have clearly shown that they need nonorganic specific preventive impact against malignant ($M_{Lg}{}^T$) growth. This impact is perhaps about to be due owing to their substance of chemicals, specifically $G_N{}^D$ (Santangelo et al. 2018).

### 25.4.2 *Cytotoxic and Anti $T_U{}^m$ Activity*

The $C_y{}^{tx}$ and antiproliferative ($P_r{}^{Lf}$) impacts of GD toward human and creature $M_{Lg}{}^T$ growth $C_e{}^L$ lines are shown in numerous examinations. In Associate in the nursing examination, Wang et al. (2007) undertake to assess the toxicity of ten $G_N{}^D$ (20(S)-PPD, 5, 12, 14, 15, 44, 84, 88, 91 and 124), detached from the product of *P.* $G^n{}_S$, toward a couple of human $M_{Lg}{}^T$ growth $C_e{}^L$ lines, together with bosom unwellness $C_e{}^L$ lines (e.g., MCF-7 $C_e{}^L$), respiratory organ $M_{Lg}{}^T$ growth $C_e{}^L$ lines (e.g., H838 $C_e{}^L$) and prostate unwellness $C_e{}^L$ lines (e.g., LNCaP and PC3 $C_e{}^L$). Amongst the $G_N{}^D$ tried, $G_N{}^D$ 20(S)-PPD, Rh2 (15), and $G_N{}^D$ 20(R)-25-OH PPD indicate significant movement of all told $C_e{}^L$ lines and are significantly the most effective inhibitors of malignancy ($M^L{}_g$) $C_e{}^L$ development and enlargement. For 20 (R)-25-OH PPD, the $IC_{50}$ esteems for many $C_e{}^L$ lines are within the scope of 10–60 metric linear unit, which is at any rate two-fold less than for any of the various $G_N^D$ tried. Each 20(S)-PPD and 20(R)-25-OH PPD distends $C_e^L$ death (apoptosis) and $C_e^L$ cycle movement during a portion of the subordinate method, whereas these impacts are less articulated for $G_N^D$ Rh2. It is eminent that 20(R)-25-OH PPD has an additional grounded impact than $G_N^D$ 20(S)-Rg3 (14) on $C_e^L$ development restraint. It has $IC_{50}$ qualities, which is 5-to 15-overlap less than for $G_N^D$ Rg3, a compound antecedently being showcased for treatment. Moreover, $G_N^D$ Rb1 (5), Rd (12), and Rg3 have much zero impact on $C_e^L$ development and enlargement. The impact on $C_e^L$ multiplication of $G_N^D$ Rh2 is furthermore discovered to be of the same size because the aglycones 20(S)-PPD and 20(S)-PPT, through the restrictive impact of $G_N^D$ 20(S)-Rh1 is ten times less. Moreover, the distance of sugars in PPD and PPT aglycone structures seems to decrease the strength to actuate caspase ($C_A^{sp}$)-mediated $C_e^L$ death as PPD and PPT are found to instigate $C_A^{sp}$-mediated $C_e^L$ death to the next degree than GD Rh2, though Rh1 failed to incite $C_A^{sp}$-mediated $C_e^L$ death. This shows that matters of sugar moieties at C-3 or C-6 may boot assume employment within the anti-$T_U^m$ impact of $G_N^D$. It projects the anti-$P_r^{Lf}$ impacts of $G_N^D$. Here, completely different bioactive compounds destitute GD 65 on the capability of GD to collaborate with $C_e^L$ layer capacities based on their hydrophobic nature. This is often likewise as per the structure–action relationship on the anti-$P_r^{Lf}$ impacts of $G_N^D$ and, therefore, the upgraded action watched for unsaturated fat conjugate $G_N^D$ (Bilia and Bergonzi 2019). $G_N^D$ of the 20 (S)-PPD family is the best-contemplated gathering of $G_N^D$ with relevancy anti-$T_U^m$ impact of that $G_N^D$ Rh2, and it could be a standout amongst alternative thought-about $G_N^D$ (Rani et al. 2014). $G_N^D$ Rh2 has been looked as if it would stifle enlargement in numerous human unwellness $C_e^L$, together with bosom, colorectal, prostate, hepatic, intestinal, melanoma, and creature $C_e^L$ lines (Li et al. 2019). The anti-$P_r^{Lf}$ impact of Rh2 provides off a sway of being connected to its capability to actuate $C_A^{sp}$-mediated $C_e^L$ death further as by capturing $C_e^L$ cycle movement. As an example, Rh2 has been accounted for to actuate $C_A^{sp}$-3 enzyme, a stimulating organic compound engaged with $C_A^{sp}$-mediated $C_e^L$ death and to capture $C_e^L$ cycle movement at the G1 section of MCF-7 human bosom $M^L{}_g$ $C_e^L$, SK-HEP-1 $M_{Lg}^T$ hepatoma $C_e^L$ (Rani et al. 2016), and B16-BL6 $M_{Lg}^T$ melanoma $C_e^L$. Rh2 also can restrain $T_U^m$ development in vivo of clean mice bearing human female internal reproductive organ unwellness $C_e^L$ (Lee et al. 2018). The anti-$P_r^{Lf}$ impacts toward $M^L{}_g$ $C_e^L$ of alternative PPD $G_N^D$, as an example, Rg3, Rg5 (72), Rs3 (18) and Rs 4 to boot seem to be as a result of their capability to instigate $C_A^{sp}$-mediated $C_e^L$ death and to irritate standard $C_e^L$ cycle occasions. It is a very fact that the anti-$P_r^{Lf}$ impacts of $G_N^D$ (Ryoo et al. 2019), together with PPD and PPT, toward urinary organ proximal tube-shaped structure $C_e^L$ can be as a result of a change of c-fos and c-jun quality articulation (Medina-Franco 2019).

### 25.4.3 Inhibition of $T_U{}^m$ $C_e{}^L$ Invasion and $M_a{}^T{}_s$

The counteractive action of $M^L{}_g$ $M_a{}^T{}_s$ is critical, and therefore there should be an improvement in the guess of $C_N{}^r$ patients. The first trademark venture of $M^L{}_g$ $M_a{}^T{}_s$ is the $N_e{}^{PL}$ $C_e{}^L$ intrusion of encompassing tissues and vasculature. Kitagawa Associate in nursing partners designed up an intrusion model for evaluating $N_e{}^{PL}$ $C_e{}^L$ attack capability in vitro (Pal and Saha 2019). During this model, $N_e{}^{PL}$ $C_e{}^L$ are seeded on a vital refined monolayer of host $C_e{}^L$, for instance, mesothelium or epithelial tissue $C_e{}^L$. The $N_e{}^{PL}$ $C_e{}^L$ infiltrate the monolayer and develop and structure $N_e{}^{PL}$ $C_e{}^L$ states beneath the monolayer. The limit of the entrance of $N_e{}^{PL}$ $C_e{}^L$ in vitro relates well thereupon of in vivo implantation into guinea pigs. Afterward, the in vitro model permits concentrating on the impacts of drugs on $N_e{}^{PL}$ $C_e{}^L$ attack. By utilizing this in vitro model, over ten GD are tried for the hindrance of $N_e{}^{PL}$ $C_e{}^L$ attack and $M_a{}^T{}_s$. GD 20(R)-Rg3 (42) has been ascertained to be associate in nursing intense matter of attack of a couple of $N_e{}^{PL}$ $C_e{}^L$ as well as haptonema (MM1), skin $C_N{}^r$ (B16FE7), human tiny respiratory organ $M_{Lg}{}^T$ neoplastic disease ($D_E{}^a{}_S$) (OC10), and human exocrine gland glandular $C_N{}^r$ (PSN-1) $C_e{}^L$ Whereas GD Rb2 (7), 20(R)-Rg2 (111), and 20(S)-Rg3 (14) have simply indicated lowest repressing action on $N_e{}^{PL}$ $C_e{}^L$ intrusion. Neither GD Rc (10), Re (84), Rh1 (91), Rh2 (15), nor 20(R)-Rh1 (112) were found to possess any impact within the model. As shown by Azuma and Mochizuki (1994) and Mochizuki et al. (1995), the enantiomers 20(S)- and 20(R)-Rg3 appear to possess an impact on $N_e{}^{PL}$ $M_a{}^T{}_s$ development as exhibited in vitro on 2 extremely $M_{Lg}{}^T$ $T_U{}^m$ $C_e{}^L$, B16-BL6 skin $C_N{}^r$ and colon 26-M3. $M_{Lg}{}^T$ neoplastic $D_E{}^a{}_S$, and in vivo by $N_e{}^{PL}$ immunization of B16-BL6 skin $C_N{}^r$ in mice. In any case, the impacts of 20(S)- and 20(R)-Rg3 against pneumonic $M_a{}^T{}_s$ in vitro and in vivo appear, by all accounts, to look as one thing else, with 20(S)-Rg3 demonstrating the weakest impact in vivo and therefore the most grounded impact in vitro contrasted and 20(R)-Rg3.

### 25.4.4 Inhibition of $T_U{}^m$ Angiogenesis

Angiogenesis is a physiological process involving the growth of new blood vessels from preexisting vessels and is considered a normal process in growth and development, as well as in wound healing. However, this is also a fundamental step in the transition of $T_U{}^m$ from a dormant state to a state where the $T_U{}^m$ $C_e{}^L$ proliferate ($M_{Lg}{}^T$ state). Inhibition of angiogenesis, therefore, prevents $T_U{}^m$ growth, proliferation, and secondary $M_a{}^T{}_s$ and is essential for the prevention and treatment of $C_N{}^r$ ( Folkman et al. 1995). Only a few studies on the angio-suppressive effects of GD have been performed, and they mainly concern the $G_N{}^D$ Rb2 (7) and 20(R)-Rg3 (42). Sato et al. (1994) study the effect of $G_N{}^D$ Rb2 on angiogenesis and $M_a{}^T{}_s$ produced by B16-BL6 melanoma $C_e{}^L$ in syngeneic mice. Intravenous administration of $G_N{}^D$ Rb2 on day 1, 3, or 7 after $T_U{}^m$ inoculation results in a remarkable reduction in the number of vessels oriented toward the $T_U{}^m$ mass, but do not cause significant inhibition of $T_U{}^m$ growth. The angio-suppressive effect is dose-dependent in the ranges of 10–50 mg/mouse.

In contrast, intratumorally or oral administration of $G_N{}^D$ Rb2 causes a marked inhibition of both neovascularization and $T_U{}^m$ growth. $G_N{}^D$ Rb2 does not affect the growth of rat lung endothelial $C_e{}^L$. However, it inhibits in a dose-dependent fashion the invasion of rat lung endothelial $C_e{}^L$ into the reconstituted basement membrane (Matrigel), which is considered to be an essential event in $T_U{}^m$ neovascularization. Multiple administrations of GD Rb2 after the *i.v.* inoculation of B16-BL6 melanoma $C_e{}^L$ results in significant inhibition of lung $M_a{}^T{}_s$ as compared with that of the untreated control. The results suggest that the inhibition of $T_U{}^m$-associated angiogenesis by $G_N{}^D$ Rb2 may partly contribute to the inhibition of lung $T_U{}^m$ $M_a{}^T{}_s$. Yue et al. (2006) examine the ability of GD 20(R)-Rg3 to interfere with the various steps of $T_U{}^m$ angiogenesis. GD 20(R)-Rg3 is, for example, found to inhibit the proliferation of human umbilical vein endothelial $C_e{}^L$ (HUVEC) with an $IC_{50}$ value of 10 nM. GD 20(R)-Rg3 also dose-dependently suppresses the capillary tube formation of HUVEC on the Matrigel from 1 to 1000 nM in the presence or absence of 20 ng/ml vascular endothelial growth factor (VEGF). The $T_U{}^m$ angio-suppressive effects and the inhibiting effect of $M_a{}^T{}_s$ of GD Rb2 and 20(R)-Rg3 are probably related to their inhibitive effect on the release of VEGF from $T_U{}^m$ $C_e{}^L$.

### 25.4.5 Immunomodulatory Effects

The Immuno ($I_m{}^M$) modulatory activities of $G_N{}^D$ square measure are closely associated with their anti-carcinogenic, anti-inflammatory ($I_n{}^F{}_l$) and anti-allergic activities. The immune responses square measure is controlled by T helper (Th) $C_e{}^L$ and may broadly categorize into cellular mediated responses ($C_e{}^L$-mediated immunity) mediated by Th1 $C_e{}^L$, macrophages, and protein (antibody-mediated immunity) responses directed by Th2 $C_e{}^L$. The $C_e{}^L$ square measure is concerned with activation and directional different immune $C_e{}^L$ like $C_y{}^{tx}$.

T $C_e{}^L$ and natural killer (NK) $C_e{}^L$, and thence square measure is significantly necessary within the system. The event and differentiation of Th $C_e{}^L$ square sure strictly regulated by antigen-presenting nerve fiber $C_e{}^L$ (DCs). DCs that generate Th1 responses could also be achieved to forestall or treat pathological conditions that square measure caused by infections and $M_{Lg}{}^T$ disorders via secretion of sort one cytokines($C_y{}^T{}_k$) like interferon-g (IFN-g) and interleukin-2 (IL-2) to facilitate T-$C_e{}^L$-mediated toxicity. In distinction, DCs that generate Th2 responses could also be wont to forestall or treat conditions within which Th1 responses square measure disturbed, for instance, contact allergic reaction and response disorders, by secretion of the sort a pair of $C_y{}^T{}_k$, like IL-4 and IL-10, to assist B $C_e{}^L$ to secrete protecting antibodies (Takei et al. 2004). Therefore, any compound capable of modulating or operating particularly phagocyte activation by making assembly of small and enormous lymphocytes ($L_y{}^M{}_T$) (e.g., NK, T, and B $C_e{}^L$) becomes very important within the interference and treatment of $T_U{}^m$, infectious agents, and chronic $I_n{}^F{}_l$ $D_E{}^a{}_S$ (e.g. Autoimmune disorder, asthma, and atherosclerosis). It is renowned that numerous $G^n{}_S$ species have different $I_m{}^M$ modulatory activities in which the most active elements is square measure $G_N{}^D$. Yu et al. (2005) investigate numerous PPT-type $G_N{}^D$ isolated from *P.* $G^n{}_S$ leaves (20(S)-PPT, $P^{NX}$ atriol (20(S)-PT), F1 (80), Re (84), Rg1 (88), Rh1 (92), and a pair of 0(R)-Rh1 (112)) for his ability to modulate sort one differentially and sort 2 $C_y{}^T{}_k$ productions from murine splenocytes. GD F1 and Rg1 are found to influence a pair of $C_y{}^T{}_k$ production through regulation of the expression. For instance, in IL-4, GD Rh1 and 20(R)-Rh1 influence one $C_y{}^T{}_k$ production by regulation of the assembly of IL-12 and thereby influence the expression of IFN-g and T-bet. The latter being a particular Th1 transcriptional subject, is thought to initiate the development of Th1 and inhibit differentiation of Th2. The results clearly show that PPT-type $G_N{}^D$ have different $I_m{}^M$ modulatory effects together with each immune-stimulatory and $I_m{}^M$ logical disorder effects. This can be additionally in accordance with a study of Cho et al. (2002). World Health Organization finds that the $G_N{}^D$ Rb1 (5), Rb2 (7), Re (84), and Rg1 modulate WBC proliferation elicited by T $L_y{}^M{}_T$ mitogens [e.g., concanavalin A] and therefore the lymph $C_e{}^L$ agent, lipopolysaccharide $G_N{}^D$ sixty-nine (LPS), yet acts as protein IL-2, a potent trigger of WBC proliferation. $G_N{}^D$ Rb1 and Re considerably increase Con A-induced WBC proliferation, whereas Rg1 does not affect the proliferation. On the opposite hand, Rb2 powerfully blocks the mitogen-induced WBC proliferation with $IC_{50}$ values around twenty-one. This clearly shows that $G_N{}^D$ Rb2 may be a potent $I_m{}^M$ logical disorder agent. $G_N{}^D$ Rb2 and Rb1 have no restrictive effects on the proliferation of IL-2- aroused CD8þ T $C_e{}^L$, whereas Re and Rg1 show robust restrictive effects with $IC_{50}$ values of 57 and 64.7 mM, severally. These results clearly indicate that $G_N{}^D$ could modulate WBC proliferation. $G_N{}^D$ of *P. noto* $G^n{}_S$ and *P.* $G^n{}_S$, like Rb1, Rb2, and Rg1 have additionally shown to powerfully suppress the assembly of TNF-α in macrophages treated with LPS.

## *25.4.6 Anti-inflammatory Activity*

Furthermore, these $G_N{}^D$ also seem to suppress the production of other $I_n{}^F{}_l$ $C_y{}^T{}_k$, such as IL-6 and IL-1b, and hence demonstrate that widely distributed $G_N{}^D$ possesses anti-$I_n{}^F{}_l$ and $I_m{}^M$ suppressive properties in vitro. The activation of macrophages and hence the production of various types of $L_y{}^M{}_T$ has been shown to be essential for the prevention and treatment of $T_U{}^m$ and infectious $D_E{}^a{}_S$. $G_N{}^D$ Rg1 has been reported to have mainly $I_m{}^M$ modulatory effects that increase both humoral and $C_e{}^L$-mediated immunities by enhancing the activity of Th $C_e{}^L$ and NK $C_e{}^L$ responsive to given antigens. Furthermore, it has been reported that maturation of DCs is promoted by metabolized $G_N{}^D$ such as compound K. The anti-$I_n{}^F{}_l$ and anti-allergic properties of $G_N{}^D$ are more or less directly linked to their immune-stimulatory and anti-carcinogenic effects as well as in $D_E{}^a{}_S$ where $I_n{}^F{}_l$

conditions play a significant role such as in atherosclerosis and neuro ($N_E^{ur}$) degenerative $D_E^a{}_S$. Allergic $D_E^a{}_S$ of type 1, such as asthma, allergic rhinitis, atopic dermatitis, and food allergy, afflicts up to 20% of the human population in many countries. Allergen reactivity in these allergic $D_E^a{}_S$ is based on $I_m^M$ globin E (IgE)-mediated pharmacological processes in a variety of $C_e^L$ populations, in particular basophils and mast $C_e^L$. Degranulation of basophils and mast $C_e^L$ with antigen cross-linked IgE releases histamine, prostaglandins, leukotrienes, and $C_y^T{}_k$ affecting macrophages, $L_y^M{}_T$, eosinophils, and neutrophils, causing tissue injuries and $I_n^F{}_l$ $D_E^a{}_S$. $C_y^T{}_k$ and/or bacterial LPS induce nitric oxide synthase and cyclooxygenase-2 expression in, for example, macrophages and hence the production of nitric oxide and prostaglandins, respectively. Sustained production of NO and PGs has been implicated in the pathogenesis of $I_n^F{}_l$ $D_E^a{}_S$ and $C_N^r$ (Zhang et al. 2019). Several $G_N^D$ have shown to reduce the expression of iNOS and COX-2 and to inhibit the production of NO and PGs in macrophages as well as the inhibition of nuclear factor (NF)-kB transcription factor, which regulates iNOS and COX-2 gene expression. $G_N^D$ Rh1 (92) and Rh2 (15) and $G_N^D$ 20(S)-PPT, a metabolite of, for example, Rh1 or Rg1 and compound K, a metabolite of, for example, $G_N^D$ Rb1, have been reported to inhibit the production of NO and PGE2 and to inhibit the activation of NF-kB, in LPS-stimulated murine macrophages (RAW 264.7 $C_e^L$) (Kwon et al. 2018). The inhibition of NF-kB and COX-2 expression has also been demonstrated for compound K in mouse ear edema induced by the prototype $T_U^m$ promoter 12-O-tetradecanylphorbol-13-acetate (Moon et al. 2018). The results suggest that these $G_N^D$ can inhibit NO and PGs production by regulation of the signal transduction related to the activation of NF-kB. The anti-$I_n^F{}_l$ effects of $G_N^D$ have also been demonstrated in microglial $C_e^L$, which are resident macrophages of the CNS. It is found that the PPDs $G_N^D$ Rb2, Rd, and the PPTs $G_N^D$ Rg1, Re are able to inhibit LPS-induced NO formation and TNF-a production due to the inhibition of NF-kB in N9 microglial $C_e^L$. Thus, these $G_N^D$ may be used in the prevention or treatment of $I_n^F{}_l$ $D_E^a{}_S$, such as allergic inflammation and $N_E^{ur}$ logical $D_E^a{}_S$ (e.g., Alzheimer's and Parkinson's $D_E^a{}_S$) as well as $C_N^r$. The anti-allergic effect of $G_N^D$ has been studied in vitro and in vivo on rodent peritoneal mast $C_e^L$ and on IgE-induced passive cutaneous anaphylaxis (PCA), the latter being a model for study of type 1 sensitivity reactions. GD Rb1, Rc, Rd, F2, and Rh1 have been shown to inhibit histamine and/or leukotriene release from peritoneal mast $C_e^L$, whereas GD Rh1, Rh2, and compound K have been shown to be potent inhibitors of the PCA reaction in rodents (Chen et al. 2019). The inhibitory activity of Rh1, Rh2, and compound K on the PCA reaction is found to be more potent than the commercial anti-allergic drug disodium cromoglycate. These $G_N^D$ furthermore show a membrane-stabilizing effect, and it has been suggested that this membrane-stabilizing effect, which may prevent membrane perturbations, is the leading cause of their anti-allergic activity.

### 25.4.7 Antistress Activity

Antistress impact of $G_N^D$ complete $S_p^N$, $G_N^D$ Rg3, and Rb1 toward immobilization stress has likewise been exhibited by researches on the cerebrum level of endogenous polyamines, which are fundamental for $C_e^L$ development, multiplication, recovery, separation of the mind and outstanding pressure boosts markers. In this examination, it is discovered that $G_N^D$ Rg3 and Rb1 hinder the action of the catalyst ornithine decarboxylase, associated with the digestion and catabolism of polyamines and constricting the degrees of the polyamine putrescine. Along these lines, $G_N^D$ Rg3 and Rb1 may assume a $N_E^{ur}$ protective job in the immobilization-focused on the mind (Tam et al. 2018). Impacts on the CNS by various $G_N^D$ species have been appeared to have both stimulatory and inhibitory consequences and may adjust $N_E^{ur}$ transmission. $G_N^D$, and specifically $G_N^D$ Rb1, Rg1, and Re appear to assume an outstanding job in these impacts (Zheng et al. 2018).

### 25.4.8 Memory, Learning, and $N_E^{ur}$ Protection

Focal cholinergic ($C^h{}_L{}^n$) frameworks have been embroiled in intercession learning and memory forms. Since scopolamine is a $C^h{}_L{}^n$ receptor opponent, the exhibition debilitated by scopolamine may bring about the brokenness of focal $C^h{}_L{}^n$ components and subsequently may bring about memory shortages. Results from creature studies have demonstrated that $G_N^D$ Rb1, Rg1, and Re can forestall (scopolamine-incited memory shortages). These ameliorative impacts of Rg1 and Re have been demonstrated to be firmly identified with an expansion of choline acetyltransferase movement in the average septum of youthful and mature rodents (Majid 2019). GDRb1 and Rg1 have likewise demonstrated to be prepared to do in part, turning around scopolamine-instigated amnesia by improving $C^h{}_L{}^n$ movement and using halfway $N_E^{ur}$ trophic and $N_E^{ur}$ protective impacts.

Furthermore, it has been shown that $G_N^D$ Rb1 expands the take-up of $C^h{}_L{}^n$ system in focal $C^h{}_L{}^n$ nerve ending and encourages the arrival of acetylcholine from hippocampal cuts. These outcomes obviously recommend that $G_N^D$ may encourage learning and improving the fundamental synaptic transmission just as nerve development (Metwaly et al. 2019). $G_N^D$ additionally appears to have a $N_E^{ur}$ protective impact where nerve development likewise assumes a significant job. $G_N^D$ on in vitro investigations appeared to build survival of refined $N_E^{ur}$ nal $C_e^L$ and upgrade the outgrowth of neuritis. For instance, $G_N^D$ Rb1 has appeared to build the neurite outgrowth of refined cerebral cortex neurons and animate neurite outgrowth of PC12 $C_e^L$ without nerve development factor. The capacity of $G_N^D$ to recover $N_E^{ur}$ nal systems has additionally been exhibited in SK-N-SH $C_e^L$ for PPD-type $S_p^N$. For example, $G_N^D$ & noto $G_N^D$ R4 & Fa, while PPT, ocotillol, and oleanolic corrosive sort $S_p^N$ had no impact on neurite outgrowth (Wu et al. 2019). This plainly demonstrates that some $G_N{}^D$ can broaden axons and dendrites in neurons that may make up and fix harmed organizers, as found in the dementia brain (Szczuka et al. 2019).

### *25.4.9 Anti-diabetic Activity*

It has been demonstrated that the anti-diabetic GD 77 impacts of GDRb1 and 20(S)-PPT are likely identified with their capacity to enact peroxisome PPAR gamma (PPARG). PPARG is an individual component obtained from the atomic receptor of ligand-initiated translation factors that direct the declaration of critical qualities engaged with lipid and glucose digestion and adipocyte separation (Wang et al. 2019). PPARG is fundamentally communicated in fat tissue, and the enactment of PPARG improves the capacity of adipocytes to store lipids, in this way lessening lipotoxicity in muscle and liver. The qualities communicated by the initiation of PPARG depend to a great extent on the sort of actuating ligand present as they select an alternate arrangement of cofactors (Park et al. 2019). Consequently, the transcriptional reaction of the PPARG results in cofactors that lead to expanded lipid stockpiling and diminished vitality use. For example, transcriptional factor-2 or enlistment of cofactors lead to expanding insulin-animated glucose take-up and positive guideline of glucose digestion and vitality consumption (e.g., the steroid receptor coactivator-1).The enactment of PPARG causes body-wide lipid repartitioning by expanding the triglyceride content in fat tissue and bringing down free unsaturated fats triglycerides available for use. Liver and muscle, along these lines, improve insulin sensitives (Barman et al. 2019). A few examinations have shown that the hypoglycemic impact of $G^n{}_S$ items depends both on the $G_N{}^D$ content and the 78 large P. Christensen profile, plainly shows that the anti-diabetic impact of $G^n{}_S$ relies upon the convergence of single GD and hence, it is inferred that not all GD has anti-diabetic impacts (Truong et al. 2019).

## 25.5 Conclusions

$G_N{}^D$ is particularly $T_r{}^I d$ interestingly present in $P^{NX}$ species. In light of the restorative significance of GD, examinations on their compound structure, therapeutic exercises, $B_{syt}$ proteins & generation improvement have gained much consideration. The $B_{syt}$ $P_t{}^W$ of $G^n{}_S$ $S_p{}^N$ imparts normal and expanded chemicals to those of $T_r{}^I d$ in natural botanical systems. Until this point in time, there are more than 150 known diverse $G_N{}^D$ with different numbers, linkage positions, and sorts of the sugar moiety, and a large portion of them are oleanonic in nature. Current proof recommends that quality development coming out from genome and additionally quality duplication pursued by quality obsession through the affirmative determination of sub-and neo-functionalized homologs, may give the basic hereditary base to $G_N^D$ $B_{syt}$ and $P_L^t$ adjustment and species radiation. The mechanism(s) driving the development of $G_N^D$, just as the occasions hidden the expansion of $G_N^D$ in $P^{NX}$ species, still cannot seem to be explained. Another zone needing further research identifies the fundamental blend of the different $G^n{}_S$ $S_p^N$ inside explicit tissues. $G^n{}_S$ $S_p^N$ are assumed to go as resistance atoms in $P_L^t$ pressure and pathogen associations. The pharmacological viability of $G_N^D$ depends on their auxiliary premise, particularly their hydroxyl gatherings and sugar moieties, associating with layer lipids. Later on, more understanding is found in the basic restorative impacts of $G_N^D$. For example, crosstalk with hormone flagging $P_t^W$ will enable auxiliary alterations to accomplish improved beneficial exercises and capacities. The tedious and work serious development of $G^n{}_S$ in the field has driven bioengineering approaches. For example, the culture of tissues and $C_e^L$ compound elicitation during creation, transgenic $P_L^t$, and designed yeast frameworks are used to improve $G_N^D$ generation. Especially, transgenic $P_L^t$ over-communicating qualities associated with $G_N^D$ amalgamation, for example, HMGR1, SS, CYPs, and DDS have huge expanded $G_N^D$ yields. As of late, built yeast $C_e^L$ communicating $G_N^D$ delivering catalysts, bring about the creation of PPD, PPT and oleanolic corrosive just as compound K. These achievements give a modest and proficient mechanical stage for the production of $G_N^D$ for clinical applications. Distinguishing proof of extra practical catalysts for biosynthesizing $G_N^D$ will prompt more methodologies for proficient and huge scale generation of $G_N^D$ variations.

## References

Bai L, Gao J, Wei F, Zhao J, Wang D, Wei J (2018) Therapeutic potential of ginsenosides as an adjuvant treatment for diabetes. Front Pharmacol 9:423–438

Barman S, Pal M, Majumder M, Bagchi A, Raha A, Mukherjee P (2019) Activity of ginseng on central nervous system. Pharma Innovation 8(5):420–422

Chen L-x, Qi Z, Shao Z-j, Li S-s, Qi Y-l, Gao K, Liu S-x, Li Z, Sun Y-s, Li P-y (2019) Study on antidepressant activity of pseudo-ginsenoside HQ on depression-like behavior in mice. Molecules 24(5):870–883

Darsandhari S, Shrestha B, Pandey RP, Lee S, Jung HJ, Kim YJ, Sohng JK (2019) Enzymatically synthesized ginsenoside exhibits antiproliferative activity in various cancer cell lines. Res J Appl Sci 9(5):893–910

Dou T-Y, Chen J, Qian X-K, Li K, Ge G-B (2019) Biotransformation of glycoginsenosides to intermediate products and aglycones using a hemicellulosome produced by *Cellulosimicrobium cellulan*. Appl, Biochem, Microbiol 55(2):117–122

Fu Y (2019) Biotransformation of ginsenoside Rb1 to Gyp-XVII and minor ginsenoside Rg3 by endophytic bacterium *Flavobacterium* sp. GE 32 isolated from *Panax ginseng*. Lett Appl Microbiol 68(2):134–141

He Y, Hu Z, Li A, Zhu Z, Yang N, Ying Z, He J, Wang C, Yin S, Cheng S (2019) Recent advances in biotransformation of saponins. Molecules 24(13):2365–2378

Hu Y, Wang N, Yan X, Yuan Y, Luo F, Jiang Z, Zhou Y (2019) Ginsenoside Re impacts on biotransformation products of ginsenoside Rb1 by *Cellulosimicrobium cellulans* sp. 21 and its mechanisms. Process Biochem 77:57–62

Karra A, Konstantinou M, Tzortziou M, Tsialtas I, Kalousi F, Garagounis C, Hayes J, Psarra A-M (2019) Potential dissociative glucocorticoid receptor activity for protopanaxadiol and protopanaxatriol. Int J Mol Sci 20(1):94–102

Kim HW, Han SH, Lee SW, Choi HS, Suh HJ, Hong KB (2019a) Enzymatic hydrolysis increases ginsenoside content in Korean red ginseng (*Panax ginseng* CA Meyer) and their biotransformation under hydrostatic pressure. J Sci Food Agric 99(15):6806–6813

Kim Y-S, Kim D-Y, Kang DW, Park C-S (2019b) Hydrolysis of the outer β-(1, 2)-D-glucose linkage at the C-3 position of ginsenosides by a commercial β-galactosidase and its use in the production of minor ginsenosides. Biocatal Biotransform 37(1):53–58

Kwon H-J, Lee H, Choi G-E, Kwon SJ, Song AY, Kim SJ, Choi WS, Hwang S-H, Kim SC, Kim HS (2018) Ginsenoside F1 promotes cytotoxic activity of NK cells via insulin-like growth factor-1-dependent mechanism. Front Immunol 16(4):114–120

Lee JH, Min DS, Lee CW, Song KH, Kim YS, Kim HP (2018) Ginsenosides from Korean red ginseng ameliorate lung inflammatory responses: inhibition of the MAPKs/NF-κB/c-Fos pathways. J Ginseng Res 42(4):476–484

Lee DG, Lee JS, Kim K-T, Kim HY, Lee S (2019) Analysis of major ginsenosides in various ginseng samples. Appl Biol Chem 62(1):87–91

Li G, Zhang N, Geng F, Liu G, Liu B, Lei X, Li G, Chen X (2019) High-throughput metabolomics and ingenuity pathway approach reveals the pharmacological effect and targets of Ginsenoside Rg1 in Alzheimer's disease mice. Sci Rep 9(1):1–11

Liu Z, Li J-X, Wang C-Z, Zhang D-L, Wen X, Ruan C-C, Li Y, Yuan C-S (2019) Microbial conversion of protopanaxadiol-type ginsenosides by the edible and medicinal mushroom *Schizophyllum commune*: a green biotransformation strategy. ACS Omega 4(8):13114–13123

Majid A (2019) *Panax ginseng*—a review. Univ Thi-Qar J Science 7(1):96–102

Medina-Franco JL (2019) New approaches for the discovery of pharmacologically-active natural compounds. Molecules 18(2):115–120

Metwaly AM, Lianlian Z, Luqi H, Deqiang D (2019) Black ginseng and its saponins: preparation, phytochemistry and pharmacological effects. Molecules 24(10):1856–1860

Moon SS, Lee HJ, Mathiyalagan R, Kim YJ, Yang DU, Lee DY, Min JW, Jimenez Z, Yang DC (2018) Synthesis of a novel α-glucosyl ginsenoside F1 by cyclodextrin glucanotransferase and its in vitro cosmetic applications. Biomolecules 8(4):142–147

Nguyen NH, Nguyen CT (2019) Pharmacological effects of ginseng on infectious diseases. Inflam Pharmacol 12(9):1–13

Nimse SB, Pal D (2015) Free radicals, natural antioxidants, and their reaction mechanisms. RSC Adv 5(35):27986–8006

Pal D, Saha S (2019) Chondroitin: a natural biomarker with immense biomedical applications. RSC Adv 9(48):28061–28077

Park SJ, Nam J, Ahn CW, Kim Y (2019) Anti-diabetic properties of different fractions of Korean red ginseng. J Ethnopharmacol 236(2):220–230

Rani P, Pal D, Hegde RR, Hashim SR (2014) Anticancer, anti-inflammatory, and analgesic activities of synthesized 2-(substituted phenoxy) acetamide derivatives. Biomed Res. Int 16 (9):1–9

Rani P, Pal D, Rama Hegde R, Riaz Hashim S (2016) Leuckart synthesis and pharmacological assessment of novel acetamide derivatives. Anti-Cancer Agent in Med Chem 16(7):898–906

Rahimi S, Kim J, Mijakovic I, Jung KH, Choi G, Kim SC, Kim YJ (2019) Triterpenoid-biosynthetic UDP-glycosyltransferases from plants. Biotechnol Adv 37 (7):107394–107399

Ryoo N, Rahman MA, Hwang H, Ko SK, Nah S-Y, Kim H-C, Rhim H (2019) Ginsenoside Rk1 is a novel inhibitor of NMDA receptors in cultured rat hippocampal neurons. J Ginseng Res 13 (6):715–720

Santangelo R, Silvestrini A, Mancuso C (2018) Ginsenosides, catechins, quercetin and gut microbiota: current evidence of challenging interactions. Food Chem. Toxicol 123(1):42–49

Senbagalakshmi P, Muthukrishnan S, Jebasingh T, Kumar TS, Rao MV, Kumar TS, Rao MV (2019) Squalene, biosynthesis and its role in production of bioactive compounds, a proper scientific challenge. JETIR 6(2):1–22

Szczuka D, Nowak A, Zakłos-Szyda M, Kochan E, Szymańska G, Motyl I, Blasiak J (2019) American ginseng (*Panax quinquefolium* L.) as a source of bioactive phytochemicals with pro-health properties. Nutrients 11(5):1041–1049

Tam DNH, Truong DH, Nguyen TTH, Tran L, Nguyen HD, eldin Shamandy B, Le TMH, Tran DK, Sayed D, Mizukami S (2018) Ginsenoside Rh1: a systematic review of its pharmacological properties. Planta Med 84(3):139–152

Tang QY, Chen G, Song WL, Fan W, Wei KH, He SM, Zhang GH, Tang JR, Li Y, Lin Y, Yang SC(2019) Transcriptome analysis of panax zingiberensis identifies genes encoding oleanolic acid glucuronosyl transferase involved in the biosynthesis of oleanane-type ginsenosides. Planta 249(2):393–406

Takei M, Tachikawa E, Hasegawa H, Lee JJ (2004) Dendritic cells maturation promoted by M1 and M4, end products of steroidal ginseng saponins metabolized in digestive tracts, drive a potent Th1 polarization. Biochem Pharmacol 68(3):441–452

Truong V-L, Bak MJ, Jeong W-S (2019) Chemopreventive activity of red ginseng oil in a mouse model of azoxymethane/dextran sulfate sodium-induced inflammation-associated colon carcinogenesis. J. Med. Food 22(6):578–586

Wang D-D, Jin Y, Wang C, Kim Y-J, Perez ZEJ, Baek NI, Mathiyalagan R, Markus J, Yang D-C (2018) Rare ginsenoside Ia synthesized from F1 by cloning and overexpression of the UDP-glycosyltransferase gene from *Bacillus subtilis*: synthesis, characterization, and in vitro melanogenesis inhibition activity in BL6B16 cells. J Ginseng Res 42(1):42–49

Wang C, Liu J, Deng J, Wang J, Weng W, Chu H, Meng Q (2019) Advances in the chemistry, pharmacological diversity and metabolism of 20 (R)-ginseng saponins. J Ginseng Res 44 (1):14–23

Wang CZ, Aung HH, Ni M, Wu JA, Tong R, Wicks S, He TC, Yuan CS (2007) Red American ginseng: ginsenoside constituents and anti-proliferative activities of heat-processed panax quinquefolius roots. Planta Med 73(7):669–674

Wu F, Zhu J, Li G, Wang J, Veeraraghavan VP, Krishna Mohan S, Zhang Q (2019) Biologically synthesized green gold nanoparticles from Siberian ginseng induce growth-inhibitory effect on melanoma cells (B16). Artif Cells Nano Med Biotechnol 47(1):3297–3305

Xue L, He Z, Bi X, Xu W, Wei T, Wu S, Hu S (2019) Transcriptomic profiling reveals MEP pathway contributing to ginsenoside biosynthesis in panax ginseng. BMC Genomics 20 (1):383–390

Yue PY, Wong DY, Wu PK, Leung PY, Mak NK, Yeung HW, Liu L, Cai Z, Jiang ZH, Fan TP, WongRN (2006) The angiosupressive effects of 20 (R)-ginsenoside Rg3. Biochem Pharmacol 72(4):437–445

Yu KW, Murthy HN, Hahn EJ, Paek KY (2005) Ginsenoside production by hairy root cultures of panax ginseng: influence of temperature and light quality. Biochem Eng J 23(1):53–56

Yang D, Yang X, Yan H, Fan B, Dai J, Song J, Lei Y, Guo N (2019) UPLC-MS/MS determination of twelve ginsenosides in Shenfu Tang and Dushen Tang. Int J Anal Chem 44(1):14–23

Zhang Q, Wang X, Lv L, Su G, Zhao Y (2019) Antineoplastic activity, structural modification, synthesis and structure-activity relationship of dammarane-type ginsenosides, an overview. Curr Org Chem 23(5):503–516

Zheng M, Xin Y, Li Y, Xu F, Xi X, Guo H, Cui X, Cao H, Zhang X, Han C (2018) Ginsenosides: a potential neuroprotective agent. Biomed Res Int 13(6):518–526

Zuo T, Sun Y, Xue W, Zhang Y, Wang G, Zhang X, Xu R, Zhang S (2019) Study on interaction between ginsenosides Rg_1, Rb_1 and Ro and bovine serum albumin. Zhongguo Zhong Yao Za Zhi 44(12):2559–2565

The manufacturer's authorised representative in the EU is Springer Nature Customer Service Centre GmbH, Europaplatz 3, 69115 Heidelberg, Germany. If you have any concerns regarding our products, please contact ProductSafety@springernature.com

Printed and bound by CPI Group (UK) Ltd, Croydon, CR0 4YY
22/09/2026
02221048-0018